Lecture Notes in Physics

The Lecture Notes in Physics

The series Lecture Notes in Physics (LNP), founded in 1969, reports new developments in physics research and teaching – quickly and informally, but with a high quality and the explicit aim to summarize and communicate current knowledge in an accessible way. Books published in this series are conceived as bridging material between advanced graduate textbooks and the forefront of research and to serve three purposes:

- to be a compact and modern up-to-date source of reference on a well-defined topic
- to serve as an accessible introduction to the field to postgraduate students and nonspecialist researchers from related areas
- to be a source of advanced teaching material for specialized seminars, courses and schools

Both monographs and multi-author volumes will be considered for publication. Edited volumes should, however, consist of a very limited number of contributions only. Proceedings will not be considered for LNP.

Volumes published in LNP are disseminated both in print and in electronic formats, the electronic archive being available at springerlink.com. The series content is indexed, abstracted and referenced by many abstracting and information services, bibliographic networks, subscription agencies, library networks, and consortia.

Proposals should be sent to a member of the Editorial Board, or directly to the managing editor at Springer:

Christian Caron
Springer Heidelberg
Physics Editorial Department I
Tiergartenstrasse 17
69121 Heidelberg / Germany
christian.caron@springer.com

H. Fehske
R. Schneider
A. Weiße (Eds.)

Computational Many-Particle Physics

Springer

Editors

Holger Fehske
Alexander Weiße
Universität Greifswald
Institut für Physik
Felix-Hausdorff-Str. 6
17489 Greifswald,
Germany
holger.fehske@physik.uni-greifswald
weisse@physik.uni-greifswald.de

Ralf Schneider
Max-Planck-Institut für Plasmaphysik
Wendelsteinstr. 1
17491 Greifswald, Germany
ralf.schneider@ipp.mpg.de

H. Fehske, R. Schneider and A. Weiße (Eds.), *Computational Many-Particle Physics*, Lect. Notes Phys. 739 (Springer, Berlin Heidelberg 2008), DOI 10.1007/ 978-3-540-74686-7

ISSN 0075-8450
ISBN 978-3-642-09414-9 e-ISBN 978-3-540-74686-7

Springer is a part of Springer Science+Business Media
springer.com

Softcover reprint of the hardcover 1st edition 2008

Cover design: eStudio Calamar S.L., F. Steinen-Broo, Pau/Girona, Spain

Preface

Many-particle physics is at work whenever we delve into the rich phenomenology of the real world, or into laboratory experiments. Nevertheless, our physical description of nature is mostly built upon single-particle theories. For instance, Kepler's laws provide a basic understanding of our solar system, many features of the periodic table can be understood from the solution of a single hydrogen atom, and even complicated microprocessors with an unbearable number of electrons floating through millions of transistors can be developed based on the effective single-particle models of semiconductor physics. These approaches are successful because quite often interactions affect physical systems in a perturbative way. Classical perturbation theory yields corrections to a planet's orbit due to other planets, quantum chemistry relies on various approximation schemes to deal with complicated atoms and small molecules, and solid state theory uses weakly interacting quasiparticles as elementary excitations. This fortunate situation changes, however, when we try to understand more complex or strongly interacting systems, or when we try to explore the nature of matter itself. Condensates of cold bosonic atoms, for example, show subtle many-particle effects, strongly correlated fermions may give rise to high-temperature superconductivity, and the way quarks build up elementary particles (hadronization) is a highly non-trivial few-body problem. Another example are quantum computers, which many scientist envision as a replacement for our present-day microprocessors, and which exploit the entanglement property of quantum many-particle states. Last but not least, we mention the complexity of fusion plasmas, which some day may help feeding our ever-growing hunger for new energy resources. Unfortunately, even the most sophisticated analytical approaches largely fail to describe such systems. Hence, at present, unbiased numerical investigations provide the most reliable tool to address these problems. This is the point where the expert use of large-scale computers comes into play.

The increasing importance of computational many-particle physics calls for a comprehensive introduction into this rapidly developing field suitable for graduate students and young researchers. Therefore, we decided to organize a summer school on "Computational Many-Particle Physics" in September 2006, during the 550th anniversary of the University Greifswald. Generously sponsored by the Wilhelm and Else Heraeus Foundation and hosted by the Max-Planck-Institute for Plasma Physics and the Institute for Physics, we brought together more than 40 students and 20 distinguished scientists working on such diverse fields as fusion plasmas, statistical physics, solid state theory and high performance computing. The present Lecture Notes summarize and extend the material showcased over a 2-week period of tightly scheduled tutorials, seminars and exercises. The emphasis is on a very pedagogical and systematic introduction to various numerical concepts and techniques, with the hope that the reader may quickly start to program himself. The spectrum of the numerical methods presented is very broad, covering classical as well as quantum few- and many-particle systems. The trade-off between the number of particles, the complexity of the underlying microscopic models and the importance of the interactions determine the choice of the appropriate numerical approach. Therefore, we arranged the book along the algorithms and techniques employed, rather than on the physics applications, which we think is more natural for a book on numerical methods.

We start with methods for classical many-particle systems. Here, molecular dynamics approaches trace the motion of individual particles, kinetic approaches work with the distribution functions of particles and momenta, while hybrid approaches combine both concepts. A prominent example is the particle-in-cell method typically applied to model plasmas, where the time evolution of distribution functions is approximated by the dynamics of pseudo-particles, representing thousands or millions of real particles. Of course, at a certain length scale the quantum nature of the particles becomes important. As an attempt to close the gap between classical and quantum systems, we outline a number of semi-classical (Wigner-function, Boltzmann- and Vlasov-equation based) approaches, which in particular address transport properties. The concept of Monte Carlo sampling is equally important for classical, statistical and quantum physical problems. The corresponding chapters therefore account for a substantial part of the book and introduce the major stochastic approaches in application to very different physical situations. Focussing on solids and their properties, we continue with ab initio approaches to the electronic structure problem, where band structure effects are taken into account with full detail, but Coulomb interactions and the resulting correlations are treated approximately. Dynamical mean field theories and cluster approaches aim at improving the description of correlations and bridge the gap to an exact numerical treatment of basic microscopic models. Exact diagonalization of finite systems gives access to their ground-state, spectral and thermodynamic properties. Since these methods work with the full many-particle Hamiltonian, the study of a decent number of particles or larger system sizes is a challenging task, and there is a strong demand to circumvent these limitations. Along this line the density matrix renormalization group represents a clever technique to restrict the many-particle Hilbert space to the physically most important subset. Finally, all the discussed methods heavily rely on the use of powerful computers, and the book would be incomplete without two detailed chapters on parallel programming and optimization techniques for high performance computing.

Of course, the preparation of such a comprehensive book would have been impossible without support from many colleagues and sponsors. First of all, we thank the lecturers and authors for their engagement, enthusiasm and patience. We are greatly indebted to Milena Pfafferott and Andrea Pulss for their assistance during the editorial work and the fine-tuning of the articles. Jutta Gauger, Beate Kemnitz, Thomas Meyer and Gerald Schubert did an invaluable job in the organization of the summer school. Finally, we acknowledge financial support from the Wilhelm and Else Heraeus foundation, the Deutsche Forschungsgemeinschaft through SFB 652 and TR 24 and the Helmholtz-Gemeinschaft through COMAS.

Greifswald,
July 2007

Holger Fehske
Ralf Schneider
Alexander Weiße

Contents

Part IX The Density Matrix Renormalisation Group: Concepts and Applications

Part I

Molecular Dynamics

1 Introduction to Molecular Dynamics

Ralf Schneider, Amit Raj Sharma, and Abha Rai

Max-Planck-Institut für Plasmaphysik, Teilinstitut Greifswald,
17491 Greifswald, Germany

Molecular dynamics is the science of simulating the time dependent behavior of a system of particles. The time evolution of the set of interacting atoms is followed by integrating their equation of motion with boundary conditions appropriate for the geometry or symmetry of the system. Molecular dynamics generate information at the microscopic level, which are: atomic positions, velocities. In order to calculate the microscopic behavior of a system from the laws of classical mechanics, MD requires, as an input, a description of the interaction potential (or force field). The quality of the results of an MD simulation depends on the accuracy of the description of inter-particle interaction potential. This choice depends very strongly on application. Thus MD technique acts as a *computational microscope*. This microscopic information is then converted to the macroscopic observable like pressure, temperature, heat capacity and stress tensor etc. using statistical mechanics. Molecular dynamic techniques have been widely used by almost all the branches of science. Namely, determination of reaction rates in chemistry, solid state structures, surfaces and defects formation in material science, protein folding in biochemistry and so on. Recent applications employing common force fields include an exploration of protein folding pathways in solution [1], structural and dynamical properties of ion channels [2, 3]. The disadvantage of a model force-field is that a system is restricted to a single molecular connectivity. This prohibits force field models from describing chemical processes involving bond breaking and forming. An alternative approach is the combination of classical dynamics with electronic structure: internuclear forces are computed *on the fly* from an electronic structure calculation as a MD simulation proceeds [4, 5]. This method, known as ab initio molecular dynamics, requires no input potential model and is capable of describing chemical events, although it has high computational overhead.

1.1 Basic Approach

The essential elements for a molecular dynamics simulation are (i) the interaction potential (i.e., potential energy) for the particles, from which the forces can be calculated, and (ii) the equations of motion governing the dynamics of the particles. We follow the laws of classical mechanics, mainly Newton's law

$$\boldsymbol{F}_i = m_i \boldsymbol{a}_i \ , \tag{1.1}$$

R. Schneider et al.: *Introduction to Molecular Dynamics*, Lect. Notes Phys. **739**, 3–40 (2008)
DOI 10.1007/978-3-540-74686-7_1

for each atom i in a system constituted by N atoms. Here, m_i is the atom mass, $\boldsymbol{a}_i$ its acceleration and $\boldsymbol{F}_i$ the force acting upon it due to the interactions with the other atoms. Equivalently one can solve classical Hamiltonian equation of motion

$$\dot{\boldsymbol{p}}_i = -\frac{\partial H}{\partial \boldsymbol{r}_i} , \tag{1.2}$$

$$\dot{\boldsymbol{r}}_i = \frac{\partial H}{\partial \boldsymbol{p}_i} , \tag{1.3}$$

where $\boldsymbol{p}_i$ and $\boldsymbol{r}_i$ are the momentum and position co-ordinates for the ith atom. H, the Hamiltonian, which is defined as a function of position and momenta, is given by

$$H(\boldsymbol{p}_i, \boldsymbol{r}_i) = \sum_{i=1}^{N} \frac{\boldsymbol{p}_i^2}{2m_i} + V(\boldsymbol{r}_i) . \tag{1.4}$$

The force on an atom can be calculated as the derivative of energy with respect to the change in the atom's position

$$\boldsymbol{F}_i = m_i \boldsymbol{a}_i = -\nabla_i V = -\frac{\mathrm{d}E}{\mathrm{d}\boldsymbol{r}_i} . \tag{1.5}$$

Knowledge of the atomic forces and masses can then be used to solve for the positions of each atom along a series of extremely small time steps (on the order of femtoseconds). The velocities are calculated from the accelerations

$$\boldsymbol{a}_i = \frac{\mathrm{d}\boldsymbol{v}_i}{\mathrm{d}t} . \tag{1.6}$$

Finally, the positions are calculated from the velocities

$$\boldsymbol{v}_i = \frac{\mathrm{d}\boldsymbol{r}_i}{\mathrm{d}t} . \tag{1.7}$$

To summarize the procedure, at each step, the forces on the atoms are computed and combined with the current positions and velocities to generate new positions and velocities a short time ahead. The force acting on each atom is assumed to be constant during the time interval. The atoms are then moved to the new positions, an updated set of forces is computed and new dynamics cycle goes on.

Usually molecular dynamics simulations scale by either $O(N \log N)$ or $O(N)$, with N as the number of atoms. This makes simulations with macroscopic number of atoms or molecules ($\sim 10^{23}$) impossible to handle with MD. Therefore, statistical mechanics is used to extract the macroscopic information from the microscopic information provided by MD.

Two important properties of the equations of motion should be noted. One is that they are time reversible, i.e., they take the same form when the transformation $t \to -t$ is made. The consequence of time reversal symmetry is that the microscopic physics is independent of the direction of the flow of time. Therefore, in contrast to the Monte Carlo method, molecular dynamics is a *deterministic* technique: Given an initial set of positions and velocities, the subsequent time evolution is *in principle* [6] completely determined from its current state. Molecular dynamics calculates the real dynamics, i.e. behavior of the system, from which the time average of the system's properties can be calculated. The second important property of the equations of motion is that they conserve the Hamiltonian. This can be easily seen by computing the time derivative of H and substituting (1.2) and (1.3) for the time derivatives of position and momentum

$$\frac{\mathrm{d}H}{\mathrm{d}t} = \sum_{i=1}^{N} \left[\frac{\partial H}{\partial \boldsymbol{r}_i} \dot{\boldsymbol{r}}_i + \frac{\partial H}{\partial \boldsymbol{p}_i} \dot{\boldsymbol{p}}_i \right] = \sum_{i=1}^{N} \left[\frac{\partial H}{\partial \boldsymbol{r}_i} \frac{\partial H}{\partial \boldsymbol{p}_i} - \frac{\partial H}{\partial \boldsymbol{p}_i} \frac{\partial H}{\partial \boldsymbol{r}_i} \right] = 0\,. \tag{1.8}$$

The conservation of the Hamiltonian is equivalent to the conservation of the total energy of the system and provides an important link between molecular dynamics and statistical mechanics.

1.1.1 Statistical Ensemble

Statistical mechanics connects the microscopic details of a system to physical observables such as equilibrium thermodynamic properties, transport coefficients, and spectra. Statistical mechanics is based on the Gibbs ensemble concept. That is, many individual microscopic configurations of a very large system lead to the same macroscopic properties, implying that it is not necessary to know the precise detailed motion of every particle in a system in order to predict its properties. It is sufficient to simply average over a large number of identical systems, each in a different microscopic configuration; i.e., the macroscopic observables of a system are formulated in terms of ensemble averages. Statistical ensembles are usually characterized by fixed values of thermodynamic variables such as energy E, temperature T, pressure P, volume V, particle number N or chemical potential μ. One fundamental ensemble is called the *micro-canonical* ensemble and is characterized by constant particle number N, constant volume V and constant total energy E, and is denoted as the NVE ensemble. Other examples include the *canonical* or NVT ensemble, the *isothermal-isobaric* or NPT ensemble, and the *grand canonical* or μVT ensemble. The thermodynamic variables that characterize an ensemble can be regarded as experimental control parameters that specify the conditions under which an experiment is performed.

Now consider a system of N particles occupying a container of volume V and evolving under Hamilton's equations of motion. According to (1.8), the Hamiltonian will be a constant E, equal to the total energy of the system. In addition, the number of particles and the volume are assumed to be fixed. Therefore, a dynamical trajectory of this system will generate a series of classical states having constant N, V, and E, corresponding to a micro-canonical ensemble. If the dynamics generates all possible states having a fixed N, V, and E, then an average over this trajectory will yield the same result as an average in a micro-canonical ensemble. The energy conservation condition, $H(\boldsymbol{p},\boldsymbol{r}) = E$, which imposes a restriction on the classical microscopic states accessible to the system, defines a hyper-surface in the phase space called the constant energy surface. A system evolving according to Hamilton's equations of motion will remain on this surface. The assumption that a system, given an infinite amount of time, will cover the entire constant energy hyper-surface is known as the ergodic hypothesis. Thus, under the ergodic hypothesis, averages over a trajectory of a system obeying Hamilton's equations are equivalent to averages over the micro-canonical ensemble.

1.2 Macroscopic Parameters

Statistical mechanics provides a link between the macroscopic properties of matter (like temperature, pressure, etc.) and the microscopic properties (like positions, velocities, individual kinetic and potential energies) of atoms and molecules that constitute it. These macroscopic properties reflect the time average behavior of the atoms at equilibrium (i.e. in one of the many possible degenerate minimum energy states accessible to the system). Often even in an NVE simulation one does some simple tricks to control temperature and/or pressure. This give something of a NVT or NVP and NVE hybrid. However temperature and pressure fluctuate, and the system does not behave as a true NVT or NVP ensemble in the thermodynamic sense. But on average temperature and pressure have the desired value. In true NVT or NPT (non-Hamiltonian) algorithms it is possible to have T and P have exactly the desired value, and the simulation directly corresponds to the thermodynamic ensembles.

At the start of the MD simulation the atomic positions and velocities have to be initialized. In the case of crystalline solids the starting positions will be defined by the crystal symmetry and positions of atoms within the unit cell of the crystal. The unit cell is then repeated to fill up the desired dimensions of the system. Realistic atomic displacements from crystal lattice sites can also be derived using the Debye model. For amorphous solids the particles can be randomly distributed within the desired dimensions making sure that there exists a minimum distance between the atoms so that strong local forces do not exist in the system.

The initial velocities are set by assuming a Maxwell-Boltzmann distribution for velocities along the three dimensions. This is done by using Gaussian distributed random numbers multiplied by a mean square velocity given by $\sqrt{2k_{\mathrm{B}}T/m}$ in each of the three directions and making sure that the system has total momentum equal to zero. Generally speaking, if sensible (tailored to avoid large impulsive forces) position and velocity distributions are chosen, particle positions at equilibrium relax to oscillating around the minimum energy locations of the potential Φ. A Maxwellian distribution of velocities is naturally obtained in the simulation.

Therefore the initial temperature and total energy of the system has been fixed. The temperature is fixed by the velocity distribution. The total energy of the system is given by

$$E_{\text{tot}} = (KE)_{\text{tot}} + (PE)_{\text{tot}} , \tag{1.9}$$

where $(KE)_{\text{tot}}$ is the total kinetic energy in the system given by

$$(KE)_{\text{tot}} = \sum_{i=1}^{N} \frac{1}{2} m (v_{x,i}^2 + v_{y,i}^2 + v_{z,i}^2) \tag{1.10}$$

and $(PE)_{\text{tot}}$ is the total potential energy of the system given by

$$(PE)_{\text{tot}} = \sum_{i=1}^{N} \Phi_i(r_i) \tag{1.11}$$

with $v_{x,y,z}$ being the velocities, r being the positions of atoms, and i being the index that sums over all the atoms N in the system. $\Phi_i(r_i)$ is the potential energy of the ith atom due to all other atoms in the system.

1.2.1 Temperature Scaling

In equilibrium simulations, especially if long-range interactions are involved and a potential truncated at a cut-off radius is used, an unavoidable slow drift occur that need correction. A possible trivial temperature scaling is to force the system temperature to be exactly T during every time step. This can be a rather severe perturbation of the atom motion especially if there are only a few atoms. Better methods to control temperature and pressure are discussed in [7, 8, 9] and will be shortly summarized in the following.

The Berendsen method [7] is essentially a direct scaling, but softened with a time constant. Let T_0 be the desired temperature, Δt is the time step of the simulation and τ_T be the time constant for temperature control. In the Berendsen temperature control scheme, all velocities are scaled at each time step by a factor λ given by

$$\lambda = \sqrt{1 + \frac{\Delta t}{\tau_T}\left(\frac{T_o}{T} - 1\right)} , \tag{1.12}$$

τ_T has to be greater than Δt. According to Berendsen [7] if $\tau_T > 100\Delta t$ then the system has natural fluctuations about the average.

1.2.2 Pressure Scaling

The Berendsen pressure control is implemented by changing all atom positions, and the system cell size during the simulation. If the desired pressure is P_0 and τ_P is the time constant for pressure control, which should be typically greater than $100\Delta t$, the scaling factor μ is given by:

$$\mu = \left[1 - \frac{\beta \Delta t}{\tau_P}(P_o - P)\right]^{1/3} , \tag{1.13}$$

where β is the isothermal compressibility of the system ($= 1/\text{bulk modulus}$) and P is the current pressure. The change in all atom positions and the system size is given by

$$\boldsymbol{r}(t + \delta t) = \mu \boldsymbol{r}(t) \; , \tag{1.14}$$

$$\boldsymbol{S}(t + \delta t) = \mu \boldsymbol{S}(t) \; , \tag{1.15}$$

and the volume of the system also changes by

$$\boldsymbol{V}(t + \delta t) = \mu^3 \boldsymbol{V}(t) \; . \tag{1.16}$$

This type of temperature and pressure scaling should be done after the solution of the equations of motions gives realistic fluctuations in temperature and pressure for a system in equilibrium and when large values of τ_T and τ_P are chosen.

1.2.3 Time Scale Dilemma

Design of a molecular dynamics simulation can often encounter limits of computational power. The simulation's time duration is dependent on the time length of each time-step, between which forces are recalculated. The time-step must be chosen small enough to avoid discretization errors, and the number of time-steps, and thus simulation time, must be chosen large enough to capture the effect being modeled without taking an extraordinary period of time i.e. smaller than the vibrational frequency of the system. The length of the simulation should be large enough that the system goes through all possible phase space points in the ensemble. As a rule of thumb: the atoms should not move more than $1/20$ of the nearest neighbor distance in the chosen time step. There exists a wide range of time scales over which specific processes occur and one need to resolve vibrations at these scales, for example, bond vibrations (femtosecond), collective vibrations (picosecond) and protein foldings (millisecond to microsecond). The integration time step which is determined by the fastest varying force is of the order femtoseconds. This limits the accessible time scale by MD simulations from picoseconds to several nanoseconds. So, no matter how many processors (how powerful the computer is) one can only reach several picoseconds in time because time cannot be parallelize [10]. As a consequence of time scale dilemma, slower mechanisms like MD has limited accessibility to handle diffusion. This can only be overcome using multi-scale models.

1.3 Inter-Atomic Potentials

1.3.1 Pair Potentials

For pair potentials, the total potential energy of a system can be calculated from the sum of energy contributions from pairs of atoms and it depends only on the distance between atoms. One example of a pair potential is the *Lennard-Jones potential* [11] (also known as the 6–12 potential). Other examples of pair potential are Coulomb potential, Morse potential [12] etc. Lennard-Jones potential is the most commonly used form

$$V(r)^{\mathrm{LJ}} = 4\epsilon\left[\left(\frac{\sigma}{r}\right)^{12} - \left(\frac{\sigma}{r}\right)^{6}\right], \tag{1.17}$$

where ϵ is the cohesive energy well depth and σ is the equilibrium distance. The $(\sigma/r)^{12}$ term describes the repulsive force due to overlapping of electron orbitals (Pauli repulsion) and does not have a true physical motivation, other than that the exponent must be larger than 6 to get a potential well. One often uses 12 because it can be calculated efficiently(square of 6). The term $(\sigma/r)^6$ describes the attractive force (Van der Waals) and can be derived classically by considering how two charged spheres induce dipole-dipole interactions into each other. This potential was used in the earliest studies of the properties of liquid argon [13, 14]. LJ potentials are not a good choice for very small r ($r \lesssim 0.1$ nm) since the true interaction is $\sim (1/r)\exp(-r)$ and not $1/r^{12}$.

Typical simulation sizes in molecular dynamics simulation are very small up to 1000 atoms. As a consequence, most of the extensive quantities are small in magnitude when measured in macroscopic units. There are two possibilities to overcome this problem: Either one should work with atomic-scale units (ps, amu, nm) or to make all the observable quantities dimensionless with respect to their characteristic values. The second approach is more popular. The scaling is done with the model parameters e.g size σ, energy ϵ, mass m. So the common recipe is, one chooses a value for one atom/molecule pair potential arbitrarily (ϵ) and then other model parameters (say energy E) are given in terms of this reference value ($E^* = E/\epsilon$). The other parameters are also calculated similarly. For example, dimensionless distance ($r^* = r/\sigma$), energy($E^* = E/\epsilon$), temperature ($T^* = kT/\epsilon$), time ($t^* = t/[\sigma(m/\epsilon)^{1/2}]$), force ($F* = F\sigma/\epsilon$), diffusion coefficient ($D^* = D/[\sigma(\epsilon/m)^{1/2}]$) and so on.

Now if we write the LJ potential in dimensionless form

$$V^*(r^*)^{\mathrm{LJ}} = 4\left[\left(\frac{1}{r^*}\right)^{12} - \left(\frac{1}{r^*}\right)^{6}\right]. \tag{1.18}$$

We see that it is parameter independent, consequently all the properties must also be parameter independent. If a potential only has a couple of parameters then this scaling has a lot of advantages. Namely, potential evaluation can be really efficient in reduced units and as the results are always the same, so the results can be transferred to different systems with straight forward scaling by using the model parameters σ, ϵ and m. This is equivalent to selecting unit value for the parameters and it is convenient to report system properties in this form e.g $P^*(\rho^*)$.

1.3.2 Molecular Interaction Models

To describe atomic interactions in molecules more complex than a dimer a pair potential is not enough. Since molecules are bonded by covalent bonds, at least angular terms are needed, and in many cases many more complicated terms as well. For instance, in carbon chains the difference between *single* and *double* bonds often is important, and for this at least a four-body term is needed.

To describe complex molecules a large set of inter-atomic potentials (often also called force fields) have been developed by chemists, physicists and biochemists. At least when force fields are used to describe atom motion inside molecules and interactions between molecules (but not actual chemical reactions) the term molecular mechanics is often used.

The total energy of a molecule can be given as

$$E = E_{\text{bond}} + E_{\text{angle}} + E_{\text{torsion}} + E_{\text{oop}} + E_{\text{cross}} + E_{\text{nonbond}} \; . \tag{1.19}$$

Where:

- E_{bond} describes the energy change related to a change of bond length, and thus is simply a pair potential V_2.
- E_{angle} describes the energy change associated with a change in the bond angle, i.e. is a three-body potential V_3.
- E_{torsion} describes the torsion, i.e. energy associated with the rotation between two parts of a molecule relative to each other.
- E_{oop} describes *out-of-plane* interactions, i.e. the energy change when one part of a molecule is out of the plane with another.
- E_{cross} are cross terms between the other interaction terms.
- E_{nonbond} describes interaction energies which are not associated with covalent bonding. Could be e.g. ionic or van-der-Waal-terms.

In the following we describe the terms, using notation more common on chemistry rather than the physics notation used earlier.

1.3.2.1 The Term E_{bond}

This term describes the energy change associated with the bond length. It is a simple pair potential, and could be e.g. a Morse or LJ potential. At its simplest, it is purely harmonic, i.e.

$$E_{\text{bond}} = \sum_{\text{bonds}} \frac{1}{2} k_b (b - b_0)^2 \; , \tag{1.20}$$

where b is the bond length. If we write this term instead as

$$E_i = \sum_j \frac{1}{2} k (r_{ij} - r_0)^2 \; , \tag{1.21}$$

we see that this is essentially the same thing as the pair potentials dealt with earlier. So this is essentially the same thing as approximating the bond as a string with the string constant k. Although the approximation is very simple, it can be good enough in problems where we are always close to equilibrium, since any smooth potential well can always be to the first order approximated by a harmonic well. But harmonic potentials obviously can not describe large displacements of atoms or bond breaking reasonably. In solids, the harmonic approximation corresponds to the elastic regime, i.e. the one where stress is directly proportional to the strain (Hooke's law).

To improve on the bond model beyond the elastic regime, one can add higher-order terms to it, e.g.

$$E_{\text{bond}} = \sum_{\text{bonds}} K_2(b - b_0)^2 + K_3(b - b_0)^3 + K_4(b - b_0)^4 \, . \tag{1.22}$$

This way also larger strain can be described, but this still does not describe bond breaking (dissociation).

Also the Morse potential

$$E_{\text{bond}} = \sum_{\text{bonds}} D_b\{1 - \mathrm{e}^{-a(b-b_0)}\}^2 \tag{1.23}$$

is much used to describe bond energies. It is good in that it tends to zero when b tends to infinity so it can describe bond breaking. But on the other hand it never goes fully to zero, which is not quite realistic either as in reality a covalent bond does break essentially completely at some inter-atomic distance.

1.3.2.2 Angular Terms E_{angle}

The angular terms describe the energy change associated with two bonds forming an angle with each other. Most kinds of covalent bonds have some angle which is most favored by them – for sp^3 hybridized bonds it is $\sim 109°$, for sp^2 120° and so on. Like for bond lengths, the easiest way to describe bond angles is to use a harmonic term like

$$E_{\text{angle}} = \sum_{\theta} H_\theta(\theta - \theta_0)^2 \, , \tag{1.24}$$

where θ_0 is the equilibrium angle and H_θ a constant which describes the angular dependence well.

This may work well up to 10° or so, but for larger angles additional terms are needed. A typical means for improvement is the third-order terms and so forth, for instance

$$E_{\text{angle}} = \sum_{\theta} H_2(\theta - \theta_0)^2 + H_3(\theta - \theta_0)^3 \, . \tag{1.25}$$

1.3.2.3 Torsional Terms E_{torsion}

The bond and angular terms were already familiar from the potentials for solids. In the physics and chemistry of molecules there are many important effects which can not be described solely with these terms. The most fundamental of these is probably torsion. By this, the rotations of one part of a molecule with respect to another is meant. A simple example is the rotation of two parts of the ethane molecule C_2H_6 around the central C-C carbon bond.

Torsional forces can be caused by e.g. dipole-dipole-interactions and bond conjugation. If the angle between two parts is described by an angle ϕ, it is clear that the function f which describes the rotation should have the property $f(\phi) = f(\phi + 2\pi)$, because it is possible to do a full rotation around the central bond and return to the initial state. The trigonometric functions sine and cosine of course fulfill this requirement, so it is natural to describe the torsional energy with a few terms in a Fourier series

$$E_{\text{torsion}} = V_1(1 + \cos(\phi)) + V_2(1 + \cos(2\phi)) + V_3(1 + \cos(3\phi)) . \tag{1.26}$$

The first part of the torsional term V_1 is often interpreted to be related to dipole-dipole interactions, V_2 to bond conjugation and V_3 to steric energy.

1.3.2.4 Out-of-Plane Terms E_{oop}

With the out-of-plane-terms one describes the energy which in (some cases) is associated with the displacement of atoms out of the plane in which they should be. This is relevant in some (parts of) molecules where atoms are known to lie all in the same plane. The functional form can be rather simple

$$E_{\text{oop}} = \sum_{\chi} H_\chi \chi^2 , \tag{1.27}$$

where χ is the displacement out of the plane.

1.3.2.5 Cross Terms E_{cross}

The cross-terms are functions which contain several of the above-mentioned quantities. They could e.g. describe how a stretched bond has a weaker angular dependence than a normal one. Or they can describe the relations between two displacements, an angle and a torsion and so on.

1.3.2.6 Non-Bonding Terms E_{nonbond}

With the non-bonding terms all effects which affect the energy of a molecule but are not covalent bonds are meant. These are e.g. van-der-Waals-terms, electrostatic Coulomb interactions and hydrogen bonds. For this terms one could thus further divide

$$E_{\text{nonbond}} = E_{\text{vdW}} + E_{\text{Coulomb}} + E_{\text{hbond}} . \tag{1.28}$$

The van der Waals term is often a simple Lennard-Jones-potential, and E_{Coulomb} a Coulomb potential for some, usually fractional, charges q_i.

1.3.3 Reactive Potentials

Most of the potential functions used in MD simulations are intended for modeling physical processes, not chemical reactions. The formation and breaking of chemical bonds are inherently quantum mechanical processes, and are often studied using first-principles methods. Nevertheless, classical potentials do exist that can empirically model changes in covalent bonding.

One successful method for treating covalent bonding interactions in computer simulations is the Tersoff-type potential [15, 16, 17, 18]. Unlike traditional molecular mechanics force fields [19, 20, 21, 22, 23, 24, 25, 26], the Tersoff model allows for the formation and dissociation of covalent chemical bonds during a simulation. Many-body terms reflecting the local coordination environment of each atom are used to modify the strength of more conventional pairwise terms. With this approach, individual atoms are not constrained to remain attached to specific neighbors, or to maintain a particular hybridization state or coordination number. Models of this sort, despite being purely classical, can provide a realistic description of covalent bonding processes in non-electrostatic systems. Potentials of this type have been developed to treat systems containing silicon [16], carbon [17, 27], germanium [18], oxygen [27], or hydrogen [27], as well as heterogeneous systems containing various combinations of these species [18, 28, 29, 30, 31].

One particularly successful example of a Tersoff potential is the reactive empirical bond-order (REBO) potential developed by Brenner [30, 31, 32, 33]. This model uses a Tersoff-style potential to describe the covalent bonding interactions in carbon and hydrocarbon systems. Originally developed for use in simulating the chemical vapor deposition of diamond [30], the REBO potential has been extended to provide more accurate treatment of the energetic, elastic, and vibrational properties of solid carbon and small hydrocarbons [33]. This potential has been used to model many different materials and processes, including fullerenes [32], carbon nanotubes [34], amorphous carbon [35], and the tribology and tribochemistry of diamond interfaces [36, 37, 38, 39, 40, 41, 42].

The REBO potential is not appropriate for studying every hydrocarbon system, however. In particular, the absence of dispersion and non-bonded repulsion terms makes the potential poorly suited for any system with significant intermolecular interactions. This is the case for many important hydrocarbon systems, including liquids and thin films, as well as some solid-state materials such as graphite and fullerenes. Even covalent materials such as diamond can benefit from a treatment including non-bonded interactions. The bulk phase is dominated by covalent interactions, but longer-range forces become quite important when studying interfacial systems [27].

Various attempts have been made previously to combine non-bonded interactions with the Tersoff or REBO potentials in a way that preserves the reactive capabilities of the model [43, 44, 45]. One such improvement of the Tersoff potential was presented by Kai Nordlund et al. [46] which retains the good description of the covalent bonding and yet also describes accurately both the short-range replusive part of the potential and the long-range bonding between graphite planes. One way to do this is to simply reduce the repulsive barrier associated with the Lennard-Jones or other potential [47], although this results in barriers which are too large for radical species and too small for saturated compounds. Another alternative, taken by Nyden et al. [44], is to allow bonds to dissociate with a Morse potential [12], and explicitly check for recombination reactions between dissociated radicals. This approach has been used to model thermal decomposition of polymers [44], but is not general enough to treat arbitrary reactions in hydrocarbons, such as addition across unsaturated bonds. Another method, used by Che et al. [45] is to reduce the repulsive non-bonded interactions based on the covalent interaction energy, rather than the distance. This method can help eliminate non-bonded interactions during bond dissociations, but will again tend to overestimate barriers in association reactions.

1.4 Numerical Integration Techniques

The potential energy is a function of the atomic positions (3N) of all the atoms in the system. Due to the complicated nature of this function, there is no analytical solution to the equations of motion and these equation must be solved numerically.

Numerous numerical algorithms have been developed for integrating the equations of motion. We list several here.

(i) Verlet algorithm [14],
(ii) Leap-frog algorithm [48],
(iii) Velocity Verlet [49],
(iv) Beeman's algorithm [50] and
(v) Symplectic reversible integrators [51, 52].

In choosing which algorithm to use, one considers the following criteria:

(i) The algorithm should conserve energy and momentum and is reversible. When $\delta t \to -\delta t$ the system should go back to original state.
(ii) It should be computationally efficient.
(iii) It should permit a long time step for integration.
(iv) Only one force evaluation per time step (important for complex potential).

1.4.1 Verlet's Algorithm

The most widely used finite-difference method is a third-order Störmer algorithm first used by Verlet [14] and widely known as the Verlet's method. It is derived from the two Taylor expansion

$$\boldsymbol{r}(t+\delta t) = \boldsymbol{r}(t) + \delta t\, \boldsymbol{v}(t) + \frac{1}{2}\delta t^2\, \boldsymbol{a}(t) + \frac{1}{3!}\delta t^3\, \dot{\boldsymbol{a}}(t) + O(\delta t^4)\,, \quad (1.29)$$

$$\boldsymbol{r}(t-\delta t) = \boldsymbol{r}(t) - \delta t\, \boldsymbol{v}(t) + \frac{1}{2}\delta t^2\, \boldsymbol{a}(t) - \frac{1}{3!}\delta t^3\, \dot{\boldsymbol{a}}(t) + O(\delta t^4)\,, \quad (1.30)$$

summing the above two equations eliminates the odd-order terms. Rearranging gives

$$r(t+\delta t) + r(t-\delta t) = 2r(t) + \delta t^2 a(t) , \tag{1.31}$$
$$r(t+\delta t) = 2r(t) - r(t-\delta t) + \delta t^2\, a(t) + O(\delta t^4) . \tag{1.32}$$

Notice that the position vector r at time $t+\delta t$ is calculated from position vector at time t and $t-\delta t$, this makes the Verlet's algorithm a two-step method. Therefore it is not self-starting, initial positions $r(0)$ and velocities $v(0)$ are not sufficient to begin a calculation. Also the velocities are missing from the above equation and can be calculated from

$$v(t) = \frac{r(t+\delta t) - r(t-\delta t)}{2\delta t} . \tag{1.33}$$

In its original form it treats velocity as less important than positions. This approach is conflicting for ergodic system. The phase space trajectory depends equally on positions and velocities.

The local error (error per iteration) in position of the Verlet integrator is $O(\delta t^4)$ and local error in velocity is $O(\delta t^2)$. However the global error in position is $O(\delta t^2)$ and the global error in velocity is $O(\delta t^2)$.

Because the velocity is determined in a non-cumulative way from the positions in the Verlet integrator, the global error in velocity is also $O(\delta t^2)$. In molecular dynamics simulations, the global error is typically far more important than the local error, and the Verlet integrator is therefore known as a second-order integrator.

1.4.2 General Predictor-Corrector Algorithms

Predictor-corrector methods are composed of three steps: prediction, evaluation and correction. Starting from the current position $r(t)$ and velocity $v(t)$, the numerical steps are as follows.

(i) Predict the position $r(t+\delta t)$ and velocity $v(t+\delta t)$ at the end of the next step.
(ii) Evaluate the forces by taking the gradient of the potential at $\delta t + t$ using the predicted position. The difference in the calculated acceleration (this step) and the predicted acceleration (step 1) constitutes an error signal.
(iii) The error signal is used to correct the predictions using some combination of the predicted and previous values of position and velocity.

Using a Taylor series expansion to predict the system configuration at time $(t+\delta t)$ one gets

$$\begin{aligned} r(t+\delta t) &= r(t) + \delta t\, v(t) + \frac{1}{2}\delta t^2\, a(t) + \frac{1}{3!}\delta t^3\, b(t) + \dots , \\ v(t+\delta t) &= v(t) + \delta t\, a(t) + \frac{1}{2}\delta t^2\, b(t) + \dots , \\ a(t+\delta t) &= a(t) + \delta t\, b(t) + \dots , \\ b(t+\delta t) &= b(t) + \dots , \end{aligned} \tag{1.34}$$

where b is the time derivative of the acceleration a and is known at time t.

If the Taylor expansions are truncated, so that only the terms shown explicitly in (1.34) are left, then the quantities can be called the predicted values $\boldsymbol{r}^p$, $\boldsymbol{v}^p$, $\boldsymbol{a}^p$ and $\boldsymbol{b}^p$. The force is computed by taking the gradient of potential at the predicted position $\boldsymbol{r}^p$, and new acceleration value is computed. Since the predicted values are not based on physics the re-calculated acceleration is different from the predicted acceleration $\boldsymbol{a}^p$ (acceleration in (1.34)). The difference between the two values is called the error signal or error

$$\Delta\boldsymbol{a}(t+\delta t) = \boldsymbol{a}^c(t+\delta t) - \boldsymbol{a}^p(t+\delta t) \ . \tag{1.35}$$

This error signal is used to correct all predicted quantities in (1.34)

$$\begin{aligned} \boldsymbol{r}^c(t+\delta t) &= \boldsymbol{r}^p(t+\delta t) + c_0\Delta\boldsymbol{a}(t+\delta t) \ , \\ \boldsymbol{v}^c(t+\delta t) &= \boldsymbol{v}^p(t+\delta t) + c_1\Delta\boldsymbol{a}(t+\delta t) \ , \\ \boldsymbol{a}^c(t+\delta t) &= \boldsymbol{a}^p(t+\delta t) + c_2\Delta\boldsymbol{a}(t+\delta t) \ , \\ \boldsymbol{b}^c(t+\delta t) &= \boldsymbol{b}^p(t+\delta t) + c_3\Delta\boldsymbol{a}(t+\delta t) \ . \end{aligned} \tag{1.36}$$

All the corrected quantities are proportional to the error signal, and the proportional coefficients are determined to maximize the stability of the calculation. These corrected values are now better approximations of the true quantities, and are used to predict the quantities in the next iteration. The best choice for these coefficients depends on the order of both the differential equations and the Taylor series [53]. These coefficients are computed based on the order of the algorithm being used in the simulation. In addition, the accuracy of the numerical integrator algorithms also depends on the time step size, which is typically on the order of fractions of femtoseconds (10^{-15} s). Thus, the simulation as a whole is able to describe only short-time scale phenomena that last on the order of pico- (10^{-12}) up to nano-seconds (10^{-9} s).

1.4.3 Leap-Frog

In this algorithm, the velocities are first calculated at time $t + 1/2\delta t$; these are used to calculate the positions, $\boldsymbol{r}$, at time $t + \delta t$. In this way, the velocities leap over the positions, then the positions leap over the velocities. The advantage of this algorithm is that the velocities are explicitly calculated, however, the disadvantage is that they are not calculated at the same time as the positions. The velocities at time t can be approximated by the relationship

$$\boldsymbol{v}(t) = \frac{1}{2}\left[\boldsymbol{v}\left(t - \frac{1}{2}\delta t\right) + \boldsymbol{v}\left(t + \frac{1}{2}\delta t\right)\right] \ . \tag{1.37}$$

Therefore:

$$\boldsymbol{r}(t+\delta t) = \boldsymbol{r}(t) + \boldsymbol{v}\left(t + \frac{1}{2}\delta t\right)\delta t \ , \tag{1.38}$$

$$\boldsymbol{v}\left(t + \frac{1}{2}\delta t\right) = \boldsymbol{v}\left(t - \frac{1}{2}\delta t\right) + \boldsymbol{a}(t)\delta t \ . \tag{1.39}$$

1.4.4 Velocity Verlet

One starts with the following equations

$$\boldsymbol{r}(t+\delta t)=\boldsymbol{r}(t)+\delta t\,\boldsymbol{v}(t)+\frac{1}{2}\delta t^2\,\boldsymbol{a}(t)+\dots\,, \tag{1.40}$$

$$\boldsymbol{v}(t+\delta t)=\boldsymbol{v}(t)+\frac{1}{2}\delta t\left[\boldsymbol{a}(t)+\boldsymbol{a}(t+\delta t)\right]\,. \tag{1.41}$$

Each integration cycle consists of the following step:

(i) Calculate the velocities at mid-step using

$$\boldsymbol{v}\left(t+\frac{\delta t}{2}\right)=\boldsymbol{v}(t)+\frac{1}{2}\delta t\,\boldsymbol{a}(t)\,. \tag{1.42}$$

(ii) Calculate $\boldsymbol{r}(t+\delta t)$

$$\boldsymbol{r}(t+\delta t)=\boldsymbol{r}(t)+\boldsymbol{v}\left(t+\frac{\delta t}{2}\right)\delta t\,. \tag{1.43}$$

(iii) Calculate $\boldsymbol{a}(t+\delta t)$ from the potential.
(iv) Update the velocity using

$$\boldsymbol{v}(t+\delta t)=\boldsymbol{v}\left(t+\frac{\delta t}{2}\right)+\frac{1}{2}\delta t\,\boldsymbol{a}(t+\delta t)\,. \tag{1.44}$$

1.4.5 Beeman's Algorithm

The advantage of this algorithm is that it provides a more accurate expression for the velocities and better energy conservation. The disadvantage is that the more complex expressions make the calculation more expensive

$$\boldsymbol{r}(t+\delta t)=\boldsymbol{r}(t)+\delta t\,\boldsymbol{v}(t)+\frac{2}{3}\delta t^2\,\boldsymbol{a}(t)-\frac{1}{6}\delta t^2\,\boldsymbol{a}(t-\delta t)\,. \tag{1.45}$$

The predicted velocity is given by

$$\boldsymbol{v}(t+\delta t)=\boldsymbol{v}(t)+\frac{3}{2}\delta t\,\boldsymbol{a}(t)-\frac{1}{2}\delta t\,\boldsymbol{a}(t-\delta t)\,. \tag{1.46}$$

The acceleration is based on the predicted velocity

$$\boldsymbol{a}(t+\delta t)=F\left(\{\boldsymbol{r_i}(t+\delta t),\boldsymbol{v_i}(t+\delta t)\},\ i=1,2...n\right)\,, \tag{1.47}$$

where $\boldsymbol{v_i}$ is the predicted velocity from the previous equation. The corrected velocity is given by

$$\boldsymbol{v}(t+\delta t)=\boldsymbol{v}(t)+\frac{1}{3}\delta t\,\boldsymbol{a}(t+\delta t)+\frac{5}{6}\delta t\,\boldsymbol{a}(t)-\frac{1}{6}\delta t\,\boldsymbol{a}(t-\delta t)\,. \tag{1.48}$$

1.4.6 Gear Algorithm

The fifth-order Gear predictor-corrector method [53] predicts the molecular position $\boldsymbol{r}_i$ at time $t+\delta t$ using fifth-order Taylor series based on position and their derivatives at time t. It is particularly useful for stiff differential equations.

1.4.7 Symplectic Integrators

Symplectic integrators are designed for the numerical solution of Hamiltonian's equation of motion. They preserve Poincaré invariants when integrating classical trajectories (see [54] and earlier references therein). The Hamiltonian which is slightly perturbed from the original value is conserved. This approach has the big advantage, that it guarantees and preserves conservation laws.

1.5 Analysis of MD Runs

In this section we will describe how the output of MD simulations (positions and velocities) are analysed to get the physical quantities of interest.

1.5.1 Ergodic Hypothesis

To calculate a physical quantity A in molecular dynamics, it is calculated as the time average of A

$$\langle A\rangle_{\text{time}} = \lim_{\mathcal{T}\to\infty}\frac{1}{\mathcal{T}}\int_{t=0}^{\mathcal{T}} A\left(\boldsymbol{p}^N(t),\boldsymbol{r}^N(t)\right)\,\mathrm{d}t \approx \frac{1}{M}\sum_{t=1}^{M} A\left(\boldsymbol{p}^N,\boldsymbol{r}^N\right)\,, \tag{1.49}$$

where t is the simulation time, M is the number of time steps in the simulation and $A(\boldsymbol{p}^N,\ \boldsymbol{r}^N)$ is the instantaneous value of A. This integral is generally extremely difficult to calculate because one must calculate all possible states of the system.

In statistical mechanics experimental observables are assumed to be ensemble averages

$$\langle A\rangle_{\text{ensemble}} = \iint \mathrm{d}\boldsymbol{p}^N \mathrm{d}\boldsymbol{r}^N A\left(\boldsymbol{p}^N,\boldsymbol{r}^N\right)\rho\left(\boldsymbol{p}^N,\boldsymbol{r}^N\right)\,, \tag{1.50}$$

where $A(\boldsymbol{p}^N,\ \boldsymbol{r}^N)$ is the observable of interest, $\rho(\boldsymbol{p}^N,\ \boldsymbol{r}^N)$ is the probability density of the ensemble. The integration is carried over all possible values of position $\boldsymbol{r}$ and momenta $\boldsymbol{p}$. The *ergodic hypothesis*, which states that the time average equals the ensemble average

$$\langle A\rangle_{\text{time}} = \langle A\rangle_{\text{ensemble}}\,, \tag{1.51}$$

The basic idea is that if one allows the system to evolve in time indefinitely, then the system will eventually pass through all possible states. One goal, therefore, of a molecular dynamics simulation is to generate enough representative conformations such that this equality is satisfied. If this is the case, experimentally relevant information concerning structural, dynamic and thermodynamic properties may then be calculated using a feasible amount of computer resources. Because the simulations are of fixed duration, one must be certain to sample a sufficient amount of phase space.

1.5.2 Standard Diagnostics

There are a number of different physical quantities which one may be interested in. For a liquid, these may be liquid structure factors, transport coefficients (eg. diffusion coefficient, viscosity or thermal conductivity) etc. For solids, these may be crystal structure, adsorption of molecules on surface, melting behaviour etc. Here, we will consider the diagnostics methods to calculate internal energy, pressure tensor, self-diffusion coefficient and pair distribution function. More details are described in [55, 56].

1.5.2.1 Energy

The energy is the simplest and most straightforward quantity to calculate. From all pair of atoms (i, j), one calculates their separation $\boldsymbol{r}_{ij}$. These are then substituted into the chosen form of potential $U(r)$. The energy has contributions from both potential and kinetic terms. The kinetic energy should be calculated after the momenta p have been updated, i.e., after the force routine has been called. The kinetic energy can then be calculated, and then added to the potential energy

$$\langle E \rangle = \langle H \rangle = \langle K \rangle + \langle U \rangle = \left\langle \sum_i \frac{|p_i|^2}{2m_i} \right\rangle + \langle U(r) \rangle \,. \tag{1.52}$$

$U(r)$ is obtained directly from the potential energy calculations. For calculating avarage temperature

$$E_{\text{kin}} = \langle K \rangle = \frac{3}{2} N k_B T \;\Rightarrow\; T = \frac{1}{3N k_{\text{B}}} \left\langle \sum_{i=1}^{N} \frac{|p_i|^2}{m_i} \right\rangle \,. \tag{1.53}$$

1.5.2.2 Pressure

Pressure is a second rank tensor. For inhomogeneous systems, one calculates this tensor by finding the force across potential surfaces [57]. However, for homogeneous systems, it is not the most efficient method and one uses the virial theorem to calculate the configurational part of the pressure tensor, and then add that to the kinetic part. For the derivation of the virial theorem one can refer to [58]. The full expression for the pressure tensor of a homogeneous system of particles is given as

$$P(r,t) = \frac{1}{V}\left[\sum_{i=1}^{N} m_i v_i(t) v_i(t) + \sum_{i=1}^{N}\sum_{j>i}^{N} r_{ij}(t) F_{ij}(t)|_{r_i(t)=r}\right] , \quad (1.54)$$

where V is the volume, m_i, v_i are the mass and velocity of particle i respectively. The first term represents the kinetic contribution and the second term represents the configurational part of the pressure tensor. It is clear that the interaction between the pairs is calculated just once. Note that the above equation is valid for atomic systems at equilibrium, system of molecules require some modifications to be made, as do non-equilibrium systems.

1.5.2.3 Pair Correlation Function

The static properties of the system e.g. structure, energy, pressure etc. are obtained from the pair (or radial) correlation function. Pair correlation function, $g(r)$, gives the information on the structure of the material. It gives the probability of locating pairs of atoms separated by a distance r, relative to that for a completely random distribution at the same density (i.e. the ideal gas). For a crystal, it exhibits a sequence of peaks at positions corresponding to shells around a given system. For amorphous materials and liquid, $g(r)$ exhibits its major peak close to the average atomic separation of neighboring atoms, and oscillates with less pronounced peaks at larger distances. The magnitude of the peaks usually decays exponentially with distance as $g(r) \to 1$. In most cases, $g(r)$ vanishes below a certain distance where atomic repulsion is strong enough to prevent pairs of atoms from getting too close.

It is defined as

$$g(r) = \frac{V}{N^2}\left\langle \sum_{i=1}^{N}\sum_{j\neq i}^{N} \delta(r - r_{ij}) \right\rangle . \quad (1.55)$$

In a computer simulation, the delta function is replaced by a function that is finite (say, given a value 1) over a small range of separations, and a histogram is accumulated ove time of all pair separations that fall within this range. $g(r)$ is effectively a measure of structural properties, but is particularly important because all thermodynamic quantities may be expressed as some function of it [56, 59].

1.5.2.4 Time Correlation Function

The dynamic and transport properties of the system are obtained from time correlation functions. Any Transport coefficient K can be calculated using generalized Einstein and Green-Kurbo Formulas [60]

$$K(t) = \lim_{t\to\infty} \frac{\left\langle [A(t) - A(0)]^2 \right\rangle}{2t} = \int_0^\infty d\tau \left\langle \dot{A}(\tau)\dot{A}(0) \right\rangle . \quad (1.56)$$

If one wants to calculate the self diffusion coefficient then $A(t) = \boldsymbol{r}_i(t)$ is the atom position at time t and $\dot{A} = \boldsymbol{v}_i(t)$ is the velocity of the atom. For calculating the shear viscosity, $A(t) = \sum m_i \boldsymbol{r}_i(t)\boldsymbol{v}_i(t)$ and $\dot{A} = \boldsymbol{\sigma}_{\alpha\beta}$. Other transport quantities can also be calculated similarly. If we compare the value of $A(t)$ with its value at zero time, $A(0)$ the two values will be correlated at sufficiently short times, but at longer times the value of $A(t)$ will have no correlation with its value at $t = 0$. Information on relevant dynamical processes is contained in the time decay of $K(t)$. Time correlation function can be related to the experimental spectra by a fourier transformation.

1.5.2.5 Diffusion Coefficients

As discussed above, we obtain diffusion coefficient using the Einstein relation

$$D = \lim_{t \to \infty} \frac{\langle [r(t) - r(0)]^2 \rangle}{2dt}, \tag{1.57}$$

where D is the diffusion coefficient, d is the dimensionality of the system and $r(t)$ is the position of atom at time t. Angle brackets represents averaging over all possible time origins (see [56] for more information). This is proportional to the slope of the mean square displacement of a single particle undergoing Brownian motion at the long time limit.

Warrier et al. [61] analysed the diffusion of hydrogen atoms in porous graphite. They found, that different length scales for jumps are present in the system. J. Klafter et al. [62] talk about random walk that are *sub-diffusive* (wherein the trajectory results in a mean square displacement that shows slower-than-linear growth with time), and *super-diffusive* (wherein the trajectory results in a mean square displacement that shows faster-than-linear growth with time). Such random walks are called Lévy flights and can show up super-diffusive behaviour with infinite variance and their trajectories show self-similar patterns characteristics of fractals.

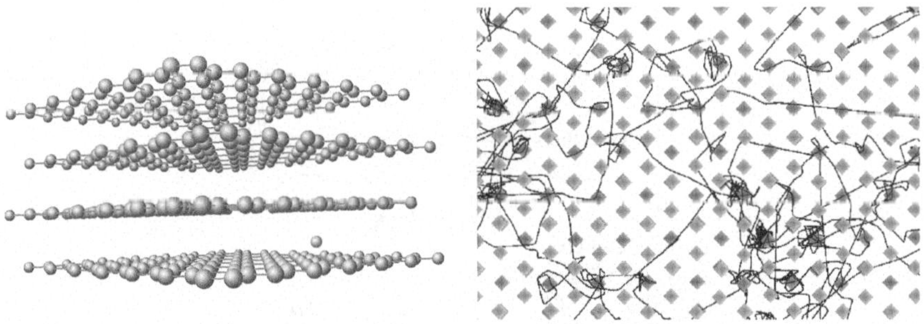

Fig. 1.1. Left: One hydrogen atom in a carbon lattice. **Right**: Diffusion paths at 900 K for a hydrogen atom in graphite. Small frequent jumps and rare large jumps are visible

1.5.3 Multi-Scale Modeling

Multi-Scale modeling is the field of solving physical problems which have important features at multiple scales, particularly multiple spatial and temporal scales. As an example, the problem of protein folding has multiple time scales. While the time scale for the vibration of the covalent bonds is of the order of femtoseconds (10^{-15} s), folding time for proteins may very well be of the order of seconds. Well-known examples of problems with multiple length scales include turbulent flows, mass distribution in the universe, and vortical structures on the weather map [63]. In addition, different physical laws may be required to describe the system at different scales. Take the example of fluids. At the macroscale (meters or millimeters), fluids are accurately described by the density, velocity and temperature fields, which obey the continuum Navier-Stokes equations. On the scale of mean free path, it is necessary to use kinetic theory (Boltzmann equations) to get a more detailed description in the terms of the one-particle phase-space distribution function. At the nanometer scale, molecular dynamics in the form of Newton's law has to be used to give the actual position and velocity of each individual atom that makes up the fluid. If a liquid such as water is used as the solvent for protein folding, then the electronic structure of the water molecules becomes important and these are described by Schrödinger's equation in quantum mechanics. The boundaries between different levels of theories may vary, depending on the system being studied, but the overall trend described above is generally valid. At each finer scale a more detailed theory has to be used, giving rise to more detailed information on the system. Warrier et al.

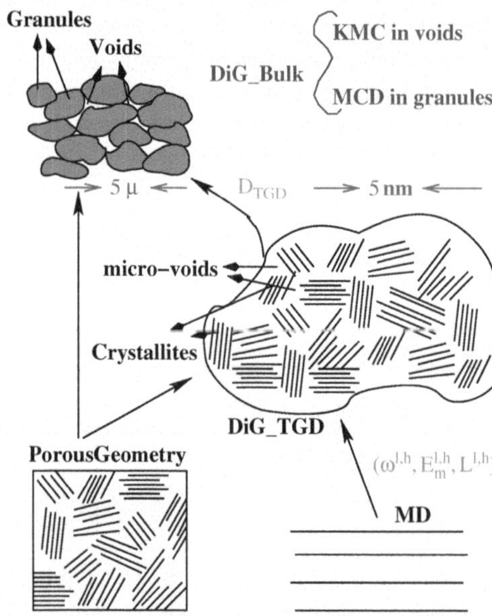

Fig. 1.2. Multi-scale modeling approach for diffusion of hydrogen in porous graphite

[61] have done a multi-scale analysis of the diffusion of hydrogen isotope in porous graphite. They used the insight gained from microscopic models (consisting of a few hundreds of atoms over a time scale of a few picoseconds and length scale of nanometersusing MD) into modeling the hydrogen isotope reactions and transports at meso-scale (trans-granular diffusion, with length scales of few microns) and further into the macro-scale (typically a centimeter over a time scale of milliseconds). Therefore a multi-scale (both in length and time) approach to modeling plasma surface interaction is necessary. The figure below explains the multi-scale modeling approach clearly.

1.6 From Classical to Quantum-Mechanical MD

Classical molecular dynamics using predefined potentials is well established as a powerful tool to investigate many-body condensed matter systems. The broadness, diversity, and level of sophistication of this technique is documented in several monographs as well as proceedings of conferences and scientific schools [56, 64, 65, 66, 67, 68, 69]. At the very heart of any molecular dynamics scheme is the question of how to describe, that is in practice how to approximate, the interatomic interactions. The traditional route followed in molecular dynamics is to determine these potentials in advance. Typically, the full interaction is broken up into two-body, three-body and many-body contributions, long-range and short-range terms etc., which have to be represented by suitable functional forms, discussed under the inter-atomic potentials section of this article. After decades of intense research, very elaborate interaction models including the non-trivial aspect to represent them analytically were devised [70, 71, 72].

Despite overwhelming success of the pre-calculated potentials, the *fixed model potential* implies serious drawbacks. Among the most delicate ones are systems where

(i) many different atom or molecule types give rise to a myriad of different interatomic interactions that have to be parameterized and/or
(ii) the electronic structure and thus the bonding pattern changes qualitatively in the course of the simulation.

These systems can be called *chemically complex*.

The reign of traditional molecular dynamics and electronic structure methods was greatly extended by the family of techniques that is called here *ab initio* molecular dynamics. Other names that are currently in use are for instance Car-Parrinello, Hellmann-Feynman, First principles, quantum chemical, on-the-fly, direct, potential-free, quantum, etc. molecular dynamics. The basic idea underlying every ab initio molecular dynamics method is to compute the forces acting on the nuclei from electronic structure calculations that are performed *on-the-fly* as the molecular dynamics trajectory is generated. In this way, the electronic variables are not integrated out before-hand, but are considered as active degrees of freedom. This implies that, given a suitable approximate solution of the many-electron problem, also *chemically complex* systems can be handled by molecular dynamics. But this also implies that the approximation is shifted from the level of selecting the model potential to the level of selecting a particular approximation for solving the Schrödinger equation.

1.7 Ab Initio MD

In this approach, a global potential energy surface is constructed in a first step either empirically or based on electronic structure calculations. In a second step, the dynamical evolution of the nuclei is generated by using classical mechanics, quantum mechanics or semi/quasiclassical approximations of various sorts.

Suppose that a useful trajectory consists of about 10^M molecular dynamics steps, i.e. 10^M electronic structure calculations are needed to generate one trajectory. Furthermore, it is assumed that 10^n independent trajectories are necessary in order to average over different initial conditions so that 10^{M+n} *ab initio* molecular dynamics steps are required in total. Finally, it is assumed that each single-point electronic structure calculation needed to devise the global potential energy surface and one ab initio molecular dynamics time step requires roughly the same amount of CPU time. Based on this truly simplistic order of magnitude estimate, the advantage of *ab initio* molecular dynamics vs. calculations relying on the computation of a global potential energy surface amounts to about $10^{3N+6+M+n}$. The crucial point is that for a given statistical accuracy (that is for M and n fixed and independent on N) and for a given electronic structure method, the computational advantage of *on-the-fly* approaches grows like 10^N with system size. Of course, considerable progress has been achieved in trajectory calculations by carefully selecting the discretization points and reducing their number, choosing sophisticated representations and internal coordinates, exploiting symmetry etc. but basically the scaling 10^N with the number of nuclei remains a problem. Other strategies consist for instance in reducing the number of active degrees of freedom by constraining certain internal coordinates, representing less important ones by a (harmonic) bath or friction, or building up the global potential energy surface in terms of few-body fragments. All these approaches, however, invoke approximations beyond the ones of the electronic structure method itself. Finally, it is evident that the computational advantage of the *on-the-fly* approaches diminish as more and more trajectories are needed for a given (small) system. For instance extensive averaging over many different initial conditions is required in order to calculate quantitatively scattering or reactive cross sections.

A variety of powerful *ab initio* molecular dynamics codes have been developed, few of them listed here are CASTEP [73], CP-PAW [74], fhi98md [75], NWChem [76], VASP [77], GAUSSIAN [78], MOLPRO [79] and ABINIT [80, 81].

1.8 Car-Parrinello Molecular Dynamics

The basic idea of the Car-Parrinello [4] approach can be viewed to exploit the quantum-mechanical adiabatic time-scale separation of fast electronic and slow nuclear motion by transforming that into classical-mechanical adiabatic energy-scale separation in the framework of dynamical systems theory. In order to achieve this goal the two-component quantum/classical problem is mapped onto a two-component purely classical problem with two separate energy scales at the expense of loosing the explicit time-dependence of the quantum subsystem dynamics.

Car and Parrinello postulated the following class of Lagrangians [4] to serve this purpose

$$L_{\mathrm{CP}} = \underbrace{\sum_I \frac{1}{2} M_I \dot{R}_I^2 + \sum_i \frac{1}{2}\mu_i \langle \dot{\psi}_i | \dot{\psi}_i \rangle}_{\text{normal kinetic energy}} - \underbrace{\langle \Psi_0 | H_e | \Psi_0 \rangle}_{\text{potential energy}} + \underbrace{\text{constraints}}_{\text{orthonormality}} , \quad (1.58)$$

where μ_i $(= \mu)$ are the *fictitious masses* or inertia parameters assigned to the orbital degrees of freedom; the units of the mass parameter μ are energy times a squared time for reasons of dimensionality. ψ_i are regarded as classical fields, M_I are the ionic masses. The potential energy in the Car-Parrinello Lagrangian can be written as

$$\langle \Psi_0 | H_e | \Psi_0 \rangle = E_{\mathrm{KS}} \left[\{\psi_i\}, \boldsymbol{R}_I \right] , \quad (1.59)$$

E_{KS} is the LDA-KS energy functional. Within the pseudopotential implementation of the local density approximation (LDA) in the Kohn-Sham (KS) scheme, the ionic potential energy corresponding to the electron in the ground state can be found by minimizing the KS total-energy functional $E_{\mathrm{KS}} \left[\{\psi_i\}, \boldsymbol{R}_I \right]$ with respect to the one-particle wavefunction $\psi_i(\boldsymbol{r})$ describing the valence-electron density subject to orthonormalization constraints. The explicit expression of E_{KS} in terms of orthonormal one-particle orbitals $\psi_i(\boldsymbol{r})$ is

$$\begin{aligned} &E_{\mathrm{KS}} \left[\{\psi_i(\boldsymbol{r})\}, \{\boldsymbol{R}_I\} \right] \\ &= \sum_i f_i \int \psi_i^*(\boldsymbol{r}) \left(-\frac{1}{2} \nabla^2 \right) \psi_i(\boldsymbol{r}) \mathrm{d}\boldsymbol{r} + \frac{1}{2} \iint \frac{\rho(\boldsymbol{r}_1)\rho(\boldsymbol{r}_2)}{|\boldsymbol{r}_1 - \boldsymbol{r}_2|} \mathrm{d}\boldsymbol{r}_1 \mathrm{d}\boldsymbol{r}_2 \\ &+ \int \epsilon_{\mathrm{XC}} \left(\rho(\boldsymbol{r}) \right) \rho(\boldsymbol{r}) \mathrm{d}\boldsymbol{r} + E_{\mathrm{eI}} \left(\left[\psi_i(\boldsymbol{r}) \right], \{\boldsymbol{R}_I\} \right) + U_I^0 \left(\{\boldsymbol{R}_I\} \right) . \end{aligned} \quad (1.60)$$

The terms on the right-hand side of the previous equation are, respectively, the electronic kinetic energy, the electrostatic Hartree term, the integral of the LDA exchange and correlation energy density ϵ_{XC}, the electron-ion pseudopotential interaction, and the ion-ion interaction potential energy. The electronic density $\rho(\boldsymbol{r})$ is given by

$$\rho(\boldsymbol{r}) = \sum_i f_i |\psi_i(\boldsymbol{r})|^2 , \quad (1.61)$$

where f_i are occupation numbers.

The corresponding Newtonian equations of motion are obtained from the associated Euler-Lagrange equations

$$\frac{\mathrm{d}}{\mathrm{d}t}\frac{\partial L}{\partial \dot{\boldsymbol{R}}_I} = \frac{\partial L}{\partial \boldsymbol{R}_I}\,, \tag{1.62}$$

$$\frac{\mathrm{d}}{\mathrm{d}t}\frac{\delta L}{\delta \dot{\psi}_i^*} = \frac{\delta L}{\delta \psi_i^*} \tag{1.63}$$

like in classical mechanics, but here for both the nuclear positions and the orbitals; note $\psi_i^* = \langle\psi_i|$ and that the constraints are holonomic (which can be expressed in the form $f(\boldsymbol{r}_1, \boldsymbol{r}_2, ..., t) = 0$). Following this route of ideas, generic Car-Parrinello equations of motion are found to be of the form

$$M_I \ddot{\boldsymbol{R}}_I(t) = -\frac{\partial}{\partial \boldsymbol{R}_I}\langle\Psi_0|H_e|\Psi_0\rangle + \frac{\partial}{\partial \boldsymbol{R}_I}\{\text{constraints}\}\,, \tag{1.64}$$

$$\mu_i \ddot{\psi}_i(t) = -\frac{\delta}{\delta \psi_i^*}\langle\Psi_0|H_e|\Psi_0\rangle + \frac{\delta}{\delta \psi_i^*}\{\text{constraints}\}\,. \tag{1.65}$$

Note that the constraints within the total wavefunction lead to *constraint forces* in the equations of motion. Note also that these constraints might be a function of both the set of orbitals $\{\psi_i\}$ and the nuclear positions $\{\boldsymbol{R}_I\}$. These dependencies have to be taken into account properly in deriving the Car-Parrinello equations following from (1.58) using (1.62) and (1.63).

According to the Car-Parrinello equations of motion, the nuclei evolve in time at a certain (instantaneous) physical temperature $\propto \sum_I M_I \dot{\boldsymbol{R}}_I^2$, whereas a *fictitious temperature* $\propto \sum_i \mu_i \langle\dot{\psi}_i|\dot{\psi}_i\rangle$ is associated to the electronic degrees of freedom. In this terminology, *low electronic temperature* or *cold electrons* means that the electronic subsystem is close to its instantaneous minimum energy $min_{\{\psi_i\}}\langle\Psi_0|H_e|\Psi_0\rangle$ i.e. close to the exact Born-Oppenheimer (BO) surface. Thus, a ground-state wavefunction optimized for the initial configuration of the nuclei will stay close to its ground state also during time evolution if it is kept at a sufficiently low temperature. The remaining task is to separate in practice nuclear and electronic motion such that the fast electronic subsystem stays cold also for long times but still follows the slow nuclear motion adiabatically (or instantaneously). Simultaneously, the nuclei are nevertheless kept at a much higher temperature. This can be achieved in nonlinear classical dynamics via decoupling of the two subsystems and (quasi-)adiabatic time evolution. This is possible if the power spectra stemming from both dynamics do not have substantial overlap in the frequency domain so that energy transfer from the *hot nuclei* to the *cold electrons* becomes practically impossible on the relevant time scales. This amounts in other words to imposing and maintaining a metastability condition in a complex dynamical system for sufficiently long times.

The Hamiltonian or conserved energy is the constant of motion (like classical MD, with relative variations smaller than 10^{-6} and with no drift), which serves as an extremely sensitive check of the molecular dynamics algorithm. Contrary to that the electronic energy displays a simple oscillation pattern due to the simplicity of the phonon modes. Most importantly, the fictitious kinetic energy of the electrons is found to perform bound oscillations around a constant, i.e. the electrons *do not heat up* systematically in the presence of the hot nuclei.

As we have seen above, Car-Parrinello method gives physical results even if the orbitals are not at the BO surface, provided that the electronic and ionic degrees of freedom remain adiabatically separated and electrons remain close to the BO surface. Loss of adiabacity would mean that there is transfer of energy from *hot nuclei* to *cold electron* and Car-Parrinello MD deviates from BO surface.

1.8.1 Adiabaticity

The metastable two-temperature regime setup in the CP dynamics is extremely efficient at approximating the constraints of maintaining the electronic energy functional at the minimum without explicit minimization. At the beginning of the numerical simulation, the electronic subsystem is in an initial state which is very close to the minimum of the energy surface. When the ions start moving, their motion causes a change in the instantaneous position of the minimum in the electronic parameter space. The electrons experience restoring forces and start moving. If they start from a neighborhood of a stable equilibrium position, there will be range of initial velocities such that a regime of small oscillations is originated.

A simple harmonic analysis of the frequency spectrum of the orbital classical fields close to the minimum defining the ground state yields [82]

$$\omega_{ij} = \left(\frac{2(\epsilon_i - \epsilon_j)}{\mu}\right)^{1/2}, \tag{1.66}$$

where ϵ_j and ϵ_i are the eigen values of occupied and unoccupied orbitals, respectively. The analytic estimate for the lowest possible electronic frequency

$$\omega_e^{\min} \propto \left(\frac{E_{\text{gap}}}{\mu}\right)^{1/2} \tag{1.67}$$

shows that this frequency increases like the square root of the electronic energy difference E_{gap} between the lowest unoccupied and the highest occupied orbital. On the other hand it increases similarly for a decreasing fictitious mass parameter μ. Since the parameters E_{gap} and the maximum phonon frequency ($\omega_n^{\max}$) are dictated by physics, the only parameter in our hands to control adiabatic separation is the fictitious mass, which is therefore also called *adiabaticity parameter*. However, decreasing μ not only shifts the electronic spectrum upwards on the frequency scale, but also stretches the entire frequency spectrum according to (1.66). This leads to an increase of the maximum frequency according to

$$\omega_e^{\max} \propto \left(\frac{E_{\text{cut}}}{\mu}\right)^{1/2}, \tag{1.68}$$

where E_{cut} is the largest kinetic energy in an expansion of the wavefunction in terms of a plane wave basis set. Limitation to decrease arbitrarily kicks in due to the maximum length of the molecular dynamics time step $\Delta t^{\max}$ that can be used. The time step is inversely proportional to the highest frequency in the system, which is $\omega_e^{\max}$ and thus the relation

$$\Delta t^{\max} \propto \left(\frac{\mu}{E_{\text{cut}}} \right)^{1/2} . \tag{1.69}$$

In the limit, when, electronic gap is very small or even vanishes $E_{\text{gap}} \rightarrow 0$ as is the case for metallic systems, all the above-given arguments break down due to the occurrence of zero-frequency electronic modes in the power spectrum according to (1.67), which necessarily overlap with the phonon spectrum. It has been shown that the coupling of separate Nosé-Hoover thermostats [68, 69, 83] to the nuclear and electronic subsystem can maintain adiabaticity by counterbalancing the energy flow from ions to electrons so that the electrons stay *cool* [84]; see [85] for a similar idea to restore adiabaticity. Although this method is demonstrated to work in practice [86], this ad hoc cure is not entirely satisfactory from both a theoretical and practical point of view so that the well-controlled Born-Oppenheimer approach is recommended for strongly metallic systems.

1.9 Potential Energy Surface

In the past two decades, or so, there have been dramatic improvements in both the accuracy and efficiency of high-level electronic structure calculations [87, 88, 89, 90]. These advances, along with the increasing speed of modern computers have made possible very high-quality ab initio calculations for small polyatomic systems [91, 92]. For three- and four-atom systems, calculations with errors less than 1 kcal/mol are feasible. Gradients and Hessians are also becoming widely available. However, many uses of this vast supply of data require that it be re-expressed with a suitable local or global representation as a potential energy surface (PES). Since the inception of quantum mechanics, considerable effort has been devoted to finding better ways of utilizing ab initio data and/or experimental data to construct PES. The earliest and most common methods involve least-squares fitting to empirical or semi-empirical functional forms [71, 93, 94]. This approach is mature and well understood, although sophisticated schemes involving complex functional forms continue to evolve. During the past decade, generic multivariate interpolation techniques have gathered attention as alternatives to complicated functional forms [95, 96, 97, 98, 99]. The goal of these methods is to produce a general framework for constructing PESs that will reduce the effort and expertise required to turn high-level calculations into usable surfaces.

Another solution is to skip the surface construction step entirely and to use the ab initio results directly in dynamical studies [100, 101, 102]. However, such direct dynamics techniques are inherently classical trajectory approaches and require tens of ab initio calculations for dynamically significant trajectories, and thus, this approach is limited by the available electronic structure calculation techniques. Its application has been restricted to cases in which modest potential quality seems sufficient and in which discrete spectral lines or state-selected dynamics are not required, as in rate constant calculations based on classical trajectories [103] or in transition state theory [104, 105]. In contrast, the highest-accuracy ab initio calculations can take hours or more of computer time, even for small systems. Another obstacle for *on-the-fly* calculations of ab initio energies is the failure or non-convergence of the ab initio method. One frequently comes across this problem when the nuclear configurations are in a state for which the selected ab initio method fails. This is seen in particular for dissociating molecules. The absence of ab initio energy on the surface can be treated as *hole* in the surface and can be corrected on the pre-calculated surface. Moreover, carefully adding the ab initio fragment data for the dissociating molecule allows to study reaction dynamics on high quality surface. Thus, the construction of accurate analytic representations of PES is a necessary step in full quantum spectroscopic and dynamics studies.

The number of high-level ab initio data points currently needed for adequate sampling of dynamically significant regions typically ranges from several hundred to several thousand points for tri- and tetra-atomic systems. Methods that use derivatives typically use fewer configurations; however, the number of pieces of information is typically in the same range [106, 107, 108, 109, 110, 111, 112, 113].

In constructing the PES the prescribed functional form must be carefully crafted so that it

(i) does not introduce arbitrary features,
(ii) achieves the required smoothness,
(iii) preserves any necessary permutation symmetry, and
(iv) agrees with any known asymptotic form of the underlying PES.

An analytic fit that has a residual significantly larger than the error in the high level ab initio calculations is only marginally more useful than if a lower-level calculation is employed. High-quality ab initio calculations demand representations that preserve their level of accuracy.

One such method named, Reproducing kernel Hilbert space (RKHS) was introduced by Hollebeek et al. [114]. Several other examples of carefully crafted analytic representations are listed in [114].

1.10 Advanced Numerical Methods

A system consisting of N particles in which the particles interact through forces with a cutoff distance R_c, each particle feels the forces from $N_c \propto \rho R_c^3$ neighbors. CPU time required to advance the system one time step δt is proportional to the number of forces calculated, $NN_c/2$. Clearly the simulation time grows as the cube of the cutoff distance. A frequently encountered problem in molecular dynamics is how to treat the long times that are required to simulate condensed systems consisting of particles interacting through long range forces. Standard methods require the calculation of the forces at every time step. Because each particle interacts with all particles within the interaction range of the potential the longer the range of the potential the larger the number forces that must be calculated at each time step.

1.10.1 Ewald Summation Method

The Ewald summation is the method of choice to compute electrostatic interactions in systems with periodic boundary conditions [56]. It avoids all problems associated with the use of a cut-off radius and there is no need for switching or shifting functions. Lennard-Jones interactions are calculated normally; due to their shorter range the errors are normally negligible. The Ewald sum consists of a short-range term that is computed in normal space (r-part) and a second term, the k-sum, that is calculated in Fourier-space (k-space). A parameter, usually labeled κ or η, controls the relationship between the two parts. Its value should be chosen so that the r-part interaction between a pair of particles is zero at the cut-off distance, which is still used although it is more a formal parameter in Ewald summation. The more one dampens the r-part (and thus shortens the computer time required for its calculation), the more costly the calculation of the k-sum becomes. Even highly optimized computer codes for the Ewald sum are, therefore, slower than cut-off based methods. If one does not make an error in the choice of η (κ) vs. the cut-off distance and includes enough terms in the k-sum, the calculation of the electrostatic energy using the Ewald summation is exact.

1.10.1.1 Minimum Image

The simulation region or *cell* is effectively replicated in all spatial directions, so that particles leaving the cell reappear at the opposite boundary. For systems governed by a short-ranged potential – say Lennard-Jones or hard spheres – it is sufficient to take just the neighbouring simulation volumes into account, leading to the *minimum-image* configuration shown in Fig. 1.3.

The potential seen by the particle at $\boldsymbol{r}_i$ is summed over all other particles $\boldsymbol{r}_j$, or their periodic images $(\boldsymbol{r}_j \pm \boldsymbol{n})$, where $\boldsymbol{n} = (i_{\hat{x}}, i_{\hat{y}}, i_{\hat{z}})L$, with $i_\alpha = 0, \pm1, \pm2, \pm3 ... \pm \infty$ whichever is closest. L denotes the length of the simulation box. More typically, this list is further restricted to particles lying within a sphere centred on r_i^6. For long-range potentials, this arrangement is inadequate because the contributions from more distant images at $2L, 3L$ etc., are no longer negligible.

1.10.1.2 Ewald Summation

One is faced with the challenge of arranging the terms in the potential energy equation so that the contribution from oppositely charged pairs of charges cancel and the summation series converges, and preferably as fast as possible.

A way to achieve this is to add image cells radially outwards from the origin as shown in Fig. 1.4 (this is to build up sets of images contained within successively larger spheres surrounding the simulation region).

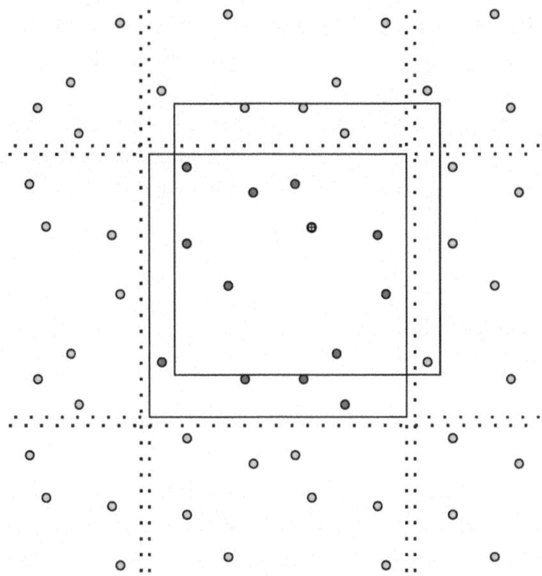

Fig. 1.3. Periodic boundary conditions for simulation region (centre, *dark-shaded particles* at positions $\boldsymbol{r}_j$), showing *minimum-image* box for reference ion $\oplus$ at position $\boldsymbol{r}_i$ containing nearest periodic images (*lightshaded particles* at positions $\boldsymbol{r}_j \pm \boldsymbol{n}$)

For the above scheme the potential at r_i due to charges at r_j and image cells is

$$V_s(\boldsymbol{r}_i) = \sum_{\boldsymbol{n}}^{i}{}' \sum_{j=1}^{N} \frac{q_j}{|\boldsymbol{r}_{ij} + \boldsymbol{n}|} , \tag{1.70}$$

where $\boldsymbol{r}_{ij} = \boldsymbol{r}_i - \boldsymbol{r}_j$, $\boldsymbol{n}$ and i_α is same as above. The prime in the summation over $\boldsymbol{n}$ indicates that the term $j = i$ is omitted for the primary cell $\boldsymbol{n} = 0$. Taking the image cells in the order perscriped by Fig. 1.4 ensures that the sum in (1.70) converges

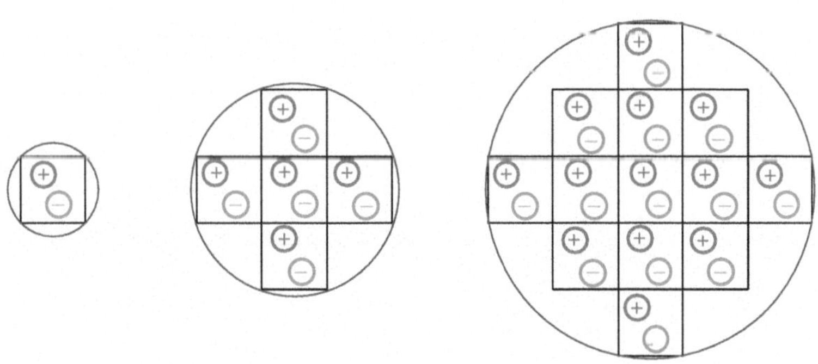

Fig. 1.4. Constructing a convergent sum over periodic images (adapted from Allen & Tildesley)

to the correct value, but only slowly. The summation over the boxes as shown in (1.70) is computionally expensive for N-body problem. The $O(N^2)$ is turned into a $N_{\text{box}} \times N^2$ operation problem.

Ewald's idea got around this problem by recasting the potential equation into sum of two rapidly converging series, one in real space and one in the reciprocal k-space. Consider the simple Gaussian distribution originally used by Ewald himself

$$\sigma(r) = \frac{\alpha^3}{\pi^{3/2}}\, \mathrm{e}^{-\alpha^2 r^2} , \tag{1.71}$$

which is normalized such that

$$\int_0^\infty \sigma(r) \mathrm{d}r = 1 . \tag{1.72}$$

Note that α determines the height and width of the effective size of the charges (called spreading function). To obtain the real-space term depicted in Fig. 1.5, we just subtract the lattice sum for the smeared out charges fom the original point-charge sum, thus

$$\begin{aligned} V_r(\boldsymbol{r}_i) &= \sum_{\boldsymbol{n}}{}' \sum_{j=1}^{N} \frac{q_j}{|\boldsymbol{r}_{ij} + \boldsymbol{n}|} \left[1 - \int_0^\infty \sigma(r - r_{ij}) \mathrm{d}^3 r \right] \\ &= \sum_{\boldsymbol{n}}{}' \sum_{j} q_j \left[\frac{1}{|\boldsymbol{r}_{ij} + \boldsymbol{n}|} - \frac{4\alpha^3}{\pi^{1/2} |\boldsymbol{r}_{ij} + \boldsymbol{n}|} \int_0^{|\boldsymbol{r}_{ij}+\boldsymbol{n}|} r^2\, \mathrm{e}^{-\alpha^2 r^2} \mathrm{d}r \right. \\ &\quad \left. - \frac{4\alpha^3}{\pi^{1/2}} \int_{|\boldsymbol{r}_{ij}+\boldsymbol{n}|}^{\infty} r\, \mathrm{e}^{-\alpha^2 r^2} \mathrm{d}r \right] . \end{aligned} \tag{1.73}$$

The second term in the above equation can be integrated by parts to give an error function

$$\mathrm{erfc}(x) = 1 - \frac{2}{\pi^{1/2}} \int_0^x \mathrm{e}^{-t^2} \mathrm{d}t , \tag{1.74}$$

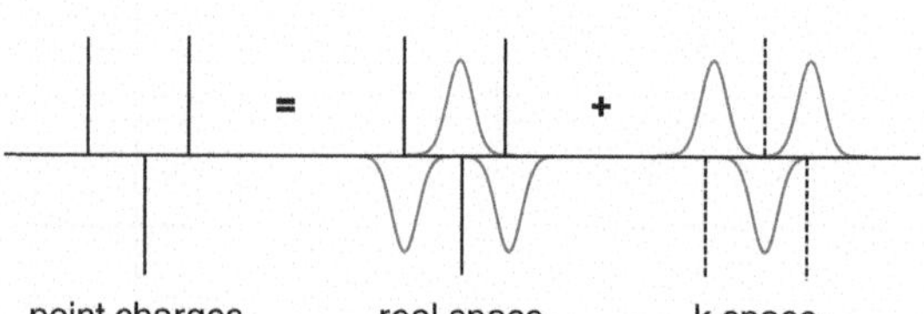

Fig. 1.5. Splitting the sum for point charges into two rapidly convergent series for Gaussian-shaped charges

plus a term which exactly cancels the third term. This gives

$$V_s(\boldsymbol{r}_i) = \sum_{\boldsymbol{n}}{}' \sum_{j=1}^{N} q_j \frac{\operatorname{erfc}(\alpha|\boldsymbol{r}_{ij}+\boldsymbol{n}|)}{|\boldsymbol{r}_{ij}+\boldsymbol{n}|} . \tag{1.75}$$

Now for the reciprocal-space sum, consider the charge density of the whole lattice at some arbitrary position r

$$\rho(r) = \sum_j q_j \delta(r - r_j) . \tag{1.76}$$

Since the lattice is periodic, we can express this quivalently as a Fourier sum

$$\rho(r) = L^{-3} \sum_j \sum_{\boldsymbol{k}} f(\boldsymbol{k}) \mathrm{e}^{-\mathrm{i}\boldsymbol{k}\cdot\boldsymbol{r})} , \tag{1.77}$$

where $\boldsymbol{k} = 2\pi/(L(i_{\hat{x}}, i_{\hat{y}}, i_{\hat{z}}))$; $i_\alpha = 0, 1, 2, \ldots$ etc. and

$$f(\boldsymbol{k}) = \int_{L^3} \rho(r)\, \mathrm{e}^{\mathrm{i}\boldsymbol{k}\cdot\boldsymbol{r}} \mathrm{d}^3 r , \tag{1.78}$$

where the integration is restricted to the unit cell volume $V = L^3$. Substituting $\rho(r)$ from (1.76) into (1.78) and making use of standard identity picks out modes corresponding to the point charges

$$f(\boldsymbol{k}) - \sum_j q_j\, \mathrm{e}^{\mathrm{i}\boldsymbol{k}\cdot\boldsymbol{r}_j} . \tag{1.79}$$

The smeared charge distribution is

$$\rho'(r) = \sum_j q_j \sigma(r - r_j) = \int_{L^3} \rho(r - r')\, \sigma(r') \mathrm{d}^3 r' . \tag{1.80}$$

This is the convolution of function $\rho(r)$ with function $\sigma(r)$, which can be expressed in Fourier space as

$$\rho'(r) = \frac{1}{L^3} \sum_{\boldsymbol{k}}{}' f(\boldsymbol{k}) \phi(\boldsymbol{k}, \alpha) \mathrm{e}^{-\mathrm{i}\boldsymbol{k}\cdot\boldsymbol{r}} , \tag{1.81}$$

where $\phi(\boldsymbol{k}, \alpha)$ is the Fourier transform of the charge-smearing function $\sigma(r)$, i.e.

$$\phi(\boldsymbol{k}, \alpha) = \mathrm{e}^{-|\boldsymbol{k}|^2/(4\alpha^2)} . \tag{1.82}$$

The potential due to the smeared charges in k-space at the reference position r_i is

$$V_k(r_i) = \int_0^\infty \frac{\rho'(r_i + r)}{r} \mathrm{d}r = \frac{1}{L^3} \sum_{\boldsymbol{k}}{}' f(\boldsymbol{k}) \phi(\boldsymbol{k}, \alpha)\, \mathrm{e}^{-\mathrm{i}\boldsymbol{k}\cdot\boldsymbol{r}} \int_0^\infty \frac{\mathrm{e}^{-\mathrm{i}\boldsymbol{k}\cdot\boldsymbol{r}}}{r} \mathrm{d}^3 r . \tag{1.83}$$

The integral on the right of the above expression is $4\pi/k^2$. Combining this with earlier results from (1.79) and (1.82) we get

$$V_k(r_i) = \frac{4\pi}{L^3} \sum_{\boldsymbol{k}}{}' \sum_j q_j \mathrm{e}^{\mathrm{i}\boldsymbol{k}\cdot(\boldsymbol{r}_j - \boldsymbol{r}_i)} \frac{\mathrm{e}^{-|\boldsymbol{k}|^2/(4\alpha^2)}}{|\boldsymbol{k}|^2} \,. \tag{1.84}$$

This potential includes an unphysical *self-term* corresponding to a smeared out charge centered at r_i, which needs to be subtracted off:

$$\begin{aligned} V_s(r_i) &= q_i \int_0^\infty \sigma(r)\mathrm{d}^3 r \\ &= \frac{4\pi q_i \alpha^3}{\pi^{3/2}} \, i \int_0^\infty r\, \mathrm{e}^{-\alpha^2 r^2} \mathrm{d}^3 r \\ &= \frac{2\alpha}{\pi^{1/2}} q_i \,. \end{aligned} \tag{1.85}$$

Adding the partial sum given by (1.75), (1.84) and (1.85) we obtain the Ewald sum

$$\begin{aligned} V_E(\boldsymbol{r}_i) = &\sum_{\boldsymbol{n}}{}' \sum_{j=1}^{N} q_j \frac{\operatorname{erfc}(\alpha|\boldsymbol{r}_{ij} + \boldsymbol{n}|)}{|\boldsymbol{r}_{ij} + \boldsymbol{n}|} \\ &+ \frac{4\pi}{L^3} \sum_{\boldsymbol{k}\neq 0} \sum_j q_j \,\mathrm{e}^{-|k|^2/(4\alpha^2)}\, \mathrm{e}^{\mathrm{i}\boldsymbol{k}\cdot(\boldsymbol{r}_j - \boldsymbol{r}_i)} - \frac{2\alpha}{\pi^{1/2}} q_i \end{aligned} \tag{1.86}$$

and the force on charge i is given by

$$\begin{aligned} \boldsymbol{f}_i &= -\nabla_{\boldsymbol{r}_i} U \\ &= \underbrace{\frac{q_i}{4\pi\epsilon_0} \sum_{\boldsymbol{n}} \sum_{j=1, j\neq i}^{N} q_j \left[\frac{\operatorname{erfc}(\alpha|\boldsymbol{r}_{ij} + \boldsymbol{n}|)}{|\boldsymbol{r}_{ij} + \boldsymbol{n}|} + \frac{2\alpha}{\sqrt{\pi}} \,\mathrm{e}^{-\alpha^2 |\boldsymbol{r}_{ij} + \boldsymbol{n}|^2} \right] \frac{\boldsymbol{r}_{ij} + \boldsymbol{n}}{|\boldsymbol{r}_{ij} + \boldsymbol{n}|}}_{\text{Real-space term}} \\ &\quad + \underbrace{\frac{2}{\epsilon_0 V} \sum_{\boldsymbol{k}>0} q_i \frac{\boldsymbol{k}}{k^2} \mathrm{e}^{-k^2/(4\alpha^2)} \Big[\sin(\boldsymbol{k}\cdot\boldsymbol{r}_i) \sum_{j=1}^{N} q_j \cos(\boldsymbol{k}\cdot\boldsymbol{r}_j) - \cos(\boldsymbol{k}\cdot\boldsymbol{r}_i) \sum_{j=1}^{N} q_j \sin(\boldsymbol{k}\cdot\boldsymbol{r}_j) \Big]}_{\text{Reciprocal-space term}} \\ &\quad \underbrace{- \frac{q_i}{6\epsilon_0 V} \sum_{j=1}^{N} q_j \boldsymbol{r}_j}_{\text{Surface dipole term}} \,. \end{aligned} \tag{1.87}$$

One needs an additional correction for the intra-molecular self-energy

$$-\frac{1}{4\pi\epsilon_0}\sum_{n=1}^{M}\sum_{\kappa=1}^{N_m}\sum_{\lambda=\kappa+1}^{N_m} q_{n\kappa}q_{n\lambda}\frac{\mathrm{erf}(\alpha|\boldsymbol{r}_{\kappa\lambda}|)}{|\boldsymbol{r}_{\kappa\lambda}|}, \tag{1.88}$$

whose derivative is absent from the equation for the forces (1.87). This term corrects for interactions between charges on the same molecule which are implicitly included in the reciprocal space sum, but are not required in the rigid-molecule model. Although the site forces $\boldsymbol{f}_i$, do include unwanted terms, these sum to zero in the evaluation of the molecular center-of-mass forces and torques (by the conservation laws for linear and angular momentum).

Both, the real- and reciprocal-space series (the sums over $\boldsymbol{n}$ and $\boldsymbol{k}$) converge fairly rapidly so that only a few terms are need to be evaluated. One defines the cut-off distances r_c and k_c so that only terms with $|\boldsymbol{r}_{ij}+\boldsymbol{n}| < r_c$ and $|\boldsymbol{k}| < k_c$ are included. The parameter α determines how rapidly the terms decrease and the values of r_c and k_c needed to achieve a given accuracy.

For a fixed α and accuracy the number of terms in the real-space sum is proportional to the total number of sites, N but the cost of the reciprocal-space sum increases as N^2. An overall scaling of $N^{3/2}$ may be achieved if α varies with N. This is discussed in detail in an excellent article by D. Fincham [115]. The optimal value of α is

$$\alpha = \sqrt{\pi}\left(\frac{t_R}{t_F}\frac{N}{V^2}\right)^{\frac{1}{6}}, \tag{1.89}$$

where t_R and t_F are the execution times needed to evaluate a single term in the real- and reciprocal-space sums respectively. If we require that the sums converge to an accuracy of $\epsilon = \exp(-p)$ the cutoffs are then given by

$$r_c = \frac{\sqrt{p}}{\alpha}, \tag{1.90}$$

$$k_c = 2\alpha\sqrt{p}\,. \tag{1.91}$$

A representative value of t_R/t_F has been established as 5.5. Though this will vary on different processors and for different potentials its value is not critical since it enters the equations as a sixth root.

It must be emphasized that the r_c is used as a cutoff for the short-ranged potentials as well as for the electrostatic part. The value chosen above does not take the nature of the non-electrostatic part of the potential into account.

1.10.1.3 Uniform Sheet Correction

In a periodic system the electrostatic energy is finite only if the total electric charge of the MD cell is zero. The reciprocal space sum for $\boldsymbol{k} = 0$ takes the form

$$\frac{1}{k^2} e^{-k^2/(4\alpha^2)} \left| \sum_{i=1}^{N} q_i \right|^2 , \tag{1.92}$$

which is zero in the case of electro-neutrality but infinite otherwise. Its omission is physically equivalent to adding a uniform jelly of charge which exactly neutralizes the unbalanced point charges. But though the form of the reciprocal space sum is unaffected by the uniform charge jelly the real-space sum is not. The real-space part of the interaction of the jelly with each point charge as well as the self-energy of the jelly itself must be included giving

$$-\frac{1}{8\epsilon_0 V \alpha^2} \left| \sum_{i=1}^{N} q_i \right|^2 . \tag{1.93}$$

1.10.1.4 Surface Dipole Term

This term accounts for different periodic boundary conditions. It was suggested by De Leeuw, Perram and Smith [116, 117, 118] in order to accurately model dipolar systems and is necessary in any calculation of a dielectric constant

$$+ \left[\frac{1}{6\epsilon_0 V} \left| \sum_{i=1}^{N} q_i \boldsymbol{r}_i \right|^2 \right] . \tag{1.94}$$

Consider a near-spherical cluster of MD cells. The *infinite* result for any property is the limit of its *cluster* value as the size of the cluster tends to infinity. However, this value is non-unique and depends on the dielectric constant, ϵ_s of the physical medium surrounding the cluster. If this medium is conductive ($\epsilon_s = \infty$) the dipole moment of the cluster is neutralized by image charges, whereas in a vacuum ($\epsilon_s = 1$) it remains. It is trivial to show that in that case the dipole moment per unit volume (or per MD cell) does not decrease with the size of the cluster. This term is then just the dipole energy, and ought to be used in any calculation of the dielectric constant of a dipolar molecular system.

1.10.2 Multipole Methods

There is a large number of N-body problems for which periodic boundaries are completely inappropriate, for example: galaxy dynamics, electron-beam transport, large proteins [119], and any number of problems with complex geometries. Two new approaches were put forward in the mid-1980's, the first from Appel [120] and Barnes & Hut [121], who proposed $O(N \log N)$-schemes based on hierarchical grouping of distant particles; the second from Greengard & Rohklin [122] with an $O(N)$ (better than $O(N \log N)$) solution with rounding-error accuracy. These two methods are known today as the *hierarchical tree algorithm* and the *Fast Multipole Method* (FMM) respectively – have revolutionized N-body simulation in a much broader sense than the specialized periodic methods discussed earlier. They offer a generic means of accelerating the computation of many-particle systems governed by central, long-range potentials.

References

1. Y. Duan, L. Wang, P. Kollman, P. Natl. Acad. Sci. USA **95**, 9897 (1998)
2. Q. Zhong, P. Moore, D. Newns, M. Klein, FEBS Lett. **427**, 267 (1998)
3. Q. Zhong, Q. Jiang, P. Moore, D. Newns, M. Klein, Biophys. J. **74**, 3 (1998)
4. R. Car, M. Parrinello, Phys. Rev. Lett. **55**, 2471 (1985)
5. G. Galli, M. Parrinello, in *Proceedings of the NATO Advanced Study Institute on Computer Simulation in Material Schience: Interatomic Potentiols, Simulation Techniques and Applications, Aussois, France, 25 March - 5 April 1991*, Vol. 3, ed. by M. Meyer, V. Pontikis (Kluwer Academic Publishers, Dordrecht, The Netherlands, 1991), Vol. 3, pp. 283–304
6. D. Heermann, *Computer Simulation Methods* (Springer, Berlin Heidelberg New York, 1986)
7. H. Berendsen, J. Postma, W. van Gunsteren, A. DiNola, J. Haak, J. Chem. Phys. **81**, 3684 (1984)
8. H. Andersen, J. Chem. Phys. **72**, 2384 (1980)
9. W. Hoover, Phys. Rev. A **31**, 1695 (1985)
10. A. Voter, F. Montalenti, T. Germann, Annu. Rev. Mater. Res. **32**, 321 (2002)
11. J. Lennard-Jones, P. Roy. Soc. Lond. **43**, 461 (1931)
12. P. Morse, Phys. Rev. **34**, 57 (1929)
13. A. Rahman, Phys. Rev. **136**, A405 (1964)
14. L. Verlet, Phys. Rev. **159**, 98 (1967)
15. J. Tersoff, Phys. Rev. Lett. **56**, 632 (1986)
16. J. Tersoff, Phys. Rev. B **37**, 6991 (1988)
17. J. Tersoff, Phys. Rev. Lett. **61**, 2879 (1988)
18. J. Tersoff, Phys. Rev. B **39**, 5566 (1989)
19. W. Jorgensen, J. Madura, C. Swenson, J. Am. Chem. Soc. **106**, 6638 (1984)
20. N. Allinger, K. Chen, J. Lii, J. Comput. Chem. **17**, 642 (1996)
21. W. Jorgensen, D. Maxwell, J. Tiradorives, J. Am. Chem. Soc. **118**, 11225 (1996)
22. W. Cornell, P. Cieplak, C. Bayly, I. Gould, K. Merz, D. Ferguson, D. Spellmeyer, T. Fox, J. Caldwell, P. Kollman, J. Am. Chem. Soc. **118**, 2309 (1996)
23. T. Halgren, J. Comput. Chem. **17**, 490 (1996)
24. S. Nath, F. Escobedo, J. de Pablo, J. Chem. Phys. **108**, 9905 (1998)
25. M. Martin, J. Siepmann, J. Phys. Chem. B **102**, 2569 (1998)
26. H. Sun, J. Phys. Chem. B **102**, 7338 (1998)
27. D. Brenner, Mat. Res. Soc. Symp. Proc. **141**, 59 (1989)
28. M. Ramana Murty, H. Atwater, Phys. Rev. B **51**, 4889 (1995)
29. A. Dyson, P. Smith, Surf. Sci. **355**, 140 (1996)
30. D. Brenner, Phys. Rev. B **42**, 9458 (1990)
31. D. Brenner, Phys. Rev. B **46**, 1948 (1992)
32. D. Brenner, J. Harrison, C. White, R. Colton, Thin Solid Films **206**, 220 (1991)
33. D. Brenner, K. Tupper, S. Sinnott, R. Colton, J. Harrison, Abstr. Pap. Am. Chem. S. **207**, 166 (1994)

34. J. Harrison, S. Stuart, D. Robertson, C. White, J. Phys. Chem. B **101**, 9682 (1997)
35. S. Sinnott, R. Colton, C. White, O. Shenderova, D. Brenner, J. Harrison, J. Vac. Sci. Technol. A **15**, 936 (1997)
36. J. Harrison, C. White, R. Colton, D. Brenner, Phys. Rev. B **46**, 9700 (1992)
37. J. Harrison, R. Colton, C. White, D. Brenner, Wear **168**, 127 (1993)
38. J. Harrison, C. White, R. Colton, D. Brenner, J. Phys. Chem. **97**, 6573 (1993)
39. J. Harrison, D. Brenner, J. Am. Chem. Soc. **116**, 10399 (1994)
40. J. Harrison, C. White, R. Colton, D. Brenner, Thin Solid Films **260**, 205 (1995)
41. M. Perry, J. Harrison, Langmuir **12**, 4552 (1996)
42. D. Allara, A. Parikh, E. Judge, J. Chem. Phys. **100**, 1761 (1994)
43. R. Smith, K. Beardmore, Thin Solid Films **272**, 255 (1996)
44. M. Nyden, T. Coley, S. Mumby, Polym. Eng. Sci **37**, 1496 (1997)
45. J. Che, T. Cagin, W. Goddard, Theor. Chem. Acc. **102**, 346 (1999)
46. K. Nordlund, J. Keinonen, T. Mattila, Phys. Rev. Lett. **77**, 699 (1996)
47. S. Stuart, B. Berne, J. Phys. Chem. **100**, 11934 (1996)
48. R. Hockney, J. Eastwood, *Computer Simulation Using Particles* (McGraw-Hill, New-York, USA, 1981)
49. W. Swope, H. Andersen, P. Berens, K. Wilson, J. Chem. Phys. **76**, 637 (1982)
50. D. Beeman, J. Comput. Phys. **20**, 130 (1976)
51. G. Martyna, M. Tuckerman, J. Chem. Phys. **102**, 8071 (1995)
52. M. Tuckerman, B. Berne, G. Martyna, J. Chem. Phys. **97**, 1990 (1992)
53. C. Gear, *Numerical Initial Value Problems in Ordinary Differential Equations (Chap. 9)* (Prentice Hall, Englewood Cliffs, NJ, USA, 1971)
54. H. Yoshida, Phys. Lett. A **150**, 262 (1990)
55. D. Frenkel, B. Smit, *Understanding Molecular Simulation: From Algorithms to Applications* (Academic Press, San Diego, 1996)
56. M. Allen, D. Tildesley, *Computer simulation of liquids* (Clarendon Press, Oxford, 1987)
57. B. Todd, D. Evans, P. Daivis, Phys. Rev. E **52**, 1627 (1995)
58. J. Irving, J. Kirkwood, J. Chem. Phys. **18**, 817 (1950)
59. D. McQuarrie, *Statistical Mechanics* (Harper and Row, New York, 1976)
60. D. Frenkel, B. Smit, *Understanding Molecular Simulation: From Algorithms to Applications* (Academic Press, San Diego, 2002)
61. M. Warrier, R. Schneider, E. Salonen, K. Nordlund, Contrib. Plasma Phys. **44**, 307 (2004)
62. J. Klafter, M. Shlesinger, G. Zumofen, Phys. Today **2**, 33 (1996)
63. E. Weinan, B. Engquist, Not. Am. Math. Soc **50**, 1062 (2003)
64. B. Berne, G. Ciccotti, C. D.F. (eds.), *Classical and Quantum Dynamics in Condensed Phase Simulations* (World Scientific Publishing Company, Singapore, Singapore, 1998)
65. K. Binder, G. Ciccotti (eds.), *Monte Carlo and Molecular Dynamics of Condensed Matter Systems* (Editrice Compositori, Bologna, Italy, 1996)
66. G. Ciccotti, D. Frenkel, I. McDonald, *Simulation of Liquids and Solids* (North Holland, Amsterdam, 1987)
67. R. Esser, P. Grassberger, J. Grotendorst, M. Lewerenz (eds.), *Molecular Dynamics on Parallel Computers* (World Scientific Publishing Company, Singapore, Singapore, 1999)
68. D. Frenkel, B. Smit, *Understanding Molecular Simulations: From Algorithms to Applications* (Academic Press, San Diego, 2005)
69. R. Haberlandt, S. Fritzsche, G. Peinel, K. Heinzinger, *Molekulardynamik - Grundlagen und Anwendungen* (H.-L.Vörtler, Lehrbuch, Vieweg, Wiesbaden, 1995)

70. G. Gray, K. Gubbins, *Theory of Molecular Fluids* (Clarendon Press, Oxford, 1984)
71. G. Schatz, Rev. Mod. Phys. **61**, 669 (1989)
72. M. Sprik, in *NATO ASI Series C*, Vol. 397, ed. by M. Allen, D. Tildesley (Kluwer Academic Publishers, Dordrecht, The Netherlands, 1993), Vol. 397, pp. 211–259
73. M. Segall, P. Lindan, M. Probert, C. Pickard, P. Hasnip, S. Clark, M. Payne, J. Phys-Condens. Mat. **14**, 2717 (2002)
74. P. Blöchl, Phys. Rev. B **50**, 17953 (1994)
75. M. Bockstedte, A. Kley, J. Neugebauer, M. Scheffler, Comput. Phys. Commun. **107**, 187 (1997)
76. R. Kendall, E. Apra, D. Bernholdt, E. Bylaska, M. Dupuis, G. Fann, R. Harrison, J. Ju, J. Nichols, J. Nieplocha, T. Straatsma, T. Windus, A. Wong, Comput. Phys. Commun. **128**, 260 (2000)
77. G. Kresse, J. Furthmuller, Phys. Rev. B **54**, 11169 (1996)
78. M. Frisch, G. Trucks, H. Schlegel, G. Scuseria, M. Robb, J. Cheeseman, J. Montgomery, Jr., T. Vreven, K. Kudin, J. Burant, J. Millam, S. Iyengar, J. Tomasi, V. Barone, B. Mennucci, M. Cossi, G. Scalmani, N. Rega, G. Petersson, H. Nakatsuji, M. Hada, M. Ehara, K. Toyota, R. Fukuda, J. Hasegawa, M. Ishida, T. Nakajima, Y. Honda, O. Kitao, H. Nakai, M. Klene, X. Li, J. Knox, H. Hratchian, J. Cross, V. Bakken, C. Adamo, J. Jaramillo, R. Gomperts, R. Stratmann, O. Yazyev, A. Austin, R. Cammi, C. Pomelli, J. Ochterski, P. Ayala, K. Morokuma, G. Voth, P. Salvador, J. Dannenberg, V. Zakrzewski, S. Dapprich, A. Daniels, M. Strain, O. Farkas, D. Malick, A. Rabuck, K. Raghavachari, J. Foresman, J. Ortiz, Q. Cui, A. Baboul, S. Clifford, J. Cioslowski, B. Stefanov, G. Liu, A. Liashenko, P. Piskorz, I. Komaromi, R. Martin, D. Fox, T. Keith, C. Al-Laham, M.A.and Peng, A. Nanayakkara, M. Challacombe, P. Gill, B. Johnson, W. Chen, M. Wong, C. Gonzalez, J. Pople, *Gaussian 03 User's Reference, Revision C.02.* Gaussian, Inc., Wallingford, CT, 2004
79. H.J. Werner, P. Knowles, R. Lindh, R. Manby, M. Schütz, P. Celani, T. Korona, G. Rauhut, R. Amos, A. Bernhardsson, A. Berning, D. Cooper, M. Deegan, A. Dobbyn, F. Eckert, C. Hampel, G. Hetzer, A. Lloyd, S. McNicholas, W. Meyer, M. Mura, A. Nicklass, P. Palmieri, R. Pitzer, U. Schumann, H. Stoll, A. Stone, R. Tarroni, T. Thorsteinsson, *MOLPRO, Version 2006.1.* Cardiff, UK (2006). A package of ab initio programs, see http://www.molpro.net
80. X. Gonze, J.M. Beuken, R. Caracas, F. Detraux, M. Fuchs, G. Rignanese, L. Sindic, M. Verstraete, G. Zerah, F. Jollet, M. Torrent, A. Roy, M. Mikami, P. Ghosez, J. Raty, D. Allan, Comp. Mater. Sci. **25**, 478 (2002)
81. X. Gonze, G. Rignanese, M. Verstraete, J. Beuken, Y. Pouillon, R. Caracas, F. Jollet, M. Torrent, G. Zerah, M. Mikami, P. Ghosez, M. Veithen, J. Raty, V. Olevanov, F. Bruneval, L. Reining, R. Godby, G. Onida, D. Hamann, D. Allan, Z. Kristallogr. **220**, 558 (2005)
82. G. Pastore, E. Smargiassi, F. Buda, Phys. Rev. A **44**, 6334 (1991)
83. M. Allen, D. Tildesley, *Computer simulation of liquids* (Clarendon Press: Oxford, 1990)
84. P.E. Blöchl, M. Parrinello, Phys. Rev. B **45**, 9413 (1992)
85. E. Fois, A. Selloni, M. Parrinello, R. Car, J. Phys. Chem. **92**, 3268 (1988)
86. A. Pasquarello, K. Laasonen, R. Car, C. Lee, D. Vanderbilt, Phys. Rev. Lett. **69**, 1982 (1992)
87. Y. Yamaguchi, Y. Osamura, J. Goddard, H. Schaefer, *A New Dimension to Quantum Chemistry: Analytic Derivative Methods in Ab Initio Molecular Electronic Structure Theory* (Oxford University Press, New York, 1994)
88. W. Hehre, L. Radom, P. Schleyer, J. Pople, *Ab initio molecular orbital theory* (Wiley, New York, 1986)

89. M. Headgordon, J. Phys. Chem. **100**, 13213 (1996)
90. W. Kohn, A. Becke, R. Parr, J. Phys. Chem. **100**, 12974 (1996)
91. T. Dunning, *Advances in Molecular and Electronic Structure Theory*, Vol. 1 (Jai Press, Greenwich, CT, 1990)
92. B. Jeziorski, R. Moszynski, K. Szalewicz, Chem. Rev. **94**, 1887 (1994)
93. J. Murrell, S. Carter, S. Farantos, P. Huxley, A. Varandas, *Molecular Potential Energy Functions* (John Wiley and Sons Ltd, New York, 1984)
94. D. Truhlar, R. Steckler, M. Gordon, Chem. Rev. **87**, 217 (1987)
95. J. Ischtwan, M. Collins, J. Chem. Phys. **100**, 8080 (1994)
96. M. Collins, Adv. Chem. Phys. **93**, 389 (1996)
97. T.S. Ho, H. Rabitz, J. Chem. Phys. **104**, 2584 (1996)
98. T. Hollebeek, T.S. Ho, H. Rabitz, J. Chem. Phys. **106**, 7223 (1997)
99. T.S. Ho, H. Rabitz, in *Fashioning a Model: Optimization Methods in Chemical Physics*, ed. by A. Ernesti, J. Hutson, N. Wright (1998), pp. 28–34
100. T. Helgaker, E. Uggerud, H. Jensen, Chem. Phys. Lett. **173**, 145 (1990)
101. W. Chen, W. Hase, H. Schlegel, Chem. Phys. Lett. **228**, 436 (1994)
102. R. Steckler, G. Thurman, J. Watts, R. Bartlett, J. Chem. Phys. **106**, 3926 (1997)
103. A. Varandas, P. Abreu, Chem. Phys. Lett. **293**, 261 (1998)
104. Y. Chuang, D. Truhlar, J. Phys. Chem. A **101**, 3808 (1997)
105. J. Corchado, J. Espinosa-Garcia, O. Roberto-Neto, Y. Chuang, D. Truhlar, J. Phys. Chem. A **102**, 4899 (1998)
106. M. Jordan, K. Thompson, M. Collins, J. Chem. Phys. **102**, 5647 (1995)
107. M. Jordan, K. Thompson, M. Collins, J. Chem. Phys. **103**, 9669 (1995)
108. M. Jordan, M. Collins, J. Chem. Phys. **104**, 4600 (1996)
109. K. Thompson, M. Collins, J. Chem. Soc. Faraday T. **93**, 871 (1997)
110. K. Thompson, M. Jordan, M. Collins, J. Chem. Phys. **108**, 564 (1998)
111. K. Thompson, M. Jordan, M. Collins, J. Chem. Phys. **108**, 8302 (1998)
112. T. Ishida, G. Schatz, J. Chem. Phys. **107**, 3558 (1997)
113. I. Takata, T. Taketsugu, K. Hirao, M. Gordon, J. Chem. Phys. **109**, 4281 (1998)
114. T. Hollebeek, T.S. Ho, H. Rabitz, Annu. Rev. Phys. Chem. **50**, 537 (1999)
115. D. Fincham, Mol. Simulat. **13**, 1 (1994)
116. S. Deleeuw, J. Perram, E. Smith, P. Roy. Soc. Lond. A Mat. **373**, 27 (1980)
117. S. Deleeuw, J. Perram, E. Smith, P. Roy. Soc. Lond. A Mat. **373**, 57 (1980)
118. S. Deleeuw, J. Perram, E. Smith, P. Roy. Soc. Lond. A Mat. **388**, 177 (1983)
119. T. Schlick, R. Skeel, A. Brunger, L. Kale, J. Board, J. Hermans, K. Schulten, J. Comput. Phys. **151**, 9 (1999)
120. A. Appel, Siam J. Sci. Stat. Comp. **6**, 85 (1985)
121. J. Barnes, P. Hut, Nature **324**, 446 (1986)
122. L. Greengard, V. Rokhlin, J. Comput. Phys. **73**, 325 (1987)

2 Wigner Function Quantum Molecular Dynamics

V. S. Filinov[1,2], M. Bonitz[2], A. Filinov[2], and V. O. Golubnychiy[2]

[1] Institute for High-Energy Density, Russian Academy of Sciences, Moscow 127412, Russia
[2] Institut für Theoretische Physik und Astrophysik, Christian-Albrechts-Universtität, 24098 Kiel, Germany

Classical molecular dynamics (MD) is a well established and powerful tool in various fields of science, e.g. chemistry, plasma physics, cluster physics and condensed matter physics. Objects of investigation are few-body systems and many-body systems as well. The broadness and level of sophistication of this technique is documented in many monographs and reviews, see for example [1, 2]. Here we discuss the extension of MD to quantum systems (QMD). There have been many attempts in this direction which differ from each other, depending on the type of system under consideration. One variety of QMD has been developed for condensed matter systems. This approach is reviewed e.g. in [3] and will not be discussed here. In this contribution we deal with unbound electrons as they occur in gases, fluids or plasmas. Here, a quite successful strategy is to replace classical point particles by wave packets [3, 4, 5, 6]. This method, however, struggles with problems related to the dispersion of such a wave packet and difficulties to properly describe strong electron-ion interaction and bound-state formation. We try to avoid these restrictions by an alternative approach: We start the discussion of quantum dynamics by a general consideration of quantum distribution functions.

2.1 Quantum Distribution Functions

There exists a variety of different representations of quantum mechanics including the so-called *Wigner representation* which involves a class of functions depending on coordinates and momenta. In the classical limit, the Wigner distribution function f_W turns into the phase space distribution f known from classical statistical mechanics. In contrast to f, the Wigner function may be non-positive as a consequence of the coordinate-momentum (Heisenberg) uncertainty. This will lead to a modification of the particle trajectories which is discussed in Sect. 2.3. An important property of the distribution functions is that they can be used to compute the expectation value of an arbitrary physical observable $\langle A \rangle$, defined by the operator $\widehat{A}(\widehat{p}, \widehat{q})$ [7]

$$\langle A \rangle(t) = \int \mathrm{d}p\, \mathrm{d}q\, A_W(p,q)\, f_W(p,q,t)\,, \qquad 1 = \int \mathrm{d}p\, \mathrm{d}q\, f_W(q,p,t)\,, \qquad (2.1)$$

where $A_W(p,q)$ is a scalar function. For simplicity we considered the one-dimensional (1D) case; the generalization to higher dimensions and N particles

V. S. Filinov et al.: *Wigner Function Quantum Molecular Dynamics*, Lect. Notes Phys. **739**, 41–60 (2008)
DOI 10.1007/978-3-540-74686-7_2

is straightforward by re-defining the coordinate and momentum as vectors, $q = \{\boldsymbol{q_1}, \ldots, \boldsymbol{q_N}\}, p = \{\boldsymbol{p_1}, \ldots, \boldsymbol{p_N}\}$. f_{W} is defined via the nonequilibrium N-particle density operator $\widehat{\rho}$ in coordinate representation (i.e. the density matrix),

$$f_{\mathrm{W}}(p,q,t) = \frac{1}{2\pi\hbar} \int \mathrm{d}\nu \, \left\langle q + \frac{\nu}{2} \, |\widehat{\rho}| \, q - \frac{\nu}{2} \right\rangle \mathrm{e}^{-\mathrm{i}\nu p} \,, \tag{2.2}$$

and $A_{\mathrm{W}}(p,q)$ is analogously defined from the coordinate representation of $\widehat{A}$.

We now consider the time evolution of the wave function under the influence of a general Hamiltonian of the form

$$\widehat{H} = \sum_{j=1}^{N} \frac{\widehat{p}_i^2}{2m} + \sum_{i=1}^{N} \widetilde{V}(q_i) + \sum_{i<j} V(q_i, q_j) \,, \tag{2.3}$$

where $\widetilde{V}(q_i)$ and $V(q_i, q_j)$ denote an external and an interaction potential, respectively. The equation of motion for f_{W} has the form [8, 7] (see also Sect. 2.3)

$$\frac{\partial f_{\mathrm{W}}}{\partial t} + \frac{\boldsymbol{p}}{m} \cdot \boldsymbol{\nabla}_q f_{\mathrm{W}} = \int\limits_{-\infty}^{\infty} \mathrm{d}s \; f_{\mathrm{W}}\,(p-s,q,t) \; \widetilde{\omega}\,(s,q,t) \,, \tag{2.4}$$

where the function

$$\widetilde{\omega}\,(s,q,t) = \frac{2}{\pi\hbar^2} \int \mathrm{d}q' V\,(q-q',t) \sin\left(\frac{2sq'}{\hbar}\right) \tag{2.5}$$

takes into account the non-local contribution of the potential energy in the quantum case. Equivalently, expanding the integral around $q' = 0$, (2.4) can be rewritten by an infinite sum of local potential terms

$$\frac{\partial f_{\mathrm{W}}}{\partial t} + \frac{p}{m}\frac{\partial f_{\mathrm{W}}}{\partial q} = \sum_{n=0}^{\infty} \frac{(\hbar/(2\mathrm{i}))^{2n}}{(2n+1)!} \left(\frac{\partial^{2n+1} V}{\partial q^{2n+1}}, \frac{\partial^{2n+1} f_{\mathrm{W}}}{\partial p^{2n+1}} \right) \,, \tag{2.6}$$

where $(\partial^{2n+1} V/\partial q^{2n+1}, \partial^{2n+1} f_{\mathrm{W}}/\partial p^{2n+1})$ denotes the scalar product of two vectors which for an N-particle system contain $3N$ components.

If the potential does not contain terms higher than second order in q, i.e. $\partial^n V/\partial q^n|_{n\geq 3} = 0$, (2.6) reduces to the classical Liouville equation for the distribution function f:

$$\frac{\partial f}{\partial t} + \frac{p}{m}\frac{\partial f}{\partial q} = \frac{\partial V}{\partial q}\frac{\partial f}{\partial p} \,. \tag{2.7}$$

The Wigner function must satisfy a number of conditions [9], therefore, the initial function $f_{\mathrm{W}}(q,p,0)$ cannot be chosen arbitrarily. Even if $f_{\mathrm{W}}(q,p,t)$ satisfies the classical equation (2.7) it nevertheless describes the evolution of a quantum distribution because a properly chosen initial function $f_{\mathrm{W}}(q,p,0)$ contains, in general, all powers of $\hbar$. In particular, the uncertainty principle holds for averages of operators calculated with $f_W(q, p, 0)$ and $f_W(q, p, t)$.

One can rewrite (2.6) in a form analogous to the classical Liouville equation (2.7) by replacing V by a new effective potential V_{eff} defined as

$$\frac{\partial V_{\text{eff}}}{\partial q}\frac{\partial f_W}{\partial p} = \frac{\partial V}{\partial q}\frac{\partial f_W}{\partial p} - \frac{\hbar^2}{24}\frac{\partial^3 V}{\partial q^3}\frac{\partial^3 f_W}{\partial p^3} + \cdots . \tag{2.8}$$

Equation (2.7) can be efficiently solved with the *method of characteristics*, see e.g. [10]. This is the basis of our QMD approach where an ensemble of classical (Wigner) trajectories is used to solve (numerically) the quantum Wigner-Liouville equation (2.4) which will be discussed in Sect. 2.3. The time-dependence of the trajectories is given by the classical equations of motion

$$\frac{\partial q}{\partial t} = \frac{p}{m}, \qquad \frac{\partial p}{\partial t} = -\frac{\partial V_{\text{eff}}(p, q, t)}{\partial q} . \tag{2.9}$$

Of course, a direct solution of (2.9) with the definition (2.8) is only useful if the series is rapidly converging and there is only a small number of non-zero terms.

Clearly there is a principle difficulty with this approach if the series of terms with the potential derivatives is not converging. This is the case, e.g., for a Coulomb potential (at zero distance). There are at least three solutions to this problem. The first one is to solve the Wigner-Liouville equation by Monte Carlo (MC) techniques [11, 12, 13, 14], which is discussed below in Sect. 2.3. The second one is to replace the original potential on the r.h.s. of (2.8) by some model potential having a finite number of nonzero derivatives, see e.g. [15]. The third approach is to perform a suitable average of V_{eff}, e.g. over a thermal ensemble of particles. This has been done both for external potentials and also for two particle interaction. The use of an effective quantum pair potential in classical MD is discussed in Chap. 1.

2.2 Semiclassical Molecular Dynamics

2.2.1 Quantum Pair Potentials

In order to obtain an effective pair potential which is finite at zero interparticle distance, we consider (2.4) for two particles. Assuming further thermodynamic equilibrium with a given temperature $k_BT = 1/\beta$, spatial homogeneity and neglecting three-particle correlations, one can solve for the two-particle Wigner function $f_{W,12} = F_{12}^{\text{eq}}(r_1, p_1, r_2, p_2, \beta) \approx F_{12}^{\text{eq}}(r_1 - r_2, p_1, p_2, \beta)$.

This is now rewritten as in the canonical case [7], $F_{12}^{\text{eq}}(r_1 - r_2, p_1, p_2, \beta) \equiv F_1^{\text{eq}}(p_1, \beta) F_2^{\text{eq}}(p_2, \beta) \exp(-\beta V_{12}^{\text{qp}})$, which defines the desired quantum pair potential V_{12}^{qp}.

The first solution for V_{12}^{qp} was found by Kelbg in the limit of weak coupling [16, 17, 18]. It has the form of (2.10) with $\gamma_{ij} \to 1$, for details and references see [10, 19]. The Kelbg potential, or slightly modified versions, is widely used in numerical simulations of dense plasmas [4, 5, 20, 21, 22]. It is finite at zero distance which correctly captures basic quantum diffraction effects preventing any divergence. However, the absolute value at $r = 0$ is incorrect which has lead to the derivation of further improved potentials, see [10, 19, 23] and references therein. Here we use the *improved Kelbg potential* (IKP),

$$\Phi(r_{ij}, \beta) = \frac{q_i q_j}{r_{ij}} \left\{ 1 - e^{-r_{ij}^2/\lambda_{ij}^2} + \sqrt{\pi} \frac{r_{ij}}{\lambda_{ij}\gamma_{ij}} \left(1 - \operatorname{erf}\left[\gamma_{ij} \frac{r_{ij}}{\lambda_{ij}} \right] \right) \right\}, \quad (2.10)$$

where $r_{ij} = |\boldsymbol{r}_{ij}|$, $x_{ij} = r_{ij}/\lambda_{ij}$, $\lambda_{ij}^2 = \hbar^2\beta/(2\mu_{ij})$ and $\mu_{ij}^{-1} = m_i^{-1} + m_j^{-1}$, which contains additional free parameters γ_{ij} that can be obtained from a fit to the exact solution of the two-particle problem [19].

2.2.2 Molecular Dynamics Simulations

We have performed extensive MD simulations of dense partially ionized hydrogen in thermodynamic equilibrium using different IKP for electrons with different spin projections. To properly account for the long-range character of the potentials, we used periodic boundary conditions with the standard Ewald procedure, see Chap. 1. The number of electrons and protons was $N = 200$. For our MD simulations we use standard Runge-Kutta or Verlet algorithms (see Chap. 1) to solve Newton's equations (2.9), where V_{eff} is replaced by the IKP. Because of the temperature dependence of the IKP we applied a temperature scaling at every time step for all components separately (for protons and two sorts of electrons) to guarantee a constant temperature of all components in our equilibrium simulations. In each simulation the system was equilibrated for at least 10^4 MD steps, only after this the observables have been computed.

In Fig. 2.1 we show the internal energy per atom as a function of temperature for two densities and compare it to path integral Monte Carlo (PIMC) results [19, 24]. The density is given by the Brueckner parameter $r_s = \bar{r}/a_B$, where $\bar{r}$ is the average interparticle distance and a_B denotes the Bohr radius. For high temperatures and weak coupling, $\Gamma = e^2/(\bar{r}k_B T) < 1$ for the fully ionized plasma, the two simulations coincide within the limits of statistical errors. If we use the original Kelbg potential, at temperatures below 300 000 K (approximately two times the binding energy), the MD results start to strongly deviate from the PIMC results. In contrast the IKP fully agrees with the PIMC data even at temperatures far below the hydrogen binding energy (1 Ry), where the plasma is dominated by atoms, which is a remarkable extension of semi-classical MD into the theoretically very difficult regime of moderate coupling, moderate degeneracy and partial ionization.

Interestingly, even bound states can be analyzed in our simulations by following the electron trajectories. At $T < 1$ Ry, we observe an increasing number of electrons undergoing strong deflection (large-angle scattering) on protons and eventually performing quasi-bound trajectories. Most of these electrons remain bound only for a few classical orbits and then leave the proton again. Averaged over a long time, our simulations are able to reveal the degree of ionization of the plasma. For temperatures below approximately 50 000 K, which is close to the binding energy of hydrogen molecules, the simulations cannot be applied. Although we clearly observe

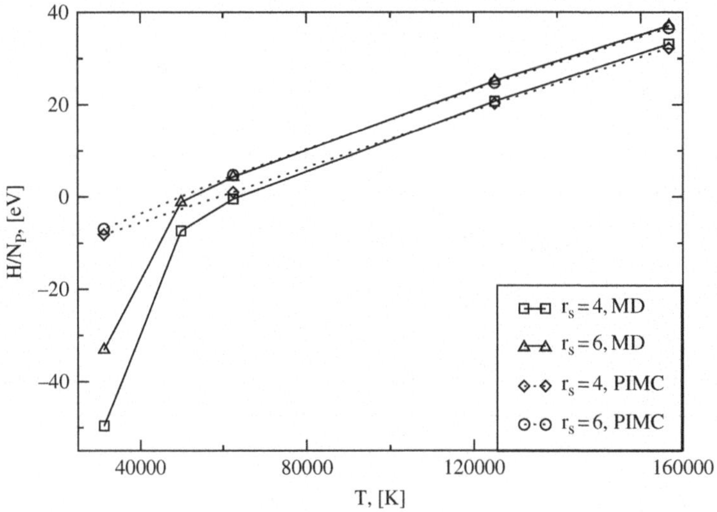

Fig. 2.1. Internal energy per hydrogen atom at $r_s = 4$ and $r_s = 6$ versus temperature, MD results are compared to restricted PIMC simulations [19, 24]

molecule formation (see below), there also appear clusters of several molecules which is unphysical under the present conditions and is caused by the approximate two-particle treatment of quantum effects in the IKP. This turns out to be the reason for the too small energy at low temperatures (see Fig. 2.1).

Let us now turn to a more detailed analysis of the spatial configuration of the particles. In Fig. 2.2 the pair distribution functions of all particle species with the same charge are plotted at two densities. Consider first the case of $T = 125\,000\,\mathrm{K}$ (upper panels). For both densities all functions agree qualitatively showing a depletion at zero distance due to Coulomb repulsion. Besides, there are differences which arise from the spin properties. Electrons with the same spin show a *Coulomb hole* around $r = 0$ which is broader than the one of the protons due to the Pauli principle with additional repulsion of electrons with the same spin projection. This trend is reversed at low temperatures (see middle panel), which is due to the formation of hydrogen atoms and molecules. In this case, electrons, i.e., their classical trajectories, are spread out around the protons giving rise to an increased probability of close encounters of two electrons belonging to different atoms compared to two protons.

Now, let us compare electrons with parallel and electrons with anti-parallel spins. In all cases, we observe a significantly increased probability to find two electrons with opposite spin at distances below one Bohr radius, which is due to the missing Pauli repulsion. This trend increases when the temperature is lowered because of increasing quantum effects. Before analyzing the lowest temperature in Fig. 2.2, let us consider the electron-proton (e-p) distributions. Multiplying these functions by r^2 gives essentially the radial probability density $W_{ep}(r) = r^2 g_{ep}(r)$,

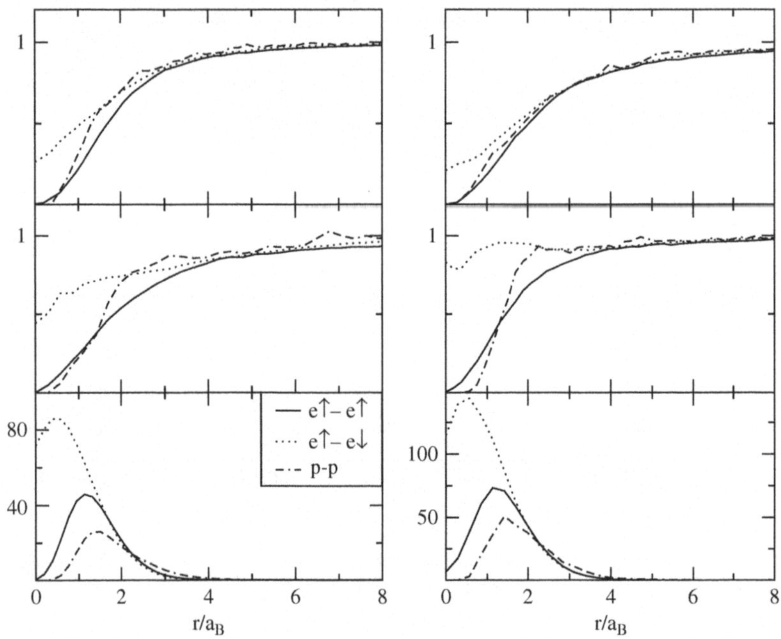

Fig. 2.2. Electron-electron (e-e) and proton-proton (p-p) pair distribution functions for a correlated hydrogen plasma with $r_s = 4$ (**left row**) and $r_s = 6$ (**right row**) for $T = 125\,000$ K, 61 250 K and 31 250 K (**from top to bottom**) [19]

which is plotted in Fig. 2.3. At low temperatures this function converges to the ground state probability density of the hydrogen atom $W_{ep}(r) = r^2|\psi|^2_{1s}(r)$ influenced by the surrounding plasma. Here, lowering of the temperature leads towards the formation of a shoulder around $1.4a_B$ for $r_s = 4$ and $1.2a_B$ for $r_s = 6$ which is due to the formation of hydrogen atoms; this is confirmed by the corresponding quasi-bound electron trajectories. At this temperature, the observed most probable electron distance is slightly larger than one a_B as in the atom hydrogen ground state. Of course, classical MD cannot yield quantization of the bound electron motion, but it correctly reproduces (via averaging over the trajectories) the statistical properties of the atoms, such as the probability density averaged over the energy spectrum.

At 62 500 K and $r_s = 6$ (right middle part of Fig. 2.2) the simulations show a first weak signature of molecule formation – see the maximum of the p-p distribution function around $r = 2a_B$ and the maximum of the distribution function of electrons with anti-parallel spins around $r = 1.5a_B$. Upon further lowering of the temperature by a factor of two (lower panel of Fig. 2.2) the p-p functions exhibit a clear peak very close to $r = 1.4a_B$, the theoretical p-p separation in H_2. At the same time, also the e-e functions have a clear peak around $r = 0.5a_B$, the two electrons are concentrated between the protons. In contrast, in the case of parallel spins, no molecules are formed, the most probable electron distance is around $r = 1.2a_B$.

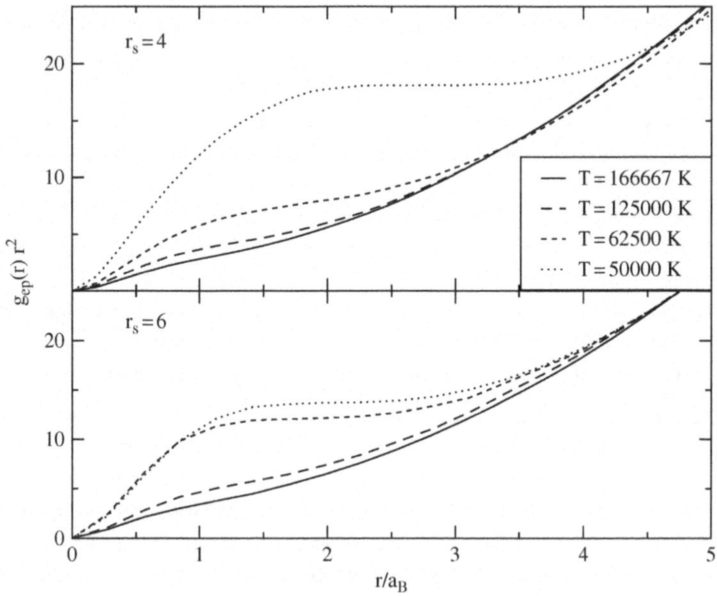

Fig. 2.3. Electron-proton (e-p) pair distribution functions multiplied by r^2 as function of e-p distance at r_s= 4 (**top**) and r_s = 6 (**bottom**) at four temperatures [19]

2.2.3 Molecular Dynamics Results for Dynamical Quantities

We now extend the analysis to the dynamical properties of a hydrogen plasma in equilibrium using the fluctuation-dissipation theorem. The time-dependent microscopic density of plasma species α is defined as

$$\rho^\alpha(\boldsymbol{r},t) = \sum_{i=1}^{N_\alpha} \delta[\boldsymbol{r} - \boldsymbol{r}_i^\alpha(t)] \,, \tag{2.11}$$

with the Fourier components

$$\rho^\alpha(\boldsymbol{k},t) = \sum_{i=1}^{N_\alpha} \mathrm{e}^{\mathrm{i}\boldsymbol{k}\cdot\boldsymbol{r}_i^\alpha(t)} \,, \tag{2.12}$$

where $\boldsymbol{r}_i^\alpha(t)$ denotes the trajectory of particle i obtained in the simulation. We now define the three partial density-density time correlation functions (DDCF) between sorts α and η as

$$A^{\alpha\eta}(\boldsymbol{k},t) = \frac{1}{N_\alpha + N_\eta} \left\langle \rho^\alpha(\boldsymbol{k},t)\rho^\eta(-\boldsymbol{k},0) \right\rangle \,, \tag{2.13}$$

where, due to isotropy, $\boldsymbol{k} = k$. Here $\langle \rho^\alpha(\boldsymbol{k},t)\rho^\eta(-\boldsymbol{k},0)\rangle$ denotes averaging along the trajectories by shifting the time interval and keeping the difference equal to t.

Note also, that $A^{\alpha\eta}(\boldsymbol{k},t) = A^{\eta\alpha}(\boldsymbol{k},t)$ for all pairs α and η. In addition to the spin-resolved electron functions we can also consider the spin averaged correlation function $A(\boldsymbol{k},t) = A^{\uparrow\uparrow}(\boldsymbol{k},t) + A^{\downarrow\uparrow}(\boldsymbol{k},t)$.

We have performed a series of simulation runs of equilibrium fluctuations in hydrogen plasmas with coupling parameters Γ and electron degeneracy parameters $\chi_e = \rho\Lambda_e^3$ with the electron de Broglie wavelength $\Lambda_e = \hbar/\sqrt{2\pi m_e k_B T}$ ranging from zero (classical system) to one (quantum or degenerate system). The electron DDCF for $\Gamma = 1$ and $\chi_e = 1$ are plotted in Fig. 2.4 for four values of the dimensionless wavenumber $q = k\bar{r}$. The correlation functions ($\uparrow\uparrow$ and $\downarrow\uparrow$) have two characteristic features – a highly damped, high-frequency part and a weakly damped low-frequency tail. The latter originates from slow ionic motion whereas the high-frequency part is related to oscillations with frequencies close to the electron plasma frequency ω_{pl}. On the other hand, the time scale of the ion motion is determined by the ion plasma frequency $\omega_{pl}^i = \sqrt{4\pi\rho_i Z_i^2 e^2/m_i}$, the ratio of the two time scales is $\sqrt{m_i/m_e} \approx 43$. The slow proton oscillations are clearly seen in the proton DDCF, shown in Fig. 2.5. To resolve the proton oscillations the whole simulation (including the electron dynamics) has to extend over several proton plasma periods $T_p = 2\pi/\omega_{pl}^i$ thereby resolving the fast electronic motions as well, which sets the numerical limitation of the calculation.

The temporal Fourier transform of the DDCF yields another very important quantity – the dynamic structure factor, $S_{\alpha,\eta}(\omega,q)$, which allows one to analyze, e.g., the dispersion of the coupled electron and proton oscillations. Fig. 2.6 shows

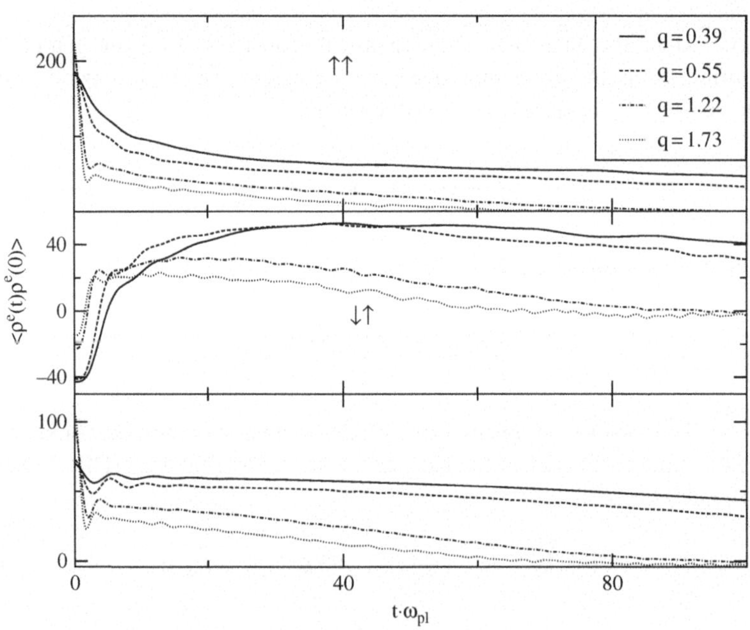

Fig. 2.4. Electron DDCF (2.13) multiplied by $(N_e^{\uparrow} + N_e^{\downarrow})$ for $\Gamma = 1$ and $\chi_e = 1$ for four wave vectors. **Upper (middle) panel**: Correlation functions for parallel (antiparallel) spins. **Bottom**: Spin-averaged function [25]

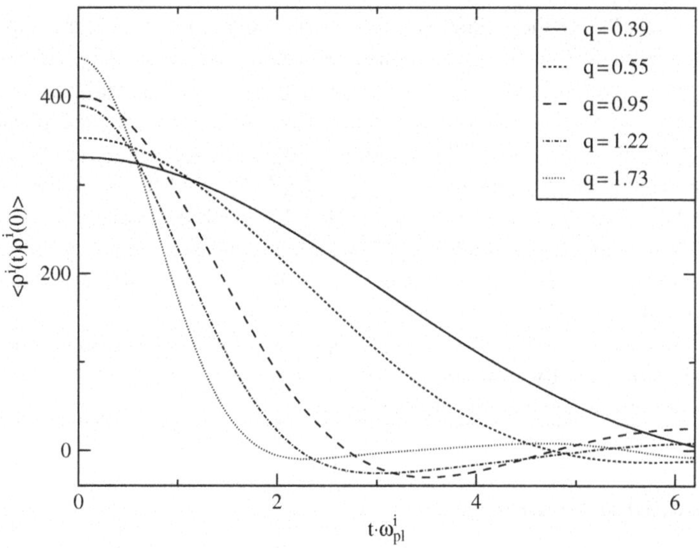

Fig. 2.5. Proton DDCF (2.13) for $\Gamma = 1$ and $\chi_e = 1$ for five wave vectors (in units of $1/\bar{r}$) [25]

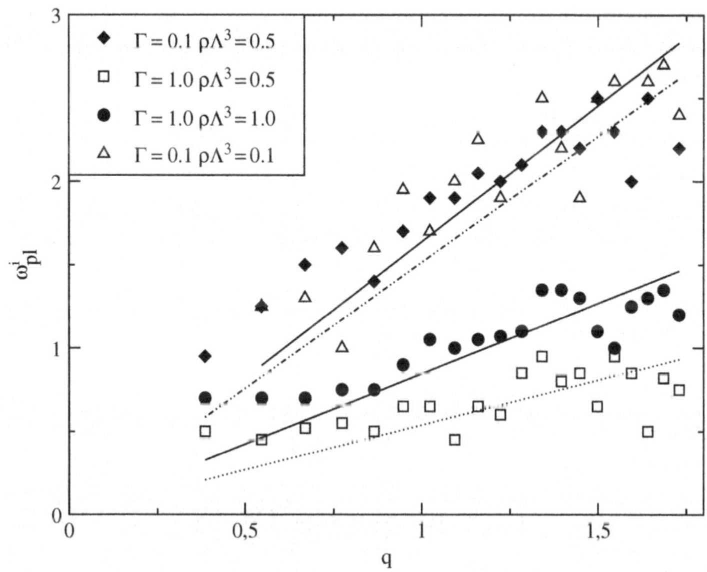

Fig. 2.6. Ion-acoustic wave dispersion in a dense hydrogen plasma. Lines correspond to weighted linear fits to the MD data (*symbols*). The scatter of the data is due to the limited particle number N and simulation time and can be systematically reduced. Also, smaller q-values require larger N [25]

dispersion results for the collective proton oscillations, for the electron modes see [22, 24], which follow from the peak positions of $S_{ii}(\omega, q)$. Fig. 2.6 shows the peak frequency versus wave number, i.e. the dispersion of longitudinal ion-acoustic waves, $\omega(q) = v_{\rm MD}\, q$, where $v_{\rm MD}$ denotes our MD result for the phase velocity. This can be compared to the familiar analytical expression for an ideal two-temperature ($T_e \gg T_i$) plasma $v_s = \sqrt{Z_i k_B T_e/m_i}$, where v_s is the ion sound velocity. We observe deviations of about 10% for weak degeneracy $\chi_e < 0.5$, and about 10% for large degeneracy $\chi_e \geq 1$, which are due to nonideality (correlations) and quantum effects, directly included in our simulations. For further details on this method see [6, 24, 25].

Thus semiclassical MD is a powerful approach to correlated quantum plasmas. Thermodynamic and dynamic properties are accurately computed if accurate quantum pair potentials, such as the IKP, are used.

2.3 Quantum Dynamics

Let us now discuss the method of Wigner trajectories in more detail. As we have seen, the Wigner function W (to avoid confusion, in this section we rename $f_{\rm W} \to W$) in (2.2) is the Fourier transform of the non-diagonal elements of the density matrix which, for a pure state, is $\rho\left(q + \frac{\nu}{2}, q - \frac{\nu}{2}\right) = \psi(q + \frac{\nu}{2}, t)\psi^*(q - \frac{\nu}{2}, t)$, where the $N-$particle wave functions satisfy the Schrödinger equation with an initial condition

$$\mathrm{i}\hbar \frac{\partial \psi}{\partial t} = \widehat{H}\psi\,, \qquad \psi\,(t_0) = \psi^0\,(q)\,, \tag{2.14}$$

which contains the Hamiltonian (2.3); recall that q is a vector of dimension Nd. By taking the time derivative of W in (2.2) and substituting $\partial\psi/\partial t$ in the l.h.s of the Schrödinger equation we recover (2.4), after integrating by parts. For convenience, on both sides we add the contribution of the classical force, $\boldsymbol{F}(\boldsymbol{q}) = -\boldsymbol{\nabla}_q V(\boldsymbol{q})$, which leads to a new function ω which differs from $\widetilde{\omega}$ in (2.4) by an additional term, the last term in (2.16),

$$\frac{\partial W}{\partial t} + \frac{\boldsymbol{p}}{m} \cdot \boldsymbol{\nabla}_q W + \boldsymbol{F}(\boldsymbol{q}) \cdot \boldsymbol{\nabla}_p W = \int\limits_{-\infty}^{\infty} \mathrm{d}s\, W\,(p - s, q, t)\;\omega\,(s, q, t)\;, \tag{2.15}$$

$$\omega\,(s, q, t) = \frac{2}{(\pi\hbar^2)^{Nd}} \int \mathrm{d}q' V\,(q - q', t) \sin\left(\frac{2sq'}{\hbar}\right) + \boldsymbol{F}\,(q) \cdot \boldsymbol{\nabla}_s \delta\,(s)\;. \tag{2.16}$$

In the classical limit ($\hbar \to 0$), the r.h.s of (2.15) vanishes and we obtain the classical Liouville equation

$$\frac{\partial W}{\partial t} + \frac{\boldsymbol{p}}{m} \cdot \boldsymbol{\nabla}_q W + \boldsymbol{F}\,(\boldsymbol{q}) \cdot \boldsymbol{\nabla}_p W = 0\,. \tag{2.17}$$

The solution of (2.17) is known and can be expressed by the Green function [9]

$$G(p,q,t;p_0,q_0,t_0) = \delta\left[p - \overline{p}(t;t_0,p_0,q_0)\right]\delta\left[q - \overline{q}(t;t_0,p_0,q_0)\right] \,, \tag{2.18}$$

where $\overline{p}(\tau)$ and $\overline{q}(\tau)$ are the phase space trajectories of all particles, which are the solutions of Hamilton's equations with the initial conditions at $\tau = t_0 = 0$,

$$\begin{aligned} \frac{\mathrm{d}\bar{q}}{\mathrm{d}\tau} &= \frac{\bar{p}(\tau)}{m}; & \bar{q}(0) &= q_0, \\ \frac{\mathrm{d}\bar{p}}{\mathrm{d}\tau} &= \boldsymbol{F}(\bar{q}(\tau)); & \bar{p}(0) &= p_0. \end{aligned} \tag{2.19}$$

Using the Green function, the time-dependent solution of the classical Liouville equation takes the form

$$W(p,q,t) = \int \mathrm{d}p_0\, \mathrm{d}q_0\; G(p,q,t;p_0,q_0,0)\; W_0(p_0,q_0) \,. \tag{2.20}$$

With this result, it is now possible to construct a solution also for the quantum case. To this end we note that it is straightforward to convert (2.15) into an integral equation

$$\begin{aligned} W(p,q,t) = &\int \mathrm{d}p_0\, \mathrm{d}q_0\; G(p,q,t;p_0,q_0,0)\; W_0(p_0,q_0) \\ &+ \int_0^t \mathrm{d}t_1 \int \mathrm{d}p_1\, \mathrm{d}q_1\; G(p,q,t;p_1,q_1,t_1) \\ &\times \int_{-\infty}^{\infty} \mathrm{d}s_1\, \omega(s_1,q_1,t_1)\; W(p_1 - s_1,q_1,t_1) \,, \end{aligned} \tag{2.21}$$

which is exact and can be solved efficiently by iteration [10, 11]. The idea is to replace the unknown function W under the integral in (2.21) by an approximation. The first approximation is obtained by solving (2.21) to lowest order, i.e. by neglecting the integral term completely. This gives the first order result for W which can again be substituted for W in the integral in (2.21) and so on. This way we can systematically derive improved approximations for W. The procedure leads to a series of terms of the following general form,

$$\begin{aligned} W(p,q,t) = \; & W^{(0)}(p,q,t) + W^{(1)}(p,q,t) + \int_0^t \mathrm{d}t_1 \int \mathrm{d}1\; G(p,q,t;1,t_1) \\ &\times \int_0^{t_1} \mathrm{d}t_2 \int \mathrm{d}2\; G(p_1 - s_1,q_1,t_1;2,t_2) \\ &\times \int_{-\infty}^{\infty} \mathrm{d}s_2\, \omega(s_2,q_2,t_2)\; W(p_2 - s_2,q_2,t_2) \,, \end{aligned} \tag{2.22}$$

where we have introduced the notations $n \equiv q_n, p_n$, $\mathrm{d}n \equiv \mathrm{d}q_n \mathrm{d}p_n$ and

$$W^{(0)}(p,q,t) = \int \mathrm{d}0\, G(p,q,t;0,0) W_0(0)\,,$$

$$W^{(1)}(p,q,t) = \int_0^t \mathrm{d}t_1 \int_{-\infty}^{\infty} \mathrm{d}1\, G(p,q,t;1,t_1) \int_{-\infty}^{\infty} \mathrm{d}s_1\, \omega(s_1,q_1,t_1) \times \int \mathrm{d}0\, G(p_1 - s_1, q_1, t_1; 0, 0) W_0(0)\,. \tag{2.23}$$

The terms $W^{(0)}$ and $W^{(1)}$ are the first of an infinite series. To shorten the notation, all higher order terms are again summed up giving rise to the last term in (2.22). Below we will give also the third term, $W^{(2)}$, but first we discuss the physical interpretation of each contribution.

$W^{(0)}(p,q,t)$, as it follows from the Green function $G(p,q,t;p_0,q_0,0)$, describes the propagation of the Wigner function along the classical characteristics, i.e., the solutions of Hamilton's equations (2.19) in the time interval $[0,t]$. It is worth mentioning, that this first term describes both classical and quantum effects, due to the fact that the initial Wigner function $W_0(p_0,q_0)$, in general, contains all powers of Planck's constant $\hbar$ contained in the initial state wave functions. These are quantum diffraction and spin effects, depending on the quality of the initial function.

The second and third terms on the r.h.s. of (2.22) describe additional quantum corrections to the time evolution of $W(p,q,t)$ arising from non-classical time propagation, in particular, the Heisenberg uncertainty principle. Let us consider the term $W^{(1)}(p,q,t)$ in more detail. It was first proposed in [11]. Later on it was demonstrated that the multiple integral (2.23) can be calculated stochastically by Monte Carlo techniques [12, 13, 14]. For this we need to generate an ensemble of trajectories in phase space. To each trajectory we ascribe a specific weight, which gives its contribution to (2.23). For example, let us consider a trajectory which starts at point $\{p_0, q_0, \tau = 0\}$. This trajectory acquires a weight equal to the value $W_0(p_0,q_0)$. Up to the time $\tau = t_1$ the trajectory is defined by the Green function $G(p_1 - s_1, q_1, t_1; p_0, q_0, 0)$. At $\tau = t_1$, as it follows from (2.23), the weight of this trajectory must be multiplied by the factor $\omega(s_1,q_1,t_1)$, and simultaneously a perturbation in momentum takes place: $(p_1 - s_1) \rightarrow p_1$. As a result the trajectory becomes discontinuous in momentum space, but continuous in the coordinate space. Obviously this is a manifestation of the Heisenberg uncertainty of coordinates and momenta. Now the trajectory consists of two parts – two classical trajectories which are the solutions of (2.19), which are separated, at $\tau = t_1$ by a momentum jump of magnitude s_1. What about the value s_1 of the jump and the time moment t_1? Both appear under integrals with a certain probability. To sample this probability adequately, a statistical ensemble of trajectories should be generated, further the point in time t_1 must be chosen randomly in the interval $[0,t]$, and the momentum jump s_1 randomly in the interval $[-\infty, +\infty]$. Finally, also different starting points $\{p_0, q_0\}$ of trajectories at $\tau = 0$ must be considered due to the integration $\int \mathrm{d}p_0 \mathrm{d}q_0$. Considering a sufficiently large number of trajectories of such type we can accurately calculate $W^{(1)}(p,q,t)$ – the first correction to the classical evolution of the quantum distribution function $W^{(0)}(p,q,t)$.

Let us now take into account the third term in (2.22). We substitute, instead of $W(p_2 - s_2, q_2, t_2)$, its integral representation, using (2.21). As a result we get for this term

$$\begin{aligned} W^{(2)}(p,q,t) = &\int_0^t \mathrm{d}t_1 \int \mathrm{d}1 G(p,q,t;1,t_1) \int_{-\infty}^{\infty} \mathrm{d}s_1\, \omega(s_1,q_1,t_1) \\ &\times \int_0^{t_1} \mathrm{d}t_2 \int \mathrm{d}2\, G(p_1 - s_1, q_1, t_1; 2, t_2) \int_{-\infty}^{\infty} \mathrm{d}s_2\, \omega(s_2,q_2,t_2) \\ &\times \int \mathrm{d}0 G(p_2 - s_2, q_2, t_2; 0, 0)\, W_0(0)\,. \end{aligned} \tag{2.24}$$

If we apply the stochastic interpretation of the integrals, as we did above for $W^{(1)}(p,q,t)$, this term can be analogously calculated using an ensemble of classical trajectories with *two* momentum jumps taking place at time moments $\tau = t_1$ and $\tau = t_2$, and with a weight function multiplied by the factors $\omega(s_1,q_1,t_1)$ and $\omega(s_2,q_2,t_2)$, respectively.

Applying the above procedure several times, we can get the higher order correction terms. As a result, $W(p,q,t)$ will be expressed as an iteration series, with each term of the series representing a contribution of trajectories of a definite topological type – with one, two, three, etc. momentum jumps. In Fig. 2.7 we show an example of trajectories contributing to the terms $W^{(0)}$, $W^{(1)}$ and $W^{(2)}$.

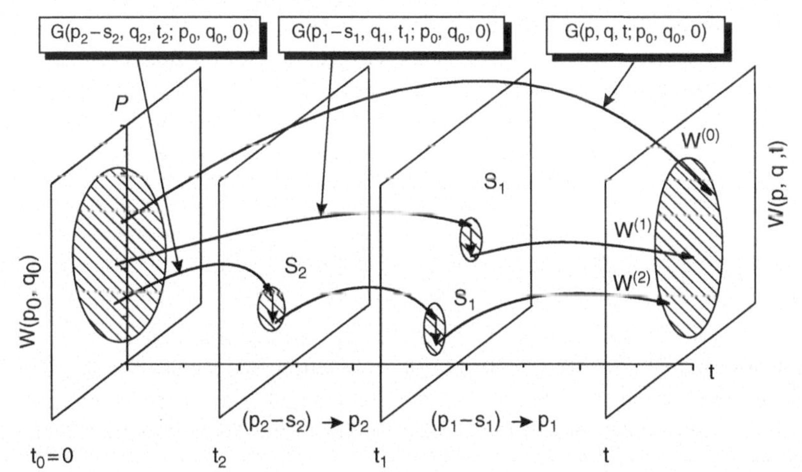

Fig. 2.7. Illustration of the iteration series. Three types of trajectories are shown: Without (*top curve*), with one (*middle*) and with two (*lower*) momentum jumps

As was noted in Sect. 2.1 the Wigner function allows us to compute the quantum-mechanical expectation value of an arbitrary one-particle operator $\widehat{A}$. Using the idea of iteration series (2.22), we obtain an iteration series also for the expectation value

$$\langle\widehat{A}\rangle(t) = \int \mathrm{d}p\mathrm{d}q\; A(p,q)W(p,q,t) = \langle\widehat{A}\rangle^{(0)}(t) + \langle\widehat{A}\rangle^{(1)}(t) + \dots\,, \qquad (2.25)$$

where different terms correspond to different terms in the series for W. The series (2.25) maybe computed much more efficiently than the one for W since the result does not depend on coordinates and momenta anymore.

Certainly, in the iteration series it is possible to take into account only a finite number of terms and contributions of a limited number of trajectories. Interestingly, it is not necessary to compute the individual terms iteratively. Instead, all relevant terms can be calculated simultaneously using the basic concepts of MC methods [26]. An important task of the MC procedure will be to generate stochastically the trajectories which give the dominant contribution to the result, for details see [10].

2.4 Time Correlation Functions in the Canonical Ensemble

So far we have considered the dynamics of pure states where the density matrix ρ, which is the matrix representation of the density operator $\widehat{\rho}$, is defined by a single wave function ψ. However, at finite temperature ρ is, in general, defined by an incoherent superposition of wave functions (mixed states). Here we consider the canonical ensemble as the most common one. Time correlation functions $C_{FA}(t) = \langle F(0)A(t)\rangle$ are among the most important quantities in statistical physics which describe transport properties, such as diffusion, dielectric properties, chemical reaction rates, equilibrium or non-equilibrium optical properties. An example has already been considered in Sect. 2.2 – the density-density auto-correlation function (2.13). Here we use a more general expression for the quantum correlation function of two quantities A and F given by the operators $\widehat{F}$ and $\widehat{A}$. In the canonical ensemble the averaging is performed by a trace with the canonical density operator $\widehat{\rho}^{\mathrm{eq}} = Z^{-1}\exp(-\beta\widehat{H})$, with $\beta = 1/k_{\mathrm{B}}T$, and the correlation function has the form [27]

$$C_{FA}(t) = \frac{1}{Z}\mathrm{Tr}\left(\widehat{F}\,\mathrm{e}^{\mathrm{i}\widehat{H}t_\beta^*}\,\widehat{A}\,\mathrm{e}^{-\mathrm{i}\widehat{H}t_\beta}\right)\,, \qquad (2.26)$$

where $\widehat{H}$ is the Hamiltonian (2.3), t_β is a complex time argument $t_\beta = t - \mathrm{i}\beta/2$ which absorbs $\widehat{\rho}^{\mathrm{eq}}$, $Z = \mathrm{Tr}\widehat{\rho}^{\mathrm{eq}}$ is the partition function, and we use $\hbar = 1$.

The time correlation function can now be computed by first writing (2.26) in coordinate representation and then transforming to the Wigner picture, using the Weyl representation of $\widehat{F}$ and $\widehat{A}$,

$$
\begin{aligned}
C_{FA}(t) &= \frac{1}{Z}\int \mathrm{d}q_1\mathrm{d}q_2\mathrm{d}q_3\mathrm{d}q_4 \,\left\langle q_1|\widehat{F}|q_2\right\rangle \left\langle q_2|\mathrm{e}^{\mathrm{i}\widehat{H}t^*_\beta}|q_3\right\rangle \\
&\quad\times \left\langle q_3|\widehat{A}|q_4\right\rangle \left\langle q_4|\mathrm{e}^{-\mathrm{i}\widehat{H}t_\beta}|q_1\right\rangle \\
&= \int \mathrm{d}p_1\mathrm{d}q_1\mathrm{d}p_2\mathrm{d}q_2 \; F(p_1,q_1)\; A(p_2,q_2)\; W(p_1,q_1;p_2,q_2;t;\beta)\,,
\end{aligned}
\tag{2.27}
$$

where $W(p_1, q_1; p_2, q_2; t; \beta)$ is now a generalization of the Wigner function which is defined as double Fourier transformation of the product of two non-diagonal matrix elements of the density operator

$$
\begin{aligned}
W(p_1,q_1;p_2,q_2;t;\beta) &= \frac{1}{Z(2\pi)^{2Nd}}\int \mathrm{d}\xi_1\mathrm{d}\xi_2 \; \mathrm{e}^{\mathrm{i}p_1\xi_1}\;\mathrm{e}^{\mathrm{i}p_2\xi_2} \\
&\times \left\langle q_1-\frac{\xi_1}{2}\right|\mathrm{e}^{\mathrm{i}\widehat{H}t^*_\beta}\left| q_2+\frac{\xi_2}{2}\right\rangle \left\langle q_2-\frac{\xi_2}{2}\right|\mathrm{e}^{-\mathrm{i}\widehat{H}t_\beta}\left| q_1+\frac{\xi_1}{2}\right\rangle .
\end{aligned}
\tag{2.28}
$$

Calculating the partial time derivatives of the function W it can be shown that the function W satisfies a system of two Wigner-Liouville equations [12, 13]

$$
\begin{aligned}
&\frac{\partial W}{\partial t} + \frac{\boldsymbol{p}_1}{m}\cdot \boldsymbol{\nabla}_{q_1} W + \boldsymbol{F}(\boldsymbol{q}_1)\cdot \boldsymbol{\nabla}_{p_1} W = I_1\,, \\
&\frac{\partial W}{\partial t} + \frac{\boldsymbol{p}_2}{m}\cdot \boldsymbol{\nabla}_{q_2} W + \boldsymbol{F}(\boldsymbol{q}_2)\cdot \boldsymbol{\nabla}_{p_2} W = I_2\,,
\end{aligned}
\tag{2.29}
$$

where on the r.h.s. we have two collision integrals

$$
\begin{aligned}
I_1 &= \int_{-\infty}^{\infty} \mathrm{d}s_1 \; W\left(p_1 - s_1, q_1; p_2, q_2; t; \beta\right)\, \omega\left(s_1, q_1, t\right)\,, \\
I_2 &= \int_{-\infty}^{\infty} \mathrm{d}s_2 \; W\left(p_1, q_1; p_2 - s_2, q_2; t; \beta\right)\, \omega\left(s_2, q_2, t\right)\,,
\end{aligned}
\tag{2.30}
$$

and the function $\omega(s, q, t)$ is defined in the same way as in the microcanonical ensemble, see (2.16).

2.4.1 Initial Conditions for the Wigner-Liouville Equation

Using (2.28) at $t = 0$, we find that the initial value of the Wigner function is given by the integral

$$
\begin{aligned}
W_0(1;2;0;\beta) &= \frac{1}{Z(2\pi)^{2Nd}}\int \mathrm{d}\xi_1\mathrm{d}\xi_2 \; \mathrm{e}^{\mathrm{i}p_1\xi_1}\;\mathrm{e}^{\mathrm{i}p_2\xi_2} \\
&\times \left\langle q_1-\frac{\xi_1}{2}\right|\mathrm{e}^{-\beta\widehat{H}/2}\left| q_2+\frac{\xi_2}{2}\right\rangle \left\langle q_2-\frac{\xi_2}{2}\right|\mathrm{e}^{-\beta\widehat{H}/2}\left| q_1+\frac{\xi_1}{2}\right\rangle
\end{aligned}
\tag{2.31}
$$

with $1 = q_1, p_1$ and $2 = q_2, p_2$.

Let us now exploit the group property of the density operator $\widehat{\rho}$ and the high temperature approximation for the matrix elements of $\langle q'|\widehat{\rho}|q\rangle$ (see Chap. 13)

$$
\begin{aligned}
\mathrm{e}^{-\beta\widehat{H}} &= \left[\mathrm{e}^{-\beta/M\widehat{H}}\right]^M \\
\left\langle q' \left| \mathrm{e}^{-\beta/(2M)\widehat{H}} \right| q'' \right\rangle &\approx \left\langle q' \left| \mathrm{e}^{-\beta/(2M)\widehat{K}} \right| q'' \right\rangle \left\langle q' \left| \mathrm{e}^{-\beta/(2M)\widehat{U}} \right| q'' \right\rangle . \qquad (2.32)
\end{aligned}
$$

Then we obtain

$$
\begin{aligned}
W_0(1;2;0;\beta) \approx & \frac{1}{Z(2\pi\hbar)^{2Nd}} \int \mathrm{d}q_1' \ldots \mathrm{d}q_M' \, \mathrm{d}q_1'' \ldots \mathrm{d}q_M'' \mathrm{e}^{-\sum_{m=2}^M K_m - \sum_{m=1}^M U_m} \\
& \times \int \mathrm{d}\xi_1 \mathrm{e}^{\mathrm{i}p_1\xi_1/\hbar} \left\langle q_M' \left| \mathrm{e}^{-\beta\widehat{K}/(2M)} \right| q_1 + \frac{\xi_1}{2} \right\rangle \left\langle q_1 - \frac{\xi_1}{2} \left| \mathrm{e}^{-\beta\widehat{K}/(2M)} \right| q_1'' \right\rangle \\
& \times \int \mathrm{d}\xi_2 \mathrm{e}^{\mathrm{i}p_2\xi_2/\hbar} \left\langle q_M'' \left| \mathrm{e}^{-\beta\widehat{K}/(2M)} \right| q_2 + \frac{\xi_2}{2} \right\rangle \left\langle q_2 - \frac{\xi_2}{2} \left| \mathrm{e}^{-\beta\widehat{K}/(2M)} \right| q_1' \right\rangle ,
\end{aligned}
\qquad (2.33)
$$

where $K_m = (\pi/\lambda_M^2)\left[(q_m' - q_{m-1}')^2 + (q_m'' - q_{m-1}'')^2\right]$ and $U_m = (\beta/(2M)) [U(q_m') + U(q_m'')]$. Here we have assumed that $M \gg 1$, and $\lambda_M^2 = 2\pi\hbar^2\beta/(mM)$ denotes the thermal de Broglie wave length corresponding to the inverse temperature $\beta/(2M)$. A direct calculation of the last two factors in (2.33) gives

$$
\begin{aligned}
& \int \mathrm{d}\xi_1 \mathrm{e}^{\mathrm{i}p_1\xi_1/\hbar} \left\langle q_M' \left| \mathrm{e}^{-\beta\widehat{K}/(2M)} \right| q_1 + \frac{\xi_1}{2} \right\rangle \left\langle q_1 - \frac{\xi_1}{2} \left| \mathrm{e}^{-\beta\widehat{K}/(2M)} \right| q_1'' \right\rangle \\
& = \left\langle q_M' \left| \mathrm{e}^{-\beta\widehat{K}/(2M)} \right| q \right\rangle \phi(p; q_M', q_1) \left\langle q \left| \mathrm{e}^{-\beta\widehat{K}/(2M)} \right| q_1 \right\rangle ,
\end{aligned}
\qquad (2.34)
$$

where

$$
\phi\left(p; q_M', q_1\right) = (2\lambda_M^2)^{Nd/2} \mathrm{e}^{-(p\lambda_M/\hbar + \mathrm{i}\pi(q'-q'')/\lambda_M)^2/(2\pi)} \qquad (2.35)
$$

The final result for the Wigner function at $t = 0$ can be written as

$$
\begin{aligned}
W(1;2;0;\beta) \approx & \int \mathrm{d}q_1' \ldots \mathrm{d}q_M' \, \mathrm{d}q_1'' \ldots \mathrm{d}q_M'' \, \Psi(1;2;q_1' \ldots q_M'; q_1'' \ldots q_M''; 0; \beta) \\
& \times \phi(p_2; q_M', q_1'') \, \phi(p_1; q_M'', q_1') ,
\end{aligned}
\qquad (2.36)
$$

where

$$
\Psi(p_1, q_1; p_2, q_2; q_1' \ldots q_M'; q_1'' \ldots q_M''; \beta) = \frac{1}{Z} \mathrm{e}^{-\sum_{m=1}^{M+1} K_m - \sum_{m=1}^M U_m} . \qquad (2.37)
$$

Here we have introduced the notation $\{q_0' \equiv q_1; q_0'' \equiv q_2\}$ and $\{q_{M+1}' \equiv q_2; q_{M+1}'' \equiv q_1\}$. Fig. 2.8 illustrates the simulation idea. Two closed loops with the set of points

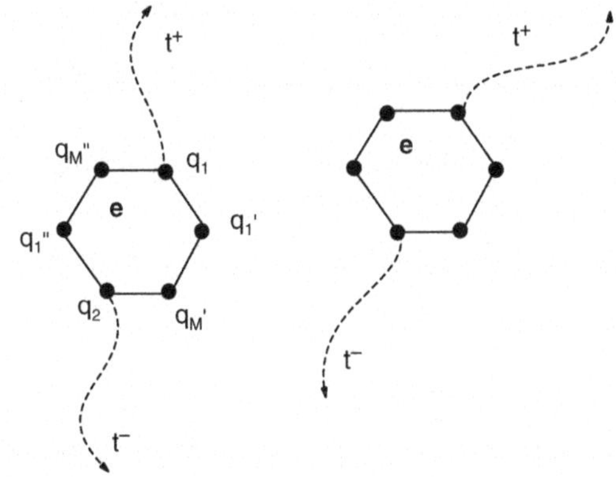

Fig. 2.8. Two closed loops illustrating the path integral representation of two electrons in the density matrices in (2.33). Two special points, (p_1, q_1) and (p_2, q_2), are starting points for two dynamical trajectories propagating forward and backward in time

show the path integral representation of the density matrices in (2.33). The left chain of points, i.e. $\{q_1, q_1', \ldots, q_M', q_2, q_1'', \ldots, q_M''\}$ characterizes the path of a single quantum particle. The chain has two special points (p_1, q_1) and (p_2, q_2). As it follows from (2.28) and (2.29) these points are the original points for the Wigner function, the additional arguments arise from the path integral representation. As we show in the next section, we can consider these points as starting points for two dynamical trajectories propagating forward and backward in time, i.e. $t \to t^+$ and $t \to t^-$. The Hamilton equations for the trajectories are defined in the next section.

2.4.2 Integral Equations

The solution follows the scheme explained before. The only difference is that we now have to propagate two trajectories instead of one,

$$
\begin{aligned}
\frac{\mathrm{d}\bar{q}_1}{\mathrm{d}\tau} &= \frac{\bar{p}_1(\tau)}{2m}\,, && \bar{q}_1(0) = q_1^0\,, \\
\frac{\mathrm{d}\bar{p}_1}{\mathrm{d}\tau} &= \frac{1}{2}\boldsymbol{F}[\bar{q}_1(\tau)]\,, && \bar{p}_1(0) = p_1^0\,, \\
\frac{\mathrm{d}\bar{q}_2}{\mathrm{d}\tau} &= -\frac{\bar{p}_2(\tau)}{2m}\,, && \bar{q}_2(0) = q_2^0\,, \\
\frac{\mathrm{d}\bar{p}_2}{\mathrm{d}\tau} &= -\frac{1}{2}\boldsymbol{F}[\bar{q}_2(\tau)], && \bar{p}_2(0) = p_2^0\,.
\end{aligned}
\tag{2.38}
$$

The first (second) trajectory propagates forward (backward). Let us substitute expressions for $\boldsymbol{F}[\bar{q}_1(\tau)], \bar{p}_1(\tau), \boldsymbol{F}[\bar{q}_2(\tau)]$ and $\bar{p}_2(\tau)$ from (2.38) into (2.29) and subtract the second equation from the first. As a result, on the l.h.s. we obtain a full differential of the Wigner function. After multiplication by the factor $1/2$ and integration over time, the integral equation for the Wigner function takes the form

$$
\begin{aligned}
W(p_1,q_1;p_2,q_2;t;\beta) &= \int \mathrm{d}p_1^0\mathrm{d}q_1^0\mathrm{d}p_2^0\mathrm{d}q_2^0 \\
&\times G(p_1,q_1,p_2,q_2,t;p_1^0,q_1^0,p_2^0,q_2^0,0)W(p_1^0,q_1^0;p_2^0,q_2^0;0;\beta) \\
&+ \int_0^t \mathrm{d}\tau \int \mathrm{d}p_1^1\mathrm{d}q_1^1\,\mathrm{d}p_2^1\mathrm{d}q_2^1\; G(p_1,q_1,p_2,q_2,t;p_1^1,q_1^1,p_2^1,q_2^1,\tau) \\
&\times \int_{-\infty}^{\infty} \mathrm{d}s\,\mathrm{d}\eta\; \vartheta(s,q_1^1;\eta,q_2^1;\tau)\; W(p_1^1-s,q_1^1;p_2^1-\eta,q_2^1;\tau;\beta)\,, \qquad (2.39)
\end{aligned}
$$

where $\vartheta(s,q_1^1;\eta,q_2^1;\tau) = [\omega(s,q_1^1)\delta(\eta) - \omega(\eta,q_2^1)\delta(s)]/2$. The dynamical Green function G is defined as $G(p_1,q_1,p_2,q_2,t;p_1^0,q_1^0,p_2^0,q_2^0,0)=\delta[p_1-\bar{p_1}(\tau;p_1^0,q_1^0,0)]$ $\delta[q_1 - \bar{q_1}(\tau;p_1^0,q_1^0,0)]\delta[p_2 - \bar{p_2}(\tau;p_2^0,q_2^0,0)]\delta[q_2 - \bar{q_2}(\tau;p_2^0,q_2^0,0)]$. Let us denote the first term on the r.h.s. of (2.39) as $W^{(0)}(p_1,q_1;p_2,q_2;t;\beta)$. This term represents the Wigner function of the initial state propagating along classical trajectories (characteristics – solutions of (2.38)). Using the approach applied for the microcanonical ensemble, we obtain expressions for $W^{(1)}(p_1,q_1;p_2,q_2;t;\beta)$, $W^{(2)}(p_1,q_1;p_2,q_2;t;\beta),\ldots$ and represent $W(p_1,q_1;p_2,q_2;t;\beta)$ as iteration series. In this case, we can calculate this also with an ensemble of trajectories using the quantum dynamics MC approach described in [28]. As a result the expression for the time correlation function (2.27) can be rewritten as

$$
\begin{aligned}
C_{FA}(t) &= \int \mathrm{d}p_1\mathrm{d}q_1\,\mathrm{d}p_2\mathrm{d}q_2 F(p_1,q_1)A(p_2,q_2)W(p_1,q_1;p_2,q_2;t;\beta) \\
&= \Big(\phi(P)|W^{(0)}(P;\beta)\Big) + \sum_{i=1}^{\infty}\Big(\phi(P)|W^{(i)}(P;\beta)\Big)\,, \qquad (2.40)
\end{aligned}
$$

where $\big(\phi(P)|W^{(i)}(P;\beta)\big)$ denotes the integral in the phase space $\{p_1,q_1,p_2,q_2\}$ (now we consider a $2N$-particle system), and $\phi(P) = F(p_1,q_1)A(p_2,q_2)$.

An illustrative example for the calculations of the time correlation functions C_{FA} is the momentum-momentum autocorrelation function $C_{PP}(t)$ for a 1D system of interacting electrons in an array of fixed random scatterers at finite temperature [28]. This system is of high interest because at zero temperature it shows Anderson localization if e-e interaction is neglected. It is a long standing question what the effect of e-e interaction on localization will be. The present method is, in principle, well suited to answer this question. In [28] the first applications of the method to an 1D system at finite temperature have been presented showing that Coulomb e-e interaction has the trend to enhance the mobility of localized electrons [10, 28].

2.5 Discussion

We have presented a general idea how to extend the powerful method of molecular dynamics to quantum systems. First, we discussed semi-classical MD, i.e., classical

MD with accurate quantum pair potentials. This method is very efficient and allows to compute thermodynamic properties of partially ionized plasmas for temperatures above the molecule binding energy (i.e. as long as three and four particle correlations can be neglected). Further, frequency dependent quantities, e.g., the plasmon spectrum, are computed correctly for $\omega < \omega_{pl}$. Further progress is possible if more general quantum potentials are derived.

In the second part, we considered methods for a rigorous solution of the quantum Wigner-Liouville equation for the N-particle Wigner function. Results were derived for both, a pure quantum state and a mixed state (canonical ensemble). Although this method is by now well formulated, it is still very costly in terms of CPU time, so that practical applications are only starting to emerge. Yet, we expect that, due to its first principle character, Wigner function QMD will become increasingly important for a large variety of complex many-body problems.

This work is supported by the Deutsche Forschungsgmeinschaft through SFB TR 24 and in part by Award No. Y2-P-11-02 of the U.S. Civilian Research and Development Foundation for the Independent States of the Former Soviet Union (CRDF) and of Ministry of Education and Science of Russian Federation, and RF President Grant NS-3683.2006.2 for governmental support of leading scientific schools.

References

1. M. Allen, D. Tildesley, *Computer Simulations of Liquids* (Clarendon Press, Oxford, 1987)
2. D. Frenkel, B. Smit, *Understanding Molecular Simulations: From Algorithms to Applications* (Academic Press, Fribourg, 2002)
3. H. Feldmeier, J. Schnack, Rev. Mod. Phys. **72**, 655 (2000)
4. D. Klakow, C. Toepffer, P.G. Reinhard, Phys. Lett. A **192**, 55 (1994)
5. D. Klakow, C. Toepffer, P.G. Reinhard, J. Chem. Phys. **101**(12), 10766 (1994)
6. G. Zwicknagel, T. Pschiwul, J. Phys. A: Math. General **39**, 4359 (2006)
7. M. Bonitz, *Quantum Kinetic Theory* (B.G. Teubner, Stuttgart/Leipzig, 1998)
8. E. Wigner, Phys. Rev. **40**, 749 (1932)
9. V. Tatarsky, Sov. Phys. Usp. **26**(4), 311 (1983)
10. M. Bonitz, D. Semkat (eds.), *Introduction to Computational Methods in Many Body Physics* (Princeton: Rinton Press, 2006)
11. V. Filinov, Y. Medvedev, V. Kamskyi, Mol. Phys. **85**(4), 711 (1995)
12. V. Filinov, Mol. Phys. **88**(6), 1517 (1996)
13. V. Filinov, Y. Lozovik, A. Filinov, E. Zakharov, A. Oparin, Phys. Scripta **58**, 297 (1998)
14. Y. Lozovik, A. Filinov, Sov. Phys. JETP - USSR **88**, 1026 (1999)
15. Y. Lozovik, A. Filinov, A. Arkhipov, Phys. Rev. E **67**, 026707 (2003)
16. G. Kelbg, Ann. Physik **467**(3–4), 219 (1963)
17. G. Kelbg, Ann. Physik **467**(7–8), 354 (1964)
18. G. Kelbg, Ann. Physik **469**(7–8), 394 (1964)
19. A. Filinov, V. Golubnychiy, M. Bonitz, W. Ebeling, J. Dufty, Phys. Rev. E **70**, 046411 (2004); W. Ebeling, A. Filinov, M. Bonitz, V. Filinov, T. Pohl, J. Phys. A: Math. Gen. **39**, 4309 (2006)

20. J. Hansen, I. McDonald, Phys. Rev. A **23**, 2041 (1981)
21. V. Filinov, M. Bonitz, W. Ebeling, V. Fortov, Plasma Phys. Contr. Fusion **43**, 743 (2001); V.S. Filinov, V.E. Fortov, M. Bonitz, D. Kremp, Physics Lett. A **274**, 228 (2000); V.S. Filinov, V.E. Fortov, M. Bonitz, P.R. Levashov, JETP Letters **74**, 384 (2001) [Pis'ma V ZhETF **74**, 422 (2001)]; M. Bonitz, V.S. Filinov, V.E. Fortov, P.R. Levashov, H. Fehske, Phys. Rev. Lett. **95**, 235006 (2005)
22. V. Golubnychiy, M. Bonitz, D. Kremp, M. Schlanges, Phys. Rev. E **64**, 016409 (2001)
23. H. Wagenknecht, W. Ebeling, A. Förster, Contrib. Plasma Phys. **41**, 15 (2001)
24. V. Golubnychiy, Molecular dynamics simulations of strongly correlated mesoscopic and macroscopic coulomb systems. Ph.D. thesis, Kiel University (2004)
25. M. Bonitz, A. Filinov, V. Golubnychiy, T. Bornath, W. Kraeft, Contrib. Plasma Phys. **5–6**, 450 (2005)
26. I. Sobol, *The Monte Carlo Method (Popular Lectures in Mathematics)* (University of Chicago Press, 1975)
27. D. Zubarev, *Nonequilibrium Statistical Thermodynamics* (Plenum Press, New York/London, 1974)
28. V. Filinov, P. Thomas, I. Varga, T. Meier, M. Bonitz, V. Fortov, S. Koch, Phys. Rev. B **65**, 165124 (2002)

Part II

Classical Monte Carlo

3 The Monte Carlo Method, an Introduction

Detlev Reiter

Institut für Energieforschung - Plasmaphysik, Forschungszentrum Jülich GmbH, 52425 Jülich, Germany

This chapter presents the basic principles of stochastic algorithms, usually called Monte Carlo methods. After some historical notes, the generation of random numbers is discussed. Then, as a first non-trivial example, the concept is applied to the evaluation of integrals. More involved problems will be discussed in the two subsequent chapters of this part.

3.1 What is a Monte Carlo Calculation?

In an early lecture note (around 1960, but see also [1]) one of the pioneers of the Monte Carlo technique M.H. Kalos, quotes the two "definitions": (i) A last resort when doing numerical integration, and (ii) a way of wastefully using computer time. Today common assumptions characterize it as a numerical method involving random numbers in a significant way.

In a certain sense any large computer calculation has random aspects, due to roundoff errors. Also deterministic molecular dynamics (MD) calculations, in which the interaction of a large number of moving particles is followed by integrating Newton's equation can have random results, due to randomly chosen initial conditions and/or due to the large number of particles. These are usually excluded from that definition of Monte Carlo techniques. The involvement of randomness in Monte Carlo methods is rendered more precisely to mean deliberate use of random numbers in a calculation which has the structure of a stochastic process.

Two major areas of application are in statistical mechanics (many particle systems) and in linear kinetic (particle) transport theory. The first type of calculations are embodied in a very specific sampling technique, and are not discussed here. The second, for example traffic flow, finance, genetics, but in particular neutronics, radiation transport, cosmic rays, neutral and charged particle transport in plasmas, etc., rely on a study of many interesting stochastic processes by imitating the random processes directly on the computer. Although intuition, and the resulting high transparency of the procedure, is an important ingredient in this type of stochastic analysis, a sound mathematical basis also exists. This allows rigorous mathematical proofs to be given that certain methods actually provide solutions to certain generic mathematical equations, e.g., to Fredholm Integral Equations of second kind in case of transport problems. Distinct from most numerical schemes, in these stochastic methods error estimates are provided by the method itself rather than requiring

D. Reiter: *The Monte Carlo Method, an Introduction*, Lect. Notes Phys. **739**, 63–78 (2008)
DOI 10.1007/978-3-540-74686-7_3

additional considerations. Some mathematical background and basic statistical results are also needed to analyze results of Monte Carlo simulations, for estimation of errors and for obtaining more economical approaches beyond simple simulation of nature.

There exists a vast amount of introductory literature, from a basic text-book level up to monographs focussing on very specialized applications (see, for example, [1, 2, 3, 4]), probably now hundreds of web-based lecture notes and uncounted journal articles. This present introductory chapter certainly duplicates most, if not all of that material. Our aims are to introduce the terminology, and to convey the message that Monte Carlo Methods do have a solid basis in measure theory (with the theory of probability as special case thereof). Strict mathematical proofs of convergence of the method to the exact solution exist, but also, and distinct from most numerical concepts, implementation can be be strongly guided by intuition and retain a high transparency even in very complex situations.

We will, after some short historical remarks below, start with introducing the concepts of random events, of error estimates and unbiased procedures for estimation. Practical implementations of Monte Carlo techniques rely on our ability to draw random number from any probability law we wish. Only a few, most basic facts and concepts in this regard will be repeated here in Sect. 3.2. In Sect. 3.3 the central limit theorem and variance reduction techniques will be demonstrated *at work* using the generic example for Monte Carlo methods: Integration by stochastic sampling. The relation of this very general but intuitively clear and transparent application to the mathematically and statistically more involved transport problems (Monte Carlo particle simulation) will be frequently used as guidance here and will be discussed in more detail in Chap. 5.

3.1.1 Historical Notes

Monte Carlo concepts fall into the branch of experimental mathematics. In ordinary mathematics conclusions are deduced from postulates (Deduction). In experimental mathematics conclusions are inferred from observations (Induction). Monte Carlo methods comprise that branch of experimental mathematics, which is concerned with experiments on random events (mainly random numbers). Monte Carlo methods can be of probabilistic or deterministic type.

Usually the first reference to the Monte Carlo Method is the famous needle experiment of Compte de Buffon (1733), a French biologist (1707–1788), Fig. 3.1. Buffon pointed out that if a needle of length L is tossed on a plane with parallel lines a distance D apart $(D > L)$, it has probability $p = 2L/(\pi D)$ to fall such that is crosses one of the lines. Later, also Laplace suggested this procedure to determine π by counting the number of crosses n in N repetitions of the experiment. Then

$$\frac{n}{N} = \frac{2L}{\pi D} \Rightarrow \pi \approx \frac{2L}{D} \cdot \frac{n}{N} . \tag{3.1}$$

This historical use of Monte Carlo has all key features of the method:

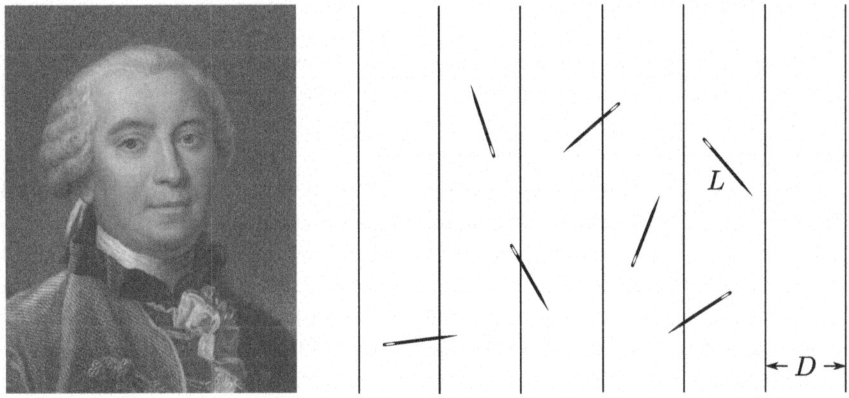

Fig. 3.1. Buffon's needles: What is the probability p, that a needle (length L), which falls randomly on a sheet, crosses one of the lines (distance D)? (**Left**: ©Copyright 1998–2003: The Regents of the University of California)

- *Convergence:* About $N = 100\ 000$ trials are needed for only two digits after the comma. Convergence is slow, but foolproof.
- *Transparency:* The method is intuitively understandable, even without any mathematical reasoning.
- *Error estimates, optimization:* Error estimates and optimal choice of L, D are provided by theory of probability. (Binomial distribution, statistical variance as 2nd central moment etc.).

Modern use Monte Carlo techniques, in the age of digital computers, was initiated by the pioneering work of John von Neumann and Stanislaw Ulam in thermonuclear weapon development. They are also credited for having coined the phrase *Monte Carlo*.

Many monographs on Monte Carlo Methods start with an introduction to measure theory and in particular to elementary probability theory. Although we will introduce and use the proper mathematical vocabulary too, we will, with respect to purely mathematical aspects, refer to those and largely rely upon the intuitive meaning. We refer in particular to the classic monograph by Hammersley and Handscomb [2]. This book provides a short and very readable overview of Monte Carlo. Remarkably, the theoretical foundations today remain rather similar to those from 1964, when this book was first published. Just the applications are far more sophisticated today. The illustrative examples on Monte Carlo integration and some of the advanced techniques in this present introduction will be based upon this text[1].

[1] A pdf-file of that book, which is out of print since long, can be downloaded from the internet, e.g., http://www.eirene.de/html/textbooks.html.

3.1.2 The Basic Principle

The principle is to find (estimate) mean values, i.e. expectation values, I of some system components. If a deterministic problem is to be solved, one first has to invent a stochastic system such that a mean value (= expectation value) coincides with the desired solution I of the deterministic problem.

In any case: I is a single numerical quantity of interest (not an entire functional dependence), and one might always think of I as some definite integral.

The simple intuitive interpretations are given below, but in abstract mathematical terms this stochastic model is given by the probability space (Ω, σ, p, X). Ω is a set of elementary (random) events ω, the σ-field is a set of subsets of Ω to which the measurable function p assigns a value (the probability) from the interval [0, 1], such that the Kolmogoroff axioms for a probability are fulfilled. X is a random variable on Ω, assigning a (usually real) number (or vector) to each random event, e.g.: $X(\omega) \rightarrow \mathbb{R}$, such that $I = E(X)$, the expected value of X.

The expectation value $E(X)$ and variance $\sigma^2(X)$ are defined as the first moment and second central moment, respectively, and, unless otherwise stated, we assume that they both exist

$$E(X) := \int_{\Omega} \mathrm{d}p\, X\ ,$$

$$\sigma^2(X) := \int_{\Omega} \mathrm{d}p\, (X - E(X))^2\ . \tag{3.2}$$

Note that $E(X) = E_p(X)$, $\sigma^2(X) = \sigma_p^2(X)$, i.e., the moments of X of course depend upon the probability measure p.

A stochastic approximation to I is then obtained by producing an independent sequence of random events $\omega_i, i = 1, \ldots N$ according to probability law p and evaluating

$$E(X_N) = I_N = \frac{1}{N}\sum_{i=1}^{N} X(\omega_i)\ . \tag{3.3}$$

The estimator I_N is just the arithmetic mean of many (N) outcomes of the random experiment.

Even without any of this abstract mathematical background it is intuitively clear (see examples below) that I_N will converge to $E(X)$, hence to I by construction, as the number of samples N is increased. However the laws of large numbers and the central limit theorems of probability theory not only provide sound mathematical proofs that this Monte Carlo procedure is exact (unbiased) but also that it converges: $I_N \rightarrow I$ for $N \rightarrow \infty$, albeit slowly (with $1/\sqrt{N}$). In particular the central limit theorem of probability theory[2] asserts that the probability distribution of I_N, for large enough N, converges to a Gaussian distribution, with mean value $I = E(X)$ and

[2] See any textbook on Monte Carlo, or Probability Theory.

variance $\sigma^2(I_N) = \sigma^2(X)/N$. Hence the typical results from statistical error analysis under Gaussian distribution laws apply, e.g., also the resulting confidence levels. It is, therefore, common practice in Monte Carlo applications to quote results as

$$I \approx I_N \pm \sigma(I_N) \quad \text{or} \quad I \approx I_N \pm 2 \cdot \sigma(I_N)\,, \tag{3.4}$$

which have confidence levels of about 66% and 95%, respectively.

Of course, in applications the variance $\sigma^2(X)$ is usually even more difficult to compute than the mean value $E(X)$. It is therefore replaced by the empirical variance

$$\begin{aligned} s^2 &= \frac{1}{N-1}\sum_{i=1}^{N}[X(\omega_i) - E(X_N)]^2 \\ &= \frac{1}{N-1}\left(\sum_{i=1}^{N} X^2(\omega_i) - \frac{1}{N}\left[\sum_{i=1}^{N} X(\omega_i)\right]^2\right) \end{aligned} \tag{3.5}$$

and one has also, under the assumptions made, for large sample size N

$$\begin{aligned} s^2 &\to \sigma^2\,, \\ \sigma^2(I_N) &\approx s_N^2 = \frac{1}{N}\,s^2\,. \end{aligned} \tag{3.6}$$

Hence, for large enough N, in the Gaussian based error estimates (3.4) σ can safely be replaced by s_N, at least for large sample size $N \gtrsim 100$. In the opposite case $N \lesssim 100$ Student's t-distribution should be employed in error analysis instead.

3.2 Random Number Generation

The Monte Carlo method rests on our ability to produce random numbers drawn from any particular probability distribution $F(x)$, or, if it exists, from the probability density function (pdf) $f(x)$, with $F(x) = \int_{-\infty}^{x} \mathrm{d}t\; f(t)$.

Examples are wetting of a surface by rain (uniform distribution), radioactive decay (Poisson distribution), or the distribution of velocities of molecules in a gas (Gaussian distribution). In general, the probability law must be known, either from theory, experiment or plausibility.

A theorem from measure theory states:

Theorem 1. *Each (probability) measure μ can be decomposed into a weighted sum $\mu = p_1\,\mu_c + p_2\,\mu_d + p_3\,\mu_a$ of three parts:*

(i) Part one has a continuous distribution (with probability density $f(x)$).
(ii) Part two has a discrete distribution.
(iii) Part three is a pathological contribution.

The third part is required for the abstract mathematical case only (general measurable spaces, σ-algebras, ...), but it does not occur in practical Monte Carlo applications. This means, for any distribution law arising in an application we can obtain random numbers in two steps: First a random decision (based on the two remaining weighting factors p_1, p_2) whether the continuous or the discrete distribution is to be sampled, and second then generating a random number from the chosen distribution μ_c or μ_d. We will show below that for both cases, continuous and discrete distributions, general procedures for random number generation exist, at least in principle. We refer to the standard reference on the production of nonuniform random numbers [5]. This book deals with the myriad number of ways to transform the uniform random numbers into anything else one might want. Also the first section (pp. 1–193) of [6] is a very comprehensive introduction to random number generation.

3.2.1 Uniform Random Numbers

Uniform random numbers are the basis for generation of random numbers with all other distribution laws. A random variable is uniformly distributed on an interval [a, b], if the distribution density f is

$$f(x) = \frac{1}{b-a}\chi_{[a,b]} , \qquad x \in \mathbb{R} \tag{3.7}$$

with $\chi_{[a,\ b]} = 1$ if x is in the interval $[a,\ b]$ and $f(x) = 0$ elsewhere.

The classical method to generate uniform random numbers on $[0, 1]$ is by so called linear congruential random number generators, which are defined by the recursion

$$\xi_{n+1} = [a\ \xi_n\ +\ b] \quad \text{mod}\ m \tag{3.8}$$

Here a is a magic multiplicand, m if often chosen to be the largest integer representable on the machine ($m = 2^{32}$, etc.), and b should be prime to m. Proofs for particular choices of (large) parameters a and m that the generator achieves the largest possible period of $m - 1$ different random numbers are quite cumbersome. Optimal parameter choices are typically found experimentally, see again [6]. The finite periodicity limits precision only in very large calculations, e.g. on modern massively parallel computing systems. A rather subtle issue is also *independence* of an entire sequence of random numbers (loc.cit.).

3.2.2 Non-uniform Random Numbers

As stated above we only need to consider discrete distributions and continuous distributions.

Finding random numbers with a given discrete distribution is trivial: Let a discrete distribution, with k elementary outcomes labelled by natural numbers $\{0, 1, 2, \ldots, k\}$, be given by

$$
\begin{aligned}
P(X = i) &= p_i \geq 0 \,, \\
\sum_{i=0}^{k} p_i &= 1 \,, \\
F(i) = P(X \leq i) &= \sum_{j=0}^{i} p_i
\end{aligned}
\tag{3.9}
$$

with $P(X = i)$ the probability of event i. F is the (cumulative) distribution. Let ξ be a uniform random number on [0, 1], then the random variable X with $X = i$ if $F(i-1) < \xi \leq F(i)$ is distributed according to F.

3.2.2.1 Inversion Method

The inversion method provides random samples z from a distribution F by converting uniform random numbers ξ. This is simply done by setting

$$z := \min\{x | F(x) \geq \xi\} \sim F \,. \tag{3.10}$$

If F is strictly monotonous, then $z = F^{-1}(\xi)$. For example, if $f(x)$ is the distribution density function (pdf) to be generated, then first find the cumulative function $F(x) = \int_{-\infty}^{x} \mathrm{d}t \; f(t)$, pick a uniform random number ξ on [0, 1] and set $\xi = F(z)$ and finally invert this to find random number z, which is then distributed according to $f(x)$.

The same transformation rules as for any density function apply also for a pdf. Hence the general strategy is: Try to transform a given pdf $f(x)$ to another distribution $\widetilde{f}$, such that the inverse of the new cumulative distribution $\widetilde{F}$ is explicitly known. Then apply the method of inversion and transform back. Figure 3.2 illustrates the method of inversion for the normal (Gaussian) distribution

$$
\begin{aligned}
\phi(x) &= \frac{1}{\sqrt{2\pi}} \mathrm{e}^{-x^2/2} \,, \\
\Phi(x) &= \int_{-\infty}^{x} \mathrm{d}t \; \phi(t) = \frac{1}{2}\left[1 + \mathrm{erf}\left(\frac{x}{\sqrt{2}}\right)\right] \,.
\end{aligned}
\tag{3.11}
$$

Unfortunately, the Gaussian error function $\mathrm{erf}(x)$ and hence $\Phi(x)$ cannot be inverted in closed form. We will show how to generate Gaussian random numbers, even without numerical inversion, further below.

From this procedure follows directly the natural and best format for storing (also multi-dimensional) tabulated data for random sampling in Monte Carlo applications: Form the inverse cumulative distribution function $F^{-1}(x)$ (i.e.: the quantile function) and store this for x uniformly spaced in [0, 1]. Then take ξ from a uniform distribution on [0, 1] and find $F^{-1}(\xi)$ by interpolation in this table.

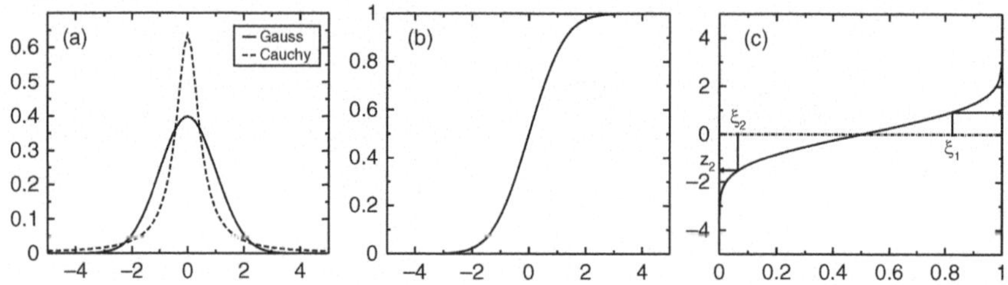

Fig. 3.2. (**a**) Comparison of Cauchy (*dashed line*) and normal distribution (*solid line*). (**b**) Cumulative distribution function $\Phi(x)$ of normal distribution (3.11), (**c**) Inverse cumulative distribution of normal distribution $\Phi^{-1}(\xi)$. Uniform random numbers ξ_1, ξ_2 (abscissa) are converted to random numbers z_1, z_2 from a normal distribution (ordinate)

3.2.2.2 Rejection

Another general method for generating non-uniform random numbers is the rejection method (J.v. Neumann, 1947). This method is *always* applicable, although it may sometimes be rather inefficient. For distributions with finite support, i.e., $f(x) \neq 0$ only on a finite domain M (say, $M = [a,\ b]$), find the maximum c of $f(x)$, sample a random pair (ξ_1, ξ_2) with ξ_1 uniform on M and ξ_2 uniform on [0, c]. If $\xi_2 \leq f(\xi_1)$, accept ξ_1. Otherwise reject this pair and pick a new pair. Repeat this procedure until a pair is accepted. Clearly, the efficiency of this method (e.g. measured as average number of accepted random pairs to number of pairs produced) may be quite poor, in particular if the distribution $f(x)$ has sharp maxima.

A more general, and sometimes more efficient rejection method, working even on infinite sampling domains M, results if one finds a second distribution $g(x)$ and a numerical constant c such that $f(x) \leq c \cdot g(x)$. Again find a pair (z_1, z_2) of random numbers, however with z_1 not sampled uniformly on M but from distribution $g(z)$ instead. z_2 is uniform on the interval $[0, c]$. The random variable z_1 is accepted if $z_2 \leq f(z_1)/g(z_1)$. Otherwise a new pair (z_1, z_2) is generated. See Chap. 5 for an important application in particle simulation.

3.2.2.3 Examples

3.2.2.3.1 Inversion

Important examples in which the inversion method can be applied are, e.g., the exponential distribution (of the mean free flight length of radiation in matter), the cosine distribution of polar emission angles against surface normals, the surfaces crossing Maxwellian flux distribution $f(v_\perp) \propto v_\perp f_{\text{Maxw}}(v_\perp)$ of normal velocity components of gas molecules with Maxwellian velocity distribution (f_{Maxw}). We explicitly illustrate the inversion method here for the Cauchy distribution: The Cauchy distribution, see Fig. 3.2(a), in physical applications also called Lorentz distribution, is an example of a distribution function that has no moments. It arises often in radiation transfer, e.g., as line-shapes of naturally- or Stark broadened lines or in other resonance phenomena

$$f_C(x) = \frac{c}{\pi} \frac{1}{(x-b)^2 + c^2} \,. \tag{3.12}$$

Here b is the median (line shift), and c is the half width at half maximum (HWHM). Generating random number with a Cauchy distribution is usúally done by inversion. First transform to a standardized Cauchy, by $s = (x - b)/c$. The cumulative distribution is then given as

$$F_C(x) = \frac{1}{\pi} \int_{-\infty}^{x} \frac{1}{s^2+1} \,\mathrm{d}s = \frac{1}{2} + \frac{1}{\pi} \arctan\left(\frac{x-b}{c}\right) \,. \tag{3.13}$$

Therefore the random number $z = b + c \cdot \tan\{\pi(\xi - 1/2)\}$, with ξ a uniformly distributed random number on [0, 1], has a Cauchy (b,c) distribution.

3.2.2.3.2 The Box Muller Method for Gaussian Random Numbers

Because the Gaussian error function cannot be inverted in closed form, the following combination of transformation, rejection and inversion method is typically applied: Not one, but two independent normally distributed random numbers (z_1, z_2) are produced by first transforming random variables Z_1, Z_2 from cartesian to polar coordinates R, Φ. The angle Φ is then uniform in $[0, 2\pi]$. Only $\cos(\Phi)$ and $\sin(\Phi)$ are needed, and a rejection method (comparing a unit circle and a surrounding square) can be used for them. The variable R has, due to the Jacobian of the transformation, a Gaussian flux distribution (see above) rather than a Gaussian itself, and this can be directly generated by the Method of Inversion. Transforming back $Z_1 = R \cdot \cos(\Phi)$ and $Z_2 = R \cdot \sin(\Phi)$ provides a pair of independent Gaussian random numbers.

3.3 Integration by Monte Carlo

Integration by Monte Carlo is a stochastic method for the deterministic problem of finding an integral, which in sufficiently complex high dimensional situations can be competitive or even superior to numerical methods.

Let's consider the source rate of particles (likewise, of momentum, heat, etc.) in a macroscopic system (e.g., a fluid flow), in which these particles (microscopic objects) are ruled by a kinetic, i.e. microscopic (Boltzmann) equation. Examples are chemical sources (particle, momentum, energy) in plasma chemistry, or radiative heat source in case of radiation transfer theory.

Such terms then read

$$I = \int_V \mathrm{d}x \; g(x) f(x) := \int_V \mathrm{d}f \; g(x) \,. \tag{3.14}$$

Here f is the one particle distribution (density) function $f(\boldsymbol{r}, \boldsymbol{v}, i, t)$ or $f(x)$, where the state x of the relevant phase-space may, e.g., be characterized by a position vector $\boldsymbol{r}$, a velocity vector $\boldsymbol{v}$, the time t, i.e. continuous variables, and further a discrete chemical species index i, also for example for internal quantal states. $g(x)$ is again some weighting function determined by the particular moment of interest. In mathematical terms one would refer to this as Lebesgue-Stieltjes Integral of measurabel function $g(x)$ with respect to (probability) measure defined by distribution density $f(x)$.

We will discuss Integration by Monte Carlo using the example from [2]: Let the integration domain V be the unit interval $[0, 1]$, $f(x)$ the uniform distribution on $[0, 1]$ (i.e.: $f(x) = 1$ on $[0, 1]$, and $f(x) = 0$ elsewhere) and $g(x) = (\exp(x) - 1)/(\mathrm{e} - 1)$. Clearly,

$$I = \int_0^1 \frac{\mathrm{e}^x - 1}{\mathrm{e} - 1} \mathrm{d}x = 0.4180\ldots . \tag{3.15}$$

We will now integrate this same function by Monte Carlo. Our first method does not require any theory, but instead, inspired by Buffon's needle experiment, we will just use pairs ξ_1, ξ_2 of independent uniform random numbers and compare the known area (the unit square $[0, 1] \times [0, 1]$) with the unknown area I, which is the area underneath function $g(x)$, in [0,1]. I.e., we count a *hit* if the point defined by the pair of random numbers is under the curve $g(x)$, and a *miss* otherwise.

As can clearly be seen on Fig. 3.3 the ratio of hits to total number of samples converges to the exact values of the integral, as expected, and also the statistical error, indicated as empirical standard deviation s_N, (3.5) scales with $1/\sqrt{N}$ as expected.

Of course such a Monte Carlo integration method is patently foolish. By this method we have, in principle, replaced the single integral over function g by a double integral over the area between abscissa and function $g(x)$. The conventional text-book method (crude Monte Carlo) can be obtained from this one by the observation that once the first random number ξ_1 of the pair is known, we do not have to rely upon ξ_2 to decide about counting zero or one. Given ξ_1, then an one will be counted with probability p: $p = g(\xi_1)$. Hence instead we can use that (conditional) expected value p of the binomial distribution $b(1, p)$ directly. This is, admittedly, a quite obscure explanation for something really trivial. But it is also the underlying idea behind a powerful variance reducing Monte Carlo technique known under different names in different areas of application: Conditional expectation estimator (in neutron shielding), [4], averaging transformation (transfer theory, mainly in Russian literature), [7], or energy partitioning method in radiative heat transfer [8].

This method is opposite to randomization: We have replaced a sampled result (zero or one) by its expectation value. In our particular example we have carried out one of the two integrations analytically, conditional on the outcome ξ_1. The second random number is not needed at all in this particular trivial case but this not the relevant point. What is important also in general terms is that one (generally: some) of the two (generally: many) integrals has been done analytically, and only the remaining ones by random sampling. The general rule is: Always try to do as many integrations analytically or numerically and resort to Monte Carlo only for the rest. In particle transport theory this concept will lead to powerful hybrid methods combining information gained analytically (or numerically) and stochastically, bridging

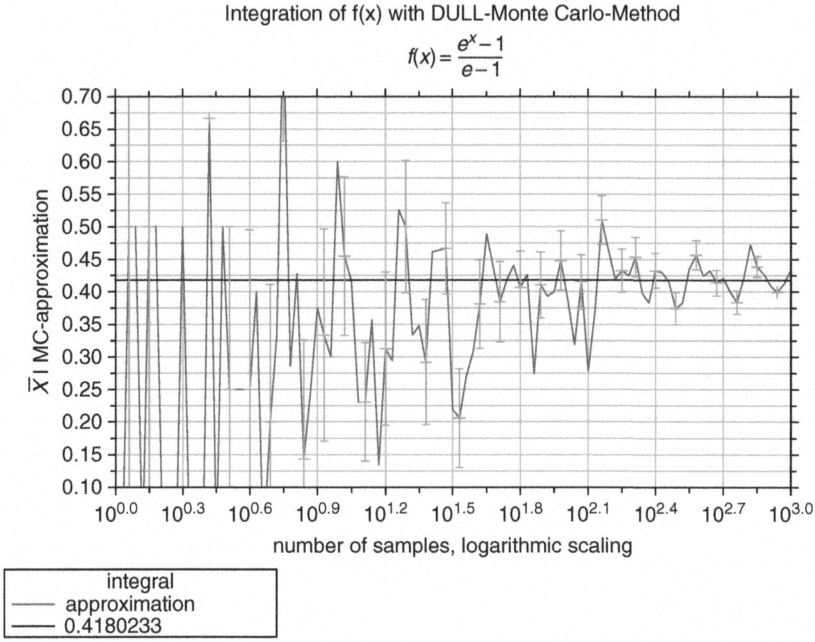

Fig. 3.3. Evaluating Integral of $(\exp(x) - 1)/(\mathrm{e} - 1)$ on [0,1], method: hit or miss Monte Carlo

continuously the gap between stochastic and numerical methods. Sometimes, however, these resulting methods may loose their transparency.

In this crude Monte Carlo integration I is obtained as estimated mean value (expectation value) of function $g(x)$ with respect to the uniform probability distribution $f(x)$ on [0, 1], $I = E_f(g)$, see remark after (3.2). Also indicated in Fig. 3.4 is again the empirical standard deviation s_N, which, as expected, is significantly smaller than with the hit or miss method.

Note that although this method has certainly a smaller statistical error per sample, the efficiency gain of one over the other method has also to account for the extra labor involved in evaluating the smoother estimator (which is hardly measurable in this trivial example chosen here).

In general Monte Carlo terminology one would refer to the uniform distribution $f(x)$ as the underlying stochastic law, according to which random samples X are produced. The random variable $g(X)$ is called estimator, score, or response function.

3.3.0.4 Importance sampling

We are now in the position to explain the famous Monte Carlo concept of importance sampling for improving the statistical performance of Monte Carlo methods. Distinct from the conditional expectation technique discussed above, in which the

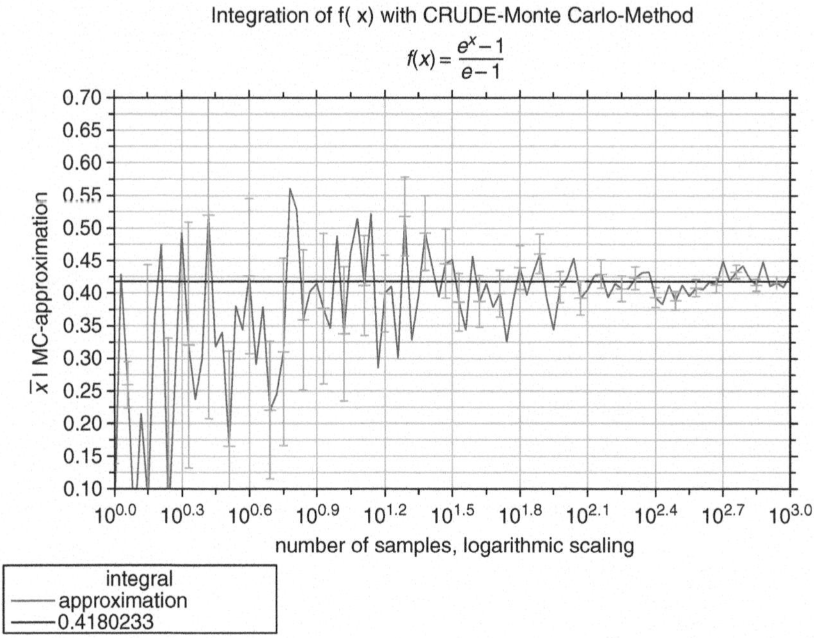

Fig. 3.4. Evaluating Integral of $(\exp(x) - 1)/(\mathrm{e} - 1)$ on [0,1], method: crude Monte Carlo

statistical noise is reduced solely by modifying (smoothing) the estimator $g(x)$, in importance sampling the underlying random variable (or random process) $f(x)$ is altered to another one, $\widetilde{f}(x)$ in order to achieve variance reduction. A compensating weight correction factor in introduced in the estimator to maintain the same mean value $I = E(g(X))$

$$I = \int\limits_V g(x) \cdot f(x) \mathrm{d}x = \int\limits_V g(x) \frac{f(x)}{\widetilde{f}(x)} \widetilde{f}(x) \mathrm{d}x = \int\limits_V \widetilde{g}(x) \widetilde{f}(x) \mathrm{d}x \ . \tag{3.16}$$

Hence we have $\widetilde{g}(x) = g(x)f(x)/\widetilde{f}(x)$. The name of this method, importance sampling originates from the special techniques often used to find optimal biassing schemes (i.e.: $\widetilde{f}(x)$) of the random process, in particular in transfer theory. A more general, but also somewhat imprecise terminology would refer to this concept as non-analog Monte Carlo, as compared to the analog Monte Carlo scheme. In the latter the underlying probability distribution law is directly taken from the application, whereas in the former one uses a different distribution, motivated by practical, economical or other reasons, and statistical weights to compensate this.

As seen from (3.16), the value of I is independent of how the integrant is decomposed into a product of a probability density and a response function, but the variances, $\sigma_f^2(g)$ and $\sigma_{\widetilde{f}}^2(\widetilde{g})$, certainly can be different.

Let's take, again, our example, to illustrate the concept: In order to reduce the variance $\sigma^2_{\widetilde{f}}(\widetilde{g})$ of $\widetilde{g}$ with respect to probability law $\widetilde{f}$ we should try to make $\widetilde{g}$ as constant as possible on [0,1]. The Taylor expansion of our particular function $g(x)$ indicates that the ratio $\widetilde{g}(x) = g(x)/x$ should be more constant than $g(x)$ itself. Hence we try $\widetilde{f}(x) \propto x$, i.e., $\widetilde{f}(x) = 2x$ so that $\widetilde{f}(x)$ is normalized to one on [0,1].

Our importance sampling procedure to evaluate I now proceeds as follows: Draw random numbers $\widetilde{\xi}$ from $\widetilde{f}(x)$. By the method of inversion, this is done by setting $\widetilde{\xi} = \sqrt{\xi}$, with ξ a uniform random number on [0,1]. Then, again, form the arithmetic average of many (N) random variables $\widetilde{g}(\widetilde{\xi})$. Figure 3.5 shows the result of such an integration, again vs. N. Clearly the convergence is (i) to the correct value, (ii) still only $\propto 1/\sqrt(N)$, but (iii) the error bars s_N are much smaller than in both previously discussed Monte Carlo integration methods.

Again, it needs to be pointed out that the efficiency of the procedure is neither determined by the variance, not by N per CPU-time, but only by the figure of merit: variance per CPU time. And hence, importance sampling, more generally, non-analog sampling, can go both ways in Monte Carlo. Its performance has to be assessed on a case by case basis.

As a general observation, one should note that in non-analog Monte Carlo schemes the error assessment simply based upon the empirical variance, and error bars obtained from the central limit theorem, can be less reliable than in analog simulations. Although the variance may be decreased by a clever importance sampling

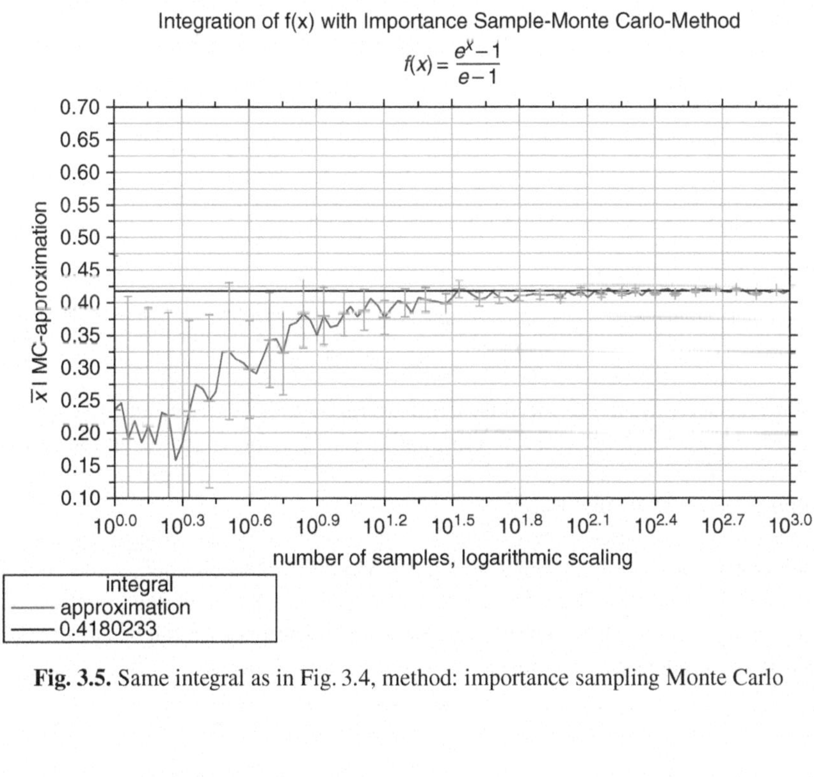

Fig. 3.5. Same integral as in Fig. 3.4, method: importance sampling Monte Carlo

method, the variance of the variance may increase, thus invalidating conventional error bar estimates, see [9].

As in the case of conditional expectation Monte Carlo we can design an extreme case of importance sampling with zero statistical error after only one sample: Let us set $\tilde{f}(x) = g(x)/I$, hence: $\tilde{g}(x) = I = \text{const.}$ Monte Carlo integration proceeds by sampling from this distribution $\tilde{f}(x)$ which, in case of our particular example can be done by the rejection technique. Then, independent of the sampling, I is scored. Unfortunately we needed the knowledge of the final result I already to design this perfect zero variance scheme.

3.3.0.5 δf Monte Carlo

Finally we use our simple integral to illustrate the concept of the δf Monte Carlo method, which is widely used in kinetic particle transport simulations. Starting point is the idea to split the unknown parameter into a large known *nearby* quantity and small unknown perturbation. In particle simulations this can also be the single particle distribution function $f(x)$ solving some kinetic equation or moments of this pdf. In near equilibrium situations we have

$$f(x) = f_{\text{equil}}(x) + \delta f(x) \tag{3.17}$$

with, for example, the Maxwellian equilibrium distribution f_{equil} and a small perturbation δf. It can then be advantageous to solve, by Monte Carlo sampling, only for δf rather than for the full distribution.

So let us consider our integral again, and write, accordingly, $I = I_0 + \delta I$ with I_0 the known part

$$I_0 = \frac{1}{\mathrm{e}-1}\int_0^1 \mathrm{d}x\ g_0(x) = \frac{1}{\mathrm{e}-1}\int_0^1 \mathrm{d}x\ \left(x + \frac{1}{2}x^2\right) = \frac{2}{3}\frac{1}{\mathrm{e}-1} \tag{3.18}$$

and δI the rest. Clearly,

$$\delta I = \int_0^1 \mathrm{d}x\ \frac{\mathrm{e}^x - 1 - x - x^2/2}{\mathrm{e}-1}\ . \tag{3.19}$$

Figure 3.6 shows the result of the estimate for I, with I_0 known and δI evaluated by crude Monte Carlo. Clearly by eliminating a large, known, contribution to I the relative errors of the estimates for any given sample size N are greatly reduced as compared to previous methods.

This method is also related to the so called correlation sampling technique, in which one would evaluate both I and I_0 by Monte Carlo techniques, but using the same random numbers. Both estimates are then positively correlated and the statistical precision of the Monte Carlo estimate for the difference δI can be substantially better than in independent estimates of I and I_0 or of I alone.

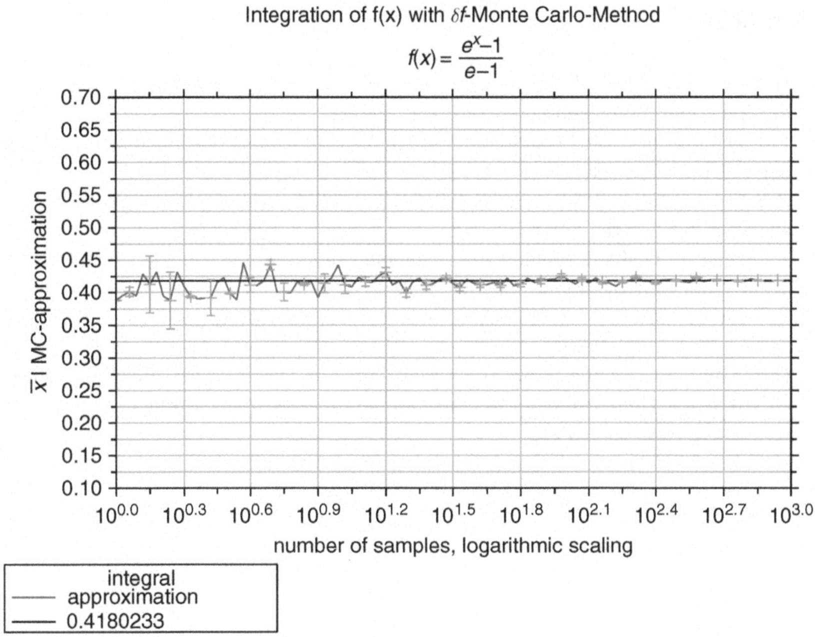

Fig. 3.6. Same integral as in Fig. 3.4, method: δf Monte Carlo

3.4 Summary

The purpose of this introduction was to show that random numbers can be generated from any given probability density distribution, and that Monte Carlo Methods can be regarded as stochastic (rather than numerical) procedures for integration. Monte Carlo consists of inventing a random game such that the expected value of a proper random variable is exactly equal to the parameter which is to be computed. Averaging over repeated independent Monte Carlo samples from that game converges (in the proper measure theoretical sense) to the desired solution.

The additional complication arising in many particle physics applications and in transfer theory is due to one fact only: Distinct from the material in this present chapter the sampling distribution $f(x)$ is sometimes not known explicitly. Instead it will be given only implicitly as solution of a, usually, very complicate equation (e.g.: the Boltzmann equation, the Fokker-Planck equation, etc.). We will see that this extra complication can be dealt with by sampling from certain stochastic processes (generating particle trajectories), rather than sampling from a given pdf, see Chap. 5. But the rest: Estimation of multi-dimensional integrals, the unbiased nature of the method, proof of convergence, error bars, variance reduction methods, remain essentially the same as in this present introduction.

References

1. H. Kalos, P.A. Whitlock, *Monte Carlo Methods, Vol. 1: Basics* (Wiley-Interscience Publications, John Wiley and Sons, New York, 1986)
2. J.M. Hammersley, D.C. Handscomb, *Monte Carlo Methods* (Chapman and Hall, London & New York, 1964)
3. R.Y. Rubenstein, in *Wiley Series in Probability and Mathematical Statistics* (John Wiley and Sons, New York, 1981)
4. J. Spanier, E. Gelbard, *Monte Carlo Principles and Neutron Transport Problems* (Addison Wesley Publication Company, 1969)
5. L. Devroye, *Non-Uniform Random Variate Generation* (Springer-Verlag, Berlin Heidelberg New York, 1986)
6. D.E. Knuth, in *Seminumerical Algorithms*, Vol. 2 (Addison Wesley, Reading, 1998)
7. G. Mikhailov, *Optimization of Weighted Monte Carlo Methods* (Springer Verlag, Berlin Heidelberg New York, 1992)
8. A. Wang, M.F. Modest, J. Quant. Spectrosc. R. A **104**, 288 (2007)
9. K. Noack, Ann. nucl. Energy **18**(6), 309 (1991)

4 Monte Carlo Methods in Classical Statistical Physics

Wolfhard Janke

Institut für Theoretische Physik and Centre for Theoretical Sciences, Universität Leipzig, 04009 Leipzig, Germany

The purpose of this chapter is to give a brief introduction to Monte Carlo simulations of classical statistical physics systems and their statistical analysis. To set the general theoretical frame, first some properties of phase transitions and simple models describing them are briefly recalled, before the concept of importance sampling Monte Carlo methods is introduced. The basic idea is illustrated by a few standard local update algorithms (Metropolis, heat-bath, Glauber). Then methods for the statistical analysis of the thus generated data are discussed. Special attention is payed to the choice of estimators, autocorrelation times and statistical error analysis. This is necessary for a quantitative description of the phenomenon of critical slowing down at continuous phase transitions. For illustration purposes, only the two-dimensional Ising model will be needed. To overcome the slowing-down problem, non-local cluster algorithms have been developed which will be described next. Then the general tool of reweighting techniques will be explained which is extremely important for finite-size scaling studies. This will be demonstrated in some detail by the sample study presented in the next section, where also methods for estimating spatial correlation functions will be discussed. The reweighting idea is also important for a deeper understanding of so-called generalized ensemble methods which may be viewed as dynamical reweighting algorithms. After first discussing simulated and parallel tempering methods, finally also the alternative approach using multicanonical ensembles and the Wang-Landau recursion are briefly outlined.

4.1 Introduction

Classical statistical physics is a well understood subject which poses, however, many difficult problems when a concrete solution for interacting systems is sought. In almost all non-trivial applications, analytical methods can only provide approximate answers. Numerical computer simulations are, therefore, an important complementary method on our way to a deeper understanding of complex physical systems such as (spin) glasses and disordered magnets or of biologically motivated problems such as protein folding. Quantum statistical problems in condensed matter or the broad field of elementary particle physics and quantum gravity are other major applications which, after suitable mappings, also rely on classical simulation techniques.

W. Janke: *Monte Carlo Methods in Classical Statistical Physics*, Lect. Notes Phys. **739**, 79–140 (2008)
DOI 10.1007/978-3-540-74686-7_4

In these lecture notes we shall confine ourselves to a survey of computer simulations based on Markov chain Monte Carlo methods which realize the importance sampling idea. Still, not all aspects can be discussed in these notes in detail, and for further reading the reader is referred to recent textbooks [1, 2, 3, 4], where some of the material is presented in more depth. For illustration purposes, here we shall focus on the simplest spin models, the Ising and Potts models. From a theoretical point of view, also spin systems are still of current interest since they provide the possibility to compare completely different approaches such as field theory, series expansions, and simulations. They are also the ideal testing ground for general concepts such as universality, scaling or finite-size scaling, where even today some new features can still be discovered. And last but not least, they have found a revival in slightly disguised form in quantum gravity and and random network theory, where they serve as idealized matter fields on Feynman diagrams or fluctuating graphs.

This chapter is organized as follows. In Sect. 4.2, first the definition of the standard Ising model is recalled and the most important observables (specific heat, magnetization, susceptibility, correlation functions, ...) are briefly discussed. Next some characteristic properties of phase transitions, their scaling properties, the definition of critical exponents and finite-size scaling are briefly summarized. In Sect. 4.3, the basic method underlying all importance sampling Monte Carlo simulations is described and some properties of local update algorithms (Metropolis, heat-bath, Glauber) are discussed. The following Sect. 4.4 is devoted to non-local cluster algorithms which in some cases can dramatically speed up the simulations. A fairly detailed account of the initial non-equilibrium period and ageing phenomena as well as statistical error analysis in equilibrium is given in Sect. 4.5. Here temporal correlation effects are discussed, which explain the problems with critical slowing down at a continuous phase transition and exponentially large flipping times at a first-order transition. In Sect. 4.6, we discuss reweighting techniques which are particularly important for finite-size scaling studies. A worked out example of such a study is presented in the following Sect. 4.7. Finally, more refined generalized ensemble simulation methods are briefly outlined in Sect. 4.8, focusing on simulated and parallel tempering, the multicanonical ensemble and the Wang-Landau recursion. The lecture notes close in Sect. 4.9 with a few concluding remarks.

4.2 Statistical Physics Primer

To set the scenery for the simulation methods discussed below, we need to briefly recall a few basic concepts of statistical physics [5, 6, 7, 8]. In these lecture notes we will only consider classical systems and mainly focus on the canonical ensemble where the partition function is generically given as

$$\mathcal{Z} = \sum_{\text{states}} \mathrm{e}^{-\beta\mathcal{H}} = \mathrm{e}^{-\beta\mathcal{F}} , \tag{4.1}$$

with the summation running over all possible states of the system. The state space may be continuous or discrete. As usual $\beta \equiv 1/k_{\mathrm{B}}T$ denotes the inverse temperature fixed by an external heat bath, k_{B} is Boltzmann's constant, $\mathcal{H}$ is the Hamiltonian of the system, encoding the details of the interactions which may be short-, medium-, or long-ranged, and $\mathcal{F}$ is the free energy. Expectation values denoted by angular brackets $\langle \ldots \rangle$ then follow as

$$\langle \mathcal{O} \rangle = \sum_{\text{states}} \mathcal{O} \mathrm{e}^{-\beta \mathcal{H}} / \mathcal{Z} \,, \tag{4.2}$$

where $\mathcal{O}$ stands symbolically for any observable, e.g., the energy $E \equiv \mathcal{H}$.

As we will see in the next section, the most elementary Monte Carlo simulation method (Metropolis algorithm) can, in principle, cope with all conceivable variants of this quite general formulation. Close to a phase transition, however, this basic algorithm tends to become very time consuming and for accurate quantitative results one needs to employ more refined methods. Most of them are much more specific and take advantage of certain properties of the model under study. One still quite broad class of systems are lattice models, where one assumes that the degrees of freedom live on the sites or/and links of a D-dimensional lattice. These are often taken to be hypercubic, but more complicated regular lattice types (e.g., triangular (T), body-centered cubic (BCC), face-centered cubic (FCC), etc.) and even random lattices do not cause problems in principle. The degrees of freedom may be continuous or discrete field variables such as a gauge field or the height variable of a crystal surface, continuous or discrete spins such as the three-dimensional unit vectors of the classical Heisenberg model or the ± 1 valued spins of the Ising model, or arrow configurations along the links of the lattice such as in Baxter's vertex models, to give only a few popular examples.

To be specific and to keep the discussion as simple as possible, most simulation methods will be illustrated with the minimalistic Ising model [9, 10] where

$$\mathcal{H} = -J \sum_{\langle ij \rangle} \sigma_i \sigma_j - h \sum_i \sigma_i \tag{4.3}$$

with $\sigma_i = \pm 1$. Here J is a coupling constant which is positive for a ferromagnet ($J > 0$) and negative for an anti-ferromagnet ($J < 0$), h is an external magnetic field, and the symbol $\langle ij \rangle$ indicates that the lattice sum is restricted to run only over all nearest-neighbor pairs. In the examples discussed below, usually periodic boundary conditions are applied. And to ease the notation, we will always assume units in which $k_{\mathrm{B}} = 1$ and $J = 1$.

Basic observables are the internal energy per site, $u = U/V$, with $U = -\mathrm{d} \ln \mathcal{Z} / \mathrm{d}\beta \equiv \langle \mathcal{H} \rangle$, and the specific heat

$$C = \frac{\mathrm{d}u}{\mathrm{d}T} = \beta^2 \frac{\langle E^2 \rangle - \langle E \rangle^2}{V} = \beta^2 V \left(\langle e^2 \rangle - \langle e \rangle^2 \right) \,, \tag{4.4}$$

where we have set $\mathcal{H} \equiv E = eV$ with V denoting the number of lattice sites, i.e., the lattice volume. The magnetization per site $m = M/V$ and the susceptibility χ are defined as

$$M = \frac{1}{\beta}\frac{\mathrm{d}\ln \mathcal{Z}}{\mathrm{d}h} = V\langle\mu\rangle \;, \qquad \mu = \frac{1}{V}\sum_i \sigma_i \;, \tag{4.5}$$

and

$$\chi = \beta V \left(\langle\mu^2\rangle - \langle\mu\rangle^2\right) \,. \tag{4.6}$$

The correlation between spins σ_i and σ_j at sites labeled by i and j can be measured by considering correlation functions like the two-point spin-spin correlation $G(i,j)$, which is defined as

$$G(\boldsymbol{r}) = G(i,j) = \langle\sigma_i\sigma_j\rangle - \langle\sigma_i\rangle\langle\sigma_j\rangle \;, \tag{4.7}$$

where $\boldsymbol{r} = \boldsymbol{r}_i - \boldsymbol{r}_j$ (assuming translational invariance). Away from criticality and at large distances $|\boldsymbol{r}| \gg 1$ (where we have assumed a lattice spacing $a = 1$), $G(\boldsymbol{r})$ decays exponentially

$$G(\boldsymbol{r}) \sim |\boldsymbol{r}|^{\kappa}\, \mathrm{e}^{-|\boldsymbol{r}|/\xi} \;, \tag{4.8}$$

where ξ is the correlation length and the exponent κ of the power-law prefactor depends in general on the dimension and on whether one studies the ordered or disordered phase. Some model (and simulation) specific details of the latter observables and further important quantities like various magnetization cumulants will be discussed later when dealing with concrete applications.

The Ising model is the paradigm model for systems exhibiting a continuous (or, roughly speaking, second-order) phase transition from an ordered low-temperature to a disordered high-temperature phase at some critical temperature T_c when the temperature is varied. In two dimensions (2D), the thermodynamic limit of this model in zero external field has been solved exactly by Onsager [11], and also for finite $L_x \times L_y$ lattices the exact partition function is straightforward to compute [12, 13]. For infinite lattices, even the correlation length is known in arbitrary lattice directions [14, 15]. The exact magnetization for $h = 0$, apparently already known to Onsager [16][1], was first derived by Yang [17, 18], and the susceptibility is known to very high precision [19, 20], albeit still not exactly. In 3D no exact solutions are available, but analytical and numerical results from various methods give a consistent and very precise picture.

The most characteristic feature of a second-order phase transition is the divergence of the correlation length at T_c. As a consequence thermal fluctuations are equally important on *all* length scales, and one therefore expects power-law singularities in thermodynamic functions. The leading divergence of the correlation length is usually parameterized in the high-temperature phase as

$$\xi = \xi_{0+}|1 - T/T_c|^{-\nu} + \ldots \quad (T \geq T_c) \;, \tag{4.9}$$

where the ... indicate sub-leading analytical as well as confluent corrections. This defines the critical exponent $\nu > 0$ and the critical amplitude ξ_{0+} on the high-temperature side of the transition. In the low-temperature phase one expects a similar behavior

[1] See also the historical remarks in Refs. [14, 15].

$$\xi = \xi_{0-}(1 - T/T_c)^{-\nu} + \dots \qquad (T \leq T_c)\,, \tag{4.10}$$

with the same critical exponent ν but a different critical amplitude $\xi_{0-} \neq \xi_{0+}$.

An important consequence of the divergence of the correlation length is that qualitative properties of second-order phase transitions should not depend on short-distance details of the Hamiltonian. This is the basis of the universality hypothesis [21] which means that all (short-ranged) systems with the same symmetries and same dimensionality should exhibit similar singularities governed by one and the same set of critical exponents. For the amplitudes this is not true, but certain amplitude ratios are also universal.

The singularities of the specific heat, magnetization (for $T < T_c$), and susceptibility are similarly parameterized by the critical exponents α, β and γ, respectively,

$$\begin{aligned} C &= C_{\text{reg}} + C_0|1 - T/T_c|^{-\alpha} + \dots\,, \\ m &= m_0(1 - T/T_c)^{\beta} + \dots\,, \\ \chi &= \chi_0|1 - T/T_c|^{-\gamma} + \dots\,, \end{aligned} \tag{4.11}$$

where C_{reg} is a regular background term, and the amplitudes are again in general different on the two sides of the transition. Right at the critical temperature T_c, two further exponents δ and η are defined through

$$\begin{aligned} m &\propto h^{1/\delta}\,, \\ G(\boldsymbol{r}) &\propto r^{-D+2-\eta}\,. \end{aligned} \tag{4.12}$$

In the 1960's, Rushbrooke [22], Griffiths [23], Josephson [24, 25] and Fisher [26] showed that these six critical exponents are related via four inequalities. Subsequent experimental evidence indicated that these relations were in fact equalities, and they are now firmly established and fundamentally important in the theory of critical phenomena. With D representing the dimensionality of the system, the scaling relations are

$$\begin{aligned} D\nu &= 2 - \alpha && \text{(Josephson's law)}\,, \\ 2\beta + \gamma &= 2 - \alpha && \text{(Rushbrooke's law)}\,, \\ \beta(\delta - 1) &= \gamma && \text{(Griffiths' law)}\,, \\ \nu(2 - \eta) &= \gamma && \text{(Fisher's law)}\,. \end{aligned} \tag{4.13}$$

In the conventional scaling scenario, Rushbrooke's and Griffiths' laws can be deduced from the Widom scaling hypothesis that the Helmholtz free energy is a homogeneous function [27, 28]. Widom scaling and the remaining two laws can in turn be derived from the Kadanoff block-spin construction [29] and ultimately from that of the renormalization group (RG) [30]. Josephson's law can also be derived from the hyperscaling hypothesis, namely that the free energy behaves near criticality as

Table 4.1. Critical exponents of the Ising model in two (2D) and three (3D) dimensions. All 2D exponents are exactly known [31, 32], while for the 3D Ising model the world-average for ν and γ calculated in [33] is quoted. The other exponents follow from the hyperscaling relation $\alpha = 2 - D\nu$, and the scaling relations $\beta = (2 - \alpha - \gamma)/2$, $\delta = \gamma/\beta + 1$, and $\eta = 2 - \gamma/\nu$

dimension	ν	α	β	γ	δ	η
$D = 2$	1	0 (log)	1/8	7/4	15	1/4
$D = 3$	0.630 05(18)	0.109 85	0.326 48	1.237 17(28)	4.7894	0.036 39

the inverse correlation volume: $f_\infty(t) \sim \xi_\infty^{-D}(t)$. Twice differentiating this relation one recovers Josephson's law (4.13). The critical exponents for the 2D and 3D Ising model [31, 32, 33] are collected in Table 4.1.

In any numerical simulation study, the system size is necessarily finite. While the correlation length may still become very large, it is therefore always finite. This implies that also the divergences in other quantities are rounded and shifted [34, 35, 36, 37]. How this happens is described by finite-size scaling (FSS) theory, which in a nut-shell may be explained as follows: Near T_c the role of ξ is taken over by the linear size L of the system. By rewriting (4.9) or (4.10) and replacing $\xi \to L$

$$|1 - T/T_c| \propto \xi^{-1/\nu} \longrightarrow L^{-1/\nu} , \tag{4.14}$$

it is easy to see that the scaling laws (4.11) are replaced by the FSS Ansätze,

$$\begin{aligned} C &= C_{\text{reg}} + aL^{\alpha/\nu} + \dots , \\ m &\propto L^{-\beta/\nu} + \dots , \\ \chi &\propto L^{\gamma/\nu} + \dots . \end{aligned} \tag{4.15}$$

As a mnemonic rule, a critical exponent x of the temperature scaling law is replaced by $-x/\nu$ in the corresponding FSS law. In general these scaling laws are valid in a neighborhood of T_c as long as the scaling variable

$$x = (1 - T/T_c)L^{1/\nu} \tag{4.16}$$

is kept fixed [34, 35, 36, 37]. This implies for the locations T_{max} of the (finite) maxima of thermodynamic quantities such as the specific heat or susceptibility, an FSS behavior of the form

$$T_{\text{max}} = T_c(1 - x_{\text{max}}L^{-1/\nu} + \dots) . \tag{4.17}$$

In this more general formulation the scaling law for, e.g., the susceptibility reads

$$\chi(T, L) = L^{\gamma/\nu} f(x) , \tag{4.18}$$

where $f(x)$ is a scaling function. By plotting $\chi(T, L)/L^{\gamma/\nu}$ versus the scaling variable x, one thus expects that the data for different T and L fall onto a master curve described by $f(x)$. This is a nice visual method for demonstrating the scaling properties.

Similar considerations for first-order phase transitions [38, 39, 40, 41] show that here the δ-function like singularities in the thermodynamic limit, originating from phase coexistence, are also smeared out for finite systems [42, 43, 44, 45, 46]. They are replaced by narrow peaks whose height (width) grows proportional to the volume (1/volume) with a displacement of the peak location from the infinite-volume limit proportional to 1/volume [47, 48, 49, 50, 51, 52].

4.3 The Monte Carlo Method

Let us now discuss how the expectation values in (4.2) can be estimated in a Monte Carlo simulation. For any reasonable system size, a direct summation of the partition function is impossible, since already for the minimalistic Ising model with only two possible states per site the number of terms would be enormous: For a moderate 20×20 lattice, the state space consists already of $2^{400}\approx 10^{120}$ different spin configurations.[2] Also a naive random sampling of the spin configurations does not work. Here the problem is that the relevant region in the high-dimensional phase space is relatively narrow and hence too rarely hit by random sampling. The solution to this problem is known since long under the name importance sampling [53].

4.3.1 Importance Sampling

The basic idea of importance sampling is to set up a suitable Markov chain that draws configurations not at random but according to their Boltzmann weight

$$\mathcal{P}^{\mathrm{eq}}(\{\sigma_i\}) = \frac{\mathrm{e}^{-\beta\mathcal{H}(\{\sigma_i\})}}{\mathcal{Z}} . \tag{4.19}$$

A Markov chain defines stochastic rules for transitions from one state to another subject to the condition that the probability for the new configuration only depends on the preceding state but not on the history of the whole trajectory in state space, i.e., it is almost local in time. Symbolically this can be written as

$$\ldots \xrightarrow{W} \{\sigma_i\} \xrightarrow{W} \{\sigma_i\}' \xrightarrow{W} \{\sigma_i\}'' \xrightarrow{W} \ldots , \tag{4.20}$$

where the transition probability W has to satisfy the following conditions:

(i) $W(\{\sigma_i\} \longrightarrow \{\sigma_i\}') \geq 0$ for all $\{\sigma_i\}, \{\sigma_i\}'$,
(ii) $\sum_{\{\sigma_i\}'} W(\{\sigma_i\} \longrightarrow \{\sigma_i\}') = 1$ for all $\{\sigma_i\}$,
(iii) $\sum_{\{\sigma_i\}} W(\{\sigma_i\} \longrightarrow \{\sigma_i\}')\mathcal{P}^{\mathrm{eq}}(\{\sigma_i\}) = \mathcal{P}^{\mathrm{eq}}(\{\sigma_i\}')$ for all $\{\sigma_i\}'$.

[2] This number should be compared with the estimated number of protons in the Universe which is about 10^{80}.

From condition (iii) we see that the desired Boltzmann distribution $P^{\rm eq}$ is a fixed point of W (eigenvector of W with unit eigenvalue). A somewhat simpler sufficient condition is detailed balance,

$$\mathcal{P}^{\rm eq}(\{\sigma_i\})\, W(\{\sigma_i\} \longrightarrow \{\sigma_i\}') = \mathcal{P}^{\rm eq}(\{\sigma_i\}')\, W(\{\sigma_i\}' \longrightarrow \{\sigma_i\}) \,. \tag{4.21}$$

By summing over $\{\sigma_i\}$ and using condition (ii), the more general condition (iii) follows. After an initial equilibration period (cf. Sect. 4.5.1), expectation values can be estimated as an arithmetic mean over the Markov chain of length N, e.g.,

$$E = \langle \mathcal{H} \rangle = \sum_{\{\sigma_i\}} \mathcal{H}(\{\sigma_i\})\mathcal{P}^{\rm eq}(\{\sigma_i\}) \approx \frac{1}{N}\sum_{j=1}^{N} \mathcal{H}(\{\sigma_i\}_j) \,, \tag{4.22}$$

where $\{\sigma_i\}_j$ denotes the spin configuration at "time" j. A more detailed exposition of the mathematical concepts underlying any Markov chain Monte Carlo algorithm can be found in many textbooks and reviews [1, 2, 3, 4, 34, 54, 55].

4.3.2 Local Update Algorithms

The Markov chain conditions (i)–(iii) are still quite general and can be satisfied by many different concrete update rules. In a rough classification one distinguishes between local and non-local algorithms. Local update algorithms discussed in this subsection are conceptually much simpler and, as the main merit, quite universally applicable. The main drawback is their relatively poor performance close to second-order phase transitions where the spins or fields of a typical configuration are strongly correlated over large spatial distances. Here non-local update algorithms based on multigrid methods or in particular self-adaptive cluster algorithms discussed later in Sect. 4.4 perform much better.

4.3.2.1 Metropolis Algorithm

The most flexible update rule is the classic Metropolis algorithm [56], which is applicable in practically all cases (lattice/off-lattice, discrete/continuous, short-range/long-range interactions, ...). Here one proposes an update for a single degree of freedom (spin, field, ...) and accepts this proposal with probability

$$W(\{\sigma_i\}_{\rm old} \longrightarrow \{\sigma_i\}_{\rm new}) = \begin{cases} 1 & E_{\rm new} < E_{\rm old} \\ {\rm e}^{-\beta(E_{\rm new}-E_{\rm old})} & E_{\rm new} \geq E_{\rm old} \end{cases} \,, \tag{4.23}$$

where $E_{\rm old}$ and $E_{\rm new}$ denote the energy of the old and new spin configuration $\{\sigma_i\}_{\rm old}$ and $\{\sigma_i\}_{\rm new}$, respectively, where $\{\sigma_i\}_{\rm new}$ differs from $\{\sigma_i\}_{\rm old}$ only locally by one modified degree of freedom at, say, $i = i_0$. More compactly, this may also be written as

$$W(\{\sigma_i\}_{\rm old} \longrightarrow \{\sigma_i\}_{\rm new}) = \min\{1, {\rm e}^{-\beta\Delta E}\} \,, \tag{4.24}$$

where $\Delta E = E_{\rm new} - E_{\rm old}$. If the proposed update lowers the energy, it is always accepted. On the other hand, when the new configuration has a higher energy, the update has still to be accepted with a certain probability in order to ensure the proper treatment of entropic contributions – in thermal equilibrium, it is the free energy $F = U - TS$ which has to be minimized and not the energy. Only in the limit of zero temperature, $\beta \to \infty$, the acceptance probability for this case tends to zero and the Metropolis method degenerates to a minimization algorithm for the energy functional. With some additional refinements, this is the basis for the simulated annealing technique [57], which is often applied to hard optimization and minimization problems.

The verification of the detailed balance condition (4.21) is straightforward. If $E_{\rm new} < E_{\rm old}$, then the l.h.s. of (4.21) becomes $\exp(-\beta E_{\rm old}) \times 1 = \exp(-\beta E_{\rm old})$. On the r.h.s. we have to take into account that the reverse move would increase the energy, $E_{\rm old} > E_{\rm new}$ (with $E_{\rm old}$ now playing the role of the new energy), such that now the second line of (4.23) with $E_{\rm old}$ and $E_{\rm new}$ interchanged is relevant. This gives $\exp(-\beta E_{\rm new}) \times \exp(-\beta(E_{\rm old} - E_{\rm new})) = \exp(-\beta E_{\rm old})$ on the r.h.s. of (4.21), completing the demonstration of detailed balance. In the opposite case with $E_{\rm new} < E_{\rm old}$, a similar reasoning leads to $\exp(-\beta E_{\rm old}) \times \exp(-\beta(E_{\rm new} - E_{\rm old})) = \exp(-\beta E_{\rm new}) = \exp(-\beta E_{\rm new}) \times 1$. Admittedly, this proof looks a bit like a tautology. To uncover its non-trivial content, it is a useful exercise to replace the r.h.s. of the Metropolis rule (4.23) by some general function $f(E_{\rm new} - E_{\rm old})$ and repeat the above steps [58].

Finally a few remarks on the practical implementation of the Metropolis method are in order. To decide whether a proposed update should be accepted or not, one draws a uniformly distributed random number $r \in [0, 1)$, and if $W \leq r$, the new state is accepted. Otherwise one keeps the old configuration and continues with the next spin. In computer simulations, random numbers are generated by means of pseudo-random number generators (RNGs), which produce (more or less) uniformly distributed numbers whose values are very hard to predict – by using some deterministic rule (see [59] and references therein). In other words, given a finite sequence of subsequent pseudo-random numbers, it should be (almost) impossible to predict the next one or to even guess the deterministic rule underlying their generation. The goodness of an RNG is thus measured by the difficulty to derive its underlying deterministic rule. Related requirements are the absence of trends (correlations) and a very long period. Furthermore, an RNG should be portable among different computer platforms and, very importantly, it should yield reproducible results for testing purposes. The design of RNGs is a science in itself, and many things can go wrong with them. As a recommendation one should better not experiment too much with some fancy RNG one has picked up somewhere from the Web, say, but rely on well-tested and well-documented routines.

There are many different ways how the degrees of freedom to be updated can be chosen. They may be picked at random or according to a random permutation, which can be updated every now and then. But also a simple fixed lexicographical (sequential) order is permissible.[3] In lattice models one may also update first all odd and then all even sites, which is the usual choice in vectorized codes. A so-called sweep is completed when on the average[4] for all degrees of freedom an update was proposed. The qualitative behavior of the update algorithm is not sensitive to these details, but its quantitative performance does depend on the choice of the update scheme.

4.3.2.2 Heat-Bath Algorithm

This algorithm is only applicable to lattice models and at least in its most straightforward form only to discrete degrees of freedom with a few allowed states. The new value σ'_{i_0} at site i_0 is determined by testing all its possible states in the heat-bath of its (fixed) neighbors (e.g., four on a square lattice and six on a simple-cubic lattice with nearest-neighbor interactions):

$$W(\{\sigma_i\}_{\text{old}} \longrightarrow \{\sigma_i\}_{\text{new}}) = \frac{\mathrm{e}^{-\beta\mathcal{H}(\{\sigma_i\}_{\text{new}})}}{\sum_{\sigma_{i_0}} \mathrm{e}^{-\beta\mathcal{H}(\{\sigma_i\}_{\text{old}})}} = \frac{\mathrm{e}^{-\beta\sigma'_{i_0} S_{i_0}}}{\sum_{\sigma_{i_0}} \mathrm{e}^{-\beta\sigma_{i_0} S_{i_0}}}, \tag{4.25}$$

where $S_{i_0} = -\sum_j \sigma_j - h$ is an effective spin or field collecting all neighboring spins (in their old states) interacting with the spin at site i_0 and h is the external magnetic field. Note that this decomposition also works in the case of vectors ($\sigma_i \to \boldsymbol{\sigma_i}$, $h \to \boldsymbol{h}$, $S_{i_0} \to \boldsymbol{S_{i_0}}$), interacting via the usual dot product ($\sigma'_{i_0} S_{i_0} \to \boldsymbol{\sigma'_{i_0}} \cdot \boldsymbol{S_{i_0}}$). As the last equality in (4.25) shows, all other contributions to the energy not involving σ'_{i_0} cancel due to the ratio in (4.25), so that for the update at each site i_0 only a small number of computations is necessary (e.g, about four for a square and six for a simple-cubic lattice of arbitrary size). Detailed balance (4.21) is obviously satisfied since

$$\mathrm{e}^{-\beta\mathcal{H}(\{\sigma_i\}_{\text{old}})} \frac{\mathrm{e}^{-\beta\mathcal{H}(\{\sigma_i\}_{\text{new}})}}{\sum_{\sigma_{i_0}} \mathrm{e}^{-\beta\mathcal{H}(\{\sigma_i\}_{\text{new}})}} = \mathrm{e}^{-\beta\mathcal{H}(\{\sigma_i\}_{\text{new}})} \frac{\mathrm{e}^{-\beta\mathcal{H}(\{\sigma_i\}_{\text{old}})}}{\sum_{\sigma_{i_0}} \mathrm{e}^{-\beta\mathcal{H}(\{\sigma_i\}_{\text{old}})}}. \tag{4.26}$$

How is the probability (4.25) realized in practice? Due to the summation over all local states, special tricks are necessary when each degree of freedom can take many different states, and only in special cases the heat-bath method can be efficiently generalized to continuous degrees of freedom. In many applications, however, the admissible local states of σ_{i_0} can be labeled by a small number of integers, say $n = 1, \ldots, N$. Since the probability in (4.25) is normalized to unity, the sequence $(P_1, P_2, \ldots, P_n, \ldots, P_N)$ decomposes the unit interval into segments of length $P_n = \exp(-\beta n S_{i_0})/\sum_{k=1}^{N} \exp(-\beta k S_{i_0})$. If one now draws a random number $R \in [0, 1)$ and compares the accumulated probabilities $\sum_{k=1}^{n} P_k$ with R, then the new state n_0 is given as the smallest integer for which $\sum_{k=1}^{n_0} P_k \geq R$. Clearly, for a large number of possible local states, the determination of n_0 can become quite time-consuming (in particular, if many small P_n are at the beginning

[3] Some special care is necessary, however, for one-dimensional spin chains.

[4] This is only relevant when the random update order is chosen.

of the sequence, in which case a clever permutation of the P_n-list can help a lot). The order of updating the individual variables can be chosen as for the Metropolis algorithm (random, sequential, ...).

In the special case of the Ising model with only two states per spin, $\sigma_i = \pm 1$, (4.25) reads explicitly as

$$W(\{\sigma_i\}_{\text{old}} \longrightarrow \{\sigma_i\}_{\text{new}}) = \frac{e^{-\beta\sigma'_{i_0} S_{i_0}}}{e^{\beta S_{i_0}} + e^{-\beta S_{i_0}}} \,. \tag{4.27}$$

And since $\Delta E = E_{\text{new}} - E_{\text{old}} = (\sigma'_{i_0} - \sigma_{i_0}) S_{i_0}$, the probability for a spin flip, $\sigma'_{i_0} = -\sigma_{i_0}$, becomes [58]

$$W_{\sigma_{i_0} \to -\sigma_{i_0}} = \frac{e^{-\beta \Delta E/2}}{e^{\beta \Delta E/2} + e^{-\beta \Delta E/2}} \,. \tag{4.28}$$

The acceptance ratio (4.28) is plotted in Fig. 4.1 as a function of ΔE for various (inverse) temperatures and compared with the corresponding ratio (4.24) of the Metropolis algorithm. As we shall see in the next paragraph, for the Ising model, the Glauber and heat-bath algorithm are identical.

4.3.2.3 Glauber Algorithm

The Glauber update prescription [60] is conceptually similar to the Metropolis algorithm in that also here a local update proposal is accepted with a certain probability

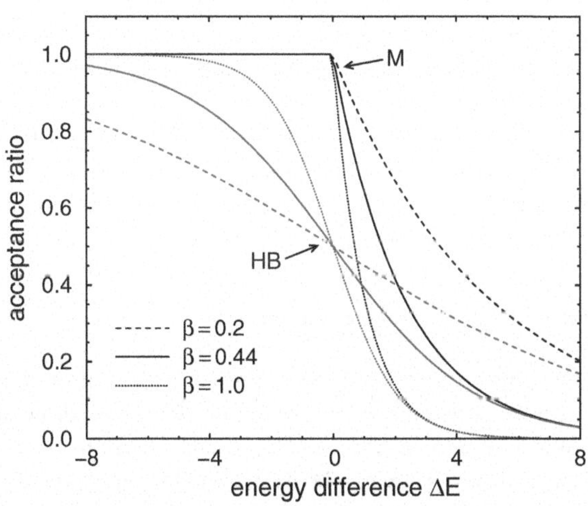

Fig. 4.1. Comparison of the acceptance ratio for a spin flip with the heat-bath (HB) (or Glauber) and Metropolis (M) algorithm in the Ising model for three different inverse temperatures β. Note that for all values of ΔE and temperature, the Metropolis acceptance ratio is higher than that of the heat-bath algorithm

or otherwise rejected. For the Ising model with spins $\sigma_i = \pm 1$ the acceptance probability can be written as

$$W_{\sigma_{i_0} \to -\sigma_{i_0}} = \frac{1}{2} \left[1 + \sigma_{i_0} \tanh\left(\beta S_{i_0}\right)\right] , \tag{4.29}$$

where as before $\sigma_{i_0} S_{i_0}$ with $S_{i_0} = -\sum_j \sigma_j - h$ is the energy of the i_0^{th} spin in the current old state.

Due to the point symmetry of the hyperbolic tangent, one may rewrite σ_{i_0} tanh (βS_{i_0}) as $\tanh\left(\sigma_{i_0} \beta S_{i_0}\right)$. And since as before $\Delta E = E_{\text{new}} - E_{\text{old}} = -2\sigma_{i_0} S_{i_0}$, (4.29) becomes

$$W_{\sigma_{i_0} \to -\sigma_{i_0}} = \frac{1}{2} \left[1 - \tanh\left(\beta \Delta E/2\right)\right] , \tag{4.30}$$

showing explicitly that the acceptance probability only depends on the total energy change as in the Metropolis case. In this form it is thus possible to generalize the Glauber update rule from the Ising model with only two states per spin to any general model that can be simulated with the Metropolis procedure. Also detailed balance is straightforward to prove. Finally by using trivial identities for hyperbolic functions, (4.30) can be further recast to read

$$\begin{aligned} W_{\sigma_{i_0} \to -\sigma_{i_0}} &= \frac{1}{2} \left[\frac{\cosh(\beta \Delta E/2) - \sinh(\beta \Delta E/2)}{\cosh(\beta \Delta E/2)} \right] \\ &= \frac{\mathrm{e}^{-\beta \Delta E/2}}{\mathrm{e}^{\beta \Delta E/2} + \mathrm{e}^{-\beta \Delta E/2}} , \end{aligned} \tag{4.31}$$

which is just the flip probability (4.28) of the heat-bath algorithm for the Ising model, i.e., heat-bath updates for the special case of a 2-state model and the Glauber update algorithm are identical. In the general case with more than two states per spin, however, this is not the case.

The Glauber (or equivalently heat-bath) update algorithm for the Ising model is also theoretically of interest since in this case the dynamics of the Markov chain can be calculated analytically for a one-dimensional system [60]. For two and higher dimensions, however, no exact solutions are known.

4.3.3 Performance of Local Update Algorithms

Local update algorithms are applicable to a very wide class of models and the computer codes are usually quite simple and very fast. The main drawback are temporal correlations of the generated Markov chain which tend to become huge in the vicinity of phase transitions. They can be determined by analysis of autocorrelation functions

$$A(k) = \frac{\langle \mathcal{O}_i \mathcal{O}_{i+k} \rangle - \langle \mathcal{O}_i \rangle \langle \mathcal{O}_i \rangle}{\langle \mathcal{O}_i^2 \rangle - \langle \mathcal{O}_i \rangle \langle \mathcal{O}_i \rangle} , \tag{4.32}$$

where $\mathcal{O}$ denotes any measurable quantity, for example the energy or magnetization. More details and how temporal correlations enter into the statistical error analysis will be discussed in Sect. 4.5.2.3. For large time separations k, $A(k)$ decays exponentially ($a = \text{const}$)

$$A(k) \xrightarrow{k\to\infty} a e^{-k/\tau_{\mathcal{O},\text{exp}}} , \tag{4.33}$$

which defines the exponential autocorrelation time $\tau_{\mathcal{O},\text{exp}}$. At smaller distances usually also other modes contribute and $A(k)$ behaves no longer purely exponentially.

This is illustrated in Fig. 4.2 for the 2D Ising model on a rather small 16×16 square lattice with periodic boundary conditions at the infinite-volume critical point $\beta_c = \ln(1+\sqrt{2})/2 = 0.440\,686\,793\ldots$. The spins were updated in sequential order by proposing always to flip a spin and accepting or rejecting this proposal according to (4.23). The raw data of the simulation are collected in a time-series file, storing 1 000 000 measurements of the energy and magnetization taken after each sweep over the lattice, after discarding (quite generously) the first 200 000 sweeps for equilibrating the system from a disordered start configuration. The last 1 000 sweeps of the time evolution of the energy are shown in Fig. 4.2(a). Using the complete time series the autocorrelation function was computed according to (4.32) which is shown in Fig. 4.2(b). On the linear-log scale of the inset we clearly see the asymptotic linear behavior of $\ln A(k)$. A linear fit of the form (4.33), $\ln A(k) = \ln a - k/\tau_{e,\text{exp}}$, in the range $10 \le k \le 40$ yields an estimate for the exponential autocorrelation time of $\tau_{e,\text{exp}} \approx 11.3$. In the small k behavior of $A(k)$ we observe an initial fast drop, corresponding to faster relaxing modes, before the asymptotic behavior sets in. This is the generic behavior of autocorrelation functions in realistic models where the small-k deviations are, in fact, often much more pronounced than for the 2D Ising model.

Close to a critical point, in the infinite-volume limit, the autocorrelation time typically scales as

$$\tau_{\mathcal{O},\text{exp}} \propto \xi^z , \tag{4.34}$$

where $z \ge 0$ is the so-called dynamical critical exponent. Since the spatial correlation length $\xi \propto |T-T_c|^{-\nu} \to \infty$ when $T \to T_c$, also the autocorrelation time $\tau_{\mathcal{O},\text{exp}}$

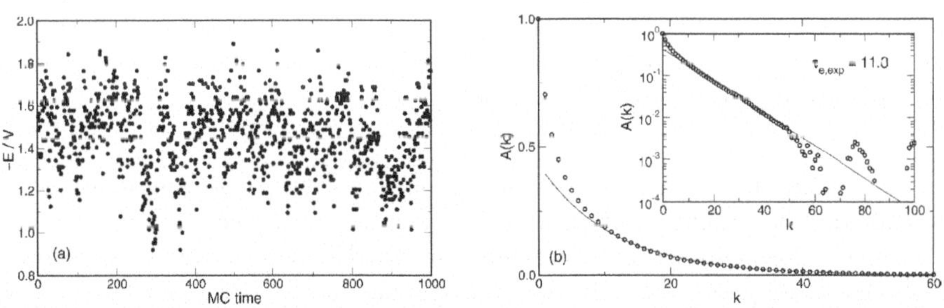

Fig. 4.2. (**a**) Part of the time evolution of the energy $e = E/V$ for the 2D Ising model on a 16×16 lattice at $\beta_c = \ln(1+\sqrt{2})/2 = 0.440\,686\,793\ldots$ and (**b**) the resulting autocorrelation function. In the inset the same data are plotted on a logarithmic scale, revealing a fast initial drop for very small k and noisy behavior for large k. The solid lines show a fit to the ansatz $A(k) = a\exp(-k/\tau_{e,\text{exp}})$ in the range $10 \le k \le 40$ with $\tau_{e,\text{exp}} = 11.3$ and $a = 0.432$

diverges when the critical point is approached, $\tau_{\mathcal{O},\text{exp}} \propto |T - T_c|^{-\nu z}$. This leads to the phenomenon of critical slowing down at a continuous phase transition. This is not in the first place a numerical artefact, but can also be observed experimentally for instance in critical opalescence, see Fig. 1.6 in [5]. The reason is that local spin-flip Monte Carlo dynamics (or diffusion dynamics in a lattice-gas picture) describes at least qualitatively the true physical dynamics of a system in contact with a heat-bath (which, in principle, enters stochastic elements also in molecular dynamics simulations). In a finite system, the correlation length ξ is limited by the linear system size L, and similar to the reasoning in (4.14) and (4.15), the scaling law (4.34) becomes

$$\tau_{\mathcal{O},\text{exp}} \propto L^z \,. \tag{4.35}$$

For local dynamics, the critical slowing down effect is quite pronounced since the dynamical critical exponent takes a rather large value around

$$z \approx 2 \,, \tag{4.36}$$

which is only weakly dependent on the dimensionality and can be understood by a simple random-walk or diffusion argument in energy space. Non-local update algorithms such as multigrid schemes or in particular the cluster methods discussed in the next section can reduce the value of the dynamical critical exponent z significantly, albeit in a strongly model-dependent fashion.

At a first-order phase transition, a completely different mechanism leads to an even more severe slowing-down problem [47]. Here, the password is phase coexistence. A finite system close to the (pseudo-) transition point can flip between the coexisting pure phases by crossing a two-phase region. Relative to the weight of the pure phases, this region of state space is strongly suppressed by an additional Boltzmann factor $\exp(-2\sigma L^{d-1})$, where σ denotes the interface tension between the coexisting phases, L^{d-1} is the (projected) area of the interface and the factor two accounts for periodic boundary conditions, which enforce always an even number of interfaces for simple topological reasons. The time spent for crossing this highly suppressed rare-event region scales proportional to the inverse of this interfacial Boltzmann factor, implying that the autocorrelation time increases exponentially with the system size,

$$\tau_{\mathcal{O},\text{exp}} \propto e^{2\sigma L^{d-1}} \,. \tag{4.37}$$

In the literature, this behavior is sometimes termed supercritical slowing down, even though, strictly speaking, nothing is critical at a first-order phase transition. Since this type of slowing-down problem is directly related to the shape of the probability distribution, it appears for all types of update algorithms, i.e., in contrast to the situation at a second-order transition, here it cannot be cured by employing multigrid or cluster techniques. It can be overcome, however, at least in part by means of tempering and multicanonical methods also briefly discussed at the end of these notes in Sect. 4.8.

4.4 Cluster Algorithms

In this section we first concentrate on the problem of critical slowing down at a second-order phase transition which is caused by very large spatial correlations, reflecting that excitations become equally important on all length scales. It is therefore intuitively clear that some sort of non-local updates should be able to alleviate this problem. While it was realized since long that whole clusters or droplets of spins should play a central role in such an update, it took until 1987 before Swendsen and Wang [61] proposed a legitimate cluster update procedure first for q-state Potts models [62] with

$$\mathcal{H}_{\text{Potts}} = -J \sum_{\langle ij \rangle} \delta_{\sigma_i,\sigma_j} , \tag{4.38}$$

where $\sigma_i = 1, \ldots, q$. For $q = 2$ (and a trivial rescaling) the Ising model (4.3) is recovered. Soon after this discovery, Wolff [63] introduced the so-called single-cluster variant and developed a generalization to O(n)-symmetric spin models. By now cluster update algorithms have been constructed for many other models as well [64]. However, since in all constructions some model specific properties enter in a crucial way, they are still far less general applicable than local update algorithms of the Metropolis type. We therefore first concentrate again on the Ising model where (as for more general Potts models) the prescription for a cluster-update algorithm can be easily read off from the equivalent Fortuin-Kasteleyn representation [65, 66, 67, 68]

$$Z = \sum_{\{\sigma_i\}} \mathrm{e}^{\beta \sum_{\langle ij \rangle} \sigma_i \sigma_j} \tag{4.39}$$

$$= \sum_{\{\sigma_i\}} \prod_{\langle ij \rangle} \mathrm{e}^{\beta} \left[(1-p) + p \delta_{\sigma_i,\sigma_j} \right] \tag{4.40}$$

$$= \sum_{\{\sigma_i\}} \sum_{\{n_{ij}\}} \prod_{\langle ij \rangle} \mathrm{e}^{\beta} \left[(1-p) \delta_{n_{ij},0} + p \delta_{\sigma_i,\sigma_j} \delta_{n_{ij},1} \right] \tag{4.41}$$

with

$$p = 1 - \mathrm{e}^{-2\beta} . \tag{4.42}$$

Here the n_{ij} are bond occupation variables which can take the values $n_{ij} = 0$ or $n_{ij} = 1$, interpreted as deleted or active bonds. The representation (4.40) in the second line follows from the observation that the product $\sigma_i \sigma_j$ of two Ising spins can only take the two values ± 1, so that $\exp(\beta \sigma_i \sigma_j) = x + y \delta_{\sigma_i \sigma_j}$ can easily be solved for x and y. And in the third line (4.41) we made use of the trivial (but clever) identity $a + b = \sum_{n=0}^{1} (a \delta_{n,0} + b \delta_{n,1})$.

4.4.1 Swendsen-Wang Cluster

According to (4.41) a cluster update sweep consists of two alternating steps. First, updates of the bond variables n_{ij} for given spins, and second updates of the spins σ_i for a given bond configuration. In practice one proceeds as follows:

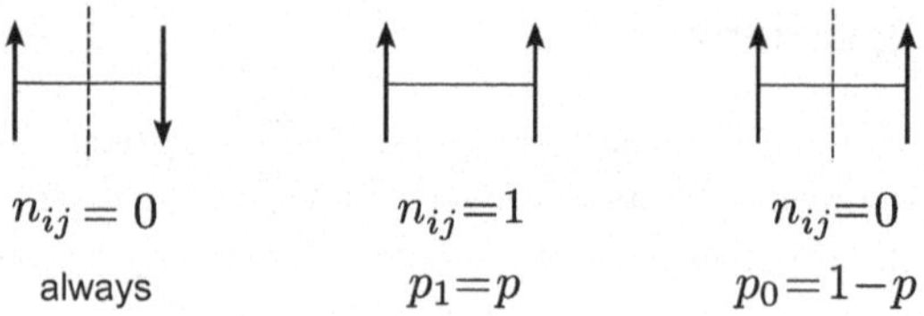

Fig. 4.3. Illustration of the bond variable update. The bond between unlike spins is always deleted as indicated by the *dashed line*. A bond between like spins is only active with probability $p = 1 - \exp(-2\beta)$. Only at zero temperature ($\beta \to \infty$) stochastic and geometrical clusters coincide

(i) Set $n_{ij} = 0$ if $\sigma_i \neq \sigma_j$, or assign values $n_{ij} = 1$ and 0 with probability p and $1 - p$, respectively, if $\sigma_i = \sigma_j$, cp. Fig. 4.3.
(ii) Identify clusters of spins that are connected by active bonds ($n_{ij} = 1$).
(iii) Draw a random value ± 1 independently for each cluster (including one-site clusters), which is then assigned to all spins in a cluster.

Technically the cluster identification part is the most complicated step, but there are by now quite a few efficient algorithms available which can even be used on parallel computers. Vectorization, on the other hand, is only partially possible.

Notice the difference between the just defined stochastic clusters and geometrical clusters whose boundaries are defined by drawing lines through bonds between unlike spins. In fact, since in the stochastic cluster definition also bonds between like spins are deleted with probability $p_0 = 1 - p = \exp(-2\beta)$, stochastic clusters are smaller than geometrical clusters. Only at zero temperature ($\beta \to \infty$) p_0 approaches zero and the two cluster definitions coincide.

As described above, the cluster algorithm is referred to as Swendsen-Wang (SW) or multiple-cluster update [61]. The distinguishing point is that the *whole* lattice is decomposed into stochastic clusters whose spins are assigned a random value $+1$ or -1. In one sweep one thus attempts to update all spins of the lattice.

4.4.2 Wolff Cluster

Shortly after the original discovery of cluster algorithms, Wolff [63] proposed a somewhat simpler variant in which only a single cluster is flipped at a time. This variant is therefore sometimes also called single-cluster algorithm. Here one chooses a lattice site at random, constructs only the cluster connected with this site, and then flips all spins of this cluster. In principle, one could also here choose for all spins in the updated cluster a new value $+1$ or -1 at random, but then nothing at all would be changed if one hits the current value of the spins. Typical configuration plots before and after the cluster flip are shown in Fig. 4.4, which also nicely illustrates the difference between stochastic and geometrical clusters already stressed in the last paragraph. The upper right plot clearly shows that, due to the randomly distributed inactive bonds between like spins, the stochastic cluster is much smaller than the underlying black geometrical cluster which connects all neighboring like spins.

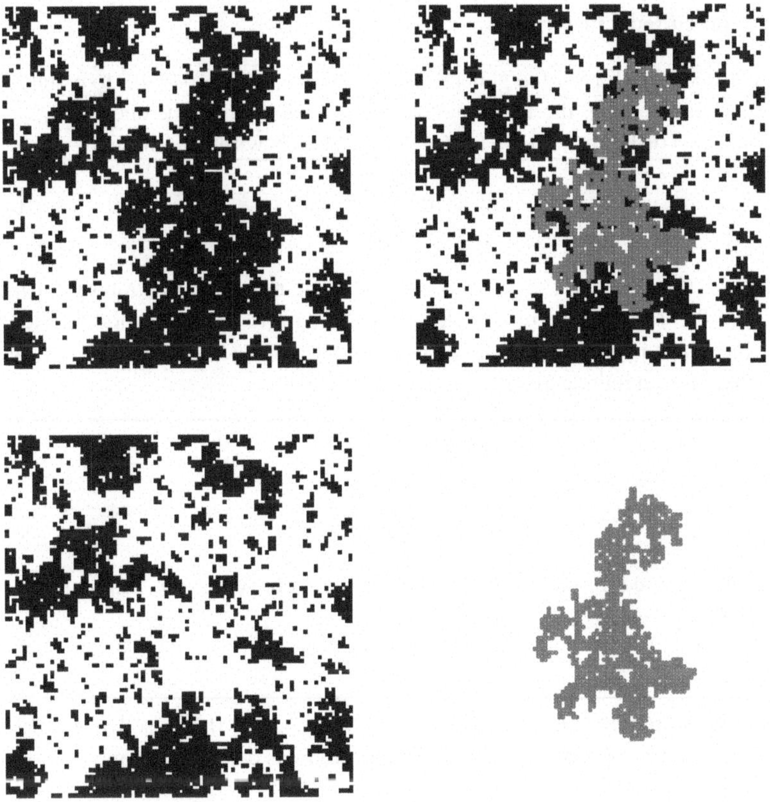

Fig. 4.4. Illustration of the Wolff cluster update, using actual simulation results for the 2D Ising model at $0.97\beta_c$ on a 100×100 lattice. **Upper left**: Initial configuration. **Upper right**: The stochastic cluster is marked. Note how it is embedded in the larger geometric cluster connecting all neighboring like (black) spins. **Lower left**: Final configuration after flipping the spins in the cluster. **Lower right**: The flipped cluster

In the single-cluster variant some care is necessary with the definition of the unit of time since the number of flipped spins varies from cluster to cluster. It also depends crucially on temperature since the average cluster size automatically adapts to the correlation length. With $\langle|C|\rangle$ denoting the average cluster size, a sweep is usually defined to consist of $V/\langle|C|\rangle$ single cluster steps, assuring that on the average V spins are flipped in one sweep. With this definition, autocorrelation times are directly comparable with results from the Swendsen-Wang or Metropolis algorithm. Apart from being somewhat easier to program, Wolff's single-cluster variant is usually even more efficient than the Swendsen-Wang multiple-cluster algorithm, especially in 3D. The reason is that with the single-cluster method, on the average, larger clusters are flipped.

4.4.3 Embedded Clusters

While it is straightforward to generalize the derivation (4.39)–(4.42) to q-state Potts models (because as in the Ising model each contribution to the energy, $\delta_{\sigma_i \sigma_j}$, can take only two different values), for O(n) spin models with Hamiltonian

$$H = -J \sum_{\langle ij \rangle} \boldsymbol{\sigma}_i \cdot \boldsymbol{\sigma}_j \,, \tag{4.43}$$

with $\boldsymbol{\sigma}_i = (\sigma_{i,1}, \sigma_{i,2}, \ldots, \sigma_{i,n})$ and $|\boldsymbol{\sigma}_i| = 1$, one needs a new strategy for $n \geq 2$ [63, 69, 70, 71] (the case $n = 1$ degenerates again to the Ising model). Here the basic idea is to isolate Ising degrees of freedom by projecting the spins $\boldsymbol{\sigma}_i$ onto a randomly chosen unit vector $\boldsymbol{r}$

$$\begin{aligned} \boldsymbol{\sigma}_i &= \boldsymbol{\sigma}_i^{\parallel} + \boldsymbol{\sigma}_i^{\perp} \,, \\ \boldsymbol{\sigma}_i^{\parallel} &= \epsilon \, |\boldsymbol{\sigma}_i \cdot \boldsymbol{r}| \, \boldsymbol{r} \,, \\ \epsilon &= \text{sign}(\boldsymbol{\sigma}_i \cdot \boldsymbol{r}) \,. \end{aligned} \tag{4.44}$$

If this is inserted in the original Hamiltonian one ends up with an effective Hamiltonian

$$H = -\sum_{\langle ij \rangle} J_{ij} \epsilon_i \epsilon_j + \text{const} \,, \tag{4.45}$$

with positive random couplings $J_{ij} = J|\boldsymbol{\sigma}_i \cdot \boldsymbol{r}||\boldsymbol{\sigma}_j \cdot \boldsymbol{r}| \geq 0$, whose Ising degrees of freedom ϵ_i can be updated with a cluster algorithm as described above.

4.4.4 Performance of Cluster Algorithms

The advantage of cluster algorithms is most pronounced close to criticality where excitations on all length scales occur. A convenient performance measure is thus the dynamical critical exponent z (even though one should always check that the proportionality constant in $\tau \propto L^z$ is not exceedingly large, but this is definitely not the case here [72]). Some results on z are collected in Table 4.2, which allow us to conclude:

(i) Compared to local algorithms with $z \approx 2$, z is dramatically reduced for both cluster variants in 2D and 3D [73, 74, 75].
(ii) In 2D, Swendsen-Wang and Wolff cluster updates are equally efficient, while in 3D, the Wolff update is clearly favorable.
(iii) In 2D, the scaling with system size can hardly be distinguished from a very weak logarithmic scaling. Note that this is consistent with the Li-Sokal bound [76, 77] for the Swendsen-Wang cluster algorithm of $\tau_{\text{SW}} \geq C\,(= C_0 + A \ln L$ for the 2D Ising model), implying $z_{\text{SW}} \geq \alpha/\nu\,(= 0$ for the 2D Ising model).
(iv) Different observables (e.g., energy E and magnetization M) may yield quite different values for z when defined via the scaling behavior of the integrated autocorrelation time discussed below in Sect. 4.5.2.3.

Table 4.2. Dynamical critical exponents z for the 2D and 3D Ising model ($\tau \propto L^z$). The subscripts indicate the observables and method used (exp resp. int: exponential resp. integrated autocorrelation time, rel: relaxation, dam: damage spreading)

algorithm	2D	3D	observable	authors
Metropolis	2.1667(5)	–	$z_{M,\mathrm{exp}}$	Nightingale and Blöte [78, 79]
	–	2.032(4)	z_{dam}	Grassberger [80, 81]
	–	2.055(10)	$z_{M,\mathrm{rel}}$	Ito et al. [82]
Swendsen-Wang cluster	0.35(1)	0.75(1)	$z_{E,\mathrm{exp}}$	Swendsen and Wang [61]
	0.27(2)	0.50(3)	$z_{E,\mathrm{int}}$	Wolff [72]
	0.20(2)	0.50(3)	$z_{\chi,\mathrm{int}}$	Wolff [72]
	$0(\log L)$	–	$z_{M,\mathrm{exp}}$	Heermann and Burkitt [83]
	0.25(5)	–	$z_{M,\mathrm{rel}}$	Tamayo [84]
Wolff cluster	0.26(2)	0.28(2)	$z_{E,\mathrm{int}}$	Wolff [72]
	0.13(2)	0.14(2)	$z_{\chi,\mathrm{int}}$	Wolff [72]
	0.25(5)	0.3(1)	$z_{E,\mathrm{rel}}$	Ito and Kohring [85]

4.4.5 Improved Estimators

The intimate relationship of cluster algorithms with the correlated percolation representation of Fortuin and Kasteleyn leads to another quite important improvement which is not directly related with the dynamical properties discussed so far. Within the percolation picture, it is quite natural to introduce alternative estimators (measurement prescriptions) for most standard quantities which turn out to be so-called improved estimators. By this one means measurement prescriptions that yield the same expectation value as the standard ones but have a smaller statistical variance which helps to reduce the statistical errors. Suppose we want to measure the expectation value $\langle \mathcal{O} \rangle$ of an observable $\mathcal{O}$. Then any estimator $\widehat{\mathcal{O}}$ satisfying $\langle \widehat{\mathcal{O}} \rangle = \langle \mathcal{O} \rangle$ is permissible. This does not determine $\widehat{\mathcal{O}}$ uniquely since there are infinitely many other possible choices $\widehat{\mathcal{O}}' = \widehat{\mathcal{O}} + \widehat{\mathcal{X}}$, as long as the added estimator $\widehat{\mathcal{X}}$ has zero expectation $\langle \widehat{\mathcal{X}} \rangle = 0$. The variance of the estimator $\widehat{\mathcal{O}}'$, however, can be quite different and is not necessarily related to any physical quantity (contrary to the standard mean-value estimator of the energy, for instance, whose variance is proportional to the specific heat). It is exactly this freedom in the choice of $\widehat{\mathcal{O}}$ which allows the construction of improved estimators.

For the single-cluster algorithm an improved cluster estimator for the spin-spin correlation function in the high-temperature phase $G(\boldsymbol{x}_i - \boldsymbol{x}_j) \equiv \langle \boldsymbol{\sigma}_i \cdot \boldsymbol{\sigma}_j \rangle$ is given by [71]

$$\widehat{G}(\boldsymbol{x}_i - \boldsymbol{x}_j) = n\frac{V}{|C|}(\boldsymbol{r}\cdot\boldsymbol{\sigma}_i)\,(\boldsymbol{r}\cdot\boldsymbol{\sigma}_j)\,\Theta_C(\boldsymbol{x}_i)\Theta_C(\boldsymbol{x}_j)\,, \tag{4.46}$$

where $\boldsymbol{r}$ is the normal of the mirror plane used in the construction of the cluster of size $|C|$ and $\Theta_C(\boldsymbol{x})$ is its characteristic function ($=1$ if $\boldsymbol{x}\in C$ and zero otherwise). In the Ising case ($n=1$), this simplifies to

$$\widehat{G}(\boldsymbol{x}_i - \boldsymbol{x}_j) = \frac{V}{|C|}\Theta_C(\boldsymbol{x}_i)\Theta_C(\boldsymbol{x}_j)\,, \tag{4.47}$$

i.e., to the test whether the two sites $\boldsymbol{x}_i$ and $\boldsymbol{x}_j$ belong to same stochastic cluster or not. Only in the former case, the average over clusters is incremented by one, otherwise nothing is added. This implies that $\widehat{G}(\boldsymbol{x}_i - \boldsymbol{x}_j)$ is strictly positive which is not the case for the standard estimator $\boldsymbol{\sigma}_i\cdot\boldsymbol{\sigma}_j$, where ± 1 contributions have to average to a positive value. It is therefore at least intuitively clear that the cluster (or percolation) estimator has a smaller variance and is thus indeed an improved estimator, in particular for large separations $|\boldsymbol{x}_i - \boldsymbol{x}_j|$.

For the Fourier transform $\widetilde{G}(\boldsymbol{k}) = \sum_{\boldsymbol{x}} G(\boldsymbol{x})\exp(-i\boldsymbol{k}\cdot\boldsymbol{x})$, (4.46) implies the improved estimator

$$\widehat{\widetilde{G}}(\boldsymbol{k}) = \frac{n}{|C|}\left[\left(\sum_{i\in C}\boldsymbol{r}\cdot\boldsymbol{\sigma}_i\cos\boldsymbol{k}\boldsymbol{x}_i\right)^2 + \left(\sum_{i\in C}\boldsymbol{r}\cdot\boldsymbol{\sigma}_i\sin\boldsymbol{k}\boldsymbol{x}_i\right)^2\right]\,, \tag{4.48}$$

which, for $\boldsymbol{k}=\boldsymbol{0}$, reduces to an improved estimator for the susceptibility $\chi' = \beta V\langle m^2\rangle$ in the high-temperature phase

$$\widehat{\widetilde{G}}(\boldsymbol{0}) = \widehat{\chi}'/\beta = \frac{n}{|C|}\left(\sum_{i\in C}\boldsymbol{r}\cdot\boldsymbol{\sigma}_i\right)^2\,. \tag{4.49}$$

For the Ising model ($n=1$) this reduces to $\chi'/\beta = \langle |C|\rangle$, i.e., the improved estimator of the susceptibility is just the average cluster size of the single-cluster update algorithm. For the XY ($n=2$) and Heisenberg ($n=3$) model one finds empirically that in two as well as in three dimensions $\langle |C|\rangle \approx 0.81\chi'/\beta$ for $n=2$ [69, 86] and $\langle |C|\rangle \approx 0.75\chi'/\beta$ for $n=3$ [71, 87], respectively.

Close to criticality, the average cluster size becomes large, growing $\propto \chi' \propto L^{\gamma/\nu} \simeq L^2$ (since $\gamma/\nu = 2-\eta$ with η usually small) and the advantage of cluster estimators diminishes. In fact, in particular for short-range quantities such as the energy (the next-neighbor correlation) it may even degenerate into a depraved or deteriorated estimator, while long-range quantities such as $G(\boldsymbol{x}_i - \boldsymbol{x}_j)$ for large distances $|\boldsymbol{x}_i - \boldsymbol{x}_j|$ usually still profit from it. A significant reduction of variance by means of the estimators (4.46)–(4.49) can, however, always be expected outside the FSS region where the average cluster size is small compared to the volume of the system.

Finally it is worth pointing out that at least for 2D Potts models also the geometrical clusters do encode critical properties – albeit those of different but related (tricritical) models [88, 89, 90, 91, 92][5].

[5] See also the extensive list of references to earlier work given therein.

4.5 Statistical Analysis of Monte Carlo Data

4.5.1 Initial Non-Equilibrium Period and Ageing

When introducing the importance sampling technique in Sect. 4.3.1 it was already indicated in (4.22) that within Markov chain Monte Carlo simulations, the expectation value $\langle \mathcal{O} \rangle$ of some quantity $\mathcal{O}$, for instance the energy, can be estimated as arithmetic mean

$$\langle \mathcal{O} \rangle = \sum_{\{\sigma_i\}} \mathcal{O}(\{\sigma_i\}) P^{\text{eq}}(\{\sigma_i\}) \approx \overline{\mathcal{O}} = \frac{1}{N} \sum_{j=1}^{N} \mathcal{O}_j \,, \tag{4.50}$$

where $\mathcal{O}_j = \mathcal{O}(\{\sigma_i\}_j)$ is the measured value for the jth configuration and N is the number of measurement sweeps. Also a warning was given that this is only valid after a sufficiently long equilibration period without measurements, which is needed by the system to approach equilibrium after starting the Markov chain in an arbitrary initial configuration.

This initial equilibration or thermalization period, in general, is a non-trivial non-equilibrium process which is of interest in its own right and no simple general recipe determining how long one should wait before starting measurements can be given. Long suspected to be a consequence of the slow dynamics of glassy systems only, the phenomenon of ageing for example has also been found in the phase-ordering kinetics of simple ferromagnets such as the Ising model. To study this effect numerically, one only needs the methods introduced so far since most theoretical concepts assume a local spin-flip dynamics as realized by one of the three local update algorithms discussed above. Similarly to the concept of universality classes in equilibrium, all three algorithms should yield qualitatively similar results, being representatives of what is commonly referred to as dynamical Glauber universality class.

Let us assume that we pick as the initial configuration of the Markov chain a completely disordered state. If the simulation is run at a temperature $T > T_c$, equilibration will, in fact, be fast and nothing spectacular happens. If, however, we choose instead to perform the simulation right at T_c or at a temperature $T < T_c$, the situation is quite different. In the latter two cases one speaks of a quench, since now the starting configuration is in a statistical sense far away from a typical equilibrium configuration at temperature T. This is easiest to understand for temperatures $T \ll T_c$, where the typical equilibrium state consists of homogeneously ordered configurations. After the quench, local regions of parallel spins start forming domains or clusters, and the non-equilibrium dynamics of the system is governed by the movement of the domain walls. In order to minimize their surface energy, the domains grow and straighten their surface. A typical time evolution for the 2D Ising model is illustrated in Fig. 4.5, showing spin configurations after a quench to $T < T_c$, starting from an initially completely disordered state.

This leads to a growth law for the typical correlation length scale of the form $\xi \sim t^{1/z}$, where t is the time (measured in units of sweeps) elapsed since the quench

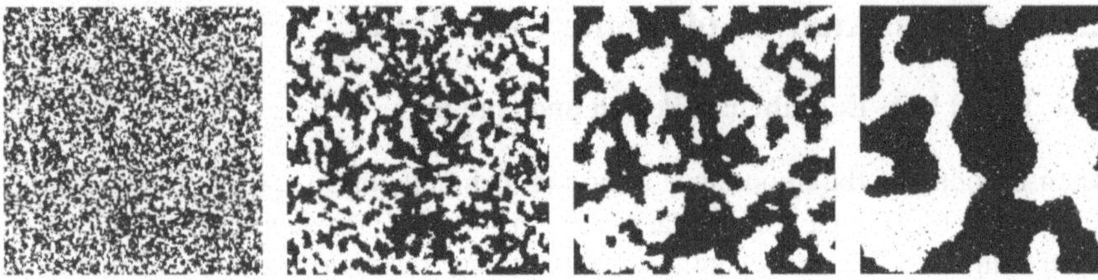

Fig. 4.5. Phase-ordering with progressing Monte Carlo time (from left to right) of an initially disordered spin configuration for the 2D Ising model at $T = 1.5 \approx 0.66\,T_c$ [93]

and z is the dynamical critical exponent already introduced in Sect. 4.3.2. In the case of a simple ferromagnet like the Ising- or q-state Potts model with a non-conserved scalar order parameter, below T_c the dynamical exponent can be found exactly as $z = 2$ [94], according to diffusion or random-walk arguments. Right at the transition temperature, critical dynamics (for a recent review, see [95]) plays the central role and the dynamical exponent of, e.g., the 2D Ising model takes the somewhat larger non-trivial value $z \approx 2.17$ [78, 79] cf. Table 4.2. To equilibrate the whole system, ξ must approach the system size L, so that the typical relaxation time for equilibration scales as

$$\tau_{\text{relax}} \sim L^z \ . \tag{4.51}$$

Note that this implies in the infinite-volume limit $L \to \infty$ that true equilibrium can never be reached.

Since $1/z < 1$, the relaxation process after the quench happens on a growing time scale. This can be revealed most clearly by measurements of two-time quantities $f(t, s)$ with $t > s$, which no longer transform time-translation invariantly as they would do for small perturbations in equilibrium, where f would be a function of the time difference $t - s$ only. Instead, in phase-ordering kinetics, two-time quantities depend non-trivially on the ratio t/s of the two times. The dependence of the relaxation on the so-called waiting time s is the notional origin of ageing: Older samples respond more slowly.

For the most commonly considered two-time quantities, dynamical scaling forms can be theoretically predicted (for recent reviews see, e.g., [96, 97]). Well studied are the two-time autocorrelation function (here in q-state Potts model notation)

$$C(t, s) = \frac{1}{q-1}\left(\frac{q}{V}\sum_{i=1}^{V}\left[\delta_{\sigma_i(t),\sigma_i(s)}\right]_{\text{av}} - 1\right) = s^{-b} f_C(t/s) \ , \tag{4.52}$$

with the asymptotic behavior $f_C(x) \to x^{-\lambda_C/z}$ $(x \gg 1)$, and the two-time response function

$$R(t, s) = \left.\frac{\delta\left[\sigma_i(t)\right]_{\text{av}}}{\delta h_i(s)}\right|_{h=0} = s^{-1-a} f_R(t/s) \ , \tag{4.53}$$

where $f_R(x) \to x^{-\lambda_R/z}$ $(x \gg 1)$. Here $h(s)$ is the amplitude of a small spatially random external field which is switched off after the waiting time s and $[\ldots]_{\text{av}}$

denotes an average over different random initial configurations (and random fields in (4.53)). In phase-ordering kinetics after a quench to $T < T_c$, in general $b = 0$ (and $z = 2$) [94], but all other exponents depend on the dimensionality of the considered system. In the simplest case of the Ising model in two dimensions, it is commonly accepted that $\lambda_C = \lambda_R = 5/4$. The value of the remaining exponent a, however, is more controversial [98, 99], with strong claims for $a = 1/z = 1/2$ [96, 100], but also $a = 1/4$ [101, 102] has been conjectured. In computer simulation studies the two-time response function is rather difficult to handle and it is more convenient to consider the integrated response or thermoremanent magnetization (TRM) [103],

$$\rho(t,s) = T\int_0^s du R(t,u) = \frac{T}{h} M_{\mathrm{TRM}}(t,s) \,. \tag{4.54}$$

By extending dynamical scaling to local scale invariance (LSI) in analogy to conformal invariance [104], even explicit expressions of the scaling functions $f_C(x)$ and $f_R(x)$ have been predicted [105] (for a recent review, see [106]). For the 2D and 3D Ising model, extensive numerical tests of the LSI predictions have been performed by Henkel, Pleimling and collaborators [107, 108, 109], showing a very good agreement with the almost parameter-free analytical expressions. Recently this could be confirmed also for more general q-state Potts models with $q = 3$ and $q = 8$ in two dimensions [93, 110].

If one is primarily interested in equilibrium properties of the considered statistical system, there is, of course, no need to study the initial equilibration period in such a great detail. It is, however, advisable to watch the time evolution of the system and to make sure that no apparent trends are still visible when starting the measurements. If estimates of the autocorrelation or relaxation time are available, a good a priori estimate is to wait at least about 20 $\tau_{\mathcal{O},\mathrm{exp}}$. Finally, as a (not further justified) rule of thumb, most practicers of Monte Carlo simulations spend at least about 10% of the total computing time on the equilibration or thermalization period.

4.5.2 Statistical Errors and Autocorrelation Times

4.5.2.1 Estimators

As already indicated in (4.50), conceptually it is important to distinguish between the expectation value $\langle \mathcal{O} \rangle$ and the mean value $\overline{\mathcal{O}}$, which is an estimator for the former. While $\langle \mathcal{O} \rangle$ is an ordinary number and represents the exact result (which is usually unknown, of course), the estimator $\overline{\mathcal{O}}$ is still a random number which for finite N fluctuates around the theoretically expected value. Of course, from a single Monte Carlo simulation with N measurements, we obtain only a single number for $\overline{\mathcal{O}}$ at the end of the day. To estimate the statistical uncertainty due to the fluctuations, i.e., the statistical error bar, it seems at first sight that one would have to repeat the whole simulation many times. Fortunately, this is not necessary since one can estimate the variance of $\overline{\mathcal{O}}$,

$$\sigma^2_{\overline{\mathcal{O}}} = \langle [\overline{\mathcal{O}} - \langle \overline{\mathcal{O}} \rangle]^2 \rangle = \langle \overline{\mathcal{O}}^2 \rangle - \langle \overline{\mathcal{O}} \rangle^2 \,, \tag{4.55}$$

from the statistical properties of individual measurements $\mathcal{O}_i, i = 1, \ldots, N$, in a single Monte Carlo run.

4.5.2.2 Uncorrelated Measurements and Central-Limit Theorem

Inserting (4.50) into (4.55) gives

$$\sigma^2_{\overline{\mathcal{O}}} = \langle \overline{\mathcal{O}}^2 \rangle - \langle \overline{\mathcal{O}} \rangle^2 = \frac{1}{N^2} \sum_{i,j=1}^{N} \langle \mathcal{O}_i \mathcal{O}_j \rangle - \frac{1}{N^2} \sum_{i,j=1}^{N} \langle \mathcal{O}_i \rangle \langle \mathcal{O}_j \rangle \,, \tag{4.56}$$

and by collecting diagonal and off-diagonal terms one arrives at [111]

$$\sigma^2_{\overline{\mathcal{O}}} = \frac{1}{N^2} \sum_{i=1}^{N} \left(\langle \mathcal{O}_i^2 \rangle - \langle \mathcal{O}_i \rangle^2 \right) + \frac{1}{N^2} \sum_{i \neq j}^{N} \left(\langle \mathcal{O}_i \mathcal{O}_j \rangle - \langle \mathcal{O}_i \rangle \langle \mathcal{O}_j \rangle \right) \,. \tag{4.57}$$

Assuming equilibrium, the individual variances $\sigma^2_{\mathcal{O}_i} = \langle \mathcal{O}_i^2 \rangle - \langle \mathcal{O}_i \rangle^2$ do not depend on "time" i, such that the first term gives $\sigma^2_{\mathcal{O}_i}/N$. The second term with $\langle \mathcal{O}_i \mathcal{O}_j \rangle - \langle \mathcal{O}_i \rangle \langle \mathcal{O}_j \rangle = \langle (\mathcal{O}_i - \langle \mathcal{O}_i \rangle)(\mathcal{O}_j - \langle \mathcal{O}_j \rangle) \rangle$ records the correlations between measurements at times i and j. For completely uncorrelated data (which is, of course, an unrealistic assumption for importance sampling Monte Carlo simulations), the second term would vanish and (4.57) simplifies to

$$\epsilon^2_{\overline{\mathcal{O}}} \equiv \sigma^2_{\overline{\mathcal{O}}} = \sigma^2_{\mathcal{O}_i}/N \,. \tag{4.58}$$

This result is true for any distribution $\mathcal{P}(\mathcal{O}_i)$. In particular, for the energy or magnetization, distributions of the individual measurements are often plotted as physically directly relevant (N independent) histograms (see, e.g., Fig. 4.8(b) below) whose squared width ($= \sigma^2_{\mathcal{O}_i}$) is proportional to the specific heat or susceptibility, respectively.

Whatever form the distribution $\mathcal{P}(\mathcal{O}_i)$ assumes (which, in fact, is often close to Gaussian because the $\mathcal{O}_i$ are usually already lattice averages over many degrees of freedom), by the central limit theorem the distribution of the mean value is Gaussian, at least for uncorrelated data in the asymptotic limit of large N. The variance of the mean, $\sigma^2_{\overline{\mathcal{O}}}$, is the squared width of this (N dependent) distribution which is usually taken as the one-sigma squared error, $\epsilon^2_{\overline{\mathcal{O}}} \equiv \sigma^2_{\overline{\mathcal{O}}}$, and quoted together with the mean value $\overline{\mathcal{O}}$. Under the assumption of a Gaussian distribution for the mean, the interpretation is that about 68% of all simulations under the same conditions would yield a mean value in the range $[\overline{\mathcal{O}} - \sigma_{\overline{\mathcal{O}}}, \overline{\mathcal{O}} + \sigma_{\overline{\mathcal{O}}}]$ [113]. For a two-sigma interval which also is sometimes used, this percentage goes up to about 95.4%, and for a three-sigma interval which is rarely quoted, the confidence level is higher than 99.7%.

4.5.2.3 Correlated Measurements and Autocorrelation Times

For correlated data the second term in (4.57) does not vanish and things become more involved [114, 115, 116]. Using the symmetry $i \leftrightarrow j$ to reduce the summation $\sum_{i \neq j}^{N}$ to $2\sum_{i=1}^{N}\sum_{j=i+1}^{N}$, reordering the summation, and using time-translation invariance in equilibrium, one finally obtains [111]

$$\sigma_{\overline{\mathcal{O}}}^2 = \frac{1}{N}\left[\sigma_{\mathcal{O}_i}^2 + 2\sum_{k=1}^{N}\left(\langle\mathcal{O}_1\mathcal{O}_{1+k}\rangle - \langle\mathcal{O}_1\rangle\langle\mathcal{O}_{1+k}\rangle\right)\left(1 - \frac{k}{N}\right)\right] , \qquad (4.59)$$

where, due to the last factor $(1 - k/N)$, the $k = N$ term may be trivially kept in the summation. Factoring out $\sigma_{\mathcal{O}_i}^2$, this can be written as

$$\epsilon_{\overline{\mathcal{O}}}^2 \equiv \sigma_{\overline{\mathcal{O}}}^2 = \frac{\sigma_{\mathcal{O}_i}^2}{N} 2\tau'_{\mathcal{O},\mathrm{int}} , \qquad (4.60)$$

where we have introduced the (proper) integrated autocorrelation time

$$\tau'_{\mathcal{O},\mathrm{int}} = \frac{1}{2} + \sum_{k=1}^{N} A(k)\left(1 - \frac{k}{N}\right) , \qquad (4.61)$$

with

$$A(k) \equiv \frac{\langle\mathcal{O}_1\mathcal{O}_{1+k}\rangle - \langle\mathcal{O}_1\rangle\langle\mathcal{O}_{1+k}\rangle}{\sigma_{\mathcal{O}_i}^2} \xrightarrow{k\to\infty} a e^{-k/\tau_{\mathcal{O},\mathrm{exp}}} \qquad (4.62)$$

being the normalized autocorrelation function ($A(0) = 1$) already introduced in (4.32). Since in any meaningful simulation study $N \gg \tau_{\mathcal{O},\mathrm{exp}}$, $A(k)$ in (4.61) is already exponentially small before the correction term in parentheses becomes important. For simplicity this correction is hence usually omitted (as is the prime of $\tau'_{\mathcal{O},\mathrm{int}}$ in (4.61)) and one employs the following definition for the integrated autocorrelation time

$$\tau_{\mathcal{O},\mathrm{int}} = \frac{1}{2} + \sum_{k=1}^{N} A(k) . \qquad (4.63)$$

The notion "integrated" derives from the fact that this may be interpreted as a trapezoidal discretization of the (approximate) integral $\tau_{\mathcal{O},\mathrm{int}} \approx \int_0^N \mathrm{d}k A(k)$. Notice that, in general, $\tau_{\mathcal{O},\mathrm{int}}$ (and also $\tau'_{\mathcal{O},\mathrm{int}}$) is different from $\tau_{\mathcal{O},\mathrm{exp}}$. In fact, one can show [117] that $\tau_{\mathcal{O},\mathrm{int}} \leq \tau_{\mathcal{O},\mathrm{exp}}$ in realistic models. Only if $A(k)$ is a pure exponential, the two autocorrelation times, $\tau_{\mathcal{O},\mathrm{int}}$ and $\tau_{\mathcal{O},\mathrm{exp}}$, coincide (up to minor corrections for small $\tau_{\mathcal{O},\mathrm{int}}$ [58, 111]).

As far as the accuracy of Monte Carlo data is concerned, the important point of (4.60) is that due to temporal correlations of the measurements the statistical error $\epsilon_{\overline{\mathcal{O}}} \equiv \mathcal{O} \Rightarrow \sqrt{\sigma_{\overline{\mathcal{O}}}^2}$ on the Monte Carlo estimator $\overline{\mathcal{O}}$ is enhanced by a factor of $\sqrt{2\tau_{\mathcal{O},\mathrm{int}}}$. This can be rephrased by writing the statistical error similar to the uncorrelated case as $\epsilon_{\overline{\mathcal{O}}} = \sqrt{\sigma_{\mathcal{O}_j}^2/N_{\mathrm{eff}}}$, but now with a parameter

$$N_{\text{eff}} = N/2\tau_{\mathcal{O},\text{int}} \leq N \;, \tag{4.64}$$

describing the effective statistics. This shows more clearly that only every $2\tau_{\mathcal{O},\text{int}}$ iterations the measurements are approximately uncorrelated and gives a better idea of the relevant effective size of the statistical sample. In view of the scaling behavior of the autocorrelation time in (4.34), (4.35) or (4.37), it is obvious that without extra care this effective sample size may become very small close to a continuous or first-order phase transition, respectively.

4.5.2.4 Bias

A too small effective sample size does not only affect the error bars, but for some quantities even the mean values can be severely underestimated. This happens for so-called biased estimators, as is for instance the case for the specific heat and susceptibility. The specific heat can be computed as $C = \beta^2 V\left(\langle e^2\rangle - \langle e\rangle^2\right) = \beta^2 V \sigma_{e_i}^2$, with the standard estimator for the variance

$$\widehat{\sigma}_{e_i}^2 = \overline{e^2} - \overline{e}^2 = \overline{(e-\overline{e})^2} = \frac{1}{N}\sum_{i=1}^{N}\left(e_i - \overline{e}\right)^2 \;. \tag{4.65}$$

Subtracting and adding $\langle\overline{e}\rangle^2$, one finds for the expectation value

$$\langle\widehat{\sigma}_{e_i}^2\rangle = \langle\overline{e^2} - \overline{e}^2\rangle = \left(\langle\overline{e^2}\rangle - \langle\overline{e}\rangle^2\right) - \left(\langle\overline{e}^2\rangle - \langle\overline{e}\rangle^2\right) = \sigma_{e_i}^2 + \sigma_{\overline{e}}^2 \;. \tag{4.66}$$

Using (4.60) this gives

$$\langle\widehat{\sigma}_{e_i}^2\rangle = \sigma_{e_i}^2\left(1 - \frac{2\tau_{e,\text{int}}}{N}\right) = \sigma_{e_i}^2\left(1 - \frac{1}{N_{\text{eff}}}\right) \neq \sigma_{e_i}^2 \;. \tag{4.67}$$

The estimator $\widehat{\sigma}_{e_i}^2$ in (4.65) thus systematically underestimates the true value by a term of the order of $\tau_{e,\text{int}}/N$. Such an estimator is called weakly biased (weakly because the statistical error $\propto 1/\sqrt{N}$ is asymptotically larger than the systematic bias; for medium or small N, however, also prefactors need to be carefully considered).

We thus see that for large autocorrelation times or equivalently small effective statistics N_{eff}, the bias may be quite large. Since $\tau_{e,\text{int}}$ scales quite strongly with the system size for local update algorithms, some care is necessary when choosing the run time N. Otherwise the FSS of the specific heat or susceptibility and thus the determination of the static critical exponent α/ν or γ/ν could be completely spoiled by the temporal correlations [118]! Any serious simulation study should therefore provide at least a rough order-of-magnitude estimate of autocorrelation times.

4.5.3 Numerical Estimation of Autocorrelation Times

The above considerations show that not only for the error estimation but also for the computation of static quantities themselves, it is important to have control over autocorrelations. Unfortunately, it is very difficult to give reliable a priori estimates, and an accurate numerical analysis is often too time consuming. As a rough estimate it is about ten times harder to get precise information on dynamic quantities than on static quantities like critical exponents. A (weakly biased) estimator $\widehat{A}(k)$ for the autocorrelation function is obtained as usual by replacing in (4.32) the expectation values (ordinary numbers) by mean values (random variables), e.g., $\langle \mathcal{O}_i \mathcal{O}_{i+k} \rangle$ by $\overline{\mathcal{O}_i \mathcal{O}_{i+k}}$. With increasing separation k the relative variance of $\widehat{A}(k)$ diverges rapidly. To get at least an idea of the order of magnitude of $\tau_{\mathcal{O},\mathrm{int}}$ and thus the correct error estimate (4.60), it is useful to record the running autocorrelation time estimator

$$\widehat{\tau}_{\mathcal{O},\mathrm{int}}(k_{\mathrm{max}}) = \frac{1}{2} + \sum_{k=1}^{k_{\mathrm{max}}} \widehat{A}(k) \, , \tag{4.68}$$

which approaches $\tau_{\mathcal{O},\mathrm{int}}$ in the limit of large k_{max} where, however, its statistical error increases rapidly. As an example, Fig. 4.6(a) shows results for the 2D Ising model from an analysis of the same raw data as in Fig. 4.2.

As a compromise between systematic and statistical errors, an often employed procedure is to determine the upper limit k_{max} self-consistently by cutting off the summation once $k_{\mathrm{max}} \geq 6\, \widehat{\tau}_{\mathcal{O},\mathrm{int}}(k_{\mathrm{max}})$, where $A(k) \approx \mathrm{e}^{-6} \approx 10^{-3}$. In this case an a priori error estimate is available [116, 119, 120]

$$\epsilon_{\tau_{\mathcal{O},\mathrm{int}}} = \tau_{\mathcal{O},\mathrm{int}} \sqrt{\frac{2(2k_{\mathrm{max}}+1)}{N}} \approx \tau_{\mathcal{O},\mathrm{int}} \sqrt{\frac{12}{N_{\mathrm{eff}}}} \, . \tag{4.69}$$

For a 5% relative accuracy one thus needs at least $N_{\mathrm{eff}} \approx 5\,000$ or $N \approx 10\,000\, \tau_{\mathcal{O},\mathrm{int}}$ measurements. For an order of magnitude estimate consider the 2D Ising model on a square lattice with $L = 100$ simulated with a local update algorithm. Close to criticality, the integrated autocorrelation time for this example is of the order of $L^z \approx L^2 \approx 100^2$ (ignoring an priori unknown prefactor of order unity which

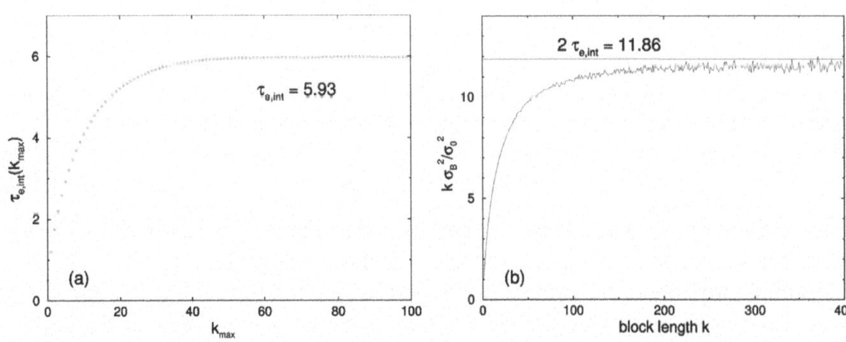

Fig. 4.6. (**a**) Integrated autocorrelation time approaching $\tau_{e,\mathrm{int}} \approx 5.93$ for large upper cutoff k_{max} and (**b**) binning analysis for the energy of the 2D Ising model on a 16×16 lattice at β_c, using the same data as in Fig. 4.2. The *horizontal line* in (**b**) shows $2\tau_{e,\mathrm{int}}$ with $\tau_{e,\mathrm{int}}$ read off from (**a**)

depends on the considered quantity), implying $N \approx 10^8$. Since in each sweep L^2 spins have to be updated and assuming that each spin update takes about $0.1\,\mu$sec, we end up with a total time estimate of about $10^5\,\text{s} \approx 1$ CPU-day to achieve this accuracy.

An alternative is to approximate the tail end of $A(k)$ by a single exponential as in (4.33). Summing up the small k part exactly, one finds [121]

$$\tau_{\mathcal{O},\text{int}}(k_{\text{max}}) = \tau_{\mathcal{O},\text{int}} - c\,\mathrm{e}^{-k_{\text{max}}/\tau_{\mathcal{O},\text{exp}}}\,, \tag{4.70}$$

where c is a constant. The latter expression may be used for a numerical estimate of both the exponential and integrated autocorrelation times [121].

4.5.4 Binning Analysis

It should be clear by now that ignoring autocorrelation effects can lead to severe underestimates of statistical errors. Applying the full machinery of autocorrelation analysis discussed above, however, is often too cumbersome. On a day by day basis the following binning analysis is much more convenient (though somewhat less accurate). By grouping the N original time-series data into N_B non-overlapping bins or blocks of length k (such that[6] $N = N_B k$), one forms a new, shorter time series of block averages

$$\mathcal{O}_j^{(B)} \equiv \frac{1}{k}\sum_{i=1}^{k} \mathcal{O}_{(j-1)k+i} \tag{4.71}$$

with $j = 1, \ldots, N_B$, which by choosing the block length $k \gg \tau$ are almost uncorrelated and can thus be analyzed by standard means. The mean value over all block averages obviously satisfies $\overline{\mathcal{O}^{(B)}} = \overline{\mathcal{O}}$ and their variance can be computed according to the standard (unbiased) estimator, leading to the squared statistical error of the mean value

$$\epsilon_{\overline{\mathcal{O}}}^2 \equiv \sigma_{\overline{\mathcal{O}}}^2 = \sigma_B^2/N_B = \frac{1}{N_B(N_B-1)}\sum_{j=1}^{N_B}(\mathcal{O}_j^{(B)} - \overline{\mathcal{O}^{(B)}})^2\,. \tag{4.72}$$

By comparing with (4.60) we see that $\sigma_B^2/N_B = 2\tau_{\mathcal{O},\text{int}}\sigma_{\mathcal{O}_i}^2/N$. Recalling the definition of the block length $k = N/N_B$, this shows that one may also use

$$2\tau_{\mathcal{O},\text{int}} = k\sigma_B^2/\sigma_{\mathcal{O}_i}^2 \tag{4.73}$$

for the estimation of $\tau_{\mathcal{O},\text{int}}$. This is demonstrated in Fig. 4.6(b). Estimates of $\tau_{\mathcal{O},\text{int}}$ obtained in this way are often referred to as blocking τ or binning τ.

A simple toy model (bivariate time series), where the behavior of the blocking τ and also of $\tau_{\mathcal{O},\text{int}}(k_{\text{max}})$ for finite k resp. k_{max} can be worked out exactly, is discussed in [58]. These analytic formulas are very useful for validating the computer implementations.

---

[6] Here we assume that N was chosen cleverly. Otherwise one has to discard some of the data and redefine N.

4.5.5 Jackknife Analysis

Even if the data are completely uncorrelated in time, one still has to handle the problem of error estimation for quantities that are not directly measured in the simulation but are computed as a non-linear combination of basic observables. This problem can either be solved by error propagation or by using the Jackknife method [122, 123], where instead of considering rather small blocks of length k and their fluctuations as in the binning method, one forms N_B large Jackknife blocks $\mathcal{O}_j^{(J)}$ containing all data but the jth block of the previous binning method,

$$\mathcal{O}_j^{(J)} = \frac{N\overline{\mathcal{O}} - k\mathcal{O}_j^{(B)}}{N-k} \tag{4.74}$$

with $j = 1, \ldots, N_B$, cf. the schematic sketch in Fig. 4.7.

Each of the Jackknife blocks thus consists of $N - k$ data, i.e., it contains almost as many data as the original time series. When non-linear combinations of basic variables are estimated, the bias is hence comparable to that of the total data set (typically $1/(N-k)$ compared to $1/N$). The N_B Jackknife blocks are, of course, trivially correlated because one and the same original data enter in $N_B - 1$ different Jackknife blocks. This trivial correlation caused by re-using the original data over and over again has nothing to do with temporal correlations. As a consequence, the Jacknife block variance σ_J^2 will be much smaller than the variance estimated in the binning method. Because of the trivial nature of the correlations, however, this reduction can be corrected by multiplying σ_J^2 with a factor $(N_B - 1)^2$, leading to

$$\epsilon_{\overline{\mathcal{O}}}^2 \equiv \sigma_{\overline{\mathcal{O}}}^2 = \frac{N_B - 1}{N_B} \sum_{j=1}^{N_B} (\mathcal{O}_j^{(J)} - \overline{\mathcal{O}^{(J)}})^2 \,. \tag{4.75}$$

To summarize this section, any realization of a Markov chain Monte Carlo update algorithm is characterized by autocorrelation times which enter directly into the statistical errors of Monte Carlo estimates. Since temporal correlations always increase the statistical errors, it is thus a very important issue to develop Monte Carlo

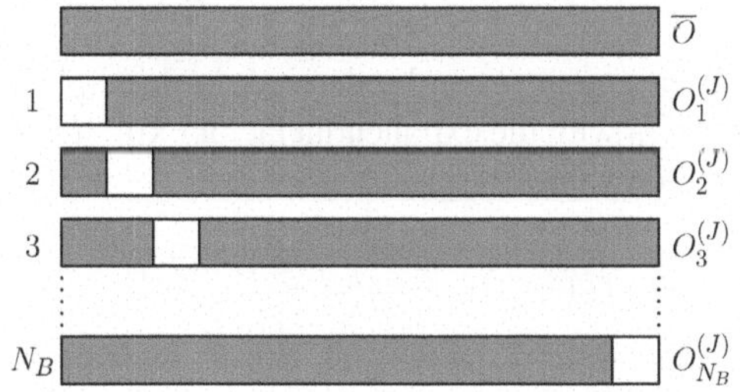

Fig. 4.7. Schematic sketch of the organization of Jackknife blocks. The *grey part* of the N data points is used for calculating the total and the Jackknife block averages. The white blocks enter into the more conventional binning analysis using non-overlapping blocks

update algorithms that keep autocorrelation times as small as possible. This is the reason why cluster and other non-local algorithms are so important.

4.6 Reweighting Techniques

The physics underlying reweighting techniques [124, 125] is extremely simple and the basic idea has been known since long (see the list of references in [125]), but their power in practice has been realized only relatively late in 1988. The important observation by Ferrenberg and Swendsen [124, 125] was that the best performance is achieved near criticality where histograms are usually broad. In this sense reweighting techniques are complementary to improved estimators, which usually perform best off criticality.

4.6.1 Single-Histogram Technique

The single-histogram reweighting technique [124] is based on the following very simple observation. If we denote the number of states (spin configurations) that have the same energy E by $\Omega(E)$, the partition function at the simulation point $\beta_0 = 1/k_\mathrm{B}T_0$ can always be written as[7]

$$Z(\beta_0) = \sum_{\{s\}} \mathrm{e}^{-\beta_0 H(\{s\})} = \sum_E \Omega(E)\mathrm{e}^{-\beta_0 E} \propto \sum_E P_{\beta_0}(E) \,, \tag{4.76}$$

where we have introduced the unnormalized energy histogram (density)

$$P_{\beta_0}(E) \propto \Omega(E)\mathrm{e}^{-\beta_0 E} \,. \tag{4.77}$$

If we would normalize $P_{\beta_0}(E)$ to unit area, the r.h.s. would have to be divided by $\sum_E P_{\beta_0}(E) = Z(\beta_0)$, but the normalization will be unimportant in what follows. Let us assume we have performed a Monte Carlo simulation at inverse temperature β_0 and thus know $P_{\beta_0}(E)$. It is then easy to see that

$$P_\beta(E) \propto \Omega(E)\mathrm{e}^{-\beta E} = \Omega(E)\mathrm{e}^{-\beta_0 E}\mathrm{e}^{-(\beta-\beta_0)E} \propto P_{\beta_0}(E)\mathrm{e}^{-(\beta-\beta_0)E} \,, \tag{4.78}$$

i.e., the histogram at any point β can be derived, in principle, by reweighting the simulated histogram at β_0 with the exponential factor $\exp[-(\beta-\beta_0)E]$. Notice that in reweighted expectation values

$$\langle f(E)\rangle(\beta) = \frac{\sum_E f(E)P_\beta(E)}{\sum_E P_\beta(E)} \,, \tag{4.79}$$

the normalization of $P_\beta(E)$ indeed cancels. This gives, for instance, the energy $\langle e\rangle(\beta) = \langle E\rangle(\beta)/V$ and the specific heat $C(\beta) = \beta^2 V[\langle e^2\rangle(\beta) - \langle e\rangle(\beta)^2]$, in principle, as a continuous function of β from a single Monte Carlo simulation at β_0, where $V = L^D$ is the system size.

As an example of this reweighting procedure, using actual Swendsen-Wang cluster simulation data (with 5 000 sweeps for equilibration and 50 000 sweeps for measurements) of the 2D Ising model at $\beta_0 = \beta_c = \ln(1 + \sqrt{2})/2 = 0.440\,686\ldots$ on a 16×16 lattice with periodic boundary conditions, the specific heat $C(\beta)$ is shown in Fig. 4.8(a) and compared with the curve obtained from the exact Kaufman solution [12, 13] for finite $L_x \times L_y$ lattices. This clearly demonstrates that, in practice, the β-range over which reweighting can be trusted is limited. The reason for this limitation are unavoidable statistical errors in the numerical determination of P_{β_0} using a Monte Carlo simulation. In the tails of the histograms the relative statistical errors are largest, and the tails are exactly the regions that contribute most when multiplying $P_{\beta_0}(E)$ with the exponential reweighting factor to obtain $P_\beta(E)$ for β-values far off the simulation point β_0. This is illustrated in Fig. 4.8(b) where the simulated histogram at $\beta_0 = \beta_c$ is shown together with the reweighted histograms at $\beta = 0.375 \approx \beta_0 - 0.065$ and $\beta = 0.475 \approx \beta_0 + 0.035$, respectively. For the 2D Ising model the quality of the reweighted histograms can be judged by comparing with the curves obtained from Beale's [112] exact expression for $\Omega(E)$.

4.6.1.1 Reweighting Range

As a rule of thumb, the range over which reweighting should produce accurate results can be estimated by requiring that the peak location of the reweighted his-

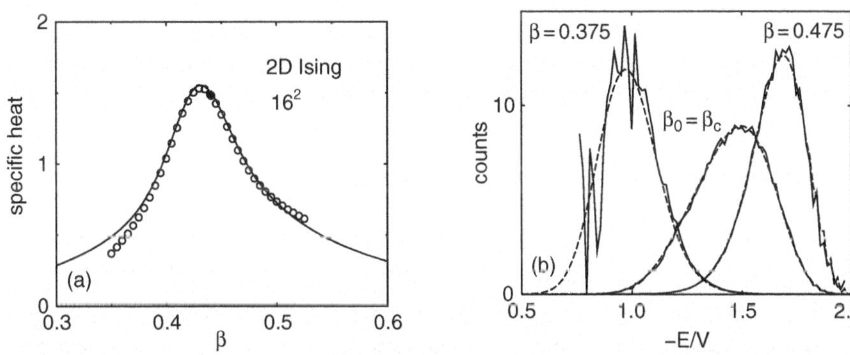

Fig. 4.8. **(a)** The specific heat of the 2D Ising model on a 16×16 square lattice computed by reweighting from a single Monte Carlo simulation at $\beta_0 = \beta_c$, marked by the filled data symbol. The *continuous line* shows for comparison the exact solution of Kaufman [12, 13]. **(b)** The corresponding energy histogram at β_0, and reweighted to $\beta = 0.375$ and $\beta = 0.475$. The *dashed lines* show for comparison the exact histograms obtained from Beale's expression [112]

[7] For simplicity we consider here only models with discrete energies. If the energy varies continuously, sums have to be replaced by integrals, etc. Also lattice size dependences are suppressed to keep the notation short.

togram should not exceed the energy value at which the input histogram had decreased to about one half or one third of its maximum value. In most applications this range is wide enough to locate from a single simulation, e.g., the specific-heat maximum by employing a standard maximization subroutine to the continuous function $C(\beta)$. This is by far more convenient, accurate and faster than the traditional way of performing many simulations close to the peak of $C(\beta)$ and trying to determine the maximum by spline or least-squares fits.

For an analytical estimate of the reweighting range we now require that the peak of the reweighted histogram is within the width $\langle e\rangle(T_0) \pm \Delta e(T_0)$ of the input histogram (where a Gaussian histogram would have decreased to $\exp(-1/2) \approx 0.61$ of its the maximum value)

$$|\langle e\rangle(T) - \langle e\rangle(T_0)| \leq \Delta e(T_0) , \tag{4.80}$$

where we have made use of the fact that for a not too asymmetric histogram $P_{\beta_0}(E)$ the maximum location approximately coincides with $\langle e\rangle(T_0)$. Recalling that the half width Δe of a histogram is related to the specific heat via $(\Delta e)^2 \equiv \langle(e - \langle e\rangle)^2\rangle = \langle e^2\rangle - \langle e\rangle^2 = C(\beta_0)/\beta_0^2 V$ and using the Taylor expansion $\langle e\rangle(T) = \langle e\rangle(T_0) + C(T_0)(T - T_0) + \ldots$, this can be written as $C(T_0)|T - T_0| \leq T_0\sqrt{C(T_0)/V}$ or

$$\frac{|T - T_0|}{T_0} \leq \frac{1}{\sqrt{VC(T_0)}} . \tag{4.81}$$

Since $C(T_0)$ is known from the input histogram this is quite a general estimate of the reweighting range. For the example in Fig. 4.8 with $V = 16\times16$, $\beta_0 = \beta_c \approx 0.44$ and $C(T_0) \approx 1.5$, this estimate yields $|\beta - \beta_0|/\beta_0 \approx |T - T_0|/T_0 \leq 0.04$, i.e., $|\beta - \beta_0| \leq 0.02$ or $0.42 \leq \beta \leq 0.46$. By comparison with the exact solution we see that this is indeed a fairly conservative estimate of the reliable reweighting range.

If we only want to know the scaling behavior with system size $V = L^D$, we can go one step further by considering three generic cases:

(i) *Off-criticality*, where $C(T_0) \approx$ const, such that

$$\frac{|T - T_0|}{T_0} \propto V^{-1/2} = L^{-D/2} . \tag{4.82}$$

(ii) *Criticality*, where $C(T_0) \simeq a_1 + a_2 L^{\alpha/\nu}$, with a_1 and a_2 being constants, and α and ν denoting the standard critical exponents of the specific heat and correlation length, respectively. For $\alpha > 0$, the leading scaling behavior becomes $|T - T_0|/T_0 \propto L^{-D/2}L^{-\alpha/2\nu}$. Assuming hyperscaling ($\alpha = 2 - D\nu$) to be valid, this simplifies to

$$\frac{|T - T_0|}{T_0} \propto L^{-1/\nu} , \tag{4.83}$$

i.e., the typical scaling behavior of pseudo-transition temperatures in the finite-size scaling regime of a second-order phase transition [126]. For $\alpha < 0$, $C(T_0)$ approaches asymptotically a constant and the leading scaling behavior of the reweighting range is as in the off-critical case.

(iii) *First-order transitions*, where $C(T_0) \propto V$. This yields

$$\frac{|T - T_0|}{T_0} \propto V^{-1} = L^{-D} , \tag{4.84}$$

which is again the typical finite-size scaling behavior of pseudo-transition temperatures close to a first-order phase transition [47].

4.6.1.2 Reweighting of Non-Conjugate Observables

If we also want to reweight other quantities such as the magnetization $\langle m \rangle$ we have to go one step further. The conceptually simplest way would be to store two-dimensional histograms $P_{\beta_0}(E, M)$ where $M = Vm$ is the total magnetization. We could then proceed in close analogy to the preceding case, and even reweighting to non-zero magnetic field h would be possible, which enters via the Boltzmann factor $\exp(\beta h \sum_i s_i) = \exp(\beta h M)$. However, the storage requirements may be quite high (of the order of V^2), and it is often preferable to proceed in the following way. For any function $g(M)$, e.g., $g(M) = M^k$, we can write

$$\begin{aligned}\langle g(M) \rangle &= \sum_{\{s\}} g(M(\{s\}))\mathrm{e}^{-\beta_0 H}/Z(\beta_0) \\ &= \sum_{E,M} \Omega(E, M) g(M)\mathrm{e}^{-\beta_0 E}/Z(\beta_0) \\ &= \sum_{E} \frac{\sum_M \Omega(E, M) g(M)}{\sum_M \Omega(E, M)} \sum_M \Omega(E, M)\mathrm{e}^{-\beta_0 E}/Z(\beta_0) .\end{aligned} \tag{4.85}$$

Recalling that $\sum_M \Omega(E, M)\mathrm{e}^{-\beta_0 E}/Z(\beta_0) = \Omega(E)\mathrm{e}^{-\beta_0 E}/Z(\beta_0) = P_{\beta_0}(E)$ and defining the microcanonical expectation value of $g(M)$ at fixed energy E (sometimes denoted as a list)

$$\langle\langle g(M) \rangle\rangle(E) \equiv \frac{\sum_M \Omega(E, M) g(M)}{\sum_M \Omega(E, M)} , \tag{4.86}$$

we arrive at

$$\langle g(M) \rangle = \sum_E \langle\langle g(M) \rangle\rangle(E) P_{\beta_0}(E) . \tag{4.87}$$

Identifying $\langle\langle g(M) \rangle\rangle(E)$ with $f(E)$ in (4.79), the actual reweighting procedure is precisely as before. An example for computing $\langle\langle |M| \rangle\rangle(E)$ and $\langle\langle M^2 \rangle\rangle(E)$ using the data of Fig. 4.8 is shown in Fig. 4.9. Mixed quantities, e.g. $\langle E^k M^l \rangle$, can be treated similarly. One caveat of this method is that one has to decide beforehand which lists $\langle\langle g(M) \rangle\rangle(E)$ one wants to store during the simulation, e.g., which powers k in $\langle\langle M^k \rangle\rangle(E)$ are relevant.

An alternative and more flexible method is based on time series. Suppose we have performed a Monte Carlo simulation at β_0 and stored the time series of N

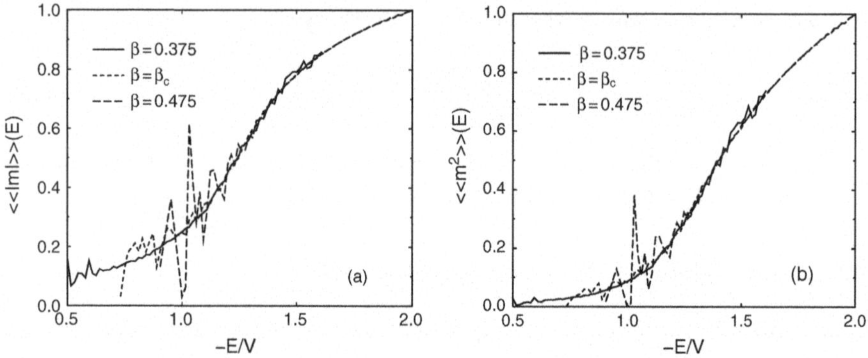

Fig. 4.9. Microcanonical expectation values for **(a)** the absolute magnetization and **(b)** the magnetization squared obtained from the 2D Ising model simulations shown in Fig. 4.8

measurements $E_1, E_2, \ldots, E_N$ and $M_1, M_2, \ldots, M_N$. Then the most general expectation values at another inverse temperature β can simply be obtained from

$$\langle\langle f(E,M)\rangle\rangle = \frac{\sum_{i=1}^{N} f(E_i,M_i)\mathrm{e}^{-(\beta-\beta_0)E_i}}{\sum_{i=1}^{N} \mathrm{e}^{-(\beta-\beta_0)E_i}} , \tag{4.88}$$

i.e., in particular all moments $\langle E^k M^l \rangle$ can be computed. Notice that this can also be written as

$$\langle\langle f(E,M)\rangle\rangle = \frac{\langle f(E,M)\mathrm{e}^{-(\beta-\beta_0)E}\rangle_0}{\langle \mathrm{e}^{-(\beta-\beta_0)E}\rangle_0} , \tag{4.89}$$

where the subscript zero refers to expectation values taken at β_0. Another very important advantage of the last formulation is that it works without any systematic discretization error also for continuously distributed energies and magnetizations.

As nowadays hard-disk space is no real limitation anymore, it is advisable to store time series in any case. This guarantees the greatest flexibility in the data analysis. As far as the memory requirement of the actual reweighting code is concerned, however, the method of choice is sometimes not so clear. Using directly histograms and lists, one typically has to store about $(6-8)\,V$ data, while working directly with the time series one needs $2N$ computer words. The cheaper solution (also in terms of CPU time) thus obviously depends on both, the system size V and the run length N. It is hence sometimes faster to generate from the time series first histograms and the required lists and then proceed with reweighting the latter quantities.

4.6.2 Multi-Histogram Technique

The basic idea of the multi-histogram technique [127] can be summarized as follows:

(i) Perform m Monte Carlo simulations at $\beta_1, \beta_2, \ldots, \beta_m$ with N_i, $i = 1, \ldots, m$, measurements.

(ii) Reweight all runs to a common reference point β_0.
(iii) Combine at β_0 all information by computing error weighted averages.
(iv) Reweight the combined histogram to any other β.

Here we shall assume that the histograms $P_{\beta_i}(E)$ are naturally normalized $\sum_E P_{\beta_i}(E) = N_i$, such that the statistical errors for each of the histograms $P_{\beta_i}(E)$ are approximately given by $\sqrt{P_{\beta_i}(E)}$. By choosing as reference point $\beta_0 = 0$ and working out the error weighted combined histogram one ends up with

$$\Omega(E) = \frac{\sum_{i=1}^m P_{\beta_i}(E)}{\sum_{i=1}^m N_i Z_i^{-1} \mathrm{e}^{-\beta_i E}} , \tag{4.90}$$

where the unknown partition function values $Z_i \equiv Z(\beta_i)$ are determined self-consistently from

$$Z_i = \sum_E \Omega(E) \mathrm{e}^{-\beta_i E} = \sum_E \mathrm{e}^{-\beta_i E} \frac{\sum_{k=1}^m P_{\beta_k}(E)}{\sum_{k=1}^m N_k Z_k^{-1} \mathrm{e}^{-\beta_k E}} , \tag{4.91}$$

up to an unimportant overall constant. A good starting point for the recursion is to fix, say, $Z_1 = 1$ and use single histogram reweighting to get an estimate of $Z_2/Z_1 = \exp[-(\widehat{F}_2 - \widehat{F}_1)]$, where $\widehat{F}_i = \beta_i F(\beta_i)$. Once Z_2 is determined, the same procedure can be applied to estimate Z_3 and so on. In the limit of infinite statistics, this would already yield the solution of (4.91). In realistic simulations the statistics is of course limited and the (very few) remaining recursions average this uncertainty to get a self-consistent set of Z_i. In order to work in practice, the histograms at neighboring β-values must have sufficient overlap, i.e., the spacings of the simulation points must be chosen according to the estimates (4.82)–(4.84).

Multiple-histogram reweighting has been widely applied in many different applications. Some problems of this method are that autocorrelations cannot properly be taken into account when computing the error weighted average (which is still correct but no longer optimized), the procedure for computing mixed quantities such as $\langle E^k M^l \rangle$ is difficult to justify (even though it does work as an ad hoc prescription quite well), and the statistical error analysis becomes quite cumbersome.

As an alternative one may compute by reweighting from each of the m simulations all quantities of interest as a function of β, including their statistical error bars which now also should take care of autocorrelations as discussed in Sect. 4.5.2.3. In this way one obtains, at each β-value, m estimates, e.g. $e_1(\beta) \pm \Delta e_1, e_2(\beta) \pm \Delta e_2, \ldots, e_m(\beta) \pm \Delta e_m$, which may be optimally combined according to their error bars to give $e(\beta) \pm \Delta e$. If the relative error $\Delta e/e(\beta)$ is minimized, this leads to [87]

$$e(\beta) = \left(\frac{e_1(\beta)}{(\Delta e_1)^2} + \frac{e_2(\beta)}{(\Delta e_2)^2} + \ldots + \frac{e_m(\beta)}{(\Delta e_m)^2} \right) (\Delta e)^2 , \tag{4.92}$$

with

$$\frac{1}{(\Delta e)^2} = \frac{1}{(\Delta e_1)^2} + \frac{1}{(\Delta e_2)^2} + \ldots + \frac{1}{(\Delta e_m)^2} . \tag{4.93}$$

Notice that in this way the average for each quantity can be individually optimized.

4.7 Finite-Size Scaling Analysis

Equipped with the various technical tools discussed above, the purpose of this section is to outline a typical FSS analysis of Monte Carlo simulations of second-order phase transitions. The described procedure is generally applicable but to keep the notation short, all formulas are formulated for Ising like systems. For instance for O(n) symmetric models, m should be replaced by $\boldsymbol{m}$ etc. The main results of such studies are usually estimates of the critical temperature and the critical exponents characterizing the universality class of the transition.

4.7.1 General Framework

To facilitate most flexibility in the analysis, it is advisable to store during data production the time series of measurements. Standard quantities are the energy and magnetization, but depending on the model at hand it may be useful to record also other observables. In this way the full dynamical information can be extracted still after the actual simulation runs and error estimation can be easily performed. For example it is no problem to experiment with the size and number of Jackknife bins. Since a reasonable choice depends on the a priori unknown autocorrelation time, it is quite cumbersome to do a reliable error analysis on the flight during the simulation. Furthermore, basing data reweighting on time-series data is more efficient since histograms, if needed or more convenient, can still be produced from this data but working in the reverse direction is obviously impossible.

For some models it is sufficient to perform for each lattice size a single long run at some coupling β_0 close to the critical point β_c. This is, however, not always the case and also depends on the observables of interest. In this more general case, one may use several simulation points β_i and combine the results by the multi-histogram reweighting method or may apply a very recently developed finite-size adapted generalized ensemble method [128]. In both situations, one can compute from the time series of the energies $e = E/V$ (if E happens to be integer valued, this should be stored of course) by reweighting the internal energy $\langle e \rangle(\beta)$, the specific heat $C(\beta) = \beta^2 V \left(\langle e^2 \rangle - \langle e \rangle^2\right)$, and for instance also the energetic fourth-order parameter

$$V(\beta) = 1 - \frac{\langle e^4 \rangle}{3\langle e^2 \rangle^2} \tag{4.94}$$

as a function of temperature. Similarly, from measurements of the magnetization $m = M/V$ one can derive the temperature variation of the mean magnetization[8] $m(\beta) = \langle |m| \rangle$, the susceptibility $\chi(\beta) = \beta V \left(\langle m^2 \rangle - \langle |m| \rangle^2\right)$ (or $\chi'(\beta) = \beta V \langle m^2 \rangle$ for $\beta \leq \beta_c$), the magnetic cumulants (Binder parameters)

[8] Notice that here and in the following formulas, $|m|$ is used instead of m as would follow from the formal definition (4.5) of the zero-field magnetization $m(\beta) = 1/(V\beta) \lim_{h\to 0} \partial \ln \mathcal{Z}(\beta, h)/\partial h$. The reason is that for a symmetric model on finite lattices one obtains $\langle m \rangle(\beta) = 0$ for all temperatures due to symmetry. Only in the proper infinite-volume limit, that is $\lim_{h\to 0} \lim_{V\to\infty}$, spontaneous symmetry breaking can occur below T_c. In a simulation on finite lattices, this is reflected by a symmetric double-peak

$$U_2(\beta) = 1 - \frac{\langle m^2 \rangle}{3\langle |m| \rangle^2} ,$$
$$U_4(\beta) = 1 - \frac{\langle m^4 \rangle}{3\langle m^2 \rangle^2} , \tag{4.95}$$

and their slopes

$$\begin{aligned}\frac{\mathrm{d}U_2(\beta)}{\mathrm{d}\beta} &= \frac{V}{3\langle |m| \rangle^2} \left[\langle m^2 \rangle \langle e \rangle - 2 \frac{\langle m^2 \rangle \langle |m| e \rangle}{\langle |m| \rangle} + \langle m^2 e \rangle \right] \\ &= V(1 - U_2) \left[\langle e \rangle - 2 \frac{\langle |m| e \rangle}{\langle |m| \rangle} + \frac{\langle m^2 e \rangle}{\langle m^2 \rangle} \right] , \\ \frac{\mathrm{d}U_4(\beta)}{\mathrm{d}\beta} &= V(1 - U_4) \left[\langle e \rangle - 2 \frac{\langle m^2 e \rangle}{\langle m^2 \rangle} + \frac{\langle m^4 e \rangle}{\langle m^4 \rangle} \right] . \end{aligned} \tag{4.96}$$

Further quantities with a useful FSS behavior are the derivatives of the magnetization,

$$\begin{aligned}\frac{\mathrm{d}\langle |m| \rangle}{\mathrm{d}\beta} &= V \left(\langle |m| e \rangle - \langle |m| \rangle \langle e \rangle \right) , \\ \frac{\mathrm{d}\ln \langle |m| \rangle}{\mathrm{d}\beta} &= V \left(\frac{\langle |m| e \rangle}{\langle |m| \rangle} - \langle e \rangle \right) , \\ \frac{\mathrm{d}\ln \langle m^2 \rangle}{\mathrm{d}\beta} &= V \left(\frac{\langle m^2 e \rangle}{\langle m^2 \rangle} - \langle e \rangle \right) . \end{aligned} \tag{4.97}$$

These latter five quantities are good examples for expectation values containing both, powers of e and m.

In the infinite-volume limit most of these quantities exhibit singularities at the transition point. As already discussed in Sect. 4.2, in finite systems the singularities are smeared out and the standard observables scale in the critical region according to

$$C = C_{\text{reg}} + L^{\alpha/\nu} f_C(x)[1 + \ldots] , \tag{4.98}$$
$$\langle |m| \rangle = L^{-\beta/\nu} f_m(x)[1 + \ldots] , \tag{4.99}$$
$$\chi = L^{\gamma/\nu} f_\chi(x)[1 + \ldots] , \tag{4.100}$$

where C_{reg} is a regular background term, α, ν, β (in the exponent of L) and γ are the usual critical exponents, and $f_i(x)$ are FSS functions with

structure of the magnetization distribution (provided the runs are long enough). By averaging m one thus gets zero by symmetry, while the peak locations $\pm m_0(L)$ are close to the spontaneous magnetization and the average of $|m|$ is a good estimator. Things become more involved for slightly asymmetric models, where this recipe would produce a systematic error and thus cannot be employed. For strongly asymmetric models, on the other hand, one peak clearly dominates and the average of m can usually be measured without too many problems.

$$x = (\beta - \beta_c)L^{1/\nu} \tag{4.101}$$

being the scaling variable (do not confuse the unfortunate double-meaning of β – here $\beta = 1/k_B T$). The brackets $[1+\ldots]$ indicate corrections-to-scaling terms which become unimportant for sufficiently large system sizes L.

A particular role play the magnetic cumulants or Binder parameters U_2 and U_4 which scale according to

$$U_{2p} = f_{U_{2p}}(x)[1 + \ldots] \,, \tag{4.102}$$

i.e., for constant scaling variable x, they take approximately the same value for all lattice sizes, in particular $U_{2p}^* \equiv f_{U_{2p}}(0)$ at β_c. Their curves as function of temperature for different L hence cross around (β_c, U_{2p}^*) (with slopes $\propto L^{1/\nu}$), apart from corrections-to-scaling collected in $[1 + \ldots]$ which explain small systematic deviations. From a determination of this crossing point, one thus obtains a basically unbiased estimate of β_c, the critical exponent ν, and U_{2p}^*. Note that in contrast to the truly universal critical exponents, U_{2p}^* is only weakly universal. By this one means that the infinite-volume limit of such quantities does depend in particular on the boundary conditions and geometrical shape of the considered lattice, e.g., on the aspect ratio $r = L_y/L_x$ [129, 130, 131, 132, 133, 134, 135, 136].

Differentiating U_{2p} with respect to β, one picks up an extra power of L from the scaling function, $\mathrm{d}U_{2p}/\mathrm{d}\beta = (\mathrm{d}x/\mathrm{d}\beta)f'_{U_{2p}} = L^{1/\nu}f'_{U_{2p}}$. This leads to

$$\frac{\mathrm{d}U_{2p}}{\mathrm{d}\beta} = L^{1/\nu}f'_{U_{2p}}(x)[1 + \ldots] \,, \tag{4.103}$$

and similarly for the magnetization derivatives

$$\frac{\mathrm{d}\langle|m|\rangle}{\mathrm{d}\beta} = L^{(1-\beta)/\nu}f'_m(x)[1 + \ldots] \,, \tag{4.104}$$

$$\frac{\mathrm{d}\ln\langle|m|^p\rangle}{\mathrm{d}\beta} = L^{1/\nu}f_{dm_p}(x)[1 + \ldots] \,. \tag{4.105}$$

By applying standard reweighting techniques to the time-series data one first determines the temperature dependence of $C(\beta)$, $\chi(\beta)$, ..., in the neighborhood of the simulation point $\beta_0 \approx \beta_c$ (a reasonably good guess of β_0 can usually be obtained quite easily from a few short test runs). It should be stressed that in a serious study, by estimating the valid reweighting range, one should at any rate make sure that no systematic errors crept in by this procedure (which may be easily overlooked if one works too mechanically). Once the temperature dependence is known, one can determine the maxima, e.g., $C_{\text{max}}(\beta_{\text{max}_C}) \equiv \max_\beta C(\beta)$, by applying standard extremization routines: When reweighting is implemented as a subroutine, for instance $C(\beta)$ can be handled as a normal function with a continuously varying argument β, i.e., no interpolation or discretization error is involved when iterating towards the maximum. The locations of the maxima of C, χ, $\mathrm{d}U_2/\mathrm{d}\beta$, $\mathrm{d}U_4/\mathrm{d}\beta$, $\mathrm{d}\langle|m|\rangle/\mathrm{d}\beta$, $\mathrm{d}\ln\langle|m|\rangle/\mathrm{d}\beta$, and $\mathrm{d}\ln\langle m^2\rangle/\mathrm{d}\beta$ provide us with seven sequences of pseudo-transition points $\beta_{\max_i}(L)$ which all should scale according to $\beta_{\max_i}(L) = \beta_c + a_i L^{-1/\nu} + \ldots$. In other words, the scaling variable $x = (\beta_{\max_i}(L) - \beta_c)L^{1/\nu} = a_i + \ldots$ should be constant, if we neglect the small higher-order corrections indicated by $\ldots$.

Notice that while the precise estimates of a_i do depend on the value of ν, the qualitative conclusion that $x \approx$ const for each of the $\beta_{\max_i}(L)$ sequences does not require any a priori knowledge of ν or β_c. Using this information one thus has several possibilities to extract unbiased estimates of the critical exponents ν, α/ν, β/ν, and γ/ν from least-squares fits assuming the FSS behaviors (4.98)–(4.105).

4.7.2 A Practical Recipe

The typical procedure of an FSS analysis then proceeds as follows:

(i) Estimate the critical exponent ν by least-square fits to the scaling behavior (4.103) and (4.105). For this one may consider directly the maxima of $\mathrm{d}U_{2p}/\mathrm{d}\beta$ and $\mathrm{d}\ln\langle|m|^p\rangle/\mathrm{d}\beta$, $p = 1, 2$, or work with any other FSS sequence $\beta_{\max_i}(L)$.

Remarks: Considering only the asymptotic behavior, e.g., $\mathrm{d}\ln\langle|m|\rangle/\mathrm{d}\beta = aL^{1/\nu}$, and taking the logarithm, $\ln(\mathrm{d}\ln\langle|m|\rangle/\mathrm{d}\beta) = c + (1/\nu)\ln(L)$, one ends up with a linear two-parameter fit yielding estimates for the constant $c = \ln(a)$ and the exponent $1/\nu$. For small lattice sizes the asymptotic ansatz is, of course, not justified. Taking into account the (effective) correction term $[1 + bL^{-w}]$ would result in a non-linear four-parameter fit for a, b, $1/\nu$ and w. Even if we would fix w to some theoretically expected value (as is sometimes done), we would be still left with a non-linear fit which is usually much harder to control than a linear fit (where only a set of linear equations with a unique solution has to be solved, whereas a non-linear fit involves a numerical minimization of the χ^2-function, possessing possibly several local minima). The alternative method is to use the linear fit ansatz and to discard successively more and more small lattice sizes until the χ^2 per degree of freedom or the goodness-of-fit Q [113] has reached an acceptable value and does not show any further trend. Of course, all this relies heavily on correct estimates of the statistical error bars on the original data for $\mathrm{d}\ln\langle|m|\rangle/\mathrm{d}\beta$.

Furthermore, when combining the various fit results for ν to a final value, some care is necessary with the final statistical error estimate on ν, since the various fits for determining ν are of course correlated (since they use the data from one and the same simulation).

(ii) Once ν is estimated one can use the scaling form $\beta_{\max_i}(L) = \beta_c + a_i L^{-1/\nu} + \ldots$ to extract β_c and a_i. As a useful check, one should repeat these fits at the error margins of ν, but usually this dependence turns out to be very weak.

Remark: Regarding the β_c fit alone, the uncertainty in the proper value of ν looks like a kind of systematic error or bias, whose origin, however, is also of statistical nature occurring in the first step.

(iii) As a useful cross-check one can determine β_c also from the Binder parameter crossings. For a first rough estimate, this is a very convenient and fast method.

Remarks: As a rule of thumb, an accuracy of about 3–4 digits for β_c can be obtained with this method without any elaborate infinite-volume extrapolations – the crossing points lie usually much closer to β_c than the various maxima locations. For high precision, however, it is quite cumbersome to control the necessary extrapolations and often more accurate estimates can be obtained by considering the scaling of the maxima locations. Also, error estimates of crossing points involve the data for two different lattice sizes which tends to be quite unhandy.

(iv) Next, similarly to ν, the ratios of critical exponents α/ν, β/ν, and γ/ν can be obtained from fits to (4.98)–(4.100), and (4.104). Again the maxima of these quantities or any of the FSS sequences $\beta_{\max_i}$ can be used. What concerns the fitting procedure the same remarks apply as for ν.

Remarks: The specific heat C usually plays a special role in that the exponent α is difficult to determine. The reason is that α is usually relatively small (3D Ising model: $\alpha \approx 0.1$), may be zero (logarithmic divergence as in the 2D Ising model) or even negative (as for instance in the 3D XY and Heisenberg models). In all these cases, the constant background contribution C_{reg} in (4.98) becomes important, which enforces a non-linear three-parameter fit with the just described problems. Also for the susceptibility χ, a regular background term cannot be excluded, but it is usually much less important since $\gamma \gg \alpha$. Therefore, in (4.99), (4.100), and (4.104), similar to the fits for ν, one may take the logarithm and work with much more stable linear fits.

(v) As a final step one may re-check the FSS behavior of C, χ, $\mathrm{d}U_2/\mathrm{d}\beta, \ldots$ at the numerically determined estimate of β_c. These fits should be repeated also at $\beta_c \pm \Delta\beta_c$ in order to estimate by how much the uncertainty in β_c propagates into the thus determined exponent estimates.

Remark: In (the pretty rare) cases where β_c is known exactly (e.g., through self-duality), this latter option is by far the most accurate one. This is the reason, why for such models numerically estimated critical exponents are usually also much more precise.

4.7.3 Finite-Size Scaling at Work – An Example

The purpose of this subsection is to illustrate the above outlined recipe with actual data from recent simulations of a 2D Ising model with next-nearest neighbor interactions [137]. The Hamiltonian has the form

$$\mathcal{H} = -J \sum_{\langle i,j \rangle} \sigma_i \sigma_j - J_{\text{d}} \sum_{(k,l)} \sigma_k \sigma_l \,, \tag{4.106}$$

where the spins can take the values $\sigma_i = \pm 1$, J denotes the nearest-neighbor (nn) coupling and J_{d} is the next-nearest-neighbor (nnn) coupling along the two diagonals of a square lattice. The corresponding pairs of spins are denoted by the brackets $\langle i, j \rangle$ and (k, l), respectively. In [137] we restricted ourselves to that region of the phase diagram where the ground states show ferromagnetic order ($J \geq 0$, $J_\mathrm{d} \geq -J/2$), and always assumed periodic boundary conditions. Absorbing the nn coupling J into the inverse temperature β (i.e., formally putting $J = 1$), the remaining second parameter is the coupling-constant ratio $\alpha = J_\mathrm{d}/J$. In the following we will concentrate on the case $\alpha = 0.5$ [138]. The linear size of the lattices varies from $L = 10, 20, 40, \ldots$ up to 640. All simulations are performed with the single-cluster algorithm which is straightforward to adapt to nnn interactions by setting bonds also along the diagonals. Similarly to the standard nn model, the integrated autocorrelation time close to criticality is found for $\alpha = 1$ [137] to scale only weakly with lattice size: $\tau_{\mathrm{e,int}} \propto L^z$ with $z = 0.134(3)$.

Another example following closely the lines sketched above is provided by a Monte Carlo study of the 3D Ising model, albeit not on a regular but on Poissonnian random lattices of Voronoi-Delaunay type [139]. The random lattices are treated as quenched disorder in the local coordination numbers and hence necessitate an additional average over many realizations (in the study described in [139], for each lattice size 96 independent realizations were used). This introduces in all FSS formulas additional disorder averages which complicate some aspects of the analysis. The general concept of FSS analysis, however, does not depend on this special feature and it may be worthwhile to consult [139] for a supplementary example.

4.7.3.1 Critical Exponent ν

Having recorded the times series of the energy and magnetization, all quantities of the preceding paragraph can be computed in the FSS region. The scaling behavior of the maxima of $\mathrm{d}\ln\langle|m|^p\rangle/\mathrm{d}\beta$ and $\mathrm{d}U_{2p}/\mathrm{d}\beta$ for $p = 1$ and $p = 2$ is shown in

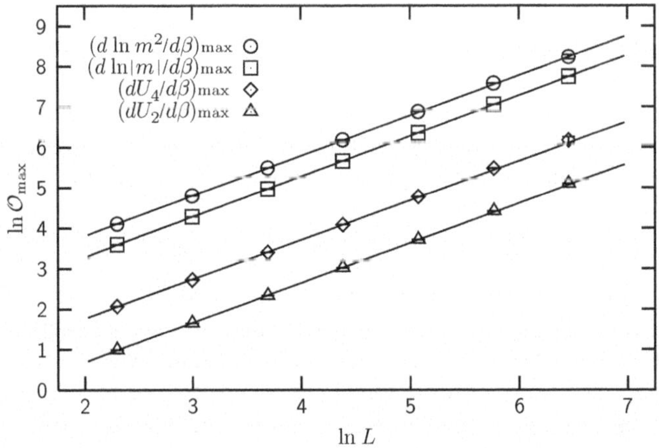

Fig. 4.10. FSS of the maxima of $\mathrm{d}\ln\langle|m|^p\rangle/\mathrm{d}\beta$ and $\mathrm{d}U_{2p}/\mathrm{d}\beta$ for $p = 1$ and $p = 2$ of the 2D nnn Ising model (4.106) with $\alpha = J_\mathrm{d}/J = 0.5$ and fits to extract $1/\nu$

Table 4.3. Fit results for the correlation length critical exponent ν of the 2D nnn Ising model with $\alpha = J_d/J = 0.5$, and the weighted average of the four estimates. Also listed are the χ^2 per degree of freedom, χ^2/d.o.f., and the goodness-of-fit parameter Q [113]

	$d\ln\langle\lvert m\rvert\rangle/d\beta$	$d\ln\langle m^2\rangle/d\beta$	$dU_2/d\beta$	$dU_4/d\beta$	weighted av.
ν	1.0031(17)	1.0034(21)	1.0027(24)	1.0025(44)	1.0031(11)
χ^2/d.o.f	0.98	0.60	2.02	0.49	
Q	0.37	0.55	0.13	0.61	

the log-log plot of Fig. 4.10. From the parameters of the four linear fits over the data points with $L_{\rm min} > 40$ collected in Table 4.3, we obtain a weighted average of $\nu = 1.003\,1 \pm 0.001\,1$.

As the more detailed analysis in [139] clearly shows, considering all $4\times7 = 28$ possible FSS sequences (the four observables shown in Fig. 4.10 evaluated at the seven different $\beta_{\max_i}$ sequences) does not significantly improve the precision of the final estimate. The reason are the quite strong correlations between most of these 28 estimates. In a really large-scale simulation such a more detailed analysis may still be valuable, however, since it potentially helps to detect systematic trends which otherwise may remain unnoticed. Also here the weighted average is clearly dominated by the result from the $d\ln\langle\lvert m\rvert\rangle/d\beta$ fit, and correlations between the first and second pair of estimates are obvious. Therefore, to account for these correlations at least heuristically, we usually quote in our investigations the weighted average, but take the smallest contributing error estimate (here thus from the $d\ln\langle\lvert m\rvert\rangle/d\beta$ fit). This recipe then gives from Table 4.3 the final result

$$\nu = 1.003\,1 \pm 0.001\,7\,, \tag{4.107}$$

in good agreement with the 2D Ising universality class (cf. Table 4.1).

4.7.3.2 Critical Coupling $\boldsymbol{\beta_c}$

Fixing the critical exponent ν at the numerically determined estimate (or in the present context at the exactly known value $\nu = 1$), it is now straightforward to obtain estimates of the critical coupling β_c from linear least-squares fits to

$$\beta_{\max_i} = \beta_c + a_i L^{-1/\nu}\,, \tag{4.108}$$

where $\beta_{\max_i}$ are the pseudo-transition points discussed earlier. Depending on the quantity considered, here we found a significant improvement of the fit quality if the smallest lattice sizes were excluded. This is illustrated in Table 4.4, where detailed results for various fit ranges are compiled.

As final result we quote the weighted average of the five estimates and again the smallest contributing error bar,

$$\beta_c = 0.262\,817\,4 \pm 0.000\,0017\,. \tag{4.109}$$

Table 4.4. FSS fits of the pseudo-transition points $\beta_{\max} = \beta_c + aL^{-1/\nu}$ of the nnn model (4.106) with $\alpha = 0.5$ for varying fit ranges, assuming $\nu = 1$. Here n is the number of data points, $L_{\min}$ denotes the smallest lattice size considered, and Q is the standard goodness-of-fit parameter. The selected fit ranges used for the final average are high-lighted in boldface. The last line labeled HTS gives a high-temperature series estimate [140] for comparison

observables	n	$L_{\min}$	β_c	Q
$\beta_{\max}^{C}$	7	10	0.262 699(13)	0.00
	6	20	0.262 766(15)	0.03
	5	**40**	**0.262 799(18)**	**0.88**
	4	80	0.262 807(22)	0.89
$\beta_{\text{inf}}^{\vert m\vert}$	7	10	0.262 8706(36)	0.00
	6	20	0.262 8398(40)	0.00
	5	**40**	**0.262 8272(47)**	**0.16**
	4	80	0.262 8212(58)	0.38
$\beta_{\max}^{\chi}$	7	10	0.262 8253(12)	0.00
	6	20	0.262 8195(13)	0.00
	5	40	0.262 8178(14)	0.09
	4	**80**	**0.262 8153(17)**	**0.66**
$\beta_{\text{inf}}^{\ln\vert m\vert}$	7	10	0.262 8437(62)	0.00
	6	**20**	**0.262 8243(68)**	**0.24**
	5	40	0.262 8183(77)	0.42
	4	80	0.262 8099(97)	0.70
$\beta_{\text{inf}}^{\ln m^2}$	7	10	0.262 8684(94)	0.00
	6	**20**	**0.262 837(11)**	**0.43**
	5	40	0.262 837(13)	0.57
	4	80	0.262 818(17)	0.55
average			0.262 8204(144)	
weighted average			0.262 8174(16)	
final			0.262 8174(17)	
HTS (Oitmaa [140])			0.262 808	

The corrections to the asymptotic FSS behavior can be also visually inspected in Fig. 4.11, where the Monte Carlo data and fits are compared. One immediately notices a systematic trend that the $L = 10$ data deviate from the linear behavior. For larger L, however, the deviations are already so small that only a quantitative judgement in terms of the χ^2 per degree of freedom or goodness-of-fit parameter Q of the fits [113] can lead to a sensible conclusion.

4.7.3.3 Binder Parameters U_2 and U_4

Following our general recipe sketched above, the Binder parameter $U_4(L)$ is shown in Fig. 4.12 as a function of temperature. Even though the temperature range is much

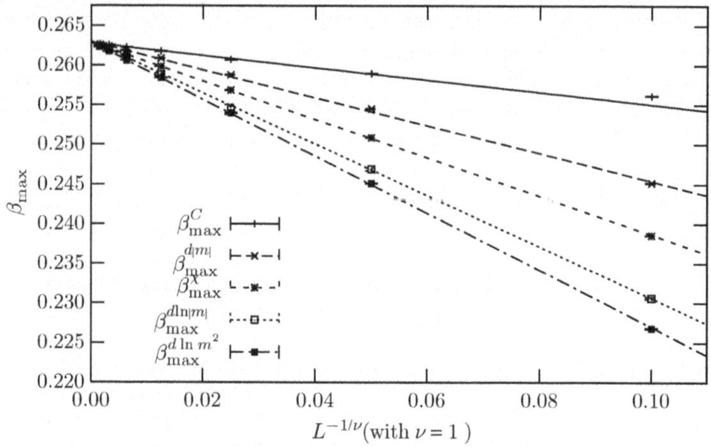

Fig. 4.11. FSS fits of the pseudo-transition points $\beta_{\max_i}$ with $\nu = 1.0$ fixed of the 2D nnn Ising model (4.106) with $\alpha = J_d/J = 0.5$. The error weighted average of the FSS extrapolations yields $\beta_c = 0.262\ 817\ 4(16)$, cf. Table 4.4 for details

smaller than in the $\beta_{\max_i}$ plot of Fig. 4.11, a clear-cut crossing point can be observed. Already from the crossing of the two curves for the very modestly sized lattices with $L = 10$ and $L = 20$ (which can be obtained in a few minutes of computing time), one can read off that $\beta_c \approx 0.262\ 8$. This clearly demonstrates the power of this method, although it should be stressed that the precision is exceptionally good for this model.

On the scale of Fig. 4.12 one reads off that $U_4^* \approx 0.61$. Performing an extrapolation (on a very fine scale) to infinite size at $\beta = \beta_c$, one obtains the more accurate

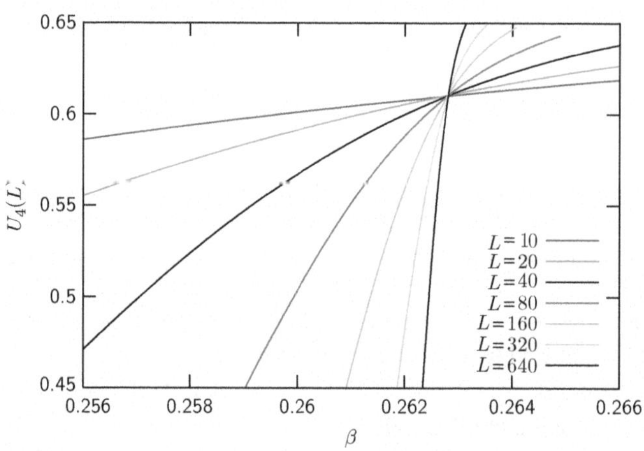

Fig. 4.12. Fourth-order Binder parameter U_4, exhibiting a sharp crossing point around $(\beta_c, U_4^*) \approx (0.262\ 82, 0.61)$. Note the much smaller temperature scale compared to Fig. 4.11

estimate of $U_4^* = 0.610\ 8(1)$. This result for the 2D nnn Ising model with $\alpha = 0.5$ is in perfect agreement with the very precisely known value for the standard square lattice nn Ising model with periodic boundary conditions from extrapolating exact transfer-matrix data for $L \leq 17$ [129], $U_4^* = 0.610\ 690\ 1(5)$, and a numerical evaluation of an exact expression [130], $U_4^* = 0.610\ 692(2)$. This illustrates the expected universality of U_4^* (and also U_2^*) for general isotropic interactions (e.g., also for $\alpha = 1$ one finds the same result within error bars [137]), as long as boundary conditions, lattice shapes etc. are the same. As emphasized already in Sect. 4.7.1, the cumulants are, however, only weakly universal in the sense that they do depend sensitively on the anisotropy of interactions, boundary conditions and lattice shapes (aspect ratios) [131, 132, 133, 134, 135, 136].

4.7.3.4 Critical Exponent γ

The exponent ratio γ/ν can be obtained from fits to the FSS behavior (4.100) of the susceptibility. By monitoring the quality of the fits, using all data starting from $L = 10$ is justified. The fits collected in Table 4.5 all have $Q \geq 0.15$.

Still it is fairly obvious, that the two fits with $Q < 0.2$ have some problems. Discarding them in the averages, one obtains from the weighted average (and again quoting the smallest contributing error estimate to heuristically take into account the correlations among the individual fits)

Table 4.5. Fit results for the critical exponents γ/ν, β/ν, and $(1-\beta)/\nu$. The fits for γ/ν and $(1-\beta)/\nu$ take all lattices with $L \geq 10$ into account while the fits for β/ν start at $L = 20$

at $K_{\max}$ of	γ/ν	Q	β/ν	Q	$(1-\beta)/\nu$	Q
C	1.7574(28)	0.87	0.12856(38)	0.00	0.8889(13)	0.00
χ	1.7407(10)	0.16	0.12480(32)	0.45	0.8710(24)	0.93
$\mathrm{d}U_4/\mathrm{d}\beta$	1.7700(50)	0.40	0.12481(39)	0.51	0.9154(99)	0.38
$\mathrm{d}U_2/\mathrm{d}\beta$	1.7417(12)	0.42	0.12562(32)	0.02	0.8815(35)	0.39
$\mathrm{d}\langle\lvert m\rvert\rangle/\mathrm{d}\beta$	1.7356(11)	0.19	0.12191(33)	0.00	0.8760(15)	0.82
$\mathrm{d}\ln\langle\lvert m\rvert\rangle/\mathrm{d}\beta$	1.7520(20)	0.62	0.12407(34)	0.02	0.8923(49)	0.57
$\mathrm{d}\ln\langle m^2\rangle/\mathrm{d}\beta$	1.7630(32)	0.76	0.12363(37)	0.01	0.9047(68)	0.81
average	1.7515(49)	≥ 0	0.12477(78)	≥ 0	0.8900(60)	≥ 0
weighted av.	1.7423(06)		0.12468(13)		0.8822(09)	
final	1.7423(10)		0.12468(32)		0.8822(13)	
average	1.7568(49)	≥ 0.2	0.12483(32)	≥ 0.02	0.8901(71)	≥ 0.3
weighted av.	1.7477(09)		0.12485(17)		0.8775(11)	
final	1.7477(12)		0.12485(32)		0.8775(15)	
exact	1.75		0.125		0.875	

$$\gamma/\nu = 1.747\,7 \pm 0.001\,2 \tag{4.110}$$

to be compared with the exact result $7/4 = 1.75$. For the critical exponent η, the estimate (4.110) implies $\eta = 2 - \gamma/\nu = 0.252\ 3 \pm 0.001\ 2$, and, by inserting our value of $\nu = 1.003\ 1(17)$, one obtains $\gamma = 1.753\ 1 \pm 0.004\ 2$. Here and in the following we are quite conversative and always quote the maximal error, i.e., $\max\{(O_1 + \epsilon_1)(O_2 + \epsilon_2) - O_1 O_2, O_1 O_2 - (O_1 - \epsilon_1)(O_2 - \epsilon_2)\}$.

4.7.3.5 Critical Exponent β

The exponent ratio β/ν can be obtained either from the FSS behavior of $\langle|m|\rangle$ or $\mathrm{d}\langle|m|\rangle/\mathrm{d}\beta$, (4.99) or (4.104). In the first case, Table 4.5 shows that most β_{max_i} sequences yield poor Q values (≤ 0.1) even if the $L = 10$ lattice data is discarded. If one averages only the fits with $Q \geq 0.02$, the final result is

$$\beta/\nu = 0.124\,85 \pm 0.000\,32\ , \tag{4.111}$$

and, by using our estimate (4.107) for ν, $\beta = 0.125\ 23 \pm 0.000\ 54$, in very good agreement with the exact result $\beta/\nu = \beta = 1/8 = 0.125\ 00$ for the 2D Ising universality class. Assuming hyperscaling to be valid, the estimate (4.111) implies $\gamma/\nu = D - 2\beta/\nu = 1.750\ 30(64)$.

From the Q values in Table 4.5 one can conclude that the FSS of $\mathrm{d}\langle|m|\rangle/\mathrm{d}\beta$ is somewhat better behaved, so that one can keep again all lattice sizes $L \geq 10$ in the fits. By discarding only the fit for the β_{max_C} sequence, which has an exceptionally small Q value, one arrives at

$$(1 - \beta)/\nu = 0.877\,5 \pm 0.001\,5\ , \tag{4.112}$$

so that by inserting our estimate (4.107) for ν, $\beta/\nu = 0.119\ 4 \pm 0.003\ 2$, and finally $\beta = 0.119\ 8 \pm 0.003\ 0$.

4.7.3.6 Critical Exponent α

Due to the regular background term C_{reg} in the FSS behavior (4.98), the specific heat is usually among the most difficult quantities to analyze [141]. In the present example the critical exponent α is expected to be zero, as can be verified by using the hyperscaling relation $\alpha = 2 - D\nu = -0.0062(34)$. In such a situation it may be useful to test at least the consistency of a linear two-parameter fit with α/ν kept fixed. In the present case with $\alpha = 0$, this amounts to the special form $C = C_{\mathrm{reg}} + a\ln(L)$. As can be inspected in Fig. 4.13, the expected linear behavior is, in fact, satisfied over the whole range of lattice sizes.

To conclude this example analysis [138], it should be stressed that no particular care was taken to arrive at high-precision estimates for the critical exponents since in the original work [137] primarily the critical coupling was of interest. In applications aiming also at accurate exponent estimates, one may experiment more elaborately

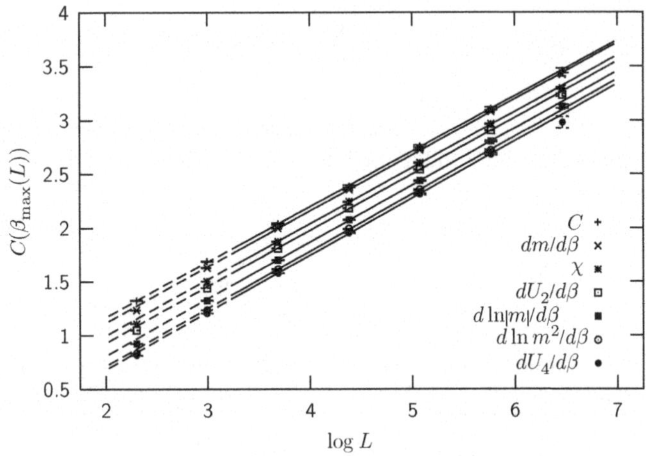

Fig. 4.13. FSS behavior of the specific heat evaluated at the various $\beta_{\max_i}$ sequences, assuming $\alpha = 0$, i.e., a logarithmic scaling $\propto \ln L$

with the fit ranges and averaging procedures. If (small) inconsistencies happen to persist, it is in particular also wise to re-check the extent of the reliable reweighting range, which often turns out to be the source of trouble in the first place (... which we have not seriously attempted to exclude in this example analysis).

4.7.4 Spatial Correlation Length

Since critical phenomena are intimately connected with diverging spatial correlations, it is in many applications important to also estimate the correlation length. In the high-temperature phase down to the critical point, we have $\langle \sigma_i \rangle = 0$ and the two-point correlation function (4.7) simplifies to

$$G(\boldsymbol{r}_i - \boldsymbol{r}_j) = \langle \sigma_i \sigma_j \rangle \, . \tag{4.113}$$

By summing over all lattice points one obtains the susceptibility (without β prefactor)

$$\begin{aligned} \chi'/\beta &= \frac{1}{V} \sum_{\boldsymbol{r}_i, \boldsymbol{r}_j} G(\boldsymbol{r}_i - \boldsymbol{r}_j) = \sum_{\boldsymbol{r}} G(\boldsymbol{r}) \\ &= V \left\langle \left(\frac{1}{V} \sum_{\boldsymbol{r}} \sigma_i \right)^2 \right\rangle = V \langle m^2 \rangle \, . \end{aligned} \tag{4.114}$$

Recall that above T_c, $\langle m \rangle = 0$. On D-dimensional periodic lattices with edge lengths $L_1 = L_2 = \ldots = L$, the correlation function can be decomposed into Fourier modes

$$G(\boldsymbol{r}_i - \boldsymbol{r}_j) = \frac{1}{V} \sum_{n_1, n_2, \ldots = 0}^{L-1} \widehat{G}(\boldsymbol{k}) \mathrm{e}^{i\boldsymbol{k}\cdot(\boldsymbol{r}_i - \boldsymbol{r}_j)} , \tag{4.115}$$

where $\boldsymbol{k} \equiv (2\pi/L)(n_1, n_2, \ldots)$ are the discrete lattice momenta. In the high-temperature phase the amplitudes for long-wavelength modes ($|\boldsymbol{k}| \to 0$) are effectively given by

$$\widehat{G}(\boldsymbol{k}) = a \left[\sum_{i=1}^{D} 2(1 - \cos k_i) + m^2 \right]^{-1} \overset{|\boldsymbol{k}| \to 0}{\approx} a \frac{1}{\boldsymbol{k}^2 + m^2} , \tag{4.116}$$

with β dependent prefactor a and mass parameter m. Inserting this into (4.114), one finds for large distances $|\boldsymbol{r}| \gg 1$ (but $|\boldsymbol{r}| \ll L/2$ for finite periodic lattices)

$$G(\boldsymbol{r}) \propto |\boldsymbol{r}|^{-(D-1)/2} \mathrm{e}^{-m|\boldsymbol{r}|} \tag{4.117}$$

with ($|\boldsymbol{r}| \gg 1$), so that the inverse mass can be identified as the correlation length $\xi \equiv 1/m$.

4.7.4.1 Zero-Momentum Projected Correlation Function

In order to avoid the power-like prefactor in (4.117) and to increase effectively the statistics one actually measures in most applications a so-called projected (zero-momentum) correlation function defined by ($\boldsymbol{r} = (x_1, x_2, \ldots)$)

$$\begin{aligned} g(x_1 - x_1') &= \frac{1}{L^{D-1}} \sum_{x_2, x_3, \ldots = 1}^{L} \sum_{x_2', x_3', \ldots = 1}^{L} G(\boldsymbol{r}_i - \boldsymbol{r}_j) \\ &= L^{D-1} \left\langle \left[\frac{1}{L^{D-1}} \sum_{x_2, x_3, \ldots = 1}^{L} \sigma_{x_1, x_2, x_3, \ldots} \right] \left[\frac{1}{L^{D-1}} \sum_{x_2', x_3', \ldots = 1}^{L} \sigma_{x_1', x_2', x_3', \ldots} \right] \right\rangle , \end{aligned} \tag{4.118}$$

i.e., the correlations of line magnetizations $L^{-1} \sum_{x_2=1}^{L} \sigma_{x_1, x_2}$ for 2D systems or surface magnetizations $L^{-2} \sum_{x_2, x_3 = 1}^{L} \sigma_{x_1, x_2, x_3}$ for 3D systems at x_1 and x_1'. Notice that in all dimensions

$$\chi'/\beta = \frac{1}{2}\, g(0) + \sum_{i=1}^{L-1} g(i) + \frac{1}{2}\, g(L) \tag{4.119}$$

is given by the trapezoidal approximation to the area $\int_0^L g(x) \mathrm{d}x$ under the projected correlation function $g(x)$. Applying the summations in (4.118) to the Fourier decomposition of $G(\boldsymbol{r}_i - \boldsymbol{r}_j)$ and using

$$\frac{1}{L^{D-1}} \sum_{x_2,x_3,\ldots=1}^{L} \mathrm{e}^{ik_2x_2+ik_3x_3+\ldots} = \delta_{k_2,0}\,\delta_{k_3,0}\cdots\,, \tag{4.120}$$

it is easy to see that

$$g(x_1-x_1') = \frac{a}{L}\sum_{n_1=0}^{L-1} \frac{\mathrm{e}^{ik_1(x_1-x_1')}}{2(1-\cos k_1)+m^2} \tag{4.121}$$

is the one-dimensional version of (4.115) and (4.116), since all but one momentum component are projected to zero in (4.120). This can be evaluated exactly as

$$\begin{aligned} g(x) &= \frac{a}{2\sinh m^*}\,\frac{\cosh[m^*(L/2-x)]}{\sinh(m^*L/2)} \\ &= \frac{a}{2\sinh m^*}\left[\mathrm{e}^{-m^*x} + \frac{2\mathrm{e}^{-m^*L}}{1-\mathrm{e}^{-m^*L}}\cosh(m^*x)\right]\,, \end{aligned} \tag{4.122}$$

with m and m^* related by

$$\begin{aligned} \frac{m}{2} &= \sinh\left(\frac{m^*}{2}\right)\,, \\ \frac{m^*}{2} &= \ln\left[\frac{m}{2} + \sqrt{\left(\frac{m}{2}\right)^2+1}\right]\,. \end{aligned} \tag{4.123}$$

For $\xi > 10$ $(m < 0.1)$ the difference between ξ and $\xi^* \equiv 1/m^*$ is completely negligible, $(\xi^*-\xi)/\xi < 0.042\%$. Notice that there is no x-dependent prefactor in (4.122). And also note that $G(\boldsymbol{r})$ computed for $\boldsymbol{r}$ along one of the coordinate axes is a truly D-dimensional correlation function (albeit along some special direction), exhibiting the $\boldsymbol{r}^{(D-1)/2}$ prefactor of (4.117).

Figure 4.14 shows as an example $g(x)$ for the standard nn Ising model at $T = 2.5 \approx 1.1\ T_c$ on a 50×50 square lattice. By fitting the Monte Carlo data to the cosh-form (4.122), $m^* = 0.167\,9$ is obtained or $\xi^* = 5.957$. Inserting this value into (4.123), one obtains $\xi = 1/m = 5.950$. This is in very good agreement (at a 0.1-0.2% level) with the exactly known correlation length (of the two-dimensional correlation function) along one of the two main coordinate axes, $\xi_{||}^{(\mathrm{ex})} = -1/(\ln(\tanh(\beta))+2\beta) = 5.962\,376\,984\ldots$ [14, 15].

4.7.4.2 Second-Moment Correlation Length

Alternatively, one may also measure directly the Fourier amplitudes

$$\widehat{G}(\boldsymbol{k}) = \sum_{\boldsymbol{r}} G(\boldsymbol{r})\mathrm{e}^{-i\boldsymbol{k}\cdot\boldsymbol{r}} = \frac{1}{V}\langle|\widehat{\sigma}(\boldsymbol{k})|^2\rangle\,, \tag{4.124}$$

for a few long-wavelength modes $\widehat{\sigma}(\boldsymbol{k}) = \sum_{\boldsymbol{r}} \sigma(\boldsymbol{r})\mathrm{e}^{i\boldsymbol{k}\cdot\boldsymbol{r}}$, where the normalization is chosen such that $\widehat{G}(\boldsymbol{0}) = \chi'/\beta$. From (4.116) we read off that

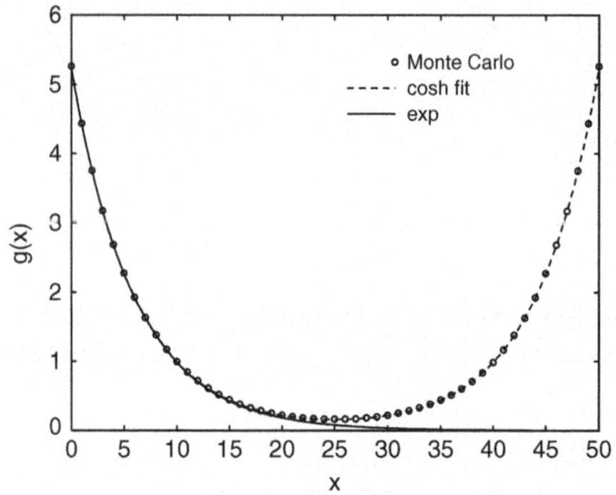

Fig. 4.14. Zero momentum projected correlation function $g(x)$ for the standard 2D nn Ising model at $T = 2.5 > T_c$. Also shown is a fit with the cosh-ansatz (4.122), yielding $m^* = 0.167\,9$ or $\xi^* = 5.957$, and the exponential approximation $\propto \exp(-m^* x)$

$$\widehat{G}(\boldsymbol{k})^{-1} = \frac{1}{a}\left(\sum_{i=1}^{D} 2(1-\cos k_i) + m^2\right) \equiv c_1 \kappa^2 + c_0 \; , \tag{4.125}$$

with $c_1 = 1/a$ and $c_0 = m^2/a$, so that the squared correlation length

$$\xi^2 = 1/m^2 = c_1/c_0 \tag{4.126}$$

can be extracted from a linear fit of $\widehat{G}(\boldsymbol{k})^{-1}$ versus $\kappa^2 = \sum_{i=1}^{D} 2(1-\cos k_i) \approx \boldsymbol{k}^2$. In 2D, for instance, one may use $\boldsymbol{k} = 2\pi\boldsymbol{n}/L$ with $\boldsymbol{n} = (0,0), (1,0), (1,1), (2,0)$, and $(2,1)$, as done for the example in Fig. 4.15, which shows Monte Carlo data for $\widehat{G}(\boldsymbol{k})^{-1}$ from the same run used for Fig. 4.14 and a fit with (4.125). From the parameters c_1 and c_0 one then obtains $\xi = \sqrt{c_1/c_0} = 5.953$.

Even the simplest expression, using only $\boldsymbol{k} = \boldsymbol{0}$ and $\boldsymbol{k} = \boldsymbol{1} = (2\pi/L)(1,0,0,\ldots)$ and involving *no* fit at all, can be used:

$$\xi = \frac{1}{2\sin(\pi/L)}\left[\frac{\widehat{G}(\boldsymbol{0})}{\widehat{G}(\boldsymbol{1})} - 1\right]^{1/2} . \tag{4.127}$$

This quantity, which is comparatively easy to measure in a Monte Carlo simulation, is usually referred to as second-moment correlation length. In the 2D Ising example, with $\widehat{G}(\boldsymbol{0}) = 62.66$ and $\widehat{G}(\boldsymbol{1}) = 1.768$ (cp. Fig. 4.15) and $L = 50$, (4.127) evaluates to $\xi = 5.965$, again in good agreement with the exact result for $\xi_{||}^{(\mathrm{ex})}$. Finally note that the Fourier method gives directly ξ (and not ξ^*).

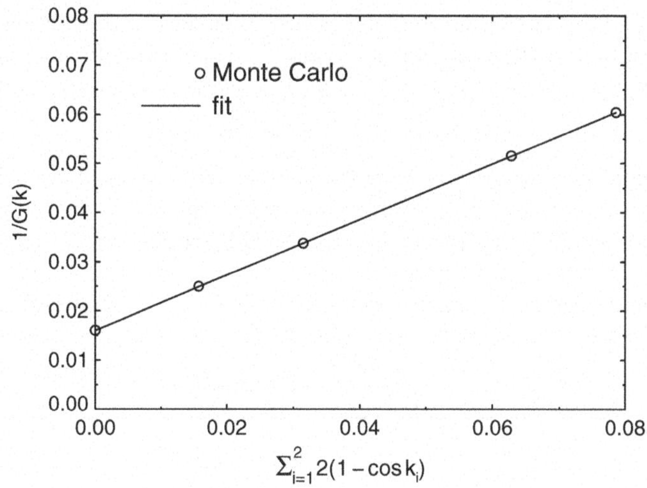

Fig. 4.15. Inverse long-wavelength Fourier components $\widehat{G}(\boldsymbol{k})^{-1}$ versus squared lattice momenta $\kappa^2 \equiv \sum_{i=1}^{2} 2(1 - \cos k_i) \approx \boldsymbol{k}^2$ for the 2D Ising model at $T = 2.5 > T_c$. The fit (4.125), $c_1\kappa^2 + c_0$, gives $c_1 = 0.565\ 5$ and $c_0 = 0.015\ 96$, and hence by (4.126), $\xi = \sqrt{c_1/c_0} = 5.953$

4.8 Generalized Ensemble Methods

All Monte Carlo methods described so far assumed a conventional canonical ensemble where the probability distribution of configurations is governed by a Boltzmann factor $\propto \exp(-\beta E)$. A simulation at some inverse temperature β_0 then covers a certain range of configuration space but not all (recall the discussion of the reweighting range). In principle a broader range can be achieved by patching several simulations at different temperatures using the multi-histogram method. Loosely speaking generalized ensemble methods aim at replacing this static patching by a single simulation in an appropriately defined generalized ensemble. The purpose of this section is to give at least a brief survey of the available methods.

4.8.1 Simulated Tempering

One approach are tempering methods which may be characterized as dynamical multi-histogramming. Similarly to the static reweighting approach, in simulated as well as in parallel tempering one considers m simulation points $\beta_1 < \beta_2 < \ldots < \beta_m$ which here, however, are combined already during the simulation in a specific, dynamical way.

In simulated tempering simulations [142, 143] one starts from a joint partition function (expanded ensemble)

$$\mathcal{Z}_{\mathrm{ST}} = \sum_{i=1}^{m} \mathrm{e}^{g_i} \sum_{\{\sigma\}} \mathrm{e}^{-\beta_i \mathcal{H}(\{\sigma\})}\,, \tag{4.128}$$

where $g_i = \beta_i f(\beta_i)$ and the inverse temperature β is treated as an additional dynamical degree of freedom that can take the values $\beta_1, \ldots, \beta_m$. Employing a Metropolis algorithm, a proposed move from $\beta = \beta_i$ to β_j is accepted with probability

$$W = \min\left[1, \exp[-(\beta_j - \beta_i)\mathcal{H}(\{\sigma\})] + g_j - g_i\right] . \tag{4.129}$$

Similar to multi-histogram reweighting (and also to multicanonical simulations), the free-energy parameters g_i are a priori unknown and have to be adjusted iteratively. To assure a reasonable acceptance rate for the β-update moves (usually between neighboring β_i-values), the histograms at β_i and β_{i+1}, $i = 1, \ldots, m-1$, must overlap. An estimate for a suitable spacing $\delta\beta = \beta_{i+1} - \beta_i$ of the simulation points β_i is hence immediately given by the results (4.82)–(4.84) for the reweighting range,

$$\delta\beta \propto \begin{cases} L^{-D/2} & \text{off-critical} \\ L^{-1/\nu} & \text{critical} \\ L^{-D} & \text{first-order} \end{cases} . \tag{4.130}$$

Overall the simulated tempering method shows some similarities to the avoiding-rare-events variant of multicanonical simulations briefly discussed in the next subsection.

4.8.2 Parallel Tempering

In parallel tempering (replica exchange Monte Carlo, multiple Markov chain Monte Carlo) simulations [144, 145, 146], the starting point is a product of partition functions (extended ensemble),

$$\mathcal{Z}_{\text{PT}} = \prod_{i=1}^{m} \mathcal{Z}(\beta_i) = \prod_{i=1}^{m} \sum_{\{\sigma\}_i} \mathrm{e}^{-\beta_i \mathcal{H}(\{\sigma\}_i)} , \tag{4.131}$$

and all m systems at different simulation points $\beta_1 < \beta_2 < \ldots < \beta_m$ are simulated in parallel, using any legitimate update algorithm (Metropolis, cluster,...). This freedom in the choice of update algorithm is a big advantage of the parallel tempering method. After a certain number of sweeps, exchanges of the current configurations $\{\sigma\}_i$ and $\{\sigma\}_j$ are attempted (equivalently, the β_i may be exchanged, as is done in most implementations). Adapting the Metropolis criterion (4.24) to the present situation, the proposed exchange will be accepted with probability $W = \min(1, \mathrm{e}^{\Delta})$, where

$$\Delta = (\beta_j - \beta_i)\left[E(\{\sigma\}_j) - E(\{\sigma\}_i)\right] . \tag{4.132}$$

To assure a reasonable acceptance rate, usually only nearest-neighbor exchanges ($j = i \pm 1$) are attempted and the β_i should again be spaced with the $\delta\beta$ given in (4.130). In most applications, the smallest inverse temperature β_1 is chosen in the high-temperature phase where the autocorrelation time is expected to be very short and the system decorrelates rapidly. Conceptually this approach follows again the avoiding-rare-events strategy.

Notice that in parallel tempering no free-energy parameters have to be adjusted. The method is thus very flexible and moreover can be almost trivially parallelized.

4.8.3 Multicanonical Ensembles

To conclude this introduction to simulation techniques, at least a very brief outline of multicanonical ensembles shall be given. For more details, in particular on practical implementations, see the recent reviews [4, 147, 148, 149, 150]. Similarly to the tempering methods of the last section, multicanonical simulations may also be interpreted as a dynamical multi-histogram reweighting method. This interpretation is stressed by the notation used in the original papers by Berg and Neuhaus [151, 152] and explains the name multicanonical. At the same time, this method may also be viewed as a specific realization of non-Boltzmann sampling [153] which has been known since long to be a legitimate alternative to the more standard Monte Carlo approaches [154]. The practical significance of non-Boltzmann sampling was first realized in the so-called umbrella-sampling method [155, 156, 157], but it took many years before the introduction of the multicanonical ensemble [151, 152] turned non-Boltzmann sampling into a widely appreciated practical tool in computer simulation studies of phase transitions. Once the feasibility of such a generalized ensemble approach was realized, many related methods and further refinements were developed.

Conceptually the method can be divided into two main strategies. The first strategy can be best described as avoiding rare events which is close in spirit to the alternative tempering methods. In this variant one tries to connect the important parts of phase space by easy paths which go around suppressed rare-event regions which hence cannot be studied directly. The second approach is based on enhancing the probability of rare event states, which is for example the typical strategy for dealing with the highly suppressed mixed-phase region of first-order phase transitions [47, 150]. This allows a direct study of properties of the rare-event states such as, e.g., interface tensions or more generally free energy barriers, which would be very difficult (or practically impossible) with canonical simulations and also with the tempering methods described in Sects. 4.8.1 and 4.8.2.

In general the idea is as follows. With $\{\sigma\}$ representing generically the degrees of freedom (discrete spins or continuous field variables), the canonical Boltzmann distribution

$$\mathcal{P}_{\mathrm{can}}(\{\sigma\}) \propto \mathrm{e}^{-\beta\mathcal{H}(\{\sigma\})} \tag{4.133}$$

is replaced by an auxiliary multicanonical distribution

$$\mathcal{P}_{\mathrm{muca}}(\{\sigma\}) \propto W(Q(\{\sigma\}))\mathrm{e}^{-\beta\mathcal{H}(\{\sigma\})} \equiv \mathrm{e}^{-\beta\mathcal{H}_{\mathrm{muca}}(\{\sigma\})}\,, \tag{4.134}$$

introducing a multicanonical weight factor $W(Q)$ where Q stands for any macroscopic observable such as the energy or magnetization. This defines formally $\mathcal{H}_{\mathrm{muca}} = \mathcal{H} - (1/\beta)\ln W(Q)$ which may be interpreted as an effective multicanonical Hamiltonian. The Monte Carlo sampling can then be implemented as usual by comparing $\mathcal{H}_{\mathrm{muca}}$ before and after a proposed update of $\{\sigma\}$, and canonical expectation values can be recovered exactly by inverse reweighting

$$\langle \mathcal{O}\rangle_{\mathrm{can}} = \langle \mathcal{O}W^{-1}(Q)\rangle_{\mathrm{muca}}/\langle W^{-1}(Q)\rangle_{\mathrm{muca}} \tag{4.135}$$

similarly to (4.89). The goal is now to find a suitable weight factor W such that the dynamics of the multicanonical simulation profits most.

To be specific, let us assume in the following that the relevant macroscopic observable is the energy E itself. This is for instance the case at a temperature driven first-order phase transition, where the canonical energy distribution $P_{\text{can}}(E)$ develops a characteristic double-peak structure [47]. As an illustration, simulation data for the 2D seven-state Potts model [158] are shown in Fig. 4.16. With increasing system size, the region between the two peaks becomes more and more suppressed ($\propto \exp(-2\sigma_{od}L^{D-1})$ where σ_{od} is the (reduced) interface tension, L^{D-1} the cross-section of a D-dimensional system, and the factor two accounts for the fact that with the usually employed periodic boundary condition at least two interfaces are present due to topological reasons) and the autocorrelation time thus grows exponentially with the system size L. In the literature, this is sometimes termed supercritical slowing down (even though nothing is critical here). Given such a situation, one usually adjusts $W = W(E)$ such that the multicanonical distribution $P_{\text{muca}}(E)$ is approximately constant between the two peaks of $P_{\text{can}}(E)$, thus aiming at a random-walk (pseudo-) dynamics of the Monte Carlo process, cf. Fig. 4.16.

The crucial non-trivial point is, of course, *how* this can be achieved. On a piece of paper, $W(E) \propto 1/P_{\text{can}}(E)$ – but we do not know $P_{\text{can}}(E)$ (otherwise there would be little need for the simulation ...). The solution of this problem is a recursive computation. Starting with the canonical distribution, or some initial guess based on results for already simulated smaller systems together with finite-size scaling

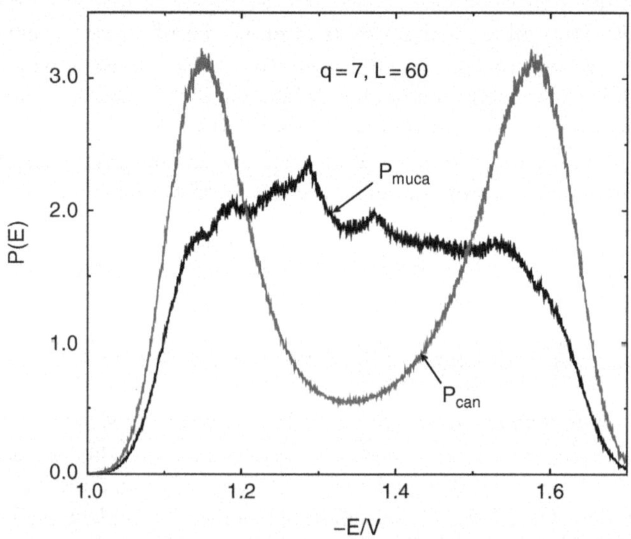

Fig. 4.16. The canonical energy density $P_{\text{can}}(E)$ of the 2D 7-state Potts model on a 60×60 lattice at inverse temperature $\beta_{\text{eqh},L}$, where the two peaks are of equal height, together with the multicanonical energy density $P_{\text{muca}}(E)$, which is approximately constant between the two peaks

extrapolations, one performs a relatively short simulation to get an improved estimate of the canonical distribution. When this is inverted one obtains a new estimate of the multicanonical weight factor, which then is used in the next iteration and so on. In this naive variant only the simulation data of the last iteration are used in the construction of the improved weight factor.

A more sophisticated recursion, in which the updated weight factor, or more conveniently the ratio $R(E) = W(E+\Delta E)/W(E)$, is computed from *all* available data accumulated so far, works as follows [159]:

(i) Perform a simulation with $R_n(E)$ to obtain the nth histogram $H_n(E)$.
(ii) Compute the statistical weight of the nth run:

$$p(E) = H_n(E)H_n(E+\Delta E)/[H_n(E)+H_n(E+\Delta E)] \,. \tag{4.136}$$

(iii) Accumulate statistics:

$$\begin{aligned} p_{n+1}(E) &= p_n(E) + p(E) \,, \\ \kappa(E) &= p(E)/p_{n+1}(E) \,. \end{aligned} \tag{4.137}$$

(iv) Update weight ratios:

$$R_{n+1}(E) = R_n(E)\,[H_n(E)/H_n(E+\Delta E)]^{\kappa(E)} \,. \tag{4.138}$$

Go to (i).

The recursion is initialized with $p_0(E) = 0$. To derive this recursion one assumes that (unnormalized) histogram entries $H_n(E)$ have an a priori statistical error $\sqrt{H_n(E)}$ and (quite crudely) that all data are uncorrelated. Due to the accumulation of statistics, this procedure is rather insensitive to the length of the nth run in the first step and has proved to be rather stable and efficient in practice.

In most applications local update algorithms have been employed, but for certain classes of models also non-local multigrid methods [119, 120, 160, 161] are applicable [121, 162]. A combination with non-local cluster update algorithms, on the other hand, is not straightforward. Only by making direct use of the random-cluster representation as a starting point, a multibondic variant [163, 164, 165] has been developed. For a recent application to improved finite-size scaling studies of second-order phase transitions, see [128]. If P_{muca} was completely flat and the Monte Carlo update moves would perform an ideal random walk, one would expect that after V^2 local updates the system has travelled on average a distance V in total energy. Since one lattice sweep consists of V local updates, the autocorrelation time should scale in this idealized picture as $\tau \propto V$. Numerical tests for various models with a first-order phase transition have shown that in practice the data are at best consistent with a behavior $\tau \propto V^\alpha$, with $\alpha \geq 1$. While for the temperature-driven transitions of 2D Potts models the multibondic variant seems to saturate the bound [163, 164, 165], employing local update algorithms, typical fit results are $\alpha \approx 1.1$–1.3, and due to the limited accuracy of the data even a weak exponential growth law cannot really be excluded.

In fact, at least for the field-driven first-order transition of the 2D Ising model below T_c, where one works with the magnetization instead of the energy (sometimes called multimagnetical simulations), it has been demonstrated recently [166] that even for a perfectly flat multicanonical distribution there are two hidden free energy barriers (in directions orthogonal to the magnetization) which lead to an exponential growth of τ with lattice size, which is albeit much weaker than the leading supercritical slowing down of the canonical simulation. Physically the two barriers are related to the nucleation of a large droplet of the wrong phase (say down-spins in the background of up-spins) [167, 168, 169, 170, 171, 172, 173] and the transition of this large, more or less spherical droplet to the strip phase (coexisting strips of down- and up-spins, separated by two straight interfaces) around $m = 0$ [174].

4.8.4 Wang-Landau Recursion

Another more recently proposed method deals directly with estimators $\Omega(E)$ of the density of states [175, 176]. By flipping spins randomly, the transition probability from energy level E_1 to E_2 is

$$p(E_1 \to E_2) = \min\left[\frac{\Omega(E_1)}{\Omega(E_2)}, 1\right] . \tag{4.139}$$

Each time an energy level is visited, the estimator is multiplicatively updated

$$\Omega(E) \to f\,\Omega(E) , \tag{4.140}$$

where initially $\Omega(E) = 1$ and $f = f_0 = \mathrm{e}^1$. Once the accumulated energy histogram is sufficiently flat, the factor f is refined

$$f_{n+1} = \sqrt{f_n} \tag{4.141}$$

with $n = 0, 1, \ldots$, and the energy histogram reset to zero until some small value such as $f = \mathrm{e}^{10^{-8}} \approx 1.000\,000\,01$ is reached.

For the 2D Ising model this procedure converges very rapidly towards the exactly known density of states, and also for other applications a fast convergence has been reported. Since the procedure is known to violate detailed balance, however, some care is necessary in setting up a proper protocol of the recursion. Most authors who employ the obtained density of states directly to extract canonical expectation values by standard reweighting argue that, once f is close enough to unity, systematic deviations become negligible. While this claim can be verified empirically for the 2D Ising model (where exact results are available for judgement), possible systematic deviations are difficult to assess in the general case. A safe way would be to consider the recursion (4.139)–(4.141) as an alternative method to determine the multicanonical weights, and then to perform a usual multicanonical simulation based on them. As emphasized earlier, any deviations of multicanonical weights from their optimal shape do not show up in the final canonical expectation values; they rather only influence the dynamics of the multicanonical simulations.

4.9 Concluding Remarks

The intention of these lecture notes was to give an elementary introduction to the concepts of modern Markov chain Monte Carlo simulations and to illustrate their usefulness by applications to the very simple Ising lattice spin model. The basic Monte Carlo methods employing local update rules are straightforward to generalize to all models with discrete degrees of freedom and, with small restrictions, also to all models with continuous variables and off-lattice systems. Non-local cluster update methods are much more efficient but also more specialized. Some generalizations to Potts and O(n) symmetric spin models have been indicated and also further models may be efficiently simulated by this method, but there is no guarantee that for a given model a cluster update procedure can be developed. The statistical error analysis is obviously completely general, and also the example finite-size scaling analysis can be taken as a guideline for any model exhibiting a second-order phase transition. Finally, reweighting techniques and generalized ensemble ideas such as tempering methods, the multicanonical ensemble and Wang-Landau sampling can be adapted to almost every statistical physics problem at hand once the relevant macroscopic observables are identified.

Acknowledgements

Many people have influenced these lecture notes with their advice, discussions, questions, and active contributions. In particular I wish to thank Michael Bachmann, Bertrand Berche, Pierre-Emmanuel Berche, Bernd A. Berg, Alain Billoire, Kurt Binder, Elmar Bittner, Christophe Chatelain, Thomas Haase, Malte Henkel, Desmond A. Johnston, Christoph Junghans, Ralph Kenna, David P. Landau, Eric Lorenz, Thomas Neuhaus, Andreas Nußbaumer, Michel Pleimling, Adriaan Schakel, and Martin Weigel for sharing their insight and knowledge with me. Special thanks go to Elmar Bittner for his help with the sample finite-size scaling analysis.

This work was partially supported by the Deutsche Forschungsgemeinschaft (DFG) under grants JA 483/22-1 and JA 483/23-1, the EU RTN-Network "ENRAGE": *Random Geometry and Random Matrices: From Quantum Gravity to Econophysics* under grant MRTN-CT-2004-005616, and the JUMP computer time grants hlz10, hlz11, and hlz12 of NIC at Forschungszentrum Jülich.

References

1. M. Newman, G. Barkema, *Monte Carlo Methods in Statistical Physics* (Clarendon Press, Oxford, 1999)
2. D. Landau, K. Binder, *Monte Carlo Simulations in Statistical Physics* (Cambridge University Press, Cambridge, 2000)
3. K. Binder, D. Heermann, *Monte Carlo Simulations in Statistical Physics: An Introduction*, 4th edn. (Springer, Berlin, 2002)

4. B. Berg, *Markov Chain Monte Carlo Simulations and Their Statistical Analysis* (World Scientific, Singapore, 2004)
5. H. Stanley, *Introduction to Phase Transitions and Critical Phenomena* (Oxford Press, Oxford, 1979)
6. J. Binney, N. Dowrick, A. Fisher, M. Newman, *The Theory of Critical Phenomena* (Oxford University Press, Oxford, 1992)
7. D. Lavis, G. Bell, *Statistical Mechanics of Lattice Systems 2* (Springer, Berlin, 1999)
8. C. Domb, J. Lebowitz (eds.), *Phase Transitions and Critical Phenomena* (Academic Press, New York, 1976)
9. W. Lenz, Phys. Z. **21**, 613 (1920)
10. E. Ising, Phys. Z. **31**, 253 (1925)
11. L. Onsager, Phys. Rev. **65**, 117 (1944)
12. B. Kaufman, Phys. Rev. **76**, 1232 (1949)
13. A. Ferdinand, M. Fisher, Phys. Rev. **185**, 832 (1969)
14. B. McCoy, T. Wu, *The Two-Dimensional Ising Model* (Harvard University Press, Cambridge, 1973)
15. R. Baxter, *Exactly Solved Models in Statistical Mechanics* (Academic Press, New York, 1982)
16. L. Onsager, Nuovo Cimento **6**, 261 (1949)
17. C. Yang, Phys. Rev. **85**, 808 (1952)
18. C. Chang, Phys. Rev. **88**, 1422 (1952)
19. W. Orrick, B. Nickel, A. Guttmann, J. Perk, Phys. Rev. Lett. **86**, 4120 (2001)
20. W. Orrick, B. Nickel, A. Guttmann, J. Perk, J. Stat. Phys. **102**, 795 (2001)
21. R. Griffiths, Phys. Rev. Lett. **24**, 1479 (1970)
22. G. Rushbrooke, J. Chem. Phys. **39**, 842 (1963)
23. R. Griffiths, Phys. Rev. Lett. **14**, 623 (1965)
24. B. Josephson, Proc. Phys. Soc. **92**, 269 (1967)
25. B. Josephson, Proc. Phys. Soc. **92**, 276 (1967)
26. M. Fisher, Phys. Rev. **180**, 594 (1969)
27. L. Widom, J. Chem. Phys. **43**, 3892 (1965)
28. L. Widom, J. Chem. Phys. **43**, 3898 (1965)
29. L. Kadanoff, Physics **2**, 263 (1966)
30. K. Wilson, J. Kogut, Phys. Rep. **C12**, 75 (1974)
31. F. Wu, Rev. Mod. Phys. **54**, 235 (1982)
32. F. Wu, Rev. Mod. Phys. **55**, 315(E) (1983)
33. M. Weigel, W. Janke, Phys. Rev. **B62**, 6343 (2000)
34. K. Binder, in *Monte Carlo Methods in Statistical Physics* ed. by K. Binder (Springer, Berlin, 1979), p. 1
35. M. Barber, in *Phase Transitions and Critical Phenomena*, Vol. 8 ed. by C. Domb, J. Lebowitz (Academic Press, New York, 1983), p. 146
36. V. Privman (ed.), *Finite-Size Scaling and Numerical Simulations of Statistical Systems* (World Scientific, Singapore, 1990)
37. K. Binder: in Computational Methods in Field Theory, Schladming Lecture Notes, eds. H. Gausterer, C.B. Lang (Springer, Berlin, 1992), P. 59
38. J. Gunton, M. Miguel, P. Sahni, in *Phase Transitions and Critical Phenomena*, Vol. 8, ed. by C. Domb, J. Lebowitz (Academic Press, New York, 1983)
39. K. Binder, Rep. Prog. Phys. **50**, 783 (1987)
40. H. Herrmann, W. Janke, F. Karsch (eds.), *Dynamics of First Order Phase Transitions* (World Scientific, Singapore, 1992)

41. W. Janke, in *Computer Simulations in Condensed Matter Physics*, Vol. VII, ed. by D. Landau, K. Mon, H.B. Schüttler (Springer, Berlin, 1994), p. 29
42. M. Fisher, A. Berker, Phys. Rev. B **26**, 2507 (1982)
43. V. Privman, M. Fisher, J. Stat. Phys. **33**, 385 (1983)
44. K. Binder, D. Landau, Phys. Rev. B **30**, 1477 (1984)
45. M. Challa, D. Landau, K. Binder, Phys. Rev. B **34**, 1841 (1986)
46. V. Privman, J. Rudnik, J. Stat. Phys. **60**, 551 (1990)
47. W. Janke, in *Computer Simulations of Surfaces and Interfaces, NATO Science Series, II. Mathematics, Physics and Chemistry* Vol. 114, ed. by B. Dünweg, D. Landau, A. Milchev (Kluwer, Dordrecht, 2003), pp. 111–135
48. C. Borgs, R. Kotecký J. Stat. Phys. **61**, 79 (1990)
49. J. Lee, J. Kosterlitz Phys. Rev. Lett. **65**, 137 (1990)
50. C. Borgs, R. Kotecký, S. Miracle-Solé, J. Stat. Phys. **62**, 529 (1991)
51. C. Borgs, W. Janke, Phys. Rev. Lett. **68**, 1738 (1992)
52. W. Janke, Phys. Rev. B **47**, 14757 (1993)
53. J. Hammersley, D. Handscomb, *Monte Carlo Methods* (Chapman and Hall, London, New York, 1964)
54. D. Heermann, *Computer Simulation Methods in Theoretical Physics*, 2nd edn. (Springer, Berlin, 1990)
55. K. Binder (ed.), *The Monte Carlo Method in Condensed Matter Physics* (Springer, Berlin, 1992)
56. N. Metropolis, A. Rosenbluth, M. Rosenbluth, A. Teller, E. Teller, J. Chem. Phys. **21**, 1087 (1953)
57. S. Kirkpatrick, C. Gelatt. Jr., M. Vecchi, Science **220**, 671 (1983)
58. W. Janke, in *Ageing and the Glass Transition – Summer School, University of Luxembourg, September 2005, Lecture Notes in Physics*, Vol. 716, ed. by M. Henkel, M. Pleimling, R. Sanctuary (Springer, Berlin, Heidelberg, 2007), *Lecture Notes in Physics*, pp. 207–260
59. W. Janke, in *Proceedings of the Euro Winter School Quantum Simulations of Complex Many-Body Systems: From Theory to Algorithms, NIC Series*, Vol. 10, ed. by J. Grotendorst, D. Marx, A. Muramatsu (John von Neumann Institute for Computing, Jülich, 2002), pp. 447–458
60. R. Glauber, J. Math. Phys. **4**, 294 (1963)
61. R. Swendsen, J.S. Wang, Phys. Rev. Lett. **58**, 86 (1987)
62. R. Potts, Proc. Camb. Phil. Soc. **48**, 106 (1952)
63. U. Wolff, Phys. Rev. Lett. **62**, 361 (1989)
64. W. Janke, Mathematics and Computers in Simulations **47**, 329 (1998)
65. P. Kasteleyn, C. Fortuin, J. Phys. Soc. Japan **26**, 11 (1969)
66. C. Fortuin, P. Kasteleyn, Physica **57**, 536 (1972)
67. C. Fortuin, Physica **58**, 393 (1972)
68. C. Fortuin, Physica **59**, 545 (1972)
69. U. Wolff, Nucl. Phys. **B322**, 759 (1989)
70. M. Hasenbusch, Nucl. Phys. **B333**, 581 (1990)
71. U. Wolff, Nucl. Phys. **B334**, 581 (1990)
72. U. Wolff, Phys. Lett. A **228**, 379 (1989)
73. C. Baillie, Int. J. Mod. Phys. C **1**, 91 (1990)
74. M. Hasenbusch, S. Meyer, Phys. Lett. B **241**, 238 (1990)
75. R. Swendsen, J.S. Wang, A. Ferrenberg, in *The Monte Carlo Method in Condensed Matter Physics* ed. by K. Binder (Springer, Berlin, 1992)

76. X.L. Li, A. Sokal, Phys. Rev. Lett. **63**, 827 (1989)
77. X.L. Li, A. Sokal, Phys. Rev. Lett. **67**, 1482 (1991)
78. M. Nightingale, H. Blöte, Phys. Rev. Lett. **76**, 4548 (1996)
79. M. Nightingale, H. Blöte, Phys. Rev. B **62**, 1089 (2000)
80. P. Grassberger, Physica A **214**, 547 (1995)
81. P. Grassberger, Physica A **217**, 227 (E) (1995)
82. N. Ito, K. Hukushima, K. Ogawa, Y. Ozeki, J. Phys. Soc. Japan **69**, 1931 (2000)
83. D. Heermann, A. Burkitt, Physica A **162**, 210 (1990)
84. P. Tamayo, Physica A **201**, 543 (1993)
85. N. Ito, G. Kohring, Physica A **201**, 547 (1993)
86. W. Janke, Phys. Lett. A **148**, 306 (1990)
87. C. Holm, W. Janke, Phys. Rev. B **48**, 936 (1993)
88. W. Janke, A. Schakel, Nucl. Phys. **B700**, 385 (2004)
89. W. Janke, A. Schakel, Comp. Phys. Comm. **169**, 222 (2005)
90. W. Janke, A. Schakel, Phys. Rev. E **71**, 036703 (2005)
91. W. Janke, A. Schakel, Phys. Rev. Lett. **95**, 135702 (2005)
92. W. Janke, A. Schakel, in *Order, Disorder and Criticality: Advanced Problems of Phase Transition Theory*, Vol. 2, ed. by Y. Holovatch (World Scientific, Singapore, 2007), pp. 123–180
93. E. Lorenz, Ageing phenomena in phase-ordering kinetics in Potts models. Diploma thesis, Universität Leipzig (2005). www.physik.uni-leipzig.de/~lorenz/ diplom.pdf
94. A. Rutenberg, A. Bray, Phys. Rev. E **51**, 5499 (1995)
95. P. Calabrese, A. Gambassi, J. Phys. A **38**, R133 (2005)
96. C. Godrèche, J.M. Luck, J. Phys.: Condens. Matter **14**, 1589 (2002)
97. L.F. Cugliandolo: in Slow Relaxation and Non Equilibrium Dynamics in Condensed Matter, Les Houches Lectures, eds. J.-L. Barrat, J. Dalibard, J. Kurchan, M.V. Feigel'man (Springer, Berlin, 2003)
98. F. Corberi, E. Lippiello, M. Zannetti, Phys. Rev. Lett. **90**, 099601 (2003)
99. M. Henkel, M. Pleimling, Phys. Rev. Lett. **90**, 099602 (2003)
100. L. Berthier, J. Barrat, J. Kurchan, Europhys. J. B **11**, 635 (1999)
101. F. Corberi, E. Lippiello, M. Zannetti, Europhys. J. B **24**, 359 (2001)
102. F. Corberi, E. Lippiello, M. Zannetti, Phys. Rev. E **65**, 046136 (2003)
103. A. Barrat, Phys. Rev. E **57**, 3629 (1998)
104. M. Henkel, *Conformal Invariance and Critical Phenomena* (Springer, Berlin, 1999)
105. M. Henkel, M. Pleimling, C. Godrèche, J.M. Luck, Phys. Rev. Lett. **87**, 265701 (2001)
106. M. Henkel, Nucl. Phys. **B641**, 405 (2002)
107. M. Henkel, M. Paessens, M. Pleimling, Europhys. Lett. **62**, 664 (2003)
108. M. Henkel, M. Pleimling, Phys. Rev. E **68**, 065101 (R) (2003)
109. M. Henkel, A. Picone, M. Pleimling, Europhys. Lett. **68**, 191 (2004)
110. E. Lorenz, W. Janke, Europhys. Lett. **77**, 10003 (2007)
111. W. Janke, in *Proceedings of the Euro Winter School Quantum Simulations of Complex Many-Body Systems: From Theory to Algorithms*, *NIC Series*, Vol. 10, ed. by J. Grotendorst, D. Marx, A. Muramatsu (John von Neumann Institute for Computing, Jülich, 2002), pp. 423–445
112. P. Beale, Phys. Rev. Lett. **76**, 78 (1996)
113. W. Press, S. Teukolsky, W. Vetterling, B. Flannery, *Numerical Recipes in Fortran 77 – The Art of Scientific Computing*, 2nd edn. (Cambridge University Press, Cambridge, 1999)

114. M. Priestley, *Spectral Analysis and Time Series*, Vol. 2 (Academic, London, 1981). Chaps. 5–7
115. T. Anderson, *The Statistical Analysis of Time Series* (Wiley, New York, 1971)
116. N. Madras, A. Sokal, J. Stat. Phys. **50**, 109 (1988)
117. A. Sokal, L. Thomas, J. Stat. Phys. **54**, 797 (1989)
118. A. Ferrenberg, D. Landau, K. Binder, J. Stat. Phys. **63**, 867 (1991)
119. A. Sokal, *Monte Carlo Methods in Statistical Mechanics: Foundations and New Algorithms* (Cours de Troisième Cycle de la Physique en Suisse Romande, Lausanne, 1989)
120. A. Sokal, in *Quantum Fields on the Computer*, ed. by M. Creutz (World Scientific, Singapore, 1992), p. 211
121. W. Janke, T. Sauer, J. Stat. Phys. **78**, 759 (1995)
122. B. Efron, *The Jackknife, the Bootstrap and Other Resampling Plans* (Society for Industrial and Applied Mathematics [SIAM], Philadelphia, 1982)
123. R. Miller, Biometrika **61**, 1 (1974)
124. A. Ferrenberg, R. Swendsen, Phys. Rev. Lett. **61**, 2635 (1988)
125. A. Ferrenberg, R. Swendsen, Phys. Rev. Lett. **63**, 1658(E) (1989)
126. N. Wilding, in *Computer Simulations of Surfaces and Interfaces, NATO Science Series, II. Mathematics, Physics and Chemistry* Vol. 114, ed. by B. Dünweg, D. Landau, A. Milchev (Kluwer, Dordrecht, 2003), pp. 161–171
127. A. Ferrenberg, R. Swendsen, Phys. Rev. Lett. **63**, 1195 (1989)
128. B. Berg, W. Janke, Phys. Rev. Lett. **98**, 040602 (2007)
129. G. Kamieniarz, H. Blöte, J. Phys. A **26**, 201 (1993)
130. J. Salas, A. Sokal, J. Stat. Phys. **98**, 551 (2000)
131. X. Chen, V. Dohm, Phys. Rev. E **70**, 056136 (2004)
132. V. Dohm, J. Phys. A **39**, L259 (2006)
133. W. Selke, L. Shchur, J. Phys. A **38**, L739 (2005)
134. M. Schulte, C. Drope, Int. J. Mod. Phys. C **16**, 1217 (2005)
135. M. Sumour, D. Stauffer, M. Shabat, A. El-Astal, Physica A **368**, 96 (2006)
136. W. Selke, Europhys. J. B **51**, 223 (2006); preprint `http://arxiv.org/abs/cond-mat/0701515`
137. A. Nußbaumer, E. Bittner, W. Janke, Europhys. Lett. **78**, 16004 (2007)
138. E. Bittner, W. Janke, The pain of example analyses – a (self-)critical discussion. Unpublished results
139. W. Janke, R. Villanova, Phys. Rev. B **66**, 134208 (2002)
140. J. Oitmaa, J. Phys. A **14**, 1159 (1981)
141. C. Holm, W. Janke, Phys. Rev. Lett. **78**, 2265 (1997)
142. E. Marinari, G. Parisi, Europhys. Lett. **19**, 451 (1992)
143. A. Lyubartsev, A. Martsinovski, S. Shevkunov, P. Vorontsov-Velyaminov, J. Chem. Phys. **96**, 1776 (1992)
144. C. Geyer, in *Proceedings of the 23rd Symposium on the Interface*, ed. by E. Keramidas (Interface Foundation, Fairfax, Virginia, 1991), pp. 156–163
145. C. Geyer, E. Thompson, J. Am. Stat. Assoc. **90**, 909 (1995)
146. K. Hukushima, K. Nemoto, J. Phys. Soc. Japan **65**, 1604 (1996)
147. B. Berg, Fields Inst. Comm. **26**, 1 (2000)
148. B. Berg, Comp. Phys. Comm. **104**, 52 (2002)
149. W. Janke, Physica A **254**, 164 (1998)
150. W. Janke, in *Computer Simulations of Surfaces and Interfaces, NATO Science Series, II. Mathematics, Physics and Chemistry –Proceedings of the NATO Advanced Study Institute, Albena, Bulgaria, 9–20 September 2002*, Vol. 114, ed. by B. Dünweg, D. Landau, A. Milchev (Kluwer, Dordrecht, 2003)

151. B. Berg, T. Neuhaus, Phys. Lett. B **267**, 249 (1991)
152. B. Berg, T. Neuhaus, Phys. Rev. Lett. **68**, 9 (1992)
153. W. Janke, Int. J. Mod. Phys. C **3**, 1137 (1992)
154. K. Binder: in Phase Transitions and Critical Phenomena, Vol. 5b, eds. C. Domb, M.S. Green (Academic Press, New York, 1976), p. 1
155. G. Torrie, J. Valleau, Chem. Phys. Lett. **28**, 578 (1974)
156. G. Torrie, J. Valleau, J. Comp. Phys. **23**, 187 (1977)
157. G. Torrie, J. Valleau, J. Chem. Phys. **66**, 1402 (1977)
158. W. Janke, B. Berg, M. Katoot, Nucl. Phys. **B382**, 649 (1992)
159. B. Berg, W. Janke. Unpublished notes
160. J. Goodman, A. Sokal, Phys. Rev. Lett. **56**, 1015 (1986)
161. J. Goodman, A. Sokal, Phys. Rev. D **40**, 2035 (1989)
162. W. Janke, T. Sauer, Phys. Rev. E **49**, 3475 (1994)
163. W. Janke, S. Kappler, Nucl. Phys. (proc suppl.) **B42**, 876 (1995)
164. W. Janke, S. Kappler, Phys. Rev. Lett. **74**, 212 (1995)
165. M. Carroll, W. Janke, S. Kappler, J. Stat. Phys. **90**, 1277 (1998)
166. T. Neuhaus, J. Hager, J. Stat. Phys. **113**, 47 (2003)
167. K. Binder, M. Kalos, J. Stat. Phys. **22**, 363 (1980)
168. H. Furukawa, K. Binder, Phys. Rev. A **26**, 556 (1982)
169. M. Biskup, L. Chayes, R. Kotecký, Europhys. Lett. **60**, 21 (2002)
170. M. Biskup, L. Chayes, R. Kotecký, Comm. Math. Phys. **242**, 137 (2003)
171. M. Biskup, L. Chayes, R. Kotecký, J. Stat. Phys. **116**, 175 (2003)
172. K. Binder, Physica A **319**, 99 (2003)
173. A. Nußbaumer, E. Bittner, T. Neuhaus, W. Janke, Europhys. Lett. **75**, 716 (2006)
174. K. Leung, R. Zia, J. Phys. A **23**, 4593 (1990)
175. F. Wang, D. Landau, Phys. Rev. Lett. **86**, 2050 (2001)
176. F. Wang, D. Landau, Phys. Rev. E **64**, 056101 (2001)

5 The Monte Carlo Method for Particle Transport Problems

Detlev Reiter

Institut für Energieforschung - Plasmaphysik, Forschungszentrum Jülich GmbH, 52425 Jülich, Germany

5.1 Transport Problems and Stochastic Processes

The concept of the natural stochastic process for a transport problem will be introduced. Basic methods of estimating macroscopic quantities (e.g. particle densities, energy densities, collision rates), from the sample pathes of random walks will be discussed. A key distinction is made between processes with continuous sample pathes (diffusion processes, e.g. Brownian motion) and jump processes (Markov-chains, e.g. radiation or particle transport). Continuous path processes are controlled by a Fokker-Planck type equation, whereas discontinuous jump processes lead to Fredholm integral equations of the 2nd kind, with the linear form of the Boltzmann equation as prototypical example. A practically important example of the first type in plasma physics is the diffusive (in velocity space) charged test particle transport in a prescribed bath of electrons and ions (due to the long range nature of the Coulomb interaction). Other important examples are quantum-mechanical diffusion Monte Carlo (DiffMC), but also smooth particle hydrodynamics, a convection-diffusion in real space of fluid parcels simulated by a Monte Carlo concept, which is an alternative and complementary to the conventional discretization tools from computational fluid dynamics. Other important applications are in the fields of population dynamics, investment finance, turbulent diffusion, and many more. In the second category fall the neutron-, radiation shielding transport problems, recycling and neutral particle transport in plasmas, cosmic ray shower simulations and many more. The transition from the second to the first type of transport problems is provided by diffusion approximations, which are frequently applied in many different fields of physics and computational science. Monte Carlo methods for such continuous processes involve both stochastic and numerical concepts. The computation of trajectories requires not only random sampling but also a time discretization, because the (continuous) trajectories of a diffusion process cannot be simulated on a digital device. They must be approximated by trajectories of a discontinuous jump process. There exist Monte Carlo counterparts to the explicit Euler scheme, as well as higher order concepts such as the famous Milstein scheme, see the textbook by P.E. Kloeden and E. Platen [1] for an excellent and comprehensive introduction into that topic. Due to this finite discretization in time one may think of the underlying diffusion process as being approximated by a discontinuous (jump-) process. After having sorted out the numerical issues (Taylor expansions, here referred to as Ito-Taylor expansions because the underlying equation is a stochastic differential equation)

D. Reiter: *The Monte Carlo Method for Particle Transport Problems*, Lect. Notes Phys. **739**, 141–158 (2008)
DOI 10.1007/978-3-540-74686-7_5

the Monte Carlo method then basically simulates an approximating integral equation of a jump process. A very clear discussion on the approximation of diffusion processes by jump-processes (i.e., opposite to the direction usually used in physical arguments to derive Fokker Planck equations) is, e.g., given in the monograph by C.W. Gardiner [2]. Once this approximation is done, the Monte Carlo procedures for Fokker-Planck equations and for Boltzmann equations become analog. We shall, therefore, only discuss discontinuous jump processes from now on, hence only Monte Carlo methods for solving Fredholm integral equations. We will follow a similar strategy as in the introductory chapter on Monte Carlo methods in these lecture notes, Chap. 3: Although we will try to make explicit the underlying mathematical basis of the method, we strongly build on the key advantage of Monte Carlo methods over numerical concepts: The important role of intuition to guide the derivation of the algorithm, which consequently retains a high level of transparency.

5.1.1 Historical Notes

The first Monte Carlo computer simulations have been carried out within the US atomic bomb project (Manhatten project), under the leadership of John von Neumann (Fig. 5.1, left) and Stan Ulam.

Neutron migration in material was simulated by a cascade of decisions based on non-uniform random numbers (Fig. 5.1, right): At the start of a neutron velocity and position was sampled. Then the mean free flight distance (from an exponential distribution) was determined, leading to the decision: Collision or transit through the medium? If transit, the neutron was moved and new free flight distance is sampled.

(a)

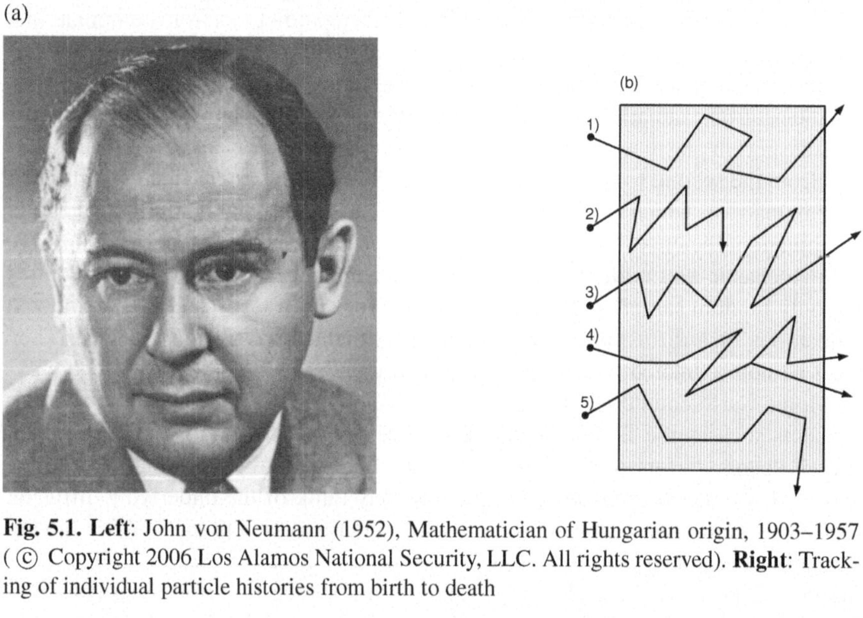

Fig. 5.1. Left: John von Neumann (1952), Mathematician of Hungarian origin, 1903–1957 (© Copyright 2006 Los Alamos National Security, LLC. All rights reserved). **Right**: Tracking of individual particle histories from birth to death

If collision: Again a random decision is made: Scattering, fission or absorption? If scattering: Sample new velocity. If fission: How many new neutrons (at which velocities)? If absorption: Stop.

This strategy allowed complex geometries to be modelled, a continuous energy representation of data and the calculation accuracy is limited only by statistics and data uncertainties, but no numerical approximations are made.

The sound mathematical basis of this procedure is based on the relation between the probability to find a particular random walk from this procedure, the evaluation of random variables (estimators) along these trajectories and the Neumann series for expanding the solution to Fredholm integral equations of the second kind. Most of this material was published shortly after World War II, in a large number of papers and later also monographs, see [3, 4, 5] for excellent overviews.

5.2 The Transport Equation: Fredholm Integral Equation of Second Kind

As it has been discussed in Chap. 3 Monte Carlo simulation can always be viewed as integration procedure of a certain function g with respect to a probability distribution f. This is formally done by making connection to the mathematical definition of an expectation value of a random function $G(X)$ with respect to a distribution law F

$$I = \int_V \mathrm{d}x \; g(x) f(x) := \int_V \mathrm{d}F \; G(X) \; . \tag{5.1}$$

The new element in this present chapter on particle transport is the fact that the distribution law f may not be known explicitly anymore. Hence direct random sampling from f is not possible. Common to all Monte Carlo applications to transport theory is that f is given only implicity, as solution of a governing kinetic equation. This kinetic equation can be a differential equation (diffusive transport, Fokker-Planck type differential equations, i.e., very soft interactions causing only small changes), an integral equation (ballistic transport, Boltzmann type integral equations, hard interactions, causing discontinuous jumps), or of mixed type. We refer to the historic papers on this relation between analytic properties of trajectories of random walks and corresponding differential and integral equations by W. Feller [6], and references therein. As will be discussed next the key idea is then to generate an entire random walk (Markov chain) rather than sequences of independent random numbers.

It is worth noting that also a second very wide class of Monte Carlo applications, namely those to problems in statistical mechanics (chapter 4), is based upon a similar idea: Ensemble averages (very high dimensional integrals) are found there by generating a random walk in the Gibb's phase-space, rather than explicitly considering the underlying many-body distribution law itself. Because of this similarity of the concept with the historically earlier developed neutron transport applications also this procedure was then named Monte Carlo method, see Metropolis [7].

5.2.1 The Generic Transport Equation

To introduce the terminology, we briefly recall the basic definitions and principles of a Monte Carlo linear transport model, following the lead of many textbooks on Monte Carlo methods for computing neutron transport (see e.g., Spanier and Gelbard, [3]). We begin with the linear transport equation for the dependent variable ψ (see below), written as integral equation (linear non-homogeneous Fredholm integral equation (FIE) of 2nd kind). This equation reads

$$\begin{aligned} \psi(x) &= S(x) + \int \mathrm{d}x' K(x' \to x)\psi(x') \, , \\ c(x') &= \int \mathrm{d}x' K(x' \to x) \, . \end{aligned} \tag{5.2}$$

Distinct from standard terminology in (analytic) transport theory we do not discuss analytic properties of the various terms in this equation, but, instead, point out their probabilistic interpretation, as needed for a Monte-Carlo solution of that equation.

One may view the function $\psi(x)$ as probability to find state x in a relevant phase space (think of a particle distribution function in kinetic theory). There is an (un-collided) contribution directly from an external source S to $\psi(x)$, as well as a contribution from previous states x', from which the objects are transferred to x with probability $K(x' \to x)$. The normalization function $c(x')$ is to be interpreted as mean number of objects emerging from one transition event, given that one object went into that event at point x' (collision).

The quantity to be computed is (compare with (5.1))

$$I_g(x) = \int \mathrm{d}x \; g(x)\psi(x) \, . \tag{5.3}$$

In Sect. 5.3 we will interpret the generic FIE (5.2) of transport theory with the particular Boltzmann equation for dilute gases in physics, because this serves then as guidance of intuition for all our further discussions. The objects will be interpreted as particles, events will be collisions with a host (background) medium.

5.3 The Boltzmann Equation

We now make connection between the generic transport equation (5.2) and the most famous and important transport equation in science: The Boltzmann equation for dilute gases: The phase space is then the space of all relevant independent variables (co-ordinates) of a single particle and the dependent quantity of interest ψ is then the one particle distribution function $f(\boldsymbol{r}, \boldsymbol{v}, i, t)$, $f(\boldsymbol{r}, E, \boldsymbol{\Omega}, i, t)$, or $f(x)$ where the state x is characterized by a position vector $\boldsymbol{r}$, a velocity vector $\boldsymbol{v}$, a chemical species index i and the time t, etc.

In radiation transfer applications instead of $\boldsymbol{v}$ one utilizes the kinetic energy E, (or frequency, or wavelength) and the unit (speed) vector $\boldsymbol{\Omega} = \boldsymbol{v}/|\boldsymbol{v}|$ in the direction of particle motion. The number density $n_i(\boldsymbol{r})$ for species i at a point $\boldsymbol{r}$ then is

$$n_i(\boldsymbol{r},t) = \int \mathrm{d}^3 v \, f(\boldsymbol{r},\boldsymbol{v},i,t) \,, \tag{5.4}$$

which is a special case of a moment (response) as defined in (5.3), with $g = \delta^3(\boldsymbol{r} - \boldsymbol{r'})\delta_{i,j}$. I.e., g can contain appropriate delta function to select a particular species and point in space. Any other macroscopic (and also microscopic) quantity of interest, such as fluxes, energy density, momentum transfer rates, etc. averaged over subdomains (including single points) of phase space, can readily be seen to be covered by the general expression (5.3).

As already discussed we start by assuming that events (here collisions) lead to discontinuous trajectories, at least in some of the phase space variables. Further: Lets consider only one specific particle species i_0 from now on, omitting this species index. We assume that there are only collisions (events) of this species i_0 with only one further species (labelled b), and that exactly one particle of each of these species will also be present after the collision event. I.e., inelastic and chemical reactions are excluded, for keeping notation simple (but they are further discussed below). The familiar Boltzmann equation [8] for the distribution function f for this species i_0 reads

$$\begin{aligned} &\left[\frac{\partial}{\partial t} + \boldsymbol{v} \cdot \nabla_{\boldsymbol{r}} + \frac{\boldsymbol{F}(\boldsymbol{r},\boldsymbol{v},t)}{m} \cdot \nabla_{\boldsymbol{v}}\right] f(\boldsymbol{r},\boldsymbol{v},t) \\ &= \iiint \sigma(\boldsymbol{v}',\boldsymbol{v}';\boldsymbol{v},\boldsymbol{v})|\boldsymbol{v}' - \boldsymbol{v}'| f(\boldsymbol{v}') f_b(\boldsymbol{v}') \\ &\quad - \iiint \sigma(\boldsymbol{v},\boldsymbol{v};\boldsymbol{v}',\boldsymbol{v}')|\boldsymbol{v} - \boldsymbol{v}| f(\boldsymbol{v}) f_b(\boldsymbol{v}) \,. \end{aligned} \tag{5.5}$$

Integrations are over the velocities $\boldsymbol{v}', \boldsymbol{v}$ and $\boldsymbol{v}'$. Here $\sigma(\boldsymbol{v}',\boldsymbol{v}';\boldsymbol{v},\boldsymbol{v})$ is the cross section for a binary particle collision process defined such that the conservation laws for total energy and momentum are fulfilled. The first two arguments in σ, namely the velocities $\boldsymbol{v}', \boldsymbol{v}'$ in the first integral, correspond to the species i_0 and b, respectively, prior to a collision. These are turned into the post collision velocities $\boldsymbol{v}, \boldsymbol{v}$, again for species i_0 and b, respectively. The first integral, therefore, describes transitions $(\boldsymbol{v}',\boldsymbol{v}' \to \boldsymbol{v},\boldsymbol{v})$ into the velocity space interval $[\boldsymbol{v}, \boldsymbol{v} + \mathrm{d}\boldsymbol{v}]$ for species i_0, and the second integral describes loss from that interval for this species. Furthermore, m is the particle mass and $\boldsymbol{F}(\boldsymbol{r},\boldsymbol{v},t)$ is the volume force field. The right hand side is the collision integral $\delta f/(\delta t)|_b$. If there are more than just one possible type of collision partners, then the collision integral has to be replaced by a sum of collision integrals over all collision partners b, including, possibly, $b = i_0$ (self collisions)

$$\frac{\delta f}{\delta t} = \sum_b \left.\frac{\delta f}{\delta t}\right|_b \,. \tag{5.6}$$

This is readily generalized to the semi-classical Boltzmann equation for chemical reactions (including, for example, vibrational relaxation or exchange of internal energy as special cases) symbolized as $i_0 + j_0 \leftrightarrow i_1 + j_1$. These species indices label both the chemical species and/or the internal quantum state. In this case the sum in the collision integral is over j_0, i_1 and j_1 and the cross sections in the corresponding collision integrals $\sigma_{i_0 j_0}^{i_1 j_1}(\boldsymbol{v}, \boldsymbol{v}, \boldsymbol{v}', \boldsymbol{v}')$ are differential for scattering at a certain solid angle and post collision energies with simultaneous transition from (i_0, j_0) to (i_1, j_1). Further generalizations to include particle splitting, absorption or fragmentation into more than two post collision products are straight forward, but can more conveniently be formulated in the C-collision kernel formulation used below.

All these collision operators are bi-linear in the distribution functions. The first term on the right hand side is due to scattering into the element $\mathrm{d}\boldsymbol{v}$ of velocity space and we shall abbreviate it by defining the collision kernel (redistribution function) C as a proper integral over pre- and post collision velocities of species b-particles:

$$\left.\frac{\delta f}{\delta t}\right|_{\text{gain}} = \int \mathrm{d}^3 v' \, C(\boldsymbol{v}' \to \boldsymbol{v}) |\boldsymbol{v}'| f(\boldsymbol{v}') \,. \tag{5.7}$$

Despite its simple physical content (transition probability from $\boldsymbol{v}'$ to $\boldsymbol{v}$, given a collision at $\boldsymbol{r}$) the collision kernel C can be a quite complicated integral, as it involves not only multiple differential cross sections, but also, possibly, particle multiplication factors, e.g. in case of fission by neutron impact, dissociation of molecules by electron impact, or stimulated photon emission from excited atoms. It can also include absorption, in which case the post collision state must be an extra limbo state outside the phase-space considered. Due to both particle multiplication and/or absorption the collision kernel C is not normalized to one, generally.

The second term on the right hand side is much simpler, because the function $f(\boldsymbol{v})$ can be taken out of the integral. We even take the product $|\boldsymbol{v}| \cdot f$ before the integral. The remaining integral is then just the total macroscopic cross section Σ_t, i.e., the inverse local mean free path (dimension: 1/length). It is solely defined by total cross sections and independent of particle multiplication factors, since we only consider binary collisions (exactly two pre-collision partners always).

This term is then often taken on the left hand side of the Boltzmann equation with a positive sign, in the form

$$\left.\frac{\delta f}{\delta t}\right|_{\text{loss}} = \Sigma_t(\boldsymbol{r}, \boldsymbol{v}) |\boldsymbol{v}| f(\boldsymbol{v}) \,. \tag{5.8}$$

With these formal substitutions the Boltzmann equation takes a form which is often more convenient, in particular in linear transport theory

$$\begin{aligned} &\left[\frac{\partial}{\partial t} + \boldsymbol{v} \cdot \nabla_{\boldsymbol{r}} + \frac{\boldsymbol{F}(\boldsymbol{r}, \boldsymbol{v}, t)}{m} \cdot \nabla_{\boldsymbol{v}}\right] f(\boldsymbol{r}, \boldsymbol{v}, t) + \Sigma_t(\boldsymbol{r}, \boldsymbol{v}) |\boldsymbol{v}| f(\boldsymbol{v}) \\ &= \int \mathrm{d}^3 v' \, C(\boldsymbol{v}' \to \boldsymbol{v}) |\boldsymbol{v}'| f(\boldsymbol{v}') + Q(\boldsymbol{r}, \boldsymbol{v}, t) \,. \end{aligned} \tag{5.9}$$

In this equation an external source term Q has also been added, for completeness.

5.3.1 The Linear Boltzmann Equation

If the distributions f_b of collision partners b are assumed to be given, then the kernel C does not dependent on the dependent quantities f. Also the extinction coefficient Σ_t is independent of the dependent variable $f = f_{i_0}$, and the out-scattering loss term (last term on left hand side) just describes the loss of particle flux of i_0 particles due to any kind of interaction of them with the host medium. Equation (5.9) above becomes a linear integro-differential equation. If the characteristic time constants for the considered transport phenomena are very short compared to those for evolution of the macroscopic background medium one can then neglect explicit time dependence.

If the particles travel on straight lines between collisions i.e., with no forces acting on them: $\boldsymbol{F} = \mathbf{0}$, then the scalar transport flux (angular flux) Φ, where

$$\Phi(x) = |\boldsymbol{v}| \cdot f(\boldsymbol{r}, \boldsymbol{v}, i) \,, \tag{5.10}$$

is sometimes used in preference to $f(x)$ as dependent variable. Alternatively, in computational domains with non-vanishing collisionality (i.e., if $\Sigma_t(x) \neq 0$ everywhere) the (pre-) collision density Ψ is used, i.e.,

$$\Psi(x) = \Sigma_t(x)\Phi(x) \,, \tag{5.11}$$

where, again, the macroscopic cross section Σ_t is the total inverse local mean free path (dimension: 1/length). This cross section can be written as a sum $\Sigma_t = \sum \Sigma_k$ over macroscopic cross sections for the different types (identified by the index k) of collision processes.

With these simplifications the transport equation takes the well known form in linear transport theory (e.g., neutronics, radiation transfer, cosmic rays, etc.)

$$\begin{aligned} &\frac{\boldsymbol{v}}{|\boldsymbol{v}|} \cdot \nabla_{\boldsymbol{r}}\Phi(\boldsymbol{r}, \boldsymbol{v}) + \Sigma_t(\boldsymbol{r}, \boldsymbol{v})\Phi(\boldsymbol{r}, \boldsymbol{v}) \\ &= Q(\boldsymbol{r}, \boldsymbol{v}) + \int \mathrm{d}\boldsymbol{v}'\Phi(\boldsymbol{r}, \boldsymbol{v}')\Sigma_t(\boldsymbol{r}, \boldsymbol{v}') \cdot C(\boldsymbol{r}, \boldsymbol{v}' \to \boldsymbol{v}) \,. \end{aligned} \tag{5.12}$$

5.4 The Linear Integral Equation for the Collision Density

We will now give these algebraically very complex equations a very simple stochastic interpretation. We first note that by formally integrating the characteristics for (5.12) the same transport equation can also be written in our generic form (Fredholm IE) for Monte Carlo transport simulations (5.2), which we express for the (pre-) collision density (distribution density of particles in phase space going into a collision, per unit time)

$$\Psi(x) = S(x) + \int \mathrm{d}x'\Psi(x')K(x' \to x) \,. \tag{5.13}$$

In order to see this, define the Green's function $G(\boldsymbol{v}, i; \boldsymbol{r}' \to \boldsymbol{r})$. This is the solution to (5.12), but with the right hand side replaced by a delta-point source at $x = (\boldsymbol{r}', \boldsymbol{v}, i)$. For this let, again, $\boldsymbol{\Omega}$ denote the unit vector in the direction of particle flight, and let $\boldsymbol{\Omega}'$ and $\boldsymbol{\Omega}''$ be two further unit vectors such that these three vectors form an ortho-normal basis at the point $\boldsymbol{r}'$. The Green's function G then reads as follows

$$\begin{aligned} &G(\boldsymbol{v}, i; \boldsymbol{r}' \to \boldsymbol{r}) \\ &= \mathrm{e}^{-\int_0^{\boldsymbol{\Omega}(\boldsymbol{r}-\boldsymbol{r}')} \mathrm{d}s \Sigma_t(\boldsymbol{r}'+s\boldsymbol{\Omega})} \delta(\boldsymbol{\Omega}'(\boldsymbol{r}-\boldsymbol{r}')) \\ &\times \delta(\boldsymbol{\Omega}''(\boldsymbol{r}-\boldsymbol{r}')) H(\boldsymbol{\Omega}(\boldsymbol{r}-\boldsymbol{r}')) \end{aligned} \tag{5.14}$$

with $H(x) = 0$ if $x \leq 0$, and $H(x) = 1$ if $x > 0$, the Heaviside step function. Thus, G is closely related to the distribution density $T(l)$ for the distance l for a free flight starting from $\boldsymbol{r}'$ to the next point of collision $\boldsymbol{r} = \boldsymbol{r}' + l \cdot \boldsymbol{\Omega}$. The integral

$$\alpha(\boldsymbol{r}', \boldsymbol{r}) = \mathrm{e}^{-\int_0^{\boldsymbol{\Omega}(\boldsymbol{r}-\boldsymbol{r}')} \mathrm{d}s \Sigma_t(\boldsymbol{r}'+s\boldsymbol{\Omega})} \tag{5.15}$$

in (5.14) is well known to characterize the optical thickness of the medium in linear transport theory.

Multiplying (5.12) with that Green's function and integrating over initial variables $\boldsymbol{r}'$ turns this integro-differential equation into an integral equation for the flux Φ, which (almost) has the required generic form

$$\begin{aligned} \Phi(x) &= \int \mathrm{d}x' Q(x') G(x' \to x) \\ &+ \int \mathrm{d}x' \Phi(x') \Sigma(x') G(x' \to x) C(x' \to x) \\ &= \int \mathrm{d}x' Q(x') \frac{1}{\Sigma(x)} T(x' \to x) \\ &+ \int \mathrm{d}x' \Phi(x') \frac{\Sigma(x')}{\Sigma(x)} T(x' \to x) C(x' \to x) \,. \end{aligned} \tag{5.16}$$

Here we have introduced the transport kernel $T(x \to x') = \Sigma(x') G(x \to x')$, which will play the role of the distribution of free flight length between two collision events.

Multiplying this equation with $\Sigma(x)$ and using the definition for the pre-collision density: $\Psi = \Sigma\Phi$ yields exactly the generic equation (5.13). The source term S in this equation is now seen to be $S = \int QT \mathrm{d}x'$, i.e. it is the contribution to Ψ directly from source Q, then transported (free flight) to the first point of collision with T. It is the density of (un-collided) particles going into their first collision. The kernel $K(x \to x')$ is now identified as $K = CT$: a particle going into a collision at x' is first collided by sampling from C, then transported to the next collision at x with operator T. The once collided contribution (particles going into their second collision) is $\int QTCT \mathrm{d}x'$. The twice collided contribution of particles going into their third collision is consequently: $\int QTCTCT \mathrm{d}x'$, and so on.

This separation by generations of particles is just the intuitive particle interpretation of the Neumann series expansion solving the generic integral equation (5.13). This equation also has the general form of the backward integral equation of a Markovian jump-process [6] and it is therefore particularly well suited for a Monte Carlo method of solution. A direct intuitive interpretation of the integral equation as given above is already sufficient to understand the Monte Carlo method of solution.

In (5.13) x' and x are the states at two successive collisions (jumps). The integral $\int \mathrm{d}x'$ is to be understood as an integral over physical space and over velocity space and a summation over all discrete species indices. The transition kernel K is usually decomposed, in our context, into a collision- and a transport (free flight) kernel, i.e., into C and T, respectively, where

$$K(\boldsymbol{r}', \boldsymbol{v}', i' \rightarrow \boldsymbol{r}, \boldsymbol{v}, i) = C(\boldsymbol{r}'; \boldsymbol{v}', i' \rightarrow \boldsymbol{v}, i) T(\boldsymbol{v}, i; \boldsymbol{r}' \rightarrow \boldsymbol{r}) . \tag{5.17}$$

The kernel C is (excluding normalization) the conditional distribution for new coordinates ($\boldsymbol{v}$,i) given that a particle of species i' and with velocity $\boldsymbol{v}'$ has undergone a collision at position $\boldsymbol{r}'$. This kernel can further be decomposed into

$$C(\boldsymbol{r}', \boldsymbol{v}', i' \rightarrow \boldsymbol{v}, i) = \sum_k p_k \, C_k(\boldsymbol{r}'; \boldsymbol{v}', i' \rightarrow \boldsymbol{v}, i) \ , \qquad p_k = \frac{\Sigma_k}{\Sigma_t} \tag{5.18}$$

with summation over the index k for the different types of collision processes under consideration and p_k defined as the (conditional) probability for a collision to be of type k. The normalizing factor

$$c_k(x') = \sum_i \int \mathrm{d}^3 v \, C_k(\boldsymbol{r}', \boldsymbol{v}', i' \rightarrow \boldsymbol{v}, i) \ , \qquad \mathcal{C}_k = \frac{1}{c_k} C_k \tag{5.19}$$

gives the mean number of secondaries for this collision process. The normalized function $\mathcal{C}_k$ then is a conditional probability density. The particle absorption process can conveniently be described by adding an absorbing state x_a to the μ-space (generally referred to as one-point compactification of this space in the language of mathematical topology). This limbo state, once it is reached, is never left again if the kernels T or C are employed as transition probabilities.

The Green's function G and similarly the kernel $T(\boldsymbol{r}' \rightarrow \boldsymbol{r}) := \Sigma G(\boldsymbol{r}' \rightarrow \boldsymbol{r})$ describes the motion of the test particles between the collision events. It is the probability distribution of the mean free flight length l between events. In more compact notation

$$T(\boldsymbol{v}', l) = \Sigma_t(\boldsymbol{v}', \boldsymbol{r}) \mathrm{e}^{-\int_{\boldsymbol{r}'}^{\boldsymbol{r}} \mathrm{d}s \, \Sigma_t(\boldsymbol{v}', s)} . \tag{5.20}$$

As the problem is linear, the source Q arising in the inhomogeneous part can be normalized to one and, thus, Q can be regarded as a distribution density in phase space for the primary birth points of particles.

Also a secondary birth point distribution (or post collision density) χ of particles emerging from a collision event (or directly from the source Q) is sometimes defined and used as dependent variable, instead of Ψ, ϕ or f

$$\chi(x) = Q(x) + \int dx' \Psi(x') C(x' \to x). \tag{5.21}$$

Comparing this with the previous definitions and equations one easily sees that

$$\phi(x) = \int dx' \chi(x') G(x' \to x) \,, \tag{5.22}$$

$$\chi(x) = Q(x) + \int dx' \chi(x') T(x' \to x) C(x' \to x) \,. \tag{5.23}$$

This equation too has exactly the same form as our generic equation for Ψ. But now the inhomogeneous part is directly the physical source Q, and the order of C and T is reversed in the transport kernel. But this is also obvious: $\chi(x)$ is the emerging collision density (per unit time), hence for the next higher generation of emerging particles first the free flight (T) and then the scattering (C) must be applied.

As already mentioned, a detailed knowledge of Φ, Ψ or χ is often not required, and the output of Monte Carlo simulations are responses R, defined by

$$R = \langle \Psi | g_c \rangle = \int dx \Psi(x) g_c(x) \left(= \langle \Phi | g_t \rangle = \int dx \Phi(x) g_t(x) \right) \,, \tag{5.24}$$

where $g_c(x)$, $g_t(x)$ are given detector functions.

For example all terms in computational micro-macro models, in which microscopic transport (of some species) is coupled to macroscopic (fluid) transport of some other species, can be written in this way [9].

5.5 Monte Carlo Solution

It can be shown that a unique solution $\Psi(x)$ (or, equivalently, $\chi(x), \phi(x)$) exists subject to appropriate boundary conditions and under only mild restrictions (basically on the constants c_k and p_a) to ensure that the particle generation process stays sub-critical. And a stochastic (Monte Carlo) solution to the generic equation (5.9) is now straight forward, because it is formulated in probabilistic terms as follows. Let's, for example, take (5.13): A discrete Markov chain is defined using QT as an initial distribution. I.e. sample a birth point from the physical source Q and transport to first collision with T. This amounts to sampling from S, the inhomogeneous part.

Then employ $K = CT$ as a transition probability. Histories ω^n from this stochastic process are generated: $\omega^n = (x_0, x_1, x_2, \dots, x_n)$, where $x_j = x_a$ for all $j \geq n$ and $x_i \neq x_a$ for all $i < n$ with x_n being the first state after transition into the absorbing state x_a. x_0 denotes the initial state distributed as described by Q. Note that the length n of the chain ω^n is a random variable itself. A random sampling procedure to generate such chains is carried out in Monte Carlo codes by converting machine generated (pseudo-) random numbers $\xi_{i_1}, \xi_{i_2}, \dots$ into random numbers with the distributions Q, T and C. Having computed N (typically several thousand to several million) chains ω_i, $i = 1, 2, \dots, N$, the responses R are estimated as the arithmetic mean of functions (statistics, or estimators) $X(\omega)$, i.e.,

$$R \approx \widetilde{R} = \frac{1}{N} \sum_{i=1}^{N} X(\omega_i) \,. \tag{5.25}$$

One possible choice for $X(\omega)$ is the so called collision estimator X_c,

$$X_c(\omega_i^n) = \sum_{l=1}^{n} g_c(x_l) \prod_{j=1}^{l-1} \frac{c(x_j)}{(1 - p_a(x_j))} \,. \tag{5.26}$$

This estimator evaluates the response function g at the points of collisions along the random walks, starting at the first collision. The factors in the product account for particle absorption and multiplication.

For example: If $g = 1$ and $c = p_{sc}$, i.e., no particle multiplication, then this estimator simply counts collisions. It is then also intuitively clear that the response R_g is just the collision density averaged over phase space (or a sub-domain of phase space, if $g = 0$ outside that sub-domain).

But it can be shown rigourously that the statistical expectation E(X_c) produces

$$R = E(X_c) = \int \mathrm{d}(\omega) X_c(\omega) h(\omega) \tag{5.27}$$

with $h(\omega)$ being the probability density for finding a chain ω from the Markov process defined above. This means: X_c is, indeed, an unbiased (correct) estimator for response R.

5.5.1 Outline of a Proof

We now sketch the idea of the proof. We will refer to the construction of a Markov chain by directly employing the terms Q, T and C in the integral equation (5.13) as analog and the resulting procedure as analog Monte Carlo. Note that this means that possible physical particle splitting events (fission processes, cascading of ray showers, dissociation of molecules) have been eliminated already, and this is corrected for by the weight factors p_a and c which result from normalization of the scattering kernel C. Hence the analog Markov process is not a branching process anymore, even if the underlying physical process was a branching process.

In order to cover variance reducing methods already in this proof, we also consider another, non-analog, equation, of exactly the same type

$$\widetilde{\psi}(x) = \widetilde{S}(x) + \int \mathrm{d}x' \widetilde{K}(x \to x') \widetilde{\psi}(x') \tag{5.28}$$

and we use (5.28) to construct a random walk process, rather than (5.13).

If (5.28) = (5.13) we speak of an analog Monte Carlo game, otherwise of non-analog Monte Carlo: Variance reduction is then possible by making clever choices for the non-analog process, as already discussed under the topic importance sampling in the introductory chapter before. For the initial distribution of the Markov chain we set

$$f_1(x) = \widetilde{S}(x) . \tag{5.29}$$

The transition probability is defined by

$$f_{2/1}(x_1 \to x_2) = \widetilde{p}_{\mathrm{a}}(x_1)\widetilde{q}(x_2) + \widetilde{p}_{\mathrm{sc}}(x_1)\frac{\widetilde{K}(x_1 \to x_2)}{\widetilde{c}(x_1)} , \tag{5.30}$$

$\widetilde{p}_{\mathrm{a}}$ is, again, the absorption probability, $\widetilde{p}_{\mathrm{sc}}$ is the scattering probability ($= 1 - \widetilde{p}_{\mathrm{a}}$), and $\widetilde{q}(x)$ is (an entirely irrelevant) distribution, formally needed after transition into the limbo state *absorbed particle.*

The probability for finding a particular chain $(x_1, ..., x_k)$, ending with absorption in x_k, is given by the product

$$h(x_1, \ldots, x_k) = f_1(x_1) \prod_{j=1}^{k-1} f_{2/1}(x_j \to x_{j+1}) . \tag{5.31}$$

We now define the estimator for the non-analog Monte Carlo process (with the analog estimator X, (5.26) as special case)

$$\widetilde{X}(w) = X(w) \underbrace{\frac{S(x_1)}{\widetilde{S}(x_1)} \prod_{j=1}^{k=1} \frac{K(x_j \to x_{j+1})\, p_{sc}(x_j)\, \widetilde{c}(x_j)\, p_a(x_k)}{\widetilde{K}(x_j \to x_{j+1})\, \widetilde{p}_{sc}(x_j)\, c(x_j)\, \widetilde{p}_a(x_k)}}_{(5.13) \neq (5.28)} . \tag{5.32}$$

The Monte Carlo method for solving a Fredholm IE by this random walk and with this estimating method is exact, because:

Theorem 1. *If K is subcritical, i.e., the absorption p_a is strong enough compared to particle multiplication c, and if some measure-theoretical conditions are fulfilled as well, namely $\widetilde{p} = 0 \Rightarrow p = 0$ (Radon Nikodym) for any non-analog probability $\widetilde{p}$ and corresponding analog probability p in the Markov chain, then*

$$\begin{aligned} E(\widetilde{X}(w)) &= I_g(\psi) \\ &= \int \mathrm{d}x\ S(x)g(x) \\ &\quad + \iint \mathrm{d}x'\ \mathrm{d}x\ S(x')K(x' \to x)g(x) \\ &\quad + \iiint \mathrm{d}x''\ \mathrm{d}x'\ \mathrm{d}x\ S(x'')K(x'' \to x')K(x' \to x)g(x) \\ &\quad + \ldots \end{aligned} \tag{5.33}$$

(v. Neumann Series)

Outline of a Proof: One calculates the expectation value of the estimator $\widetilde{X}$ by multiplying the probability $h(\omega)$ to find a particular random walk ω with the value of the estimator for that history: $\widetilde{X}(\omega)$, and then integrates over all possible random walks, those with length one, plus those with length two, etc., summing over all possible lengths k of random walks

$$\begin{aligned}
&E(\widetilde{X}(w)) \\
&= \sum_{k=1}^{\infty} \iint \ldots \int \mathrm{d}x_1 \ldots \mathrm{d}x_k \widetilde{X}(x_1, \ldots, x_k) h(x_1, \ldots, x_k) \\
&= \ldots \quad \text{(after same lengthy algebra)} \quad \ldots \\
&= \sum_{i=1}^{\infty} \iint \ldots \int S(x_1) \prod_{j=1}^{i-1} K(x_j \to x_{j+1}) g(x_i) N_{i,k}(x_i, \ldots, x_{i+k}) \qquad (5.34)
\end{aligned}$$

with

$$\begin{aligned}
N_{i,k} &= 1 - \lim_{k\to\infty} \begin{pmatrix} \text{Probability that a chain, which starts at } x_i \\ \text{will not end at one of the next } k \text{ events.} \end{pmatrix} \\
&= 1 \quad \text{(because K is subcritical)}\,, \qquad (5.35)
\end{aligned}$$

hence: $E(\widetilde{X}(w)) = I_g(\psi) = \langle g|\psi\rangle$, by convergence of the v. Neumann series of the FIE.

5.5.2 Other Estimators

Other estimators (track-length type estimators) are employed frequently. These estimators are unbiased as well but have higher moments (e.g. variance) different from those of X_c. Instead of evaluating the detector function g_c(x) at the points of collisions x_l as X_c does, they involve line integrals of g_t(x) along the trajectories, e.g.,

$$X_t(\omega_i^n) = \sum_{l=0}^{n-1} \left\{ \int_{x_l}^{x_{l+1}} \mathrm{d}s\; g_t(s) \right\} \prod_{j=1}^{l-1} \frac{c(x_j)}{(1 - p_a(x_j))}\,, \qquad (5.36)$$

again with $R = E(X_t) = E(X_c)$. See (5.24) for the definition of response functions g_c and g_t.

It can be seen (see also [3]), that the collision estimator, written not for the pre-collision density Ψ but for the post-collision density χ (integral equation (5.23)) results in a track-length type conditional expectation estimator X_e: This conditional expectation estimator reads

$$X_e(\omega_i^n) = \sum_{l=0}^{n-1} \left\{ \int_{x_l}^{x_{\text{end}}} \mathrm{d}s\; g_t(s)\, \mathrm{e}^{-\int_0^s \mathrm{d}s' \Sigma_t(s')} \right\} \prod_{j=1}^{l-1} \frac{c(x_j)}{(1 - p_a(x_j))}\,. \qquad (5.37)$$

Here x_{end} is the nearest point on a boundary along the test flight originating in x_l.

The proof is identical to the one given above for the collision estimator, but using (5.23) for χ as starting point instead (which has the identical mathematical form), and the definition of the flux ϕ expressed in terms of χ and the Green's function G in (5.22).

With this proof for the estimator X_e, as special case of a collision estimator after an averaging transformation of the FIE, also the track-length estimator X_t is proofed to be unbiased for the same response. Because the exponential in X_e is just the sampling distribution for the flight length between collisions, X_t results from X_e by randomization: Rather than evaluating the integral over $g_t \exp(\dots)$ in X_e, one samples the next collision point from this exponential distribution and evaluates only g_t until this point. This is exactly what the track-length estimator X_t does.

This estimator X_e is related to X_t by extending the line integration, which is restricted to the path from x_l to x_{l+1} in formula (5.36), to the line segment from x_l to x_{end}. I.e., the line integration (scoring) may be extended into a region beyond the next point of collision, into which the generated history would not necessarily reach. X_e is especially useful for deep penetration problems. Furthermore, for a point source Q and a purely absorbing host medium its variance is exactly zero: This Monte Carlo scheme then has turned into a purely analytic or numerical concept. See also the similar discussions on zero variance estimators in the introductory chapter before.

5.6 Some Special Sampling Techniques

5.6.1 Sampling from Collision Kernel C

Methods for random number generation from the collision kernel C (i.e., sampling the post collision velocity after a collision) are largely case dependent. Usually first a discrete random number is used to determine the type of collision process k, next one finds post collision parameters and weight from kernel C_k, see (5.19). In case of scattering, one frequently encountered sampling distribution is given by the following consideration: Take a classical Monte Carlo test-particle, velocity $\boldsymbol{v}_0$, traveling in a host medium of other particles, which have a known velocity distribution f_b, often: $f_b = f_{\text{Maxw}}(\boldsymbol{v}_b)$, a Maxwellian, with a given temperature T_b. Given that a collision point has been found (after sampling from the transport kernel T), the task is to find (sample) the velocity $\boldsymbol{v}_c$ of the collision partner. Once both $\boldsymbol{v}_0$ and $\boldsymbol{v}_c$ are known (and the masses of the particles involved), the new velocities can be calculated from the collision kinetics (e.g., classical orbits, or using differential cross sections, etc.).

The distribution of velocities of the collision partners going into a collision at this point in phase space is given as

$$f_c(\boldsymbol{v}_c) = \frac{\sigma(v_{\text{rel}}) v_{\text{rel}} f_b(\boldsymbol{v}_c)}{\langle \sigma v \rangle (\boldsymbol{v}_0, T_b)} \, . \tag{5.38}$$

Here $v_{rel} = |\boldsymbol{v}_0 - \boldsymbol{v}_c|)$ and the f_b-averaged rate coefficient $c = \langle \sigma(v_{\text{rel}}) v_{\text{rel}} \rangle_{f_b}$ is the normalization constant.

Sampling this velocity $\boldsymbol{v_c}$ now proceeds as (see introductory chapter before, Sect. 3.2.2.2):

- Choose g as f_b, and set $M = \max\{\sigma(v)v\}/c$.
- Sample $\boldsymbol{z_1} = \boldsymbol{v_c}$ from f_b, and form the ratio: $f_c(\boldsymbol{v_c})/g(\boldsymbol{v_c}) = \sigma(v_{\text{rel}})v_{\text{rel}}/c$.
- Compare Mz_2 with this ratio, or, what is the same: Mcz_2 with $\sigma \cdot |\boldsymbol{v_0} - \boldsymbol{v_c}|$, i.e. the normalization c cancels out. Accept $\boldsymbol{v_c}$ if $Mcz_2 \leq \sigma|\boldsymbol{v_0} - \boldsymbol{v_c}|$, otherwise repeat this procedure.

Note that although this sampling from the collision kernel C by rejection is possible without knowing the normalization (i.e., the rate coefficient), this same rate coefficient still enters in the transport kernel T, and there it is needed indeed to find the free flight distance of the particles.

5.6.2 Sampling from Transport Kernel T: Null Collisions

We now discuss one special sampling method for the transport kernel T, which is known under various different names in Monte Carlo literature: Null collisions (in PIC simulations), pseudo collisions (in fusion plasma applications) or delta-events (in neutron transport).

Lets take l as coordinate along the flight starting from $\boldsymbol{r}$, remove all irrelevant parameters, and assume that the mean free path $\lambda = 1/\Sigma_t$ is independent of the spatial co-ordinate $\boldsymbol{r'}$, along the trajectory under consideration. Then the transport kernel T, see (5.20), is simply given by the exponential distribution

$$T(\boldsymbol{v}, l) = \Sigma_t e^{-\Sigma_t \cdot l} \tag{5.39}$$

and the flight distance l can directly be sampled by the method of inversion of the chapter before, Sect. 3.2.2.1.

If, however, the parameters of the host medium are varying along the flight path (either continuously or, in a grid, from cell to cell) then it may sometimes be computationally advantageous to modify the collision rate, such that the mean free path remains constant along a flight path. I.e., one replaces $\Sigma_t(\boldsymbol{r}, \boldsymbol{v})$ by $\Sigma_t^*(\boldsymbol{v})$ with

$$\Sigma_t^* = \Sigma_t(\boldsymbol{v}, \boldsymbol{r}) + \Sigma_\delta(\boldsymbol{v}, \boldsymbol{r}) \tag{5.40}$$

with $\Sigma_t^* = \text{const}$.

According to the discussions above for non-analog methods then statistical weights T/T^* would appear in the estimators each time a particle is pushed a distance l^* sampled from the transport kernel T^*, in which Σ_t has been replaced by Σ_t^*. The following trick allows to avoid these weight corrections: Tactically assume that the modification of Σ_t results from an additional, artificial isotope in the background medium, we call it the δ-isotope.

Out-scattering by this isotope leads to an additional loss term $\Sigma_\delta \Phi$ on the left hand side of the transport equation (5.12), i.e. now to a total loss term $\Sigma_t^* \Phi$ there. On the left hand side, in the collision integral, we also now add the same artificial collision density

$$\Sigma_\delta(\boldsymbol{r},\boldsymbol{v})\Phi(\boldsymbol{r},\boldsymbol{v}) = \int \mathrm{d}\boldsymbol{v}'\Phi(\boldsymbol{r},\boldsymbol{v}')\Sigma_\delta(\boldsymbol{r},\boldsymbol{v}')\delta(\boldsymbol{v}'-\boldsymbol{v}) \,. \tag{5.41}$$

Clearly, by adding this on both sides of the equation the solution Φ is not altered. But C is modified to become C^*

$$C \to C^* = \frac{\Sigma_t}{\Sigma_t^*}C + \frac{\Sigma_\delta}{\Sigma_t^*}\delta(\boldsymbol{v}'-\boldsymbol{v}) \,. \tag{5.42}$$

Rather than applying weight corrections T/T^* we now need to sample from the non-analog kernel collision C^*. But this is trivial: A first random number is used to decide if the collision is real or with the δ-isotope. In the second case the scattering is actually a null event: The flight continues without any change in velocity, due to the delta distribution for post collision velocities in the δ-scattering kernel.

Note, that typically $\Sigma_\delta \geq 0$, i.e., the mean free path in the simulation is reduced. More general δ-scattering operators, also allowing for negative values of Σ_δ, i.e., increased mean free pathes, have also been derived [10]. They seem not to be in use very much. Although they are unbiased (correct), they require negative weight corrections.

5.7 An Illustrative Example

We close this chapter by considering one example: A linear transport problem with all features discussed in this chapter is, for example, given by (dilute) neutral particle transport from the surrounding vacuum chamber into the hot (fully ionized) hydrogen plasma in magnetic fusion devices (here: a tokamak). Atoms and molecules are formed by recombination and surface erosion of charged particles at these surfaces. They penetrate into the plasma, were molecules dissociate (branching process), atoms and molecules scatter (the former mainly via resonant charge exchange, the latter elastically) and they both are ionized or pumped at certain wall segments (absorption). A view into the machine is shown in Fig. 5.2, left.

The area of main plasma surface interaction (defining the location of the external source distribution S for neutrals) are the leading edges (upper and lower) of the toroidally (almost) symmetric belt limiter. 45 typical trajectories (random walks), computed in 3D space with analog sampling, are also shown in Fig. 5.2, right. The 3D trajectories have been projected into one poloidal section of the torus for this plot. Densities of atoms, see Fig. 5.3, left and molecules, see Fig. 5.3, right.

The shading has been done with respect to the logarithm of the density, because at TEXTOR (and even more so in larger tokamaks) the neutral gas density drops by many orders of magnitude from the edge to the core region. Still its density is an important quantity, for example for interpretation of charge exchange recombination spectroscopy. Various non-analog methods, together with the conditional expectation estimating technique, are usually applied to obtain statistically reliable results.

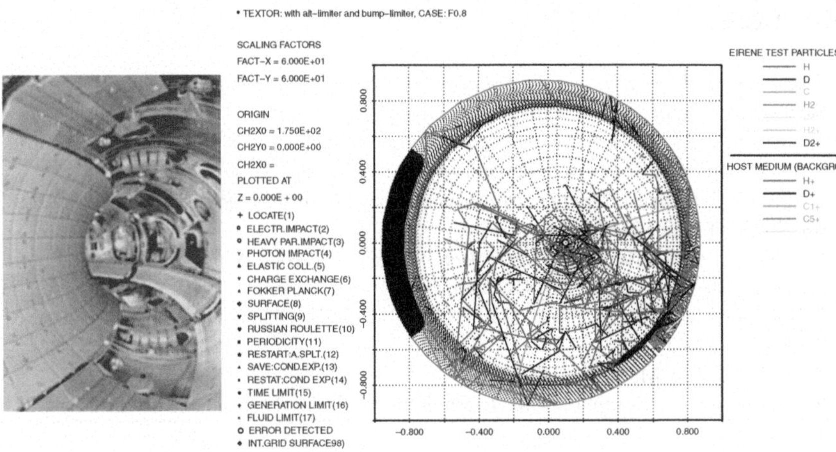

Fig. 5.2. Left: Inside view of TEXTOR Tokamak, FZ-Jülich. Major and minor radius of torus: 1.75 m and 0.5 m, respectively. **Right**: 45 typical Monte Carlo trajectories (atoms and molecules). Analog sampling, Host medium: hydrogen plasma, central electron density: $4{\cdot}10^{19}\,\mathrm{m}^{-3}$, central plasma temperature: 1.5 keV

As can be seen the molecular density is compressed underneath the toroidal limiter blade (bright area). This is also the location of the pump-ducts. The atoms penetrate deeper into the plasma, for the TEXTOR conditions shown here the density typically drops from $10^{18}\,\mathrm{m}^{-3}$ at the outer edge to $10^{13}\,\mathrm{m}^{-3}$ in the plasma center. More details on the particular application of Monte Carlo transport methods to neutral particle transport in fusion plasmas can be found at the URL: `www.eirene.de`.

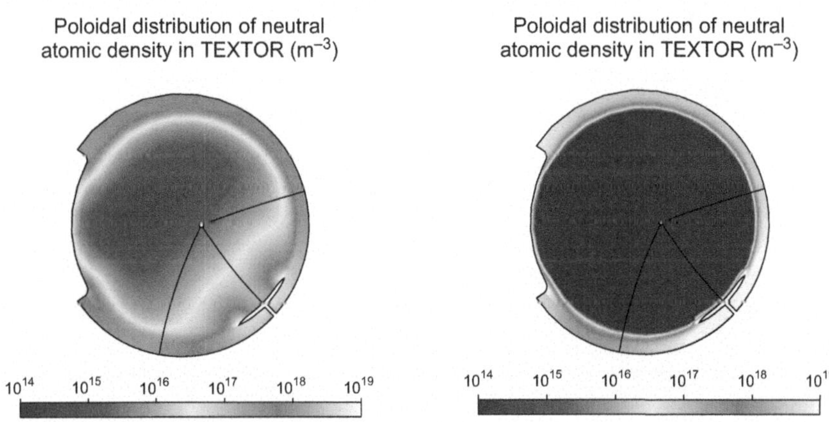

Fig. 5.3. Neutral particle density in TEXTOR, poloidal distribution. Monte Carlo solution, with track-length estimator. **Left**: atom density. **Right**: molecule density. Shading according to logarithmic scale for density, density range: $10^{14} - 10^{18}\,\mathrm{m}^{-3}$

References

1. P. Kloeden, E. Platen, *Numerical Solution of Stochastic Differential Equations*. Springer Series: Applications of Mathematics (Springer Verlag, 1995)
2. C. Gardiner, *Handbook of Stochastic Methods: for Physics, Chemistry and the Natural Sciences*. Springer Series in Synergetics (Springer Verlag, 2004)
3. J. Spanier, E. Gelbard, *Monte Carlo Principles and Neutron Transport Problems* (Addi son Wesley Publication Company, 1969)
4. H. Kalos, P. Whitlock, *Monte Carlo Methods*, Vol. I: Basics (Wiley-Interscience Publications, John Wiley and Sons, New York, 1986)
5. J. Hammersley, D. Handscomb, *Monte Carlo Methods* (Chapman and Hall, London & New York, 1964)
6. W. Feller, T. Am. Math. Soc. **48** (1940)
7. N. Metropolis, A. Rosenbluth, M. Rosenbluth, A. Teller, E. Teller, J. Chem. Phys. **21**, 1087 (1953)
8. C. Cercignani, *The Boltzmann Equation and its Applications*. Springer Series: Applied Mathematical Sciences (Springer Verlag, 1975)
9. D. Reiter, J. Nucl. Mater. **196–198** (1992)
10. L. Carter, E. Cashwell, W. Taylor, Nucl. Sci. Eng. **48** (1972)

Part III

Kinetic Modelling

6 The Particle-in-Cell Method

David Tskhakaya

Association Euratom-OEAW, Institute of Theoretical Physics, A-6020 Innsbruck, Austria
Andronikashvili Institute of Physics, 380077 Tbilisi, Georgia

Probably the first Particle-in-Cell (PIC) simulations have been made at late 1950s by Buneman [1] and Dawson [2] who simulated the motion of 100–1 000 particles including interaction between them. Our day PIC codes simulate 10^5–10^{10} particles and represent a powerful tool for kinetic plasma studies. They are used practically in all branches of plasma physics modelling laboratory, as well as astrophysical plasma. PIC codes have a number of advantages: They represent so-called lowest codes, i.e. the number of assumptions made in the physical model is reduced to the minimum, they can simulate high-dimensional cases and can tackle complicated atomic and plasma-surface interactions. The prize for these advantages is a long simulation time: Some simulations can take up to 10^4 hours of CPU. As a result, they require a high level of optimization and are usually designed for professional use.

With this chapter we aim at introducing the reader to the basics of the PIC simulation technique. It is based mainly on available literature cited below, but includes some original unpublished material, too. For the interested reader I can recommend two classical monographs, [3] and [4], and the papers [5, 6] describing new developments in this field (see also references cited in the text).

The chapter is organized as follows. The main PIC features are discussed in Sect. 6.1. In Sect. 6.2 we consider solvers of equations of motion used in PIC and discuss their accuracy and stability aspects. Initialization of particle distribution, boundary effects and particle sources are described in Sect. 6.3. In Sects. 6.4 and 6.5 we show how plasma macro-parameters are calculated and discuss solvers of Maxwell's equations. Particle collisions are considered in Sect. 6.6. Final remarks are given in Sect. 6.7.

6.1 General Remarks

The idea of the PIC simulation is trivial: The code simulates the motion of plasma particles and calculates all macro-quantities (like density, current density and so on) from the position and velocity of these particles. The macro-force acting on the particles is calculated from the field equations. The name "Particle-in-Cell" comes from the way of assigning macro-quantities to the simulation particles. In general, any numerical simulation model, which simultaneously solves equations of motion of N particles

D. Tskhakaya: *The Particle-in-Cell Method*, Lect. Notes Phys. **739**, 161–189 (2008)
DOI 10.1007/978-3-540-74686-7_6

$$\frac{\mathrm{d}\boldsymbol{X}_i}{\mathrm{d}t} = \boldsymbol{V}_i \qquad \text{and} \qquad \frac{\mathrm{d}\boldsymbol{V}_i}{\mathrm{d}t} = \boldsymbol{F}_i\left(t, \boldsymbol{X}_i, \boldsymbol{V}_i, A\right) \tag{6.1}$$

for $i = 1, \ldots, N$ and of macro fields $A = L_1(B)$, with the prescribed rule of calculation of macro quantities $B = L_2\left(\boldsymbol{X}_1, \boldsymbol{V}_1, \ldots, \boldsymbol{X}_N, \boldsymbol{V}_N\right)$ from the particle position and velocity can be called a PIC simulation. Here $\boldsymbol{X}_i$ and $\boldsymbol{V}_i$ are the generalized (multi-dimensional) coordinate and velocity of the particle i. A and B are macro fields acting on particles and some macro-quantities associated with particles, respectively. L_1 and L_2 are some operators and $\boldsymbol{F}_i$ is the force acting on a particle i. As one can see, PIC simulations have much broader applications then just plasma physics. On the other hand, inside the plasma community PIC codes are usually associated with codes solving the equation of motion of particles with the Newton-Lorentz's force (for simplicity we consider an unrelativistic case)

$$\frac{\mathrm{d}\boldsymbol{X}_i}{\mathrm{d}t} = \boldsymbol{V}_i \qquad \text{and} \qquad \frac{\mathrm{d}\boldsymbol{V}_i}{\mathrm{d}t} = \frac{e_i}{m_i}\left(\boldsymbol{E}\left(\boldsymbol{X}_i\right) + \boldsymbol{V}_i \times \boldsymbol{B}\left(\boldsymbol{X}_i\right)\right) \tag{6.2}$$

for $i = 1, \ldots, N$ and the Maxwell's equations

$$\begin{aligned} &\nabla \boldsymbol{D} = \rho\left(\boldsymbol{r}, t\right), && \frac{\partial \boldsymbol{B}}{\partial t} = -\boldsymbol{\nabla} \times \boldsymbol{E}, && \boldsymbol{D} = \varepsilon \boldsymbol{E}, \\ &\nabla \boldsymbol{B} = 0, && \frac{\partial \boldsymbol{D}}{\partial t} = \boldsymbol{\nabla} \times \boldsymbol{H} - \boldsymbol{J}\left(\boldsymbol{r}, t\right), && \boldsymbol{B} = \mu \boldsymbol{H}, \end{aligned} \tag{6.3}$$

together with the prescribed rule of calculation of ρ and $\boldsymbol{J}$

$$\rho = \rho\left(\boldsymbol{X}_1, \boldsymbol{V}_1, \ldots, \boldsymbol{X}_N, \boldsymbol{V}_N\right), \tag{6.4}$$

$$\boldsymbol{J} = \boldsymbol{J}\left(\boldsymbol{X}_1, \boldsymbol{V}_1, \ldots, \boldsymbol{X}_N, \boldsymbol{V}_N\right). \tag{6.5}$$

Here ρ and $\boldsymbol{J}$ are the charge and current densities and ε and μ the permittivity and permeability of the medium, respectively. Below we will follow this definition of the PIC codes.

PIC codes usually are classified depending on dimensionality of the code and on the set of Maxwell's equations used. The codes solving a whole set of Maxwell's equations are called electromagnetic codes, contrary electrostatic ones solve just the Poisson equation. E.g., the XPDP1 code represents a 1D3V electrostatic code, which means that it is 1D in usual space and 3D in velocity space, and solves only the electrostatic field from the Poisson equation [7]. Some advanced codes are able to switch between different dimensionality and coordinate system, and use electrostatic, or electro-magnetic models (e.g. the XOOPIC code [8]).

A simplified scheme of the PIC simulation is given in Fig. 6.1. Below we consider each part of it separately.

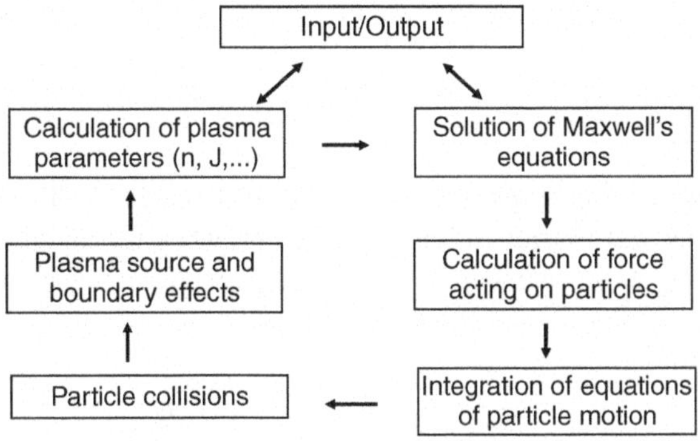

Fig. 6.1. Scheme of the PIC simulation

6.2 Integration of Equations of Particle Motion

6.2.1 Description of Particle Movers

During PIC simulation the trajectory of all particles is followed, which requires solution of the equations of motion for each of them. This part of the code is frequently called "particle mover".

A few words about the simulation particles itself. The number of particles in real plasma is extremely large and exceeds by orders of magnitude a maximum possible number of particles, which can be handled by the best supercomputers. Hence, during a PIC simulation it is usually assumed that one simulation particle consists of many physical particles. Because the ratio charge/mass is invariant to this transformation, this superparticle follows the same trajectory as the corresponding plasma particle. One has to note that for 1D and 2D models this transformation can be easily avoided by choosing of sufficiently small simulated volume, so that the number of real plasma particles can be chosen arbitrary.

As we will see below, the number of simulated particles is defined by a set of physical and numerical restrictions, and usually it is extremely large ($> 10^5$). As a result, the main requirements to the particle mover are the high accuracy and speed. One of such solvers represents the so called leap-frog method (see [3] and [4]), which we will consider in detail.

As in other numerical codes the time in PIC is divided into discrete time moments, in other words the time is grided. This means that physical quantities are calculated only at given time moments. Usually, the time step, Δt, between the nearest time moments is constant, so that the simulated time moments can be given via following expression: $t \rightarrow t_k = t_0 + k\Delta t$ and $A(t) \rightarrow A_k = A(t = t_k)$ with $k = 0, 1, 2, \ldots$, where t is the time, t_0 the initial moment and A denotes any physical quantity. The leap-frog method calculates particle velocity not at usual time steps t_k, but between them $t_{k+1/2} = t_0 + (k + 1/2)\,\Delta t$. In this way equations become time centred, so that they are sufficiently accurate and require relatively short calculation time

$$\frac{\boldsymbol{X}_{k+1} - \boldsymbol{X}_k}{\Delta t} = \boldsymbol{V}_{k+1/2} \,,$$
$$\frac{\boldsymbol{V}_{k+1/2} - \boldsymbol{V}_{k-1/2}}{\Delta t} = \frac{e}{m}\left(\boldsymbol{E}_k + \frac{\boldsymbol{V}_{k+1/2} + \boldsymbol{V}_{k-1/2}}{2} \times \boldsymbol{B}_k\right) . \tag{6.6}$$

The leap-frog scheme is an explicit solver, i.e. it depends on old forces from the previous time step k. Contrary to implicit schemes, when for calculation of particle velocity a new filed (at time step $k+1$) is used, explicit solvers are simpler and faster, but their stability requires a smaller time step Δt.

By substituting

$$\boldsymbol{V}_{k\pm 1/2} = \boldsymbol{V}_k \pm \frac{\Delta t}{2}\boldsymbol{V}'_k + \frac{\Delta t^2}{8}\boldsymbol{V}''_k \pm \frac{1}{6}\left(\frac{\Delta t}{2}\right)^3 \boldsymbol{V}'''_k + \ldots \,,$$
$$\boldsymbol{X}_{k+1} = \boldsymbol{X}_k + \Delta t \boldsymbol{V}_k + \frac{\Delta t^2}{2}\boldsymbol{V}'_k + \frac{\Delta t^3}{6}\boldsymbol{V}''_k + ... \tag{6.7}$$

into (6.6) we obtain the order of the error $\sim \Delta t^2$. It satisfies a general requirement for the scaling of numerical accuracy $\Delta t^{a>1}$. In order to understand this requirement we recall that for a fixed simulated time the number of simulated time steps scales as $N_t \sim \Delta t^{-1}$. Then, after N_t time steps an accumulated total error will scale as $N_t \Delta t^a \sim \Delta t^{a-1}$, where Δt^a is the scale of the error during one step. Thus, only $a > 1$ can guarantee, that the accuracy increases with decreasing Δt.

There exist different methods of solution of finite-difference equations (see (6.6)). Below we consider the Boris method (see [3]), which is frequently used in PIC codes

$$\boldsymbol{X}_{k+1} = \boldsymbol{X}_k + \Delta t \boldsymbol{V}_{k+1/2} \qquad \text{and} \qquad \boldsymbol{V}_{k+1/2} = \boldsymbol{u}_+ + q\boldsymbol{E}_k \tag{6.8}$$

with $\boldsymbol{u}_+ = \boldsymbol{u}_- + (\boldsymbol{u}_- + (\boldsymbol{u}_- \times \boldsymbol{h})) \times \boldsymbol{s}$, $\boldsymbol{u}_- = \boldsymbol{V}_{k-1/2} + q\boldsymbol{E}_k$, $\boldsymbol{h} = q\boldsymbol{B}_k$, $\boldsymbol{s} = 2\boldsymbol{h}/(1+h^2)$ and $q = \Delta t/(2(e/m))$. Although these equations look very simple, their solution represent the most time consuming part of PIC, because it is done for each particle separately. As a result, the optimization of the particle mover can significantly reduce the simulation time.

In general, the Boris method requires 39 operations (18 adds and 21 multiplies), assuming that $\boldsymbol{B}$ is constant and $\boldsymbol{h}$, $\boldsymbol{s}$ and q are calculated only once at the beginning of simulation. But if $\boldsymbol{B}$ has one or two components, then the number of operations can be significantly reduced. E.g., if $\boldsymbol{B} \parallel \boldsymbol{z}$ and $\boldsymbol{E} \parallel \boldsymbol{x}$ then (6.8) can be reduced to the following ones

$$\boldsymbol{X}_{k+1} = \boldsymbol{X}_k + \Delta t \boldsymbol{V}_{k+1/2} \,,$$
$$V^x_{k+1/2} = u^x_- + \left(V^y_{k+1/2} + V^y_{k-1/2}\right) h + qE^x_k \,,$$
$$V^y_{k+1/2} = V^y_{k-1/2}\left(1 - sh\right) - u^x_- s \tag{6.9}$$

with $u^x_- = V^x_{k-1/2} + qE^x_k$. They require just 17 operations (8 multiplies and 9 adds), which can save up to 50% of the CPU time. Some advanced PIC codes include a subroutine for searching the fastest solver for a given simulation setup, which significantly decreases the CPU time.

6.2.2 Accuracy and Stability of the Particle Mover

In order to find correct simulation parameters one has to know the absolute accuracy and corresponding stability conditions for the particle mover. They are different for different movers and the example considered below is applied just to the Boris scheme.

First of all let us consider the accuracy of a Larmor rotation. By assuming $\boldsymbol{V}_{k-1/2} \perp \boldsymbol{B}$ we can define the rotation angle during the time Δt from

$$\cos(\omega \Delta t) = \frac{\boldsymbol{V}_{k+1/2}\boldsymbol{V}_{k-1/2}}{V_{k-1/2}^2} . \tag{6.10}$$

On the other hand, substituting (6.8) into (6.10) we obtain

$$\frac{\boldsymbol{V}_{k+1/2}\boldsymbol{V}_{k-1/2}}{V_{k-1/2}^2} = \frac{1-\frac{(\Delta t \Omega)^2}{4}}{1+\frac{(\Delta t \Omega)^2}{4}} \tag{6.11}$$

with $\Omega = eB/m$, so that for a small Δt we get $\omega = \Omega(1-(\Delta t\Omega)^2/12)+\dots$. E.g., for a 1% accuracy the following condition has to be satisfied: $\Delta t\Omega \leq 0.35$.

In order to formulate the general stability condition some complicated calculations are required (see [4]). Below we present simple estimates of the stability criteria for the (explicit) particle mover.

Let us consider the equation of a linear harmonic oscillator

$$\frac{\mathrm{d}^2 X}{\mathrm{d}t^2} = -\omega_0^2 X , \tag{6.12}$$

having the following analytic solution

$$X = A\mathrm{e}^{-\mathrm{i}\omega_0 t} , \tag{6.13}$$

where A is an arbitrary imaginary number. The corresponding leap-frog equations take the following form

$$\frac{X_{k+1} - 2X_k + X_{k-1}}{\Delta t^2} = -\omega_0^2 X_k . \tag{6.14}$$

We assume that the solution has a form similar to (6.13), $X_k = A\exp(-\mathrm{i}\omega t_k)$. After substitution into (6.14) and performing simple transformations we find

$$\sin\left(\frac{\omega\Delta t}{2}\right) = \pm\frac{\omega_0 \Delta t}{2} . \tag{6.15}$$

Hence, for a stable solution $\mathrm{Im}(\omega) \leq 0$ the condition

$$\omega_0 \Delta t < 2 \tag{6.16}$$

is required. PIC often use a much more restrictive condition

$$\omega_0 \Delta t \leq 0.2 \,, \tag{6.17}$$

giving sufficiently accurate results. Interesting to note that this number has been derived few decades ago when the number of simulation time steps was typically of the order of $N_t \sim 10^4$. From (6.15) we obtain $\omega = \omega_0(1 - (\omega_0 \Delta t)^2/24) + \ldots$. Hence, a cumulative phase error after N_t steps should be $\Delta(\omega \Delta t) \approx (N_t(\omega_0 \Delta t)^3)/24$. Assuming $N_t = 10^4$ and $\Delta\,(\omega \Delta t) < \pi$ we obtain the condition (6.17). Although modern simulations contain much larger number of time steps up to $N_t = 10^7$, this condition still can work surprisingly well.

The restrictions on Δt described above can require the simulation of unacceptably large number of time steps. In order to avoid these restrictions different implicit schemes have been introduced: $\boldsymbol{V}_{k+1/2} = \boldsymbol{F}(\boldsymbol{E}_{k+1}, ...)$. The difference from the explicit scheme is that for the calculation of the velocity a new field is used, which is given at the next time moment.

One of examples of an implicit particle mover represents the so called 1 scheme (see [9])

$$\begin{aligned} \frac{\boldsymbol{X}_{k+1} - \boldsymbol{X}_k}{\Delta t} &= \boldsymbol{V}_{k+1/2} \,, \\ \frac{\boldsymbol{V}_{k+1/2} - \boldsymbol{V}_{k-1/2}}{\Delta t} &= \frac{e}{m} \left(\frac{\boldsymbol{E}_{k+1}(x_{k+1}) + \boldsymbol{E}_{k-1}}{2} \right. \\ &\quad \left. + \frac{\boldsymbol{V}_{k+1/2} + \boldsymbol{V}_{k-1/2}}{2} \times \boldsymbol{B}_k \right) . \end{aligned} \tag{6.18}$$

It can be shown that for a harmonic oscillator (see (6.12))

$$V,\, X \sim \frac{1}{(\omega_0 \Delta t)^{2/3}} \,, \tag{6.19}$$

if $\omega_0 \Delta t \gg 1$. Hence, the corresponding oscillations are heavily damped and the solver (see (6.18)) can filter unwanted oscillations. As a result, the condition (6.16) can be neglected.

6.3 Plasma Source and Boundary Effects

6.3.1 Boundary Effects

From the physics point of view, the boundary conditions for the simulated particles are relatively easy to formulate: Particles can be absorbed at boundaries, or injected from there with any distribution. On the other hand, an accurate numerical implementation of particle boundary conditions can be tricky. The problem is that (i) the velocity and position of particles are shifted in time ($\Delta t/2$), and (ii) the velocity of

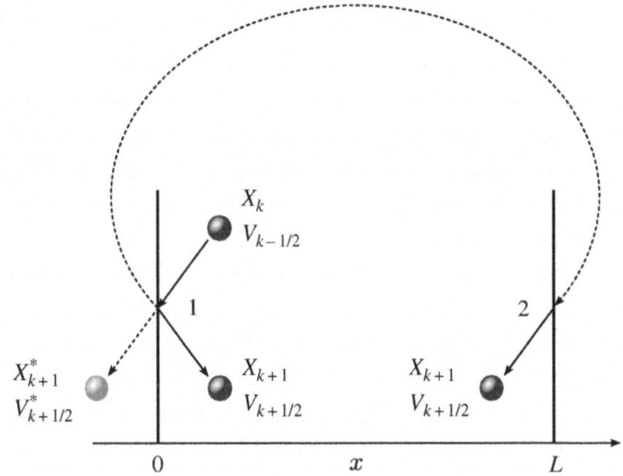

Fig. 6.2. Particle reflection (1) and reinjection (2) at the boundaries. X^*_{k+1} and $V^*_{k+1/2}$ denote the virtual position and velocity of a particle if there would be no boundary

particles are known at discrete time steps, while a particle can cross the boundary at any moment between these steps.

In unbounded plasma simulation particles are usually reflected at the boundaries, or reinjected from the opposite side (see Fig. 6.2). A frequently used reflection model, so called specular reflection, is given as

$$X^{\text{refl}}_{k+1} = -X_{k+1} \qquad \text{and} \qquad V^{x,\text{refl}}_{k+1/2} = -V^x_{k+1/2} \,. \tag{6.20}$$

Here, the boundary is assumed to be located at $x = 0$ (see Fig. 6.2). The specular reflection represents the simplest reflection model, but due to relatively low accuracy it can cause artificial effects. Let us estimate the accuracy of reflection (see (6.20)). The exact time when particle reaches a boundary and the corresponding velocity can be written as

$$\begin{aligned} t_0 &= t_k + \left| \frac{X_k}{V^x_{k-1/2}} \right| , \\ V_0 &= V^x_{k-1/2} + \left| \frac{X_k}{V^x_{k-1/2}} \right| \frac{e}{m} E^x_k \,. \end{aligned} \tag{6.21}$$

Accordingly, the velocity after the reflection will be

$$\begin{aligned} V^{x,\text{refl}}_{k+1/2} &= -V_0 + \left(\Delta t - \left| \frac{X_k}{V^x_{k-1/2}} \right| \right) \frac{e}{m} E^x_k \\ &= -V^x_{k-1/2} + \left(\Delta t - 2 \left| \frac{X_k}{V^x_{k-1/2}} \right| \right) \frac{e}{m} E^x_k \,. \end{aligned} \tag{6.22}$$

The second term on the right hand side of (6.22) represents the error made during the specular reflection, which can cause an artificial particle acceleration and heating.

Particle reinjection is applied usually when the fields satisfy periodic boundary conditions. The reinjection is given by $X_{k+1}^{\text{reinj}} = L - X_{k+1}$ and $V_{k+1/2}^{x,\text{reinj}} = V_{k+1/2}^{x}$, where $x = L$ denotes the opposite boundary. If the fields are not periodic, then this expression has to be modified. Otherwise a significant numerical error can arise.

The PIC codes simulating bounded plasmas are usually modeling particle absorption and injection at the wall, and some of them are able to tackle complicated plasma-surface interactions too.

Numerically, particle absorption is the most trivial operation and done by removing of the particle from memory. Contrary to this, for particle injection complicated numerical models can be required. When a new particle is injected it has to be taken into account that the initial coordinate and velocity are known at the same time, while the leap-frog scheme uses a time shifted values of them. In most cases the number of particles injected per time step is much smaller than the number of particles near the boundary, hence, the PIC code use simple injection models. For example, an old version of the XPDP1 code (see [7]) has used $\boldsymbol{V}_{k+1/2} = \boldsymbol{V} + e\Delta t\,(R - 0.5)\,\boldsymbol{E}_k/m$ and $X_{k+1} = R\Delta t V_{k+1/2}^{x}$, which assumes that particle has been injected at time $t_0 = t_{k+1} - R\Delta t$ with R being an uniformly distributed number between 0 and 1. $\boldsymbol{V}$ is the velocity obtained from a given injection distribution function (usually the Maxwellian one). The BIT1 code [10] uses a more simpler injection routine

$$\boldsymbol{V}_{k+1/2} = \boldsymbol{V} \qquad \text{and} \qquad X_{k+1} = R\Delta t V_{k+1/2}^{x}\,, \tag{6.23}$$

which is independent of the field at the boundary and hence, insensitive to a possible field error there. Description of higher order schemes can be found in [11].

Strictly speaking, the plasma-surface interaction processes can not be attributed to a classical PIC method, but probably all advanced PIC codes simulating bounded plasma contain elements of Monte-Carlo techniques [12]. A general scheme of plasma-surface interactions implemented in PIC codes is given below.

When a primary particle is absorbed at the wall, it can cause the emission of a secondary particle (a special case is reflection of the same particle). In general the emission probability F depends on the surface properties and primary particle energy ϵ and incidence angle α. Accordingly, the PIC code calculates $F(\epsilon, \alpha)$ and compares it to a random number R, uniformly distributed between 0 and 1. If $F > R$ then a secondary particle is injected. The velocity of a secondary particle is calculated according to a prescribed distribution $f_{\text{sev}}(\boldsymbol{V})$. Some codes allow multiple secondary particle injection, including as a special case the thermal emission. The functions F and f_{sev} are obtained from different sources on surface and solid physics.

6.3.2 Particle Loading

The particles in a PIC simulation appear, either by initial loading, or via particle injection from the boundary and at a volumetric source. In any case the corresponding velocities have to be calculated from a given distribution function $f(\boldsymbol{V})$. Important to note, that there is a significant difference between volumetric particle loading and particle injection from the wall. In the first case the particle velocity is calculated directly from $f(\boldsymbol{V})$. Contrary to this, the velocity of particles injected from the wall has to be calculated according to $V^x f(\boldsymbol{V})$, where V^x is the component of the velocity normal to the wall. This becomes clear if we recall that the injection distribution function is the probability that particles having a distribution $f(\boldsymbol{V})$ will cross the boundary with a given velocity $\boldsymbol{V}$. For simplicity we do not distinguish below these two functions denoting them $f(\boldsymbol{V})$.

There exist two possibilities of calculation of velocities according to a given distribution:

(i) The most effective way for a 1D case is to use a cumulative distribution function

$$F(V) = \frac{\int_{V_{\min}}^{V} f(V')\, \mathrm{d}\boldsymbol{V}'}{\int_{V_{\min}}^{V_{\max}} f(V')\, \mathrm{d}\boldsymbol{V}'} \tag{6.24}$$

with $F(V_{\min}) = 0$ and $F(V_{\max}) = 1$, representing a probability that the velocity of particle lays between $V_{\min}$ and V. By equating this function to a sequence of uniformly distributed numbers U (or to random numbers R) between 0 and 1 and inverting it, we produce a sequence of V with the distribution $f(V)$ [3]:

$$F^{-1}(U) = V\,. \tag{6.25}$$

The same method can be applied to multi-dimensional cases which can be effectively reduced to 1D, e.g., by variable separation: $f(\boldsymbol{V}) = f_1(V^x)\, f_2(V^y)\, f_3(V^z)$. Often inversion of (6.25) can be done analytically, otherwise it is done numerically.

As an example we consider the injection of Maxwell-distributed particles: $f(V) \sim V \exp(-V^2/(2V_{\mathrm{T}}^2))$. According to (6.24) and (6.25) we get

$$F(V) = 1 - \mathrm{e}^{-V^2/(2V_{\mathrm{T}}^2)} \quad \text{and} \quad V = V_{\mathrm{T}}\sqrt{-2\ln(1-U)}\,. \tag{6.26}$$

(ii) Another possibility is to use two sets of random numbers R_1 and R_2 (for sim plicity we consider a 1D case) $V = V_{\min} + R_1(V_{\max} - V_{\min})$, if $f(V)/(f_{\max}) > R_2$ use V, else try once more. This method requires random number generators of high level and it is time consuming. As a result, it is usually used when the method considered above can not be applied (e.g. for complicated multi-dimensional $f(\boldsymbol{V})$).

In advanced codes these distributions are generated and saved at the beginning of a simulation, so that later no further calculations are required except getting $\boldsymbol{V}$ from the memory. The same methods are used for spatial distributions $f(\boldsymbol{X})$, too.

As it was mentioned above, required velocity distributions can be generated by set of either ordered numbers U or by random numbers R, which are uniformly distributed between 0 and 1. A proper choice of these numbers is not a trivial task and depends on the simulated system; e.g., using of random numbers can cause some noise. In addition, numerically generated random numbers in reality represent pseudo-random numbers, which can correlate and cause some unwanted effects. Contrary to this, the distributions generated by a set of ordered numbers, e.g. $U = (i + 0.5)/N$, $i = 1, \ldots, N-1$, are less noisy. On the other hand, in this case the generated distributions represent a multi-beam distribution, which sometimes can cause a beam instability [3].

6.4 Calculation of Plasma Parameters and Fields Acting on Particles

6.4.1 Particle Weighting

All numerical schemes considered up to now can be applied not only to PIC, but to any test particle simulation too. In order to simulate a real plasma one has to self-consistently obtain the force acting on particles, i.e. to calculate particle and current densities and solve Maxwell's equations. The part of the code calculating macro quantities associated with particles (n, $\boldsymbol{J}$, ...) is called "particle weighting".

For a numerical solution of field equations it is necessary to grid the space: $\boldsymbol{x} \to \boldsymbol{x_i}$ with $\boldsymbol{i} = 0, \ldots, \boldsymbol{N}_g$. Here $\boldsymbol{x}$ is a general 3D coordinate and $\boldsymbol{N}_g$ number of grid cells (e.g. for 3D Cartesian coordinates $\boldsymbol{N}_g = \left(N_g^x,\ N_g^y,\ N_g^z\right)$). Accordingly, the plasma parameters are known at these grid points: $A(\boldsymbol{x}) \to A_{\boldsymbol{i}} = A(\boldsymbol{x} = \boldsymbol{x}_i)$. The number of simulation particles at grid points is relatively low, so that one can not use an analytic approach of point particles, which is valid only when the number of these particles is very large. The solution is to associate macro parameters to each of the simulation particle. In other words to assume that particles have some shape $S(\boldsymbol{x} - \boldsymbol{X})$, where $\boldsymbol{X}$ and $\boldsymbol{x}$ denote the particle position and observation point. Accordingly, the distribution moments at the grid point $\boldsymbol{i}$ associated with the particle "j" can be defined as

$$A_{\boldsymbol{i}}^m = a_j^m S(\boldsymbol{x_i} - \boldsymbol{X}_j) \ , \tag{6.27}$$

where $A_{\boldsymbol{i}}^0 = n_{\boldsymbol{i}}$, $A_{\boldsymbol{i}}^1 = n_{\boldsymbol{i}} \boldsymbol{V}_{\boldsymbol{i}}$, $A_{\boldsymbol{i}}^2 = n_{\boldsymbol{i}} V_{\boldsymbol{i}}^2$ etc. and $a_j^0 = 1/V_g$, $a_j^1 = \boldsymbol{V}^j/V_g$, $a_j^2 = (V^j)^2/V_g$ etc. V_g is the volume occupied by the grid cell. The total distribution moments at a given grid point are expressed as

$$A_{\boldsymbol{i}}^m = \sum_{j=1}^{N} a_j^m S(\boldsymbol{x_i} - \boldsymbol{X}_j) \ . \tag{6.28}$$

Stability and simulation speed of PIC simulations strongly depend on the choice of the shape function $S(\boldsymbol{x})$. It has to satisfy a number of conditions. The first two conditions correspond to space isotropy

$$S(x) = S(-x) \ , \tag{6.29}$$

and charge conservation

$$\sum_i S(\boldsymbol{x}_i - \boldsymbol{X}) = 1 \,. \tag{6.30}$$

The rest of the conditions can be obtained requiring an increasing accuracy of the weighting scheme. In order to derive them let us consider a potential generated at the point x by a unit charge located at the point X, $G(x - X)$. In other words $G(x - X)$ is the Green's function (for simplicity we consider a 1D case). Introducing the weighting scheme we can write the potential generated by some particle located at X as

$$\phi(x) = e \sum_{i=1}^{m} S(x_i - X)\, G(x - x_i) \,, \tag{6.31}$$

here e is the particle charge and m the number of nearest grid points with assigned charge. Expanding $G(x - x_i)$ near $(x - X)$ we get

$$\begin{aligned}
\phi(x) &= e \sum_{i=1}^{m} S(x_i - X)\, G(x - X) \\
&\quad + e \sum_{i=1}^{m} S(x_i - X) \sum_{n=1}^{\infty} \frac{(X - x_i)^n}{n!} \frac{\mathrm{d}^n G(x - X)}{\mathrm{d}x^n} \\
&= eG(x - X) + \delta\phi(x) \,, \\
\delta\phi(x) &= e \sum_{n=1}^{\infty} \frac{1}{n!} \frac{\mathrm{d}^n G(x - X)}{\mathrm{d}x^n} \sum_{i=1}^{m} S(x_i - X)(X - x_i)^n \,.
\end{aligned} \tag{6.32}$$

The first term on the right hand side of expression (6.32) represents a physical potential, while $\delta\phi$ is an unphysical part of it introduced by weighting. It is obvious to require this term to be as small as possible. This can be done by requiring

$$\sum_{i=1}^{m} S(x_i - X)(x_i - X)^n = 0 \tag{6.33}$$

with $n = 1, \ldots, n_{\max} - 1$. Substituting the expression (6.33) into (6.32) we get

$$\begin{aligned}
\delta\phi(x) &= \sum_{n=n_{\max}}^{\infty} \frac{1}{n!} \frac{\mathrm{d}^n G(x - X)}{\mathrm{d}x^n} \sum_{i=1}^{m} S(x_i - X)(X - x_i)^n \\
&\sim G(x - X) \sum_{i=1}^{m} S(x_i - X) \sum_{n=n_{\max}}^{\infty} \frac{(X - x_i)^n}{n!\,(x - X)^n} \,.
\end{aligned} \tag{6.34}$$

Thus, at large distance from the particle ($|X - x_i| < |x - X|$) $\delta\phi(x)$ decreases with increasing $n_{\max}$.

The shape functions can be directly constructed from the conditions (6.29), (6.30) and (6.33). The later two represent algebraic equations for $S(x_i - X)$. Hence, the number of conditions (6.33), which can be satisfied depends on the maximum number of nearest grid points m to which the particle is weighted. A simplest shape function assigns density to the nearest grid point ($m = 1$) and satisfies just the first two conditions (6.29) and (6.30). For a 1D Cartesian coordinates it is given as $S^0(x) = 1$, if $|x| < \Delta x/2$, otherwise $S^0(x) = 0$, where Δx is the size of spatial grid. This weighting is called zero order or NGP weighting and was used in first PIC codes (see Fig. 6.3). Although the NGP scheme requires less CPU time, it is relatively noisy and probably not in use any more. The next, first order weighting scheme assigns density to two nearest grid points ($m = 2$) and given as $S^1(x) = 1 - |x|/(\Delta x)$, if $|x| < \Delta x$, otherwise $S^1(x) = 0$. Often it is called a cloud in cell (CIC) scheme. It satisfies one more condition in (6.33), with $n_{\max} = 1$, and a more accurate than the NGP scheme. Probably the CIC represents the most commonly used weighting scheme. Generalization to multi-dimensional Cartesian coordinates is trivial: $S(\boldsymbol{x}) = S(x)\,S(y)\,S(z)$. The higher order schemes (see [4]) can increase the accuracy of simulation (when other parameters are fixed), but require significantly longer CPU time.

For completeness I note, that some authors often use another definition of the particle shape $D(x)$ (e.g. see [4])

$$S(x_i - x) = \int_{x_i-\Delta x/2}^{x_i+\Delta x/2} D(x' - x)\, \mathrm{d}x' \,. \tag{6.35}$$

The meaning of this expression is that the density at the grid point x_i assigned by the particle located at the point x represents the average of the particle real shape $D(x' - x)$ over the area $[x_i - \Delta x/2; x_i + \Delta x/2]$. For the nearest grid point and

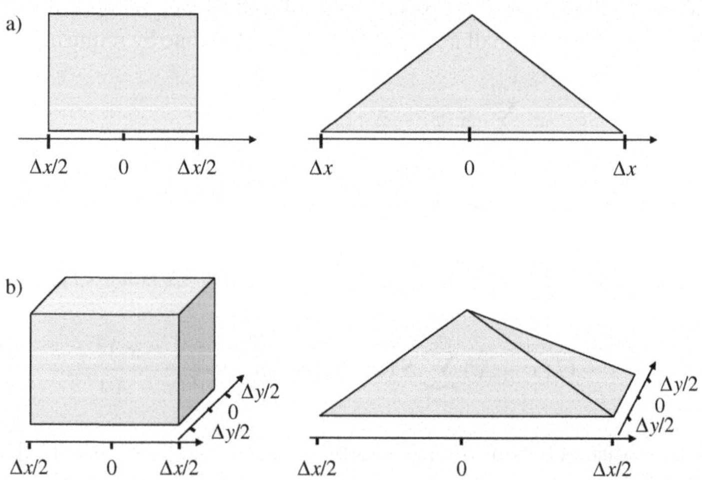

Fig. 6.3. Particle shapes for the NGP (**left**) and linear (**right**) weightings in 1D (**a**) and 2D (**b**) cases

linear weightings $D(x) = \delta(x)$ and $D(x) = H(\Delta x/2 - |x|)$, respectively. Here $H(x)$ is the step-function: $H(x) = 1$, if $x > 0$, else $H(x) = 0$.

6.4.2 Field Weighting

After the calculation of charge and current densities the code solves the Maxwell's equations (cf. Fig. 6.1) and delivers fields at the grid points $\boldsymbol{i} = 0, \ldots, \boldsymbol{N}_g$. These fields can not be used directly for the calculation of force acting on particles, which are located at any point and not necessarily at the grid points. Calculation of fields at any point is done in a similar way as charge assignment and called field weighting. So, we have $\boldsymbol{E}_i$ and $\boldsymbol{B}_i$ and want to calculate $\boldsymbol{E}(\boldsymbol{x})$ and $\boldsymbol{B}(\boldsymbol{x})$ at any point $\boldsymbol{x}$. This interpolation should conserve momentum, which can be done by requiring that the following conditions are satisfied:

(i) Weighting schemes for the field and particles are same

$$\boldsymbol{E}(\boldsymbol{x}) = \sum_{\boldsymbol{i}} \boldsymbol{E}_i \boldsymbol{S}(\boldsymbol{x}_i - \boldsymbol{x}) \ . \tag{6.36}$$

(ii) The field solver has a correct space symmetry, i.e. formally the field can be expressed in the following form (for simplicity we consider the 1D case)

$$E_i = \sum_k g_{ik} \rho_k \tag{6.37}$$

with $g_{ik} = -g_{ki}$, where ρ_k is the charge density at the grid point k. In order to understand this condition better, let us consider a 1D electrostatic system. By integrating the Poisson equation we obtain

$$E(x) = \frac{1}{2\varepsilon_0} \left(\int_a^x \rho \, \mathrm{d}x - \int_x^b \rho \, \mathrm{d}x \right) + E_b + E_a \ , \tag{6.38}$$

where a and b define boundaries of the system. Assuming that either a and b are sufficiently far and $E_{a,b} = \rho_{a,b} = 0$, or the system (potential) is periodic $E_b = -E_a$, $\rho_b = \rho_a$, we obtain

$$\begin{aligned} E(x_i) &= \frac{1}{2\varepsilon_0} \left(\int_a^{x_i} \rho \, \mathrm{d}x - \int_{x_i}^b \rho \, \mathrm{d}x \right) \\ &= \frac{\Delta x}{4\varepsilon_0} \left(\sum_{k=1}^{i-1} (\rho_k + \rho_{k+1}) - \sum_{k=i}^{N_g - 1} (\rho_k + \rho_{k+1}) \right) \\ &= \frac{\Delta x}{4\varepsilon_0} \sum_{k=1}^{N_g} g_{ik} \rho_k \end{aligned} \tag{6.39}$$

with $N_g \to \infty$, $\Delta x = [b, a]/N_g$ and

$$g_{ik} = \begin{cases} 2 & \text{if } i > k \\ -2 & \text{if } i < k \\ 0 & \text{if } i = k \end{cases} . \tag{6.40}$$

Thus, the condition (6.37) is satisfied.

Let us check different conservation constraints.

(i) The self-force of the particle located at the point x can be calculated as follows

$$\begin{aligned} F_{\text{self}} &= e \sum_i E_i S(x_i - x) = e \sum_{i,\,k} g_{ik} S(x_i - x)\, \rho_k \\ &= \frac{e^2}{V_g} \sum_{i,\,k} g_{ik} S(x_i - x)\, S(x_i - x) = (i \leftrightarrow k) \\ &= -\frac{e^2}{V_g} \sum_{i,\,k} g_{ik} S(x_i - x)\, S(x_i - x) = -F_{\text{self}} = 0 , \end{aligned} \tag{6.41}$$

(ii) The two-particle interaction force is given as

$$\begin{aligned} F_{12} &= e_1 E_2(x_1) = e_1 \sum_i E_{2,i} S(x_i - x_1) \\ &= \frac{e_1 e_2}{V_g} \sum_{i,\,k} g_{ik} S(x_i - x_1)\, S(x_k - x_2) \\ &= -\frac{e_1 e_2}{V_g} \sum_{i,\,k} g_{ki} S(x_i - x_1)\, S(x_k - x_2) \\ &= -e_2 \sum_{i,\,k} g_{ki} S(x_k - x_2)\, \rho_{1,k} = -e_2 E_1(x_2) \\ &= -F_{21} . \end{aligned} \tag{6.42}$$

Here, E_p denotes the electric field generated by the particle p.

(iii) Momentum conservation

$$\begin{aligned} \frac{\mathrm{d}\boldsymbol{P}}{\mathrm{d}t} &= \boldsymbol{F} = \sum_{p=1}^{N} e_p \left(\boldsymbol{E}(\boldsymbol{x}_p) + \boldsymbol{V}_p \times \boldsymbol{B}(\boldsymbol{x}_p) \right) \\ &= \sum_{p=1}^{N} e_p \sum_i \boldsymbol{E}_i S(\boldsymbol{x}_i - \boldsymbol{x}_p) + \sum_{p=1}^{N} e_p \boldsymbol{V}_p \times \sum_i \boldsymbol{B}_i S(\boldsymbol{x}_i - \boldsymbol{x}_p) \\ &= \sum_i \boldsymbol{E}_i \sum_{p=1}^{N} e_p S(\boldsymbol{x}_i - \boldsymbol{x}_p) - \sum_i \boldsymbol{B}_i \times \sum_{p=1}^{N} e_p \boldsymbol{V}_p S(\boldsymbol{x}_i - \boldsymbol{x}_p) \\ &= V_g \sum_i \left(\rho_i \boldsymbol{E}_i + \boldsymbol{J}_i \times \boldsymbol{B}_i \right) . \end{aligned} \tag{6.43}$$

Representing fields as a sum of external and internal components $\boldsymbol{E_i} = \boldsymbol{E_i^{\text{ext}}} + \boldsymbol{E_i^{\text{int}}}$ and $\boldsymbol{B_i} = \boldsymbol{B_i^{\text{ext}}}$, where $\boldsymbol{E_i^{\text{int}}}$ is given in expression (6.36), after some trivial transformations we finally obtain the equation of momentum conservation

$$\frac{\mathrm{d}\boldsymbol{P}}{\mathrm{d}t} = V_g \sum_{\boldsymbol{i}} \left(\rho_{\boldsymbol{i}} \boldsymbol{E_i^{\text{ext}}} + \boldsymbol{J_i} \times \boldsymbol{B_i} \right) . \tag{6.44}$$

As we see, the conditions (6.36) and (6.37) guarantee that during the force weighting the momentum is conserved and the inter-particle forces are calculated in a proper way. It has to be noted that:

(i) We neglected contribution of an internal magnetic field $\boldsymbol{B}^{\text{int}}$.
(ii) The momentum conserving schemes considered above does not necessarily conserve the energy too (for energy conserving schemes see [3] and [4]).
(iii) The condition (6.37) is not satisfied in general for coordinate systems with nonuniform grids, causing the self-force and incorrect inter-particle forces.

For exampple, if we introduce a nonuniform grid $\Delta x^i = \Delta x \alpha^i$ with $\alpha^i \neq \alpha^{j \neq i}$, in expression (6.39) we obtain

$$E\left(x_i\right) = \frac{\Delta x}{4\varepsilon_0} \sum_{k=1}^{N_g} g_{ik} \rho_k \tag{6.45}$$

with

$$g_{ik} = \begin{cases} \alpha^k + \alpha^{k-1} & \text{if } i > k \\ -\left(\alpha^k + \alpha^{k-1}\right) & \text{if } i < k, N_g \to \infty, \ \Delta x = \frac{[b,a]}{\sum_{i=1}^{N_g} \alpha^i} \\ \alpha^{i-1} - \alpha^i & \text{if } i = k \end{cases} , \tag{6.46}$$

so that $g_{ki} \neq -g_{ik}$.

6.5 Solution of Maxwell's Equations

6.5.1 General Remarks

Numerical solution of Maxwell's equations is a continuously developing independent direction in numerical plasma physics (e.g., see [13]). Field solvers in general can be divided into three groups:

(i) Mesh-relaxation methods, when the solution is initially guessed and then systematically adjusted until the solution is obtained with required accuracy;
(ii) Matrix methods, when Maxwell's equations are reduced to a set of linear finite difference equations and solved by some matrix method, and
(iii) Methods using the so called fast Fourier transform (FFT) and solving equations in Fourier space.

According to the type of the equations to be solved the field solvers can be explicit or implicit. E.g., the explicit solver of the Poisson equation solves the usual Poisson equation

$$\nabla \left[\varepsilon \left(\boldsymbol{x} \right) \nabla \varphi \left(\boldsymbol{x}, t \right) \right] = -\rho \left(\boldsymbol{x}, t \right) \ , \tag{6.47}$$

while an implicit one solves the following equation

$$\nabla \left[\left(1 + \eta \left(\boldsymbol{x} \right) \right) \varepsilon \left(\boldsymbol{x} \right) \varphi \left(\boldsymbol{x}, t \right) \right] = -\rho \left(\boldsymbol{x}, t \right) \ . \tag{6.48}$$

Here $\eta\left(\boldsymbol{x}\right)$ is the implicit numerical factor, which arises due to the fact that in its implicit formulation a new position (and hence ρ) of particle is calculated from a new field given at the same moment.

As an example we consider some matrix methods, which are frequently used in different codes. For a general overview of different solvers the interested reader can use [3] or [4].

6.5.2 Electrostatic Case, Solution of Poisson Equation

Let us consider the Poisson equation in a Cartesian coordinate system

$$\left(\frac{\partial^2}{\partial x^2} + \frac{\partial^2}{\partial y^2} + \frac{\partial^2}{\partial z^2} \right) \varphi \left(\boldsymbol{r} \right) = -\frac{1}{\varepsilon_0} \rho \left(\boldsymbol{r} \right) \ , \tag{6.49}$$

and formulate the corresponding finite difference equations. For this we use the transformation

$$\frac{\partial^2}{\partial x^2} \varphi \Rightarrow \frac{a \varphi_{i+1} + b \varphi_i + c \varphi_{i-1}}{\Delta x^2} \ . \tag{6.50}$$

Other components are treated in a similar way. Our aim is to choose the constants a, b and c, so that the error will be smallest. From the symmetry constraint we can write $a = c$. Then by expanding $\varphi_{i\pm1}$ at $x = x_i$

$$\begin{aligned} \varphi_{i\pm1} &= \varphi_i \pm \Delta x \left(\varphi_i \right)' + \frac{\Delta x^2}{2} \left(\varphi_i \right)'' \pm \frac{\Delta x^3}{6} \left(\varphi_i \right)''' + \frac{\Delta x^4}{24} \left(\varphi_i \right)^{(4)} \ldots \ , \\ \left(\varphi_i \right)^{(k)} &= \left. \frac{\partial^k}{\partial x^k} \varphi \right|_{x = x_i} \ , \end{aligned} \tag{6.51}$$

and substituting in (6.50) we obtain

$$a \varphi_{i+1} + b \varphi_i + c \varphi_{i-1} = \varphi_i \left(2a + b \right) + \left(\varphi_i \right)'' a \Delta x^2 + \left(\varphi_i \right)^{(4)} a \frac{\Delta x^4}{12} + \ldots \ . \tag{6.52}$$

Hence, by choosing $a = 1$ and $b = -2a = -2$ we get

$$\left(\frac{\partial^2 \varphi}{\partial x^2} \right)_{x = x_i} - \frac{\varphi_{i+1} - 2\varphi_i + \varphi_{i-1}}{\Delta x^2} = \frac{\Delta x^2}{12} \left(\varphi_i \right)^{(4)} + O\left(\Delta x^4 \right) \ . \tag{6.53}$$

Hence, the finite difference equation (6.50) with $b = -2$ and $a = c = 1$ has second order accuracy ($\sim \Delta x^2$). Usually this accuracy is sufficient, otherwise one can consider a more accurate scheme

$$\frac{\partial^2}{\partial x^2}\varphi \Rightarrow \frac{a\varphi_{i+2} + b\varphi_{i+1} + c\varphi_i + d\varphi_{i-1} + e\varphi_{i-2}}{\Delta x^2}\,. \tag{6.54}$$

6.5.2.1 1D Case: Bounded Plasma with External Circuit

An excellent example of an 1D Poisson solver has been introduced in [7]. The solver is applied to an 1D bounded plasma between two electrodes and solves Poisson and external circuit equations simultaneously. Later, this solver has been applied to a 2D plasma model [14]. Below we consider an simplified version of this solver assuming that the external circuit consists of a voltage (or current) source $V(t)$ ($I(t)$) and a capacitor C (see Fig. 6.4)).

The Poisson equation for a 1D plasma is given as

$$\varphi_{i+1} - 2\varphi_i + \varphi_{i-1} = -\frac{\Delta x^2}{\varepsilon_0}\rho_i\,. \tag{6.55}$$

It is a second order equation, so that we need two boundary conditions for the solution. The first one can be a potential at the right-hand-side (rhs) wall:

$$\varphi_{Ng} = 0. \tag{6.56}$$

The second condition can be formulated at the left-hand-side (lhs) wall:

$$\frac{\varphi_0 - \varphi_1}{\Delta x} = E(x = \frac{\Delta x}{2}) = E_0 + \frac{1}{\varepsilon_0}\int_0^{\Delta x/2} \rho\, \mathrm{d}x \approx E_0 + \frac{\Delta x}{2\varepsilon_0}\rho_0\,. \tag{6.57}$$

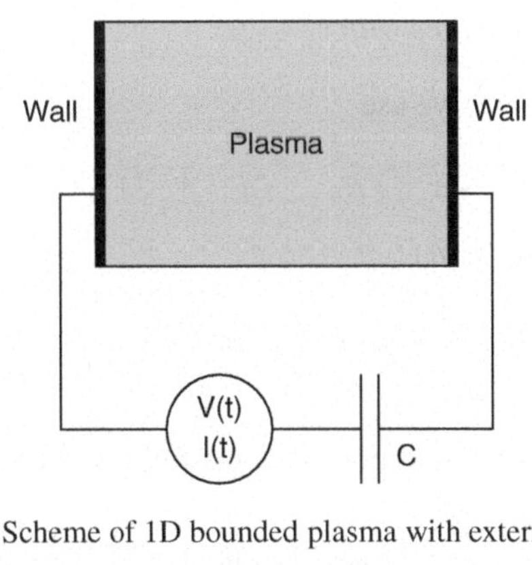

Fig. 6.4. Scheme of 1D bounded plasma with external circuit

Recalling that E_0 is the electric field at the l.h.s. wall, we can write $E_0 = \sigma_{\text{lhs}}/\varepsilon_0$, where σ_{lhs} is the surface charge density there. Hence, the second boundary condition can be formulated as

$$\varphi_0 - \varphi_1 = \frac{\Delta x}{\varepsilon_0}\left(\sigma_{\text{lhs}} + \frac{\Delta x}{2}\rho_0\right) . \tag{6.58}$$

In order to calculate σ_{lhs} we have to employ the circuit equation.

6.5.2.1.1 Voltage Driven Source with Finite C

In this case charge conservation at the l.h.s. wall can be written as

$$\sigma_{\text{lhs}}(t) = \sigma_{\text{lhs}}(t - \Delta t) + \frac{Q_{\text{pl}} + Q_{\text{ci}}}{S} , \tag{6.59}$$

where Q_{pl} and Q_{ci} are the charge deposited during Δt time on the lhs wall by the plasma and the external circuit, respectively. S is the area of the wall surface. Q_{pl} can be calculated by counting the charge of the plasma particles absorbed at the lhs wall, and Q_{ci} can be given as $Q_{\text{ci}} = Q^{\text{c}}(t) - Q^{\text{c}}(t - \Delta t)$, where Q^{c} is the charge at the capacitor. Q^{c} can be calculated using the Kirchhoff's law

$$\frac{Q^{\text{c}}}{C} = V(t) + \varphi_{N_g} - \varphi_0 = V(t) - \varphi_0 . \tag{6.60}$$

Substituting the expressions (6.59) and (6.60) into (6.58) we obtain

$$\begin{aligned} &\varphi_0\left(1 + \frac{C}{S}\frac{\Delta x}{\varepsilon_0}\right) - \varphi_1 \\ &= \frac{\Delta x}{\varepsilon_0}\left(\frac{Q_{\text{pl}} + C\left(V(t) - V(t - \Delta t) + \varphi_0(t - \Delta t)\right)}{S}\right. \\ &\left. + \sigma_{\text{lhs}}(t - \Delta t) + \frac{\Delta x}{2}\rho_0\right) . \end{aligned} \tag{6.61}$$

6.5.2.1.2 Voltage Driven Source with $C \to \infty$

In this case in spite of (6.58) we use

$$\varphi_0 - \varphi_{Ng} = \varphi_0 = V(t) . \tag{6.62}$$

6.5.2.1.3 Open Circuit ($C = 0$)

In this case we write

$$\sigma_{\text{lhs}}(t) = \sigma_{\text{lhs}}(t - \Delta t) + \frac{Q_{\text{pl}}}{S} , \tag{6.63}$$

so that the second boundary condition takes the following form

$$\varphi_0 - \varphi_1 = \frac{\Delta x}{\varepsilon_0}\left(\sigma_{\text{lhs}}(t - \Delta t) + \frac{Q_{\text{pl}}}{S} + \frac{\Delta x}{2}\rho_0\right) . \tag{6.64}$$

6.5.2.1.4 Current Driven Source

In this case Q_{ci} can be directly calculate from the expression $Q_{ci} = \Delta t I(t)$. Then the second boundary condition can be given as

$$\varphi_0 - \varphi_1 = \frac{\Delta x}{\varepsilon_0}\left(\sigma_{\text{lhs}}\left(t-\Delta t\right) + \frac{Q_{\text{pl}} + \Delta t I\left(t\right)}{S} + \frac{\Delta x}{2}\rho_0\right) . \tag{6.65}$$

Combining equations (6.55), (6.56) and (6.61)–(6.65) we can write the set of difference equations in the following matrix form

$$\begin{pmatrix} a & b & 0 & \dots\dots\dots & 0 \\ c & -2 & 1 & 0 \;\dots\dots & 0 \\ 0 & 1 & -2 & 1 \;\; 0 \;\dots & 0 \\ & \ddots & \ddots & \ddots & \\ 0 \dots & 0 & 1 & -2 \;\; 1 & 0 \\ 0 \dots\dots & & 0 & 1 \;\; -2 & 1 \\ 0 \dots\dots\dots & & & 0 \;\; 1 & -2 \end{pmatrix} \begin{pmatrix} \varphi_0 \\ \varphi_1 \\ \varphi_2 \\ \vdots \\ \varphi_{Ng-2} \\ \varphi_{Ng-1} \end{pmatrix} = -\frac{\Delta x^2}{\varepsilon_0} \begin{pmatrix} d/\Delta x \\ \rho_1 + e \\ \rho_2 \\ \vdots \\ \rho_{Ng-2} \\ \rho_{Ng-1} \end{pmatrix} . \tag{6.66}$$

Here, for the cases

(i) Voltage driven source or open circuit: $a = -1 - C\Delta x/S\varepsilon_0$, $b = 1$, $c = 1$, $d = \sigma_{\text{lhs}}(t-\Delta t) + (Q_{\text{pl}} + C(V(t) - V(t-\Delta t) + \varphi_0(t-\Delta t)))/S + \Delta x/2\rho^0$ and $e = 0$.
(ii) Short circuit ($C \to \infty$): $a = b = c = d = 0$ and $e = \varepsilon_0/(\Delta x^2)V(t)$.
(iii) Current driven source: $a = -1$, $b = 1$, $c = 1$, $d = \sigma_{\text{lhs}}(t - \Delta t) + (Q_{\text{pl}} + \Delta t I(t))/S + \Delta x/2\rho^0$ and $e = 0$.

The matrix (6.66) can be solved by standard inverse matrix solvers (e.g., see [15]).

6.5.2.2 2D Case: Generalization of the 1D Solver

This 1D solver can be generalized for a 2D case. The main difference between the 1D and 2D cases represent the decomposition of the field and the boundary conditions at internal objects introduced in 2D (for details see [14]).

Field decomposition is given by

$$\varphi\left(t,x,y\right) = \varphi^{\text{pl}}\left(t,x,y\right) + \varphi^{\text{con}}\left(t\right)\varphi^{\text{vac}}\left(x,y\right) . \tag{6.67}$$

Here φ^{pl} is the plasma field with the zero boundary conditions

$$\Delta\varphi^{\text{pl}}\left(t,x,y\right) = -\frac{1}{\varepsilon_0}\rho\left(x,y\right) , \qquad \varphi^{\text{pl}}\big|_b = 0 , \tag{6.68}$$

where φ^{vac} is the vacuum field with the unit boundary conditions

$$\Delta\varphi^{\text{vac}}(x, y) = 0\,, \qquad \varphi^{\text{vac}}\big|_b = 1\,. \tag{6.69}$$

$\varphi^{\text{con}}(t)$ is the field at conductors, which is either calculated self-consistently (for electrodes), or prescribed (e.g., at the wall). The symbol $|_b$ denotes a plasma boundary.

It's easy to see that φ in (6.67) represents an exact solution of the Poisson equation with the given boundary conditions. The advantage of this decomposition is that

(i) the vacuum field has to be calculated just once and
(ii) the Poisson equation (6.68) with the zero boundary conditions is easier to solve, than one with a general boundary conditions.

As a result, the field decomposition can save a lot of CPU time.

The equation of the plasma field (6.68) is reduced to a set of finite difference equations

$$\frac{\varphi_{i+1,j} - 2\varphi_{ij} + \varphi_{i-1,j}}{\Delta x^2} + \frac{\varphi_{i,j+1} - 2\varphi_{ij} + \varphi_{i,j-1}}{\Delta y^2} = -\frac{1}{\varepsilon_0}\rho_{ij} \tag{6.70}$$

with $\varphi|_b = 0$, which can be solved by matrix method. In a similar way the Laplace equation (6.69) can be solved for the vacuum field.

The corresponding boundary conditions at the wall of internal objects are calculated using Gauss' law

$$\oint \varepsilon \boldsymbol{E}\, \mathrm{d}\boldsymbol{S} = \int \rho\, \mathrm{d}V + \oint \sigma\, \mathrm{d}S\,, \tag{6.71}$$

which in a finite volume representation can be written as

$$\begin{aligned} &\Delta y \Delta z \left(\varepsilon_{i+1/2,j} E_{i+1/2,j} - \varepsilon_{i-1/2,j} E_{i-1/2,j}\right) \\ &\qquad + \Delta x \Delta z \left(\varepsilon_{i,j+1/2} E_{i,j+1/2} - \varepsilon_{i,j-1/2} E_{i,j-1/2}\right) \\ &= \rho_{ij} \Delta V_{ij} + \sigma_{ij} \Delta S_{ij}\,. \end{aligned} \tag{6.72}$$

Here ΔV_{ij} and ΔS_{ij} are the volume and area associated with the given grid point i, j. The electric fields entering in this equation are calculated according to the following expressions

$$\begin{aligned} E_{i\pm1/2,j} &= \pm\frac{\varphi_{i,j} \quad \varphi_{i\pm1,j}}{\Delta x}\,, \\ E_{i,j\pm1/2} &= \pm\frac{\varphi_{i,j} - \varphi_{i,j\pm1}}{\Delta x}\,. \end{aligned} \tag{6.73}$$

Calculation of the potential at the plasma boundary $\varphi^{\text{con}}(t)$ consists in general of three parts. The potential at the outer wall is fixed and usually chosen as 0. The potential at the electrodes, which are connected to an external circuit is done in a similar way as for the 1D case considered above. For calculation of the potential at the internal object equation (6.72) is solved. We note that the later task is case dependent and not a trivial one, e.g., the solution depends on the object shape or material (conductor or dielectric). For further details see [14].

6.5.2.3 2D Case: Cartesian/Fourier Solver

The number of operations to be performed by a matrix solver (per dimension) scales as $\sim N_g^2$ and drastically increases with N_g. This number can be significantly reduced by using a fast Fourier Transform (FFT) solver, which scales as $\sim N_g \ln N_g$ (see [15]). This scaling can be significantly improved by using different optimizations. One example when FFT solvers can be applied is a 2D plasma, which is bounded in one direction and unbounded or periodic in the other one. In this case one can apply a discrete Fourier transform along the periodic direction

$$A_{ij} = \frac{1}{2\pi} \sum_{k=0}^{Ny-1} A_i^k \mathrm{e}^{-\mathrm{i}2\pi j/N_y k} \tag{6.74}$$

with $A = \varphi, \rho$. By substituting this expression into (6.70) we obtain

$$\varphi_{i+1}^k - 2\left(1 + 2\left(\frac{\Delta x}{\Delta y}\sin\left(\frac{\pi k}{N_y}\right)\right)^2\right)\varphi_i^k + \varphi_{i-1}^k = -\frac{\Delta x^2}{\varepsilon_0}\rho_i^k \,. \tag{6.75}$$

It is easy to see that (6.75) is similar to the one for the 1D model considered above and can be solved in the same way. The main difference are the boundary conditions along the x-axis. E.g., if the plasma is bounded between two conducting walls, then $\varphi_0^k = \varphi_{N_g}^k = 0$ if $k > 0$, and for the $k = 0$-component we have exactly the same equation as for 1D with the same boundary condition.

6.5.3 Electromagnetic Case

For sufficiently strong fields and/or very fast processes it is necessary to solve the complete set of Maxwell's equations (6.3). It is obvious that corresponding solvers are more complicated than ones considered above. Correspondingly a detailed description of them is out of the scope of this work. Here we present just one of possible schemes, which is implemented in the XOOPIC code [8].

In order to ensure high speed and accuracy it is convenient to introduce a leap-frog scheme also for the fields. The leap frog scheme is applied to the space coordinates too, which means that electric and magnetic fields are shifted in time by $\Delta t/2$, and different components of them are shifted in space by $\Delta x/2$. In other words:

(i) $\boldsymbol{E}$ is defined at $t = n\Delta t$ and $\boldsymbol{B}$ and $\boldsymbol{J}$ at $t = (n + 1/2)\Delta t$ time moments.
(ii) "i" components of the electric field and current density are defined at the points $x_i + \Delta_i/2$, x_k and x_j, and same component of the magnetic field at x_i, $x_k + \Delta_k/2$ and $x_j + \Delta_j/2$. Here x_s and Δ_s for $s = i, k, j$ denote the grid point and grid size along the s-axis. i, k and j denote the indices of the right-handed Cartesian coordinate system.

As a result the finite-differenced Ampere's and Faraday's laws in Cartesian coordinates can be written as

$$\frac{D^{i,t}_{i+1/2,k,j} - D^{i,t-\Delta t}_{i+1/2,k,j}}{\Delta t}$$
$$= \frac{H^{j,t-\Delta t/2}_{i+1/2,k+1/2,j} - H^{j,t-\Delta t/2}_{i+1/2,k-1/2,j}}{\Delta x_k}$$
$$- \frac{H^{k,t-\Delta t/2}_{i+1/2,k,j+1/2} - H^{k,t-\Delta t/2}_{i+1/2,k,j-1/2}}{\Delta x_j} - J^{i,t-\Delta t/2}_{i+1/2,k,j} , \tag{6.76}$$

$$\frac{B^{i,t+\Delta t/2}_{i,k+1/2,j+1/2} - B^{i,t-\Delta t/2}_{i,k+1/2,j+1/2}}{\Delta t}$$
$$= \frac{D^{k,t}_{i,k+1/2,j+1} - D^{k,t}_{i,k+1/2,j}}{\Delta x_j} - \frac{D^{j,t}_{i,k+1,j+1/2} - D^{j,t}_{i,k,j+1/2}}{\Delta x_k} . \tag{6.77}$$

The solver works in the following way. The equations $\nabla \boldsymbol{D} = \rho$ and $\nabla \boldsymbol{B} = 0$ prescribe the initial electromagnetic fields. They remain satisfied due to Ampere's and Faraday's law, which are solved from the finite difference equations (6.76) and (6.77). The corresponding boundary conditions strongly depend on the simulated plasma model. E.g., at the wall representing an ideal conductor $E_{\|} = B_{\perp} = 0$, where $\|$ and $\perp$ denote the components parallel and normal to the wall, respectively.

As one can see, the components of the electromagnetic field obtained from (6.76) and (6.77) are defined at different time moments and spatial points than for the particle mover. Moreover, the current density obtained from the particle position is not defined at $t = (n + 1/2)\,\Delta t$ as it is required for Ampere's law (6.76). Hence, it is necessary to additionally couple the particle and field solvers [3].

It is useful to derive a general stability criteria for the electromagnetic case. For this we consider electromagnetic waves in vacuum

$$\boldsymbol{A} = \boldsymbol{A}_0 \mathrm{e}^{\mathrm{i}\boldsymbol{k}\boldsymbol{x} - \omega t} , \tag{6.78}$$

with $\boldsymbol{A} = \boldsymbol{E}, \boldsymbol{B}$. After substitution of (6.78) into field equations (6.76) and (6.77) and trivial transformations we obtain

$$\left(\frac{\sin(\omega t/2)}{c\Delta t}\right)^2 = \sum_{i=1}^{3} \left(\frac{\sin(k_i x_i/2)}{\Delta x_i}\right)^2 , \tag{6.79}$$

where $c = \sqrt{1/\varepsilon_0 \mu_0}$ is the speed of light. It is obvious that the solution is stable (i.e. Im$\omega < 0$) if

$$(c\Delta t)^2 < \left(\sum_{i=1}^{3} \frac{1}{\Delta x_i^2}\right)^{-1} . \tag{6.80}$$

Often, this so called Courant condition requires unnecessary small time step for the particle mover. In order to relax it one can introduce separate time steps for field and particles. This procedure is called "sub-cycling" [3].

The routines described above namely: The field solver, the particle mover with proper boundary conditions and the particle source, weighting of particles and fields represent a complete PIC code in its classical understanding. Starting from 1970s a number of PIC codes include different models of particle collisions. Today the majority of PIC codes include at least some kind of collision operator, which have to be attributed to a PIC technique. These operators are usually based on statistical methods and correspondingly are called Monte Carlo (MC) models. Often different authors use the name PIC-MC code. The MC simulations represent an independent branch in numerical physics and the interested reader can find more on MC method in corresponding literature (e.g., see Part II). Below we consider the main features of the MC models used in PIC codes.

6.6 Particle Collisions

6.6.1 Coulomb Collisions

The forces acting on the particles in a classical PIC scheme correspond to macro fields, so that the simulated plasma is assumed to be collisionless. In order to simulate a collisional plasma it is necessary to implement corresponding routines. Moreover, the field solver is organized in such a way that self-forces are excluded, hence, the field generated by a particle inside the grid cell decreases with decreasing distance from this particle. As a result, inter-particle forces inside grid cells are underestimated (see Fig. 6.5). Hence, they can be (at least partially) compensated by introducing the Coulomb collision operator.

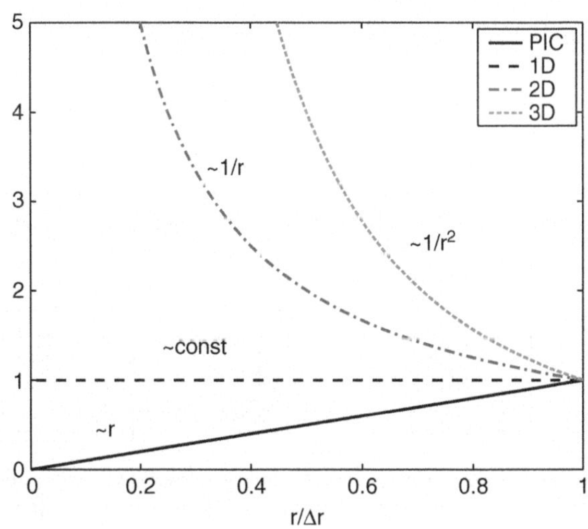

Fig. 6.5. Inter-particle forces inside grid cell

The first codes simulating Coulomb collisions were the particle-particle codes simulating the exact interaction between each particle pair. Of course this method, which scales as N^2 can not be used in contemporary PIC simulations. Later different MC models have been developed.

The simplest linear model assumes that the particle distribution is near to a Maxwellian and calculates an average force acting on particles due to collisions [16]. Although this is the fastest operator it probably can not be used for most of kinetic plasma simulations, when particle distributions are far from the Maxwellian. A nonlinear analogue of this model has been introduced in [17]. Here, the exact collision inter-particle inter-particle is obtained from the particle velocity distribution function. Unfortunately, the number of particles required for building up a sufficiently accurate velocity distribution is extremely large (see [18]), which makes it practically impossible to simulate large systems.

Most of nonlinear Coulomb collision operators used in our day PIC codes are based on the binary collision model introduced in [19]. In this model each particle inside a cell is collided with one particle from the same cell. This collision operator conserves energy and momentum and it is sufficiently accurate. The main idea is based on the fact that there is no need to consider Coulomb interaction between two particles separated by a distance larger than the Debye radius λ_D (e.g., see [20]). Since a typical size of the PIC cell is of the order of λ_D, the interaction between the particles in different cells can be neglected. This method consists of the following three steps (see Fig. 6.6):

(i) First, all particles are grouped according to the cells where they are located;
(ii) Then these particles are paired in a random way, so that one particle has only one partner;
(iii) Finally, the paired particles are (statistically) collided.

The latter is not trivial and we consider it in some detail.

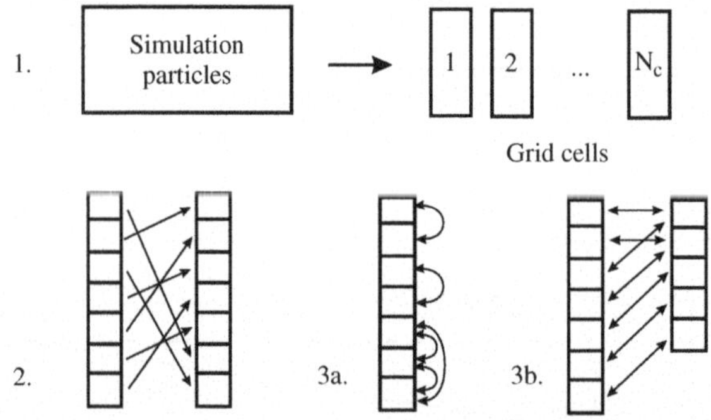

Fig. 6.6. Binary collision model from [19]. **1**. Grouping of particles in the cells; **2**. Randomly changing the particle order inside the cells; **3a**. Colliding particles of the same type; **3b**. Colliding particles of different types

According to the momentum and energy conservation constrains, we can express the after-collision velocities of particles $\boldsymbol{V}_1'$, $\boldsymbol{V}_2'$ via their before-collision values $\boldsymbol{V}_1$ and $\boldsymbol{V}_2$ [21]

$$\boldsymbol{V}_1' = \boldsymbol{V}_1 + \frac{m_2}{m_1 + m_2}\Delta\boldsymbol{V} \quad \text{and} \quad \boldsymbol{V}_2' = \boldsymbol{V}_2 - \frac{m_1}{m_1 + m_2}\Delta\boldsymbol{V} \tag{6.81}$$

with $\Delta\boldsymbol{V} = \boldsymbol{V}' - \boldsymbol{V}$, $\boldsymbol{V} = \boldsymbol{V}_2 - \boldsymbol{V}_1$, $\boldsymbol{V}' = \boldsymbol{V}_2' - \boldsymbol{V}_1'$ and $V'^2 = V^2$. As we see the calculation can be reduced to the scattering of the relative velocity $\boldsymbol{V}$

$$\Delta\boldsymbol{V} = \left(\widehat{O}\,(\chi,\psi) - 1\right)\boldsymbol{V}\,, \tag{6.82}$$

where $\widehat{O}\,(\alpha,\beta)$ is the matrix corresponding to the rotation on angles α and β (see [19]). χ and ψ represent the scattering and azimuthal angles.

The scattering angle χ is calculated from a corresponding statistical distribution. By using the Fokker-Plank collision operator one can show (see [22]) that during the time Δt_c the scattering angle has the following Gaussian distribution

$$\begin{aligned} P\,(\chi) &= \frac{\chi}{\langle\chi^2\rangle_{\Delta t_c}}\mathrm{e}^{-\chi^2/(2\langle\chi^2\rangle_{\Delta t_c})}\,, \\ \langle\chi^2\rangle_{\Delta t_c} &\equiv \frac{e_1^2 e_2^2}{2\pi\varepsilon_0^2}\,\frac{n\Delta t_c\Lambda}{\mu^2 V^3}\,. \end{aligned} \tag{6.83}$$

Here $e_{1,2}$ and $\mu = m_1 m_2/\,(m_1{+}m_2)$ denote the charge and reduced mass of the collided particles, respectively. n and Λ are the density and the Landau logarithm [20], respectively. The distribution (6.83) can be inverted to get

$$\chi = \sqrt{-2\,\langle\chi^2\rangle_t \ln R_1}\,. \tag{6.84}$$

Correspondingly, the azimuthal angle ψ is chosen randomly between 0 and 2π

$$\psi = 2\pi R_2\,. \tag{6.85}$$

R_1 and R_2 are random numbers between 0 and 1.

Finally, the routine for two-particle collision is reduced to the calculation of expressions (6.81), (6.82), (6.84), and (6.85).

The Coulomb interaction is a long range interaction, when a cumulative effect of many light collisions with small scattering angle represents the main contribution to the collisionality. Accordingly, the time step for the Coulomb collisions Δt_c should be sufficiently small: $\langle\chi^2\rangle_{\Delta t}\,(V = V_T) \ll 1$. It is more convenient to formulate this condition in the equivalent following form

$$\nu_c\Delta t_c \ll 1 \qquad \text{and} \qquad \nu_c = \frac{e_1^2 e_2^2}{2\pi\varepsilon_0^2}\,\frac{n\Lambda}{\mu^2 V_T^3}\,, \tag{6.86}$$

where ν_c is the characteristic relaxation time for the given Coulomb collisions [23] and V_T is the thermal velocity of the fastest collided particle species. Although usually $\Delta t_c \gg \Delta t$, the binary collision operator is the most time consuming part of the PIC code. Recently, in order to speed up the collisional plasma simulations a number of updated versions of this operator have been developed (e.g., see [6, 24] and [25]).

6.6.2 Charged-Neutral Particle Collisions

Under realistic conditions the plasma contains different neutral particles, which suffer collisions with the plasma particles. The corresponding collision models used in PIC codes can be divided in two different schemes: Direct Monte-Carlo and null-collision models.

The direct Monte-Carlo model is a common MC scheme when all particles carry information about their collision probability. In this scheme all particles have to be analyzed for a collision probability. Hence, the direct MC requires some additional memory storage and sufficiently large amount of the CPU time.

The null collision method (see [26] and [27]) requires a smaller number of particles to be sampled and it is relatively faster. It uses the fact that in each simulation time step only a small fraction of charged particles suffer collisions with the neutrals. Hence, there is no necessity to analyze all particles. As a first step the maximum collision probability is calculated for each charged particle species

$$P_{\max} = \left(1 - \mathrm{e}^{-\sigma n \Delta s}\right)_{\max} = 1 - \mathrm{e}^{-(\sigma V)_{\max} n_{\max} \Delta t} , \tag{6.87}$$

where $\sigma = \sum \sigma_i(V)$ and n are the total collision cross-section, i.e. the sum of cross-sections σ_i for all possible collision types and the neutral density, respectively. $\Delta s = V\Delta t$ is the distance, which the particle travels per Δt time. Accordingly, the maximum number of particles which can suffer a collision per Δt time is given as $N_{nc} = P_{\max} N \ll N$. As a result only N_{nc} particle per time step have to be analyzed. These N_{nc} particles are randomly chosen, e.g., by using the expression $i = R_j N$ with $j = 1, \ldots, N_{nc}$, where i is the index of the particle to be sampled and R_j are the random numbers between zero and one. The sampling procedure itself includes the calculation of the collision probability of a sampled particle and choosing which kind of collision it should suffer (if any). For this a random number R is compared to the corresponding relative collision probabilities: if

$$R \leq \frac{P_1}{P_{\max}} = \frac{1 - \mathrm{e}^{-\sigma_1 V n \Delta t}}{P_{\max}} \approx \frac{n\sigma_1(V)V}{(\sigma V)_{\max} n_{\max}} , \tag{6.88}$$

a type one collision takes place; else if

$$R \leq \frac{P_1 + P_2}{P_{\max}} \approx \frac{nV(\sigma_1(V) + \sigma_2(V))}{(\sigma V)_{\max} n_{\max}} , \tag{6.89}$$

a type two collision takes place, and so on. If

$$R > \frac{\sum P_i}{P_{\max}} \approx \frac{nV\sum \sigma_i(V)}{(\sigma V)_{\max} n_{\max}} \tag{6.90}$$

no collision takes place.

The difference between the nonlinear and linear null collision methods is the way how the collided neutral particle is treated. In the linear models the neutral velocity is picked up from the prescribed distribution (usually the Maxwellian distribution with the given density and temperature profiles). Contrary to this, in the nonlinear case the motion of neutral particles is resolved in the simulation, and the collided ones are randomly chosen from the same cells, where the colliding charged-particle are.

When the collision partners and corresponding collision types are chosen, the collision itself takes place. Each collision type needs a separate consideration, so that here we discuss the general principle.

The easiest collisions are the ion-neutral charge-exchange collisions. In this case the collision is reduced to an exchange of velocities

$$\boldsymbol{V}_1' = \boldsymbol{V}_2 \qquad \text{and} \qquad \boldsymbol{V}_2' = \boldsymbol{V}_1 \, . \tag{6.91}$$

The recombination collisions are also easy to implement. In this case the collided particles are removed from the simulation and the newly born particle, i.e. the recombination product, has the velocity derived from the momentum conservation

$$\boldsymbol{V}_{\text{new}} = \frac{m_1 \boldsymbol{V}_1 + m_2 \boldsymbol{V}_2}{m_{\text{new}}} \, . \tag{6.92}$$

The elastic collisions are treated in a similar way as the Coulomb collisions using (6.81). The scattering angle depends on the given atomic data. E.g., often it is assumed that the scattering is isotropic

$$\cos\chi = 1 - 2R \, . \tag{6.93}$$

In order to save computational time during the electron-neutral elastic collisions the neutrals are assumed to be at rest. Accordingly, in spite of resolving (6.81) a simplified expression is used for the calculation of the after-collision electron velocity

$$V_{\text{e}}' \approx V_{\text{e}} \sqrt{1 - \frac{2m_{\text{e}}}{M_{\text{n}}} \left(1 - \cos\chi\right)} \, . \tag{6.94}$$

Excitation collisions are done in a similar way as the elastic ones, just before the scattering the threshold energy E_{th} is subtracted from the charged particle energy

$$\boldsymbol{V} \Rightarrow \boldsymbol{V}' = \boldsymbol{V} \sqrt{1 - \frac{E_{\text{th}}}{E}} \Rightarrow \text{scattering} \Rightarrow \boldsymbol{V}'' \, . \tag{6.95}$$

Important to note is that one has to take care on the proper coordinate system, e.g., in (6.95) the first transform should be done in a reference system, where the collided neutral is at rest.

Implementation of inelastic collisions when secondary particles are produced is case dependent. E.g., in electron-neutral ionization collisions, first the neutral particle is removed from the simulation and a secondary electron-ion pair is born. The velocity of this ion is equal to the neutral particle velocity. The velocity of electrons is calculated in the following way. First, the ionization energy is subtracted from the primary electron energy and then the rest is divided between the primary and secondary electrons. This division is done according to given atomic data. After finding these energies the electrons are scattered on the angles χ_{prim} and χ_{sec}.

In a similar way the neutral-neutral and inelastic charged-charged particle collisions can be treated.

6.7 Final Remarks

The material presented above represents just the basics of PIC codes. Nowadays PIC codes use different optimizations including paralleling of the code and memory optimizations (see [5, 6, 7, 8, 9, 10, 11, 12, 13, 14, 15, 16, 17, 18, 19, 20, 21, 22, 23, 24, 25, 26, 27, 28]), developing of more efficient collision operators (see [6, 24, 25, 29]) and grid-less solvers (see [30] and [31]).

It has to be noted, that in the present work we do not consider the analytic theory of the PIC simulation, which can provide useful information on possible numerical oscillation modes and can help to better understand number of conditions to be satisfied in PIC simulation. The interested reader can find the corresponding material in [3] and [4].

References

1. O. Buneman, Phys. Rev. **115**(3), 503 (1959)
2. J. Dawson, Phys. Fluids **5**(4), 445 (1962)
3. C. Birdsall, A. Langdon, *Plasma Physics Via Computer Simulation* (McGraw-Hill, New York, 1985)
4. R. Hockney, J. Eastwood, *Computer Simulation Using Particles* (IOP, Bristol and New York, 1989)
5. V. Decyk, Comput. Phys. Commun. **87**(1-2), 87 (1995)
6. D. Tskhakaya, R. Schneider, J. of Comp. Phys. **225**(1), 829–839 (2007)
7. J. Verboncoeur, M. Alves, V. Vahedi, C. Birdsall, J. Comput. Phys. **104**(2), 321 (1993)
8. J. Verboncoeur, A. Langdon, N. Gladd, Comput. Phys. Commun. **87**(1–2), 199 (1995)
9. D. Barnes, T. Kamimura, J.N. Le Boeuf, T. Tajima, J. Comput. Phys. **52**(3), 480 (1983)
10. D. Tskhakaya, S. Kuhn, Contrib. Plasma Phys. **42**(2–4), 302 (2002)
11. K. Cartwright, J. Verboncoeur, C. Birdsall, J. Comput. Phys. **162**(2), 483 (2000)
12. D. Tskhakaya, S. Kuhn, Plasma Phys. Contr. F. **47**, A327 (2005)
13. F. F. Collino, T. Fouquet, P. Joly, J. Comput. Phys. **211**(1), 9 (2006)
14. V. Vahedi, G. DiPeso, J. Comput. Phys. **13**(1), 149 (1997)
15. W. Press, S. Teukolsky, W. Vetterling, B. Flannery, *Numerical Recipes in C* (Cambridge University Press, Cambridge, New York Port Chester, Melbourne, Sydney, 2002)
16. A. Bergmann, Contrib. Plasma Phys. **38**, 231 (1998)
17. O. Batishchev, X. Xu, J. Byers, R. Cohen, S. Krasheninnikov, T. Rognlien, D. Sigmar, Phys. Plasmas **3**(9), 3386 (1996)
18. O. Batishchev, S. Krasheninnikov, P. Catto, A. Batishcheva, D. Sigmar, X. Xu, J. Byers, T. Rognlien, R. Cohen, M. Shoucri, I. Shkarofskii, Phys. Plasmas **4**(5), 1672 (1997)
19. T. Takizuka, H. Abe, J. Comput. Phys. **25**(3), 205 (1977)

20. N. Krall, A. Trivelpiece, *Principles of Plasma Physics* (San Francisco Press, Inc., Box 6800, San Francisco, 1986)
21. L. Landau, E. Lifshitz, *Course of TheoreticalPhysics*, vol. 1, Mechanics (Pergamon Press, Oxford-London-Paris, 1960)
22. R. Shanny, J. Dawson, J. Greene, Phys. Fluids **10**(6), 1281 (1967)
23. D. Book, *NRL Plasma formulary* (Naval Research Laboratory, Washington D.C., 1978)
24. K. Nanbu, Phys. Rev. E **55**(4), 4642 (1997)
25. A. Bobylev, K. Nanbu, Phys. Rev. E **61**(4), 4576 (2000)
26. C. Birdsall, IEEE T. Plasma Sci. **19**(2), 65 (1991)
27. V. Vahedi, M. Surendra, Comput. Phys. Commun. **87**, 179 (1995)
28. K. Bowers, J. Comput. Phys. **173**(2), 393 (2001)
29. K. Matyash, R. Schneider, A. Bergmann, W. Jacob, U. Fantz, P. Pecher, J. Nucl. Mater. **313-316**, 434 (2003)
30. A. Christlieb, R. Krasny, J. Verboncoeur, IEEE T. Plasma Sci. **32**(2), 384 (2004)
31. A. Christlieb, R. Krasny, J. Verboncoeur, Comput. Phys. Commun. **164**(1–3), 306 (2004)

7 Gyrokinetic and Gyrofluid Theory and Simulation of Magnetized Plasmas

Richard D. Sydora

Department of Physics, University of Alberta, Edmonton, Alberta, Canada T6G 2G7

Charged particle dynamics in slowly varying electromagnetic fields leads to a guiding center formalism in which the particle motion can be described as the sum of a fast gyromotion about the guiding center and a slower drift velocity. Collective oscillations in the magnetized plasma, with frequency below the cyclotron frequency can be effectively studied using this approach since the detailed particle gyration and associated fast time scale does not have to be followed. It is possible to retain the gyro-averaged effects of particle cyclotron motion and include their influence on the self-consistent electric and magnetic fluctuations. Some of the physical properties of these gyrokinetic plasmas in the discrete and continuum limit are presented along with the particle simulation approach. Applications of the simulation model to current-driven and current gradient-driven instabilities are used to illustrate the techniques.

7.1 Introduction

Magnetized plasmas contain a wide range of time and space scales that span many orders of magnitude. This makes realistic simulations of time-dependent phenomena very difficult and capturing all the scales within a single calculation in still beyond reach of our present computational capabilties. Charged particle motion in time-varying, nonuniform electric and magnetic fields, in the presence of collective effects and collisions is very complex. This complexity arises because the inter-particle forces have both a short and long range nature. For the short range, the cross-section of Coulomb collisions strongly decreases with increasing energy of the interacting particles and for lower densities. Therefore, the mean free path of the charged particles in such physical systems as high temperature magnetically-confined fusion plasmas or in low density space and astrophysical plasmas becomes enormous; hundreds to thousands of meters or kilometers. The particle trajectories become more influenced by the electromagnetic forces which are determined by external sources and internal processes. An external source could be a magnetic field which is necessarily confined to a finite volume and is generally curved and inhomogeneous. The Lorentz force that acts of the particles binds them to the magnetic field and forces them to follow the field lines. The internal processes created by collective plasma motions have a range of scales and these also modify the trajectories leading to cross-field or anomalous plasma transport.

R. D. Sydora: *Gyrokinetic and Gyrofluid Theory and Simulation of Magnetized Plasmas*, Lect. Notes Phys. **739**, 191–219 (2008)
DOI 10.1007/978-3-540-74686-7_7

In this chapter we are concerned with collective plasma effects which reside in the low frequency range $\omega < \Omega_i$, where $\Omega_i = eB/m_i$ is the ion cyclotron frequency. This is motivated by the experimental observation [1, 2, 3] that the dominant contribution to low frequency microturbulence in magnetically confined plasmas originate from temporal and spatial scales that are associated with the drift frequency $\omega \simeq \Omega_i(\rho_s/L_\perp)$, where $\rho_s = \sqrt{m_i T_e}/(eB) = \sqrt{T_e/T_i}\ \rho_i$ and ρ_i is the thermal ion gyroradius defined as $\rho_i = v_{ti}/\Omega_i$ with ion thermal velocity $v_{ti} = \sqrt{T_i/m_i}$. Since a typical scale separation between ρ_s and $L_\perp$ in experiment is $\rho_s/L_\perp \sim 10^{-3}$–$10^{-2}$, this makes ω/Ω_i of this order and therefore kinetic simulations using the complete set of Vlasov-Maxwell equations or particle simulations based on the Lorentz-Newton and Maxwell's equations quite impractical.

Another important experimental indication of important physical scales, particularly relevant to convective transport in inhomogeneous magnetized plasmas, is the observed peaks in the wavenumber spectra around $k_\perp\rho_s \simeq 0.2 - 0.5$ in density fluctuation measurements [4, 5]. Therefore, the electric and magnetic fields associated with these fluctuations must include finite-gyroradius effects. The observed characteristics of low frequency turbulent fluctuations suggest an ordering $\omega/\Omega_i \sim \rho_i/L_\perp \sim O(\epsilon)$ and $k_\perp\rho_i \sim O(1)$, which helps in deriving reduced kinetic equations for the evolution of the phase space distribution function that removes all dependence on gyrophase. Thus, the detailed cyclotron time scale does not have to be explicitly followed. Analytical orbit averaging has been used to derive energy and phase space preserving drift-kinetic and gyrokinetic equations of motion.

Gyrokinetic theory was originally developed in the 1960's as an extension to guiding center theory [6] to include the finite gyroradius effects on low frequency, short perpendicular wavelength electrostatic fluctuations in general magnetic geometry [7, 8]. In 1978 Catto [9] develops an important approach for gyrokinetic equations by first transforming the particle coordinates to the guiding center variables in the Vlasov equation (or collisionless Boltzmann equation) before performing the gyrophase averaging. This key result then allowed for a more consistent development of the linear theory [10, 11], an early formulation of nonlinear gyrokinetic theory [12] and a gyrokinetic particle simulation model [13]. In the early 1980's two important advances in guiding center theory occur. First, Boozer [14] develops particle drift motion in magnetic coordinates which greatly simplifies the analysis of orbits in complex geometry and second, Littlejohn [15] develops guiding center theory based on action variational and Lie perturbation methods to obtain phase space conserving equations. This was soon followed by an extension of the method to gyrokinetic theory [16]. In the late 1980's Hahm [17], Brizard [18] and co-workers extend the methodology to general magnetic geometry. There is an excellent recent review on the rigorous perturbation approach using action variational methods [19]. Improved numerical algorithms for performing the gyrophase averaging [20] in 2D were also made in the late 1980's as well as the first 3D gyrokinetic simulations [21, 22]. In the 1990's 3D gyrokinetic particle simulations with general magnetic geometry advanced with the rapid growth in massively parallel computational facilities [23, 24, 25]. Recently 3D toroidal geometry simulations have made it possible to study turbulent fluctuations in magnetically confined fusion plasmas from about the scale size of the ion gyroradius (typically a few millimeters) up to the minor radius of the cross-section (about 0.5–1 m). The anomalous transport coefficients obtained from the models, such as the ion heat diffusivity are well within the experimental range [26]. The anomalous electron thermal diffusivity is not well understood and there are indications that fluctuations scales near the electron gyroradius need to be included [27].

The basic gyrokinetic equations can also be used to formulate reduced or continuum equations representing the time evolution of moments such as density, current and pressure [16, 28, 29, 30, 31]. The two-fluid equations can be used to capture the different dynamics of electrons and ions parallel and perpendicular to the magnetic field and the coupling between both species by electric and magnetic field interactions. These, so-called gyrofluid models are able to incorporate the finite gyroradius, ion polarization drift and coupling to sound waves. There are many computational advantages in using these continuum-based models in addition to efficiency, such as the clear indentification of important fluid-type nonlinearities, inclusion of sources and handling of more collisional regimes.

In this chapter, we present some of the key steps in development of gyrokinetic and gyrofluid models starting with single particle motion in a magnetic field. Once the basic transformation from gyro-center to gyrophase averaged coordinates is established, it is possible to construct a kinetic theory or self-consistent gyrokinetic Vlasov-Poisson-Ampere equations which form the basis of an N-body particle simulation model. The set of equations possess an energy invariant which can be used to precisely monitor the exchange of energy among the fields and particles and the system is inherently phase space conserving. The moments of the fundamental gyrokinetic equations lead to a set of gyrofluid partial differential equations which form the basis of magnetized fluid simulations in the low frequency regime and some of the steps in the derivation are presented. The elements of the gyrokinetic particle simulation approach are discussed along with the fundamental normal modes and equilibrium statistical properties of the model. To illustrate the techniques, the results from a couple of example simulations in simpler geometry (slab or Cartesian) are presented and are related to current-driven and current gradient-driven microinstabilities as a potential source of low frequency turbulence in laboratory and space plasmas.

7.2 Single Particle Dynamics

Since the distribution function is conserved along particle trajectories in collisionless kinetics, it is useful to first discuss single particle dynamics. We begin by examining the classical treatment of charged particle motion based on the Newton-Lorentz form and then use this to motivate the guiding center transformation [6, 32] and drift dynamics. Then, the gyro-drift formulation is presented where the resultant equations contain only the slower time scales of interest such as the transit and drift time scales but retain the gyro-averaged effects of particle cyclotron motion. This leads to computationally efficient methods for the N-body dynamics. To obtain the gyro-drift equations we utilize the more modern approach using action-variational Lie perturbuation methods applied to single particle motion [15, 16] under the influence of strong ambient magnetic field and electromagnetic perturbation. This preserves the Hamiltonian structure of the system under coordinate system changes.

7.2.1 Full Particle and Drift Motion

When the dominant force acting on the individual particles in a plasma is electromagnetic, the equations of motion for a particle with mass m and charge q in the electromagnetic fields $\boldsymbol{E}(\boldsymbol{r},t), \boldsymbol{B}(\boldsymbol{r},t)$ are

$$\frac{\mathrm{d}\boldsymbol{x}}{\mathrm{d}t} = \boldsymbol{v} , \tag{7.1}$$

$$\frac{\mathrm{d}\boldsymbol{v}}{\mathrm{d}t} = \frac{q}{m}(\boldsymbol{E} + \boldsymbol{v} \times \boldsymbol{B}) . \tag{7.2}$$

Each of the N plasma particles satisfy such equations and the solution of the $6N$ equations are the particle trajectories. These trajectories determine the local charge and current density

$$\rho(\boldsymbol{r},t) = \sum_j q_j \delta(\boldsymbol{r} - \boldsymbol{r}_j(t)) ,$$
$$\boldsymbol{J}(\boldsymbol{r},t) = \sum_j \boldsymbol{v}_j q_j \delta(\boldsymbol{r} - \boldsymbol{r}_j(t)) , \tag{7.3}$$

which become the sources in Maxwell's equations.

For the situation with a strong ambient magnetic field, the particle motion can be effectively described as a fast gyro-motion about a slowly moving guiding center or gyro-center. When the electromagnetic fields are slowly varying, on a time scale longer than the gyro-period, we may treat the gyro-phase as an ignorable coordinate and effectively utilize the adiabatic invariant associated with the wide separation of time scales. This can reduce the dynamical phase space variables from six to four. Before a more formal presentation of gyro-drift dynamics, it is useful to present a simplified example of drift motion to motivate the guiding center transformation.

If we consider the case where electric and magnetic fields are constant is space and time, and orthogonal $\boldsymbol{E} \perp \boldsymbol{B}$, then the forces in the direction of the magnetic field vanish so that the motion along the field line is uniform. In other words, $v_{\|} = v_{\|}(t = 0)$, where $v_{\|} = \boldsymbol{v} \cdot \boldsymbol{b}$ is the particle velocity along the homogeneous and uniform magnetic field and $\boldsymbol{b} = \boldsymbol{B}/B$ is the unit vector in the direction of $\boldsymbol{B}$. For the motion perpendicular to the field line, we introduce a velocity $\boldsymbol{u}$ which is related to $\boldsymbol{v}$ by

$$\boldsymbol{v} = \boldsymbol{u} + \boldsymbol{v}_{\mathrm{E}} + v_{\|}\boldsymbol{b} , \tag{7.4}$$

where we define $\boldsymbol{v}_{\mathrm{E}} = \boldsymbol{E} \times \boldsymbol{B}/B^2$ describing uniform motion in a direction perpendicular to $\boldsymbol{E}$ and $\boldsymbol{B}$, assumed constant. If we substitute (7.4) into (7.2) we obtain

$$\frac{\mathrm{d}\boldsymbol{u}}{\mathrm{d}t} = \frac{q}{m}\boldsymbol{u} \times \boldsymbol{B} , \tag{7.5}$$

which has the harmonic solution

$$\boldsymbol{u} = u(\boldsymbol{e}_1 \cos(\theta) - \boldsymbol{e}_2 \sin(\theta)) , \tag{7.6}$$

where $\theta = \Omega_{\mathrm{c}} t - \theta_0$ is the phase angle of gyro-rotation and $\Omega_{\mathrm{c}} = qB/m$ is the gyrofrequency. The position of the charge particle gyrating along a circle centered at position $\boldsymbol{R}$ on a field line with constant velocity u is

$$\boldsymbol{\rho} = \rho(\boldsymbol{e}_1 \sin(\theta) + \boldsymbol{e}_2 \cos(\theta)) \tag{7.7}$$

and $\rho = u/\Omega_{\mathrm{c}}$ is the gyroradius. Particles with opposite sign of charge move along the gyro-orbits in opposite directions. The position of the particle with respect to the gyro-center and gyroradius is $\boldsymbol{r}(t) = \boldsymbol{R}(t) + \boldsymbol{\rho}(t)$ and this is shown schematically in Fig. 7.1. The gyro-center moves with a velocity

$$\frac{\mathrm{d}\boldsymbol{R}}{\mathrm{d}t} = \frac{1}{T_{\mathrm{c}}} \int_0^{T_{\mathrm{c}}} \boldsymbol{v} \, \mathrm{d}t = v_{\parallel}\boldsymbol{b} + \boldsymbol{v}_{\mathrm{E}} , \tag{7.8}$$

where $T_{\mathrm{c}} = 2\pi/\Omega_{\mathrm{c}}$ is the gyroperiod.

The particle trajectory can be described as a helix around the gyro-center due to the combined ballistic parallel and perpendicular gyrational motion and this helix slowly drifts with velocity $\boldsymbol{v}_{\mathrm{E}}$ in the cross-field direction. To lowest order, this slow drift is identical for electrons and ions because it is independent of mass and charge. Therefore, there is no current density associated with it. An adiabatic invariant can

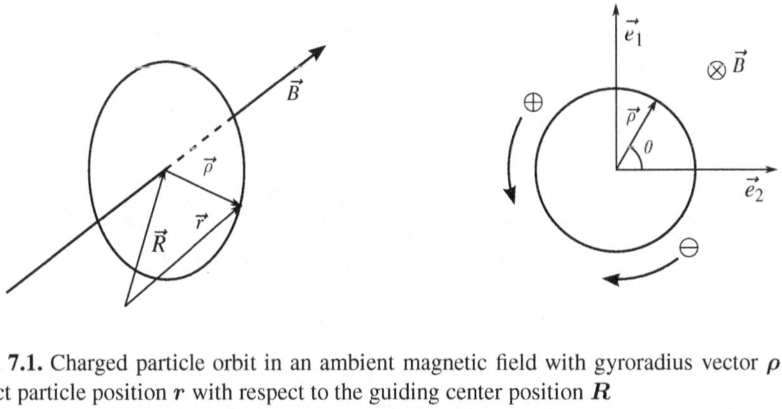

Fig. 7.1. Charged particle orbit in an ambient magnetic field with gyroradius vector $\boldsymbol{\rho}$ and exact particle position $\boldsymbol{r}$ with respect to the guiding center position $\boldsymbol{R}$

also be derived from this time scale separation. If we consider the current associated with the particle gyro-motion $I = q/T_c$, the flux enclosed by the current loop is the magnetic moment $\mu = I\pi\rho^2$ which is therefore

$$\mu = q\frac{\Omega_c}{2\pi}\pi\rho^2 = \frac{mu^2}{2B} \,. \tag{7.9}$$

We next consider the electromagnetic fields that are no longer constant but vary in space and time. Computational advantages over following exact particle motion arise when an ordering of the multiple scales is applied. If we introduce a characteristic length L and time τ over which the fields vary and these satisfy

$$\frac{\rho}{L} \ll 1 \,, \qquad \frac{1}{\Omega_c \tau} \ll 1 \,, \tag{7.10}$$

then we can extend the basic drift dynamics equations. Since Ω_c^{-1} is proportional to m/q this may be adopted as a smallness parameter and allows us to obtain the inertial corrections to the drift motion and magnetic moment. To lowest order we have shown that the gyro-center position is given by a vector version of the simple uniform field result. However, the exact gyro-center position is more complicated. We may re-write the original Lorentz-Newton equations (7.1) and (7.2), in terms of guiding center coordinates $(\boldsymbol{R}, v_\perp, v_\parallel, \phi)$, making use of cylindrical coordinates

$$\boldsymbol{u} = v_\parallel \boldsymbol{b} + v_\perp(\boldsymbol{e}_1 \cos(\phi) + \boldsymbol{e}_2 \sin(\phi)) = \boldsymbol{v} - \boldsymbol{v}_E \tag{7.11}$$

and $\boldsymbol{\rho} = \boldsymbol{b} \times \boldsymbol{u}/\Omega_c$, $\boldsymbol{R} = \boldsymbol{r} - \boldsymbol{\rho}$, with local orthogonal unit vectors $(\boldsymbol{e}_1, \boldsymbol{e}_2, \boldsymbol{b})$ satisfying $\boldsymbol{e}_\perp = \boldsymbol{e}_1 \cos(\phi) + \boldsymbol{e}_2 \sin(\phi)$ and $\boldsymbol{e}_\phi = \boldsymbol{b} \times \boldsymbol{e}_\perp = -\boldsymbol{e}_1 \sin(\phi) + \boldsymbol{e}_2 \cos(\phi)$. These transformation relations lead to the gyro-center equations of motion

$$\frac{\mathrm{d}\boldsymbol{R}}{\mathrm{d}t} = v_\parallel \boldsymbol{b} + \boldsymbol{v}_E + \frac{1}{\Omega_c}\boldsymbol{b} \times \frac{\mathrm{d}\boldsymbol{v}_E}{\mathrm{d}t} + \boldsymbol{u} \times \frac{\mathrm{d}}{\mathrm{d}t}\left(\frac{\boldsymbol{b}}{\Omega_c}\right) , \tag{7.12}$$

$$\frac{\mathrm{d}v_\perp}{\mathrm{d}t} = -\boldsymbol{e}_\perp \cdot \left(v_\parallel \frac{\mathrm{d}\boldsymbol{b}}{\mathrm{d}t} + \frac{\mathrm{d}\boldsymbol{v}_E}{\mathrm{d}t}\right) , \tag{7.13}$$

$$\frac{\mathrm{d}v_\parallel}{\mathrm{d}t} = \frac{q}{m}E_\parallel + \boldsymbol{v}_E \cdot \frac{\mathrm{d}\boldsymbol{b}}{\mathrm{d}t} + v_\perp \boldsymbol{e}_\perp \cdot \frac{\mathrm{d}\boldsymbol{b}}{\mathrm{d}t} , \tag{7.14}$$

$$\frac{\mathrm{d}\phi}{\mathrm{d}t} = -\Omega_c - \boldsymbol{e}_2 \cdot \frac{\mathrm{d}\boldsymbol{e}_1}{\mathrm{d}t} - \frac{1}{v_\perp}\boldsymbol{e}_\phi \cdot \left(v_\parallel \frac{\mathrm{d}\boldsymbol{b}}{\mathrm{d}t} + \frac{\mathrm{d}\boldsymbol{v}_E}{\mathrm{d}t}\right) , \tag{7.15}$$

where the total time dervative is taken along the particle trajectory $\mathrm{d}/\mathrm{d}t = \partial/\partial t + (v_\parallel \boldsymbol{b} + \boldsymbol{v}_E + v_\perp \boldsymbol{e}_\perp) \cdot \boldsymbol{\nabla}$. In order to express the fields as a function of the guiding center position, $\boldsymbol{R}$, we use a Taylor expansion around this point $\boldsymbol{b}(\boldsymbol{r}, t) = \boldsymbol{b}(\boldsymbol{R}, t) + \boldsymbol{\rho} \cdot \boldsymbol{\nabla}\boldsymbol{b}(\boldsymbol{R}, t) + \ldots$.

From these transformed equations we note several important points.

(i) The ordering we established indicates that the fastest timescale is gyromotion since $\mathrm{d}\phi/\mathrm{d}t \simeq -\Omega_c$ is the largest term.

(ii) $E_{\|}$ must be small because the q/m factor in front of it is large. If $E_{\|}$ were large, the particles would accelerate on the time scale of gyration and in opposite directions. This would create a charge separation and generate electric fields on the shortest time scale, violating our initial assumption of slowly varying fields.
(iii) The parallel and perpendicular velocities are slowly varying and may be considered constant on the fast gyrofrequency time scale.
(iv) Lastly, the inertial corrections to the gyro-center motion are obtained such as the polarization drift, the third term on the right hand side of (7.12), which is associated with time varying electric fields and is much larger for the ions due to its proportionality to mass.

It is possible to continue working with these equations and derive gyrophase independent drift motion. We will not go through the detailed steps here but state the result [32] for the guiding center drift velocity to first order in the parameter m/q is

$$\begin{aligned}\frac{\mathrm{d}\boldsymbol{R}}{\mathrm{d}t} &= \left(v_{\|} + \frac{v_{\perp}^2}{2\Omega_{\mathrm{c}}}\boldsymbol{b}\cdot\boldsymbol{\nabla}\times\boldsymbol{b}\right)\boldsymbol{b} + \boldsymbol{v}_{\mathrm{E}} + \frac{v_{\perp}^2}{2\Omega_{\mathrm{c}}}\boldsymbol{b}\times\frac{\boldsymbol{\nabla}B}{B} \\ &\quad + \frac{\boldsymbol{b}}{\Omega_{\mathrm{c}}}\times\left(v_{\|}\frac{\mathrm{d}\boldsymbol{b}}{\mathrm{d}t'} + \frac{\mathrm{d}\boldsymbol{v}_{\mathrm{E}}}{\mathrm{d}t'}\right),\end{aligned} \tag{7.16}$$

where $\mathrm{d}/\mathrm{d}t' = \partial/\partial t + (v_{\|}\boldsymbol{b} + \boldsymbol{v}_E)\cdot\boldsymbol{\nabla}$ and all fields are taken at the guiding center position. The first term on the right hand side of (7.16) is the particle transit motion along the field, the second term is the $\boldsymbol{E}\times\boldsymbol{B}$ drift motion and the third term is the gradient-B drift. This gradient-B drift arises because, as the particle gyrates in the inhomogeneous field, it periodically experiences stronger and weaker field strengths along its gyro-orbit, leading to a net drift motion in the direction perpendicular to $\boldsymbol{B}$ and $\boldsymbol{\nabla}B$. The final two terms in (7.16) are the curvature drift and polarization drift effects, respectively.

7.2.2 Gyro-Drift Particle Motion

If the electromagnetic field variations occur on the spatial scale of the gyro-orbit then we must include the gyrophase averaging effects in the equations of motion. In order to determine the proper gyrophase-averaged equations of motion we must make a transformation from gyro-center to gyrophase-averaged coordinates. A particularly powerful way to obtain these equations is to use a combination of one-form mechanics and Lie perturbation theory [19]. The one-form or Lagrangian completely determines the equations of motion which are obtained by the least action principle. Lie perturbation theory can be used to generate equations of motion that are gyrophase independent to any desired order. In this formalism the particle dynamics transforms simply under coordinate change since a one-form transforms covariantly. The coordinate transformations are general and need not be canonical and are guaranteed to be phase space preserving.

To illustrate the procedures, we consider straight magnetic fields and neglect the curvature and gradient-B drift effects. The single particle Lagrangian can be written in terms of canonical variables $(\boldsymbol{P},\boldsymbol{Q})$

$$L\mathrm{d}t = \boldsymbol{P} \cdot \mathrm{d}\boldsymbol{Q} - H(\boldsymbol{P}, \boldsymbol{Q})\mathrm{d}t \; , \tag{7.17}$$

where for single charged-particle motion in a fixed magnetic field has the Hamiltonian

$$H = \frac{1}{2m}(\boldsymbol{P} - q\boldsymbol{A}_0(\boldsymbol{Q}))^2 + q\Phi(\boldsymbol{Q}, t) \; , \tag{7.18}$$

where $\boldsymbol{A}_0$ is the vector potential of the background magnetic field and the canonical variables are

$$\begin{aligned} \boldsymbol{P}(t) &= m\boldsymbol{v}(t) + q\boldsymbol{A}_0(\boldsymbol{r}(t)) \; , \\ \boldsymbol{Q}(t) &= \boldsymbol{r}(t) \; . \end{aligned} \tag{7.19}$$

The Lagrangian can be written as the one-form of Poincaré-Cartan [33]

$$\gamma = \gamma_\mu \mathrm{d}z^\mu = \gamma_i \mathrm{d}z^i - h\mathrm{d}t \; , \tag{7.20}$$

where $\boldsymbol{z} = \boldsymbol{z}(\boldsymbol{P}, \boldsymbol{Q}, t)$ is the phase space coordinate system and in the summation convention $\mu = 0, ..., 6$ and $i = 1, ..., 6$, with $z^0 = t$, $\gamma_i = \boldsymbol{P} \cdot \partial \boldsymbol{Q}/\partial z^i$, and $h = H - \boldsymbol{P} \cdot \partial \boldsymbol{Q}/\partial t$. Therefore, the one-form can be written as

$$\gamma = (q\boldsymbol{A}_0(\boldsymbol{r}) + m\boldsymbol{v}) \cdot \mathrm{d}\boldsymbol{r} - \left(\frac{mv^2}{2} + q\Phi(\boldsymbol{r}, t)\right) \mathrm{d}t \tag{7.21}$$

and $\boldsymbol{r}$ and $\boldsymbol{v}$ can be obtained from the canonical variables $\boldsymbol{Q}$ and $\boldsymbol{P}$. $\boldsymbol{A}_0$ is the vector potential of the background magnetic field. The action associated with the one-form is given by

$$S = \int_{t_0}^{t_f} \gamma_\mu \frac{\mathrm{d}z^\mu}{\mathrm{d}t} \mathrm{d}t \tag{7.22}$$

and mimimization of S with respect to variations in z^μ leads to the Euler-Lagrange equations

$$\left(\frac{\partial \gamma_\mu}{\partial z^\nu} - \frac{\partial \gamma_\nu}{\partial z^\mu}\right) \frac{\mathrm{d}z^\mu}{\mathrm{d}t} = 0 \; . \tag{7.23}$$

As in standard perturbation theory, we separate the one-form γ into the sum of an easily solvable part of the motion γ_0 and a perturbation γ_1, which in our case will contain the gyrophase dependence. The one-form in gyro-center coordinates $\boldsymbol{z} = (\boldsymbol{R}, v_\parallel, \widehat{\mu}, \theta; t)$ with $\widehat{\mu} = mv_\perp^2/2\Omega_\mathrm{c}$, $\theta = \tan^{-1}(\boldsymbol{v} \cdot \boldsymbol{e}_2/\boldsymbol{v} \cdot \boldsymbol{e}_1)$ and $(\boldsymbol{e}_2, \boldsymbol{e}_1)$ are the unit vectors in (x, y), is

$$\begin{aligned} \gamma_0 &= q\boldsymbol{A}_0 \cdot \mathrm{d}\boldsymbol{R}_\perp + mv_z \mathrm{d}R_z + \widehat{\mu}\mathrm{d}\theta - h_0 \mathrm{d}t \; , \\ \gamma_1 &= -h_1 \mathrm{d}t \; , \end{aligned} \tag{7.24}$$

where $h_0 = \widehat{\mu}\Omega_\mathrm{c} + mv_z^2/2$ is the zeroth order Hamiltonian and $h_1 = q\Phi(\boldsymbol{R} + \boldsymbol{\rho}, t)$, $\boldsymbol{\rho} = \boldsymbol{b} \times \boldsymbol{v}/\Omega_\mathrm{c}$ and Φ is the electric potential. An Euler-Lagrange equation of motion can be derived from this one-form and it describes the fast periodic gyromotion about θ. The θ-dependent non-secular perturbation is removed by transforming the fundamental one-form to gyrophase-averaged coordinates $\bar{Z} = (\bar{\boldsymbol{R}}, \bar{v}_z, \bar{\widehat{\mu}}, \bar{\theta}; t)$. This is accomplished by using the Lie transform which gives the fundamental one-form in the new coordinate system as [16, 34]

$$\begin{aligned}\bar{\Gamma}_0 &= \gamma_0 \\ \bar{\Gamma}_1 &= \mathrm{d}S_1 - L_1\gamma_0 + \gamma_1\end{aligned} \tag{7.25}$$

with $(L_1\gamma_0)_\nu = g_1^\mu(\partial_\nu\gamma_{0\mu} - \partial_\mu\gamma_{0\nu})$. The generating function S_1 is

$$S_1 = \frac{q}{\Omega_\mathrm{c}} \int \mathrm{d}\theta' \widetilde{\Phi} \tag{7.26}$$

and is related to the difference between the gyro-center and gyro-averaged potential $\widetilde{\Phi} = \Phi - \langle\Phi\rangle_\theta$ with $\langle\Phi\rangle_\theta = 1/(2\pi)\int_0^{2\pi} \Phi(\boldsymbol{R} + \boldsymbol{\rho}, t)\mathrm{d}\theta$.

The generator, g_1^μ, is also obtained within the low frequency ordering as

$$\begin{aligned}g_1^{\boldsymbol{R}} &= \frac{1}{qB}\boldsymbol{\nabla}_{\boldsymbol{R}} S_1 \times \boldsymbol{b}\,, \\ g_1^{v_z} &= \frac{1}{m}\boldsymbol{b}\cdot\boldsymbol{\nabla}_{\boldsymbol{R}} S_1\,, \\ g_1^{\widehat{\mu}} &= \frac{\partial S_1}{\partial\theta}\,, \\ g_1^{\theta} &= -\frac{\partial S_1}{\partial\widehat{\mu}}\,.\end{aligned} \tag{7.27}$$

The fundamental one-form in the gyro-averaged coordinates becomes

$$\bar{\Gamma} = q\boldsymbol{A}_0\cdot\mathrm{d}\bar{\boldsymbol{R}}_\perp + m\bar{v}_z\mathrm{d}\bar{R}_z + \bar{\widehat{\mu}}\mathrm{d}\bar{\theta} - \bar{h}\mathrm{d}t \tag{7.28}$$

with gyrophase-averaged Hamiltonian

$$\bar{h} = \bar{\widehat{\mu}}\Omega_\mathrm{c} + \frac{1}{2}m\bar{v}_z^2 + q\langle\Phi\rangle_{\bar{\theta}}\,. \tag{7.29}$$

By taking the variation of $\bar{\Gamma}$ we obtain the Euler-Lagrange equation for the particle motion in gyro-averaged coordinates, to first order

$$\begin{aligned}\frac{\mathrm{d}\bar{\boldsymbol{R}}}{\mathrm{d}t} &= \bar{v}_z\boldsymbol{b} + \frac{\boldsymbol{b}\times\boldsymbol{\nabla}_{\bar{\boldsymbol{R}}}\langle\Phi\rangle_{\bar{\theta}}}{B}\,, \\ \frac{\mathrm{d}\bar{v}_z}{\mathrm{d}t} &= -\frac{q}{m}\boldsymbol{b}\cdot\boldsymbol{\nabla}_{\bar{\boldsymbol{R}}}\langle\Phi\rangle_{\bar{\theta}}\,, \\ \frac{\mathrm{d}\bar{\widehat{\mu}}}{\mathrm{d}t} &= 0\,, \\ \frac{\mathrm{d}\bar{\theta}}{\mathrm{d}t} &= \Omega_\mathrm{c} + q\frac{\partial\langle\Phi\rangle_{\bar{\theta}}}{\partial\bar{\widehat{\mu}}}\,.\end{aligned} \tag{7.30}$$

It is straightforward to generate the higher order corrections to the gyro-averaged drift motion, parallel acceleration and magnetic moment using this formalism [16].

7.3 Continuum Gyrokinetics

7.3.1 Gyrokinetic Vlasov Equation

The single particle gyro-drift dynamics can be used to obtain a Vlasov equation for the gyro-averaged particle distribution function $\bar{F}(\bar{Z})$. The gyrokinetic Vlasov equation for species α is

$$\frac{\partial \bar{F}_\alpha}{\partial t} + \frac{\mathrm{d}\bar{\boldsymbol{R}}_\alpha}{\mathrm{d}t} \cdot \frac{\partial \bar{F}_\alpha}{\partial \bar{\boldsymbol{R}}} + \frac{\mathrm{d}\bar{v}_{z\alpha}}{\mathrm{d}t}\frac{\partial \bar{F}_\alpha}{\partial \bar{v}_z} = 0 \tag{7.31}$$

and for the ions becomes

$$\frac{\partial \bar{F}_i}{\partial t} + \left(\bar{v}_z \boldsymbol{b} + \frac{\boldsymbol{b} \times \boldsymbol{\nabla}_{\bar{\boldsymbol{R}}}\langle \Phi \rangle_{\bar{\theta}}}{B}\right) \cdot \frac{\partial \bar{F}_i}{\partial \bar{\boldsymbol{R}}} - \frac{e}{m_\mathrm{i}} \boldsymbol{b} \cdot \boldsymbol{\nabla}_{\bar{\boldsymbol{R}}} \langle \Phi \rangle_{\bar{\theta}} \frac{\partial \bar{F}_i}{\partial \bar{v}_z} = 0\,. \tag{7.32}$$

For electrons with the smaller gyroradius $\rho_\mathrm{e} \ll \rho_\mathrm{i}$, leads to the drift-kinetic equation

$$\frac{\partial \bar{F}_\mathrm{e}}{\partial t} + \left(\bar{v}_z \boldsymbol{b} + \frac{\boldsymbol{b} \times \boldsymbol{\nabla}_{\bar{\boldsymbol{R}}}\Phi}{B}\right) \cdot \frac{\partial \bar{F}_\mathrm{e}}{\partial \bar{\boldsymbol{R}}} + \frac{e}{m_\mathrm{e}} \boldsymbol{b} \cdot \boldsymbol{\nabla}_{\bar{\boldsymbol{R}}} \Phi \frac{\partial \bar{F}_\mathrm{e}}{\partial \bar{v}_z} = 0\,. \tag{7.33}$$

The particle trajectories correspond to the characteristics of the gyrokinetic Vlasov partial differential equation and is the basis for the gyrokinetic particle-in-cell simulation model discussed in Sect. 7.5.

7.3.2 Field Equations

In order to obtain the self-consistent electric potential for the gyrokinetic Vlasov equation, we must consider the density response in real space $\boldsymbol{r}$ and not in $\boldsymbol{R}$. The Lie transform can help us relate the distribution function in the gyro-averaged coordinates $\bar{F}$ to the gyro-center coordinates f. To first order

$$f(\boldsymbol{R}, v_z, \widehat{\mu}, \theta; t) \simeq (1 + g^\mu \partial_\mu)\bar{F}(\bar{\boldsymbol{R}}, \bar{v}_z, \bar{\widehat{\mu}}, \bar{\theta}; t) \tag{7.34}$$

and the particle density in real space becomes

$$\begin{aligned} n(\boldsymbol{r}, t) &= \int f(\boldsymbol{R}, v_z, \widehat{\mu}, \theta; t)\delta(\boldsymbol{R} - \boldsymbol{r} + \boldsymbol{\rho}) J \mathrm{d}^6 z \\ &\simeq \int \bar{F}(\bar{\boldsymbol{R}}, \bar{v}_z, \bar{\widehat{\mu}}, t)\delta(\bar{\boldsymbol{R}} - \boldsymbol{r} + \bar{\boldsymbol{\rho}})\bar{J}\mathrm{d}^6\bar{Z} \\ &\quad + \int \Bigg[\left(\frac{\boldsymbol{\nabla}_{\bar{\boldsymbol{R}}} S_1 \times \boldsymbol{b}}{B} - \frac{1}{m}\frac{\partial S_1}{\partial \bar{v}_z}\boldsymbol{b} \right) \cdot \boldsymbol{\nabla}_{\bar{\boldsymbol{R}}} \bar{F} \\ &\quad + \frac{1}{m} \boldsymbol{b} \cdot \boldsymbol{\nabla}_{\bar{\boldsymbol{R}}} S_1 \frac{\partial \bar{F}}{\partial \bar{v}_z} + \frac{\partial S_1}{\partial \bar{\theta}}\frac{\partial \bar{F}}{\partial \bar{\widehat{\mu}}} \Bigg] \delta(\bar{\boldsymbol{R}} - \boldsymbol{r} + \bar{\boldsymbol{\rho}})\bar{J}\mathrm{d}^6\bar{Z}\,, \end{aligned} \tag{7.35}$$

where J and $\bar{J}$ are the Jacobians of the gyro-center and gyro-averaged coordinates, respectively. In evaluating the second term, we can linearize the distribution about a local Maxwellian defined as

$$\bar{F}_{\mathrm{M}}(\bar{\boldsymbol{R}}, \bar{v}_z, \bar{\bar{\mu}}) = \frac{n_0(\bar{\boldsymbol{R}})}{(2\pi T(\bar{\boldsymbol{R}})/m)^{3/2}} e^{-(m\bar{v}_z^2/2+\bar{\bar{\mu}}\Omega_{\mathrm{c}})/T(\bar{\boldsymbol{R}})} \tag{7.36}$$

and use the ordering $\rho/L \ll 1$, where L is the density and temperature gradient scale variation, to show that the leading order term is $(\partial S_1/\partial\bar{\theta})(\partial \bar{F}_{\mathrm{M}}/\partial\bar{\bar{\mu}}) = q/(\Omega_{\mathrm{c}})$ $(\Phi - \langle\Phi\rangle_{\bar{\theta}})(\partial \bar{F}_{\mathrm{M}}/\partial\bar{\bar{\mu}})$ and the particle density simplifies to

$$n(\boldsymbol{r},t) = \int \left[\bar{F}(\bar{\boldsymbol{R}}, \bar{v}_z, \bar{\bar{\mu}}, t) + \frac{q}{\Omega_{\mathrm{c}}}(\Phi - \langle\Phi\rangle_{\bar{\theta}})\frac{\partial \bar{F}_{\mathrm{M}}}{\partial\bar{\bar{\mu}}}\right] \delta(\bar{\boldsymbol{R}} - \boldsymbol{r} + \bar{\boldsymbol{\rho}})\bar{J}\mathrm{d}^6\bar{Z}\,. \tag{7.37}$$

This expression can be used to construct a Poisson equation by taking the difference between the electron and ion number densities

$$\begin{aligned} &\frac{e}{\Omega_{\mathrm{i}}}\int (\Phi - \langle\Phi\rangle_{\bar{\theta}})\frac{\partial \bar{F}_{\mathrm{M}}}{\partial\bar{\bar{\mu}}}\delta(\bar{\boldsymbol{R}} - \boldsymbol{r} + \bar{\boldsymbol{\rho}}_{\mathrm{i}})\bar{J}_i\mathrm{d}^6\bar{Z} \\ &= \int \bar{F}_{\mathrm{i}}\delta(\bar{\boldsymbol{R}} - \boldsymbol{r} + \bar{\boldsymbol{\rho}}_{\mathrm{i}})\bar{J}_i\mathrm{d}^6\bar{Z} - \int \bar{F}_{\mathrm{e}}\delta(\bar{\boldsymbol{R}} - \boldsymbol{r})\bar{J}_{\mathrm{e}}\mathrm{d}^6\bar{Z}\,, \end{aligned} \tag{7.38}$$

where the small gyroradius limit for the electrons $\rho_{\mathrm{e}} \to 0$ has been taken. Using F_{m} from (7.36), the gyrokinetic Poisson equation becomes

$$\frac{\tau}{\lambda_{\mathrm{De}}^2}(\Phi - \widetilde{\Phi}) = 4\pi e(\bar{n}_{\mathrm{i}} - n_{\mathrm{e}}) \tag{7.39}$$

with $\tau = T_{\mathrm{e}}/T_{\mathrm{i}}$, $\lambda_{\mathrm{De}}^2 = T_{\mathrm{e}}/(4\pi n_{\mathrm{e}}e^2)$ and

$$\widetilde{\Phi}(\boldsymbol{r}) = \left\langle \int \langle\Phi\rangle_{\bar{\theta}}(\bar{\boldsymbol{R}})\bar{F}_{\mathrm{m}}\delta(\bar{\boldsymbol{R}} - \boldsymbol{r} + \bar{\boldsymbol{\rho}})\,\mathrm{d}\bar{\boldsymbol{R}}\,\mathrm{d}\bar{\bar{\mu}}\,\mathrm{d}\bar{v}_z\right\rangle_{\bar{\theta}}, \tag{7.40}$$

which physically, is the electrostatic potential viewed by the gyrocenter.

The electrostatic potential can be written in an operator form, convenient for numerical computation, by using a Fourier representation. The electrostatic potential in the particle or laboratory coordinates is

$$\Phi(\boldsymbol{r}) = \sum_{\boldsymbol{k}} \Phi_{\boldsymbol{k}} \mathrm{e}^{\mathrm{i}\boldsymbol{k}\cdot\boldsymbol{r}} \tag{7.41}$$

and the gyro-averaged potential is

$$\langle\Phi(\bar{\boldsymbol{R}})\rangle_{\bar{\theta}} = \sum_{\boldsymbol{k}} \Phi_{\boldsymbol{k}} J_0\left(\frac{k_\perp v_\perp}{\Omega_{\mathrm{i}}}\right)\mathrm{e}^{\mathrm{i}\boldsymbol{k}\cdot\bar{\boldsymbol{R}}} \tag{7.42}$$

using the Bessel function relation $J_0(k_\perp v_\perp/\Omega_{\mathrm{i}}) = \int_0^{2\pi}\exp(\pm\mathrm{i}\boldsymbol{k}\cdot\bar{\boldsymbol{\rho}})\mathrm{d}\bar{\theta}/(2\pi)$. The J_0 represents the difference between the potential at the guiding center and the averaged potential. The gyrocenter electrostatic potential transformed to the particle coordinates involves a double gyrophase average and assuming the Maxwellian distribution for $\bar{F}_{\text{m}}$ we obtain

$$\widetilde{\Phi}(\boldsymbol{r}) = \sum_{\boldsymbol{k}} \Phi_{\boldsymbol{k}} \Gamma_0(b) \mathrm{e}^{\mathrm{i}\boldsymbol{k}\cdot\boldsymbol{r}} , \tag{7.43}$$

where $\Gamma_0(b) = \int J_0^2(k_\perp v_\perp/\Omega_{\text{i}})\bar{F}_{\text{m}}(\bar{\mu})d\bar{\mu} = I_0(b)\exp(-b)$. I_0 is the modified Bessel function with argument $b = k_\perp^2 \rho_{\text{i}}^2$ with ion thermal gyroradius $\rho_{\text{i}} = v_{\text{thi}}/\Omega_{\text{i}}$ and $v_{\text{thi}} = (T_{\text{i}}/m_{\text{i}})^{1/2}$. The Poisson equation (7.39) in this operator form is therefore expressed as

$$\frac{\tau}{\lambda_{\text{De}}^2}(1-\Gamma_0)\Phi = 4\pi e(\bar{n}_{\text{i}} - n_{\text{e}}) \tag{7.44}$$

and makes the $1-\Gamma_0$ operator easy to invert.

Returning to (7.38), we can expand the delta functions about $\bar{\boldsymbol{R}} - \boldsymbol{r}$ for the ions and the left-hand-side of the Poisson equation becomes

$$(\nabla_\perp \cdot \frac{\omega_{\text{pi}}^2}{\Omega_{\text{i}}^2} \nabla_\perp)\Phi = -4\pi e(\bar{n}_{\text{i}} - n_{\text{e}}) \tag{7.45}$$

and the ion plasma frequency is $\omega_{\text{pi}} = (4\pi n e^2/m_{\text{i}})^{1/2}$. This can also be obtained from (7.44) in the long wavelength limit $b \ll 1$, where $1 - \Gamma_0(b) \simeq b$ and the operator becomes $\tau b/\lambda_{\text{De}}^2 = k_\perp^2 \tau \rho_{\text{i}}^2/\lambda_{\text{De}}^2 = k_\perp^2 \omega_{\text{pi}}^2/\Omega_{\text{i}}^2$.

The operator on the left hand side of (7.45) represents the shielding effect due to the ion polarization field. It is the lowest order contribution to the density fluctuations provided by the polarization drift. We can obtain it heuristically by considering the polarization drift from (7.12)

$$\boldsymbol{v}_p = \frac{1}{\Omega_{\text{c}} B} \frac{\partial \boldsymbol{E}}{\partial t} = \frac{m}{eB^2} \frac{\partial \boldsymbol{E}}{\partial t} , \tag{7.46}$$

which gives a polarization current density

$$\boldsymbol{J}_p = e n_{io} \boldsymbol{v}_{\text{pi}} - e n_{eo} \boldsymbol{v}_{\text{pe}} \simeq \frac{n_{io} m_{\text{i}}}{eB^2} \frac{\partial \boldsymbol{E}_\perp}{\partial t} , \tag{7.47}$$

that is dominated by the ions. Using the continuity equation and integrating, the polarization density is

$$n_p = \frac{4\pi m_{\text{i}}}{eB^2} \boldsymbol{\nabla}_\perp \cdot (n_{io} \boldsymbol{\nabla}_\perp \Phi) = \frac{1}{e} \boldsymbol{\nabla}_\perp \cdot \left(\frac{\omega_{\text{pi}}^2}{\Omega_{\text{i}}^2} \boldsymbol{\nabla}_\perp \Phi \right) . \tag{7.48}$$

Equations (7.32), (7.33) and (7.39) form the basis for electrostatic gyrokinetic simulation of low frequency magnetized plasma dynamics.

In addition to electrostatic perturbations, it is also possible to self-consistently generate magnetic perturbations via currents that are parallel to the ambient magnetic field. The currents are induced via inductive electric fields $E_z = -\partial A_z/\partial t$, if the magnetic field is in the z-direction. From Ampere's law, the parallel current causes magnetic perturbations that are primarily perpendicular to the main field with $\delta B_\perp < B_{oz}$ and this is termed a field line bending effect. The perpendicular magnetic field perturbations can be expressed as a vector potential $\boldsymbol{\nabla} \times (A_z \boldsymbol{b}) = \boldsymbol{\nabla} A_z \times \boldsymbol{b}$. The compressional magnetic field perturbations can also be included by determining the perpendicular currents and higher β plasmas may be studied, where β characterizes the ratio of the plasma pressure to magnetic field pressure. Magnetically confined plasmas with high β are of contemporary interest in fusion, space and astrophysical plasmas.

We proceed to outline the gyrokinetic Vlasov-Poisson-Ampere system of equations in the low β regime where parallel currents are important. We assume the currents are sufficiently weak and density fluctuations small such that the gyrokinetic equations satisfy the ordering

$$\begin{aligned} \frac{\omega}{\Omega_\mathrm{i}} \simeq \frac{\rho_\mathrm{i}}{L} \simeq \frac{e\Phi}{T_\mathrm{e}} \simeq \frac{\delta B_\perp}{B_0} \simeq O(\epsilon) , \\ k_\perp \rho_\mathrm{i} \simeq O(1) , \end{aligned} \tag{7.49}$$

where we keep the finite gyroradius effects for the ions. If we introduce a canonical momentum

$$p_z = v_z + \frac{q}{m} A_z \tag{7.50}$$

into the one-form gyro-center Hamiltonian, we have

$$\begin{aligned} h_0 &= \hat{\mu}\Omega_\mathrm{c} + m p_z^2/2 , \\ h_1 &= -q p_z A_z(\boldsymbol{R}+\boldsymbol{\rho}, t) + q\Phi(\boldsymbol{R}+\boldsymbol{\rho}, t) \end{aligned} \tag{7.51}$$

and the gyrophase-averaged Hamiltonian can be derived using the Lie perturbation method as

$$\bar{h} = \bar{\hat{\mu}}\Omega_\mathrm{c} + \frac{1}{2} m \bar{p}_z^2 + q\langle\Phi\rangle_{\bar{\theta}} - q\bar{p}_z \langle A_z\rangle_{\bar{\theta}} . \tag{7.52}$$

From this result it is possible to derive the gyrokinetic Vlasov equation including magnetic perturbations from parallel currents [35]

$$\frac{\partial \bar{F}_\mathrm{i}}{\partial t} + \left(\bar{v}_z \boldsymbol{b}^* + \frac{\boldsymbol{b} \times \boldsymbol{\nabla}_{\bar{\boldsymbol{R}}} \langle\Psi\rangle_{\bar{\theta}}}{B} \right) \cdot \frac{\partial \bar{F}_\mathrm{i}}{\partial \bar{\boldsymbol{R}}} - \frac{e}{m_\mathrm{i}} \boldsymbol{b}^* \cdot \boldsymbol{\nabla}_{\bar{\boldsymbol{R}}} \langle\Psi\rangle_{\bar{\theta}} \frac{\partial \bar{F}_\mathrm{i}}{\partial \bar{p}_z} = 0 , \tag{7.53}$$

where $\langle\Psi\rangle_{\bar{\theta}}$ is a generalized potential defined by $\langle\Psi\rangle_{\bar{\theta}} = \langle\Phi\rangle_{\bar{\theta}} - \bar{v}_z \langle A_z\rangle_{\bar{\theta}}$. Ampere's law expressed in terms of the parallel vector potential becomes

$$\begin{aligned} \nabla_\perp^2 A_z = -4\pi e \Bigg[& \int (\bar{p}_z - \frac{e}{m_\mathrm{i}} A_z) \bar{F}_\mathrm{i} \delta(\bar{\boldsymbol{R}} - \boldsymbol{r} + \bar{\boldsymbol{\rho}}_\mathrm{i}) \bar{J}_i \mathrm{d}^6 \bar{Z} \\ & - \int (\bar{p}_z + \frac{e}{m_\mathrm{e}} A_z) \bar{F}_\mathrm{e} \bar{J}_\mathrm{e} \mathrm{d}^6 \bar{Z} \Bigg] . \end{aligned} \tag{7.54}$$

The canonical momentum formulation does not explicitly contain the induction electric field but is present when one transforms the distribution function momentum characteristic back to its evolution along the velocity characteristic. The cost of removing the explicit induction electric field in the characteristics is that we must solve (7.54) as a nonlinear elliptic partial differential equation for the vector potential.

7.3.3 Energy Conservation

In this section the total energy conservation of the gyrokinetic Vlasov-Poisson system is derived in the transformed corrdinates. By using the fundamental conservation law in the Hamiltonian system

$$\int H(\boldsymbol{z},t)\frac{\partial f(\boldsymbol{z},t)}{\partial t} J\mathrm{d}^6\boldsymbol{z} = 0 \tag{7.55}$$

we can apply this relation for each species in (7.32) and (7.33) to obtain the total energy as

$$\begin{aligned}&\tfrac{\mathrm{d}}{\mathrm{d}t}\left(\int \tfrac{1}{2}m_\mathrm{i}\bar{v}_z^2\bar{F}_\mathrm{i}\bar{J}_\mathrm{i}\mathrm{d}^6\bar{Z} + \int \tfrac{1}{2}m_\mathrm{e}\bar{v}_z^2\bar{F}_\mathrm{e}\bar{J}_\mathrm{e}\mathrm{d}^6\bar{Z}\right)\\ &\quad + \int e\langle\Phi\rangle_{\bar{\theta}}\tfrac{\partial\bar{F}_\mathrm{i}}{\partial t}\bar{J}_\mathrm{i}\mathrm{d}^6\bar{Z} + \int e\Phi\tfrac{\partial\bar{F}_\mathrm{e}}{\partial t}\bar{J}_\mathrm{e}\mathrm{d}^6\bar{Z} = 0\end{aligned} \tag{7.56}$$

without the adiabatic invariant, $\bar{\bar{\mu}}$, contribution. Using the gyrokinetic Poisson equation (7.45), the field energy contribution is

$$\begin{aligned}&\int e\langle\Phi\rangle_{\bar{\theta}}\frac{\partial\bar{F}_\mathrm{i}}{\partial t}\bar{J}_\mathrm{i}\mathrm{d}^6\bar{Z} + \int e\Phi\frac{\partial\bar{F}_\mathrm{e}}{\partial t}\bar{J}_\mathrm{e}\mathrm{d}^6\bar{Z}\\ &\quad = \frac{1}{4\pi}\int \Phi\frac{\partial}{\partial t}\left[\left(\nabla_\perp\cdot\frac{\omega_\mathrm{pi}^2}{\Omega_\mathrm{i}^2}\nabla_\perp\right)\Phi\right]\mathrm{d}^3\boldsymbol{r}\\ &\quad = \frac{\mathrm{d}}{\mathrm{d}t}\frac{1}{8\pi}\int\left[\frac{\omega_\mathrm{pi}^2}{\Omega_\mathrm{i}^2}|\nabla_\perp\Phi|^2\right]\mathrm{d}^3\boldsymbol{r}\,,\end{aligned} \tag{7.57}$$

where the right-hand-side is the ion polarization drift field energy. The total energy invariant is therefore

$$\begin{aligned}&\tfrac{\mathrm{d}}{\mathrm{d}t}\left(\int \tfrac{1}{2}m_\mathrm{i}\bar{v}_z^2\bar{F}_\mathrm{i}\bar{J}_\mathrm{i}\mathrm{d}^6\bar{Z} + \int \tfrac{1}{2}m_\mathrm{e}\bar{v}_z^2\bar{F}_\mathrm{e}\bar{J}_\mathrm{e}\mathrm{d}^6\bar{Z}\right)\\ &\quad + \tfrac{\mathrm{d}}{\mathrm{d}t}\tfrac{1}{8\pi}\int\left[\tfrac{\omega_\mathrm{pi}^2}{\Omega_\mathrm{i}^2}|\nabla_\perp\Phi|^2\right]\mathrm{d}^3\boldsymbol{r} = 0\,.\end{aligned} \tag{7.58}$$

7.4 Gyrofluid Model

It is now possible to construct moments of the gyrokinetic Vlasov equation to obtain gyrofluid equations. When the underlying distribution functions remain close to a Maxwellian it is possible to close the system with a few lower order moments.

We will present this case here, although it is possible to include higher order moments and linear wave-particle resonance effects such as Landau damping. These are known as gyro-Landau fluid closure methods [29]. The moment equations allow for a computationally efficient way to investigate nonlinearity in low frequency magnetized plasmas. However, if the velocity space nonlinearities become significant, the closure methods may require too many higher order moments and the simulations become impractical. Therefore, it is important to carefully compare the results of kinetic simulations and fluid closure approaches to be sure important physical effects are not left out.

We now derive a three-field gyrofluid model $(\Phi, A_z, p_\mathrm{e})$ based on the gyrokinetic Vlasov-Poisson-Ampere system discussed previously. We will first work in the long wavelength limit $(k_\perp \rho_\mathrm{i} < 1)$ and neglect the finite ion temperature effects $(T_\mathrm{i} = 0)$ but include the ion polarization drift or polarization shielding in the gyrokinetic Poisson equation. The gyrokinetic Vlasov equation, now in gyro-center coordinates, is

$$\frac{\partial f}{\partial t} + \left(v_\| \boldsymbol{b}^* + \frac{\boldsymbol{b} \times \nabla \Phi}{B_0} \right) \cdot \frac{\partial f}{\partial \boldsymbol{r}} + \frac{q}{m} \left(-\boldsymbol{b}^* \cdot \boldsymbol{\nabla} \Phi - \frac{\partial A_z}{\partial t} \right) \frac{\partial f}{\partial v_\|} = 0 \tag{7.59}$$

for each species and where the unit vector along the magnetic field becomes tilted to

$$\boldsymbol{b}^* = \boldsymbol{b} + \frac{\boldsymbol{\nabla} A_z \times \boldsymbol{b}}{B_0} \tag{7.60}$$

in the parallel velocity representation. The field equations are

$$\frac{\omega_\mathrm{pi}^2}{\Omega_\mathrm{i}^2} \nabla_\perp^2 \Phi = 4\pi e (n_\mathrm{e} - n_\mathrm{i}) \, , \tag{7.61}$$

$$\nabla_\perp^2 A_z = -4\pi (J_\mathrm{e} + J_\mathrm{i}) \, . \tag{7.62}$$

We first form the density and current moments for each species

$$\begin{aligned} n &= \int \mathrm{d}v_\| f(\boldsymbol{r}, v_\|) \, , \\ J &= q \int \mathrm{d}v_\| v_\| f(\boldsymbol{r}, v_\|) = qnv \, , \end{aligned} \tag{7.63}$$

where v is the fluid velocity along the magnetic field. By integration of (7.59) over velocity space we obtain the continuity equation for each species as

$$\frac{\mathrm{d}n}{\mathrm{d}t} + \boldsymbol{b}^* \cdot \boldsymbol{\nabla}(nv) = 0 \, . \tag{7.64}$$

Taking the convective derivative $(\mathrm{d}/\mathrm{d}t)$, defined by

$$\frac{\mathrm{d}}{\mathrm{d}t} = \frac{\partial}{\partial t} + \frac{\boldsymbol{b} \times \boldsymbol{\nabla} \Phi}{B_0} \cdot \boldsymbol{\nabla} \tag{7.65}$$

of (7.61) and using the continuity equation (7.64) with (7.62), an equation for the vorticity evolution is obtained

$$\frac{\mathrm{d}}{\mathrm{d}t}\nabla_\perp^2 \Phi = -\frac{\Omega_\mathrm{i}^2}{\omega_\mathrm{pi}^2}\boldsymbol{b}^* \cdot \nabla(\nabla_\perp^2 A_z) \,. \tag{7.66}$$

For the second field equation, which involves A_z, we assume the ions move only with the $\boldsymbol{E} \times \boldsymbol{B}$ drift and we set $v_i = 0$, which is equivalent to neglecting the coupling to ion sound waves. For electrons, we take the velocity moment of (7.59) which gives

$$\frac{\mathrm{d}J_\mathrm{e}}{\mathrm{d}t} - \frac{e}{m_\mathrm{e}}\boldsymbol{b}^* \cdot \boldsymbol{\nabla} p_\mathrm{e} = \frac{n_0 e^2}{m_\mathrm{e}}\left(-\boldsymbol{b}^* \cdot \boldsymbol{\nabla}\Phi - \frac{\partial A_z}{\partial t}\right) , \tag{7.67}$$

where the field-aligned electron pressure is $p_\mathrm{e} = m_\mathrm{e} \int \mathrm{d}v_\parallel (v_\parallel - v_\mathrm{e})^2 f_\mathrm{e}(\boldsymbol{r}, v_\parallel)$. This moment equation should be recognized as a type of Ohm's law. Using the simplest closure relation, $p_\mathrm{e} = nT_\mathrm{e}$ and $T_\mathrm{e} = const$, the field-aligned pressure gradient is related to the vorticity by

$$\nabla p_\mathrm{e} = T_\mathrm{e} \nabla n_\mathrm{e} = \frac{1}{4\pi e}\frac{\omega_\mathrm{pi}^2}{\Omega_\mathrm{i}^2}\nabla(\nabla_\perp^2 \Phi) \,, \tag{7.68}$$

where (7.61) was used. From (7.67), (7.62) and (7.68) we obtain

$$\frac{\partial A_z}{\partial t} = -\boldsymbol{b}^* \cdot \nabla \Phi + d_\mathrm{e}^2 \frac{\mathrm{d}}{\mathrm{d}t}(\nabla_\perp^2 A_z) + \rho_\mathrm{s}^2 \boldsymbol{b}^* \cdot \boldsymbol{\nabla}(\nabla_\perp^2 \Phi) \,, \tag{7.69}$$

where the $d_\mathrm{e} = c/\omega_\mathrm{pe}$ is the collisionless electron skin depth and $\rho_\mathrm{s} = c_\mathrm{s}/\Omega_\mathrm{i}$ is the ion sound radius with $c_\mathrm{s} = (T_\mathrm{e}/m_\mathrm{i})^{1/2}$ determined by the electron temperature. It is related to the ion gyroradius by $\rho_\mathrm{s} = \sqrt{T_\mathrm{e}/T_\mathrm{i}}\rho_\mathrm{i}$. If we define $\widetilde{A}_z = A_z - d_\mathrm{e}^2 \nabla_\perp^2 A_z$ the Ohm's law can be re-written as

$$\frac{\partial \widetilde{A}_z}{\partial t} = -\boldsymbol{b}^* \cdot \left[\boldsymbol{\nabla}\Phi + \frac{\boldsymbol{\nabla}\Phi \times \boldsymbol{\nabla}\widetilde{A}_z}{B_0} + \rho_\mathrm{s}^2 \boldsymbol{\nabla}(\nabla_\perp^2 \Phi) + \frac{\boldsymbol{\nabla}(\nabla_\perp^2 \Phi) \times \boldsymbol{\nabla}\widetilde{A}_z}{B_0}\right] . \tag{7.70}$$

The continuity equation for the electron density completes the three-field gyrofluid model and is coupled to both the vorticity evolution and Ohm's law

$$\frac{\partial n_\mathrm{e}}{\partial t} + \frac{\boldsymbol{b} \times \boldsymbol{\nabla}\Phi}{B_0} \cdot \boldsymbol{\nabla} n_\mathrm{e} = -\frac{1}{4\pi e}\boldsymbol{b}^* \cdot \boldsymbol{\nabla}(\nabla_\perp^2 A_z) \,. \tag{7.71}$$

The finite ion temperature effects can be incorporated by using the form of the gyrokinetic Poisson equation

$$\frac{T_\mathrm{e}}{T_\mathrm{i}\lambda_\mathrm{De}^2}(1 - \Gamma_0)\Phi = -4\pi e(n_\mathrm{e} - \bar{n}_\mathrm{i}) \,, \tag{7.72}$$

where $\bar{n}_\mathrm{i}(\boldsymbol{r}, t) = \int 2\pi v_\perp \mathrm{d}v_\perp \mathrm{d}v_\parallel J_0 f_\mathrm{i}(\boldsymbol{r}, v_\parallel, v_\perp, t)$. The ion continuity equation is obtained by taking the first moment of the ion gyrokinetic equation

$$\begin{aligned} &\frac{\partial f_\mathrm{i}}{\partial t} + \left(v_\| \boldsymbol{b}^* + \frac{\boldsymbol{b} \times \nabla (J_0 \Phi)}{B_0} \right) \cdot \frac{\partial f_\mathrm{i}}{\partial \boldsymbol{r}} \\ &\quad + \frac{e}{m_\mathrm{i}} \left[-\boldsymbol{b}^* \cdot \boldsymbol{\nabla} (J_0 \Phi) - \frac{\partial (J_0 A_z)}{\partial t} \right] \frac{\partial f_\mathrm{i}}{\partial v_\|} = 0 \,, \end{aligned} \tag{7.73}$$

where J_0 represents the gyrophase-averaging effect. We therefore obtain

$$\frac{\mathrm{d}\bar{n}_\mathrm{i}}{\mathrm{d}t} + (1 - \Gamma_0) \frac{\boldsymbol{b} \times \boldsymbol{\nabla} \Phi}{B_0} \cdot \boldsymbol{\nabla} n_\mathrm{i} = 0 \,, \tag{7.74}$$

where the dominant term for the gyro-averaged drift is retained. The finite-temperature vorticity equation is obtained by operating the convective derivative on (7.72) and again using the continuity and Ampere's equation gives

$$\frac{1}{\rho_\mathrm{i}^2} \frac{\mathrm{d}}{\mathrm{d}t} ((1 - \Gamma_0) \Phi) = \frac{\Omega_\mathrm{i}^2}{\omega_\mathrm{pi}^2} \boldsymbol{b}^* \cdot \nabla (\nabla_\perp^2 A_z) - \frac{T_\mathrm{i}}{e n_{eo}} \frac{(1 - \Gamma_0)}{\rho_\mathrm{i}^2} \frac{\boldsymbol{b} \times \boldsymbol{\nabla} \Phi}{B_0} \cdot \boldsymbol{\nabla} n_\mathrm{i} \,. \tag{7.75}$$

By using a Padé approximation to the Γ_0 operator, expressed in Fourier space as

$$1 - \Gamma_0 (k_\perp^2 \rho_\mathrm{i}^2) \simeq \frac{k_\perp^2 \rho_\mathrm{i}^2}{1 + k_\perp^2 \rho_\mathrm{i}^2} \tag{7.76}$$

and multiplying the vorticity equation on both sides by $1 + k_\perp^2 \rho_\mathrm{i}^2$, then replacing $k_\perp^2$ with the Laplacian $\nabla_\perp^2$ we finally obtain

$$\frac{\mathrm{d}}{\mathrm{d}t} (\nabla_\perp^2 \Phi) = \frac{\Omega_\mathrm{i}^2}{\omega_\mathrm{pi}^2} (1 - \rho_\mathrm{i}^2 \nabla_\perp^2) \boldsymbol{b}^* \cdot \nabla (\nabla_\perp^2 A_z) + \frac{T_\mathrm{i}}{e n_{eo}} \frac{\boldsymbol{b} \times \boldsymbol{\nabla} n_\mathrm{i}}{B_0} \cdot \boldsymbol{\nabla} (\nabla_\perp^2 \Phi) \,. \tag{7.77}$$

The multi-field gyrofluid models can be extended to include nonuniform electron and ion temperature as well as the parallel ion velocity. This leads to extended four-field [36, 37, 38] and five-field ($\Phi, A_z, p_\mathrm{e}, p_\mathrm{i}, v_i$) models.

7.5 Gyrokinetic Particle Simulation Model

The equation for the gyrophase-averaged distribution in a collisionless plasma can be put in a form that conserves phase space density along the characteristics. Expressed in this way, the particle simulation method can be used since it is equivalent to the method of characteristics. The equations for these characteristics, or each particle j, from (7.53) are

$$\begin{aligned} \frac{\mathrm{d}\boldsymbol{R}_j}{\mathrm{d}t} &= \left[v_z \boldsymbol{b} + \frac{\boldsymbol{b} \times \boldsymbol{\nabla}_{\bar{\boldsymbol{R}}} \langle \Psi \rangle_\theta}{B} \right]_j \,, \\ \frac{\mathrm{d}p_{zj}}{\mathrm{d}t} &= -\frac{q}{m} \left[\boldsymbol{b} \cdot \boldsymbol{\nabla}_{\bar{\boldsymbol{R}}} \langle \Psi \rangle_\theta \right]_j \,, \end{aligned} \tag{7.78}$$

where $\langle \Psi \rangle_\theta$ is a generalized potential defined as $\langle \Psi \rangle_\theta = \langle \Phi \rangle_\theta - v_z \langle A_z \rangle_\theta$. The gyrophase averages in the particle equations of motion are performed in real space by noticing that

$$\langle \Phi(\boldsymbol{R}) \rangle_\theta = \sum_{\boldsymbol{k}} \Phi_{\boldsymbol{k}} J_0 \left(\frac{k_\perp v_\perp}{\Omega_\mathrm{i}} \right) \mathrm{e}^{\mathrm{i}\boldsymbol{k}\cdot\boldsymbol{R}} = \left\langle \int \Phi(\boldsymbol{r}) \delta(\boldsymbol{r} - \boldsymbol{R} - \boldsymbol{\rho}) \right\rangle_\theta \tag{7.79}$$

since $J_0 = \langle \exp(\mathrm{i}\boldsymbol{k} \cdot \boldsymbol{\rho}) \rangle_\theta$ and $\langle \ldots \rangle_\theta \equiv \oint \mathrm{d}\theta / 2\pi$. Therefore, an algorithm can be developed by considering each particle as a uniformly charged ring centered at $\boldsymbol{R}_j$ and with radius $\boldsymbol{\rho}_j$. The number of points on this ring is related to the accuracy of computing J_0 to the desired order in $k_\perp \rho_\mathrm{i}$. For example, if we are to accurately represent $J_0(k_\perp \rho_\mathrm{i})$ for $k_\perp \rho_\mathrm{i} \leq 1$ then at least four points on the averaging ring are required [20]. The discrete gyro-averaging operation for M points is therefore

$$\langle \Phi(\boldsymbol{R}_j) \rangle_\theta = \frac{1}{M} \sum_{i=1}^{M} \Phi(\boldsymbol{R}_j + \boldsymbol{\rho}_j(\theta_i)) \tag{7.80}$$

and $\boldsymbol{\rho}_j = |\rho_j| (\boldsymbol{e}_1 \sin(\theta) + \boldsymbol{e}_2 \cos(\theta))$. The particle equations of motion (7.78), are finite-differenced in time and standard, second order predictor-corrector methods can be used to evolve the discrete N-particle distribution [39]

$$F(\boldsymbol{R}, p_z, \mu, t) = \sum_{i=1}^{N} \delta(\boldsymbol{R} - \boldsymbol{R}_i(t)) \delta(p_z - p_{zi}(t)) \delta(\mu - \mu_i) \, . \tag{7.81}$$

Each particle is initially assigned a guiding center position, a parallel velocity and a magnetic moment, from which a gyroradius can be computed.

Using this discrete particle distribution, the electrostatic and vector potentials are obtained from the gyrokinetic Poisson and Ampere's equations

$$\frac{\tau}{\lambda_\mathrm{De}^2}(1 - \Gamma_0)\Phi = 4\pi e (\bar{n}_\mathrm{i} - n_\mathrm{e}) \, , \tag{7.82}$$

$$\nabla_\perp^2 A_z = -4\pi (\bar{J}_{zi} + J_{ze}) \, , \tag{7.83}$$

where

$$\bar{n}_\mathrm{i}(\boldsymbol{r}) = \left\langle \int F_\mathrm{i}(\boldsymbol{R}, v_z, \mu) \delta(\boldsymbol{R} - \boldsymbol{r} + \boldsymbol{\rho}_i) \mathrm{d}\boldsymbol{R} \mathrm{d}\mu \mathrm{d}v_z \right\rangle_\theta \, ,$$

$$\bar{J}_{zi}(\boldsymbol{r}) = e \left\langle \int v_z F_\mathrm{i}(\boldsymbol{R}, v_z, \mu) \delta(\boldsymbol{R} - \boldsymbol{r} + \boldsymbol{\rho}_i) \mathrm{d}\boldsymbol{R} \mathrm{d}\mu \mathrm{d}v_z \right\rangle_\theta \tag{7.84}$$

and $v_z = p_z - (e/m_\mathrm{i}) A_z$. The electrons are not gyrophase-averaged and considered to be drift-kinetic. The gyrophased-averaged ion number density (and current density) can also be obtained numerically by using the ring-averaged of the N-particle distribution function and this gives

$$\bar{n}_\mathrm{i}(\boldsymbol{r}) = \frac{1}{M} \sum_{i=1}^{M} \left[\sum_j \delta(\boldsymbol{R}_j + \boldsymbol{\rho}_j(\theta_i)) \right] \, . \tag{7.85}$$

The charge density is obtained at discrete grid points and therefore an interpolation must be made. The delta functions are then replaced by interpolating functions which may be of low order, such as the nearest-grid-point (NGP) method, or higher order, such as the second order quadratic spline method [40]. Once the charge and current densities are formed at the grid points, the field equations may be solved by inverting the elliptic-type operators on the left-hand-side of (7.82) and (7.83). This can be done efficiently using Fast Fourier Transform (FFT) methods. Non-periodic boundary conditions may be incorporated by employing sine or cosine transforms. (7.83) for the vector potential is a nonlinear equation since the right-hand-side depends on the vector potential. Therefore, an iterative procedure must be used to converge the solution.

In order to determine the accuracy and energetics of the simulation plasma the conservation properties must be carefully examined. The total energy invariant for the gyrokinetic Vlasov-Poisson-Ampere system is used to determine the accuracy of the simulation results and is given by [35]

$$
\begin{aligned}
E_T = & \sum_j m_{\mathrm{e}} \left[\mu_{ej} B + \frac{(p_{z_{ej}} + e/m_{\mathrm{e}} A_z)^2}{2} \right] \\
& + \sum_j m_{\mathrm{i}} \left[\mu_{ij} B + \frac{(p_{z_{ij}} - e/m_{\mathrm{i}} \langle A_z \rangle_\theta)^2}{2} \right] \\
& + \frac{1}{8\pi} \int (|\boldsymbol{E}|^2 + |\boldsymbol{B}|^2) \mathrm{d}^3 \boldsymbol{r} ,
\end{aligned} \tag{7.86}
$$

where $\boldsymbol{E}$ is the electric field determined by the gradient of the electric potential in (7.82) and $\boldsymbol{B}$ is the magnetic field obtained from (7.83) and $\boldsymbol{B} = \boldsymbol{\nabla} A_z \times \widehat{\boldsymbol{b}}$.

7.5.1 Normal Modes and Fluctuation-Dissipation Properties

Linearization of the ion gyrokinetic equation, the electron drift-kinetic equation and combined with the Fourier transform of the gyrokinetic Poisson and Ampere equations, one obtains the following dispersion relation

$$
\omega^2 = \frac{k_\parallel^2 v_{\mathrm{A}}^2}{1 + k_\perp^2 d_{\mathrm{e}}^2} \left[\frac{k_\perp^2 \rho_{\mathrm{i}}^2}{1 - \Gamma_0(k_\perp^2 \rho_{\mathrm{i}}^2)} + k_\perp^2 \rho_{\mathrm{s}}^2 \right] , \tag{7.87}
$$

where ω is the real frequency. This fundamental normal mode in gyrokinetic plasmas is known as the kinetic shear Alfven wave [41] and in the long wavelength limit, $k_\perp^2 \rho_{\mathrm{i}}^2 < 1$ has the form

$$
\omega^2 = k_\parallel^2 v_{\mathrm{A}}^2 \left[\frac{1 + k_\perp^2 \rho_{\mathrm{s}}^2}{1 + k_\perp^2 d_{\mathrm{e}}^2} \right] \tag{7.88}
$$

since $1 - \Gamma_0(k_\perp^2 \rho_{\mathrm{i}}^2) \simeq k_\perp^2 \rho_{\mathrm{i}}^2$. This is the highest frequency which must be resolved in the simulation and therefore the condition $\omega \Delta t \leq 0.1$ must be satisfied. Another time step restriction arises from the electron transit motion and hence $k_\parallel (v_{\parallel e})_{\max} \Delta t < 1$.

Another important test of the simulation model is the equipartition of thermal fluctuation energy for different wavelengths. The basic fluctuation-dissipation properties of the magnetoinductive gyrokinetic model have been investigated theoretically [42] and the results are summarized here. The thermal fluctuation spectrum for electrostatic modes is determined by using the fluctuation-dissipation theorem

$$\frac{E_k^2}{8\pi} = -T \int \frac{\mathrm{d}\omega}{2\pi\omega} \mathrm{Im}(D^{-1}(\omega, k)) \tag{7.89}$$

and $D(\omega, k)$ is the linear dielectric response. Integration with respect to ω gives

$$\frac{E_k^2}{8\pi} = \frac{T}{2}\left[1 - \frac{1}{\mathrm{Re}(D(0,k))}\right] . \tag{7.90}$$

Applying this result to the gyrokinetic plasma dielectric, obtained by linearization of the gyrokinetic Vlasov-Poisson-Ampere system, we obtain

$$\frac{E_k^2}{8\pi} = \frac{T_\mathrm{e}}{2}\left[\frac{1}{1 + k^2\lambda_\mathrm{De}^2} - \frac{1 - \Gamma_0(k_\perp^2\rho_\mathrm{i}^2)}{(1 - \Gamma_0(k_\perp^2\rho_\mathrm{i}^2)) + k^2\lambda_\mathrm{Di}^2}\right] . \tag{7.91}$$

In the limit where $k^2\lambda_\mathrm{De}^2 < 1$, $k_\perp^2\rho_\mathrm{i}^2 < 1$ and $\rho_\mathrm{i} > \lambda_\mathrm{Di}$, we obtain

$$\frac{E_k^2}{8\pi} = \frac{T_\mathrm{e}}{2}\frac{\lambda_\mathrm{Di}^2}{\rho_\mathrm{i}^2} = \frac{T_\mathrm{e}}{2}\frac{\lambda_\mathrm{De}^2}{\rho_\mathrm{s}^2} \tag{7.92}$$

with $\rho_\mathrm{s} = \sqrt{T_\mathrm{e}/T_\mathrm{i}}\rho_\mathrm{i}$ and $\lambda_\mathrm{Di}^2/\rho_\mathrm{i}^2 = \Omega_\mathrm{i}^2/\omega_\mathrm{pi}^2$. Since the gyroradius scale is typically larger than the Debye length scale ($\rho_\mathrm{s} > \lambda_\mathrm{De}$), the fluctuation level for the gyrokinetic plasmas is much reduced compared to the fully kinetic plasmas containing high frequency space charge dominated fluctuations determined by [43]

$$\frac{E_k^2}{8\pi} = \frac{T_\mathrm{e}/2}{1 + k^2\lambda_\mathrm{De}^2} . \tag{7.93}$$

For the magnetic fluctuations which arise from the fluctuating currents, the thermal fluctuation spectrum is

$$\frac{B_{\perp k}^2}{8\pi} = \frac{T_\mathrm{e}/2}{1 + k_\perp^2 d_\mathrm{e}^2} \tag{7.94}$$

and $d_\mathrm{e} = c/\omega_\mathrm{pe}$.

7.6 Gyrokinetic Particle Simulation Model Applications

7.6.1 Current-driven Kinetic Alfven Wave Instability

As we have seen in the previous section the fundamental normal mode of gyrokinetic plasmas is the kinetic shear Alfven wave (KSAW). The particle inertia and finite gyroradius effects make this mode highly dispersive for short wavelengths.

Finite amplitude KSAW's can be excited by a variety of sources such as nonuniform background plasma parameters, energetic particle beams or finite currents [44]. These free energy sources have relevance to the generation of low frequency turbulence, small scale turbulence in near-Earth space plasmas and laboratory plasmas [45, 46].

In this section we consider the KSAW excitation in the presence of a uniform field-aligned equilibrium current $J_0 = -en_0v_d$ in a uniform Maxwellian plasma. In this system the parallel component of the perturbed electron current density in the wave electric field E_z is

$$J_{ez} = \sigma_z E_z = \frac{\mathrm{i}\omega Z'(\zeta_\mathrm{e})}{8\pi k_z^2 \lambda_\mathrm{De}^2} E_z \,, \tag{7.95}$$

where σ_z is the collisionless conductivity, Z' is the derivative of the standard plasma dispersion function $Z(\zeta_\mathrm{e}) = \pi^{-1/2} \int_{-\infty}^{\infty} \mathrm{d}v \exp(-v^2)/(v - \zeta_\mathrm{e})$ with argument $\zeta_\mathrm{e} = (\omega - k_z v_d)/\sqrt{2} k_z v_\mathrm{te}$. For $\zeta_\mathrm{e} \ll 1$

$$J_{ez} = -\frac{\mathrm{i}\omega}{4\pi k_z^2 \lambda_\mathrm{De}^2} \left(1 + \mathrm{i}\sqrt{\frac{\pi}{2}} \frac{(\omega - k_z v_d)}{k_z v_\mathrm{te}} \right) E_z \,. \tag{7.96}$$

When the condition $\omega < k_z v_d$ is satisfied (the drift speed is greater than the phase velocity of the KSAW), the phase shift between the current and electric field is such that the perturbations are amplified. Above the threshold drift speed, the phase velocity of the wave encounters a positive slope in the electron distribution function and this leads to an inverse Landau damping or a growth. The threshold electron drift velocity for instability is determined by

$$v_d > \frac{\omega}{k_z} \simeq v_\mathrm{A} \sqrt{\frac{1 + k_\perp^2 \rho_\mathrm{s}^2}{1 + k_\perp^2 d_\mathrm{e}^2}} \,, \tag{7.97}$$

where the kinetic Alfven wave frequency $\omega \simeq k_z v_\mathrm{A} (1 + k_\perp^2 \rho_\mathrm{s}^2)^{1/2} / (1 + k_\perp^2 d_\mathrm{e}^2)^{1/2}$ is assumed.

For the gyrokinetic simulations with self-consistent electric and magnetic fields, we initialize the electron distribution function as a shifted Maxwellian to produce a uniform current J_0 along an ambient magnetic field $\boldsymbol{B} = B_0 \hat{\boldsymbol{z}}$. This is shown in Fig. 7.2. The parameters used are: System size $L_x \times L_y = 15\rho_\mathrm{s} \times 7.5\rho_\mathrm{s}$, $m_\mathrm{i}/m_\mathrm{e} = 1837$, $v_d = 2.2 v_\mathrm{te} = 1.8 v_\mathrm{A}$, $\rho_\mathrm{s} = 8.5\Delta$, and $d_\mathrm{e} = 8\Delta$.

The time evolution of the total magnetic, electrostatic and kinetic energies are presented in Fig. 7.3 and reveal the linear instability growth and saturation. The total energy conservation is preserved to less than about one percent.

The electron kinetic energy decreases as the unstable kinetic Alfven wave electrostatic and magnetic energy components grow. The saturation of the fluctuations proceeds in two stages, first, the shorter wavelengths and then transfer some of their energy to more stable longer wavelengths which amplify and then saturate. In addition to the wave energy transfer to other more stable wavelengths, the mean electron drift current is reduced. This is illustrated in Fig. 7.2 where the final electron distribution taken at the end of the simulation is shown together with the initial one. There

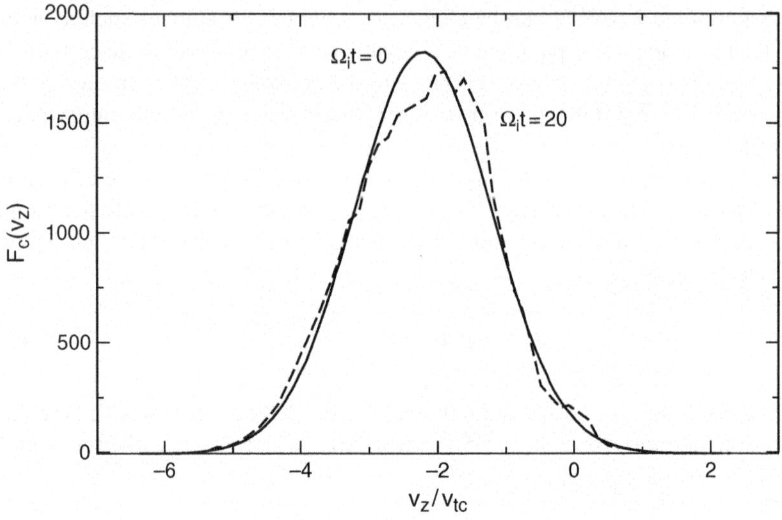

Fig. 7.2. Parallel electron distribution function taken at the initial time and final time levels

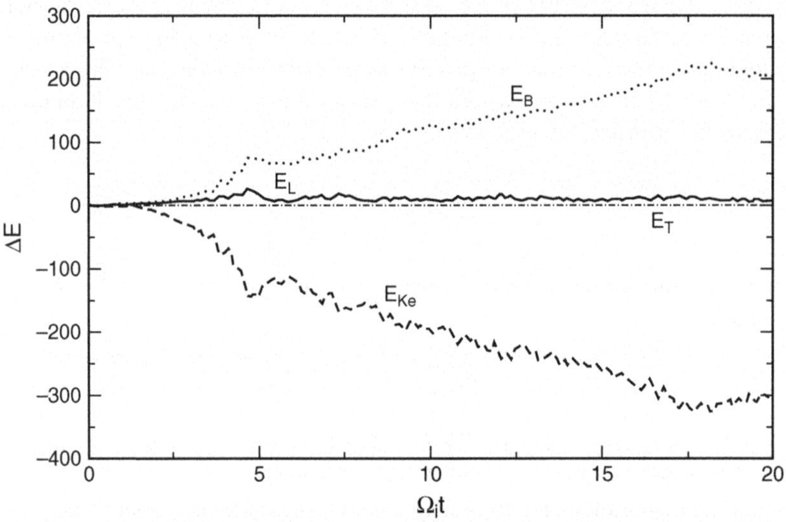

Fig. 7.3. Temporal evolution of the energy change ΔE from the initial value of the electrostatic E_{L}, magnetic E_{B}, electron kinetic E_{Ke} and total energy E_{T} for the current-driven kinetic shear Alfven wave instability

is a net slowing down of the parallel electron distribution with very weak thermal change.

The spatial distribution of the electric potential fluctuations is given in Fig. 7.4 at the saturation phase of instability. The electric potential vortices have a mean scale size of about 2–3 ρ_s and are roughly isotropic with $k_x\rho_s \simeq k_y\rho_s$. The electron density fluctuations averaged over the y-direction are also shown and reach a maximum level of $\delta n/n_0 \simeq 0.05$. There is also some indication of smaller scales ($\sim \rho_s$) being modulated by larger scales ($\sim 5\rho_s$). A more complete analysis of these results will be presented elsewhere, but one can see the large amount of information which can be obtained in the nonlinear regime of such models. The fluctuation spectra can be compared to experiment and assist in their interpretation.

7.6.2 Microtearing Instability

In this section we consider electric and magnetic fluctuations which arise from nonuniform currents. Spatially localized currents can lead to regions of anti-parallel magnetic fields which can break and reconnect via a microtearing instability to form magnetic islands. Small-scale magnetic islands have been proposed as means of inducing spontaneous symmetry breaking of perfectly nested flux surfaces in magnetically-confined toroidal plasmas [47] and in certain space plasma environments [48]. A consequence of this is the generation of substantial anomalous electron thermal transport, particularly when these islands interact radially [49, 50]. Small-scale magnetic turbulence can also act as a negative effective resistivity on large-scale magnetic field perturbations which can lead to amplification of the large-scale fields. Furthermore, sources of small-scale magnetic turbulence can produce

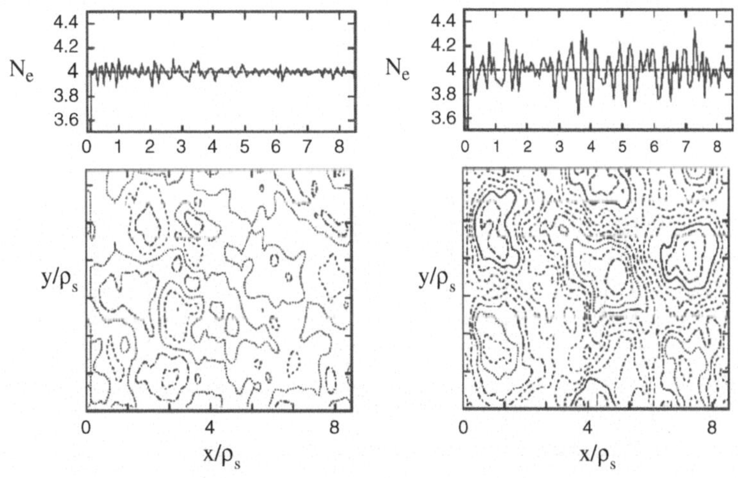

Fig. 7.4. Electron density profile (y-averaged) and electrostatic potential taken at the pre-growth ($\Omega_i t = 2$) (**left panel**) and saturation phase ($\Omega_i t = 20$) (**right panel**)

anomalous electron viscosity and enhanced current diffusion which could lead to self-sustained turbulence [51].

There has been theoretical work on the kinetic theory of magnetic island growth in the linearly unstable regime of collisionless tearing [52, 53] as well as some early particle simulations with a full particle magnetoinductive model [54]. More recently, gyrokinetic particle simulations have been applied to the collisionless tearing instability dynamics in uniform and nonuniform plasmas [55, 56].

For the results here, we use a plasma slab with sheared magnetic field $\boldsymbol{B} = B_z\widehat{z}+B_y(x)\widehat{y}$ and $|B_z| \gg |B_y|$. The shear field $B_y(x)$ is produced by a nonuniform sheet current assumed to vary only in the x-direction. It has the form $J_z(x) = -en_0v_{dz}\exp(-(x-L_x/2)^2/a^2)$ where v_{dz} is the electron drift velocity in the z-direction and this is shown in Fig. 7.5. The sheared $B_y(x)$ field has anti-parallel field lines across the middle of the simulation domain as displayed in the same figure. The initial density and temperature profiles are taken to be uniform. The boundary conditions are periodic in the y-direction and the vector potential A_z and electrostatic potential are set to be zero at the boundaries in the x-direction. The particles are specularly reflected at these boundaries.

This equilibrium serves an excellent test case because the linear growth rate and saturated island width are well-known [52, 53]. The equilibrium sheared magnetic field $B_y(x)$ can also be represented by a vector potential $A_{zo}(x)$ and the perturbed magnetic field by a vector potential $\widetilde{A}_z(x,y)$ through

$$\widetilde{\boldsymbol{B}}_\perp = \boldsymbol{\nabla}\times(\widetilde{A}_z\widehat{\boldsymbol{z}}) = \widetilde{B}_x\widehat{\boldsymbol{x}} + \widetilde{B}_y\widehat{\boldsymbol{y}} \,. \tag{7.98}$$

Since $A_z(x,y) = A_{zo} + \widetilde{A}_z$ and $\widetilde{A}_z(x,y) = \widehat{A}_z\cos(k_y y)$, it is possible to show the width of the magnetic island W is related to the amplitude of the perturbed vector potential by

$$W = \sqrt{\frac{\widehat{A}_z L_s}{B_z}}\,, \tag{7.99}$$

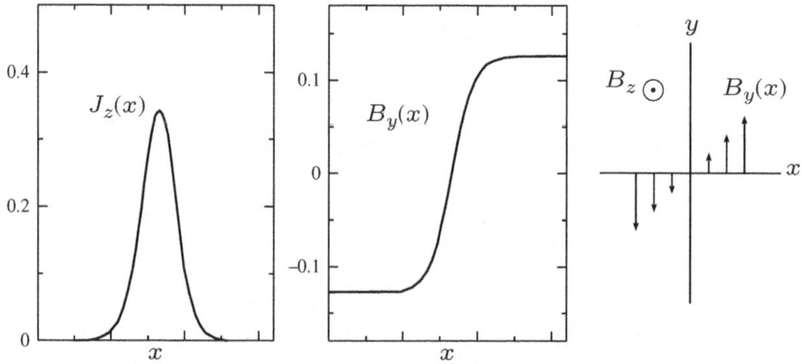

Fig. 7.5. Initial current profile, magnetic field $B_y(x)$, and schematic of the ambient magnetic direction B_z and sheared anti-parallel B_y component

where L_s is the shear scale length of the magnetic field defined as $L_s = B_z/|\mathrm{d}B_y(x)/\mathrm{d}x|$. The linear growth rate of the tearing mode has been derived from the electron drift kinetic equation in the collisionless regime as

$$\gamma_k \simeq (k_{\|} v_{\mathrm{te}})(\Delta' d_{\mathrm{e}}) \simeq \left(\frac{k_y v_{\mathrm{te}} d_{\mathrm{e}}}{L_s}\right)(\Delta' d_{\mathrm{e}}) , \tag{7.100}$$

where $d_{\mathrm{e}} = c/\omega_{\mathrm{pe}}$, v_{te} is the electron thermal velocity, and Δ' is the jump derivative of the perturbed vector potential across the anti-parallel field reversal region

$$\Delta' = \frac{[\partial \widetilde{A}_z/(\partial x)]_{-\Delta}^{\Delta}}{\widetilde{A}_z(0)} , \tag{7.101}$$

where Δ is the singular layer width and for the Gaussian current profile, the jump derivative $\Delta' \simeq 1/a$, where a is the current channel width. Therefore, under the assumption of a uniform plasma, the condition $\Delta' > 0$ is required for growing microtearing modes. The saturation level for the unstable collisionless mode is predicted to be

$$W^{\max} \simeq \frac{\Delta' d_{\mathrm{e}}^2}{2G} , \tag{7.102}$$

where G is a numerical constant with value $G = 0.41$.

The nonlinear evolution of the collisionless microtearing mode is investigated using the 2D gyrokinetic particle simulation model described earlier and the parameters used were: System size $L_x \times L_y = 16d_{\mathrm{e}} \times 16d_{\mathrm{e}}$, $d_{\mathrm{e}} = c/\omega_{\mathrm{pe}} = 8\Delta$, $\rho_{\mathrm{i}} = 4\Delta$, $T_{\mathrm{e}}/T_{\mathrm{i}} = 1$, $m_{\mathrm{i}}/m_{\mathrm{e}} = 1837$, and current layer width $2a = 10\Delta$. The y-direction is periodic and the discrete wavenumbers in this direction are given by $k_y = 2\pi m/L_y, m = 0, 1, ..., L_y/2 - 1$.

Fig. 7.6(a) displays the magnetic island half-width time evolution for the most unstable wavelength. The final saturation level is $W^{\mathrm{sat}} \simeq 1.2 d_{\mathrm{e}}$ which is comparable to the theoretical estimate of $W^{\max} \simeq (d_{\mathrm{e}}/2aG)d_{\mathrm{e}} \simeq 1.9 d_{\mathrm{e}}$. The simulation results are below the predicted value mainly due to the difference between the simulation value of Δ' which changes during the evolution of the current profile; the theory assumes it is constant.

The vector potential $A_z(x, y)$ and electrostatic potential $\Phi(x, y)$ at fixed time level are presented in Fig. 7.6(b), just prior to saturation. The magnetic island with X-point and O-point are clearly visible in the vector potential and the electrostatic potential pattern has a quadrapolar structure near the X-point of the island. After the island grows, electrons are trapped in the singular layer and undergo bounce motion with frequency $\omega_b = k_y v_{\mathrm{te}} W/2L_s$, where W is the magnetic island width. This can be seen in Fig. 7.6(a), where electron trapping oscillations appear after saturation and the period is consistent with $T_b = 2\pi/\omega_b$. When the saturated island is evolved for a long time period a nonlinear electron distibution function emerges and consists of trapped and untrapped electron orbits.

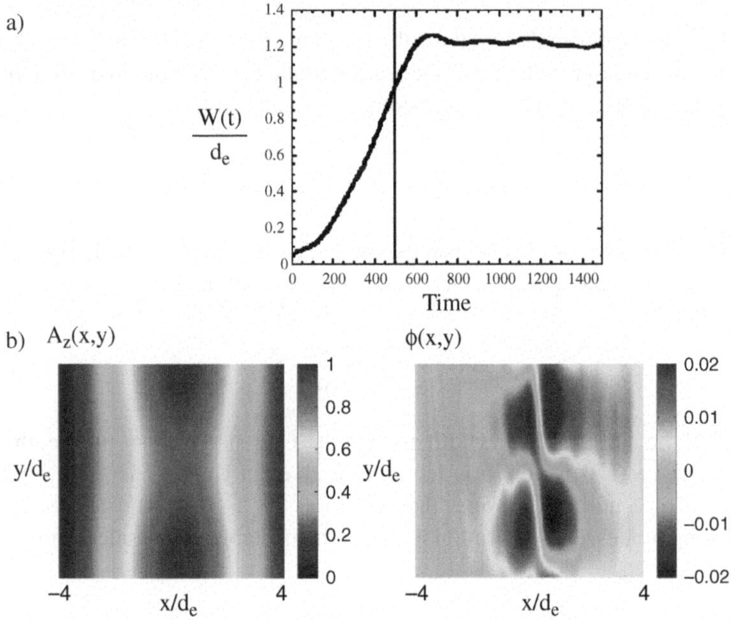

Fig. 7.6. (**a**) Magnetic island width W, temporal evolution and (**b**) vector potential $A_z(x, y)$, and electrostatic potential $\Phi(x, y)$, at time level prior to island width saturation

Fig. 7.7 shows the current profile at the initial time and near saturation. A double-peaked structure forms near the field reversal region and is related to the quasilinear changes induced by the magnetic island formation. The current becomes redistributed to the outer regions of the island.

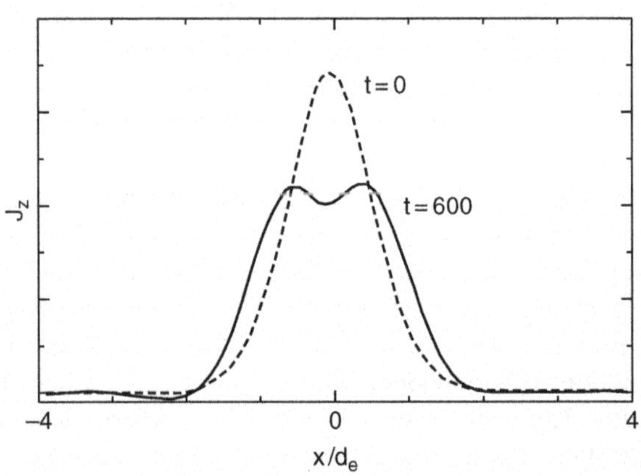

Fig. 7.7. Initial and final electron current profiles for the microtearing instability

7.7 Summary

The multiple scale nature of plasmas present inherent difficulties in the simulation of low frequency ($\omega < \Omega_i$) electromagnetic fluctuations in magnetized plasmas. In full particle simulation models based on the Vlasov-Maxwell system, the main problem is the high frequency space charge waves characterized by the electron plasma frequency ω_{pe} and electron Debye length λ_{De}, which impose severe time step and spatial resolution restrictions. Their presence gives rise to very high noise levels which can mask the evolution of low frequency quasi-neutral-type ($n_e \approx n_i$) waves whose equilibrium fluctuation energy can be orders of magnitude lower.

Beginning with single particle dynamics in an ambient magnetic field, a gyrophase averaging procedure can be developed to remove the gyrophase dependence on drift motion and thus eliminate the fast gyro-motion and associated high frequency cyclotron waves while retaining finite gyroradius effects. The methods of action variational and Lie perturbation methods can be used to derive gyro-drift equation of motion to any order and retain the phase space conserving properties in the change of variables from gyro-center to gyrophase-averaged coordinates.

Using the single particle equations of motion as characteristics of the gyrokinetic Vlasov equation, it is possible to formulate a self-consistent system of equations including a Poisson and Ampere equation for the electrostatic and magnetic potential from which the electric and magnetic fields are formed. The gyrokinetic Poisson equation physically describes the ion polarization drift effects without the need to include them explicity in the equations of motion. The gyrokinetic Vlasov-Poission-Ampere system satisfies particle and energy conservation.

By integration over the phase space, moment equations can be formed to describe continuum gyrofluids. In some cases the magnetized plasma dynamics can be modeled by just a few of the lowest order moments, resulting in computationally efficient simulations without the problems of statistical noise as in the discrete formulation. It should also be mentioned that continuum gyrokinetic Vlasov-Poission-Ampere equations are being used for turbulent transport simulations in inhomogeneous plasmas [27]. These require very large computing resources because of the large number of grid points required in the high dimensional phase space.

Gyrokinetic particle simulations have been extensively developed in recent years for the study of low frequency microturbulence in inhomogeneous plasmas. These models have the advantage of allowing one to formulate parallel algorithms for implemention on massively parallel computing architectures and simulations with over one billion particles are now feasible. The advance of low noise techniques, where only the perturbed part of the distribution function is represented by particles, has also allowed for more clearer delineation of the linear growth and saturation phase of instabilities [39, 56] as well as larger scale simulations.

The author would like to thank the organizers of the Heraeus Summer School, Prof. H. Fehske, Dr. R. Schneider and Dr. A. Weiße for their support and the hospitality of the Max-Planck-Institute for Plasma Physics, Greifswald, Germany. This research was supported in part by a grant from the Natural Sciences and Engineering Research Council (NSERC) of Canada.

References

1. J. Hugill, Nucl. Fusion **23**, 331 (1983)
2. P. Liewer, Nucl. Fusion **25**, 543 (1985)
3. F. Wagner, U. Stroth, Plasma Phys. Contr. F. **35**, 1321 (1993)
4. R. Fonck, G. Cosby, R. Durst, S. Paul, N. Bretz, S. Scott, E. Synakowski, G. Taylor, Phys. Rev. Lett. **70**, 3736 (1993)
5. G. McKee, C. Petty, R. Waltz, C. Fenzi, R. Fonck, J. Kinsey, T. Luce, K. Burrell, D. Baker, E. Doyle, X. Garbet, R. Moyer, C. Rettig, T. Rhodes, D. Ross, G. Staebler, R. Sydora, M. Wade, Nucl. Fusion **41**, 1235 (2001)
6. T. Northrop, *Adiabatic Motion of Charged Particles* (Wiley, New York, 1963)
7. P. Rutherford, E. Frieman, Phys. Fluids **11**, 569 (1968)
8. J. Taylor, R. Hastie, Plasma Phys. **20**, 479 (1968)
9. P. Catto, Plasma Phys. **20**, 719 (1978)
10. T. Antonsen, B. Lane, Phys. Fluids **23**, 1205 (1980)
11. P. Catto, W. Tang, D. Baldwin, Plasma Phys. **23**, 639 (1981)
12. E. Frieman, L. Chen, Phys. Fluids **25**, 502 (1982)
13. W. Lee, Phys. Fluids **26**, 556 (1983)
14. A. Boozer, Phys. Fluids **23**, 904 (1980)
15. R. Littlejohn, Phys. Fluids **24**, 1730 (1981)
16. D. Dubin, J. Krommes, C. Oberman, W. Lee, Phys. Fluids **26**, 3524 (1983)
17. T. Hahm, Phys. Fluids **31** (1988)
18. A. Brizard, J. Plasma Phys. **41**, 541 (1989)
19. A. Brizard, T. Hahm, Rev. Mod. Phys. **79**, 421 (2007)
20. W. Lee, J. Comput. Phys. **72**, 243 (1987)
21. R. Sydora, T. Hahm, W. Lee, J. Dawson, Phys. Rev. Lett. **64**, 2015 (1990)
22. R. Sydora, Phys. Fluids **B2**, 1455 (1990)
23. S. Parker, W. Lee, R. Santoro, Phys. Rev. Lett. **71**, 2042 (1993)
24. R. Sydora, V. Decyk, J. Dawson, Plasma Phys. Contr. F. **12**, A281 (1996)
25. Z. Lin, T. Hahm, W. Lee, W. Tang, R. White, Science **281**, 1835 (1998)
26. A. Dimits, G. Bateman, M. Beer, B. Cohen, W. Dorland, G. Hammett, C. Kim, J. Kinsey, M. Kotschenreuther, A. Kritz, L. Lao, J. Mandrekas, W. Nevins, S. Parker, A. Redd, D. Shumaker, R. Sydora, J. Weiland, Phys. Plasm. **7**, 969 (2000)
27. W. Nevins, J. Candy, S. Cowley, T. Dannert, A. Dimits, W. Dorland, C. Estrada-Mila, G. Hammett, F. Jenko, M. Pueschel, D. Shumaker, Phys. Plasm. **13**, 122306 (2006)
28. A. Brizard, Phys. Fluids **B4**, 1213 (1992)
29. W. Dorland, G. Hammett, Phys. Fluids **B5**, 812 (1993)
30. M. Beer, G. Hammett, Phys. Plasmas **3**, 812 (1996)
31. D. Strintzi, B. Scott, Phys. Plasmas **11**, 5452 (2004)
32. K. Miyamoto, *Plasma Physics for Nuclear Fusion* (MIT Press, Cambridge, MA, 1989)
33. V. Arnold, *Mathematical Methods of Classical Mechanics* (Springer-Verlag, New York, 1989)
34. J. Cary, R. Littlejohn, Ann. Phys. **151**, 1 (1983)
35. T. Hahm, W. Lee, A. Brizard, Phys. Fluids **31**, 1940 (1988)
36. R. Hazeltine, C. Hsu, P. Morrison, Phys. Fluids **30**, 3204 (1987)
37. A. Aydemir, Phys. Fluids **B4**, 3469 (1992)
38. B. Scott, Plasma Phys. Contr. F. **45**, A385 (2003)
39. H. Naitou, K. Tsuda, W. Lee, R. Sydora, Phys. Plasmas **2**, 4257 (1995)

40. C. Birdsall, A. Langdon, *Plasma Physics via Computer Simulation* (McGraw-Hill, New York, 1985)
41. A. Hasegawa, L. Chen, Phys. Fluids **19**, 1924 (1976)
42. J. Krommes, Phys. Fluids **B5**, 2405 (1993)
43. J. Dawson, Rev. Mod. Phys. **55**, 403 (1983)
44. A. Hasegawa, P. Indian Acad. Sci. A **86**, 151 (1977)
45. K. Stasiewicz, P. Bellan, C. Chaston, C. Kletzing, R. Lysak, J. Maggs, O. Pokhotelov, C. Seyler, P. Shukla, L. Stenflo, A. Streltsov, J.E. Wahlund, Space Sci. Rev. **92**, 423 (2000)
46. D. Leneman, W. Gekelman, J. Maggs, Phys. Rev. Lett. **82**, 2673 (1999)
47. B. Kadomtsev, Nucl. Fusion **31**, 1301 (1991)
48. A. Galeev, L. Zelenyi, JETP Lett. **29**, 614 (1979)
49. A. Rechester, M. Rosenbluth, Phys. Rev. Lett. **40**, 88 (1978)
50. P. Rebut, M. Hugon, Comments Plasma Phys. Contr. F. **33**, 1085 (1991)
51. M. Yagi, S.I. Itoh, K. Itoh, A. Fukuyama, M. Azumi, Phys. Plasmas **2**, 4140 (1995)
52. J. Drake, Y. Lee, Phys. Rev. Lett. **39**, 453 (1977)
53. J. Drake, Y. Lee, Phys. Fluids **20**, 1341 (1977)
54. I. Katanuma, T. Kamimura, Phys. Fluids **23**, 2500 (1980)
55. R. Sydora, Phys. Plasmas **8**, 1929 (2001)
56. W. Wan, Y. Chen, S. Parker, Phys. Plasmas **12**, 012311 (2005)

Part IV

Semiclassical Approaches

8 Boltzmann Transport in Condensed Matter

Franz Xaver Bronold

Institut für Physik, Universität Greifswald, 17487 Greifswald, Germany

This chapter presents numerical methods for the solution of Boltzmann equations as applied to the analysis of transport and relaxation phenomena in condensed matter systems.

8.1 Boltzmann Equation for Quasiparticles

Besides the traditional approaches, such as variational methods or the expansion of the distribution function in a symmetry adapted orthonormal set of functions, stochastic methods, either based on the sampling of the distribution function by superparticles, the particle-in-cell Monte Carlo collision (PIC-MCC) approach, or the direct simulation of the master equation underlying the Boltzmann description, the ensemble Monte Carlo methods, are discussed at a tutorial level. Expansion methods are most appropriate for the solution of Boltzmann equations which occur in the Fermi liquid based description of transport in normal metals and superconductors. Stochastic methods, on the other hand, are particularly useful for the solution of relaxation and transport problems in semiconductors and electronic devices.

8.1.1 Preliminary Remarks

The Boltzmann equation (BE) is of central importance for the description of transport processes in many-particle systems. Boltzmann introduced this equation in the second half of the 19th century to study irreversibility in gases from a statistical mechanics point of view. Envisaging the molecules of the gas to perform free flights, which are occasionally interrupted by mutual collisions, he obtained the well-known equation [1]

$$\frac{\partial g}{\partial t} + \boldsymbol{v} \cdot \nabla_{\boldsymbol{r}} g + \frac{\boldsymbol{F}}{M} \cdot \nabla_{\boldsymbol{v}} g = \left(\frac{\partial g}{\partial t} \right)_c , \tag{8.1}$$

where $g(\boldsymbol{r}, \boldsymbol{v}, t)$ is the velocity distribution function, M is the mass of the gas molecules, $\boldsymbol{F}$ is the external force, and the r.h.s. is the collision integral. With this equation Boltzmann could not only prove his famous H-theorem, which contains a definition of entropy in terms of the velocity distribution function and states that

F. X. Bronold: *Boltzmann Transport in Condensed Matter*, Lect. Notes Phys. **739**, 223–254 (2008)
DOI 10.1007/978-3-540-74686-7_8

for irreversible processes entropy has to increase, he could also calculate transport properties of the gas, for instance, its heat conductivity or its viscosity.

In the original form, the BE holds only for dilute, neutral gases with a short range interaction, for which $nR^3 \ll 1$, where n is the density of the gas and R is the range of the interaction potential. However, it has also been applied to physical systems, for which, at first sight, the condition $nR^3 \ll 1$ is not satisfied. For instance, the kinetic description of plasmas in laboratory gas discharges or interstellar clouds is usually based on a BE, although $R \to \infty$ for the Coulomb interaction. Thus, $nR^3 \ll 1$ cannot be satisfied, for any density. A careful study of the Coulomb collision integral showed, however, that the bare Coulomb interaction has to be replaced by the screened one, resulting in a modified BE, the Lenard-Balescu equation [2, 3], which can then indeed be employed for the theoretical analysis of plasmas [4, 5].

In the temperature and density range of interest, ionized gases are, from a statistical point of view, classical systems. The technical problems due to the Coulomb interaction not withstanding, it is therefore clear that a BE, which obviously belongs to the realm of classical statistical mechanics, can be in principle formulated for a plasma.

The BE has also been successfully applied to condensed matter, in particular, to quantum fluids, metals, and semiconductors [6, 7, 8], whose microscopic description has to be quantum-mechanical. Hence, transport properties of these systems should be calculated quantum-statistically, using a quantum-kinetic equation, instead of a BE [9, 10, 11]. In addition, naively, one would not expect the condition $nR^3 \ll 1$ to be satisfied. The densities of condensed matter are too high. A profound quantum-mechanical analysis in the first half of the 20th century [12, 13, 14] revealed, however, that the carriers in condensed matter are not the tightly bound, dense building blocks but physical excitation which, at a phenomenological level, resemble a dilute gas of *quasiparticles* for which a BE or a closely related kinetic equation can indeed be formulated.

The quasiparticle concept opens the door for Boltzmann transport in condensed matter, see Fig. 8.1. Depending on the physical system, electrons or ion cores in a solid, normal or superfluid/superconducting quantum fluids etc., various types of quasiparticles can be defined quantum-mechanically, whose kinetics can then be modelled by an appropriate semi-classical BE. Its mathematical structure, and with it the solution strategy, is essentially independent of the physical context. Below, we restrict ourselves to the transport resulting from electronic quasiparticles in a crystal. Further examples of semi-classical quasiparticle transport can be found in the excellent textbook by Smith and Jensen [7].

8.1.2 Electronic Quasiparticles in a Crystal

To facilitate a qualitative understanding of the quasiparticle concept as applied to electrons, we recall the quantum mechanics of a single electron in a crystal. Writing the electrons' wave function in Bloch form $\psi_{n\boldsymbol{k}}(\boldsymbol{r}) = \mathrm{e}^{\mathrm{i}\boldsymbol{k}\cdot\boldsymbol{r}} u_{n\boldsymbol{k}}(\boldsymbol{r})/\sqrt{V}$ with

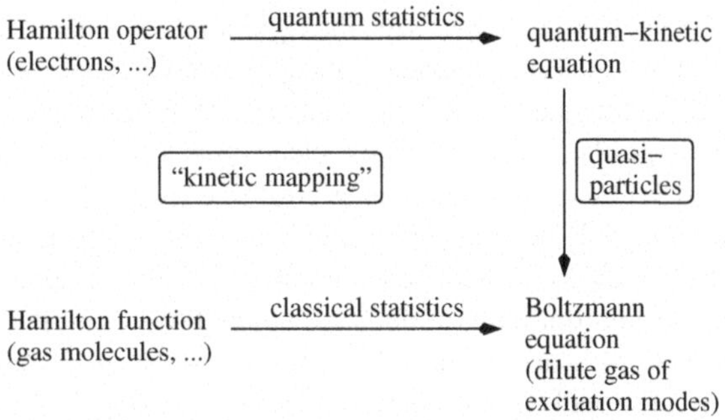

Fig. 8.1. This cartoon puts the content of this chapter into perspective. For neutral or ionized gases, the BE and its range of validity, can be directly derived from the Hamilton function for the classical gas molecules. In that case, the BE determines the distribution function for the constituents of the physical system under consideration. In the context of condensed matter, however, the BE describes the distribution function for the excitation modes (quasiparticles) and not for the constituents (electrons, ion cores, ...) although the BE has to be obtained – by quantum-statistical means – from the constituents' Hamilton operator. The definition of quasiparticles is absolutely vital for setting-up a BE. It effectively maps, as far as the kinetics is concerned, the quantum-mechanical many-particle system of the constituents to a semi-classical gas of excitation modes

an appropriately normalized, lattice periodic Bloch function[1] $u_{n\boldsymbol{k}}$, the one-particle Schrödinger equation, which determines the quasiparticle energies $E_n(\boldsymbol{k})$ with n the band index and $\boldsymbol{k}$ the wave vector in the first Brillouin zone, is given by

$$\left[\frac{\hbar^2(\nabla + \mathrm{i}\boldsymbol{k})^2}{2m_e} + V(\boldsymbol{r})\right] u_{n\boldsymbol{k}}(\boldsymbol{r}) = E_n(\boldsymbol{k}) u_{n\boldsymbol{k}}(\boldsymbol{r}) - \int \mathrm{d}\boldsymbol{r}' \Sigma(\boldsymbol{r} - \boldsymbol{r}', E_n(\boldsymbol{k})) \mathrm{e}^{\mathrm{i}\boldsymbol{k}\cdot(\boldsymbol{r}'-\boldsymbol{r})} u_{n\boldsymbol{k}}(\boldsymbol{r}') , \quad (8.2)$$

where we separated the lattice-periodic potential $V(\boldsymbol{r})$ originating from the rigidly arranged ion cores from the energy dependent potential $\Sigma(\boldsymbol{r}, E)$ (self-energy) which arises from the coupling to other charge carriers as well as to the ion cores' deviations from the equilibrium positions (phonons).

Let us first consider (8.2) for $\Sigma = 0$. An electron moving through the crystal experiences then only the lattice periodic potential V. It gives rise to extremely strong scattering, with a mean free path of the order of the lattice constant, which could never be treated in the framework of a BE. However, this scattering is not random. It originates from the periodic array of the ion cores and leads to the formation of bare energy bands. Within these bands, the motion of the electron is coherent, but with a dispersion which differs from the dispersion of a free electron. Because of

[1] The Bloch functions are orthonormal when integrated over a unit cell v_{cell} : $v_{\text{cell}}^{-1} \int_{\text{cell}} \mathrm{d}\boldsymbol{r}\ u_{n\boldsymbol{k}}(\boldsymbol{r})^* u_{n'\boldsymbol{k}'}(\boldsymbol{r}) = \delta_{nn'}\delta_{\boldsymbol{k},\boldsymbol{k}'}$.

this difference, the electron no longer sees the rigid array of ion cores. Its mean free path exceeds therefore the lattice constant, and a BE may become feasible.

However, in a crystal there is not only one electron but many, and the lattice of ion cores is not rigid but dynamic. Thus, electron-electron and electron-phonon interaction have to be taken into account giving rise to $\Sigma \neq 0$. As a result, the Schrödinger equation (8.2) becomes an implicit eigenvalue problem for the renormalized energy bands $E_n(\boldsymbol{k})$ which may contain a real and an imaginary part. For the purpose of the discussion, we assume Σ to be real. Physically, the dressing of the electron incorporated in Σ arises from the fact that the electron attracts positively charged ion cores and repels other electrons, as visualized in Fig. 8.2. The former gives rise to a lattice distortion around the considered electron and the latter leads to a depletion of electrons around it[2].

While coherent scattering on the periodic array of ion cores transforms bare electrons into band electrons, which is favorable for a BE description, residual interactions turn band electrons into dressed quasiparticles, which may be detrimental to it, because the dressing is energy and momentum dependent. Quasiparticles are therefore complex objects. Nevertheless, they are characterized by a dispersion, carry a negative elementary charge, and obey Fermi statistics, very much like band electrons. Provided they are well-defined, which means that the imaginary part of Σ, which we neglected so far in our discussion, is small compared to the real part of Σ, a BE may be thus also possible for them. However, the justification of the

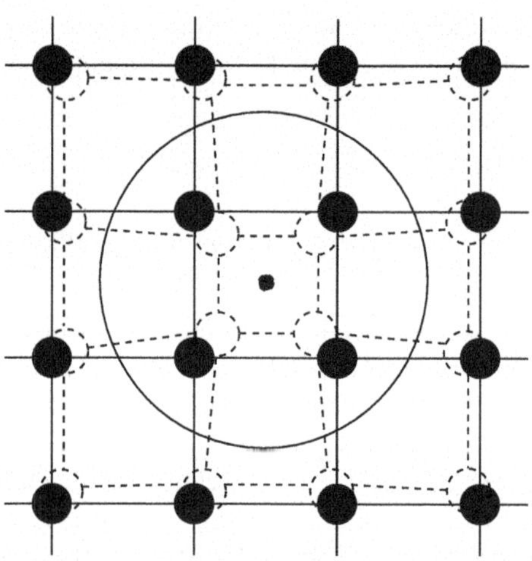

Fig. 8.2. Graphical representation of the many-body effects contained in the selfenergy Σ. The lattice distortion (*dashed lines*) and the depletion region (*large solid circle*) turn bare band electrons (*visualized by the small bullet*) into quasiparticles which carry the lattice distortion and the depletion region with it when they move through the crystal

[2] Here, exchange effects are also important, because electrons are fermions. At a technical level, the depletion region is encoded in the Coulomb hole and the screened exchange selfenergy.

quasiparticle BE will be subtle. Indeed, there are no pre-canned recipes for this task. Each physical situation has to be analyzed separately, using quantum-statistical techniques [9, 10, 11].

For standard metals [15] and superconductors [16, 17, 18, 19], for instance, the BE for quasiparticles can be rigorously derived from basic Hamiltonians, provided the quasiparticles are well-defined. The main reason is the separation of energy scales [20]: A high-energy scale set by the Fermi momentum k_F and a low-energy scale given by the width Δk of the thermal layer around the Fermi surface, see Fig. 8.3. The latter also defines the wavelength $1/\Delta k$ of the quasiparticles responsible for transport. Because of the separation of scales, an ab initio calculation of transport coefficients is possible along the lines put forward by Rainer [21] for the calculation of transition temperatures in superconductors, which is a closely related problem, see also [22].

For semiconductors, on the other hand, a BE for quasiparticles can only be rigorously derived when they are degenerate, that is, heavily doped and thus metal-like; the scales are then again well separated. However, when the doping is small, or in moderately optically excited semiconductors, the electrons are non-degenerate. Thus, neither a Fermi energy nor a transport energy scale can be defined. In that case, a BE for quasiparticles is very hard to justify from first principles [23], despite the empirical success the BE has also in these situations.

8.1.3 Heuristic Derivation of the Quasiparticle Boltzmann Equation

Since we will be mainly concerned with computational techniques developed for the solution of the quasiparticle BE, we skip the lengthy quantum-statistical

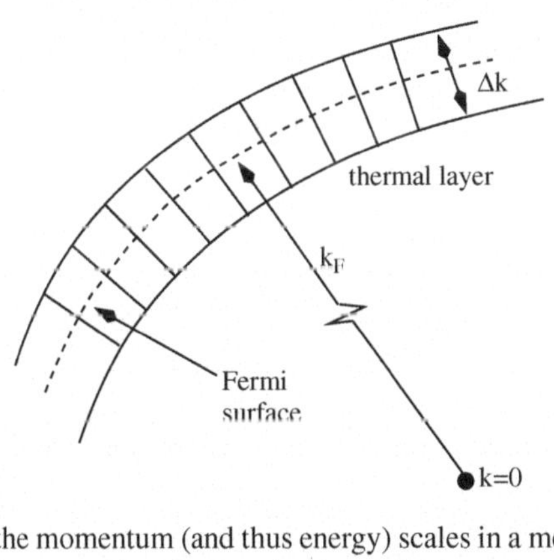

Fig. 8.3. Separation of the momentum (and thus energy) scales in a metal. The Fermi momentum k_F sets the high-energy scale, whereas the thermal smearing-out of the Fermi surface, Δk, gives the scale relevant for transport. Using quantum-statistical methods, a correlation function called g^K, which is closely related to the distribution function g can be systematically expanded in $\Delta k/k_F \sim kT/E_F$. If the quasiparticles have long enough lifetimes, g^K reduces, in leading order, to g and satisfies a BE [20]

derivation and simply summarize the necessary conditions a many-particle system has to satisfy to be suitable for a BE-based analysis:

(i) The quasiparticles should be well-defined objects. The time they spend in a state with energy $E_n(\boldsymbol{k})$, that is, the average time before scattering between quasiparticles, with material imperfections, or with phonons[3] takes place, should be large compared to $\hbar/E_n(\boldsymbol{k})$.
(ii) The coupling to the scatterer (impurities, quasiparticles, and phonons) should be weak, that is, the collision integral in the quasiparticle BE should be calculable perturbatively.
(iii) External electro-magnetic fields and temperature gradients should be weak enough so that their effect during a collision can be neglected. In addition, their spatio-temporal variation has to be small within a mean free path or a mean free time.

When conditions (i)–(iii) are satisfied, a BE exists for quasiparticles. This is the regime of Boltzmann transport in condensed matter.

The structure of the BE for quasiparticles can be guessed heuristically. In analogy to the BE (8.1), it should be an *equation of motion* for the quasiparticle distribution function in the nth band, $g_n(\boldsymbol{r}, \boldsymbol{k}, t)$. Notice, instead of the velocity $\boldsymbol{v}$, the momentum $\boldsymbol{k}$ is now used, because $\boldsymbol{k}$ is the relevant one-particle quantum number in a given band. The distribution function contains both $\boldsymbol{r}$ and $\boldsymbol{k}$ as independent variables, whereas the uncertainty principle prohibits a simultaneous measurement of the two. Thus, by necessity, the quasiparticle BE has to be a rather subtle semi-classical equation.

Within the semi-classical framework, quasiparticles perform classical intra-band free flights and occasionally scatter on imperfections, phonons, or other quasiparticles, which then leads to transitions between momentum states and possibly between bands. Accordingly, the equation of motion which governs the free flight of a quasiparticle in an electro-magnetic field specified by a vector potential $\boldsymbol{A}$ and a scalar potential U can be derived from the classical Hamiltonian [24]

$$\mathcal{H} = E_n\left(\frac{1}{\hbar}\boldsymbol{p} + \frac{e}{\hbar c}\boldsymbol{A}(\boldsymbol{r})\right) - eU(\boldsymbol{r}) \, , \tag{8.3}$$

where $E_n(\boldsymbol{k})$ is the band energy obtained from the solution of the Schrödinger equation (8.2), $\hbar\boldsymbol{k}$ is the kinetic momentum, and $\boldsymbol{p}$ and $\boldsymbol{r}$ are, respectively, the canonical momentum and coordinate of the quasiparticle. From the Hamilton equations it then follows, that a quasiparticle in the nth band centered at $\boldsymbol{r}$ and $\boldsymbol{k}$ in phase space has to move according to

[3] Naturally, phonons comprising the lattice distortion accounted for in the definition of quasiparticles do not lead to scattering. But there is a residual electron-phonon interaction which induces transitions between different quasiparticle states.

$$\left(\frac{\mathrm{d}\boldsymbol{r}}{\mathrm{d}t}\right)_n = \boldsymbol{v}_n(\boldsymbol{k}) = \frac{1}{\hbar}\nabla_{\boldsymbol{k}} E_n(\boldsymbol{k}) \,,$$
$$\left(\hbar\frac{\mathrm{d}\boldsymbol{k}}{\mathrm{d}t}\right)_n = \boldsymbol{F}_n = -e\left(\boldsymbol{E} + \frac{1}{c}\boldsymbol{v}_n(\boldsymbol{k}) \times \boldsymbol{B}\right) \tag{8.4}$$

with $\boldsymbol{E} = -\nabla_{\boldsymbol{r}} U$ and $\boldsymbol{B} = \nabla_{\boldsymbol{r}} \times \boldsymbol{A}$, which immediately leads to the quasiparticle BE

$$\frac{\partial g_n}{\partial t} + \boldsymbol{v}_n \cdot \nabla_{\boldsymbol{r}} g_n - \frac{e}{\hbar}\left(\boldsymbol{E} + \frac{1}{c}\boldsymbol{v}_n \times \boldsymbol{B}\right) \cdot \nabla_{\boldsymbol{k}} g_n = \left(\frac{\partial g_n}{\partial t}\right)_c \,, \tag{8.5}$$

when the time evolutions of the distribution function due to streaming (l.h.s.) and scattering (r.h.s.) are balanced.

Suppressing the variables $\boldsymbol{r}$ and t, the general structure of the collision integral is

$$\left(\frac{\partial g_n}{\partial t}\right)_c = \sum_{n'\boldsymbol{k}'} \{S_{n'\boldsymbol{k}',n\boldsymbol{k}} g_{n'}(\boldsymbol{k}')[1 - g_n(\boldsymbol{k})] - S_{n\boldsymbol{k},n'\boldsymbol{k}'} g_n(\boldsymbol{k})[1 - g_{n'}(\boldsymbol{k}')]\} \tag{8.6}$$

with $S_{n'\boldsymbol{k}',n\boldsymbol{k}}$ the probability for scattering from the quasiparticle state $n'\boldsymbol{k}'$ to the quasiparticle state $n\boldsymbol{k}$, which has to be determined from the quantum mechanics of scattering. Its particular form depends on the scattering process (see below). The collision integral consists of two terms: The term proportional to $1 - g_n(\boldsymbol{k})$ accounts for scattering-in (gain) processes, whereas the term proportional to $g_n(\boldsymbol{k})$ takes scattering-out (loss) processes into account. Note, for non-degenerate quasiparticles[4], $g_n(\boldsymbol{k}) \ll 1$ and the Pauli-blocking factor $1 - g_n(\boldsymbol{k})$ reduces to unity.

Some of the numerical techniques we will discuss below are tailored for the solution of the steady-state, spatially uniform, linearized quasiparticle BE, applicable to situations, where the external fields are weak and the system is close to thermal equilibrium. This equation can be obtained from the full BE (8.5) through an expansion around thermal equilibrium. In the absence of magnetic fields and for a single band it reads [6, 7]

$$-e\boldsymbol{E} \cdot \boldsymbol{v}\frac{\partial f}{\partial E}\bigg|_{E(\boldsymbol{k})} = \left(Cg^{(1)}\right)(\boldsymbol{k}) \,, \tag{8.7}$$

where the r.h.s. symbolizes the linearized collision integral and $g^{(1)} = g - f$ is the deviation of the distribution function from the Fermi function $f(E) = [\exp(E/k_{\mathrm{B}}T) + 1]^{-1}$. Here T is the temperature and E measures the energy from the chemical potential. With the help of the detailed balance condition

$$S_{\boldsymbol{k}',\boldsymbol{k}} f(E(\boldsymbol{k}'))[1 - f(E(\boldsymbol{k}))] = S_{\boldsymbol{k},\boldsymbol{k}'} f(E(\boldsymbol{k}))[1 - f(E(\boldsymbol{k}'))] \,, \tag{8.8}$$

the linearized collision integral becomes[5]

[4] Quasiparticles with mass m^* are non-degenerate when $n\lambda_{dB}^3 \ll 1$, where n is the density and $\lambda_{dB} = \sqrt{h^2/2\pi m^* k_{\mathrm{B}}T}$ is the de Broglie wavelength of the quasiparticles.

[5] Recall, we suppress in the collision integral the variables $\boldsymbol{r}$ and t.

$$\left(Cg^{(1)}\right)(\boldsymbol{k}) = \sum_{\boldsymbol{k}'} S_{\boldsymbol{k},\boldsymbol{k}'} \left\{ g^{(1)}(\boldsymbol{k}') \frac{f(E(\boldsymbol{k}))}{f(E(\boldsymbol{k}'))} - g^{(1)}(\boldsymbol{k}) \frac{1-f(E(\boldsymbol{k}'))}{1-f(E(\boldsymbol{k}))} \right\}$$
$$= \sum_{\boldsymbol{k}'} \widetilde{Q}_{\boldsymbol{k},\boldsymbol{k}'} g^{(1)}(\boldsymbol{k}') , \tag{8.9}$$

where in the second line we defined a matrix

$$\widetilde{Q}_{\boldsymbol{k},\boldsymbol{k}'} = S_{\boldsymbol{k},\boldsymbol{k}'} \frac{f(E(\boldsymbol{k}))}{f(E(\boldsymbol{k}'))} - \sum_{\boldsymbol{k}''} S_{\boldsymbol{k}',\boldsymbol{k}''} \frac{1-f(E(\boldsymbol{k}''))}{1-f(E(\boldsymbol{k}'))} \delta_{\boldsymbol{k}\boldsymbol{k}'} . \tag{8.10}$$

8.2 Techniques for the Solution of the Boltzmann Equation

In the previous section we phenomenologically derived the BE for quasiparticles[6] and listed the necessary conditions for Boltzmann transport in condensed matter. The standard way to analyze transport processes in condensed matter consists then of three main steps:

(i) Determine appropriate interaction mechanisms for the quasiparticles responsible for the transport phenomenon under consideration and calculate the relevant scattering probabilities. This step requires quasiparticle energies and wave functions. For electronic quasiparticles, they have to be obtained from (8.2) using, for instance, the ab initio methods described in Chap. 14.
(ii) Write down the BE in terms of the external driving terms and scattering probabilities and solve it.
(iii) Calculate the relevant currents for a confrontation with experiments. For weak external fields, i.e., in the linear regime, use the currents to extract transport coefficients (electric and thermal conductivity, mobility etc.) which may be more suitable for a comparison with experiments.

The calculation of the currents or transport coefficients is straightforward provided the solution of the BE is known. Solving the BE is, however, a serious task. It is a complicated non-linear integro-differential equation, which in almost all cases of interest cannot be solved analytically. Noteworthy exceptions are the linearized BE for a classical gas with an interaction potential $U(r) \sim r^{-4}$ [7] and the linearized BE for an isotropic Fermi liquid at very low temperatures, where particle-particle Coulomb scattering dominates [25, 26].

Usually the BE has to be solved with a computer. Two groups of techniques can be roughly distinguished. The first group consists of *classical* numerical techniques for the solution of integro-differential equations (approximating differentials by finite differences, integrals by sums, and using numerical routines for manipulating the resulting algebraic equations). They are mostly applied to situations where external fields and temperature gradients are weak and the BE can be linearized around the local thermal equilibrium. In principle, however, they can be also used to solve

[6] From now on, BE refers to quasiparticle BE.

the full BE. With an eye on the calculation of the electric conductivity of metals and the calculation of hot-electron distributions in semiconductors, we will describe two such methods: Numerical iteration [27, 28, 29, 30, 31, 32, 33, 34, 35] and *algebraization* through an expansion of the distribution function in terms of a set of basis functions [36, 37, 38, 39, 40, 41, 42, 43].

The second group consists of Monte Carlo techniques for the direct simulation of the stochastic motion of quasiparticles, whose distribution function is governed by the BE. These techniques are the most popular ones currently used because the concepts they invoke are easy to grasp and straightforward to implement on a computer. In addition, Monte Carlo techniques can be applied to far-off-equilibrium situations and are thus ideally suited for studying hot-electron transport in semiconductor devices which is of particular importance for the micro-electronics industry. Below, we will present two different Monte Carlo approaches. The first approach, which evolved into a design tool for electronic circuit engineers, samples the phase space of the quasiparticles by monitoring the time evolution of a single test-particle [44, 45, 46, 47, 48, 49]. Whereas the second approach generates the time evolution of N-electron configurations in a discretized momentum space [50]. This is particularly useful for degenerate electrons, where Pauli-blocking is important.

8.2.1 Numerical Iteration

8.2.1.1 Spatially Uniform, Steady-State BE with Linearized Collision Integral

Based on the linearized BE (8.7), numerical iteration has been extensively used for calculating steady-state transport coefficients for metals in uniform external fields [27, 28, 29, 30]. In contrast to the full BE, the linearized BE is not an integro-differential equation but an inhomogeneous integral equation to which an iterative approach can be directly applied. As an illustration, we consider the calculation of the electric conductivity tensor $\boldsymbol{\sigma}$.

To set up the iteration scheme, $g^{(1)}$ is written in a form which anticipates that $g^{(1)}$ will change rapidly in the vicinity of the Fermi surface, see Fig. 8.3, while it will be a rather smooth function elsewhere. The relaxation time approximation [6, 7] suggests for $g^{(1)}$ the ansatz

$$g^{(1)}(\boldsymbol{k}) = -\frac{\partial f}{\partial E}\bigg|_{E(\boldsymbol{k})} e\boldsymbol{E}\cdot\boldsymbol{v}(\boldsymbol{k})\phi(\boldsymbol{k})\ , \tag{8.11}$$

where $E(\boldsymbol{k})$ and $\boldsymbol{v}(\boldsymbol{k})$ are, respectively, the energy measured from the chemical potential and the group velocity of the quasiparticles. In terms of the function $\phi(\boldsymbol{k})$, which can be interpreted as a generalized, $\boldsymbol{k}$-dependent relaxation time, the electric current becomes [6, 7]

$$\boldsymbol{j} = 2e\int \frac{\mathrm{d}\boldsymbol{k}}{(2\pi)^3}\, \boldsymbol{v}(\boldsymbol{k})\, g^{(1)}(\boldsymbol{k})$$

$$= -2e^2 \int \frac{d\boldsymbol{k}}{(2\pi)^3}\, \phi(\boldsymbol{k})\, \frac{\partial f}{\partial E}\bigg|_{E(\boldsymbol{k})} \boldsymbol{v}(\boldsymbol{k}) : \boldsymbol{v}(\boldsymbol{k})\boldsymbol{E}$$
$$= \boldsymbol{\sigma}\boldsymbol{E}\ , \qquad (8.12)$$

from which we can read off the electric conductivity tensor

$$\boldsymbol{\sigma} = -2e^2 \int \frac{d\boldsymbol{k}}{(2\pi)^3}\, \phi(\boldsymbol{k})\, \frac{\partial f}{\partial E}\bigg|_{E(\boldsymbol{k})} \boldsymbol{v}(\boldsymbol{k}) : \boldsymbol{v}(\boldsymbol{k})\ , \qquad (8.13)$$

where : denotes the tensor product and the factor two comes from the spin. Note, although the particular structure of (8.11) is inspired by the relaxation time approximation, the iterative approach goes beyond it, because it does not replace the linearized collision integral by $-g^{(1)}/\tau$, where τ is the relaxation time, but keeps it fully intact. In addition, it is also more general than variational approaches [6, 7] because the function $\phi(\boldsymbol{k})$ is left unspecified.

To proceed, we insert (8.11) into (8.7). Using the collision integral in the form (8.9) and defining

$$X(\boldsymbol{k};\boldsymbol{E}) = -e\boldsymbol{E}\cdot\boldsymbol{v}(\boldsymbol{k})\ , \qquad (8.14)$$

we obtain

$$X(\boldsymbol{k};\boldsymbol{E})\frac{\partial f}{\partial E}\bigg|_{E(\boldsymbol{k})} = \sum_{\boldsymbol{k}'} S_{\boldsymbol{k},\boldsymbol{k}'}\left[\frac{1-f(E(\boldsymbol{k}'))}{1-f(E(\boldsymbol{k}))}\frac{\partial f}{\partial E}\bigg|_{E(\boldsymbol{k})} X(\boldsymbol{k};\boldsymbol{E})\phi(\boldsymbol{k}) \right.$$
$$\left. - \frac{f(E(\boldsymbol{k}))}{f(E(\boldsymbol{k}'))}\frac{\partial f}{\partial E}\bigg|_{E(\boldsymbol{k}')} X(\boldsymbol{k}';\boldsymbol{E})\phi(\boldsymbol{k}')\right]\ , \qquad (8.15)$$

which can be simplified to

$$\phi(\boldsymbol{k}) = \frac{1 + \sum_{\boldsymbol{k}'} S_{\boldsymbol{k},\boldsymbol{k}'} \frac{[1-f(E(\boldsymbol{k}'))]X(\boldsymbol{k}';\boldsymbol{E})}{[1-f(E(\boldsymbol{k}))]X(\boldsymbol{k};\boldsymbol{E})}\phi(\boldsymbol{k}')}{\sum_{\boldsymbol{k}'} S_{\boldsymbol{k},\boldsymbol{k}'}\frac{1-f(E(\boldsymbol{k}'))}{1-f(E(\boldsymbol{k}))}}\ , \qquad (8.16)$$

when we recall the identity

$$\frac{\partial f}{\partial E}\bigg|_{E(\boldsymbol{k})} = \frac{1}{k_B T} f(E(\boldsymbol{k}))[1-f(E(\boldsymbol{k}))]\ . \qquad (8.17)$$

Notice that the precise form of the single band scattering probability $S_{\boldsymbol{k},\boldsymbol{k}'}$ is immaterial for the iteration procedure which can thus handle all three major scattering processes: Elastic electron-impurity, inelastic electron-phonon, and electron-electron scattering.

Equation (8.16) is an inhomogeneous integral equation suitable for iteration: Starting with $\phi^{(0)} = 0$ (thermal equilibrium), a sequence of functions $\phi^{(i)}$, $i \geq 1$, can be successively generated, which comes with increasing i arbitrarily close to the exact solution, provided the process converges. Convergence is only guaranteed when the kernel is positive and continuous. This is not necessarily the case, but it can be enforced when selfscattering processes are included, see below.

The iteration process needs as an input the scattering probability. The most important scattering processes affecting the electric conductivity of metals are electron-impurity and electron-phonon scattering. The former determines the conductivity at low temperatures whereas the latter at high temperatures[7]. In our notation, these two scattering probabilities are given by [28]

$$S^{\rm imp}_{\boldsymbol{k},\boldsymbol{k}'} = \frac{2\pi}{\hbar}|M^{\rm imp}(\cos\theta_{\boldsymbol{k}\boldsymbol{k}'})|^2\delta(E(\boldsymbol{k}')-E(\boldsymbol{k}))\ , \tag{8.18}$$

$$S^{\rm ph}_{\boldsymbol{k},\boldsymbol{k}'} = \frac{2\pi}{\hbar}\sum_{\boldsymbol{q}\lambda}\sum_{\boldsymbol{Q}_i}\left|M^{\rm ph}_\lambda(\boldsymbol{k}'-\boldsymbol{k})\right|^2\{N_{\lambda\boldsymbol{q}}\delta(E(\boldsymbol{k}')-E(\boldsymbol{k})-\hbar\omega_{\lambda\boldsymbol{q}})\,\delta_{\boldsymbol{k}'-\boldsymbol{k},\boldsymbol{q}+\boldsymbol{Q}_i}$$
$$+[1+N_{\lambda\boldsymbol{q}}]\,\delta(E(\boldsymbol{k}')-E(\boldsymbol{k})+\hbar\omega_{\lambda\boldsymbol{q}})\delta_{\boldsymbol{k}-\boldsymbol{k}',\boldsymbol{q}+\boldsymbol{Q}_i}\}\ , \tag{8.19}$$

where $M^{\rm imp}(\cos\theta_{\boldsymbol{k}\boldsymbol{k}'})$ is the electron-impurity coupling which depends on the angle $\theta_{\boldsymbol{k}\boldsymbol{k}'}$ between $\boldsymbol{k}$ and $\boldsymbol{k}'$ (isotropic elastic scattering), $M^{\rm ph}_\lambda(\boldsymbol{k}'-\boldsymbol{k})$ is the electron-phonon coupling, and $N_{\lambda\boldsymbol{q}} = [\exp(\hbar\omega_{\lambda\boldsymbol{q}})-1]^{-1}$ is the equilibrium distribution function for phonons with frequency $\omega_{\lambda\boldsymbol{q}}$; $\boldsymbol{q}$, λ, and $\boldsymbol{Q}_i$ are the phonon wave-vector, the phonon polarization, and the ith reciprocal lattice vector, respectively. The coupling functions are material specific and can be found in the literature [6, 7, 8].

In order to obtain a numerically feasible integral equation, (8.18) and (8.19) are inserted into (8.16) and the momentum sums are converted into integrals. The integral over $\boldsymbol{k}'$ is then transformed into an integral over constant energy surfaces using[8]

$$\sum_{\boldsymbol{k}'} \to \int\frac{{\rm d}\boldsymbol{k}'}{(2\pi)^3} \to \frac{1}{(2\pi)^3}\int{\rm d}E(\boldsymbol{k}')\int\frac{{\rm d}\Omega(\boldsymbol{k}')}{\hbar|\boldsymbol{v}(\boldsymbol{k}')|}\ , \tag{8.20}$$

where ${\rm d}\Omega(\boldsymbol{k}')$ is the surface element on the energy surface $E(\boldsymbol{k}')$. The δ-functions appearing in the scattering probabilities (8.18) and (8.19) are then utilized to eliminate some of the integrations thereby reducing the dimensionality of (8.16). For isotropic bands $\phi(\boldsymbol{k}) \to \phi(E(\boldsymbol{k}))$, and one ends up with an one-dimensional integral equation which can be readily solved by iteration. For more details see [27, 28, 29, 30].

8.2.1.2 Spatially Uniform BE with the Full Collision Integral

The iterative approach can be also applied to the full BE. This is of particular interest for the calculation of distribution functions for electrons in strong external fields [31, 32, 33, 34, 35]. In that case, however, the BE has to be first converted into an integral equation. This is always possible because the free streaming term in (8.5) has the form of a total differential which can be integrated along its characteristics.

---

[7] Electron-electron scattering does not affect the electric conductivity, as long as normal processes are only taken into account. Umklapp processes, on the other hand, contribute to the conductivity, but the matrix elements are usually very small.

[8] The volume is put equal to one.

As an illustration, we consider a spatially uniform, non-degenerate electron system[9] in a single band and an arbitrarily strong electric field. Then we can write the BE as

$$\left\{\frac{\partial}{\partial t} - \frac{e}{\hbar}\boldsymbol{E}\cdot\nabla_{\boldsymbol{k}} + \lambda_{\boldsymbol{k}} + S_{\boldsymbol{k}}\right\} g(\boldsymbol{k},t) = \sum_{\boldsymbol{k}'} [S_{\boldsymbol{k}',\boldsymbol{k}} + S_{\boldsymbol{k}}\delta_{\boldsymbol{k},\boldsymbol{k}'}] g(\boldsymbol{k}',t) \; , \quad (8.21)$$

where we introduced the scattering-out rate

$$\lambda_{\boldsymbol{k}} = \sum_{\boldsymbol{k}'} S_{\boldsymbol{k},\boldsymbol{k}'} \; , \quad (8.22)$$

and added on both sides of the equation a selfscattering term $S_{\boldsymbol{k}} g(\boldsymbol{k},t)$, which has no physical significance, but is later needed to simplify the kernel of the integral equation.

To transform (8.21) into an integral equation, we introduce path variables $\boldsymbol{k}^* = \boldsymbol{k} + e/\hbar\boldsymbol{E}t^*$ and $t^* = t$ which describe the collisionless motion of the electrons along the characteristics of the differential operator [31, 45][10]. In terms of these variables, (8.21) can be written as

$$\frac{\mathrm{d}}{\mathrm{d}t^*}\left\{g(\boldsymbol{k}(\boldsymbol{k}^*,t^*),t^*)\,\mathrm{e}^{\int_0^{t^*}\mathrm{d}y\widetilde{\boldsymbol{\lambda}}_{\boldsymbol{k}(\boldsymbol{k}^*,y)}}\right\} = \mathrm{e}^{\int_0^{t^*}\mathrm{d}y\widetilde{\boldsymbol{\lambda}}_{\boldsymbol{k}(\boldsymbol{k}^*,y)}} \sum_{\boldsymbol{k}'} \widetilde{S}_{\boldsymbol{k}',\boldsymbol{k}(\boldsymbol{k}^*,t^*)} g(\boldsymbol{k}',t^*) \quad (8.23)$$

with $\widetilde{\boldsymbol{\lambda}}_{\boldsymbol{k}} = \lambda_{\boldsymbol{k}} + S_{\boldsymbol{k}}$ and $\widetilde{S}_{\boldsymbol{k}',\boldsymbol{k}} = S_{\boldsymbol{k}',\boldsymbol{k}} + S_{\boldsymbol{k}}\delta_{\boldsymbol{k}\boldsymbol{k}'}$. Integrating this equation from t_1^* to $t_2^* > t_1^*$ and setting $\boldsymbol{k} = \boldsymbol{k}^* - e\boldsymbol{E}t_2^*/\hbar$, $t = t_2^*$, and $t' = t_1^*$ yields

$$\begin{aligned} g(\boldsymbol{k},t) =& g\left(\boldsymbol{k} + \frac{e\boldsymbol{E}}{\hbar}(t-t'),t'\right)\mathrm{e}^{-\int_{t'}^{t}\mathrm{d}y\widetilde{\boldsymbol{\lambda}}_{\boldsymbol{k}+e\boldsymbol{E}(t-y)/\hbar}} \\ &+ \int_{t'}^{t}\mathrm{d}t^*\sum_{\boldsymbol{k}'}\mathrm{e}^{-\int_{t^*}^{t}\mathrm{d}y\widetilde{\boldsymbol{\lambda}}_{\boldsymbol{k}+e\boldsymbol{E}(t-y)/\hbar}}\,\widetilde{S}_{\boldsymbol{k}',\boldsymbol{k}+e\boldsymbol{E}(t-t^*)/\hbar}\,g(\boldsymbol{k}',t^*) \; . \end{aligned} \quad (8.24)$$

This equation, which is an integral representation of the BE (8.21), can be further simplified when we consider the physical content of the terms on the r.h.s. The first term denotes the contribution to $g(\boldsymbol{k},t)$ originating from electrons which were in state $\boldsymbol{k} + e\boldsymbol{E}(t-t')/\hbar$ at time t' and drifted to the state $\boldsymbol{k}$ at time t without being scattered. The second term, on the other hand, denotes the contribution of electrons which were scattered from any state $\boldsymbol{k}'$ to the new state $\boldsymbol{k} + e\boldsymbol{E}(t-t^*)/\hbar$ at any time t^* between t and t' and arrive at the state $\boldsymbol{k}$ at time t without being scattered. In both terms, the time t' is arbitrary. It can be any time. The only requirement is that $t' < t$. We can thus take the convenient limit $t' \to -\infty$, in which case the first term on the r.h.s. of (8.24) vanishes because $g(\boldsymbol{k},t)$ vanishes for $\boldsymbol{k} \to \infty$. The integral representation of (8.21) reduces therefore to

[9] No Pauli blocking, i.e. $1 - g \to 1$ in (8.6).

[10] A generalization of the procedure to spatially non-uniform situations is conceivable, but will not be discussed here.

$$g(\boldsymbol{k},t) = \int\limits_{-\infty}^{t} \mathrm{d}t^* \sum_{\boldsymbol{k}'} \mathrm{e}^{-\int_{t^*}^{t} \mathrm{d}y \widetilde{\lambda}_{\boldsymbol{k}+e\boldsymbol{E}(t-y)/\hbar}} \widetilde{S}_{\boldsymbol{k}',\boldsymbol{k}+e\boldsymbol{E}(t-t^*)/\hbar}\, g(\boldsymbol{k}',t^*)\ . \quad (8.25)$$

This form of the BE is not yet particularly useful, because the time integral in the exponent contains an integrand which almost always cannot be integrated exactly. Even if it can, the result would be a complicated function, unsuited for fast numerical manipulations. It is at this point, where the selfscattering term, which we artificially added on both sides of the BE, can be used to dramatically simplify the integral equation, as was first noticed by Rees [32]. Since the selfscattering rate $S_{\boldsymbol{k}}$ is completely unspecified, we can use it to enforce a particularly simple form of $\widetilde{\lambda}_{\boldsymbol{k}}$. An obvious choice is

$$\widetilde{\lambda}_{\boldsymbol{k}} = \lambda_{\boldsymbol{k}} + S_{\boldsymbol{k}} = \Gamma \equiv \text{const} \quad (8.26)$$

with a constant $\Gamma > \sup \lambda_{\boldsymbol{k}}$ in order to maintain the physical desirable interpretation of $S_{\boldsymbol{k}} = \Gamma - \lambda_{\boldsymbol{k}}$ in terms of a selfscattering rate, which, of course, has to be positive. With (8.26), (8.25) reduces after a re-labeling of the time integration variable to

$$g(\boldsymbol{k},t) = \int\limits_{0}^{\infty} \mathrm{d}\tau \sum_{\boldsymbol{k}'} \mathrm{e}^{-\Gamma\tau} \widetilde{S}_{\boldsymbol{k}',\boldsymbol{k}+e\boldsymbol{E}\tau/\hbar}\, g(\boldsymbol{k}',t-\tau)\ . \quad (8.27)$$

This form of the uniform BE is well-suited for an iterative solution [32, 34]. The parameter Γ turns out to be crucial. It not only eliminates a complicated integration but it also enforces a positive, continuous kernel which is necessary for the iteration procedure to converge [33].

From a numerical point of view, integral equations are less prone to numerical errors than differential equations. It can be therefore expected that an iteration based solution of (8.27) is numerically more robust than a numerical treatment of the BE in integro-differential form. Another nice property of the iterative approach is that it processes the whole distribution function which is available any time during the calculation. This is particularly useful for degenerate electrons, where Pauli-blocking affects electron-electron and electron-phonon scattering rates. In the simplest, and thus most efficient, particle-based Monte Carlo simulations, in contrast, the distribution function is only available at the end of the simulation, see Sect. 8.2.3.

At first sight the dimensionality of the integral equation[11] seems to ruin any efficient numerical treatment of (8.27). This is however not necessarily so. The time integration, for instance, is a convolution and can be eliminated by a Laplace transform ($t \leftrightarrow s$). In the Laplace domain, (8.27) contains s only as a parameter not as an integration variable. The efficiency of the method depends then on the efficiency with which the remaining $\boldsymbol{k}$-integration can be performed. For realistic band structures and scattering processes this may be time consuming. However, it is always possible to express $g(\boldsymbol{k},s)$ in a symmetry adopted set of basis functions, thereby converting (8.27) into a set of integral equations with lower dimensionality.

[11] Three momentum variables and one time variable.

Hammar [35], for instance, used Legendre polynomials to expand the angle dependence of the distribution function and obtained an extremely fast algorithm for the calculation of hot-electron distributions in $p - Ge$ and $n - GaAs$. In addition, it is conceivable to construct approximate kernels, which are numerically easier to handle.

8.2.2 Expansion into an Orthonormal Set of Basis Functions

Another technique which is often used to solve the BE is the expansion of the one-particle distribution function in terms of an orthonormal set of basis functions. The BE reduces then to a set of integro-differential equations with less independent variables. The method leads thus to a partial algebraization of the BE.

A typical example from plasma physics is the Lorentz ansatz for the distribution function

$$g(\boldsymbol{r}, \boldsymbol{v}, t) = g^{(0)}(\boldsymbol{r}, v, t) + \boldsymbol{g}^{(1)}(\boldsymbol{r}, v, t) \cdot \frac{\boldsymbol{v}}{v} + ... \tag{8.28}$$

which comprises the first two terms of the expansion of the velocity space angle dependence of $g(\boldsymbol{r}, \boldsymbol{v}, t)$ in terms of spherical harmonics. In particular, calculations of transport coefficients for gas discharges are based on this expansion, see, for instance, the review by Winkler [5]. The simplification arises here from the fact that the expansion coefficients $g^{(i)}$ with $i = 0, 1, ...$ are independent of the angle variables in velocity space. Hence, the equations determining $g^{(i)}$, which are obtained from the BE by inserting (8.28) and averaging over the velocity space angles, have two less independent variables. Symmetries of the discharge can be used to further reduce the number of independent variables. This depends, however, on the details of the discharge and thus on the particular form of the BE.

Naturally, the expansion method is most useful for the spatially-uniform, steady-state, linearized BE, where the expansion coefficients depend at most on the magnitude of the velocity and the collision integral is linear in the expansion coefficients. As an example from condensed matter physics, we discuss here the expansion method developed by Allen [37] and Pinski [38]. It is an adaptation of (8.28) to quasiparticles and has been applied to various metals and alloys [39, 40, 41, 42, 43]. In that case, the *algebraization* is even complete leading to a linear set of algebraic equations for the expansion coefficients, which are just constants. In addition, the Allen-Pinski expansion typifies an ab initio approach because it is usually furnished with first-principle electronic structure data.

Having in mind the calculation of the electric conductivity of a metal, we again start from the linearized BE (8.7), but now with the linearized collision integral written in the form (8.9). Defining the function ϕ by

$$g^{(1)}(\boldsymbol{k}) = -\left.\frac{\partial f}{\partial E}\right|_{E(\boldsymbol{k})} \phi(\boldsymbol{k}) , \tag{8.29}$$

which deviates slightly from the definition used in the previous subsection, and assuming the electric field to be in x-direction, the electric current in x-direction becomes, see (8.12),

$$j_x = -2e \sum_{\boldsymbol{k}} v_x(\boldsymbol{k}) \frac{\partial f}{\partial E}\bigg|_{E(\boldsymbol{k})} \phi(\boldsymbol{k}) \tag{8.30}$$

with ϕ satisfying the linearized BE in the form

$$-eE_x v_x(\boldsymbol{k}) \frac{\partial f}{\partial E}\bigg|_{E(\boldsymbol{k})} = \sum_{\boldsymbol{k}'} Q_{\boldsymbol{k}\boldsymbol{k}'} \phi(\boldsymbol{k}') \,, \tag{8.31}$$

where the kernel of the collision integral is given by

$$Q_{\boldsymbol{k}\boldsymbol{k}'} = -\frac{\partial f}{\partial E}\bigg|_{E(\boldsymbol{k})} \widetilde{Q}_{\boldsymbol{k}\boldsymbol{k}'} \tag{8.32}$$

with $\widetilde{Q}_{\boldsymbol{k}\boldsymbol{k}'}$ defined in (8.10).

Although the iterative approach described in the previous subsection could be used to calculate $\phi(\boldsymbol{k})$ from (8.31), this is not very efficient, in particular, for metals with a complicated Fermi surface, where the required numerical integrations can be rather subtle and time-consuming. Allen [37] suggested therefore to expand $\phi(\boldsymbol{k})$ in a complete set of functions, which takes the symmetry of the crystal and thus the topology of the Fermi surface into account, and to transform (8.31) to a symmetry-adapted matrix representation which is then solved by matrix inversion. For that purpose Allen [37] introduced a particular biorthogonal product basis, consisting of Fermi surface harmonics $F_J(\boldsymbol{k})$ and energy polynomials $\sigma_n(E(\boldsymbol{k}))$.

The strength of Allen's approach stems form the mathematical properties of the basis functions. The Fermi surface harmonics $F_J(\boldsymbol{k})$ with $J = X, Y, Z, X^2, ...$ are polynomials of the three Cartesian components of the velocity $\boldsymbol{v}(\boldsymbol{k})$. They are periodic functions in $\boldsymbol{k}$-space and orthogonal when integrated over the Fermi surface [36]. More precisely

$$\sum_{\boldsymbol{k}} F_J(\boldsymbol{k}) F_{J'}(\boldsymbol{k}) \delta(E(\boldsymbol{k}) - E) = N(E) \delta_{JJ'} \tag{8.33}$$

with $N(E) = \sum_{\boldsymbol{k}} \delta(E(\boldsymbol{k}) - E)$ the single-spin density of states at energy E; recall, we put the volume equal to one. Fermi surface harmonics are useful for the description of variations of $\phi(\boldsymbol{k})$ on the energy shell, which may be anisotropic and even consisting of various unconnected pieces. For spherical energy shells, the $F_J(\boldsymbol{k})$ reduce to spherical harmonics $Y_{lm}(\hat{\boldsymbol{k}})$ with $\hat{\boldsymbol{k}}$ the unit vector in direction of $\boldsymbol{k}$. For general topologies they have to be constructed on a computer. The Fermi surface harmonics transform as basis functions of the irreducible point group of the crystal for which they are constructed. This is a particularly useful property, because it leads to a block-diagonal matrix representation for the BE (8.7). For single sheet, cubic symmetry energy surfaces, the lowest order Fermi surface harmonics are $F_J(\boldsymbol{k}) = v_J(\boldsymbol{k})/v(E(\boldsymbol{k}))$ with $J = X, Y, Z$ and a normalization factor $v(E) = [N(E)^{-1} \sum_{\boldsymbol{k}} v_x(\boldsymbol{k})^2 \delta(E(\boldsymbol{k}) - E)]^{1/2}$, which is the root-mean-square velocity at the energy surface E. Further details about Fermi surface harmonics, in particular, the construction principle for arbitrary energy surfaces, can be found in [36].

The energy polynomials $\sigma_n(E)$ are n^{th} order polynomials in $E/(k_B T)$, which are orthogonal with respect to the weight function $-\partial f/\partial E$ with

$$\int \mathrm{d}E \left(-\frac{\partial f}{\partial E}\right) \sigma_n(E)\sigma_{n'}(E) = \delta_{nn'} \; . \tag{8.34}$$

They will be used to describe variations perpendicular to the Fermi surface. The first two polynomials are $\sigma_0(E) = 1$ and $\sigma_1(E) = \sqrt{3}E/(\pi k_B T)$. Higher order ones have to be again constructed on a computer, using the recursion relation given by Allen [37]. As pointed out by Pinski [38], another possible choice for the energy polynomials, which may lead to faster convergence in some cases, is $\sigma_n(E) = \sqrt{2n+1}P_n(\tanh[E/(2k_B T)])$, where $P_n(E)$ is the nth order Legendre polynomial.

Allen used the functions $F_J(\boldsymbol{k})$ and $\sigma_n(E(\boldsymbol{k}))$ to define two complete sets of functions which are biorthogonal. With the proper normalization, they are given by

$$\chi_{Jn}(\boldsymbol{k}) = \frac{F_J(\boldsymbol{k})\sigma_n(E(\boldsymbol{k}))}{N(E(\boldsymbol{k}))v(E(\boldsymbol{k}))} \; ,$$
$$\xi_{Jn}(\boldsymbol{k}) = -F_J(\boldsymbol{k})\sigma_n(E(\boldsymbol{k}))v(E(\boldsymbol{k}))\frac{\partial f}{\partial E}\bigg|_{E(\boldsymbol{k})} \tag{8.35}$$

with $N(E)$ and $v(E)$, respectively, the single-spin density of states and the root-mean-square velocity at energy E (see above). With the help of (8.33) and (8.34), it is straightforward to show that $\chi_{Jn}(\boldsymbol{k})$ and $\xi_{Jn}(\boldsymbol{k})$ satisfy the biorthogonality conditions

$$\sum_{\boldsymbol{k}} \chi_{Jn}(\boldsymbol{k})\xi_{J'n'}(\boldsymbol{k}) = \delta_{JJ'}\delta_{nn'} \; ,$$
$$\sum_{Jn} \chi_{Jn}(\boldsymbol{k})\xi_{Jn}(\boldsymbol{k}') = \delta_{\boldsymbol{k}\boldsymbol{k}'} \; . \tag{8.36}$$

Any function of $\boldsymbol{k}$ can be either expanded in terms of the functions $\chi_{Jn}(\boldsymbol{k})$ or in terms of the functions $\xi_{Jn}(\boldsymbol{k})$. The functions χ_{Jn} are most convenient for expanding functions which are smooth in energy. Since in (8.29) we split-off the factor $-\partial_E f$, we expect $\phi(\boldsymbol{k})$ to exhibit this property and thus write

$$\phi(\boldsymbol{k}) = \sum_{Jn} \phi_{Jn}\chi_{Jn}(\boldsymbol{k}) \; . \tag{8.37}$$

The functions ξ_{Jn}, on the other hand, vary strongly in the vicinity of the Fermi surface. They are used at intermediate steps to express functions which peak at the Fermi energy.

We are now able to rewrite (8.31). Using the definition of ξ_{Jn}, the l.h.s. immediately becomes

$$\text{l.h.s. of (8.31)} = eE_x\xi_{X0}(\boldsymbol{k}) \; . \tag{8.38}$$

For the r.h.s., we find

$$\begin{aligned}\text{r.h.s. of (8.31)} &= \sum_{k'k''} Q_{kk'}\delta_{k'k''}\phi(k'') \\ &= \sum_{J'n'}\sum_{k'} Q_{kk'}\chi_{J'n'}(k')\sum_{k''}\xi_{J'n'}(k'')\phi(k'') \\ &= \sum_{J'n'}\sum_{k'} Q_{kk'}\chi_{J'n'}(k')\phi_{J'n'} ,\end{aligned} \tag{8.39}$$

where in the second line we expressed the Kronecker delta via (8.36) and in the third line we used the inverse of (8.37)

$$\phi_{Jn} = \sum_{k} \xi_{Jn}(k)\phi(k) . \tag{8.40}$$

Multiplying (8.38) and (8.39) from the left with $\chi_{Jn}(k)$ and summing over all k leads to the final result

$$E_x\delta_{n0}\delta_{JX} = \sum_{J'n'} Q_{Jn,J'n'}\phi_{J'n'} \tag{8.41}$$

with $Q_{Jn,J'n'} = \sum_{kk'} \chi_{Jn}(k)Q_{kk'}\chi_{J'n'}(k')$.

Equation (8.41) is the symmetry-adapted matrix representation of the linearized BE (8.31). Its solution gives the expansion coefficients ϕ_{Jn}. To complete the calculation, we have to express the electric current j_x in terms of these coefficients. Using in (8.30) the definition for $\xi_{X0}(k)$ and the biorthogonality condition (8.36), we obtain

$$j_x = 2e\phi_{X0} , \tag{8.42}$$

which with (8.41) yields

$$\sigma_{xx} = 2e^2[Q^{-1}]_{X0,X0} . \tag{8.43}$$

Thus, in Allen's basis, the xx-component of the electric conductivity tensor is just the upper-most left matrix element of the inverse of the matrix which represents the linearized collision integral. Remember, because of the symmetry of the basis functions, this matrix is block-diagonal. The numerical inversion is therefore expected to be fast.

The numerical bottleneck is the calculation of the matrix elements $Q_{Jn,J'n'}$. They depend on the symmetry of the metal and, of course, on the scattering processes. For realistic band structures, this leads to rather involved expressions, which, fortunately, are amenable to some simplifications arising from the fact that in metals $k_B T/E_F \ll 1$, where E_F is the Fermi energy. The k-integration can thus be restricted to the thermal layer with width Δk, see Fig. 8.3. For explicit expressions, we refer to the literature [37, 38, 39, 40, 41, 42, 43]. Although Allen's approach is not straightforward to implement, it has the advantage that it can handle complicated Fermi surfaces in a transparent manner. In practice, the matrix elements $Q_{Jn,J'n'}$ are expressed in terms of generalized coupling functions which can be either directly obtained from experiments or from ab initio band structure calculations. Allen's method of solving the linearized BE is therefore geared towards an ab initio calculation of transport coefficients for metals.

8.2.3 Particle-Based Monte Carlo Simulation

The most widely accepted method for solving the electron transport problem in semiconductors is the particle-based Monte Carlo simulation [44, 45, 46, 47, 48, 49]. In its general form, it simulates the stochastic motion of a finite number of test-particles and is equivalent to the solution of the BE. This technique has been, for instance, used to simulate field-effect transistors from a microscopic point of view, starting from the band structures and scattering processes of the semiconducting materials transistors are made off.

A Monte Carlo simulation of Boltzmann transport is a one-to-one realization of Boltzmann's original idea of free flights, occasionally interrupted by random scattering events, as being responsible for the macroscopic transport properties of the gas under consideration; here, the electrons in a semiconductor. The approach relies only on general concepts of probability theory, and not on specialized mathematical techniques. Because of the minimum of mathematical analysis, realistic band structures, scattering probabilities, and geometries can be straightforwardly incorporated into a Monte Carlo code. However, the method has some problems to account for Pauli-blocking in degenerate electron systems. It has thus not been applied to degenerate semiconductors, metals, or quantum fluids.

8.2.3.1 Spatially Uniform, Steady-State BE with the Full Collision Integral

First, we focus on the simplest situation: Steady-state transport in a spatially uniform, non-degenerate semiconductor. For the techniques described in the previous Subsections, this situation is already rather demanding, in particular, when a realistic electronic structure is used, which leads to involved $\boldsymbol{k}$-summations. The Monte Carlo simulation, on the other hand, requires no $\boldsymbol{k}$-summations. Moreover, in that particular case, transport coefficients for electrons can be calculated rather easily because ergodicity guarantees that the simulation of a single test-particle is sufficient to sample the whole phase space.

Instead of attempting to solve

$$\left\{-\frac{e}{\hbar}\boldsymbol{E}\cdot\nabla_{\boldsymbol{k}} + \lambda_{\boldsymbol{k}} + S_{\boldsymbol{k}}\right\} g(\boldsymbol{k},t) = \sum_{\boldsymbol{k}'}\left[S_{\boldsymbol{k}',\boldsymbol{k}} + S_{\boldsymbol{k}}\delta_{\boldsymbol{k}\boldsymbol{k}'}\right] g(\boldsymbol{k}',t)\,, \tag{8.44}$$

which is the BE appropriate for steady-state transport in a single band of a semiconductor subject to an uniform electric field, the Monte Carlo approach simulates the motion of a single test-electron in momentum space. For that purpose, it generates a large number of free flights, where the test-electron drifts freely in the electric field until it suffers one of the possible scattering processes, see Fig. 8.4. The technique uses random variables to represent the duration of the free flight, to select the type of scattering which terminates the free flight, and to determine the momentum of the test-electron after the scattering event, which is then used as the initial momentum for the next free flight. The steady-state does not depend on the initial condition for the first free flight. Any convenient choice can therefore be made.

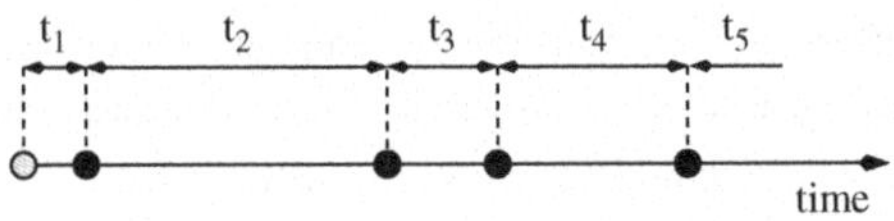

Fig. 8.4. Schematic representation of the particle-based Monte Carlo simulation of steady-state Boltzmann transport in spatially uniform solids. A single test-particle suffices here because ergodicity guarantees that the whole phase space is sampled. The test-particle performs free flights in the external field, randomly interrupted by one of the possible scattering processes (*black bullets*). The simulation consists of a finite number of free flights, starting from an arbitrary initial condition (*grey bullet*). For each free flight the simulation uses random numbers to generate its duration t_i, to select the terminating scattering process, and to determine the test-particles' momentum after the scattering event, which then serves as the initial momentum for the next free flight

Because of ergodicity, the ensemble average of a single-particle observable $O(\boldsymbol{k})$ is equal to the time average of this observable. Splitting the total simulation time t_s into a finite number of free flights with duration t_i we obtain

$$\langle O \rangle = \langle O \rangle_{t_s} = \sum_i \frac{1}{t_s} \int_0^{t_i} \mathrm{d}t \; O(\boldsymbol{k}(t)) \;, \tag{8.45}$$

where O can be, for instance, the energy of the electron $E(\boldsymbol{k})$ or its velocity $\boldsymbol{v}(\boldsymbol{k}) = \hbar^{-1}\nabla_{\boldsymbol{k}} E(\boldsymbol{k})$. Note, for each free flight, the time integration starts all over again from zero. The test-electron has no memory, reminiscent of the Markovian property of the BE.

The probability distributions for the random variables used in the Monte Carlo simulation are given in terms of the electric field and the transition probabilities for the various scattering processes. For realistic band structures the distributions can be quite complicated, in particular, the distribution of the duration of the free flights. Special techniques have to be used to relate the random variables needed in the simulation to the uniformly distributed random variables generated by a computer.

Let us first consider the distribution of the duration of the free flights. The probability for the test-electron to suffer the next collision in the time interval $\mathrm{d}t$ centered around t is given by

$$P(t)\mathrm{d}t = \widetilde{\lambda}_{\boldsymbol{k}(t)} \mathrm{e}^{-\int_0^t \mathrm{d}t' \widetilde{\lambda}_{\boldsymbol{k}(t')}} \mathrm{d}t \;, \tag{8.46}$$

where $\boldsymbol{k}(t) = \boldsymbol{k}_0 - (e/\hbar)\boldsymbol{E}t$, with $\boldsymbol{k}_0$ the arbitrary wave vector from which the first free flight started, and $\widetilde{\lambda}_{\boldsymbol{k}}$ the total transition rate from state $\boldsymbol{k}$ due to all scattering processes, including selfscattering. In Sect. 8.2.1, we introduced selfscattering in order to simplify the integral representation of the BE. But it is also very useful in the present context because, using again the choice (8.26), it leads to $\widetilde{\lambda} = \Gamma$ and thus to

$$P(t)\mathrm{d}t = \Gamma \mathrm{e}^{-\Gamma t} \mathrm{d}t \,. \tag{8.47}$$

Note, without selfscattering, we should have integrated over $\lambda_{\boldsymbol{k}(t)}$, comprising only real scattering events. For realistic band structures, this could have been done only numerically, and would have lead to a rather complicated $P(t)\mathrm{d}t$, useless for further numerical processing.

To relate the random variable t to a random variable $R \in [0,1]$ with a uniform distribution, we consider the cumulant of $P(t)$,

$$c(t) = \int_0^t \mathrm{d}t' P(t') = 1 - \mathrm{e}^{-\Gamma t} , \tag{8.48}$$

which for a random value of t is a random variable R, uniformly distributed between $[0,1]$ because $c(0) = 0$ and $c(\infty) = 1$. Thus, $c(t) = R$. Inverting (8.48) and introducing a new random variable $R_1 = 1 - R$, which is also uniformly distributed in $[0,1]$, the duration of the free flights can be easily generated by

$$t = -\frac{1}{\Gamma} \ln R_1 . \tag{8.49}$$

Having determined the duration of a free flight, a scattering event, responsible for the end of the free flight, has to be chosen. As mentioned before, we assume that $\widetilde{\lambda}_{\boldsymbol{k}}$ is the total transition rate from state $\boldsymbol{k}$, which is now the terminating momentum of the free flight, including selfscattering, that is, $\widetilde{\lambda}_{\boldsymbol{k}} = \sum_{p=1}^{N} \lambda_{\boldsymbol{k}}^{p} + S_{\boldsymbol{k}} = \sum_{p=1}^{N+1} \lambda_{\boldsymbol{k}}^{p}$, where N is the total number of real scattering processes and $\lambda_{\boldsymbol{k}}^{N+1} = S_{\boldsymbol{k}}$. From definition (8.26) follows $\widetilde{\lambda}_{\boldsymbol{k}} = \Gamma$. One way of choosing the terminating scattering process is therefore to generate a random variable $R_2 \in [0,1]$ and to form partial sums until

$$\frac{1}{\Gamma} \sum_{p=1}^{m-1} \lambda_{\boldsymbol{k}}^{p} < R_2 \leq \frac{1}{\Gamma} \sum_{p=1}^{m} \lambda_{\boldsymbol{k}}^{p} \tag{8.50}$$

is satisfied. The mth process is then the terminating one.

Equation (8.50) contains selfscattering and real scattering processes. In the former the momentum is conserved, that is, $\boldsymbol{k} = \boldsymbol{k}'$, with $\boldsymbol{k}'$ the momentum after the scattering event. For real scattering processes, however, $\boldsymbol{k}'$ is a random variable which has to be determined from the transition probabilities of the various scattering processes whose precise forms in turn depend on the material. We describe therefore only the basic strategy for choosing the momentum of the test-electron after the scattering event, assuming electron-impurity and electron-phonon scattering to be responsible for the end of the free flight, see Fig. 8.5.

From the momentum $\boldsymbol{k}$ the test-electron has at the end of the free flight, we obtain its energy $E(\boldsymbol{k})$ before the scattering event. Energy conservation can then be used to determine the energy E' of the test-electron after the scattering event. For electron-impurity scattering $E' = E(\boldsymbol{k})$ (elastic scattering) whereas for electron-phonon scattering $E' = E(\boldsymbol{k}) \pm \hbar\omega$, where for simplicity we assumed dispersionless phonons with energy $\hbar\omega$. The upper and lower sign corresponds, respectively, to phonon absorption and emission, which are treated here as separate processes. In most cases, the part of the semiconductors' band structure relevant for transport can

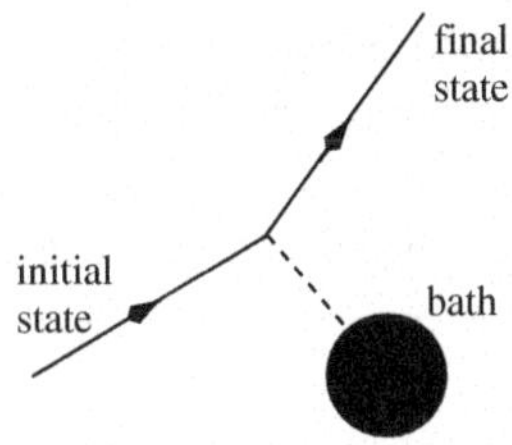

Fig. 8.5. Illustration of the scattering event in the test-particle-based Monte Carlo simulation. The test-particle scatters off a generalized bath representing impurities and phonons. For elastic scattering, the test-electron gains or loses only momentum, whereas for inelastic scattering it can also transfer or receive energy

be assumed to be isotropic. The magnitude of the final state momentum, $k' = |\boldsymbol{k}'|$ can thus be obtained from $E' = E(k')$. The orientation of the vector $\boldsymbol{k}'$, however, is still unspecified.

When the scattering process is randomizing, all momentum states on the final state energy surface are equally probable. Measuring the orientation of $\boldsymbol{k}'$ in polar coordinates, with $\boldsymbol{k}$ pointing in the z-direction, the probability[12] for $\boldsymbol{k}'$ to be given by $k'_x = k' \cos\phi' \sin\theta'$, $k'_y = k' \sin\phi' \sin\theta'$, and $k'_z = k' \cos\theta'$, with $0 \leq \phi' \leq 2\pi$ and $0 \leq \theta' \leq \pi$, is $P(\phi', \theta') = (1/4\pi)\sin\theta'$. To relate the random variables ϕ' and θ' to two uniformly distributed random variables in the interval $[0, 1]$, we first condition the two-variable probability $P(\phi', \theta')$ in the form

$$P(\phi', \theta') = P_1(\phi') P_2(\theta'|\phi') \,, \tag{8.51}$$

where $P_1(\phi') = \int_0^\pi \mathrm{d}\theta' P(\phi', \theta') = 1/(2\pi)$ is the marginal probability for ϕ' and $P_2(\theta'|\phi') = P(\phi', \theta')/(P_1(\phi')) = \sin\theta'/2$ is the conditional probability for θ' given ϕ'. We then apply cumulants (as described above) separately to P_1 and P_2 and obtain

$$\begin{aligned} \phi' &= 2\pi R_3 \,, \\ \cos\theta' &= 1 - 2R_4 \end{aligned} \tag{8.52}$$

with R_3 and R_4 uniformly distributed random variables in the interval $[0, 1]$.

For non-randomizing scattering processes, the probability for the angles is proportional to the transition rate written in the polar coordinates introduced above. Hence, for given k and k', the properly normalized function $P(\phi', \theta'; k, k') = (\sin\theta'/4\pi) S(k, k', \phi', \theta')$ gives the probability for the azimuth ϕ' and the polar angle θ', both depending therefore on k and k'. Applying again the method of conditioning, which can be applied to any two-variable probability, together with the cumulants, the random variables $\phi'(k, k')$ and $\theta'(k, k')$ can be again expressed in terms of uniformly distributed random variables in the interval $[0, 1]$.

The simulation consists of a finite number of free flights of random duration and random initial conditions. Average single particle properties, in particular the

[12] Strictly speaking, it is the probability density.

drift velocity, can then be obtained from (8.45). Assuming the electric field to be in z-direction, the drift will be also along the z-axis, that is, only the z-component of the electron momentum will be changed due to the field. Hence, writing (8.45) with $O = v_z(k_z(t))$ and integrating with respect to k_z instead of t, the drift velocity is given by [44]

$$\langle v_z \rangle = \frac{1}{K} \sum_{\text{flights}} \int_{k_{z,i}}^{k_{z,f}} \frac{1}{\hbar} \frac{\partial E}{\partial k_z} \, \mathrm{d}k_z = \frac{1}{\hbar K} \sum_{\text{flights}} (E_f - E_i) \,, \tag{8.53}$$

where the sum goes over all free flights, $k_{z,i}$ and $k_{z,f}$ denote the z-component of the initial and final momentum of the respective free flights, and K is the total length of the $\boldsymbol{k}$-space trajectory.

In some cases, the distribution function $g(\boldsymbol{k})$ may be also of interest. In order to determine $g(\boldsymbol{k})$ from the motion of a single test-particle a grid is set up in momentum space at the beginning of the simulation. During the simulation the fraction of the total time the test-electron spends in each cell is then recorded and taken as a measure for the distribution function. This rule results from an application of (8.45). Indeed, using $O(\boldsymbol{k}(t)) = n_i(\boldsymbol{k}(t))$ with $n_i(\boldsymbol{k}(t)) = 1$ when the test-particle is in cell i and zero otherwise, gives $g(\boldsymbol{k}_i) \equiv \langle n_i \rangle = \Delta t_i / t_s$, with Δt_i the time spend in cell i. Averaged single particle quantities for the steady-state could then be also obtained from the sum

$$\langle O \rangle = \sum_{\boldsymbol{k}} O(\boldsymbol{k}) g(\boldsymbol{k}) \,, \tag{8.54}$$

but for a reasonable accuracy the grid in momentum space has to be very fine. It is therefore more convenient to calculate $\langle O \rangle$ directly from (8.45).

It is instructive to demonstrate that the Monte Carlo procedure just outlined is indeed equivalent to solving the steady-state BE (8.44). The equivalence proof has been given by Fawcett and coauthors [44] and we follow closely their treatment. The starting point is the definition of a function $P_n(\boldsymbol{k}_0, \boldsymbol{k}, t)$ which is the probability that the test-electron will have momentum $\boldsymbol{k}$ at time t during the nth free flight when it started at $t = 0$ with momentum $\boldsymbol{k}_0$. The explicit time dependence must be retained because the electron can pass through the momentum state $\boldsymbol{k}$ any time during the nth free flight. This probability satisfies an integral equation

$$P_n(\boldsymbol{k}_0, \boldsymbol{k}, t) = \sum_{\boldsymbol{k}', \boldsymbol{k}''} \int_0^t \mathrm{d}t' P_{n-1}(\boldsymbol{k}_0, \boldsymbol{k}', t) \widetilde{S}_{\boldsymbol{k}', \boldsymbol{k}''} \times \mathrm{e}^{-\int_0^{t-t'} \mathrm{d}t'' \widetilde{\lambda}_{\boldsymbol{k}'' - e\boldsymbol{E}t''/\hbar}} \delta_{\boldsymbol{k}, \boldsymbol{k}'' - e\boldsymbol{E}(t-t')/\hbar} \,, \tag{8.55}$$

whose r.h.s. consists of three probabilities which are integrated over. The first one $P_{n-1}(\boldsymbol{k}_0, \boldsymbol{k}', t)$ is the probability that the test-electron passes through some momentum state $\boldsymbol{k}'$ during the $(n-1)^{th}$ free flight, the second $\widetilde{S}_{\boldsymbol{k}', \boldsymbol{k}''}$ is the probability that it will be scattered from state $\boldsymbol{k}'$ to state $\boldsymbol{k}''$, whereas the exponential factor is the probability that it will not be scattered while drifting from $\boldsymbol{k}''$ to $\boldsymbol{k}$ during the nth free flight. The Kronecker-δ ensures that the test-electron follows the trajectory appropriate for the applied electric field $\boldsymbol{E}$. The Monte Carlo simulation generates realizations of the random variable $\boldsymbol{k}(t)$ in accordance to the probability $P_n(\boldsymbol{k}_0, \boldsymbol{k}, t)$.

Integrating in (8.55) over $\boldsymbol{k}''$ and t' and substituting $\tau = t - t'$ and $y = \tau - t''$ yields an equation,

$$P_n(\boldsymbol{k}_0, \boldsymbol{k}, t) = \sum_{\boldsymbol{k}'} \int_0^t \mathrm{d}\tau P_{n-1}(\boldsymbol{k}_0, \boldsymbol{k}', t - \tau) \widetilde{S}_{\boldsymbol{k}', \boldsymbol{k}+e\boldsymbol{E}\tau/\hbar}\, \mathrm{e}^{-\int_0^\tau \mathrm{d}y \widetilde{\boldsymbol{\lambda}}_{\boldsymbol{k}+e\boldsymbol{E}y/\hbar}} , \tag{8.56}$$

which is a disguised BE. To make the connection with the BE more explicit, we consider the count at $\boldsymbol{k}$ obtained after N collisions

$$C_N(\boldsymbol{k}_0, \boldsymbol{k}) = \lim_{t_s \to \infty} \sum_{n=1}^{N} \frac{1}{t_s} \int_0^{t_s} \mathrm{d}t P_n(\boldsymbol{k}_0, \boldsymbol{k}, t) . \tag{8.57}$$

This number is provided by the Monte Carlo procedure and at the same time it can be identified with $g(\boldsymbol{k})$ for $N \gg 1$. Thus, $C_N(\boldsymbol{k}_0, \boldsymbol{k})$ is the bridge, which will carry us from the test-particle Monte Carlo simulation to the traditional BE.

We now perform a series of mathematical manipulations at the end of which we will have obtained the steady-state, spatially-uniform BE (8.44). Inserting (8.56) into the definition (8.57), applying on both sides $-(e/\hbar)\boldsymbol{E} \cdot \nabla_{\boldsymbol{k}}$ from the left, and using the two identities

$$\begin{aligned} -\frac{e\boldsymbol{E}}{\hbar} \cdot \nabla_{\boldsymbol{k}} \widetilde{S}_{\boldsymbol{k}', \boldsymbol{k}+e\boldsymbol{E}\tau/\hbar} &= -\frac{\partial}{\partial \tau} \widetilde{S}_{\boldsymbol{k}', \boldsymbol{k}+e\boldsymbol{E}\tau/\hbar} , \\ -\frac{e\boldsymbol{E}}{\hbar} \cdot \nabla_{\boldsymbol{k}} \mathrm{e}^{-\int_0^\tau \mathrm{d}y \widetilde{\boldsymbol{\lambda}}_{\boldsymbol{k}+e\boldsymbol{E}y/\hbar}} &= -\left(\frac{\partial}{\partial \tau} + \widetilde{\lambda}_{\boldsymbol{k}}\right) \mathrm{e}^{-\int_0^\tau \mathrm{d}y' \widetilde{\boldsymbol{\lambda}}_{\boldsymbol{k}+e\boldsymbol{E}y/\hbar}} \end{aligned} \tag{8.58}$$

gives

$$\begin{aligned} -\frac{e\boldsymbol{E}}{\hbar} \cdot \nabla_{\boldsymbol{k}} C_N(\boldsymbol{k}_0, \boldsymbol{k}) &= -\widetilde{\lambda}_{\boldsymbol{k}} C_N(\boldsymbol{k}_0, \boldsymbol{k}) \\ &- \lim_{t_s \to \infty} \sum_{n=1}^{N} \frac{1}{t_s} \sum_{\boldsymbol{k}'} \int_0^{t_s} \mathrm{d}\tau P_{n-1}(\boldsymbol{k}_0, \boldsymbol{k}', t - \tau) \frac{\partial}{\partial \tau} \left[\widetilde{S}_{\boldsymbol{k}', \boldsymbol{k}+e\boldsymbol{E}\tau/\hbar} \mathrm{e}^{-\int_0^\tau \mathrm{d}y \widetilde{\boldsymbol{\lambda}}_{\boldsymbol{k}+e\boldsymbol{E}y/\hbar}}\right] , \end{aligned} \tag{8.59}$$

where we used definition (8.57) once more to obtain the first term on the r.h.s. This equation can be rewritten into

$$
\begin{aligned}
-\frac{e\boldsymbol{E}}{\hbar}\cdot\nabla_{\boldsymbol{k}}C_N(\boldsymbol{k}_0,\boldsymbol{k}) &= -\widetilde{\lambda}_{\boldsymbol{k}}C_N(\boldsymbol{k}_0,\boldsymbol{k})\\
&- \lim_{t_s\to\infty}\sum_{n=1}^{N}\frac{1}{t_s}\int_0^{t_s}\mathrm{d}t\sum_{\boldsymbol{k}'}P_{n-1}(\boldsymbol{k}_0,\boldsymbol{k}',0)\widetilde{S}_{\boldsymbol{k}',\boldsymbol{k}+e\boldsymbol{E}t/\hbar}\mathrm{e}^{-\int_0^t \mathrm{d}y\widetilde{\lambda}_{\boldsymbol{k}+e\boldsymbol{E}y/\hbar}}\\
&+ \lim_{t_s\to\infty}\sum_{n=1}^{N}\frac{1}{t_s}\int_0^{t_s}\mathrm{d}t\sum_{\boldsymbol{k}'}P_{n-1}(\boldsymbol{k}_0,\boldsymbol{k}',t)\widetilde{S}_{\boldsymbol{k}',\boldsymbol{k}} - \lim_{t_s\to\infty}\sum_{n=1}^{N}\frac{1}{t_s}\int_0^{t_s}\mathrm{d}t\\
&\times\sum_{\boldsymbol{k}'}\int_0^t \mathrm{d}\tau\frac{\partial}{\partial t}\left(P_{n-1}(\boldsymbol{k}_0,\boldsymbol{k}',t-\tau)\widetilde{S}_{\boldsymbol{k}',\boldsymbol{k}+e\boldsymbol{E}\tau/\hbar}\mathrm{e}^{-\int_0^\tau \mathrm{d}y\widetilde{\lambda}_{\boldsymbol{k}+e\boldsymbol{E}y/\hbar}}\right), \quad (8.60)
\end{aligned}
$$

when the τ-integration is carried out by parts and $\partial_\tau P_{n-1} = -\partial_t P_{n-1}$ is used. Pulling now in the fourth term on the r.h.s. the differential operator ∂_t in front of the τ-integral produces two terms, one of which cancels with the second term on the r.h.s. and the other vanishes in the limit $t_s \to \infty$. As a result, only the first and third term on the r.h.s. of (8.60) remain. Using finally in the third term again the definition (8.57) yields

$$
-\frac{e\boldsymbol{E}}{\hbar}\cdot\nabla_{\boldsymbol{k}}C_N(\boldsymbol{k}_0,\boldsymbol{k}) + \widetilde{\lambda}_{\boldsymbol{k}}C_N(\boldsymbol{k}_0,\boldsymbol{k}) = \sum_{\boldsymbol{k}'}C_{N-1}(\boldsymbol{k}_0,\boldsymbol{k}')\widetilde{S}_{\boldsymbol{k}',\boldsymbol{k}}\,, \quad (8.61)
$$

which, recalling the definitions of $\widetilde{\lambda}_{\boldsymbol{k}}$ and $\widetilde{S}_{\boldsymbol{k}',\boldsymbol{k}}$ and using $C_{N-1}(\boldsymbol{k}_0,\boldsymbol{k}) \to C_N(\boldsymbol{k}_0,\boldsymbol{k}) \to g(\boldsymbol{k})$ for $N \to \infty$, is identical to the BE (8.44). Thus, the simulation of a large number of free flights, each one terminating in a random scattering event, indeed simulates the steady-state BE for a spatially uniform semiconductor.

8.2.3.2 Spatially Non-Uniform BE with the Full Collision Integral

For spatially non-uniform situations[13], typical for semiconductor devices, the simulation of a single test-particle is not enough (see Fig. 8.6). With a single test-particle, for instance, it is impossible to represent the source term of the Poisson equation. However, this equation needs to be solved in conjunction with the BE to obtain the self-consistent electric field responsible for space-charge effects which, in turn, determine the current-voltage characteristics of electronic devices.

Instead of a single test particle it is necessary to simulate an ensemble of test-particles for prescribed boundary conditions for the Poisson equation and the BE, where the latter have to be translated into boundary conditions for the test-particles. The boundary conditions for the Poisson equation are straightforward; Dirichlet condition, i.e., fixed potentials, at the electrodes and Neumann condition, i.e., zero electric field, at the remaining boundaries. But the boundary conditions for the test-particles, which need to be consistent with the ones for the Poisson equation, can be rather subtle, resulting in sophisticated particle injection and reflection strategies,

[13] The same holds for time-dependent situations.

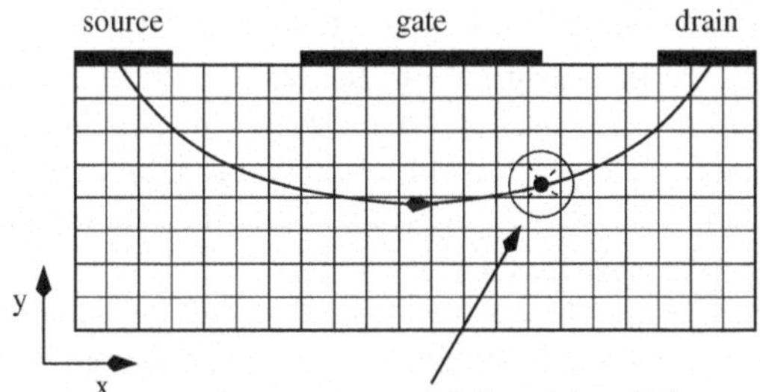

Fig. 8.6. Schematic representation of the cross section of a metal-semiconductor field-effect transistor (MESFET). Shown is a typical discretization of the two-dimensional simulation volume, a representative electron trajectory, and the contacting through a source, a gate, and a drain electrode. The BE has to be solved together with the Poisson equation; for both equations boundary conditions are required, see Sect. 8.2.3.2. Since the Poisson equation is grid-bound, whereas the BE is not, charge assignment and force interpolation symbolized by the thin lines inside the circle are required for the simultaneous solution of the two

in particular, when the doping profile of the semiconductor structure is taken into account. An authoritative discussion of the boundary conditions, as well as other aspects of device modeling, can be found in the textbook by Jacoboni and Lugli [47].

Conceptually, the Monte Carlo simulation for semiconductor devices resembles the particle-in-cell simulations for plasmas described in Chap. 6, and we refer there for technical details. In particular, the techniques for the solution of the Poisson equation and the particle weighting and force interpolation required for the coupling of the grid-free electron kinetics (simulation of the BE) with the grid-bound electric field (solution of the Poisson equation) are identical. In addition, except of the differences which arise from the particular electric contacting of the simulation volume, the implementation of particle injection and reflection (boundary conditions for the test-particles) are also basically the same. The only differences are that the test-particles have to be of course propagated during a free flight according to $\mathrm{d}\boldsymbol{k}/\mathrm{d}t = -(e/\hbar)\boldsymbol{E}$ and $\mathrm{d}\boldsymbol{r}/\mathrm{d}t = \hbar^{-1}\nabla_{\boldsymbol{k}}E(\boldsymbol{k})$ and that the scattering processes are the ones appropriate for semiconductors: Electron-impurity scattering, electron-phonon scattering, and, in some cases, electron-electron scattering.

In this generalized form, the particle-based Monte Carlo simulation has become the standard tool for analyzing Boltzmann transport of electrons in semiconductors. In combination with ab initio band structure data, including scattering rates, it is by now an indispensable tool for electronics engineers optimizing the performance of semiconductor devices [46, 47, 48, 49].

8.2.4 Ensemble-Based Monte Carlo Simulation

The test-particle-based Monte Carlo algorithm described in the previous Subsection cannot be applied to degenerate electron systems where the final state of the scattering may be blocked by the Pauli principle. Mathematically, the Pauli-blocking is encoded in the collision integral (8.6) through the factor $1 - g_n(\boldsymbol{r}, \boldsymbol{k}, t)$. It depends therefore on the one-particle distribution function which in the test-particle-based Monte Carlo algorithm is only available at the end of the simulation. In principle, the distribution function from a previous run could be used, but this requires additional book-keeping, which, if nothing else, demonstrates that the algorithm presented in the previous Subsection loses much of its simplicity.

An alternative method, which is most suitable for degenerate Fermi systems is the ensemble-based Monte Carlo simulation. There are various ways to simulate an ensemble. We describe here a simple approach, applicable to a spatially homogeneous electron system. It is based on the master equation for the probability $P_\nu(t)$ that at time t the many-particle system is in configuration $\nu = (n_{\boldsymbol{k}_1}, n_{\boldsymbol{k}_2}, ...)$. For fermions, $n_{\boldsymbol{k}_i} = 0$ when the momentum state $\boldsymbol{k}_i$ is empty and $n_{\boldsymbol{k}_i} = 1$ when the state is occupied. The one-particle distribution function, which is the solution of the corresponding BE, is then given by an ensemble average

$$g(\boldsymbol{k}, t) = \sum_\nu P_\nu(t) n_{\boldsymbol{k}} \, . \tag{8.62}$$

The algorithm has been developed by El-Sayed and coworkers and we closely follow their treatment [50]. The purpose of the algorithm is to simulate electron relaxation in a two-dimensional, homogeneous degenerate electron gas, with electron-electron scattering as the only scattering process. Such a situation can be realized, for instance, in the conduction band of a highly optically excited semiconductor quantum well at low enough temperatures. It is straightforward to take other scattering processes into account. Inhomogeneous situations, typical for device modeling, can be in principle also treated but it requires a major overhaul of the approach which we will not discuss.

Taking only direct electron-electron scattering into account, the force-free Boltzmann equation for a homogeneous, two-dimensional electron gas reads[14]

$$\frac{\partial g_{\boldsymbol{k}}}{\partial t} = 2 \sum_{\boldsymbol{p},\boldsymbol{k}'\boldsymbol{p}'} W_{\boldsymbol{k}\boldsymbol{p};\boldsymbol{k}'\boldsymbol{p}'} \left([1 - g_{\boldsymbol{k}}][1 - g_{\boldsymbol{p}}] g_{\boldsymbol{k}'} g_{\boldsymbol{p}'} - g_{\boldsymbol{k}} g_{\boldsymbol{p}} [1 - g_{\boldsymbol{k}'}][1 - g_{\boldsymbol{p}'}]\right) \tag{8.63}$$

with

$$W_{\boldsymbol{k}\boldsymbol{p};\boldsymbol{k}'\boldsymbol{p}'} = \frac{2\pi}{\hbar} \left| V(|\boldsymbol{k} - \boldsymbol{k}'|) \right|^2 \delta_{\boldsymbol{k}+\boldsymbol{p};\boldsymbol{k}'+\boldsymbol{p}'} \delta\left(E(\boldsymbol{k}) + E(\boldsymbol{p}) - E(\boldsymbol{k}') - E(\boldsymbol{p}')\right) \tag{8.64}$$

and $V(q) = 2\pi e^2/[\epsilon_0(q + q_s)]$ the statically screened Coulomb interaction in two dimensions, $V = L^2$ is again put to one. The factor two in front of the sum in (8.63) comes from the electron spin. As indicated above, the simulation of this equation via the test-particle-based Monte Carlo technique is complicated because the Pauli blocking factors depend on the (instantaneous) distribution function. The ensemble Monte Carlo method proposed by El-Sayed and coworkers [50] simulates therefore

[14] Notice the slight change in our notation: $g(\boldsymbol{k}) \to g_{\boldsymbol{k}}$.

the master equation underlying the Boltzmann description. This equation determines the time evolution of the probability for the occurrence of a whole configuration in momentum space,

$$\frac{\mathrm{d}P_\nu}{\mathrm{d}t} = -\frac{P_\nu(t)}{\tau_\nu} + \sum_{\nu'} W_{\nu'\nu} P_{\nu'}(t) \,. \tag{8.65}$$

Here

$$\frac{1}{\tau_\nu} = \sum_{\nu'} W_{\nu\nu'} \tag{8.66}$$

is the lifetime of the configuration ν, and $W_{\nu,\nu'}$ is the transition rate from configuration ν to ν'. Specifically for electron-electron scattering,

$$W_{\nu\nu'} = \frac{1}{2} \sum_{\boldsymbol{k}\boldsymbol{p}\boldsymbol{k}'\boldsymbol{p}'} W_{\boldsymbol{k}\boldsymbol{p};\boldsymbol{k}'\boldsymbol{p}'} n_{\boldsymbol{k}} n_{\boldsymbol{p}} [1 - n_{\boldsymbol{k}'}][1 - n_{\boldsymbol{p}'}] D^{\nu\nu'}_{\boldsymbol{k}\boldsymbol{p};\boldsymbol{k}'\boldsymbol{p}'} \tag{8.67}$$

with $D^{\nu\nu'}_{\boldsymbol{k}\boldsymbol{p};\boldsymbol{k}'\boldsymbol{p}'} = \delta_{n'_{\boldsymbol{k}} n_{\boldsymbol{k}}-1} \delta_{n'_{\boldsymbol{p}} n_{\boldsymbol{p}}-1} \delta_{n'_{\boldsymbol{k}'} n_{\boldsymbol{k}'}+1} \delta_{n'_{\boldsymbol{p}'} n_{\boldsymbol{p}'}+1} \prod_{\boldsymbol{q} \neq \boldsymbol{k},\boldsymbol{p},\boldsymbol{k}',\boldsymbol{p}'} \delta_{n'_{\boldsymbol{q}} n_{\boldsymbol{q}}}$.

The crucial point of the method is that the sampling of the configurations can be done in discrete time steps τ_ν. The master equation (8.65) then simplifies to

$$P_\nu(t + \tau_\nu) = \sum_{\nu'} \Pi_{\nu'\nu} P_{\nu'}(t) \tag{8.68}$$

with

$$\Pi_{\nu'\nu} = \tau_\nu W_{\nu'\nu} \tag{8.69}$$

the transition probability from configuration ν to configuration ν'[15]. Thus, when the system was at time t in the configuration ν_0, that is $P_\nu(t) = \delta_{\nu\nu_0}$, then the probability to find the system at time $t + \tau_\nu$ in the configuration ν is $P_\nu(t + \tau_\nu) = \Pi_{\nu_0\nu}$. In a simulation the new configuration can be therefore chosen according to the probability $\Pi_{\nu_0\nu}$.

However, there is a main drawback. In order to determine τ_ν form (8.66) a high-dimensional, configuration-dependent integral has to be numerically calculated before the time propagation can be made. Clearly, this is not very efficient. To overcome the problem, the selfscattering method is used again, but now at the level of the master equation, where selfscattering events can be also easily introduced because (8.65) is unchanged, when a diagonal element is added to the transition rate. It is therefore possible to work with a modified transition rate

$$W^s_{\nu\nu'} = W_{\nu\nu'} + W_\nu \delta_{\nu\nu'} \,, \tag{8.70}$$

giving rise to (cp. with (8.66))

---

[15] The normalization required for the interpretation of $\Pi_{\nu'\nu}$ in terms of a probability is a consequence of the detailed balance $W_{\nu\nu'} = W_{\nu'\nu}$ which holds for energy conserving processes.

$$\frac{1}{\tau_\nu^s} = \frac{1}{\tau_\nu} + W_\nu \, . \tag{8.71}$$

The diagonal elements of the modified transition probability $\Pi^s_{\nu_0\nu}$, that is, (8.69) with $\tau_\nu \to \tau_\nu^s$ and $W_{\nu_0\nu} \to W^s_{\nu_0\nu}$, are now finite. There is thus a finite probability to find the system at time $t + \tau_\nu^s$ still in the configuration ν_0, in other words, there is a finite probability for selfscattering $\nu \to \nu$.

Allowing for selfscattering provides us with the flexibility we need to speed up the simulation. Imagine τ_ν has a lower bound τ^s. Then, we can always add

$$W_\nu = \frac{1}{\tau^s} - \frac{1}{\tau_\nu} > 0 \tag{8.72}$$

to the transition rate which, when inserted in (8.71), leads to $\tau_\nu^s = \tau^s$. The sampling time step can be therefore chosen configuration independent, before the sampling starts. In addition, from the fact that τ^s is a lower bound to τ_ν follows $1/\tau_\nu \leq 1/\tau^s$. Thus, $1/\tau^s$ can be easily obtained from (8.66) using an approximate integrand which obeys or even enforces this inequality. In particular, using

$$n_{\boldsymbol{k}} n_{\boldsymbol{p}} \leq n_{\boldsymbol{k}} n_{\boldsymbol{p}} [1 - n_{\boldsymbol{k}'}][1 - n_{\boldsymbol{p}'}] \tag{8.73}$$

in (8.66) leads to

$$\frac{1}{\tau^s} = \frac{\gamma}{2} N(N-1) \, , \tag{8.74}$$

where $N = \sum_{\boldsymbol{k}} n_{\boldsymbol{k}}$ is the total number of electrons and $\gamma = \sup_{\boldsymbol{k},\boldsymbol{p}} \gamma_{\boldsymbol{kp}}$ with $\gamma_{\boldsymbol{kp}} = \sum_{\boldsymbol{k}'\boldsymbol{p}'} W_{\boldsymbol{kp};\boldsymbol{k}'\boldsymbol{p}'}$.

We now have to work out the modified transition probability $\Pi^s_{\nu_0\nu} = \tau_\nu^s W^s_{\nu_0\nu} = \tau^s W^s_{\nu_0\nu}$. Following El-Sayed and coworkers [50], we consider a configuration ν_1 which differs from the configuration ν_0 only in the occupancy of the four momentum states $\boldsymbol{k}_1$, $\boldsymbol{p}_1$, $\boldsymbol{k}_1'$, and $\boldsymbol{p}_1'$. Then

$$\Pi^s_{\nu_0\nu_i} = P^{(1)}(\boldsymbol{k}_1, \boldsymbol{p}_1) \cdot P^{(2)}_{\boldsymbol{k}_1,\boldsymbol{p}_1}(\boldsymbol{k}_1', \boldsymbol{p}_1') \cdot P^{(3)}_{\boldsymbol{k}_1\boldsymbol{p}_1;\boldsymbol{k}_1'\boldsymbol{p}_1'}(\nu_i) \tag{8.75}$$

with

$$P^{(1)}(\boldsymbol{k}_1, \boldsymbol{p}_1) = \frac{n_{\boldsymbol{k}_1}}{N} \frac{n_{\boldsymbol{p}_1}}{N-1} \tag{8.76}$$

the probability for the electrons with momentum $\boldsymbol{k}_1$ and $\boldsymbol{p}_1$ to be the scatterer,

$$P^{(2)}_{\boldsymbol{k}_1,\boldsymbol{p}_1}(\boldsymbol{k}_1', \boldsymbol{p}_1') = \frac{W_{\boldsymbol{k}_1\boldsymbol{p}_1;\boldsymbol{k}_1'\boldsymbol{p}_1'}}{\gamma_{\boldsymbol{k}_1\boldsymbol{p}_1}} \tag{8.77}$$

the probability that the two electrons with $\boldsymbol{k}_1$ and $\boldsymbol{p}_1$ are scattered into momentum states $\boldsymbol{k}_1'$ and $\boldsymbol{p}_1'$, respectively, and

$$P^{(3)}_{\boldsymbol{k}_1\boldsymbol{p}_1;\boldsymbol{k}_1'\boldsymbol{p}_1'}(\nu_i) = \begin{cases} \frac{\gamma_{\boldsymbol{k}_1\boldsymbol{p}_1}}{\gamma}(1 - n_{\boldsymbol{k}_1'})(1 - n_{\boldsymbol{p}_1'}) & i = 1 \\ 1 - \frac{\gamma_{\boldsymbol{k}_1\boldsymbol{p}_1}}{\gamma}(1 - n_{\boldsymbol{k}_1'})(1 - n_{\boldsymbol{p}_1'}) & i = 0 \end{cases} \tag{8.78}$$

the probability for the selected momentum states to perform a real ($i = 1$) or a selfscattering ($i = 0$) event, respectively. Note, the factor $(1 - n_{\boldsymbol{k}_1'})(1 - n_{\boldsymbol{p}_1'})$ guarantees that real scattering events occur only when the final momentum states are empty. All three probabilities are normalized to unity when summed over the domain of the independent variables in the brackets.

In order to implement the ensemble-based Monte Carlo simulation, the momentum space is discretized into a large number of cells which can be either occupied or empty (see Fig. 8.7). A configuration is then specified by the occupancies of all cells. The temporal evolution of the configurations proceeds in discrete time steps τ^s and is controlled by the probability $\Pi^s_{\nu'\nu}$. The basic structure of the algorithm is thus as follows: First, the initial distribution $g_{\boldsymbol{k}}(t = 0)$ is sampled to create the initial configuration ν_0, which is then propagated in time in the following manner:

(i) Increment the time by τ^s.
(ii) Choose at random two initial momentum states, $\boldsymbol{k}_1$ and $\boldsymbol{p}_1$, and two final momentum states, $\boldsymbol{k}_1'$ and $\boldsymbol{p}_1'$.
(iii) Perform the selfscattering test consisting of two inquiries:
First, check whether the chosen momentum states are legitimate by asking whether $R_1 > P^{(1)}(\boldsymbol{k}_1, \boldsymbol{p}_1)$ and $R_2 > P^{(2)}_{\boldsymbol{k}_1,\boldsymbol{p}_1}(\boldsymbol{k}_1', \boldsymbol{p}_1')$, with $R_1, R_2 \in [0, 1]$ two uniformly distributed random numbers. Second, determine whether the final states are empty or not. In the former case, a real scattering event takes place provided $R_3 > P^{(3)}_{\boldsymbol{k}_1\boldsymbol{p}_1;\boldsymbol{k}_1'\boldsymbol{p}_1'}(\nu_1)$, with $R_3 \in [0, 1]$ again an uniformly distributed random variable, whereas in the latter selfscattering occurs.
(iv) Generate the new configuration ν_1, which is the old configuration ν_0 with the occupancies $n_{\boldsymbol{k}_1}$, $n_{\boldsymbol{p}_1}$, $n_{\boldsymbol{k}_1'}$, and $n_{\boldsymbol{p}_1'}$ changed in accordance to the outcome of the selfscattering test.

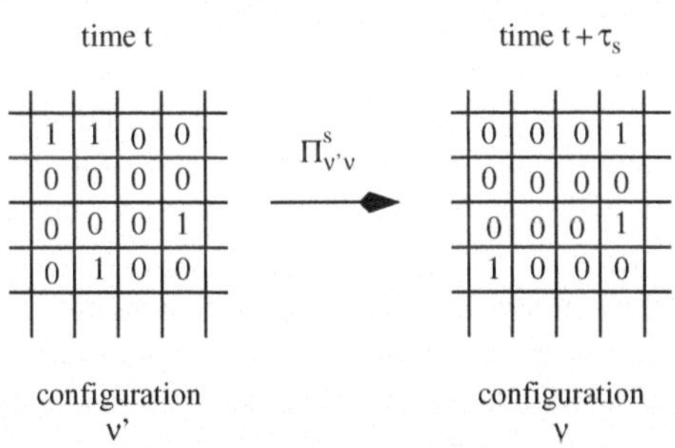

Fig. 8.7. Schematic representation of the ensemble-based Monte Carlo simulation. A sufficiently large part of the two-dimensional momentum space is discretized into small cells. Each cell with size $\Delta k_x \Delta k_y$ is labelled by its central momentum $\boldsymbol{k}_i$, $i = 1, 2, ..., M$ with M the total number of cells. An ensemble of $N < M$ electrons occupies the cells: $n(\boldsymbol{k}_i) = 1$, when an electron is in cell i, and $n(\boldsymbol{k}_i) = 0$ otherwise; $\sum_i n(\boldsymbol{k}_i) = N$. The occupancies of all cells constitute a configuration ν. During the simulation a sequence of configurations is generated stochastically whereby the transition probability from configuration ν' at time t to configuration ν at time $t + \tau_s$ is $\Pi^s_{\nu'\nu}$

The algorithm is terminated after a pre-fixed number of real scattering processes took place. The diagnostics and the calculation of moments can be done on the fly, for instance, before the time is incremented from t to $t + \tau^s$.

Provided the momentum space is appropriately discretized, the algorithm is very fast. A typical run to study, for instance, the relaxation of an initial non-equilibrium Gaussian electron distribution, $g_{\boldsymbol{k}}(t = 0) = C \exp[(E(\boldsymbol{k}) - E_0)/\sigma)^2]$, where C is a constant to fix the electron density and E_0 and σ are the position and width of the Gaussian profile, respectively, took for $N = 5000$ on the Intel 80368 processor available to El-Sayed and coworkers only 10–20 minutes [50]. With the increased computing power available now, this kind of ensemble Monte Carlo simulation is extremely fast. It should be therefore a useful tool for the simulation of dense, degenerate Fermi systems in non-equilibrium, such as, highly optically excited semiconductors, metal clusters in strong laser fields (see Chap. 9), or nuclear matter in heavy ion collisions.

8.3 Conclusions

In this section, we discussed Boltzmann transport in condensed matter, focusing on the conditions, which need to be satisfied for a BE to be applicable to the quasiparticles in a crystal, and on computational tools to solve the quasiparticle BE. Although the quasiparticle BE cannot always be rigorously derived from first principles, it provides in most cases a surprisingly accurate description of transport processes in condensed matter. Most of semiconductor device engineering, for instance, is based on a quasiparticle BE, despite the lack of a satisfying microscopic derivation.

We presented various strategies for the numerical solution of the quasiparticle BE. For the steady-state, spatially uniform, linearized BE, usually employed for the calculation of transport coefficients for metals, we discussed numerical iteration and the expansion of the one-particle distribution function in terms of a symmetry-adapted set of basis functions. In the context of condensed matter, Fermi surface harmonics are here particularly useful because they adequately describe the topology of the Fermi surface, which may be anisotropic, or even consisting of unconnected pieces in momentum space.

As far as the numerical solution of the time-dependent, nonlinear BE is concerned, we discussed iteration and Monte Carlo simulation. Both approaches have been used in the past to calculate hot electron distributions in strongly biased semiconductors. Iteration is here based on the integral representation of the BE. The approach is mathematically very elegant although its potential has not been fully exploited. By far the most popular method for the numerical solution of the BE is the Monte Carlo simulation. It has the virtue of an intuitively obvious approach, requiring a minimum of preparatory mathematical analysis, before the computer generates the solution. In addition, it requires no $\boldsymbol{k}$-summations, which makes the incorporation of realistic band structures particularly easy. We discussed two Monte Carlo algorithms. In the first, particle-based algorithm, a single test-particle is used to build-up the stationary distribution function for electrons in a semiconductor subject to an uniform electric field. This approach, appropriately modified for time-dependent and spatially inhomogeneous settings, has become a design tool for the electronic circuit engineer, indicating its power, flexibility, and practical importance. The second approach propagates an ensemble of N electrons in discrete time steps through a discretized momentum space and is particularly useful for spatially homogeneous, degenerate electron systems with a pronounced Fermi statistics.

There are of course situations where the BE cannot be applied to condensed matter. In particular, transport properties of liquids, amorphous solids, and strongly correlated systems (transition metals, Kondo insulators etc.) cannot be described within the framework of a BE. The mean free path of quasiparticles, if they can be defined, is too short in this type of condensed matter and the separation of the quasiparticles' motion into free flights, with a few randomly occurring scattering events, is impossible. However, provided external fields and temperature gradients are weak, transport processes can be alternatively studied within linear response theory (Kubo formalism [51]). This approach relates linear transport coefficients to thermodynamic correlation functions, which can then be calculated, for instance, with the methods outlined in Chap. 19, Sect. 19.2.2. The Kubo formalism is also applicable when the mean free path is short. It is therefore the method of choice for the calculation of transport coefficients in situations where the BE cannot be used.

References

1. L.W. Boltzmann, Ber. Wien. Akad. **66**, 275 (1872)
2. A. Lenard, Ann. Phys. (New York) **10**, 390 (1960)
3. R. Balescu, Phys. Fluids **3**, 52 (1960)
4. G. Ecker, *Theory of fully ionized plasmas* (Academic Press, New York, 1972)
5. R. Winkler, in *Advances in Atomic, Molecular, and Optical Physics*, Vol. 43, ed. by B. Bederson, H. Walther (Academic Press, New York, 2000), p. 19
6. J.M. Ziman, *Electrons and Phonons* (Oxford University Press, Oxford, 1960)
7. H. Smith, H.H. Jensen, *Transport Phenomena* (Clarendon Press, Oxford, 1989)
8. L.M. Roth, in *Handbook on Semiconductors Completely Revised Edition*, Vol. 1, ed. by P.T. Landsberg (Elsevier Science Publishers, Amsterdam, 1992), p. 489
9. L.P. Kadanoff, G. Baym, *Quantum Statistical Mechanics* (W. A. Benjamin, Inc., New York, 1962)
10. L.V. Keldysh, Sov. Phys. JETP **20**, 1018 (1965)
11. E.M. Lifshitz, L.P. Pitaevskii, *Physical Kinetics* (Pergamon Press, New York, 1981)
12. F. Bloch, Zeitschrift f. Physik **52**, 555 (1928)
13. R. Peierls, Ann. d. Physik **3**, 1055 (1929)
14. L.D. Landau, Sov. Phys. JETP **3**, 920 (1958)
15. R.E. Prange, L.P. Kadanoff, Phys. Rev. **134A**, 566 (1964)
16. G. Eilenberger, Zeitschrift f. Physik **214**, 195 (1968)
17. A.I. Larkin, Y.N. Ovchinnikov, Sov. Phys. JETP **28**, 1200 (1969)
18. G. Eliashberg, Sov. Phys. JETP **34**, 668 (1972)
19. A.I. Larkin, Y.N. Ovchinnikov, Sov. Phys. JETP **41**, 960 (1976)

20. D. Rainer, in *Recent Progress in Many-Body Theories*, Vol. 4, ed. by E. Schachinger, H. Mitter, H. Sormann (Plenum Press, New York and London, 1995), p. 9
21. D. Rainer, Prog. Low Temp. Phys. **10**, 371 (1986)
22. B. Nicklaus, *Quasi-particle picture and ab-initio band structure of electrons in solids: Transport and superconducting properties (in German)* (PhD Thesis, Universität Bayreuth, 1993)
23. V. Špička, P. Lipavský, Phys. Rev. B **52**, 14615 (1995)
24. I.M. Lifshitz, M.I. Kaganov, Sov. Phys. Usp. **2**, 831 (1960)
25. H.H. Jensen, H. Smith, J.W. Wilkins, Phys. Lett. **27B**, 532 (1968)
26. J. Sykes, G.A. Brooker, Ann. Phys. **56**, 1 (1970)
27. K. Takegahara, S. Wang, J. Phys. F: Metal Phys. **7**, L293 (1977)
28. C.R. Leavens, J. Phys. F: Metal Phys. **7**, 163 (1977)
29. H.L. Engquist, Phys. Rev. B **21**, 2067 (1980)
30. H.L. Engquist, G. Grimvall, Phys. Rev. B **21**, 2072 (1980)
31. H. Budd, Phys. Rev. **158**, 798 (1967)
32. H.D. Rees, J. Phys. Chem. Solids **30**, 643 (1969)
33. M.O. Vassel, J. Math. Phys. **11**, 408 (1970)
34. H.D. Rees, J. Phys. C: Solid State Phys. **5**, 641 (1972)
35. C. Hammar, J. Phys. C: Solid State Phys. **6**, 70 (1973)
36. P.B. Allen, Phys. Rev. B **13**, 1416 (1976)
37. P.B. Allen, Phys. Rev. B **17**, 3725 (1978)
38. F.J. Pinski, Phys. Rev. B **21**, 4380 (1980)
39. F.J. Pinski, P.B. Allen, W.H. Butler, Phys. Rev. B **23**, 5080 (1981)
40. T.P. Beaulac, P.B. Allen, F.J. Pinski, Phys. Rev. B **26**, 1549 (1982)
41. I. Mertig, E. Mrosan, J. Phys. F: Metal Phys. **12**, 3031 (1982)
42. T. Vojta, I. Mertig, R. Zeller, Phys. Rev. B **46**, 15761 (1992)
43. W.W. Schulz, P.B. Allen, Phys. Rev. B **52**, 7994 (1995)
44. W. Fawcett, A.D. Boardman, S. Swain, J. Phys. Chem. Solids **31**, 1963 (1970)
45. C. Jacoboni, L. Reggiani, Rev. Mod. Phys. **55**, 645 (1983)
46. M.V. Fischetti, S.E. Laux, Phys. Rev. B **38**, 9721 (1988)
47. C. Jacoboni, P. Lugli, *The Monte Carlo Method for Semiconductor Device Simulation* (Springer-Verlag, Wien, 1989)
48. C. Moglestue, Rep. Prog. Phys. **53**, 1333 (1990)
49. M.V. Fischetti, S.E. Laux, P.M. Solomon, A. Kumar, J. Comp. Electr. **3**, 287 (2004)
50. K. El-Sayed, T. Wicht, H. Haug, L. Bányai, Z. Phys. B **86**, 345 (1992)
51. R. Kubo, J. Phys. Soc. Japan **12**, 570 (1957)

9 Semiclassical Description of Quantum Many-Particle Dynamics in Strong Laser Fields

Thomas Fennel and Jörg Köhn

Institut für Physik, Universität Rostock, 18051 Rostock, Germany

Semiclassical kinetic methods provide an approximate description of the dynamics in quantum many-particle systems without directly refering to their wavefunction. This is desired for problems that resist an explicit quantum mechanical treatment, such as the highly nonlinear laser excitation of finite fermionic systems. The central idea behind the semiclassical approach is the approximation of quantum dynamics with effective transport equations of classical structure. Methods from classical many-particle theory then allow for an efficient solution. This strategy does not imply that all quantum effects are neglected. In fact, important features, such as the Pauli principle and exchange-correlation effects, can reasonably be taken into account using suitable initial conditions and effective potentials. Therefore, semiclassical models can be very convenient tools to study nonlinear processes in three-dimensional quantum systems without symmetry restrictions.

This lecture provides a guided tour through the basics of static and time-dependent semiclassical modelling of fermionic systems. A quick derivation of the semiclassical equations of motion is presented in terms of the density matrix formalism, which leads to an effective Vlasov equation. The test particle approach and the particle-mesh technique allow for an efficient numerical solution of the semiclassical problem. A consistent ground-state theory is provided by an extended Thomas-Fermi model. To give a practical example, we apply the described methods to simple-metal clusters. We discuss the semiclassical cluster ground state and show how optical properties in linear response can be calculated efficiently by the real-time method. Finally, typical aspects of the ionization dynamics of simple-metal clusters are adressed for highly nonlinear femtosecond laser excitation.

9.1 Semiclassical Many-Particle Dynamics in Mean-Field Approximation

As a starting point we describe a formal way to derive the semiclassical approximation to quantum many-particle dynamics. We consider a system of N interacting electrons in a time-dependent external potential $V_{\text{ext}}(\boldsymbol{r}, t)$ and in absence of a magnetic field. The exact evolution of the corresponding antisymmetric N-particle wavefunction $\Psi(\boldsymbol{r}_1 \ldots \boldsymbol{r}_N, t)$ is given by the time-dependent Schrödinger equation

T. Fennel and J. Köhn: *Semiclassical Description of Quantum Many-Particle Dynamics in Strong Laser Fields*, Lect. Notes Phys. **739**, 255–273 (2008)
DOI 10.1007/978-3-540-74686-7_9

$$\mathrm{i}\hbar\frac{\partial}{\partial t}\Psi = \Big[\sum_{i=1}^{N}\left(\frac{-\hbar^2}{2m}\nabla^2_{\boldsymbol{r}_i} + V_{\mathrm{ext}}(\boldsymbol{r}_i)\right) + \sum_{i<j} V_{\mathrm{ee}}(\underbrace{|\boldsymbol{r}_i - \boldsymbol{r}_j|}_{r_{ij}})\Big]\Psi\,, \tag{9.1}$$

where $V_{\mathrm{ee}} = e^2/(4\pi\epsilon_0 r_{ij})$ is the Coulomb potential and the full expression in square brackets is the Hamilton operator.

9.1.1 Density Matrix

For useful approximations of (9.1) it is convenient to reformulate the problem in terms of the density matrix [1, 2], which is defined as

$$\widetilde{\rho}(\boldsymbol{r}_1 \ldots \boldsymbol{r}_N, \boldsymbol{r}'_1 \ldots \boldsymbol{r}'_N, t) = \Psi^*(\boldsymbol{r}_1 \ldots \boldsymbol{r}_N, t)\Psi(\boldsymbol{r}'_1 \ldots \boldsymbol{r}'_N, t)\,. \tag{9.2}$$

The density matrix has a similar interpretation as the density operator in quantum statistics. For example, the probability of finding the system in a state with one electron at each $\boldsymbol{r}_i$ is given by the diagonal elements (primed coordinates set equal to the corresponding unprimed ones). The evolution of $\widetilde{\rho}$ follows from (9.1) and reads

$$\begin{aligned} -\mathrm{i}\hbar\frac{\partial}{\partial t}\widetilde{\rho} = &\Bigg[\sum_{i=1}^{N}\left(\frac{-\hbar^2}{2m}(\nabla^2_{\boldsymbol{r}_i} - \nabla^2_{\boldsymbol{r}'_i}) + V_{\mathrm{ext}}(\boldsymbol{r}_i) - V_{\mathrm{ext}}(\boldsymbol{r}'_i)\right) \\ &+ \sum_{i<j}^{N}(V_{\mathrm{ee}}(r_{ij}) - V_{\mathrm{ee}}(r'_{ij}))\Bigg]\widetilde{\rho}(\boldsymbol{r}_1 \ldots \boldsymbol{r}_N, \boldsymbol{r}'_1 \ldots \boldsymbol{r}'_N, t)\,. \end{aligned} \tag{9.3}$$

So far this does not seem to simplify the problem, since the number of variables has doubled. The strategy becomes more transparent after introducing the reduced k-particle density matrices

$$\begin{aligned} &\rho^{(k)}(\boldsymbol{r}_1 \ldots \boldsymbol{r}_k, \boldsymbol{r}'_1 \ldots \boldsymbol{r}'_k, t) \\ &\quad = \frac{N!}{(N-k)!}\int \widetilde{\rho}(\boldsymbol{r}_1 \ldots \boldsymbol{r}_N, \boldsymbol{r}'_1 \ldots \boldsymbol{r}'_k, \boldsymbol{r}_{k+1} \ldots \boldsymbol{r}_N, t)\mathrm{d}^3\boldsymbol{r}_{k+1} \ldots \mathrm{d}^3\boldsymbol{r}_N \end{aligned} \tag{9.4}$$

by writing $\boldsymbol{r}'_i = \boldsymbol{r}_i$ for all but k spacial coordinates, and integrating over these $N-k$ variables. To derive the equation of motion for the reduced k-particle density matrix, insert (9.4) into (9.3) and integrate in the same way over all but k coordinates. The terms $(\nabla^2_{\boldsymbol{r}_i} - \nabla^2_{\boldsymbol{r}'_i})$ and $(V_{\mathrm{ext}}(\boldsymbol{r}_i) - V_{\mathrm{ext}}(\boldsymbol{r}'_i))$ vanish if the ith coordinate is integrated out. Also, interaction terms $(V_{\mathrm{ee}}(r_{ij}) - V_{\mathrm{ee}}(r'_{ij}))$ cancel, when both primed coordinates are equal to the unprimed ones. Then for the one-body density matrix follows

$$\begin{aligned} &-\mathrm{i}\hbar\frac{\partial}{\partial t}\rho^{(1)}(\boldsymbol{r}, \boldsymbol{r}') \\ &\quad = \left(\frac{-\hbar^2}{2m}(\nabla^2_{\boldsymbol{r}} - \nabla^2_{\boldsymbol{r}'}) + V_{\mathrm{ext}}(\boldsymbol{r}) - V_{\mathrm{ext}}(\boldsymbol{r}')\right)\rho^{(1)}(\boldsymbol{r}, \boldsymbol{r}') \\ &\qquad + \int (V_{\mathrm{ee}}(|\boldsymbol{r} - \boldsymbol{r}_2|) - V_{\mathrm{ee}}(|\boldsymbol{r}' - \boldsymbol{r}_2|)\rho^{(2)}(\boldsymbol{r}, \boldsymbol{r}_2, \boldsymbol{r}', \boldsymbol{r}_2)\mathrm{d}^3\boldsymbol{r}_2\,. \end{aligned} \tag{9.5}$$

The first term on the right hand side contains all single particle contributions, while the second interaction term describes two-body effects and depends on the next higher matrix $\rho^{(2)}$. Similarly, the evolution of $\rho^{(2)}$ requires prior knowledge of the three-body density matrix $\rho^{(3)}$. Thus, the exact reformulation results in a series of coupled equations of motion for the reduced density matrices $\rho^{(k)}$, representing the quantum counterpart to the famous BBGKY[1] hierarchy, known from classical statistical mechanics. For a useful approximation this series must be truncated at some level. Let us keep only (9.5) and close this equation by an approximation for $\rho^{(2)}$. A simple approach is a product of one-body density matrices (Hartree approximation)

$$\rho^{(2)}(\boldsymbol{r}_1, \boldsymbol{r}_2, \boldsymbol{r}_1', \boldsymbol{r}_2') = \rho^{(1)}(\boldsymbol{r}_1, \boldsymbol{r}_1')\rho^{(1)}(\boldsymbol{r}_2, \boldsymbol{r}_2') \,. \tag{9.6}$$

Now the integral in (9.5) can be carried out and allows to include the interaction terms in an effective field according to

$$-\mathrm{i}\hbar\frac{\partial}{\partial t}\rho^{(1)}(\boldsymbol{r}, \boldsymbol{r}') = \left(\frac{-\hbar^2}{2m}(\nabla_{\boldsymbol{r}}^2 - \nabla_{\boldsymbol{r}'}^2) + V_{\text{eff}}(\boldsymbol{r}) - V_{\text{eff}}(\boldsymbol{r}')\right)\rho^{(1)}(\boldsymbol{r}, \boldsymbol{r}') \,, \tag{9.7}$$

with the effective potential

$$V_{\text{eff}}(\boldsymbol{r}) = V_{\text{ext}}(\boldsymbol{r}) + \frac{e^2}{4\pi\epsilon_0}\int \frac{1}{|\boldsymbol{r} - \boldsymbol{r}''|}\underbrace{\rho^{(1)}(\boldsymbol{r}'', \boldsymbol{r}'')}_{=n(\boldsymbol{r}'')}\,\mathrm{d}^3\boldsymbol{r}'' \,. \tag{9.8}$$

The second term in (9.8) is just the classical Hartree potential resulting from the total electron density of the system $n(\boldsymbol{r}'')$. Thus we have found a closed mean-field approximation to the dynamics of the one-body density matrix.

9.1.2 Wigner Function and Vlasov Equation

To obtain the equation of motion for the one-body density matrix $\rho^{(1)}$ in the form of a classical kinetic equation for a phase-space distribution function, we switch to the Wigner representation [3], which reads

$$f_{\text{W}}(\boldsymbol{r}, \boldsymbol{p}, t) = \frac{1}{(2\pi\hbar)^3}\int e^{\mathrm{i}\boldsymbol{p}\cdot\boldsymbol{q}/\hbar}\,\rho^{(1)}\left(\boldsymbol{r} + \frac{\boldsymbol{q}}{2}, \boldsymbol{r} - \frac{\boldsymbol{q}}{2}\right)\mathrm{d}^3\boldsymbol{q} \,. \tag{9.9}$$

The Wigner function f_{W} contains the same information as the one-body density matrix (it is just a Fourier transform). Like an ordinary phase-space distribution, f_{W} is a function of momentum and space, but it can be negative in some regions. However, single particle observables can be calculated from f_{W} in the same way as from a classical distribution function. For example, particle or current densities are given by

$$\begin{aligned} n(\boldsymbol{r}) &= \int f_{\text{W}}(\boldsymbol{r}, \boldsymbol{p})\mathrm{d}^3\boldsymbol{p} \,, \\ \boldsymbol{j}(\boldsymbol{r}) &= -e\int \frac{\boldsymbol{p}}{m} f_{\text{W}}(\boldsymbol{r}, \boldsymbol{p})\mathrm{d}^3\boldsymbol{p} \,. \end{aligned} \tag{9.10}$$

[1] Born, Bogoliubov, Green, Kirkwood and Yvon.

The dynamics of f_{W} follows from (9.7) as

$$-\mathrm{i}\hbar\,\tfrac{\partial}{\partial t}f_{\mathrm{W}}(\boldsymbol{r},\boldsymbol{p},t) = \tfrac{1}{(2\pi\hbar)^3}\int \mathrm{e}^{\mathrm{i}\boldsymbol{p}\cdot\boldsymbol{q}/\hbar}\Bigg[\tfrac{-\hbar^2}{2m}\big(\nabla^2_{\boldsymbol{r}+\frac{\boldsymbol{q}}{2}} - \nabla^2_{\boldsymbol{r}-\frac{\boldsymbol{q}}{2}}\big)$$

$$+V_{\mathrm{eff}}(\boldsymbol{r}+\tfrac{\boldsymbol{q}}{2}) - V_{\mathrm{eff}}(\boldsymbol{r}-\tfrac{\boldsymbol{q}}{2})\Bigg]\rho^{(1)}(\boldsymbol{r}+\tfrac{\boldsymbol{q}}{2},\boldsymbol{r}-\tfrac{\boldsymbol{q}}{2})\mathrm{d}^3\boldsymbol{q}\,. \qquad (9.11)$$

Using the identity $\nabla^2_{\boldsymbol{r}+\boldsymbol{q}/2} - \nabla^2_{\boldsymbol{r}-\boldsymbol{q}/2} = 2\nabla_{\boldsymbol{r}}\nabla_{\boldsymbol{q}}$ and the Taylor expansion of the potential,

$$V_{\mathrm{eff}}(\boldsymbol{r}+\boldsymbol{s}) = \mathrm{e}^{\boldsymbol{s}\cdot\nabla_{\boldsymbol{r}}}V_{\mathrm{eff}}(\boldsymbol{r})\,, \qquad (9.12)$$

Equation (9.11) can be rewritten as

$$\frac{\partial}{\partial t}f_{\mathrm{W}} + \frac{\boldsymbol{p}}{m}\nabla_{\boldsymbol{r}}f_{\mathrm{W}} - \frac{2}{\hbar}f_{\mathrm{W}}\sin\left(\frac{\hbar}{2}\overleftarrow{\nabla}_{\boldsymbol{p}}\cdot\overrightarrow{\nabla}_{\boldsymbol{r}}\right)V_{\mathrm{eff}}(\boldsymbol{r}) = 0\,. \qquad (9.13)$$

In the third term, the operator $\nabla_{\boldsymbol{p}}$ acts on the distribution, whereas $\nabla_{\boldsymbol{r}}$ acts on the potential, as indicated by the arrows on top. Now, a Taylor series of the sine can be seen as a formal expansion in orders of $\hbar$, which is the so-called Wigner expansion. The semiclassical approximation is obtained in the limit $\hbar \to 0$, where only the first term of the expansion contributes ($\sin(x) \approx x$), provided the potential as well as the distribution are sufficiently smooth [4]. In this case we obtain the desired classical appearance of the equation of motion

$$\frac{\partial}{\partial t}f(\boldsymbol{r},\boldsymbol{p},t) + \frac{\boldsymbol{p}}{m}\nabla_{\boldsymbol{r}}f(\boldsymbol{r},\boldsymbol{p},t) - \nabla_{\boldsymbol{p}}f(\boldsymbol{r},\boldsymbol{p},t)\nabla_{\boldsymbol{r}}V_{\mathrm{eff}}(\boldsymbol{r},t) = 0\,, \qquad (9.14)$$

where the change of the distribution function in time is expressed by a drift contribution and a field term. Equation (9.14) is a second order approximation to the exact evolution of the Wigner function. The next higher order correction is of third order, and proportional to $\hbar^2$ [5]. We have dropped the index of the distribution function and, from now on, denote (9.14) only as Vlasov equation, assuming its validity for the classical as well as for the approximate semiclassical case. To calculate the time dependence of $f(\boldsymbol{r},\boldsymbol{p},t)$, (9.14) has to be solved self-consistently with the effective potential

$$V_{\mathrm{eff}}(\boldsymbol{r},t) = V_{\mathrm{ext}}(\boldsymbol{r}) + \frac{e^2}{4\pi\epsilon_0}\int\frac{n(\boldsymbol{r}',t)}{|\boldsymbol{r}-\boldsymbol{r}'|}\mathrm{d}^3\boldsymbol{r}'\,. \qquad (9.15)$$

The described approximations have simplified the equations of motion to the pure classical level and we have lost quantum effects as tunnelling, interference and exchange. However, some quantum corrections to the dynamics can be included without loosing the classical appearance of the problem. The Hartree-Fock approximation instead of (9.6) yields an exchange contribution in the interaction term, which can be treated in local density approximation (LDA). This adds an extra term to V_{eff}, i.e., the exchange potential

$$V_x(\boldsymbol{r}) = -\frac{e^2}{4\pi^2\epsilon_0}(3\pi^2 n(\boldsymbol{r}))^{1/3} , \tag{9.16}$$

which corresponds to the Dirac exchange energy [6]. Similarly, correlation effects can be introduced in terms of a local potential, as from [7, 8]. A semiclassical formulation of the Pauli principle will be discussed later in Sect. 9.2.1. Without symmetry restrictions, the direct solution of the Vlasov equation requires to evolve a six-dimensional function in phase space, which is numerically unfavorable. An efficient practical solution is offered by the test particle method described in the next section, which, however, requires a non-negative distribution function. This can be achieved either by smoothing out the rapid oscillations of the Wigner function to remove their negative values, or by using a suitable approximation to the initial state of the distribution function. Once the distribution function is continuously differentiable and non-negative, it remains non-negative upon propagation according to the Vlasov equation.

9.1.3 Test Particle Method

Many techniques behind semiclassical kinetic methods are inherited form nuclear physics, such is the test particle method [4]. The idea of the test particle method is to sample the continuous distribution function with a swarm of fractional particles and to map the dynamics into classical equations of motion for the discrete samples. A straightforward way of representation is

$$f(\boldsymbol{r},\boldsymbol{p},t) = \frac{1}{N_s}\sum_i^{N_{pp}} g_r(\boldsymbol{r}-\boldsymbol{r}_i(t))g_p(\boldsymbol{p}-\boldsymbol{p}_i(t)) \tag{9.17}$$

with the positions $\boldsymbol{r}_i$ and the momenta $\boldsymbol{p}_i$ of the test particles and the smooth weighting functions g_r and g_p in coordinate and momentum space. The parameter N_s sets the number of test particles per physical particle and defines the total number of test particles $N_{pp} = N\,N_s$. One possible choice for the weighting are normalized Gaussians

$$g(\boldsymbol{x}) = \frac{1}{\pi^{3/2}d^3}e^{-x^2/d^2} , \tag{9.18}$$

where d is a numerical smoothing parameter. For reasons that will become evident in a moment, lets define the single-particle Hamiltonian of a test particle as

$$h_i^{pp}(\boldsymbol{r}_i,\boldsymbol{p}_i) = \left[\frac{p_i^2}{2m} + \int V_{\text{eff}}(\boldsymbol{r})g_r(\boldsymbol{r}-\boldsymbol{r}_i)\mathrm{d}^3\boldsymbol{r}\right] , \tag{9.19}$$

and assume classical motion for the test particles according to

$$\begin{aligned}
\dot{\boldsymbol{r}}_i &= \frac{\partial h_i^{pp}}{\partial \boldsymbol{p}_i} = \frac{\boldsymbol{p}_i}{m} ,\\
\dot{\boldsymbol{p}}_i &= -\frac{\partial h_i^{pp}}{\partial \boldsymbol{r}_i} = \underbrace{-\int V_{\text{eff}}(\boldsymbol{r})\nabla_{\boldsymbol{r}_i} g_r(\boldsymbol{r}-\boldsymbol{r}_i)\mathrm{d}^3\boldsymbol{r}}_{\boldsymbol{f}_i} .
\end{aligned} \tag{9.20}$$

To see that this is a reasonable approximation to the Vlasov propagation, insert (9.17) in (9.14). In the limit $d \to 0$ (i.e. a weighting with δ-functions) we recover (9.20) exactly. For the semiclassical treatment a smooth phase-space distribution is essential to suppress the tendency of classical thermalization, which requires a finite width of the test particles in practice [9]. This, however, is not necessarily a shortcoming, since the width parameter can be used to define a semiclassical version of uncertainty. [2]

The test particle method has also a statistical meaning, if the distribution function is interpreted as an statistical ensemble containing N_s possible realizations of the systems. During propagation all observables, including the effective potential, are taken as instantaneous averages over all realizations and therefore describe a statistical mean value.

With the test particle representation the semiclassical dynamics is mapped onto the classical propagation of test particles in a self-consistent potential. This is equivalent to standard PIC simulations, except that in our case many fractional test particles (typically 10^2–10^5) represent one physical particle, whereas in PIC one test particle represents many real particles. However, in practice the numerical simulation of a large number of test particles requires the same efficient methods, such as the particle-mesh technique.

9.1.4 Particle-Mesh Technique

In general, the calculation of the forces is the most expensive part in simulations of the dynamics of interacting particles, since forces depend on all pairs of particles. This leads to the known N^2-scaling in direct particle-particle simulations. The strategy behind the particle-mesh technique is to use a gridded potential in coordinate space and to approximate the forces by finite differences [11]. In our case, even the numerical approximation of derivatives drops out, since we can express the forces directly as a convolution of the potential and the analytically know gradient of the weighting function, see (9.20). Now, for high particle numbers the particle-mesh treatment is obviously advantageous to the direct force calculations, if the numerical method to calculate the potential scales better than N^2. For our problem this is possible, as we discuss in a moment. For a Coulomb-coupled system the force calculation using the particle-mesh technique consists of three steps:

(i) Inject all particles to a grid for the charge density.
(ii) Find the potential by solving Poisson's equation on the grid.
(iii) Compute forces for all particles from the potential.

For our semiclassical problem we just add local potentials resulting from ions, external laser fields and the approximated exchange-correlation effects to the potential obtained from step (ii). The only demanding task is the solution of the Poisson equation on a grid. A common way is the solution in frequency space, which results in $N \log(N)$ scaling due to discrete Fourier transforms. The discrete Fourier

[2] This is related to the Husimi picture, see [10].

representation introduces periodic boundary conditions, which makes it appealing for periodic systems. If this is undesirable, additional effort is needed to subtract supercell contributions (see [12]). For non-periodic systems, such as nanoparticles, it is even possible to realize order N scaling in the potential calculation by using iterative multigrid methods [13, 14]. This, however, requires an initialization of the potential at the boundary (i.e. the surface of the simulation box), which can be done approximately, e.g., by a multipole expansion. We will use this multigrid method for the application to metal clusters given in Sect. 9.3.

9.2 Semiclassical Ground State

So far we have discussed only the semiclassical approximation to the dynamics and have indicated strategies for its efficient numerical solution. What is still needed is an appropriate ground state theory that provides a consistent initial condition to the propagation, i.e., an appropriate initial test particle distribution. A useful semiclassical approximation to the ground state can be derived from the theory of Fermi gases and the Thomas-Fermi model.

9.2.1 Homogenous Fermi Gas

As the most simple model of a fermionic many-particle system, the infinite Fermi gas assumes noninteracting particles. The corresponding solutions of the stationary Schrödinger equation are eigenfunctions of the kinetic energy operator, i.e., plane waves. Restriction to a fixed volume L^3 with periodic boundary conditions yields the density of states in k-space for the Fermi gas with paired spins as

$$g(\boldsymbol{k}) = \frac{2L^3}{(2\pi)^3} . \tag{9.21}$$

The occupation number of each state in k-space is given from the Fermi-Dirac distribution

$$f_{\mathrm{FD}}(\epsilon(\boldsymbol{k}) - \mu) = \frac{1}{1 + \mathrm{e}^{(\epsilon(\boldsymbol{k})-\mu)/(k_{\mathrm{B}}T)}} , \tag{9.22}$$

with the single-particle energy $\epsilon(\boldsymbol{k}) = \hbar^2 k^2/(2m)$ and the chemical potential μ. For a given chemical potential μ the number of particles we find in the volume L^3 is given by

$$N(\mu) = \int \frac{2L^3}{(2\pi)^3} f_{\mathrm{FD}}(\epsilon(\boldsymbol{k}) - \mu)\mathrm{d}^3\boldsymbol{k} . \tag{9.23}$$

By substituting $\boldsymbol{p} = \hbar\boldsymbol{k}$ and dividing by the volume L^3 we can write the particle density as an integral over momentum space,

$$n(\mu) = \int f(\boldsymbol{p})\mathrm{d}^3\boldsymbol{p} , \tag{9.24}$$

where $f(\boldsymbol{p}) = 2f_{\mathrm{FD}}(\epsilon(\boldsymbol{p}) - \mu)/(2\pi\hbar)^3$ and $f(\boldsymbol{p})$ is the momentum distribution of the Fermi gas. The Pauli principle appears here implicity as an upper limit of the distribution function according to $f(\boldsymbol{p}) \leq 2/(2\pi\hbar)^3$, which we can use for semiclassical considerations as it stands. At zero temperature all states are fully occupied up to the chemical potential, i.e., the distribution becomes a step function

$$f^{T=0}(\boldsymbol{p}) = \frac{2}{(2\pi\hbar)^3}\Theta(\mu - \epsilon(\boldsymbol{p})) \,. \tag{9.25}$$

It is now straightforward to find the zero point kinetic energy density as a function of the particle density

$$u_{\mathrm{kin}}(n) = \frac{3}{10}\frac{\hbar^2(3\pi^2)^{2/3}}{m}n^{5/3}, \tag{9.26}$$

where m is the mass of the fermions. Now (9.26) can be used to find approximate solutions to realistic problems, such as the electron distribution in an atom. This leads to the Thomas-Fermi approximation.

9.2.2 Thomas-Fermi Approximation

The original Thomas-Fermi theory was developed to describe the electronic structure of heavy atoms at zero temperature and leads to a problem with spherical symmetry [15, 16]. Here we consider a more general form that can be derived from a variational principle and contains the original form as a special case. The central idea is to describe electrons in an external potential as a Fermi gas at zero temperature by using LDA. Then, the total energy can be written in terms of the total electron density $n(\boldsymbol{r})$ as

$$E_{\mathrm{tot}}[n(\boldsymbol{r})] = \int \Bigg[u_{\mathrm{kin}}(n(\boldsymbol{r})) + V_{\mathrm{ext}}(\boldsymbol{r})n(\boldsymbol{r}) + \frac{1}{2}\int \frac{e^2}{4\pi\epsilon_0}\frac{n(\boldsymbol{r})n(\boldsymbol{r}')}{|\boldsymbol{r}-\boldsymbol{r}'|}\mathrm{d}^3\boldsymbol{r}' \Bigg] \mathrm{d}^3\boldsymbol{r} \,, \tag{9.27}$$

where the terms in square brackets describe the approximate kinetic energy density taken from Fermi gas, the interaction with an external potential and the electron-electron interaction. Obviously, there is a spurious self-interaction, since an electron interacts with its own contribution to the total electron density $n(\boldsymbol{r})$, but we assume this error to be small for systems with many electrons. To find the density with minimal energy, i.e., the ground state, we solve the variational problem

$$\frac{\delta}{\delta n}\left(E_{\mathrm{tot}}[n] - \mu \int n(\boldsymbol{r})\mathrm{d}^3\boldsymbol{r} \right) = 0 \,, \tag{9.28}$$

where μ has the meaning of a Lagrange multiplier to fix the total number of electrons. The interpretation of (9.28) is the following: The (extremal) energy must remain unchanged for any infinitesimal change of the density by $\delta n(\boldsymbol{r})$. This leads to the condition

$$n(\boldsymbol{r}) = \frac{(2m)^{3/2}}{3\pi^2\hbar^2}\left[\mu - V_{\text{eff}}(\boldsymbol{r})\right]^{3/2} , \tag{9.29}$$

with the effective potential from (9.15). Thus, in our notation, the Thomas-Fermi ground state is defined by a pair of self-consistent equations. For systems with spherical symmetry the problem can be reduced to a single one-dimensional nonlinear differential equation using Poisson's equation. If the external potential V_{ext} is a Coulomb potential, as for a nucleus, this yields the famous Thomas-Fermi equation [15, 16]. However, we consider the unrestricted case.

As the most simple version of density functional theory (DFT), the Thomas-Fermi approximation provides a reasonable parameter-free description of heavy atoms, but has serious shortcomings. For example, molecules are predicted to be completely unstable within Thomas-Fermi theory [17], since exchange effects are neglected [18]. In addition, the predicted values of the first atomic ionization potentials are far too small. To cure these problems it was suggested by Dirac to treat exchange effects in the same way as the kinetic energy [6], i.e., by approximating the exchange energy locally with the Hartree-Fock result from the Fermi gas. This adds the exchange energy density in LDA,

$$u_x(n(\boldsymbol{r})) = -\frac{3e^2}{16\pi^2\epsilon_0}(3\pi^2)^{1/3}n^{4/3}(\boldsymbol{r}) , \tag{9.30}$$

to the integrand of (9.27). The solution of the variational problem is similar, but yields an additional term in the effective potential. This is the LDA exchange potential we have already seen in (9.16).

However, the solution of this extended Thomas-Fermi-Dirac model can lead to unphysical jumps in the electron density in some cases. Quantum mechanics avoids sharp density jumps, since the large gradient of the corresponding wavefunction would result in a very high kinetic energy. Fortunately, we can take advantage of the test particle representation here, since the weighting functions introduce an artificial smoothing of the density.

9.2.3 Generalized Thomas-Fermi Approximation for Test Particles

Now we describe necessary modifications of the Thomas-Fermi model to obtain a ground state theory that is consistent with the constraints of the test particle representation of (9.17). Obviously, we need to know the distribution of the discrete test particles, i.e., the position and momenta of their centers. Therefore we define a density $n_\delta(\boldsymbol{r})$ that gives the number density of test particles centers in space. Furthermore, we assume the local momentum distribution of the test particles to be the same as in a Fermi gas at zero temperature, which means, that all states are fully occupied up to the local Fermi momentum. This allows to use the Fermi gas result from (9.26) to approximate the kinetic energy density, but now as a function of the test particle density according to

$$u_{\text{kin}}(\boldsymbol{r}) = \frac{3}{10}\frac{\hbar^2}{m}(3\pi^2)n_\delta^{5/3}(\boldsymbol{r}) \,. \tag{9.31}$$

From the test particle density we find the effective real-space electron density $n_{\text{eff}}(\boldsymbol{r})$ after convolution with the corresponding weighting function as

$$n_{\text{eff}}(\boldsymbol{r}) = \int n_\delta(\boldsymbol{r}')g_r(\boldsymbol{r}-\boldsymbol{r}')\mathrm{d}^3\boldsymbol{r}' \,. \tag{9.32}$$

The effective density can then be used to describe all contributions to the potential energy density due to external fields, Coulomb interactions between electrons, and exchange. For simplicity, we restrict the derivation to an external potential and the classical Coulomb energy, leading to

$$u_{\text{pot}}[n_{\text{eff}}](\boldsymbol{r}) = V_{\text{ext}}(\boldsymbol{r})n_{\text{eff}}(\boldsymbol{r}) + \frac{1}{2}\int \frac{e^2}{4\pi\epsilon_0}\frac{n_{\text{eff}}(\boldsymbol{r})n_{\text{eff}}(\boldsymbol{r}'')}{|\boldsymbol{r}-\boldsymbol{r}''|}\mathrm{d}^3\boldsymbol{r}'' \,. \tag{9.33}$$

The dependence on the test particle density is implicit, because it was used to define the effective density. After integrating the kinetic and potential energy densities and introducing a Lagrange multiplier we find the variational problem for the minimal total energy,

$$\frac{\delta}{n_\delta}\left(\int \left[u_{\text{kin}}(n_\delta(\boldsymbol{r})) + u_{\text{pot}}[n_{\text{eff}}](\boldsymbol{r}) - \mu\, n_\delta(\boldsymbol{r})\right]\mathrm{d}^3\boldsymbol{r}\right) = 0 \,. \tag{9.34}$$

Since the varied quantity is the test particle density n_δ, the variation of the first and last term under the integral is straightforward and analogous to the previous section. The treatment of the potential energy term is more difficult, since it is a functional of the effective density. Application of the chain rule for the functional derivative yields

$$\begin{aligned}&\int \left[\frac{1}{2}\frac{\hbar^2}{m}(3\pi^2)n_\delta^{2/3}(\boldsymbol{r}) - \mu\right]\delta n_\delta(\boldsymbol{r})\mathrm{d}^3\boldsymbol{r}\\ &+\int\!\!\int \underbrace{\int \frac{\delta u_{\text{pot}}[n_{\text{eff}}](\boldsymbol{r})}{\delta n_{\text{eff}}(\boldsymbol{r}')}\mathrm{d}^3\boldsymbol{r}}_{V_{\text{eff}}(\boldsymbol{r}')}\underbrace{\frac{\delta n_{\text{eff}}(\boldsymbol{r}')}{\delta n_\delta(\boldsymbol{r}'')}}_{g_r(\boldsymbol{r}'-\boldsymbol{r}'')}\mathrm{d}^3\boldsymbol{r}'\delta n_\delta(\boldsymbol{r}'')\mathrm{d}^3\boldsymbol{r}'' = 0 \,.\end{aligned} \tag{9.35}$$

In the second term on the right hand side, the integration over $\mathrm{d}^3\boldsymbol{r}$ yields the effective potential to our potential energy density from (9.33). It has the simple and familiar form

$$V_{\text{eff}}(\boldsymbol{r}) = V_{\text{ext}}(\boldsymbol{r}) + \frac{e^2}{4\pi\epsilon_0}\int \frac{n_{\text{eff}}(\boldsymbol{r}')}{|\boldsymbol{r}-\boldsymbol{r}'|}\mathrm{d}^3\boldsymbol{r}' \,. \tag{9.36}$$

With this definition we perform the $\mathrm{d}^3\boldsymbol{r}'$-integration in (9.35) and introduce the smoothed test particle potential

$$V_\delta(\boldsymbol{r}) = \int V_{\mathrm{eff}}(\boldsymbol{r}')g(\boldsymbol{r}' - \boldsymbol{r})\mathrm{d}^3\boldsymbol{r}' \,. \tag{9.37}$$

Inserting this into (9.35) and renaming of $\boldsymbol{r}''$ to $\boldsymbol{r}$ results

$$\int \underbrace{\left[\frac{1}{2}\frac{\hbar^2}{m}(3\pi^2)n_\delta^{2/3}(\boldsymbol{r}) + V_\delta(\boldsymbol{r}) - \mu\right]}_{\equiv 0} \delta n_\delta(\boldsymbol{r})\mathrm{d}^3\boldsymbol{r} = 0 \,. \tag{9.38}$$

Since the integrand must vanish to fulfill this equation for arbitrary δn_δ we find the condition for extremal energy after solving for n_δ, which reads

$$n_\delta(\boldsymbol{r}) = \frac{(2m)^{3/2}}{3\pi^2\hbar^2}\left[\mu - V_\delta(\boldsymbol{r})\right]^{3/2} \,. \tag{9.39}$$

It is not surprising that the structure of this equation is analog to (9.29), but here we describe the test particle density n_δ as a function of the test particle potential V_δ. For density weighting with delta functions, where $n_{\mathrm{eff}} = n_\delta$ and therefore $V_\delta = V_{\mathrm{eff}}$, we recover (9.29) as a limiting case.

For a given external potential the determination of the test particle ground state density n_δ requires to solve (9.32), (9.36), (9.37) and (9.39) self-consistently. Further quantum corrections due to exchange and correlation effects can be easily incorporated, if they are treated in LDA. Therefore, the corresponding potentials, such as that of (9.16) for the LDA exchange, are just added to the effective potential in (9.36) as a fuction of the effective density. Once the test particle density is known, the positions of numerical test particles can be generated by simple Monte-Carlo sampling of $n_\delta(\boldsymbol{r})$. The local momenta are sampled according to the assumed homogeneous occupation of the local Fermi sphere up to the local Fermi momentum

$$p_\delta^{\max}(\boldsymbol{r}) = \left(3\pi^2\hbar^3 n_\delta(\boldsymbol{r})\right)^{1/3} \,. \tag{9.40}$$

For sufficiently fine sampling ($N_s \gg 1$) and a finite width of the weighting functions d_r the semiclassically initialized system is numerically stable upon the propagation described in Sect. 9.1.3. In practice, the parameters N_s and d_r are chosen to provide the required level of long-term stability of the model, i.e., sufficient suppression of spurious classical thermalization (see [9, 14]).

9.3 Application to Simple-Metal Clusters

To give a practical example, this section describes the application of the semiclassical method to the dynamics of simple-metal clusters in intense infrared femtosecond laser fields. In many respects the properties of simple metals, such as their stability and optical response, are governed by delocalized valence electrons that can be reasonably approximated as a Fermi gas. This is the major justification for the applicability of the semiclassical method.

As we want to resolve the structure of the systems, the dynamics and potentials of the ions (nuclei plus strongly bound electrons) have to be taken into account. As the contributions of deeper bound electrons are assumed to be less important, it is convenient to resolve only the valance electrons explicitly, while their interaction with core electrons and nuclei is described by pseudopotentials. This is also a common strategy in time-dependent density functional theory. Here, we consider sodium clusters where each atom contributes one active valence electron to the model explicitly. For all results discussed in this section the exchange-correlation potential from [7] and Gaussian density weighting ($d_r = 1.15$ Å) have been used.

9.3.1 Ground State with Atomic Pseudopotentials

For the alkaline metals, where the singly charged ion has a closed-shell electronic structure, it is sufficient to model the ion as an effective charge distribution with spherical symmetry. A convenient form is a sum of Gaussians according to

$$\rho_{\mathrm{ion}}(r) = \sum_{n=1}^{k} c_n \frac{e}{\pi^{3/2} a_n^3} \, \mathrm{e}^{-r^2/a_n^2} \,, \tag{9.41}$$

where c_n and a_n are the charge and the width of each Gaussian. The corresponding potential of an electron at position $\boldsymbol{r}$ in the field of a pseudo-ion at position $\boldsymbol{R}$ is

$$V_{\mathrm{e}\leftrightarrow\mathrm{ion}}(\boldsymbol{r}, \boldsymbol{R}) = -e \sum_{n=1}^{k} c_n \frac{e}{4\pi\epsilon_0} \frac{\mathrm{erf}\,(|\boldsymbol{r}_e - \boldsymbol{R}|/a_n)}{|\boldsymbol{r}_e - \boldsymbol{R}|} \,, \tag{9.42}$$

where $\mathrm{erf}(x)$ is the error function. The parameters a_n and c_n can be optimized so that the model reproduces central properties of the described element, such as ionization potential and polarizability [14]. Examples for the semiclassical prediction on the basis of optimized pseudopotential with two Gaussians are given in Table 9.1, illustrating the reasonable agreement with experimental values. The sum over the pseudopotential of all ions at positions $\boldsymbol{R}_i$ then provides the external potential for the electronic problem

$$V_{\mathrm{ext}}(\boldsymbol{r}) = \sum_i V_{\mathrm{e}\leftrightarrow\mathrm{ion}}(\boldsymbol{r}, \boldsymbol{R}_i) \,. \tag{9.43}$$

Table 9.1. Semiclassically calculated atomic properties for the sodium atom using a two-Gaussian pseudo-potential compared [14]

		model	reference
ionization potential	[eV]	5.30	5.13
polarizability	[$\text{Å}^3/(4\pi\epsilon_0)$]	21.9	23.6

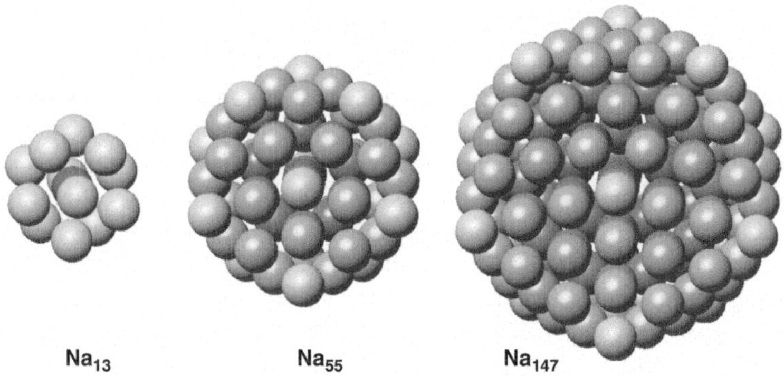

Fig. 9.1. Icosahedral structure of small sodium clusters predicted by the semiclassical ground state theory. Since electronic shells effects are not resolved in the semiclassical theory, the ground state geometries are biased by geometric packing effects [14]

Using the parameters from the optimized atomic problem, the total energy of the full ground state can be minimized with respect to the ionic coordinates to find the cluster geometry, e.g., be simulated annealing. Plots of optimized geometries for three cluster sizes are shown in Fig. 9.1. It should be noted, that the semiclassical theory is biased by geometric packing effects and ignores electronic shell closures. Nevertheless, the results are surprisingly close to DFT calculations [19], except for very small particle numbers.

Having obtained the initial state of the considered system (ionic structure plus test particle distribution), the time-dependent response can be calculated by direct numerical propagation for various external perturbations, e.g., due to a laser field or collisions with charged ions. However, this treatment is inefficient for a systematic characterization of the system, since all possible scenarios would require a separate calculation. In the limit of small excitations, where the response can be assumed to be almost linear and allows mode decomposition, it is possible to extract the full spectrum out of a single numerical calculation, as we discuss here in terms of the optical response.

9.3.2 Optical Response in the Linear Regime: Real-Time Method

In dipole approximation ($d \ll \lambda$), the linear optical response of a finite and isotropic system to an external electric field is fully characterized by its complex dynamic polarizability $\alpha(\omega)$. This quantity relates the spectral amplitudes of the induced dipole moment $\boldsymbol{p}(\omega)$ linearly to those of a driving external field $\boldsymbol{E}(\omega)$ by

$$\boldsymbol{p}(\omega) = \alpha(\omega)\boldsymbol{E}(\omega)\,. \tag{9.44}$$

As the dipole moment must be real in the time domain, it is required that $\alpha(\omega) = \alpha^*(-\omega)$. The knowledge of $\alpha(\omega)$ enables to calculate important optical properties of the system, as, e.g., the light absorption cross section from

$$\sigma(\omega) = \frac{\omega}{c\epsilon_0} \mathrm{Im}[\alpha(\omega)] \,. \tag{9.45}$$

A convenient way to calculate $\alpha(\omega)$ for a finite system (on the basis of a time-based numerical model) is offered by the real-time method [20], as it requires only a single simulation run. The idea behind is to excite all modes of the system at once and to extract their spectral weights from a simple Fourier transform of the response in the time domain. To see this, assume an external field oriented in z-direction, having constant spectral amplitudes for all frequencies $E_z(\omega) = f/(2\pi)$. The corresponding field in the time domain[3] is $E_z(t) = f\delta(t)$ and has the meaning of an impulsive force, instantaneously changing the velocity of all charged particles by $\Delta v_z = qf/m$ at time $t = 0$, where q is the charge and m the particle mass. In practice, only electrons are considered to be kicked, as ions are basically unaffected due to their higher mass. The impulsive perturbation leads to an excitation of all possible optical modes of the system in proportion to their excitation strength. The resulting dipole moment in the time domain, $p_z(t)$, which can be easily recorded from a simulation, is just a weighted superposition of harmonic oscillations. Their amplitudes characterize the corresponding optical activity of the investigated system. Now, the Fourier transform of the time-dependent dipole moment, if we assume it is a continuous function and use (9.44), turns out to be directly proportional to the polarizability according to

$$\alpha(\omega) = \frac{2\pi}{f} p_z(\omega) \,. \tag{9.46}$$

A numerical simulation, of course, requires to sample the evolution of the dipole moment by a finite number of data points $p_z(t_n)$. Assuming an even number of points N and a fixed time step Δt, a discrete Fourier transform provides an array for the polarizability at N discrete values ω_k from

$$\alpha(\omega_k) = \frac{\Delta t}{f_z} \sum_{n=0}^{N-1} p_z(t_n) \mathrm{e}^{-2\pi \mathrm{i} nk/N} \tag{9.47}$$

with $\omega_k = k\Delta\omega$, $\Delta\omega = 2\pi/(N\Delta t)$ and $k = -N/2, \ldots, N/2$. This form has the advantage that the spectrum can be calculated with Fast Fourier Transform. Due to the mentioned symmetry properties of $\alpha(\omega)$ it is sufficient to use only the values for positive frequencies. Obviously, the timestep and the number of iterations are directly related to the bandwidth and the resolution of the spectrum and must therefore be chosen adequately for a given problem. Also, the magnitude of the field impulse f is a sensitive parameter, as it must be small enough to remain in the linear response regime. A simple cross-check is to vary the value of f, as the resulting $\alpha(\omega)$ must be independent of the strength of the perturbation.

As an example, in Fig. 9.2 the real-time method is applied to icosahedral Na_{147}. Starting from the ground state, an initial velocity offset is introduced to all electrons and the system is propagated in time. The resulting dipole signal is given in (a).

[3] We use $g(\omega) = \frac{1}{2\pi}\int_{-\infty}^{\infty} g(t)\mathrm{e}^{-\mathrm{i}\omega t}\mathrm{d}t$ and $g(t) = \int_{-\infty}^{\infty} g(\omega)\mathrm{e}^{\mathrm{i}\omega t}\mathrm{d}\omega$.

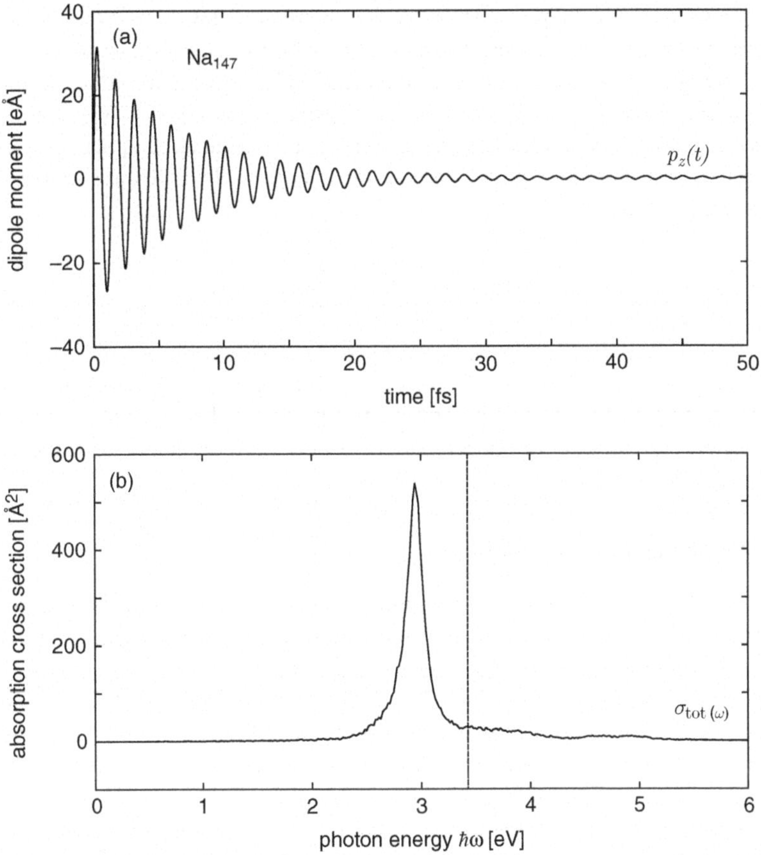

Fig. 9.2. Calculated linear optical response of Na_{147}: (**a**) Evolution of the dipole signal in the time domain, $p_z(t)$, recorded from the semiclassical model, after giving all electrons a constant velocity offset of $\Delta v_z = -1$ Å/fs. (**b**) Corresponding total light absorption cross section $\sigma(\omega)$ by using the polarizability obtained from a Fourier transform of the dipole moment. The dominant peak at $\hbar\omega = 2.95$ eV corresponds to the plasmon resonance of the nanoparticle. A significant red-shift with respect to the classical value of the Mie plasmon (vertical line in (**b**)) is predicted [14]

From its Fourier transform we obtain the polarizability, and the absorption cross section, see (b), follows directly from (9.45). The optical spectrum is dominated by a strong peak, i.e., the plasmon resonance of the metallic nanoparticle. Sharp transitions through single particle-hole excitations are absent, as discrete electronic states are not resolved within the semiclassical treatment. However, the predicted response is reasonable and surprisingly close to results obtained from orbital based quantum mechanical approaches such as the time-dependent density functional theory [19, 20]. This is due to the fact that the response of simple metals is dominated by collective effects, which are well covered in the semiclassical treatment. An interesting feature of the plasmon resonance in small metal particles is the significant red-shift with respect to the classical value, which is a clear quantum effect. The magnitude of the semiclassically predicted shift is in agreement with experimental observations. It can be explained by the non-zero electron density outside the cluster surface, often referred to as *spill-out*. In a classical metallic sphere, where the density makes a sharp step at the surface, the energy of the collective dipole mode (Mie plasmon) reads

$$\omega_{\text{Mie}} = \left(\frac{e^2 n_i}{3\epsilon_0 m_e} \right)^{1/2} , \tag{9.48}$$

where n_i is the number density of ionic charges. As bulk sodium[4] is an almost ideal metal, the prediction of (9.48) of $\hbar\omega_{\text{Mie}} = 3.41\,\text{eV}$ gives a good estimate for the macroscopic limit, see vertical line in Fig. 9.2(b). The red-shift of plasmon in case of a cluster is a function of particle size and decreases gradually with increasing particle size.

9.3.3 Nonlinear Laser Excitations

So far, we have gone a long way without considering truly nonlinear scenarios. Therefore, let us finally discuss an application of the semiclassical treatment to metal clusters in ultrashort intense laser pulses. On the basis of the calculated optical absorption spectrum, cf. Fig. 9.2, a high absorption cross section is expected for laser excitations close to the plasmon resonance. For laser photon energies far away from the resonance, only a weak response is predicted. This is true, but only within the linear regime. In intense laser pulses (say $I \gg 10^{10}\,\text{W/cm}^2$) the system is changing rapidly during the interaction process, due to laser heating or the emission of electrons, which results in transient optical properties. As observed in many experiments, the cluster response is very sensitive to the temporal shape of the laser field, leading to strong variations in the numbers and energies of emitted electrons, ions and photons. The mechanisms behind these phenomena are a fascinating aspect of clusters in intense fields. However, full quantum mechanical treatment is unfeasible and simplified approaches are necessary for a theoretical description. In case of metal clusters, the semiclassical method is a useful compromise, providing valuable insight into the dynamics of nonlinear laser-cluster interactions. This is demonstrated below by analyzing the origin of maximum cluster ionization at optimal delay of dual pulses.

We consider the excitation of Na_{55} by a sequence of two linearly polarized 50 fs laser pulses of moderate intensity ($I_0 = 4 \times 10^{12}\,\text{W/cm}^2$), having a variable delay Δt and a photon energy of $\hbar\omega = 1.54\,\text{eV}$ (Titanium-Sapphire laser at 800 nm). This means, the system is probed well below its collective mode in the ground state, in accordance to the typical situation in experiments on simple-metal clusters. A set of simulations for various pulse delays, say $\Delta t = 0 \ldots 1\,\text{ps}$, will specify a characteristic optimal pulse separation, resulting in maximal total ionization. This behavior

[4] $n_i = 2.53 \times 10^{22}\,\text{cm}^{-3}$.

has also been observed in measurements [21]. To identify the mechanism underlying this effect, Fig. 9.3 shows a set of time-dependent observables from the simulation with the optimal delay. For the given laser parameters this is $\Delta t_{\mathrm{opt}} \approx 250\,\mathrm{fs}$.

Let us first concentrate on the impact of the leading pulse. The almost vanishing phase lag between the laser and the dipole moment (b) is a marker for low energy absorption from the first pulse, as the system is excited far off the resonance. Remember, this is what we know from a driven oscillator. Only a small amplitude of the dipole moment (a) and weak ionization of the cluster (c) is induced by the leading pulse. However, the cluster is excited strong enough to become unstable, as can be seen from the increasing radius (d). There are two important mechanism driving the expansion, i.e., the Coulomb pressure due to total cluster charge and a hydrodynamic contribution, resulting from the heated electron gas. Now, if we

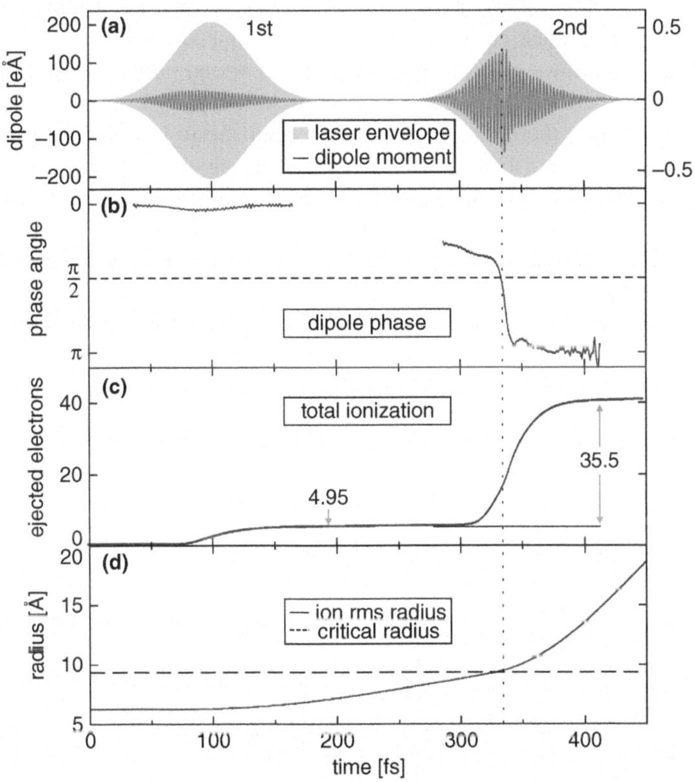

Fig. 9.3. Response of Na_{55} for dual pulse laser excitation with $I = 4 \times 10^{12}\,\mathrm{W/cm^2}$, $\hbar\omega = 1.54\,\mathrm{eV}$ (800 nm), and an optical delay of 250 fs. Shown are the envelope of the laser field (*grey*) and the corresponding electron dipole amplitude (**a**), the phase angle between the laser field and the dipole signal (**b**), the total cluster ionization (**c**), and the root-mean-square radius of the ion distribution (**d**). Note that the dipole phase angle passes $\pi/2$ as the rms-radius is close to the critical value R_c (*dotted line*) [21]

inspect the impact of the second pulse, much larger amplitudes in the dipole moment are found and the ionization is increased by a factor of seven. This is a significant difference to the first excitation step, although the pulses are identical. The enhancement can be explained by a dynamic plasmon resonance. A clear hint to a collective resonance phenomena over quasi-static field effects is a transient phase lag of $\pi/2$. The critical cluster radius for frequency matching can be estimated from the simple classical plasmon formula in (9.48), cf. Fig. 9.3(d). The cluster radius passes this critical value right at the time where the system absorbs energy most efficiently and therefore emits many electrons. This effect is called plasmon-enhanced ionization. Connected to the plasmon enhancement is an efficient non-thermal electron acceleration mechanism, discussed in [22].

The optimal pulse delay calculated within the semiclassical model is of the same order of magnitude as values obtained from corresponding experiments. It is, of course, only an approximation to the real behavior, but has a number of advantages over purely classical MD techniques. The introduction of exchange and correlation effects allows to start from a stable and bound ground state. An initial Fermi-Dirac distribution is stable, as the mean-field test particle approach removes binary collisions and, therefore, unphysical thermalization to a Boltzmann distribution. However, as described above, the treatment neglects collisions if the system becomes highly excited. This shortcoming can be removed by introducing a Ühling-Uhlenbeck collision term [23]. However, this is beyond the scope of this contribution. For further reading about the semiclassical method and a comparison to quantum mechanical models we refer to [24].

The authors gratefully acknowledge financial support by the Deutsche Forschungsgemeinschaft within the Sonderforschungsbereich 652. Computer time was provided by the High Performance Computing Center for North Germany (HLRN).

References

1. A. Messiah, *Quantum Mechanics* (North-Holland, 1976)
2. G. Bertsch, in *Many-Body Dynamics of Heavy-Ion Collisions*, ed. by R.B. et al. (North-Holland, Amsterdam, 1978)
3. E. Wigner, Phys. Rep. **40**, 739 (1932)
4. G. Bertsch, S.D. Gupta, Phys. Rep. **160**, 189 (1988)
5. A. Smerzi, Phys. Rev. Lett. **76**, 559 (1996)
6. P. Dirac, Proc. Cambridge Philos. Soc. **26**, 376 (1930)
7. O. Gunnarsson, B. Lundquist, Phys. Rev. B **13**, 4274 (1976)
8. J. Perdew, A. Zunger, Phys. Rev. B **23**, 5048 (1981)
9. C. Jarzynski, G. Bertsch, Phys. Rev. C **53**, 1028 (1995)
10. A. Domps, P. L'Eplattenier, P. Reinhard, E. Suraud, Ann. Phys. (Leipzig) **6**, 455 (1997)
11. R. Hockney, J. Eastwood, *Computer simulation using particles* (McGraw-Hill Book Company, 1981)
12. A. Castro, A. Rubio, M.J. Stott, Can. J. Phys. **81**, 1151 (2003)
13. T. Beck, Rev. Mod. Phys. **72**, 1041 (2000)
14. T. Fennel, G. Bertsch, K.H. Meiwes-Broer, Eur. Phys. J. D **29**, 367 (2004)

15. L.H. Thomas, Proc. Cambridge Philos. Soc. **23**, 542 (1927)
16. E. Fermi, Z. Phys. **48**, 73 (1928)
17. E. Teller, Rev. Mod. Phys. **34**, 627 (1962)
18. E.H. Lieb, Rev. Mod. Phys. **348**, 553 (1976)
19. C. Legrand, E. Suraud, P. Reinhard, J. Phys. B. **39**, 2481 (2006)
20. Y. Yabana, G. Bertsch, Phys. Rev. B **15**(7), 3108 (1996)
21. T. Döppner, T. Fennel, T. Diederich, J. Tiggesbäumker, K.H. Meiwes-Broer, Phys. Rev. Lett. **94**, 13401 (2005)
22. T. Fennel, T. Döppner, J. Passig, C. Schaal, J. Tiggesbäumker, K.H. Meiwes-Broer, Phys. Rev. Lett. **98**, 143401 (2007)
23. A. Domps, P.G. Reinhard, E. Suraud, Ann. Phys **280**, 211 (2000)
24. P. Reinhard, E. Suraud, *Introduction to Cluster Dynamics* (Wiley-VCH, Berlin, 2004)

Part V

Quantum Monte Carlo

10 World-line and Determinantal Quantum Monte Carlo Methods for Spins, Phonons and Electrons

F.F. Assaad[1] and H.G. Evertz[2]

[1] Institut für Theoretische Physik und Astrophysik, Universität Würzburg, 97074 Würzburg, Germany
[2] Institut für Theoretische Physik und Computational Physics, Technische Universität Graz, A-8010 Graz, Austria

In this chapter we will concentrate primarily on world-line methods with loop updates, for spins and also for spin-phonon systems, as well as on the auxiliary field quantum Monte Carlo (QMC) method. Both approaches are based on a path integral formulation of the partition function which maps a d-dimensional quantum system onto a $d + 1$ dimensional classical system. The additional dimension is nothing but the imaginary time. World-line based approaches for quantum spin systems offer a simple realization of the mapping from quantum to classical, and allow for new approaches to phonons, as recently developed. Auxiliary field QMC methods provide access to fermionic systems both at finite temperature and in the ground state. An important example is the Hirsch-Fye approach that allows for an efficient simulation of impurity models, such as the Kondo and Anderson models, and is widely used in the domain of dynamical mean field theories (DMFT).

10.1 Introduction

The correlated electron problem remains one of the central challenges in solid state physics. Given the complexity of the problem, numerical simulations provide an essential source of information to test ideas and develop intuition. In particular for a given model describing a particular material we would ultimately like to be able to carry out efficient numerical simulations so as to provide *exact* results on thermodynamic, dynamical, transport and ground-state properties. If the model shows a continuous quantum phase transition we would like to characterize it by computing the critical exponents. Without restriction on the type of model, this is an extremely challenging goal.

There are however a set of problems for which numerical techniques have provided invaluable insight and will continue to do so. Here we list a few which are exact, capable of reaching large system sizes (the computational effort scales as a power of the volume), and provide ground-state, dynamical as well as thermodynamic quantities: (i) Density matrix renormalization group applied to general one-dimensional (1D) systems [1, 2], (ii) world-line based QMC methods such as the loop algorithm [3, 4] or directed loops [5] applied to non-frustrated spin systems in arbitrary dimensions or to 1D electron-models on bipartite lattices, and (iii) auxiliary field QMC methods [6]. The latter method is capable of handling a class of

F.F. Assaad and H.G. Evertz: *World-line and Determinantal Quantum Monte Carlo Methods for Spins, Phonons and Electrons*, Lect. Notes Phys. **739**, 277–356 (2008)
DOI 10.1007/978-3-540-74686-7_10

models with spin and charge degrees of freedom in dimensions larger than unity. This class contains fermionic lattice models with attractive interactions (e.g. attractive Hubbard model), models invariant under particle-hole transformation, as well as impurity problems modelled by Kondo or Anderson Hamiltonians.

In this lecture we first introduce the world-line approach, exemplarily for the 1D XXZ-chain, see Sect. 10.2. In Sect. 10.3, we discuss world-line representations of $\exp(-\beta H)$ without Trotter-time discretization errors (where $\beta = 1/(k_\mathrm{B}T)$), including the stochastic series expansion (SSE). We emphasize that the issue of such a representation of $\exp(-\beta H)$ is largely independent of the Monte Carlo algorithm used to update the world lines. In Sect. 10.4 we explain the loop algorithm from an operator point of view, and discuss some applications and generalizations. Sect. 10.5 discusses ways to treat coupled systems of spins and phonons, exemplified for 1D spin-Peierls transitions. It includes a new method which allows the simulation of arbitrary bare phonon dispersions [7]. In Sect. 10.6 we describe the basic formulation of the auxiliary field QMC method. This includes the formulation of the partition function, the measurement of equal-time and time-displaced correlation functions as well as general conditions under which one can show the absence of negative sign problem. In Sect. 10.7 we concentrate on the implementation of the auxiliary field method for lattice problems. Here, the emphasis is placed on numerical stabilization of the algorithm. Sect. 10.8 concentrates on the Hirsch-Fye formulation of the algorithm. This formulation is appropriate for general impurity models, and is extensively used in the framework of dynamical mean-field theories and their generalization to cluster methods. Recently, more efficient continuous time algorithms for the impurity problem (diagrammatic determinantal QMC methods) have been introduced [8, 9]. Finally in Sect. 10.9 we briefly provide a short and necessarily biased overview of applications of auxiliary field methods.

10.2 Discrete Imaginary Time World Lines for the XXZ Spin Chain

The attractive feature of the world-line approach [10] is its simplicity. Here, we will concentrate on the 1D XXZ spin chain. The algorithm relies on a mapping of the 1D XXZ quantum spin chain to the six vertex model [11]. The classical model may then be solved exactly as in the case of the six vertex model [12] or simulated very efficiently by means of cluster Monte Carlo methods [3, 4]. The latter approach has proved to be extremely efficient for the investigation of non-frustrated quantum spin systems [13] in arbitrary dimensions. The efficiency lies in the fact that (i) the computational time scales as the volume of the classical system so that very large system sizes may be achieved, and (ii) the autocorrelation times are small.

A related method, applicable to more models, are directed loops [5, 14]. A short introduction is provided in [15]. For a general short overview of advanced world-line QMC methods see [16]. Longer reviews are provided in [4, 17].

Fermions can also be represented by world lines. For spinless fermions in any dimension the same representation as for the XXZ spin model results, albeit with additional signs corresponding to the exchange of fermions. The world-line approach will allow us to acquire some insight into the resulting sign problem. This is a major open issue in QMC methods applied to correlated systems. When it occurs the computational effort scales exponentially with system size and inverse temperature. Recent attempts in the form of novel concepts to tackle correlated electron systems are reviewed in [18, 19].

Finally, at the end of this section, we will discuss extensions of the world-line approach to tackle the problem of the dynamics of a single-hole in non-frustrated quantum magnets.

10.2.1 Basic Formulation

To illustrate the world-line QMC method, we concentrate on the XXZ quantum spin chain. This model is defined as

$$H = J_x \sum_i \left(S_i^x S_{i+1}^x + S_i^y S_{i+1}^y \right) + J_z \sum_i S_i^z S_{i+1}^z \,, \tag{10.1}$$

where $\boldsymbol{S}_i$ are spin $1/2$ operators on site i and hence satisfy the commutation rules

$$\left[S_i^\eta, S_j^\nu \right] = \mathrm{i}\epsilon^{\eta,\nu,\gamma} S_i^\gamma \delta_{i,j} \,. \tag{10.2}$$

In the above, $\epsilon^{\eta,\nu,\gamma}$ is the antisymmetric tensor and the sum over repeated indices is understood. We impose periodic boundary conditions

$$\boldsymbol{S}_{i+L} = \boldsymbol{S}_i \,, \tag{10.3}$$

where L denotes the length of the chain.

A representation of the above commutation relations is achieved with the Pauli spin matrices. For a single spin-$1/2$ degree of freedom, we can set

$$S^x = \frac{1}{2}\begin{pmatrix} 0 & 1 \\ 1 & 0 \end{pmatrix} , \quad S^y = \frac{1}{2}\begin{pmatrix} 0 & -\mathrm{i} \\ \mathrm{i} & 0 \end{pmatrix} , \quad S^z = \frac{1}{2}\begin{pmatrix} 1 & 0 \\ 0 & -1 \end{pmatrix} , \tag{10.4}$$

and the corresponding Hilbert space $\mathcal{H}_{1/2}$ is spanned by the two state vectors

$$|\uparrow\rangle = \begin{pmatrix} 1 \\ 0 \end{pmatrix} , \quad |\downarrow\rangle = \begin{pmatrix} 0 \\ 1 \end{pmatrix} . \tag{10.5}$$

It is convenient to define the raising S^+ and lowering S^- operators

$$S^+ = S^x + \mathrm{i}S^y \,, \quad S^- = S^x - \mathrm{i}S^y \,, \tag{10.6}$$

such that

$$\begin{aligned} S^-|\downarrow\rangle &= S^+|\uparrow\rangle = 0 \,, \\ S^-|\uparrow\rangle &= |\downarrow\rangle \,, \\ S^+|\downarrow\rangle &= |\uparrow\rangle \,. \end{aligned} \tag{10.7}$$

The Hilbert space of the L-site chain $\mathcal{H}_L$ is given by the tensor product of L spin $1/2$ Hilbert spaces. $\mathcal{H}_L$ contains 2^L state vectors which we will denote by

$$|\boldsymbol{\sigma}\rangle = |\sigma_1, \sigma_2, \ldots, \sigma_L\rangle \tag{10.8}$$

with $\sigma_i = \uparrow$ or $\downarrow$. A representation of the unit operator in $\mathcal{H}_L$ is given by

$$1 = \sum_{\boldsymbol{\sigma}} |\boldsymbol{\sigma}\rangle\langle\boldsymbol{\sigma}| \,. \tag{10.9}$$

We can easily solve the two-site problem

$$H_{\text{two sites}} = \underbrace{J_x \left(S_1^x S_2^x + S_1^y S_2^y\right)}_{\equiv \frac{1}{2}(S_1^+ S_2^- + S_1^- S_2^+)} + J_z S_1^z S_2^z \,. \tag{10.10}$$

The eigenstates of the above Hamiltonian are nothing but the singlet and three triplet states

$$\begin{aligned}
H_{\text{two sites}} \frac{1}{\sqrt{2}}\left(|\uparrow,\downarrow\rangle - |\downarrow,\uparrow\rangle\right) &= \left(-\frac{J_z}{4} - \frac{J_x}{2}\right)\frac{1}{\sqrt{2}}\left(|\uparrow,\downarrow\rangle - |\downarrow,\uparrow\rangle\right) \,, \\
H_{\text{two sites}} \frac{1}{\sqrt{2}}\left(|\uparrow,\downarrow\rangle + |\downarrow,\uparrow\rangle\right) &= \left(-\frac{J_z}{4} + \frac{J_x}{2}\right)\frac{1}{\sqrt{2}}\left(|\uparrow,\downarrow\rangle + |\downarrow,\uparrow\rangle\right) \,, \\
H_{\text{two sites}} |\uparrow,\uparrow\rangle &= \frac{J_z}{4} |\uparrow,\uparrow\rangle \,, \\
H_{\text{two sites}} |\downarrow,\downarrow\rangle &= \frac{J_z}{4} |\downarrow,\downarrow\rangle \,.
\end{aligned} \tag{10.11}$$

The basic idea of this original world-line approach is to split the XXZ Hamiltonian into a set of independent two-site problems. The way to achieve this decoupling is with the use of a path integral and the Trotter decomposition. First we write

$$H = \underbrace{\sum_n H^{(2n+1)}}_{H_1} + \underbrace{\sum_n H^{(2n+2)}}_{H_2} \tag{10.12}$$

with $H^{(i)} = J_x \left(S_i^x S_{i+1}^x + S_i^y S_{i+1}^y\right) + J_z S_i^z S_{i+1}^z$. One may verify that H_1 and H_2 are sums of commuting (i.e. independent) two-site problems. Hence, on their own H_1 and H_2 are trivially solvable problems. However, H is not. To use this fact, we split the imaginary propagation $\exp(-\beta H)$ into successive infinitesimal propagations of H_1 and H_2. Here β corresponds to the inverse temperature. This is achieved with the Trotter decomposition introduced in detail in Sect. 10.A. The partition function is then given by

$$\begin{aligned}
\mathrm{Tr}\left[\mathrm{e}^{-\beta H}\right] &= \mathrm{Tr}\left[(\mathrm{e}^{-\Delta\tau H})^m\right] = \mathrm{Tr}\left[(\mathrm{e}^{-\Delta\tau H_1}\mathrm{e}^{-\Delta\tau H_2})^m\right] + \mathcal{O}(\Delta\tau^2) \\
&= \sum_{\boldsymbol{\sigma}_1 \ldots \boldsymbol{\sigma}_{2m}} \langle\boldsymbol{\sigma}_1|\mathrm{e}^{-\Delta\tau H_1}|\boldsymbol{\sigma}_{2m}\rangle \ldots \langle\boldsymbol{\sigma}_3|\mathrm{e}^{-\Delta\tau H_1}|\boldsymbol{\sigma}_2\rangle\langle\boldsymbol{\sigma}_2|\mathrm{e}^{-\Delta\tau H_2}|\boldsymbol{\sigma}_1\rangle + \mathcal{O}(\Delta\tau^2) \,,
\end{aligned} \tag{10.13}$$

where $m\Delta\tau = \beta$. In the last equality we have inserted the unit operator between each infinitesimal imaginary time propagation. For each set of states $|\boldsymbol{\sigma}_1\rangle \ldots |\boldsymbol{\sigma}_{2m}\rangle$ with non-vanishing contribution to the partition function we have a simple graphical representation in terms of world lines which track the evolution of the spins in space and imaginary time. An example of a world-line configuration is shown in Fig. 10.1.

Hence the partition function may be written as the sum of over all world-line configurations w, each world-line configuration having an appropriate weight $\Omega(w)$

$$Z = \sum_w \Omega(w)$$
$$\Omega(w) = \langle \boldsymbol{\sigma}_1 | \mathrm{e}^{-\Delta\tau H_1} | \boldsymbol{\sigma}_{2m}\rangle \ldots \langle \boldsymbol{\sigma}_3 | \mathrm{e}^{-\Delta\tau H_1} | \boldsymbol{\sigma}_2\rangle \langle \boldsymbol{\sigma}_2 | \mathrm{e}^{-\Delta\tau H_2} | \boldsymbol{\sigma}_1\rangle \,, \tag{10.14}$$

where w defines the states $|\boldsymbol{\sigma}_1\rangle \ldots |\boldsymbol{\sigma}_{2m}\rangle$.

Our task is now to compute the weight $\Omega(w)$ for a given world-line configuration w. Let us concentrate on the matrix element $\langle \boldsymbol{\sigma}_{\tau+1} | \exp(-\Delta\tau H_2) | \boldsymbol{\sigma}_\tau\rangle$. Since H_2 is a sum of independent two site problems, we have

$$\langle \boldsymbol{\sigma}_{\tau+1} | \mathrm{e}^{-\Delta\tau H_2} | \boldsymbol{\sigma}_\tau\rangle = \prod_{i=1}^{L/2} \langle \sigma_{2i,\tau+1}, \sigma_{2i+1,\tau+1} | \mathrm{e}^{-\Delta\tau H^{(2i)}} | \sigma_{2i,\tau}, \sigma_{2i+1,\tau}\rangle \,. \tag{10.15}$$

Hence, the calculation of the weight reduces to solving the two-site problem, see (10.10). We can compute, for example, the spin-flip matrix element

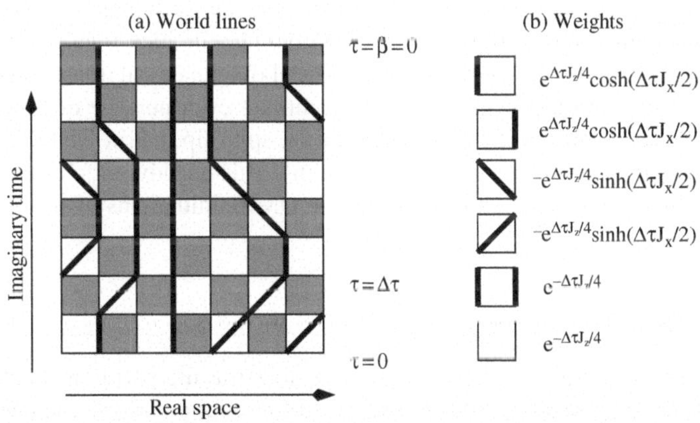

Fig. 10.1. (**a**) World-line configuration for the XXZ model of (10.1). Here, $m = 4$ and the system size is $L = 8$. The *bold lines* follow the time evolution of the up spins and empty sites, with respect to the world lines, correspond to the down spins. A full time step $\Delta\tau$ corresponds to the propagation with H_1 followed by H_2. Periodic boundary conditions are chosen in the spatial direction. In the time direction, periodic boundary conditions follow from the fact that we are evaluating a trace. (**b**) The weights for a given world-line configuration is the product of the weights of plaquettes listed in the figure. Note that, although the spin-flip processes come with a minus sign, the overall weight for the world-line configuration is positive since each world-line configuration contains an even number of spin flips

$$
\begin{aligned}
&\langle \downarrow, \uparrow | \mathrm{e}^{-\Delta\tau H_{\text{two sites}}} | \uparrow, \downarrow \rangle \\
&= \frac{1}{\sqrt{2}} \langle \downarrow, \uparrow | \mathrm{e}^{-\Delta\tau H_{\text{two sites}}} \left(\frac{1}{\sqrt{2}} (| \uparrow, \downarrow\rangle - | \downarrow, \uparrow\rangle) + \frac{1}{\sqrt{2}} (| \uparrow, \downarrow\rangle + | \downarrow, \uparrow\rangle) \right) \\
&= \frac{1}{\sqrt{2}} \langle \downarrow, \uparrow | \Big(\mathrm{e}^{-\Delta\tau(-J_z/4 - J_x/2)} \frac{1}{\sqrt{2}} (| \uparrow, \downarrow\rangle - | \downarrow, \uparrow\rangle) \\
&\quad + \mathrm{e}^{-\Delta\tau(-J_z/4 + J_x/2)} \frac{1}{\sqrt{2}} (| \uparrow, \downarrow\rangle + | \downarrow, \uparrow\rangle) \Big) \\
&= -\mathrm{e}^{\Delta\tau J_z/4} \sinh\left(\frac{\Delta\tau J_x}{2} \right) . \qquad (10.16)
\end{aligned}
$$

The other five matrix elements are listed in Fig. 10.1 and may be computed in the same manner.

We are now faced with a problem, namely that the spin-flip matrix elements are negative. However, for non-frustrated spin systems, we can show that the overall sign of the world-line configuration is positive. To prove this statement consider a bipartite lattice in arbitrary dimensions. A bipartite lattice may be split into two sub-lattices, A and B, such that the nearest neighbors of sub-lattice A belong to sub-lattice B and vice-versa. A non-frustrated spin system on a bipartite lattice has solely spin-spin interactions between two lattice sites belonging to different sub-lattices. For example, in our 1D case, the even sites correspond to say sub-lattice A and the odd sites to sub-lattice B. Under those conditions we can carry out the canonical transformation (i.e. the commutation rules remain invariant) $S_i^x \to f(i) S_i^x$, $S_i^y \to f(i) S_i^y$, and $S_i^z \to S_i^z$, where $f(i) = 1\,(-1)$ if i belongs to sublattice A (B). Under this transformation, the matrix element J_x in the Hamiltonian transforms to $-J_x$, which renders all matrix elements positive. The above canonical transformation just tells us that the spin-flip matrix element occurs an even number of times in any world-line configuration. The minus sign in the spin-flip matrix element may not be omitted in the case of frustrated spin systems. This negative sign leads to a sign problem which up to date inhibits large scale QMC simulations of frustrated spin systems.

10.2.2 Observables

In the previous section, we have shown how to write the partition function of a non-frustrated spin system as a sum over world-line configurations, each world-line configuration having a positive weight. Our task is now to compute observables

$$
\langle O \rangle = \frac{\mathrm{Tr}\left[\mathrm{e}^{-\beta H} O\right]}{\mathrm{Tr}\left[\mathrm{e}^{-\beta H}\right]} = \frac{\sum_w \Omega(w) O(w)}{\sum_w \Omega(w)} , \qquad (10.17)
$$

where $\Omega(w)$ corresponds to the weight of a given world-line configuration as obtained through multiplication of the weights of the individual plaquettes listed in Fig. 10.1 and $O(w)$ corresponds to the value of the observable for the given world-line configuration.

One of the major drawbacks of the world-line algorithm used to be that one could not measure arbitrary observables. In particular, the correlation functions such as $S_i^+ S_j^-$ which introduce a cut in a world-line configuration are not accessible with continuous world lines and local updates. This problem disappears in the loop algorithm and also with worms and directed loops, as will be discussed later. Here we will concentrate on observables which locally conserve the z-component of spin, specifically the total energy as well as the spin-stiffness.

10.2.2.1 Energy and Spin-Spin Correlations

Neglecting the systematic error originating from the Trotter decomposition, the expectation value of the energy is given by

$$\begin{aligned}\langle H\rangle &= \frac{1}{Z}\mathrm{Tr}\left[\left(\mathrm{e}^{-\Delta\tau H_1}\mathrm{e}^{-\Delta\tau H_2}\right)^m (H_1+H_2)\right]\\ &= \frac{1}{Z}\mathrm{Tr}\Big[\left(\mathrm{e}^{-\Delta\tau H_1}\mathrm{e}^{-\Delta\tau H_2}\right)^{m-1}\Big(\mathrm{e}^{-\Delta\tau H_1}H_1\mathrm{e}^{-\Delta\tau H_2}\\ &\quad+\mathrm{e}^{-\Delta\tau H_1}\mathrm{e}^{-\Delta\tau H_2}H_2\Big)\Big]\,. \end{aligned}\tag{10.18}$$

To obtain the last equation, we have used the cyclic properties of the trace: $\mathrm{Tr}\left[AB\right] = \mathrm{Tr}\left[BA\right]$. Inserting the unit operator $1 = \sum_{\boldsymbol{\sigma}} |\boldsymbol{\sigma}\rangle\langle\boldsymbol{\sigma}|$ at each imaginary time interval yields

$$\begin{aligned}\langle H\rangle &= \frac{1}{Z}\sum_{\boldsymbol{\sigma}_1,\dots\boldsymbol{\sigma}_{2m}}\left[\langle\boldsymbol{\sigma}_1|\mathrm{e}^{-\Delta\tau H_1}|\boldsymbol{\sigma}_{2m}\rangle\dots\langle\boldsymbol{\sigma}_3|\mathrm{e}^{-\Delta\tau H_1}|\boldsymbol{\sigma}_2\rangle\langle\boldsymbol{\sigma}_2|\mathrm{e}^{-\Delta\tau H_2}H_2|\boldsymbol{\sigma}_1\rangle\right.\\ &\quad\left.+\langle\boldsymbol{\sigma}_1|\mathrm{e}^{-\Delta\tau H_1}|\boldsymbol{\sigma}_{2m}\rangle\dots\langle\boldsymbol{\sigma}_3|\mathrm{e}^{-\Delta\tau H_1}H_1|\boldsymbol{\sigma}_2\rangle\langle\boldsymbol{\sigma}_2|\mathrm{e}^{-\Delta\tau H_2}|\boldsymbol{\sigma}_1\rangle\right]\\ &= \frac{1}{Z}\sum_{\boldsymbol{\sigma}_1,\dots\boldsymbol{\sigma}_{2m}}\langle\boldsymbol{\sigma}_1|\mathrm{e}^{-\Delta\tau H_1}|\boldsymbol{\sigma}_{2m}\rangle\dots\langle\boldsymbol{\sigma}_3|\mathrm{e}^{-\Delta\tau H_1}|\boldsymbol{\sigma}_2\rangle\langle\boldsymbol{\sigma}_2|\mathrm{e}^{-\Delta\tau H_2}|\boldsymbol{\sigma}_1\rangle\\ &\quad\times\left[\frac{\langle\boldsymbol{\sigma}_3|\mathrm{e}^{-\Delta\tau H_1}H_1|\boldsymbol{\sigma}_2\rangle}{\langle\boldsymbol{\sigma}_3|\mathrm{e}^{-\Delta\tau H_1}|\boldsymbol{\sigma}_2\rangle}+\frac{\langle\boldsymbol{\sigma}_2|\mathrm{e}^{-\Delta\tau H_2}H_2|\boldsymbol{\sigma}_1\rangle}{\langle\boldsymbol{\sigma}_2|\mathrm{e}^{-\Delta\tau H_2}|\boldsymbol{\sigma}_1\rangle}\right]\\ &= \frac{\sum_w \Omega(w)E(w)}{\sum_w \Omega(w)}\end{aligned}\tag{10.19}$$

with

$$E(w) = -\frac{\partial}{\partial\Delta\tau}\left[\ln\langle\boldsymbol{\sigma}_2|\mathrm{e}^{-\Delta\tau H_2}|\boldsymbol{\sigma}_1\rangle + \ln\langle\boldsymbol{\sigma}_3|\mathrm{e}^{-\Delta\tau H_1}|\boldsymbol{\sigma}_2\rangle\right]\,. \tag{10.20}$$

We can of course measure the energy on arbitrary time slices. Averaging over all the time slices to reduce the fluctuations yields the form

$$E(w) = -\frac{1}{m}\frac{\partial}{\partial\Delta\tau}\ln\Omega(w)\,. \tag{10.21}$$

Hence the energy of a world-line configuration is nothing but the logarithmic derivative of its weight. This can also be obtained more directly by taking the derivative of (10.14).

Observables O which locally conserve the z-component of the spin are easy to compute. If we decide to measure on time slice τ then $O|\boldsymbol{\sigma}_\tau\rangle = O(w)|\boldsymbol{\sigma}_\tau\rangle$. An example of such an observable is the correlation function $O = S_i^z S_j^z$.

10.2.2.2 Spin Stiffness (Superfluid Density)

The spin stiffness probes the sensitivity of the system under a twist – in spin space – of the boundary condition along one lattice direction. If long-range spin order is present, the free energy in the thermodynamic limit will acquire a dependence on the twist. If on the other hand the system is disordered, the free energy is insensitive to the twist. The spin stiffness hence probes for long range or quasi long-range spin ordering. It is identical to the superfluid density when viewing spin systems in terms of hard-core bosons. To define the spin stiffness, we consider the Heisenberg model on a d-dimensional hyper-cubic lattice of linear length L:

$$H = J \sum_{\langle \boldsymbol{i},\boldsymbol{j} \rangle} \widetilde{\boldsymbol{S}}_{\boldsymbol{i}} \cdot \widetilde{\boldsymbol{S}}_{\boldsymbol{j}} \ . \tag{10.22}$$

We impose twisted boundary condition in say the x-direction,

$$\widetilde{\boldsymbol{S}}_{\boldsymbol{i}+L\boldsymbol{e}_x} = R(\mathbf{e},\phi)\widetilde{\boldsymbol{S}}_{\boldsymbol{i}} \ . \tag{10.23}$$

where $R(\mathbf{e},\phi)$ is an $SO(3)$ rotation around the axis $\mathbf{e}$ with angle ϕ. In the other lattice directions, we consider periodic boundary conditions. The spin stiffness is then defined as

$$\rho_s = \frac{1}{L^{d-2}} \frac{-1}{\beta} \ln Z(\phi) \bigg|_{\phi=0} \ , \tag{10.24}$$

where $Z(\phi)$ is the partition function in the presence of the twist in the boundary condition, and β corresponds to the inverse temperature.

Under the canonical transformation

$$\boldsymbol{S}_{\boldsymbol{i}} = R(\mathbf{e}, -\frac{\phi}{L}\boldsymbol{i}\cdot\boldsymbol{e}_x)\widetilde{\boldsymbol{S}_{\boldsymbol{i}}} \tag{10.25}$$

the twist may be eliminated from the boundary condition

$$\begin{aligned} \boldsymbol{S}_{\boldsymbol{i}+L\boldsymbol{e}_x} &= R\left[\mathbf{e}, -\frac{\phi}{L}(\boldsymbol{i}+L\boldsymbol{e}_x)\cdot\boldsymbol{e}_x\right] \widetilde{\boldsymbol{S}}_{\boldsymbol{i}+L\boldsymbol{e}_x} \\ &= R\left[\mathbf{e}, -\frac{\phi}{L}(\boldsymbol{i}+L\boldsymbol{e}_x)\cdot\boldsymbol{e}_x\right] R(\mathbf{e},\phi)\widetilde{\boldsymbol{S}}_{\boldsymbol{i}} \\ &= R\left[\mathbf{e}, -\frac{\phi}{L}\boldsymbol{i}\cdot\boldsymbol{e}_x\right] \widetilde{\boldsymbol{S}}_{\boldsymbol{i}} = \boldsymbol{S}_{\boldsymbol{i}} \end{aligned} \tag{10.26}$$

to appear explicitly in the Hamiltonian

$$
\begin{aligned}
H &= J\sum_{\langle \boldsymbol{i},\boldsymbol{j}\rangle}\left[R(\boldsymbol{e},-\frac{\phi}{L}\boldsymbol{i}\cdot\boldsymbol{e}_x)\boldsymbol{S}_{\boldsymbol{i}}\right]\cdot\left[R(\boldsymbol{e},-\frac{\phi}{L}\boldsymbol{j}\cdot\boldsymbol{e}_x)\boldsymbol{S}_{\boldsymbol{j}}\right] \\
&= J\sum_{\langle \boldsymbol{i},\boldsymbol{j}\rangle}\boldsymbol{S}_{\boldsymbol{i}}\cdot R\left[\boldsymbol{e},\frac{\phi}{L}(\boldsymbol{i}-\boldsymbol{j})\cdot\boldsymbol{e}_x\right]\boldsymbol{S}_{\boldsymbol{j}} \\
&= J\sum_{\boldsymbol{i}}\boldsymbol{S}_{\boldsymbol{i}}\cdot R(\boldsymbol{e},-\frac{\phi a}{L})\boldsymbol{S}_{\boldsymbol{i}+\boldsymbol{a}_x} + J\sum_{\boldsymbol{i},\boldsymbol{a}\neq\boldsymbol{a}_x}\boldsymbol{S}_{\boldsymbol{i}}\cdot\boldsymbol{S}_{\boldsymbol{i}+\boldsymbol{a}}\,.
\end{aligned}
\tag{10.27}
$$

Setting the rotation axis $\boldsymbol{e}$ to $\boldsymbol{e}_z$ such that

$$
R\left(\boldsymbol{e},-\frac{\phi a}{L}\right) = \begin{pmatrix} \cos(\phi a/L) & \sin(\phi a/L) & 0 \\ -\sin(\phi a/L) & \cos(\phi a/L) & 0 \\ 0 & 0 & 1 \end{pmatrix}
\tag{10.28}
$$

the Hamiltonian may be written as

$$
\begin{aligned}
H = J\sum_{\boldsymbol{i}}&\left[S^z_{\boldsymbol{i}}S^z_{\boldsymbol{i}+\boldsymbol{a}_x} + \frac{1}{2}(\mathrm{e}^{\mathrm{i}\phi a/L}S^+_{\boldsymbol{i}}S^-_{\boldsymbol{i}+\boldsymbol{a}_x} + \mathrm{e}^{-\mathrm{i}\phi a/L}S^-_{\boldsymbol{i}}S^+_{\boldsymbol{i}+\boldsymbol{a}_x})\right] \\
&+J\sum_{\boldsymbol{i},\boldsymbol{a}\neq\boldsymbol{a}_x}\left[S^z_{\boldsymbol{i}}S^z_{\boldsymbol{i}+\boldsymbol{a}} + \frac{1}{2}(S^+_{\boldsymbol{i}}S^-_{\boldsymbol{i}+\boldsymbol{a}} + S^-_{\boldsymbol{i}}S^+_{\boldsymbol{i}+\boldsymbol{a}})\right]\,.
\end{aligned}
\tag{10.29}
$$

In the spirit of the world-line algorithm, we write the partition function as

$$
Z(\phi) = \sum_w \underbrace{\prod_p W(S_p(w),\phi)}_{\Omega(w,\phi)}\,.
\tag{10.30}
$$

The sum runs over all world-line configurations w and the weight of the world-line configuration, $\Omega(w)$, is given by the product of the individual plaquette weights $W(S_p(w),\phi)$ in the space-time lattice. $S_p(w)$ denotes the spin configuration on plaquette p in the world-line configuration w.

Since at $\phi - 0$ time reversal symmetry holds, the spin current

$$
j_s = -\frac{1}{\beta}\frac{\partial}{\partial\phi}\ln Z(\phi)\bigg|_{\phi=0}
\tag{10.31}
$$

vanishes and the spin stiffness reads

$$
\rho_s = \frac{1}{Z}\sum_w \Omega(w)\rho_s(w)
\tag{10.32}
$$

where

$$\rho_s(w) = -\frac{1}{\beta L^{d-2}}\left(\sum_p \frac{\frac{\partial^2}{\partial\phi^2} W(S_p(w),\phi)\big|_{\phi=0}}{W(S_p(w))} + \sum_{p\neq q} \frac{\frac{\partial}{\partial\phi} W(S_p(w),\phi)\big|_{\phi=0}}{W(S_p(w))}\frac{\frac{\partial}{\partial\phi} W(S_q(w),\phi)\big|_{\phi=0}}{W(S_q(w))}\right) \tag{10.33}$$

It is instructive to compute the spin stiffness in the limit $\Delta\tau \to 0$ since in this limit ρ_s is nothing but the average of the square of the total spatial winding number of the world lines. Let $\sigma_{1,p}, \sigma_{2,p}, \sigma_{3,p}$ and $\sigma_{4,p}$ correspond to the spin configuration S_p and $\boldsymbol{i}_p$, $\boldsymbol{j}_p$ to the two real space points associated to the plaquette p such that

$$\begin{aligned}
&\lim_{\Delta\tau\to 0} \frac{\frac{\partial^2}{\partial\phi^2} W(S_p(w),\phi)\Big|_{\phi=0}}{W(S_p(w))} \\
&= \lim_{\Delta\tau\to 0} -\frac{\Delta\tau J}{2}\left[\frac{\mathrm{i}\boldsymbol{e}_x\cdot(\boldsymbol{j}_p-\boldsymbol{i}_p)}{L}\right]^2 \frac{\langle \sigma_{1,p},\sigma_{2,p}|S^+_{\boldsymbol{i}_p}S^-_{\boldsymbol{j}_p}+S^-_{\boldsymbol{i}_p}S^+_{\boldsymbol{j}_p}|\sigma_{3,p},\sigma_{4,p}\rangle}{\langle \sigma_{1,p},\sigma_{2,p}|1-\Delta\tau H_{\boldsymbol{i}_p,\boldsymbol{j}_p}|\sigma_{3,p},\sigma_{4,p}\rangle} \\
&= \left[\frac{\mathrm{i}\boldsymbol{e}_x\cdot(\boldsymbol{j}_p-\boldsymbol{i}_p)}{L}\langle \sigma_{1,p},\sigma_{2,p}|S^+_{\boldsymbol{i}_p}S^-_{\boldsymbol{j}_p}+S^-_{\boldsymbol{i}_p}S^+_{\boldsymbol{j}_p}|\sigma_{3,p},\sigma_{4,p}\rangle\right]^2 .
\end{aligned} \tag{10.34}$$

In the last line have used the fact that $\langle \sigma_{1,p},\sigma_{2,p}|S^+_{\boldsymbol{i}_p}S^-_{\boldsymbol{j}_p}+S^-_{\boldsymbol{i}_p}S^+_{\boldsymbol{j}_p}|\sigma_{3,p},\sigma_{4,p}\rangle = 1$ if there is a spin-flip process on plaquette p and zero otherwise. Similarly, we have:

$$\begin{aligned}
&\lim_{\Delta\tau\to 0} \frac{\frac{\partial}{\partial\phi} W(S_p(w),\phi)\Big|_{\phi=0}}{W(S_p(w))} \\
&= \lim_{\Delta\tau\to 0} -\frac{\Delta\tau J}{2}\frac{\mathrm{i}\boldsymbol{e}_x\cdot(\boldsymbol{j}_p-\boldsymbol{i}_p)}{L} \frac{\langle \sigma_{1,p},\sigma_{2,p}|(S^+_{\boldsymbol{i}_p}S^-_{\boldsymbol{j}_p}-S^-_{\boldsymbol{i}_p}S^+_{\boldsymbol{j}_p})|\sigma_{3,p},\sigma_{4,p}\rangle}{\langle \sigma_{1,p},\sigma_{2,p}|1-\Delta\tau H_{\boldsymbol{i}_p,\boldsymbol{j}_p}|\sigma_{3,p},\sigma_{4,p}\rangle} \\
&= \frac{\mathrm{i}\boldsymbol{e}_x\cdot(\boldsymbol{j}_p-\boldsymbol{i}_p)}{L}\langle \sigma_{1,p},\sigma_{2,p}|S^+_{\boldsymbol{i}_p}S^-_{\boldsymbol{j}_p}-S^-_{\boldsymbol{i}_p}S^+_{\boldsymbol{j}_p}|\sigma_{3,p},\sigma_{4,p}\rangle .
\end{aligned} \tag{10.35}$$

Since $\langle \sigma_{1,p},\sigma_{2,p}|S^+_{\boldsymbol{i}_p}S^-_{\boldsymbol{j}_p}-S^-_{\boldsymbol{i}_p}S^+_{\boldsymbol{j}_p}|\sigma_{3,p},\sigma_{4,p}\rangle = \pm 1$ if there is a spin-flip process on plaquette p and zero otherwise the identity

$$\lim_{\Delta\tau\to 0} \frac{\frac{\partial^2}{\partial\phi^2} W(S_p(w),\phi)\Big|_{\phi=0}}{W(S_p(w))} = \left(\lim_{\Delta\tau\to 0} \frac{\frac{\partial}{\partial\phi} W(S_p(w),\phi)\Big|_{\phi=0}}{W(S_p(w))}\right)^2 \tag{10.36}$$

holds. Hence, one can rewrite the spin stiffness as

$$\rho_s(w) = \frac{1}{\beta L^d}(W_x(w))^2 , \tag{10.37}$$

where the winding number along the x-lattice direction W_x is given by

$$W_x(w) = \sum_p \boldsymbol{e}_x \cdot (\boldsymbol{j}_p - \boldsymbol{i}_p)\langle \sigma_{1,p}, \sigma_{2,p}| S^+_{\boldsymbol{i}_p} S^-_{\boldsymbol{j}_p} - S^-_{\boldsymbol{i}_p} S^+_{\boldsymbol{j}_p} |\sigma_{3,p}, \sigma_{4,p}\rangle \ . \quad (10.38)$$

10.2.3 Updating Schemes

The problem is now cast into one which may be solved with classical Monte Carlo methods where we need to generate a Markov chain through the space of world-line configurations. Along the chain the world-line configuration w, occurs on average with normalized probability $\Omega(w)$. There are many ways of generating the Markov chain. Here we will first discuss a local updating scheme and its limitations. We will then turn our attention to a more powerful updating scheme which is known under the name of loop algorithm.

10.2.3.1 Local Updates

Local updates deform a world-line configuration locally. As shown in Fig. 10.2 one randomly chooses a shaded plaquette and, if possible, shifts a world line from one side of the shaded plaquette to the other. This move is local and only involves the four plaquettes surrounding the shaded one. It is then easy to calculate the ratio of weights of the new to old world-line configurations and accept the move according to a Metropolis criterion. As it stands, the above described local move is not ergodic. For example, the z-component of spin is conserved. This problem can be circumvented by considering a move which changes a single down world line into an up one and vice-versa. However, such a global move will have very low acceptance probability at large β.

Combined, both types of moves are ergodic but only in the case of open boundary conditions in space. The algorithm is not ergodic if periodic or anti-periodic boundary conditions are chosen. Consider a starting configuration with zero winding (i.e. $W_x(w) = 0$). The reader will readily convince himself that with local updates, it will never be possible to generate a configuration with $W_x(w) \neq 0$. Hence, for example, a spin stiffness may not be measured within the world line algorithm with local updates. However, one should note that violation of ergodicity lies in

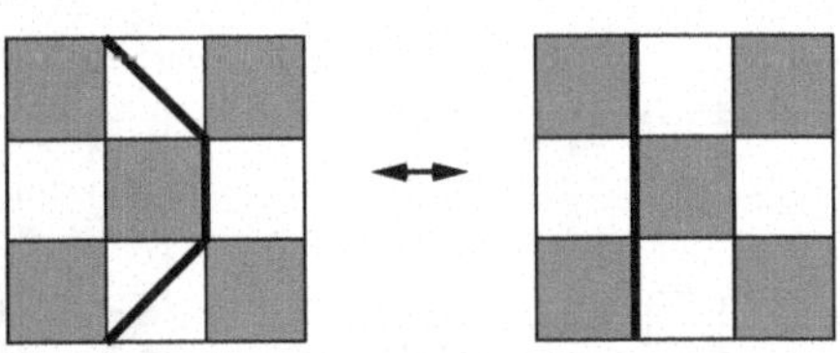

Fig. 10.2. Local updates. A *shaded plaquette* is chosen randomly and a Word Line is shifted from left to right or vice versa across the shaded plaquette

the choice of the boundary condition. Since bulk properties are boundary independent in the thermodynamic limit, the algorithm will yield the correct results in the thermodynamic limit [20].

Different local updates without such problems have been invented in recent years, namely *worms* and *directed loops*. They work by allowing a partial world line, and iteratively changing the position of its ends until it closes again. They will be discussed in Sect. 10.4.5.

10.2.3.2 Loop Updates

To introduce loop updates, it is useful to first map the XXZ model onto the six vertex model of statistical mechanics.

10.2.3.2.1 Equivalence to the Six Vertex Model

That the XXZ quantum spin chain is equivalent to the classical 2D six vertex model follows from a one to one mapping of a world-line configuration to one of the six vertex model. The identification of single plaquettes is shown in Fig. 10.3(a). The world-line configuration of Fig. 10.1 is plotted in the language of the six vertex mode in Fig. 10.3(b). The vertex model lies on a 45 degrees rotated lattice denoted by bullets in Fig. 10.3(b). At each vertex (bullets in Fig. 10.3) the number of incoming arrows equals the number of outgoing arrows. In the case of the XYZ chain, source and drain terms have to be added, yielding the eight vertex model.

The identification of the XXZ model to the six vertex model gives us an intuitive picture of loop updates [3]. Consider the world-line configuration in Fig. 10.4(a) and its corresponding vertex formulation (Fig. 10.4(b)). One can pick a site at random and follow the arrows of the vertex configuration. At each plaquette there are two possible arrow paths to follow. One is chosen, appropriately, and one follows the arrows to arrive to the next plaquette. The procedure is then repeated until one returns to the starting point. Such a loop is shown in Fig. 10.4(c). Along the loop,

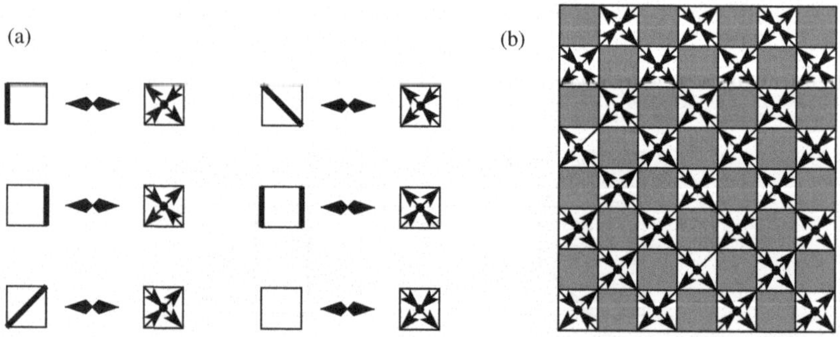

Fig. 10.3. (**a**) Identification of world-line configurations on plaquettes with the vertices of the six vertex model. (**b**) The world-line configuration of Fig. 10.1 in the language of the six vertex model

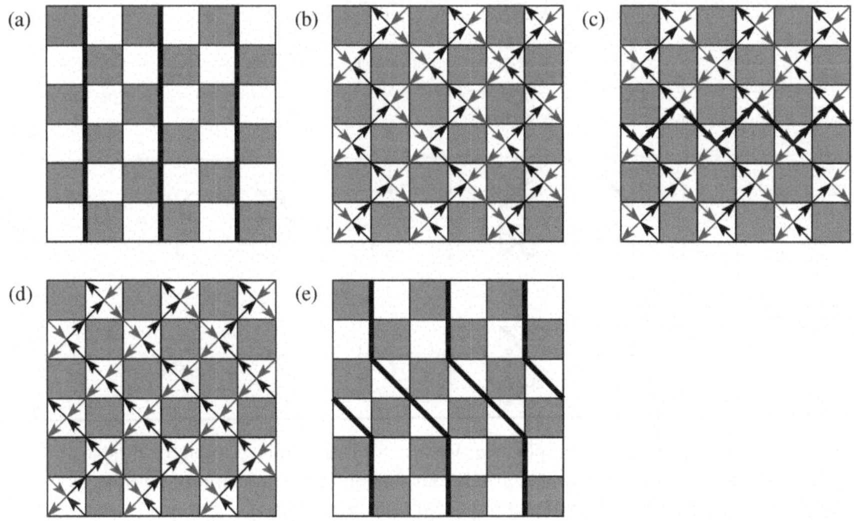

Fig. 10.4. Example of a loop update

changing the direction of the arrows generates another valid vertex configuration, see Fig. 10.4(d). The corresponding world-line configuration (after flipping the loop) is shown in Fig. 10.4(e). As apparent, this is a global update which in this example changes the winding number. This was not achievable with local moves.

10.2.3.2.2 Loop Updates

In the previous paragraph we have seen how to build a loop. Flipping the loop has the potential of generating large-scale changes in the world-line configuration and hence allows us to move quickly in the space of world lines. There is however a potential problem. If the loops were constructed at random, then the acceptance rate for flipping a loop would be extremely small and loop updates would not lead to an efficient algorithm. The loop algorithm sets up rules to build the loop such that it can be flipped without any additional acceptance step for the XXZ model.

To do so, additional variables are introduced, which specify for each plaquette the direction which a loop should take there, Fig. 10.5. These specifications, called breakups or plaquette-graphs, are completely analogous to the Fortuin-Kasteleyn bond-variables of the Swendsen-Wang cluster algorithm, discussed in Chap. 4. They can also be thought of as parts of the Hamilton operator, as discussed in Sect. 10.4. Note that when a breakup has been specified for every plaquette, this then graphically determines a complete decomposition of the vertex-lattice into a set of loops (see also below). The loop algorithm is a cluster algorithm mapping from such sets of loops to world-line configurations and back to new sets of loops. In contrast, directed loops are a local method not associated with such graphs.

Which plaquette-graphs are possible? For each plaquette and associated vertex (spin-configuration) there are several possible choices of plaquette-graphs which

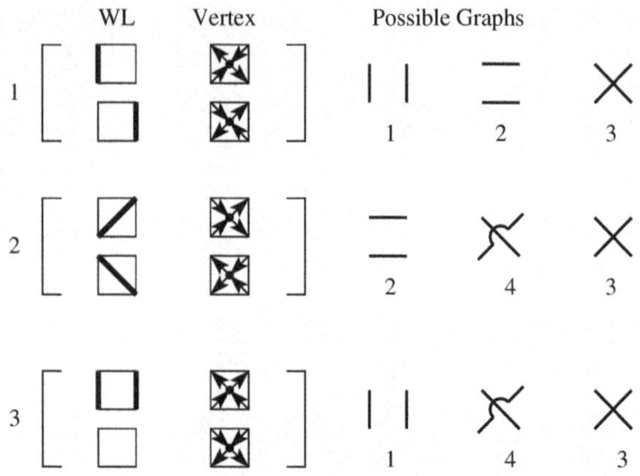

Fig. 10.5. Possible plaquette-graphs for vertex configurations. Graph one is a vertical breakup, graph two a horizontal one, graph four is diagonal. Plaquette-graph three is called frozen; it corresponds to the combined flip of all four arrows

are compatible with the fact that the arrow direction may not change in the construction of the loop. Figure 10.5 illustrates this. Given the vertex configurations one in Fig. 10.5 one can follow the arrows vertically (graph one) or horizontally (graph two). There is also the possibility to flip all the spins of the vertex. This corresponds to graph three in Fig. 10.5. The plaquette-graph configuration defines the loops along which one will propose to flip the orientation of the arrows of the vertex model.

In order to find appropriate probabilities for choosing the breakups, we need to find weights $W(S,G)$ for each of the combinations of spin configuration S on a plaquette and plaquette-graph G shown in Fig. 10.5. We require that

$$\sum_G W(S,G) = W(S) \, , \tag{10.39}$$

where $W(S)$ is the weight of the vertex S, i.e. we subdivide the weight of each spin-configuration on a vertex onto the possible graphs, for example graphs one, four and three if $S = 3$, see Fig. 10.5.

Starting from a vertex configuration S on a plaquette we choose an allowed plaquette-graph with probability

$$P(S \to (S,G)) = \frac{W(S,G)}{W(S)} \, . \tag{10.40}$$

for every vertex-plaquette of the lattice. We then have a configuration of vertices and plaquette-graphs. When a plaquette-graph has been chosen for every plaquette, the combined lines subdivide the lattice into a set of loops. To achieve a constant acceptance rate for the flip of each loop, we require that for a given plaquette-graph G

$$W(S,G) = W(S',G) \, , \tag{10.41}$$

where S' is obtained from S by flipping the arrows of the vertex configuration according the rules of the graph G. That is for $G = 1$ in Fig. 10.5 we require $W(S = 1, G = 1) = W(S = 3, G = 1)$. This equation can be satisfied with weights $W(S,G) = V(G)$ when S and G are compatible and $W(S,G) = 0$ otherwise.

When choosing the heat-bath algorithm for flipping, the probability of flipping the arrows along the loop is given by

$$P\left((S,G) \to (S',G)\right) = \frac{W(S',G)/(W(S,G))}{1 + W(S',G)/(W(S,G))} = \frac{1}{2} \,. \tag{10.42}$$

Thus each loop is flipped with probability 1/2. This generates a new, highly independent, world-line configuration. The previous plaquette-graphs are then discarded, and another update starts with the choice of new plaquette-graphs according to (10.40).

With (10.41) and (10.42) the detailed balance in the space of graphs and spins

$$W(S,G)P[(S,G) \to (S',G)] = W(S',G)P[(S',G) \to (S,G)] \tag{10.43}$$

is trivially satisfied. Detailed balance in the space of spins follows from:

$$\begin{aligned}
&W\left(S\right) P\left(S \to S'\right) \\
&= W(S) \sum_G P[S \to (S,G)] P[(S,G) \to (S',G)] \\
&= \sum_G W\left(S\right) \frac{W\left(S,G\right)}{W(S)} P[(S,G) \to (S',G)] \\
&= \sum_G W(S') \frac{W(S',G)}{W(S')} P[(S',G) \to (S,G)] = W(S') P\left(S' \to S\right) \,.
\end{aligned} \tag{10.44}$$

This completes the formal description of the algorithm. We will now illustrate the algorithm in the case of the isotropic Heisenberg model ($J = J_x = J_z$) since this turns out to be a particularly simple case. Equations (10.39) and (10.41) lead to

$$\begin{aligned}
e^{\Delta\tau J/4} \cosh(\Delta\tau J/2) &\equiv W_1 = W_{1,1} + W_{1,2} + W_{1,3} \\
e^{\Delta\tau J/4} \sinh(\Delta\tau J/2) &\equiv W_2 = W_{2,2} + W_{2,4} + W_{2,3} \\
e^{-\Delta\tau J/4} &\equiv W_3 = W_{3,1} + W_{3,4} + W_{3,3}
\end{aligned} \tag{10.45}$$

with $W_{3,1} = W_{1,1}$, $W_{1,2} = W_{2,2}$ and $W_{2,4} = W_{3,4}$. Here we adopt the notation $W_{i,j} = W(S = i, G = j)$ and $W_i = W(S = i)$. To satisfy the above equations for the special case of the Heisenberg model, we can set $W_{\bullet,3} = W_{\bullet,4} = 0$ and thereby consider only the graphs $G = 1$ and $G = 2$. The reader will readily verify that the equations

$$\begin{aligned}
e^{\Delta\tau J/4} \cosh(\Delta\tau J/2) &\equiv W_1 = W_{1,1} + W_{1,2} \\
e^{\Delta\tau J/4} \sinh(\Delta\tau J/2) &\equiv W_2 = W_{2,2} = W_{1,2} \\
e^{-\Delta\tau J/4} &\equiv W_3 = W_{1,1} = W_{3,1}
\end{aligned} \tag{10.46}$$

are satisfied. We will then only have vertical and horizontal breakups. The probability of choosing a horizontal breakup is $\tanh(\Delta\tau J/2)$ on an antiferromagnetic plaquette (i.e. type one), it is unity on type two, and zero on a ferromagnetic plaquette (type three).

Further aspects of the loop algorithm will be discussed in Sect. 10.3.

10.2.4 The Sign Problem in the World-Line Approach

The QMC approach is often plagued by the so-called sign problem. Since the origin of this problem is easily understood in the framework of the world-line algorithm we will briefly discuss it in this section on a specific model. Consider spinless electrons on an L-site linear chain

$$H = -t \sum_i c_i^\dagger \left(c_{i+1} + c_{i+2} \right) + \text{H.c.} \tag{10.47}$$

with $\{c_i^\dagger, c_j^\dagger\} = \{c_i, c_j\} = 0, \{c_i^\dagger, c_j\} = \delta_{i,j}$. Here, we consider periodic boundary conditions, $c_{i+L} = c_i$ and $t > 0$.

The world-line representation of spinless fermions is basically the same as that of spin-$1/2$ degrees of freedom (which themselves are equivalent to so-called hard-core bosons) on any lattice. For fermions, world lines stand for occupied locations in space-time. Additional signs occur when fermion world lines wind around each other, as we will now discuss.

For the above Hamiltonian it is convenient to split it into a set of independent four site problems

$$H = \underbrace{\sum_{n=0}^{L/4-1} H^{(4n+1)}}_{H_1} + \underbrace{\sum_{n=0}^{L/4-1} H^{(4n+3)}}_{H_2} \tag{10.48}$$

with $H^{(i)} = -t c_i^\dagger (c_{i+1}/2 + c_{i+2}) - t c_{i+1}^\dagger (c_{i+2} + c_{i+3}) - t c_{i+2}^\dagger c_{i+3}/2 + \text{H.c.}$. With this decomposition one obtains the graphical representation of Fig. 10.6 [21].

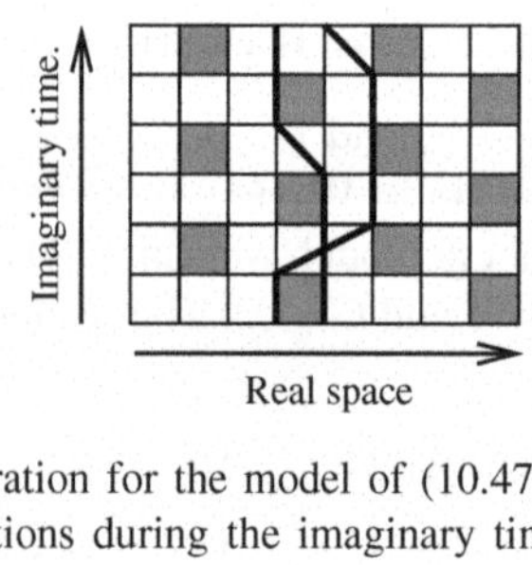

Fig. 10.6. World-line configuration for the model of (10.47). Here $m = 3$. Since the two electrons exchange their positions during the imaginary time propagation, this world-line configuration has a negative weight

The sign problem occurs due to the fact that the weights $\Omega(w)$ are not necessarily positive. An example is shown in Fig. 10.6. In this case the origin of negative signs lies in Fermi statistics. Negative weights cannot be interpreted as probabilities. To circumvent the problem, one decides to carry out the sampling with an auxiliary probability distribution

$$\overline{Pr}(\omega) = \frac{|\Omega(w)|}{\sum_w |\Omega(w)|} , \tag{10.49}$$

which in the limit of small values of $\Delta\tau$ corresponds to the partition function of the Hamiltonian of (10.47) but with fermions replaced by hard-core bosons. Thus, we can now evaluate (10.17) with

$$\langle O \rangle = \frac{\sum_w \overline{Pr}(\omega)\mathrm{sign}(w)O(w)}{\sum_w \overline{Pr}(\omega)\mathrm{sign}(w)} , \tag{10.50}$$

where both the numerator and denominator are evaluated with MC methods. Let us first consider the denominator

$$\langle \mathrm{sign} \rangle = \sum_w \overline{Pr}(\omega)\mathrm{sign}(w) = \frac{\sum_w \Omega(w)}{\sum_w |\Omega(w)|} = \frac{\mathrm{Tr}\left[\mathrm{e}^{-\beta H}\right]}{\mathrm{Tr}\left[\mathrm{e}^{-\beta H_B}\right]} . \tag{10.51}$$

Here, H_B corresponds to the Hamiltonian of (10.47) but with fermions replaced by hard-core bosons. In the limit of large inverse temperatures β the partition functions is dominated by the ground state. Thus in this limit

$$\langle \mathrm{sign} \rangle \sim \mathrm{e}^{-\beta\left(E_0 - E_0^B\right)} = \mathrm{e}^{-\beta L \Delta} , \tag{10.52}$$

where $\Delta = \left(E_0 - E_0^B\right)/L$ is an intensive, in general positive, quantity. The above equation corresponds to the sign problem. When the temperature is small or system size large, the average sign becomes exponentially small. Hence, the observable $\langle O \rangle$ is given by the quotient of two exponentially small values which are determined stochastically. Since $\langle sign \rangle$ is the average of values ± 1, its variance is extremely large. When the error-bars become comparable to the average sign, uncontrolled fluctuations in the evaluation of $\langle O \rangle$ will occur. Two comments are in order:

(i) In this simple example the sign problem occurs due to Fermi statistics. However, sign problems occur equally in frustrated spin-1/2 systems which are nothing but hard-core boson models. Note that replacing the fermions by hard-core bosons in (10.47) and considering hopping matrix elements of different signs between nearest and next-nearest neighbors will generate a sign problem in the above formulation.
(ii) The sign problem is formulation dependent.

In the world-line algorithm, we decide to work in real space. Had we chosen Fourier space, the Hamiltonian would have been diagonal and hence no sign problem would have occurred. In the auxiliary field approach discussed in the next section the sign problem would not occur for this non-interacting problem since one body operators are treated exactly. That is, the sum over all world lines is carried out exactly in that approach.

10.2.5 Single-Hole Dynamics in non Frustrated Quantum Magnets

In this section we describe generalizations of the loop algorithm which allow one to investigate the physics of single-hole motion in non-frustrated quantum magnets [22, 23, 24].

The Hamiltonian we will consider is the t-J model defined as

$$H_{t-J} = \mathcal{P}\Big(-t \sum_{\langle \boldsymbol{i},\boldsymbol{j}\rangle,\sigma} c^{\dagger}_{\boldsymbol{i},\sigma} c_{\boldsymbol{j},\sigma} + \text{H.c.} + J \sum_{\langle \boldsymbol{i},\boldsymbol{j}\rangle} \boldsymbol{S_i} \cdot \boldsymbol{S_j} - \frac{1}{4} n_{\boldsymbol{i}} n_{\boldsymbol{j}} \Big) \mathcal{P} \,. \quad (10.53)$$

The t-J model lives a Hilbert space where double occupancy on a site is excluded. In the above, this constraint is taken care of by the projection

$$\mathcal{P} = \prod_{\boldsymbol{i}} (1 - n_{\boldsymbol{i},\uparrow} n_{\boldsymbol{i},\downarrow}) \,, \quad (10.54)$$

which filters out all states with double occupancy.

To access the single-hole problem, we carry out a canonical transformation to rewrite the fermionic operators, $c^{\dagger}_{\boldsymbol{i},\sigma}$, in terms of spinless fermions and spin $1/2$ degrees of freedom. On a given site the product space of a spinless fermion and a spin $1/2$ degree of freedom consists of four states

$$|n, \sigma\rangle \equiv |n\rangle \otimes |\sigma\rangle \quad (10.55)$$

with $n = 0, 1$ and $\sigma = \uparrow, \downarrow$, on which the fermionic

$$\{f^{\dagger}, f\} = 1, \{f^{\dagger}, f^{\dagger}\} = \{f^{\dagger}, f^{\dagger}\} = 0 \,, \quad (10.56)$$

and spin $1/2$ operators,

$$\left[\sigma^{\alpha}, \sigma^{\beta}\right] = 2\mathrm{i} \sum_{\gamma} \epsilon^{\alpha,\beta,\gamma} \sigma^{\gamma} \quad (10.57)$$

act.

We can identify the four fermionic states on a given site to the four states in the product space of spinless fermions and spins as:

$$\begin{aligned} |\uparrow\rangle = c^{\dagger}_{\uparrow}|0\rangle &\leftrightarrow |1, \uparrow\rangle = |\text{vac}\rangle \,, \\ |\downarrow\rangle = c^{\dagger}_{\downarrow}|0\rangle &\leftrightarrow |1, \downarrow\rangle = \sigma^{-}|\text{vac}\rangle \,, \\ |0\rangle &\leftrightarrow |0, \uparrow\rangle = f^{\dagger}|\text{vac}\rangle \,, \\ |\downarrow\uparrow\rangle = c^{\dagger}_{\downarrow} c^{\dagger}_{\uparrow}|0\rangle &\leftrightarrow |0, \downarrow\rangle = f^{\dagger}\sigma^{-}|\text{vac}\rangle \,. \end{aligned} \quad (10.58)$$

Here $\sigma^- = (\sigma^x - \mathrm{i}\sigma^y)/2$ and $\sigma^+ = (\sigma^x + \mathrm{i}\sigma^y)/2$. The fermionic operators $(c_\sigma^\dagger)$ are identified as

$$
\begin{aligned}
c_\uparrow^\dagger &\leftrightarrow \sigma^{z,+} f - \sigma^{z,-} f^\dagger , \\
c_\downarrow^\dagger &\leftrightarrow \sigma^- \left(f^\dagger + f\right) .
\end{aligned}
\tag{10.59}
$$

with $\sigma^{z,\pm} = (1 \pm \sigma^z)/2$. Under the above canonical transformation the t-J model reads

$$
\begin{aligned}
\widetilde{H}_{t-J} &= \widetilde{\mathcal{P}}\Big(t \sum_{\langle i,j\rangle} [f_j^\dagger f_i \widetilde{P}_{i,j} + \mathrm{H.c.}] + \frac{J}{2} \sum_{\langle i,j\rangle} (\widetilde{P}_{i,j} - 1)\widetilde{\Delta}_{i,j}\Big)\widetilde{\mathcal{P}} , \\
\widetilde{P}_{i,j} &= \frac{1}{2}\left(\boldsymbol{\sigma}_i \cdot \boldsymbol{\sigma}_j + 1\right) , \\
\widetilde{\Delta}_{i,j} &= 1 - f_i^\dagger f_i - f_j^\dagger f_j , \\
\widetilde{\mathcal{P}} &= \prod_i \left(1 - f_i^\dagger f_i \sigma_i^- \sigma_i^+\right) .
\end{aligned}
\tag{10.60}
$$

We can check the validity of the above expression by considering the two-site problem $H_{t-J}^{(i,j)}$. Applying the Hamiltonian on the four states in the projected Hilbert space with two electrons gives

$$
\begin{aligned}
H_{t-J}^{(i,j)} \mid\uparrow\rangle_i \otimes \mid\uparrow\rangle_j &= 0 , \\
H_{t-J}^{(i,j)} \mid\downarrow\rangle_i \otimes \mid\downarrow\rangle_j &= 0 , \\
H_{t-J}^{(i,j)} \mid\uparrow\rangle_i \otimes \mid\downarrow\rangle_j &= \mathcal{P}\Big(- t|0\rangle_i \otimes \mid\uparrow\downarrow\rangle_j - t\mid\uparrow\downarrow\rangle_i \otimes |0\rangle_j \\
&\quad - \frac{J}{2}\mid\uparrow\rangle_i \otimes \mid\downarrow\rangle_j + \frac{J}{2}\mid\downarrow\rangle_i \otimes \mid\uparrow\rangle_j\Big) \\
&= -\frac{J}{2}\mid\uparrow\rangle_i \otimes \mid\downarrow\rangle_j + \frac{J}{2}\mid\downarrow\rangle_i \otimes \mid\uparrow\rangle_j , \\
H_{t-J}^{(i,j)} \mid\downarrow\rangle_i \otimes \mid\uparrow\rangle_j &= \mathcal{P}\Big(- t|0\rangle_i \otimes \mid\downarrow\uparrow\rangle_j - t\mid\downarrow\uparrow\rangle_i \otimes |0\rangle_j \\
&\quad - \frac{J}{2}\mid\downarrow\rangle_i \otimes \mid\uparrow\rangle_j + \frac{J}{2}\mid\uparrow\rangle_i \otimes \mid\downarrow\rangle_j\Big) \\
&= -\frac{J}{2}\mid\downarrow\rangle_i \otimes \mid\uparrow\rangle_j + \frac{J}{2}\mid\uparrow\rangle_i \otimes \mid\downarrow\rangle_j .
\end{aligned}
\tag{10.61}
$$

As apparent, starting from a state in the projected Hilbert space the kinetic energy term generates states with double occupancy which have to be projected out. In other words the projection operator does not commute with the kinetic energy. We can now check that one obtains the same result in the representation in terms of spinless fermions and spins. The above equations respectively read

$$
\begin{aligned}
\widetilde{H}^{(i,j)}_{t-J}\ |1,\uparrow\rangle_i \otimes |1,\uparrow\rangle_j &= 0\,, \\
\widetilde{H}^{(i,j)}_{t-J}\ |1,\downarrow\rangle_i \otimes |1,\downarrow\rangle_j &= 0\,, \\
\widetilde{H}^{(i,j)}_{t-J}\ |1,\uparrow\rangle_i \otimes |1,\downarrow\rangle_j &= \widetilde{\mathcal{P}}\Big(-\frac{J}{2}|1,\uparrow\rangle_i \otimes |1,\downarrow\rangle_j + \frac{J}{2}|1,\downarrow\rangle_i \otimes |1,\uparrow\rangle_j\Big) \\
&= -\frac{J}{2}|1,\uparrow\rangle_i \otimes |1,\downarrow\rangle_j + \frac{J}{2}|1,\downarrow\rangle_i \otimes |1,\uparrow\rangle_j\,, \\
\widetilde{H}^{(i,j)}_{t-J}\ |1,\downarrow\rangle_i \otimes |1,\uparrow\rangle_j &= \widetilde{\mathcal{P}}\Big(-\frac{J}{2}|1,\downarrow\rangle_i \otimes |1,\uparrow\rangle_j + \frac{J}{2}|1,\uparrow\rangle_i \otimes |1,\downarrow\rangle_j\Big) \\
&= -\frac{J}{2}|1,\downarrow\rangle_i \otimes |1,\uparrow\rangle_j + \frac{J}{2}|1,\uparrow\rangle_i \otimes |1,\downarrow\rangle_j\,,
\end{aligned}
\tag{10.62}
$$

which confirms that the matrix elements of $\widetilde{H}^{(i,j)}_{t-J}$ are identical to those of $H^{(i,j)}_{t-J}$. The reader can readily carry out the calculation in the one and zero particle Hilbert spaces to see that: $\langle\eta|H^{(i,j)}_{t-J}|\nu\rangle = \langle\widetilde{\eta}|\widetilde{H}^{(i,j)}_{t-J}|\widetilde{\nu}\rangle$, where $|\nu\rangle$ ($|\eta\rangle$) and $|\widetilde{\nu}\rangle$ ($|\eta\rangle$) correspond to the same states but in the two different representations. Since the t-J model may be written as a sum of two-sites terms, the above is equivalent to

$$
\langle\eta|H_{t-J}|\nu\rangle = \langle\widetilde{\eta}|\widetilde{H}_{t-J}|\widetilde{\nu}\rangle\,. \tag{10.63}
$$

In the representation of (10.61) the t-J model has two important properties which facilitate numerical simulations:

(i) As apparent from (10.62) the application of the Hamiltonian (without projection) on a state in the projected Hilbert space does not generate states with double occupancy. Hence, the projection operation commutes with the Hamiltonian in this representation. The reader can confirm that this is a statement which holds in the full Hilbert space. This leads to the relation

$$
\Big[t\sum_{\langle i,j\rangle}[f_j^\dagger f_i\widetilde{P}_{i,j} + \text{H.c.}] + \frac{J}{2}\sum_{\langle i,j\rangle}(\widetilde{P}_{i,j}-1)\widetilde{\Delta}_{i,j}\,,\ \widetilde{P}\Big] = 0\,, \tag{10.64}
$$

which states that the projection operator is a conserved quantity.

(ii) The Hamiltonian is bilinear in the spinless fermion operators. This has the consequence that for a fixed spin configuration the spinless fermion may be integrated out.

We now use those two properties to study the problem of single-hole dynamics in un-frustrated quantum magnets. Single-hole dynamics is determined by the Green function. In this section we will define it as

$$
G(\boldsymbol{i}-\boldsymbol{j},\tau) = \langle c^\dagger_{\boldsymbol{i},\uparrow}(\tau)c_{\boldsymbol{j},\uparrow}\rangle = \frac{1}{Z}\text{Tr}\left[\mathrm{e}^{-(\beta-\tau)H}c^\dagger_{\boldsymbol{i},\uparrow}\mathrm{e}^{-\tau H}c_{\boldsymbol{j},\uparrow}\right]\,, \tag{10.65}
$$

where the trace runs over the Hilbert space with no holes. In the representation of (10.61) the above equation reads

$$
G(\boldsymbol{r},\tau) = \langle\sigma^{z,+}_{\boldsymbol{i}}(\tau)f_{\boldsymbol{i}}(\tau)\sigma^{z,+}_{\boldsymbol{j}}f^\dagger_{\boldsymbol{j}}\rangle\,. \tag{10.66}
$$

To use the world-line formulation to the present problem, we introduce the unit operator in the Hilbert space with no holes

$$1 = \sum_{\sigma} |\boldsymbol{v}, \boldsymbol{\sigma}\rangle\langle \boldsymbol{v}, \boldsymbol{\sigma}| \, , \quad |\boldsymbol{v}, \boldsymbol{\sigma}\rangle = |1, \sigma_1\rangle_1 \otimes |1, \sigma_2\rangle_2 \otimes \ldots \otimes |1, \sigma_N\rangle_N \, , \quad (10.67)$$

as well as the unit operator in the Hilbert space with a single hole

$$1 = \sum_{\boldsymbol{r}, \boldsymbol{\sigma}} |\boldsymbol{r}, \boldsymbol{\sigma}\rangle\langle \boldsymbol{v}, \boldsymbol{r}| \, , \quad |\boldsymbol{r}, \boldsymbol{\sigma}\rangle = \sigma_{\boldsymbol{r}}^{z,+} f_{\boldsymbol{r}}^{\dagger} |\boldsymbol{v}, \boldsymbol{\sigma}\rangle \, . \quad (10.68)$$

In the above, $\boldsymbol{r}$ denotes a lattice site and N corresponds to the number of lattice sites. In the definition of the single hole-states, the operator $\sigma_{\boldsymbol{r}}^{z,+}$ guarantees that we will never generate a doubly occupied state on site $\boldsymbol{r}$ (i.e. $|0, \downarrow\rangle$).

The Green function may now be written as

$$\begin{aligned} G(\boldsymbol{i} - \boldsymbol{j}, \tau) &= \frac{1}{Z} \sum_{\boldsymbol{\sigma}_1} \langle \boldsymbol{v}, \boldsymbol{\sigma}_1 | \left(\mathrm{e}^{-\Delta\tau H_1} \mathrm{e}^{-\Delta\tau H_2} \right)^{m - n_\tau} \sigma_{\boldsymbol{i}}^{z,+} f_{\boldsymbol{i}} \\ &\times \left(\mathrm{e}^{-\Delta\tau H_1} \mathrm{e}^{-\Delta\tau H_2} \right)^{n_\tau} \sigma_{\boldsymbol{j}}^{z,+} f_{\boldsymbol{j}}^{\dagger} |\boldsymbol{v}, \boldsymbol{\sigma}_1\rangle \\ &= \frac{1}{Z} \sum_{\substack{\boldsymbol{\sigma}_1 \ldots \boldsymbol{\sigma}_{2m} \\ \boldsymbol{r}_2 \ldots \boldsymbol{r}_{2n_\tau}}} \langle \boldsymbol{v}, \boldsymbol{\sigma}_1 | \mathrm{e}^{-\Delta\tau H_1} | v, \boldsymbol{\sigma}_{2m} \rangle \\ &\times \langle \boldsymbol{v}, \boldsymbol{\sigma}_{2m-1} | \mathrm{e}^{-\Delta\tau H_2} | \boldsymbol{v}, \boldsymbol{\sigma}_{2m-2} \rangle \cdots \langle \boldsymbol{v}, \boldsymbol{\sigma}_{2n_\tau+1} | \sigma_{\boldsymbol{i}}^{z,+} f_{\boldsymbol{i}} \mathrm{e}^{-\Delta\tau H_1} | \boldsymbol{r}_{2n_\tau}, \boldsymbol{\sigma}_{2n_\tau} \rangle \\ &\times \langle \boldsymbol{r}_{2n_\tau}, \boldsymbol{\sigma}_{2n_\tau} | x \mathrm{e}^{-\Delta\tau H_2} | \boldsymbol{r}_{2n_\tau - 1}, \boldsymbol{\sigma}_{2n_\tau - 1} \rangle \cdots \langle \boldsymbol{r}_2, \boldsymbol{\sigma}_2 | \mathrm{e}^{-\Delta\tau H_2} \sigma_{\boldsymbol{j}}^{z,+} f_{\boldsymbol{j}}^{\dagger} | \boldsymbol{v}, \boldsymbol{\sigma}_1 \rangle \\ &= \frac{\sum_w \Omega(w) G_w(\boldsymbol{i} - \boldsymbol{j}, \tau)}{\sum_w \Omega(w)} \, . \end{aligned} \quad (10.69)$$

The following comments are in order:

(i) We have neglected the controlled systematic error of order $(\Delta\tau)^2$.
(ii) $n_\tau \Delta\tau = \tau$ and $m\Delta\tau = \beta$.
(iii) w denotes a world-line configuration defined by the set of spin states $|\sigma_1\rangle \ldots |\sigma_{2m}\rangle$. The Boltzmann weight of this state is given by

$$\Omega(w) = \langle \boldsymbol{v}, \boldsymbol{\sigma}_1 | \mathrm{e}^{-\Delta\tau H_1} | \boldsymbol{v}, \boldsymbol{\sigma}_{2m} \rangle \ldots \langle \boldsymbol{v}, \boldsymbol{\sigma}_2 | \mathrm{e}^{-\Delta\tau H_2} | \boldsymbol{v}, \boldsymbol{\sigma}_1 \rangle \quad (10.70)$$

such that $Z = \sum_w \Omega(w)$ in the partition function of the Heisenberg model.
(iv) The Green function for a given world-line configuration (w) reads

$$\begin{aligned} G_w(\boldsymbol{i} - \boldsymbol{j}, \tau) = \sum_{\boldsymbol{r}_{2n_\tau} \ldots \boldsymbol{r}_2} & \frac{\langle \boldsymbol{v}, \boldsymbol{\sigma}_{2n_\tau+1} | \sigma_{\boldsymbol{i}}^{z,+} f_{\boldsymbol{i}} \mathrm{e}^{-\Delta\tau H_1} | \boldsymbol{r}_{2n_\tau}, \boldsymbol{\sigma}_{2n_\tau} \rangle \cdots}{\langle \boldsymbol{v}, \boldsymbol{\sigma}_{2n_\tau+1} | \mathrm{e}^{-\Delta\tau H_1} | \boldsymbol{v}, \boldsymbol{\sigma}_{2n_\tau} \rangle \cdots} \\ & \times \frac{\cdots \langle \boldsymbol{r}_2, \boldsymbol{\sigma}_2 | \mathrm{e}^{-\Delta\tau H_2} \sigma_{\boldsymbol{j}}^{z,+} f_{\boldsymbol{j}}^{\dagger} | \boldsymbol{v}, \boldsymbol{\sigma}_1 \rangle}{\cdots \langle \boldsymbol{v}, \boldsymbol{\sigma}_2 | \mathrm{e}^{-\Delta\tau H_2} | \boldsymbol{v}, \boldsymbol{\sigma}_1 \rangle} \, . \end{aligned} \quad (10.71)$$

Defining

$$
\begin{aligned}
[A_1(\sigma_2,\sigma_1)]_{\boldsymbol{r},\boldsymbol{j}} &= \frac{\langle v,\sigma_2|f_{\boldsymbol{r}}\sigma_{\boldsymbol{r}}^{z,+}\mathrm{e}^{-\Delta\tau H_1}\sigma_{\boldsymbol{j}}^{z,+}f_{\boldsymbol{j}}^{\dagger}|v,\sigma_1\rangle}{\langle v,\sigma_2|\mathrm{e}^{-\Delta\tau H_1}|v,\sigma_1\rangle}\,,\\
[A_2(\sigma_2,\sigma_1)]_{\boldsymbol{r},\boldsymbol{j}} &= \frac{\langle v,\sigma_2|f_{\boldsymbol{r}}\sigma_{\boldsymbol{r}}^{z,+}\mathrm{e}^{-\Delta\tau H_2}\sigma_{\boldsymbol{j}}^{z,+}f_{\boldsymbol{j}}^{\dagger}|v,\sigma_1\rangle}{\langle v,\sigma_2|\mathrm{e}^{-\Delta\tau H_2}|v,\sigma_1\rangle}
\end{aligned}
\tag{10.72}
$$

and since the single-hole states are given by $|\boldsymbol{r},\boldsymbol{\sigma}\rangle = \sigma_{\boldsymbol{r}}^{z,+}f_{\boldsymbol{r}}^{\dagger}|\boldsymbol{v},\boldsymbol{\sigma}\rangle$, the Green function for a given world-line configuration is given by

$$
\begin{aligned}
G_w(\boldsymbol{i}-\boldsymbol{j},\tau) = \big[A_1(\sigma_{2n_\tau+1},\sigma_{2n_\tau})A_2(\sigma_{2n_\tau},\sigma_{2n_\tau+1})\cdots\\
\cdots A_1(\sigma_3,\sigma_2)A_2(\sigma_2,\sigma_1)\big]_{\boldsymbol{i},\boldsymbol{j}}\,.
\end{aligned}
\tag{10.73}
$$

We are now left with the task of computing the matrix A. Since H_2 is a sum of commuting bond Hamiltonians (H_b) $[A_1(\boldsymbol{\sigma}_3,\boldsymbol{\sigma}_2)]_{\boldsymbol{i},\boldsymbol{j}}$ does not vanish only if $\boldsymbol{i}$ and $\boldsymbol{j}$ belong to the same bond $\tilde{b}$. In particular, denoting the two-spin configuration on bond b by $\boldsymbol{\sigma}_{1,b},\boldsymbol{\sigma}_{2,b}$ we have

$$
\begin{aligned}
&A_2(\boldsymbol{\sigma}_2,\boldsymbol{\sigma}_1)_{\boldsymbol{i},\boldsymbol{j}}\\
&= \frac{\left[\prod_{b\neq\tilde{b}}\langle\boldsymbol{v},\boldsymbol{\sigma}_{2,b}|\mathrm{e}^{-\Delta\tau H_b}|\boldsymbol{v},\boldsymbol{\sigma}_{1,b}\rangle\right]\langle\boldsymbol{v},\boldsymbol{\sigma}_{2,\tilde{b}}|\sigma_{\boldsymbol{i}}^{z,+}f_{\boldsymbol{i}}\mathrm{e}^{-\Delta\tau H_{\tilde{b}}}\sigma_{\boldsymbol{j}}^{z,+}f_{\boldsymbol{j}}^{\dagger}|\boldsymbol{v},\boldsymbol{\sigma}_{1,\tilde{b}}\rangle}{\prod_b\langle\boldsymbol{v},\boldsymbol{\sigma}_{b,2}|\mathrm{e}^{-\Delta\tau H_b}|\boldsymbol{v},\boldsymbol{\sigma}_{b,1}\rangle}\\
&= \frac{\langle\boldsymbol{v},\boldsymbol{\sigma}_{2,\tilde{b}}|\sigma_{\boldsymbol{i}}^{z,+}f_{\boldsymbol{i}}\mathrm{e}^{-\Delta\tau H_{\tilde{b}}}\sigma_{\boldsymbol{j}}^{z,+}f_{\boldsymbol{j}}^{\dagger}|\boldsymbol{v},\boldsymbol{\sigma}_{1,\tilde{b}}\rangle}{\langle\boldsymbol{v},\boldsymbol{\sigma}_{2,\tilde{b}}|\mathrm{e}^{-\Delta\tau H_{\tilde{b}}}|\boldsymbol{v},\boldsymbol{\sigma}_{1,\tilde{b}}\rangle}\,.
\end{aligned}
\tag{10.74}
$$

Omitting the bond index, the above quantity is given by

$$
\begin{aligned}
A\left(\boldsymbol{\sigma}_2=\uparrow_{\boldsymbol{i}},\uparrow_{\boldsymbol{j}},\boldsymbol{\sigma}_1=\uparrow_{\boldsymbol{i}},\uparrow_{\boldsymbol{j}}\right) &= \begin{pmatrix}\cosh(-\Delta\tau t) & \sinh(-\Delta\tau t)\\ \sinh(-\Delta\tau t) & \cosh(-\Delta\tau t)\end{pmatrix}_{\boldsymbol{ij}}\\
A\left(\boldsymbol{\sigma}_2=\downarrow_{\boldsymbol{i}},\downarrow_{\boldsymbol{j}},\boldsymbol{\sigma}_1=\downarrow_{\boldsymbol{i}},\downarrow_{\boldsymbol{j}}\right) &= \begin{pmatrix}0 & 0\\ 0 & 0\end{pmatrix}_{\boldsymbol{ij}}\\
A\left(\boldsymbol{\sigma}_2=\downarrow_{\boldsymbol{i}},\uparrow_{\boldsymbol{j}},\boldsymbol{\sigma}_1=\downarrow_{\boldsymbol{i}},\uparrow_{\boldsymbol{j}}\right) &= \begin{pmatrix}0 & 0\\ 0 & \frac{\cosh(-\Delta\tau t)}{\mathrm{e}^{\Delta\tau J/2}\cosh(\Delta\tau J/2)}\end{pmatrix}_{\boldsymbol{ij}}\\
A\left(\boldsymbol{\sigma}_2=\uparrow_{\boldsymbol{i}},\downarrow_{\boldsymbol{j}},\boldsymbol{\sigma}_1=\uparrow_{\boldsymbol{i}},\downarrow_{\boldsymbol{j}}\right) &= \begin{pmatrix}\frac{\cosh(-\Delta\tau t)}{\mathrm{e}^{\Delta\tau J/2}\cosh(\Delta\tau J/2)} & 0\\ 0 & 0\end{pmatrix}_{\boldsymbol{ij}}\\
A\left(\boldsymbol{\sigma}_2=\downarrow_{\boldsymbol{i}},\uparrow_{\boldsymbol{j}},\boldsymbol{\sigma}_1=\uparrow_{\boldsymbol{i}},\downarrow_{\boldsymbol{j}}\right) &= \begin{pmatrix}0 & 0\\ \frac{\sinh(-\Delta\tau t)}{-\mathrm{e}^{\Delta\tau J/2}\sinh(\Delta\tau J/2)} & 0\end{pmatrix}_{\boldsymbol{ij}}\\
A\left(\boldsymbol{\sigma}_2=\uparrow_{\boldsymbol{i}},\downarrow_{\boldsymbol{j}},\boldsymbol{\sigma}_1=\downarrow_{\boldsymbol{i}},\uparrow_{\boldsymbol{j}}\right) &= \begin{pmatrix}0 & \frac{\sinh(-\Delta\tau t)}{-\mathrm{e}^{\Delta\tau J/2}\sinh(\Delta\tau J/2)}\\ 0 & 0\end{pmatrix}_{\boldsymbol{ij}}
\end{aligned}
\tag{10.75}
$$

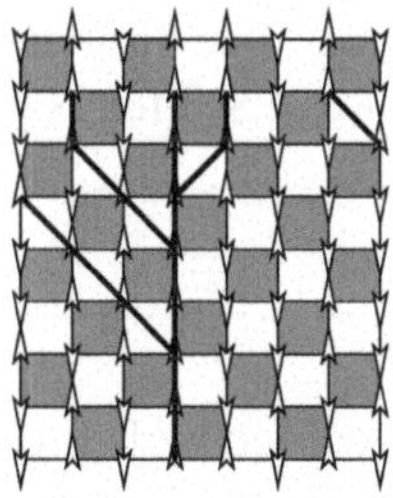

Fig. 10.7. Graphical representation of the propagation of a hole in a given world-line or spin configuration. The *solid lines* denotes the possible routes taken by the hole through the spin configuration. One will notice that due to the constraint which inhibits the states $|0, \downarrow\rangle$ the hole motion tracks the up spins

The possible paths the hole follows for a given spin configuration is shown in Fig. 10.7. With the above construction, a loop algorithm for a given non-frustrated spin system in arbitrary dimensions may be quickly generalized to tackle the important problem of single-hole dynamics in quantum magnets.

10.3 World-Line Representations without Discretization Error

The Trotter discretization of imaginary time which was used in the preceding section is conceptually easy. It was historically the first approach, but has some notable disadvantages:

- In order to obtain reliable results, one has to perform calculations at several different small values of $\Delta\tau$ and to extrapolate to $\Delta\tau = 0$.
- In practice, this extrapolation is often skipped, and instead a *small* value like $\Delta\tau = 1/32$ or $1/20$ (or even larger) is used, which implies unknown systematic discretization errors.
- Small values of $\Delta\tau$ imply a large number $L = \beta/\Delta\tau$ of time slices, so that the computer time needed for each sweep through the lattice increases like $1/\Delta\tau$. In addition, the correlation length in imaginary time, measured in time slices, grows like $1/\Delta\tau$, so that autocorrelation times for local algorithms typically grow with another factor of $(1/\Delta\tau)^2$.

Fortunately, it has been found in recent years, independently by a number of authors, that one can overcome the Trotter discretization error entirely. We will describe the most common representations: Continuous imaginary time and the stochastic series expansion.

Note that such representations of $\exp(-\beta H)$ are all world-line like. They are almost independent of the algorithm used to update the world-line configurations! That is, there are local and loop-updates both in imaginary time and in the SSE representation.

A number of other methods without time discretization errors have been developed in recent years in different contexts. See for example [8, 9, 19, 25, 26, 27, 28, 29] and Chaps. 11 and 12.

10.3.1 Limit of Continuous Time

In the context of QMC, it was first realized by Beard and Wiese [30] that the limit $\Delta\tau \to 0$ can be explicitly taken within the loop algorithm. Actually this applies to any model with a discrete state space, see Sect. 10.3.3. Let us look again at the isotropic Heisenberg AF, (10.1) with $J = J_z = J_x$. There are then only vertical and horizontal breakups in the loop algorithm.

To lowest order in $\Delta\tau$, the probability for a horizontal breakup is $J\Delta\tau/2$, proportional to $\Delta\tau$, and the probability for a vertical breakup is $1 - J\Delta\tau/2$. This is like a discrete Poisson process: The event of a horizontal breakup occurs with probability $J\Delta\tau/2$. Note that the vertical breakup does not change the world-line configuration; it is equivalent to the identity operator, see also Sect. 10.4. In the limit $\Delta\tau \to 0$ the Poisson process becomes a Poisson distribution in continuous imaginary time, with probability density $J/2$ for a horizontal breakup.

In continuous imaginary time there are no plaquettes anymore. Instead, configurations are specified by the space and time coordinates of the events, together with the local spin values. On average, there will be about one event per unit of βJ on each lattice bond. Therefore the storage requirements are reduced by $O(1/\Delta\tau)$! The events are best stored as linked lists, i.e. for each event on a bond there should be pointers to the events closest in imaginary time, for both sites of the bond.

Monte Carlo Loop updates are implemented quite differently for the multi-loop and for the single-loop variant, respectively. For multi-loop updates, i.e. the construction and flip of loops for every space-time site of the lattice, one first constructs a stochastic loop decomposition of the world-line configuration. To do so, horizontal breakups are put on the lattice with constant probability density in imaginary time for each bond, but only in time regions where they are compatible with the world-line configuration, i.e. where the spins are antiferromagnetic. Horizontal breakups must also be put wherever a world-line jumps to another site. The linked list has to be updated or reconstructed. The configuration of breakups is equivalent to a configuration of loops, obtained by vertically connecting the horizontal breakups (see Sect. 10.4). These implicitly given loops then have to be flipped with some constant probability, usually $1/2$. To do so, one can for example go to each stored event (breakup) and find, and possibly flip, the one or two loops through this breakup, unless these loop(s) have already been treated.

In single-loop-updates only one single loop is constructed and then always flipped. Here it is better to make the breakup-decisions during loop construction, see also Sect. 10.4.1). One starts at a randomly chosen space-time site (i, t_0). The loop is constructed piece by piece. It thus has a tail and a moving head. The partial loop can be called a worm (cf. Sect. 10.4.5). The loop points into the present spin-direction, say upwards in time.

For each lattice bond $\langle ij \rangle$ at the present site, the smallest of the following times is determined:

(i) The time at which the neighboring spin changes;
(ii) If the bond is antiferromagnetic, the present time t_0 plus a decay time generated with uniform probability density;
(iii) The time at which the spin at site i changes.

The loop head is moved to the smallest of all these times, t_1. Existing breakups between t_0 and t_1 are removed. If t_1 corresponds to case (ii) or (i), a breakup is inserted there, and the loop head follows it, i.e. it moves to the neighboring site and changes direction in imaginary time. Then the construction described in the present paragraph repeats.

It finishes when the loop has closed. All spins along the loop can then be flipped.

10.3.2 Stochastic Series Expansion (SSE)

The stochastic series expansion (SSE), invented by A. Sandvik [31, 32, 33] is another *representation* of $\exp(-\beta H)$ without discretization error. Note that it is *not* directly connected to any particular MC-update. Most update methods can (with some adjustments) be applied either in imaginary time or in the SSE representation.

Let the Hamiltonian be a sum of operators defined on lattice bonds

$$H = -\sum_{b}^{m_b} H_b \tag{10.76}$$

like in the nearest-neighbor Heisenberg model. The operators H_b need to be non-branching, in some basis, i.e. for each basis state $|i\rangle$, $H_b|i\rangle$ is proportional to a single basis state. All diagonal matrix elements of these operators need to be positive in order to avoid a sign problem. For the XXZ Heisenberg model one can for example use the bond operators $(S_i^+ S_j^- + S_i^- S_j^+)/2$ and $1/4 - S_i^z S_j^z$ for each bond $\langle ij \rangle$. We write the series expansion

$$\begin{aligned}
\exp(-\beta H) &= \sum_n \frac{\beta^n}{n!} (-H)^n \\
&= \sum_n \frac{\beta^n}{n!} (H_1 + H_2 + \ldots)(H_1 + H_2 + \ldots) \ldots \\
&= \sum_n \frac{\beta^n}{n!} \sum_{S_n} H_{i_1} H_{i_2} H_{i_3} \ldots ,
\end{aligned}$$

where $\sum_{S_n}$ extends over all sequences $(i_1, i_2, ..., i_n)$ of indices $i_\alpha \in \{1, 2, \ldots, m_b\}$ labelling the operators H_b. When we compute the trace $\mathrm{Tr}[\exp(-\beta H)] = \sum_i \langle i| \exp(-\beta H)|i\rangle$, the initial state $|i\rangle$ is modified in turn by each of the H_b, each time resulting in another basis state. For the XXZ-model and spin-S^z basis states, a world-line like configuration results again, but with a discrete timelike index

$\alpha = 1, 2, \ldots, n$, and only one event per value of the index. The remaining matrix elements can be evaluated easily. With suitable normalizations of the operators H_b, they can usually be made to be unity. They are zero for operator configurations which are not possible, e.g. not compatible with periodic world lines, which will thus not be produced in the Monte Carlo. Spins at sites not connected by any operator to other sites can be summed over immediately.

Note that, in contrast to imaginary time, now *diagonal* operators $S_i^z S_j^z$ occur explicitly, since the exponential factor weighing neighboring world lines has also been expanded in a power series. Thus, SSE needs more operators on average than imaginary time for a given accuracy.

The average length $\langle n \rangle$ of the operator sequence is β times the average total energy (as can be seen from $\partial \log Z/\partial\beta$) and its variance is related to the specific heat. Therefore in any finite length simulation, only a finite value of n of order $\beta\langle -H\rangle$ will occur, so that we get results without discretization error, despite the finiteness of n.

It is convenient to pad the sum in (10.77) with unit operators $\mathbf{1}$ in order to have an operator string of constant length N. For details see [31, 32, 33].

Updates in the SSE representation usually proceed in two steps. First, a diagonal update is performed, for which a switch between diagonal parts of the Hamiltonian, e.g. $S_i^z S_j^z$, and unit operators $\mathbf{1}$ is proposed. This kind of update does not change the shape of world lines. Second, non-diagonal updates are proposed, e.g. local updates analogous to the local updates of world lines in imaginary time, see Sect. 10.2. Loop updates are somewhat different, see Sect. 10.4.

10.3.3 Unified Picture: Interaction Representation

All previous representations, namely discrete and continuous imaginary time, as well as SSE, follow easily from the interaction representation of $\exp(-\beta H)$ [5, 16, 34, 35, 36, 37].

Let $H = H_0 - V$ with H_0 diagonal in the chosen basis. Then the interaction representation is

$$Z = \mathrm{Tr} \sum_{n=0}^{\infty} \mathrm{e}^{(-\beta H_0)} \int_0^{\beta} \mathrm{d}\tau_n \ldots \int_0^{\tau_3} \mathrm{d}\tau_2 \int_0^{\tau_2} \mathrm{d}\tau_1 V(\tau_1) \ldots V(\tau_n) \,, \tag{10.77}$$

where $V(\tau) = \exp(H_0\tau) V \exp(-H_0\tau)$. When the system size and β are finite, this is a convergent expansion.

Indeed, in the form of (10.77), this is already the continuous imaginary time representation of $\exp(-\beta H)$! When the time integrals are approximated by discrete sums, then the discrete time representation results.

The SSE representation can be obtained in the special case that one chooses $H_0 = 0$ and $V = -H = \sum_b^{m_b} H_b$. Then $H(\tau)$ does not depend on τ and the time integrals can be performed

$$\int_0^\beta d\tau_n \ldots \int_0^{\tau_2} d\tau_1 = \frac{\beta^n}{n!} \tag{10.78}$$

and we end up with the *ordered* sequence of operators $H_1 \ldots H_n$ of the SSE representation.

This unified picture has turned out to be very useful [7], by providing a stochastic mapping between SSE and continuous time. Starting with a continuous time configuration, one can just drop the specific times of operators to get to an SSE configuration. Starting with an SSE configuration of n ordered operators, one can draw n times between zero and β uniformly at random, sort them, and assign them to the operators, keeping their order intact. This mapping is useful in order to measure dynamical Greens functions during a simulation that uses the SSE representation. In SSE such a measurement is very costly [38], while in imaginary time it can be done efficiently with FFT.

Interestingly, for the usual representation of the Heisenberg model (10.10)

$$H_{ij} = \frac{1}{2}\left(S_i^+ S_j^- + S_i^- S_j^+\right) + S_i^z S_j^z \,, \tag{10.79}$$

the interaction representation immediately provides the continuous time limit of the discrete time world-line representation, independently of any loops.

One can see the essence of the continuous time limit by looking at the exponential of some operator O with a finite discrete spectrum (state space)

$$\begin{aligned} e^{-\beta(1-JO)} &= \left(e^{(JO-1)\Delta\tau}\right)^{\beta/\Delta\tau} \\ &= \lim_{\Delta\tau\to 0} \left((1-\Delta\tau)\mathbf{1} + \Delta\tau\, J\, O\right)^{\beta/\Delta\tau} \end{aligned} \tag{10.80}$$

The term in brackets can be interpreted as a Poisson process: With probability $\Delta\tau\, J$ choose O, else choose $\mathbf{1}$. Its limit $\Delta\tau \to 0$ is a Poisson distribution in continuous imaginary time, i.e. the operator O occurs with a constant probability density J in imaginary time.

10.4 Loop Operator Representation of the Heisenberg Model

At the root of the loop algorithm there is a representation of the model in terms of loop-operators [4], akin to the Fortuin-Kasteleyn representation of the Ising model [39, 40], and analogous to the Swendsen-Wang algorithm [41, 42], see also Chap. 4. The bond operator of the spin $1/2$ Heisenberg antiferromagnet, with a suitable constant added, is a singlet projection operator

$$-\boldsymbol{S}_i \boldsymbol{S}_j + \frac{1}{4} = \frac{1}{\sqrt{2}}\left(|\uparrow\downarrow\rangle - |\downarrow\uparrow\rangle\right) \frac{1}{\sqrt{2}}\left(\langle\uparrow\downarrow| - \langle\downarrow\uparrow|\right) . \tag{10.81}$$

On a bipartite lattice, the minus signs can be removed by rotating the operators $S^{x,y} \to -S^{x,y}$ on one of the two sublattices. We now denote the operator (10.81) pictorially in terms of contributing spin-configurations, as an operator acting towards a spin configuration at the bottom and producing a new spin configuration on the top. There are four contributing configurations

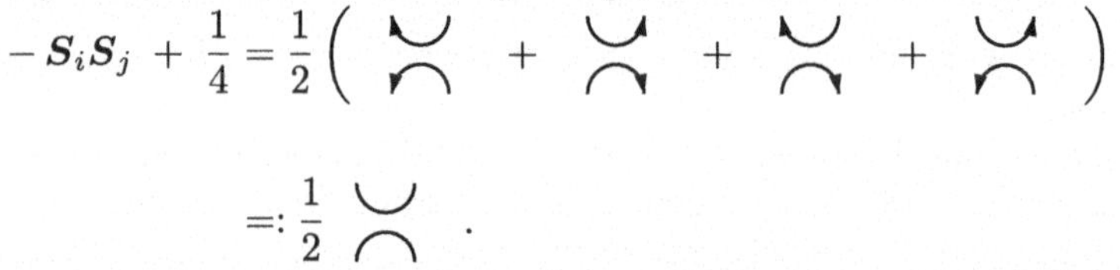

(10.82)

These are just the configurations compatible with the horizontal breakup of the loop algorithm. The horizontal breakup can thus be interpreted as an operator projecting onto a spin singlet. The partition function of the Heisenberg model is then

$$Z = \mathrm{Tr}\,\mathrm{e}^{-\beta H} \sim \mathrm{Tr}\,\mathrm{e}^{\beta J \sum_{\langle ij \rangle} \frac{1}{2}} \, . \tag{10.83}$$

From (10.77) or (10.80) we see that $\exp(-\beta H)$ then corresponds to a Poisson distribution of horizontal breakups (singlet projection operators) with density $J/2$ in imaginary time, on each lattice bond. One instance of such a distribution is shown in Fig. 10.8 on the left.

Taking the trace means to sum over all spin states on the bottom, with periodic boundary conditions in imaginary time. Between operators, the spin states cannot change. The operators can therefore be connected by lines, on which the spin direction does not change. The operator configuration, see Fig. 10.8 (left), therefore implies a configuration of loops, Fig. 10.8 (middle left). A horizontal breakup stands for a sum over two spin directions on each of its half-circles. On each loop the spin direction stays constant along the lines. Thus each loop contributes two states to the partition function. We arrive at the loop representation of the Heisenberg antiferromagnet [4, 43, 44]

$$Z = \int_0^\beta \begin{pmatrix} \text{Poisson distribution of horizontal} \\ \text{breakups with density } J/2 \end{pmatrix} 2^{\text{number of loops}} \, . \tag{10.84}$$

When $J_x \neq J_z$, similar loop representations result [4]. The loop-algorithm moves back and forth between the world-line representation and the operator

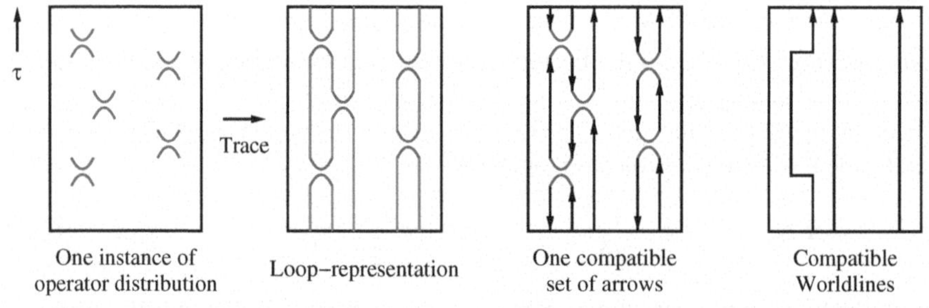

Fig. 10.8. Loop operator representation of the Heisenberg model and of the loop algorithm

representation. From a loop (-operator) configuration we get to a compatible world-line configuration by choosing one direction for each loop, see Fig. 10.8 (middle right and right). We get back to a new operator configuration by choosing one with Poisson probability, and with the constraint that it must be compatible to the current world-line configuration (i.e. operators can only appear where world lines are antiferromagnetic, and they must appear where a world-line jumps).

In the SSE representation, loop updates require only a so-called diagonal update, namely a switch between unit operators and breakups. Once the breakups are defined, the loops just have to be found and flipped. Since there is no second stochastic non-diagonal update step, this has been called, somewhat misleading, a deterministic loop update [45, 46].

10.4.1 Single Loop Updates

An alternative to the multi-loop method just sketched is to construct and flip only a single loop at a time. This is also a valid Monte Carlo method. One could imagine that all breakups and thus all loops were actually constructed, but only a single one of them flipped, see also Sect. 10.3.1. For each update, one starts with a randomly chosen space-time site and follows the spin arrow direction from there. One then constructs just the one loop to which this spin belongs, performing the breakup-decisions on the fly, i.e. the decisions on whether to move vertically in time or to put a horizontal breakup on a neighboring bond and to move there. During this construction, or afterwards, all spins on the loop are flipped. Note that the insertion of a horizontal breakup (Heisenberg spin singlet projection operator) at some place (plaquette in case of discrete time) already determines the path of the loop when and if it should return to the same place again: Either it completes then, or it will take the other half-circle of the horizontal breakup. This behavior is different from the worms and directed loops discussed later.

On average, a single loop constructed this way will be bigger than in the multi-loop variant, since the initial site will on average be more likely on a big loop than on a small one. This usually results in smaller autocorrelation times.

10.4.2 Projector Monte Carlo in Valence Bond Basis

The fact that a horizontal breakup is a singlet projection operator is also at the root of a recent efficient Projector Monte Carlo method [47] for the antiferromagnetic Heisenberg model. Indeed, a cut through a loop configuration, see Fig. 10.8 (middle left) at some imaginary time τ provides a spin state in which each pair of sites that belongs to the same loop is in a spin singlet state.

In the limit of large enough projection time and on a bipartite lattice, all sites will be in such a singlet with probability one. The state is then called an RVB state (resonating valence bond). This is an alternative way to see the famous Lieb-Mattis theorem, namely that the ground state of the Heisenberg antiferromagnet is a global spin singlet.

When one wants to investigate only the ground state, it is sufficient to restrict configurations to an RVB basis, also called valence bond basis [47].

10.4.3 Improved Estimators

The spin directions on different loops are independent. Therefore the contribution of a given loop configuration to the spin Greens function $\langle S^z(x,t)\, S^z(x',t')\rangle$ averages to zero when (x,t) and (x',t') are on different loops, whereas it gets four identical contributions when they are on the same loop [4]. Thus this Greens function can be measured within the loop representation, and it is particularly simple there. For the Heisenberg AF and at momentum π, this Greens function only takes the values zero and one: It is one when (x,t) and (x',t') are on the same loop, and zero otherwise. Thus its variance is smaller than that of $S^z(x,t)\, S^z(x',t')$ in spin representation, which takes values $+1$ and -1. Observables in loop representation such as this Greens function are therefore called improved estimators.

We also see that the Greens function corresponds directly to the space-time size of the loops: These are the physically correlated objects of the model, in the same sense that Fortuin-Kasteleyn clusters are the physically correlated objects of the Ising model [39, 40, 42].

In the loop representation one can also easily measure the off-diagonal Greens function $\langle S^+(x,t)\, S^-(x',t')\rangle$. It is virtually inaccessible in the spin world-line representation with standard local updates, since contributing configurations would require partial world lines, which do not occur there. However, in loop representation, $S^+(x,t)\, S^-(x',t')$ does get a contribution whenever (x,t) and (x',t') are located on the same loop [4]. For the spin-isotropic Heisenberg model, the estimator in loop representation is identical to that of the diagonal correlation function $\langle S^z(x,t)\, S^z(x',t')\rangle$.

10.4.4 Simulations on Infinite Size Lattice

One intriguing application of improved estimators is the possibility to do simulations on an *infinite* size lattice and/or at zero temperature whenever $\langle \boldsymbol{S}(x,t)\, \boldsymbol{S}^z(x',t')\rangle$ goes to zero at infinite distance in space and/or imaginary time, i.e. in an unbroken phase [48].

The idea is to perform single-loop-updates, all starting at the same space-time site (the "origin") instead of at a random point. The lattice of spins is assumed to be infinite in size, but only a finite portion will be needed.

Since the correlation functions go to zero, the size of each single loop will be finite. For a correlation length ξ and gap Δ it will reach spatial distances r with probability $\sim \exp(-r/\xi)$ and temporal distances τ with probability $\sim \exp(-\tau\Delta)$. The maximum distance reached will therefore be finite for any finite number of loops constructed. With each loop flip, the spin configuration is updated. It will eventually equilibrate in the region of space-time that was visited by loops often enough. The updated region is sketched schematically in Fig. 10.9. Since there is no boundary to this region, the physics of the infinite size lattice is simulated. Its properties can be measured in this region, especially the two-point Greens function, which is directly available from the loops.

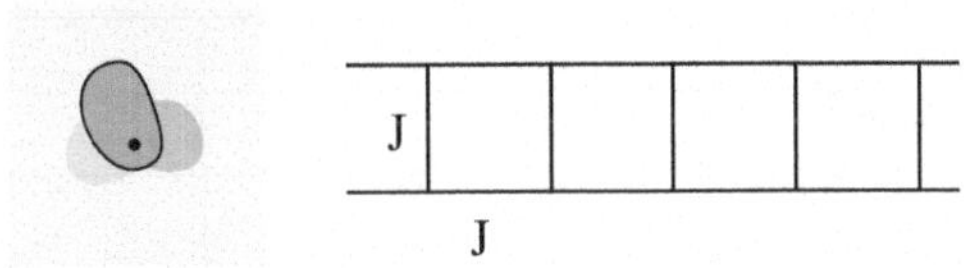

Fig. 10.9. Left: Sketch of regions updated with subsequent loops on an infinite lattice. **Right**: Heisenberg spin ladder with two legs

As an example, let us look at simulations of a Heisenberg spin ladder with $N = 2$ and with $N = 4$ legs, illustrated in Fig. 10.9. The behavior of the infinite size system usually has to be extracted by finite-size scaling from results like those for $L = 10$ and $L = 20$ in Fig. 10.10. Here they result directly, with an effort that here amounted to a few hours on a workstation, similar to a finite lattice simulation at $L = 40$. The asymptotic behavior is exponential, with a correlation length that can directly be measured from the Greens function with high precision.

Similarly, one can measure Greens functions in imaginary time, illustrated in Fig. 10.11, and directly extract the spin gap with high precision from a linear fit to $\log G(q = \pi, \tau)$. The Greens function can be translated to real frequency with the Maximum Entropy technique, resulting in the spectrum shown in Fig. 10.11 on the right.

10.4.5 Worms and Directed Loops

A generalization of single loop updates is provided by worms and directed loops [5, 14, 15, 16, 17, 35]. They are applicable to any model with a world-line like representation. At the same time, they are not cluster algorithms, so that objects like improved estimators are not available.

A single loop (or worm) is constructed iteratively in space-time. The worm-head is a priori allowed to move in any direction, including back-tracking. Each proposal for such a move is accepted or rejected with (e.g.) Metropolis probability. Thus only local updates are needed.

In contrast to the single-loop update of the loop-algorithm, the movement of the worm head is not determined by previous decisions when it crosses its own track.

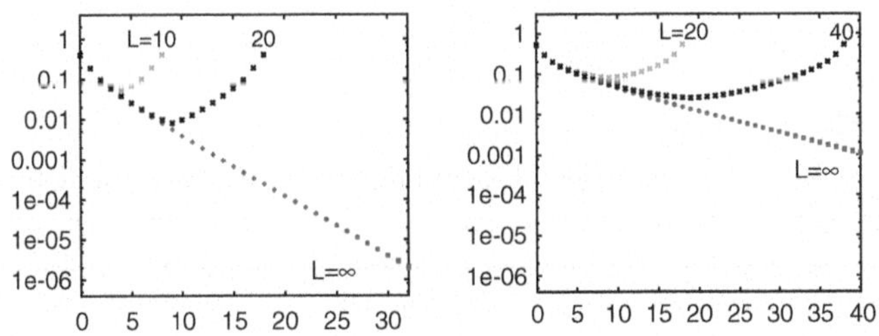

Fig. 10.10. Spatial correlation function of Heisenberg ladders at $\beta = \infty$, for finite systems of finite L and, independently, of $L = \infty$

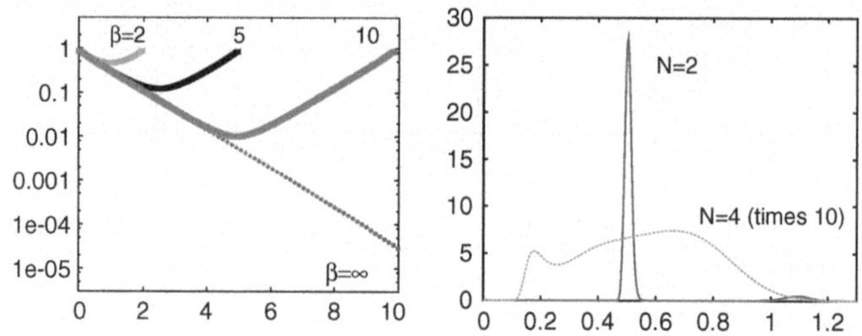

Fig. 10.11. Left: Temporal correlation function (Greens function) of Heisenberg ladders at $L = \infty$, at finite inverse temperatures $\beta = 2, 5, 10$ and, independently, at $\beta = \infty$. **Right**: Real frequency spectrum obtained by Maximum Entropy continuation

The worm algorithm and directed loops differ in details of the updates. Note that, like the loop-algorithm, they also allow the measurement of off-diagonal two-point functions and the change of topological quantum numbers like the number of particles or the spatial winding. In a suitably chosen version of directed loops, single-loop updates of the loop algorithm become a special case. For more information on worms and directed loops we refer to [5, 14, 15, 16, 17, 35].

10.5 Spin-Phonon Simulations

As an example of world-line Monte Carlo calculations we shall discuss recent investigations of the spin-Peierls transition in 1D [7]. Our discussion will also include a new way to simulate phonons which is suitable for any bare phonon dispersion $\omega(q)$.

The model consists of an 1D Heisenberg chain coupled to phonons

$$H = J \sum_{i=1}^{N} \boldsymbol{S}_i \boldsymbol{S}_{i+1} \underbrace{\begin{Bmatrix} 1+g \quad x_i & \text{bond phonons} \\ 1+g\,(x_i - x_{i+1}) & \text{site phonons} \end{Bmatrix}}_{f(\{x_i\})}$$

$$+ \underbrace{\frac{1}{2} \sum_q p_q^2 + \omega^2(q)\, x_q^2}_{H_{\text{ph}}} \; . \tag{10.85}$$

At $T = 0$ there is a quantum phase transition of the Kosterlitz-Thouless type at a critical coupling g_c to a dimerized phase. In this phase the spin-interaction $\boldsymbol{S}_i \boldsymbol{S}_{i+1}$ as well as the phonon coordinate x_i (resp. $x_i - x_{i+1}$) is larger on every second lattice bond, and a spin-gap develops, initially exponentially small [36, 37, 49].

Some of the interesting issues are, see Fig. 10.12:

(i) Does g_c depend on the bare phonon dispersion $\omega(q)$?

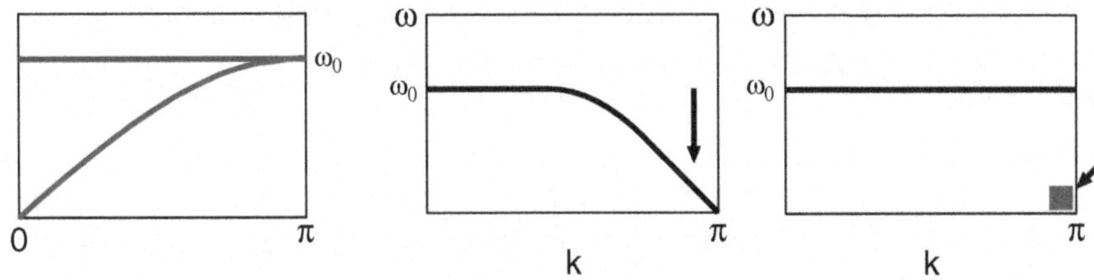

Fig. 10.12. Issues for the spin-Peierls transition. **Left**: Einstein (optical) and acoustical bare phonon dispersions. **Middle**: Softening scenario. **Right**: Central peak scenario

(ii) Is the phonon spectrum beyond the transition softened (i.e. the bare phonon spectrum moves to lower frequency, down to zero at momentum π), or does it have a separate central peak?

10.5.1 Bond Phonons with Einstein Dispersion $\omega(q) = \omega_0$

These phonons are the easiest to treat by QMC. In order to make the quantum phonons amenable to numerical treatment, one can express them with the basic Feynman path integral for each x_i (see Chap. 11), by introducing discrete Trotter times τ_j, inserting complete sets of states $x_i(\tau_j)$ and evaluating the resulting matrix elements to $O(\Delta\tau)$. A simple QMC for the phonon degrees of freedom can then be done with local updates of the phonon world lines $x_i(\tau)$.

A similar approach is possible in second quantization, by inserting complete sets of occupation number eigenstates $n_i(\tau_j)$ at the Trotter times τ_j. Again, one can perform QMC with local updates on the occupation number states [36, 37]. The discrete Trotter time can be avoided here, either with continuous time or with SSE [31, 32, 33].

Such local updates suffer from the usual difficulties of long autocorrelation times, which occur especially close to and beyond the phase transition. They can be alleviated by using parallel tempering [50, 51] (or simulated tempering [52]) (see Chap. 4). In this approach, simulations at many different couplings g (originally: at many temperatures) are run in parallel. Occasionally, a swap of configurations at neighboring g is proposed. It is accepted with Metropolis probability. The goal of this strategy is to have approximately a random walk of configurations in the space of couplings g. Configurations at high g can then equilibrate by first moving to low g, where the Monte Carlo is efficient, and then back to high g. The proper choice of couplings (and of re-weighting factors in case of simulated tempering) depends on the physics of the system and is sometimes cumbersome. It can, however, be automated [7] efficiently by measuring the distributions of energies during an initial run.

The results discussed below were obtained using loop updates for spins and local updates in second quantization for phonons, in SSE representation, similar to [36, 37], with additional automated tempering. Spectra were obtained by mapping the SSE configurations to continuous imaginary time, as explained in Sect. 10.3.3, and measuring Greens functions there using FFT.

The location of the phase transition is best determined through the finite size dependence of a staggered susceptibility, of spins, spin-dimers, or phonons. For spins it reads

$$\chi_S(\pi) = \frac{1}{N}\sum_{n,m}(-1)^m \int_0^\beta \mathrm{d}\tau \left\langle S_n^z(\tau) S_{n+m}^z(0)\right\rangle . \tag{10.86}$$

At the phase transition, $\chi_S(\pi)$ is directly proportional to the system size N, whereas above g_c there are additional logarithmic corrections. Below g_c it is proportional to $\ln N$ for any $g > 0$, i.e. there is a non-extensive central peak in the phonon spectrum for any finite spin-phonon coupling.

The phonon spectra exhibit drastic changes at the phase transition. Figure 10.13 shows that the value of ω_0 determines their qualitative behavior: At $\omega_0 = J$ the central peak becomes extensive and develops a linear branch at the phase transition, which shows the spin-wave velocity. At $\omega_0 = 0.25\,\mathrm{J}$ the behavior is completely different: The bare Einstein dispersion has softened and has joined the previously non-extensive central peak. Thus both the central peak scenario and the softening scenario occur, depending on the size of ω_0.

Note that large system sizes and low temperature are essential to get the correct spectra. The finite size gap of a finite system is of order $1/N$. When $1/N$ is larger than about $\omega_0/10$ (!), then there are drastic finite size effects in the phonon spectrum [7].

At very large values of g, the spin gap Δ_S becomes sizeable. The system enters an adiabatic regime when $\Delta_S > O(\omega_0)$ [49]. For the couplings investigated here, it is always diabatic.

10.5.2 Phonons with arbitrary dispersion $\omega(q)$

Phonons other than those treated in Sect. 10.5.1 have in the past posed great difficulties for QMC. Site phonons have a coupling

$$(1 + g(x_i - x_{i+1}))\, \boldsymbol{S}_i \boldsymbol{S}_{i+1} , \tag{10.87}$$

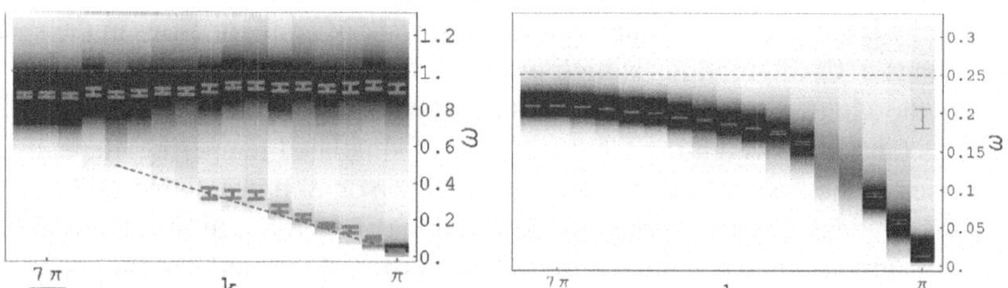

Fig. 10.13. Spectra of phonon coordinates x_i above the phase transition for bond phonons. **Left**: $\omega_0 = 1\,\mathrm{J}$, just above the phase transition. **Right**: $\omega_0 = 0.25\,\mathrm{J}$ at $g = 0.3 > g_c \simeq 0.23$. Lattice size $L = 256$ and $\beta = 512$

which causes a sign problem when second phonon quantization is used. In first quantization, phonon updates are very slow. This is even worse in case of acoustical phonons, which have a zero mode at $q = 0$. Indeed, no efficient QMC method has been available for arbitrary phonon dispersions.

Let us now discuss a new method [7] which overcomes all these difficulties. We use the interaction representation with the pure phonon Hamiltonian as the diagonal part and the spin interaction (10.87) as the interaction part which is expanded. The partition function then reads

$$Z = \mathrm{Tr}_s \sum_{n=0}^{\infty} \sum_{S} \int_0^{\beta} \mathrm{d}\tau_n \ldots \int_0^{\tau_2} \mathrm{d}\tau_1 \int \mathcal{D}x \underbrace{\prod_{l=0}^{n} f(\{x_l\}) S[l]}_{\text{spin operator sequence}} \underbrace{\mathrm{e}^{-\int_0^\beta \mathrm{d}\tau\, H_{\mathrm{ph}}(\{x(\tau)\})}}_{\text{phonon path integral}} . \tag{10.88}$$

Here $S[l]$ is a spin operator like $\boldsymbol{S}_i \boldsymbol{S}_{i+1}$. The spin-phonon coupling $f(\{x(\tau)\})$ is to be evaluated at the space-time location where the spin operators act.

For a given sequence of spin operators we now construct a Monte Carlo phonon update. The effective action S_{eff} for the phonons contains $\log(f(\{x(\tau)\})$. It is therefore not bilinear and cannot be integrated directly. However, for purposes of a Monte Carlo update, we can pretend for a moment that the coupling was $f^{\mathrm{prop}}(x) := \exp(gx)$ instead of $f(x) = 1 + gx$. Then $S_{\mathrm{eff}}^{\mathrm{prop}}$ is bilinear. For a given sequence of spin operators, we can diagonalize $S_{\mathrm{eff}}^{\mathrm{prop}}$ in momentum space and Matsubara frequencies. This results in independent Gaussian distributions of phonon coordinates in the diagonalized basis. We can then generate a new, completely independent phonon configuration by taking one sample from this distribution. In order to achieve a correct Monte Carlo update for the actual model, we take this sample as a Monte Carlo proposal and accept or reject it with Metropolis probability for the actual model, see (10.88).

The acceptance probability will depend on the difference between S_{eff} and $S_{\mathrm{eff}}^{\mathrm{prop}}$, and thus on the typical phonon extensions. In order to achieve high acceptance rates it is advantageous to change phonon configurations only in part of the complete (q, ω_n) space for each update proposal. These parts need to be smaller close to the physically important region ($q = \pi,\ \omega = 0$).

Given a phonon-configuration, the effective model for the spins is a Heisenberg antiferromagnet with couplings that vary in space-time. It can be simulated efficiently with the loop-algorithm, modified for the fact that probabilities are now not constant in imaginary time, but depend on the phonon coordinates.

The approach just sketched works for site phonons as well as for bond phonons. Remarkably, any bare phonon dispersion $\omega(q)$ can be used, since it just appears in the Gaussian effective phonon action. Measurements of phonon properties are easy, since the configurations are directly available in (q, ω_n) space.

Let us now briefly discuss some recent results [7] for site phonons. Their bare dispersion is acoustical, i.e. gapless at $q = 0$. In a recent letter [53] it was concluded that for this model, the critical coupling is $g_c = 0$, i.e. the system supposedly orders

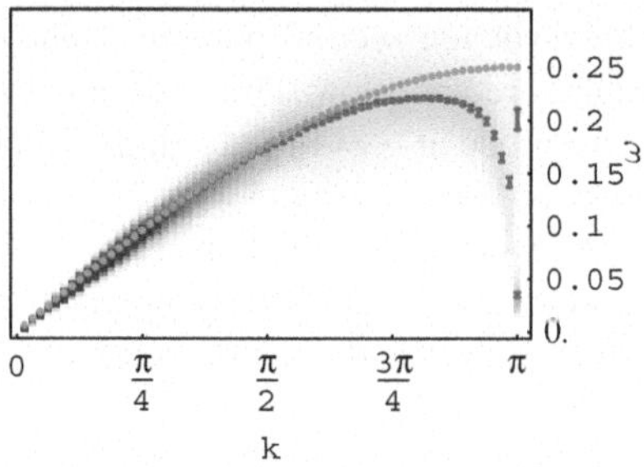

Fig. 10.14. Spectrum of phonon coordinates x_i for acoustical site phonons, at the phase transition

at any finite coupling. However, it turns out that this conclusion was based on an incorrect scaling assumption [7].

QMC examination of the spin susceptibility $\chi_S(\pi)$ on lattices up to length 256 revealed that the critical coupling is actually finite, and almost identical to that of dispersionless bond phonons with the same $\omega_0(\pi)$.

The phonon dispersion slightly above the phase transition is shown, together with the bare dispersion, in Fig. 10.14.

One can see clearly that in this case of small $\omega_0(\pi) = 0.25J$ there is again phonon softening. The spin-Peierls phase transition only affects phonons with momenta close to π. The soft bare dispersion at $q = 0$ is not affected at all. Indeed, the bare dispersion at small momenta has no influence on the phase transition [7].

10.6 Auxiliary Field Quantum Monte Carlo Methods

In the present and following sections, we will review the basic concepts involved in the formulation of various forms of auxiliary field QMC algorithms for fermionic systems. Auxiliary field methods are based on a Hubbard-Stratonovich (HS) decomposition of the two-body interaction term thereby yielding a functional integral expression

$$\mathrm{Tr}\left[e^{-\beta(H-\mu N)}\right] = \int d\Phi(i,\tau) e^{-S[\phi(i,\tau)]} \tag{10.89}$$

for the partition function. Here, i runs over all lattice sites and τ from 0 to β. For a fixed HS field $\Phi(i,\tau)$, one has to compute the action $S[\Phi(i,\tau)]$, corresponding to a problem of non-interacting electrons in an external space and imaginary time dependent field. The required computational effort depends on the formulation of the algorithm. In the Blankenbecler-Scalapino-Sugar (BSS) [6] approach for lattice models such as the Hubbard Hamiltonian, it scales as βN^3 where N corresponds to the number of lattice sites. In the Hirsch-Fye approach [54], appropriate for impurity problems it scales as $(\beta N_{\mathrm{imp}})^3$ where N_{imp} corresponds to the number of correlated sites. Having solved for a fixed HS field, we have to sum over all possible fields. This is done stochastically with the Monte Carlo method.

In comparison to the loop and SSE approaches, auxiliary field methods are slow. Recall that the computational effort for loop and SSE approaches – in the absence of a sign problem – scales as $N\beta$. However, the attractive point of the auxiliary field approach lies in the fact that the sign problem is absent in many non-trivial cases where the loop and SSE methods fail.

10.6.1 Basic Formulation

For simplicity, we will concentrate on the Hubbard model. Applications to different models such as the Kondo lattice or $SU(N)$ Hubbard Heisenberg models can be found in [55, 56]. The Hubbard model we consider reads

$$H = H_t + H_U \tag{10.90}$$

with $H_t = -t\sum_{\langle i,j\rangle,\sigma} c^\dagger_{i,\sigma} c_{j,\sigma}$ and $H_U = U\sum_i (n_{i,\uparrow} - 1/2)(n_{i,\downarrow} - 1/2)$.

If one is interested in ground-state properties, it is convenient to use the projector quantum Monte Carlo (PQMC) algorithm [57, 58, 59]. The ground-state expectation value of an observable O is obtained by projecting a trial wave function $|\Psi_T\rangle$ along the imaginary time axis

$$\frac{\langle\Psi_0|O|\Psi_0\rangle}{\langle\Psi_0|\Psi_0\rangle} = \lim_{\Theta\to\infty} \frac{\langle\Psi_T|\mathrm{e}^{-\Theta H} O \mathrm{e}^{-\Theta H}|\Psi_T\rangle}{\langle\Psi_T|\mathrm{e}^{-2\Theta H}|\Psi_T\rangle} . \tag{10.91}$$

The above equation is readily verified by writing $|\Psi_T\rangle = \sum_n |\Psi_n\rangle\langle\Psi_n|\Psi_0\rangle$ with $H|\Psi_n\rangle = E_n|\Psi_n\rangle$. Under the assumptions that $\langle\Psi_T|\Psi_0\rangle \neq 0$ and that the ground state is non-degenerate the right hand side of the above equation reads:

$$\lim_{\Theta\to\infty} \frac{\sum_{n,m}\langle\Psi_T|\Psi_n\rangle\langle\Psi_m|\Psi_T\rangle \mathrm{e}^{-\Theta(E_n - E_m - 2E_0)}\langle\Psi_n|O|\Psi_m\rangle}{\sum_n |\langle\Psi_T|\Psi_n\rangle|^2 \mathrm{e}^{-2\Theta(E_n - E_0)}} = \frac{\langle\Psi_0|O|\Psi_0\rangle}{\langle\Psi_0|\Psi_0\rangle} . \tag{10.92}$$

Finite-temperature properties in the grand-canonical ensemble are obtained by evaluating

$$\langle O\rangle = \frac{\mathrm{Tr}\left[\mathrm{e}^{-\beta(H-\mu N)} O\right]}{\mathrm{Tr}\left[\mathrm{e}^{-\beta(H-\mu N)}\right]} , \tag{10.93}$$

where the trace runs over the Fock space and μ is the chemical potential. The algorithm based on (10.93) will be referred to as finite-temperature QMC (FTQMC) method [60, 61]. Comparison of both algorithms is shown in Fig. 10.15 for the Hubbard model. At half-filling, the ground state is insulating so that charge fluctuations are absent in the low temperature limit on finite lattices. Hence, in this limit both grand-canonical and canonical approaches yield identical results. It is however clear that if one is interested solely in ground-state properties the PQMC is more efficient. This lies in the choice of the trial wave function which is chosen to be a spin singlet.

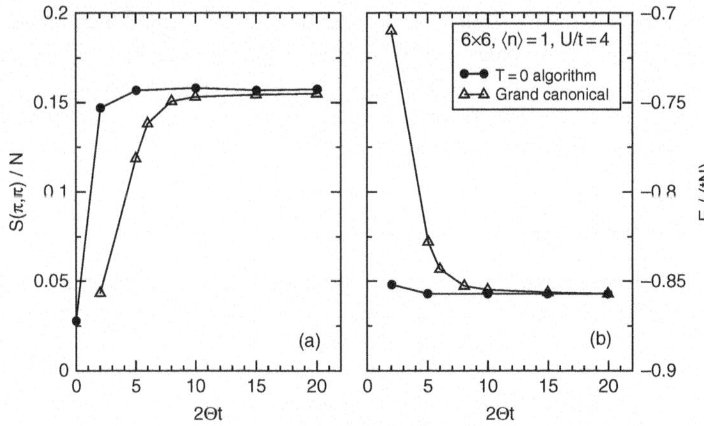

Fig. 10.15. Fourier transform of the spin-spin correlation functions at $\boldsymbol{Q} = (\pi, \pi)$ (**a**) and energy (**b**) for the half-filled Hubbard model (10.90). •: PQMC algorithm. △: FTQMC algorithm at $\beta = 2\Theta$

10.6.2 Formulation of the Partition Function

In the world-line approach, one uses the Trotter decomposition (see App. 10.A) to split the Hamiltonian into a set of two-site problems. In the auxiliary field approach, we use the Trotter decomposition to separate the single-body Hamiltonian H_0 from the two-body interaction term in the imaginary time propagation

$$Z = \mathrm{Tr}\left[\mathrm{e}^{-\beta(H-\mu N)}\right] = \mathrm{Tr}\left[\left(\mathrm{e}^{-\Delta_\tau H_U}\mathrm{e}^{-\Delta_\tau H_t}\right)^m\right] + \mathcal{O}\left(\Delta_\tau^2\right) , \tag{10.94}$$

where we have included the chemical potential in a redefinition of H_t. In the above $m\Delta_\tau = \beta$, and the systematic error of order Δ_τ^2 will be omitted in the following. At each infinitesimal time step, we use the HS decomposition of (10.236) (see App. 10.B) to decouple the Hubbard interaction

$$\mathrm{e}^{-\Delta_\tau U\sum_i (n_{i,\uparrow}-1/2)(n_{i,\downarrow}-1/2)} = C \sum_{s_1,\dots,s_N=\pm 1} \mathrm{e}^{\alpha\sum_i s_i(n_{i,\uparrow}-n_{i,\downarrow})} . \tag{10.95}$$

where $\cosh(\alpha) = \exp\left(\Delta_\tau U/2\right)$ and on an N-site lattice, the constant $C = \exp\left(\Delta_\tau UN/4\right)/2^N$.

To simplify the notation we introduce the index $x = (\boldsymbol{i}, \sigma)$ to define

$$H_t = \sum_{x,y} c_x^\dagger T_{x,y} c_y \equiv \boldsymbol{c}^\dagger T \boldsymbol{c} ,$$

$$\alpha \sum_{\boldsymbol{i}} s_{\boldsymbol{i}}\left(n_{\boldsymbol{i},\uparrow} - n_{\boldsymbol{i},\downarrow}\right) = \sum_{x,y} c_x^\dagger V(\boldsymbol{s})_{x,y} c_y \equiv \boldsymbol{c}^\dagger V(\boldsymbol{s})\boldsymbol{c} . \tag{10.96}$$

We will furthermore define the imaginary time propagators

$$U_{\boldsymbol{s}}(\tau_2,\tau_1) = \prod_{n=n_1+1}^{n_2} \mathrm{e}^{\boldsymbol{c}^\dagger V(\boldsymbol{s}_n)\boldsymbol{c}} \mathrm{e}^{-\Delta_\tau \boldsymbol{c}^\dagger T \boldsymbol{c}} ,$$
$$B_{\boldsymbol{s}}(\tau_2,\tau_1) = \prod_{n=n_1+1}^{n_2} \mathrm{e}^{V(\boldsymbol{s}_n)} \mathrm{e}^{-\Delta_\tau T} , \tag{10.97}$$

where $n_1\Delta_\tau = \tau_1$ and $n_2\Delta_\tau = \tau_2$.

Using the results of App. 10.C we can now write the partition function as

$$Z = C^m \sum_{\boldsymbol{s}_1,\dots,\boldsymbol{s}_m} \mathrm{Tr}\left[U_{\boldsymbol{s}}(\beta,0)\right] = C^m \sum_{\boldsymbol{s}_1,\dots,\boldsymbol{s}_m} \det\left[1 + B_{\boldsymbol{s}}(\beta,0)\right] . \tag{10.98}$$

For the PQMC algorithm, we will require the trial wave function to be a Slater determinant characterized by the rectangular matrix P (see App. 10.C)

$$|\Psi_T\rangle = \prod_{y=1}^{N_p} \left(\sum_x c_x^\dagger P_{x,y} \right) |0\rangle = \prod_{y=1}^{N_p} \left(\boldsymbol{c}^\dagger P\right)_y |0\rangle . \tag{10.99}$$

Hence,

$$\langle\Psi_T|\mathrm{e}^{-2\Theta H}|\Psi_T\rangle = C^m \sum_{\boldsymbol{s}_1,\dots,\boldsymbol{s}_m} \det\left[P^\dagger B_{\boldsymbol{s}}(2\Theta,0)P\right] , \tag{10.100}$$

where for the PQMC $m\Delta_\tau = 2\Theta$.

10.6.3 Observables and Wick's Theorem

One of the big advantages of the auxiliary field approach is the ability of measuring arbitrary observables. This is based on the fact that for a given Hubbard-Stratonovich field we have to solve a problem of non-interacting fermions subject to this time and space dependent field. This leads to the validity of Wick's theorem. In this section, we will concentrate on equal-time observables, show how to compute Green functions, and finally demonstrate the validity of Wick's theorem.

10.6.3.1 PQMC

In the PQMC algorithm we compute

$$\frac{\langle\Psi_T|\mathrm{e}^{-\Theta H} O \mathrm{e}^{-\Theta H}|\Psi_T\rangle}{\langle\Psi_T|\mathrm{e}^{-2\Theta H}|\Psi_T\rangle} = \sum_{\boldsymbol{s}} \mathrm{P}_{\boldsymbol{s}} \langle O\rangle_{\boldsymbol{s}} + O(\Delta_\tau^2) . \tag{10.101}$$

For each lattice site $\boldsymbol{i}$, time slice n, we have introduced an independent HS field $\boldsymbol{s} = \{s_{\boldsymbol{i},n}\}$ and

$$\mathrm{P}_{\boldsymbol{s}} = \frac{\det\left(P^\dagger B_{\boldsymbol{s}}(2\Theta,0)P\right)}{\sum_{\boldsymbol{s}} \det\left(P^\dagger B_{\boldsymbol{s}}(2\Theta,0)P\right)} ,$$

$$\langle O \rangle_{\boldsymbol{s}} = \frac{\langle \Psi_T | U_{\boldsymbol{s}}(2\Theta,\Theta) O U_{\boldsymbol{s}}(\Theta,0) | \Psi_T \rangle}{\langle \Psi_T | U_{\boldsymbol{s}}(2\Theta,0) | \Psi_T \rangle} . \tag{10.102}$$

We start by computing the equal-time Green function $O = c_x c_y^\dagger = \delta_{x,y} - \mathbf{c}^\dagger A^{(y,x)} \mathbf{c}$ with $A^{(y,x)}_{x_1,x_2} = \delta_{x_1,y}\delta_{x_2,x}$. Inserting a source term, we obtain

$$\begin{aligned}
&\langle c_x c_y^\dagger \rangle_{\boldsymbol{s}} \\
&= \delta_{x,y} - \frac{\partial}{\partial \eta} \ln \langle \Psi_T | U_{\boldsymbol{s}}(2\Theta,\Theta) \mathrm{e}^{\eta \mathbf{c}^\dagger A^{(y,x)} \mathbf{c}} U_{\boldsymbol{s}}(\Theta,0) | \Psi_T \rangle \,|_{\eta=0} \\
&= \delta_{x,y} - \frac{\partial}{\partial \eta} \ln\det\left(P^\dagger B_{\boldsymbol{s}}(2\Theta,\Theta)\mathrm{e}^{\eta A^{(y,x)}} B_{\boldsymbol{s}}(\Theta,0) P\right) |_{\eta=0} \\
&= \delta_{x,y} - \frac{\partial}{\partial \eta} \mathrm{Tr}\ln\left(P^\dagger B_{\boldsymbol{s}}(2\Theta,\Theta)\mathrm{e}^{\eta A^{(y,x)}} B_{\boldsymbol{s}}(\Theta,0) P\right) |_{\eta=0} \\
&= \delta_{x,y} - \mathrm{Tr}\left[\left(P^\dagger B_{\boldsymbol{s}}(2\Theta,0)P\right)^{-1} P^\dagger B_{\boldsymbol{s}}(2\Theta,\Theta) A^{(y,x)} B_{\boldsymbol{s}}(\Theta,0) P\right] ,
\end{aligned} \tag{10.103}$$

$$\left(1 - B_{\boldsymbol{s}}(\Theta,0) P \left(P^\dagger B_{\boldsymbol{s}}(2\Theta,0)P\right)^{-1} P^\dagger B_{\boldsymbol{s}}(2\Theta,\Theta)\right)_{x,y} \equiv \left(G_{\boldsymbol{s}}(\Theta)\right)_{x,y} . \tag{10.104}$$

We have used (10.245), (10.248) to derive the third equality. The attentive reader will have noticed that (10.245) was shown to be valid only in the case of Hermitian or anti-Hermitian matrices which is certainly not the case of $A^{(y,x)}$. However, since only terms of order η are relevant in the calculation, we may replace $\exp(\eta A)$ by $\exp(\eta(A + A^\dagger)/2)\exp(\eta(A - A^\dagger)/2)$ which is exact up to order η^2. For the latter form, one may use (10.245). To obtain the fourth equality we have used the relation $\det A = \exp(\mathrm{Tr}\ln A)$.

We now show that any multi-point correlation function decouples into a sum of products of the above defined Green functions. First, we define the cumulants

$$\begin{aligned}
&\langle\langle O_n \dots O_1 \rangle\rangle_{\boldsymbol{s}} \\
&= \left.\frac{\partial^n \ln \langle \Psi_T | U_{\boldsymbol{s}}(2\Theta,\Theta) \mathrm{e}^{\eta_n O_n} \dots \mathrm{e}^{\eta_1 O_1} U_{\boldsymbol{s}}(\Theta,0) | \Psi_T \rangle}{\partial \eta_n \dots \partial \eta_1}\right|_{\eta_1 \dots \eta_n = 0}
\end{aligned} \tag{10.105}$$

with $O_i = \mathbf{c}^\dagger A^{(i)} \mathbf{c}$. Differentiating the above definition we obtain

$$\begin{aligned}
\langle\langle O_1 \rangle\rangle_{\boldsymbol{s}} &= \langle O_1 \rangle_{\boldsymbol{s}} \\
\langle\langle O_2 O_1 \rangle\rangle_{\boldsymbol{s}} &= \langle O_2 O_1 \rangle_{\boldsymbol{s}} - \langle O_2 \rangle_{\boldsymbol{s}} \langle O_1 \rangle_{\boldsymbol{s}} \\
\langle\langle O_3 O_2 O_1 \rangle\rangle_{\boldsymbol{s}} &= \langle O_3 O_2 O_1 \rangle_{\boldsymbol{s}} \\
&\quad - \langle O_3 \rangle_{\boldsymbol{s}} \langle\langle O_2 O_1 \rangle\rangle_{\boldsymbol{s}} - \langle O_2 \rangle_{\boldsymbol{s}} \langle\langle O_3 O_1 \rangle\rangle_{\boldsymbol{s}} \\
&\quad - \langle O_1 \rangle_{\boldsymbol{s}} \langle\langle O_3 O_2 \rangle\rangle_{\boldsymbol{s}} - \langle O_1 \rangle_{\boldsymbol{s}} \langle O_2 \rangle_{\boldsymbol{s}} \langle O_3 \rangle_{\boldsymbol{s}} .
\end{aligned} \tag{10.106}$$

The following rule, which may be proven by induction, emerges

$$\begin{aligned}\langle O_n \dots O_1 \rangle_s = \langle\langle O_n \dots O_1 \rangle\rangle_s + \sum_{j=1}^{n} \langle\langle O_n \dots \widehat{O_j} \dots O_1 \rangle\rangle_s \langle\langle O_j \rangle\rangle_s \\ + \sum_{j>i} \langle\langle O_n \dots \widehat{O_j} \dots \widehat{O_i} \dots O_1 \rangle\rangle_s \\ \langle\langle O_j O_i \rangle\rangle_s + \dots + \langle\langle O_n \rangle\rangle_s \dots \langle\langle O_1 \rangle\rangle_s \,, \end{aligned} \quad (10.107)$$

where $\widehat{O_j}$ means that the operator O_j has been omitted from the product [62].

The cumulant may now be computed order by order. We concentrate on the form $\langle\langle c^\dagger_{x_n} c_{y_n} \dots c^\dagger_{x_1} c_{y_1} \rangle\rangle$ so that $A^{(i)}_{x,y} = \delta_{x,x_i}\delta_{y,y_i}$. To simplify the notation we introduce the quantities

$$\begin{aligned} B^{\rangle} &= B_s(\Theta,0)P \,, \\ B^{\langle} &= P^\dagger B_s(2\Theta,\Theta) \,. \end{aligned} \quad (10.108)$$

We have already computed $\langle\langle O_1 \rangle\rangle_s$, see (10.103),

$$\langle\langle O_1 \rangle\rangle_s = \langle\langle c^\dagger_{x_1} c_{y_1} \rangle\rangle = \mathrm{Tr}\left((1 - G_s(\Theta))A^{(1)} \right) = (1 - G_s(\Theta))_{y_1,x_1} \,. \quad (10.109)$$

For $n = 2$ we have

$$\begin{aligned} \langle\langle O_2 O_1 \rangle\rangle_s &= \langle\langle c^\dagger_{x_2} c_{y_2} c^\dagger_{x_1} c_{y_1} \rangle\rangle_s \\ &= \frac{\partial^2}{\partial\eta_2 \partial\eta_1} \mathrm{Tr} \ln \left(P^\dagger B_s(2\Theta,\Theta) e^{\eta_2 A^{(2)}} e^{\eta_1 A^{(1)}} B_s(\Theta,0) P \right) \Big|_{\eta_2,\eta_1=0} \\ &= \frac{\partial}{\partial\eta_2} \mathrm{Tr} \left[\left(B^{\langle} e^{\eta_2 A^{(2)}} B^{\rangle} \right)^{-1} B^{\langle} e^{\eta_2 A^{(2)}} A^{(1)} B^{\rangle} \right] \Big|_{\eta_2=0} \\ &= -\mathrm{Tr} \left[\left(B^{\langle} B^{\rangle} \right)^{-1} B^{\langle} A^{(2)} B^{\rangle} \left(B^{\langle} B^{\rangle} \right)^{-1} B^{\langle} A^{(1)} B^{\rangle} \right] \\ &\quad + \mathrm{Tr} \left[\left(B^{\langle} B^{\rangle} \right)^{-1} B^{\langle} A^{(2)} A^{(1)} B^{\rangle} \right] \\ &= \mathrm{Tr} \left(\overline{G_s(\Theta)} A^{(2)} G_s(\Theta) A^{(1)} \right) \\ &= \langle c^\dagger_{x_2} c_{y_1} \rangle_s \langle c_{y_2} c^\dagger_{x_1} \rangle_s \end{aligned} \quad (10.110)$$

with $\overline{G} = 1 - G$. To derive the above, we have used the cyclic properties of the trace as well as the relation $G = 1 - B^{\rangle} \left(B^{\langle} B^{\rangle} \right)^{-1} B^{\langle}$. Note that for a matrix $A(\eta)$, $(\partial/\partial\eta) A^{-1}(\eta) = -A^{-1}(\eta)[(\partial/\partial\eta)A(\eta)]A^{-1}(\eta)$. There is a simple rule to obtain the third cumulant given the second. In the above expression for the second cumulant, one replaces $B^{\langle}$ with $B^{\langle} \exp(\eta_3 A^{(3)})$. This amounts in redefining the Green function as $G(\eta_3) = 1 - B^{\rangle}(B^{\langle} \exp(\eta_3 A^{(3)}) B^{\rangle})^{-1} B^{\langle} \exp(\eta_3 A^{(3)})$. Thus,

$$\begin{aligned}
\langle\langle O_3 O_2 O_1 \rangle\rangle_s &= \langle\langle c^\dagger_{x_3} c_{y_3} c^\dagger_{x_2} c_{y_2} c^\dagger_{x_1} c_{y_1} \rangle\rangle_s \\
&= \frac{\partial}{\partial \eta_3} \mathrm{Tr}\left(\overline{G_s(\Theta, \eta_3)} A^{(2)} G_s(\Theta, \eta_3) A^{(1)} \right)\Big|_{\eta_3 = 0} \\
&= \mathrm{Tr}\left(\overline{G_s(\Theta)} A^{(3)} G_s(\Theta) A^{(2)} G_s(\Theta) A^{(1)} \right) \\
&\quad - \mathrm{Tr}\left(\overline{G_s(\Theta)} A^{(3)} G_s(\Theta) A^{(1)} \overline{G_s(\Theta)} A^{(2)} \right) \\
&= \langle c^\dagger_{x_3} c_{y_1} \rangle_s \langle c_{y_3} c^\dagger_{x_2} \rangle_s \langle c_{y_2} c^\dagger_{x_1} \rangle_s \\
&\quad - \langle c^\dagger_{x_3} c_{y_2} \rangle_s \langle c_{y_3} c^\dagger_{x_1} \rangle_s \langle c^\dagger_{x_2} c_{y_1} \rangle_s
\end{aligned} \tag{10.111}$$

since

$$\frac{\partial}{\partial \eta_3} G_s(\Theta, \eta_3)|_{\eta_3=0} = -\overline{G_s(\Theta)} A^{(3)} G_s(\Theta) = -\frac{\partial}{\partial \eta_3} \overline{G_s(\Theta, \eta_3)}|_{\eta_3=0} \,. \tag{10.112}$$

Clearly the same procedure may be applied to obtain the n+1th cumulant given the nth one. It is also clear that the nth cumulant is a sum of products of Green functions. Thus with (10.107) we have shown that any multi-point correlation function may be reduced into a sum of products of Green functions: Wicks theorem. Useful relations include

$$\langle c^\dagger_{x_2} c_{y_2} c^\dagger_{x_1} c_{y_1} \rangle_s = \langle c^\dagger_{x_2} c_{y_1} \rangle_s \langle c_{y_2} c^\dagger_{x_1} \rangle_s + \langle c^\dagger_{x_2} c_{y_2} \rangle_s \langle c^\dagger_{x_1} c_{y_1} \rangle_s \,. \tag{10.113}$$

10.6.3.2 FTQMC

For the FTQMC we wish to evaluate

$$\frac{\mathrm{Tr}\left[\mathrm{e}^{-\beta H} O \right]}{\mathrm{Tr}\left[\mathrm{e}^{-\beta H} \right]} = \sum_s \mathrm{P}_s \langle O \rangle_s + O(\Delta_\tau^2) \,. \tag{10.114}$$

where

$$\begin{aligned}
\mathrm{P}_s &= \frac{\det\left(1 + B_s(\beta, 0)\right)}{\sum_s \det\left(1 + B_s(\beta, 0)\right)} \,, \\
\langle O \rangle_s &= \frac{\mathrm{Tr}\left[U_s(\beta, \tau) O U_s(\tau, 0) \right]}{\mathrm{Tr}\left[U_s(\beta, 0) \right]} \,.
\end{aligned} \tag{10.115}$$

Here, we measure the observable on time slice τ. Single-body observables, $O = \boldsymbol{c}^\dagger A \boldsymbol{c}$ are evaluated as

$$\begin{aligned}\langle O\rangle_{\boldsymbol{s}} &= \frac{\partial}{\partial\eta}\ln\mathrm{Tr}\left[U_{\boldsymbol{s}}(\beta,\tau)\mathrm{e}^{\eta O}U_{\boldsymbol{s}}(\tau,0)\right]\Big|_{\eta=0} \\ &= \frac{\partial}{\partial\eta}\ln\det\left[1+B_{\boldsymbol{s}}(\beta,\tau)\mathrm{e}^{\eta A}B_{\boldsymbol{s}}(\tau,0)\right]\Big|_{\eta=0} \\ &= \frac{\partial}{\partial\eta}\mathrm{Tr}\ln\left[1+B_{\boldsymbol{s}}(\beta,\tau)\mathrm{e}^{\eta A}B_{\boldsymbol{s}}(\tau,0)\right]\Big|_{\eta=0} \\ &= \mathrm{Tr}\left[B_{\boldsymbol{s}}(\tau,0)(1+B_{\boldsymbol{s}}(\beta,0))^{-1}B_{\boldsymbol{s}}(\beta,\tau)A\right] \\ &= \mathrm{Tr}\left[\left(1-(1+B_{\boldsymbol{s}}(\tau,0)B_{\boldsymbol{s}}(\beta,\tau))^{-1}\right)A\right] . \end{aligned} \tag{10.116}$$

In particular the Green function is given by

$$\langle c_x c_y^\dagger\rangle_{\boldsymbol{s}} = (1+B_{\boldsymbol{s}}(\tau,0)B_{\boldsymbol{s}}(\beta,\tau))^{-1}_{x,y} . \tag{10.117}$$

Defining the cumulants as

$$\langle\langle O_n\dots O_1\rangle\rangle_{\boldsymbol{s}} = \frac{\partial^n\ln\mathrm{Tr}\left[U_{\boldsymbol{s}}(\beta,\tau)\mathrm{e}^{\eta_n O_n}\dots\mathrm{e}^{\eta_1 O_1}U_{\boldsymbol{s}}(\tau,0)\right]}{\partial\eta_n\dots\partial\eta_1}\Bigg|_{\eta_1\dots\eta_n=0} \tag{10.118}$$

with $O_i = \boldsymbol{c}^\dagger A^{(i)}\boldsymbol{c}$, one can derive Wicks theorem in precisely the same manner as shown for the PQMC. Thus both in the PQMC and FTQMC, it suffices to compute the equal-time Green functions to evaluate any equal-time observable.

10.6.4 Imaginary Time Displaced Green Functions

Imaginary time displaced correlation yield important information. On one hand they may be used to obtain spin and charge gaps [63, 64], as well quasiparticle weights [23]. On the other hand, with the use of the Maximum Entropy method [65, 66] and generalizations thereof [67], dynamical properties such as spin and charge dynamical structure factors, optical conductivity, and single-particle spectral functions may be computed. Those quantities offer the possibility of direct comparison with experiments, such as photoemission, neutron scattering and optical measurements.

Since there is again a Wick's theorem for time displaced correlation functions, it suffices to compute the single-particle Green function for a given HS configuration. We will first start with the FTQMC and then concentrate on the PQMC.

10.6.4.1 FTQMC

For a given HS field, we wish to evaluate

$$G_{\boldsymbol{s}}(\tau_1,\tau_2)_{x,y} = \langle T c_x(\tau_1)c_y^\dagger(\tau_2)\rangle_{\boldsymbol{s}} = \begin{cases} \langle c_x(\tau_1)c_y^\dagger(\tau_2)\rangle_{\boldsymbol{s}} & \text{if } \tau_1\geq\tau_2 \\ -\langle c_y^\dagger(\tau_2)c_x(\tau_1)\rangle_{\boldsymbol{s}} & \text{if } \tau_1<\tau_2 \end{cases} , \tag{10.119}$$

where T corresponds to the time ordering. Thus for $\tau_1 > \tau_2$ $G_{\boldsymbol{s}}(\tau_1, \tau_2)_{x,y}$ reduces to

$$\begin{aligned} \langle c_x(\tau_1) c_y^\dagger(\tau_2) \rangle_{\boldsymbol{s}} &= \frac{\mathrm{Tr}\left[U_{\boldsymbol{s}}(\beta, \tau_1) c_x U_{\boldsymbol{s}}(\tau_1, \tau_2) c_y^\dagger U_{\boldsymbol{s}}(\tau_2, 0) \right]}{\mathrm{Tr}\left[U_{\boldsymbol{s}}(\beta, 0) \right]} \\ &= \frac{\mathrm{Tr}\left[U_{\boldsymbol{s}}(\beta, \tau_2) U_{\boldsymbol{s}}^{-1}(\tau_1, \tau_2) c_x U_{\boldsymbol{s}}(\tau_1, \tau_2) c_y^\dagger U_{\boldsymbol{s}}(\tau_2, 0) \right]}{\mathrm{Tr}\left[U_{\boldsymbol{s}}(\beta, 0) \right]} \,. \end{aligned} \tag{10.120}$$

Evaluating $U^{-1}(\tau_1, \tau_2) c_x U_{\boldsymbol{s}}(\tau_1, \tau_2)$ boils down to the calculation of

$$c_x(\tau) = \mathrm{e}^{\tau \boldsymbol{c}^\dagger A \boldsymbol{c}} c_x \mathrm{e}^{-\tau \boldsymbol{c}^\dagger A \boldsymbol{c}} \,, \tag{10.121}$$

where A is an arbitrary matrix. Differentiating the above with respect to τ yields

$$\frac{\partial c_x(\tau)}{\partial \tau} = \mathrm{e}^{\tau \boldsymbol{c}^\dagger A \boldsymbol{c}} [\boldsymbol{c}^\dagger A \boldsymbol{c}, c_x] \mathrm{e}^{-\tau \boldsymbol{c}^\dagger A \boldsymbol{c}} = -\sum_z A_{x,z} c_z(\tau) \,. \tag{10.122}$$

Thus,

$$c_x(\tau) = (\mathrm{e}^{-A} \boldsymbol{c})_x \,, \quad \text{and similarly } c_x^\dagger(\tau) = (\boldsymbol{c}^\dagger \mathrm{e}^{A})_x \,. \tag{10.123}$$

We can use the above equation successively to obtain

$$\begin{aligned} U_{\boldsymbol{s}}^{-1}(\tau_1, \tau_2) c_x U_{\boldsymbol{s}}(\tau_1, \tau_2) &= \left(B_{\boldsymbol{s}}(\tau_1, \tau_2) \boldsymbol{c} \right)_x \\ U_{\boldsymbol{s}}^{-1}(\tau_1, \tau_2) c_x^\dagger U_{\boldsymbol{s}}(\tau_1, \tau_2) &= \left(\boldsymbol{c}^\dagger B_{\boldsymbol{s}}^{-1}(\tau_1, \tau_2) \right)_x \,. \end{aligned} \tag{10.124}$$

Since B is a matrix and not a second quantized operator, we can pull it out of the trace in (10.120) to obtain

$$G_{\boldsymbol{s}}(\tau_1, \tau_2)_{x,y} = \langle c_x(\tau_1) c_y^\dagger(\tau_2) \rangle_{\boldsymbol{s}} = \left[B_{\boldsymbol{s}}(\tau_1, \tau_2) G_{\boldsymbol{s}}(\tau_2, \tau_2) \right]_{x,y} \tag{10.125}$$

with $\tau_1 > \tau_2$, where $G_{\boldsymbol{s}}(\tau_2)$ is the equal-time Green function computed previously. A similar calculation will yield for $\tau_2 > \tau_1$

$$\begin{aligned} G_{\boldsymbol{s}}(\tau_1, \tau_2)_{x,y} &= -\langle c_y^\dagger(\tau_2) c_x(\tau_1) \rangle_{\boldsymbol{s}} \\ &= -\left[\left(1 - G_{\boldsymbol{s}}(\tau_1, \tau_1) \right) B_{\boldsymbol{s}}^{-1}(\tau_2, \tau_1) \right]_{x,y} \,. \end{aligned} \tag{10.126}$$

The above equations imply the validity of Wick's theorem for time displaced Green functions. Any n-point correlation function at different imaginary times may be mapped onto an expression containing n-point equal-time correlation functions. The n-point equal-time correlation function may then be decomposed into a sum of products of equal-time Green functions. For example, for $\tau_1 > \tau_2$ let us compute

$$
\begin{aligned}
&\langle c_x^\dagger(\tau_1)c_x(\tau_1)c_y^\dagger(\tau_2)c_y(\tau_2)\rangle \\
&= \frac{\mathrm{Tr}\left[U(\beta,\tau_2)U^{-1}(\tau_1,\tau_2)c_x^\dagger U^{-1}(\tau_1,\tau_2)U(\tau_1,\tau_2)c_x U(\tau_1,\tau_2)c_y^\dagger c_y U(\tau_2,0)\right]}{\mathrm{Tr}\left[U(\beta,0)\right]} \\
&= \sum_{z,z_1} B^{-1}(\tau_1,\tau_2)_{z,x} B(\tau_1,\tau_2)_{x,z_1} \langle c_z^\dagger(\tau_2)c_{z_1}(\tau_2)c_y^\dagger(\tau_2)c_y(\tau_2)\rangle \\
&= \sum_{z,z_1} B^{-1}(\tau_1,\tau_2)_{z,x} B(\tau_1,\tau_2)_{x,z_1} \Big[\left(1-G(\tau_2,\tau_2)\right)_{z_1,z} \left(1-G(\tau_2,\tau_2)\right)_{y,y} \\
&\quad + \left(1-G(\tau_2,\tau_2)\right)_{y,z} G(\tau_2,\tau_2)_{z_1,y} \Big] \\
&= \left[B(\tau_1,\tau_2)\left(1-G(\tau_2,\tau_2)\right)B^{-1}(\tau_1,\tau_2)\right]_{x,x} \left[1-G(\tau_2,\tau_2)\right]_{y,y} \\
&\quad + \left[\left(1-G(\tau_2,\tau_2)\right)B^{-1}(\tau_1,\tau_2)\right]_{y,x} \left[B(\tau_1,\tau_2)G(\tau_2,\tau_2)\right]_{x,y} \\
&= \left[1-G(\tau_1,\tau_1)\right]_{x,x} \left[1-G(\tau_2,\tau_2)\right]_{y,y} - G(\tau_2,\tau_1)_{y,x} G(\tau_1,\tau_2)_{x,y} \,.
\end{aligned} \tag{10.127}
$$

In the above, we have omitted the index $\boldsymbol{s}$, used (10.126) and (10.125), Wick's theorem for equal-time n-point correlation functions as well as the identity

$$
B_{\boldsymbol{s}}(\tau_1,\tau_2)G_{\boldsymbol{s}}(\tau_2,\tau_2)B_{\boldsymbol{s}}^{-1}(\tau_1,\tau_2) = G_{\boldsymbol{s}}(\tau_1,\tau_1) \,. \tag{10.128}
$$

We conclude this Subsection, by a method proposed by Hirsch [68] to compute imaginary time displaced Green functions. This equation provides a means to circumvent numerical instabilities which we will discuss in a subsequent chapter and is the basis of the Hirsch-Fye [54] algorithm. Let β be a multiple of τ_1 and $l\tau_1 = \beta$. Using the definition $\tau_i = i\tau_1$ with $i = 1 \dots l$. Let

$$
O = \begin{pmatrix}
1 & 0 & . & 0 & B_{\boldsymbol{s}}(\tau_1,0) \\
-B_{\boldsymbol{s}}(\tau_2,\tau_1) & 1 & 0 & . & 0 \\
0 & -B_{\boldsymbol{s}}(\tau_3,\tau_2) & 1 & . & 0 \\
. & 0 & -B_{\boldsymbol{s}}(\tau_4,\tau_3) & . & . \\
. & . & 0 & . & . \\
. & . & . & . & \\
0 & . & 0 & -B_{\boldsymbol{s}}(\tau_l,\tau_{l-1}) & 1
\end{pmatrix}, \tag{10.129}
$$

and

$$
G = \begin{pmatrix}
G_{\boldsymbol{s}}(\tau_1,\tau_1) & G_{\boldsymbol{s}}(\tau_1,\tau_2) & .. & G_{\boldsymbol{s}}(\tau_1,\tau_l) \\
G_{\boldsymbol{s}}(\tau_2,\tau_1) & G_{\boldsymbol{s}}(\tau_2,\tau_2) & .. & G_{\boldsymbol{s}}(\tau_2,\tau_l) \\
. & . & .. & . \\
G_{\boldsymbol{s}}(\tau_l,\tau_1) & G_{\boldsymbol{s}}(\tau_l,\tau_2) & .. & G_{\boldsymbol{s}}(\tau_l,\tau_l)
\end{pmatrix}, \tag{10.130}
$$

then

$$
O^{-1} = G \,. \tag{10.131}
$$

The above equation is readily verified by showing that $OG = 1$. Here, we illustrate the validity of the above equation for the case $l = 2$. Using (10.126), (10.125) and (10.128), bearing in mind that in this case $\tau_2 = \beta$ and omitting the index $\boldsymbol{s}$ we have

$$G(\tau_1,\tau_1) + B(\tau_1,0)G(\tau_2,\tau_1) = \underbrace{[1 + B(\tau_1,0)B(\tau_2,\tau_1)]}_{G^{-1}(\tau_1,\tau_1)} G(\tau_1,\tau_1) = 1\,, \tag{10.132}$$

$$\begin{aligned} G(\tau_1,\tau_2) + B(\tau_1,0)G(\tau_2,\tau_2) &= -(1 - G(\tau_1,\tau_1))B^{-1}(\tau_2,\tau_1) \\ &\quad + B(\tau_1,0)B(\tau_2,\tau_1)G(\tau_1,\tau_1)B^{-1}(\tau_2,\tau_1) \\ &= \Big[- G^{-1}(\tau_1,\tau_1) + \underbrace{1 + B(\tau_1,0)B(\tau_2,\tau_1)}_{G^{-1}(\tau_1,\tau_1)} \Big] G(\tau_1,\tau_1)B^{-1}(\tau_2,\tau_1) = 0\,, \end{aligned} \tag{10.133}$$

$$-B(\tau_2,\tau_1)G(\tau_1,\tau_1) + G(\tau_2,\tau_1) = -G(\tau_2,\tau_1) + G(\tau_2,\tau_1) = 0\,, \tag{10.134}$$

and

$$\begin{aligned} &- B(\tau_2,\tau_1)G(\tau_1,\tau_2) + G(\tau_2,\tau_2) \\ &\quad = B(\tau_2,\tau_1)(1 - G(\tau_1,\tau_1))B^{-1}(\tau_2,\tau_1) + G(\tau_2,\tau_2) \\ &\quad = 1 - G(\tau_2,\tau_2) + G(\tau_2,\tau_2) = 1\,, \end{aligned} \tag{10.135}$$

so that

$$\begin{pmatrix} 1 & B(\tau_1,0) \\ -B(\tau_2,\tau_1) & 1 \end{pmatrix} \begin{pmatrix} G(\tau_1,\tau_1) & G(\tau_1,\tau_2) \\ G(\tau_2,\tau_1) & G(\tau_2,\tau_2) \end{pmatrix} = \begin{pmatrix} 1 & 0 \\ 0 & 1 \end{pmatrix}. \tag{10.136}$$

10.6.4.2 PQMC

Zero-temperature time displaced Green functions are given by

$$\begin{aligned} &G_s\left(\Theta + \frac{\tau}{2}, \Theta - \frac{\tau}{2}\right)_{x,y} \\ &= \frac{\langle \Psi_T | U_s\left(2\Theta, \Theta + \frac{\tau}{2}\right) c_x U_s\left(\Theta + \frac{\tau}{2}, \Theta - \frac{\tau}{2}\right) c_y^\dagger U_s(\Theta - \frac{\tau}{2}, 0) | \Psi_T\rangle}{\langle \Psi_T | U_s(2\Theta, 0) | \Psi_T\rangle} \\ &= \left[B_s\left(\Theta + \frac{\tau}{2}, \Theta - \frac{\tau}{2}\right) G_s\left(\Theta - \frac{\tau}{2}\right)\right]_{x,y} \end{aligned} \tag{10.137}$$

and

$$\begin{aligned} &G_s\left(\Theta - \frac{\tau}{2}, \Theta + \frac{\tau}{2}\right)_{x,y} \\ &= -\frac{\langle \Psi_T | U_s(2\Theta, \Theta + \frac{\tau}{2}) c_y^\dagger U_s(\Theta + \frac{\tau}{2}, \Theta - \frac{\tau}{2}) c_x U_s(\Theta - \frac{\tau}{2}, 0) | \Psi_T\rangle}{\langle \Psi_T | U_s(2\Theta, 0) | \Psi_T\rangle} \\ &= -\left[\left(1 - G_s\left(\Theta - \frac{\tau}{2}\right)\right) B_s^{-1}\left(\Theta + \frac{\tau}{2}, \Theta - \frac{\tau}{2}\right)\right]_{x,y}. \end{aligned} \tag{10.138}$$

Here $\tau > 0$ and we have used (10.124), as well as the equal-time Green function of (10.103). Two comments are in order.

(i) For a given value of τ the effective projection parameter is $\Theta - \tau$. Thus, before starting a simulation, one has to set the maximal value of τ which will be considered, τ_M and the effective projection parameter $\Theta - \tau_M$ should be large enough to yield the ground state within the desired precision.

(ii) In a canonical ensemble, the chemical potential is meaningless. However, when single-particle Green functions are computed it is required to set the reference energy with regards to which a particle will be added or removed. In other words, it is the chemical potential which delimits photoemission from inverse photoemission.

Thus, it is useful to have an estimate of this quantity if single-particle or pairing correlations are under investigation. For observables such as spin-spin or charge-charge time displaced correlations this complication does not come into play since they are in the particle-hole channel.

10.6.5 The Sign Problem

One of the big advantages of the auxiliary field method, is that one can use symmetries to show explicitly that the sign problem does not occur. The generic way of showing the absence of sign problem is through the factorization of the determinant. In general, particle-hole symmetry allows one to avoid the sign problem (see for example [55] for the case of the Kondo lattice, Hubbard and Periodic Anderson models). In this case, the weight decouples into the product of two determinants in the spin-up and spin-down sectors. Particle-hole symmetry locks in together the sign of both determinants such that the weight remains positive. Models with attractive interactions which couple independently to an internal symmetry with an even number of states lead to weights, for a given HS configuration, which are an even power of a single determinant. If the determinant itself is real (i.e. absence of magnetic fields), the overall weight will be positive. An example is the attractive Hubbard model. The attractive Hubbard model falls into the above class and is hence free of the sign problem.

Here we will give more general conditions under which the sign problem is absent [69]. The proof is very similar to Kramers degeneracy for time reversal symmetric Hamiltonians [70]. Let us assume the existence of an anti-unitary transformation $\mathcal{K}$ with following properties (we adopt the notation of (10.96))

$$\begin{aligned} \mathcal{K}^\dagger T \mathcal{K} &= T \, , \\ \mathcal{K}^\dagger V(\boldsymbol{s}) \mathcal{K} &= V(\boldsymbol{s}) \, , \\ \mathcal{K}^\dagger \mathcal{K} &= 1 \, , \\ \mathcal{K}^2 &= -1 \, . \end{aligned} \tag{10.139}$$

It then follows that the eigenvalues of the matrix $1 + B_{\boldsymbol{s}}(\beta, 0)$ occur in complex conjugate pairs. Hence,

$$\det\left(1 + B(\beta, 0)\right) = \prod_i |\lambda_i|^2 \tag{10.140}$$

and no sign problem occurs.

Proof. Let us first remind the reader that an anti-linear operator $\mathcal{K}$ satisfies the property $\mathcal{K}(\alpha \boldsymbol{v} + \beta \boldsymbol{u}) = \alpha^\dagger \mathcal{K}\boldsymbol{v} + \beta^\dagger \mathcal{K}\boldsymbol{u}$, where α and β are complex numbers. An anti-unitary operator, corresponding to time reversal symmetry for example, is an unitary anti-linear transformation so that the scalar product remains invariant $(\mathcal{K}\boldsymbol{v}, \mathcal{K}\boldsymbol{u}) = (\boldsymbol{v}, \boldsymbol{u})$. Let us assume that $\boldsymbol{v}$ is an eigenvector of the matrix $1 + B_{\boldsymbol{s}}(\beta, 0)$ with eigenvalue λ

$$(1 + B_{\boldsymbol{s}}(\beta, 0))\boldsymbol{v} = \lambda \boldsymbol{v} \; . \tag{10.141}$$

From (10.139) and (10.97) follows that $\mathcal{K}^\dagger \left(1 + B_{\boldsymbol{s}}(\beta, 0)\right) \mathcal{K} = 1 + B_{\boldsymbol{s}}(\beta, 0)$ such that

$$\left(1 + B_{\boldsymbol{s}}(\beta, 0)\right) \mathcal{K}\boldsymbol{v} = \lambda^\dagger \mathcal{K}\boldsymbol{v} \; . \tag{10.142}$$

Hence, $\mathcal{K}\boldsymbol{v}$ is an eigenvector with eigenvalue $\lambda^\dagger$. To complete the proof, we have to show that $\boldsymbol{v}$ and $\mathcal{K}\boldsymbol{v}$ are linearly independent

$$(\boldsymbol{v}, \mathcal{K}\boldsymbol{v}) = \left(\mathcal{K}^\dagger \boldsymbol{v}, \boldsymbol{v}\right) = \left(\mathcal{K}\mathcal{K}^\dagger \boldsymbol{v}, \mathcal{K}\boldsymbol{v}\right) = -\left(\boldsymbol{v}, \mathcal{K}\boldsymbol{v}\right) \; . \tag{10.143}$$

In the above, we have used the unitarity of $\mathcal{K}$ and the relation $\mathcal{K}^2 = -1$. Hence, since $\boldsymbol{v}$ and $\mathcal{K}\boldsymbol{v}$ are orthogonal, we are guaranteed that λ and $\lambda^\dagger$ will occur in the spectrum. In particular, if λ is real, it occurs an even number of times in the spectrum.

It is interesting to note that models which show spin-nematic phases can be shown to be free of sign problems due the above symmetry even though the factorization of the determinant is not present [71].

Clearly, the sign problem remains the central issue in Monte Carlo simulations of correlated electrons. It has been argued that there is no general solution to this problem [72]. This does not exclude the possibility of finding novel algorithms which can potentially circumvent the sign problem for a larger class of models than at present. A very interesting novel algorithm, the Gaussian Monte Carlo approach, has recently been introduced by Corney and Drummond [18, 73] and is claimed to solve the negative sign problem for a rather general class of models containing the Hubbard model on arbitrary lattices and at arbitrary dopings. As it stands, this method does not produce accurate results and the interested reader is referred to [19] for a detailed discussion of those problems.

10.6.6 Summary

In principle, we now have all the elements required to carry out a QMC simulation. The space we have to sample is that of Nm Ising spins. Here N is the number of lattice sites and m the number of imaginary time slices. For each configuration of Ising spins $\boldsymbol{s}$, we can associate a weight. For the PQMC it reads

$$W_{\boldsymbol{s}} = C^m \det\left[P^\dagger B_{\boldsymbol{s}}(2\Theta, 0) P\right] \tag{10.144}$$

and for the FTQMC

$$W_{\boldsymbol{s}} = C^m \det\left[1 + B_{\boldsymbol{s}}(\beta, 0)\right] . \tag{10.145}$$

Here we will assume that the weight is positive. A Monte Carlo simulation may now be carried out as follows.

- To generate a Markov chain we can adopt a sequential, or random, single spin flip upgrading scheme. We accept the proposed change from $\boldsymbol{s}$ to $\boldsymbol{s}'$ with probability max $(1, W_{\boldsymbol{s}'}/W_{\boldsymbol{s}})$ corresponding to a Metropolis algorithm. Since we can in principle compute the weight $W_{\boldsymbol{s}}$ at the expense of a set of matrix multiplications and estimation of a determinant we can compute the quotient $W_{\boldsymbol{s}'}/W_{\boldsymbol{s}}$. This procedure will be repeated until an independent Ising spin configuration is obtained. That is after the autocorrelation time.
- For a given Ising spin configuration, and with the help of the formulas given in the preceding section, we can compute the time displaced Green functions. Since a Wick's theorem holds for a given Hubbard Stratonovich configuration of Ising spins, we have access to all observables.
- After having measured an observable, we will return to step one so as to generate a new, independent configuration of Ising spins.

The implementation of the above program will not work due to the occurrence of numerical instabilities at low temperatures. It also leads to a very inefficient code. In the next two sections will show first to implement efficiently the algorithm. We will first concentrate on simulations for lattice models and then on the Hirsch-Fye approach which is triggered at solving impurity models.

10.7 Numerical Stabilization Schemes for Lattice Models

This section is organized as follows. We will first show how to compute the equal-time Green functions both in the finite (FTQMC) and projective (PQMC) formalisms. The equal-time Green function is the fundamental quantity on which the whole algorithm relies. On one hand and in conjunction with Wick's theorem, it allows to compute any equal-time observable. On the other hand, it determines the Monte Carlo dynamics, since the ratio of statistical weights under a single spin flip is determined by the equal-time Green function (see Sect. 10.7.2). In Sect. 10.7.3 we will show how to compute imaginary time displaced Green functions in an efficient and numerically stable manner.

10.7.1 Numerical Stabilization and Calculation of the Equal-Time Green Function

The fundamental quantity on which the entire algorithm relies is the equal-time Green function. For a given HS configuration of auxiliary fields, this quantity is given by

$$\langle c_x c_y^\dagger \rangle_{\boldsymbol{s}} = \left(1 - B_{\boldsymbol{s}}(\Theta, 0) P \left(P^\dagger B_{\boldsymbol{s}}(2\Theta, 0) P\right)^{-1} P^\dagger B_{\boldsymbol{s}}(2\Theta, \Theta)\right)_{x,y} \tag{10.146}$$

for the PQMC, see (10.103), and by

$$\langle c_x c_y^\dagger \rangle_{\boldsymbol{s}} = (1 + B_{\boldsymbol{s}}(\tau,0) B_{\boldsymbol{s}}(\beta,\tau))^{-1}_{x,y} \tag{10.147}$$

for the FTQMC, see (10.117). On finite precision machines a straightforward calculation of the Green function leads to numerical instabilities at large values of β or projection parameter Θ. To understand the sources of numerical instabilities, it is convenient to consider the PQMC. The rectangular matrix P accounting for the trial wave function is just a set of column orthonormal vectors. Typically for a Hubbard model, at weak couplings, the extremal scales in the matrix $B_{\boldsymbol{s}}(2\Theta,0)$ are determined by the kinetic energy and range from $\exp(8t\Theta)$ to $\exp(-8t\Theta)$ in the 2D case. When the set of orthonormal vectors in P are propagated, the large scales will wash out the small scales yielding a numerically ill defined inversion of the matrix $P^\dagger B_{\boldsymbol{s}}(2\Theta,0)P$. To be more precise consider a two-electron problem. The matrix P then consists of two column orthonormal vectors $\boldsymbol{v}(0)_1$ and $\boldsymbol{v}(0)_2$, which after propagation along the imaginary time axis will be dominated by the largest scales in $B_{\boldsymbol{s}}(2\Theta,0)$ so that $\boldsymbol{v}(2\Theta)_1 = \boldsymbol{v}(2\Theta)_2 + \boldsymbol{\epsilon}$, where $\boldsymbol{v}(2\Theta)_1 = B_{\boldsymbol{s}}(2\Theta,0)\boldsymbol{v}_1$. It is the information contained in $\boldsymbol{\epsilon}$ which renders the matrix $P^\dagger B_{\boldsymbol{s}}(2\Theta,0)P$ non-singular. For large values of Θ this information is lost in round-off errors.

To circumvent this problem a set of matrix decomposition techniques were developed [58, 59, 61]. Those matrix decomposition techniques are best introduced with the Gram-Schmidt orthonormalization method of N_p linearly independent vectors. At imaginary time τ, $B_{\boldsymbol{s}}(\tau,0)P \equiv B^\rangle$ is given by the N_p vectors $\boldsymbol{v}_1 \dots \boldsymbol{v}_{N_p}$. Orthogonalizing those vectors yields

$$\begin{aligned}
\boldsymbol{v}'_1 &= \boldsymbol{v}_1 \\
\boldsymbol{v}'_2 &= \boldsymbol{v}_2 - \frac{\boldsymbol{v}_2 \cdot \boldsymbol{v}'_1}{\boldsymbol{v}'_1 \cdot \boldsymbol{v}'_1}\boldsymbol{v}'_1 \\
&\vdots \\
\boldsymbol{v}'_{N_p} &= \boldsymbol{v}_{N_p} - \sum_{i=1}^{N_p-1} \frac{\boldsymbol{v}_{N_p} \cdot \boldsymbol{v}'_i}{\boldsymbol{v}'_i \cdot \boldsymbol{v}'_i}\boldsymbol{v}'_i \,.
\end{aligned} \tag{10.148}$$

Since $\boldsymbol{v}'_n$ depends only on the vectors $\boldsymbol{v}_n \dots \boldsymbol{v}_1$ we can write

$$(\boldsymbol{v}'_1, \dots, \boldsymbol{v}'_{N_p}) = (\boldsymbol{v}_1, \dots, \boldsymbol{v}_{N_p}) V_R^{-1} \,, \tag{10.149}$$

where V_R is an upper unit triangular $N_p \times N_p$ matrix, that is the diagonal matrix elements are equal to unity. One can furthermore normalize the vectors $\boldsymbol{v}'_1, \dots, \boldsymbol{v}'_{N_p}$ to obtain

$$B^\rangle \equiv (\boldsymbol{v}_1, \dots, \boldsymbol{v}_{N_p}) = \underbrace{\left(\frac{\boldsymbol{v}'_1}{|\boldsymbol{v}'_1|}, \dots, \frac{\boldsymbol{v}'_{N_p}}{|\boldsymbol{v}'_{N_p}|}\right)}_{\equiv U^\rangle} D_R V_R \,, \tag{10.150}$$

where D is a diagonal matrix containing the scales. One can repeat the procedure to obtain: $B^{\langle} \equiv P^{\dagger} B_{\boldsymbol{s}}(2\Theta, \tau) = V_L D_L U^{\langle}$. The Green function for the PQMC is now particularly easy to compute:

$$
\begin{aligned}
1 - G_{\boldsymbol{s}}(\tau) &= B^{\rangle} \left(B^{\langle} B^{\rangle} \right)^{-1} B^{\langle} \\
&= U^{\rangle} D_R V_R \left(V_L D_L U^{\langle} U^{\rangle} D_R V_R \right)^{-1} V_L D_L U^{\langle} \\
&= U^{\rangle} D_R V_R \left(D_R V_R \right)^{-1} \left(U^{\langle} U^{\rangle} \right)^{-1} \left(V_L D_L \right)^{-1} V_L D_L U^{\langle} \\
&= U^{\rangle} \left(U^{\langle} U^{\rangle} \right)^{-1} U^{\langle} \,.
\end{aligned}
\tag{10.151}
$$

Thus, in the PQMC, all scales which are at the origin of the numerical instabilities disappear from the problem when computing Green functions. Since the entire algorithm relies solely on the knowledge of the Green function, the above stabilization procedure leaves the physical results invariant. Note that although appealing, the Gram-Schmidt orthonormalization is itself unstable, and hence it is more appropriate to use singular value decompositions based on Housholder's method to obtain the above UDV-form for the B matrices [74]. In practice the frequency at which the stabilization is carried out is problem dependent. Typically, for the Hubbard model with $\Delta_\tau t = 0.125$ stabilization at every 10th time slice produces excellent accuracy.

The stabilization procedure for the finite-temperature algorithm is more subtle since scales do not drop out in the calculation of the Green function. Below, we provide two ways of computing the Green function.

The first approach relies on the identity

$$
\begin{pmatrix} A & B \\ C & D \end{pmatrix}^{-1} = \begin{pmatrix} \left(A - BD^{-1}C\right)^{-1} & \left(C - DB^{-1}A\right)^{-1} \\ \left(B - AC^{-1}D\right)^{-1} & \left(D - CA^{-1}B\right)^{-1} \end{pmatrix} , \tag{10.152}
$$

where A, B, C and D are matrices. Using the above, we obtain

$$
\begin{pmatrix} 1 & B_{\boldsymbol{s}}(\beta, \tau) \\ -B_{\boldsymbol{s}}(\tau, 0) & 1 \end{pmatrix}^{-1} = \begin{pmatrix} G_{\boldsymbol{s}}(0) & -(1 - G_{\boldsymbol{s}}(0)) B_{\boldsymbol{s}}^{-1}(\tau, 0) \\ B_{\boldsymbol{s}}(\tau, 0) G_{\boldsymbol{s}}(0) & G_{\boldsymbol{s}}(\tau) \end{pmatrix} . \tag{10.153}
$$

The diagonal terms on the right hand side of the above equation correspond to the desired equal-time Green functions. The off-diagonal terms are nothing but the time displaced Green functions, see (10.125) and (10.126). To evaluate the left hand side of the above equation, we first have to bring $B_{\boldsymbol{s}}(\tau, 0)$ and $B_{\boldsymbol{s}}(\beta, \tau)$ in UDV-forms. This has to be done step by step so as to avoid mixing large and small scales. Consider the propagation $B_{\boldsymbol{s}}(\tau, 0)$, and a time interval τ_1, with $n\tau_1 = \tau$, for which the different scales in $B_{\boldsymbol{s}}(n\tau_1, (n-1)\tau_1)$ do not exceed machine precision. Since $B_{\boldsymbol{s}}(\tau, 0) = B_{\boldsymbol{s}}(n\tau_1, (n-1)\tau_1) \dots B_{\boldsymbol{s}}(\tau_1, 0)$ we can evaluate $B_{\boldsymbol{s}}(\tau, 0)$ for $n = 2$ with

$$
B_{\boldsymbol{s}}(2\tau_1, \tau_1) \underbrace{B_{\boldsymbol{s}}(\tau_1, 0)}_{U_1 D_1 V_1} = \underbrace{\left(\left(B_{\boldsymbol{s}}(2\tau_1, \tau_1) U_1 \right) D_1 \right)}_{U_2 D_2 V} V_1 = U_2 D_2 V_2 \,, \tag{10.154}
$$

where $V_2 = VV_1$. The parenthesis determine the order in which the matrix multiplication are to be done. In all operations, mixing of scales is avoided. After the multiplication with diagonal matrix D_1 scales are again separated with the use of the singular value decomposition.

Thus, for $B_s(\tau, 0) = U_R D_R V_R$ and $B_s(\beta, \tau) = V_L D_L U_L$ we have to invert

$$\begin{aligned}
&\begin{pmatrix} I & V_L D_L U_L \\ -U_R D_R V_R & I \end{pmatrix}^{-1} \\
&= \left[\begin{pmatrix} V_L & 0 \\ 0 & U_R \end{pmatrix} \underbrace{\begin{pmatrix} (V_R V_L)^{-1} & D_L \\ -D_R & (U_L U_R)^{-1} \end{pmatrix}}_{UDV} \begin{pmatrix} V_R & 0 \\ 0 & U_L \end{pmatrix} \right]^{-1} \\
&= \left[\begin{pmatrix} V_R^{-1} & 0 \\ 0 & U_L^{-1} \end{pmatrix} V^{-1} \right] D^{-1} \left[U^{-1} \begin{pmatrix} V_L^{-1} & 0 \\ 0 & U_R^{-1} \end{pmatrix} \right] .
\end{aligned} \tag{10.155}$$

In the above, all matrix multiplications are well defined. In particular, the matrix D contains only large scales since the matrices $(V_R V_L)^{-1}$ and $(U_L U_R)^{-1}$ act as a cutoff to the exponentially small scales in D_L and D_R. This method to compute Green functions is very stable and has the advantage of producing time displaced Green functions. However, it is numerically expensive since the matrices involved are twice as big as the B matrices.

Alternative methods to compute $G_s(\tau)$ which involve matrix manipulations only of the size of B include

$$\begin{aligned}
&(1 + B_s(\tau, 0) B_s(\beta, \tau))^{-1} \\
&= (1 + U_R D_R V_R V_L D_L U_L)^{-1} \\
&= U_L^{-1} \underbrace{((U_L U_R)^{-1} + D_R (V_R V_L) D_L)}_{UDV}^{-1} U_R^{-1} \\
&= (V U_L)^{-1} D^{-1} (U_R U^{-1}) .
\end{aligned} \tag{10.156}$$

Again, $(U_L U_R)^{-1}$ acts as a cutoff to the small scales in $D_R(V_R V_L) D_L$ so that D contains only large scales.

The accuracy of both presented methods may be tested by in the following way. Given the Green function at time τ we can upgrade and wrap, see (10.128), this Green function to time slice $\tau + \tau_1$. Of course, for the time interval τ_1 the involved scales should lie within the accuracy of the computer $\sim 10^{-12}$ for double precision numbers. The Green function at time $\tau + \tau_1$ obtained thereby may be compared to the one computed from scratch using (10.155) or (10.156). For a 4×4 half-filled Hubbard model at $U/t = 4$, $\beta t = 20$, $\Delta_\tau t = 0.1$ and $\tau_1 = 10\ \Delta_\tau$ we obtain an average (maximal) difference between the matrix elements of both Green functions of 10^{-10} (10^{-6}) which is orders of magnitude smaller than the statistical uncertainty. Had we chosen $\tau_1 = 50\ \Delta_\tau$ the accuracy drops to 0.01 and 100.0 for the average and maximal differences.

10.7.2 The Monte Carlo Sampling

The Monte Carlo sampling used in the auxiliary field approach is based on a single spin-flip algorithm. Acceptance or rejection of this spin flip requires the knowledge of the ratio

$$R = \frac{\mathrm{P}_{\boldsymbol{s}'}}{\mathrm{P}_{\boldsymbol{s}}} , \tag{10.157}$$

where $\boldsymbol{s}$ and $\boldsymbol{s}'$ differ only at one point in space $\boldsymbol{i}$, and imaginary time n. For the Ising field required to decouple the Hubbard interaction, (10.236) and (10.239)

$$s'_{\boldsymbol{i}',n'} = \begin{cases} s_{\boldsymbol{i}',n'} & \text{if } \boldsymbol{i}' \neq \boldsymbol{i} \text{ and } n' \neq n \\ -s_{\boldsymbol{i},n} & \text{if } \boldsymbol{i}' = \boldsymbol{i} \text{ and } n' = n \end{cases} . \tag{10.158}$$

The calculation of R boils down to computing the ratio of two determinants

$$R = \begin{cases} \dfrac{\det\left[1 + B_{\boldsymbol{s}'}(\beta, 0)\right]}{\det\left[1 + B_{\boldsymbol{s}}(\beta, 0)\right]} & \text{for the FTQMC} \\[2ex] \dfrac{\det\left[P^\dagger B_{\boldsymbol{s}'}(2\Theta, 0) P\right]}{\det\left[P^\dagger B_{\boldsymbol{s}}(2\Theta, 0) P\right]} & \text{for the PQMC} \end{cases} . \tag{10.159}$$

For the Hubbard interaction with HS transformation of (10.236) only the matrix $V(\boldsymbol{s}_n)$ will be effected by the move. Hence, with

$$\mathrm{e}^{V(\boldsymbol{s}'_n)} = \Big[1 + \underbrace{\left(\mathrm{e}^{V(\boldsymbol{s}'_n)}\mathrm{e}^{-V(\boldsymbol{s}_n)} - 1\right)}_{\Delta}\Big]\mathrm{e}^{V(\boldsymbol{s}_n)} \tag{10.160}$$

we have

$$B_{\boldsymbol{s}'}(\bullet, 0) = B_{\boldsymbol{s}}(\bullet, \tau)\,(1 + \Delta)\, B_{\boldsymbol{s}}(\tau, 0) , \tag{10.161}$$

where the $\bullet$ stands for 2Θ or β and $\tau = n\Delta_\tau$.

For the FTQMC, the ratio is given by

$$\begin{aligned} &\frac{\det\left[1 + B_{\boldsymbol{s}}(\beta,\tau)(1+\Delta)B_{\boldsymbol{s}}(\tau,0)\right]}{\det\left[1 + B_{\boldsymbol{s}}(\beta,0)\right]} \\ &= \det\left[1 + \Delta B_{\boldsymbol{s}}(\tau,0)\left(1 + B_{\boldsymbol{s}}(\beta,0)\right)^{-1} B_{\boldsymbol{s}}(\beta,\tau)\right] \\ &= \det\left[1 + \Delta\left(1 - \left(1 + B_{\boldsymbol{s}}(\tau,0)B_{\boldsymbol{s}}(\beta,\tau)\right)^{-1}\right)\right] \\ &= \det\left[1 + \Delta\left(1 - G_{\boldsymbol{s}}(\tau)\right)\right] . \end{aligned} \tag{10.162}$$

Where the last line follows from the fact that the equal-time Green function reads $G_{\boldsymbol{s}}(\tau) = (1 + B_{\boldsymbol{s}}(\tau,0)B_{\boldsymbol{s}}(\beta,\tau))^{-1}$. Hence the ratio is uniquely determined from the knowledge of the equal-time Green function.

Let us now compute the ratio for the PQMC. Introducing the notation $B^{\langle}_{\boldsymbol{s}} = P^\dagger B_{\boldsymbol{s}}(2\Theta,\tau)$ and $B^{\rangle}_{\boldsymbol{s}} = B_{\boldsymbol{s}}(\tau,0)P$, again we have to evaluate

$$\frac{\det\left[B_s^{\langle}\left(1+\Delta^{(i)}\right)B_s^{\rangle}\right]}{\det\left[B_s^{\langle}B_s^{\rangle}\right]} = \det\left[B_s^{\langle}\left(1+\Delta^{(i)}\right)B_s^{\rangle}\left(B_s^{\langle}B_s^{\rangle}\right)^{-1}\right]$$
$$= \det\left[1+B_s^{\langle}\Delta^{(i)}B_s^{\rangle}\left(B_s^{\langle}B_s^{\rangle}\right)^{-1}\right]$$
$$= \det\left[1+\Delta^{(i)}B_s^{\rangle}\left(B_s^{\langle}B_s^{\rangle}\right)^{-1}B_s^{\langle}\right], \quad (10.163)$$

where the last equation follows from the identity $\det[1+AB] = \det[1+BA]$ for arbitrary rectangular matrices[3]. We can recognize the Green function of the PQMC $B_s^{\rangle}(B_s^{\langle}B_s^{\rangle})^{-1}B_s^{\langle} = 1 - G_s(\tau)$. The result is thus identical to that of the FTQMC provided that we replace the finite-temperature equal-time Green function with the zero-temperature one. Hence, in both algorithms, the ratio is essentially given by the equal-time Green function which, at this point, we know how to compute in a numerically stable manner.

Having calculated the ratio R for a single spin flip one may now decide stochastically within, for example, a Metropolis scheme if the move is accepted or not. In case of acceptance, we have to update the Green function since this quantity is required at the next step.

Since in general the matrix Δ has only a few non-zero entries, it is convenient to use the Sherman-Morrison formula [74] which states that

$$(A+\boldsymbol{u}\otimes\boldsymbol{v})^{-1} = (1+A^{-1}\boldsymbol{u}\otimes\boldsymbol{v})^{-1}A^{-1}$$
$$= [1 - A^{-1}\boldsymbol{u}\otimes\boldsymbol{v} + A^{-1}\boldsymbol{u}\otimes\underbrace{\boldsymbol{v}A^{-1}\boldsymbol{u}}_{\equiv\lambda}\otimes\boldsymbol{v} + A^{-1}\boldsymbol{u}\otimes\lambda^2\boldsymbol{v} - \ldots]A^{-1}$$
$$= \left[1 - A^{-1}\boldsymbol{u}\otimes\boldsymbol{v}\left(1-\lambda+\lambda^2-\ldots\right)\right]A^{-1}$$
$$= A^{-1} - \frac{(A^{-1}\boldsymbol{u})\otimes(\boldsymbol{v}A^{-1})}{1+\boldsymbol{v}\bullet A^{-1}\boldsymbol{u}}, \quad (10.164)$$

where A is an $N\times N$ matrix, $\boldsymbol{u},\boldsymbol{v}$ N-dimensional vectors with tensor product defined as $(\boldsymbol{u}\otimes\boldsymbol{v})_{x,y} = \boldsymbol{u}_x\boldsymbol{v}_y$.

To show how to use this formula for the updating of the Green function, let us first assume that matrix Δ has only one non-vanishing entry $\Delta_{x,y} = \delta_{x,z}\delta_{y,z'}\eta^{(z,z')}$. In the case of the FTQMC we will then have to compute

$$G_{s'}(\tau) = [1+(1+\Delta)B_s(\tau,0)B_s(\beta,\tau)]^{-1}$$
$$= B_s^{-1}(\beta,\tau)\left[1+B_s(\beta,\tau)(1+\Delta)B_s(\tau,0)\right]^{-1}B_s(\beta,\tau)$$
$$= B_s^{-1}(\beta,\tau)\left[1+B_s(\beta,\tau)B_s(\tau,0)+\boldsymbol{u}\otimes\boldsymbol{v}\right]^{-1}B_s(\beta,\tau) \quad (10.165)$$

where $\boldsymbol{u}_x = [B_s(\beta,\tau)]_{x,z}\,\eta^{(z,z')}$ and $\boldsymbol{v}_x = [B_s(\tau,0)]_{z',x}$.

[3] This identity may be formally proven by using the relation $\det(1+AB) = \exp(\mathrm{Tr}\log(1+AB))$, expanding the logarithm and using the cyclic properties of the trace.

Using the Sherman-Morrison formula for inverting $1+B_{\boldsymbol{s}}(\beta,\tau)B_{\boldsymbol{s}}(\tau,0)+\boldsymbol{u}\otimes\boldsymbol{v}$ yields

$$[G_{\boldsymbol{s}'}(\tau)]_{x,y} = [G_{\boldsymbol{s}}(\tau)]_{x,y} - \frac{[G_{\boldsymbol{s}}(\tau)]_{x,z}\,\eta^{(z,z')}\,[1-G_{\boldsymbol{s}}(\tau)]_{z',y}}{1+\eta^{(z,z')}\,[1-G_{\boldsymbol{s}}(\tau)]_{z',z}} . \tag{10.166}$$

Precisely the same equation holds for the PQMC provided that one replaces the finite-temperature Green function by the zero-temperature one. To show this, one will first compute

$$\begin{aligned}(B^{\langle}_{\boldsymbol{s}'}B^{\rangle}_{\boldsymbol{s}'})^{-1} &= \left(B^{\langle}_{\boldsymbol{s}}(1+\Delta)B^{\rangle}_{\boldsymbol{s}}\right)^{-1} = \left(B^{\langle}_{\boldsymbol{s}}B^{\rangle}_{\boldsymbol{s}}+\boldsymbol{u}\otimes\boldsymbol{v}\right)^{-1} \\ &= (B^{\langle}_{\boldsymbol{s}}B^{\rangle}_{\boldsymbol{s}})^{-1} - \frac{(B^{\langle}_{\boldsymbol{s}}B^{\rangle}_{\boldsymbol{s}})^{-1}\boldsymbol{u}\otimes\boldsymbol{v}(B^{\langle}_{\boldsymbol{s}}B^{\rangle}_{\boldsymbol{s}})^{-1}}{1+\eta^{(z,z')}\,[1-G^{0}_{\boldsymbol{s}}(\tau)]_{z',z}}\end{aligned} \tag{10.167}$$

with $\boldsymbol{u}_x = [B^{\langle}_{\boldsymbol{s}}]_{x,z}\eta^{(z,z')}$ and $\boldsymbol{v}_x = [B^{\rangle}_{\boldsymbol{s}}]_{z',x}$. Here x runs from $1\ldots N_p$ where N_p corresponds to the number of particles contained in the trial wave function and the zero-temperature Green function reads $G^{0}_{\boldsymbol{s}}(\tau) = 1 - B^{\rangle}_{\boldsymbol{s}}(B^{\langle}_{\boldsymbol{s}}B^{\rangle}_{\boldsymbol{s}})^{-1}B^{\langle}_{\boldsymbol{s}}$. After some straightforward algebra, one obtains

$$\begin{aligned}\left[G^{0}_{\boldsymbol{s}'}(\tau)\right]_{x,y} &= \left[1-(1+\Delta)B^{\rangle}_{\boldsymbol{s}}(B^{\langle}_{\boldsymbol{s}}(1+\Delta)B^{\rangle}_{\boldsymbol{s}})^{-1}B^{\langle}_{\boldsymbol{s}}\right]_{x,y} \\ &= \left[G^{0}_{\boldsymbol{s}}(\tau)\right]_{x,y} - \frac{\left[G^{0}_{\boldsymbol{s}}(\tau)\right]_{x,z}\eta^{(z,z')}\left[1-G^{0}_{\boldsymbol{s}}(\tau)\right]_{z',y}}{1+\eta^{(z,z')}\,[1-G^{0}_{\boldsymbol{s}}(\tau)]_{z',z}}\end{aligned} \tag{10.168}$$

In the above, we have assumed that the matrix Δ has only a single non-zero entry. In general, it is convenient to work in a basis where Δ is diagonal with n non-vanishing eigenvalues. One will then iterate the above procedure n-times to upgrade the Green function.

10.7.3 Numerical Calculation of Imaginary Time Displaced Green Functions

In Sect. 10.6.4 we introduced the time displaced Green functions both within the ground-state and finite-temperature formulations. Our aim here is to show how to compute them in a numerically stable manner. We will first start with the FTQMC and then concentrate on the PQMC.

10.7.3.1 FTQMC

For a given HS field, we wish to evaluate

$$G_{\boldsymbol{s}}(\tau_1,\tau_2)_{x,y} = \langle c_x(\tau_1)c^{\dagger}_y(\tau_2)\rangle_{\boldsymbol{s}} = B_{\boldsymbol{s}}(\tau_1,\tau_2)G_{\boldsymbol{s}}(\tau_2) \quad \tau_1>\tau_2 , \tag{10.169}$$

where $G_{\boldsymbol{s}}(\tau_1)$ is the equal-time Green function computed previously and

$$G_{\boldsymbol{s}}(\tau_1,\tau_2)_{x,y} = -\langle c^\dagger_y(\tau_2) c_x(\tau_1)\rangle_{\boldsymbol{s}}$$
$$= -\left(1 - G_{\boldsymbol{s}}(\tau_1)\right) B^{-1}_{\boldsymbol{s}}(\tau_2,\tau_1) \quad \tau_2 > \tau_1 \;, \qquad (10.170)$$

see (10.126) and (10.125).

Returning to (10.153) we see that we have already computed the time displaced Green functions $G_{\boldsymbol{s}}(0,\tau)$ and $G_{\boldsymbol{s}}(\tau,0)$ when discussing a stabilization scheme for the equal-time Green functions. However, this calculation is expensive and is done only at times $\tau = n\tau_1$ where τ_1 is time scale on which all energy scales fit well on finite precision machines. To obtain the Green functions for arbitrary values of τ one uses the relations

$$G_{\boldsymbol{s}}(0,\tau+\tau_2) = G_{\boldsymbol{s}}(0,\tau) B^{-1}_{\boldsymbol{s}}(\tau_2,\tau) \;,$$
$$G_{\boldsymbol{s}}(\tau+\tau_2,0) = B_{\boldsymbol{s}}(\tau_2,\tau) G_{\boldsymbol{s}}(\tau,0) \;, \qquad (10.171)$$

where $\tau_2 < \tau_1$.

With the above method, we have access to all time displaced Green functions $G_{\boldsymbol{s}}(0,\tau)$ and $G_{\boldsymbol{s}}(\tau,0)$. However, we do not use translation invariance in imaginary time. Clearly, using this symmetry in the calculation of time displaced quantities will reduce the fluctuations which may sometimes be desirable. A numerically expensive but elegant way of producing all time displaced Green functions relies on the inversion of the matrix O given in (10.129). Here, provided the τ_1 is small enough so that the scales involved in $B_s(\tau+\tau_1,\tau)$ fit on finite precision machines, the matrix inversion of O^{-1} is numerically stable and and yields the Green functions between arbitrary time slices $n\tau_1$ and $m\tau_1$. For $\beta/\tau_1 = l$, the matrix to inverse has the dimension l times the size of the B matrices, and is hence expensive to compute. It is worth noting that on vector machines the performance grows with growing vector size so that the above method can become attractive. Having computed the Green functions $G_{\boldsymbol{s}}(n\tau_1, m\tau_1)$ we can obtain Green functions on any two time slices by using equations of the type (10.171).

10.7.3.2 PQMC

Zero-temperature time displaced Green functions are given by

$$G_s\left(\Theta+\frac{\tau}{2},\Theta-\frac{\tau}{2}\right)_{x,y} = \left[B_{\boldsymbol{s}}\left(\Theta+\frac{\tau}{2},\Theta-\frac{\tau}{2}\right) G_{\boldsymbol{s}}\left(\Theta-\frac{\tau}{2}\right)\right]_{x,y}$$
$$G_s\left(\Theta-\frac{\tau}{2},\Theta+\frac{\tau}{2}\right)_{x,y} = -\left[\left(1-G_{\boldsymbol{s}}\left(\Theta-\frac{\tau}{2}\right)\right) B^{-1}_{\boldsymbol{s}}\left(\Theta+\frac{\tau}{2},\Theta-\frac{\tau}{2}\right)\right]_{x,y} \qquad (10.172)$$

with $\tau > 0$, see (10.138).

Before showing who to compute imaginary time displaced Green functions, we first note that a direct multiplication of the equal-time Green function with B matrices is unstable for larger values of τ. This can be understood in the framework of free electrons on a 2D square lattice

$$H = -t \sum_{<i,j>} c_i^\dagger c_j \ , \tag{10.173}$$

where the sum runs over nearest neighbors. For this Hamiltonian one has

$$\langle \Psi_0 | c_{\boldsymbol{k}}^\dagger(\tau) c_{\boldsymbol{k}} | \Psi_0 \rangle = \mathrm{e}^{\tau(\epsilon_{\boldsymbol{k}} - \mu)} \langle \Psi_0 | c_{\boldsymbol{k}}^\dagger c_{\boldsymbol{k}} | \Psi_0 \rangle \ , \tag{10.174}$$

where $\epsilon_{\boldsymbol{k}} = -2t(\cos(\boldsymbol{k}\boldsymbol{a}_x) + \cos(\boldsymbol{k}\boldsymbol{a}_y))$, $\boldsymbol{a}_x$, $\boldsymbol{a}_y$ being the lattice constants. We will assume $|\Psi_0\rangle$ to be non-degenerate. In a numerical calculation the eigenvalues and eigenvectors of the above Hamiltonian will be known up to machine precision ϵ. In the case $\epsilon_{\boldsymbol{k}} - \mu > 0$ is $\langle \Psi_0 | c_{\boldsymbol{k}}^\dagger c_{\boldsymbol{k}} | \Psi_0 \rangle \equiv 0$. However, on a finite precision machine the later quantity will take a value of the order of ϵ. When calculating $\langle \Psi_0 | c_{\boldsymbol{k}}^\dagger(\tau) c_{\boldsymbol{k}} | \Psi_0 \rangle$ this roundoff error will be blown up exponentially and the result for large values of τ will be unreliable. In (10.138) the B matrices play the role of the exponential factor $\exp\left(\tau(\epsilon_{\boldsymbol{k}} - \mu)\right)$ and the equal-time Green functions correspond to $\langle \Psi_0 | c_{\boldsymbol{k}}^\dagger c_{\boldsymbol{k}} | \Psi_0 \rangle$. In the PQMC, the stability problem is much more severe than for free electrons since the presence of the time dependent HS field mixes different scales.

An elegant and efficient method [75] to alleviate this problem rests on the observation that in the PQMC the Green function is a projector. Consider again the free electron case. For a non-degenerate ground state $\langle \Psi_0 | c_{\boldsymbol{k}}^\dagger c_{\boldsymbol{k}} | \Psi_0 \rangle = 0, 1$ so that

$$\langle \Psi_0 | c_{\boldsymbol{k}}^\dagger(\tau) c_{\boldsymbol{k}} | \Psi_0 \rangle = \left(\langle \Psi_0 | c_{\boldsymbol{k}}^\dagger c_{\boldsymbol{k}} | \Psi_0 \rangle \mathrm{e}^{\epsilon_{\boldsymbol{k}} - \mu} \right)^\tau \ . \tag{10.175}$$

The above involves only well defined numerical manipulations even in the large τ limit provided that all scales fit onto finite precision machines for a unit time interval.

The implementation of this idea in the QMC algorithm is as follows. First, one has to notice that the Green function $G_{\boldsymbol{s}}(\tau)$ is a projector

$$G_{\boldsymbol{s}}(\tau)^2 = G_{\boldsymbol{s}}(\tau) \ . \tag{10.176}$$

We have already seen that for $P^\dagger B_{\boldsymbol{s}}(2\Theta, \tau) = V_L D_L U^{\langle}$ and $B_{\boldsymbol{s}}(\tau, 0) = U^{\rangle} D_R U_R$, $G_{\boldsymbol{s}}(\tau) = 1 - U^{\rangle}(U^{\langle} U^{\rangle})^{-1} U^{\langle}$. Since $\left[U^{\rangle}(U^{\langle} U^{\rangle})^{-1} U^{\langle}\right]^2 = U^{\rangle}(U^{\langle} U^{\rangle})^{-1} U^{\langle}$ we have

$$\begin{aligned} G_{\boldsymbol{s}}^2(\tau) &= G_{\boldsymbol{s}}(\tau) \ , \\ (1 - G_{\boldsymbol{s}}(\tau))^2 &= 1 - G_{\boldsymbol{s}}(\tau) \ . \end{aligned} \tag{10.177}$$

This property implies that $G_{\boldsymbol{s}}(\tau_1, \tau_3)$ obeys a simple composition identity

$$G_{\boldsymbol{s}}(\tau_1, \tau_3) = G_{\boldsymbol{s}}(\tau_1, \tau_2) G_{\boldsymbol{s}}(\tau_2, \tau_1) \ . \tag{10.178}$$

In particular for $\tau_1 > \tau_2 > \tau_3$

$$\begin{aligned} G_{\boldsymbol{s}}(\tau_1, \tau_3) &= B_{\boldsymbol{s}}(\tau_1, \tau_3) G_{\boldsymbol{s}}^2(\tau_3) = G_{\boldsymbol{s}}(\tau_1, \tau_3) G_{\boldsymbol{s}}(\tau_3) \\ &= \underbrace{G_{\boldsymbol{s}}(\tau_1, \tau_3) B_{\boldsymbol{s}}^{-1}(\tau_2, \tau_3)}_{G_{\boldsymbol{s}}(\tau_1, \tau_2)} \underbrace{B_{\boldsymbol{s}}(\tau_2, \tau_3) G_{\boldsymbol{s}}(\tau_3)}_{G_{\boldsymbol{s}}(\tau_2, \tau_3)} \ . \end{aligned} \tag{10.179}$$

A similar proof is valid for $\tau_3 > \tau_2 > \tau_1$.

Using this composition property (10.178) we can break up a large τ interval into a set of smaller intervals of length $\tau = N\tau_1$ so that

$$G_s\left(\Theta+\frac{\tau}{2},\Theta-\frac{\tau}{2}\right)=\prod_{n=0}^{N-1} G_s\left(\Theta-\frac{\tau}{2}+[n+1]\,\tau_1,\Theta-\frac{\tau}{2}+n\tau_1\right) . \tag{10.180}$$

The above equation is the generalization of (10.175). If τ_1 is small enough each Green function in the above product is accurate and has matrix elements bounded by order unity. The matrix multiplication is then numerically well defined.

We conclude this section by comparing with a different approach to computed imaginary time correlation functions in the framework of the PQMC [63]. We consider the special case of the Kondo lattice model (see Fig. 10.16). As apparent the results are identical within error-bars. The important point however, is that the method based on (10.180) is for the considered case an order of magnitude quicker in CPU time than the method of [63].

10.7.4 Practical Implementation

In this section we first describe in detail a possible efficient implementation of the finite-temperature algorithm and then comment on the differences required for the

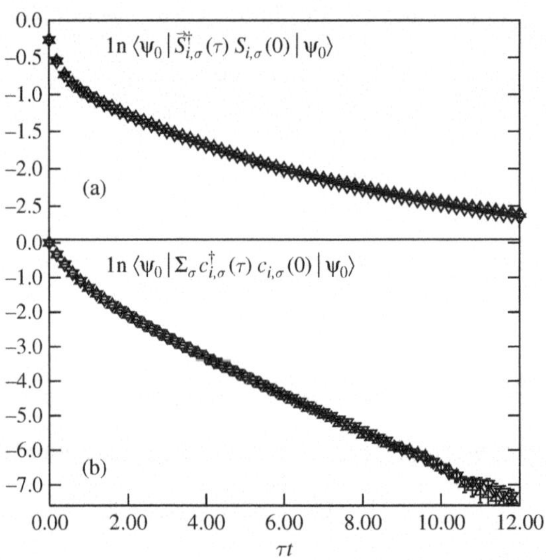

Fig. 10.16. Imaginary time displaced on-site spin-spin correlation function (**a**) and Green function (**b**). We consider a 6×6 lattice at half-filling and $J/t = 1.2$. In both (**a**) and (**b**) results obtained from (10.180) ($\triangle$) and from an alternative approach presented in [63] ($\bigtriangledown$) are plotted

implementation of the projector formalism. It is convenient to split the total imaginary time propagation β, into intervals of length τ_1 such that $n\tau_1 = \beta$. We require τ_1 to be small enough such that all scales in the matrices $B_{\boldsymbol{s}}(\tau_1, 0)$ fit into say 64 bit reals. The organization of the time slices is shown schematically in Fig. 10.17. To save computer time, we will need enough memory to store $n+1$ orthogonal matrices U, $n+1$ triangular matrices V and $n+1$ diagonal matrices D.

At the onset, we start from a randomly chosen Hubbard-Stratonovich configuration of fields $\boldsymbol{s}$. We then compute $B_{\boldsymbol{s}}(\tau_1, 0)$ carry out a singular value decomposition and store the result in U_1, D_1 and V_1. Given the UDV-decomposition of $B_{\boldsymbol{s}}(n_\tau \tau_1, 0)$, where $(1 \leq n_\tau < n)$, we compute the UDV-decomposition of $B_{\boldsymbol{s}}\left[(n_\tau + 1)\tau_1, 0\right]$ using (10.154) and store the results in $U_{n_\tau+1}$, $D_{n_\tau+1}$ and $D_{n_\tau+1}$. Hence, our storage now contains

$$U_{n_\tau} D_{n_\tau} V_{n_\tau} = B_{\boldsymbol{s}}(n_\tau \tau_1, 0) \qquad (10.181)$$

with $1 \leq n_\tau \leq n$.

At this stage we can sequentially upgrade the Hubbard Stratonovich fields from $\tau = \beta$ to $\tau = \Delta_\tau$. In doing so, we will take care of storing information to subsequently carry out a sweep from $\tau = \Delta_\tau$ to $\tau = \beta$.

10.7.4.1 From $\tau = \beta$ to $\tau = \Delta_\tau$

From the UDV-decomposition of $B_{\boldsymbol{s}}(\beta = n\tau_1, 0)$ which we read out of the storage (U_n, D_n, V_n), we compute in a numerically stable way the equal-time Green function on time slice $\tau = \beta$. Having freed the arrays U_n, D_n and V_n we set them to unity such that $B_s(\beta, n\tau_1 = \beta) \equiv 1 = V_n D_n U_n$. We can now sweep down from time slice $\tau = \beta$ to time slice $\tau = \Delta_\tau$.

Given the Green function at time $\tau = n_\tau \tau_1$ we sequentially upgrade all the Hubbard Stratonovich fields on this time slice. Each time a move is accepted, we will have to update the equal-time Green function. To move to the next time slice $\tau - \Delta_\tau$, we make use of the equation

$$G_{\boldsymbol{s}}(\tau - \Delta_\tau) = B_{\boldsymbol{s}}^{-1}(\tau, \tau - \Delta_\tau) G_{\boldsymbol{s}}(\tau) B_{\boldsymbol{s}}(\tau, \tau - \Delta_\tau) \,. \qquad (10.182)$$

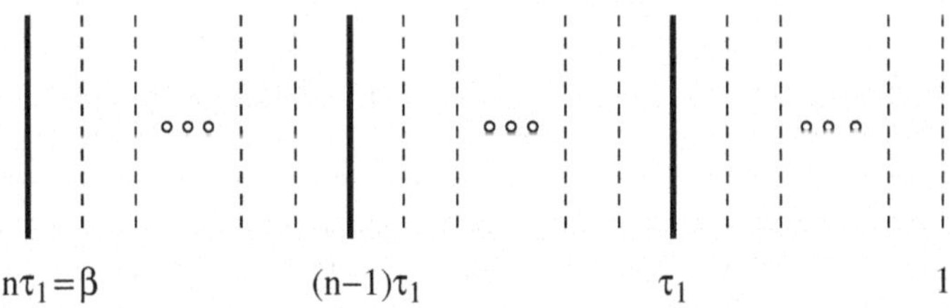

Fig. 10.17. Each line (*solid* or *dashed*) denotes a time slice separated by an imaginary time propagation Δ_τ. The solid lines correspond to time slices where we store the UDV-decomposition of the matrices $B_s(\beta, n_\tau \tau_1)$ or $B_s(n_\tau \tau_1, 0)$ depending upon the direction of the propagation ($1 \leq n_\tau \leq n$)

We will repeat the above procedure till we arrive at time slice $\tau = (n_\tau - 1)\tau_1$. At this stage, we have to recompute the equal-time Green function due to the accumulation of round-off errors and hence loss of precision. To do so, we read from the storage $U_R = U_{n_\tau - 1}$, $D_R = U_{n_\tau - 1}$ and $V_R = U_{n_\tau - 1}$ such that $B_s((n_\tau - 1)\tau_1, 0) = U_R D_R V_R$. Note that we have not yet upgraded the Hubbard Stratonovich fields involved in $B_s((n_\tau - 1)\tau_1, 0)$ so that this storage slot is still up to date. We then compute the matrix $B_s(n_\tau\tau_1, (n_\tau - 1)\tau_1)$ and read from the storage $\widetilde{V}_L = V_{n_\tau}$, $\widetilde{D}_L = V_{n_\tau}$ and $\widetilde{U}_L = V_{n_\tau}$ such that $B_s(\beta, n_\tau\tau_1) = \widetilde{V}_L \widetilde{D}_L \widetilde{U}_L$. With this information and the computed matrix $B_s(n_\tau\tau_1, (n_\tau - 1)\tau_1)$ we will calculate $B_s(\beta, (n_\tau - 1)\tau_1) = V_L D_L U_L$, see (10.154). We now store this result as $V_{n_\tau - 1} = V_L$, $D_{n_\tau - 1} = D_L$ and $U_{n_\tau - 1} = U_L$, and recompute the Green function. Note that as a cross check, one can compare both Green functions to test the numerical accuracy. Hence, we now have a fresh estimate of the Green function at time slice $\tau = (n_\tau - 1)\tau_1$ and we can iterate the procedure till we arrive at time slice Δ_τ.

Hence, in this manner, we sweep down from time slice β to time slice Δ_τ, upgrade sequentially all the Hubbard Stratonovich fields and have stored

$$B_s(\beta, n_\tau\tau_1) = V_{n_\tau} D_{n_\tau} U_{n_\tau} \tag{10.183}$$

with $0 \le n_\tau \le n$. We can now carry out a sweep from Δ_τ to β and take care of storing the information required for the sweep from β to Δ_τ.

10.7.4.2 From $\tau = \Delta_\tau$ to β

We initially set $n_\tau = 0$, read out from the storage $B_s(\beta, 0) = V_0 D_0 U_0$ and compute the Green function on time slice $\tau = 0$. This storage slot is then set to unity such that $B_s(0, 0) = U_0 D_0 V_0 \equiv 1$.

Assuming that we are on time slice $\tau = n_\tau\tau_1$, we propagate the Green function to time slice $\tau + \Delta_\tau$ with

$$G_s(\tau + \Delta_\tau) = B_s(\tau + \Delta_\tau, \tau) G_s(\tau) B_s^{-1}(\tau + \Delta_\tau, \tau) \tag{10.184}$$

and upgrade the Hubbard Stratonovich fields on time slice $\tau + \Delta_\tau$. The above procedure is repeated till we reach time slice $(n_\tau + 1)\tau_1$, where we have to recompute the Green function. To do so, we read from the storage $V_L = V_{n_\tau + 1}$ $D_L = D_{n_\tau + 1}$ and $U_L = n_\tau + 1$ such that $B_s(\beta, (n_\tau + 1)\tau_1) = V_L D_L U_L$. We then compute $B_s((n_\tau + 1)\tau_1, n_\tau\tau_1)$ and from the UDV-form of $B_s(n_\tau\tau_1, 0)$ which we obtain from the storage slot n_τ, we calculate $B_s((n_\tau + 1)\tau_1, 0) = U_R D_R V_R$. The result of the calculation is stored in slot $n_\tau + 1$, and we recompute the Green function on time slice $(n_\tau + 1)\tau_1$. We can now proceed till we reach time slice β and we will have accumulated all the information required for carrying out a sweep from β to Δ_τ.

This completes a possible implementation of the finite-temperature method. The zero-temperature method follows exactly the same logic. However, it turns out that it is more efficient to keep track of $(P^\dagger B_s(2\Theta, 0) P)^{-1}$ since (i) it is of dimension $N_p \times N_p$ in contrast to the Green function which is a $N \times N$ matrix, and (ii) it is τ independent. When Green functions are required they are computed from scratch.

10.8 The Hirsch-Fye Impurity Algorithm

As its name suggests, this algorithm is triggered at solving impurity problems such as the Kondo and Anderson models. The strong point of the algorithm is that the CPU time is independent on the volume of the system thus allowing one to carry out simulations directly in the thermodynamic limit. The price however is a β^3 scaling of the CPU time where β is the inverse temperature. Diagrammatic determinantal methods, provide an alternative approach [8, 9] to solve impurity problems. Those algorithms are formulated in continuous time and hence do not suffer from Trotter errors. The computational effort equally scales as β^3, but there is a prefactor which renders them more efficient. We will nevertheless concentrate here on the Hirsch-Fye algorithm since it is extensively used in the framework of dynamical mean-field theories [76, 77], see Chap. 16.

We will concentrate on the Anderson model defined as

$$H - \mu N = H_0 + H_U \tag{10.185}$$

with

$$\begin{aligned} H_0 &= \sum_{\boldsymbol{k},\sigma} (\epsilon(\boldsymbol{k}) - \mu)\, c^{\dagger}_{\boldsymbol{k},\sigma} c_{\boldsymbol{k},\sigma} + \frac{V}{\sqrt{N}} \sum_{\boldsymbol{k},\sigma} \left(c^{\dagger}_{\boldsymbol{k},\sigma} f_{\sigma} + f^{\dagger}_{\sigma} c_{\boldsymbol{k},\sigma} \right) \\ &\quad + \epsilon_f \sum_{\sigma} f^{\dagger}_{\sigma} f_{\sigma} \,, \\ H_U &= U(f^{\dagger}_{\uparrow} f_{\uparrow} - 1/2)(f^{\dagger}_{\downarrow} f_{\downarrow} - 1/2) \,. \end{aligned} \tag{10.186}$$

For an extensive overview of the Anderson and related Kondo model, we refer the reader to [78].

In the next section, we will review the finite-temperature formalism. Since the CPU time scales as β^3 it is expensive to obtain ground state properties, and projective formulations of Hirsch-Fye algorithm become attractive. This corresponds to the topic of Sect. 10.8.2.

10.8.1 The Finite-Temperature Hirsch-Fye Method

In Sect. 10.6 we have shown that the grand-canonical partition function may be written as

$$Z \equiv \mathrm{Tr}\left[\mathrm{e}^{-\beta(H-\mu N)}\right] = \sum_{\boldsymbol{s}} \left[\prod_{\sigma} \det\left[1 + B^{\sigma}_{m} B^{\sigma}_{m-1} \cdots B^{\sigma}_{1}\right] \right] \tag{10.187}$$

with $m\Delta_\tau = \beta$.

To define the matrices B^{σ}_{n}, we will label all the orbitals (conduction and impurity) with the index $\boldsymbol{i}$ and use the convention that $\boldsymbol{i} = 0$ denotes the f-orbital and $\boldsymbol{i} = 1 \ldots N$ the conduction orbitals. We will furthermore define the fermionic operators

$$a^\dagger_{\boldsymbol{i},\sigma} = \begin{cases} f^\dagger_\sigma & \text{if } \boldsymbol{i} = 0 \\ c^\dagger_{\boldsymbol{i},\sigma} & \text{otherwise} \end{cases}, \tag{10.188}$$

such that the non-interacting term of the Anderson takes the form

$$H_0 = \sum_\sigma H_0^\sigma, \quad H_0^\sigma = \sum_{\boldsymbol{i},\boldsymbol{j}} a^\dagger_{\boldsymbol{i},\sigma} (h_0)_{\boldsymbol{i},\boldsymbol{j}} a_{\boldsymbol{j},\sigma} \, . \tag{10.189}$$

Using the HS transformation of (10.236), the B matrices read

$$\begin{aligned} B_n^\sigma &= \mathrm{e}^{V_n^\sigma} \mathrm{e}^{-\Delta_\tau h_0} \, , \\ (V_n^\sigma)_{\boldsymbol{i},\boldsymbol{j}} &= \delta_{\boldsymbol{i},\boldsymbol{j}} \delta_{\boldsymbol{i},0} \alpha \sigma s_n \, , \\ \cosh(\alpha) &= \mathrm{e}^{\Delta_\tau U/2} \, . \end{aligned} \tag{10.190}$$

The determinant in a given spin sector may be written as

$$\det \left[1 + B_m^\sigma B_{m-1}^\sigma \dots B_1^\sigma \right] = \det O^\sigma \tag{10.191}$$

with

$$O^\sigma = \begin{pmatrix} 1 & 0 & . & . & 0 & B_1^\sigma \\ -B_2^\sigma & 1 & 0 & . & . & 0 \\ 0 & -B_3^\sigma & 1 & . & . & 0 \\ . & 0 & -B_4^\sigma & . & . & . \\ . & . & 0 & . & . & . \\ . & . & . & . & . & . \\ 0 & . & . & 0 & -B_m^\sigma & 1 \end{pmatrix} . \tag{10.192}$$

The above identity, follows by considering – omitting spin indices – the matrix $A = O - 1$. Since

$$\mathrm{Tr}\left[A^n\right] = \sum_r \delta_{n,rm} \left(-1\right)^{r(m+1)} m \mathrm{Tr}\left[(B_m \dots B_1)^r\right] \tag{10.193}$$

we obtain:

$$\begin{aligned} \det O &= \mathrm{e}^{\mathrm{Tr}\ln(1+A)} = \mathrm{e}^{\sum_{n=1}^\infty \frac{(-1)^{n+1}}{n} \mathrm{Tr}[A^n]} \\ &= \mathrm{e}^{\sum_{r=1}^\infty \frac{(-1)^{r+1}}{r} \mathrm{Tr}[(B_m \dots B_1)^r]} \\ &= \mathrm{e}^{\mathrm{Tr}\ln(1+B_m \dots B_1)} = \det\left(1 + B_m \dots B_1\right) \, . \end{aligned} \tag{10.194}$$

From (10.129) we identify

$$(O^\sigma)^{-1} \equiv g^\sigma = \begin{pmatrix} G^\sigma(1,1) & G^\sigma(1,2) & \dots & G^\sigma(1,m) \\ G^\sigma(2,1) & G^\sigma(2,2) & \dots & G^\sigma(2,m) \\ \vdots & \vdots & \dots & \vdots \\ G^\sigma(m,1) & G^\sigma(m,2) & \dots & G^\sigma(m,m) \end{pmatrix} , \tag{10.195}$$

where $G^\sigma(n_1, n_2)$ are the time displaced Green functions

$$[G^\sigma(n_1,n_2)]_{i,j} = \begin{cases} \frac{\mathrm{Tr}\left[B^\sigma_m \dots B^\sigma_{n_1+1} a_{i,\sigma} B^\sigma_{n_1} \dots B^\sigma_{n_2+1} a^\dagger_{j,\sigma} B^\sigma_{n_2} \dots B^\sigma_1\right]}{\mathrm{Tr}\left[B^\sigma_m \dots B^\sigma_1\right]} & \text{if } n_1 \geq n_2 \\ -\frac{\mathrm{Tr}\left[B^\sigma_m \dots B^\sigma_{n_2+1} a^\dagger_{j,\sigma} B^\sigma_{n_2} \dots B^\sigma_{n_1+1} a_{i,\sigma} B^\sigma_{n_1} \dots B^\sigma_1\right]}{\mathrm{Tr}\left[B^\sigma_m \dots B^\sigma_1\right]} & \text{if } n_1 < n_2 \end{cases} \tag{10.196}$$

(see (10.125) and (10.126)). The operators B^σ_n are given by

$$B^\sigma_n = \mathrm{e}^{\alpha \sigma s_n f^\dagger_\sigma f_\sigma} \mathrm{e}^{-\Delta_\tau H^\sigma_0} \, . \tag{10.197}$$

Given an HS configuration $\boldsymbol{s}$ and $\boldsymbol{s}'$ and associated matrices

$$V^\sigma = \begin{pmatrix} V^\sigma_1 & 0 & . & . & . & 0 \\ 0 & V^\sigma_2 & 0 & . & . & 0 \\ 0 & 0 & V^\sigma_3 & 0 & . & 0 \\ . & . & . & . & . & . \\ 0 & . & . & . & 0 & V^\sigma_m \end{pmatrix} \tag{10.198}$$

and V'^σ the Green functions g^σ and g'^σ satisfy the following Dyson equation

$$g^\sigma = g'^\sigma + g'^\sigma \Delta (1 - g^\sigma) \ \text{ with } \ \Delta^\sigma = (\mathrm{e}^{V'^\sigma} \mathrm{e}^{-V^\sigma} - 1) \, . \tag{10.199}$$

To demonstrate the above, we consider

$$\begin{aligned} \widetilde{O}^\sigma &= \mathrm{e}^{-V^\sigma} O^\sigma \\ &= \begin{pmatrix} \mathrm{e}^{-V^\sigma_1} & 0 & . & . & 0 & \mathrm{e}^{-\Delta_\tau h_0} \\ -\mathrm{e}^{-\Delta_\tau h_0} & \mathrm{e}^{-V^\sigma_2} & 0 & . & . & 0 \\ 0 & -\mathrm{e}^{-\Delta_\tau h_0} & \mathrm{e}^{-V^\sigma_3} & . & . & 0 \\ . & 0 & -\mathrm{e}^{-\Delta_\tau h_0} & . & . & . \\ . & . & 0 & . & . & . \\ . & . & . & . & . & . \\ 0 & . & . & 0 & -\mathrm{e}^{-\Delta_\tau h_0} & \mathrm{e}^{-V^\sigma_m} \end{pmatrix} , \end{aligned} \tag{10.200}$$

so that (omitting the spin index σ)

$$\widetilde{g} \equiv \widetilde{O}^{-1} = [\widetilde{O'} + \underbrace{\widetilde{O} - \widetilde{O'}}_{\equiv \mathrm{e}^{-V} - \mathrm{e}^{-V'}}]^{-1} = \widetilde{g'} - \widetilde{g'} \left(\mathrm{e}^{-V} - \mathrm{e}^{-V'} \right) \widetilde{g} \, . \tag{10.201}$$

The above equation follows from the identity

$$\frac{1}{(A+B)} = \frac{1}{A} - \frac{1}{A} B \frac{1}{A+B} \, . \tag{10.202}$$

Substitution, $\widetilde{g} = g \exp(V)$, leads to the Dyson equation (10.199).

The above Dyson equation is the central identity in the Hirsch-Fye algorithm: All quantities required for the algorithm may be derived directly from this equality. An implementation of the algorithm involves two steps:

10.8.1.0.1 Calculation of the impurity Green function for a given HS configuration

The starting point of the algorithm is to compute the green function for a random HS configuration of Ising spins $\boldsymbol{s}'$. We will only need the Green function for the impurity f-site. Let $x = (\tau_x, \boldsymbol{i}_x)$ with Trotter index τ_x and orbital $\boldsymbol{i}_x$. Since

$$(\mathrm{e}^{V'}\mathrm{e}^{-V} - 1)_{x,y} = (\mathrm{e}^{V'}\mathrm{e}^{\ V} - 1)_{x,x}\delta_{x,y}\delta_{\boldsymbol{i}_x,0} \tag{10.203}$$

we can use the Dyson equation only for the impurity Green function

$$g^{\sigma}_{f,f'} = g'^{\sigma}_{f,f'} + \sum_{f''} g'^{\sigma}_{f,f''} \Delta^{\sigma}_{f'',f''} (1 - g^{\sigma})_{f'',f'} \tag{10.204}$$

with indices $f \equiv (\tau, 0)$ running from $1 \dots m$. Hence, the $m \times m$ impurity Green function matrix,

$$g^{I,\sigma}_{f,f'} = g^{\sigma}_{f,f'} \tag{10.205}$$

satisfies the Dyson equation

$$g^{I,\sigma} = g'^{I,\sigma} + g'^{I,\sigma} \Delta^{I,\sigma} (1 - g^{I,\sigma}) \quad \text{with} \quad \Delta^{I,\sigma}_{f,f'} = \Delta^{\sigma}_{f,f'} \, . \tag{10.206}$$

For $V = 0$, g^I is nothing but the impurity Green function of the non-interacting Anderson model which may readily be computed. Thus using the Dyson equation, we can compute the Green function g'^I for an arbitrary HS configuration $\boldsymbol{s}'$ at the cost of an $m \times m$ matrix inversion. This involves a CPU cost scaling as m^3 or equivalently β^3.

10.8.1.0.2 Upgrading

At this point we have computed the impurity Green function for a given HS configuration $\boldsymbol{s}$. Adopting a single spin flip algorithm we will propose the configuration

$$\boldsymbol{s}'_f = \begin{cases} -\boldsymbol{s}_f & \text{if } f = f_1 \\ \boldsymbol{s}_f & \text{otherwise} \end{cases} \tag{10.207}$$

and accept it with probability

$$R = \prod_{\sigma} R^{\sigma} \tag{10.208}$$

with

$$\begin{aligned} R^{\sigma} &= \frac{\det\left[1 + B'^{\sigma}_m B'^{\sigma}_{m-1} \cdots B'^{\sigma}_1\right]}{\det\left[1 + B^{\sigma}_m B^{\sigma}_{m-1} \cdots B^{\sigma}_1\right]} = \det\left[g^{\sigma}(g'^{\sigma})^{-1}\right] \\ &= \det\left[1 + \Delta^{\sigma}\left(1 - g^{\sigma}\right)\right] \, . \end{aligned} \tag{10.209}$$

The last identity follows from the Dyson equation to express g^σ as $g^\sigma = g'^\sigma[1 + \Delta^\sigma(1 - g^\sigma)]$. Since $\boldsymbol{s}$ and $\boldsymbol{s}'$ differ only by one entry the matrix Δ^σ has one non-zero matrix element: $\Delta^\sigma_{f_1,f_1}$. Hence, $R^\sigma = 1 + \Delta^\sigma_{f_1,f_1}(1 - g^\sigma_{f_1,f_1})$. Since the impurity Green function $g^{I,\sigma}$ is at hand, we can readily compute R.

If the move (spin flip) is accepted, we will have to recalculate (upgrade) the impurity Green function. From the Dyson equation (10.206), we have

$$g'^{I,\sigma} = g^{I,\sigma}\left[1 + \Delta^{I,\sigma}\left(1 - g^{I,\sigma}\right)\right]^{-1} . \tag{10.210}$$

To compute $[1 + \Delta^{I,\sigma}(1 - g^{I,\sigma})]^{-1}$ we can use the Sherman-Morrison formula of (10.164). Setting $A = 1$, $\boldsymbol{u}_f = \Delta^{I,\sigma}_{f_1,f_1}\delta_{f_1,f}$ and $\boldsymbol{v}_f = (1 - g^{I,\sigma})_{f_1,f}$ we obtain

$$g'^{I,\sigma}_{f,f'} = g^{I,\sigma}_{f,f'} + \frac{g^{I,\sigma}_{f,f_1}\Delta^\sigma_{f_1,f_1}(g^{I,\sigma} - 1)_{f_1,f'}}{1 + (1 - g^{I,\sigma})_{f_1,f_1}\Delta^\sigma_{f_1,f_1}} . \tag{10.211}$$

Thus, the upgrading of the Green function under a single spin flip is an operation which scales as m^2. Since for a single sweep we have to visit all spins, the computational cost of a single sweep scales as m^3.

By construction, the Hirsch-Fye algorithm is free from numerical stabilization problems. For the here considered Anderson model, it has recently been shown that there is no sign problem irrespective of the conduction band electron density [79]. Clearly the attractive feature of the Hirsch-Fye impurity algorithm is that the algorithm may be formulated directly in the thermodynamic limit. This is not possible within the lattice formulation of the auxiliary field QMC method. Within this approach the dimension of the matrices scale as the total number of orbitals N, and the CPU time for a single sweep as $N^3\beta$. The Hirsch-Fye algorithm is not limited to impurity models. However, when applied to lattice models, such as the Hubbard model, it is not efficient since the CPU time will scale as $(\beta N)^3$.

To conclude this section we show a typical example of the use of the Hirsch-Fye algorithm for the Kondo model

$$H = \sum_{\boldsymbol{k},\sigma} \varepsilon(\boldsymbol{k}) c^\dagger_{\boldsymbol{k},\sigma} c_{\boldsymbol{k},\sigma} + J\boldsymbol{S}^I_c \cdot \boldsymbol{S}^I_f . \tag{10.212}$$

For the Monte Carlo formulation, the same ideas as for the lattice problem may be used for the HS decoupling of the interaction as well as to impose the constraint of no charge fluctuations on the f-sites. Figure 10.18 plots the impurity spin susceptibility

$$\chi^I = \int_0^\beta \mathrm{d}\tau \langle \boldsymbol{S}^I_f(\tau) \cdot \boldsymbol{S}^I_f \rangle \tag{10.213}$$

for various values of J/t for a half-filled conduction band. As apparent, at low energies the data collapse to the universal form $\chi^I = f\left(T/T^I_K\right)/T$ where T^I_K is the Kondo temperature [78].

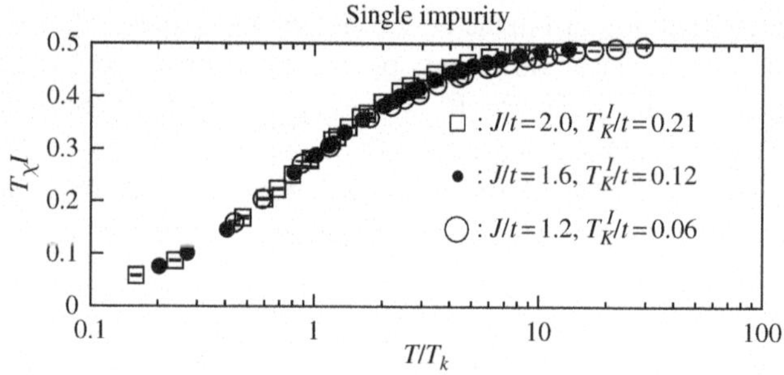

Fig. 10.18. Impurity spin susceptibility of the Kondo model as computed with the Hirsch-Fye impurity algorithm [80]

10.8.2 Ground-State Formulation

In the above finite-temperature formulation of the Hirsch-Fye algorithm, the CPU time scales as β^3 thus rendering it hard to reach the low temperature limit. Here we show how to formulate a projector version of the Hirsch-Fye algorithm. Although the CPU time will still scale as β^3 a good choice of the trial wave function may provide quicker convergence to the ground state than the finite temperature algorithm.

In the projector approach, the trial wave function $|\Psi_T\rangle$ is required to be a Slater determinant non-orthogonal to the ground state wave function. Hence, we can find an one body Hamiltonian

$$H_T = \sum_{i,j,\sigma} a^\dagger_{i,\sigma} \left(h_T\right)_{i,j} a_{j,\sigma} \ , \tag{10.214}$$

which has $|\Psi_T\rangle$ as a non-degenerate ground state. In the above, and in the context of the Anderson model, $a_{j,\sigma}$ denotes c- or f-fermionic operators. Our aim is to compute

$$\frac{\langle\Psi_T|\mathrm{e}^{-\frac{\Theta}{2}H}O\mathrm{e}^{-\frac{\Theta}{2}H}|\Psi_T\rangle}{\langle\Psi_T|\mathrm{e}^{-\Theta H}|\Psi_T\rangle} \equiv \lim_{\beta\to\infty} \frac{\mathrm{Tr}\left[\mathrm{e}^{-\frac{\Theta}{2}H}O\mathrm{e}^{-\frac{\Theta}{2}H}\mathrm{e}^{-\beta H_T}\right]}{\mathrm{Tr}\left[\mathrm{e}^{-\Theta H}\mathrm{e}^{-\beta H_T}\right]} \tag{10.215}$$

and subsequently take the limit $\Theta \to \infty$. As apparent, the above equation provides a link between the finite temperature and projection approaches. To proceed, we will consider the right hand side of the above equation and retrace the steps carried out for the standard finite-temperature formulation of the Hirsch-Fye algorithm. After Trotter decomposition and discrete Hubbard Stratonovich transformation we obtain

$$\langle\Psi_T|\mathrm{e}^{-\Theta H}|\Psi_T\rangle = \lim_{\beta\to\infty} \sum_{s} \left[\prod_\sigma \det\left[1 + B^\sigma_m B^\sigma_{m-1}\dots B^\sigma_1 \mathrm{e}^{-\beta h_T}\right]\right] \tag{10.216}$$

with $m\Delta_\tau = \Theta$. Replacing B^σ_1 by $B^\sigma_1\exp(-\beta h_T)$ in (10.192) and following the steps described for the finite-temperature version, we derive a Dyson equation (omitting spin indices) for the ground-state Green function matrix g_0

$$g_0^\sigma = g_0'^\sigma + g_0'^\sigma \Delta(1 - g_0^\sigma)\,, \quad \Delta^\sigma = (\mathrm{e}^{V'^\sigma}\mathrm{e}^{-V^\sigma} - 1)\,, \tag{10.217}$$

with

$$g_0 = \begin{pmatrix} G_0(1,1) & G_0(1,2) & \dots & G_0(1,m) \\ G_0(2,1) & G_0(2,2) & \dots & G_0(2,m) \\ . & . & \dots & . \\ G_0(m,1) & G_0(m,2) & \dots & G_0(m,m) \end{pmatrix} \tag{10.218}$$

and

$$\begin{aligned} &[G_0(n_1,n_2)]_{i,j} \\ &= \lim_{\beta\to\infty} \begin{cases} \dfrac{\mathrm{Tr}\left[B_m \dots B_{n_1+1} a_{i,\sigma} B_{n_1} \dots B_{n_2+1} a^\dagger_{j,\sigma} B_{n_2} \dots B_1 \mathrm{e}^{-\beta H_T}\right]}{\mathrm{Tr}\left[B_m \dots B_1 \mathrm{e}^{-\beta H_T}\right]} & \text{if } n_1 \geq n_2 \\ -\dfrac{\mathrm{Tr}\left[B_m \dots B_{n_2+1} a^\dagger_{j,\sigma} B_{n_2} \dots B_{n_1+1} a_{i,\sigma} B_{n_1} \dots B_1 \mathrm{e}^{-\beta H_T}\right]}{\mathrm{Tr}\left[B_m \dots B_1 \mathrm{e}^{-\beta H_T}\right]} & \text{if } n_1 < n_2 \end{cases} \\ &= \begin{cases} \dfrac{\langle \Psi_T | B_m \dots B_{n_1+1} a_{i,\sigma} B_{n_1} \dots B_{n_2+1} a^\dagger_{j,\sigma} B_{n_2} \dots B_1 | \Psi_T \rangle}{\langle \Psi_T | B_m \dots B_1 | \Psi_T \rangle} & \text{if } n_1 \geq n_2 \\ -\dfrac{\langle \Psi_T | B_m \dots B_{n_2+1} a^\dagger_{j,\sigma} B_{n_2} \dots B_{n_1+1} a_{i,\sigma} B_{n_1} \dots B_1 | \Psi_T \rangle}{\langle \Psi_T | B_m \dots B_1 | \Psi_T \rangle} & \text{if } n_1 < n_2 \end{cases} . \end{aligned} \tag{10.219}$$

As shown for the finite-temperature formulation, the simulation is entirely based on the Dyson equation. Since this equation also holds for the zero-temperature formulation precisely the same algorithm as in the finite-temperature case can be used.

In Fig. 10.19 we compare both algorithms and consider the double occupancy on the impurity site. As apparent, the ground-state formulation converges more quickly to the ground-state expectation value than the finite-temperature formulation.

The projector formulation of the Hirsch-Fye algorithm has been efficiently incorporated in the DMFT self-consistent cycle thus offering a route to compute $T = 0$ quantities within this framework [81]. Finally we note that diagrammatic determinantal methods can be extended very easily to projective schemes [82].

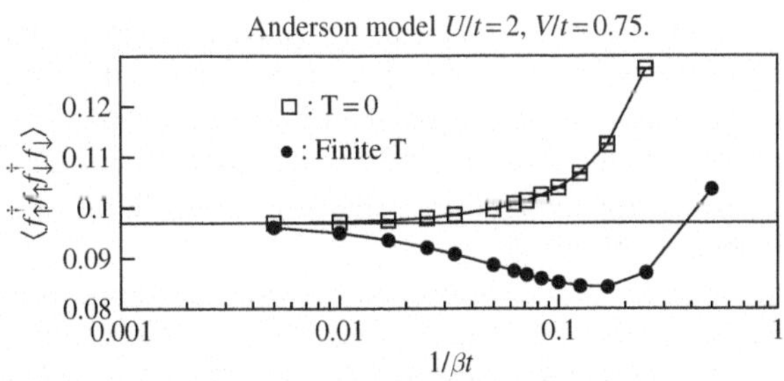

Fig. 10.19. Comparison between the zero and finite-temperature Hirsch-Fye algorithms for the symmetric Anderson model, with an 1D density of states

10.9 Selected Applications of the Auxiliary Field Method

The applications of the auxiliary field algorithms to correlated electron systems are numerous. Here we will only mention a few.

Let us start with the attractive attractive Hubbard model. This model essentially describes the electron-phonon problem in terms of the Holstein model which in the anti-adiabatic limit maps onto the attractive Hubbard model [83]. Both models are free of sign problems in arbitrary dimensions and on arbitrary lattice topologies. The salient features of those models have been investigated in detail. For instance, the crossover from long coherence length (BCS) to short coherence length superconductors. In the short coherence length limit, a liquid of preformed pairs with non-Fermi liquid character is apparent above the transition temperature [84, 85]. Furthermore, the disorder driven superfluid to insulator transition has been studied in the framework of the attractive Hubbard model [86].

Recently, a new class of models of correlated electron models showing no sign problem has been investigated [69, 87, 88, 89]. Those models have exotic ground states including phases with circulating currents [56, 89], striped phases [87] as well as possible realizations of gapless spin liquid phases [56].

A lot of the work using the BSS algorithm is centered around the repulsive Hubbard model in two dimensions, as well as the three-band Hubbard model of the CuO_2 planes in the cuprates. On the basis of Monte Carlo simulations, it is now accepted that at half band-filling those models are Mott (charge transfer for the three-band model) insulators with long-range antiferromagnetic order [61, 90, 91]. In the case of the three band Hubbard model, a minimal set of parameters were found so as to reproduce experimental findings [92]. The issue of superconductivity at low doping away from half-filling is still open. General concepts – independent on the symmetry of the pair wave function and including possible retardation effects – such as flux quantization and superfluid density have been used to attempt to answer the above question [93, 94]. Within the algorithmic limitations, no convincing sign of superconductivity has been found to date.

The nature of the doping induced metal-insulator transition in the 2D Hubbard model, has attracted considerable interest [95, 96, 97]. In particular it has been argued that the transition is driven by the divergence of the effective mass rather than by the vanishing of the number of charge carriers. The origin of such a metal insulator transition is to be found in a very flat dispersion relation around the $(\pi, 0)$ and $(0, \pi)$ points in the Brillouin zone [98, 99]. An extensive review of this topic as well as a consistent interpretation of the numerical data in terms of a hyper-scaling Ansatz may be found in [100].

Aspects of the physics of heavy fermion systems have been investigated in the framework of the 2D periodic Anderson model (PAM) [101] and of the Kondo lattice model (KLM) [55]. It is only recently that a sign free formulation of the KLM for particle-hole symmetric conduction bands has been put forward [64]. Extensive calculations both at $T = 0$ and at finite T allow to investigate the magnetic order-disorder transition triggered by the competition between the RKKY interaction and the Kondo effect [55]. Across this quantum phase transition single-hole dynamics as well as spin excitations were investigated in detail. One can show numerically that the quasiparticle residue in the vicinity of $\boldsymbol{k} = (\pi, \pi)$ tracks the Kondo scale of the corresponding single impurity problem. This statement is valid both in the magnetically ordered and disordered phases [102]. This suggest that the coherence temperature tracks the Kondo scale. Furthermore, the effect of a magnetic field on the Kondo insulating state was investigated. For the particle-hole symmetric conduction band, results show a transition from the Kondo insulator to a canted antiferromagnet [103, 104]. Finally, models with regular depletion of localized spins can be investigated [80]. Within the framework of those models, the typical form of the resistivity versus temperature can be reproduced.

The most common application of the Hirsch-Fye algorithm is in the framework of dynamical mean-field theories [77] which map the Hubbard model onto an Anderson impurity problem supplemented by a self-consistency loop. At each iteration, the Hirsch-Fye algorithm is used to solve the impurity problem at finite temperature [76] or at $T = 0$ [81]. For this particular problem, many competing methods such as DMRG [105] and NRG [106] are available. In the dynamical mean-field approximation spatial fluctuations are frozen out. To reintroduce them, one has to generalize to cluster methods such as the dynamical cluster approximation (DCA) [107] or cellular-DMFT (CDMFT) [108]. Within those approaches, the complexity of the problem to solve at each iteration is that of an N-impurity Anderson model (N corresponds to the cluster size). Generalizations of DMRG and NRG to solve this problem are difficult. On the other hand, as a function of cluster size the sign problem in the Hirsch-Fye approach becomes more and more severe but is, in many instances, still tractable. It however proves to be one of the limiting factors in achieving large cluster sizes.

10.10 Conclusion

We have discussed in details a variety of algorithms which can broadly be classified as world-line based or determinantal algorithms. For fermionic models, such as the Hubbard model, the determinantal QMC algorithm should be employed because of the reduced sign problem in this formulation. For purely 1D fermion systems and for spin models the world-line algorithms are available, which have lower autocorrelations, and better scaling because of their almost linear scaling with system size, in contrast to the cubic scaling of the determinantal algorithms.

Appendix 10.A The Trotter Decomposition

Given a Hamiltonian of the form

$$H = H_1 + H_2 \; , \tag{10.220}$$

the Trotter decomposition states that the imaginary time propagator can be split into a product of infinitesimal time propagations such that

$$e^{-\beta H} = \lim_{\Delta\tau \to 0} \left[e^{-\Delta\tau H_1} e^{-\Delta\tau H_2} \right]^m , \tag{10.221}$$

where $m\Delta\tau = \beta$. For $[H_1, H_2] \neq 0$ and finite values of the time step $\Delta\tau$ this introduces a systematic error. In many QMC algorithms we will not take the limit $\Delta\tau \to 0$, and it is important to understand the order of the systematic error produced by the above decomposition[4]. A priori, it is of the order $\Delta\tau$. However, in many non-trivial cases, the prefactor of the error of order $\Delta\tau$ vanishes [109].

For a time step $\Delta\tau$

$$e^{-\Delta\tau(H_1+H_2)} = e^{-\Delta\tau H_1} e^{-\Delta\tau H_2} - \frac{\Delta\tau^2}{2} [H_1, H_2] + \mathcal{O}(\Delta\tau^3) , \tag{10.222}$$

such that

$$e^{-\Delta\tau(H - \Delta\tau/2[H_1,H_2])} = e^{-\Delta\tau H_1} e^{-\Delta\tau H_2} + \mathcal{O}(\Delta\tau^3) . \tag{10.223}$$

We can now exponentiate both sides of the former equation to the power m

$$e^{-\beta(H - \Delta\tau/2[H_1,H_2])} = \left[e^{-\Delta\tau H_1} e^{-\Delta\tau H_2} \right]^m + \mathcal{O}(\Delta\tau^2) . \tag{10.224}$$

The systematic error is now of order $\Delta\tau^2$ since in the exponentiation, the systematic error of order $\Delta\tau^3$ occurs m times and $m\Delta\tau = \beta$.

To evaluate the left hand side of the above equation we use time dependent perturbation theory. Let $h = h_0 + h_1$, where h_1 is small in comparison to h_0. The imaginary time propagation in the interacting picture reads

$$U_I(\tau) = e^{\tau h_0} e^{-\tau h} \tag{10.225}$$

such that

$$\begin{aligned} \frac{\partial}{\partial\tau} U_I(\tau) &= e^{\tau h_0}(h_0 - h) e^{-\tau h} = -\underbrace{e^{\tau h_0} h_1 e^{-\tau h_0}}_{\equiv h_1^I(\tau)} U_I(\tau) \\ &= -h_1^I(\tau) U_I(\tau) . \end{aligned} \tag{10.226}$$

Since $U_I(0) = 1$ we can transform the differential equation to an integral one

$$U_I(\tau) = 1 - \int_0^\tau d\tau' h_1^I(\tau') U_I(\tau') = 1 - \int_0^\tau d\tau' h_1^I(\tau') + \mathcal{O}(h_1^2) . \tag{10.227}$$

Returning to (10.224) we can set $h_0 = H$, $h_1 = -\Delta\tau\,[H_1, H_2]/2$ and $\tau = \beta$ to obtain

$$\begin{aligned} \left(e^{-\Delta\tau H_1} e^{-\Delta\tau H_2}\right)^m &= e^{-\beta(H - \Delta\tau[H_1,H_2]/2)} + \mathcal{O}(\Delta\tau^2) \\ &= e^{-\beta H} + \underbrace{\frac{\Delta\tau}{2} \int_0^\beta d\tau e^{-(\beta-\tau)H} [H_1, H_2] e^{-\tau H}}_{\equiv A} + \mathcal{O}(\Delta\tau^2) . \end{aligned} \tag{10.228}$$

[4] For cases where a continuous time formulation is possible see Sect. 10.3.

In the QMC approaches with finite time steps we will compute:

$$\frac{\mathrm{Tr}\left[\left(\mathrm{e}^{-\Delta\tau H_1}\mathrm{e}^{-\Delta\tau H_2}\right)^m O\right]}{\mathrm{Tr}\left[\left(\mathrm{e}^{-\Delta\tau H_1}\mathrm{e}^{-\Delta\tau H_2}\right)^m\right]} = \frac{\mathrm{Tr}\left[\mathrm{e}^{-\beta H}O\right] + \frac{\Delta\tau}{2}\mathrm{Tr}\left[AO\right]}{\mathrm{Tr}\left[\mathrm{e}^{-\beta H}\right] + \frac{\Delta\tau}{2}\mathrm{Tr}\left[A\right]} + O(\Delta\tau^2)\,, \quad (10.229)$$

where $O = O^\dagger$ is an observable. We now show that A is an anti-Hermitian operator

$$\begin{aligned} A^\dagger &= -\int\limits_0^\beta \mathrm{d}\tau \mathrm{e}^{-\tau H}[H_1, H_2]\mathrm{e}^{-(\beta-\tau)H} \\ &= \int\limits_\beta^0 \mathrm{d}\tau' \mathrm{e}^{-(\beta-\tau')H}[H_1, H_2]\mathrm{e}^{-\tau' H} = -A\,, \end{aligned} \quad (10.230)$$

where we have carried out the substitution $\tau' = \beta - \tau$ Since A is an anti-Hermitian operator it follows that $\overline{\mathrm{Tr}\,[A]} = \mathrm{Tr}\left[A^\dagger\right] = -\mathrm{Tr}\,[A]$ as well as $\overline{\mathrm{Tr}\,[AO]} = -\mathrm{Tr}\,[AO]$. Recall that the observable O is a Hermitian operator. Thus, if O, H_1 and H_2 are simultaneously real representable in a given basis, the systematic error proportional to $\Delta\tau$ vanishes since in this case the trace is real. Hence the systematic error is of order $\Delta\tau^2$.

Clearly there are other choices of the Trotter decomposition which irrespective of the properties of H_1, H_2 and O yield systematic errors of the order $\Delta\tau^2$. For example we mention the symmetric decomposition

$$\mathrm{e}^{-\Delta\tau(H_1+H_2)} = \mathrm{e}^{-\Delta\tau H_1/2}\mathrm{e}^{-\Delta\tau H_1}\mathrm{e}^{-\Delta\tau H_2/2} + \mathcal{O}\Delta\tau^3\,. \quad (10.231)$$

However, in many cases higher order decompositions are cumbersome and numerically expensive to implement.

Appendix 10.B The Hubbard-Stratonovich Decomposition

Auxiliary field QMC methods are based on various forms of the Hubbard-Stratonovich (HS) decomposition. This transformation is not unique. The efficiency of the algorithm as well as of the sampling scheme depends substantially on the type of HS transformation one uses. In this appendix we will review some aspects of the HS transformation with emphasis on its application to the auxiliary field QMC method.

The generic HS transformation is based on the Gaussian integral

$$\int\limits_{-\infty}^{+\infty} \mathrm{d}\phi \mathrm{e}^{-(\phi+A)^2/2} = \sqrt{2\pi}\,, \quad (10.232)$$

which may be rewritten as

$$e^{A^2/2} = \frac{1}{\sqrt{2\pi}} \int_{-\infty}^{+\infty} d\phi e^{-\phi^2/2 - \phi A} . \tag{10.233}$$

Hence, if A is a one-body operator, the two-body operator $\exp(A^2/2)$, can be transformed into the integral of single-body operators interacting with a bosonic field ϕ. The importance of this identity in the Monte Carlo approach lies in the fact that for a fixed field ϕ the one-body problem is exactly solvable. The integral over the field ϕ can then be carried out with Monte Carlo methods. However, the Monte Carlo integration over a continuous field is much more cumbersome than the sum over a discrete field.

Let us consider for example the Hubbard interaction for a single site

$$H_U = U(n_\uparrow - \tfrac{1}{2})(n_\downarrow - \tfrac{1}{2}) . \tag{10.234}$$

Here, $n_\sigma = c_\sigma^\dagger c_\sigma$ where $c_\sigma^\dagger$ are spin $1/2$ fermionic operators. In the Monte Carlo approach after Trotter decomposition of the kinetic and interaction term, we will have to compute $\exp(-\Delta\tau H_U)$. Since,

$$H_U = -\frac{U}{2}\left(n_\uparrow - n_\downarrow\right)^2 + \frac{U}{4} \tag{10.235}$$

we can set $A^2 = \Delta\tau U \left(n_\uparrow - n_\downarrow\right)^2$ and use (10.233) to compute $\exp(-\Delta\tau H_U)$. There are, however, more efficient ways of carrying out the transformation which are based on the fact that the Hilbert space for a single site consists of four states $|0\rangle, |\uparrow\rangle, |\downarrow\rangle$ and $|\uparrow,\downarrow\rangle$. Let us propose the identity

$$e^{-\Delta\tau H_U} = \gamma \sum_{s=\pm 1} e^{\alpha s (n_\uparrow - n_\downarrow)} \tag{10.236}$$

and see if it is possible to find values of α and γ to satisfy it on the single site Hilbert space. Applying each state vector on both sides of the equation yields

$$\begin{aligned} e^{-\Delta\tau U/4}|0\rangle &= 2\gamma|0\rangle \\ e^{-\Delta\tau U/4}|\uparrow\downarrow\rangle &= 2\gamma|\uparrow\downarrow\rangle \\ e^{\Delta\tau U/4}|\uparrow\rangle &= 2\gamma\cosh(\alpha)|\uparrow\rangle \\ e^{\Delta\tau U/4}|\downarrow\rangle &= 2\gamma\cosh(\alpha)|\downarrow\rangle . \end{aligned} \tag{10.237}$$

Hence (10.236) is satisfied provided that

$$\begin{aligned} \gamma &= \frac{1}{2} e^{-\Delta\tau U/4} , \\ \cosh(\alpha) &= e^{\Delta\tau U/2} . \end{aligned} \tag{10.238}$$

This choice of HS transformation leads to an efficient Monte Carlo algorithm for Hubbard type models. However, as apparent it breaks $SU(2)$ spin symmetry. Since the HS field s couples to the z-component of the magnetization the spin symmetry is broken for a fixed value of the field and is restored only after summation over the field. To avoid this symmetry breaking, one can consider alternative HS transformations which couple to the density. In the same manner as above, we can show that

$$e^{-\Delta\tau H_U} = \widetilde{\gamma} \sum_{s=\pm 1} e^{i\widetilde{\alpha} s(n_\uparrow + n_\downarrow - 1)} , \qquad (10.239)$$

where $\cos(\widetilde{\alpha}) = \exp(-\Delta\tau U/2)$ and $\widetilde{\gamma} = \exp(\Delta\tau U/4)/2$. Clearly, this choice of the HS transformation conserves the $SU(2)$ spin symmetry for each realization of the field. However, this comes at the price that one needs to work with complex numbers. It turns out that when the sign problem is absent, the above choice of the HS transformation yields in general more efficient codes.

We conclude this appendix with a general discrete HS transformation which replaces (10.233). For small time steps $\Delta\tau$ we have the identity

$$e^{\Delta\tau\lambda A^2} = \sum_{l=\pm 1,\pm 2} \gamma(l) e^{\sqrt{\Delta\tau\lambda}\,\eta(l) O} + \mathcal{O}(\Delta\tau^4) , \qquad (10.240)$$

where the fields η and γ take the values

$$\begin{aligned}
\gamma(\pm 1) &= 1 + \sqrt{6}/3 , \\
\gamma(\pm 2) &= 1 - \sqrt{6}/3 , \\
\eta(\pm 1) &= \pm\sqrt{2\left(3 - \sqrt{6}\right)} , \\
\eta(\pm 2) &= \pm\sqrt{2\left(3 + \sqrt{6}\right)} .
\end{aligned} \qquad (10.241)$$

This transformation is not exact and produces an overall systematic error proportional to $\Delta\tau^3$ in the Monte Carlo estimate of an observable. However, since we already have a systematic error proportional to $\Delta\tau^2$ from the Trotter decomposition, the transformation is as good as exact. It also has the great advantage of being discrete thus allowing efficient sampling.

Appendix 10.C Slater Determinants and their Properties

In this appendix, we review the properties of Slater determinants required for the formulation of auxiliary field QMC algorithms. Consider a single-particle Hamiltonian of the form

$$H_0 = \sum_{x,y} c_x^\dagger \left[h_0\right]_{x,y} c_y , \qquad (10.242)$$

where h_0 is a Hermitian matrix, $\{c_x^\dagger, c_y\} = \delta_{x,y}$, $\{c_x^\dagger, c_y^\dagger\} = 0$, and x runs over the N_s single-particle states. Since h_0 is Hermitian, we can find an unitary matrix U such that $U^\dagger h_0 U = \lambda$, where λ is a diagonal matrix. Hence,

$$H_0 = \sum_x \lambda_{x,x} \gamma_x^\dagger \gamma_x \, ,$$
$$\gamma_x = \sum_y U_{x,y}^\dagger c_y \, ,$$
$$\gamma_x^\dagger = \sum_y c_y^\dagger U_{y,x} \, . \tag{10.243}$$

Since U is an unitary transformation the γ operators satisfy the commutation relations $\{\gamma_x^\dagger, \gamma_y\} = \delta_{x,y}$, and $\{\gamma_x^\dagger, \gamma_y^\dagger\} = 0$. An N_p-particle eigenstate of the Hamiltonian H_0 is characterized by the occupation of N_p single-particle levels, $\alpha_1 \dots \alpha_{N_p}$ and is given by

$$\gamma_{\alpha_1}^\dagger \gamma_{\alpha_2}^\dagger \cdots \gamma_{N_p}^\dagger |0\rangle = \prod_{n=1}^{N_p} \Big(\sum_x c_x^\dagger U_{x,\alpha_n} \Big) |0\rangle = \prod_{n=1}^{N_p} \left(\mathbf{c}^\dagger P \right)_n |0\rangle \, . \tag{10.244}$$

In the last equation P denotes an rectangular matrix with N_s rows and N_p columns. The last equation defines the Slater determinant. The Slater determinant is a solution of a single-particle Hamiltonian, and is completely characterized by the rectangular matrix P.

We will now concentrate on the properties of Slater determinants. The first important property is that

$$e^{\mathbf{c}^\dagger T \mathbf{c}} \prod_{n=1}^{N_p} \left(\mathbf{c}^\dagger P \right)_n |0\rangle = \prod_{n=1}^{N_p} \left(\mathbf{c}^\dagger e^T P \right)_n |0\rangle \, . \tag{10.245}$$

The propagation of a Slater determinant with a single-particle propagator $\exp[\mathbf{c}^\dagger T \mathbf{c}]$ is a Slater determinant. We will show the above under the assumption that T is a Hermitian or anti-Hermitian matrix. It is useful to go into a basis where T is diagonal $U^\dagger T U = \lambda$. U is an unitary matrix and λ a real (purely imaginary) diagonal matrix provided that T is Hermitian (anti-Hermitian). Thus we can define the fermionic operators $\boldsymbol{\gamma}^\dagger = \mathbf{c}^\dagger U$ to obtain

$$\begin{aligned}
e^{\mathbf{c}^\dagger T \mathbf{c}} \prod_{n=1}^{N_p} \left(\mathbf{c}^\dagger P \right)_n |0\rangle &= e^{\boldsymbol{\gamma}^\dagger \lambda \boldsymbol{\gamma}} \prod_{n=1}^{N_p} \left(\boldsymbol{\gamma}^\dagger U P \right)_n |0\rangle \\
&= \sum_{y_1,\dots,y_{N_p}} e^{\sum_x \lambda_{x,x} \gamma_x^\dagger \gamma_x} \gamma_{y_1}^\dagger \cdots \gamma_{y_{N_p}}^\dagger |0\rangle \left(UP \right)_{y_1,1} \cdots \left(UP \right)_{y_{N_p},N_p} \\
&= \sum_{y_1,\dots,y_{N_p}} e^{\lambda_{y_1,y_1}} \gamma_{y_1}^\dagger \cdots e^{\lambda_{y_{N_p},y_{N_p}}} \gamma_{y_{N_p}}^\dagger |0\rangle \left(UP \right)_{y_1,1} \cdots \left(UP \right)_{y_{N_p},N_p} \\
&= \prod_{n=1}^{N_p} \left(\boldsymbol{\gamma}^\dagger e^{\lambda} U P \right)_n |0\rangle = \prod_{n=1}^{N_p} \left(\mathbf{c}^\dagger U^\dagger e^{\lambda} U P \right)_n |0\rangle = \prod_{n=1}^{N_p} \left(\mathbf{c}^\dagger e^T P \right)_n |0\rangle \, .
\end{aligned} \tag{10.246}$$

The second property we will need, is the overlap of two slater determinants. Let

$$|\Psi\rangle = \prod_{n=1}^{N_p} \left(\mathbf{c}^\dagger P\right)_n |0\rangle \,,$$
$$|\widetilde{\Psi}\rangle = \prod_{n=1}^{N_p} \left(\mathbf{c}^\dagger \widetilde{P}\right)_n |0\rangle \,, \tag{10.247}$$

then

$$\langle \Psi | \widetilde{\Psi} \rangle = \det\left[P^\dagger \widetilde{P}\right] \,. \tag{10.248}$$

The above follows from

$$\begin{aligned}\langle \Psi | \widetilde{\Psi} \rangle &= \langle 0 | \prod_{n=N_p}^{1} \left(P^\dagger \mathbf{c}\right)_n \prod_{\widetilde{n}=1}^{N_p} \left(\mathbf{c}^\dagger \widetilde{P}\right)_{\widetilde{n}} |0\rangle \\ &= \sum_{\substack{y_1,\dots y_{N_p} \\ \widetilde{y}_1 \dots \widetilde{y}_{N_p}}} P^\dagger_{N_p,y_{N_p}} \dots P^\dagger_{1,y_1} \widetilde{P}_{\widetilde{y}_1,1} \dots \widetilde{P}_{\widetilde{y}_{N_p},N_p} \langle 0 | c_{y_{N_p}} \dots c_{y_1} c^\dagger_{\widetilde{y}_1} \dots c^\dagger_{\widetilde{y}_{N_p}} |0\rangle \,.\end{aligned} \tag{10.249}$$

The matrix element in the above equation does not vanish provided that all the $y_i, i : 1 \dots N_p$ take different values and that there is a permutation π , of N_p numbers such that

$$\widetilde{y}_i = y_{\pi(i)} \,. \tag{10.250}$$

Under those conditions, the matrix element is nothing but the sign of the permutation $(-1)^\pi$. Hence,

$$\begin{aligned}\langle \Psi | \widetilde{\Psi} \rangle &= \sum_{y_1,\dots y_{N_p}} |c^\dagger_{y_1} \dots c^\dagger_{y_{N_p}} |0\rangle|^2 \\ &\times \sum_{\pi \epsilon S_{N_p}} (-1)^\pi P^\dagger_{N_p,y_{N_p}} \dots P^\dagger_{1,y_1} \widetilde{P}_{y_{\pi(1)},1} \dots \widetilde{P}_{y_{\pi(N_p)},N_p} \,.\end{aligned} \tag{10.251}$$

In the above, we have explicitly included the matrix element $|c^\dagger_{y_1} \dots c^\dagger_{y_{N_p}} |0\rangle|^2$ to insure that terms in the sum with $y_i = y_j$ do not contribute since under this assumption the matrix element vanishes due to the Pauli principle. We can however omit this term since the sum over permutations will guarantee that if $y_i = y_j$ for any $i \neq j$ then $\sum_{\pi \epsilon S_{N_p}} (-1)^\pi P^\dagger_{N_p,y_{N_p}} \dots P^\dagger_{1,y_1} \widetilde{P}_{y_{\pi(1)},1} \dots \widetilde{P}_{y_{\pi(N_p)},N_p}$ vanishes. Consider for example $N_p = 2$ and $y_1 = y_2 = x$ then the sum reduces to $P^\dagger_{2,x} P^\dagger_{1,x} \widetilde{P}_{x,1} \widetilde{P}_{x,2} \sum_{\pi \epsilon S_2} (-1)^\pi = 0$ since the sum over the sign of the permutations vanishes.

With the above observation

$$\begin{aligned}\langle \Psi|\widetilde{\Psi}\rangle &= \sum_{\substack{y_1,\dots y_{N_p}\\ \pi\epsilon S_{N_p}}} (-1)^{\pi} P^{\dagger}_{N_p,y_{N_p}} \cdots P^{\dagger}_{1,y_1} \widetilde{P}_{y_{\pi(1)},1} \cdots \widetilde{P}_{y_{\pi(N_p)},N_p} \\ &= \sum_{\substack{y_1,\dots y_{N_p}\\ \pi\epsilon S_{N_p}}} (-1)^{\pi^{-1}} P^{\dagger}_{N_p,y_{N_p}} \cdots P^{\dagger}_{1,y_1} \widetilde{P}_{y_1,\pi^{-1}(1)} \cdots \widetilde{P}_{y_{N_p},\pi^{-1}(N_p)} \\ &= \sum_{\pi\epsilon S_{N_p}} (-1)^{\pi} \left(P^{\dagger}\widetilde{P}\right)_{1,\pi(1)} \cdots \left(P^{\dagger}\widetilde{P}\right)_{N_p,\pi(N_p)} = \det\left[P^{\dagger}\widetilde{P}\right] .\end{aligned} \tag{10.252}$$

Finally, we will need to establish the relation

$$\mathrm{Tr}\left[\mathrm{e}^{\mathbf{c}^{\dagger}T_1\mathbf{c}}\mathrm{e}^{\mathbf{c}^{\dagger}T_2\mathbf{c}}\dots\mathrm{e}^{\mathbf{c}^{\dagger}T_n\mathbf{c}}\right] = \det\left[1+\mathrm{e}^{T_1}\mathrm{e}^{T_2}\dots\mathrm{e}^{T_n}\right] , \tag{10.253}$$

where the trace is over the Fock space. To verify the validity of the above equation, let us set $B=\mathrm{e}^{T_1}\mathrm{e}^{T_2}\dots\mathrm{e}^{T_n}$ and $U=\mathrm{e}^{\mathbf{c}^{\dagger}T_1\mathbf{c}}\mathrm{e}^{\mathbf{c}^{\dagger}T_2\mathbf{c}}\dots\mathrm{e}^{\mathbf{c}^{\dagger}T_n\mathbf{c}}$.

$$\begin{aligned}&\det(1+B)\\ &= \sum_{\pi\epsilon S_{N_s}} (-1)^{\pi}\left(1+B_{\pi(1),1}\right)\cdots\left(1+B_{\pi(N_s),N_s}\right)\\ &= \sum_{\pi\epsilon S_{N_s}} (-1)^{\pi}\delta_{1,\pi(1)}\cdots\delta_{N_s,\pi(N_s)}\\ &\quad+\sum_{x}\sum_{\pi\epsilon S_{N_s}} (-1)^{\pi}B_{\pi(x),x}\delta_{1,\pi(1)}\cdots\widehat{\delta_{x,\pi(x)}}\cdots\delta_{N_s,\pi(N_s)}\\ &\quad+\sum_{y>x}\sum_{\pi\epsilon S_{N_s}} (-1)^{\pi}B_{\pi(x),x}B_{\pi(y),y}\\ &\quad\times\delta_{1,\pi(1)}\cdots\widehat{\delta_{x,\pi(x)}}\cdots\widehat{\delta_{y,\pi(y)}}\cdots\delta_{N_s,\pi(N_s)}\\ &\quad+\sum_{y>x>z}\sum_{\pi\epsilon S_{N_s}} (-1)^{\pi}B_{\pi(x),x}B_{\pi(y),y}B_{\pi(z),z}\\ &\quad\times\delta_{1,\pi(1)}\cdots\widehat{\delta_{x,\pi(x)}}\cdots\widehat{\delta_{y,\pi(y)}}\cdots\widehat{\delta_{z,\pi(z)}}\cdots\delta_{N_s,\pi(N_s)}+\dots .\end{aligned} \tag{10.254}$$

Here, $\widehat{\delta_{y,\pi(y)}}$ means that this term is omitted in the product $\prod_{x=1}^{N_s}\delta_{x,\pi(x)}$. To proceed, let us consider in more details the second term starting with $\sum_{y>x}$ in the last equality. Due to the δ-functions the sum over the permutation of N_s numbers reduces to two terms, namely the unit permutation and the transposition $\pi(x)=x$ and $\pi(y)=y$. Let us define the $P^{(x,y)}$ as a rectangular matrix of dimension $N_s\times 2$, with entries of the first (second) column set to one at row x (y) and zero otherwise. Hence, we can write

$$\begin{aligned}&\sum_{\pi\epsilon S_{N_s}} (-1)^{\pi}B_{\pi(x),x}B_{\pi(y),y}\delta_{1,\pi(1)}\cdots\widehat{\delta_{x,\pi(x)}}\cdots\widehat{\delta_{y,\pi(y)}}\cdots\delta_{N_s,\pi(N_s)}\\ &=\det\left[P^{(x,y),\dagger}BP^{(x,y)}\right] = \langle 0|c_x c_y U c_y^{\dagger}c_x^{\dagger}|0\rangle ,\end{aligned} \tag{10.255}$$

where in the last equation we have used the properties of (10.248) and (10.245). Repeating the same argument for different terms we obtain

$$\begin{aligned}&\det(1+B)\\&=1+\sum_x \langle 0|c_x U c_x^\dagger|0\rangle + \sum_{y>x}\langle 0|c_x c_y U c_y^\dagger c_x^\dagger|0\rangle\\&+\sum_{y>x>z}\langle 0|c_x c_y c_z U c_z^\dagger c_y^\dagger c_x^\dagger|0\rangle + \ldots = \mathrm{Tr}\,[U]\ .\end{aligned} \tag{10.256}$$

This concludes the demonstration of (10.253).

References

1. S.R. White, Physics Reports **301**, 187 (1998)
2. U. Schollwöck, Rev. Mod. Phys. **77**, 259 (2005)
3. H.G. Evertz, G. Lana, M. Marcu, Phys. Rev. Lett. **70**, 875 (1993)
4. H.G. Evertz, Adv. Phys. **52**, 1 (2003)
5. O. Syljuasen, A.W. Sandvik, Phys. Rev. E **66**, 046701 (2002)
6. R. Blankenbecler, D.J. Scalapino, R.L. Sugar, Phys. Rev. D **24**, 2278 (1981)
7. F. Michel, H.G. Evertz. URL http://arxiv.org/abs/0705.0799 and in preparation
8. A.N. Rubtsov, V.V. Savkin, A.I. Lichtenstein, Phys. Rev. B **72**, 035122 (2005)
9. P. Werner, A. Comanac, L.D. Medici, M. Troyer, A. Millis, Phys. Rev. Lett. **97**, 076405 (2006)
10. J.E. Hirsch, D.J. Scalapino, R.L. Sugar, R. Blankenbecler, Phys. Rev. B **26**, 5033 (1981)
11. M. Barma, B.S. Shastry, Phys. Rev. B **18**, 3351 (1978)
12. R.J. Baxter, *Exactly Solved Models in Statistical Mechanics* (Academic Press Limited, London, 1989)
13. M. Troyer, M. Imada, K. Ueda, J. Phys. Soc. Jpn. **66**, 2957 (1997)
14. O. Syljuasen, Phys. Rev. B **67**, 046701 (2003)
15. A.W. Sandvik, O.F. Syljuåsen, in *THE MONTE CARLO METHOD IN THE PHYSICAL SCIENCES: Celebrating the 50th Anniversary of the Metropolis Algorithm, AIP Conference Proceedings*, Vol. 690, ed. by J.E. Gubernatis (2003), pp. 299–308. URL `http://arxiv.org/abs/cond-mat/0306542`
16. M. Troyer, F. Alet, S. Trebst, S. Wessel, in *THE MONTE CARLO METHOD IN THE PHYSICAL SCIENCES: Celebrating the 50th Anniversary of the Metropolis Algorithm, AIP Conference Proceedings*, Vol. 690, ed. by J.E. Gubernatis (2003), pp. 156–169. URL `http://arxiv.org/abs/physics/0306128`
17. N. Kawashima, K. Harada, J. Phys. Soc. Jpn. **73**, 1379 (2004)
18. J.F. Corney, P.D. Drummond, Phys. Rev. Lett. **93**, 260401 (2004)
19. F.F. Assaad, P. Werner, P. Corboz, E. Gull, M. Troyer, Phys. Rev. B **72**, 224518 (2005)
20. F.F. Assaad, D. Würtz, Phys. Rev. B **44**, 2681 (1991)
21. M. Troyer, F.F. Assaad, D. Würtz, Helv. Phys. Acta. **64**, 942 (1991)
22. M. Brunner, F.F. Assaad, A. Muramatsu, Eur. Phys. J. B **16**, 209 (2000)
23. M. Brunner, F.F. Assaad, A. Muramatsu, Phys. Rev. B **62**, 12395 (2000)
24. C. Brünger, F.F. Assaad, Phys. Rev. B **74**, 205107 (2006)
25. N. Prokof'ev, B. Svistunov, Phys. Rev. Lett. **81**, 2514 (1998)
26. S. Rombouts, K. Heide, N. Jachowicz, Phys. Rev. Lett. **82**, 4155 (1999)

27. E. Burovski, A. Mishchenko, N. Prokof'ev, B. Svistunov, Phys. Rev. Lett. **87**, 186402 (2001)
28. A. Rubtsov, M. Katsnelson, A. Lichtenstein, Dual fermion approach to nonlocal correlations in the Hubbard model. URL http://arxiv.org/abs/cond-mat/0612196. Preprint
29. M. Boninsegni, N. Prokof'ev, B. Svistunov, Phys. Rev. Lett. **96**, 070601 (2006)
30. B. Beard, U. Wiese, Phys. Rev. Lett. **77**, 5130 (1996)
31. A. Sandvik, J. Kurkijärvi, Phys. Rev. B **43**, 5950 (1991)
32. A.W. Sandvik, J. Phys. A **25**, 3667 (1992)
33. A.W. Sandvik, Phys. Rev. B **56**, 11678 (1997)
34. N. Prokof'ev, B. Svistunov, I. Tupitsyn, Sov. Phys. JETP Letters 64, 911 (1996). URL http://arxiv.org/abs/cond-mat/9612091
35. N. Prokof'ev, B. Svistunov, I. Tupitsyn, Sov. Phys. JETP **87**, 310 (1998). URL http://arxiv.org/abs/cond-mat/9703200
36. A. Sandvik, R. Singh, D. Campbell, Phys. Rev. B **56**, 14510 (1997)
37. A. Sandvik, D. Campbell, Phys. Rev. Lett. **83**, 195 (1999)
38. A. Dorneich, M. Troyer, Phys. Rev. E **64**, 066701 (2001)
39. P. Kasteleyn, C. Fortuin, J. Phys. Soc. Jpn. Suppl. **26**, 11 (1969)
40. C. Fortuin, P. Kasteleyn, Physica **57**, 536 (1972)
41. R. Swendsen, J. Wang, Phys. Rev. Lett. **58**, 86 (1987)
42. A.D. Sokal, in *Quantum Fields on the Computer*, ed. by M. Creutz (World Scientific, Singapore, 1992), pp. 211–274. Available electronically via www.dbwilson.com/exact
43. M. Aizenman, B. Nachtergaele, Comm. Math. Phys. **164**, 17 (1994)
44. B. Nachtergaele, in *Probability Theory and Mathematical Statistics (Proceedings of the 6th Vilnius Conference)*, ed. by B. Grigelionis, et al. (VSP/TEV, Utrecht Tokyo Vilnius, 1994), pp. 565–590. URL http://arxiv.org/abs/cond-mat/9312012
45. A.W. Sandvik, Phys. Rev. B **59**, R14157 (1999)
46. P. Henelius, A. Sandvik, Phys. Rev. B **62**, 1102 (2000)
47. A.W. Sandvik, Phys. Rev. Lett. **95**, 207203 (2005)
48. H.G. Evertz, W. von der Linden, Phys. Rev. Lett. **86**, 5164 (2001)
49. R. Citro, E. Orignac, T. Giamarchi, Phys. Rev. B **72**, 024434 (2005)
50. K. Hukushima, K. Nemoto, J. Phys. Soc. Japan **65**, 1604 (1996)
51. K. Hukushima, H. Takayama, K. Nemoto, Int. J. Mod. Phys. C **7**, 337 (1996)
52. E. Marinari, G. Parisi, Europhys. Lett. **19**, 451 (1992)
53. W. Barford, R. Bursill, Phys. Rev. Lett. **95**, 137207 (2005)
54. J.E. Hirsch, R.M. Fye, Phys. Rev. Lett. **56**, 2521 (1986)
55. S. Capponi, F.F. Assaad, Phys. Rev. B **63**, 155114 (2001)
56. F.F. Assaad, Phys. Rev. B **71**, 075103 (2005)
57. G. Sugiyama, S. Koonin, Ann. Phys. (N.Y.) **168**, 1 (1986)
58. S. Sorella, S. Baroni, R. Car, M. Parrinello, Europhys. Lett. **8**, 663 (1989)
59. S. Sorella, E. Tosatti, S. Baroni, R. Car, M. Parrinello, Int. J. Mod. Phys. B **1**, 993 (1989)
60. J.E. Hirsch, Phys. Rev. B **31**, 4403 (1985)
61. S.R. White, D.J. Scalapino, R.L. Sugar, E.Y. Loh, J.E. Gubernatis, R.T. Scalettar, Phys. Rev. B **40**, 506 (1989)
62. A.M. Tsvelik, *Quantum Field Theory in Condensed Matter Physics* (Cambridge University press, Cambridge, 1996)
63. F.F. Assaad, M. Imada, J. Phys. Soc. Jpn. **65**, 189 (1996)
64. F.F. Assaad, Phys. Rev. Lett. **83**, 796 (1999)

65. M. Jarrell, J. Gubernatis, Phys. Rep. **269**, 133 (1996)
66. W. von der Linden, Appl. Phys. A **60**, 155 (1995)
67. K.S.D. Beach, Identifying the maximum entropy method as a special limit of stochastic analytic continuation. URL `http://arxiv.org/abs/cond-mat/0403055`. Preprint
68. J.E. Hirsch, Phys. Rev. B **38**, 12023 (1988)
69. C. Wu, S. Zhang, Phys. Rev. B **71**, 155115 (2005)
70. A. Messiah, *Quantum Mechanics.* (Dover publications, INC., Mineola, New-York, 1999)
71. S. Capponi, F.F. Assaad, Phys. Rev. B **75**, 045115 (2007)
72. M. Troyer, U. Wiese, Phys. Rev. Lett. **94**, 170201 (2005)
73. J.F. Corney, P.D. Drummond, J. Phys. A: Math. Gen. **39**, 269 (2006)
74. W.H. Press, S.A. Teukolsky, W.T. Vetterling, B.P. Flannery, *Numerical Recipes in C* (Cambridge University Press, Cambridge, 1992)
75. M. Feldbacher, F.F. Assaad, Phys. Rev. B **63**, 073105 (2001)
76. M. Jarrell, Phys. Rev. Lett. **69**, 168 (1992)
77. A. Georges, G. Kotliar, W. Krauth, M.J. Rozenberg, Rev. of. Mod. Phys. **68**, 13 (1996)
78. A.C. Hewson, *The Kondo Problem to Heavy Fermions.* Cambridge Studies in Magnetism (Cambridge Universiy Press, Cambridge, 1997)
79. J. Yoo, S. Chandrasekharan, R.K. Kaul, D. Ullmo, H.U. Baranger, J. Phys. A: Math. Gen. **38**, 10307 (2005)
80. F.F. Assaad, Phys. Rev. B **65**, 115104 (2002)
81. M. Feldbacher, F.F. Assaad, K. Held, Phys. Rev. Lett. **93**, 136405 (2004)
82. F. F. Assaad and T. Lang, Phys. Rev. B 76, 035116 (2007)
83. J.E. Hirsch, Fradkin, Phys. Rev. B **27**, 4302 (1983)
84. M. Randeria, N. Trivedi, A. Moreo, R.T. Scalettar, Phys. Rev. Lett. **69**, 2001 (1992)
85. N. Trivedi, M. Randeria, Phys. Rev. Lett. **75**, 312 (1995)
86. M. Randeria, N. Trivedi, A. Moreo, R.T. Scalettar, Phys. Rev. D **54**, R3756 (1996)
87. F.F. Assaad, V. Rousseau, F. Hébert, M. Feldbacher, G. Batrouni, Europhys. Lett. **63**, 569 (2003)
88. C. Wu, J.P. Hu, S.C. Zhang, Phys. Rev. Lett. **91**, 186402 (2003)
89. S. Capponi, C. Wu, S.C. Zhang, Phys. Rev. B **70**, 220505 (2004)
90. J.E. Hirsch, S. Tang, Phys. Rev. Lett. **62**, 591 (1989)
91. G. Dopf, A. Muramatsu, W. Hanke, Europhys. Lett. **17**, 559 (1992)
92. G. Dopf, A. Muramatsu, W. Hanke, Phys. Rev. Lett. **68**, 353 (1992)
93. F.F. Assaad, W. Hanke, D.J. Scalapino, Phys. Rev. B **50**, 12835 (1994)
94. D.J. Scalapino, S. White, S. Zhang, Phys. Rev. B **47**, 7995 (1993)
95. N. Furukawa, M. Imada, J. Phys. Soc. Jpn. **62**, 2557 (1993)
96. F.F. Assaad, M. Imada, Phys. Rev. Lett **74**, 3868 (1995)
97. F.F. Assaad, M. Imada, Phys. Rev. Lett **76**, 3176 (1996)
98. G. Dopf, J. Wagner, P. Dieterich, A. Muramatsu, W. Hanke, Phys. Rev. Lett. **68**, 2082 (1992)
99. C. Gröber, R. Eder, W. Hanke, Phys. Rev. B **62**, 4336 (2000)
100. M. Imada, A. Fujimori, Y. Tokura, Rev. Mod. Phys. **70**, 1039 (1998)
101. M. Vekic, J.W. Cannon, D.J. Scalapino, R.T. Scalettar, R.L. Sugar, Phys. Rev. Lett. **74**, 2367 (1995)
102. F.F. Assaad, Phys. Rev. B **70**, 020402 (2004)
103. I. Milat, F. Assaad, M. Sigrist, Eur. Phys. J. B **38**, 571 (2004)

104. K.S.D. Beach, P.A. Lee, P. Monthoux, Phys. Rev. Lett. **92**, 026401 (2004)
105. S. Nishimoto, F. Gebhard, E. Jeckelmann, J. Phys.: Condens. Matter **16**, 7063 (2004)
106. R. Bulla, Phys. Rev. Lett. **83**, 136 (1999)
107. M.M. M. H. Hettler, M. Jarrell, H.R. Krishnamurthy, Phys. Rev. B **61**, 12739 (2000)
108. G. Kotliar, S.Y. Savrasov, G. Pálsson, G. Biroli, Phys. Rev. Lett. **87**, 186401 (2001)
109. R.M. Fye, Phys. Rev. B **33**, 6271 (1986)

11 Autocorrelations in Quantum Monte Carlo Simulations of Electron-Phonon Models

Martin Hohenadler[1] and Thomas C. Lang[2]

[1] Theory of Condensed Matter, Cavendish Laboratory, University of Cambridge, Cambridge CB3 0HE, United Kingdom
[2] Institut für Theoretische Physik und Astrophysik, Universität Würzburg, 97074 Würzburg, Germany

The problem of autocorrelations in quantum Monte Carlo simulations of electron-phonon models is analysed for different algorithms in the framework of the Holstein model. By revisiting several cases found in the literature, it is demonstrated that neglecting autocorrelations can lead to an underestimation of statistical errors by orders of magnitude and hence to incorrect results. A modified algorithm for certain Holstein-type models, free of any autocorrelations, is discussed.

11.1 Introduction

The interaction of electrons with lattice degrees of freedom plays an important role in many materials, including conventional and high-temperature superconductors, colossal-magnetoresistance manganites, and low-dimensional nanostructures. Over more than two decades, lattice and continuum quantum Monte Carlo (QMC) simulations have proved to be a highly valuable tool to investigate the properties of coupled fermion-boson models in condensed matter theory.

Despite the recent development of other numerical methods (e.g., the density matrix renormalization group, see Part IX, QMC approaches remain in the focus of research due to their versatility. Especially in the early days of computational physics, they outperformed alternative memory-consumptive methods, and this often remains true today, e.g., in more than one dimension or at finite temperature. Apart from stand-alone applications, QMC algorithms also serve as solvers in the context of cluster methods (see Chap. 16). Finally, they represent the most reliable techniques for several classes of problems, e.g., three-dimensional (3D) spin systems (see Chap. 10).

A general introduction to the concepts common to many QMC methods has been given in Chap. 10. In this chapter, we focus on the issue of autocorrelations, which turns out to be of particular importance in the case of coupled fermion-boson models due to the different physical time scales involved, and the resulting problems in finding appropriate updating schemes. Quite disturbingly, some recent as well as early work seems to be unaware of the problem. To illustrate this point, we re-enact some specific QMC studies from the literature using the same methods, and demonstrate that statistical errors are severely underestimated if autocorrelations are neglected.

M. Hohenadler and T. C. Lang: *Autocorrelations in Quantum Monte Carlo Simulations of Electron-Phonon Models*, Lect. Notes Phys. **739**, 357–366 (2008)
DOI 10.1007/978-3-540-74686-7_11

This chapter is organized as follows. In Sect. 11.2, we introduce the model considered, and Sect. 11.3 gives a brief description of the algorithms used. Numerical evidence for the problem of autocorrelations is presented in Sect. 11.4, whereas their origin and a possible solution are the topic of Sect. 11.5. We end with our conclusions in Sect. 11.6.

11.2 Holstein Model

We consider the Holstein model which is defined by the Hamiltonian

$$H = -t \sum_{\langle i,j\rangle\sigma} c^{\dagger}_{i,\sigma} c_{j,\sigma} + \frac{\omega_0}{2} \sum_i (\widehat{p}_i^2 + \widehat{x}_i^2) - \alpha \sum_{i,\sigma} \widehat{n}_{i,\sigma} \widehat{x}_i \,. \tag{11.1}$$

Here $c^{\dagger}_{i,\sigma}$ creates an electron with spin σ at site i, and $\widehat{n}_i = \sum_\sigma \widehat{n}_{i,\sigma}$ with $\widehat{n}_{i,\sigma} = c^{\dagger}_{i,\sigma} c_{i,\sigma}$. The phonon degrees of freedom at site i are described by the momentum $\widehat{p}_i$ and coordinate (displacement) $\widehat{x}_i$ of a harmonic oscillator. The model parameters are the nearest-neighbor hopping amplitude t, the Einstein phonon frequency ω_0 and the electron-phonon coupling α. We shall also refer to the spinless Holstein model, which can be obtained from (11.1) by dropping spin indices and sums over σ. We consider D-dimensional lattices with $V = N^D$ sites and periodic boundary conditions. A useful dimensionless coupling constant is $\lambda = \alpha^2/(\omega_0 W) = 2E_{\mathrm{P}}/W$, where $W = 4tD$ and E_{P} denote the free bandwidth and the polaron binding energy, respectively.

The Holstein model provides a framework to study numerous problems associated with electron-phonon interaction, such as polaron formation, superconductivity or charge-density-wave formation. Besides, more complicated models such as the Holstein-Hubbard model share the same structure of the phonon degrees of freedom and the electron-phonon interaction, so that the following discussion in principle applies to a wider range of problems.

11.3 Numerical Methods

To set the stage for the discussion of autocorrelations, we provide here a brief summary of the most important details of the different QMC algorithms employed. For details we refer the reader to [1, 2] and Chap. 10.

11.3.1 One-Electron Method

For the one-electron case (the polaron problem), we make use of the world-line method originally proposed in [3, 4]. Dividing the imaginary-time axis $[0, \beta]$ ($\beta = (k_{\mathrm{B}}T)^{-1}$ is the inverse temperature) into intervals of length $\Delta\tau = \beta/L \ll 1$

according to the Suzuki-Trotter approximation (see Chap. 10), the result for the fermionic[3] partition function reads

$$Z_{\mathrm{f},L} = \sum_{\{\boldsymbol{r}_\tau\}} w_{\mathrm{f}}(\{\boldsymbol{r}_\tau\}), \quad w_{\mathrm{f}}(\{\boldsymbol{r}_\tau\}) = \mathrm{e}^{\sum_{\tau,\tau'=1}^{L} F(\tau-\tau')\delta_{\boldsymbol{r}_\tau,\boldsymbol{r}_{\tau'}}} \prod_{\tau=1}^{L} I(\boldsymbol{r}_{\tau+1}-\boldsymbol{r}_\tau), \tag{11.2}$$

with the fermionic weight w_{f}. The fermion world-lines, specified by a position vector $\boldsymbol{r}_\tau$ on each time slice,[4] are subject to periodic boundary conditions both in real space and imaginary time, and the sum in (11.2) is over all allowed configurations.

The retarded electron (self-)interaction due to electron-phonon coupling is described by the memory function

$$F(\tau) = \frac{\omega_0 \Delta\tau^3 \alpha^2}{4L} \sum_{\nu=0}^{L-1} \frac{\cos(2\pi\tau\nu/L)}{1-\cos(2\pi\nu/L)+(\omega_0\Delta\tau)^2/2}, \tag{11.3}$$

whereas electron hopping is manifest in the Fourier-transformed lattice propagator

$$I(\boldsymbol{r}) = \frac{1}{V}\sum_{\boldsymbol{k}} \cos(\boldsymbol{k}\cdot\boldsymbol{r})\,\mathrm{e}^{2\Delta\tau t\sum_{\zeta=1}^{D}\cos k_\zeta}. \tag{11.4}$$

The system described by the partition function (11.2) is characterized by an additional dimension (imaginary time), as well as by a complicated retarded (i.e., non-local in imaginary time) interaction. As first shown in [3], it may be simulated by means of Markov Chain MC in combination with the Metropolis algorithm [5]. The updating consists in choosing a random time slice $\tau_0 \in [1, L]$ and a random spatial component $\zeta_0 \in [1, D]$, and proposing a local change $r'_{\tau_0,\zeta_0} = r_{\tau_0,\zeta_0} \pm 1$, which is to be accepted with probability $\min[1, w_{\mathrm{f}}(\boldsymbol{r}'_\tau)/w_{\mathrm{f}}(\boldsymbol{r}_\tau)]$.

11.3.2 Many-Electron Method

A frequently used method for simulations of many-electron systems is the grand-canonical determinant QMC method [2], introduced for interacting fermions in Chap. 10. The corresponding grand-canonical Hamiltonian reads $H - \mu\sum_{i,\sigma}\widehat{n}_{i,\sigma}$, where μ denotes the chemical potential. For the Holstein model, the integration over the fermionic degrees of freedom can be done exactly, whereas for the Holstein-Hubbard model with electron-electron interaction it is done by means of MC sampling over Hubbard-Stratonovitch fields (see Chap. 10).

In the original approach of [2], the Trotter approximation to the partition function reads

[3] The bosonic part can be calculated exactly [4] and is therefore not considered.

[4] We use bold symbols to indicate the vector character of a quantity. The exact definition of the components should be clear from the context.

$$Z_L = \text{const.} \int \mathcal{D}\boldsymbol{x} \underbrace{\prod_\sigma \det\left[(1 + \prod_{\tau=1}^{L} \mathrm{e}^{-\Delta\tau K^\sigma} \mathrm{e}^{-\Delta\tau I^\sigma(\{\boldsymbol{x}_\tau\})}\right]}_{w_\mathrm{f}(\{\boldsymbol{x}_\tau\})} \underbrace{\mathrm{e}^{-\Delta\tau S_\mathrm{b}(\{\boldsymbol{x}_\tau\})}}_{w_\mathrm{b}(\{\boldsymbol{x}_\tau\})}, \tag{11.5}$$

with K^σ, I^σ denoting the matrix representations of the spin-σ component of the first respectively last term (including the minus signs) in Hamiltonian (11.1).

The bosonic action is given by

$$S_\mathrm{b}(\{\boldsymbol{x}_\tau\}) = \sum_{i=1}^{V} \sum_{\tau=1}^{L} \left[\frac{\omega_0}{2} x_{i,\tau}^2 + \frac{1}{2\omega_0 \Delta\tau^2} (x_{i,\tau} - x_{i,\tau+1})^2\right] = \sum_{i=1}^{L} \boldsymbol{x}_i^\mathrm{T} A \boldsymbol{x}_i \,. \tag{11.6}$$

Here the sampling is over all possible phonon configurations $\{\boldsymbol{x}_\tau\}$ of the bosonic degrees of freedom. In the simplest approach, we select a random time slice $\tau_0 \in [1, L]$ and a random lattice site $i_0 \in [1, V]$, and propose a modified phonon configuration $x'_{i_0,\tau_0} = x_{i_0,\tau_0} \pm \delta x$. The latter is then accepted with probability $\min[1, w_\mathrm{f}(\{\boldsymbol{x}'_\tau\}) w_\mathrm{b}(\{\boldsymbol{x}'_\tau\}) / w_\mathrm{f}(\{\boldsymbol{x}_\tau\}) w_\mathrm{b}(\{\boldsymbol{x}_\tau\})]$. The change δx is determined by requiring a reasonable acceptance rate. An improved (global) updating scheme will be discussed below.

11.4 Problem of Autocorrelations

We now come to a discussion of autocorrelations, which are analyzed using the binning and Jackknife methods introduced in Chap. 4. As shown in there, the integrated autocorrelation time $\tau_{\mathcal{O},\mathrm{int}}$ associated with an observable $\mathcal{O}$ can be estimated by plotting the statistical error as obtained from a binning/Jackknife analysis as a function of binsize k, and deducing the binsize required for "saturation". The increase of the error with increasing k demonstrates the fact that unjustified neglect of autocorrelations leads to an underestimation of errors and hence to incorrect results. Explicitly, the statistical error $\Delta\mathcal{O}$ for a given number of (correlated) measurements N_meas increases with $\tau_{\mathcal{O},\mathrm{int}}$ as $(\Delta\mathcal{O})^2 \propto 2\tau_{\mathcal{O},\mathrm{int}}/N_\mathrm{meas}$.

11.4.1 One-Electron Case

As a specific case, we consider world-line simulations of the Holstein polaron, i.e., the Hamiltonian (11.1) with a single electron. This problem has first been studied by means of QMC in [3]. The same method has also been used in [6] as well as in [7], and its generalizations represent a versatile tool for studies of more general systems with one or two fermions. In [3, 6], the authors skipped L (the number of time slices) steps between successive measurements.

We take $\lambda = 1$, close to the critical coupling where the small-polaron crossover occurs in the adiabatic regime [8], leading to the occurrence of critical slowing down

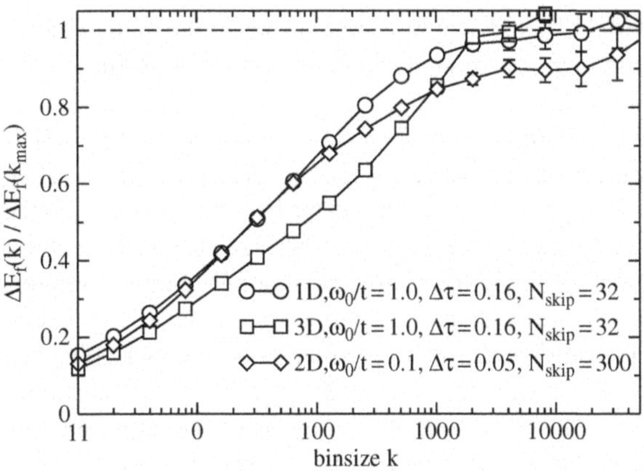

Fig. 11.1. Statistical error of the fermionic total energy E_f as a function of binsize k, normalized to the result for the maximum binsize, obtained with the world-line method [4]. Results are for the Holstein model with one electron, $N = 32$ and $\lambda = 1$

for $\omega_0/t \lesssim 1$ [7]. To compare with [4], we choose the same parameters $\beta t = 5$, $L = 32$, $N = 32$ and $\omega_0/t = 1$.

In Fig. 11.1, we show results for the statistical error of the fermionic contribution to the total energy[5] as obtained from a binning analysis with variable binsize k. The uncertainty of the binning error increases with decreasing binsize, leading to fluctuations. Note the logarithmic scale on the abscissa.

Skipping $N_{\text{skip}} = L = 32$ steps between measurements, we find for the 1D case that the statistical error increases roughly by a factor of 7, i.e., it is substantially larger than the estimate obtained from binning with $k = 1$, which corresponds to the usual procedure to calculate statistical errors from uncorrelated data.

The situation is slightly worse in three dimensions, with the real error being again about an order of magnitude larger. This may be related to the local updating, as the relative difference between two entire successive world-line configurations is smaller in higher dimensions if only a single coordinate r_{τ_0,ζ_0} is changed.

Finally, we also consider the more demanding 2D case of low temperature $\beta t = 15$ and small phonon frequency $\omega_0/t = 0.1$, which has been studied using the same method in [6, 7]. As discussed below, the smaller values of ω_0 and $\Delta\tau$ give rise to substantially longer autocorrelation times. Indeed, despite the larger number of skipped steps $N_{\text{skip}} = L = 300$, convergence of the statistical error is slower as a function of k.

An alternative algorithm free of autocorrelations, which can also be applied to the many-electron case, has been proposed in [9] and will be discussed in Sect. 11.5.

11.4.2 Many-Electron Case

It is important to point out that the occurrence of long autocorrelation times is not restricted to the one-electron case. The problem is at least as serious for the

[5] Autocorrelation times are similar for other observables.

two-electron model [7], and accurate simulations in the many-electron case turn out to be unfeasible in many cases [10] due to autocorrelations times exceeding 10^4 MC steps.

To illustrate this point, we consider two parameter sets for the Holstein model at half filling (one electron per lattice site), representative of the work in [11] and [12]. We use the finite-temperature determinant QMC method, although the results of [11] have been obtained using the projector method (see Chap. 10; autocorrelation times are usually comparable). Owing to the substantially larger computational effort as compared to one-electron calculations, we were not able to obtain converged results. Therefore, and to compare different parameters, we show in Fig. 11.2 the statistical error of the bosonic energy $E_\mathrm{b} = (\omega_0/2)\langle\sum_i(\widehat{p}_i^2 + \widehat{x}_i^2)\rangle$, normalized to the error for binsize $k = 1$. The definition of λ in terms of the coupling constant g used in [11, 12] reads $\lambda = 2g^2/(\omega_0 W)$, and we have used $N_\mathrm{skip} = 1$.

The strong increase of statistical errors as a function of binsize in Fig. 11.2 illustrates the substantial autocorrelations in such simulations. No saturation can be seen in our data even for the largest binsize $k > 10^4$ shown (cf Fig. 11.1) and, in contrast to the world-line method of Sect. 11.3, skipping thousands of steps is usually not practicable in the many-electron case. In our opinion, this suggests that reliable results for the Holstein model in the many-electron case are extremely challenging to obtain using the determinant QMC method, and the situation becomes even worse for $\omega_0/t < 1$. Similar conclusions can be drawn about the spinless Holstein model, models with phonon modes of different symmetry [10], as well as Holstein-Hubbard models with local and/or non-local Coulomb interaction [13].

Despite these difficulties, some early work [12] as well as more recent papers, e.g., [11, 14], seem to be unaware of this problem. This issue becomes even more critical if dynamical quantities such as the one-electron spectral function are

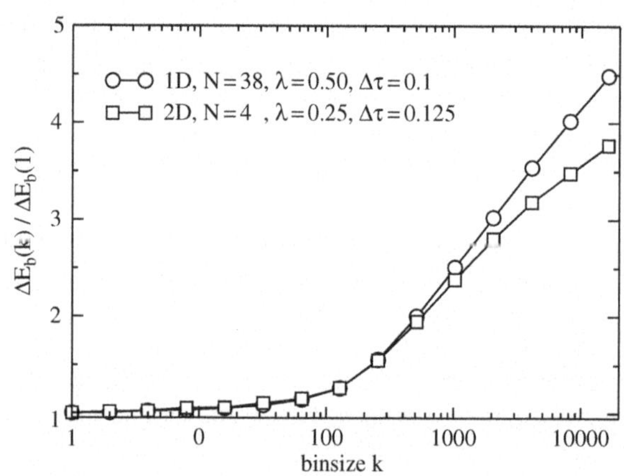

Fig. 11.2. Statistical error of the bosonic energy E_b as a function of binsize k, normalized to the result for $k = 1$, obtained with the determinant QMC method [2]. Results are for the Holstein model at half filling $n = 1$, $\beta t = 10$ and $\omega_0/t = 1$. Errorbars are not shown

calculated. The reason lies in the necessity to perform an analytical continuation to real frequencies, which turns out to be an extremely ill-conditioned problem (see Chap. 12) whose solution – obtained for example by the Maximum Entropy method [15, 16] – depends crucially on the statistical noise of the input data (the imaginary-time Green function). Any underestimation of errors, or the neglect of significant autocorrelations between measurements on different time slices can lead to incorrect results. Furthermore, meaningful statistical errors for dynamical properties are difficult to obtain, and are often not reported at all.

11.5 Origin of Autocorrelations and Principal Components

To understand the problem and to find solutions, it is instructive to look at the physical origin of autocorrelations in more detail [9]. To this end, let us consider the non-interacting limit ($t = \alpha = 0$), in which the partition function $Z_L \sim \int \mathcal{D}x\, \mathrm{e}^{-\Delta\tau S_\mathrm{b}}$. As discussed in [17], the difficulties encountered in QMC simulations, even for the simple case of a single ($N = 1$) quantum-mechanical harmonic oscillator, result from the large condition number (the ratio of largest to smallest eigenvalue) of the matrix A in the bosonic action S_b (11.6). For $\Delta\tau \ll 1$, this number scales as $(\omega_0\Delta\tau)^{-2}$ [17], causing autocorrelation times to grow quadratically with decreasing phonon frequency or increasing number of Trotter slices L. This is very unfortunate, as small phonon frequencies are frequently encountered in materials of interest, and small values of $\Delta\tau$ are desirable to control the Trotter error.

The physical reason for these correlations becomes obvious if we look at the free bosonic action (11.6), which is proportional to the energy of a given phonon configuration. The first term of S_b corresponds to the kinetic energy of the oscillators, and the second term describes a coupling in imaginary time – a pure quantum-mechanical effect. Due to this interaction, a large change of a single phonon degree of freedom, x_{i_0,τ_0} say, is very unlikely to be accepted due to the associated large energy change $\sim (\omega_0\Delta\tau)^{-1}$. However, using only small changes Δx, successive phonon configurations will be highly correlated. This behavior carries over to the interacting case with one or many electrons, as well as to more general models.

The situation is not completely obvious for the world-line algorithm, because the phonon degrees of freedom are integrated out analytically. However, the retarded self-interaction entering simulations in terms of $F(\tau)$ (11.3) gives rise to the same problem. Moreover, for large α (strong coupling), electronic hopping becomes very unlikely ($F(\tau) \sim \alpha^2$), causing the acceptance rate to approach zero and thus again giving rise to autocorrelations.[6] The situation is expected to be slightly better for the continuous-time variant of the algorithm [18] because hopping events may occur at arbitrary points in imaginary time.

[6] In the world-line algorithm, the discrete step size used for updates cannot be reduced below one lattice constant.

We now discuss a solution for the problem of autocorrelations in simulations of certain Holstein-type models. It is based on a transformation of the bosonic action to a so-called principal-component representation [9], which is obtained by writing

$$S_\mathrm{b} = \sum_{i=1}^{V} \boldsymbol{x}_i^\mathrm{T} A \boldsymbol{x}_i = \sum_{i=1}^{V} \boldsymbol{x}_i^\mathrm{T} A^{1/2} A^{1/2} \boldsymbol{x}_i = \sum_{i=1}^{V} \boldsymbol{\xi}_i^\mathrm{T} \cdot \boldsymbol{\xi}_i \,, \quad \boldsymbol{\xi}_i = A^{1/2} \boldsymbol{x}_i \,, \quad (11.7)$$

with the aforementioned $L \times L$ matrix A, and the principal components $\boldsymbol{\xi}$, in terms of which S_b becomes diagonal. Using this representation, the bosonic weight reduces to a Gaussian distribution, $w_\mathrm{b} = \exp(-\Delta\tau \sum_i \boldsymbol{\xi}_i^\mathrm{T} \cdot \boldsymbol{\xi}_i)$. For $\alpha = 0$, sampling can be done exactly in terms of the new variables $\xi_{i,\tau}$ using the Box-Muller method [19].

To further illustrate the origin of autocorrelations, as well as the transformation to principal components, we show in Fig. 11.3(a) a schematic representation of the distribution of values for a pair (p, p') of two phonon momenta (shaded area). The elongated shape originates from the strong correlations mediated by S_b, and requires a transition $A \rightarrow B$ between two points in phase space to be performed in many small steps, leading in turn to long autocorrelation times.

In contrast, the axes of the principal components ξ, ξ' in Fig. 11.3(b) lie along the axes of the ellipse, and a single MC update of ξ' is sufficient to get from A to B. Although we have sketched the more general case, the distribution after the exact transformation (11.7) – under which w_b becomes a Gaussian – is actually circular in the new variables ξ, ξ' (dashed line in Fig. 11.3(b)).

Whereas exact sampling without autocorrelations is straightforward in the non-interacting case $\alpha = 0$, the dependence of w_f on the phonon coordinates $x_{i,\tau}$ for $\alpha > 0$ does not permit a simple separation of bosonic and fermionic contributions in the updating process. Therefore, it has been proposed [9] to base the QMC algorithm on the Lang-Firsov transformed Hamiltonian, which has no explicit coupling of x to electronic degrees of freedom. To this end, it is advantageous to sample the phonon momenta p instead of x, as the former depend only weakly on the electronic degrees of freedom [9], which enables us to treat the fermionic weight w_f

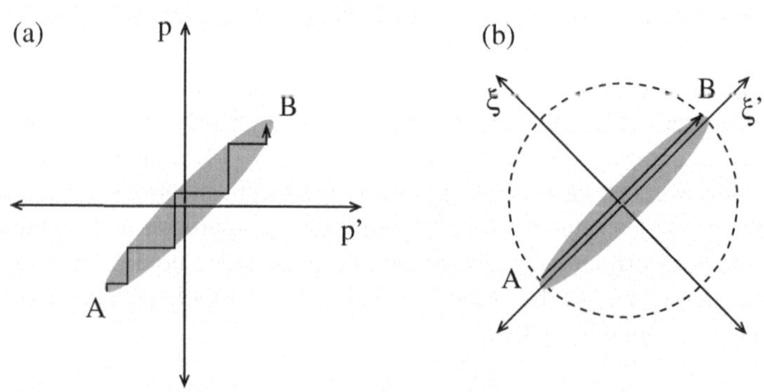

Fig. 11.3. Schematic illustration of the transformation from phonon momenta p, p' to principal components ξ, ξ' (see text)

as a part of the observables, and renders the MC sampling exact and rejection-free (every new configuration is accepted). Consequently, we avoid a warm-up phase, autocorrelations and the computationally expensive evaluation of w_{f} in the updating process.

The method outlined here has been successfully applied to the polaron [9, 20], bipolaron [21], and the (spinless) many-polaron problem [22]. Unfortunately, attempts to generalize this approach to the Holstein-Hubbard model, or the spinful Holstein model, have not been successful [13]. Although the Lang-Firsov transformation improves the sampling of phonon configurations via principal components, the complex phase in the transformed hopping term [9] induces a severe sign problem [13, 22]. Despite encouraging acceptance rates, this global updating scheme does not permit reliable statements concerning a possible decrease of autocorrelation times.

11.6 Conclusions

By revisiting several QMC studies of Holstein models carried out in the past we have illustrated the severe problem of autocorrelations in simulations of electron-phonon models, in accordance with [10]. In particular, we have shown that statistical errors can be underestimated by orders of magnitude if autocorrelations are neglected. This is particularly dangerous when calculating dynamic properties using, e.g., Maximum Entropy methods, where meaningful errorbars can usually not be obtained, introducing substantial uncertainties into the results. Long autocorrelation times can also lead to critical slowing down as well as non-ergodic sampling during finite-time MC runs – both phenomena being additional sources for underestimated statistical errors – thereby also affecting the expectation values of observables.

Similar to the infamous minus-sign problem (see Chap. 10), autocorrelations in QMC simulations seem to result from the fact that one is dealing with an ill-conditioned physical problem. As a consequence, their appearance is not restricted to the Holstein-type models considered here (see Chap. 10), or the particular QMC methods employed. Besides, autocorrelations even occur in simulations of classical systems (Chap. 4), although the problem is usually not as substantial as for coupled fermion-boson systems. This general observation strongly suggests that great care has to be taken when performing any MC simulations in order to avoid incorrect results.

Significant advances in terms of efficiency and applicability can be achieved by constructing a physically motivated global updating scheme. One such possibility has been presented here in terms of a transformation to principal components. However, a general solution to overcome the problem of autocorrelations in simulations of electron-phonon models is not yet known.

Acknowledgements

MH acknowledges financial support by the Austrian Science Fund (FWF) through the Erwin-Schrödinger Grant No J2583. We thank Pavel Kornilovitch for useful discussion.

References

1. H. de Raedt, A. Lagendijk, Phys. Rep. **127**, 233 (1985)
2. R. Blankenbecler, D.J. Scalapino, R.L. Sugar, Phys. Rev. D **24**, 2278 (1981)
3. H. De Raedt, A. Lagendijk, Phys. Rev. Lett. **49**, 1522 (1982)
4. H. De Raedt, A. Lagendijk, Phys. Rev. B **27**, 6097 (1983)
5. N. Metropolis, A. Rosenbluth, A. Teller, E. Teller, J. Chem. Phys. **21**, 1087 (1953)
6. P.E. Kornilovitch, J. Phys.: Condens. Matter **9**, 10675 (1997)
7. M. Hohenadler, H. Fehske, J. Phys.: Condens. Matter **19**, 255210 (2007)
8. H. Fehske, A. Alvermann, M. Hohenadler, G. Wellein, in *Polarons in Bulk Materials and Systems with Reduced Dimensionality*, ed. by G. Iadonisi, J. Ranninger, G. De Filippis (IOS Press, Amsterdam, Oxford, Tokio, Washington DC, 2006), Proc. Int. School of Physics "Enrico Fermi", Course CLXI, pp. 285–296
9. M. Hohenadler, H.G. Evertz, W. von der Linden, Phys. Rev. B **69**, 024301 (2004)
10. D. Eckert, Phononen im Hubbard Modell. Master's thesis, University of Würzburg (1997)
11. K. Tam, S. Tsai, D.K. Campbell, A.H. Castro Neto, Phys. Rev. B **75**, 161103 (2007)
12. P. Niyaz, J.E. Gubernatis, R.T. Scalettar, C.Y. Fong, Phys. Rev. B **48**, 16 011 (1993)
13. T.C. Lang, Dynamics and charge order in a quarter filled ladder coupled to the lattice. Master's thesis, TU Graz (2005)
14. C.E. Creffield, G. Sangiovanni, M. Capone, Eur. Phys. J. B **44**, 175 (2005)
15. W. von der Linden, Phys. Rep. **220**, 53 (1992)
16. M. Jarrell, J.E. Gubernatis, Phys. Rep. **269**, 133 (1996)
17. G.G. Batrouni, R.T. Scalettar, in *Quantum Monte Carlo Methods in Physics and Chemistry*, ed. by M.P. Nightingale, C.J. Umrigar (Kluwer Academic Publishers, Dordrecht, 1998), p. 65
18. P.E. Kornilovitch, Phys. Rev. Lett. **81**, 5382 (1998)
19. G.E.P. Box, M.E. Muller, Ann. Math. Stat. **29**, 610 (1958)
20. M. Hohenadler, H.G. Evertz, W. von der Linden, phys. stat. sol. (b) **242**, 1406 (2005)
21. M. Hohenadler, W. von der Linden, Phys. Rev. B **71**, 184309 (2005)
22. M. Hohenadler, D. Neuber, W. von der Linden, G. Wellein, J. Loos, H. Fehske, Phys. Rev. B **71**, 245111 (2005)

12 Diagrammatic Monte Carlo and Stochastic Optimization Methods for Complex Composite Objects in Macroscopic Baths

A. S. Mishchenko

CREST, Japan Science and Technology Agency, AIST, 1-1-1, Higashi, Tsukuba 305-8562, Japan
Russian Research Center Kurchatov Institute, 123182 Moscow, Russia

We give an introduction to the Diagrammatic Monte Carlo method, which provides an efficient numerical scheme for the approximation-free calculation of Matsubara Green functions and correlation functions in imaginary time. The analytic continuation from imaginary times to real frequencies is performed by a stochastic optimization procedure.

12.1 Introduction

Many physical problems can be reduced to a system of one or a few complex objects (CO) interacting with each other and with a macroscopic bosonic bath. The state of such a CO, in general, is defined by a diverse set of quantum numbers, which change when excitations of the bosonic bath are emitted and annihilated, or when two COs interact. Despite the varying physical meaning of the COs quantum numbers in different physical systems the typical Hamiltonians for a broad range of problems look very similar, and, thus, similar methods can be applied for their solution.

Historically, the most famous problem treated in the above framework is that of a polaron, i.e. of an electron coupled to phonons (see [1, 2] for an introduction). In the initial formulation a bare quasi particle (QP)[1] has no internal structure, i.e. internal quantum numbers, and it is characterized only by the translational quantum number – the momentum – which changes due to the interaction of the QP with phonons [3, 4]. Hence, in terms of the above definition, the polaron is not a CO since the quasimomentum completely defines its quantum state and there are no other quantum numbers determining the internal state of the QP. However, the polaron concept can be generalized to include additional internal degrees of freedom, which change their quantum numbers due to the interaction with the environment. Examples are the Jahn-Teller polaron, where the electron-phonon interaction changes the quantum numbers of degenerate electronic states [5, 6], and the pseudo Jahn-Teller (PJT) polaron, where electron-phonon interaction leads to transitions between electronic levels that are close in energy [7, 8]. Note, that for a CO, in addition to the quasimomentum, some internal quantum numbers are required to define the state

[1] In general, a QP is defined as an elementary excitation whose energy separation from the ground state is larger than the energy broadening due to decay.

A. S. Mishchenko: *Diagrammatic Monte Carlo and Stochastic Optimization Methods for Complex Composite Objects in Macroscopic Baths*, Lect. Notes Phys. **739**, 367–395 (2008)
DOI 10.1007/978-3-540-74686-7_12

of the system. A further generalization is a system of several COs interacting both with each other and the environment. For example, the effective interaction of two electrons through exchange by phonons can overcome the Coulomb repulsion and lead to the formation of a bound state, the bipolaron [9, 10, 11, 12]. Furthermore the attraction of a hole and an electron, both additionally coupled to lattice vibrations [13, 14, 15], may result in a variety of qualitatively different objects: Localized excitons, weakly bound pairs of a localized hole and a localized electron, etc. [16, 17].

Extending the meaning of what is called the *particle* and the *environment* and how they interact with each other, later on the polaron concept was applied to a broad variety of other phenomena. An example is the problem of a hole moving in an antiferromagnet background. Here the hole movement is accompanied by spin flips which, within spin wave approximation, are equivalent to the creation and annihilation of bosonic excitations – the magnons [18, 19]. Finally, let us consider a complex system which, in contrast to all previous examples, is not derived from a translationally invariant QP: The physics of the decoherence of a qubit, which is so notorious in the race to implement a quantum computer, can be formulated in terms of a two-level system coupled to a spectrum of bosonic excitations [20, 21]. Even though the problem has a completely different physical meaning its Hamiltonian is similar to those encountered in the examples above. Hence, the problem can solved with the same methods.

A particularly challenging class of problems are the strongly correlated systems. Here a bare CO and a bosonic bath are seldom well defined. Experimentally the response to a given momentum transfer is a broad distribution of the energy transfers, and a proper dispersion relation of the elementary excitations is hard to define or may not exist at all. One of the possible reasons for the ambiguity of the dispersion relation is the interaction of the COs with the bosonic bath, which is explicitly included in the Hamiltonian that defines the main interactions of a strongly correlated system. In practice, however, even the simplest Hamiltonians of strongly correlated systems turn out to be too complicated for a complete solution and many important interactions need to be neglected. Many studies [22, 23] of the single band Holstein-Hubbard model, for instance, neglect the influence of the other bands and the decay of phonons due to anharmonicity. On the other hand, it often happens that these neglected broadenings of the CO and the bosonic bath are known from the beginning, e.g., from perturbation theory. In this case the QPs defined by the unperturbed Hamiltonian are damped from the onset and one faces the challenge how to formulate the interactions, e.g. the coupling to the bosonic bath, in terms of damped objects.

This chapter is organized as follows: In Sect. 12.1.1 the advantages of the diagrammatic Monte Carlo method (DMC) are discussed. Sect. 12.1.2 presents the basic models for COs in correlated systems. The quantities and functions of interest, which are relevant to these models, are introduced in Sect. 12.2. The basics of the DMC method, which is capable of providing an exact numerical solutions for the problems formulated in the introduction [24, 25, 26, 27, 28, 29, 30, 31, 32, 33, 34, 35, 36, 37, 38, 39], are presented in Sect. 12.3. This section also contains tutorial algorithms for simple analytically solvable models. A novel stochastic optimization (SO) method for the analytic continuation of imaginary time functions to real frequencies [26, 31, 32], which avoids the difficulties of the popular maximum entropy method (MEM), is briefly reviewed in Sect. 12.4. Sect. 12.5 contains conclusions and perspectives.

12.1.1 Need for new Numerical Methods

Hardly any numerical method, not to speak of analytical approaches, can give approximation-free results for measurable spectral quantities of a CO, such as the optical conductivity, the angle resolved photoemission spectrum of a polaron or the damping of a qubit. There are plenty of effective methods which are either restricted to finite systems or applicable only to specific cases of macroscopic systems, such as low dimensional systems, etc. What we need is a general strategy for the whole class of problems formulated above, i.e. for a few COs in a macroscopic system of arbitrary dimension interacting with an arbitrary bath in the most general form. This implies arbitrary momentum dependence of the coupling constant of the CO to the bosonic bath which, in turn, has an arbitrary dispersion of bosonic excitations. In addition, it is important to treat the information on the damping of the CO and of the bosonic bath on the same (approximation-free) level as the interactions. Most of the standard numerical methods are based on the solution of an eigenvalue problem where all bare eigenstates have well defined energies. Therefore, any information which is not explicitly encoded in the Hamiltonian cannot be incorporated in the solution, in particular, it is not possible to describe damped QPs.

The DMC method provides an elegant way to handle all these difficulties. It relies on an exact numerical summation of the Feynman expansion for the considered correlation function, and is independent of the analytic expression for the initial bare Green functions (GFs). Hence, additional information, e.g. damping, which is not included in the bare Hamiltonian, can easily be incorporated afterwards using standard rules [40]. Note also, that there are no restrictions on the bosonic bath.

12.1.2 Models of few Interacting Complex Objects

Formulating models suitable for the DMC-SO approach I start from general polaron models. The simplest problem of a complex polaronic object, where the center-of-mass motion does not separate from the other degrees of freedom, is given by a system of two QPs,

$$H_0^{\text{par}} = \sum_{\boldsymbol{k}} \varepsilon_a(\boldsymbol{k}) a_{\boldsymbol{k}}^\dagger a_{\boldsymbol{k}} + \sum_{\boldsymbol{k}} \varepsilon_h(\boldsymbol{k}) h_{\boldsymbol{k}} h_{\boldsymbol{k}}^\dagger \,. \tag{12.1}$$

Here $a_{\boldsymbol{k}}$ and $h_{\boldsymbol{k}}$ are annihilation operators, and $\varepsilon_a(\boldsymbol{k})$ and $\varepsilon_h(\boldsymbol{k})$ are the dispersions of the QPs, which interact with each other through the instantaneous Coulomb potential $\mathcal{U}$,

$$H_{\text{a-h}} = -\frac{1}{N} \sum_{\boldsymbol{kpp'}} \mathcal{U}_{\boldsymbol{k}}(\boldsymbol{p},\boldsymbol{p}')\, a^{\dagger}_{\boldsymbol{k}+\boldsymbol{p}} h^{\dagger}_{\boldsymbol{k}-\boldsymbol{p}} h_{\boldsymbol{k}-\boldsymbol{p}'} a_{\boldsymbol{k}+\boldsymbol{p}'} \,, \tag{12.2}$$

where N is the number of lattice sites. The QPs are scattered by Q different branches of bosons,

$$H_{\text{par-bos}} = \mathrm{i} \sum_{\kappa=1}^{Q} \sum_{\boldsymbol{k},\boldsymbol{q}} (b^{\dagger}_{\boldsymbol{q},\kappa} - b_{-\boldsymbol{q},\kappa}) \Big[\gamma_{aa,\kappa}(\boldsymbol{k},\boldsymbol{q}) a^{\dagger}_{\boldsymbol{k}-\boldsymbol{q}} a_{\boldsymbol{k}} + \gamma_{hh,\kappa}(\boldsymbol{k},\boldsymbol{q}) h^{\dagger}_{\boldsymbol{k}-\boldsymbol{q}} h_{\boldsymbol{k}} + \gamma_{ah,\kappa}(\boldsymbol{k},\boldsymbol{q}) h^{\dagger}_{\boldsymbol{k}-\boldsymbol{q}} a_{\boldsymbol{k}} \Big] + h.c. \tag{12.3}$$

with $\gamma_{[aa,ah,hh],\kappa}$ are the interaction constants, which are described by the Hamiltonian

$$H_{\text{bos}} = \sum_{\kappa=1}^{Q} \sum_{\boldsymbol{q}} \omega_{\boldsymbol{q},\kappa} b^{\dagger}_{\boldsymbol{q},\kappa} b_{\boldsymbol{q},\kappa} \,. \tag{12.4}$$

In general, each QP can be a composite object with an internal degree of freedom represented by R different states

$$H_0^{\text{PJT}} = \sum_{\boldsymbol{k}} \sum_{i=1}^{R} \epsilon_i(\boldsymbol{k}) a^{\dagger}_{i,\boldsymbol{k}} a_{i,\boldsymbol{k}} \,, \tag{12.5}$$

with quantum numbers that can also be affected by the non-diagonal part of the particle-boson interaction

$$H_{\text{par-bos}} = i \sum_{\kappa}^{Q} \sum_{\boldsymbol{k},\boldsymbol{q}} \sum_{i,j=1}^{R} \gamma_{ij,\kappa}(\boldsymbol{k},\boldsymbol{q}) (b^{\dagger}_{\boldsymbol{q},\kappa} - b_{-\boldsymbol{q},\kappa}) a^{\dagger}_{i,\boldsymbol{k}-\boldsymbol{q}} a_{j,\boldsymbol{k}} + h.c. \,. \tag{12.6}$$

The complicated model (12.1)–(12.6) is still insufficient to describe a number of strongly correlated systems. Due to the coupling of the QPs (12.1) or (12.5) and the bosonic fields (12.4) to external degrees of freedom, which are missing in (12.1)–(12.6), these excitations may not be well defined. Frequently, the dispersion relation $\epsilon(\boldsymbol{k})$ of the QP measured, e.g., in angle resolved photoemission is subject to a substantial broadening, and the linewidth can become larger or comparable to the energy transfer corresponding to the peak position of the signal [41].

Theoretically this situation can be modelled with the Lehmann function of the QP [40, 42, 43],

$$L_{\boldsymbol{k}}(\omega) = \sum_{\nu} \delta\left(\omega - E_{\nu}(\boldsymbol{k})\right) \left| \langle \nu | a^{\dagger}_{\boldsymbol{k}} | \text{vac} \rangle \right|^2 \,, \tag{12.7}$$

which is normalized to unity $\int_0^{+\infty} \mathrm{d}\omega L_{\boldsymbol{k}}(\omega) = 1$ and describes the probability that a QP with momentum $\boldsymbol{k}$ has energy ω. Here $\{|\nu\rangle\}$ is a complete set of eigenstates of the Hamiltonian H in a sector of given momentum $\boldsymbol{k}$: $H|\nu(\boldsymbol{k})\rangle = E_{\nu}(\boldsymbol{k})|\nu(\boldsymbol{k})\rangle$ $(E_{\nu}(\boldsymbol{k}) \geq 0)$. Only for a non-interacting system the Lehmann function reduces to a delta function $L_{\boldsymbol{k}}^{(0)}(\omega) = \delta(\omega - \epsilon(\boldsymbol{k}))$ and thus sets up the dispersion relation $\omega = \epsilon(\boldsymbol{k})$. Note that the Lehmann function is the measurable quantity observed in angle resolved photoemission experiments [41].

Specifying the parameters of the model (12.1)–(12.6) we can study an enormous variety of physical problems: In the case of an attractive potential $\mathcal{U}(\boldsymbol{p}, \boldsymbol{k}, \boldsymbol{k}') > 0$, (12.1) and (12.2) account for an exciton with static screening [44, 45]. Besides, expressions (12.1)–(12.4) describe a bipolaron for repulsive interaction [9, 10, 11, 12] $\mathcal{U}(\boldsymbol{p}, \boldsymbol{k}, \boldsymbol{k}') < 0$ and an exciton-polaron otherwise [13, 14, 15]. The simplest model for exciton-phonon interaction, where only two ($R = 2$) lowest states of the relative electron-hole motion are relevant, e.g., for the one-dimensional charge-transfer exciton [46, 47, 48], is defined by the Hamiltonians (12.4)–(12.6). The same relations (12.4)–(12.6) describe the Jahn-Teller (all ϵ_i in Hamiltonian (12.5) are the same) and PJT polarons. The problem of a hole in an antiferromagnet within spin-wave approximation is expressed in terms of the Hamiltonians (12.4)–(12.6) with $Q = 1$ and $R = 1$ [18]. When the hole also interacts with phonons, one has to take into account one more bosonic branch and set $Q = 2$ in (12.4) and (12.6). Finally, the simplest nontrivial problem of a polaron, i.e. of a bare QP without internal structure, interacting with one phonon branch is described by the noninteracting Hamiltonian

$$H_0 = \sum_{\boldsymbol{k}} \epsilon(\boldsymbol{k}) a_{\boldsymbol{k}}^\dagger a_{\boldsymbol{k}} + \sum_{\boldsymbol{q}} \omega_{\boldsymbol{q}} b_{\boldsymbol{q}}^\dagger b_{\boldsymbol{q}} \,, \tag{12.8}$$

and the interaction term

$$H_{\text{int}} = \sum_{\boldsymbol{k},\boldsymbol{q}} V(\boldsymbol{k}, \boldsymbol{q})(b_{\boldsymbol{q}}^\dagger - b_{-\boldsymbol{q}}) a_{\boldsymbol{k}-\boldsymbol{q}}^\dagger a_{\boldsymbol{k}} + h.c. \,. \tag{12.9}$$

The simplest polaron problem, in turn, can be subdivided into continuous and lattice polaron models.

The dynamics of a dissipative two-state system, which we need to understand when operating real quantum computers [20], can be reduced to the so-called spin-boson Hamiltonian [21], where a two-level system interacts with a bosonic bath. The properties of the two-level system are determined by the tunneling matrix element Δ and the bias ϵ. The bosonic bath and the interaction are described by a set of oscillator frequencies $\{\omega_\alpha\}$ and coupling constants $\{\gamma_\alpha\}$. It is convenient to consider the two biased levels and the bosonic bath as the unperturbed system

$$H_0 = \frac{1}{2}\epsilon \left[c_1^\dagger c_1 - c_2^\dagger c_2\right] + \sum_\alpha \omega_\alpha b_\alpha^\dagger b_\alpha \,, \tag{12.10}$$

and treat the tunneling

$$H_{\text{int}}^{(1)} = \Delta \left[c_1^\dagger c_2 + c_2^\dagger c_1\right] \tag{12.11}$$

and the coupling to the bosonic bath

$$H_{\text{int}}^{(2)} = \sum_\alpha \sum_{\delta=1}^{2} \gamma_\alpha c_\delta^\dagger c_\delta \left[b_\alpha^\dagger + b_\alpha\right] \tag{12.12}$$

as a perturbation. Analytically solvable cases, where the spectral function

$$J(\omega) = \pi \sum_{\alpha} \gamma_{\alpha}^{2} \delta\left(\omega - \omega_{\alpha}\right) \tag{12.13}$$

can be approximated as $J(\omega) \sim \omega^s$, have been thoroughly studied [21]. However, as yet there is no method to obtain an answer for a general form of $J(\omega)$.

12.2 Physical Properties of Interest

In this section I discuss properties of the exciton-polaron which can be evaluated by DMC and SO methods. To obtain information on QPs it is necessary to calculate the Matsubara GFs in imaginary time representation and afterwards make an analytic continuation to real frequencies [40]. For the two-particle problem (12.1)–(12.4) the relevant quantity is the two-particle GF [27, 28]

$$G_{\boldsymbol{k}}^{\boldsymbol{pp}'}(\tau) = \langle \mathrm{vac} | a_{\boldsymbol{k}+\boldsymbol{p}'}(\tau) h_{\boldsymbol{k}-\boldsymbol{p}'}(\tau) h_{\boldsymbol{k}-\boldsymbol{p}}^{\dagger} a_{\boldsymbol{k}+\boldsymbol{p}}^{\dagger} | \mathrm{vac} \rangle \; , \tag{12.14}$$

where $h_{\boldsymbol{k}-\boldsymbol{p}}(\tau) = \exp(H\tau) h_{\boldsymbol{k}-\boldsymbol{p}} \exp(-H\tau)$, $\tau > 0$. In the case of the exciton-polaron the vacuum state $|\mathrm{vac}\rangle$ is the state with filled valence and empty conduction bands. For the bipolaron problem it is a system without particles. In the simpler case of a QP with internal two-level structure described by (12.4)–(12.6) the relevant quantity is the one-particle matrix GF [28, 34]

$$G_{\boldsymbol{k},ij}(\tau) = \langle \mathrm{vac} | a_{i,\boldsymbol{k}}(\tau) a_{j,\boldsymbol{k}}^{\dagger} | \mathrm{vac} \rangle \; , \tag{12.15}$$

with $i, j = 1, 2$. For a polaron composed of a bare QP without internal structure the matrix (12.15) reduces to the one-particle scalar GF

$$G_{\boldsymbol{k}}(\tau) = \langle \mathrm{vac} | a_{\boldsymbol{k}}(\tau) a_{\boldsymbol{k}}^{\dagger} | \mathrm{vac} \rangle \; . \tag{12.16}$$

Information about the response to a weak external perturbation, e.g. optical absorption, is contained in the current-current correlation function $\langle J_\beta(\tau) J_\delta \rangle$, where β,δ are Cartesian indices.

The Lehmann spectral representation [40, 43] of $G_{\boldsymbol{k}}(\tau)$ (12.14)–(12.16) at zero temperature reads

$$G_{\boldsymbol{k}}(\tau) = \int_0^{\infty} \mathrm{d}\omega L_{\boldsymbol{k}}(\omega) \mathrm{e}^{-\omega\tau} \; , \tag{12.17}$$

where the Lehmann function $L_{\boldsymbol{k}}(\omega)$ given in (12.7) reveals information on the ground state and the excited states. $L_{\boldsymbol{k}}(\omega)$ has poles (sharp peaks) at the energies of stable (metastable) states of the particle. For example, if there is a stable state at energy $E(\boldsymbol{k})$, the Lehmann function reads $L_{\boldsymbol{k}}(\omega) = Z^{(\boldsymbol{k})} \delta(\omega - E(\boldsymbol{k})) + \dots$, and the state with the lowest energy $E_{\mathrm{gs}}(\boldsymbol{k})$ in a sector of a given momentum $\boldsymbol{k}$ is characterized by the asymptotic behavior of the GF

$$G_{\boldsymbol{k}}\left(\tau \gg \frac{1}{E_{\text{ex}}(\boldsymbol{k}) - E_{\text{gs}}(\boldsymbol{k})}\right) \quad \longrightarrow \quad Z^{(\boldsymbol{k})} \mathrm{e}^{-E_{\text{gs}}(\boldsymbol{k})\tau} , \tag{12.18}$$

where $Z^{(\boldsymbol{k})}$ is the weight of the ground state and $E_{\text{ex}}(\boldsymbol{k})$ the energy of the first excited state of the system. Then, the ground state properties are obtained from the logarithmic plot of the GF (see Fig. 12.1).

Note that the energy and Z-factors of the lowest state in the sector of given momentum are not the only properties which can be extracted from the asymptotic behavior. For example, the analysis of the asymptotic behavior of the two-particle GF (12.14) of an exciton [27]

$$G_{\boldsymbol{k}}^{\boldsymbol{p}=\boldsymbol{p}'}(\tau \to \infty) = |\xi_{\boldsymbol{kp},\text{gs}}|^2 \mathrm{e}^{-E_{\text{gs}}(\boldsymbol{k})\tau} \tag{12.19}$$

yields absolute values for the coefficients $\xi_{\boldsymbol{kp},\text{gs}}$ of the wave function of the relative electron-hole motion for an exciton in the lowest state of the given momentum

$$\Psi_{\text{gs}}(\boldsymbol{k}) = \sum_{\boldsymbol{p}} \xi_{\boldsymbol{kp},\text{gs}} a^{\dagger}_{\boldsymbol{k+p}} h^{\dagger}_{\boldsymbol{k\text{-}p}} |\text{vac}\rangle \, . \tag{12.20}$$

Another example is the GF of a polaron. From the asymptotic behavior of the n-phonon GFs [26, 34]

$$G_{\boldsymbol{k}}(n,\tau;\boldsymbol{q}_1,\ldots,\boldsymbol{q}_n) = \langle \text{vac}| b_{\boldsymbol{q}_n}(\tau) \cdots b_{\boldsymbol{q}_1}(\tau) a_{\boldsymbol{p}}(\tau) a^{\dagger}_{\boldsymbol{p}} b^{\dagger}_{\boldsymbol{q}_1} \cdots b^{\dagger}_{\boldsymbol{q}_n} |\text{vac}\rangle \, , \tag{12.21}$$

where $\boldsymbol{p} = \boldsymbol{k} - \sum_{j=1}^{n} \boldsymbol{q}_j$, we obtain detailed information about the lowest state,

$$\Psi_{\text{gs}}(\boldsymbol{k}) = \sum_{i=1}^{R} \sum_{n=0}^{\infty} \sum_{\boldsymbol{q}_1 \ldots \boldsymbol{q}_n} \theta_i(\boldsymbol{k};\boldsymbol{q}_1,\ldots,\boldsymbol{q}_n) c^{\dagger}_{i,\boldsymbol{k}-\boldsymbol{q}_1 \ldots - \boldsymbol{q}_n} b^{\dagger}_{\boldsymbol{q}_1} \ldots b^{\dagger}_{\boldsymbol{q}_n} |\text{vac}\rangle \, , \tag{12.22}$$

like the partial n-phonon contributions

$$Z^{(\boldsymbol{k})}(n) = \sum_{i=1}^{R} \sum_{\boldsymbol{q}_1,\ldots,\boldsymbol{q}_n} |\theta_i(\boldsymbol{k};\boldsymbol{q}_1,\ldots,\boldsymbol{q}_n)|^2 \, , \tag{12.23}$$

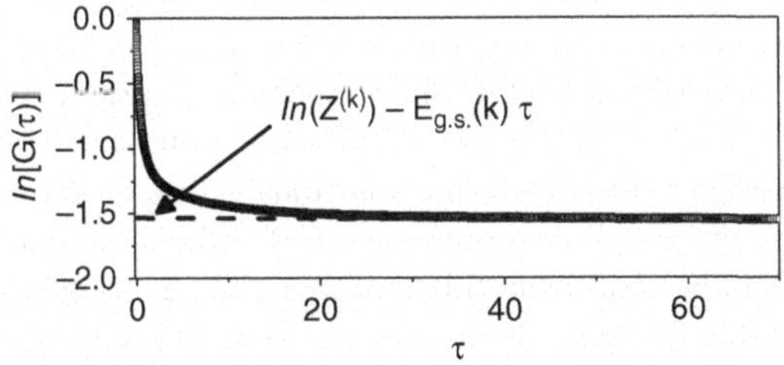

Fig. 12.1. Typical behavior of the GF of a polaron and determination of $Z^{(\boldsymbol{k})}$-factor and energy of the ground state from the fit of the linear asymptotics

which are normalized to unity $\sum_{n=0}^{\infty} Z^{(\mathbf{k})}(n) \equiv 1$, and the average number of phonons

$$\langle N \rangle \equiv \langle \Psi_{\rm gs}(\mathbf{k}) | \sum_{\mathbf{q}} b^{\dagger}_{\mathbf{q}} b_{\mathbf{q}} | \Psi_{\rm gs}(\mathbf{k}) \rangle = \sum_{n=1}^{\infty} n Z^{(\mathbf{k})}(n) \tag{12.24}$$

in the polaronic cloud.

Information on excited states can be obtained by analytic continuation of the imaginary time GF to real frequencies, which requires the solution of the Fredholm equation $G_{\mathbf{k}}(\tau) = \mathcal{F}[L_{\mathbf{k}}(\omega)]$ (12.17),

$$L_{\mathbf{k}}(\omega) = \mathcal{F}^{-1}_{\omega}[G_{\mathbf{k}}(\tau)] \,. \tag{12.25}$$

Equation (12.17) is a rather general relation between the imaginary time GF or correlator and the spectral properties of the system. The solution of

$$\mathcal{I}(\omega) = \mathcal{F}^{-1}_{\omega} \left[\sum_{\mathbf{p}\mathbf{p}'} G^{\mathbf{p}\mathbf{p}'}_{\mathbf{k}=0}(\tau) \right] , \tag{12.26}$$

for instance, yields the light absorption of excitons $\mathcal{I}(\omega)$. Moreover, the real part of the optical conductivity $\sigma_{\beta\delta}(\omega)$ can be expressed [29] in terms of the current-current correlation function $\langle J_{\beta}(\tau) J_{\delta} \rangle$ as

$$\sigma_{\beta\delta}(\omega) = \frac{\pi}{\omega} \mathcal{F}^{-1}_{\omega}[\langle J_{\beta}(\tau) J_{\delta} \rangle] \,. \tag{12.27}$$

12.3 The Diagrammatic Monte Carlo Method

Diagrammatic Monte Carlo is an algorithm for the calculation of the GF of a QP, e.g. (12.14)–(12.16), which is free of any systematic errors. Although DMC is based on a Feynman expansion of the Matsubara GF the method does not require advanced skills in the derivation of Feynman series, since most expansions have been formulated a long time ago [40, 42]. The only problem, which has not been solved, is the actual summation of the series without truncation or other approximations. To explain the general idea of the algorithm, below we start with the simplest traditional many-particle problem: The polaron without any internal structure [26]. Then, to give a feeling for the craft of building DMC algorithms, we proceed with a sequence of increasingly involved problems: Beginning with the most trivial task of simulating the bare Matsubara GF of noninteracting systems, we continue with the trivial problem of a free particle in an attractive potential. Then we consider a particle in non-retarded fields, turn to the notoriously difficult exciton problem, and, finally, end up with the simplest example of a CO with an internal quantum degree of freedom. It should be clear from this sequence that, as soon as an algorithm for the trivial problem of a free particle in an attractive field is constructed, its adaption to the complicated exciton problem is straightforward. We recommend the reader to study all the simple examples up to Sect. 12.3.6, and then return to the general formulation of the method in Sect. 12.3.1.

12.3.1 The General Concept of DMC: Green Function of a Polaron

As noted earlier, DMC is based on the Feynman expansion of the Matsubara GF in imaginary time using the interaction representation

$$G_{\boldsymbol{k}}(\tau) = \left\langle \mathrm{vac} \left| T_\tau \left(a_{\boldsymbol{k}}(\tau) a_{\boldsymbol{k}}^\dagger(0) \mathrm{e}^{-\int_0^\infty H_{\mathrm{int}}(\tau')\mathrm{d}\tau'} \right) \right| \mathrm{vac} \right\rangle_{\mathrm{con}} \tag{12.28}$$

with $\tau > 0$. Here T_τ is the imaginary time ordering operator, $|\mathrm{vac}\rangle$ is a vacuum state without particles and phonons, and H_{int} is the interaction Hamiltonian of (12.9). The exponent denotes the formal summation of a Taylor series which corresponds to multiple integrations over the internal variables $\{\tau_1', \tau_2', \ldots\}$. The operators are taken in the interaction representation $A(\tau) = \exp[\tau(H_{\mathrm{par}}+H_{\mathrm{ph}})]A\exp[-\tau(H_{\mathrm{par}}+H_{\mathrm{ph}})]$, and the index "con" denotes an expansion which contains only connected diagrams where no integral over internal time variables $\{\tau_1', \tau_2', \ldots\}$ can be factorized.

Applying the Wick theorem, a matrix element of time-ordered operators can be written as a sum of terms, each being the product of matrix elements of pairs of operators. Then the expansion (12.28) becomes an infinite series of integrals with an ever increasing number of integration variables

$$G_{\boldsymbol{k}}(\tau) = \sum_{m=0,2,4,\ldots}^{\infty} \sum_{\xi_m} \int \mathrm{d}x_1' \cdots \mathrm{d}x_m' \mathcal{D}_m^{(\xi_m)}\left(\tau; \{x_1', \ldots, x_m'\}\right) . \tag{12.29}$$

Here the index ξ_m stands for different Feynman diagrams (FDs) of the same order m because for $m > 2$ there is more than one diagram of the same order m. The zero-order term with $m = 0$ is the bare GF of the QP.

The aim of DMC is the evaluation of the series (12.29) with the help of importance sampling. Hence, we need to find a positive weight function and an update procedure to formulate something similar to the well known Metropolis algorithm [49, 50, 51]. In statistical physics the latter is used to calculate the expectation value of an observable Q, which is defined as a sum over all states μ of the system with energies E_μ, each term weighted with the Boltzmann probability, $\langle Q\rangle = Z^{-1}\sum_\mu Q_\mu \exp[-\beta E_\mu]$. Here $\beta = 1/T$ is inverse temperature and $Z = \sum_\mu \exp[-\beta E_\mu]$ the partition function. Since it is impossible to sum over all possible states μ of the macroscopic system $\{\mu\}$ the classical MC uses the concept of importance sampling, where the sum is approximated by adding only the contributions of a small but typical set of states. These states are selected such that the probability of a particular state ν equals $\mathcal{D}_\nu = Z^{-1}\exp[-\beta E_\nu]$. This can be achieved through a Markov chain $\nu \to \nu' \to \nu'' \to \ldots$ with appropriate transition probabilities between subsequent states. Within the Metropolis scheme the system is offered a new configuration ν', and the move $\nu \to \nu'$ is accepted with probability $M = \mathcal{D}_{\nu'}/\mathcal{D}_\nu$, if $M < 1$, or one otherwise. After N steps of such a stochastic (Markov) process the estimator for the observable $\langle Q\rangle$ reads

$$Q_N = \frac{1}{N} \sum_{i=1}^{N} Q_\nu \,, \tag{12.30}$$

where Q_ν is the value of Q in the state ν.

In close analogy to the weight function of classical MC, $\mathcal{D}_\nu$, the DMC method uses the weight function $\mathcal{D}_m^{(\xi_m)}(\tau; \{x'_1, \dots, x'_m\})$, which depends on the internal integration variables $\{x'_1, \dots, x'_m\}$ and the external variable τ. The term with $m = 0$ is the GF of the noninteracting QP, $G_{\boldsymbol{k}}^{(0)}(\tau)$.

For orders $m > 0$, $\mathcal{D}_m^{(\xi_m)}(\tau; \{x'_1, \dots, x'_m\})$ can be expressed as a product of GFs of noninteracting QPs, GFs of phonons, and of interaction vertices $V(\boldsymbol{k}, \boldsymbol{q})$. For the simplest case of a Hamiltonian system the expressions for the GFs are well known: They read $G_{\boldsymbol{k}}^{(0)}(\tau_2 - \tau_1) = \exp\left[-\epsilon(\boldsymbol{k})(\tau_2 - \tau_1)\right]$ with $(\tau_2 > \tau_1)$ for the QPs and $D_{\boldsymbol{q}}^{(0)}(\tau_2 - \tau_1) = \exp\left[-\omega_{\boldsymbol{q}}(\tau_2 - \tau_1)\right]$ with $(\tau_2 > \tau_1)$ for the phonons [42, 40].

An important feature, which distinguishes the DMC method from other exact numerical approaches, is the possibility to explicitly include renormalized GFs into an exact expansion without any change of the algorithm. If we know the damping of the QP caused by interactions that are not included in the Hamiltonian, we can use the renormalized GF

$$\widetilde{G}_{\boldsymbol{k}}^{(0)}(\tau) = \frac{1}{\pi} \int_{-\infty}^{\infty} d\omega e^{-\omega\tau} \frac{\mathrm{Im}\Sigma_{\mathrm{ret}}(\boldsymbol{k}, \omega)}{\left(\omega - \epsilon(\boldsymbol{k}) - \mathrm{Re}\Sigma_{\mathrm{ret}}(\boldsymbol{k}, \omega)\right)^2 + \left(\mathrm{Im}\Sigma_{\mathrm{ret}}(\boldsymbol{k}, \omega)\right)^2} \tag{12.31}$$

for our calculation, instead of bare the GF $G_{\boldsymbol{k}}^{(0)}(\tau)$. To avoid double counting the retarded self energy $\Sigma_{\mathrm{ret}}(\boldsymbol{k}, \omega)$ should contain only those interactions which are not included in the Hamiltonian treated by the DMC procedure. The rules for the evaluation of $\mathcal{D}_m^{(\xi_m)}$ do not depend on the order and topology of the FDs. In Fig. 12.2 we show examples of typical diagrams. Here GFs of noninteracting QPs $G_{\boldsymbol{k}}^{(0)}(\tau_2 - \tau_1)$, or $\widetilde{G}_{\boldsymbol{k}}^{(0)}(\tau_2 - \tau_1)$, correspond to horizontal lines, whereas noninteracting GFs of phonons $D_{\boldsymbol{q}}^{(0)}(\tau_2 - \tau_1)$, multiplied by the prefactor of the appropriate vertices $V(\boldsymbol{k}', \boldsymbol{q}) V^*(\boldsymbol{k}'', \boldsymbol{q})$, are denoted by semi-circles. $\mathcal{D}_m^{(\xi_m)}$ then is the product of all GSs occuring in a given diagram. For example, the weight of the second order term in Fig. 12.2(b) is

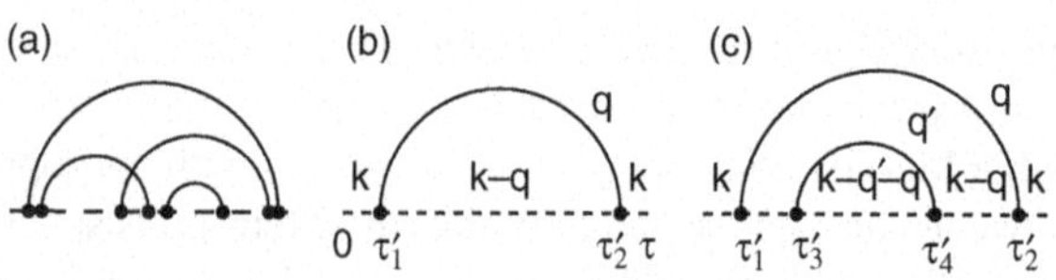

Fig. 12.2. (**a**) Typical FD contributing into expansion (12.29). (**b**) FD of the second order and (**c**) forth order

$$\mathcal{D}_2(\tau;\{\tau_2',\tau_1',\boldsymbol{q}\}) = |V(\boldsymbol{k},\boldsymbol{q})|^2\, D_{\boldsymbol{q}}^{(0)}(\tau_2'-\tau_1')G_{\boldsymbol{k}}^{(0)}(\tau_1') \\ \times G_{\boldsymbol{k}-\boldsymbol{q}}^{(0)}(\tau_2'-\tau_1')G_{\boldsymbol{k}}^{(0)}(\tau-\tau_2')\,. \qquad (12.32)$$

The DMC process is a numerical procedure which, based on the Metropolis principle [49, 50, 51] and the weight function $\mathcal{D}_m^{(\xi_m)}(\tau;\{x_1',\ldots,x_m'\})$, samples different FDs in the parameter space $(\tau, m, \xi_m, \{x_m'\})$. In parallel, it collects the statistics of the external variable τ such that the result converges to the exact GF $G_{\boldsymbol{k}}(\tau)$. Hence, within DMC the Markov process involves changes of both the internal variables and the external variable τ, as well as a switching between different orders and topologies of the FDs. The statistics of the variable τ is measured, e.g. by a histogram method.

12.3.1.1 Sampling of a Single Term of the Expansion

Even though the Markov process combines the sampling of the internal parameters of a diagram and the switching between different diagrams, it is instructive to explain these two update mechanisms separately. Let us start with the sampling of one particular diagram of weight $\mathcal{D}_m^{(\xi_m)}(\tau;\{x_1',\ldots,x_m'\})$, which has much in common with classical MC. Given a set $\{\tau;\{x_1',\ldots,x_m'\}\}$, an update $x_l^{(\text{old})} \to x_l^{(\text{new})}$ of an arbitrarily chosen parameter is suggested. This update is accepted or rejected according to the Metropolis principle. After many steps, altering all variables, the statistics of the external variable converges to the exact dependence of the term on τ. The suggestion for the new value of the parameter $x_l^{(\text{new})} = S^{-1}(R)$ is generated from a random number $R \in [0,1]$, where $S^{-1}(R)$ is the root of the integral equation

$$\int_{x_l^{(\text{min})}}^{x_l^{(\text{new})}} \mathrm{d}x' W(x') = R\,. \qquad (12.33)$$

Here $W(x')$ is a normalized distribution function $W(x_l)$ defined in the range $x_l^{(\text{min})} < x' < x_l^{(\text{max})}$. There are only two restrictions for this otherwise arbitrary function. First, the new parameters $x_l^{(\text{new})}$ must not violate the FD topology, i.e., for example, internal time τ_1' in Fig. 12.2(c) has to be in the range $[x^{(\text{min})} = 0, x^{(\text{max})} = \tau_3']$. Second, the distribution should be nonzero for the whole domain, allowed by the FD topology. This ergodicity property is crucial for the convergence of the algorithm. At each step, the update $x_l^{(\text{old})} \to x_l^{(\text{new})}$ is accepted with probability $P_{\text{acc}} = M$ (if $M < 1$) and always accepted otherwise. The ratio M has the following form

$$M = \frac{\mathcal{D}_m^{(\xi_m)}\left(\tau;\{x_1',\ldots,x_l^{(\text{new})},\ldots,x_m'\}\right)/W\left(x_l^{(\text{new})}\right)}{\mathcal{D}_m^{(\xi_m)}\left(\tau;\{x_1',\ldots,x_l^{(\text{old})},\ldots,x_m'\}\right)/W(x_l^{(\text{old})})}\,. \qquad (12.34)$$

For the uniform distribution $W = \text{const.} = (x_l^{(\max)} - x_l^{(\min)})^{-1}$, the probability of any combination of parameters is proportional to the weight function $\mathcal{D}$. However, for better convergence the distribution $W(x_l^{\text{new}})$ should be as close as possible to the actual distribution given by the function $\mathcal{D}_m^{(\xi_m)}(\{\dots, x_l^{(\text{new})}, \dots, \})$. If these two distributions coincide, $M \equiv 1$ for every update. Hence, all updates are accepted and the sampling is most effective. For example, if the distribution

$$W([\tau_4']^{(\text{new})}) = \frac{\Delta E \mathrm{e}^{-([\tau_4']^{(\text{new})} - \tau_3')\Delta E}}{1 - \mathrm{e}^{-(\tau_2' - \tau_3')\Delta E}} \tag{12.35}$$

is used to update parameter τ_4' in the FD of Fig. 12.2(c), $[\tau_4']^{(\text{new})}$ must be generated by random numbers $R \in [0, 1]$ as

$$[\tau_4']^{(\text{new})} = \tau_3' - \frac{\ln\left(1 - R(1 - \mathrm{e}^{-(\tau_2' - \tau_3')\Delta E})\right)}{\Delta E} . \tag{12.36}$$

Then, according to (12.33)–(12.35), $M \equiv 1$, and all updates are accepted. In (12.35), the distribution is normalized to unity in the constrained domain $[\tau_3', \tau_2']$, and $\Delta E = \varepsilon(\boldsymbol{k} - \boldsymbol{q} - \boldsymbol{q}') + \omega_{\boldsymbol{q}'} - \varepsilon(\boldsymbol{k} - \boldsymbol{q})$.

12.3.1.2 Switching between Diagrams of Different Order

The switching between diagrams of different order differs from the above process in that it modifies a term with a given topology. Obviously this process also changes the dimension of the parameter space. All FDs contributing to the polaron GF can be sampled with two complimentary updates. Update $\mathcal{A}$,

$$\mathcal{D}_m^{(\xi_m)}(\tau; \{x_1', \dots, x_m'\}) \underset{\mathcal{A}}{\longrightarrow} \mathcal{D}_{m+2}^{(\xi_{m+2})}(\tau; \{x_1', \dots, x_m';\ \boldsymbol{q}', \tau_3', \tau_4'\}) , \tag{12.37}$$

transforms a given FD into a higher order FD with an extra phonon arch, which connects two time points τ_3' and τ_4' by a phonon propagator of momentum $\boldsymbol{q}'$, see Fig. 12.2(c). On the opposite, update $\mathcal{B}$ performs the reciprocal transformation. Note that the ratio of the weights $\mathcal{D}_{m+2}^{(\xi_{m+2})}/\mathcal{D}_m^{(\xi_m)}$ is not dimensionless. The dimensionless Metropolis ratio

$$M = \frac{p_{\mathcal{B}}}{p_{\mathcal{A}}} \frac{\mathcal{D}_{m+2}^{(\xi_{m+2})}(\tau; \{x_1', \dots, x_m'; \boldsymbol{q}', \tau', \tau''\})}{\mathcal{D}_m^{(\xi_m)}(\tau; \{x_1', \dots, x_m'\})} \frac{1}{W(\boldsymbol{q}', \tau', \tau'')} \tag{12.38}$$

contains the normalized probability function $W(\boldsymbol{q}', \tau', \tau'')$, which is used for generating new parameters. $p_{\mathcal{A}}$ and $p_{\mathcal{B}}$ are the probabilities of selecting update $\mathcal{A}$ or $\mathcal{B}$, respectively. The context factor $p_{\mathcal{B}}/p_{\mathcal{A}}$ ensures that the probability for the occurrence of a given diagram is defined only by its weight function $\mathcal{D}$. Thus, the Metropolis ratio for the process $\mathcal{A}$ has to be divided by $p_{\mathcal{A}}$ and multiplied by $p_{\mathcal{B}}$. The context factor, of course, depends on the way the adding and removing procedure is organized. Below we describe self-balanced add/remove processes and derive the corresponding factors $p_{\mathcal{B}}/p_{\mathcal{A}}$.

Let us assume that the DMC process adds and removes lines with equal probability. To add a phonon propagator the $\mathcal{A}$-procedure randomly chooses an arbitrary electronic propagator. The value of the left end of the phonon propagator, τ_3', is selected with uniform probability $d\tau_3'/\Delta\tau$, where $\Delta\tau$ is the length of the electronic propagator considered. Then, the right end of the phonon propagator, τ_4', is seeded with (normalized) probability density $\propto d\tau_4'\bar{\omega}\exp(-\bar{\omega}(\tau_4'-\tau_3'))$, where $\bar{\omega}$ is an average frequency of the phonon spectrum. Hence, according to (12.33), the value of τ_4' is given by

$$\tau_4' = \tau_3' - \frac{1}{\bar{\omega}} \ln(R) \,. \tag{12.39}$$

If τ_4' is larger than the right end of the diagram, τ, the update is rejected. The momentum $\boldsymbol{q}'$ of the new phonon propagator is choosen from an uniform distribution over the whole Brillouin zone, $d\boldsymbol{q}'/V_{\mathrm{BZ}}$. Then, according to the rule (12.38)

$$M = \frac{p_{\mathcal{B}}}{p_{\mathcal{A}}} \frac{\mathcal{D}_{m+2}^{(\xi_{m+2})}}{\mathcal{D}_m^{(\xi_m)}} \frac{\mathrm{d}\tau_3' \, \mathrm{d}\tau_4' \, \mathrm{d}\boldsymbol{q}'/V_{\mathrm{BZ}}}{(\mathrm{d}\tau_3'/\Delta\tau)\mathrm{d}\tau_4' \, \bar{\omega} \, \mathrm{e}^{-\bar{\omega}(\tau_4'-\tau_3')} \, \mathrm{d}\boldsymbol{q}'/V_{\mathrm{BZ}}} \,. \tag{12.40}$$

The removal step $\mathcal{B}$ selects an arbitrary phonon propagator and accepts the update with the reciprocal M^{-1} of the probability, which would be used when adding the same propagator in step $\mathcal{A}$. Let us emphasize that the context factor $p_{\mathcal{B}}/p_{\mathcal{A}}$ depends on the way how the add or removal process is organized. If, for instance, the procedure addresses these processes with equal probabilities, the naive expectation that $p_{\mathcal{B}}/p_{\mathcal{A}} = 1$ is wrong. To understand this, let us consider two diagrams, $\mathcal{D}_m$ and $\mathcal{D}_{m+2}$. The diagram $\mathcal{D}_m$ contains N_{e} electron and $N_{\mathrm{ph}} = (N_{\mathrm{e}} - 1)/2$ phonon propagators. The procedure $\mathcal{A}$ transforms the diagram $\mathcal{D}_m$ to $\mathcal{D}_{m+2}$ with $N_{\mathrm{e}} + 2$ electron and $N_{\mathrm{ph}} + 1 = (N_{\mathrm{e}} + 1)/2$ phonon propagators. The procedure $\mathcal{B}$ transforms the second diagram to the first one, respectively. When procedure $\mathcal{A}$ selects an electron propagator for inserting the point τ_3' in $\mathcal{D}_m$, we have N_{e} possibilities, hence, $p_{\mathcal{A}} = 1/N_{\mathrm{e}}$. On the other hand, when the procedure $\mathcal{B}$ selects a phonon propagator for removal from $\mathcal{D}_{m+2}$, there are $N_{\mathrm{ph}} + 1 = (N_{\mathrm{e}} + 1)/2$ possibilities and $p_{\mathcal{B}} = 2/(N_{\mathrm{e}} + 1)$. Therefore, detailed balance requires a context factor of

$$\frac{p_{\mathcal{B}}}{p_{\mathcal{A}}} = \frac{N_{\mathrm{e}}}{N_{\mathrm{ph}} + 1} \,. \tag{12.41}$$

Note that this factor essentially depends on how the processes are organized. For example, if the rule of equal add and removal probability is relaxed and the add process is addressed f times more frequently than the removal process, the probability of process $\mathcal{A}$ is $p_{\mathcal{A}} = f/N_{\mathrm{e}}$ and the context factor reads $p_{\mathcal{B}}/p_{\mathcal{A}} = N_{\mathrm{e}}/[f(N_{\mathrm{ph}}+1)]$. Writing expression (12.41) I intentionally do not use the relation $N_{\mathrm{ph}} + 1 = (N_{\mathrm{e}} + 1)/2$ because it is valid only in the particular case of a polaron interacting with one phonon branch without any other terms in the interaction Hamiltonian. If the system includes interactions with other phonon branches or external potentials, the relation between the number of phonon and electron propagators does not hold, while expression (12.41) is still valid.

Note that the ratio $\mathcal{D}_{m+2}^{(\xi_{m+2})}/\mathcal{D}_m^{(\xi_m)}$ depends on the topology of the higher-order FD. When the FD in Fig. 12.2(c) is updated, e.g., from the FD in Fig. 12.2(b), the ratio has the following form

$$\frac{\mathcal{D}_{m+2}^{(\xi_{m+2})}}{\mathcal{D}_m^{(\xi_m)}} = |V(\boldsymbol{k}-\boldsymbol{q},\boldsymbol{q}')|^2 D_0(\boldsymbol{q}';\tau_4'-\tau_3')\frac{G_0(\boldsymbol{k}-\boldsymbol{q}-\boldsymbol{q}';\tau_4'-\tau_3')}{G_0(\boldsymbol{k}-\boldsymbol{q};\tau_4'-\tau_3')} . \quad (12.42)$$

12.3.1.3 General Features of DMC

Finally, let us add a few words about the general features of the DMC algorithm. Note that all updates are local, i.e. do not depend on the structure of the whole FD. Neither the rules nor the CPU time needed for the update depend on the order of the FD. The DMC method does not imply any explicit truncation of the FD's order due to the finite size of computer memory. Even for strong coupling, where the typical number of contributing phonon propagators N_{ph} is large, the memory requirements are marginal. In fact, according to the central limit theorem, the number of phonon propagators obeys a Gauss distribution centered at $\bar{N}_{\mathrm{ph}}$ with a half width of the order of $\bar{N}_{\mathrm{ph}}^{1/2}$ [52]. Hence, if memory for at least $2\bar{N}_{\mathrm{ph}}$ propagators is reserved, the diagram order hardly surpasses this limit.

12.3.2 How to Expand the Exponent

For a beginner the rules given in the previous section and thoroughly described in [26] may seem rather complicated and not easy to understand. In what follows we therefore apply the DMC method to a set of increasingly complex examples. We start with the Matsubara GF of a noninteracting particle with energy ε,

$$G^{(0)}(\tau) = \mathrm{e}^{-(\varepsilon-\mu)\tau} , \quad (12.43)$$

where μ is an artificial chemical potential. For simplicity, we chose $\mu < \varepsilon$, then G decreases with increasing τ. [2]

Let us now describe the calculation of the GF (12.43) being armed only with the rules of Sect. 12.3.1. The first thing we need is the statistics[3] of an external variable τ.

To this end we introduce a histogram with cells $\{[\xi i; \xi(i+1)]\}$ of width ξ in the range $0 < \tau < \tau_{\mathrm{max}}$, see Fig. 12.3. Initially the counters for all cells are set to zero. When the process of DMC updates is running, the counter of cell i is increased by

[2] This is not very important when such tricks as the *guiding function* [26, 53, 54, 55] are used and the statistics of the external variable is restricted to a certain finite domain. Though, in the simplest case considered here the condition $\mu < \varepsilon$ is important to keep the domain of the largest probability density of τ near the value zero.

[3] The simplest way to accumulate statistics is in a histogram, which leads to a number of systematic errors. However, one can avoid the histogram mesh and generate the exact statistics for GF [26].

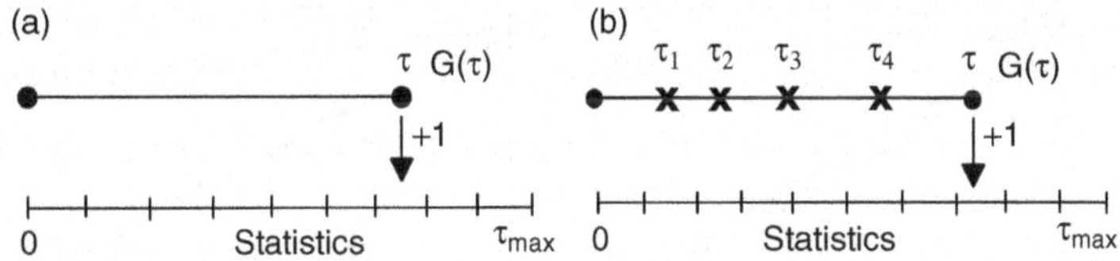

Fig. 12.3. Accumulation of statistics for (**a**) the GF of a QP and (**b**) the GF of a QP in an attractive potential

one whenever the position of the external variable τ is within the cell $\xi(i) < \tau < \xi(i+1)$, see Fig. 12.3(a).

Next, we need to initialize τ with an arbitrary value from the domain $[0, \tau_{\text{max}}]$ and set up rules for the update procedure. I suggest two methods: The "simplest" one and the "best" one.

12.3.2.1 Simple Update Method

The new external parameter τ_{new} is suggested as a shift $\tau_{\text{old}} \to \tau_{\text{new}} = \tau_{\text{old}} + \delta(R - 1/2)$ of the old value τ_{old}. The new value is generated by a random number $R \in [0, 1]$ with uniform distribution $W(x) = 1/\delta$ in the range $[\tau_{\text{old}} - \delta/2, \tau_{\text{old}} + \delta/2]$. If τ_{new} is not in the range $[0, \tau_{\text{max}}]$, the update is rejected. Otherwise, the decision to accept or reject the update is based on the Metropolis procedure with probability ratio $M = \exp\left[-(\varepsilon - \mu)(\tau_{\text{new}} - \tau_{\text{old}})\right]$.

12.3.2.2 Best Update Method

One generates τ_{new} with probability

$$W(x) = (\varepsilon - \mu)\mathrm{e}^{-(\varepsilon-\mu)x} , \tag{12.44}$$

normalized in the range $[0, +\infty]$ Then, according to the rules, one solves the equation

$$\int_0^{\tau_{\text{new}}} W(x)\mathrm{d}x = R \tag{12.45}$$

and obtains the generator for the new value

$$\tau_{\text{new}} = -\frac{1}{\varepsilon - \mu} \ln R . \tag{12.46}$$

Inserting the probability densities $W(\tau_{\text{new}})$, $W(\tau_{\text{old}})$, and the weights $D(\tau_{\text{new}})$, $D(\tau_{\text{old}})$, in the general expression (12.34) one gets $M \equiv 1$ and, hence, all updates are accepted. Note that this update is accepted even if $\tau > \tau_{\text{max}}$, though there is nothing to add to the statistics, since the external variable is out of the histogram

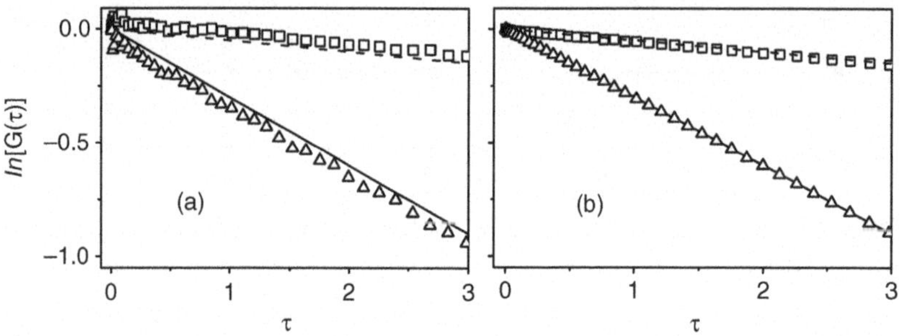

Fig. 12.4. GFs of a QP in the logarithmic scale for $\varepsilon = 0$ and $\mu = -0.3$. Solid line represents the exact GF (12.43) of a free QP $\ln[G^{(0)}(\tau)] = -(\varepsilon - \mu)\tau$. Dashed line describe the exact GF of the QP in the attractive potential (12.48) $\ln[G^{(1)}(\tau)] = -(\varepsilon - V - \mu)\tau$ for $V = 0.25$. Triangles and squares are the results of DMC method for small (**a**) and large (**b**) amount of DMC updates, respectively

range[4]. Finally, when reasonable statistics is accumulated, the data is normalized such that $G^{(0)}(\tau = 0) \equiv 1$.

In Fig. 12.4 we show the convergence of the statistics of the external variable τ (triangles) to the exact answer (solid lines). The DMC result is very close to the exact data after $\approx 10^7$ DMC updates (which means at about one second of Pentium IV CPU time) and perfectly reproduces the exact GF after $\approx 10^9$ updates (one minute CPU time).

12.3.3 Attractive Potential

If the Hamiltonian of a noninteracting system

$$H^{(0)} = \varepsilon c^\dagger c \tag{12.47}$$

is supplemented by an attractive potential

$$\widehat{H}^{(\mathrm{int})} = -|V| c^\dagger c\,, \tag{12.48}$$

the energy is renormalized as $\varepsilon \to \varepsilon - |V|$, and the exact Matsubara GF takes the form

$$G^{(1)}(\tau) = \mathrm{e}^{-(\varepsilon - |V| - \mu)\tau} \equiv \mathrm{e}^{-(\varepsilon-\mu)\tau} \sum_{n=0}^{\infty} \frac{|V\tau|^n}{n!}\,. \tag{12.49}$$

[4] One can restrict the external variable τ to the range $[0, \tau_{\max}]$ using the probability density $W(x) = [(1 - \exp(-(\varepsilon - \mu)\tau_{\max}))]^{-1}(\varepsilon - \mu)\exp(-(\varepsilon - \mu)x)$, which is normalized in the range $[0, \tau_{\max}]$. In this case one generates τ_{new} as $\tau_{\mathrm{new}} = -(\varepsilon - \mu)^{-1}\ln[1 - R[1 - \exp(-(\varepsilon - \mu)\tau_{\max})]]$. Note the similarity of the above equation with (12.36). It occurs because in both cases the distribution of the random variables is exponential and normalized in a finite range.

To get an idea of the diagrammatic expansion of DMC let us solve the problem by Feynman expansion. Since for the given problem the system is always in the Hilbert space sector with one particle $\left\langle \mathrm{vac} \left| c^\dagger c \right| \mathrm{vac} \right\rangle = 1$, we can introduce the unity operator η and consider $H^{(\mathrm{int})} = -|V|\eta$. Then, the Feynman expansion reads

$$G^{(1)}(\tau) = \left\langle \mathrm{vac} \left| T_\tau \left(c(\tau) c^\dagger(0) \mathrm{e}^{|V| \int_0^\infty \eta(\tau')\mathrm{d}\tau'} \right) \right| \mathrm{vac} \right\rangle_{\mathrm{con}} , \tag{12.50}$$

with $\tau > 0$, and the structure of diagrams is that of Fig. 12.3(b). According to the general rules, the weight of the diagram is the product of particle propagators $\ldots, \exp[(\varepsilon - \mu)(\tau_{i+1} - \tau_i)], \ldots$ and vertices $|V|$. Hence, the weight of each order-m diagram of length τ is $\mathcal{D}_m(\tau) = |V|^m \exp[(\varepsilon - \mu)\tau]$.

The GF can be calculated using three different updates: The modification of the right diagram end τ, and a pair of self-balanced updates which add/remove the vertex $|V|$, see the crosses in Fig. 12.3(b). Below I introduce the minimal set of updates sufficient to reach the numerically exact solution. Note that this set is the simplest one but not the most efficient.

Moving the external parameter τ: The value $\tau - \tau_{\mathrm{last}}$ obeys the distribution (12.44), where τ_{last} is the position of the vertex with largest imaginary time, or $\tau = 0$ when there is not a single vertex. Therefore we can use the recipes of Sect. 12.3.2 and obtain a rejection-free update method, if τ_{new} is generated through

$$\tau_{\mathrm{new}} = \tau_{\mathrm{last}} - \frac{1}{\varepsilon - \mu} \ln R \, . \tag{12.51}$$

Add or remove an interaction vertex: To add an interaction vertex one randomly chooses one particle propagator from the N_{prop} existing propagators in the FD of Fig. 12.5(a), the dashed line, for example. Then the position of the new vertex is suggested with uniform probability density $W(x) = (\tau_r - \tau_l)^{-1}$, hence, τ_{new} is chosen as $\tau_{\mathrm{new}} = \tau_l + (\tau_r - \tau_l)R$. The Metropolis ratio thus reads

$$M = \frac{N_{\mathrm{prop}}}{N_{\mathrm{vert}} + 1} |V| (\tau_r - \tau_l) \, . \tag{12.52}$$

The structure of this ratio is intentionally given in a form where all factors have a one to one correspondence with those of (12.38) and (12.40). Note the roles of the last two factors in (12.52), (12.38) and (12.40). N_{vert} is the number of vertices in the FD of Fig. 12.5(a). The first factor is the context factor $p_{\mathcal{B}}/p_{\mathcal{A}}$ of (12.38), which is necessary to self-balance add and removal processes, and whose form depends on how these processes are organized. The expression $N_{\mathrm{prop}}/(N_{\mathrm{vert}} + 1)$ accounts for

Fig. 12.5. Updates adding (**a**)→(**b**) and removing (**a**)←(**b**) an interaction vertex. The circle in (**b**) is an existing vertex of the present FD which is suggested for removal. The circle in (**a**) is a vertex suggested for the adding procedure

a process, where one of the vertices is selected randomly and then removed, τ_{i+1} in Fig. 12.5(b), for example. Note that there are $N^*_{\text{vert}} = N_{\text{vert}} + 1$ choices in the FD of Fig. 12.5(b). Hence, for a self-balanced MC process one has to divide the weight by the probability to suggest the addition of a new vertex, $p_{\mathcal{A}} = 1/N_{\text{prop}}$, and multiply by the probability to suggest the removal of the same vertex, $p_{\mathcal{B}} = 1/N^*_{\text{vert}} = 1/(N_{\text{vert}} + 1)$. This explains the factor $p_{\mathcal{B}}/p_{\mathcal{A}} = N_{\text{prop}}/(N_{\text{vert}} + 1)$ in (12.52).

The careful reader may have noticed that the context factor is equal to unity, since for the FDs in Figs. 12.3(b) and 12.5 we always find the relation $N_{\text{prop}} = N_{\text{vert}} + 1$. However, this is correct only for the specific example of the interaction with a single attractive potential. If, e.g., an interaction with phonons is added, the relation between the numbers of vertices and propagators is different, though the expression (12.52) is still correct. Hence, it seems better to stick to the correct reasoning even in this simple example, and introduce context factors which are valid in more general and complicated situations. For example, in the case of several types of interaction vertices one can introduce self-balanced updates for each type of vertices. In this case N_{prop} is the number of all propagators and N_{vert} is the number of vertices of the given type.

The Metropolis ratio for the removal procedure is constructed as the inverse of expression (12.52), which describes the adding of that same vertex which is now considered for removal,

$$M = \left(\frac{N^*_{\text{prop}} - 1}{N^*_{\text{vert}}} |V| (\tau_{i+2} - \tau_i) \right)^{-1} . \tag{12.53}$$

Here $N^*_{\text{prop}} = N_{\text{prop}} + 1$ ($N^*_{\text{vert}} = N_{\text{vert}} + 1$) is the number of propagators (vertices) in the FD of Fig. 12.5(b).

In conclusion, the general strategy is the following: We start from a bare FD without interaction vertices and with the external parameter τ in the range $\tau_{\text{min}} < \tau < \tau_{\text{max}}$, see Fig. 12.3). Then, with some probability one of the three updates, move, add, or remove is suggested. Note that with the given context factors the probabilities to address add and removal processes must be equal. One can, of course, address add and removal processes with different probabilities, but in this case the context factor $p_{\mathcal{B}}/p_{\mathcal{A}}$ need to be modified accordingly. Finally, statistics is collected as shown in Fig. 12.3, and in the end the data is normalized implying the condition $G^{(1)}(\tau = 0) \equiv 1$.

In Fig. 12.4 we show the convergence of the statistics for the external variable τ (squares) to the exact answer (dashed line). After $\approx 10^7$ DMC updates the data is very close to the exact result, and perfectly reproduces the exact GF after $\approx 3 \times 10^9$ DMC updates. Note that the integration over different orders of FDs and over the internal imaginary times of the interaction vertices requires a larger number of DMC updates, compared to the free particle.

12.3.4 Field with Internal Degrees of Freedom

It is straightforward to adapt the algorithm of the previous section to the less trivial case of the interaction

$$H_{\text{int}} = -\sum_{k,k'} |V(k,k')| c^\dagger c \,, \tag{12.54}$$

where, for simplicity, we assume that the degrees of freedom k and k' are restricted to finite domains: $k_{\text{min}} < k < k_{\text{max}}$ and $k'_{\text{min}} < k' < k'_{\text{max}}$. Then, all rules are identical to those of the previous section, except for two modifications. First, one changes $|V|$ to $|V(k,k')|$ in (12.52) and (12.53). Second, in the add-procedure one generates k and k' with uniform probability densities $k = k_{\text{min}} + (k_{\text{max}} - k_{\text{min}})R$ and $k' = k'_{\text{min}} + (k'_{\text{max}} - k'_{\text{min}})R$.

The exact result for the GF

$$G^{(2)}(\tau) = \mathrm{e}^{-(\varepsilon - \int_{k_{\text{min}}}^{k_{\text{max}}} \mathrm{d}k \int_{k'_{\text{min}}}^{k'_{\text{max}}} \mathrm{d}k' |V(k,k')| - \mu)\tau} \tag{12.55}$$

is a trivial modification of (12.49) because the state of a QP does not depend on the variables k and k'. Note that, compared to the case of a constant potential $V = \int_{k_{\text{min}}}^{k_{\text{max}}} \mathrm{d}k \int_{k'_{\text{min}}}^{k'_{\text{max}}} \mathrm{d}k' |V(k,k')|$, we need more DMC updates to converge to the exact result, because of additional integrations over the internal variables k and k'.

12.3.5 Exciton

The exciton problem is an example of a highly nontrivial two-body problem, where the center of mass motion cannot be trivially separated from the relative electron-hole motion.

However, the Feynman expansion for this two-body problem with Hamiltonian (12.1) and (12.2) can be effectively reduced to the linear class of FDs considered above. The upper panel in Fig. 12.6 presents the ladder diagrams for the two-particle GF of the exciton with momentum $\boldsymbol{k}$. The weight of each diagram is the product

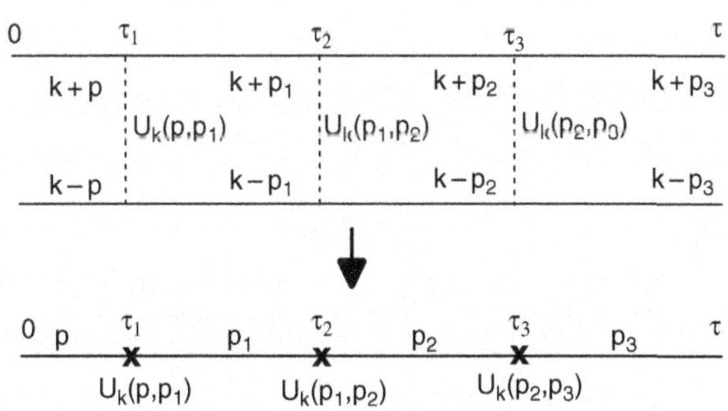

Fig. 12.6. Upper panel: Ladder diagrammatic expansion for GF of an exciton with total momentum $\boldsymbol{k}$. **Lower panel**: Equivalent one-line representation for the same class of diagrams

of the corresponding interaction vertices $U_{\boldsymbol{k}}(\boldsymbol{p},\boldsymbol{p}')$ (vertical dashed lines) and the propagators of electrons and holes with corresponding momenta (horizontal solid lines). However, for the given structure of ladder diagrams the electron and hole propagators can be combined into single propagators for the electron-hole pair. The propagator of the electron-hole pair is the product of hole and electron propagators and has the form

$$G_{\boldsymbol{k}}(\boldsymbol{p},\tau_{i+1}-\tau_i) = \mathrm{e}^{-(\epsilon_{\boldsymbol{k}}(\boldsymbol{p})-\mu)(\tau_{i+1}-\tau_i)} . \tag{12.56}$$

The energy of the electron-hole pair $\epsilon_{\boldsymbol{k}}(\boldsymbol{p}) = \varepsilon_c(\boldsymbol{k}+\boldsymbol{p}) - \varepsilon_v(\boldsymbol{k}-\boldsymbol{p})$ corresponds to the difference of the hole and electron energies with the center of mass momentum $\boldsymbol{k}$ and the relative momentum $2\boldsymbol{p}$. Then, for such a kind of Feynman expansion one can formulate an effective bare Hamiltonian

$$H^{(0)} = \sum_{\boldsymbol{p}} \epsilon_{\boldsymbol{k}}(\boldsymbol{p})\xi^\dagger_{\boldsymbol{p}}\xi_{\boldsymbol{p}} \tag{12.57}$$

and interaction term

$$H^{(\mathrm{int})} = \sum_{\boldsymbol{p}_1\boldsymbol{p}_2} U_{\boldsymbol{k}}(\boldsymbol{p}_1,\boldsymbol{p}_2)\xi^\dagger_{\boldsymbol{p}_2}\xi_{\boldsymbol{p}_1} + h.c. , \tag{12.58}$$

i.e. the expansion reduces to the line shown in the lower panel of Fig. 12.6.

12.3.5.1 Updates

The MC procedure for this series of FDs is a trivial modification of the techniques presented in previous sections. Updating the external parameter τ one needs to take into account that the distribution $W(x) = (\epsilon_{\boldsymbol{k}}(\boldsymbol{p}_3) - \mu)\exp\left[-(\epsilon_{\boldsymbol{k}}(\boldsymbol{p}_3) - \mu)x\right]$ depends on the momentum $\boldsymbol{p}_3$ of the propagator at the end of the FD. The updates, which add/remove an interaction vertex to/from the FD are similar to the previous examples. One of the N_{prop} propagators is chosen randomly and a time τ' is selected in the range $[\tau_l,\tau_r]$ with uniform probability. Then, the momentum $\boldsymbol{p}_2$ is selected with uniform probability from the Brillouin zone and attributed to the new propagator between the imaginary times τ' and τ_r, τ' is shown by circle in Fig. 12.7(a). Finally, the Metropolis ratio is very similar to that obtained for the simple potential model

$$M = \left[\frac{N_{\mathrm{prop}}}{N_{\mathrm{vert}}+1}\right] \frac{U_{\boldsymbol{k}}(\boldsymbol{p}_1,\boldsymbol{p}_2)\, U_{\boldsymbol{k}}(\boldsymbol{p}_2,\boldsymbol{p}_3)}{1/(\tau_r-\tau_l)\, U_{\boldsymbol{k}}(\boldsymbol{p}_1,\boldsymbol{p}_3)} \mathrm{e}^{-(\epsilon_{\boldsymbol{k}}(\boldsymbol{p}_2)-\epsilon_{\boldsymbol{k}}(\boldsymbol{p}_1))(\tau_r-\tau')} . \tag{12.59}$$

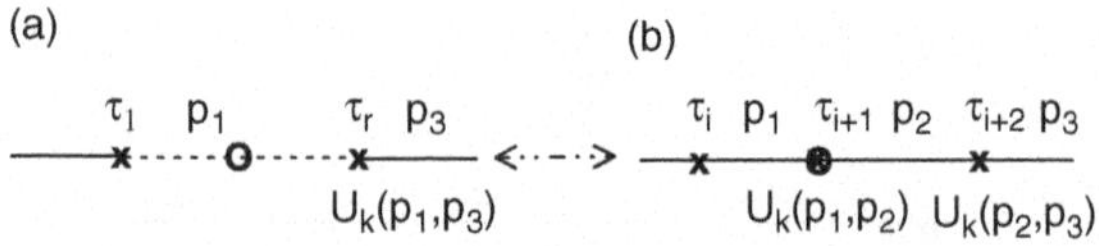

Fig. 12.7. Updates adding **(a)**→**(b)** and removing **(a)**←**(b)** an interaction vertex. See caption of Fig. 12.5

The first two factors are the same as in (12.52). The exponent takes into account the change of the FD weight due to the modification of the momentum of the electron-hole pair between τ' and τ_r, and the factor in front of the exponent appears due to the change of momentum $\boldsymbol{p}_1 \rightarrow \boldsymbol{p}_2$ in the vertex at τ_r. The ratio for the removal procedure is a straightforward modification of (12.53).

12.3.5.2 Estimators for Physical Quantities

Having calculated the Green function, let us now extract further properties of the exciton from the limit $G(\tau \rightarrow \infty)$. An eigenstate $\Psi_\nu(\boldsymbol{k})$ with energy E_ν can be written as

$$\Psi_\nu(\boldsymbol{k}) \equiv \sum_{\boldsymbol{p}} \xi_{\boldsymbol{k},\boldsymbol{p},\nu} e^\dagger_{\boldsymbol{k}+\boldsymbol{p}} h^\dagger_{\boldsymbol{k}-\boldsymbol{p}} |0\rangle \ , \tag{12.60}$$

where the amplitudes $\xi_{\boldsymbol{k},\boldsymbol{p},\nu} = \langle \nu; \boldsymbol{k} | e^\dagger_{\boldsymbol{k}+\boldsymbol{p}} h^\dagger_{\boldsymbol{k}-\boldsymbol{p}} |0\rangle$ describe the wave function of the internal motion of the exciton. In terms of exciton eigenstates we have

$$G_{\boldsymbol{k}}^{\boldsymbol{p}=\boldsymbol{p}'}(\tau) = \sum_\nu |\xi_{\boldsymbol{k},\boldsymbol{p},\nu}|^2 \mathrm{e}^{-E_\nu \tau} \ . \tag{12.61}$$

If τ is much larger than the inverse energy difference between the ground state and the first excited states, the GF projects to the ground state

$$G_{\boldsymbol{k}}^{\boldsymbol{p}=\boldsymbol{p}'}(\tau \rightarrow \infty) = |\xi_{\boldsymbol{k},\boldsymbol{p},\mathrm{gs}}|^2 \mathrm{e}^{-E_{\mathrm{gs}} \tau} \ . \tag{12.62}$$

Due to the normalization condition $\sum_{\boldsymbol{p}} |\xi_{\boldsymbol{k},\boldsymbol{p},\nu}|^2 \equiv 1$, the asymptotic behavior of the sum $\widetilde{G}_{\boldsymbol{k}} = \sum_{\boldsymbol{p}} G_{\boldsymbol{k}}^{\boldsymbol{p}=\boldsymbol{p}'}$ is especially simple: $\widetilde{G}(\tau) \rightarrow \exp(-E_{\mathrm{gs}}\tau)$.

By definition, in the limit $\tau \rightarrow \infty$, we have $G_{\boldsymbol{k}}^{\boldsymbol{p}=\boldsymbol{p}'}/\widetilde{G}_{\boldsymbol{k}} = |\xi_{\boldsymbol{k},\boldsymbol{p},\mathrm{gs}}|^2$, i.e. the distribution over the quasimomentum $\boldsymbol{p}$ is related to the wave function of internal motion, which is calculated by simulating the set of GFs $G_{\boldsymbol{k}}^{\boldsymbol{p}=\boldsymbol{p}'}$ with $\boldsymbol{p} = \boldsymbol{p}'$.

One can ask how to calculate the asymptotic behavior of $G_{\boldsymbol{k}}^{\boldsymbol{p}=\boldsymbol{p}'}(\tau)$ when, obviously, the first order diagram does not obey the condition $\boldsymbol{p} = \boldsymbol{p}'$ except for the case of the $U_{\boldsymbol{k}}(\boldsymbol{p}, \boldsymbol{p}' = \boldsymbol{p})$ vertex. Moreover, working with the function $\widetilde{G}$ we encounter a certain formal problem: The zero- and first-order diagrams with respect to $U_{\boldsymbol{k}}(\boldsymbol{p}, \boldsymbol{p}' = \boldsymbol{p})$ contain macroscopically large factors N. However, since we are only interested in the ground-state properties, we can safely omit the obstructive terms, which in a careful analysis turn out to be irrelevant in the limit $\tau \rightarrow \infty$. Therefore, in the simulation one simply starts from an arbitrary second order diagram and excludes all diagrams of order less than two.

12.3.5.3 Numeric Results

The exciton problem (12.1)–(12.2) has been studied for many years, but as yet there was no rigorous technique available for its solution. The only solvable cases are the

Frenkel small-radius limit [56] and the Wannier large-radius limit [57], but the range of validity of these two approximations was unclear.

To study the conditions for the validity of the Frenkel and Wannier approaches with DMC, we consider a three-dimensional (3D) system and assume an electron-hole spectrum with symmetric valence and conduction bands of width E_c and a direct gap E_g at zero momentum [27]. We find that for large ratio $\kappa = E_c/E_g$ ($\kappa > 30$) the exciton binding energy is in good agreement with the Wannier approximation, see Fig. 12.8(a), and the probability density of the relative electron-hole motion, see Fig. 12.8(c), corresponds to the hydrogen-like result. For smaller values of κ, however, both the binding energy and the wave function of the relative motion, see Fig. 12.8(d) deviate noticeably from the large radius results. It is quite

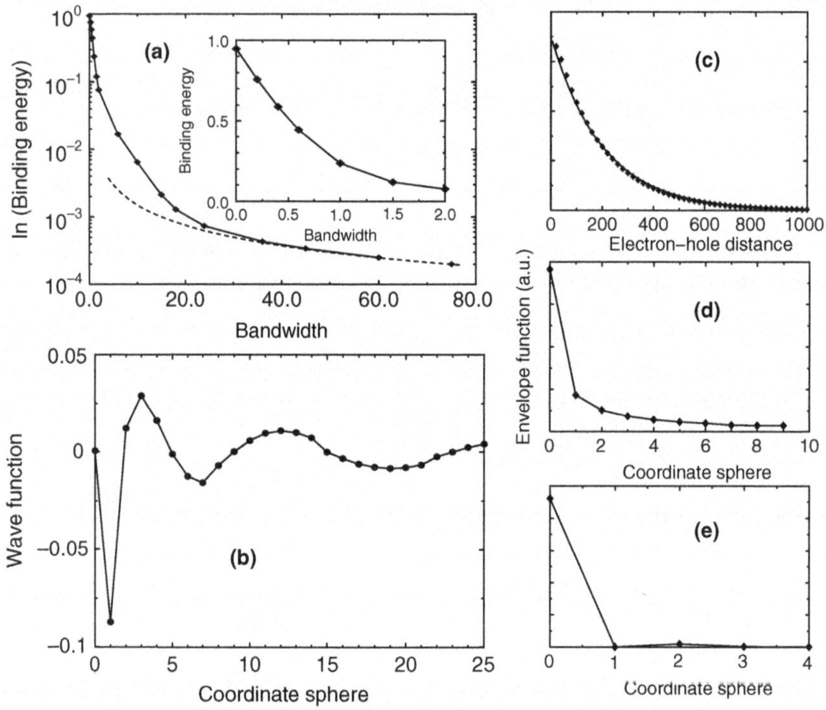

Fig. 12.8. Panel (**a**): Dependence of the exciton binding energy on the bandwidth $E_c = E_v$ for conduction and valence bands. The dashed line corresponds to the Wannier model. The solid line is the cubic spline, the derivatives at the right and left ends being fixed by the Wannier limit and perturbation theory, respectively. Inset in panel (**a**): The initial part of the plot. Panel (**b**): Wave function of internal motion in real space for the optically forbidden monopolar exciton. Panels (**c**)–(**e**): The wave function of internal motion in real space: (**c**) Wannier ($E_c = E_v = 60$); (**d**) intermediate ($E_c = E_v = 10$); (**e**) near-Frenkel ($E_c = E_v = 0.4$) regimes. The solid line in the panel (**c**) is the Wannier model result while solid lines in other panels are to guide the eyes only

surprising that we need such large valence and conduction bandwidths ($\kappa > 20$) for the Wannier approximation to be applicable.

Similarly, the range of validity of the Frenkel approach is limited as well. Even a strongly localized wave function does not guarantee good agreement between the exact binding energy and the Frenkel approximation. For $1 < \kappa < 10$ the wave function is already strongly localized, but the binding energies differ considerably. For example, at $\kappa = 0.4$, the relative motion is rather suppressed, cf. Fig. 12.8(e), but the binding energy of the Frenkel approximation is two times larger than the exact result, see inset in Fig. 12.8(a).

Another long-standing issue is the formation of charge transfer excitons in 3D systems and the appropriate modelling of mixed valence semiconductors [58]. A decade ago some of the unusual properties of SmS and SmB_6 were explained on the basis of an excitonic instability mechanism, thereby assuming a charge-transfer nature of the optically forbidden exciton [59, 60]. Although this model explained quantitatively the phonon spectra [61, 62], optical properties [63, 64], and magnetic neutron scattering data [65], its basic assumption has been criticized as being groundless [66, 67]. We have studied the excitonic wave function of mixed valence materials, starting from typical dispersions of the valence and conduction bands: An almost flat valence band is separated from a broad conduction band with its maximum in the center and minimum at the border of the Brillouin zone [27]. The results presented in Fig. 12.8(b) support the assumption of [59, 60], since the wave function of the relative motion has an almost vanishing on-site component and its maximal charge density at nearby neighbors.

12.3.6 Two-Level System

In this section we apply the diagrammatic expansion to a two-level system, see (12.10) and (12.11), which is the simplest object with an internal structure. One can generalize this example to the full problem of a two-level system in a bosonic bath (12.10)–(12.12) or to the problem of the Jahn-Teller and PJT polaron [28, 34].

While adapting the technique developed in Sect. 12.3.3, we need to take into account that the interaction switches the quantum numbers between different states of the QP with energies $\varepsilon_{1,2} = \pm\epsilon/2$. Therefore, when a new vertex is introduced into the expansion, it has to exchange the energies $\varepsilon_1 \leftrightarrow \varepsilon_2$ in all propagators situated, e.g., to the right of the vertex, see Fig. 12.9. For example, the particle propagator $G_2(\tau_{\text{last}} - \tau_{i+2}) = \exp(-(\varepsilon_2 - \mu)(\tau_{\text{last}} - \tau_{i+2}))$ of type two in Fig. 12.9(a) changes to that of type one $G_1(\tau_{\text{last}} - \tau_{i+2}) = \exp(-(\varepsilon_1 - \mu)(\tau_{\text{last}} - \tau_{i+2}))$ in Fig. 12.9(b). Therefore, the ratio for the add-propagator update in Fig. 12.9 is

Fig. 12.9. Add/remove updates changing the number of interaction vertices. Solid (*dashed*) lines correspond to propagators of the particle in a state with energy ε_1 (ε_2)

$$M = \left(\frac{N_{\text{prop}}}{N_{\text{vert}} + 1} \right) \Delta(\tau_r - \tau_l) \mathrm{e}^{-\epsilon \delta S} \ , \tag{12.63}$$

where $\delta S = (\tau_{i+2} - \tau_{i+1}) - (\tau_{last} - \tau_{i+2}) + (\tau - \tau_{last})$. Note that each additional vertex switches between the GFs $G_{11}(\tau)$ and $G_{12}(\tau)$. The statistics for $G_{11}(\tau)$ is updated when the right end of the diagram corresponds to a propagator of type 1, which is denoted by a solid line, i.e. when there is an even number of interaction vertices, and a contribution to the statistics of GF $G_{12}(\tau)$ is counted otherwise.

To realize the importance of the above remark one can take the code for an attractive potential from Sect. 12.3.3, and use it for the calculation of the GF of the degenerate two-level system. In the case of zero bias $\epsilon = 0$ the exponential factor in (12.63) is irrelevant and the DMC algorithms for both problems are equivalent. The only difference is the way how the statistics for the GFs is collected, since a diagram contributes to $G_{11}(\tau)$, $G_{12}(\tau)$, for even, odd, number of interaction vertices.

The analytic GFs for the two-level system (12.10)–(12.11) in the case of zero bias can be obtained in the following way: Diagonalization of the Hamiltonian of the two-level system (12.10)–(12.11) without coupling to bosons yields two eigenstates with energies $\pm\Delta$. Then, the GFs $G_{11}(\tau) = \langle \text{vac} | a_1(\tau) a_1^\dagger | \text{vac} \rangle$ and $G_{12}(\tau) = \langle \text{vac} | a_1(\tau) a_2^\dagger | \text{vac} \rangle$ can be obtained by a canonical transformation $a_{1,2} = 1/\sqrt{2}[a_{\text{up}} \pm a_{\text{low}}]$ of the initial creation and annihilation operators $a_{1,2}$ and $a_{1,2}^\dagger$ into the operators of the upper and lower state $a_{\text{up, low}}$ and $a_{\text{up, low}}^\dagger$. Then, taking into account that $a_{\text{up, low}}(\tau) = \exp[-(\pm\Delta - \mu)\tau] \ a_{\text{up, low}}$, one arrives at the following expressions

$$G_{11,12}(\tau) = \frac{1}{2} \left(\mathrm{e}^{-(-\Delta-\mu)\tau} \pm \mathrm{e}^{-(\Delta-\mu)\tau} \right) . \tag{12.64}$$

Comparing the DMC data with the exact GFs (12.64) we observe that the suggested strategy for the accumulation of statistics is correct, cf. Fig. 12.10.

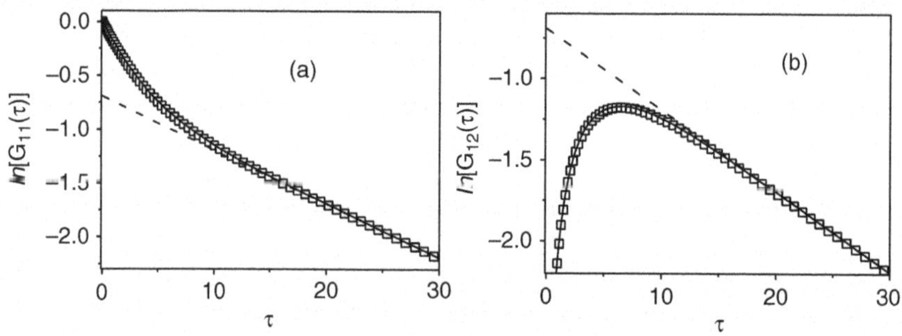

Fig. 12.10. Comparison of the DMC data (squares) for $G_{11}(\tau)$ (**a**) and $G_{12}(\tau)$ (**b**) with the solid lines corresponding to the analytic expressions (12.64) for the degenerate two-level system. The dashed line marks the asymptotics $\ln[G_{11,12}(\tau)]_{\tau\to+\infty} = \ln(1/2) - (-\Delta - \mu)\tau$ of both GFs. Calculations are done for $\mu = -0.2$ and $\Delta = 0.15$

12.4 Stochastic Optimization Method

The solution of the integral equation (12.17) is known to be an ill conditioned problem. The GF $G_{\mathbf{k}}(\tau)$ is known only with statistic errors and on a finite number of imaginary times in a finite range $[0, \tau_{\max}]$. Due to this incomplete and noisy information, there is an infinite number of approximate solutions which reproduce the GF within some range of deviation, and the problem is to choose the best one. Another problem is the saw tooth noise instability, which remained a stumbling block for decades. It occurs when the problem is solved naively, e.g. by using a least-squares approach for minimizing the deviation measure

$$D\left(\widetilde{L}_{\mathbf{k}}(\omega)\right) = \int_0^{\tau_{\max}} \left|G_{\mathbf{k}}(\tau) - \widetilde{G}_{\mathbf{k}}(\tau)\right| G_{\mathbf{k}}^{-1}(\tau)\, \mathrm{d}\tau\,. \tag{12.65}$$

Here $\widetilde{G}_{\mathbf{k}}(\tau)$ is obtained from an approximate Lehmann function $\widetilde{L}_{\mathbf{k}}(\omega)$ by applying the integral operator $\widetilde{G}_{\mathbf{k}}(\tau) = \mathcal{F}(\widetilde{L}_{\mathbf{k}}(\omega))$ in (12.17). The saw tooth instability corrupts the Lehmann function in regions where the actual Lehmann function is smooth. Fast fluctuations of the solution $\widetilde{L}_{\mathbf{k}}(\omega)$ often have much larger amplitude than the value of the actual Lehmann function $L_{\mathbf{k}}(\omega)$. Standard tools for the suppression of saw tooth noise are mostly based on the early idea of Fillips-Tikhonov regularization [68, 69, 70, 71]. In these approaches a nonlinear functional, which suppresses large derivatives of the approximate solution $\widetilde{L}_{\mathbf{k}}(\omega)$, is added to the linear deviation measure (12.65). The most popular variant of those regularization approaches is the MEM [43].

However, the typical Lehmann function of a QP in a boson field consists of δ-function peaks and a smooth incoherent continuum with a sharp border [26, 36]. Hence, suppression of high derivatives, as the general strategy of the regularization method, fails. Moreover, any specific implementation of the regularization method uses a predefined mesh in ω-space, which could be completely inappropriate for the case of sharp peaks. If the actual location of a sharp peak is between predefined discrete points, the rest of spectral density can be distorted beyond recognition. Finally, MEM assumes a Gauss distribution of statistic errors in $G_{\mathbf{k}}(\tau)$, which might be invalid in some cases [43].

Recently, a SO method, which circumvents the above mentioned difficulties, was developed [26]. The SO method is based on the calculation of a large enough number M of statistically independent non-regularized solutions $\{\widetilde{L}_{\mathbf{k}}^{(s)}(\omega), s = 1, \ldots, M\}$, whose deviation measures $D^{(s)}$ are smaller than some upper limit D_u, which depends on the statistic noise of the GF $G_{\mathbf{k}}(\tau)$. Then, using the linearity of the expressions (12.17) and (12.65), the final solution is found as the average of particular solutions $\{\widetilde{L}_{\mathbf{k}}^{(s)}(\omega)\}$

$$L_{\mathbf{k}}(\omega) = \frac{1}{M} \sum_{s=1}^{M} \widetilde{L}_{\mathbf{k}}^{(s)}(\omega)\,. \tag{12.66}$$

The particular solution $\widetilde{L}_{\mathbf{k}}^{(s)}(\omega)$ is parameterized in terms of a sum

$$\widetilde{L}_{\mathbf{k}}^{(s)}(\omega) = \sum_{t=1}^{K} \chi_{\{P_t\}}(\omega) \tag{12.67}$$

of rectangles $\{P_t\} = \{h_t, w_t, c_t\}$ with height $h_t > 0$, width $w_t > 0$, and center c_t. The configuration

$$\mathcal{C} = P_t \ , \tag{12.68}$$

with $t = 1, \ldots, K$, which satisfies the normalization condition $\sum_{t=1}^{K} h_t w_t = 1$, defines the function $\widetilde{G}_{\mathbf{k}}(\tau)$. The generation of a particular solution starts from an arbitrary initial configuration $\mathcal{C}_s^{\text{init}}$. Then, the deviation measure is optimized with a random sequence of updates, until the deviation is less than D_u. In addition to the updates, which do not change the number of terms in the sum (12.67), there are updates which increase or decrease K. Hence, since the number of elements K is not fixed, any spectral function can be reproduced with the desired accuracy.

Although each particular solution $\widetilde{L}_{\mathbf{k}}^{(s)}(\omega)$ suffers from saw tooth noise in regions where the Lehmann function is smooth, the statistical independence of each solution leads to a self-averaging of this noise in the sum (12.66). Note that the noise is suppressed without suppressing high derivatives. Hence, in contrast to regularization approaches, sharp peaks and edges are not smeared out. Moreover, the continuous parameterization (12.67) does not need a predefined mesh in ω-space, and, since the Hilbert space of solutions is sampled directly, no assumptions about the distribution of statistical errors are required.

In Fig. 12.11 we present results for an averaging over an increasing number of statistically independent particular solutions. One can notice that the spikes in the spectral analysis data disappear with increasing M. Note, that neither the general shape of the triangle, which is an artificial Lehmann function with infinite first derivatives, nor the sharp low-energy edge of the spectral density are corrupted by the SO method.

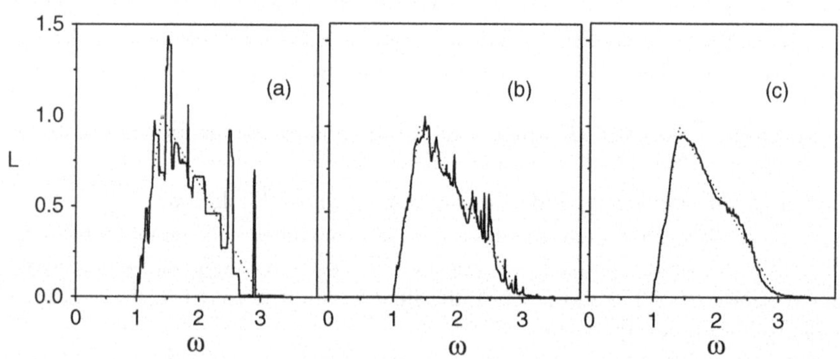

Fig. 12.11. Comparison of the actual spectral function (dashed line) with the results of a spectral analysis after averaging over (**a**) $M = 4$, (**b**) $M = 28$, and (**c**) $M = 500$ particular solutions

The SO method was successfully applied to restore the Lehmann function of a Fröhlich polaron [26], a Rashba-Pekar exciton-polaron [36], a hole-polaron in the t-J-model [35, 30], and of a many-particle spin system [72]. The calculation of the optical conductivity of the polaron by the SO method can be found in [29]. The SO method is particularly helpful when the GF's asymptotic limit, which gives information about the ground state, cannot be reached. For example, sign fluctuations of the terms in expansion (12.29) for a hole in the t-J-model lead to poor statistics at large times [35]. Nevertheless, the SO method is capable of recovering the energy and the Z-factor even when the GF is known only at small imaginary times [35].

Comparing MEM and SO, I would like to point out that SO surpasses MEM when the statistical errors of the MC data are small enough, and has no advantages otherwise. Moreover, the CPU time required for the SO procedure is two orders of magnitude larger than that necessary for MEM. However, this price is worth to be paid, since we avoid the approximations of MEM and get a result which is as close as possible to the exact solution. Moreover, the above limitation is not essential, because the CPU time required for accumulating good MC statistics is much larger than that for a SO analysis.

12.5 Conclusions and Perspectives

To summarize, the combination of diagrammatic MC and stochastic optimization methods is a powerful tool for obtaining approximation free data for few complex objects, which interact with each other and with one or several bosonic baths in a macroscopic system. In this contribution I have restricted myself to the description of the methods in application to several simple examples. This might help a beginner to write first simple DMC-SO codes. A more detailed review of the results obtained by DMC-SO can be found in [31, 32]. The numerical approach has already been used to obtain the Lehmann function [26] and the optical conductivity [29, 39] of Fröhlich polarons, Lehmann functions of the Rashba-Pekar polaron [36], a hole in an antiferromagnet [35], and a hole in an antiferromagnet interacting with optical phonons [30, 37]. In addition, the ground state properties of the pseudo Jahn-Teller polaron [34] and the exciton [27] have been studied.

These techniques can also be applied to a wide range of other problems. Among the most obvious ones are the ground state properties and the excitations of the exciton-polaron and the Holstein polaron. Particularly interesting is the case of a 3D polaron interacting with acoustic phonons, where all other numeric methods are unable to provide an exact solution. Switching from momentum to real space, i.e. calculating the on-site Lehmann functions of a polaron at and near the impurity potential, one can reveal the experimental signal observed by scanning tunneling microscopy. As yet, the interaction of a two-level system with a bosonic bath is not studied by the DMC and SO methods.

I thank N. Nagaosa, A. Sakamoto, N.V. Prokofev, B.V. Svistunov, E.A. Burovski, and H. Fehske for collaborations and discussions. This work is supported by Russian Fund of Basic Researches (RFBR) grants 04-02-17363a and 07-0200067-a.

References

1. J. Appel, *Polarons*. Solid State Physics, Vol. 21 (Academic, New York, 1968)
2. S.I. Pekar, *Untersuchungen über die Elektronentheorie der Kristalle* (Akademie Verlag, Berlin, 1954)
3. L.D. Landau, Phys. Z. Sowjetunion **3**, 664 (1933)
4. H. Fröhlich, H. Pelzer, S. Zienau, Philos. Mag. **41**, 221 (1950)
5. J. Kanamori, Appl. Phys. **31**, S14 (1960)
6. K.I. Kugel, D.I. Khomskii, Sov. Phys. Usp. **25**, 231 (1982)
7. Y. Toyozawa, J. Hermanson, Phys. Rev. Lett. **21**, 1637 (1968)
8. I.B. Bersuker, *The Jahn-Teller Effect* (IFI/Plenum, New York, 1983)
9. V.L. Vinetskii, Sov. Phys. JETP **13**, 1023 (1961)
10. P.W. Anderson, Phys. Rev. Lett. **34**, 953 (1975)
11. H. Hiramoto, Y. Toyozawa, J. Phys. Soc. Jpn. **54**, 245 (1985)
12. A. Alexandrov, J. Ranninger, Phys. Rev. B **23**, 1796 (1981)
13. H. Haken, Il Nuovo Cimento **3**, 1230 (1956)
14. F. Bassani, G. Pastori Parravicini, *Electronic States and Optical Transitions in Solids* (Pergamon, Oxford, 1975)
15. J. Pollmann, H. Büttner, Phys. Rev. B **16**, 4480 (1977)
16. A. Sumi, J. Phys. Soc. Jpn. **43**, 1286 (1977)
17. M. Ueta, H. Kanzaki, K. Kobayashi, Y. Toyozawa, E. Hanamura, *Excitonic Processes in Solids* (Springer-Verlag, Berlin, 1986)
18. C.L. Kane, P.A. Lee, N. Read, Phys. Rev. B **39**, 6880 (1989)
19. Y.A. Izyumov, Phys. Usp. **40**, 445 (1997)
20. A.J. Leggett, Science **296**, 861 (2002)
21. A.J. Leggett, S. Chakravarty, A.T. Dorsey, M.P.A. Fisher, A. Garg, W. Zwerger, Rev. Mod. Phys. **59**, 1 (1987)
22. A. Macridin, G.A. Sawatzky, M. Jarrell, Phys. Rev. B **69**, 245111 (2004)
23. H. Fehske, G. Wellein, G. Hager, A. Weiße, A.R. Bishop, Phys. Rev. B **69**, 165115 (2004)
24. N.V. Prokof'ev, B.V. Svistunov, I.S. Tupitsyn, Sov. Phys. JETP **87**, 310 (1998)
25. N.V. Prokof'ev, B.V. Svistunov, Phys. Rev. Lett. **81**, 2514 (1998)
26. A.S. Mishchenko, N.V. Prokof'ev, A. Sakamoto, B.V. Svistunov, Phys. Rev. B **62**, 6317 (2000)
27. E.A. Burovski, A.S. Mishchenko, N.V. Prokof'ev, B.V. Svistunov, Phys. Rev. Lett. **87**, 186402 (2001)
28. A.S. Mishchenko, N. Nagaosa, N.V. Prokof'ev, B.V. Svistunov, E.A. Burovski, Nonlinear Optics **29**, 257 (2002)
29. A.S. Mishchenko, N. Nagaosa, N.V. Prokof'ev, A. Sakamoto, B.V. Svistunov, Phys. Rev. Lett. **91**, 236401 (2003)
30. A.S. Mishchenko, N. Nagaosa, Phys. Rev. Lett. **93**, 036402 (2004)
31. A.S. Mishchenko, Phys. Usp. **48**, 887 (2005)
32. A.S. Mishchenko, N. Nagaosa, J. Phys. Soc. J. **75**, 011003 (2006)
33. A.S. Mishchenko, *Proceedings of the international school of physics "Enrico Fermi", Course CLXI* (IOS Press, 2006), pp. 177–206
34. A.S. Mishchenko, N. Nagaosa, Phys. Rev. Lett. **86**, 4624 (2001)
35. A.S. Mishchenko, N.V. Prokof'ev, B.V. Svistunov, Phys. Rev. B **64**, 033101 (2001)
36. A.S. Mishchenko, N. Nagaosa, N.V. Prokof'ev, A. Sakamoto, B.V. Svistunov, Phys. Rev. B **66**, 020301 (2002)
37. A.S. Mishchenko, N. Nagaosa, Phys. Rev. B **73**, 092502 (2006)

38. A.S. Mishchenko, N. Nagaosa, J. Phys. Chem. Solids **67**, 259 (2006)
39. G. De Filippis, V. Cataudella, A.S. Mishchenko, C.A. Perroni, J.T. Devreese, Phys. Rev. Lett. **96**, 136405 (2006)
40. G.D. Mahan, *Many particle physics* (Plenum Press, New York, 2000)
41. A. Damascelli, Z. Hussain, Z.X. Shen, Rev. Mod. Phys. **75**, 473 (2003)
42. A.A. Abrikosov, L.P. Gor'kov, D.I. E., *Quantum Field Theoretical Method in Statistical Physics* (Pergamon Press, Oxford, 1965)
43. M. Jarrell, J.E. Gubernatis, Phys. Rep. **269**, 133 (1996)
44. R. Knox, *Theory of Excitons* (Academic Press, New York, 1963)
45. I. Egri, Phys. Rep. **119**, 363 (1985)
46. D. Haarer, Chem. Phys. Lett. **31**, 192 (1975)
47. D. Haarer, M.R. Philpott, M. H., J. Chem. Phys. **63**, 5238 (1975)
48. A. Elschner, G. Weiser, Chem. Phys. **98**, 465 (1985)
49. N. Metropolis, A.W. Rosenbluth, M.N. Rosenbluth, A.H. Teller, E. Teller, J. Chem. Phys. **21**, 1087 (1953)
50. M.E.J. Newman, G.T. Barkema, *Carlo Methods in Statistical Physics* (Clarendon Press, Oxford, 2002)
51. D.P. Landau, K. Binder, *A Guide to Monte Carlo Simulations in Statistical Physics* (University Press, Cambridge, 2000)
52. A.W. Sandvik, J. Kurkijärvi, Phys. Rev. B **43**, 5950 (1991)
53. D.M. Ceperley, J. Comp. Phys. **51**, 404 (1983)
54. D.M. Ceperley, A.B. J., J. Chem. Phys. **81**, 5833 (1984)
55. N. Prokof'ev, B. Svistunov, I. Tupitsyn, Phys. Rev. Lett. **82**, 5092 (1999)
56. J. Frenkel, Phys. Rev. **37**, 17 (1931)
57. G.H. Wannier, Phys. Rev. **52**, 191 (1937)
58. S. Curnoe, K.A. Kikoin, Phys. Rev. B **61**, 15714 (2000)
59. K.A. Kikoin, A.S. Mishchenko, Zh. Eksp. Teor. Fiz. **94**, 237 (1988). [Sov. Phys. JETP **67**, 2309 (1988)]
60. K.A. Kikoin, A.S. Mishchenko, J. Phys.: Condens. Matter **2**, 6491 (1990)
61. P.A. Alekseev, I.A. S., B. Dorner, et.al, Europhys. Lett. **10**, 457 (1989)
62. A.S. Mishchenko, K.A. Kikoin, J. Phys.: Condens. Matter **3**, 5937 (1991)
63. G. Travaglini, P. Wachter, Phys. Rev. B **29**, 893 (1984)
64. P. Lemmens, A. Hoffman, A.S. Mishchenko, et.al, Physica B **206-207**, 371 (1995)
65. K.A. Kikoin, A.S. Mishchenko, J. Phys.: Condens. Matter **7**, 307 (1995)
66. T. Kasuya, Europhys. Lett. **26**, 277 (1994)
67. T. Kasuya, Europhys. Lett. **26**, 283 (1994)
68. A.N. Tikhonov, V.Y. Arsenin, *Solutions of Ill-Posed Problems* (Winston, Washington, 1977)
69. E. Perchik (2003). URL `http://arxiv.org/abs/math-ph/0302045`. Preprint
70. D.L. Phillips, J. Assoc. Comput. Mach. **9**, 84 (1962)
71. D.L. Tikhonov, Sov. Math. Dokl. **4**, 1035 (1963)
72. S.S. Aplesnin, J. Exp. Theor. Phys. **97**, 969 (2003)

13 Path Integral Monte Carlo Simulation of Charged Particles in Traps

Alexei Filinov, Jens Böning, and Michael Bonitz

Institut für Theoretische Physik und Astrophysik, Christian-Albrechts-Universität, 24098 Kiel, Germany

13.1 Introduction

This chapter is devoted to the computation of equilibrium (thermodynamic) properties of quantum systems. In particular, we will be interested in the situation where the interaction between particles is so strong that it cannot be treated as a small perturbation. For weakly coupled systems many efficient theoretical and computational techniques do exist. However, for strongly interacting systems such as nonideal gases or plasmas, strongly correlated electrons and so on, perturbation methods fail and alternative approaches are needed. Among them, an extremely successful one is the Path Integral Monte Carlo (PIMC) method which we are going to consider in this chapter.

13.2 Idea of Path Integral Monte Carlo

If we perform classical simulations of a system in equilibrium, we usually start from the Boltzmann-type probability distribution, $p_{\mathrm{B}} \sim \exp(-\beta U_N(R))/Z$, ($\beta = 1/k_{\mathrm{B}}T$) and then the Monte Carlo method (Part II and [1]) can be used to sample the particle coordinates $R = (\boldsymbol{r}_1, \boldsymbol{r}_2, \ldots, \boldsymbol{r}_N)$, a $3N$-dimensional vector. Now the question arises what is the appropriate probability density in the quantum case. The answer is provided by the density operator $\widehat{\rho}$. Consider a general expression for thermodynamic averages in statistical thermodynamics. The N-particle density matrix $\widehat{\rho}(\beta)$ contains the complete information about the system with the observables given by

$$\langle \widehat{O} \rangle(\beta) = \frac{\mathrm{Tr}[\widehat{O}\,\widehat{\rho}]}{\mathrm{Tr}\,[\widehat{\rho}]} = \frac{\int \mathrm{d}R\, \langle R|\widehat{O}\,\widehat{\rho}|R\rangle}{\int \mathrm{d}R\, \langle R|\widehat{\rho}(\beta)|R\rangle} = \frac{\int \mathrm{d}R\, \mathrm{d}R'\, \langle R|\widehat{O}|R'\rangle\, \langle R'|\widehat{\rho}|R\rangle}{\int \mathrm{d}R\, \langle R|\widehat{\rho}|R\rangle}\,. \tag{13.1}$$

This expression is simplified if the physical observable $\widehat{O}$ is diagonal in the coordinate representation, i.e. $\langle R'|\widehat{O}|R\rangle = \langle R|\widehat{O}|R\rangle\, \delta(R' - R)$. In this case we need only the diagonal matrix element $\langle R|\widehat{\rho}|R\rangle$.

As in the classical case we have to perform an integration over $3N$ (or more) degrees of freedom, but in contrast, now we generally do not know the analytical

A. Filinov et al.: *Path Integral Monte Carlo Simulation of Charged Particles in Traps*, Lect. Notes Phys. **739**, 397–412 (2008)
DOI 10.1007/978-3-540-74686-7_13

expression for the N-particle density operator which has to be substituted in (13.1). This problem was first overcome by Feynman [2]. The key idea was to express the unknown density operator,

$$\langle R|\widehat{\rho}(\beta)|R'\rangle \quad \text{with} \quad \widehat{\rho} = \mathrm{e}^{-\beta\widehat{H}}, \tag{13.2}$$

at a given inverse temperature β by its high-temperature asymptote which is known analytically. However, this comes at a high price: Instead of an already complicated $3N$-dimensional integral, now it expands to much higher dimensions ($3NM$), where M is an integer which in practice is chosen between $1 \leq M \leq 3000$.

13.2.1 Group Property of Density Matrix

One simple and straightforward strategy is to use the group property of the density matrix. It allows to express the density matrix at low temperatures in terms of its values at higher temperature, i.e.

$$\begin{aligned}\rho(R, R'; \beta_1 + \beta_2) &= \langle R|\mathrm{e}^{-(\beta_1+\beta_2)\widehat{H}}|R'\rangle = \int \mathrm{d}R_1\, \langle R|\mathrm{e}^{-\beta_1\widehat{H}}|R_1\rangle\langle R_1|\mathrm{e}^{-\beta_2\widehat{H}}|R'\rangle \\ &= \int \mathrm{d}R_1\, \rho(R, R_1; \beta_1)\rho(R_1, R'; \beta_2). \end{aligned} \tag{13.3}$$

Using the group property M times we find the generalization

$$\widehat{\rho} = \mathrm{e}^{-\beta\widehat{H}} = \mathrm{e}^{-\Delta\beta\widehat{H}} \ldots \mathrm{e}^{-\Delta\beta\widehat{H}}, \quad \Delta\beta = \frac{\beta}{M}\,. \tag{13.4}$$

This means that the density operator $\widehat{\rho}$ is expressed as a product of M new density operators, $\exp(-\Delta\beta\widehat{H})$, each corresponding to an M times higher temperature.

Finally, using (13.4) for any fixed end-points R and R' we can write the off-diagonal matrix element as[1]

$$\begin{aligned}\rho(R, R'; \beta) = \int & \mathrm{d}R_1\mathrm{d}R_2 \ldots \mathrm{d}R_{M-1} \\ & \rho(R, R_1; \Delta\beta)\rho(R_1, R_2; \Delta\beta) \ldots \rho(R_{M-1}, R'; \Delta\beta)\,, \end{aligned} \tag{13.5}$$

where M factors are connected by $M-1$ intermediate integrations.

13.2.2 High-Temperature Approximation

Equations (13.4) and (13.5) are correct for any finite M as long as we use exact expressions for the high-temperature N-particle density matrices, $\rho(R_{i-1}, R_i; \Delta\beta)$. Unfortunately, they are unknown, and to proceed further we need to introduce approximations.

[1] The total dimension of the integral, $(M-1)\,3N$, may be very large. The success of the method relies on highly efficient Monte Carlo integration.

The approximation we employ is based on Trotter's theorem (1959) applied to a general Hamiltonian, $\widehat{H} = \widehat{T} + \widehat{V}$, which contains both kinetic and potential energy operators, i.e.

$$\begin{aligned}\widehat{\rho} = \mathrm{e}^{-\beta(\widehat{T}+\widehat{V})} &= \lim_{M\to\infty}\left[\mathrm{e}^{-\Delta\beta\widehat{T}}\mathrm{e}^{-\Delta\beta\widehat{V}}\right]^M \\ &\approx \left[\mathrm{e}^{-\Delta\beta\widehat{T}}\mathrm{e}^{-\Delta\beta\widehat{V}}\right]^M + O\left(\mathrm{e}^{-\Delta\beta^2 M[\widehat{T},\widehat{V}]/2}\right) \\ &\approx \left[\mathrm{e}^{-\Delta\beta\widehat{T}}\mathrm{e}^{-\Delta\beta\widehat{V}}\right]^M + O\left(\frac{1}{M}\right). \end{aligned} \tag{13.6}$$

Note that $\widehat{T}$ and $\widehat{V}$ do not commute giving rise to the commutator, $[\widehat{T}, \widehat{V}]$, which is only the first term of a series[2]. Neglecting the terms $[\widehat{T}, \widehat{V}]$ gives an error of the order $O\,[1/M]$. This error can be made arbitrarily small by choosing a sufficiently large number of factors M.

Using the Trotter result (13.6), we immediately obtain an approximation for high temperatures[3]

$$\begin{aligned}\rho(R_i, R_{i+1}; \Delta\beta) &\approx \langle R_i|\mathrm{e}^{-\Delta\beta\widehat{T}}\mathrm{e}^{-\Delta\beta\widehat{V}}|R_{i+1}\rangle \\ &= \lambda_\Delta^{-3N}\mathrm{e}^{-\pi(R_i-R_{i+1})^2/\lambda_\Delta^2 - \Delta\beta V(R_i;\Delta\beta)}\,, \end{aligned} \tag{13.7}$$

where $\lambda_\Delta = \sqrt{2\pi\hbar^2\Delta\beta/m}$ is the De Broglie wavelength. Substituting (13.7) in (13.5) we get our final result for low temperatures

$$\rho(R, R'; \beta) = \int \mathrm{d}R_1 \ldots \mathrm{d}R_{M-1}\, \mathrm{e}^{-\sum\limits_{i=0}^{M-1}\pi(R_i-R_{i+1})^2/\lambda_\Delta^2}\, \mathrm{e}^{-\sum\limits_{i=0}^{M-1}\Delta\beta V(R_i)}\,, \tag{13.8}$$

with the boundary conditions: $R_0 = R$ and $R_M = R'$. Hence, we have constructed a suitable representation of the N-particle density matrix, which can be evaluated numerically with the help of a Monte Carlo algorithm.

13.2.3 Visualization of Diagonal Elements of the Density Matrix

As we can see from (13.5) and (13.8), all N particles have their own images on M different planes (or 'time slices'). We can view these images (for each particle $3M$ sets of coordinates) as a 'trajectory' or a 'path' in the configurational space. The inverse temperature argument β can be considered as an imaginary time of the path. The set of M time slices is ordered along the β-axis and separated by intervals $\Delta\beta$. In Fig. 13.1 we show typical configurations of particle trajectories which contribute to the diagonal density matrix element (13.5) with $R = R'$. The full density matrix $\rho(R, R; \beta)$ is obtained after integration over all possible path configurations with the fixed end points ($R = R'$).

[2] Double, triple and higher-order commutators have higher powers $\Delta\beta^n$ as a prefactor and can be dropped in the limit $\Delta\beta \to 0$.

[3] Other more accurate high-temperature approximations are discussed in [1, 3].

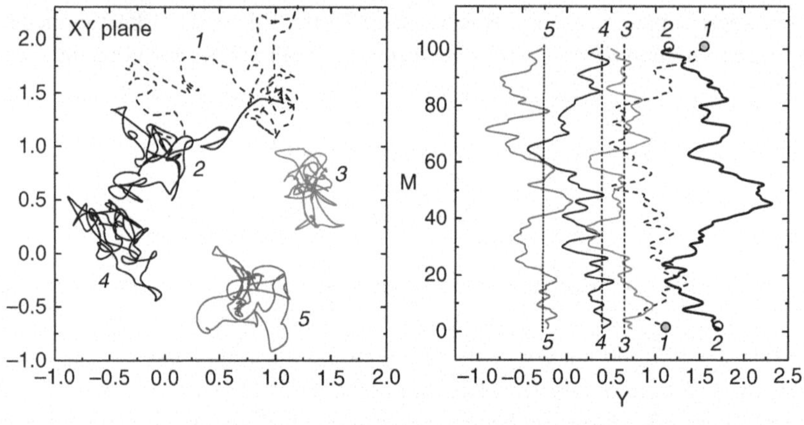

Fig. 13.1. Snapshot of 5 particles in a two-dimensional parabolic potential. Each particle is presented as a continuous path (obtained by a smooth interpolation through a discrete set of $M = 100$ points). Right panel shows how the paths are stretched along the time (β) axis. Particles 1 and 2 are in a pair exchange

If we look at the final analytical result for the high-temperature density matrix (13.8), we recognize the usual Boltzmann factor with some effective action in the exponent. This action describes two types of interaction. The first term,

$$\sum_{i=0}^{M-1} \frac{\pi}{\lambda_\Delta^2}(R_i - R_{i+1})^2 = \frac{\pi}{\lambda_\Delta^2} \sum_{j=1}^{N} \sum_{i=0}^{M-1} (\boldsymbol{r}_i^j - \boldsymbol{r}_{i+1}^j)^2 = \frac{k}{2} \sum_{j=1}^{N} \sum_{i=0}^{M-1} (\Delta r_{i,i+1}^j)^2 , \tag{13.9}$$

comes from the kinetic energy density matrices of free particles (j denotes summation over N particles, and i over M 'time slices'). This energy can be interpreted as the energy of a spring, $U = k(\Delta r)^2/2$. Changing one of the coordinates $\boldsymbol{r}_i^j$ at the time slice i is equivalent to a change of the spring energy of two nearest links, $U_i = k(\Delta r_{i-1,i}^j)^2/2$ and $U_{i+1} = k(\Delta r_{i,i+1}^j)^2/2$. These springs provide that the nearest points on the path are usually at some average distance proportional to λ_Δ. With temperature reduction the average size of the path increases with λ_Δ.

The second term $\Delta\beta V(R_i)$ in (13.8) adds interactions to the system (e.g. an external potential or inter-particle pair interaction)

$$\sum_{i=0}^{M-1} \Delta\beta V(R_i) = \Delta\beta\Big(\sum_{j=1}^{N} \sum_{i=0}^{M-1} V_{\text{ext}}(\boldsymbol{r}_i^j) + \sum_{j<k} \sum_{i=0}^{M-1} V_{\text{pair}}(\boldsymbol{r}_i^j, \boldsymbol{r}_i^k) \Big). \tag{13.10}$$

Each potential term depends only on the particle coordinates on the same time slice, i.e. $(\boldsymbol{r}_i^1, \boldsymbol{r}_i^2, \ldots, \boldsymbol{r}_i^N)$. As a result the number of pair interactions at each time slice, $N(N-1)/2$, is conserved.

In all expressions above we have considered the particles as distinguishable. The generalization to quantum particles obeying Fermi/Bose statistics is considered below, and discussed in more detail in [1, 3, 4, 5].

13.3 Basic Numerical Issues of PIMC

Having the general idea of the PIMC simulations we are ready to formulate the first list of important issues which we need to solve.

13.3.1 How to Sample Paths

It is necessary to explore the whole coordinate space for each intermediate point. This is very time consuming. To speed up convergence we move several slices (points of path) at once.

The key point is to sample a path using mid-points R_m and a consequent iteration (bisection), see Fig. 13.2(b).

With the definition: $0 < t < \beta, \tau = i_0 \Delta\beta\,[i_0 = 1, 2, 3, \ldots], R \equiv R(t), R' \equiv R(t + 2\tau), R_m \equiv R(t + \tau)$, the guiding rule to sample a mid-point R_m is

$$P(R_m) = \frac{\langle R|\mathrm{e}^{-\tau\widehat{H}}|R_m\rangle\langle R_m|\mathrm{e}^{-\tau\widehat{H}}|R'\rangle}{\langle R|\mathrm{e}^{-2\tau\widehat{H}}|R'\rangle} \approx (2\pi\sigma_\tau^2)^{-d/2}\mathrm{e}^{-(R_m-\overline{R})^2/2\sigma_\tau^2}\,, \tag{13.11}$$

where d is the spatial dimension of the system. In practice, we can neglect in the sampling distribution the potential energy and use only the ratio of the free-particle density matrices. As a result we get a Gaussian distribution with the mean $\overline{R} = (R + R')/2$ and the variance $\sigma_\tau^2 = \hbar^2\tau/2m$. This will lead to 100% acceptance of sampling for ideal systems (and close to one for a weakly interacting system).

For strongly interacting systems the overlap of the paths sampled from the free-particle distribution (13.11) results in large increase of the interaction energy and in a poor acceptance probability at the last level of the bisection sampling [1, 3]. This can be improved by using the optimized mean and the variance

$$\overline{R} = \frac{R + R'}{2} + \sigma_\tau \frac{\partial V(\overline{R})}{\partial R}\,, \quad \sigma_\tau^2 = \frac{\hbar^2\tau}{2m} + \left(\frac{\hbar^2\tau}{m}\right)^2 \Delta V(\overline{R})\,, \tag{13.12}$$

which also accounts for interaction between nearest neighbors (gradient of the potential energy).

The advantages of the bisection sampling method [1, 3] are:

- Detailed balance is satisfied at each level.
- We do not waste time on moves for which paths come close and the potential energy increases strongly (for repulsive interaction). Such configurations are rejected already at early steps.
- Computer time is spent more efficiently because we consider mainly configurations with high acceptance rate.
- The sampling of particle permutations is easy to perform.

13.3.2 Choice of the High-Temperature Density Matrix

From (13.8) we note, that the first free-particle terms can be considered as a weight over all possible random walks (Brownian random walks) in the imaginary time β with the ends points R and R'. In the limit $M \to \infty$ we directly obtain the Feynman-Kac relation

$$\rho(R, R'; \beta) = \rho_0(R, R'; \beta) \left\langle e^{-\int_0^\beta \mathrm{d}t\, V(R(t))} \right\rangle_{\mathrm{FK}} . \tag{13.13}$$

In the quasi-classical limit ($\beta \to 0$), only the classical path is important, $R_0(t) = (1 - t/\beta)\, R + tR'/\beta$, which leads to the semi-classical approximation of the high-temperature density matrix

$$\rho(R, R'; \Delta\beta) = \rho_0(R, R'; \Delta\beta) e^{-\int_0^{\Delta\beta} \mathrm{d}t\, V(R_0(t))} , \tag{13.14}$$

which is already much better compared to (13.7) with the substitution of classical (in many cases divergent) potentials.

For systems with pair interactions, in the limit of small $\Delta\beta$, the full density matrix (13.13) can be approximated by a product of pair density matrices

$$\left\langle e^{-\int_0^\beta \mathrm{d}t\, V(R(t))} \right\rangle_{\mathrm{FK}} \approx \prod_{j<k} \left\langle e^{-\int_0^\beta \mathrm{d}t\, V_{pair}[\boldsymbol{r}_j(t), \boldsymbol{r}_k(t)]} \right\rangle_{\mathrm{FK}} , \tag{13.15}$$

which is known as the pair approximation. It supposes that on the small time interval $\Delta\beta$ the correlations of two particles become independent from the surroundings. Different derivations of the effective pair potential (average on the r.h.s of (13.15)) have been proposed in the literature [6, 7, 8]. More accurate effective interaction potentials, which take into account two, three and higher order correlation effects, help to reduce the number of time slices by a factor of 10 or more.

The implementation of periodic boundary conditions leads to further modifications, see e.g. [9, 10, 11, 12, 13, 14, 15].

13.3.3 How to Calculate Physical Properties

There are different approaches for calculating expectation values of physical observables, such as the energy, momentum distribution, etc., which are called estimators. In each particular case convergence can be improved by the choice of a proper estimator. Consider, for example, the thermodynamical estimator of the internal energy

$$E = -\frac{\partial}{\partial\beta} (lnZ) = -\frac{1}{Z}\frac{\partial Z}{\partial\beta} = -\frac{1}{Z} \int \mathrm{d}R \frac{\partial \rho(R, R; \beta)}{\partial\beta} . \tag{13.16}$$

After direct substitution of (13.8), one obtains[4]

[4] This is only valid for particles with Boltzmann statistics. For fermions one has to include additional terms related to the β-derivative of the exchange determinant [1, 16, 17].

$$E = \frac{dMN}{2\beta} - \left\langle \sum_{i=0}^{M-1} \frac{Mm}{2\hbar^2\beta^2}(R_i - R_{i+1})^2 \right\rangle_{\rho(R;\beta)} + \left\langle \frac{1}{M}\sum_{i=0}^{M-1} V(R_i) \right\rangle_{\rho(R;\beta)} . \quad (13.17)$$

This form of the energy estimator has a much larger statistical variance σ_s compared to the virial estimator [18]. Since the statistical error in Monte Carlo simulations decreases as $\delta E \approx \sigma_s/\sqrt{N_{\mathrm{MC}}}$ (with N_{MC} being the number of MC-steps), with the direct estimator (13.17) one usually needs 2-4 times more MC runs to get the same accuracy as given by the virial estimator.

One of the approaches to obtain the virial estimator for the energy relies on the introduction of temperature dependent coordinates [16, 17], i.e. $\tilde{R}_i = R_0 + \lambda_\Delta \sum_{m=1}^{i} \xi^m$, $i = 1, \ldots, M-1$. Here ξ^i is a set of unit vectors, and R_0 is a set of particle coordinates at the zero time slice ($R_0 = R$). Once this has been done, the estimator takes the form [1]

$$E = \frac{dN}{2\beta} + \left\langle V(\tilde{R}) + \beta \frac{\partial V(\tilde{R})}{\partial \tilde{R}} \frac{\partial \tilde{R}}{\partial \beta} \right\rangle_{\rho(\tilde{R};\beta)} . \quad (13.18)$$

One can note at once, that for weakly interacting systems at high temperatures, the virial result (13.18) directly gives the classical kinetic energy (first term) and does not depend on the chosen number of time slices M, whereas using the direct estimator (13.17) we get this result by calculating the difference of two large terms which are diverging as $M \to \infty$.

13.3.4 Acceptance Ratio

When we try different kinds of moves in the Metropolis algorithm, it may happen that some moves will be frequently rejected or accepted. In both cases, we loose the efficiency of the algorithm. The system will be trapped in some local region of phase space for a long time (number of MC steps), and will not explore the whole space within reasonable computer time. In practice, the parameters of the moves are usually chosen to get an acceptance ratio of roughly 50%, which requires the construction of good apriori sampling distributions for the different kinds of PIMC moves (particle displacement, path deformation, permutation sampling).

A discussion of these topics, which is beyond the scope of this lecture, can be found in [1, 3].

13.3.5 Quantum Exchange – PIMC for Bosons and Fermions

Now we come to 'real' quantum particles. As we have already discussed, the properties of a system of N particles at a finite temperature T are determined by the density operator. Due to the Fermi/Bose statistics the total density matrix should be (anti)symmetric under arbitrary exchange of two identical particles (e.g. electrons, holes, with the same spin projection), i.e. we have to replace $\hat{\rho} \to \hat{\rho}^{\mathrm{A/S}}$ for fermions/bosons. As a result the full density matrix will be a superposition of all $N!$

permutations of N identical particles. Let us consider the case of two types (e,h) of particles with numbers N_e, N_h

$$\rho^{A/S}(R_e, R_h, R_e, R_h; \beta) = \frac{1}{N_e! N_h!} \sum_{P_e P_h} (\mp 1)^{P_e} (\mp 1)^{P_h} \rho(R_e, R_h, \widehat{P}_e R_e, \widehat{P}_h R_h; \beta) , \tag{13.19}$$

where $P_{e(h)}$ is the parity of a permutation (number of equivalent pair transpositions) and $\widehat{P}_{e(h)}$ the permutation operator. We directly see that for bosons all terms have a positive sign, while for fermions the sign of the prefactor alternates depending whether the permutation is even or odd.

In the last case a severe problem arises. The Metropolis algorithm gives the same distribution of permutations for both Fermi and Bose systems. The reason is that, for sampling permutations, we use the modulus of the off-diagonal density matrix, $|\rho(R, \widehat{P}R; \beta)|$ (implementation of the importance sampling in the Metropolis scheme). We find that:

- For *bosons* all permutations contribute with the same (positive) sign. Hence with the increase of the permutation statistics, accuracy in the calculation of the density matrix increases proportionally.
- For *fermions* positive and negative terms cancel almost completely (corresponding to even and odd permutations), since both are close in their absolute values. Accurate calculation of this small difference is hampered noticeably with the increase of quantum degeneracy (low T, high density). The consequences are large fluctuations in the computed averages. This is known as the fermion sign problem. It was shown [5] that the efficiency of the straightforward calculations scales like $\exp(-2N\beta\Delta F)$, where ΔF is the free energy difference per particle of the same fermionic and bosonic system, and N is the particle number.

13.3.6 Numerical Sampling of Permutations

Fermi and Bose statistics require sampling of permutations, see (13.19), in addition to the integrations in real space. From the $N!$ possibilities, we need to pick up a permutation which has a non-zero probability for a given particle configuration.

To realize a permutation we pick up two end-points $\{R_i, R_{i+i_0}\}$ along the β-axis with $i_0 = 2^{l-1}$ $(l = 1, 2, \ldots)$. Although the permutation operator $\widehat{P}$ in (13.19) acts on the last time-slice, $R_{e(h)}$, the permutation of the paths, $\{R_i, R_{i+i_0}\} \rightarrow \{R_i, \widehat{P}R_{i+i_0}\}$ can be carried out at any time slice because the operator $\widehat{P}$ commutes with the Hamiltonian. In a permutation (k permuted particles) the path coordinates between the fixed points R_i and R_{i+i_0} are removed and new paths connecting one particle to another (new k links) or a new path connecting a particle on itself (if a given particle undergoes the identity permutation) are sampled.

It is evident that a local permutation move consisting of a cyclic exchange of $k \geq 2$ neighboring particles will be more probable than a global exchange involving a macroscopic number of particles, and, in general, the probability of exchange will decrease with the increase of k. The most probable are local updates: Permutations of only few (2, 3, 4) particles. Moreover, any of the $N!$ permutations can be decomposed in a sequence of successive pair transpositions (two particle exchange), and we can explore the whole permutation space by making only local updates which have a high acceptance ratio.

In MC simulations we choose as the sampling probability of permutations

$$T(P \to P') = \frac{\rho_{\text{kin}}(R_i, \widehat{P}R_{i+i_0}; i_0\Delta\beta)}{\sum\limits_{P \in \Omega(P)} \rho_{\text{kin}}(R_i, \widehat{P}R_{i+i_0}; i_0\Delta\beta)}, \tag{13.20}$$

where we have used the product of the k one-particle density matrices

$$\rho_{\text{kin}}(R_i, \widehat{P}R_{i+i_0}, ; i_0\Delta\beta) \propto \mathrm{e}^{-\sum_{j\in k} \pi(\boldsymbol{r}_i^j - \widehat{P}\boldsymbol{r}_{i+i_0}^j)^2/(i_0\lambda_\Delta^2)} . \tag{13.21}$$

Here $\Omega(P)$ denotes the neighborhood of the current permutation P from which the permutation P' is sampled. For example, for the exchange of two particles, $\Omega(P)$ equals the number of neighbors of the given particle in the range of several De Broglie wavelengths, $\lambda(t)$, $t = i_0\Delta\beta$, because only these particles are possible candidates for the exchange.

To satisfy the detailed balance principle, we make a final decision[5] about the sampled permutation using the acceptance probability

$$A(P \to P') = \min\left[1, \frac{\sum\limits_{P \in \Omega(P)} \rho_{\text{kin}}(R_i, \widehat{P}R_{i+i_0}; i_0\Delta\beta)}{\sum\limits_{P' \in \Omega(P')} \rho_{\text{kin}}(R_i, \widehat{P}'R_{i+i_0}; i_0\Delta\beta)}\right], \tag{13.22}$$

where $\Omega(P')$ is the neighborhood of the new permutation P'. If the neighborhoods of the current and new permutation are equal, the acceptance probability is one.

As an illustration, in Fig. 13.2 we show a world line picture of five particles. Particle indices in Fig. 13.2 (a) and (c) are placed near the starting and end point of the particle trajectories. Hence, when the sequence of indices at $m = 0$ and $m = 100$ does not coincide the particles are permuted (see Figs. 13.2 (a) and (b)).

As we can see from Fig. 13.2(a) the paths of particles '1' and '2' are closed (two identical permutations), three other particles are in one cyclic exchange, and the whole permutation can be denoted as $\{1, 2, 5, 3, 4\}$ (as we can see the end of the path '3' coincides with the beginning of path '5', the end of path '4' coincides with the beginning of path '3' and the path '5' ends up at the starting position of path '4'. Now we decide to make a transposition between particles '1' and '4'. To do this we choose randomly time slices where new paths will be sampled. In our case it was $m = 17 - 33$. First, we exchange the edge points at the time slice $m = 33$, i.e. $\boldsymbol{r}_1'^{33} \equiv \boldsymbol{r}_4^{33}$ and $\boldsymbol{r}_4'^{33} \equiv \boldsymbol{r}_1^{33}$. Hence the position of the edge points is not sampled, they are a part of the unchanged trajectories. Once the initial and final points are chosen, we use the bisection algorithm to sample two paths connecting edge points, see Fig. 13.2(b).

[5] The sampled permutation can be rejected earlier when the new paths connecting R_i and $\widehat{P}R_{i+i_0}$ are sampled with the bisection algorithm [1, 3].

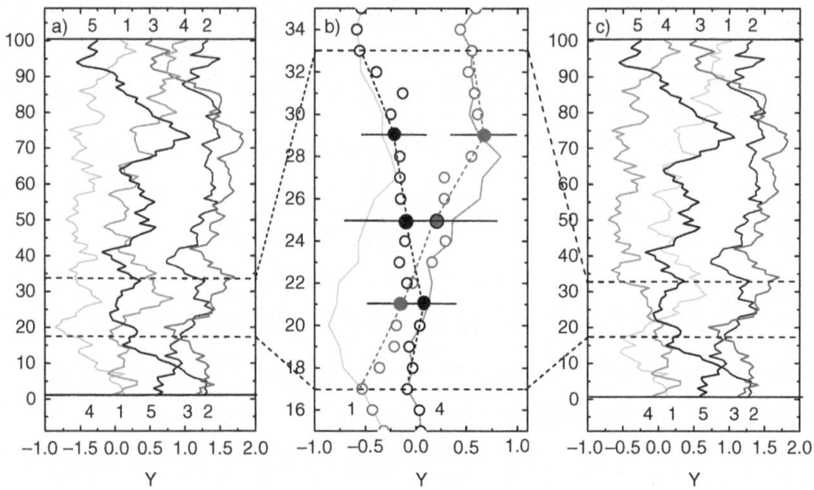

Fig. 13.2. (**a**),(**c**) The Y-coordinates of $N = 5$ identical particles as a function of the time-slice number m. Labels show particle indices. *Thick gray* and *light gray* lines show the paths of the particles '1' and '4' which are exchanged by sampling new paths at time-slices $m = 17 - 33$ (these time-slices are in the region between two *dashed lines*). (**b**) Sampling of new paths using the bisection algorithm for the particles '1' and '4'. The new paths are constrained at the time-slices $m = 17-33$. Old (new) paths are shown by lines (*circles*). The filled circles show two mid-points sampled at the level $l = 1$ (center of the interval, $m = 25$) and four other mid-points for sub-intervals $[17, 25]$ and $[25, 33]$ sampled at level $l = 2$. Open circles show final new paths for two particles obtained with the sampling at levels $l = 3, 4$ and the transposition, i.e. by exchanging the paths starting from $m = 33$ up to the end point, $m = 100$

13.4 PIMC for Degenerate Bose Systems

Currently much experimental activity is devoted to the study of ensembles of dilute gases of Bose atoms and optically excited indirect excitons in single/double well nanostructures (see e.g. [19, 20, 21] and references therein). The most exciting is certainly the possibility to observe signatures of Bose condensation and superfluidity. The essential point of these experiments is that the number of trapped atoms are limited to a few ten thousand particles and one should expect significant deviation (finite-size effects) from the macroscopic limit, leading e.g. to a 'softening' of the condensate fraction curve in the transition region and also to a shift of the critical temperature to lower values. This is particularly important for the case of a few hundreds of particles which become accessible to a direct theoretical investigation using quantum Monte Carlo approaches which allows to treat many-body correlation effects from 'first principles'.

In PIMC, as was shown by Feynman [2], the Bose statistics manifest itself as a special topology of the particle trajectories which can form macroscopically large permutation cycles. The free external parameters, like temperature, density, interaction strength, have a direct influence on these cycle distributions and, hence, on the superfluid and condensate fractions. Below we demonstrate how the latter can be easily related to the statistics of path configurations sampled by PIMC.

To be more specific in the discussion below we consider a system of trapped bosons with Coulomb interaction described by the Hamiltonian

$$\widehat{H} = \widehat{H}_0 + \sum_{i<j}^{N} \frac{e^2}{\varepsilon|\boldsymbol{r}_i - \boldsymbol{r}_j|^2}, \quad \widehat{H}_0 = \sum_{i=1}^{N} \left(-\frac{\hbar^2}{2m} \nabla^2_{\boldsymbol{r}_i} + \frac{m}{2} \omega^2 r_i^2 \right), \tag{13.23}$$

which can be also reduced to the dimensionless form (in the harmonic oscillator units)

$$\widetilde{H} = \frac{\widehat{H}}{\hbar\omega} = \frac{1}{2} \sum_{i=1}^{N} (-\nabla^2_{\widetilde{\boldsymbol{r}}_i} + \widetilde{r}_i^2) + \lambda \sum_{i<j}^{N} \frac{1}{\widetilde{r}_{ij}}, \tag{13.24}$$

using: $r \to \widetilde{r} = (r/l_0), E \to \widetilde{E} = (E/\hbar\omega)$ with $l_0^2 = \hbar/m\omega$. In this mesoscopic trapped system the density is controlled by the harmonic trap frequency ω and is characterized by the coupling parameter $\lambda = (e^2/\epsilon l_0)/(\hbar\omega) = l_0/a_B \propto \omega^{-1/2}$. In the limit $\lambda \to \infty$ the external potential vanishes, while for $\lambda \to 0$ the Coulomb interaction can be neglected (formal transition to non-interacting bosons).

13.4.1 Superfluidity

First, we consider the fraction of the superfluid (mass) density $\gamma_s = \rho_s/\rho$ which, within the Landau two-fluid model is computed from the classical and quantum momenta of inertia, I_c and I_q, according to $\gamma_s = 1 - I_q/I_c$ [22, 23]. The quantities I_c and I_q (corresponding to rotation along the Z-axis) are effectively computed in PIMC simulations [24] from the area enclosed by the particle paths $\mathbf{A}$, using

$$\frac{\rho_s}{\rho} = \frac{4m^2\langle A_z^2\rangle}{\hbar^2\beta I_{c,z}}, \quad \boldsymbol{A} = \frac{1}{2}\sum_{i=1}^{N}\sum_{k=0}^{M-1} \boldsymbol{r}_k^i \times \boldsymbol{r}_{k+1}^i,$$
$$I_{c,z} = \left\langle \sum_{i=1}^{N}\sum_{k=0}^{M-1} m_i \boldsymbol{r}_{k,\perp}^i \cdot \boldsymbol{r}_{k+1,\perp}^i \right\rangle, \tag{13.25}$$

where N is the particle number, M the number of time slices used in the path integral presentation, and $\langle \ldots \rangle$ denotes the thermal average with respect to the bosonic (symmetric) N-particle diagonal density matrix

$$\langle \ldots \rangle = \frac{1}{Z} \int\int \mathrm{d}\boldsymbol{r}_1 \mathrm{d}\boldsymbol{r}_2 \ldots \mathrm{d}\boldsymbol{r}_N \, (\ldots) \, \rho^S(\boldsymbol{r}_1, \boldsymbol{r}_2, \ldots, \boldsymbol{r}_N; \beta) \,. \tag{13.26}$$

This formula has been derived [24] for finite systems by assuming that particles are placed in an external field, e.g. in a rotating cylinder. Then one assumes that the system is put in a permanent slow rotation with the result that the normal component follows the imposed rotation while the superfluid part stays at rest. The effective moment of inertia is defined as the work required to rotate the system by a unit angle.

For macroscopic systems the path area formula (13.25) can be modified [3, 25]. Instead of a filled cylinder, one considers two cylinders with the radius R and spacing $\bar{d}$, where $\bar{d} \ll R$. Such a torus is topologically equivalent to the usual periodic boundary conditions. As a result we have: $I_c = mNR^2$ and $A_z = WR/2$, where W is the winding number, defined as the flux of paths winding around the torus and crossing any plane

$$\gamma_s = \frac{\rho_s}{\rho} = \frac{\langle W^2 \rangle}{2\lambda\beta N}, \quad \boldsymbol{W} = \sum_{i=1}^{N} \int_0^\beta \mathrm{d}t \left[\frac{\mathrm{d}\boldsymbol{r}_i(t)}{\mathrm{d}t} \right] . \tag{13.27}$$

13.4.2 Off-Diagonal Long-Range Order

The magnitude of off-diagonal long-range order is, in macroscopic systems, also directly accessible with PIMC. It is characterized by the asymptotic behavior of the single-particle off-diagonal density matrix

$$\rho(\boldsymbol{r}_1, \boldsymbol{r'}_1; \beta) = \frac{V}{Z} \int \mathrm{d}\boldsymbol{r}_2 \ldots \mathrm{d}\boldsymbol{r}_N \; \rho^{\mathrm{S}}(\boldsymbol{r}_1, \boldsymbol{r}_2, \ldots, \boldsymbol{r}_N, \boldsymbol{r'}_1, \boldsymbol{r}_2, \ldots, \boldsymbol{r}_N; \beta) \, , \tag{13.28}$$

$$n_0(\beta) = \lim_{\boldsymbol{r'}_1 \to \infty} \rho(\boldsymbol{r}_1, \boldsymbol{r'}_1; \beta) \, , \tag{13.29}$$

where n_0 is the fraction of particles in the condensate and V is the volume of the simulation cell. For a homogeneous isotropic system, $\rho(\boldsymbol{r}_1, \boldsymbol{r'}_1) = \rho(|\boldsymbol{r}_1 - \boldsymbol{r'}_1|)$ and, by taking the Fourier transform of an off-diagonal element, one obtains the momentum distribution

$$\rho(\boldsymbol{k}) = \frac{1}{(2\pi)^d} \int \mathrm{d}(\boldsymbol{r}_1 - \boldsymbol{r'}_1) \, \mathrm{e}^{-\mathrm{i}\boldsymbol{k}(\boldsymbol{r}_1 - \boldsymbol{r'}_1)} \, \rho(|\boldsymbol{r}_1 - \boldsymbol{r'}_1|; \beta) \, , \tag{13.30}$$

which shows a sharp increase at zero momentum when the temperature drops below the critical temperature T_c of Bose condensation.

Obviously, a finite trapped system of particles considered in real experiments behaves differently. The radial density is strongly inhomogeneous with the highest value at the trap center. However, these systems do represent an analog of the homogeneous macroscopic system in the angular direction (for traps with angular symmetry as in the case (13.23)). Hence, the macroscopic formulas (13.29) and (13.30) should be modified in an appropriate way and the corresponding momentum distribution, the condensate fraction and superfluidity acquire an additional dependence on the radial distance from the trap center.

As follows from (13.28), for the numerical evaluation of the single-particle density matrix one should allow that one of the N simulated particles has an open path, e.g. $\boldsymbol{r}_1 \neq \boldsymbol{r'}_1$. The paths of the other $N-1$ particles can close at their beginning (identical permutation) or at the start of another particle's path. The coordinates $\boldsymbol{r}_1, \boldsymbol{r'}_1$ are independent but their probability is given by the N-particle density matrix. In simulations we record a histogram (distribution) given by

$$\rho(\boldsymbol{r}, \boldsymbol{r}'; \beta) \propto \langle \delta(\boldsymbol{r}_1 - \boldsymbol{r})\delta(\boldsymbol{r'}_1 - \boldsymbol{r}')\rangle_W , \tag{13.31}$$

$$W = \rho^S(\boldsymbol{r}_1, \boldsymbol{r}_2, \ldots, \boldsymbol{r}_N, \boldsymbol{r'}_1, \boldsymbol{r}_2, \ldots, \boldsymbol{r}_N; \beta)/Z' , \tag{13.32}$$

(Z' is the normalization factor) which is then used to obtain the momentum distribution (13.30). The probability W is sampled using the path integral representation of ρ^S.

Recently a new method to sample the single-particle density matrix (13.28) has been proposed [26, 27]. It is based on generalization of the conventional PIMC to the grand canonical ensemble. A worm algorithm [26, 27] allows for a simultaneous sampling of both diagonal configurations contributing to the partition function and off-diagonal ones which contribute to the one-particle Matsubara Green function. The method has been recently applied to study of Bose condensation in crystalline ^{4}He and superfluidity in para-hydrogen droplets [28, 29], where high efficiency in sampling of long permutation cycles (practically unaffected by system size) and significantly improved convergence in the calculation of superfluid properties has been demonstrated.

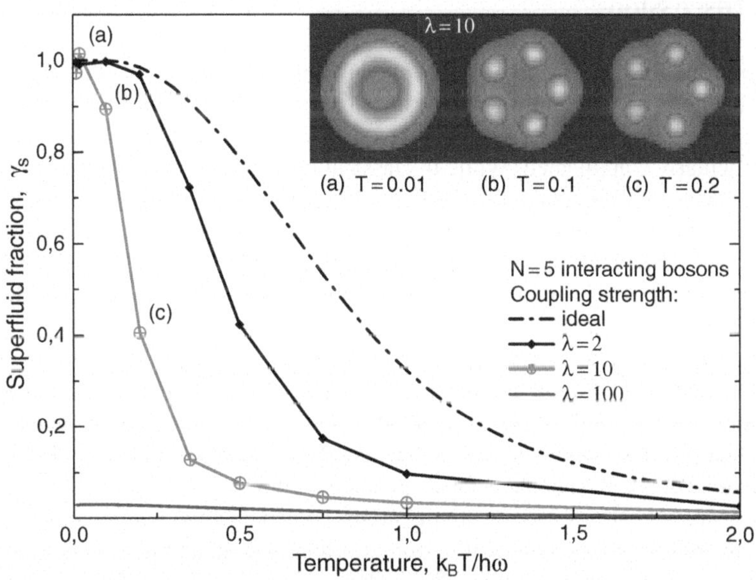

Fig. 13.3. Superfluid fraction for $N = 5$ charged bosons with Coulomb interaction in a two-dimensional harmonic trap (see the Hamiltonian (13.23)). Parameters are: $\lambda = 0, 2, 10, 100$, and temperature, $T = k_B T/\hbar\omega$. Symbols denote PIMC data (from [30]). *Dash-dotted line* displays an analytical result, $\gamma_s = 1 - I_q/I_c$, for the ideal harmonically confined bosons. The insets show the density distributions at $\lambda = 10$ for three temperatures

13.4.3 Example: Interacting Bosons

With the PIMC method it is possible to include inter-particle interactions like e.g. Coulomb repulsion (13.23) from first principles. The effective strength of the interaction can be controlled by the trap frequency and is measured by the parameter λ. As an illustration in Fig. 13.3 we present numerical results from PIMC simulations. Shown is the temperature dependence of the superfluid fraction for several values of the control parameter λ (the range $2 \leq \lambda \leq 10$ corresponds to typical particle densities in semiconductor quantum dots). The repulsive interaction causes a shift of the transition temperatures to lower values. When cooled down, the system typically forms a crystal like state in intermediate temperature regions until it melts into a ring like structure with delocalized particles, see insets in Fig. 13.3. Obviously, the latter shows a high superfluid response which is proportional to the ratio of the area enclosed by paths to the cross-section of the whole system (see (13.25)). In the ideal case, the system skips the intermediate crystal phase and directly reaches the delocalized state. In the case of dominating interaction strengths, the system stays highly localized even at absolute zero. Note, that even for the crystal phase the simulations yield a non vanishing value γ_s. This is a finite size effect because of a nonzero area ratio (13.25) (for details see [30]).

13.5 Discussion

We close this lecture with a few general comments. Quantum and classical Monte Carlo methods are currently actively developing computational tools for a basically exact treatment of many-body systems in equilibrium. Quantum simulations are particularly complicated: While in classical mechanics one only has to evaluate integrals over the Boltzmann distribution, in quantum mechanics one also needs to determine the quantum density matrix or, at low temperature, the wave function. The basis for the PIMC approach lies in the correspondence principle, which states that quantum mechanics reduces to classical mechanics in the limits of low density and high temperature.

The ability of quantum Monte Carlo methods (including PIMC) to provide an accurate treatment of quite a general class of model Hamiltonians has lead to applications in many fields of physics, including low-temperature degenerate plasmas, solid state physics, nanomaterials, collective effects in ultra-cold Bose and Fermi gases, molecules etc. This list is far from being complete.

Typical applications include neutral atoms cooled down to temperatures of several nano Kelvin, or systems with strong correlations, quantum effects in solids, melting or liquid-vapor transitions. Particularly interesting are the crystal formation of electrons or holes in bulk semiconductors [31] and quantum dots [32, 33], the superfluidity of dense ^{4}He in Vycor [28, 34], the equation of state, phase transitions and the phase diagram of hot, dense hydrogen [16, 17, 35, 36, 37].

In addition, there are calculations concerned with the superfluid transition of ^{4}He [38, 39]. Since ^{4}He is one of the simplest bosonic systems for experimentalists as well as for theoreticians, it has also been studied in inhomogeneous conditions, e.g. superfluidity in doped helium/hydrogen droplets [40]. Recently the superfluidity has been predicted also in small para-hydrogen clusters containing several tens of particles [29, 41, 42]. Interestingly, the superfluid fraction behaves in a quite non-monotic manner, depending on the cluster size.

On the other hand, electron-hole systems in semiconductors have been the source of interesting new physics for several decades. An electron and a hole can form abound state, the exciton, which is the neutral bosonic quasiparticle of a semiconductor. Formation of a Bose-Einstein condensate of excitons has been a target of many experiments [19, 20], though none have produced a clear proof of Bose condensation. Recent numerical PIMC studies [43, 44] of such systems support theoretical predictions of the possibility of exciton condensation.

As we have seen, PIMC is among the most general algorithms for quantum many-body systems. Nevertheless, a number of problems remain to be solved in the future. Among them are: a) the fermion sign problem, b) the large computational costs, which limit the simulations to system sizes of several hundred particles, c) the fast, efficient and more accurate calculation of spin and magnetic effects, d) information about excitation spectra, which could be obtained by PIMC using imaginary time correlation functions.

The field of application of PIMC is limited to systems in thermal equilibrium. The extension of this method to nonequilibrium is challenging. A possible approach is discussed in the lecture of V. Filinov et al., Chap. 2.

References

1. A. Filinov, M. Bonitz, in *Introduction to Computational Methods for Many Body Systems*, ed. by M. Bonitz, D. Semkat (Rinton Press, Princeton, 2006)
2. R. Feynman, A. Hibbs, *Quantum Mechanics and Path Integrals* (McGraw Hill, New York, 1965)
3. D. Ceperley, Rev. Mod. Phys. **67**, 279 (1995)
4. H. Kleinert, *Path Integrals in Quantum Mechanics, Statistics and Polymer Physics*, 2nd edn. (World Scientific, 1995)
5. D. Ceperley, in *Monte Carlo and Molecular Dynamics of Condensed Matter Systems*, ed. by K. Binder, G. Ciccotti (Editrice Compositori, Bologna, 1996)
6. W. Ebeling, H. Hoffmann, G. Kelbg, Contr. Plasma Phys. **7**, 233 (1967). And references therein
7. A. Filinov, V. Golubnychiy, M. Bonitz, W. Ebeling, J. Dufty, Phys. Rev. E **70**, 046411 (2004)
8. H. Kleinert, Phys. Rev. D **57**, 2264 (1998)
9. T. Gaskell, Proc. Phys. Soc. **77**, 1182 (1961)
10. T. Gaskell, Proc. Phys. Soc. **80**, 1091 (1962)
11. D. Ceperley, Phys. Rev. B **18**, 3126 (1978)
12. V. Natoli, D. Ceperley, J. Comput. Phys. **117**, 171 (1995)
13. C. Lin, F. Zong, D. Ceperley, Phys. Rev. E **64**, 016702 (2001)
14. P. Kent, R. Hood, A. Williamson, R. Needs, W. Foulkes, G. Rajagopal, Phys. Rev. B **59**, 1917 (1999)

15. A. Williamson, G. Rajagopal, R. Needs, L. Fraser, W. Foulkes, Y. Wang, M.Y. Chou, Phys. Rev. B **55**, R4851
16. V. Filinov, M. Bonitz, W. Ebeling, V. Fortov, Plasma Physics and Controlled Fusion **43**, 743 (2001)
17. V. Filinov, M. Bonitz, D. Kremp, W. Kraeft, W. Ebeling, P. Levashov, V. Fortov, Contrib. Plasma Phys. **41**, 135 (2001)
18. M. Herman, E. Bruskin, , B. Berne, J.Chem.Phys. **76**, 5150 (1982)
19. V. Negoita, D. Snoke, K. Eberl, Phys. Rev. B **60**, 2661 (1999)
20. L. Butov, Phys. Rev. Lett. **86**, 5608 (2001)
21. E. Kim, M. Chan, Nature **427**, 225 (2004)
22. L. Landau, J. Phys. USSR **5**, 71 (1941)
23. E. Andronikashvili, J. Phys. USSR **10**, 201 (1946)
24. P. Sindzingre, M. Klein, D. Ceperley, Phys. Rev. Lett. **63**, 1601 (1981)
25. E. Pollock, D. Ceperley, Phys. Rev. B **36**, 8343 (1987)
26. M. Boninsegni, N. Prokof'ev, B. Svistunov, Phys. Rev. Lett. **96**, 070601 (2006)
27. M. Boninsegni, N. Prokof'ev, B. Svistunov, Phys. Rev. E **74**, 036701 (2006)
28. M. Boninsegni, N. Prokof'ev, B. Svistunov, Phys. Rev. Lett. **96**, 105301 (2006)
29. F. Mezzacapo, M. Boninsegni, Phys. Rev. Lett. **97**, 045301 (2006)
30. J. Böning, Superfluidity in mesoscopic systems of charged bosons. Master's thesis, Christian-Albrechts Universität, Kiel (2007)
31. M. Bonitz, V. Filinov, V. Fortov, P. Levashov, H. Fehske, Phys. Rev. Lett. **95**, 23500 (2005)
32. R. Egger, W. Häusler, C. Mak, H. Grabert, Phys. Rev. Lett. **82**, 3320 (1999)
33. A. Filinov, M. Bonitz, Y. Lozovik, Phys. Rev. Lett. **86**, 3851 (2001)
34. S. Khairallah, D. Ceperley, Phys. Rev. Lett **95**, 185301 (2005)
35. W. Magro, B. Militzer, D. Ceperley, B. Bernu, C. Pierleoni, in *Strongly Coupled Coulomb Systems*, ed. by G.J. Kalman, J.M. Rommel, K. Blagoev (Plenum Press, New York, 1998)
36. B. Militzer, E. Pollock, Phys. Rev. E **61**, 3470 (2000)
37. B. Militzer, D. Ceperley, Phys.Rev. E **63**, 66404 (2001)
38. D. Ceperley, E. Pollock, Phys. Rev. Lett. **56**, 351 (1986)
39. M. Gordillo, D. Ceperley, Phys. Rev. B **58**, 6447 (1998)
40. E. Draeger, D. Ceperley, Phys. Rev. Lett. **90**, 065301 (2003)
41. P. Sindzingre, D.M. Ceperley, M.L. Klein, Phys. Rev. Lett. **67**, 1871 (1991)
42. F. Mezzacapo, M. Boninsegni, Phys. Rev. A **75**, 033201 (2007)
43. J. Shumway, D. Ceperley, Solid State Comm. **134**, 1922 (2005)
44. A. Filinov, M. Bonitz, P. Ludwig, Y. Lozovik, phys. stat. sol. (c) **3**, 2457 (2006)

Part VI

Ab-Initio Methods in Physics and Chemistry

14 Ab-Initio Approach to the Many-Electron Problem

Alexander Quandt

Institut für Physik, Universität Greifswald, 17487 Greifswald, Germany

The chemical and physical properties of solids, molecules and nanomaterials depend on a subtle interplay of the spatial arrangement of the ions and the resulting distribution and density of electrons, which provide the binding forces of the system. Predicting the structure and the properties of novel materials, e.g., nanosystems, therefore is impossible without falling back on the elementary interactions and the most accurate ab initio methods for their simulation. We give a survey of the most popular ab initio methods used by quantum chemists, and describe some important modifications that made those methods available for the study of complex nanomaterials of moderate size.

14.1 Introduction

The term *ab initio*[1] refers to a family of theoretical concepts and computational methods that literally treat the many-electron problem *from the beginning*. In other words, these methods start from the exact (non-relativistic) many-body Hamiltonian of an atomic, molecular or solid system comprising M atoms and N electrons. In a strict sense, the one and only approximation ever made will be the Born-Oppenheimer approximation [1], where one assumes that the electronic and nuclear time scales effectively decouple. Then one may freeze the nuclear degrees of freedom $R \equiv \{\boldsymbol{R}_1 \dots \boldsymbol{R}_M\}$ and solve the corresponding Schrödinger equation for a many electron wavefunction Ψ that will explicitly depend on the electronic degrees of freedom $r = \{\boldsymbol{r}_1 \dots \boldsymbol{r}_N\}$, only. Therefore in the framework of the Born-Oppenheimer approximation, the exact many-electron Hamiltonian will be (in atomic units, see [2]):

$$H(r, R) \equiv -\sum_i^N \frac{1}{2}\Delta_{\boldsymbol{r}_i} - \sum_i^N \sum_\alpha^M \frac{Z_\alpha}{|\boldsymbol{r}_i - \boldsymbol{R}_\alpha|} + \sum_{i<j}^N \frac{1}{|\boldsymbol{r}_i - \boldsymbol{r}_j|} + \sum_{\alpha<\beta}^M \frac{Z_\alpha Z_\beta}{|\boldsymbol{R}_\alpha - \boldsymbol{R}_\beta|} \, .$$

Here, the first term denotes the operator of the kinetic electronic energies, the second term refers to the various attractive electron-nucleus interactions, the third term describes the various electron-electron repulsions, and the final term describes the repulsions between the various nuclei of the system.

[1] Latin: ab, *from* + initio, ablative of initium, *beginning*.

A. Quandt: *Ab-Initio Approach to the Many-Electron Problem*, Lect. Notes Phys. **739**, 415–436 (2008)
DOI 10.1007/978-3-540-74686-7_14

In order to determine the corresponding (ground-state) many-electron wavefunction $\Psi(r, m)$, where $m \equiv \{m_1 \ldots m_N\}$ stands for a set of electronic spin variables, which is the solution of a Schrödinger equation with the Hamiltonian of (14.1), one usually falls back on a related variational principle:

$$E_0(R) = \min_{\Psi(r,m)} \frac{\langle \Psi(r, m)|H(r, R)|\Psi(r, m)\rangle}{\langle \Psi(r, m)|\Psi(r, m)\rangle} . \tag{14.1}$$

In general, the resulting electronic wavefunction $\Psi(r, m)$ will be some approximation to the real antisymmetric wavefunction of the corresponding many-electron system. And the art of ab initio will simply consist of finding ingenious ways to numerically determine an approximate wavefunctions. The ultimate goal of course must be chemical accuracy, which turns out to be one of the most formidable challenges of applied computational sciences.

But before going into the details of the numerical machineries related to quantum chemistry, it should be emphasized that textbooks like Coulson's *Valence* [3] and Pauling's *Nature of the Chemical Bond* [4]), which shaped our modern picture of the chemical bond, originate from a time, when modern ab initio methods just started to grow up, and accurate ab initio calculations beyond simple systems comprising a few electrons were just not feasible. How is it possible that these books are still valid? The answer must be sought in the general validity of an orbital picture of the chemical bond, and therefore we devoted the whole Sect. 14.2 to illustrate this important concept.

Starting from this orbital picture, one may systematically arrive at a more detailed description of the chemical bond. Along these lines, we will derive the Hartree-Fock method and some of its extensions in Sect. 14.3, and density functional theory in Sect. 14.4. We will see that the orbital picture and all of the related concepts form Sect. 14.2 remain valid, which will allow for a rather intuitive interpretation of chemical bonding in molecules and in solids.

Once one can assure chemical accuracy in determining the electronic components of a certain molecular or solid system for any given nuclear configuration, then all of the questions related to chemical stability and chemical reactivity will just boil down to a detailed analysis of the corresponding total energy hypersurfaces. Section 15.1 will give an introduction to this concept, and there we will also present some ingenious ways to step over those energy hypersurfaces, in search for chemical isomers and the products and educts of chemical reactions. Then in Sect. 15.2 we will illustrate how such concepts are applied in practice. To this end, Sect. 15.2 will contain a short recapitulation of a series of recent theoretical and experimental studies that finally lead to the discovery of a whole new class of boron based nanomaterials. Interesting enough, these discoveries were anticipated by ab initio calculations, rather than being the result of a benevolent laboratory ghost.

However, with the modelling and the prediction of novel nanostructured materials, we are already hitting the limits for the applicability of modern ab initio methods. Nevertheless the theory of the physical and chemical properties of nanomaterials and other complex materials may be extended to rather large nanosystems, by using suitably parameterized model Hamiltonians, where the parameterization will be fitted to ab initio calculations or experimental data. Some important strategies to set up model Hamiltonians will be presented in Sect. 15.3, which is based on original work by O. Gunnarsson (MPI FKF Stuttgart), who kindly made his lecture notes and various illustrations available for this purpose. In Sect. 15.3 we will also indicate how to built a consistent theoretical picture of important physical processes within complex materials by using such model Hamiltonians. Finally in Sect. 15.4 we will give a short summary and dare to make a bet on "the shape of things to come" (H.G. Wells).

The possibility to predict novel materials and to draw a consistent picture of their basic physical and chemical properties is quite remarkable, compared to the first applications of ab initio methods, which just aimed at an accurate description of simple one- and two-electron systems [5]. But the predictive power of modern ab initio methods turns out to be the product of decades of intense research, driven by at least *three* equally important developments.

First of all there was a rather dramatic increase in the hardware capacities of modern computing systems, due to a continuous down-scaling of microelectronic devices. The importance of increasing computing facilities for modern ab initio methods is directly connected to a matrix representation of the one-electron problem described in Sect. 14.2.2. In brief, our current inability to treat systems as large as complex protein structures is mainly due to the fact that we cannot store and handle matrices beyond a certain size.

The necessary increase of storage and computing facilities by a continuous down-scaling of microelectronic devices involves a number of serious technological challenges [6], and one might wonder whether Moore's law, which predicts an exponential growth of computing power, may still hold in the future, when microelectronic devices smaller than the sub-lithographic range of about 40 nm must be produced. But a closer look back into the past reveals that Moore's law was even holding long before the silicon era, see Fig. 14.1. Thus, there is some hope

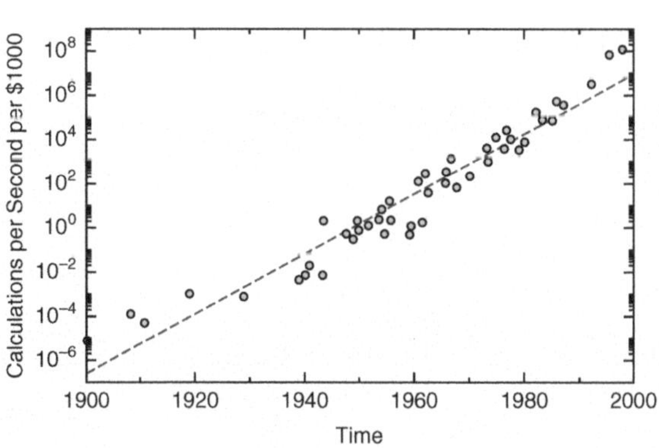

Fig. 14.1. Moore's law predicts an exponential growth of computing power, which obviously extends over various technologies (electromechanical: 1900–1935, relays: 1934–1940, vacuum tubes: 1940–1960, transistors: 1960–1970, integrated circuits: since 1970), see [7]

that similar shifts in fabrication technologies and distributed computing will extend Moore's law even into the far future [7], and that these developments will provide us with increasingly powerful computational platforms for future ab initio simulations.

A second important factor for the dramatic progress of ab initio methods were several algorithmic breakthroughs, which considerably boosted the performance of modern ab initio program packages. In order to understand the strong dependence of modern ab initio codes from the development of powerful numerical algorithms, we listed some of the most popular algorithms in Tab. 14.1. This listing was taken from a recent effort to identify the top ten algorithms of the 20th century [8]. It comes as no big surprise that the vast majority of these algorithms are actually forming key elements of modern ab initio codes, and progress along these lines implies progress in the computational performance of ab initio codes. Probably, the top ten algorithms of the 21st century will also make their way into future ab initio codes.

The third important factor for the progress of ab initio methods were theoretical and conceptional breakthroughs in applying the variational principle described by (14.1). Nowadays chemical accuracy may routinely be achieved for system sizes that imply hundreds of atoms and electrons, and these developments turned out to be so useful for our current understanding of molecular and solid systems, that the 1998 Nobel prize in Chemistry was given to some of the protagonists in the field of ab initio methods, Walter Kohn and John A. Pople. We will describe some of their achievements in Sect. 14.3 and 14.4, but in order to get a more detailed impression about their pioneering work, we recommend the study of some decent textbooks, for example [2, 9, 10, 11] or [12].

Table 14.1. Top ten algorithms of the 20th century, after [8]

Algorithm	author(s)	year
Monte Carlo method	J. v. Neumann, S. Ulam, N. Metropolis	1946
Simplex method for linear programming	G. Danzig	1947
Krylov subspace iteration method	M. Hestenes, E. Stiefel, C. Lanczos	1950
Decompositional approach to matrix computations	A. Householder	1951
Fortran optimizing compiler	J. Backus	1957
QR algorithm	J.G.F. Francis	1961
Quicksort	T. Hoare	1962
Fast Fourier transform	J. Cooley, J. Tuckey	1965
Integer relation detection algorithm	H. Ferguson, R. Forcade	1977
Fast multipole algorithm	L. Greengard, V. Rokhlin	1987

Let us finally emphasize that these lecture notes are meant to be tutorial in the first place, and to provide the reader with some sort of jump start concerning modern ab initio methods. Therefore these notes are no substitute for an extended review article about ab initio methods, and the interested reader is asked to consult further literature in order to get a more detailed picture of the vast field of modern ab initio methods. Beyond that, knowledge usually comes with practice, and we really want to encourage the reader to get some practical experience with modern ab initio methods, for example after implementing and running some of the program packages listed in Sect. 15.A.

14.2 An Orbital Approach to Chemistry

Once the variational principle of (14.1) can be solved with sufficient accuracy, the structural and chemical properties of molecular or solid systems may be predicted quite reliably. But as we said before, there was already a profound understanding of the structure and the properties of materials (see [3] or [4]) at a time when numerical calculations were just restricted to atoms, small molecules and simple solids. The basic reason why theoretical chemistry was able to advance for a long time without the help of modern ab initio simulations, is due to the famous observation by Lewis [13], that there is some sort of orbital picture underlying chemistry. Armed with this rather intuitive concept, chemists then went on and combined early orbital based theoretical concepts with experimental investigations. One of the classical examples for such a strategy is Lipscomb's discovery of multicentered bonding in the boranes [14].

Nowadays many of the rather crude or hand waving concepts that appeared in the early years of quantum chemistry and computational materials science may be put on a solid basis using ab initio calculations. For example, quantum chemists have developed schemes to analyze in great detail the nature of chemical bonding in molecular systems, based on the concept of Natural Bond Orbitals [15, 16]. For solid systems, there are similar approaches based on a tight-binding picture of the chemical bond [17, 18]. This tight-binding approach also turns out to be helpful in deriving analytical models to describe the physical properties of solids, where the essential model parameters will be taken from ab initio methods, rather than experiment (see Sects. 15.3 and [19]).

In the following, we will give a short survey of orbital theory, and introduce some important theoretical concepts related to this approach. Orbital theory will reappear in Sects. 14.3 and 14.4, where we discuss the basic ideas of modern ab initio methods.

14.2.1 The Lewis Picture of the Chemical Bond

It is very likely that already during high school, your chemistry teacher might have introduced you into the language of Lewis-diagrams, just like the ones shown in Fig. 14.2. And after some time, you might have even learned to check chemical

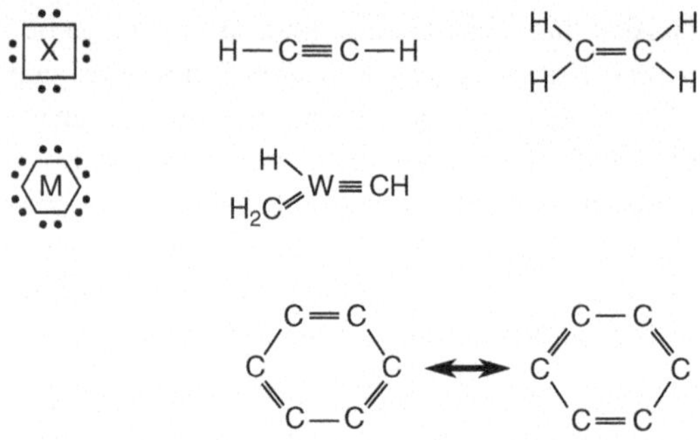

Fig. 14.2. Lewis diagrams. (**a**) Octet rule for main row elements and some examples. (**b**) Dodectet rule for d-block elements and a simple example. (**c**) Resonance effects stabilizing a π-electron system

structures by carefully counting electrons from one to eight. But it is also very likely that someone at the university finally told you that it is all rubbish. Well, the next sections will show you that even the most simple minded Lewis picture of the chemical bond is not that far off the truth.

14.2.1.1 Chemistry with a Pencil

Let us have a closer look at Fig. 14.2. Under (a), we find a rather suggestive representation of the famous octet rule, which tells you that main row elements bind over localized electrons pairs, and in a way that all main row elements involved in chemical binding may be able to completely fill up their valence shells $(s+3p)$ with shared electrons. There is a similar rule for d-block elements shown in (b), where the valence shell comprises six orbitals $(s+5d)$, which leads to a dodectet rule [16].

A single Lewis diagram of course is a very localized description of the chemical bond, and in most cases, there is additional stabilization through delocalization effects. In the classical resonance picture of the chemical bond [4], delocalized bonding will be represented by a series of resonance structures, as indicated in Fig. 14.2(c) for the well-known case of the delocalized π-electron system of benzene. The real π-electron wavefunctions will be superpositions of these resonance structures, such that all carbon-carbon bonds in Fig. 14.2(c) will turn out to be equal.

14.2.1.2 Donor-Acceptor Interactions

It is possible to translate the language of Lewis-diagrams into a purely quantum mechanical picture using ab initio methods [16]. Here we just want to argue on the basis of a simple model system. Suppose $\phi_i(\boldsymbol{r})$ to be the localized eigenfunction (orbital) of a Hermitian one-electron operator $f_{\mathrm{loc}}(\boldsymbol{r})$:

$$f_{\mathrm{loc}}(\boldsymbol{r})\phi_i(\boldsymbol{r}) = \epsilon_i\phi_i(\boldsymbol{r}) \; . \tag{14.2}$$

According to the Pauli principle, every orbital could be filled with two electrons, and therefore we may consider the doubly occupied orbital solutions of (14.2) to correspond to some localized electron pairs in the Lewis diagrams, to be located around the atomic cores. Now let us assume that the influence of the nearby ionic cores and electrons may be described by the addition of a perturbation term $f_{\text{pert}}(\boldsymbol{r})$. Then according to second order perturbation theory, we obtain the following results:

$$\phi_i^{(1)}(\boldsymbol{r}) = \sum_{j \neq i} \frac{\langle \phi_j | f_{\text{pert}} | \phi_i \rangle}{\epsilon_i - \epsilon_j} \phi_j(\boldsymbol{r}) \, , \tag{14.3}$$

$$E_i^{(1)} = \langle \phi_i | f_{\text{pert}} | \phi_i \rangle \, , \tag{14.4}$$

$$E_i^{(2)} = \sum_{j \neq i} \frac{|\langle \phi_i | f_{\text{pert}} | \phi_j \rangle|^2}{\epsilon_i - \epsilon_j} \, . \tag{14.5}$$

These equations have some interesting interpretation. First of all (14.3) tells us that under the influence of a perturbing environment, our localized orbitals will mix and form delocalized orbitals. Equation (14.4) is a simple energy shift lacking any further interpretation. But (14.4) contains a lot of chemistry. Here, the expression for the second order energy correction involves a sum of terms that become negative and rather large (i.e. bonding), whenever there is a strong interaction $\langle \phi_i | f_{\text{pert}} | \phi_j \rangle$ between orbitals i and j that are close in energy, and $\epsilon_i < \epsilon_j$. Therefore the system usually stabilizes through one or just a few bonding contributions, which correspond to a specific energetic scenario indicated in Fig. 14.3.

The latter diagram also has some chemical interpretation. The occupied orbital i of energy ϵ_i is strongly interacting with a nearby unoccupied orbital j^* of energy $\epsilon_{j^*} > \epsilon_i$. According to Lewis [13], the occupied orbital i is an electron pair donor (Lewis base), whereas the unoccupied orbital j^* is an electron pair acceptor (Lewis acid). The strong interaction between the donor orbital and the acceptor orbital leads to the formation of a delocalized bonding orbital, which is lower in energy by an amount:

$$E_{i \to j^*}^{(2)} = -2 \frac{|\langle \phi_i | f_{\text{pert}} | \phi_{j^*} \rangle|^2}{\epsilon_{j^*} - \epsilon_i} \, . \tag{14.6}$$

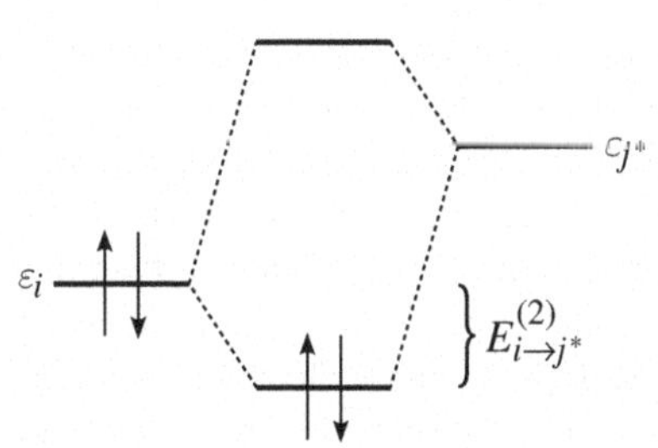

Fig. 14.3. Donor-acceptor interaction between a doubly occupied orbital i and an unoccupied orbital j^*, forming a new bonding orbital stabilized by an energy $E_{i \to j^*}^{(2)}$

This delocalized bonding orbital will be filled by the electron pair that originally occupied the localized orbital i.

Therefore, given the validity of a one-electron picture, where every (localized) electron is only slightly perturbed by a local interaction $f_{\text{pert}}(\boldsymbol{r})$ corresponding to the averaged influence of the environment, the chemical bonding will largely be dominated by donor-acceptor interactions of the type shown in Fig. 14.3. And this seems to be the standard scenario of quantum chemistry.

14.2.2 One-Electron Hamiltonians

Now that we understood the importance of the one-electron approach to chemistry, we may ask ourselves how to find a numerical solution to a general one-electron Schrödinger equation similar to (14.2). But we already know that for a proper description of the chemical bond, there must be some perturbing local interaction $f_{\text{pert}}(\boldsymbol{r})$ with the molecular or solid environment, and this interaction will lead to a mixing of the localized orbitals related to $f_{\text{loc}}(\boldsymbol{r})$, see (14.3).

Based on this knowledge, we will better forget about our previous perturbative approach. Instead, we will try to solve the one-electron Schrödinger equation for a more realistic one-particle operator $f(\boldsymbol{r}) = f_{\text{loc}}(\boldsymbol{r}) + f_{\text{pert}}(\boldsymbol{r})$. Furthermore we will not even try to determine the proper eigenstates of $f_{\text{loc}}(\boldsymbol{r})$, but instead we will expand the eigenstates $\phi_i(\boldsymbol{r})$ of $f(\boldsymbol{r})$ in a suitable set of basis functions $\varphi_\mu(\boldsymbol{r})$:

$$\phi_i(\boldsymbol{r}) = \sum_\mu C_{\mu i}\varphi_\mu(\boldsymbol{r}) \,. \tag{14.7}$$

14.2.2.1 Basis Functions

From a chemical point of view, the basis functions $\varphi(\boldsymbol{r})$ should either be chosen such that they mimic localized electronic states, for example the eigenstates of a single atom (atomic orbitals). Or the basis functions should have some important physical or chemical properties in common with the one-electron states they should describe, for example the periodicity of electron wavefunctions in a solid.

Beyond that, the basis functions also serve some numerical purpose, namely the transformation of a Schrödinger equation similar to (14.2) into a generalized matrix eigenvalue problem to be discussed below. Then the criteria must be that the numerical algebra related to these basis functions should be as simple as possible.

Consequently the basis functions φ_μ neither have to be orthogonal, nor do they really have to correspond to any known $f_{\text{loc}}(\boldsymbol{r})$. Instead, for the usual one-electron system encountered in quantum chemistry or solid state physics, it will be most important to pick a set of basis functions of the right physical shape. This chosen basis set must be large enough to mimic electrons in a realistic fashion, but at the same time small enough to keep the related matrix eigenvalue problem manageable. Some of the most popular choices for basis functions are:

$$\varphi(\boldsymbol{r}) = \mathrm{e}^{-\alpha r^2} \tag{14.8}$$

$$\varphi(\boldsymbol{r}) = x^l y^m z^n \mathrm{e}^{-\eta r} \tag{14.9}$$

$$\varphi(\boldsymbol{r}) = \frac{1}{\sqrt{V}} \mathrm{e}^{\mathrm{i}\boldsymbol{k}\cdot\boldsymbol{r}} \tag{14.10}$$

$$\varphi(\boldsymbol{r}) = \begin{cases} \frac{1}{\sqrt{V}} \sum_{\boldsymbol{G}} c_{\boldsymbol{G}}\, \mathrm{e}^{[\mathrm{i}(\boldsymbol{G}+\boldsymbol{k})\cdot\boldsymbol{r}]} & \boldsymbol{r} \in I \\ \sum_{lm} A_{lm} u_l(r) Y_{lm}(\theta, \phi) & \boldsymbol{r} \in S \end{cases} . \tag{14.11}$$

Within quantum chemistry, the most popular choices are the Gauss-type orbitals (GTO) of (14.8). Their algebra is well understood [2], and the corresponding basis sets have been optimized by generations of quantum chemists. Although the atomic states seem to be more similar to the Slater-type orbitals (STO) of (14.9), it turns out that the STOs may be well approximated by a suitable fixed linear combination of GTOs [2].

For solids, the simplest type of basis functions are the plane waves (PW) of (14.10). There are various numerical advantages in using such a basis set, in particular in the framework of Car-Parrinello molecular dynamics [20]. The algebra related to the PWs is extremely simple, and the basic numerics can be carried out quite effectively using FFT routines [21].

The basis functions of (14.11) are augmented planewaves (APW), which go back to Slater [22]. These functions are designed to meet the special bonding situations in (closely packed) solids. Inside a sphere S near the nucleus, the potential will be nearly spherically symmetric and similar to the potential of a single atom, whereas in the interstitial region I, the potential will be almost constant. Both parts of the corresponding electronic wavefunction in (14.11) have to match on the surface of S. Using some clever approximations [23], all of these requirements can be met in the framework of the linearized augmented planewave method (LAPW), where the determination of eigenstates based on APWs may again be reduced to a standard generalized matrix eigenvalue problem [21].

Other interesting basis sets are Muffin-Tin orbitals [24], Wannier functions [25], or wavelets [26].

14.2.2.2 Matrix Equations

In the last paragraph we saw that each type of basis function $\varphi_\mu(\boldsymbol{r})$ seems to require its own individual type of algebra. But in the end, the general problem of solving the one-electron Schrödinger equation

$$f(\boldsymbol{r})\phi_i(\boldsymbol{r}) = \epsilon_i \phi_i(\boldsymbol{r}) \quad \text{with} \quad \phi_i(\boldsymbol{r}) = \sum_\nu C_{\nu i} \varphi_\nu(\boldsymbol{r}) , \tag{14.12}$$

will always boil down to a generalized matrix eigenvalue problem by sandwiching (14.12) with a basis function $\varphi_\nu(\boldsymbol{r})$:

$$\sum_{\nu} F_{\mu\nu} C_{\nu i} = \epsilon_i \sum_{\nu} S_{\mu\nu} C_{\nu i}$$
$$S_{\mu\nu} \equiv \langle \varphi_\mu | \varphi_\nu \rangle$$
$$F_{\mu\nu} \equiv \langle \varphi_\mu | f | \varphi_\nu \rangle$$
$$\Longrightarrow \mathbf{FC} = \mathbf{SCE} \quad \text{(using matrix notation)} . \tag{14.13}$$

Here the matrix elements $F_{\mu\nu}$ are a measure for the interaction strength between two orbitals, and the matrix elements $S_{\mu\nu}$ are a measure for their mutual overlap. The coefficient matrix $C_{\nu i}$ and the diagonal matrix $\epsilon_i \delta_{ij}$ are to be determined by numerically solving the generalized matrix eigenvalue problem.

In principle (14.13) should be a standard numerical task. There exists a bulk of profound literature dealing with such problems (see e.g. [27]), and there are quite powerful program packages to tackle such problems, see `http://www.netlib.org/lapack/` or [28]. Nevertheless, as we will see in the next section, the generalized matrix eigenvalue problem related to the one-electron problem turns out to be rather special. And therefore even the most powerful solvers, which are designed to tackle the most general cases, may not really be the method of choice for solving this problem.

14.2.3 Some Useful Simplifications

When blindly setting up the eigenvalue problem of (14.13), we may actually be overdoing things for at least two good reasons. First of all, it is very unlikely that electrons located at opposite sides of a larger molecule would still be interacting with each other. Second, it is well known that only the valence electrons really participate in the chemical bonding, whereas the core electrons remain in orbitals that are practically indistinguishable from their atomic counterparts. In the following, we will present some general approximation schemes that will take these issues into account.

14.2.3.1 The Tight-Binding Approximation

It will be another textbook wisdom that the overlap between bonding atomic orbitals is supposed to be a measure for the strength of that bond (principle of maximum overlap). In a more scientific language, we may put it that way:

$$F_{\mu\nu} \approx K \langle \varphi_\mu | \varphi_\nu \rangle = K S_{\mu\nu} . \tag{14.14}$$

Most basis functions will decay rather quickly away from the centers where they are located, and this means that quantum chemistry is a rather near-sighted business, where the matrices $F_{\mu\nu}$ and $S_{\mu\nu}$ may be thinned out considerably. In the end (14.13) will be a rather sparse matrix eigenvalue problem, and even some (over-)simplified versions of (14.13) might be of considerable theoretical interest. Let us have a look at the following approximations:

$$\sum_{\nu} F_{\mu\nu} C_{\nu i} = \epsilon_i \sum_{\nu} S_{\mu\nu} C_{\nu i} \, , \tag{14.15}$$

with $S_{\mu\nu} = 0$ except when φ_μ and φ_ν are located within nearest-neighbor distance, or $S_{\mu\nu} = \delta_{\mu\nu}$ (Hückel-type approach), $F_{\mu\nu} = 0$ except when φ_μ and φ_ν are located within nearest-neighbor distance. These equations are the essence of the tight-binding approximation, where all the contributions to $F_{\mu\nu}$ and $S_{\mu\nu}$ are zero, except those that involve basis functions, which are located at neighboring sites. In such a case, the interactions between valence orbitals located on neighboring atoms become somewhat standardized and may be tabulated. Furthermore these tight-binding models are also a perfect starting point for a series of simple, but rather powerful analytical models in solid state physics, see Sect. 15.3 and [19].

14.2.3.2 Pseudopotentials

The idea behind the pseudopotential approach may easily be stated in a few sentences. As we already mentioned before, only the valence electrons are contributing to the chemical bond. Therefore it would be best to substitute all of the core electrons by a pseudopotential, which weakens the original potential within the core region. This would lead to a pseudo-wavefunction χ_{v} for the valence electrons, which would be much smoother inside the core region than the real valence wavefunction ϕ_{v}, which wiggles around much faster, due to some orthogonality constraints with respect to the core states ϕ_{c}, see Fig. 14.4.

Altogether we may assume that the pseudo-wavefunctions will also contain some contributions from the core states, and therefore we make the following Ansatz:

$$\begin{aligned} f\phi_{\mathrm{c}} &= \epsilon_{\mathrm{c}}\phi_{\mathrm{c}} \quad \text{(core)} \qquad f\phi_{\mathrm{v}} = \epsilon_{\mathrm{v}}\phi_{\mathrm{v}} \quad \text{(valence)} \\ \chi_{\mathrm{v}} &= \phi_{\mathrm{v}} + \sum_{\mathrm{c}} \langle \phi_{\mathrm{c}} | \chi_{\mathrm{v}} \rangle \phi_{\mathrm{c}} \qquad \text{(pseudo-wavefunction)} \, . \end{aligned} \tag{14.16}$$

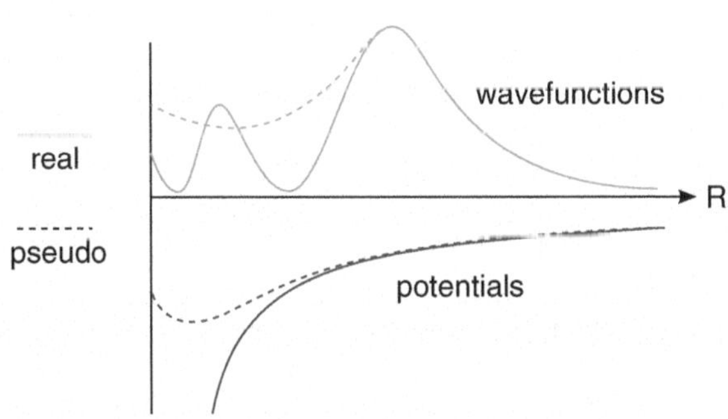

Fig. 14.4. In the framework of the pseudopotential approach, the wavefunctions of the valence electrons are substituted by smooth pseudo-wavefunctions, which implies a weakened interaction potential within the core region

The next theoretical step is to introduce some core projector P, such that

$$(1-P)\phi_c = 0 \qquad (1-P)\chi_v = \phi_v \,. \tag{14.17}$$

Then it is just a matter of simple algebra to show that:

$$\begin{aligned}
&[(1-P)f(1-P)]\,\chi_v = \epsilon_v(1-P)(1-P)\chi_v \\
&(f - Pf - fP + PfP + \epsilon_v P)\chi_v = [-\frac{1}{2}\Delta_{\boldsymbol{r}} + V_{PS}(\boldsymbol{r})]\chi_v = \epsilon_v\chi_v \,.
\end{aligned} \tag{14.18}$$

Obviously, the pseudo-wavefunctions have the same energies as the real valence electron wavefunctions, but the corresponding one-electron Hamiltonian f has been modified quite dramatically (it should be energy-dependent!). It turns out that this modified one-electron Hamiltonian may usually be written in the form of the last line in (14.18), using a simple parameterized form for the pseudopotential V_{PS} like:

$$V_{PS} \equiv \frac{Z - N_c}{r} + \frac{A}{r}e^{-\lambda r} \,. \tag{14.19}$$

This parameterization comprises the fitting parameters A and λ, Z should just be the nuclear charge, and N_c the number of core electrons. Of course, there are more sophisticated ways to construct a pseudopotential V_{PS}, in particular using planewave basis sets [21].

14.2.4 Noninteracting Many-Electron Systems

So far we only considered one-electron Hamiltonians and their corresponding orbitals. But usually a molecule or solid will contain a large number of electrons, and one might wonder about the right formalism to describe such a system. Furthermore, we have largely ignored electron spin, and now it is time to put the spin back into our formalism.

There are two things that we have to require to set up this formalism. First, the electrons must be independent and indistinguishable. Second, the corresponding many-electron wavefunction must be antisymmetric, such that:

$$\begin{aligned}
&\Psi(\boldsymbol{x}_1 \dots \boldsymbol{x}_i \dots \boldsymbol{x}_j \dots \boldsymbol{x}_N) = -\Psi(\boldsymbol{x}_1 \dots \boldsymbol{x}_j \dots \boldsymbol{x}_i \dots \boldsymbol{x}_N) \\
&\boldsymbol{x}_i = (\boldsymbol{r}_i, m_i), \quad \boldsymbol{r}_i : \text{ cartesian coordinates, } m_i : \text{ spin} \,.
\end{aligned} \tag{14.20}$$

The indistinguishability and independence of the electrons requires a many-electron Hamiltonian, which is a sum of identical one-electron operators for each electron i:

$$\begin{aligned}
H(r)\Psi(r,m) &= (\sum_i^N f(\boldsymbol{r}_i))\Psi(\boldsymbol{x}_1 \dots \boldsymbol{x}_N) = (\sum_i^N \epsilon_i)\Psi(r,m) \\
&= E\Psi(r,m) \,.
\end{aligned} \tag{14.21}$$

This Hamiltonian could be inserted into the variational principle of (14.1) under the constraint of orthonormality for a set of suitable spin orbitals

$$\{\chi_i(\boldsymbol{x}) \equiv \phi_i(\boldsymbol{r})\omega(m)\}_{i=1\ldots N} \,. \tag{14.22}$$

These spin orbitals will form a Slater determinant, which is defined as follows:

$$\Psi^{\mathrm{SD}}(r,m) \equiv \frac{1}{\sqrt{N!}} \begin{vmatrix} \chi_1(\boldsymbol{x}_1) & \chi_2(\boldsymbol{x}_1) & \ldots & \chi_N(\boldsymbol{x}_1) \\ \chi_1(\boldsymbol{x}_2) & \chi_2(\boldsymbol{x}_2) & \ldots & \chi_N(\boldsymbol{x}_2) \\ \ldots & \ldots & \ldots & \ldots \\ \chi_1(\boldsymbol{x}_N) & \chi_2(\boldsymbol{x}_N) & \ldots & \chi_N(\boldsymbol{x}_N) \end{vmatrix} \,. \tag{14.23}$$

Then, after performing a unitary transformation, we finally arrive at a one-electron Schrödinger equation for the spin orbitals similar to (14.12).

14.3 Hartree-Fock Theory

In the following two sections, we will discuss various schemes to construct realistic one-electron Hamiltonians. And starting from this basis, we will also be able to derive two of the key methods in quantum chemistry and computational materials science. The present Sect. will be devoted to Hartree-Fock (HF) theory, which was the standard workhorse of ab initio calculations until the dawn of density functional theory. Consequently the whole field of Hartree-Fock based ab initio methods is a giant one (see [2] or [9]), and here we are able to just present a rather modest and biased selection.

We will start our derivation of the Hartree-Fock equation from the pioneering work of Hartree [29]. His theory was set up in the spirit of a one-electron theory described above, and then it was Fock [30], who pointed out how to transform Hartree's theory into a real many-electron approach. Once the related one-electron Hartree-Fock equations are solved, one may set up advanced perturbative or variational schemes of increasing accuracy using the HF orbitals [9], some of them being the very benchmark methods of modern quantum chemistry. In this article, we only discuss one of these approaches called Configuration Interaction (CI) method.

14.3.1 Hartree's Method

Hartree's idea [29] was to reduce the many-electron problem of chemical bonding to a one-electron form, where every electron has its own individual wavefunction ϕ_i and energy level ϵ_i. To this end, he suggested a one-electron Schrödinger equation of the following kind:

$$-\frac{1}{2}\Delta_{\boldsymbol{r}}\, \phi_i(\boldsymbol{r}) + V(\boldsymbol{r})\phi_i(\boldsymbol{r}) = \epsilon_i\phi_i(\boldsymbol{r}) \,. \tag{14.24}$$

The first term denotes the kinetic energy operator for an electron with wavefunction ϕ_i, and the second term represents a general interaction potential for this electron.

The brilliant insight of Hartree was to assume that every electron is moving in a potential caused by the classical electrostatic interaction with the nuclei, and caused by the classical electrostatic interaction of the electron with smeared out negative electric charges that correspond to the electron density

$$\rho(\boldsymbol{r}') = \sum_i^N \phi_i^*(\boldsymbol{r}')\phi_i(\boldsymbol{r}') . \tag{14.25}$$

This is the very essence of the mean-field approach. We see that the electron density in (14.25) is obviously built from the electron wavefunctions themselves, and therefore the corresponding potential must be constructed by iterating (14.24) until one obtains a self-consistent electron density or wavefunction.[2]

In contrast to a common prejudice, the potential given by Hartree was actually the following:

$$V(\boldsymbol{r}) = -\sum_\alpha^M \frac{Z_\alpha}{|\boldsymbol{R}_\alpha - \boldsymbol{r}|} + \int \frac{\rho(\boldsymbol{r}') - \phi_i^*(\boldsymbol{r}')\phi_i(\boldsymbol{r}')}{|\boldsymbol{r} - \boldsymbol{r}'|} \mathrm{d}\boldsymbol{r}' \text{ (for electron } i) . \tag{14.26}$$

We see that Hartree obviously corrected the interaction between electron i and the electronic mean field, such that the electron i will not interact with itself, which would certainly be unphysical.

The final conceptional step of the Hartree theory was to pack the one-electron wavefunctions together to form a many-electron wavefunction. Here Hartree assumed a simple product wavefunction:

$$\Psi(r) = \prod_i^N \phi_i(\boldsymbol{r}_i) . \tag{14.27}$$

It was Fock [30], who pointed out that the many-electron wavefunction of Hartree theory should better be approximated by a single Slater determinant (see (14.23)) in order to guarantee its antisymmetry (see (14.21)). In the next section, we will see that this assumption will add another term to the mean field called exchange interaction. Finally we note that spin is obviously missing from Hartree's original theory.

14.3.2 The Hartree-Fock Method

We now derive Hartree-Fock theory, starting from the variational principle of (14.1). This will lead to a set of non-linear one-electron Schrödinger equations, similar to the Hartree theory ((14.24)–(14.26)). Then, by introducing basis functions, these equations may be transformed into a nonlinear matrix equation (Roothan equation), where we have to determine the self-consistent solution to a generalized eigenvalue problem similar to (14.13).

[2] The first "supercomputer" to carry out these calculations was Hartree's father.

14.3.2.1 The Hartree-Fock Equation

In the framework of Hartree-Fock theory, the ground-state wavefunction Ψ_0 of a N-electron systems will be approximated by a single Slater determinant Ψ_0^{SD} made of N spin orbitals χ_i (see (14.23)). But in order to determine Ψ_0^{SD}, it will be necessary to derive a suitable one-electron Schrödinger equation. We therefore plug Ψ_0^{SD} into the energy functional E_{tot} of (14.1):

$$\begin{aligned} E_{\mathrm{tot}}[\Psi_0^{\mathrm{SD}}] &= \langle \Psi_0^{\mathrm{SD}} | H | \Psi_0^{\mathrm{SD}} \rangle \;=\; E_{\mathrm{tot}}[\{\chi_i\}, R] \\ &= \sum_a [a|h|a] + \frac{1}{2} \sum_{ab} [aa|bb] - [ab|ba] + V_{nn}[R] \\ [a|h|a] &= \int \mathrm{d}\boldsymbol{x}_1 \chi_a^*(\boldsymbol{x}_1) \left(-\frac{1}{2}\Delta_{\boldsymbol{r}_1} - \sum_A \frac{Z_A}{|\boldsymbol{R}_A - \boldsymbol{r}_1|} \right) \chi_a(\boldsymbol{x}_1) \\ [ab|cd] &= \int \mathrm{d}\boldsymbol{x}_1 d\boldsymbol{x}_2 \; \chi_a^*(\boldsymbol{x}_1)\chi_b(\boldsymbol{x}_1) \left(\frac{1}{|\boldsymbol{r}_1 - \boldsymbol{r}_2|} \right) \chi_c^*(\boldsymbol{x}_2)\chi_d(\boldsymbol{x}_2) \,, \end{aligned} \tag{14.28}$$

and apply the variational principle with a twist: We vary E_{tot} with respect to the (conjugate) spin orbitals χ_a^*, under the constraint that these spin orbitals should be orthogonal. To this end we introduce a Lagrangian multiplier ϵ_{ab}. Thus (14.1) will be transformed into the following variational principle:

$$\begin{aligned} \text{Orthonormality} \;:\qquad & [a|b] = \int \mathrm{d}\boldsymbol{x}_1 \chi_a^*(\boldsymbol{x}_1)\chi_b(\boldsymbol{x}_1) = \delta_{ab} \\ \frac{\delta}{\delta\chi_a^*} L[\Psi_0] &= \frac{\delta}{\delta\chi_a^*} \left(E[\{\chi_i\}] - \sum_{ab} \epsilon_{ab}([a|b] - \delta_{ab}) \right) = 0 \,. \end{aligned} \tag{14.29}$$

This leads to a coupled set of one-electron Schrödinger equations. Finally after carrying out a unitary transformation [2], we obtain the following non-linear one-electron Schrödinger equation

$$f_{\mathrm{HF}}\left(\{\chi_i\}\right) \chi_a = \epsilon_a \chi_a \,. \tag{14.30}$$

Analogous to Hartree theory, the one-electron Hamiltonian f_{HF} is a functional of the orbitals that ought to be determined from it using (14.30), and therefore we have to iterate everything until we find a self-consistent solution. This iterative procedure can be very annoying, in particular when the system is large, and every iteration step turns out to be extremely slow. Furthermore a bad convergence also tends to slow down any further simulation step like the exploration of total energy hypersurfaces in search for new materials, for which we need the total energy E_{tot} from (14.28) at a whole series of nuclear configuration R, and as quickly as possible.

Now it turns out that a self-consistent solution to (14.30) is actually equivalent to a direct minimization of E_{tot} from (14.28) for set of trial orbitals generated by various f_{HF} of (14.30). There are quite powerful minimization techniques that actually exploit this idea, which lead to a dramatic improvement in convergence for large systems [20].

Finally we write out (14.30) explicitly

$$
\begin{aligned}
-\frac{1}{2}\Delta_{\boldsymbol{r}}\, \chi_i(\boldsymbol{x}) - \sum_{\alpha}^{M} \frac{Z_\alpha}{|\boldsymbol{R}_\alpha - \boldsymbol{r}|}\chi_i(\boldsymbol{x}) + \sum_{j\neq i}^{N} \left[\int \frac{\chi_j^*(\boldsymbol{x}')\chi_j(\boldsymbol{x}')}{|\boldsymbol{r} - \boldsymbol{r}'|} \mathrm{d}\boldsymbol{x}' \right] \chi_i(\boldsymbol{x}) \\
+ \sum_{j\neq i}^{N} \left[\int \frac{\chi_j^*(\boldsymbol{x}')\chi_i(\boldsymbol{x}')}{|\boldsymbol{r} - \boldsymbol{r}'|} \mathrm{d}\boldsymbol{x}' \right] \chi_j(\boldsymbol{x}) = \epsilon_i \chi_i(\boldsymbol{x}) \,, \qquad (14.31)
\end{aligned}
$$

and compare it to Hartree theory ((14.24)–(14.26)). Everything looks pretty much the same, except for the last term in (14.31), which is called exchange potential, for obvious reasons. And the N orbitals with the lowest energies ϵ_i will actually form the ground state in terms of a Slater determinant Ψ_0^{SD}, where we must take into account that every orbital will be occupied by two electrons of opposite spin, unless we introduce the notorious spin contamination of unrestricted Hartree-Fock theory (see [2]).

14.3.2.2 The Roothan Equation

Analogous to our procedure in Sect. 14.2.2, we expand the spin orbitals χ_i in a suitable set of basis functions φ_μ

$$
\begin{aligned}
\chi_i(\boldsymbol{x}) &= \phi_i(\boldsymbol{r})\omega_i(m) \\
\phi_i(\boldsymbol{r}) &= \sum_{\mu} C_{\mu i}\varphi_\mu(\boldsymbol{r}) \,. \qquad (14.32)
\end{aligned}
$$

And by applying the techniques from Sect. 14.2.2, (14.31) will go over into a nonlinear matrix equation called Roothan equation [2]. We explicitly write out everything in its full glory, just to stop the overconfident reader, who might be convinced that he/she will be able to write a HF program overnight:

$$
\begin{aligned}
\sum_{\nu} F_{\mu\nu} C_{\nu i} &= \epsilon_i \sum_{\nu} S_{\mu\nu} C_{\nu i} \,, \\
S_{\mu\nu} &= \int \mathrm{d}\boldsymbol{r}\; \varphi_\mu^*(\boldsymbol{r})\varphi_\nu(\boldsymbol{r}) \,, \\
F_{\mu\nu} &= T_{\mu\nu} + V_{\mu\nu}^{\mathrm{nucl}} + G_{\mu\nu} = H_{\mu\nu}^{\mathrm{core}} + G_{\mu\nu} \,, \\
T_{\mu\nu} &= \int \mathrm{d}\boldsymbol{r} \varphi_\mu^*(\boldsymbol{r})[-\frac{1}{2}\Delta_{\boldsymbol{r}}]\varphi_\nu(\boldsymbol{r}) \,, \\
V_{\mu\nu}^{\mathrm{nucl}} &= \int \mathrm{d}\boldsymbol{r} \varphi_\mu^*(\boldsymbol{r})[-\sum_{\alpha} \frac{Z_\alpha}{|\boldsymbol{r} - \boldsymbol{R}_\alpha|}]\varphi_\nu(\boldsymbol{r}) \,,
\end{aligned}
$$

$$G_{\mu\nu} = \sum_a^{N/2} \sum_{\lambda\rho} C_{\lambda a} C^*_{\rho a} [2(\mu\nu|\rho\lambda) - (\mu\lambda|\rho\nu)] \,,$$

$$(\mu\nu|\lambda\rho) = \int \mathrm{d}\boldsymbol{r}_1 \mathrm{d}\boldsymbol{r}_2 \, \varphi^*_\mu(\boldsymbol{r}_1)\varphi_\nu(\boldsymbol{r}_1) \frac{1}{|\boldsymbol{r}_1 - \boldsymbol{r}_2|} \varphi^*_\lambda(\boldsymbol{r}_2)\varphi_\rho(\boldsymbol{r}_2) \,,$$

$$E_{\text{tot}} = \frac{1}{2} \sum_{\mu\nu} [2 \sum_a^{N/2} C_{\mu a} C^*_{\nu a}] (H^{\text{core}}_{\mu\nu} + F_{\mu\nu}) + \sum_\alpha \sum_{\beta>\alpha} \frac{Z_\alpha Z_\beta}{|\boldsymbol{R}_\alpha - \boldsymbol{R}_\beta|} \,. \quad (14.33)$$

Again we notice that the operator $F_{\mu\nu}$ depends on the coefficient matrix $C_{\nu i}$ that ought to be determined from (14.33). Therefore we have to solve this equation iteratively, and schemes to accelerate such a procedure are known for a long time, see [31].

14.3.3 Beyond Hartree-Fock

The one-electron Schrödinger equation of Hartree-Fock theory will generate a complete set of orthogonal spin orbitals, which may be used to set up more than just the ground-state Slater determinant. Altogether, these Slater determinants are also forming a complete set of states within the antisymmetric part of a many-particle Hilbert space. Therefore, any many-electron wavefunction Ψ may be expanded in those Slater determinants:

$$\begin{aligned}
\Psi &\equiv c_0\Psi_0 + \sum_{a\alpha} c^\alpha_a \Psi^\alpha_a + \sum_{a<b;\alpha<\beta} c^{\alpha\beta}_{ab}\Psi^{\alpha\beta}_{ab} + \sum_{a<b<c,\alpha<\beta<\gamma} c^{\alpha\beta\gamma}_{abc}\Psi^{\alpha\beta\gamma}_{abc} + \ldots \\
\Psi_0 &= \Psi^{\text{SD}}(\chi_1 \cdots \chi_a \cdots \chi_b \cdots \chi_c \cdots \chi_N) \\
&\;\;\vdots \qquad\qquad \vdots \\
\Psi^{\alpha\beta\gamma}_{abc} &= \Psi^{\text{SD}}(\chi_1 \cdots \chi_\alpha \cdots \chi_\beta \cdots \chi_\gamma \cdots \chi_N) \,.
\end{aligned} \quad (14.34)$$

Note that we would specifiy those Slater determinants by those orbitals that are actually substituting orbitals of the ground-state Slater determinant Ψ^{SD}_0. The various expansion coefficients may be determined from the variational principle of (14.1), which corresponds to the diagonalization of a giant Hamilton matrix H [2]

$$H = \begin{pmatrix}
\langle\Psi_0|H|\Psi_0\rangle & 0 & \langle\Psi_0|H|\Psi^{\alpha\beta}_{ab}\rangle & 0 & \ldots \\
 & \langle\Psi^\alpha_a|H|\Psi^\beta_b\rangle & \langle\Psi^\alpha_a|H|\Psi^{\beta\gamma}_{bc}\rangle & \langle\Psi^\alpha_a|H|\Psi^{\beta\gamma\delta}_{bcd}\rangle & \ldots \\
 & & \langle\Psi^{\alpha\beta}_{ab}|H|\Psi^{\gamma\delta}_{cd}\rangle & \langle\Psi^{\alpha\beta}_{ab}|H|\Psi^{\gamma\delta\epsilon}_{cde}\rangle & \ldots \\
 & & & & \text{etc.}
\end{pmatrix} \quad (14.35)$$

There are actually some selection rules, which make the matrix H a little bit sparser, but usually one needs a large number of Slater determinants to really improve upon the HF method. Therefore the CI method is only applied to obtain some benchmark results for smaller systems, but there is a whole plethora of similarly accurate post-HF ab initio methods described in [2] or [9], and a lot of them are actually going back to Pople.

14.4 Density Functional Theory

The central ideas of density functional theory were written up in a landmark paper of Hohenberg and Kohn [32], who showed that the variational principle of (14.1) is equivalent to a variational principle for an energy functional $E_{\rm tot}$, which depends on the ground-state one-electron density ρ_0, rather than the many-electron wave-function Ψ_0. Lateron, Kohn and Sham [33] showed that this rather abstract concept may actually be translated into a powerful computational scheme, which involves a one-electron equation similar to the Hartree and Hartree-Fock equations discussed above.

It turns out that such a density functional scheme will be as fast as the Hartree method, but with the accuracy of post-HF methods. Therefore, with the advent of density functional theory, it suddenly became possible to simulate large molecular and solid systems, and ab initio simulations quickly developed into a whole new branch of materials science, which could actually challenge experimental research, see Sect. 15.2.

In the following, we will discuss some of the formal aspects of density functional theory, and derive the Kohn-Sham one-electron equations. We also mention some technical details that seem to be indispensable for running accurate density functional calculations. For further details, the reader is referred to a bulk of interst-ing monographs [10, 11, 12], or to a review article [34].

14.4.1 Formal Density Functional Theory

The key entity of density functional theory is the (spinless, reduced) one-electron density:

$$\rho_0(\boldsymbol{r}_1) = N \int \Psi_0^*(\boldsymbol{x}_1 \ldots \boldsymbol{x}_N)\Psi_0(\boldsymbol{x}_1 \ldots \boldsymbol{x}_N)\mathrm{d}m_1\mathrm{d}\boldsymbol{x}_2 \ldots \mathrm{d}\boldsymbol{x}_N \ , \tag{14.36}$$

which is obviously related to the ground state Ψ_0 of a many-electron system, but being much simpler. The rather intuitive assumption of density functional theory is that this one-electron density will entirely determine the properties of a many-electron system for a given nuclear configuration R. In other words, there is a one-to-one correspondence between $\rho_0(\boldsymbol{r})$ and a many-electron Hamiltonian $\mathrm{H}(r, v(R))$ with external potential $v(R)$. Hohenberg and Kohn actually gave an elementary and very popular proof for it [32], but there is also a sound mathematical proof, which fills all the remaining gaps [35].

This one-to-one relation implies the existence of a density dependent energy functional $E_{\rm v}[\rho_0(\boldsymbol{r}), R]$:

$$\begin{aligned} E_{\text{tot}}[\Psi_0(r), v(R)] &\equiv \langle \Psi_0(r)|H(r, v(R))|\Psi_0(r)\rangle \\ = E_{\text{v}}[\rho_0(\boldsymbol{r}), R] &\equiv T[\rho_0(\boldsymbol{r})] + V_{\text{v}}[\rho_0(\boldsymbol{r}), R] + V_{ee}[\rho_0(\boldsymbol{r})] + V_{\text{nn}}[R] \,. \end{aligned} \tag{14.37}$$

But $E_{\text{v}}[\rho_0(\boldsymbol{r}), R]$ is a rather abstract object, which contains a general kinetic energy functional T, a functional V_{v} that describes the interaction of the electrons with the external field $v(R)$, a functional V_{ee} describing the electron-electron repulsion, and the classical nucleus-nucleus repulsion V_{nn} that we already mentioned before. The exact form of $E_{\text{v}}[\rho_0(\boldsymbol{r}), R]$ is unknown up to now, and the art of density functional theory is to find a useful approximations to it [10].

Hohenberg and Kohn also showed that the variational principle of (14.1) for the ground-state wavefunction Ψ_0 may be transformed into a variational principle for the ground-state one-electron density ρ_0:

$$\langle \Psi_{\text{tr}}|H(v)|\Psi_{\text{tr}}\rangle = E_{\text{v}}[\rho_{tr}] \geq E_{\text{v}}[\rho_0] = \langle \Psi_0|H(v)|\Psi_0\rangle \Rightarrow \frac{\delta}{\delta\rho} E_{\text{v}}[\rho]\bigg|_{\rho_0} = 0 \,. \tag{14.38}$$

There are some mathematical subtleties related to this variational principle. In particular it is not clear up to now which types of trial densities ρ_{tr} are actually allowed in (14.38). But for this and other mathematical details, we refer the interested reader to [35] or [10]

14.4.2 The Kohn-Sham Method

So far, we could convince ourselves that there exists some abstract density functional $E_{\text{v}}[\rho_0(\boldsymbol{r}), R]$ (14.37), and an equally abstract variational principle to determine the ground-state density ρ_0 (14.38). However, Kohn and Sham showed [33] that density functional theory may be put in a form similar to Hartree or Hartree-Fock theory.

The key concept are the one-electron Hamiltonians that we discussed at great length in Sect. 14.2. Because Kohn and Sham made the assumption that the one-electron density ρ_0 should be equal to the one-electron density of a non-interacting reference system:

$$\begin{aligned} \left(-\frac{1}{2}\Delta_{\boldsymbol{r}} + V(\boldsymbol{r})\right)\phi_i(\boldsymbol{r}) &= \epsilon_i\phi_i \\ \Psi_{\text{s}} = \Psi^{\text{SD}}(\phi_1 \ldots \phi_N) &\Longrightarrow \rho_0(\boldsymbol{r}) = \sum_i |\phi_i(\boldsymbol{r})|^2 \,. \end{aligned} \tag{14.39}$$

Then ρ_0 will be made of the orbital solutions ϕ_i to (14.39), and the corresponding many-electron ground-state wavefunction Ψ_{s} will be a single Slater determinant made from the most stable orbitals, which is already pretty close to Hartree-Fock theory!

In order to arrive at a potential V similar to the Hartree or Hartree-Fock one-electron interaction potential, it is necessary to make some cosmetics and rearrange various parts of (14.37):

$$
\begin{aligned}
E_{\mathrm{el}}[\rho] &= T[\rho] + V_{\mathrm{ee}}[\rho] + V_{\mathrm{v}}[\rho] = T_{\mathrm{s}}[\rho] + V_{\mathrm{ee}}^{\mathrm{class}}[\rho] + E_{\mathrm{xc}}[\rho] + V_{\mathrm{v}}[\rho] \,, \\
T_{\mathrm{s}}[\rho] &= \sum_i \langle \phi_i | -\frac{1}{2}\Delta_{\boldsymbol{r}} | \phi_i \rangle \,, \\
E_{\mathrm{xc}}[\rho] &= T[\rho] - T_{\mathrm{s}}[\rho] + V_{\mathrm{ee}}[\rho] - V_{\mathrm{ee}}^{\mathrm{class}}[\rho] \,.
\end{aligned} \tag{14.40}
$$

The exchange correlation functional E_{xc} will become our garbage collection, containing all non-classical electron interactions, as well as corrections to the kinetic energy functional T_{s} of the non-interacting reference system. The quality of any density functional based simulation will depend quite critically on reasonable approximations for E_{xc} as a functional of ρ_0, see the next section.

With these rearrangements, we may carry out the variational principle of (14.38), where the variation with respect to ρ will go over into a variation with respect to the (conjugate) orbitals ϕ_i^*:

$$
\begin{aligned}
\frac{\delta}{\delta\phi_i^*} F[\{\phi_i\}] &= \frac{\delta}{\delta\phi_i^*}\left(E_{\mathrm{el}}[\{\phi_i\}] - \sum_{ij} \lambda_{ij} \int \phi_i^*(\boldsymbol{x})\phi_i(\boldsymbol{x})\mathrm{d}\boldsymbol{x} \right) = 0 \\
&\Rightarrow f_{\mathrm{ks}}\phi_i = \left(-\frac{1}{2}\Delta_{\boldsymbol{r}} + v_{\mathrm{s}}(\rho_0, \boldsymbol{r}, R) \right) \phi_i = \epsilon_i \phi_i \,.
\end{aligned} \tag{14.41}
$$

Thus we formally obtain the kind of non-linear one-electron Schrödinger equation that was postulated in (14.39). And again we will have to solve this equation iteratively.

We may then write out f_{ks} and compare it to the Hartree ((14.24)–(14.26)) and the Hartree-Fock (14.31) one-electron Hamiltonians:

$$
\begin{aligned}
-\frac{1}{2}\Delta_{\boldsymbol{r}}\, \phi_i(\boldsymbol{r}) - \sum_{\alpha}^{M} \frac{Z_\alpha}{|\boldsymbol{R}_\alpha - \boldsymbol{r}|}\phi_i(\boldsymbol{r}) + \left[\int \frac{\rho_0(\boldsymbol{r}')}{|\boldsymbol{r} - \boldsymbol{r}'|}\mathrm{d}\boldsymbol{r}' \right] \phi_i(\boldsymbol{r}) \\
+ \frac{\delta E_{\mathrm{xc}}}{\delta\rho}[\rho_0(\boldsymbol{r})]\phi_i(\boldsymbol{r}) = \epsilon_i\phi_i(\boldsymbol{r}) \,.
\end{aligned} \tag{14.42}
$$

The biggest difference is the last term in (14.42) called exchange-correlation potential, which was completely missing within Hartree theory. And in the framework of Hartree-Fock theory, there was a complicated orbital-dependent exchange potential taking the place of this density dependent exchange-correlation potential. In the next section, we will discuss some technical details related to (14.42) that will also concern the construction of a suitable exchange-correlation potential.

14.4.3 Some Technical Details

The obvious similarities between the Kohn-Sham scheme and the Hartree(-Fock) method will makes it easy to implement density functional theory into any existing Hartree-Fock code. To this end, we just have to re-write the Kohn-Sham equations in matrix form, using the standard procedure based on an expansion of the orbitals and the one-electron density in a suitable set of basis functions. Given a precise exchange-correlation functional E_{xc}, we will have a Hartree-like method with post-Hartree-Fock accuracy!

Various types of exchange-correlation functionals have been discussed at great length in the literature [10], but the most popular ones fall into the following classes:

$$E_{\mathrm{xc}}^{\mathrm{LDA}}[\rho] = \int \rho(\boldsymbol{r}) e_{\mathrm{gas}}[\rho(\boldsymbol{r})] \mathrm{d}\boldsymbol{r} \,,$$
$$E_{\mathrm{xc}}^{\mathrm{GA}}[\rho] = E_{\mathrm{xc}}^{\mathrm{LDA}} + \delta E_{\mathrm{xc}}[\rho(\boldsymbol{r}), |\nabla_{\boldsymbol{r}} \rho(\boldsymbol{r})|] \,. \tag{14.43}$$

The first type of exchange-correlation functional $E_{\mathrm{xc}}^{\mathrm{LDA}}$ refers to the local density approximation (LDA), and $\mathrm{e}_{\mathrm{gas}}$ is the energy density of the electron gas. These types of exchange-correlation functionals are basically some parameterized forms of the exchange-correlation functional of a homogeneous electron gas [10]. The LDA seems to be a rather poor assumption, because the electron density within a molecule or solid is usually varying noticeably, which is the opposite of a homogeneous electron gas.

But LDA works quite well, mainly due to some miraculous error compensation [34]. And as indicated in (14.43), it is also possible to obtain even better exchange-correlation functionals $E_{\mathrm{xc}}^{\mathrm{GA}}$ by determining some correction terms, which depend on the density and the density gradient [36].

Finally the reader may have noticed that we did not introduce any spin into our formalism. No need to worry, it turns out that the general formalism of density functional theory can easily be modified to meet this requirement, just by introducing an exchange-correlation functional that will depend on two different one-electron densities for different electron spins. This method is called spin-density functional theory, see [34] and [10].

References

1. M. Born, K. Huang, *Dynamical Theory of Crystal Lattices* (Oxford University Press, Oxford, 1954)
2. A. Szabo, N.S. Ostlund, *Modern Quantum Chemistry* (McGraw-Hill, New York, 1989)
3. C.A. Coulson, *Valence* (Clarendon Press, Oxford, 1952)
4. L. Pauling, *The Nature of the Chemical Bond*, 3rd edn. (Cornell University Press, Ithaca, 1960)
5. H.A. Bethe, E.E. Salpeter, *Quantum Mechanics of One- and Two-Electron Atoms* (Springer, Berlin Göttingen Heidelberg, 1957)
6. C.A. Mead, Journal of VLSI Sig. Process. **8**, 9 (1994)
7. R. Kurzweil, *The Age of Spiritual Machines* (Penguin Putnam, New York, 2000)
8. J. Dongarra, F. Sullivan, Comp. in Sci. & Eng. **2**, 22 (2000)
9. F.E. Harrisa, H.J. Monkhorst, D.L. Freeman, *Algebraic and Diagrammatic Methods in Many-Fermion Theory* (Oxford University Press, Oxford, 1989)
10. R.G. Parr, W. Yang, *Density Functional theory of Atoms and Molecules* (Oxford University Press, Oxford, 1989)

11. R.M. Dreizler, E.K.U. Gross, *Density Functional Theory* (Springer, Berlin Heidelberg, 1990)
12. N.H. March, *Electron Density Theory of Atoms and Molecules* (Academic Press Limited, New York, 1992)
13. G.N. Lewis, *Valence and the Structure of Atoms and Molecules* (The Chemical Catalog Co., New York, 1923)
14. W.N. Lipscomb, Acc. Chem. Research **6**, 257 (1973)
15. F. Weinhold, C.R. Landis, Chemistry education **2**, 91 (2001)
16. F. Weinhold, C.R. Landis, *Valence and Bonding. A Natural Bond Orbital Donor–Acceptor Perspective* (Cambridge University Press, Cambridge, 2005)
17. A.P. Sutton, *Electronic Structure of Materials* (Clarendon Press, Oxford, 1994)
18. D. Pettifor, *Bonding and Structure of Molecules and Solids* (Clarendon Press, Oxford, 1995)
19. W.A. Harrison, *Elementary Electronic Structure*, revised edn. (World Scientific, Singapore, 2004)
20. M.C. Payne, M.P. Teter, D.C. Allan, T.A. Arias, J.D. Joannopoulos, Rev. Mod. Phys. **64**, 1045 (2002)
21. D.J. Singh, *Planewaves, pseudopotentials and the LAPW method* (Kluwer Academic Publishers, Dordrecht, 1994)
22. J.C. Slater, Phys. Rev. **51**, 846 (1937)
23. O.K. Andersen, Phys. Rev.B **12**, 3060 (1975)
24. O.K. Andersen, Z. Pavlowska, O. Jepsen, Phys. Rev.B **34**, 5253 (1986)
25. N. Mazari, D. Vanderbilt, Phys. Rev.B **56**, 12847 (1997)
26. T.A. Arias, Rev. Mod. Phys. **71**, 267 (1999)
27. G.H. Golub, C.F. van Loan, *Matrix computations* (The Johns Hopkins University Press, Baltimore London, 1996)
28. E. Anderson, Z. Bai, C. Bishof, S. Blackford, J. Demmel, J. Dongarra, J.D. Croz, A. Greenbaum, S. Hammarling, A. McKenney, D. Sorensen, *LAPACK user's guide* (SIAM, Philadelphia, 1999)
29. D.R. Hartree, Proc. Camb. Phil. Soc. **24**, 111 (1928)
30. V. Fock, Z. Phys. **61**, 126 (1930)
31. P. Pulay, Chem. Phys. Lett. **73**, 393 (1980)
32. P. Hohenberg, W. Kohn, Phys. Rev. **136**, B864 (1964)
33. W. Kohn, L. Sham, Phys. Rev. **140**, A1133 (1965)
34. R.O. Jones, O. Gunnarsson, Rev. Mod. Phys. **61**, 690 (1989)
35. E.H. Lieb, Int. J. Quantum Chem. **24**, 243 (1983)
36. J. Tao, J.P. Perdew, V.N. Staroverov, G.E. Scuseria, Phys. Rev. Lett. **91**, 146401 (2003)

15 Ab-Initio Methods Applied to Structure Optimization and Microscopic Modelling

Alexander Quandt

Institut für Physik, Universität Greifswald, 17487 Greifswald, Germany

Having discussed the technical and conceptional aspects of ab initio methods, we now show how to do computational materials science based on such methods. The following Sections will be devoted to the study of energy hypersurfaces, interatomic forces and various techniques to step over these energy hypersurfaces. We will also explain how to carry out elementary, but indispensable theoretical tasks of modern materials sciences like the prediction of novel materials, or the deciphering of chemical reaction schemes and finally describe microscopic modeling based on suitable model Hamiltonians.

15.1 Exploring Energy Hypersurfaces

The exploration of energy hypersurfaces is of great importance for many branches of chemistry, chemical engineering and materials sciences. For example, a lot of research activities within modern microbiology are actually devoted to the notorious problem of protein folding, and one of the main goals is to create numerical simulation tools to uncover the basic protein folding mechanism. In Sect. 15.2 we will discuss a somewhat simpler, but nevertheless rather surprising example from basic materials sciences, which demonstrates the predictive power of modern ab initio methods. And finally we want to point out that there is a recent monograph by Wales [1], which covers most of the topics presented in this section, and many more.

15.1.1 About Energy Hypersurfaces

An energy hypersurface for a molecular or solid system is a mapping of (ab initio) total energies like the ones from (14.28), (14.33) or (14.37) as a function of the corresponding nuclear configurations $R = (\boldsymbol{R}_1, \ldots, \boldsymbol{R}_M)$

$$E_{\text{hyp}}(\boldsymbol{R}_1, \ldots, \boldsymbol{R}_M) = E_{\text{hyp}}(R) \equiv E_{\text{tot}}[\ldots, R] \, . \tag{15.1}$$

In Fig. 15.1 we made a simple sketch of such an energy hypersurface. The simplicity of this figure is slightly misleading. Normally R is a large multivector, and therefore the energy landscape may be full of stationary points. But those stationary points are of the highest chemical relevance:

A. Quandt: *Ab-Initio Methods Applied to Structure Optimization and Microscopic Modelling*, Lect. Notes Phys. **739**, 437–469 (2008)
DOI 10.1007/978-3-540-74686-7_15

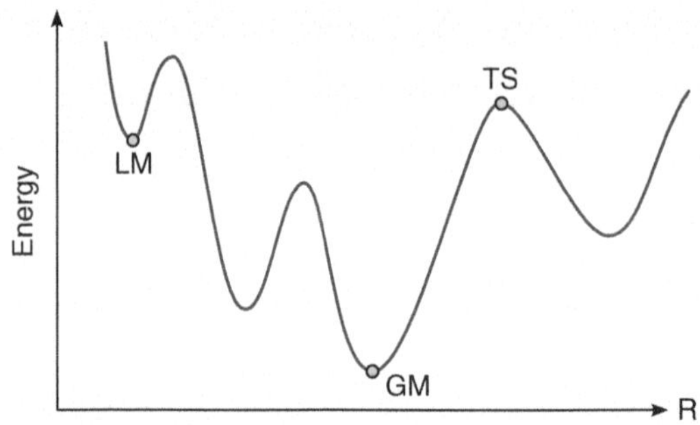

Fig. 15.1. Sketch of an energy hypersurface $E_{\text{hyp}}(R)$, indicating the location of transition states (TS), local minima (LM) and global minima (GM)

$$\left.\frac{\partial E_{\text{hyp}}}{\partial R}\right|_{R_0} = 0 \qquad \text{stationary state ,}$$

$$\left.\frac{\partial^2 E_{\text{hyp}}}{\partial^2 R}\right|_{R_0} = \begin{cases} > 0 & \text{for all coord.: isomer ,} \\ < 0 & \text{for at least one coord.: transition state .} \end{cases} \tag{15.2}$$

Among those isomers, there will usually be a large number of local minima (LM), and just one or a handful of global minima (GM). These global minima ought to be detected to make a reliable prediction of the most stable configurations of a certain system, which is a serious numerical challenge. In Sect. 15.1.3 we will present several techniques to step over energy hypersurfaces, in a way that will actually increase our chances to detect the most relevant local and global minima.

Furthermore, there may be chemical of physical processes that connect various chemically relevant minima. Here it will be of immediate chemical relevance to know the transition states that are located on a path connecting both minima. Similarly one might want to know the size of the energetic activation barriers between both minima. Unfortunately, transition states are even more difficult to detect than minima, and there is also no guarantee that the numerical search algorithms for transition states will generate any meaningful result [2].

A simple toy model will illustrate the complexity of such a task [1]. Just assume that we want to examine a large system of m mutually independent subsystems comprising N atoms. For the number of isomers n_{isomer} we find that:

$$n_{\text{isomers}}(mN) \approx n_{\text{isomers}}(N)^m \Longrightarrow n_{\text{isomers}}(N) \approx \mathrm{e}^{\alpha N} \ , \tag{15.3}$$

which means that the number of isomers is growing exponentially with the subsystem size N. For the transition states we assume that each of them is located in one subsystem, and that a transition state of the complete system with mN atoms is only occurring when one of the subsystem is in a transition state, and all of the others are in a minimum. Therefore the number of transition states n_{tstates} may be calculated as follows:

$$n_{\text{tstates}}(mN) \approx m\left(n_{\text{isomers}}(N)^{m-1}\right) n_{\text{tstates}}(N) \Longrightarrow n_{\text{tstates}}(N) \approx N\mathrm{e}^{\alpha N} \ . \tag{15.4}$$

Again this will imply exponential growth with N. Finally we see that

$$\frac{n_{\text{tstates}}}{n_{\text{isomers}}} \approx N \, , \tag{15.5}$$

i.e. the ratio of the number of transition states vs. the number of isomers grows linearly with the subsystems size N, which explains the increasing difficulty to detect transition states.

Given the complexity of a rugged energy hypersurface defined over a configuration space made of large multivectors R, we may also ask ourselves how such a complicated object might actually be visualized. The monograph of Wales [1] presents several interesting techniques like monotonic sequences, disconnectivity graphs of minimum-transition state-minimum triplets, and a network analysis of disconnectivity graphs to determine some typical scaling laws, and to prove the existence of chemically relevant hubs.

However, beyond these techniques mentioned in [1], there is a large literature concerning the graphical visualization of complex data [3], and it might actually pay off to try one of these techniques to represent and analyze energy hypersurfaces.

15.1.2 Forces

Given an energy hypersurface $E_{\text{hyp}}(R)$, the formal definition of ab initio interatomic forces is rather simple:

$$\boldsymbol{F}_k = \nabla_{\boldsymbol{R}_k} E_{\text{hyp}}(R) = \frac{\partial E_{\text{hyp}}(R)}{\partial R} \, . \tag{15.6}$$

Thus the forces on a nucleus with coordinate $\boldsymbol{R}_k$ is just the derivative of the ab initio total energy with respect to this coordinate. The forces on a whole configuration R are then forming a corresponding force multivector, as indicated in (15.6). In the following, we will consistently use this notation, and specify the $\boldsymbol{R}_k$ only when necessary.

There is a disarmingly simple force theorem by Hellmann and Feynman, and we will discuss it in the following (see [2] for a critical revision of this concept). Assume that Ψ is the *exact* (normalized) eigenstate of $H(r, R)$. Then we obtain the following result for the forces:

$$\begin{aligned}\frac{\partial E_{\text{hyp}}(R)}{\partial R} &= \langle \frac{\partial \Psi}{\partial R} | H(r,R) | \Psi \rangle + \langle \Psi | H(r,R) | \frac{\partial \Psi}{\partial R} \rangle \\ &\quad + \langle \Psi | \frac{\partial H(r,R)}{\partial R} | \Psi \rangle \\ &= \langle \Psi | \frac{\partial H(r,R)}{\partial R} | \Psi \rangle \, . \end{aligned} \tag{15.7}$$

The last line follows from the fact that for the exact (normalized) eigenstate Ψ of $H(r, R)$ we find that

$$\frac{\partial\langle\Psi|\Psi\rangle}{\partial R} = 0 = \langle\frac{\partial\Psi}{\partial R}|\Psi\rangle + \langle\Psi|\frac{\partial\Psi}{\partial R}\rangle . \tag{15.8}$$

As simple as this theorem might be, as hard it is to apply in practice! Note that we were generally composing Ψ using orbitals that might be expanded in some suitable localized basis sets, see (14.7) and (14.32). If these basis sets are not somehow following the gradient $\partial\Psi/\partial R$, there is no reason that (15.7) will reduce to the simple result of its last line (see Appendix C of [2]). The way to include a proper basis-set-following is to determine the formal changes in the orbital expansion coefficients $C_{\mu i}$ (see (14.7) and (14.32)):

$$\begin{aligned}\frac{\partial E_{\text{hyp}}(\{C_{\mu i}\}, R)}{\partial R} &= \frac{\partial \widetilde{E}_{\text{hyp}}(\{C_{\mu i}\}, R)}{\partial R} + \sum_{\mu i}\left(\frac{\partial E_{\text{hyp}}(\{C_{\mu i}\}, R)}{\partial C_{\mu i}}\right)\frac{\partial C_{\mu i}}{\partial R} \\ &= \text{an artwork} \dots .\end{aligned} \tag{15.9}$$

The tilde-sign in this equation denotes all terms that explicitly depend on R. The complex analytical artwork indicated by (15.9) for (post) Hartree-Fock methods may be found in [4], including higher derivatives.

15.1.3 Stepping over Energy Hypersurfaces

Now we want to present some methods to step over energy surfaces in order to detect isomers and transition states. Useful references are the Appendix C of [2] and [1]. Note that none of theses methods is foolproof, and you will dramatically increase your chances to become a fool, if you leave common sense and chemical intuition behind to blindly trust a numerical blackbox.

15.1.3.1 Structure Optimization

The goal of any structure optimization method is to detect a stationary point, hopefully being the most stable isomer of the system. If there is no indication where to search, one simply has to construct a reasonable starting configuration R_0. Then one usually applies one's favorite search algorithm, which will step over the energy hypersurface in a systematic fashion, and finally reveal the location of a stationary point. This procedure can be repeated with different starting configurations to achieve a certain sampling of the energy hypersurfaces. The algorithms presented in this paragraph are all local search algorithms, which at best might be able to detect some stationary points next to a chosen starting configuration. They are to be used with care.

The simplest way to step over an energy hypersurface is a steepest descent path. In such a case we will move from one configuration R_i to the next configuration R_{i+1} along a direction determined by the local forces:

$$E_{\text{hyp}}^{\text{min}} = \min_{\lambda} E_{\text{hyp}}\left(R_i - \lambda\nabla_R E_{\text{hyp}}(R_i)\right) \Rightarrow R_{i+1} = R_i - \lambda_{\text{min}}\nabla_R E_{\text{hyp}}(R_i) . \tag{15.10}$$

Here λ_{min} is the λ which minimizes E_{hyp} along the steepest descent direction. For complicated hypersurfaces, the steepest descent procedure will mainly consist of bouncing around like a drunk sailor. A more sober way of stepping over energy hypersurfaces is the famous Newton-Raphson method. When applying this method, one is permanently optimistic that for a given configuration R_i, the next configuration R_{i+1} will be a stationary point, involving the following approximations:

$$\begin{aligned} E_{\text{hyp}}(R_{i+1}) &\approx E_{\text{hyp}}(R_i) + (R_{i+1} - R_i)\nabla_R E_{\text{hyp}}(R_i) \\ &\quad + \frac{1}{2}(R_{i+1} - R_i)\underbrace{\nabla_R \otimes \nabla_R E_{\text{hyp}}(R_i)}_{\text{Hessian } H(R_i)}(R_{i+1} - R_i) \\ \nabla_R E_{\text{hyp}}(R_{i+1}) &\approx \nabla_R E_{\text{hyp}}(R_i) + H(R_i)(R_{i+1} - R_i) \equiv 0\,. \end{aligned} \tag{15.11}$$

The Hessian $H(R_i)$ involves analytical second derivatives and may be quite costly to determine. Therefore the search step

$$R_{i+1} = R_i - H^{-1}(R_i)\nabla_R E_{\text{hyp}}(R_i)\,, \tag{15.12}$$

will be by far more tedious than the determination of a steepest descent step, which involves the determination of the forces, only (see (15.10)).

There is a whole family of Quasi-Newtonian algorithms, which circumvent these conceptional difficulties by starting with an initial guess for the inverse Hessian H^{-1}, and updating the latter for every subsequent search step using the forces. The most popular algorithms of this family can be found in the Numerical Recipes [5], but there is also a simple algorithm described in the Appendix C of [2], which may easily be programmed and implemented by the reader.

We want to close this section with a little survey of the most popular structure optimization methods (see [2]):

- *Methods without gradients.* The most popular method is due to Nelder and Mead [5]. These methods should only be used, if there is really no chance to determine analytical derivatives.
- *Methods involving analytical first derivatives and numerical second derivatives.* The whole family of Quasi-Newtonian methods mentioned above falls under this category, the most prominent examples being the Davidson-Fletcher-Powell method [5], or the Broyden-Fletcher-Goldfarb-Shanno method [5]. There is a second family of methods falling into this category, which is based on the conjugate gradient method. The latter is a rather smart line search algorithm, which proceeds along conjugate directions rather than steepest descent directions. Like the steepest descent method described above, the conjugate gradient method involves analytical first derivatives, only. The most prominent examples are the conjugate gradient methods of Polak and Ribiere [5], and of Fletcher and Powell [5].
- *Methods involving analytical first and second derivatives.* These methods are usually too costly if one is only interested in the isomers of a given molecular or

solid system. However, some of the algorithms to detect transition states involve the knowledge of analytical second derivatives (see [2]). In the following paragraph we will present a simple method to detect transition states, which will involve analytical first derivatives, only.

For a detailed description and proper references we constantly referred to the Numerical Recipes [5], which really should be your first address when trying to understand and implement those methods.

15.1.3.2 Nudged Elastic Band Method

The standard setting for this method is the typical triplet setting on the energy hypersurfaces, where two isomers are connected by a transition state. We assume that both isomers are already known, which could be the educts and the products of a chemical reaction. In order to detect the transition state and the corresponding energy barrier of a reaction path connecting both isomers, one may apply the general procedure indicated in Fig. 15.2.

Between the isomers M_1 and M_2, one may choose a set of images I_i at somewhat intermediate geometries. Those images are supposed to be connected by elastic spring forces of strength k

$$F_{i,\text{spring}} = k\left(|R_{i+1} - R_i| - |R_i - R_{i-1}|\right)\widehat{t_i} \tag{15.13}$$

which will prevent them from collapsing into one single image. The $\widehat{t_i}$ is an estimate for the normalized tangent vector to the path at R_i. Note that the F_i and R_k and $\widehat{t_i}$ are all multivectors.

The total force on an image I_i is defined as:

$$F_i = F_{i,\text{spring}} - \frac{\partial E_{\text{hyp}}}{\partial R}(R_i) + \left(\frac{\partial E_{\text{hyp}}}{\partial R}(R_i) \cdot \widehat{t_i}\right)\widehat{t_i}\,. \tag{15.14}$$

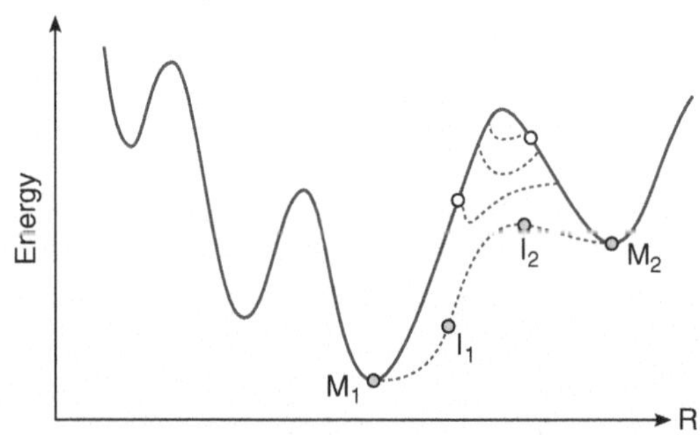

Fig. 15.2. The nudged elastic band method involves two known isomers M_1 and M_2, and a set of images I_k located between them, which interact via springs. After starting with a rather poor configuration (*white circles*), the elastic band between both isomers will slip downhill into its final position (*grey circles*), which marks the proper pathway over a transition state close to I_2

The last term in (15.14), which involves the scalar product of the multivectors, will remove the component of the chemical force along the path. This means that we now have artificial harmonic forces along the path, and the components of the real chemical forces perpendicular to them.

The forces F_i for each image are minimized using one of the algorithms with numerical second derivatives described in the last paragraph. It will correspond to an high-dimensional elastic band, that slips downhill on an energy hypersurface into the proper reaction pathway connecting two isomers, as indicated in Fig. 14.5.

15.1.3.3 Global Optimization

An isomer (=local minimum) on an energy hypersurface is usually surrounded by a catchment basin, which is defined as the volume in configuration space, from which all steepest-descent paths (see (15.10)) converge to that local minimum [1]. Once we are inside such a catchment basin, any of the local minimization methods described above will detect the corresponding isomer. The idea of the annealing/deformation methods of global optimization [1] is to deform an energy hypersurface into a simpler object that is easier to explore. In Fig. 15.3 we show such a hypersurface, which involves catchment basins, only, and which may be generated by the following transformation:

$$\widetilde{E}_{\text{hyp}}(R) = \min\{E_{\text{hyp}}(R)\} \ . \tag{15.15}$$

The operation $\min\{\}$ means optimization around R to detect a local minimum (isomer) located at $\widetilde{R}$, using one of the local structure optimization algorithms described above.

On top of that, we use the standard procedure of simulated annealing [5], the only twist being a rather large increment from the location of the current minimum

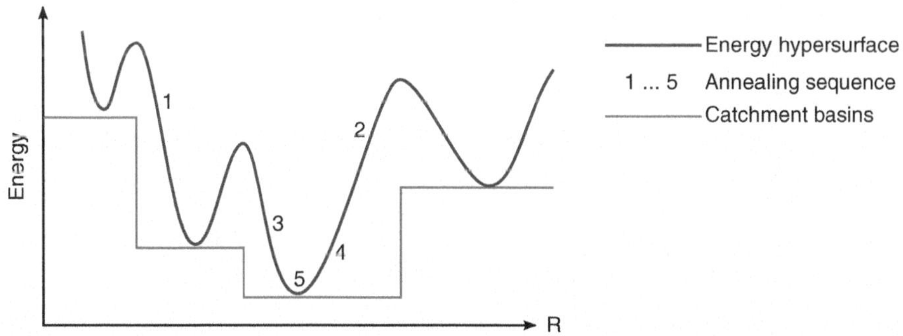

Fig. 15.3. A simulated annealing sequence will eventually be able to detect a global minimum on a rather simple energy hypersurface, but this procedure is hopelessly inaccurate for complex hypersurfaces. The basin hopping algorithm involves the transformation of an energy hypersurface into a simpler object composed of catchment basins. Such a hypersurface is much easier to sample using simulated annealing, and inside each basin, the original energy hypersurface will be sampled in search for local minima

$\widetilde{R}_i \Rightarrow R_{i+1}$ compared to standard simulated annealing runs:

$$\text{For } \widetilde{R}_i \Rightarrow R_{i+1} \text{ try } E_{\text{hyp}}(\widetilde{R}_i) \Rightarrow \widetilde{E}_{\text{hyp}}(R_{i+1}) \; ;$$
$$\text{accept } R_{i+1} \text{ if } e^{(E_{\text{hyp}}(\widetilde{R}_i)-\widetilde{E}_{\text{hyp}}(R_{i+1}))/(k_B T)} > \text{random number} \in [0,1] \; . \tag{15.16}$$

Then the search will either continue from the next local minimum $\widetilde{R}_{i+1}$, or again from the old minimum $\widetilde{R}_i$. By gradually lowering the temperature T, the search will be narrowed down on a basin, which hopefully contains the global minimum to be detected, see Fig. 15.3.

15.2 Applied Theoretical Chemistry

The term "applied theoretical chemistry" was coined by Roald Hoffmann to characterize a special blend of computational methods and the construction of general models to gain deeper insights into the chemistry of materials. The ultimate goals of applied theoretical chemistry will be new materials, new applications, and a better theoretical framework for our understanding of materials properties.

From a computational point of view, applied theoretical chemistry implies the massive use of structure optimization algorithms, preferably in combination with ab initio methods, just to be sure that one uses accurate forces (see Sect. 15.1). The numerical accuracy of ab initio forces somewhat compensates for the limited system sizes, because clear chemical trends among small systems are often transferable over medium sized to really large systems [6]. On the other hand, neither thousands nor millions of atoms interacting via unrealistic pair potentials will ever lead to any chemically relevant result.

As for the construction of models in the framework of applied theoretical chemistry, one should point out that for unknown complex systems, this will usually be an iterative rather than a straightforward process. And it will require a detailed knowledge of basic chemistry, some intuition and a good portion of common sense – or just good luck!

In the following, we will illustrate how the general procedures of applied theoretical chemistry have lead to new insights into the basic chemistry of boron. Beyond that, these ab initio simulations also predicted and anticipated the discovery of novel boron based nanomaterials, which opened a whole new field for nanotechnological applications. Further details may be found in a recent review article [7].

15.2.1 Nanotechnology and Nanomaterials

In Fig. 14.1 we saw that Moore's law was obviously holding through rather dramatic technological changes. Now even the most optimistic interpolations of Moore's law into the near future clearly predict that silicon-based computer technologies will soon hit the lithographic barrier of about 40 nm, and probably run out of steam.

These technologies will have to substituted by other technologies, and it will involve new materials, new devices and radically new concepts for the layout of future computing machines.

In order to understand the advantages and disadvantages of shrinking devices down to the nanodomain, we listed the classical scaling behavior of some key physical properties with system size L in Table 15.1. The only assumptions are that speed and electrostatic fields should be constant, and that forces are acting via surfaces, which are proportional to L^2 (continuum model [8]).

We notice that nanodevices will have some obvious advantages over microelectronic devices: They will be cheaper, they will operate at smaller voltages and higher frequencies, and they will tolerate more power. On the other hand, the resistance of nanodevices will be rather high, their capacitance will be low, they will be rather noisy and short-lived. Of course, these simple scaling laws might have to be amended due to the laws of quantum mechanics, which definitely govern the world of nanosystems [9].

Nevertheless, even under some of the unfortunate conditions listed in Table 15.1, there exists already a successful nanotechnology called biology for billions of years, offering many possibilities for reverse engineering and technological transfer to novel nanomaterials. And even if there are still a lot of nanotechnological lessons to be learned from Mother Nature, some remarkable technological breakthroughs within the last decade have shown that one does not have to be too pessimistic about the future of nanotechnology [9].

There is indeed a growing number of inorganic nanomaterials, which could become key materials for future nanoelectronics, the most prominent ones being carbon fullerenes and carbon nanotubes [10]. And although we know that "prediction is difficult, especially about the future" (N. Bohr), let us have a look at Fig. 15.4, which depicts a possible scenario for future nanoelectronics based on nanotubes. Pretty high up on the list of presents one would like to receive is a controlled layout of heterogeneous tubular networks. Furthermore one would like to have stable and noiseless junctions in between different nanotubular materials, as well as at the

Table 15.1. Scaling of various physical properties with system size L. We postulate constant speed and electrostatic fields, and assume a continuum model, where forces are acting through surfaces of size L^2. [8]

property	scaling	effect on nanosystems
mass	L^3	cheaper
frequency $\sim$ speed/length	L^{-1}	higher frequencies
power density $\sim$ force $\cdot$ speed/volume	L^{-1}	take more power
voltage $\sim$ electrostatic field $\cdot$ length	L	small voltages
resistance $\sim$ area/length	L	higher resistance
capacitance $\sim$ charge/voltage	L	small capacitance
wear life $\sim$ thickness/speed	L	short lifetime
thermal speed $\sim \sqrt{\text{thermal energy/mass}}$	$L^{-3/2}$	noisy

interfaces of nanotubular networks with the outside world. Finally those nanotubular networks will probably require some supporting substrate, or they might have to be embedded into some matrix. Therefore one would also need some detailed knowledge about the interactions between those materials and the nanotubes.

So much for the future. Let us now return to reality, which looks less promising, at least for carbon nanotubes. First of all, the chirality of carbon nanotubes, which decides about their electronic properties (semiconducting vs. metallic), may not be controlled during synthesis [10]. And despite some recent progress [11], there is no known mechanism to achieve any technologically relevant layout of nanotubular networks. Third, there seems to be no suitable binding partner for carbon to form heterogeneous networks with a certain nanoelectronic functionality. And forth the interfaces between carbon and silicon are noisy and rather unstable.

Therefore the search has long been opened to find other nanotubular materials with promising new properties [12, 13], and to achieve even more ambitious goals [12] than the ones sketched in Fig. 15.4.

15.2.2 Novel Boron Based Nanomaterials

One candidate nanotubular material has been found in a system, where nobody really expected to find nanotubes. As we will illustrate in the next paragraph, traditional boron chemistry seems to be incompatible with the existence of boron nanotubes [14]. Nevertheless in the last paragraph of this section, we will draw a radically different picture of boron chemistry [15], which has been established through a large series of numerical and experimental studies on small boron clusters and boron nanostructures [7]. The motor for this development were several theoretical studies on boron clusters and boron nanotubes [16, 17, 18], which combined ab initio structure optimization methods with a chemically motivated Aufbau principle for small boron clusters to predict new classes of nanostructured boron materials [7].

15.2.2.1 Pure Boron Chemistry in a Nutshell

The most prominent features of traditional boron chemistry are boron icosahedra, as well as a complicating bonding pattern involving 2-center and 3-center bonds,

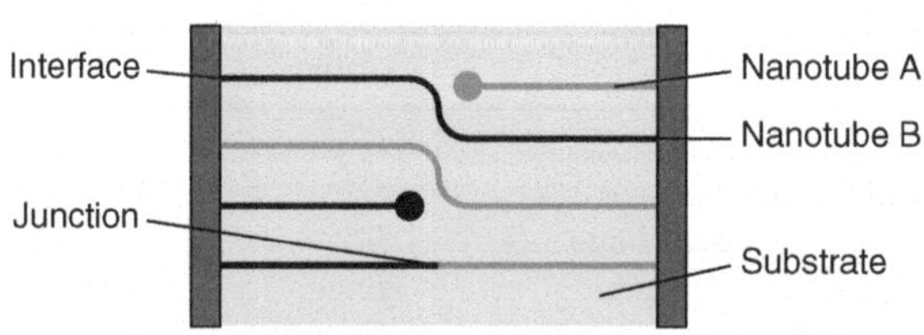

Fig. 15.4. Heterogeneous nanotubular network as a possible blueprint for future technologies. Such applications may require the controlled layout of nanotubular networks, the formation of stable and noiseless tubular heterojunctions, a detailed knowledge of tube-substrate or tube-matrix interactions, and noiseless interfaces between nanotubes and the outside world

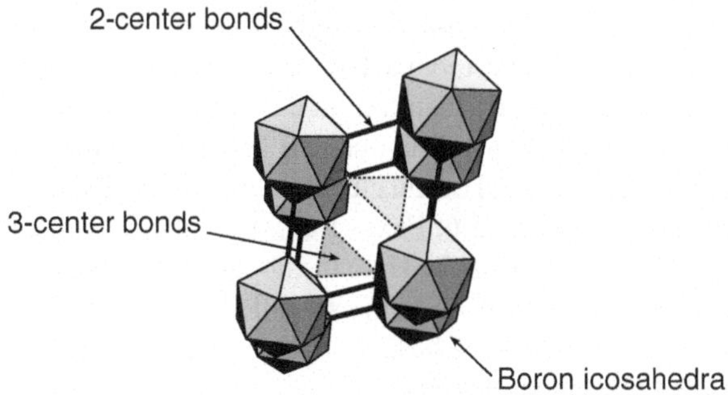

Fig. 15.5. Rhombohedral unit cell of α-boron with icosahedral boron clusters located around its vertices. The bonding is rather complicated, and it involves 2-center and 3-center bonds between boron icosahedra

see Fig. 15.5 and [14]. By the way, there is a common prejudice that icosahedral symmetry should be impossible for crystalline systems, but Fig. 15.5 is certainly the perfect counterexample.

One might ask oneself: How is it possible, that a chemical element with only five electrons will show such a complex bonding pattern? The answer is: Because it has only five electrons! Let us have a look a Table 15.2, where we showed the electronic configurations of single atoms for Be, B and C, together with their coordinations in pure solid phases.

Obviously Be and B have a smaller number of valence electrons than stable orbitals for this shell (see below), and they turn out to be rather highly coordinated. This is a general trend observed for electron deficient (ED) materials, and Pauling [14] gave the following characteristics for this kind of bonding:

(i) The ligancy of ED atoms is higher than the number of valence electrons, and even higher than the number of stable orbitals ($\mathbf{4}$:$\mathbf{1} \times (2s) + 3 \times (2p)$).
(ii) ED elements atoms cause adjacent atoms to increase their ligancy to values greater than the orbital numbers.

A typical electron deficient element is a metal like Be, but even boron, which is a semiconductor [19]), shows both characteristics. First we see from Table 15.2

Table 15.2. Electronic configuration of single atoms, and typical atomic coordinations within solid phases for the electron deficient (ED) elements Be and B, in comparison to a non ED element like C. Note the rather high atomic coordinations within the solid configurations of Be and B, which is in clear contrast to C

element	atomic config.	coordination (solids)	ED?
Be	$(1s^2)2s^2$	8 (bcc), 12 (fcc)	yes
B	$(1s^2)2s^22p^1$	6 (α-boron)	yes
C	$(1s^2)2s^22p^2$	3-4 (graphite), 4 (diamond)	no

and Fig. 15.5 that boron has a coordination higher than four. Second recent ab initio studies of B-C clusters [20] and tubular B-C heterojunctions [21] show that even carbon takes coordinations higher than four in a boron environment.

However, boron icosahedra are only one part of the story. The other part were a series of ab initio studies on small boron clusters summarized by Boustani [16]. The main results are shown in Fig. 15.6. First of all, it is quite obvious from Fig. 15.6 (a) that boron icosahedra are unstable. Here the ab initio studies clearly suggest that the stable isolated B_{12}-clusters are flat (the so-called boron flat out, see [15]). This behavior may be understood on the basis of a general aromaticity theory for boron clusters (see [7] and references therein).

Second, from the ab initio studies of small boron clusters, one may infer a general Aufbau principle for boron clusters. This Aufbau principle states that the stable boron clusters can be built from two basic units, only: The pentagonal and hexagonal pyramidal units B_6 and B_7 shown in Fig. 15.6 (b).

15.2.2.2 Boron Nanotubes and Nanowires

One of the most interesting consequences of this Aufbau principle [16] is shown in Fig. 15.7 (a): Further and further additions of hexagonal B_7 units should lead to stable nanostructures in the form of boron sheets or boron nanotubes. In the following,

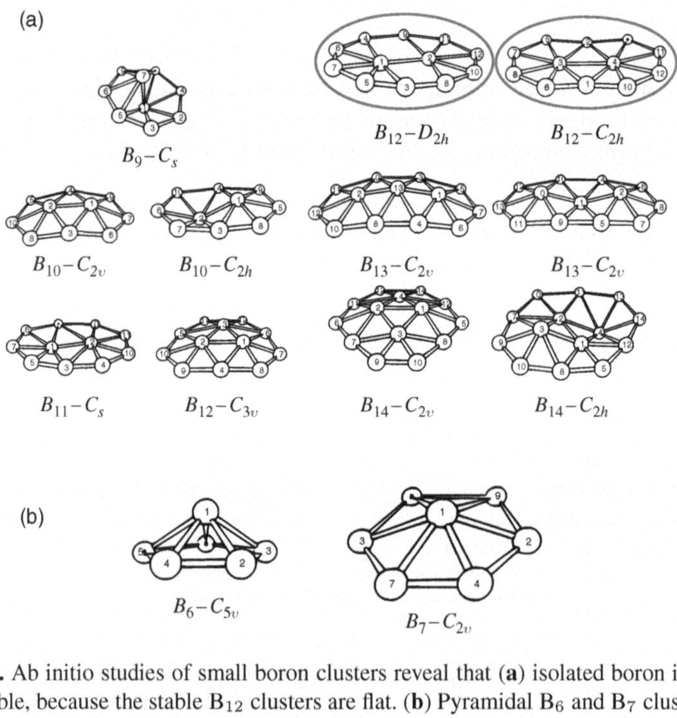

Fig. 15.6. Ab initio studies of small boron clusters reveal that (**a**) isolated boron icosahedra are unstable, because the stable B_{12} clusters are flat. (**b**) Pyramidal B_6 and B_7 clusters being the basic units of an Aufbau principle for boron clusters [16]

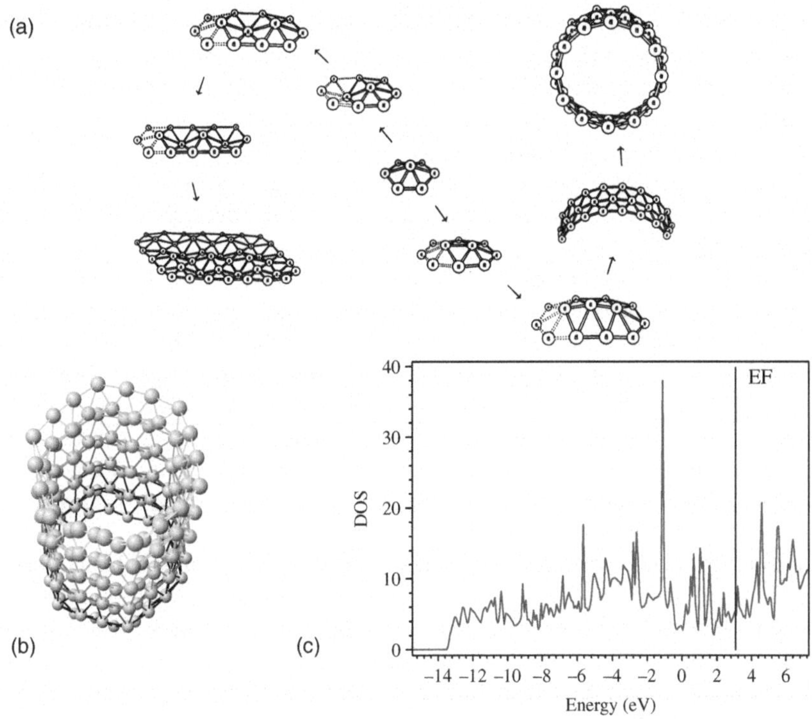

Fig. 15.7. (**a**) According to the Aufbau principle [16] one may add hexagonal B_7 units to either form stable quasiplanar structures or stable tubular structures. (**b**) Portrait of a stable boron nanotube. (**c**) Typical density of states for boron nanotubes, which should be metallic, independent of their chirality [22]

we will focus our discussion on boron nanotubes. As for the boron sheets, the interested reader must be referred to a recent article [22] dealing with structure models for stable boron sheets and their relations to boron nanotubes.

Boron nanotubes were originally postulated in [17] on the basis of an extensive ab initio study, which demonstrated the principal stability of such structures. Beyond that, an much larger class of metal-boron nanotubes was predicted in [23], which is also summarized in [7].

A proper structure model for a pure boron nanotube is shown in Fig. 15.7 (b). From a structural point of view, each boron nanotube may be characterized by a certain chirality, and one may classify them according to a scheme developed for carbon nanotubes (see [10]). When trying to determine the basic electronic properties, one finds that boron nanotubes should always be metallic [22], independent of their chirality, as shown in Fig. 15.7 (c). This is in striking contrast to carbon nanotubes, where the basic electronic properties (metallic vs. semiconducting) depend quite critically on their chirality (see [10]).

Furthermore, recent ab initio simulations of boron clusters claim that boron nanotubes should be more reactive than carbon nanotubes, and much easier to embed into a polymer matrix [22]. The same study also postulates that boron nanotubes have some non-isotropic mechanical behavior, which might be the key for a structure control of such nanosystems, in contrast to carbon, whose isotropic mechanical properties are the main obstacle for structure control! And finally another ab initio study shows that carbon and boron nanotubes should form stable junctions [21].

With all of these nice properties predicted by ab initio calculations, boron nanotubes and similar nanotubular materials could become one of the key materials to carry out the ambitious nanoelectronics program sketched in Fig. 15.4. Therefore many groups in the past have worked on the synthesis boron nanotubes (for a review, see [7]), and the first successful attempt to synthesize boron nanotubes by Ciuparu et. al. [24] is shown in Fig. 15.8 (a). Another interesting result of the increasing activities to synthesize boron nanotubes are the discoveries of novel types of boron nanowires, nanoribbons or nanobelts [7]. One nice example [25] of amorphous boron nanowires is shown in Fig. 15.8 (b).

Theoretical and experimental activities in the field of boron based nanomaterials are strongly increasing [7], and this is only partially due to their technologically interesting materials properties. Another strong motivation might be that large and pure samples of boron nanotubes would be the perfect probe for experimental studies to check our current understanding [26] of electron transport and superconductivity in quasi-one dimension.

(a)

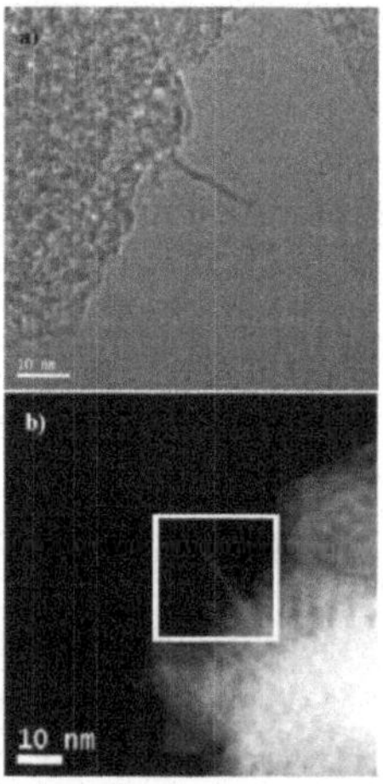

(b)

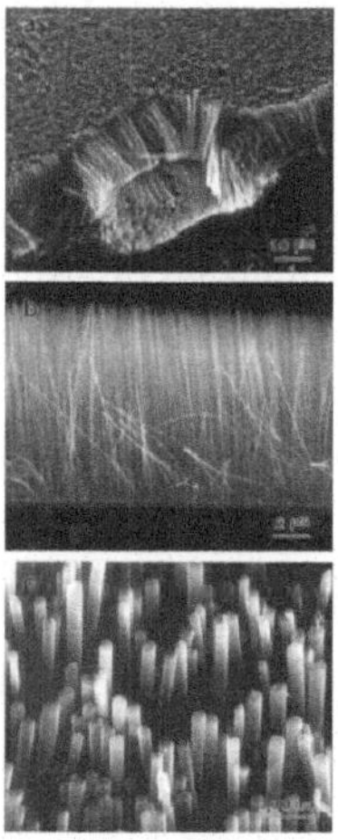

Fig. 15.8. (**a**) Boron nanotubes growing out of a template structure [24]. (**b**) Amorphous boron nanowires [25]

15.3 Model Hamiltonians

In the previous section, the prediction of novel nanomaterials and some of their basic materials properties were largely based on a subtle combination of a large number of ab initio studies and a constant refinement of the corresponding chemical models. For more complex systems such a procedure will be impossible, and in order to describe the most important physical and chemical properties of such systems, one has to fall back on model Hamiltonians. Furthermore, the assumption of independent electrons that interact with a mean field generated by the remaining electrons has to be dropped in the case of strongly correlated systems, like constrained (low-dimensional) systems or high-T_c superconductors.

In such cases, our first task will be to set up a parameterized form for these model Hamiltonians, which will comprise the most important states and the strongest interactions, only. And our second task will be the careful determination of the basic parameters using data from ab initio or experimental studies. Hopefully we will obtain a model Hamiltonian good enough to extract the basic physics and the basic chemistry of complex materials.

Of course there will always be a danger that a "lucky" combination of an over-simplified model, some unrealistic parameters and some invalid approximations will somehow produce the "correct" results. But such a dubious model Hamiltonian will be quite fatal for any advanced computational scheme, which aims at describing complicated processes in complex materials way beyond the range of the invalid approximations that ran into the parameterization of the model Hamiltonian.

Therefore it will be crucial to know how to set up a good model Hamiltonian and to find some realistic parameters for it. To this end we will first discuss some popular types of model Hamiltonians in Sect. 15.3.1, which involve a standard parameterization based on the so-called hopping and Coulomb integrals. Then in Sect. 15.3.2 and 15.3.3 we will describe some techniques to derive the corresponding model parameters from ab initio (and experimental) data, and discuss some of the necessary corrections and augmentations.

Experience shows that there is no standard procedure to derive model Hamiltonians without making any uncontrolled assumptions. Therefore one should use them very carefully, in particular antique model Hamiltonians of rather dubious origin. Mind that in the case of doubt, things are unlikely to become worse if you will sit down and try to find a better model Hamiltonian or a better parameterization, using an ab initio program, and following the basic steps described in this section.

15.3.1 How to Derive Model Hamiltonians

In order to understand the problems involved in deriving a suitable model Hamiltonian, we will switch to a general representation of a many-electron Hamiltonian in the framework of second quantization. For a more detailed description of ab initio methods within such a framework, we refer to the literature [2, 27, 28].

In the second paragraph, we will contrast the exact Hamiltonian with simple model Hamiltonians due to Anderson [29] and Hubbard [30]. This will motivate an intuitive, but less rigorous approach to many-electron problems on the basis of some suitable model Hamiltonians. Finally, in the last paragraph of this section, we will try to bridge the gap between the exact Hamiltonian and the simplest model Hamiltonians by deriving a general downfolding approach by Löwdin [31], which systematically reduces the number of degrees of freedom necessary to describe a complex system.

15.3.1.1 The Language of Second Quantization

We want to represent the electronic part of our general Hamiltonian from (14.1) in the framework of second quantization, where the fermionic degrees will be represented by a set of fermionic field operators

$$\begin{aligned} \psi(\boldsymbol{r}) &= \sum_i \phi_i(\boldsymbol{r}) c_i \ , \\ \psi^\dagger(\boldsymbol{r}) &= \sum_i \phi_i^*(\boldsymbol{r}) c_i^\dagger \end{aligned} \tag{15.17}$$

Here i runs over the labels of a complete basis, including spin. These fermionic operators have the following anticommutator relations:

$$\begin{aligned} [\psi_\sigma^\dagger(\boldsymbol{r}), \psi_{\sigma'}(\boldsymbol{r}')]_+ &= \delta_{\sigma\sigma'}\delta(\boldsymbol{r}-\boldsymbol{r}') \ , \\ [\psi_\sigma(\boldsymbol{r}), \psi_{\sigma'}(\boldsymbol{r}')]_+ &= [\psi_\sigma^\dagger(\boldsymbol{r}), \psi_{\sigma'}^\dagger(\boldsymbol{r}')]_+ = 0 \ , \end{aligned} \tag{15.18}$$

where σ and σ' label the spin components of the field operators. A single Slater determinant defined in (14.23) or (14.34) will be interpreted as a set of creation operators $c_i^\dagger$ acting on the vacuum state $|0\rangle$:

$$\Psi^{\mathrm{SD}}(\phi_1 \ldots \phi_N) = c_1^\dagger \ldots c_N^\dagger |0\rangle \ . \tag{15.19}$$

With

$$H(r,R) = \sum_i^N \left(-\frac{1}{2}\Delta_{\boldsymbol{r}_i}\right) + \sum_i^N \left(-\sum_\alpha^M \frac{Z_\alpha}{|\boldsymbol{r}_i - \boldsymbol{R}_\alpha|}\right) + \sum_{i<j}^N \frac{1}{|\boldsymbol{r}_i - \boldsymbol{r}_j|} \tag{15.20}$$

the electronic part of (14.1) may be re-written as:

$$\begin{aligned} &\int \Big(\psi^\dagger(\boldsymbol{r})\Big(-\frac{1}{2}\Delta_{\boldsymbol{r}} + \sum_\alpha^M \frac{Z_\alpha}{|\boldsymbol{r}-\boldsymbol{R}_\alpha|}\Big)\psi(\boldsymbol{r})\Big) \mathrm{d}\boldsymbol{r} \\ &\quad + \frac{1}{2}\int\int \Big(\psi^\dagger(\boldsymbol{r})\psi^\dagger(\boldsymbol{r}')\frac{1}{|\boldsymbol{r}-\boldsymbol{r}'|}\psi(\boldsymbol{r}')\psi(\boldsymbol{r})\Big)\mathrm{d}\boldsymbol{r}\mathrm{d}\boldsymbol{r}' \\ &\qquad = \sum_{ij} t_{ij} c_i^\dagger c_j + \frac{1}{2}\sum_{ijkl} v_{ijkl} c_i^\dagger c_j^\dagger c_l c_k \equiv H(c^\dagger, c) \ . \end{aligned} \tag{15.21}$$

At this point it already becomes rather clear that any transferable parameterization of a model Hamiltonian will either involve a lot of parameters $t_{\dots}$ and $v_{\dots}$, or one has to find a way to remove a lot of interaction terms and eliminate a lot of degrees of freedom. If successful, one might finally obtain a model Hamiltonian for complex systems, which contains a few adjustable parameters, only. In order to arrive at this point, we better rely on intuitive approaches, where we somehow guess the right form of the Hamiltonian. A useful model Hamiltonian will only comprise those interactions and degrees of freedom that really determine the physical or chemical properties we are interested in.

However the formally neglected interactions and degrees of freedom will not be dropped. Instead we will include them in a chosen model Hamiltonian in terms of a proper renormalization of the model parameters, but according to the following rules:

- Include implicitly, as a renormalization of the parameters, what is not included explicitly in the model.
- What is included explicitly in the model should not be included implicitly ($\Rightarrow$ no double-counting).

15.3.1.2 Simple Model Hamiltonians

In this section, we want to discuss some model Hamiltonians that are useful for our basic understanding of physical and chemical phenomena in complex materials and strongly correlated systems.

Our first task will be to derive a one-electron model Hamiltonian as discussed in Sect. 14.2.3. Such a Hamiltonian is formally given by the first part of (15.19), and it is diagonal in the orthogonal basis spanned by its eigenstates $\phi_i(\boldsymbol{x})$. Nevertheless, apart from the atomic case, these eigenstates are not very localized, and in order to to arrive at a transferable model Hamiltonian, we better try to expand the fermionic field operators $\psi^\dagger(\boldsymbol{r})$ and $\psi(\boldsymbol{r})$ in a complete basis set of (localized, atomic-like) orbitals $\varphi_\mu(\boldsymbol{r})$:

$$\begin{aligned}\psi(\boldsymbol{r}) &= \sum_\mu \varphi_\mu(\boldsymbol{r}) b_\mu \\ \psi^\dagger(\boldsymbol{r}) &= \sum_\mu \varphi^*_\mu(\boldsymbol{r}) b^\dagger_\mu \,.\end{aligned} \tag{15.22}$$

This leads to the following representation:

$$H(c^\dagger, c) = \sum_{i\sigma} \epsilon_i c^\dagger_{i\sigma} c_{i\sigma} \Rightarrow H(b^\dagger, b) = \sum_{\mu\nu\sigma} t_{\mu\nu} b^\dagger_{\mu\sigma} b_{\nu\sigma} \,. \tag{15.23}$$

In order to reduce the number of model parameters for $H(b^\dagger, b)$, we may assume that the diagonal elements of $t_{\mu\nu}$ should be close to atomic energy levels, and that the hopping terms (resonances) should extend to nearest neighbors, only. Furthermore, we may assume that these hopping terms somehow depend on the distance of the hopping centers, and the mutual orientation of the contributing orbitals. We will illustrate this point in more detail in Sect. 15.3.2.

We now discuss a number of model Hamiltonians for strongly localized systems. In such systems, the on-site electron-electron repulsion is much stronger than the resonance energies associated with the overlap of orbitals centered around different atoms. The former effect will keep the electrons as far away from each other as possible, whereas the latter effect will keep them close to each other, in order to maximize the overlap between neighboring orbitals (see Sect. 14.2.1).

The simplest model Hamiltonian for strongly correlated systems comprises two electrons distributed over two orbitals. Following [27], we denote the contributions from theses orbitals with a label l, which means ligand, and a label f, which might stand for a $4f$-electron. The corresponding orbital energies are ϵ_l and ϵ_f with $\epsilon_f < \epsilon_l$. The hybridization between the l and f orbitals, which is characterized by a parameter V, is assumed to be small such that $V \ll (\epsilon_l - \epsilon_f)$. Finally we assume that the strong repulsion between the f-orbitals should be characterized by a very large parameter $U \gg (\epsilon_l - \epsilon_f)$. Then we make the following Ansatz:

$$H = \epsilon_l \sum_\sigma l_\sigma^\dagger l_\sigma + \epsilon_f \sum_\sigma f_\sigma^\dagger f_\sigma + V \sum_\sigma (l_\sigma^\dagger f_\sigma + f_\sigma^\dagger l_\sigma) + U n_\uparrow^f n_\downarrow^f \ . \quad (15.24)$$

The operators $l^{(\dagger)}$ and $f^{(\dagger)}$ create and destroy electrons with spin σ in the corresponding l and f states, and $n_\alpha^f = f_\alpha^\dagger f_\alpha$ is the occupation number for f electrons of spin α. When $V = 0$, the electrons will just sit on their atomic sites, due to the strong repulsion U. Furthermore, a polar state, where two electrons actually sit on the same f site, might safely be excluded by assuming that $U \to \infty$. For further discussion see [27].

Next we present a popular model Hamiltonian for a magnetic impurity embedded in a metal, which is due to Anderson [29]. The basic setup for a $3d$ impurity embedded in a sp host is indicated in Fig. 15.9 (a). The conduction electrons of the periodic sp-host are noninteracting with each other. Instead they interact with

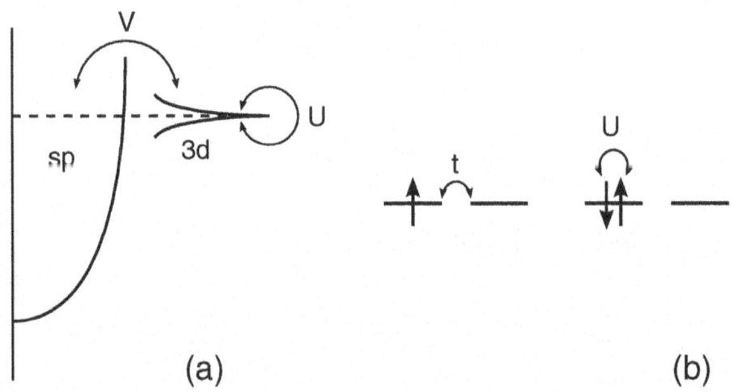

Fig. 15.9. (**a**) Anderson model for a $3d$ impurity embedded in a sp host. There is a weak hybridization V between the host and the impurity, and a strong repulsion U that effects the d-electrons, only. (**b**) Hubbard model to describe strong electron correlations in metallic compounds. We assume hopping t between different sites, and a strong on-site repulsion U

a periodic potential generated by the lattice and the mean field of the remaining electrons. Therefore they may be represented by Bloch states (see [32]):

$$\begin{aligned} \phi_{n\boldsymbol{k}}(\boldsymbol{r}) &= N\mathrm{e}^{\mathrm{i}\boldsymbol{k}\cdot\boldsymbol{r}} u_{n\boldsymbol{k}}(\boldsymbol{r}) \quad \text{with:} \quad u_{n\boldsymbol{k}}(\boldsymbol{r}+\boldsymbol{R}) = u_{n\boldsymbol{k}}(\boldsymbol{r}) \\ &\Longrightarrow \phi_{n\boldsymbol{k}}(\boldsymbol{r}+\boldsymbol{R}) = \mathrm{e}^{\mathrm{i}\boldsymbol{k}\cdot\boldsymbol{R}} \phi_{n\boldsymbol{k}}(\boldsymbol{r}) \,. \end{aligned} \tag{15.25}$$

These states are obviously composed of a plane wave times a function $u_{n\boldsymbol{k}}(\boldsymbol{r})$, which is periodic with the lattice period $\boldsymbol{r}$. They correspond to freely propagating electrons with dispersion $\epsilon_n(\boldsymbol{k})$, where n is the band index, and the continuum $\boldsymbol{k}$ that forms this band are restricted to the first Brillouin zone [32].

According to Hund's rule, there is a multiplet of electronic states distributed among the orbitals of the impurity, and a strong Coulomb repulsion U between spins of different orientation. It is best to describe these states by atomic-like localized orbitals, for example Wannier functions, which are the Fourier transformed of Bloch functions:

$$w_n(\boldsymbol{r}-\boldsymbol{R}) = N' \sum_{\boldsymbol{k}} \mathrm{e}^{-\mathrm{i}\boldsymbol{k}\cdot\boldsymbol{R}} \phi_{n\boldsymbol{k}}(\boldsymbol{r}) \,. \tag{15.26}$$

We also assume a weak hybridization V between the impurity and its host. This leads to the following model Hamiltonian:

$$\begin{aligned} H = & \sum_{\boldsymbol{k}\sigma} \epsilon(\boldsymbol{k}) c^\dagger_{\boldsymbol{k}\sigma} c_{\boldsymbol{k}\sigma} + \epsilon_{3d} \sum_m n^d_m + \frac{U}{2} \sum_{m\neq m'} n^d_m n^d_{m'} \\ & + \sum_{m\boldsymbol{k}\sigma} (V_{m\boldsymbol{k}\sigma} d^\dagger_m c_{\boldsymbol{k}\sigma} + V^*_{m\boldsymbol{k}\sigma} c^\dagger_{\boldsymbol{k}\sigma} d_m) \,. \end{aligned} \tag{15.27}$$

Here m and m' are quantum numbers that characterizes the spin up and spin down multiplets residing on the d-host. The meaning of the remaining terms should be clear from (15.24).

Finally we want to mention a model Hamiltonian due to Hubbard [30], which is used to describe strong correlations among $3d$-electrons in a transition metal (compound), as illustrated in Fig. 15.9 (b). We assume hopping t between different sites, and strong Coulomb repulsion U among spin multiplets characterized by m and m', sitting on the same site. This leads to the following Ansatz:

$$H = \sum_{ij} \sum_{mm'\sigma} t_{im,jm'} d^\dagger_{im\sigma} d_{jm'\sigma} + U \sum_i \sum_{(m\sigma)<(m'\sigma')} n_{im\sigma} n_{im'\sigma'} \,. \tag{15.28}$$

Again, the meaning of all terms should be clear from (15.24) and (15.27). For more details about this model see Chap. 18.

Note that the model Hamiltonians presented in this section are the topic of many research papers, and we will not even try to comment on the physics described by these models. Instead we want to point out that these model Hamiltonians are sometimes augmented by adding long-range Coulomb interactions, electron-phonon coupling and other effects, which might be relevant for the real system under consideration.

15.3.1.3 Downfolding Approach

We now discuss a general downfolding method due to Löwdin [31], which is also known as matrix condensation or Schur complement [33]. It is a general technique that may be applied to one-electron and many-electron Hamiltonians alike. In order to eliminate some degrees of freedom from a quantum mechanical description of a chosen system, we will partition the full Hilbert space related to the system Hamiltonian H into a model space with corresponding projection operator $P = P^2$, and the rest of that Hilbert space with projector $Q = (1 - P) = Q^2$. With a slight abuse of notation, such a partitioning may be formalized in terms of a block matrix representation of the Hamiltonian H, and a vector representation of a general state ψ from the Hilbert space related to H:

$$H \Rightarrow \begin{pmatrix} PHP & PHQ \\ QHP & QHQ \end{pmatrix}; \qquad \psi \Rightarrow \begin{pmatrix} P\psi \\ Q\psi \end{pmatrix} . \tag{15.29}$$

If we let $(H - \epsilon I)$ operate on ψ, where I denotes the identity matrix and ϵ a real number, we will obtain a new state ψ' different from zero, unless ϵ is an eigenvalue, and ψ the corresponding eigenvector:

$$\begin{pmatrix} (PHP - \epsilon PIP) & PHQ \\ QHP & (QHQ - \epsilon QIQ) \end{pmatrix} \cdot \begin{pmatrix} P\psi \\ Q\psi \end{pmatrix} = \begin{pmatrix} P\psi' \\ Q\psi' \end{pmatrix} \tag{15.30}$$

This is equivalent to the following block equations:

$$\begin{aligned} (PHP - \epsilon PIP)P\psi + (PHQ)Q\psi &= P\psi' , \\ (QHP)P\psi + (QHQ - \epsilon QIQ)Q\psi &= Q\psi' . \end{aligned} \tag{15.31}$$

If we multiply the second equation with $-(PHQ)(QHQ - \epsilon QIQ)^{-1}$ and add this to the first equation, we obtain a new set of block equations:

$$\begin{aligned} (H_{\text{red}}(\epsilon) - \epsilon PIP)P\psi &= \psi'_{\text{red}}(\epsilon) , \\ (QHP)P\psi + (QHQ - \epsilon QIQ)Q\psi &= Q\psi' , \end{aligned} \tag{15.32}$$

where H_{red} is a somewhat reduced, but ϵ-dependent matrix, the so-called Schur complement, and ψ'_{red} is a new energy-dependent component of the primed state:

$$\begin{aligned} H_{\text{red}}(\epsilon) &= PHP - PHQ\frac{1}{(QHQ - \epsilon QIQ)}QHP , \\ \psi'_{\text{red}}(\epsilon) &= P\psi' - PHQ\frac{1}{(QHQ - \epsilon QIQ)}Q\psi' . \end{aligned} \tag{15.33}$$

This corresponds to a new matrix equation equivalent to (15.30):

$$\begin{pmatrix} (H_{\text{red}}(\epsilon) - \epsilon PIP) & 0 \\ QHP & (QHQ - \epsilon QIQ) \end{pmatrix} \cdot \begin{pmatrix} P\psi \\ Q\psi \end{pmatrix} = \begin{pmatrix} \psi'_{\text{red}}(\epsilon) \\ Q\psi' \end{pmatrix} . \tag{15.34}$$

If ψ is an eigenvector of H with eigenvalue ϵ, we know from (15.30), that the new state ψ' must be a null vector. And if we plug this into (15.34), we get another eigenvalue problem for the same eigenvalue ϵ, but now the matrix is a tridiagonal block matrix, and the corresponding determinant condition simply reduces to:

$$\det(H_{\text{red}}(\epsilon) - \epsilon PIP) \cdot \det(QHQ - \epsilon QIQ) = 0$$
$$\Rightarrow \det((PHP - \epsilon PIP) - PHQ\frac{1}{(QHQ - \epsilon QIQ)}QHP) = 0\,. \quad (15.35)$$

Thus the reduced Hamiltonian $H_{\text{red}}(\epsilon)$ will have the same spectrum as the original Hamiltonian H, but only if the corresponding eigenstates ψ of H live in our model space selected by P. In practice one does not loose too much accuracy if one uses a modified reduced Hamiltonian $H_{\text{red}}(\widetilde{\epsilon})$, which depends on some suitably chosen (i.e. typical) energy $\widetilde{\epsilon}$.

15.3.2 Parameterization of Hopping Integrals

Now we will describe some successful approaches to determine hopping integrals. In the first paragraph we will present a simple, but nevertheless very accurate description of the band structure of C_{60} using a model Hamiltonian that mainly involves the knowledge of hopping integrals. And in the second paragraph, we discuss the construction and parameterization of analytical Hamiltonians, using a general downfolding technique described above.

15.3.2.1 Band Structure of C_{60}

An interesting example to illustrate the usage of model Hamiltonians are the structural and physical properties of C_{60} molecules and their related solid structures [34]. Undoped C_{60} molecules as shown in Fig. 15.10 (a) crystallize as a sc phase at temperatures below 249 K, but at room temperature the preferred structure is fcc. Within those solid phases, the C_{60} molecules interact with each other over relatively large distances, and their mutual orientation must be the result of a rather weak chemical bonding. This is certainly an interesting problem to be tackled using a suitable model Hamiltonian.

Furthermore it is possible to dope C_{60} solids with alkali atoms $A\,(B)$ = K, Rb, Cs. Those enter the solid at various tetragonal or octagonal sites inside the molecule [35], and each of them donates one extra electron, but they have little effect on the electronic states close to the Fermi energy E_{F} indicated in Fig. 15.10 (b). Also their main structural effect is limited to a mere expansion of the lattices.

For $A_{n-x}B_x\,C_{60}$ solids with $n \leq 3$ one actually observes superconductivity [36] with T_{c} in the range of 40 K, whereas for higher doping levels the solid structure becomes bct or bcc, and superconductivity disappears. Heavily doped compounds with $n = 6$ are insulating. Therefore the electronic structure of (doped) C_{60} solids will be another interesting question to be tackled using model Hamiltonians.

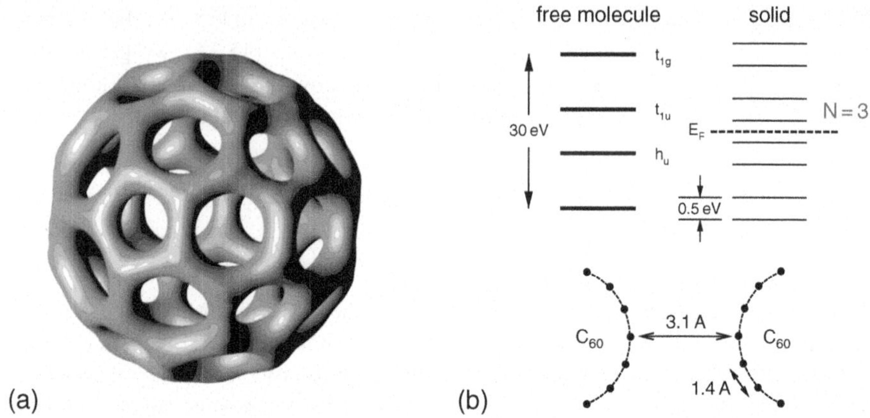

Fig. 15.10. (**a**) C_{60} molecule. (**b**) Structural details and the valence states for molecular and solid C_{60}. The valence states of the molecule broaden to some rather narrow bands of the solid, due to weak C_{60}-C_{60} bonding

Without going into the various details described in [34] and [35], we now want to illustrate how to arrive at a useful model Hamiltonian for (doped) C_{60} solids. Each carbon atom in C_{60} has one $2s$ and three $2p$ orbitals, which form three approximate sp^2 orbitals in the molecular surface pointing towards neighboring carbon atoms, and one radial p_r orbital. The sp^2 orbitals are forming strongly σ-bonding or antibonding orbitals far away from the Fermi level, and they are irrelevant for the properties that we are interested in. But the 60 $2p_r$ orbitals form weakly π-bonding or antibonding orbitals close to the Fermi level, and they point towards neighboring C_{60} molecules. These are the type of atomic orbitals that should be included in our preliminary model Hamiltonian for a C_{60} molecule, which is a simple hopping Hamiltonian similar to (15.23):

$$H = \epsilon_{2p_r} \sum_{i\sigma} c^{\dagger}_{i\sigma} c_{i\sigma} + \sum_{\langle ij \rangle \sigma} t_{ij} c^{\dagger}_{i\sigma} c_{j\sigma} \; . \tag{15.36}$$

Here i and j are running over all the 60 sites of the C_{60} molecule, and the hopping term involves nearest neighbors $\langle ij \rangle$, only. This assumption may be dropped in the case of a solid, but only for those sites that really contribute to the bonding between different C_{60} molecules [35]. Note that in the case of a solid phase, the operators $c_i^{(\dagger)}$ will refer to Bloch states $\phi_{i\boldsymbol{k}}(\boldsymbol{r})$ (see (15.25)), rather than atomic orbitals $\phi_i(\boldsymbol{r})$. In such a case, the molecular states will broaden and become subbands, as indicated in Fig. 15.10 (b).

To make our model more realistic, we assume that the hopping terms t_{ij} will depend on the mutual orientation of the atomic orbitals located at sites $\boldsymbol{R}_i$ and $\boldsymbol{R}_j$, and on the interatomic distance between them. This leads to the following Ansatz [35]:

$$t_{ij}(d_{ij}) = [V_{pp\sigma}(d_{ij}) - V_{pp\pi}(d_{ij})](\widehat{\boldsymbol{R}}_i \cdot \widehat{\boldsymbol{d}}_{ij})(\widehat{\boldsymbol{R}}_j \cdot \widehat{\boldsymbol{d}}_{ij}) + V_{pp\pi}(d_{ij})(\widehat{\boldsymbol{R}}_i \cdot \widehat{\boldsymbol{R}}_j)$$
$$V_{pp\sigma}(d_{ij}) = -4V_{pp\pi}(d_{ij}) = v_\sigma d_{ij} \mathrm{e}^{-\lambda d_{ij}} \ . \tag{15.37}$$

Here $\widehat{\boldsymbol{R}} = \boldsymbol{R}/R$ is a unit vector in the radial direction, and $\widehat{\boldsymbol{d}}_{ij} = \boldsymbol{d}_{ij}/d_{ij}$ is a unit vector between sites i and j. Thus there are three parameters ϵ_{2p_r}, v_σ and λ, which have to be determined.

In order to parameterize our model Hamiltonian, the on-site energy ϵ_{2p_r} can be taken from an atomic calculation, or from the center of gravity of the subbands generated by the $2p_r$ orbitals in the case of a solid. But as long as these levels are the only contributing orbitals, the ϵ_{2p_r} may as well be set equal to zero. The hopping terms can be fitted to ab initio data for simpler molecular carbon systems. Or they may be fitted to the ab initio bandwidth of the subbands generated by the $2p_r$ orbitals in the case of a solid [34], using the general relation between an assumed rectangular density of states of width W (Friedel model, see [37]), and the second moment M_2 of the real density of states for that band (a derivation is given in [37]):

$$\frac{W}{12} = \sqrt{M_2} = \sqrt{\frac{1}{nN} \sum_{\langle ij \rangle} t_{ij} t_{ji}} \ . \tag{15.38}$$

Here n is the number of orbitals that contribute to the band, M is the number of contributing atoms, and $\langle ij \rangle$ runs over all the orbitals in the basis, but the hopping will be restricted to nearest neighbors, only. What we basically have to do now is to count all hopping cycles of length two to the appropriate neighbors, plug this into (15.38), and match the resulting bandwidth with the width of the corresponding density of states. In practice it might be easier to simply adjust the parameter v_σ, such that ab initio bandwidths will be reproduced, and in combination with a variation of lattice sizes, the second parameter λ may be fitted quite accurately [34, 35].

But it turns out that the model Hamiltonian can be reduced even further. In alkali-doped C_{60} molecules, the important orbitals are three degenerate t_{1u} orbitals close to the Fermi level, see Fig. 15.10 (b). In the original basis of the $2p_r$ atomic orbitals $\phi_i(\boldsymbol{r})$, which are located around $\boldsymbol{R}_i$, the three t_{1u} orbitals $\phi'_m(\boldsymbol{r})$ are just:

$$\phi'_m(\boldsymbol{r}) = \sum_{i=1}^{60} c_i^m \phi_i(\boldsymbol{r}) \ . \tag{15.39}$$

The corresponding hopping terms $t_{m\mu,n\nu}$ between two t_{1u} orbitals labelled by m and n, which are associated with different C_{60} molecules located around $\boldsymbol{R}_\mu$ and $\boldsymbol{R}_\nu$, may be calculated from the basic hopping terms t_{ij} defined in (15.37):

$$t_{m\mu,n\nu} = \sum_i^{60} \sum_j^{60} c_i^m c_j^n t_{i\mu,j\nu} \ . \tag{15.40}$$

Thus we finally obtain a Hamiltonian for alkali-doped C_{60} solids, where every molecule may just be described by three t_{1u} states, instead of the $60 \times 4 = 240$ $2p_r$ orbitals:

$$H = \epsilon_{t_{1u}} \sum_{m\mu\sigma} c^{\dagger}_{m\mu\sigma} c_{m\mu\sigma} + \sum_{m\mu,n\nu,\sigma} t_{m\mu,n\nu} c^{\dagger}_{m\mu\sigma} c_{n\nu\sigma} \, . \tag{15.41}$$

Again, the hopping terms should comprise nearest neighbors only, and the basic fitting may be carried out on the basis of the general procedure described above (see (15.38)). In Fig. 15.11, we show a comparison between an ab initio band structure for an alkali-doped C_{60}-molecule and a band structure obtained using the model Hamiltonian from (15.41), which are obviously matching quite well [38]. Other important properties like the orientation of C_{60} molecules and the main features of superconductivity may also be predicted quite reliably [34, 35] using the model Hamiltonians of (15.36) and (15.41).

15.3.2.2 Analytical Hamiltonians Using Downfolding

Whenever the physical or chemical properties of a material are clearly determined by a subset of its orbital states or bands, one may apply the downfolding procedure described in the previous Sect. 15.3.1 ((15.29)–(15.35)) to arrive at simple one-band [39] or two-band [40] Hamiltonians, whose analytical treatment is indeed rather trivial. In the following, we want to repeat the essential steps to arrive at an analytical treatment of the band structure of $YBa_2Cu_3O_7$ described in [39].

The doped cuprates $YBa_2Cu_3O_{7-x}$ (YBCO) are known to be high-T_c superconductors [41], where superconductivity seems to be restricted to a doping range of $x \leq 0.7$. These materials are layered compounds made from copper-oxide planes with a planar unit cell CuO_2, as shown in Fig. 15.12. In YBCO the CuO_4 squares indicated in Fig. 15.12 are actually part of pyramidal units, which leave space for the embedding of the Ba and Y atoms. Furthermore these squares are part of some

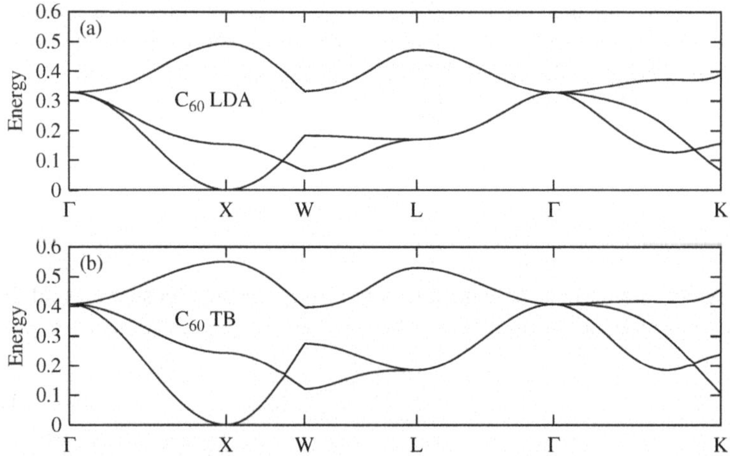

Fig. 15.11. Comparison between an ab initio band structure for RbC_{60} (above) and band structure calculation using the model Hamiltonian of (15.41) shown below [38]

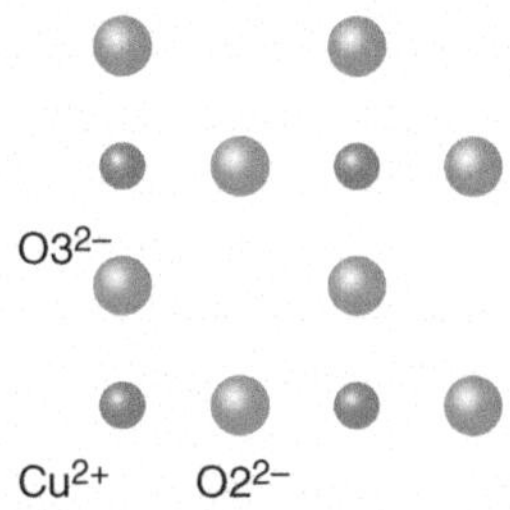

Fig. 15.12. Part of a copper-oxide plane within $YBa_2Cu_3O_7$

CuO_3 chains of edge-sharing CuO_4 squares. Those chains determine various physical properties, but the superconductivity is thought to be mediated by electrons within the copper-oxide planes.

A simple-minded electron count ($Cu^{3+}Cu_2^{2+}$) in undoped $YBa_2Cu_3O_7$ reveals that the Cu in the copper-oxide planes has the formal charge Cu^{2+}, see Fig. 15.12. But the formal charge of Cu^{2+} implies that the d-shell of the copper atom will be incomplete (d^9). The corresponding hole is mainly put into the highest antibonding state of a Cu-O bond, which is of $3d_{x^2-y^2}$ character. In such a situation, one would certainly expect a metallic behavior, but YBCO is actually a semiconductor, caused by strong correlations of the electrons within the copper-oxide planes [41]. Of course, these rather formal considerations must be amended for real YBCO materials, where doping turns out to be an essential precondition for superconductivity.

If we take all of these basic structural and electronic features into account, the band structure of YBCO may be simplified using a first downfold to remove all bands other than the ones that refer to electrons within the planes. This leads to an 8-band model Hamiltonian, which comprises the states of type $Cu_{x^2-y^2}$, $O2_x$, $O3_y$, Cu_s, Cu_{xz}, Cu_{yz}, $O2_z$ and $O3_z$. This model Hamiltonian may be parameterized following a procedure explained in the previous paragraph (see (15.38)). But in order to obtain an orthonormal model Hamiltonian, we have to modify our general downfolding procedure. To understand this, we first expand the Hamiltonian $H_{\text{red}}(\epsilon)$ of (15.33) around the Fermi energy ϵ_F [39, 40]:

$$H_{\text{red}}(\epsilon) - \epsilon \approx H_{\text{red}}(\epsilon_F) + (\epsilon - \epsilon_F)\frac{\partial H_{\text{red}}}{\partial \epsilon} - \epsilon \equiv H - \epsilon S \,. \qquad (15.42)$$

Up to first order in ϵ, the expansion will obviously lead to a generalized eigenvalue problem that we already encountered before in (14.13). Such a generalized eigenvalue problem indicates that the chosen basis functions are non-orthogonal. Therefore, in order to obtain a Hamiltonian for an orthogonal basis, we just have to make the following transformation [2]:

$$\begin{aligned}&(H - \epsilon S)C = 0\\ &\Longrightarrow S^{-\frac{1}{2}}HS^{-\frac{1}{2}}(S^{\frac{1}{2}}C) - \epsilon S^{-\frac{1}{2}}SS^{-\frac{1}{2}}(S^{\frac{1}{2}}C) = (H' - \epsilon I)C' = 0 \,. \qquad (15.43)\end{aligned}$$

We may then continue to downfold copper and oxygen bands and arrive at a 3-band model Hamiltonian, which contains the $Cu_{x^2-y^2}$, $O2_x$ and $O3_y$ bands.

The price to pay is that oxygen on-site energies will be renormalized, and that the reduced Hamiltonian will contain 2nd-nearest-neighbor $O2_x \leftrightarrow O3_y$ hopping (see Fig. 15.12), as well as 3rd-nearest-neighbor $O2_x \leftrightarrow O2_x$ and $O3_y \leftrightarrow O3_y$ hopping. In other words, hopping becomes more and more long-ranged, and downfolding remains accurate solely over a smaller and smaller energy range.

If we finally downfold the remaining oxygen bands to obtain a 1-band model Hamiltonian for the essential $Cu_{x^2-y^2}$ band, the latter will contain up to 9th-nearest-neighbor hopping integrals [39]. After all, the downfolded bands did not vanish into thin air! All the way down to the 1-band model Hamiltonian, their basic character survived in the various renormalizations of the remaining on-site and hopping terms.

Finally we want to recommend another study, which employs the downfolding procedure described in this paragraph to a much simpler and exactly solvable model for $3d$ compounds [42]. That paper also illustrates some of the techniques discussed in the following section.

15.3.3 Parameterization of Coulomb Integrals

In this section, we present a method for the parameterization of Coulomb integrals based on constrained density functional theory. The first paragraph will explain the main theoretical concepts behind such an approach. However, this procedure will not be perfect, and therefore we also added a short second paragraph, where we briefly mention some of the necessary corrections, and give some helpful references.

15.3.3.1 Constrained Density Functional Theory

We want to focus again on the Anderson Hamiltonian given in (15.27). In order to parameterize it, we may assume that the hybridization between the band electrons and the electrons of the impurity are effectively zero, and therefore the localized electrons of the impurity are somewhat decoupled from the rest of the system. Then we obtain the following model Hamiltonian:

$$\begin{aligned} H &= \epsilon_{3d} \sum_m n_m^d + \frac{U}{2} \sum_{m \neq m'} n_m^d n_{m'}^d \\ \rightarrow \quad E(n) &= \epsilon_{3d} n + \frac{U}{2} n(n-1) \end{aligned} \tag{15.44}$$

The second line follows from the fact that under the given circumstances, the occupation number $n = \sum_m \langle d_m^\dagger d_m \rangle$ is a good quantum number. We immediately see that:

$$\begin{aligned} \frac{\partial E(n)}{\partial n} &= \epsilon_{3d} + Un - \frac{U}{2} \equiv \epsilon_{3d}^{\text{DFT}} \\ \frac{\partial^2 E(n)}{\partial n^2} &= U = E(n+2) + E(n) - E(n+1) = \frac{\partial \epsilon_{3d}^{\text{DFT}}}{\partial n} \, . \end{aligned} \tag{15.45}$$

In the first line, we were setting the derivative of the energy as a functional of the occupation number of the $3d$ states equal to the corresponding one-particle state of the Kohn-Sham equations. This is known as Janak's theorem, and an elementary derivation can be found in [43]. Furthermore we see from the second line in (15.45), that U might in principle be obtained from the knowledge of total energies $E(n)$ for discrete changes of the occupation numbers n, or from the knowledge of the variation of the $3d$ Kohn-Sham eigenstate as a function of a continuous n.

To determine U for rare earth compounds, one can follow Herring [44] and assume that changes in the occupation numbers of the $4f$ states are accompanied by changes in the occupation of other localized atomic states, such that the atom as a whole will remain neutral (perfect screening). This approach was used in [45] and [46], and good agreement with experiment [47] was obtained, see Fig. 15.13. Later work [48] confirmed that perfect screening is indeed a rather useful assumption for rare earth compounds, but not for transition metal compounds.

When the perfect screening assumption is invalid, one can apply constrained density functional theory [49]. Here the idea is to fix the occupation number of a Kohn-Sham state ϕ_i to a value N_i by introducing a Lagrangian multiplier v. To this end, we formally rewrite (14.41) including this additional Lagrangian multiplier:

$$E[N_i] = \min_{\phi_k(\boldsymbol{r})} \left[F[n(\boldsymbol{r})] + v \int_{\Omega} (n_i(\boldsymbol{r}) - N_i)\, \mathrm{d}\boldsymbol{r} \right] . \tag{15.46}$$

We then obtain a set of one-particle equations similar to (14.42), but with an additional projection potential v, which acts on $\phi_i(\boldsymbol{r})$ in a restricted (atomic) domain Ω, only. All other orbitals are allowed to relax, thus describing an optimally screened excitation.

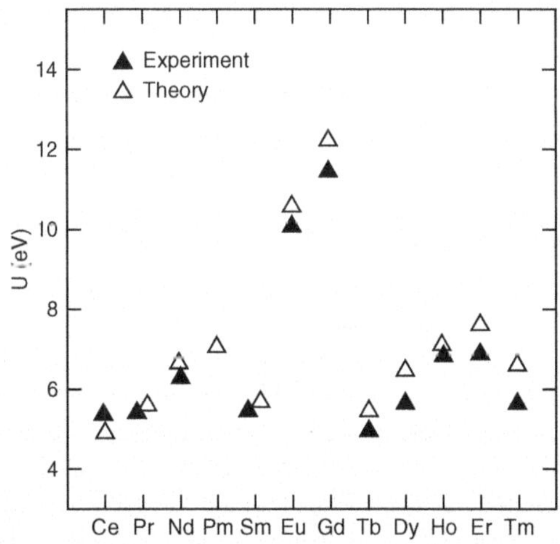

Fig. 15.13. Experimental verification [47] of an early ab initio determination of U for $4f$ elements [45, 46]

The projection potential v usually has to be adjusted by hand, in order to obtain a certain value of $n_i(\boldsymbol{r})$ within the atomic domain Ω. If we now return to our original problem of determining U for a local $3d$ impurity inside a (decoupled) host, we immediately realize from (15.45) that such a variation of v will be the key to determine U from ab initio calculations. To this end, we only have to calculate the variation of the Kohn-Sham eigenstate $\epsilon_{3d}^{\mathrm{DFT}}(N_{3d})$ with N_{3d}, or alternatively, we will have to calculate the (constrained) total energies $E[N_{3d}]$, $E[N_{3d}+1]$ and $E[N_{3d}+2]$.

Finally we want to point out that constrained density functional theory is a very general approach to obtain the parameterization of other popular model Hamiltonians. In particular, constrained density functional theory may be used to parameterize some standard magnetic model Hamiltonians [49].

15.3.3.2 Some Corrections

So far we could make a clear suggestion to determine U using constrained density functional theory, but somehow the hopping terms in the Anderson model of (15.27) have evaporated into thin air. There are various studies that actually show how to include these terms.

In [50] the authors parameterize the complete Anderson Hamiltonian of (15.27) by matching the energy hypersurfaces $E[n]$ for constrained density functional theory and for the model system, which is treated in a self-consistent mean-field fashion. The fitting procedure of the hopping terms is carried out using the techniques described in the previous section. These authors of [50] also show how to avoid double-counting by removing some of the kinetic energy contributions to U induced by constrained DFT.

In [51, 52] the authors present a technique to cut hopping within the Anderson model by removing hopping integrals from localized orbitals. This method takes advantage of the sparsity, locality and near-orthogonality of general muffin-tin orbital basis sets used with LMTO [53].

The value of U may also be taken from experiment. If we return to the problem of finding an appropriate model Hamiltonian for a free C_{60} molecule described in Sect. 15.3.2, it might be much more realistic to actually introduce some molecular U_{mol} for the mutual repulsion between two electrons in a t_{1u} orbital on the same C_{60} molecule. Then this parameter U_{mol} may be determined using constrained DFT [54]:

$$U_{\mathrm{mol}} = \frac{\partial \epsilon_{t_{1u}}}{\partial n_{t_{1u}}} \,. \tag{15.47}$$

In order to determine an U_{solid} for solid C_{60}, these authors included a dipole interaction in a self-consistent fashion [54]. When compared to the results of Auger spectroscopy [55], this model Hamiltonian gives excellent agreement between theoretical and experimental data. Therefore other authors [55] have given some detailed instructions how to determine U_{solid} from experimental data. The resulting model Hamiltonian may be further improved by including interactions between the t_{1u} states of energy $\epsilon_{t_{1u}}$ (generated and destroyed by $c_m^{\dagger}, c_m$), and some intramolecular (!) phonon modes of energy ω_μ (generated and destroyed by $b_\mu^{\dagger}, b_\mu$), see [56]:

$$H = \cdots + \epsilon_{t_{1u}} c^{\dagger}_{m} c_{m} + \omega_{\mu} b^{\dagger}_{\mu} b_{\mu} + g_{mn;\mu} (b^{\dagger}_{\mu} + b_{\mu}) c^{\dagger}_{m} c_{n} \tag{15.48}$$

where

$$|g_{mn;\mu}|^2 \sim \Delta\epsilon_{t_{1u}}(\omega_\mu) \quad \Rightarrow \quad \lambda \sim N(\epsilon_\mathrm{F})(\Delta\epsilon_{t_{1u}}(\omega_\mu))^2 \, . \tag{15.49}$$

The coupling constant $g_{mn;\mu}$ is related to the shift $\Delta\epsilon_{t_{1u}}(\omega_\mu)$ of the atomic energies $\epsilon_{t_{1u}}$, when the C_{60} molecule is distorted in the direction of the phonon mode corresponding to ω_μ. In order to determine the dimensionless coupling constant λ, we also have to know the density of states at the Fermi level, denoted by $N(\epsilon_\mathrm{F})$. All of these values may easily be extracted from ab initio data, see [56].

The Hamiltonian of (15.48) already includes the Jahn-Teller effect due to the coupling to phonons with H_g-symmetry [57]. It is also possible to include Hund's rule coupling [58], and such an augmented model Hamiltonian, which is entirely based on ab initio data, can be used to develop a consistent theory of strong superconductivity in C_{60} solids [58, 59].

In summary, it seems that for $4f$ compounds, C_{60} and high-T_c cuprates, one may determine U rather accurately. For many $3d$ compounds, the theoretically determined U turns out to be too large [60]. However, a recent study, which includes proper RPA screening, leads to largely improved results even for early $3d$ systems [61].

15.4 Summary and Outlook

We want to start our summary with a short remark about the topics that we did not treat in these lecture notes. First of all, the mathematical background of ab initio methods would be a topic too complex to be treated here, but there is a recent review article [62], which explains the main directions and some key results. Second we did not treat the whole field of ab initio molecular dynamics, but we frequently cited a rather detailed review article by Payne et al.[63], which is devoted to this topic. We highly recommend this article, in particular, as it also describes in some detail the numerical background of modern ab initio methods.

Other than that, we took the reader on a rather long and extensive trip, starting from the basic Hamiltonian of (14.1) and the basic variational principle of (14.1). We gave strong support to the one-electron picture in chemistry (Sect. 14.2), and used this concept to analyze some of the key ab initio methods in Sects. 14.3 and 14.4. In Sect. 15.1 we explained the concept of ab initio energy hypersurfaces, and discussed various ways to explore them, and in Sect. 15.2 we showed how to use ab initio methods to find new classes of nanomaterials. Finally in Sect. 15.3 we described how to set up model Hamiltonians using ab initio data, in order to understand the properties of complex or defective materials, where the full ab initio program described in Sects. 14.3–15.1 will not be applicable. Such parameterized model Hamiltonians are also central to many of the theoretical methods described in these lecture notes, and the accuracy of their parameterizations will be crucial to obtain qualitative *and* quantitative results for advanced computational methods. Which emphasizes again the importance of ab initio methods for our basic understanding of the physical and chemical properties of molecules and solids.

Right at the beginning, we pointed out that a continuing success of modern ab initio methods will not only depend on better theoretical concepts, but also on better algorithms and better computing hardware. Better theoretical concepts might imply the construction of novel basis sets, which are ideally suited to treat mesoscopic or low-dimensional systems, rather than the popular Gaussian or planewave basis sets implemented in many ab initio packages. Better algorithms could imply novel algorithms for sparse-matrix eigenvalue problems, or just some new techniques to visualize and analyze chemistry data provided by ab initio methods. And better computing hardware could imply new techniques of distributed computing, or just a brave jump into a new technology.

Whatever simulation tools the future may bring, two things will always remain: Ab initio alchemists who want to treat ab initio simulations like a black box, and new pages in the "Journal of Non Reproducible ab initio Results". Therefore we finally added a short Appendix to help you choose your alchemist's package of choice. Have fun!

Finally the author would like to thank J. Kunstmann (MPI FKF Stuttgart) for various illustrations used in Sect. 15.2.2, and O. Gunnarsson (MPI FKF Stuttgart) for his lecture notes and a number of illustrations, that were forming the basis of Sect. 15.3.

Appendix 15.A Links to Popular Ab Initio Packages

In the following, we will list the web addresses of some popular ab initio packages. This list is incomplete, and only the packages marked with an asterisk are free of charge. Some packages may already belong to the standard equipment of your local chemistry or materials science departments, or to one of the larger supercomputing centers to where you routinely submit larger computing tasks. Please check carefully before you decide to pay a lot of money.

Typical quantum chemistry packages are:

- GAUSSIAN, an all purpose gaussian based ab initio package that features virtually all methods of modern quantum chemistry (`www.gaussian.org`).
- GAMESS-UK*, a free quantum chemistry package featuring a lot of methods (`www.cfs.dl.ac.uk`), which evolved from an earlier program called GAMESS. The latter has also improved over the years (`www.msg.ameslab.gov/GAMESS`).

For solid state and materials science applications, there are a number of well-maintained packages available, most of them based on density functional theory:

- VASP, a planewave and density functional based code that is very popular among solid state physicists and materials scientists (`cms.mpi.univie.ac.at/vasp/`).

- SIESTA*, another planewave and density functional based code similar to VASP (`http://www.uam.es/departamentos/ciencias/fismateriac/siesta`).
- ABINIT*, an open source planewave and density functional based ab initio package, which is maintained by an very active newsgroup (`www.abinit.org`).
- TB-LMTO-ASA*, a density functional based ab initio package featuring Muffin-Tin-orbitals (`www.fkf.mpg.de/andersen`). It is fast, easy to handle, and may directly be used to set up analytical models, see Sect. 15.3 and [37].
- CRYSTAL, a package that contains Hartree-Fock and density functional based methods for solid systems (`www.crystal.unito.it`).

References

1. D.J. Wales, *Energy Landscapes* (Cambridge University Press, Cambridge, 2003)
2. A. Szabo, N.S. Ostlund, *Modern Quantum Chemistry* (McGraw-Hill, New York, 1989)
3. S.K. Card, J.D. MacKinlay, B. Shneiderman, *Readings in information visualization : using vision to think* (Morgan Kaufmann Publishers, San Francisco, 1999)
4. Y. Yamaguchi, Y. Osamura, J.D. Goddard, H.F.S. III, *A New Dimension to Quantum Chemistry : Analytic Derivative Methods in Ab initio Molecular Electronic Structure Theory* (Oxford University Press, Oxford, 1994)
5. W.H. Press, S.A. Teukolsky, W.T. Vetterling, B.R. Flannery, *Numerical Recipes*, Vol. 1, 2nd edn. (Cambridge University Press, Cambridge, 1992)
6. F. Weinhold, C.R. Landis, *Valence and Bonding. A Natural Bond Orbital Donor–Acceptor Perspective* (Cambridge University Press, Cambridge, 2005)
7. A. Quandt, I. Boustani, Chem. Phys. Chem **6**, 2001 (2005)
8. K.E. Drexler, *Nanosystems* (Wiley, New York, 1992)
9. E.L. Wolf, *Nanophysics and Nanotechnology* (Wiley-VCH, Weinheim, 2004)
10. M.S. Dresselhaus, G. Dresselhaus, P. Eklund, *Science of Fullerenes and Carbon Nanotubes* (Academic Press, San Diego, 1996)
11. E. Joselevich, C.M. Lieber, Nano Lett. **2**, 1137 (2002)
12. B. Halford, Chem. and Eng. News **83**, 30 (2005)
13. W. Tremel, Angew. Chem. **111**, 2311 (1999)
14. L. Pauling, *The Nature of the Chemical Bond*, 3rd edn. (Cornell University Press, Ithaca, 1960)
15. S.K. Ritter, Chem. and Eng. News **82**, 28 (2004)
16. I. Boustani, Phys. Rev. B **55**, 16426 (1997)
17. I. Boustani, A. Quandt, Europhys. Lett. **39**, 527 (1997)
18. A. Gindulyte, N. Krishnamachari, W.N. Lipscomb, L. Massa, Inorg. Chem **37**, 6546 (1998)
19. S. Lee, D.M. Bylander, L. Kleinmann, Phys. Rev. B **42**, 1316 (1990)
20. K. Exner, P. v. R. Schleyer, Science **290**, 1937 (2000)
21. J. Kunstmann, A. Quandt, J. Chem. Phys. **121**, 10680 (2004)
22. J. Kunstmann, A. Quandt, Phys. Rev. B **74**, 035413 (2006)
23. A. Quandt, A.Y. Liu, I. Boustani, Phys. Rev. B **64**, 125422 (2001)
24. D. Ciuparu, R.F. Klie, Y. Zhu, L. Pfefferle, J. Phys. Chem. B **108**, 3967 (2004)

25. L. Cao, Z. Zhang, L. Sun, C. Gao, M. He, Y. Wang, Y. Li, X. Zhang, G. Li, J. Zhang, W. Wang, Adv. Mater. **13**, 1701 (2001)
26. Y. Imry, *Introduction to mesoscopic physics*, 2nd edn. (Oxford University Press, Oxford, 2002)
27. P. Fulde, *Electron Correlations in Molecules and Solids*, 3rd edn. (Springer, Berlin Heidelberg New York, 1995)
28. F.E. Harrisa, H.J. Monkhorst, D.L. Freeman, *Algebraic and Diagrammatic Methods in Many-Fermion Theory* (Oxford University Press, Oxford, 1989)
29. P.W. Anderson, Phys. Rev. **124**, 41 (1961)
30. J. Hubbard, Proc. Roy. Soc. (London) A **276**, 238 (1963)
31. P.O. Löwdin, J. Chem. Phys. **19**, 1396 (1951)
32. N.W. Ashcroft, N.D. Mermin, *Solid State Physics* (Holt, Rinehard and Winston, Philadelphia, 1976)
33. G.H. Golub, C.F. van Loan, *Matrix computations* (The Johns Hopkins University Press, Baltimore London, 1996)
34. O. Gunnarsson, S. Satpathy, O. Jepsen, O.K. Andersen, Phys. Rev. Lett. **67**, 3002 (1991)
35. S. Satpathy, V.P. Antropov, O.K. Andersen, O. Jepsen, O. Gunnarsson, A.I. Liechtenstein, Phys. Rev. B **46**, 1773 (1992)
36. M.J. Rosseinsky, A.P. Ramizez, S.H. Glarum, D.W. Murphy, R.C. Haddon, A.F. Hebard, T.T.M. Palstra, A.R. Kortan, S.M. Zahurak, A.V. Makhija, Phys. Rev. Lett. **66**, 2830 (1991)
37. W.A. Harrison, *Elementary Electronic Structure*, revised edn. (World Scientific, Singapore, 2004)
38. O. Gunnarsson, S.C. Erwin, R.M.M. E. Koch, Phys. Rev. B **57**, 2159 (1998)
39. O.K. Andersen, A.I. Liechtenstein, O. Jepsen, F. Paulsen, J. Phys. Chem. Solids **56**, 1573 (1995)
40. O. Jepsen, O.K. Andersen, Z. Phys. B **97**, 35 (1995)
41. P.W. Anderson, *The theory of superconductivity in the high-T_c cuprates* (Princeton University Press, Princeton, 1997)
42. O. Gunnarsson, Phys. Rev. B **41**, 514 (1990)
43. R.G. Parr, W. Yang, *Density Functional theory of Atoms and Molecules* (Oxford University Press, Oxford, 1989)
44. C. Herring, *Magnetism* (Academic Press, New York, 1966)
45. J.F. Herbst, R.E. Watson, J.W. Wilkins, Phys. Rev. B **13**, 1439 (1976)
46. J.F. Herbst, R.E. Watson, J.W. Wilkins, Phys. Rev. B **17**, 3089 (1978)
47. J.K. Lang, Y. Baer, P.A. Cox, Phys. Lett. **42**, 74 (1979)
48. V.I. Anisimov, O. Gunnarsson, Phys. Rev. B **43**, 7570 (1991)
49. P.H. Dederichs, S. Blügel, R. Zeller, H. Akai, Phys. Lett. **53**, 2512 (1984)
50. M.S. Hybertsen, M. Schlüter, N.E. Christensen, Phys. Rev. B **39**, 9028 (1989)
51. A.K. McMahan, R.M. Martin, S. Satpathy, Phys. Rev. B **38**, 6650 (1988)
52. O. Gunnarsson, O.K. Andersen, J.Z. O. Jepsen, Phys. Rev. B **39**, 1708 (1989)
53. O.K. Andersen, Z. Pavlowska, O. Jepsen, Phys. Rev. B **34**, 5253 (1986)
54. V.P. Antropov, O. Gunnarsson, O. Jepsen, Phys. Rev. B **46**, 13647 (1992)
55. R.W. Lof, M.A. van Veenendaak, B. Koopmans, H.T. Jonkman, G.A. Sawatzky, Phys. Rev. Lett. **68**, 3924 (1992)
56. V.P. Antropov, O. Gunnarsson, A.I. Liechtenstein, Phys. Rev. B **48**, 7651 (1993)
57. J.E. Han, E. Koch, O. Gunnarsson, Phys. Rev. Lett. **84**, 1276 (2000)
58. M. Capone, M. Fabrizio, C. Castellani, E. Tosatti, Science **296**, 2364 (2002)
59. J.E. Han, O. Gunnarsson, V.H. Crespi, Phys. Rev. Lett. **90**, 167006 (2003)

60. V. Drchal, O. Gunnarsson, O. Jepsen, Phys. Rev. B **44**, 3518 (1991)
61. F. Aryasetiawan, K. Karlsson, O. Jepsen, U. Schönberger, Phys. Rev. B **74**, 125106 (2006)
62. C.L. Bris, P.L. Lions, Bull. Am. Math. Soc. **42**, 291 (2005)
63. M.C. Payne, M.P. Teter, D.C. Allan, T.A. Arias, J.D. Joannopoulos, Rev. Mod. Phys. **64**, 1045 (2002)

Part VII

Effective Field Approaches

16 Dynamical Mean-Field Approximation and Cluster Methods for Correlated Electron Systems

Thomas Pruschke

Institute for Theoretical Physics, University of Göttingen, 37077 Göttingen, Germany
pruschke@theorie.physik.uni-goettingen.de

Among the various approximate methods used to study many-particle systems the simplest are mean-field theories, which map the interacting lattice problem onto an effective single-site model in an effective field. Based on the assumption that one can neglect non-local fluctuations, they allow to construct a comprehensive and thermodynamically consistent description of the system and calculate various properties, for example phase diagrams. Well-known examples for successful mean-field theories are the Weiss theory for spin models or the Bardeen-Cooper-Schrieffer theory for superconductivity. In the case of interacting electrons the proper choice of the mean-field becomes important. It turns out that a static description is no longer appropriate. Instead, a *dynamical* mean-field has to be introduced, leading to a complicated effective single-site problem, a so-called quantum impurity problem.

This chapter gives an overview of the basics of dynamical mean-field theory and the techniques used to solve the effective quantum impurity problem. Some key results for models of interacting electrons, limitations as well as extensions that systematically include non-local physics are presented.

16.1 Introduction

Strongly correlated electron systems still present a major challenge for a theoretical treatment. The simplest model describing correlation effects in solids is the one-band Hubbard model [1, 2, 3]

$$H = \sum_{i,j,\sigma} t_{ij} c^\dagger_{i\sigma} c_{j\sigma} + U \sum_i c^\dagger_{i\uparrow} c_{i\uparrow} c^\dagger_{i\downarrow} c_{i\downarrow} \,, \tag{16.1}$$

where we use the standard notation of second quantization to represent the electrons for a given lattice site $\boldsymbol{R}_i$ and spin orientation σ by annihilation (creation) operators $c^{(\dagger)}_{i\sigma}$. The first term describes a tight-binding band with tunneling amplitude for the conduction electrons t_{ij}, while the second represents the local part of the Coulomb interaction. Since for this model we assume that the conduction electrons do not have further orbital degrees of freedom, this local Coulomb interaction acts only if two electrons at the same site $\boldsymbol{R}_i$ with opposite spin are present.

The complementary nature of the two terms present in the Hubbard model (16.1) – the kinetic energy or tight-binding part is diagonal in momentum representation,

T. Pruschke: *Dynamical Mean-Field Approximation and Cluster Methods for Correlated Electron Systems*, Lect. Notes Phys. **739**, 473–503 (2008)
DOI 10.1007/978-3-540-74686-7_16

the interaction part in direct space – already indicates that it will be extremely hard to solve. One can, however, get at least for filling $\langle n\rangle = 1$ (half filling) some insight into the physics of the model by a few simple arguments: In the limit $U \to 0$ we will have a simple metal. On the other hand, for $t_{ij} \to 0$, or equivalently $U \to \infty$, the system will consist of decoupled sites with localized electrons and hence represents an insulator. We thus can expect that there exists a critical value U_c, where a transition from a metal to an insulator occurs. Furthermore, from second order perturbation theory around the atomic limit [4], we find that for $|t_{ij}|/U \to 0$ the Hubbard model (16.1) maps onto a Heisenberg model

$$H = \sum_{ij} J_{ij} \boldsymbol{S}_i \cdot \boldsymbol{S}_j \,, \tag{16.2}$$

where $\boldsymbol{S}_i$ represents the spin operator at site $\boldsymbol{R}_i$ and the exchange constant is given by

$$J_{ij} = 2\frac{t_{ij}^2}{U} > 0 \,. \tag{16.3}$$

Note that this immediately implies that we will have to expect that the ground state of the model at half filling will show strong antiferromagnetic correlations.

Away from half filling $\langle n\rangle \neq 1$ the situation is much less clear. There exists a theorem by Nagaoka [5], that for $U = \infty$ and one hole in the half-filled band the ground state can be ferromagnetic due to a gain in kinetic energy; to what extent this theorem applies for a thermodynamically finite doping and finite U has not been solved completely yet. The mapping to the Heisenberg model can still be performed leading to the so-called t-J model [4], which again tells us that antiferromagnetic correlations will be at least present and possibly compete with Nagaoka's mechanism for small J_{ij} or even dominate the physics if J_{ij} is large enough. This is the realm where models like the Hubbard or t-J model are thought to describe at least qualitatively the physics of the cuprate high-T_C superconductors [6].

The energy scales present in the model are the bandwidth W of the tight-binding band and the local Coulomb parameter U. From the discussion so far it is clear that typically we will be interested in the situation $U \approx W$ or even $U \gg W$. This means, that there is either no clear-cut separation of energy scales, or the largest energy scale in the problem is given by the two-particle interactions. Thus, standard perturbation techniques using the interaction as perturbation are usually not reliable even on a qualitative level; expansions around the atomic limit, on the other hand, are extremely cumbersome [7] and suffer from non-analyticities [8] which render calculations at low temperatures meaningless.

The knowledge on correlated electrons systems in general and the Hubbard model (16.1) in particular acquired during the past decades is therefore mainly due to the development of a variety of computational techniques, for example quantum Monte Carlo (QMC), exact diagonalization (ED), and the density-matrix renormalization group (see Parts V, VIII and IX). Since these methods – including modern developments – have been covered in great detail during this school, I will not discuss them again at this point but refer the reader to the corresponding chapters in this book. The aspect interesting here is that basically all of them are restricted to low-dimensional systems: For ED, calculations in $D > 2$ are impossible due to the size of the Hilbert space, and in case of the DMRG the way the method is constructed restricts it basically to $D = 1$. QMC in principle can be applied to any system; however, the sign problem introduces a severe limitation to the range of applicability regarding system size, temperature or interaction strengths.

In particular the restriction to finite and usually also small systems make a reliable discussion of several aspects of the physics of correlated electron systems very hard. Typically, one expects these materials to show a rather large variety of ordered phases, ranging from different magnetic phases with and without orbital or charge ordering to superconducting phases with properties which typically cannot be accounted for in standard weak-coupling theory [9]. Moreover, metal-insulator transitions driven by correlation effects are expected [9], which are connected to a small energy scale of the electronic system. Both aspects only become visible in a macroscopically large system: For small finite lattices phase transitions into ordered states cannot appear, and an identification of such phases requires a thorough finite-size scaling, which usually is not possible. Furthermore, finite systems typically have finite-size gaps scaling with the inverse system size, which means that small low-energy scales appearing in correlated electron materials cannot be identified.

These restrictions motivate the question, if there exists a – possibly approximate – method that does not suffer from restrictions on temperature and model parameters but nevertheless works in the thermodynamic limit and thus allows for phase transitions and possibly very small low-energy scales dynamically generated due to the correlations. Such methods are the subject of this contribution.

In Sect. 16.2.1 I will motivate them on a very basic level using the concept of the mean-field theory well-known from statistical physics, and extend this concept for the Hubbard model in Sect. 16.2.2, obtaining the so-called dynamical mean-field theory (DMFT). As we will learn in Sect. 16.2.2.4, the DMFT still constitutes a non-trivial many-particle problem and I will thus briefly discuss techniques available to solve the equations of the DMFT. Following some selected results for the Hubbard model in Sect. 16.2.3 I will touch a recent development to use DMFT in material science in Sect. 16.2.4. Section 16.3 of this contribution will deal with extensions of the DMFT, which will be motivated in Sect. 16.3.1. The actual algorithms and their computational aspects will be discussed in Sects. 16.3.2 and 16.3.3. Some selected results for the Hubbard model in Sect. 16.3.4 will finish this chapter.

16.2 Mean-Field Theory for Correlated Electron Systems

16.2.1 Classical Mean-Field Theory for the Heisenberg Model

In the introduction to the Hubbard model we already encountered the Heisenberg model (16.2) as low-energy limit for $U \to \infty$. Regarding its solvability, this model shares some more features with the Hubbard model in that it poses a computational rather hard problem in $D \geq 2$. However, there exists a very simple approximate theory which nevertheless describes the properties of the Heisenberg model at least qualitatively correct, the Weiss mean-field theory [10]. As we all learned in the course on statistical physics, the basic idea of this approach is the approximate replacement

$$\boldsymbol{S}_i \cdot \boldsymbol{S}_j \approx \boldsymbol{S}_i \cdot \langle \boldsymbol{S}_j \rangle_{\text{MFT}} + \langle \boldsymbol{S}_i \rangle_{\text{MFT}} \cdot \boldsymbol{S}_j \tag{16.4}$$

$$H \approx \sum_i H^{(i)}_{\text{MFT}} = 2 \sum_{ij} J_{ij} \boldsymbol{S}_i \cdot \langle \boldsymbol{S}_j \rangle_{\text{MFT}} \tag{16.5}$$

where $\langle \ldots \rangle_{\text{MFT}}$ stands for the thermodynamic average with respect to the mean-field Hamiltonian (16.5) and we dropped a for the present discussion unimportant term $\langle \boldsymbol{S}_i \rangle_{\text{MFT}} \cdot \langle \boldsymbol{S}_j \rangle_{\text{MFT}}$. If we define an effective magnetic field or Weiss field according to

$$\boldsymbol{B}_{i,\text{MF}} := 2 \sum_{j \neq i} J_{ij} \langle \boldsymbol{S}_j \rangle_{\text{MF}} \ , \tag{16.6}$$

we may write

$$H^{(i)}_{\text{MF}} = \boldsymbol{S}_i \cdot \boldsymbol{B}_{i,\text{MF}} \ . \tag{16.7}$$

This replacement is visualized in Fig. 16.1. The form (16.7) also explains the name assigned to the theory: The Hamiltonian (16.2) is approximated by a single spin in an effective magnetic field, the mean-field, given by the average over the surrounding spins. Note that this treatment does not make any reference to the system size, i.e. it is also valid in the thermodynamic limit.

The fact, that the mean-field $\boldsymbol{B}_{i,\text{MF}}$ is determined by $\langle \boldsymbol{S}_j \rangle_{\text{MFT}}$ immediately leads to a self-consistency condition for the latter

$$\langle \boldsymbol{S}_i \rangle_{\text{MFT}} = \mathcal{F}\left[\langle \boldsymbol{S}_j \rangle_{\text{MFT}} \right] \ . \tag{16.8}$$

The precise form of the functional will in general depend on the detailed structure of the Hamiltonian (16.2). For a simple cubic lattice and nearest-neighbor exchange

$$J_{ij} = \begin{cases} J & \text{for } i, j \text{ nearest neighbors} \\ 0 & \text{otherwise} \end{cases} \tag{16.9}$$

one finds the well-known result

$$\langle S^z_i \rangle_{\text{MFT}} = -\frac{1}{2} \tanh\left(\frac{4DJ \langle S^z_j \rangle_{\text{MFT}}}{k_{\text{B}} T} \right) \ , \tag{16.10}$$

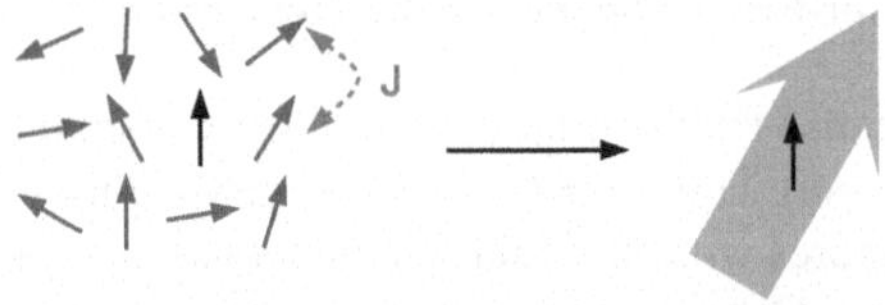

Fig. 16.1. Sketch for the mean-field theory of the Heisenberg model

where we put the quantization axis into the z direction and assumed the same value $\langle S_j^z\rangle_{\text{MFT}}$ for all $2D$ nearest neighbors. For $k_{\text{B}}T < |J^*|$, where $J^* := 2DJ$, this equation has a solution $|\langle S_i^z\rangle_{\text{MFT}}| \neq 0$, i.e. the system undergoes a phase transition to an ordered state (antiferromagnetic for $J > 0$ and ferromagnetic for $J < 0$).

As we know, this mean-field treatment yields the qualitatively correct phase diagram for the Heisenberg model in $D = 3$, but fails in dimensions $D \leq 2$ and close to the phase transition for $D = 3$. The reason is that one has neglected the fluctuations $\delta \boldsymbol{S}_i = \boldsymbol{S}_i - \langle \boldsymbol{S}_i \rangle$ of the neighboring spins. Under what conditions does that approximation become exact? The answer is given in [10]: The mean-field approximation becomes exact in the formal limit $D \to \infty$, provided on keeps $J^* = 2DJ$ constant. In this limit, each spin has $2D \to \infty$ nearest neighbors (for a simple cubic lattice). Assuming ergodicity of the system, one finds that the phase space average realized by the sum over nearest neighbors becomes equal to the ensemble average, i.e.

$$\frac{1}{2D}\sum_{j \neq i} \boldsymbol{S}_j \overset{D\to\infty}{=} \frac{1}{2D}\sum_{j \neq i} \langle \boldsymbol{S}_j \rangle + \mathcal{O}(\frac{1}{D}) \,. \tag{16.11}$$

The requirement J^* =const. finally is necessary, because otherwise the energy density $\langle H\rangle/N$ would either be zero or infinity, and the resulting model would be trivial.

For the Heisenberg model (16.2) one can even show that $D > 3$ is already sufficient to make the mean-field treatment exact, which explains why this approximation can yield a rather accurate description for magnets in $D = 3$.

Obviously, the above argument based on the limit $D = \infty$ is rather general and can be applied to other models to define a proper mean-field theory. For example, applied to disorder models, one obtains the coherent potential approximation (CPA), where the disorder is replaced by a coherent local scattering potential, which has to be determined self-consistently via the disorder average. A more detailed discussion of the capabilities and shortcomings of this mean-field theory is given in Chap. 17. Here, we want to use the limit $D \to \infty$ to construct a mean-field theory for models like the Hubbard model (16.1).

16.2.2 The Dynamical Mean-Field Theory

16.2.2.1 A First Attempt

Guided by the previous section, the most obvious possibility to construct something like a mean-field theory for the Hubbard model (16.1) is to approximate the two-particle interaction term as

$$c_{i\uparrow}^\dagger c_{i\uparrow} c_{i\downarrow}^\dagger c_{i\downarrow} \to c_{i\uparrow}^\dagger c_{i\uparrow} \langle c_{i\downarrow}^\dagger c_{i\downarrow}\rangle + \langle c_{i\uparrow}^\dagger c_{i\uparrow}\rangle c_{i\downarrow}^\dagger c_{i\downarrow} \,. \tag{16.12}$$

This approximation, which is also known as Hartree approximation, is discussed extensively in standard books on many particle theory (for example [11]). Without going into the details, one can immediately state some serious defects for half filling $\langle n\rangle = 1$:

- It leads to a metallic solution for arbitrarily large U, in contradiction to the expectations based on fundamental arguments.
- One does find an antiferromagnetically ordered phase, but for large U the Néel temperature T_N →const. instead of the expectation $T_\mathrm{N} \propto 1/U$ based on the mapping (16.3) to the antiferromagnetic Heisenberg model.

What goes wrong here? The answer is quite simple: The Hartree factorization neglects *local* charge fluctuations

$$\delta n_{i\sigma} = c^\dagger_{i\sigma} c_{i_\sigma} - \langle c^\dagger_{i\sigma} c_{i_\sigma} \rangle \tag{16.13}$$

which however are of order one. Thus, an argument rendering this approximation exact in a nontrivial limit is missing here.

16.2.2.2 The Limit $D \to \infty$

As we have observed in Sect. 16.2.1, the proper way to set up a mean-field theory is to consider the limit $D \to \infty$. Again, this limit has to be introduced such that the energy density $\langle H \rangle / N$ remains finite. As far as the interaction term in (16.1) is concerned, no problem arises, because it is purely local and thus does not care about dimensionality. The critical part is obviously the kinetic energy

$$\frac{1}{N} \langle H_\mathrm{kin} \rangle = \frac{1}{N} \sum_{i,j} \sum_{\sigma} t_{ij} \langle c^\dagger_{i\sigma} c_{j\sigma} \rangle \ . \tag{16.14}$$

To keep the notation simple, I will concentrate on a simple cubic lattice with nearest-neighbor hopping

$$t_{ij} = \begin{cases} -t & \text{for } \boldsymbol{R}_i \text{ and } \boldsymbol{R}_j \text{ nearest neighbors} \\ 0 & \text{otherwise} \end{cases} \tag{16.15}$$

in the following. Starting at a site $\boldsymbol{R}_i$, one has to apply H_kin to move an electron to or from site $\boldsymbol{R}_j$ in the nearest-neighbor shell, i.e. $\langle c^\dagger_{i\sigma} c_{j\sigma} \rangle \propto t$ and consequently

$$\frac{1}{N} \langle H_\mathrm{kin} \rangle = \frac{1}{N} \sum_{i,j} \sum_{\sigma} t_{ij} \langle c^\dagger_{i\sigma} c_{j\sigma} \rangle \propto -2Dt^2 \ , \tag{16.16}$$

where the factor $2D$ arises because we have to sum over the $2D$ nearest neighbors [12]. Thus, in order to obtain a finite result in the limit $D \to \infty$, it has to be performed such that $Dt^2 = t^*$ =const. or $t = t^*/\sqrt{D}$ [12].

What are the consequences of this scaling? To find an answer to this question, let us consider the quantity directly related to $\langle c^\dagger_{i\sigma} c_{j\sigma} \rangle$, namely the single-particle Green function [11]

$$G_{\boldsymbol{k}\sigma}(z) = \frac{1}{z + \mu - \epsilon_{\boldsymbol{k}} - \Sigma_{\boldsymbol{k}\sigma}(z)} \ , \tag{16.17}$$

where the kinetic term enters as dispersion $\epsilon_{\boldsymbol{k}}$, obtained from the Fourier transform of t_{ij}, and the two-particle interaction leads to the self-energy $\Sigma_{\boldsymbol{k}\sigma}(z)$, which can, for example, be obtained from a perturbation series using Feynman diagrams [11]. For the following argument it is useful, to discuss the perturbation expansion in real space, i.e. we study now $\Sigma_{ij,\sigma}(z)$. If we represent the Green function $G^{(0)}_{ij,\sigma}(z)$ for $U = 0$ by a (directed) full line and the two-particle interaction U by a dashed line, the first few terms of the Feynman perturbation series read

$$\Sigma_{ij,\sigma}(z) =$$

$$\delta_{ij} +$$

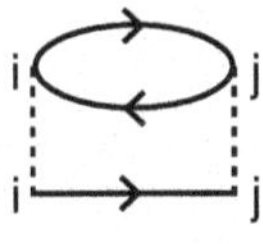

$$+ \dots \qquad (16.18)$$

The first, purely local term, evaluates to $U\langle n_{i\bar{\sigma}}\rangle$, i.e. it is precisely the Hartree approximation which we found not sufficient to reproduce at least the fundamental expectations. Let us now turn to the second term. To discuss it further, we need the important property $G^{(0)}_{ij,\sigma}(z) \propto t^{d(i,j)}$, where $d(i,j)$ is the "taxi-cab metric", i.e. the smallest number of steps to go from site $\boldsymbol{R}_i$ to site $\boldsymbol{R}_j$. Inserting the scaling $t = t^*/\sqrt{D}$, we find

$$G^{(0)}_{ij,\sigma}(z) \propto \left(\frac{1}{\sqrt{D}}\right)^{d(i,j)} . \qquad (16.19)$$

When we insert this scaling property into the second-order term in the expansion (16.2.2.2), we obtain for j being a nearest neighbor of i

$$\Sigma_{ij,\sigma}(z) - U\langle n_{i\bar{\sigma}}\rangle\delta_{ij} =$$

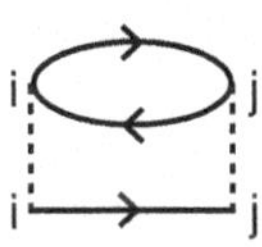

$$+ \dots \propto \left(\frac{1}{\sqrt{D}^3}\right)^{d(i,j)} \propto \frac{1}{\sqrt{D}^3} . \qquad (16.20)$$

A closer inspection [12] yields an additional factor D on the right-hand side of the equation, and we finally arrive at the scaling behavior

$$\Sigma_{ij,\sigma}(z) - U\langle n_{i\bar{\sigma}}\rangle\delta_{ij} \propto \frac{1}{\sqrt{D}} \overset{D\to\infty}{\to} 0 \qquad (16.21)$$

for the non-local part of the one-particle self-energy. Note that the local contributions $\Sigma_{ii,\sigma}(z)$ stay finite, i.e.

$$\lim_{D\to\infty} \Sigma_{ij,\sigma}(z) = \Sigma_\sigma(z)\delta_{ij} , \qquad (16.22)$$

which in momentum space translates into a $\boldsymbol{k}$-independent self-energy and hence

$$G_{\boldsymbol{k}\sigma}(z) = \frac{1}{z + \mu - \epsilon_{\boldsymbol{k}} - \Sigma_\sigma(z)} . \qquad (16.23)$$

The finding that in the limit $D \to \infty$ renormalizations due to local two-particle interactions become purely local, is rather interesting and helpful in its own right. For example, one can use it to set up a perturbation theory which is, due to the missing spatial degrees of freedom, much more easy to handle. Within this framework, the Hubbard model and related models were studied by several groups studying low-energy dynamics and transport properties [13, 14, 15] or the phase diagram at weak coupling $U \to 0$ [16, 17, 18]. There is, however, an additional way one can make use of the locality of the self-energy, which directly leads to the theory nowadays called dynamical mean-field theory (DMFT).

16.2.2.3 Mean-Field Theory for the Hubbard Model

The fundamental observation underlying the DMFT, namely that one can use the locality of the self-energy to map the lattice model onto an effective impurity problem, was first made by Brandt and Mielsch [19]. For the actual derivation of the DMFT equations for the Hubbard model one can use several different techniques. I will here present the one based on a comparison of perturbation expansions [20]. A more rigorous derivation can for example be found in the review by Georges et al. [21]. Let us begin by calculating the local Green function $G_{ii,\sigma}(z)$, which can be obtained from $G_{\boldsymbol{k}\sigma}(z)$ by summing over all $\boldsymbol{k}$, i.e.

$$G_{ii,\sigma}(z) = \frac{1}{N} \sum_{\boldsymbol{k}} \frac{1}{z + \mu - \epsilon_{\boldsymbol{k}} - \Sigma_\sigma(z)} \,. \tag{16.24}$$

Since $\boldsymbol{k}$ appears only in the dispersion, we can rewrite the $\boldsymbol{k}$-sum as integral over the density of states (DOS) of the model with $U = 0$

$$\rho^{(0)}(\epsilon) = \frac{1}{N} \sum_{\boldsymbol{k}} \delta(\epsilon - \epsilon_{\boldsymbol{k}}) \tag{16.25}$$

as

$$G_{ii,\sigma}(z) = \int \mathrm{d}\epsilon \frac{\rho^{(0)}(\epsilon)}{z + \mu - \epsilon - \Sigma_\sigma(z)} = G_{ii}^{(0)}(z + \mu - \Sigma_\sigma(z)) \,, \tag{16.26}$$

where

$$G_{ii}^{(0)}(\zeta) = \int \mathrm{d}\epsilon \frac{\rho^{(0)}(\epsilon)}{\zeta - \epsilon} \tag{16.27}$$

is the local Green function for $U = 0$. Note that due to the analytic properties of $\Sigma_\sigma(z)$ the relation $\mathrm{sign}\,\{\mathrm{Im}\,[z + \mu - \Sigma_\sigma(z)]\} = \mathrm{sign}\,\mathrm{Im}\,z$ always holds.

Now we can make use of well-known properties of quantities like $G_{ii}^{(0)}(z)$ which can be represented as Hilbert transform of a positive semi-definite function like the DOS $\rho^{(0)}(\epsilon)$ (see for example [11]), namely they can quite generally be written as

$$G_{ii}^{(0)}(\zeta) = \frac{1}{\zeta - \tilde{\Delta}(\zeta)} \,, \tag{16.28}$$

where $\widetilde{\Delta}(\zeta)$ is completely determined by $\rho^{(0)}(\epsilon)$. If we define $\mathcal{G}_\sigma(z)^{-1} := z + \mu - \Delta_\sigma(z)$, where $\Delta_\sigma(z) := \widetilde{\Delta}(z + \mu - \Sigma_\sigma(z))$, we can write the Green function for $U > 0$ as

$$G_{ii,\sigma}(\zeta) = \frac{1}{z + \mu - \Delta_\sigma(z) - \Sigma_\sigma(z)} = \frac{1}{\mathcal{G}_\sigma(z)^{-1} - \Sigma_\sigma(z)} \,. \tag{16.29}$$

Let us now assume that we switch off U at site $\boldsymbol{R}_i$ only. Then $\mathcal{G}_\sigma(z)$ can be viewed as non-interacting Green function of an impurity model with a perturbation series

$\Sigma_\sigma(z) =$ 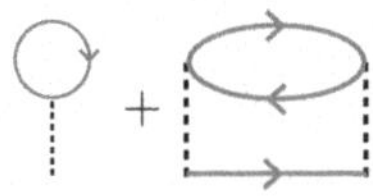 $+ \ldots$ (16.30)

for the self-energy. The full line now represents $\mathcal{G}_\sigma(z)$, but the dashed line visualizes still the same two-particle interaction as in (16.2.2.3). Looking into the literature, for example into the book by Hewson [22], one realizes that this is precisely the perturbation expansion for the so-called single impurity Anderson model (SIAM) [23]

$$H = \sum_{\boldsymbol{k}\sigma} \varepsilon_{\boldsymbol{k}} \alpha^\dagger_{\boldsymbol{k}\sigma} \alpha_{\boldsymbol{k}\sigma} + \varepsilon_f \sum_\sigma c^\dagger_\sigma c_\sigma + U c^\dagger_\uparrow c_\uparrow c^\dagger_\downarrow c_\downarrow + \frac{1}{\sqrt{N}} \sum_{\boldsymbol{k}\sigma} \left(\alpha^\dagger_{\boldsymbol{k}\sigma} c_\sigma + \text{H.c.} \right) , \tag{16.31}$$

which has been studied extensively in the context of moment formation in solids.

Obviously, the quantity $\mathcal{G}_\sigma(z)$ – or equivalently $\Delta_\sigma(z)$ – takes the role of the Weiss field in the MFT for the Heisenberg model. However, in contrast to the MFT for the Heisenberg model, where we ended up with an effective Hamiltonian of a single spin in a static field, we now have an effective local problem which is coupled to a dynamical field, hence the name DMFT. Instead of Weiss field, $\mathcal{G}_\sigma(z)$ or $\Delta_\sigma(z)$ are called effective medium in the context of the DMFT.

The missing link to complete the mean-field equations is the self-consistency condition which relates the Weiss field $\mathcal{G}_\sigma(z)$ with the solution of the effective impurity problem. This reads

$$G_{ii,\sigma}(z) = \int \mathrm{d}\epsilon \frac{\rho^{(0)}(\epsilon)}{z + \mu - \epsilon - \Sigma_\sigma(z)} \stackrel{!}{=} G_\sigma^{\rm SIAM}(z) \,. \tag{16.32}$$

Thus, $\Sigma_\sigma(z)$ has to be chosen such that the local Green function of the Hubbard model is identical to the Green function of a fictitious SIAM with non-interacting Green function

$$\mathcal{G}_\sigma(z) = \frac{1}{G_{ii,\sigma}(z)^{-1} + \Sigma_\sigma(z)} \,. \tag{16.33}$$

The resulting flow-chart for the iterative procedure to solve the Hubbard model

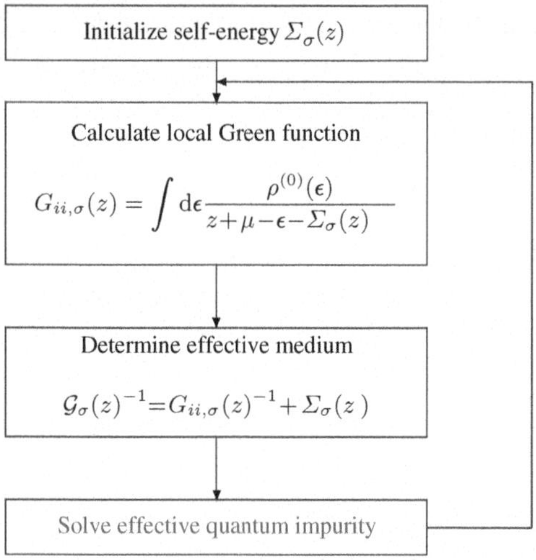

Fig. 16.2. The self-consistency loop for the DMFT for the Hubbard model

with the DMFT is shown in Fig. 16.2. The only unknown in it is the box at the bottom saying "solve effective quantum impurity problem". What the notion quantum impurity stands for and how the SIAM can be solved will be discussed next.

16.2.2.4 DMFT as Quantum Impurity Problem

Up to now all considerations have been purely analytical. Since this is a book on computational aspects in many-body systems, one may wonder how this theory fits into this field. The simple answer lies in the solution of the SIAM necessary to complete the DMFT loop. Although the Hamiltonian (16.31) of the SIAM looks comparatively simple, it is already an extremely challenging model. It has been set up in the early 1960's, but a reliable solution for both thermodynamic and dynamic quantities at arbitrary model parameters and temperature became possible only in the late 1980's. It belongs to a class of models nowadays called quantum-impurity models, where a small set of interacting quantum degrees of freedom are coupled to a continuum of non-interacting quantum states. The physical properties of these models comprise the well-known Kondo effect [22], quantum phase transitions and non-Fermi liquid behavior. The actual difficulty in solving these models is that they typically have a ground state that in the thermodynamical limit is orthogonal to the one of the possible reference systems – e.g. $U = 0$ or $V_{\boldsymbol{k}} = 0$ for the SIAM – and thus cannot be treated properly with perturbation expansions. A characteristic signature of this non-orthogonality is the appearance of an exponentially small energy scale. Both aspects together make it extremely hard to solve the models even numerically, because any representation by a finite system makes it impossible to resolve such energy scales.

Besides the numerical renormalization group discussed in the next paragraph, one of the first computational techniques used to solve quantum impurity problems for finite temperatures was quantum Monte Carlo based on the Hirsch-Fye algorithm [21, 24]. This algorithm and its application to e.g. Hubbard model has already been discussed extensively in Chap. 10. For these models, the short-ranged interaction and hopping allow for a substantial reduction of the computational effort and a rather efficient code. For quantum impurity problems, however, the orthogonality catastrophe mentioned above leads to long-ranged correlations in imaginary time. Consequently, when we denote with L the number of time slices in the simulation, the code scales with L^3 (instead of $L \ln L$ for lattice models [25]). Thus, although the algorithm does not show a sign problem for quantum impurity problems, the computational effort increases very strongly with decreasing temperature and also increasing local interaction. As a result, the quantum Monte Carlo based on the Hirsch-Fye algorithm is severely limited in the temperatures and interaction parameters accessible. For those interested, a rather extensive discussion of the algorithm and its application to the DMFT can be found in the reviews by Georges et al. and Maier et al. [21, 24]. Note that with quantum Monte Carlo one is generically restricted to finite temperature, although within the projector quantum Monte Carlo the ground state properties can be accessed in some cases, too [26].

A rather clever method to handle quantum-impurity systems comprising such a huge range of energy scales was invented by Wilson in the early 1970's [27], namely the numerical renormalization group (NRG). In this approach, the continuum of states is mapped onto a discrete set, however with exponentially decreasing energy scales. This trick allows to solve models like the SIAM for arbitrary model parameters and temperatures. A detailed account of this method is beyond the scope of this contribution but can be found in a recent review by Bulla et al. [28]. Here, the interesting aspect is the actual implementation. One introduces a discretization parameter $\Lambda > 1$ and divides the energy axis into intervals $[\Lambda^{-(n+1)}, \Lambda^{-n}]$, $n = 0, 1, \ldots$, for both positive and negative energies. After some manipulations [22, 28, 29, 30] one arrives at a representation

$$H \approx H_{\text{imp}} + \sum_{n=0}^{\infty} \sum_{\sigma} \left(\varepsilon_n \alpha_{n\sigma}^{\dagger} \alpha_{n\sigma} + t_n \alpha_{n-1\sigma}^{\dagger} \alpha_{n\sigma} + \text{H.c.} \right) , \tag{16.34}$$

where H_{imp} is the local part of the quantum impurity Hamiltonian. To keep the notation short, I represented the impurity degrees of freedom by the operators $\alpha_{-1,\sigma}^{(\dagger)}$. The quantities ε_n and t_n have the property, that they behave like $\varepsilon_n \propto \Lambda^{-n/2}$ and $t_n \propto \Lambda^{-n/2}$ for large n. The calculation now proceeds as follows: Starting from the impurity degrees of freedom ($n = -1$) with the Hamiltonian $H_{-1} \equiv H_{\text{imp}}$, one successively adds site after site of the semi-infinite chain, generating a sequence of Hamiltonians

$$\begin{aligned} H_{N+1} = \sqrt{\Lambda} H_N + \sum_{\sigma} \Big(& \sqrt{\Lambda}^{N+1} \varepsilon_{N+1} \alpha_{N+1,\sigma}^{\dagger} \alpha_{N+1,\sigma} \\ & + \sqrt{\Lambda}^{N+1} t_{N+1} \alpha_{n\sigma}^{\dagger} \alpha_{n\sigma} + \text{H.c.} \Big) . \end{aligned} \tag{16.35}$$

The factors Λ in the mapping ensure that at each step N the lowest energy eigenvalues are always of order one. Since for the chain parameters $\Lambda^{(N+1)/2}t_{N+1} \to 1$ holds, the high energy states of the Hamiltonian at step N will not significantly contribute to the low-energy states at step $N+1$ and one discards them. This truncation restricts the size of the Hilbert space at each step sufficiently that the usual exponential growth is suppressed and one can actually repeat the procedure up to almost arbitrarily large chains.

At each step N, one then has to diagonalize H_N, generating all eigenvalues and eigenvectors. The eigenvectors are needed to calculate matrix elements for the next step by a unitary transformation of the matrix elements from the previous step. Since this involves two matrix multiplications, the numerical effort (together with the diagonalization) scales with the third power of the dimension of the Hilbert space. Invoking symmetries of the system, like e.g. charge and spin conservation, one can reduce the Hamilton matrix at each step to a block structure. This block structure on the one hand allows for an efficient parallelization and use of SMP machines (for example with OpenMP). On the other hand, the size of the individual blocks is much smaller than the actual size of the Hilbert space. For example, with 1000 states kept in the truncation one has a dimension of the order of 200 for the largest subblock. The use of the block structure thus considerably reduces the computational effort necessary at each step.

Moreover, one can identify each chain length N with a temperature or energy scale $\Lambda^{-N/2}$ and can thus approach arbitrarily low temperatures and energies. With presently available workstations the computational effort of solving the effective impurity model for DMFT calculations at $T = 0$ then reduces to a few minutes using on the order of $10 \ldots 100$ MB of memory.

Unfortunately, an extension of Wilson's NRG to more complex quantum impurity models including e.g. orbital degrees of freedom or multi-impurity systems (needed for example for the solution of cluster mean-field theories, see Sect. 16.3) is not possible beyond four impurity degrees of freedom (where the consumption of computer resources increases to order of days computation time with $\sim 20-30$ GB memory usage), because the step "construct Hamilton matrix of step $N+1$ from Hamilton matrix of step N" increases the size again exponentially with respect to the number of impurity degrees of freedom. For a compensation, one has to increase the number of truncated states in each step appropriately. However, this procedure breaks down when one starts to truncate states that contribute significantly to the low-energy properties of the Hamiltonian at step $N+1$. In this situation, one is left with quantum Monte Carlo algorithms as only possible solver at $T > 0$. At $T = 0$, there exists presently not yet a reliable tool to solve quantum impurity models with substantially more that two impurity degrees of freedom (spin degeneracy). First attempts to use the density matrix renormalization group method to solve quantum impurity problems can for example be found in [31, 32, 33].

16.2.3 DMFT Results for the Hubbard Model

In the following sections selected results for the Hubbard model within the DMFT will be discussed. I restrict the presentation to the case of a particle-hole symmetric non-interacting system, i.e. the simple-cubic lattice with nearest-neighbor hopping according to (16.15). More general situations including next nearest-neighbor hopping have also been studied and results can for example be found in [21, 34]. Moreover, the DMFT also allows for a consistent calculation of two-particle properties like susceptibilities and also transport quantities. A detailed discussion of the aspects of these calculations go well beyond the scope of this article and the interested reader is referred to the extensive literature on these subjects [21, 24, 35].

16.2.3.1 General Structures of the Green Function in DMFT

The fundamental quantity we calculate in the DMFT is the local single-particle Green function $G_{ii,\sigma}(z)$. Its imaginary part $N_\sigma(\omega) := -\operatorname{Im} G_{ii,\sigma}(\omega + i0^+)/\pi$ is called DOS of the interacting system. The generic result for this quantity for a typical value of $U = \mathcal{O}(W)$ at $T = 0$, where W is the bandwidth of the dispersion $\epsilon_{\boldsymbol{k}}$, is shown in Fig. 16.3 [21, 35]. One can identify three characteristic structures: Two broad peaks below and above the Fermi energy $\omega = 0$, which describe the incoherent charge excitations. They are separated by the energy U and referred to as lower Hubbard band (LHB) and upper Hubbard band (UHB), respectively. In addition a rather sharp resonance exists at the Fermi energy, which is a result of coherent quasiparticles in the sense of Landau's Fermi liquid theory.

This interpretation becomes more apparent when one looks at the single-particle self-energy, shown as inset to Fig. 16.3. The region close to the Fermi energy is characterized by a behavior $\operatorname{Re} \Sigma_\sigma(\omega + i0^+) \propto \omega$ and $\operatorname{Im} \Sigma_\sigma(\omega + i0^+) \propto \omega^2$. Both

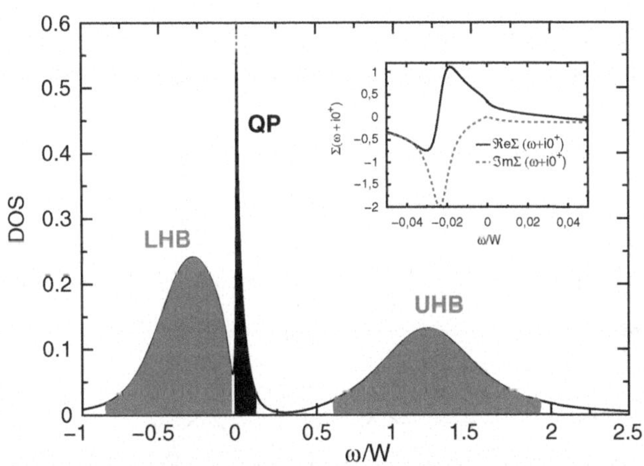

Fig. 16.3. Generic DMFT result for the DOS of the Hubbard model at $T = 0$. Model parameters are $U/W = 1.5$ and $\langle n \rangle = 0.97$. W denotes the bandwidth of the dispersion $\epsilon_{\boldsymbol{k}}$. The inset shows the corresponding self-energy $\Sigma_\sigma(\omega + i0^+)$ in the region about the Fermi energy. One nicely sees the parabolic maximum in the imaginary part and the linear real part as $\omega \to 0$

are features characteristic for a Fermi liquid. The slope of the real part determines the quasiparticle renormalization factor or effective mass of the quasiparticles.

16.2.3.2 The Mott-Hubbard Metal-Insulator Transition

One particular feature we expect for the Hubbard model is the occurrence of a metal-insulator transition (MIT) in the half-filled case $\langle n \rangle = 1$. As already mentioned, this particular property can serve as a test for the quality of the approximation used to study the model. That the expected MIT indeed appears in the DMFT has first been noticed by Jarrell [20] and was subsequently studied in great detail [21]. The MIT shows up in the DOS as vanishing of the quasiparticle peak with increasing U. An example for this behavior can be seen in Fig. 16.4. The full curve is the result of a calculation with a value of $U < W$, the dashed obtained with $U > 1.5\,W \approx U_{\mathrm{c}}$. For the latter, the quasiparticle peak at $\omega = 0$ has vanished, i.e. we have $N(\omega = 0) = 0$. Since the DOS at the Fermi level determines all properties of a Fermi system, in particular the transport, we can conclude from this result that for $U > U_{\mathrm{c}}$ the conductivity will be zero, hence the system is an insulator. One can now perform a series of calculations for different values of U and temperatures T to obtain the phase diagram for this MIT (see e.g. [36] and references therein). The result is shown in Fig. 16.5. As an unexpected feature of this MIT one finds that there exists a hysteresis region, i.e. starting from a metal and increasing U leads to a different $U_{\mathrm{c},2}$ as starting from the insulator at large U and decreasing U. The coexistence region terminates in a second-order critical end point, which has the properties of the liquid-gas transition [37, 38]. At $T = 0$, the transition is also second order and characterized by a continuously vanishing Drude weight in the optical conductivity [21], or equivalently a continuously vanishing quasiparticle renormalization factor [39]. Interestingly, the actual critical line falls almost onto the

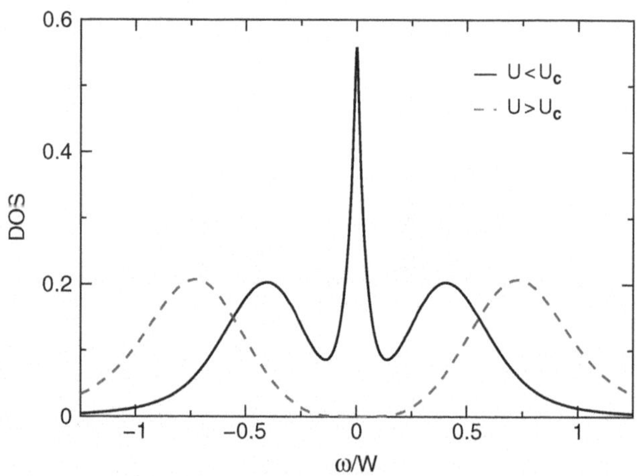

Fig. 16.4. Variation of DOS across the MIT

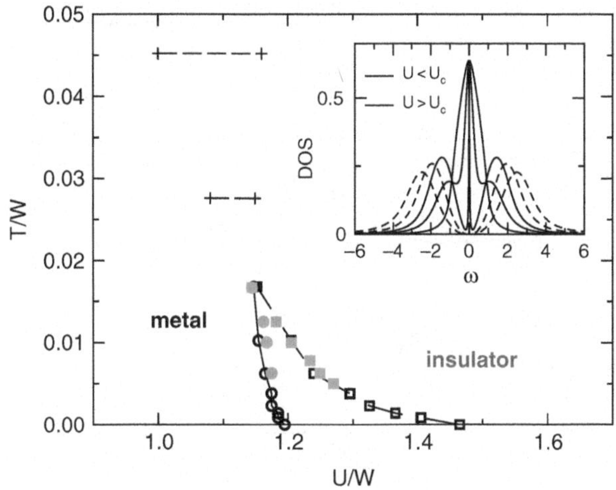

Fig. 16.5. Paramagnetic phase diagram for the Hubbard model at half filling. The transition between metal and insulator shows a hysteresis denoted by the *two critical lines*. The inset shows the behavior of the DOS as U increases. Figure taken from [36]

upper transition [40]. Finally, for temperatures larger than the upper critical point the MIT turns into a crossover.

16.2.3.3 Magnetic Properties

Up to now we have discussed the paramagnetic phase of the Hubbard model. What about the magnetic properties? Does the DMFT in particular cure the failure of the Hartree approximation, where T_N became constant when $U \to \infty$?

Investigations of magnetic properties can be done in two ways. First, one can calculate the static magnetic susceptibility and search for its divergence. This will give besides the transition temperature also the proper wave vector of the magnetic order [21, 41]. For the NRG another method is better suited and yields furthermore also information about the single-particle properties and hence transport properties in the antiferromagnetic phase [21, 34]: One starts the calculation with a small symmetry breaking magnetic field, which will be switched off after the first DMFT iteration. As result, the system will converge either to a paramagnetic state or a state with finite polarization. The apparent disadvantage is, that only certain commensurate magnetic structures can be studied, such as the ferromagnet or the Néel antiferromagnet.

For half filling, the result of such a calculation for the Néel structure at $T = 0$ is shown in Fig. 16.6. Quite generally, we expect a stable antiferromagnetic phase at arbitrarily small values of U with an exponentially small Néel temperature [11, 16]. Indeed we find that the Néel antiferromagnet is the stable solution for all values of U at $T = 0$ [42].

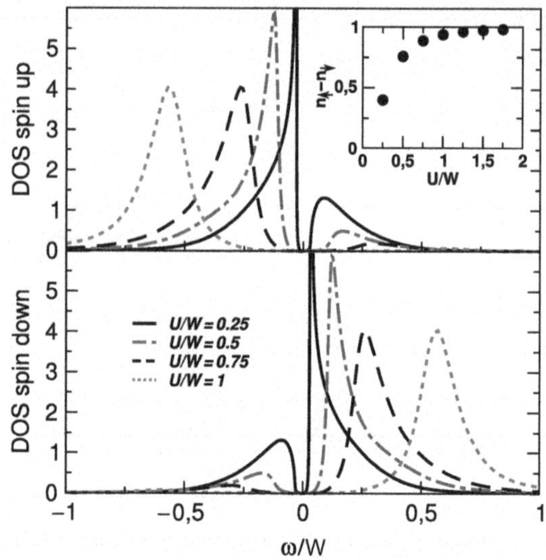

Fig. 16.6. DOS for spin up and spin down in the antiferromagnetic phase at half filling and $T = 0$. The inset shows the magnetization as function of U

The next question concerns the magnetic phase diagram, in particular the dependence of the Néel temperature T_{N} on U. To this end one has to perform a rather large number of DMFT calculations systematically varying T and U. The result of such a survey are the circles denoting the DMFT values for $T_{\mathrm{N}}(U)$ in the phase diagram in Fig. 16.7. The dotted line is a fit that for small U behaves $\propto \exp(-\alpha/U)$, predicted by weak-coupling theory, while for large U a decay like $1/U$ is reached. Thus, the DMFT indeed reproduces the correct U dependence in both limits $U \to 0$ and $U \to \infty$.

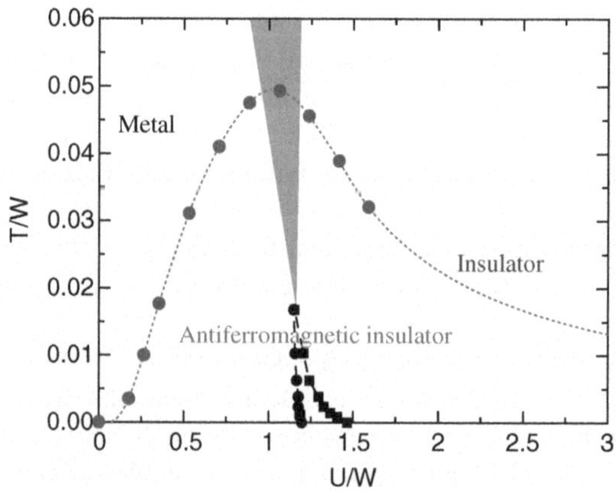

Fig. 16.7. Phase diagram for the Néel state at half filling. In addition the paramagnetic MIT phase lines are included

Another observation is that the phase diagram is completely dominated by the Néel phase. It even encompasses the MIT phase, whose phase lines are given by the squares in Fig. 16.7. Note, that the Néel phase is an insulator, too. Thus, there is the obviously interesting question if an additional transition occurs within the antiferromagnetic insulator from what is called Slater insulator, driven by bandstructure effects, and Mott-Hubbard insulator, driven by correlation effects. Up to now, no hard evidence for such a transition could be found [34, 43].

Last, but not least, one may wonder how the magnetic phase diagram develops away from half filling. Here, two interesting conjectures are known. Weak coupling theory predicts, that for small U the Néel state remains stable up to a certain filling $\langle n_c \rangle < 1$, but shows phase separation [18]. In the limit $U \to \infty$, on the other hand, one expects a ferromagnetic phase to appear, which is driven by kinetic energy gain instead of an effective exchange interaction [5]. Here, the full power of the NRG as solver for the quantum impurity problem can be seen. There are no restrictions regarding the value of U or the temperature T. Consequently, one can scan the whole phase space of the Hubbard model to obtain the U-δ phase diagram at $T = 0$ shown in Fig. 16.8. The quantity $\delta = 1 - \langle n \rangle$ denotes the doping and the vertical axis has been rescaled according to $U/(W + U)$ in order to show the results for the whole interval $U \in [0, \infty)$. One indeed finds an extended region of antiferromagnetism (AFM) for finite doping, which in addition shows phase separation (PS) for values $U < W$. For larger U, the actual magnetic structure could not be resolved yet [42]. At very large U the antiferromagnet is replaced by an extended island of Nagaoka type ferromagnetism (FM) extending out to $\approx 30\%$ doping [42]. These examples show that the DMFT is indeed capable of reproducing at least qualitatively the rather

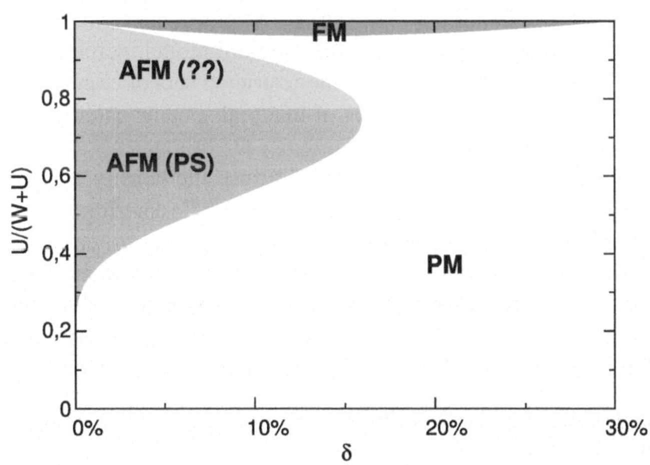

Fig. 16.8. Magnetic phase diagram of the Hubbard model for $T = 0$. The vertical axis has been rescaled as $U/(W+U)$ to encompass the whole interval $[0, \infty)$. The abbreviations mean paramagnetic metal (PM), antiferromagnet (AFM), phase separation (PS) and ferromagnet (FM). $\delta = 1 - \langle n \rangle$ denotes the doping

complex physical properties of the Hubbard model (16.1). Moreover, the values for transition temperatures obtained are strongly reduced as compared to a Stoner or Hartree approximation [11], thus illuminating the importance of local dynamical fluctuations due to the two-particle interaction respected by the DMFT.

16.2.4 Further Application: Combining First Principles with DMFT

The finding, that the DMFT for the Hubbard model, besides properly reproducing all expected features at least qualitatively, also leads to a variety of non-trivial novel aspects of the physics of this comparatively simple model [21, 35], rather early triggered the expectation, that this theory can also be a reasonable ansatz to study real 3D materials. This idea was further supported by several experimental results on transition metal compound suggesting that the metallic state can be described as a Fermi liquid with effective masses larger than the ones predicted by bandstructure theory [9]. Moreover, with increasing resolution of photoemission experiments, structures could be resolved that very much looked like the ubiquitous lower Hubbard band and quasiparticle peak found in DMFT, for example in the series (Sr,Ca)VO_3 [44, 45, 46, 47]. It was thus quite reasonable, to try to describe such materials within a Hubbard model [48, 49]. However, the explanation of the experiments required an unphysical variation of the value of U across the series.

The explanation for the failure lies in the orbital degrees of freedom neglected in the Hubbard model (16.1) but definitely present in transition metal ions. Thus, a development of quantum impurity solvers for models including orbital degrees of freedom started [50, 51, 52]. At the same time it became clear, that the number of adjustable parameters in a multi-orbital Hubbard model increases dramatically with the degrees of freedom. In view of the restricted sets of experiments that one can describe within the DMFT, the idea of material specific calculations with this method actually appears rather ridiculous.

The idea which solved that problem was to use the density functional theory (DFT) [53, 54] to generate the dispersion relation $\epsilon_{\boldsymbol{k}}^{mm'}$ entering the multi-orbital Hubbard model [55, 56]. Moreover, within the so-called constrained DFT [57] even a calculation of Coulomb parameters is possible. Thus equipped, a material-specific many-body theory for transition metal oxides and even lanthanides became possible, nowadays called LDA+DMFT [58, 59, 60, 61]. The scheme basically works as follows [58, 61]:

- For a given material, calculate the band structure using DFT with local density approximation [54].
- Identify the states where local correlations are important and downfold the bandstructure to these states to obtain a Hamilton matrix $\mathbf{H}(\boldsymbol{k})$ describing the dispersion of these states. If necessary, include other states overlapping with the correlated orbitals (for example oxygen $2p$ for transition metal oxides).

– From a constrained DFT calculation, obtain the Coulomb parameters for the correlated orbitals.
– Perform a DMFT calculations using the expression

$$\mathbf{G}_{ii,\sigma}(z) = \frac{1}{N}\sum_{\boldsymbol{k}} \frac{1}{z+\mu-\mathbf{H}(\boldsymbol{k})-\boldsymbol{\Sigma}_\sigma(z)} \tag{16.36}$$

for the local Green function, which now can be a matrix in the orbital indices taken into account. Note that the self-energy can be a matrix, too.
– If desired, use the result of the DMFT to modify the potential entering the DFT and repeat from the first step until self-consistency is achieved [56].

As an example for the results obtained in such a parameter-free calculation I present the DOS for (Sr,Ca)VO_3 obtained with the LDA+DMFT scheme compared to photoemission experiments [62] in Fig. 16.9. Apparently, both the position of the structures and the weight are obtained with rather good accuracy. From these calculations one can now infer that the structures seen are indeed the lower Hubbard band originating from the $3d$ levels, here situated at about $-2\,\mathrm{eV}$, and a quasiparticle peak describing the coherent states in the system.

This example shows that the DMFT is indeed a rather powerful tool to study 3D materials where local correlations dominate the physical properties. There is, however, not a simple rule of thumb which can tell us when this approach is indeed applicable and when correlations beyond the DMFT may become important. Even in seemingly rather simple systems non-local correlations can be important and modify the dominant effects of the local interactions in a subtle way [63].

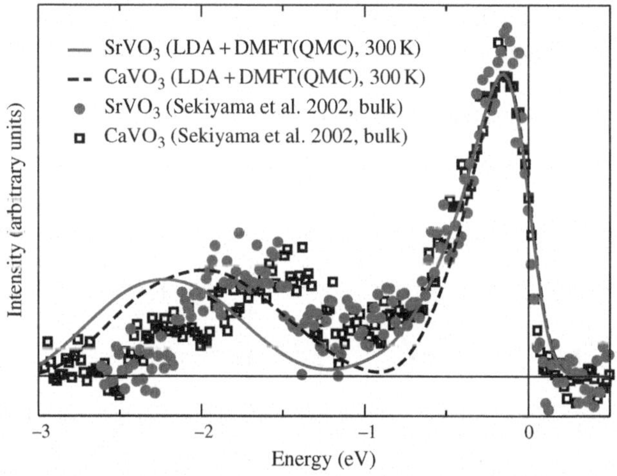

Fig. 16.9. DOS for (Sr,Ca)VO_3 obtained from a parameter free LDA+DMFT calculation (*full lines*) compared to results from photoemission experiments (*symbols*). Taken from [62]

16.3 Extending the DMFT: Effective Cluster Theories

16.3.1 Questions Beyond the DMFT

The DMFT has turned out to be a rather successful theory to describe properties of strongly correlated electron systems in three dimensions sufficiently far away from e.g. magnetic phase transitions. Its strength lies in the fact that it correctly includes the local dynamics induced by the local two-particle interactions. It is, on the other hand, well-known that in one or two dimensions or in the vicinity of a transition to a state with long-range order the physics is rather dominated by non-local dynamics, e.g. spin waves for materials showing magnetic order. Such features are of course beyond the scope of the DMFT.

As a particular example let us take a look at the qualitative properties of the Hubbard model in $D = 2$ on a square lattice at and close to half filling. As we already know, the model has strong antiferromagnetic correlations for intermediate and strong U, leading to a phase transition to a Néel state at finite T_{N} in $D = 3$. However, in $D = 2$ the theorem by Mermin and Wagner [64] inhibits a true phase transition at finite T, only the ground state may show long-range order. Nevertheless, the non-local spin correlations exist and can become strong at low temperature [6]. In particular, a snapshot of the system will increasingly look like the Néel state, at least in a certain vicinity of a given lattice site.

Such a short-range order in both time and space can have profound effects for example on the photoemission spectrum. In a true Néel ordered state the broken translational symmetry leads to a reduced Brillouin zone and hence to a folding back of the bandstructure, as depicted in Fig. 16.10(a). At the boundary of this so-called magnetic Brillouin zone, a crossing of the dispersions occurs, which will be split by interactions and leads to the gap in the DOS and the insulating behavior of the Néel antiferromagnet. When we suppress the long-range order but still allow for short-range correlations, the behavior in Fig. 16.10(b) may occur. There is no true broken translational symmetry, hence the actual dispersion will not change. However, the

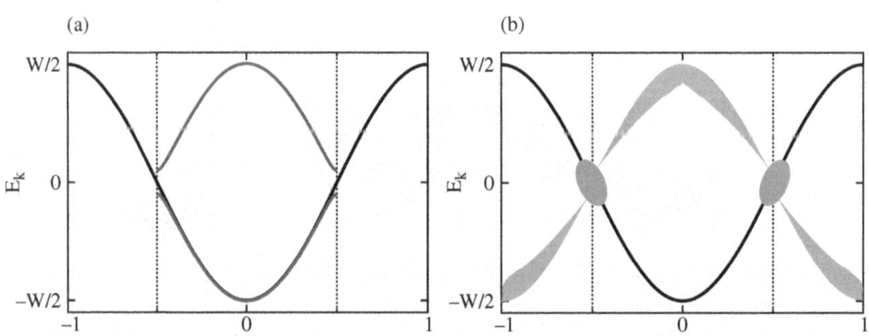

Fig. 16.10. Sketch of the effect of long-range Néel order (**a**) vs. strong short-ranged correlations (**b**) on the single-particle properties of the Hubbard model in $D = 2$

system "feels" the ordered state on certain time and length scales, which leads to broadened structures at the position of the back-folded bands (shadow bands) in the spectral function [65, 66]. Furthermore, the tendency to form a gap at the crossing points at the boundary of the magnetic Brillouin zone can lead to a suppression of spectral weight at these points (pseudo-gaps) [65].

The paradigm for such a behavior surely are the high-T_C superconductors, but other low-dimensional materials show similar features, too.

16.3.2 From the Impurity to Clusters

Let us in the beginning state the minimum requirements, that a theory extending the DMFT to include non-local correlation should fulfill: It should

- work in thermodynamic limit,
- treat local dynamics exactly,
- include short-ranged dynamical fluctuations in a systematic and possibly non-perturbative way,
- be complementarity to finite-system calculations
- and of course remain computationally manageable.

It is of course tempting, to try and start from the DMFT as an approximation that already properly includes local dynamics and add the non-local physics somehow. Since the DMFT becomes exact in the limit $D \to \infty$, an expansion in powers of $1/D$ may seem appropriate [67]. However, while such approaches work well for wave functions, their extension to the DMFT suffer from so-called self-avoiding random walk problems, and no proper resummation has been successful yet.

A more pragmatic approach tries to add the non-local fluctuations by hand [68, 69], but here the problem of possible overcounting of processes arises. Moreover, the type of fluctuations included is strongly biased and the way one includes them relies on convergence of the perturbation series.

In yet another idea one extends the DMFT by including two-particle fluctuations locally [70]. In this way, one can indeed observe effects like pseudo-gap formation in the large-U Hubbard model [71], but cannot obtain any $\boldsymbol{k}$-dependence in the spectral function, because the renormalizations are still purely local.

The most successful compromise that fulfills all of the previously stated requirements is based on the concept of clusters. There, the basic idea is to replace the impurity of the DMFT by a small cluster embedded in a medium representing the remaining infinite lattice. In this way, one tries to combine the advantages of finite-system calculations, i.e. the proper treatment of local and at least short-ranged correlations, with the properties of the DMFT, viz the introduction of the thermodynamic limit via the Weiss field. The schematic representation of this approach is shown in Fig. 16.11. This idea is not new, but has been tried in the context of disordered systems before [72], and also in various ways for correlated electron models [24]. A rather successful implementation is the cluster perturbation theory, discussed in Chap. 19. A recent review discussing these previous attempts and their problems is given in [24].

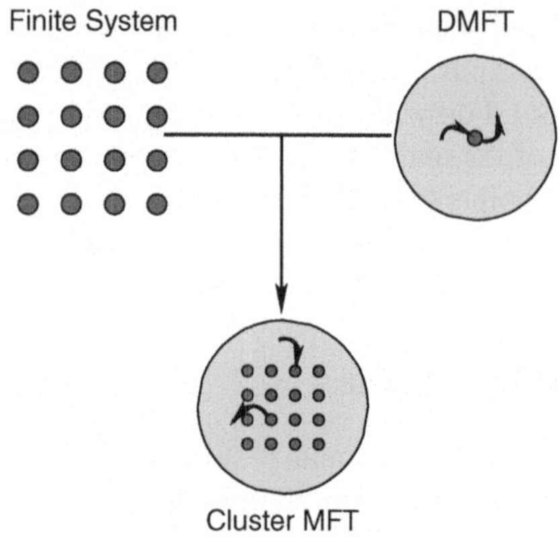

Fig. 16.11. Schematic picture of the idea of a cluster based MFT

16.3.3 Implementing the Algorithm

The implementation of the concept of a cluster MFT is straightforward and will be discussed here using the so-called DCA [24] as example. The other methods basically follow the same strategy, but differ in the details.

The DCA is an extension of the DMFT in $\boldsymbol{k}$-space. Starting from the observation that for short-ranged fluctuations one expects that $\boldsymbol{k}$-dependencies of certain quantities like the single-particle self-energy will be weak, one coarse-grains their $\boldsymbol{k}$-dependence by introducing a suitable set of N_c momenta $\boldsymbol{K}$ in the first Brillouin zone (see Fig. 16.12 with $N_c = 4$ as example). The $\boldsymbol{k}$-dependence of the single-particle self-energy $\Sigma_\sigma(\boldsymbol{k}, z)$ is then approximated according to $\Sigma_\sigma(\boldsymbol{k}, z) \approx \Sigma_\sigma(\boldsymbol{K}, z)$. This means, that one effectively reduces the resolution in real space to length scales $\Delta R \sim \pi/\Delta K$, where ΔK is a measure of the difference of individual

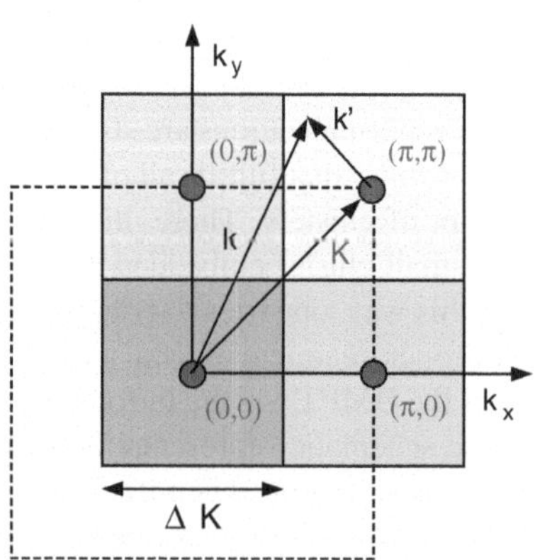

Fig. 16.12. Tiling the first Brillouin zone in the DCA

$\boldsymbol{K}$-vectors in the coarse-grained Brillouin zone. Consequently, we can expect to treat non-local correlations up to this length scale correctly.

The next step now is to integrate out the remaining $\boldsymbol{k}$-vectors in the sectors around each $\boldsymbol{K}$-point. If we do this for the single-particle Green function, we obtain a quantity

$$\bar{G}_\sigma(\boldsymbol{K}, z) := \frac{N_c}{N} \sum_{\boldsymbol{k}'} \frac{1}{z + \mu - \epsilon_{\boldsymbol{K}+\boldsymbol{k}'} - \Sigma_\sigma(\boldsymbol{K}, z)} \tag{16.37}$$

we will call the effective cluster Green function. Obviously, the quantity $\bar{G}_\sigma(\boldsymbol{K}, z)$ describes an effective periodic cluster model. The procedure now follows precisely the ideas of the DMFT. Switching off the interaction in the effective cluster leads to an effective non-interacting system described by a Green function

$$\bar{\mathcal{G}}_\sigma(\boldsymbol{K}, z) = \frac{1}{\bar{G}_\sigma(\boldsymbol{K}, z)^{-1} + \Sigma_\sigma(\boldsymbol{K}, z)} \tag{16.38}$$

and a self-consistency loop depicted in Fig. 16.13.

As in the DMFT, the problematic step is the last box, i.e. the solution of the effective quantum cluster problem. Note that although we started the construction from a cluster, the presence of the energy-dependent medium $\bar{\mathcal{G}}_\sigma(\boldsymbol{K}, z)$ renders this problem again a very complicated many-body problem, just like the effective quantum impurity problem in the DMFT. However, the situation here is even worse, because the dynamical degrees of freedom represented by this medium mean that

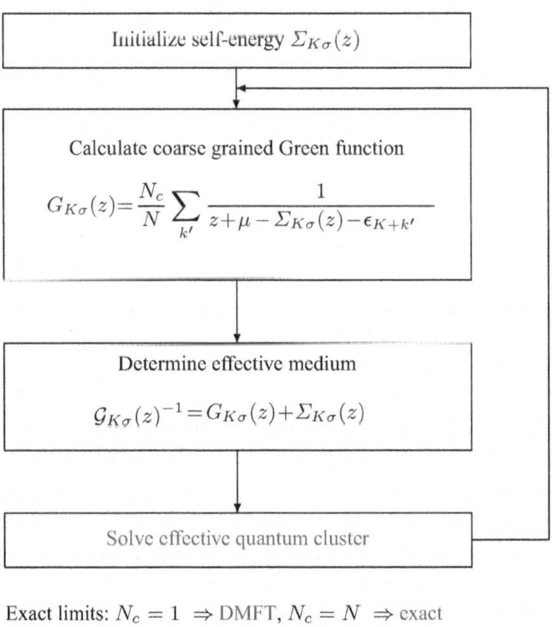

Fig. 16.13. Flow-diagram of the algorithm for the DCA

even for clusters as small as $N_c = 4$, the effective system to solve has infinitely many degrees of freedom. For example the NRG, which is so successful for the Hubbard model in the DMFT, will suffer from a huge increase of the Hilbert space (4^{N_c}) in each step, which makes the method useless. Up to now the only reasonable technique is quantum Monte Carlo (QMC), and most of the results presented in the next section will be based on QMC simulations.

Before we move to the presentation of some results for the Hubbard model, let me make some general comments on the method. First, while the concept of a cluster MFT seems to be a natural extension of the DMFT, it lacks a similar justification by an exact limit. The best one can do is view the cluster MFT as interpolation scheme between the DMFT and the real lattice, systematically including longer ranged correlations. Moreover, the use of a finite cluster introduces the problem of boundary conditions (BC). In a real space implementation [73] one has to use open BC and thus has broken translational invariance. As a consequence, $\boldsymbol{k}$ is not a good quantum number any more and one has to work out averaging procedures to recover the desired diagonality in $\boldsymbol{k}$-space. The DCA implements periodic BC, but introduces patches in the Brillouin zone, where $\Sigma_\sigma(\boldsymbol{K}, z)$ is constant. As result, one obtains a histogram of self-energy values and must use a fitting procedure to recover a smooth function $\Sigma_\sigma(\boldsymbol{k}, z)$, if desired.

Another potential problem can be causality [72]. In early attempts to set up cluster approaches, one typically ran into the problem that spectral functions could become negative. It has been shown, however, that the different implementations of the cluster MFT are manifestly causal [24].

Last but not least one may wonder how one can implement non-local two-particle interactions in this scheme, for example nearest-neighbor Coulomb interaction or the exchange interaction in models like the t-J model. In the DMFT, these interactions reduce to their mean-field description [74]. For cluster mean-field theories, they should in fact be treated similarly to the single-particle hopping. One then is faced with the requirement, to not only solve for dynamic single-particle properties in the presence of the effective bath, but also set up a similar scheme for the two-particle quantities of the effective cluster [24]. In this respect the cluster MFT acquire a structure similar to the so-called EDMT proposed by Q. Si et al. [70].

16.3.4 Results for the Hubbard Model

In the following I present some selected results obtained with the DCA for the Hubbard model in $D = 2$ on a square lattice. If not mentioned otherwise, we will again use the nearest-neighbor hopping (16.15). A much wider overview can be found in the review [24].

The first obvious question to ask is how the cluster MFT will modify the single-particle properties of the Hubbard model. As mentioned, the Mermin-Wagner theorem states that no long-range magnetic order can occur, but from the discussion in the beginning of this chapter we expect at least the presence of strong non-local spin fluctuations which should lead to precursors of the ordering at $T = 0$ in the physical quantities. In Fig. 16.14 the results of calculations for half filling and $U = W/2$ with

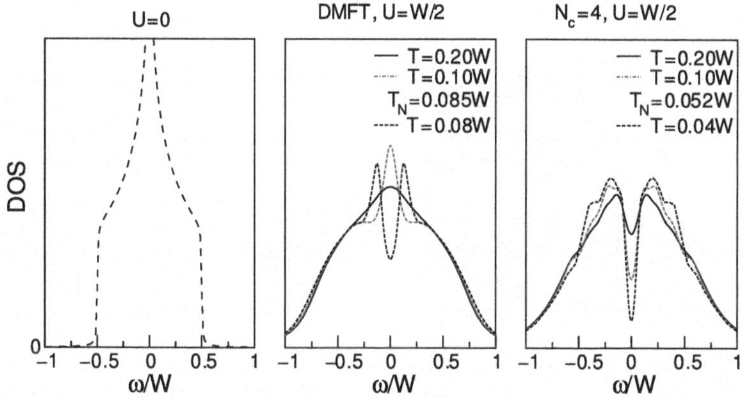

Fig. 16.14. DOS for the $2D$ Hubbard model at half filling and $U = W/2$ for different temperature using the DMFT (**middle panel**) and the DCA with $N_c = 4$ (**right panel**). The left panel shows the bare DOS at $U = 0$ for comparison. Figure taken from [75]

the DMFT (middle panel) and the DCA with a cluster size of $N_c = 4$ (right panel) for different temperatures are shown. For comparison the bare DOS is included in the left panel. In the DMFT, one obtains a phase transition into the Néel state at some $T_\mathrm{N} > 0$. For $T > T_\mathrm{N}$, the DOS shows the ubiquitous three-peak structure, while for $T < T_\mathrm{N}$ a gap appears in the DOS. No precursor of the transition can be seen. The DCA, on the other hand, already shows a pronounced pseudo-gap even at elevated temperatures, which becomes deeper with decreasing temperatures. This reflects the expected presence of spin fluctuations. Since the DCA still represents a MFT, a phase transition will eventually occur here, too. However, the corresponding transition temperature is reduced from its DMFT value and the DOS seemingly varies smoothly from $T > T_\mathrm{N}$ to $T < T_\mathrm{N}$ here.

The influence of spin fluctuations close to half filling can also be seen in the spectral functions $A(\boldsymbol{k}, \omega) = -\operatorname{Im} mG(\boldsymbol{k}, \omega + i0^+)/\pi$, which are plotted along high-symmetry directions of the first Brillouin zone of the square lattice (see Fig. 16.16) in Fig. 16.15. The calculations were done with $N_c = 16$ at a temperature $T = W/30$ at $U = W$ using a Hirsch-Fye QMC algorithm and maximum entropy to obtain the real-frequency spectra from the QMC imaginary time date [24, 77]. In the calculation an additional next-nearest neighbor hopping $t' = 0.05\,W$ was included. For $\langle n \rangle = 0.8$ (left panel of Fig. 16.15) nice quasiparticles along the non-interacting Fermi surface (base-line in the spectra) can be seen and the imaginary part of the self-energy (plot in the left corner of the panel) has a nice parabolic extremum at $\omega = 0$ and is basically $\boldsymbol{k}$-independent. Thus, in this parameter regime the DMFT can be a reliable approximation, at least as far as single-particle properties in the paramagnetic phase are concerned. For $\langle n \rangle = 0.95$ (right panel in Fig. 16.15), on the other hand, quasiparticles are found along the diagonal of the Brillouin zone (cold spot), while the structures are strongly overdamped in the region $\boldsymbol{k} \approx (0, \pi)$ (hot spot). The notion hot spot comes from the observation, that in this region the

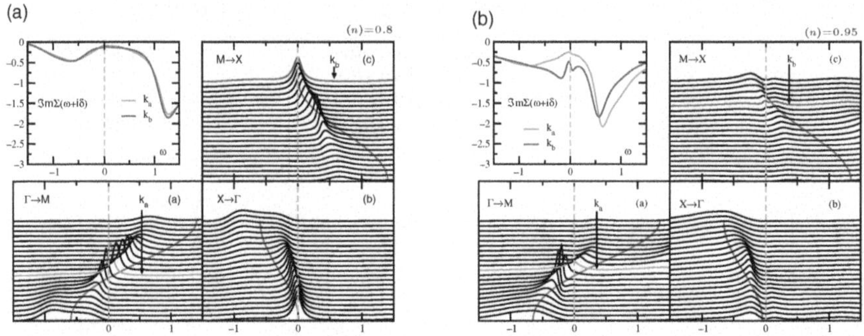

Fig. 16.15. Spectral functions along high-symmetry directions of the first Brillouin zone of the square lattice (see Fig. 16.16) obtained from a DCA calculation with $N_c = 16$ for different fillings $\langle n \rangle = 0.8$ (**left panel**) and $\langle n \rangle = 0.95$ (**right panel**). The figures in the left corners show the imaginary part of the self-energy at special $\boldsymbol{k}$-points indicated by the arrows in the spectra. The model parameters are $U = W$ and $T = W/30$. Figure taken from [76]

Fermi surface can be connected with the reciprocal lattice vector $\boldsymbol{Q}$ describing the antiferromagnetic ordering (see Fig. 16.16) (nesting). Obviously, these $\boldsymbol{k}$-points will be particularly susceptible to spin fluctuations and acquire additional damping due to the coupling to those modes.

Finally, one may wonder what the DCA can do for 3D systems. As example, I show results of a calculation of the Néel temperature for the 3D Hubbard model at half filling in Fig. 16.17. The figure includes several curves: The one labelled "Weiss" is obtained from a Weiss mean-field treatment of the antiferromagnetic Heisenberg model with an exchange coupling $J \sim t^2/U$ according to (16.3). The one called "Heisenberg" represents a full calculation for the 3D Heisenberg model with this exchange coupling, "SOPT" denotes a second-order perturbation theory calculation for the Hubbard model, "Staudt" recent QMC results [79] and finally "DMFT" and "DCA" the values for T_{N} obtained from DMFT and DCA respectively. Obviously, the DCA results in a substantial reduction of T_{N} as compared to the DMFT, leading to the correct values for all U. As expected, the DMFT overestimates T_{N} as usual for a mean-field theory, but, as we already know, is otherwise consistent with the anticipated behavior at both small and large U on the mean-field level.

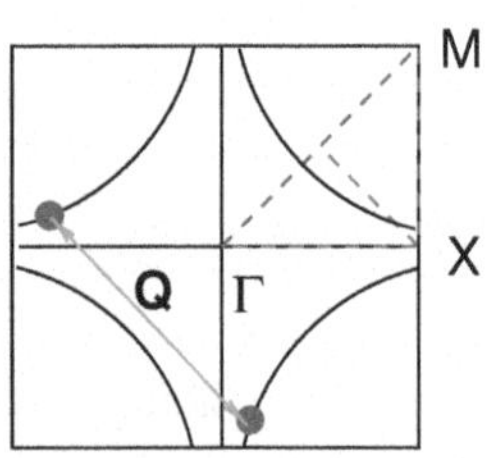

Fig. 16.16. First Brillouin zone of the square lattice

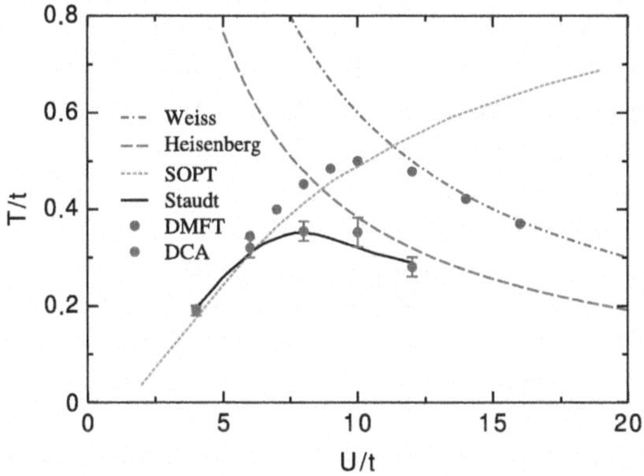

Fig. 16.17. Néel temperature as function of U for the 3D Hubbard model at half filling. For the different curves see text. Taken from [78]

Note that for the QMC results and the DCA a finite size scaling has been performed, where for the DCA lattices up to 32 sites were included, i.e. substantially smaller that in [79].

16.4 Conclusions

Starting from the Weiss mean-field theory for the Heisenberg model, we have developed a proper mean-field theory for correlated fermionic lattice models with local interactions. In contrast to the mean-field theory for the Heisenberg model, the fundamental quantity in this so-called dynamical mean-field theory is the single-particle Green function, and the effective local problem turned out to be a quantum-impurity model. Quantum impurity models are notoriously hard to solve, even with advanced computational techniques. As a special example, we discussed the numerical renormalization group approach in some detail.

As we have seen, the dynamical mean-field theory allows to calculate a variety of static and dynamic properties for correlated lattice models like the Hubbard model and its relatives. In contrast to the Hartree-Fock treatment, dynamical renormalizations lead to non-trivial phenomena like a Fermi liquid with strongly enhanced Fermi liquid parameters, a paramagnetic metal-insulator transition and magnetic properties that correctly describe the crossover from weak-coupling Slater antiferromagnetism to Heisenberg physics and Nagaoka ferromagnetism as $U \to \infty$.

In combination with density functional theory, which allows to determine model parameters for a given material ab initio, a particularly interesting novel approach to a parameter-free and material-specific calculation of properties of correlated materials arises. Several applications have demonstrated the power of this method, which can even lead to a quantitative agreement between theory and experiment.

Thus, is the DMFT an all-in-one tool, suitable for every purpose? Definitely not. We also learned that we have to pay a price for the gain: The DMFT completely neglects *non-local* fluctuations. This means, for example, that it does not care for the dimensionality of the system and will in particular lead to phase transitions even in nominally one-dimensional problems. Furthermore, even in three dimensions one cannot realize ordered states with non-local order parameters – e.g. d-wave superconductivity. Thus, for low-dimensional system or in the vicinity of phase transitions, the DMFT surely is not a good approach.

These deficiencies can be cured at least in part by extending the notion of *local* to also include clusters in addition to single lattice sites. One then arrives at extensions of the DMFT like the cluster dynamical mean-field theory or the dynamical cluster approximation. These theories allow to incorporate at least spatially short-ranged fluctuations into the calculations. We have learned that these extensions indeed lead to new phenomena, like formation of pseudo-gaps in the one-particle spectra and the appearance of new ordered phases with non-local order parameters. Cluster theories also lead to further renormalizations of transition temperatures or, with large enough clusters, lead to a suppression of phase transitions in low-dimensional systems, in accordance with e.g. the Mermin-Wagner theorem.

Again one has to pay a price for this gain, namely a tremendously increased computational effort. For this reason, calculations are up to now possible only for comparatively high temperatures and only moderate values for the interaction parameters. For the same reason, while the DMFT can also be applied to realistic materials with additional orbital degrees of freedom, cluster mean-field extensions are presently restricted to single-orbital models. Also, questions concerning critical properties of phase transitions are out of reach.

Another phenomenon frequently occurring in correlated electron systems, which cannot be handled by both theories, are quantum phase transitions. This class of phenomena typically involves long-ranged two-particle fluctuations and very low temperatures, which are of course beyond the scope of any computational resource presently available.

The roadmap for further developments and investigations is thus obvious. We need more efficient algorithms to calculate dynamical properties of complex quantum impurity systems, preferably at low temperatures and $T = 0$. First steps into this direction have already been taken through the development of new Monte Carlo algorithms [80, 81] which show much better performance than the conventional Hirsch-Fye algorithm and are also sign-problem free [82].

With more efficient algorithms also new possibilities for studies of properties of correlated electron systems arise: Studies of f-electron systems (heavy Fermions) with DFT+DMFT or even DFT+cluster mean-field theories; low-temperature properties of one- or two-dimensional correlated electron systems with large interaction parameter; critical properties and properties in the vicinity of quantum phase transitions.

This collection of examples shows that, although the DMFT and its cluster extensions are already well established, the list of possible applications and improvements is large and entering into the field by no means without possible reward.

References

1. J. Hubbard, Proc. Roy. Soc. London A **276**, 238 (1963)
2. J. Kanamori, Prog. Theor. Phys. **30**, 275 (1963)
3. M.C. Gutzwiller, Phys. Rev. Lett. **10**, 159 (1963)
4. P. Fulde, *Electron Correlations in Molecules and Solids*. Springer Series in Solid-State Sciences (Springer Verlag, Berlin/Heidelberg/New York, 1991)
5. Y. Nagaoka, Phys. Rev. **147**, 392 (1966)
6. E. Dagotto, Rev. Mod. Phys. **66**, 763 (1994)
7. N. Grewe, H. Keiter, Phys. Rev. B **24**, 4420 (1981)
8. N.E. Bickers, Rev. Mod. Phys. **59**, 845 (1987)
9. M. Imada, A. Fujimori, Y. Tokura, Rev. Mod. Phys. **70**, 1039 (1998)
10. C. Itzykson, J.M. Drouffe, *Statistical Field Theory Vol. I & II* (Cambridge University Press, Cambridge, 1989)
11. J. Negele, H. Orland, *Quantum Many-Particle Physics* (Addison-Wesley, 1988)
12. W. Metzner, D. Vollhardt, Phys. Rev. Lett. **62**, 324 (1989)
13. H. Schweitzer, G. Czycholl, Z. Phys. B **77**, 327 (1990)
14. H. Schweitzer, G. Czycholl, Phys. Rev. Lett. **67**, 3724 (1991)
15. B. Menge, E. Müller-Hartmann, Z. Phys. B **82**, 237 (1991)
16. P.G.J. van Dongen, Phys. Rev. Lett. **67**, 757 (1991)
17. P.G.J. van Dongen, Phys. Rev. B **50**, 14016 (1994)
18. P.G.J. van Dongen, Phys. Rev. B **54**, 1584 (1996)
19. U.B. und C. Mielsch, Z. Phys. B **82**, 37 (1991)
20. M. Jarrell, Phys. Rev. Lett. **69**, 168 (1992)
21. A. Georges, G. Kotliar, W. Krauth, M.J. Rozenberg, Rev. Mod. Phys. **68**, 13 (1996)
22. A.C. Hewson, *The Kondo Problem to Heavy Fermions*. Cambridge Studies in Magnetism (Cambridge University Press, Cambridge, 1993)
23. P.W. Anderson, Phys. Rev. **124**, 41 (1961)
24. T.A. Maier, M. Jarrell, T. Pruschke, M. Hettler, Rev. Mod. Phys. **77**, 1027 (2005)
25. R. Blankenbecler, D.J. Scalapino, R.L. Sugar, Phys. Rev. D **24**, 2278 (1981)
26. M. Feldbacher, K. Held, F. Asaad, Phys. Rev. Lett. **93**, 136405 (2004)
27. K.G. Wilson, Rev. Mod. Phys. **47**, 773 (1975)
28. R. Bulla, T. Costi, T. Pruschke, (2007). URL `http://arxiv.org/abs/cond-mat/0701105`. Preprint
29. H.R. Krishnamurthy, J.W. Wilkins, K.G. Wilson, Phys. Rev. B **21**, 1003 (1980)
30. H.R. Krishnamurthy, J.W. Wilkins, K.G. Wilson, Phys. Rev. B **21**, 1044 (1980)
31. S. Nishimoto, E. Jeckelmann, J. Phys.: Condens. Matter **16**, 613 (2004)
32. S. Nishimoto, T. Pruschke, R.M. Noack, J. Phys.: Condens. Matter **18**, 981 (2006)
33. C. Raas, G.S. Uhrig, F.B. Anders, Phys. Rev. B **69**, R041102 (2004)
34. T. Pruschke, Prog. Theor. Phys. Suppl. **160**, 274 (2005)
35. T. Pruschke, M. Jarrell, J.K. Freericks, Adv. in Phys. **44**, 187 (1995)
36. R. Bulla, T.A. Costi, D. Vollhardt, Phys. Rev. B **64**, 045103 (2001)
37. G. Moeller, Q. Si, G. Kotliar, M. Rozenberg, D.S. Fisher, Phys. Rev. Lett. **74**, 2082 (1995)
38. G. Kotliar, E. Lange, , M.J. Rozenberg, Phys. Rev. Lett. **84**, 5180 (2000)
39. R. Bulla, Phys. Rev. Lett. **83**, 136 (1999)
40. N.H. Tong, S.Q. Shen, F.C. Pu, Phys. Rev. B **64**, 235109 (2001)
41. M. Jarrell, T. Pruschke, Z. Phys. B **90**, 187 (1993)
42. R. Zitzler, T. Pruschke, R. Bulla, Eur. Phys. J. B **27**, 473 (2002)
43. T. Pruschke, R. Zitzler, J. Phys.: Condens. Matter **15**, 7867 (2003)

44. Y. Aiura, F. Iga, Y. Nishihara, H. Ohnuki, H. Kato, Phys. Rev. B **47**, 6732 (1993)
45. K. Morikawa, T. Mizokawa, K. Kobayashi, A. Fujimori, H. Eisaki, S. Uchida, F. Iga, Y. Nishihara, Phys. Rev. B **52**, 13711 (1995)
46. K. Maiti, D.D. Sarma, M.J. Rozenberg, I.H. Inoue, H. Makino, O. Goto, M. Pedio, R. Cimino, Europhys. Lett. **55**, 246 (2001)
47. I.H. Inoue, C. Bergemann, I. Hase, S.R. Julian, Phys. Rev. Lett. **88**, 236403 (2002)
48. M.J. Rozenberg, G. Kotliar, H. Kajueter, G.A. Thomas, D.H. Rapkine, J.M. Honig, P. Metcalf, Phys. Rev. Lett. **75**, 105 (1995)
49. M.J. Rozenberg, I.H. Inoue, H. Makino, F. Iga, Y. Nishihara, Phys. Rev. Lett. **76**, 4781 (1996)
50. M.J. Rozenberg, Phys. Rev. B **55**, R4855 (1997)
51. J.E. Han, M. Jarrell, D.L. Cox, Phys. Rev. B **58**, 4199 (1998)
52. K. Held, D. Vollhardt, Eur. Phys. J. B **5**, 473 (1998)
53. O.K. Andersen, Phys. Rev. B **12**, 3060 (1975)
54. R.O. Jones, O. Gunnarsson, Rev. Mod. Phys. **61**, 689 (1989)
55. V.I. Anisimov, A.I. Poteryaev, M.A. Korotin, A.O. Anokhin, G. Kotliar, J. Phys.: Condens. Matter **9**, 7359 (1997)
56. V.I. Anisimov, D.E. Kondakov, A.V. Kozhevnikov, I.A. Nekrasov, Z.V. Pchelkina, J.W. Allen, S.K. Mo, H.D. Kim, P. Metcalf, S. Suga, A. Sekiyama, G. Keller, I. Leonov, X. Ren, D. Vollhardt, Phys. Rev. B **71**, 125119 (2005)
57. V.I. Anisimov, O. Gunnarsson, Phys. Rev. B **43**, 7570 (1991)
58. K. Held, I.A. Nekrasov, G. Keller, V. Eyert, N. Blümer, A.K. McMahan, R.T. Scalettar, T. Pruschke, V.I. Anisimov, D. Vollhardt, in *Quantum Simulations of Complex Many-Body Systems: From Theory to Algorithms*, *NIC Series*, vol. 10, ed. by J. Grotendorst, D. Marks, A. Muramatsu (2002), *NIC Series*, vol. 10, pp. 175–209
59. K. Held, I.A. Nekrasov, N. Blümer, V.I. Anisimov, D. Vollhardt, Int. J. Mod. Phys. **15**, 2611 (2001)
60. K. Held, I.A. Nekrasov, G. Keller, V. Eyert, N. Blümer, A.K. McMahan, R.T. Scalettar, T. Pruschke, V.I. Anisimov, D. Vollhardt,
61. G. Kotliar, S.Y. Savrasov, K. Haule, V.S. Oudovenko, O. Parcollet, C.A. Marianetti, Rev. Mod. Phys. **78**, 865 (2006)
62. A. Sekiyama, H. Fujiwara, S. Imada, S. Suga, H. Eisaki, S.I. Uchida, K. Takegahara, H. Harima, Y. Saitoh, I.A. Nekrasov, G. Keller, D.E. Kondakov, A.V. Kozhevnikov, T. Pruschke, K. Held, D. Vollhardt, V.I. Anisimov, Phys. Rev. Lett. **93**, 156402 (2004)
63. A.I. Poteryaev, A.I. Lichtenstein, G. Kotliar, Phys. Rev. Lett. **93**, 086401 (2004)
64. A. Gelfert, W. Nolting, Journal of Physics: Condensed Matter **13**, R505 (2001)
65. A. Kampf, J.R. Schrieffer, Phys. Rev. B **41**, 6399 (1990)
66. A.P. Kampf, J.R. Schrieffer, Phys. Rev. B **42**, 7967 (1990)
67. F. Gebhard, Phys. Rev. B **41**, 9452 (1990)
68. T. Obermeier, T. Pruschke, J. Keller, Physica B **230–232**, 892 (1997)
69. M.V. Sadovskii, I.A. Nekrasov, E.Z. Kuchinskii, T. Pruschke, V.I. Anisimov, Phys. Rev. B **72**, 155105 (2005)
70. J.L. Smith, Q. Si, Phys. Rev. B **61**, 5184 (2000)
71. K. Haule, A. Rosch, J. Kroha, P. Wölfle, Phys. Rev. Lett. **89**, 236402 (2002)
72. A. Gonis, *Green Functions for Ordered and Disordered Systems*. Studies in Mathematical Physics (North-Holland, Amsterdam, 1992)
73. G. Kotliar, S.Y. Savrasov, G. Pallson, G. Biroli, Phys. Rev. Lett. **87**, 186401 (2001)
74. E. Müller-Hartmann, Z. Phys. **B 74**, 507 (1989)
75. T. Maier, M. Jarrell, T. Pruschke, J. Keller, Eur. Phys. J. B **13**, 613 (2000)

76. T.A. Maier, T. Pruschke, M. Jarrell, Phys. Rev. B **66**, 075102 (2002)
77. M. Jarrell, J.E. Gubernatis, Physics Reports **269**, 133 (1996)
78. P.R.C. Kent, M. Jarrell, T.A. Maier, T. Pruschke, Phys. Rev. B **72**, 060411 (2005)
79. R. Staudt, M. Dzierzawa, A. Muramatsu, Eur. Phys. J. B **17**, 411 (2000)
80. A.N. Rubtsov, V.V. Savkin, A.I. Lichtenstein, Phys. Rev. B **72**, 035122 (2005)
81. P. Werner, A.J. Millis, Phys. Rev. B **74**, 155107 (2006)
82. E. Gull, P. Werner, A.J. Millis, M. Troyer, (2006). URL `http://arxiv.org/abs/cond-mat/0609438`. Preprint

17 Local Distribution Approach

Andreas Alvermann and Holger Fehske

Institut für Physik, Universität Greifswald, 17487 Greifswald, Germany

In this contribution we describe a stochastic approach to the analysis of random spatial fluctuations and accompanying correlation phenomena like Anderson localization. We first elucidate the basic conceptual ideas which motivate the use of distributions of local Green functions in this approach, and then present details of the technique and its implementation. We illustrate its application by examples taken from the field of disordered solids. The inclusion of interaction by means of dynamical mean-field theory then is a possible starting point for a unified treatment of disorder and interaction.

17.1 Introduction

Any theory of condensed matter – at least a proper quantum mechanical one – has to include spatial and temporal fluctuations, and the correlations that develop between these. Fluctuations in time naturally arise in any interacting system, where a particle can exchange energy with the rest of the system. In a number of situations spatial fluctuations are equally important. As we learn in the Born-Oppenheimer approximation [1], electrons in a solid see the ions mainly through a static potential. In disordered systems spatial fluctuations then arise from an irregular arrangement of the ions. Even for a regular crystal, at finite temperature ions are elongated from their equilibrium positions, and the ionic potential fluctuates in space. On a technical level, the Hubbard-Stratonovich transformation [2, 3] shows how an interacting fermion system can be mapped onto a non-interacting one coupled to auxiliary fields which fluctuate in space (and time).

In traditional mean-field descriptions, such as the Weiss theory of magnetism, fluctuations are at best approximately described, if not neglected at all. As a major improvement the dynamical mean-field theory (DMFT) [4] for a detailed explanation and a list of references we refer the reader to Chap. 16 – includes fluctuations and correlations in time by establishing a self-consistent theory for a local but energy-dependent interaction self-energy. In the course of the DMFT construction, which is based on the limit of infinite dimension ($d = \infty$), spatial fluctuations are averaged out. A natural question is whether one can set up a kind of mean-field theory which accounts for fluctuations and correlations in space. This contribution will try to explain that an affirmative answer can be found if one adopts a viewpoint which has been first advocated for by P. W. Anderson in developing his theory of

A. Alvermann and H. Fehske: *Local Distribution Approach*, Lect. Notes Phys. **739**, 505–526 (2008)
DOI 10.1007/978-3-540-74686-7_17

localization in disordered systems [5]: To take the stochastic nature of spatial fluctuations serious. Then quantities like the density of states become site-dependent random quantities, and one has to deal with their distribution instead of some averages.

In this tutorial we are going to describe an approach resting on this stochastic viewpoint. This approach employs the distribution of the local density of states as the quantity of interest, and is accordingly denoted as local distribution (LD) approach. We will explain how to turn this approach into a working method, and apply it to two important examples of disordered non-interacting systems. In the discussion of the results we will relate it to a method based on averages, the coherent potential approximation (CPA) [6]. Then we outline how to combine the stochastic approach with DMFT to address both interaction and disorder. Anderson localization of a Holstein polaron serves as a particular example in this context. Finally, we take a short look how to cast the Holstein model at finite temperature into a stochastic framework. There is one word of warning to the reader: What we are going to explain is a fully worked out machinery only to a lesser degree, but constitutes an original way of thinking which has yet found some applications. This tutorial will hopefully serve the purpose to get the reader accustomed to the fundamental concepts of a stochastic approach to spatial fluctuations, and to convince him that the stochastic viewpoint is essential for an appropriate treatment.

17.1.1 Basic Concepts

We can present the basic ideas best if we concentrate on disordered systems, where spatial fluctuations are explicitly imposed.[1] In a substitutionally disordered system, like a doped semiconductor or an alloy, disorder primarily manifests through site-dependent random potentials ϵ_i. A model to describe electron motion in such a disordered crystal is given by

$$H = \sum_i \epsilon_i c_i^\dagger c_i - t \sum_{\langle i,j \rangle} c_i^\dagger c_j \; . \tag{17.1}$$

In this Hamiltonian, $c_i^{(\dagger)}$ denote fermionic operators for tight-binding electrons on a crystal lattice, and the ϵ_i account for local potentials arising from the ions composing the crystal. Note that this is a model of non-interacting fermions whose non-trivial properties arise from the randomness present in ϵ_i. Due to randomness, the ϵ_i are not fixed to some concrete values, but only their range of possible values is specified by a probability distribution $p(\epsilon_i)$. Two examples, which will be discussed below in detail, are the binary alloy with $p(\epsilon_i) = c_A \delta(\epsilon_i + \Delta/2) + (1 - c_A)\delta(\epsilon_i - \Delta/2)$, and the Anderson model of localization $p(\epsilon_i) = (1/\gamma)\, \Theta(\gamma/2 - |\epsilon_i|)$ (see (17.9) and (17.10)).

A material of certain composition corresponds to some $p(\epsilon_i)$, while any single specimen of this material is described by choosing values for ϵ_i according to $p(\epsilon_i)$.

[1] For reviews on the interesting physics of disordered systems we refer the reader to [7, 8].

Any $p(\epsilon_i)$ therefore defines many Hamiltonians (17.1), one for each concrete choice of all $\{\epsilon_i\}$. Any experiment is carried out on a single specimen, i.e. one of these Hamiltonians, while in generally we want to describe common properties of all Hamiltonians defined by $p(\epsilon_i)$. How then is the *typical* behavior for some $p(\epsilon_i)$ related to the *specific* behavior for fixed ϵ_i? For any finite system, there is a small chance to find untypical values for $\{\epsilon_i\}$. For the binary alloy (see (17.9) below) for example, there is a finite probability $c_A^N + c_B^N$ to have all ϵ_i equal on N sites – which gives an ordered system with untypical behavior for the disordered one. In a crystal with many sites however, this probability is vanishingly small: In this sense any disordered specimen is typical for the material class.[2]

In a disordered system translational invariance is broken. In contrast to the description of ordered systems we then employ quantities that depend on position, like the local density of states (LDOS) $\rho_i(\omega)$. The LDOS counts the number of states at a certain energy ω at lattice site i. It is related to the local Green function $G_{ii}(\omega) = \langle i|(\omega - H)^{-1}|i\rangle$ by

$$\rho_i(\omega) = -\operatorname{Im} G_{ii}(\omega)/\pi \,. \tag{17.2}$$

From the LDOS the density of states (DOS) $\rho(\omega)$ is obtained as the average over the crystal volume, $\rho(\omega) = \frac{1}{N}\sum_i \rho_i(\omega)$ for an N-site lattice. The LDOS generally contains more information than the DOS. Only in absence of disorder, $\rho_i(\omega) = \rho(\omega)$ for all i. But with disorder, $\rho_i(\omega)$ fluctuates through the system. The important point we will discuss later is that it would be wrong to say that the LDOS fluctuates *around* the DOS. In generally, LDOS fluctuations can render the concept of an averaged DOS to described the system in whole almost useless.

A tool to measure the LDOS in the laboratory is scanning tunneling microscopy (STM). In STM, the tunneling current between a tip and the surface of a specimen is measured. The tunneling current is, in a suitable approximation, proportional to $\rho_i(\omega)$, convoluted with some apparatus function which accounts for the finite energy resolution of the STM device. For a given applied voltage STM can therefore produce a spatially resolved picture of the LDOS. Note that due to the finite energy resolution several states contribute to the picture of $\rho_i(\omega)$: One always measures the typical behavior of some eigenstates of the Hamiltonian in the vicinity of ω.

What could not be done with STM, can be done by numerical techniques: To measure the LDOS even inside a three dimensional cube (Fig. 17.1). The computer first generates $N = L^3$ values for the ϵ_i in (17.1) using a random number generator, and then calculates the LDOS for L^2 sites in a quadratic slice of the cube using e.g. the kernel polynomial method (KPM) (see Chap. 19 in this book). Taking this

[2] The critical reader might note that this is not the more difficult question whether all quantities are self-averaging, that is mean and typical values coincide for large system sizes. The latter is true if the distribution of a quantity is sharp or at least peaked at the mean value. As e.g. the distribution $P[\rho_i(\omega)]$ of the local density of states shows, this is in general not the case. Whether it is true for the conductivity is a different question. The distribution of a quantity itself is nevertheless always typical.

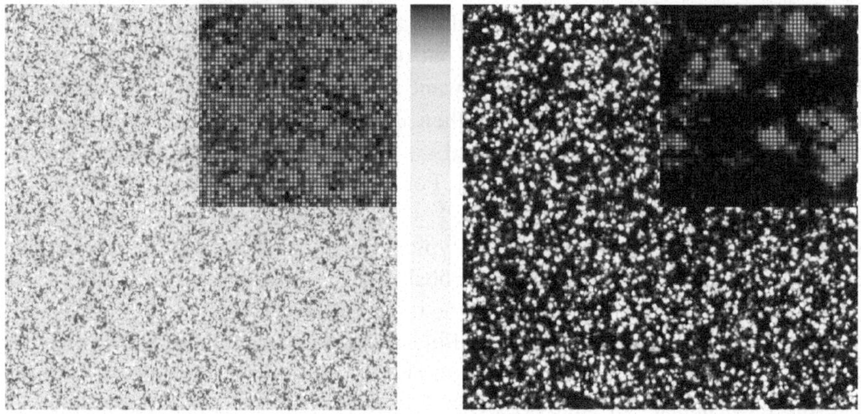

Fig. 17.1. LDOS $\rho_i(\omega)$ for a disordered cube of $N = L^3$, $L = 512$, sites. The values of ϵ_i were obtained according to the disorder distribution (17.10) of the Anderson model, with $\gamma/t = 10.0$, the calculation has been performed for periodic boundary conditions to avoid boundary effects. The pictures show a slice of L^2 sites, the value of $\rho_i(\omega)$ is color-coded, from black for very small to white for very large values (see color bar in the middle). In the upper right edge the 50^2 sites in the upper left edge of the picture are shown in magnification. Left: At energy $\omega/t = 0.0$, the LDOS is comparable throughout the crystal. Right: At $\omega/t = 7.69$, the LDOS is concentrated in finite, separated regions of the crystal. Evidently, the character of states is very different depending on energy. This indicates the existence of a phase transition, the so-called Anderson localization, which we will discuss in Sect. 17.2.1

picture,[3] one should easily accept that the site-dependence of the LDOS constitutes an eminent aspect of disordered systems. Apparently, the DOS is not significant for the different structure of the LDOS: On average, both LDOS pictures in Fig. 17.1 would look the same.

To account for the difference, we have to describe the fluctuations of the LDOS. Then, both LDOS pictures look different: The right one has strong fluctuations, most values being small but some very large, while in the left picture values are equally distributed in some range, and extreme values are rare. To quantify this behavior we can understand the LDOS with its different values at different sites as a statistical quantity, whose fluctuations are described by a distribution $P[\rho_i(\omega)]$. To construct the distribution from the explicit knowledge of the LDOS, we had to count how often the LDOS takes a value in a certain range. By this counting we would obtain $P[\rho_i(\omega)]$ as a histogram. Then, we could also recover the DOS as an (arithmetic) average

$$\rho(\omega) = \int_0^\infty \rho_i \, P[\rho_i(\omega)] \, \mathrm{d}\rho_i \; . \tag{17.3}$$

[3] With respect to the previous footnote, for $N = 512^3$ sites we expect that the LDOS shows typical behavior. Indeed, for two different sets of randomly generated values for the ϵ_i, the two pictures for the LDOS look qualitatively the same.

To find $P[\rho_i(\omega)]$ not only for one specific Hamiltonian out of the many given by (17.1) for a certain $p(\epsilon_i)$, we had to repeat this counting for many different choices of the ϵ_i until we get the typical form of $P[\rho_i(\omega)]$, which then no longer depends on concrete values of the ϵ_i but only on the disorder distribution $p(\epsilon_i)$. The aim of the stochastic approach is to construct this distribution at once.

Let us rethink the concept of the LDOS distribution, which we so far have introduced merely as a way of reorganizing information obtained from a calculation that does not mention distributions at all. To adopt the stochastic viewpoint entirely we must convince ourselves that distributions of observables are inherent in the definition of the model (17.1). Clearly, the Green function depends on all values $\{\epsilon_i\}$. Each of the values $G_{ii}(\omega; \{\epsilon_i\})$ occurs with the probability of the realization $\{\epsilon_i\}$, which is in turn given by the distribution $p(\epsilon_i)$. That is: The Green function by itself is a random variable right from the beginning, and we must deal with its distribution $P[G_{ii}(\omega)]$. As we will see this point of view is essential for the very understanding of disorder physics. We can now precisely formulate the task to be solved: To determine $P[G_{ii}(\omega)]$ from $p(\epsilon_i)$.

The distribution $P[G_{ii}(\omega)]$ has two important properties. First, though it clearly depends e.g. on energy ω, it does not depend on the lattice site i – remember, any value $G_{ii}(\omega; \{\epsilon_i\})$ for given $\{\epsilon_i\}$ does –, since due to the definition of model (17.1) each lattice site is equivalent. On the level of distributions we recover translational invariance which is otherwise lost. We keep the subscript i just to indicate a local Green function. Second, ergodicity implies a two-fold meaning of $P[G_{ii}(\omega)]$: It gives either the probability for a Green function value at a fixed lattice site but all possible $\{\epsilon_i\}$, or the probability for all lattice sites in a typical realization $\{\epsilon_i\}$. As we stated above, for an infinite lattice we get typical realizations almost surely.

17.1.2 Local Distribution Approach

We have yet advocated many times for using the distribution of the LDOS (or a Green function) instead of its average, the DOS. We now establish a scheme that provides us directly with the distribution for an infinite lattice. Since it is entirely formulated in terms of distributions of local Green functions, we call it local distribution (LD) approach.

For an arbitrary lattice, both the free DOS $\rho^0(\omega)$ and the connectivity K, i.e. the number of nearest neighbors, enter the LD equations. Compared to theories in the limit $d = \infty$, we have the additional parameter K. Since it is a bit tedious to establish the equations in the general case, we restrict to the case of a Bethe lattice (see Fig. 17.2) where we get simple equations straightforwardly, as has been first realized in [9]. As a byproduct, we obtain exact equations in this case. All principal physical features are retained despite this simplification, as we will demonstrate below.

The local Green function $G_{ii}(\omega)$ can always be expanded as

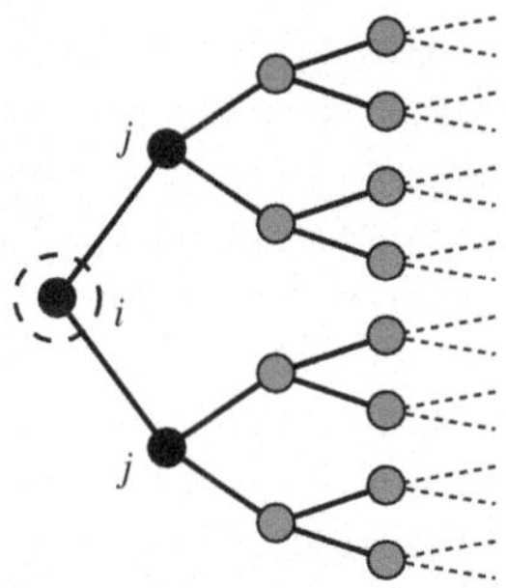

Fig. 17.2. Part of the half-infinite Bethe lattice for $K = 2$. The Bethe lattice is an infinite loop-free connected graph, where each site is connected to $K+1$ different sites. Cutting one edge, we obtain the half-infinite Bethe lattice (or Bethe tree) as shown here. The relevance of Bethe lattices originates from the fact that a number of approximations become exact there – like the LD approach. However, the precise structure of the Bethe lattice is not as relevant for the LD approach as it may seem: In principle, only the free DOS is of importance. Especially simple equations are obtained for the Bethe lattice since the inverse Hilbert transform for the Bethe DOS is a simple, algebraic function

$$G_{ii}(\omega) = \left[\omega - \epsilon_i - t^2 \sum_{j,k=1}^{K} G_{jk}^{(i)}(\omega)\right]^{-1} . \tag{17.4}$$

Here, j, k run over all K neighbors of i, and the superscript (i) indicates that $G_{jk}^{(i)}(\omega)$ has to be calculated with site i removed from the lattice. On the Bethe lattice, no path connects different sites j, k adjacent to i once i has been removed. Accordingly, (17.4) simplifies to

$$G_i(\omega) = \left[\omega - \epsilon_i - t^2 \sum_{j=1}^{K} G_j(\omega)\right]^{-1} , \tag{17.5}$$

where $G_j(\omega)$ denotes the Green function $G_{jj}^{(i)}(\omega)$ where the site i to the left of j is removed (see Fig. 17.2).

Equation (17.5) contains only Green functions of the same type. Hence it is, in the absence of disorder ($\epsilon_i = 0$ for all i), a closed equation for the local Green function $G_i(\omega) = G_j(\omega)$. Solving that quadratic equation, we find the free Green function for the Bethe lattice with corresponding semi-circular density of states,

$$G_i^0(\omega) = \frac{8}{W^2}\left(\omega - \sqrt{\omega^2 - \frac{W^2}{4}}\right) , \tag{17.6}$$

$$\rho^0(\omega) = \frac{8}{\pi W^2}\sqrt{\frac{W^2}{4} - \omega^2} \qquad |\omega| \leq \frac{W}{2} , \tag{17.7}$$

where $W = 4t\sqrt{K}$ is the bandwidth. Note that the DOS does not depend on K if W is fixed. In the limit $d = \infty$, for $K \to \infty$, the scaling $t \propto \widetilde{t}/\sqrt{K}$ keeps the bandwidth constant (cf. Chap. 16).

With disorder, the solution of (17.5) is less simple. Then, $G_i(\omega) \neq G_j(\omega)$, and (17.5) encodes an infinite set of coupled equations, depending on an infinite number of parameters $\{\epsilon_i\}$. The site-dependence of $G_i(\omega)$ prevents a closed equation for the local Green function, and hence a simple solution of the problem. But let us look at (17.5) once more from the stochastic viewpoint. We already know that the Green functions in this equation are random variables. We therefore find that (17.5) determines one random variable $G_i(\omega)$ from $K+1$ random variables ϵ_i and $G_j(\omega)$, $j = 1, \ldots, K$. We also know that $P[G_i(\omega)] = P[G_j(\omega)]$ for all j. Moreover the K Green functions $G_j(\omega)$ which appear on the r.h.s. of (17.5) are independently distributed. These two properties amount to read (17.5) as a self-consistency or fix-point equation for one random variable $G_i(\omega)$: It determines $G_i(\omega)$ on the l.h.s of (17.5) from K copies of $G_i(\omega)$ on the r.h.s. The on-site energy ϵ_i enters the equation as the source of randomness, parameterized by $p(\epsilon_i)$.

To explicitly state this essential point of the LD approach: By the stochastic reinterpretation of (17.5), the infinite set of equations for values of $G_i(\omega)$ turns out to be a single equation for the stochastic variable $G_i(\omega)$ (i.e., for its distribution $P[G_i(\omega)]$), with only one parameter $p(\epsilon_i)$. This amounts to a solution for $P[G_i(\omega)]$) entirely in terms of distributions, as provided by the sampling procedure described below.

For any finite K, (17.5) is a closed equation for the distribution of the random variable $G_i(\omega)$, which cannot be reduced to an equation for a single value like the average of $G_i(\omega)$. In the limit $d \to \infty$ however, spatial fluctuations are averaged out, and (17.5) should simplify to one for averages then. Indeed, with the scaling $t \propto \widetilde{t}/\sqrt{K}$ for $K \to \infty$, the r.h.s. of (17.5) contains a sum of K summands multiplied with $1/K$. Hence this sum becomes an average for $K \to \infty$ according to the law of large numbers. Integrating over ϵ_i gives an average also on the l.h.s., and we obtain the equation

$$G^{\mathrm{ave}}(\omega) = \int \left[\omega - \epsilon - \frac{W^2}{16} G^{\mathrm{ave}}(\omega)\right]^{-1} p(\epsilon)\, \mathrm{d}\epsilon \tag{17.8}$$

for the disorder averaged Green function $G^{\mathrm{ave}}(\omega)$. This equation is just the self-consistency equation of the so-called coherent potential approximation (CPA) for the Bethe lattice[4]. Since (17.5) is exact we have, for the special case of the Bethe lattice, proven that the CPA becomes exact for $K \to \infty$.

17.1.3 Monte Carlo Solution of the Stochastic Fix-Point Equation

It remains to solve the stochastic self-consistency equation (17.5) for $P[G_i(\omega)]$. We employ a sampling technique which is related to the Gibbs sampling method. Here

[4] For an extensive review on CPA see [6].

we have to deal with infinitely many random variables instead of finitely many as in standard Gibbs sampling.

Generally, the sampling solves any stochastic fix-point equation of the form $x = F[x, \dots, x, \epsilon]$, where x and ϵ are random variables, $F[x_1, \dots, x_K, \epsilon]$ is a function[5] that takes K values x_i of x and one value of ϵ. The distribution $p(\epsilon)$ of the external variable ϵ is known a priori. Obviously (17.5) is of that form, with $F[G_1, \dots, G_K, \epsilon_i]$ given by the r.h.s. of the equation. Then, an implicit equation has to be solved: If one already knew the solution $P[x]$ one would obtain it again by means of $F[x, \dots, x, \epsilon]$. Note the difference to the prominent Monte Carlo technique of importance sampling: While the latter one performs an integral with respect to a given known distribution, we have to construct the distribution from scratch. For that purpose we need to represent the distribution, which is conveniently done by a sample with a certain number N_s of entries x_i. Each entry will, as soon as the solution to the fix-point equation is obtained, be a possible value of x, and the fraction of entries in a certain range does determine $P[x]$. To read off $P[x]$ from the sample, we therefore construct a histogram by counting the appearances of entries in specified intervals; to build up a sample to $P[x]$ we throw N_s dice, weighted with $P[x]$, and take the N_s outcomes as sample entries. We note that any permutation of the sample still represents the same distribution.

The algorithm shown below solves the stochastic fix-point equation like one is tempted to solve any fix-point equation: By iteration. Starting with initial random values the sample is repeatedly updated until convergence is obtained. Then the distribution represented by the sample is a fix-point of the equation. To examine the following algorithm closely is a good way to comprehend the interpretation of (17.5) as a stochastic self-consistency equation.

```
input: distribution p(e), functional F, sample size Ns
output: sample and distribution for P[x]

(1) initialize sample S[i] with random data
(2) for i=1,Ns
   (2a) find random value for e
        using a random number generator for p(e)
   (2b) find random indices j[1],...,j[K] within 1,...,Ns
   (2c) calculate new value for S[i]=F[S[j[1]],...,S[j[K]],e]
(3) if notConverged goto 2

(4) construct distribution P[x] from S[i] as histogramm
(4') calculate averages of P[x] by summing over S[i]
```

We remind ourselves that convergence of the sample does not mean convergence of its entries but of the distribution represented. In practice, we may check this by comparison of some moments extracted of the distributions before and after each update `(2)`. In principle, convergence of the sampling algorithm cannot be

[5] For the equation to make sense, one requires $F[x_{\sigma 1}, \dots, x_{\sigma K}, \epsilon] = F[x_1, \dots, x_K, \epsilon]$ for all permutations σ.

guaranteed, but depends on convergence of the non-stochastic equation obtained for $\epsilon = 0$, and the properties of the fix-point distribution. Two examples for the convergence of the sampling algorithm (Figs. 17.3, 17.4) illustrate this dependence. First, take the non-stochastic fix-point equation $x = f(x)$ with $f(x) = x^3 - 1.25 \cdot x$. This equation has three fix-points $x_1 = 0$, $x_{2/3} = \pm 1.5$. All fix-points are instable, since the slope $|f'(x_i)|$, $i = 1, 2, 3$, is larger than 1. Direct iteration of $f(x)$ converges to the stable two-cycle ± 0.5, but misses the instable fix-point x_1. The usual trick to avoid two-cycles, namely rewriting the equation as $x = (x + f(x))/2$, results for starting values in $(-1.5, 1.5)$ in convergence to x_1, where the slope is negative. To check convergence of the sampling algorithm the fix-point equation is rewritten as a stochastic equation $x = F[x_1, \ldots, x_k] = f(\sum_{i=1}^{K} x_i/K)$, with identical fix-points. If we initialize the sample with values in $[0, 1]$ – any subset of $(-1.5, 1.5)$ would work – the distribution of x constructed in the sampling converges to a δ-peak at the fix-point $x_1 = 0$ of the original equation (see Fig. 17.3). As for this example, convergence of the sampling algorithm is generally better than for direct iteration of the original equation. This result should imply good convergence for (17.5). For $p(\epsilon_i) = \delta(\epsilon_i)$, i.e. without disorder, already direct iteration converges to the Green function $G^0(\omega)$ of the Bethe lattice, and sampling of the distribution is therefore expected to converge fairly well. Nevertheless, as the second example shows, convergence may worsen for the full stochastic equation even if it is pretty good for the non-stochastic one. For the second example we apply the sampling algorithm to the solution of (17.5) with disorder, i.e. $p(\epsilon_i) \neq \delta(\epsilon_i)$. While in the previous example the convergence to the fix-point distribution is very regular, the ϵ_i provide an explicit source of randomness which leads to fluctuations in the sample during

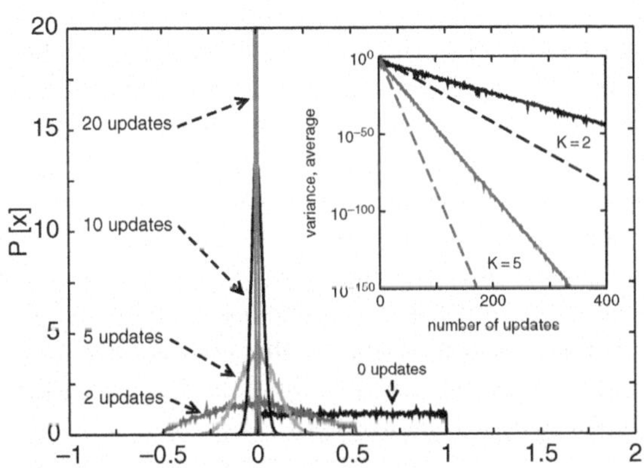

Fig. 17.3. Convergence of a distribution within the sampling algorithm. Solving the equation $x = f(x)$ with $f(x) = x^3 - 1.25 \cdot x$ as a stochastic equation with $K = 2$. The picture shows the distribution $P[x]$ of x after some updates of a sample with $N_s = 5 \times 10^4$ entries; the inset displays the arithmetic average (solid line) and variance (dashed line) of the sample for $K = 2$ and $K = 5$. The distribution converges to a δ-distribution at the fix-point $x_0 = 0$, although $|f'(x_0)| = 1.25 > 1$

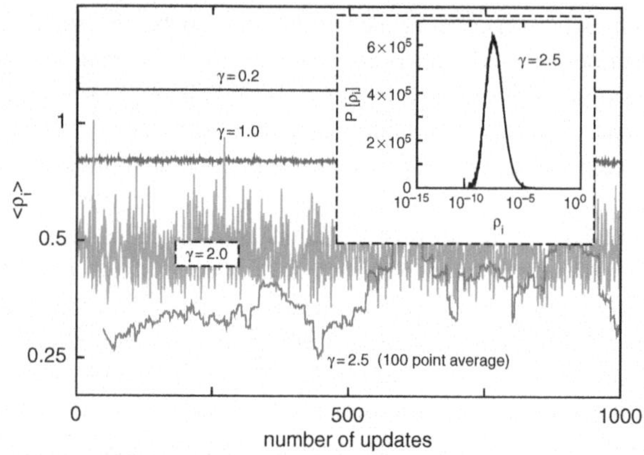

Fig. 17.4. Convergence of a distribution within the sampling algorithm. Fluctuations of the average $\langle \rho_i \rangle$ of the LDOS distribution $P[\rho_i(\omega)]$ to (17.5) during updates of a sample with $N_s = 5 \times 10^4$ entries. The disorder distribution $p(\epsilon_i)$ is taken from (17.10), and $\omega = 0$. The curves to $\gamma = 0.2$ and $\gamma = 1.0$ correspond to the distributions shown in Fig. 17.8. For $\gamma = 2.5$, the average of 100 consecutive updates is shown instead of $\langle \rho_i \rangle$ (the fluctuations of $\langle \rho_i \rangle$ would fill the picture). The inset displays $P[\rho_i(\omega)]$ for $\gamma = 2.5$. Note the logarithmic abscissa

sampling. In Fig. 17.4 we show the fluctuations of the average of the LDOS distribution $P[\rho_i(\omega)]$. The larger γ in this example, i.e. the larger the variance of ϵ_i, the stronger fluctuations are. This is not an artifact of the algorithm, but results unavoidably from the properties of the fix-point distribution. As the inset in Fig. 17.4 shows, the fix-point distribution has extremely large variance. Resolving this equation by a sample with a finite number of entries results in typical large fluctuations associated with the statistics of rare events. We will see below, that the strength of fluctuations may even diverge, which signals a phase transition (here, the Anderson transition from extended to localized states, see Sect. 17.2.1). With strong fluctuations, the algorithm does not converge even in an approximate sense, and a single sample is not a good representation of the distribution. To sample the full distribution we then have to use a large number of consecutive samples obtained in update step (2).

Note that convergence in the first example has been faster for $K = 5$ than for $K = 2$. For (17.5) this observation implies that convergence becomes better with increasing K – just as one comes close to the limit $K = \infty$, where the stochastic equation can be replaced by one for averages.

17.2 Applications of the LD Approach

After the construction of the LD approach and the explanation of the Monte Carlo sampling we shall now discuss some results of the LD approach. In addition to the examples given here we also refer the reader to [10, 11, 12, 13]. For all examples, we set $K = 2$ in (17.5), and measure energies in units of the bandwidth W (if we fix $W = 1, t = 1/\sqrt{32}$ for the $K = 2$-Bethe lattice).

17.2.1 Non-Interacting Disordered Systems

Let us begin with two examples of non-interacting disordered systems [14]. The first example is the binary alloy model, which describes a solid composed of two atomic species A, B. The on-site energies are distributed as

$$p(\epsilon_i) = c_A \delta(\epsilon_i + \Delta/2) + (1 - c_A)\delta(\epsilon_i - \Delta/2) , \tag{17.9}$$

where Δ is the separation of the energy levels of A,B atoms, and c_A ($c_B = 1 - c_A$) is the concentration of A (B) atoms.

At a first glance we should expect, for $\Delta > W$, two bands in the DOS centered at $\pm\Delta/2$, with weight c_A and $1 - c_A$ respectively. Indeed this is what we get by the CPA, if we solve (17.8). If we compare to the result obtained from the stochastic approach, solving (17.5) by the sampling algorithm, we find that the averaged CPA description misses important features of the alloy (see Fig. 17.5). Remember that the stochastic approach is exact in this situation: The DOS shown gives the true picture of the system.

Why does CPA fail in this case? Physically, the electron motion is strongly affected by multiple scattering on finite clusters of either A or B atoms, whereby the DOS develops a rich structure. The most prominent peaks in the DOS can be directly attributed to small clusters, as indicated in Fig. 17.5. For the parameters chosen here, the concentration c_A is below the classical percolation threshold, hence all clusters of the minority species A are finite. This is the origin of the fragmentation

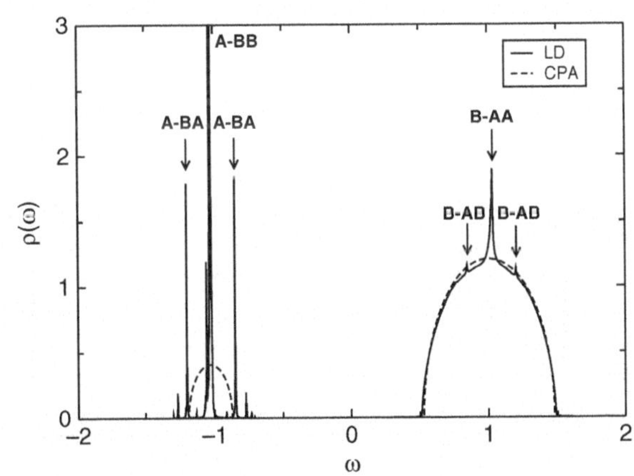

Fig. 17.5. DOS $\rho(\omega)$ for the binary alloy model, with $\Delta = 2.0$, $c_A = 0.1$. The picture shows both CPA and LD results. To resolve the δ-peaks in the minority band, the LD curve has been broadened by including an artificial imaginary part $\eta = 10^{-3}$ in the energy $\omega + i\eta$. Arrows mark contributions from small finite clusters of atoms. Figure taken from [14]

of the minority A-band. CPA, being constructed in the limit $K \to \infty$, averages over spatial fluctuations and does therefore not properly account for multiple scattering. From the stochastic viewpoint, this is manifest in the LDOS distribution $P[\rho_i(\omega)]$ (see Fig. 17.6), which cannot be represented by a single value. Especially it is not senseful to replace $P[\rho_i(\omega)]$ by $\rho(\omega)$ as in the CPA.

The second example we consider is the Anderson model of localization, which assumes a box distribution of on-site energies

$$p(\epsilon_i) = \frac{1}{\gamma}\Theta(\frac{\gamma}{2} - |\epsilon_i|) , \tag{17.10}$$

with $\gamma \geq 0$ as the strength of disorder. In contrast to the binary alloy with its discrete distribution, the DOS in the Anderson model is well described by CPA, expect for some details at the band edges (see Fig. 17.7). But, invisible in the DOS, the character of states is different towards the band edges and in the band center, as could already be anticipated from Fig. 17.1. While states in the band center resemble distorted Bloch waves, which extend through the whole crystal, states towards the band edge have appreciable weight only in finite (separated) regions of the crystal. An electron in such a state is not itinerant any more, hence the state is called localized in contrast to extended Bloch-like states. As localized states do not contribute to the electrical conductivity, Anderson localization is a mechanism to drive a metal into an insulator as a result of disorder. While for interaction-driven metal-insulator transitions like the Mott or Peierls transition a gap in the DOS opens at the transition, the DOS stays finite at the Anderson transition from localized to extended states. It is only the conductivity which drops to zero.

Guided by our discussion of Fig. 17.1, one expects that localized and extended states can be distinguished by means of the LDOS distribution. Fig. 17.8 shows

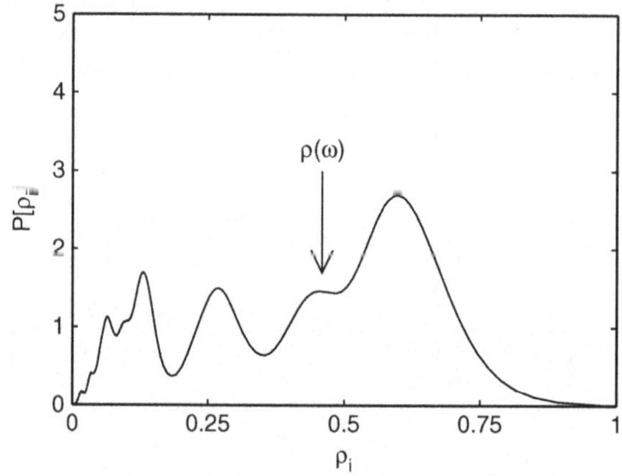

Fig. 17.6. LDOS distribution $P[\rho_i(\omega)]$ for the binary alloy model at $\omega = 0.0$, with $\Delta = 0.3$, $c_A = 0.1$. The arrow marks the DOS $\rho(\omega)$. Evidently a single value cannot represent the distribution in any sense

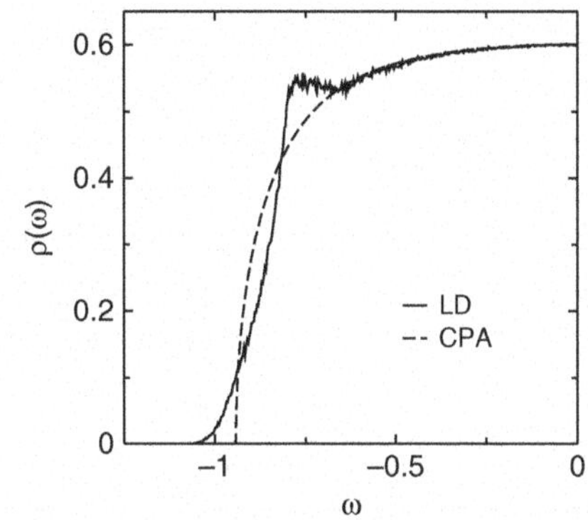

Fig. 17.7. DOS $\rho(\omega)$ for the Anderson model, at $\gamma = 1.5$. The picture shows both CPA and LD results. Since $\rho(-\omega) = \rho(\omega)$, only one half of the figure is shown. Note the sharp band edge within the CPA approximation, and the smooth *Lifshitz* tails in the LD result. These tails result from the exponentially few (localized) states at sites with large $|\epsilon_i|$ which are not resolved within CPA

$P[\rho_i(\omega)]$ for weak and moderate disorder. For weak disorder, the distribution resembles a Gaussian peaked at the (averaged) DOS $\rho(\omega)$. With increasing disorder, as fluctuations of the LDOS grow, the distribution becomes increasingly broad and asymmetric. The DOS is then not representative for the distribution anymore. With even increasing disorder, the distribution becomes singular at the transition to localized states: All but infinitesimally small weight resides at $\rho_i = 0$. This singularity in $P[\rho_i(\omega)]$ has to be accessed via analytical continuation of a Green function to the real axis, as is depicted in Fig. 17.9 Although the distribution becomes singular at the localization transition, the DOS is nevertheless still finite due to negligible

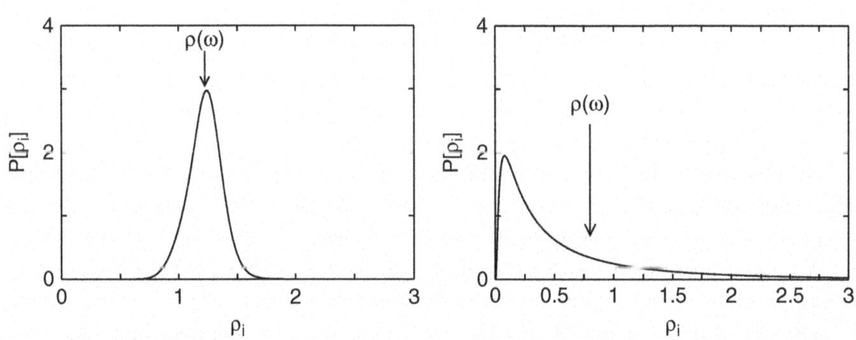

Fig. 17.8. LDOS distribution $P[\rho_i(\omega)]$ for the Anderson model, in the band center $\omega = 0$. The arrows mark the DOS $\rho(\omega)$. Left: For weak disorder $\gamma = 0.2$, the distribution is peaked at the $\rho(\omega)$. Right: Already for moderate disorder $\gamma = 1.0$, the DOS is not significant for the distribution, which is very skew and broad. Compare this to the distribution shown in Fig. 17.4 for even stronger disorder

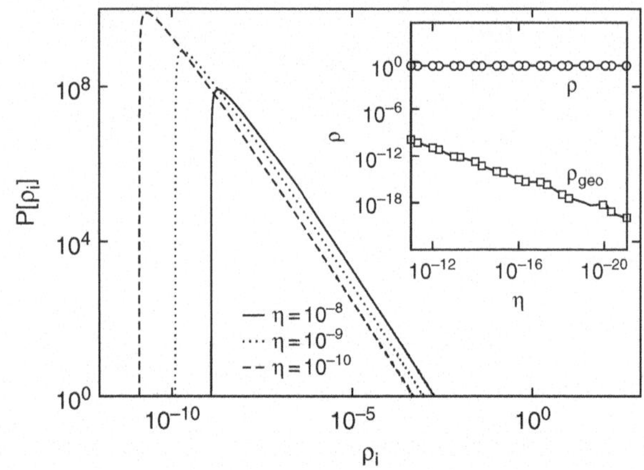

Fig. 17.9. The figure shows, how $P[\rho_i(\omega)]$ for localized states in the Anderson model depends on the imaginary part η in the energy argument of the Green function $G_i(\omega + i\eta)$. For $\eta \to 0$, numerically performing analytical continuation to the real axis, the DOS $\rho(\omega)$ stays finite, but a typical moment, like the geometrically averaged LDOS $\rho_{\rm geo}(\omega)$, goes to zero

weight at infinitely large values of $\rho_i(\omega)$. Anderson localization therefore manifests itself in the full distribution $P[\rho_i(\omega)]$ but not in an averaged value like $\rho(\omega)$. As for the binary alloy, a description in averages is prevented by the pronounced spatial fluctuations which constitute localization.

To obtain the phase diagram of the Anderson model (Fig. 17.10) which shows the transition line between localized and extended states – the so-called mobility edge –, we employ the above criterion based on the LDOS distribution. Since the DOS does not indicate Anderson localization, the phase diagram could not be obtained from CPA. Indeed, as should be apparent from our discussion, CPA misses localization at all. Looking at the distributions, we can expect CPA to describe the system well only for small disorder and away from the band edges, when $P[\rho_i(\omega)]$ is peaked at $\rho_i(\omega)$ (left panel in Fig. 17.8). There, an electron propagates diffusively, and correlations in the electron motion are weak. There is however no simple way to extend an averaged theory like CPA to all disorder strengths or energies.

We should mention that localization also occurs in the binary alloy model. For small enough c_A and large Δ (see Fig. 17.11 for the DOS), when all A-clusters are finite and scattering on interlying B-atoms is strong, one expects that all states in the A-band are localized. Tunneling processes between separated A-clusters may nevertheless give rise to extended states. For the parameters in Fig. 17.5 states in the A-band are localized. The DOS then consists of a series of δ-peaks which had to be broadened with some finite η to be seen in the picture. Note that for the Anderson model the δ-peaks densely fill the energy regime of localized states – a so-called Dirac comb or dense pure point spectrum in mathematical terms. The DOS of the Anderson model is therefore smooth, while for the binary alloy the tendency to gap formation prevails. This is a precursor of percolative behavior for $\Delta \to \infty$.

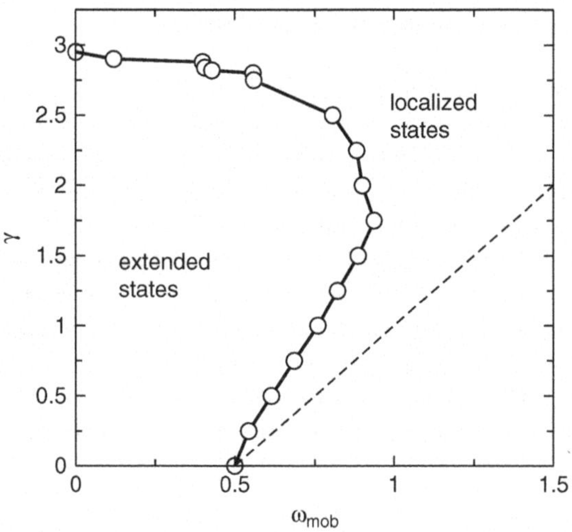

Fig. 17.10. Phase diagram of the Anderson model. Shown is the mobility edge ω_{mob} vs. γ. The dashed line shows the exact band edge $\omega = (W + \gamma)/2$. The trajectory is symmetric under $\omega_{\mathrm{mob}} \mapsto -\omega_{\mathrm{mob}}$. Note that for small γ, ω_{mob} grows before it tends to zero when γ approaches the critical value for complete localization (so-called re-entrant behavior)

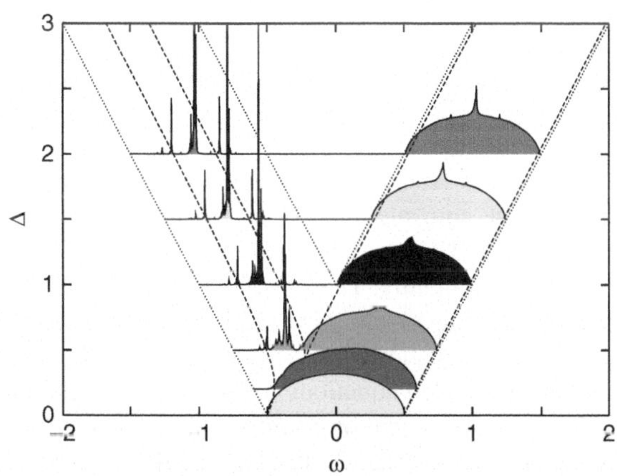

Fig. 17.11. Part of a phase diagram of the binary alloy model for concentration $c_A = 0.1$, showing the DOS for various Δ. The dashed curves show the CPA band edges, the dotted lines mark $\omega = \pm\Delta/2 \pm W/2$. Figure taken from [14]

17.2.2 Extension to Interacting Disordered Systems

Within DMFT, interaction properties are subsumed in a $\boldsymbol{k}$-independent self-energy $\Sigma(\omega)$. Transforming to real space, we get a local self-energy $\Sigma_{ii}(\omega)$, which however does not depend on the lattice site i for ordered systems. The LD approach rests on the observation, that in the presence of disorder previously site-independent quantities become site-dependent. This applies not only to the local Green function $G_{ii}(\omega)$ but also to the self-energy $\Sigma_{ii}(\omega)$. In order to extend the LD approach to interacting systems – or the DMFT to disordered ones – one has to introduce a site-dependent self-energy $\Sigma_{ii}(\omega)$, which can be understood as a random variable like the Green function $G_{ii}(\omega)$ [13].

Remember that in DMFT $\Sigma_{ii}(\omega)$ with $G_{ii}(\omega) = G^0_{ii}(\omega - \Sigma_{ii}(\omega))$ is obtained from the solution of an Anderson impurity problem, in dependence on a local propagator

$$\mathcal{G}_{ii}(\omega) = \left[G_{ii}(\omega)^{-1} + \Sigma_{ii}(\omega)\right]^{-1} , \tag{17.11}$$

which excludes interaction at site i. Formally, $\Sigma_{ii}(\omega)$ is a functional of $\mathcal{G}_{ii}(\omega)$,

$$\Sigma_{ii}(\omega) = \Sigma_{ii}[\mathcal{G}_{ii}(\omega)] , \tag{17.12}$$

whose explicit form is not known in most cases. For the Bethe lattice with its semi-circular DOS, simple expressions for $\mathcal{G}_i(\omega)$ and $G_i(\omega)$ exist, namely

$$\begin{aligned} \mathcal{G}_i(\omega) &= \left[\omega - \epsilon_i - t^2 \sum_{j=1}^{K} G_j(\omega)\right]^{-1} , \\ G_i(\omega) &= \left[\mathcal{G}_i(\omega)^{-1} - \Sigma_i(\omega)\right]^{-1} \\ &= \left[\omega - \epsilon_i - \Sigma_i(\omega) - t^2 \sum_{j=1}^{K} \mathcal{G}_j(\omega)\right]^{-1} , \end{aligned} \tag{17.13}$$

– this is of course just the equivalent to (17.5) – while the complexity of (17.12) does not reduce a single bit. Clearly, with the Green function $G_{ii}(\omega)$ also the self-energy $\Sigma_{ii}(\omega)$ is a random quantity. The Equations (17.11)–(17.13) therefore have the same status in an interacting system as (17.5) has without interaction: They form stochastic self-consistency equations for $\Sigma_{ii}(\omega)$ and $G_{ii}(\omega)$. Again, what would be an infinite number of coupled equations for self-energies and Green functions, reduces to few self-consistency equations if reformulated by means of distributions.

Solving these equations by Monte Carlo sampling the impurity problem (17.12) has to be solved in each update step `(2c)`. This constitutes the main part of the high computational complexity of the combined LD+DMFT approach. While in DMFT one has to solve the impurity problem some times till convergence, it has to be solved here repeatedly for each entry of the sample. The computational effort is therefore at least N_s times larger than in DMFT.

In few cases the DMFT impurity problem can be solved exactly. With an explicit solution for (17.12) at hand, the numerical effort to perform the sampling of

$G_i(\omega)$ can be handled. One example is the single polaron Holstein model [15] with Hamiltonian

$$H = -t \sum_{\langle i,j \rangle} c_i^\dagger c_j - \sqrt{\varepsilon_p \omega_0} \sum_i (b_i^\dagger + b_i) c_i^\dagger c_i + \omega_0 \sum_i b_i^\dagger b_i \;, \tag{17.14}$$

where an electron is coupled to optical phonons of energy ω_0. For this model, $\Sigma_i(\omega)$ is obtained as an infinite continued fraction [16]

$$\Sigma_i(\omega) = \cfrac{1\varepsilon_p \omega_0}{[\mathcal{G}_i(\omega - 1\omega_0)]^{-1} - \cfrac{2\varepsilon_p \omega_0}{[\mathcal{G}_i(\omega - 2\omega_0)]^{-1} - \cfrac{3\varepsilon_p \omega_0}{\dots}}} \;. \tag{17.15}$$

The continued fraction is an expansion in terms of the maximal number of virtual phonons that are excited at the same time. Evidently, this expansion is non-perturbative, and contains diagrams of arbitrary order at any truncation depth of the fraction.

To give an impression of the physical content of the Holstein model, we show in Fig. 17.12 the DOS $\rho(\omega)$ in the anti-adiabatic (i.e. for large ω_0) strong coupling regime as obtained from a DMFT calculation based on (17.15). This picture illustrates the formation of a new quasi-particle which is a compound object of an

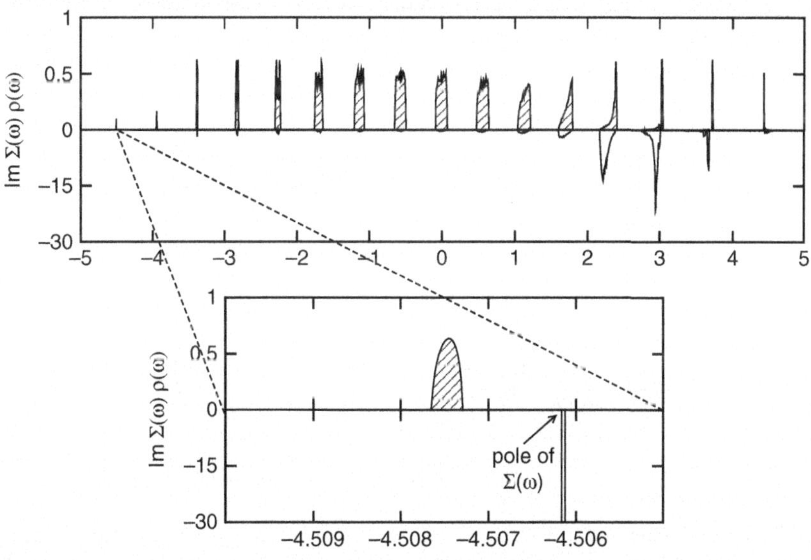

Fig. 17.12. The Holstein polaron at strong coupling and large phonon frequency. We show the DOS for the Holstein model with $\omega_0/W = 0.5625$, $\varepsilon_p/W = 4.5$. The center of the lowest sub-band is located nearly at $-\varepsilon_p$ (the polaron shift), and bands are separated by ω_0. The bandwidth of the lowest sub-band, which is shown in detail in the lower panel, is $W_{\text{sub}} = 3.45 \times 10^{-4} W$

electron with a surrounding cloud of phonons. This so-called small polaron is characterized by an extremely large mass resulting in a narrow quasi-particle band (in Fig. 17.12 the effective mass of the polaron is four orders of magnitude larger than the free electron mass). Note that, while the lowest polaron is fully coherent, as an effect of inelastic electron-phonon interaction higher bands are incoherent. Accordingly, the imaginary part of the self-energy is finite. The reader should be aware that the properties of the polaron intimately depend on the parameter values. Here we do by no means provide a general picture of polaron physics. For detailed discussions see e.g. [17, 18, 19], for a DMFT study of small polarons [20].

If the Hamiltonians (17.1) and (17.14) are combined, we obtain a model to study possible effects of Anderson localization of a polaron. Like for the polaron itself, the physics of polaron localization is diverse and complicated. A general discussion, as partly given in [21], is far beyond the scope of this tutorial. For the parameters used in Fig. 17.12 however, the polaron in its lowest band is a small and heavy quasi-particle with infinite lifetime. We therefore expect that disorder affects this quasi-particle like a free electron, but with the mass of the polaron. We can scrutinize this expectation within the LD+DMFT approach, which provides the mobility edge trajectory for the lowest sub-band (Fig. 17.13). Rescaling the trajectory properly it perfectly matches the trajectory of the Anderson model in Fig. 17.10. As a fundamental observation we note that the critical disorder for complete localization of all states in the polaron sub-band is renormalized by W_{sub}/W as compared to the free electron: In any real material such a polaron would be localized for almost arbitrarily small disorder.

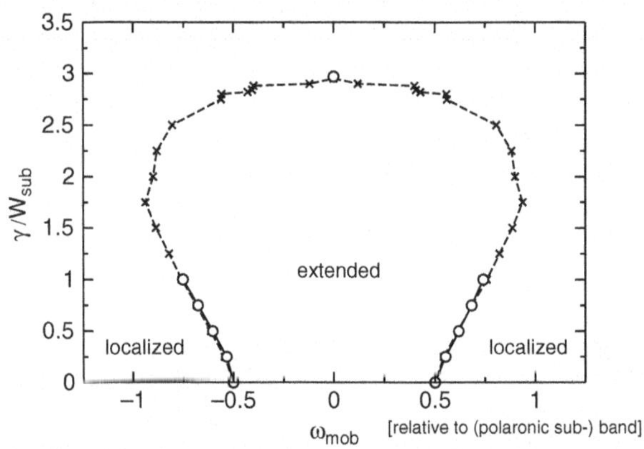

Fig. 17.13. Phase diagram for Anderson localization of a Holstein polaron at strong coupling and large phonon frequency. As in the previous figure, $\omega_0/W = 0.5625$, $\varepsilon_p/W = 4.5$. Shown is the mobility edge for the lowest polaronic sub-band (*circles*) in comparison to the Anderson model for a free electron (*crosses*). γ and ω_{mob} is rescaled to the respective bandwidth. The energy scale of both curves accordingly differs by almost four orders of magnitude, as $W_{\mathrm{sub}} = 3.45 \times 10^{-4} W$

To avoid any misconception we like to point out that polaron localization in this example shows up as renormalization of e.g. the critical disorder only for the reason that the polaron here is a small quasi-particle with infinite lifetime. If the polaron would extend over some lattice sites, disorder would affect the structure of the polaron itself, instead of affecting the polaron as a compound entity. Likewise, if the quasi-particle lifetime is finite localization is weakened since the motion of the particle is incoherent then. This is already expressed by the fact that Anderson localization can only be detected in the limit $\eta \to 0$, as discussed before. Incidentally, the band fragmentation we saw in the minority band of the binary alloy is not destroyed by damping since finite gaps exist there. Remember that the latter one shows up in the DOS for finite η, while the former one shows up only in the distribution for $\eta \to 0$.

17.2.3 The Holstein Model at Finite Temperature

In the previous section we addressed the Holstein model at zero temperature, and imposed spatial fluctuations by disorder. But even without disorder, the physics of the Holstein model (17.14) may be strongly influenced by static scattering off spatial fluctuations. As mentioned in the introduction, this is the case for heavy ions, i.e. small oscillator frequency ω_0, when ions act as static scatterers to first order. If at finite temperature ions are displaced from their equilibrium positions, the concomitant random potential acts as a static disorder potential to first order (see also [19]).

Let us consider the limit of large ionic mass M, keeping the spring constant $k_s = M\omega_0^2$ of the harmonic oscillator $\omega_0 b_i^\dagger b_i$ constant. This limit, the so-called adiabatic limit $\omega_0 \to 0$ of small phonon frequency, is opposite to the regime of large phonon frequency to which the example from the previous section (Fig. 17.12) belongs. In the limit $\omega_0 \to 0$ ions are nearly classical particles. Classical states in the context of the harmonic oscillator can be constructed as coherent states $|\alpha\rangle$. Remember that a coherent states is a Gaussian wavepacket centered at $\bar{X}_i^\alpha = \langle\alpha|X_i|\alpha\rangle = \sqrt{2/(M\omega_0)}\,\mathrm{Re}\,\alpha$, with the position operator $X_i = \sqrt{1/(2M\omega_0)}(b_i + b_i^\dagger)$.

It is not difficult to convince oneself, that the thermal (Boltzmann) trace over boson eigenstates $|n\rangle$ can be expressed as an integral over coherent states:

$$\mathrm{Tr}_\beta[\ldots] = \frac{1}{2}\sum_{n=0}^{\infty} \mathrm{e}^{-\beta n\omega_0}\langle n|\ldots|n\rangle = \frac{\mathrm{e}^{\beta\omega_0}-1}{\pi}\int \mathrm{d}^2\alpha\, \mathrm{e}^{(1-\exp(\beta\omega_0))|\alpha|^2}\langle\alpha|\ldots|\alpha\rangle. \tag{17.16}$$

In the spirit of Monte Carlo integration the complex plane integral $\int \mathrm{d}^2\alpha\ldots$ has a stochastic counterpart: The integral value is obtained by sampling the expectation value $\langle\alpha|\ldots|\alpha\rangle$ for a complex random variable α with Gaussian probability density $\propto \exp[(1-\exp(\beta\omega_0))|\alpha|^2]$. This results in a stochastic interpretation of the Holstein model at finite temperature. The random part of the model is the initial state of the bosonic subspace, given by random coherent states $|\alpha_i\rangle$ at site i according to the specific distribution for α_i. The bosonic vacuum at $T = 0$ is therefore replaced by a fluctuating vacuum, where the strength of fluctuations depends on T.

From a local point of view as in the previous section, we need the Green function $G_i^\alpha(\omega)$, which in contrast to the Holstein model at $T = 0$ is not evaluated in the bosonic vacuum but within a certain coherent state $|\alpha_i\rangle$. The Green function is given by

$$G_i^\alpha(\omega) = \left[\mathcal{G}_i(\omega)^{-1} - \sqrt{2\varepsilon_p k_s}\bar{X}_i^\alpha - \Sigma_i^\alpha(\omega)\right]^{-1} . \tag{17.17}$$

This expression is of the same type as (17.13), with a static disorder contribution given by the random variable $\bar{X}_i^\alpha$, and a self-energy contribution $\Sigma_i^\alpha(\omega)$ accounting for finite lifetime effects, i.e. finite ω_0. Note that $\sqrt{\varepsilon_p}\bar{X}_i^\alpha$, being an effect of interaction, enters $G_i^\alpha(\omega)$ but not $\mathcal{G}_i(\omega)$. $\bar{X}_i^\alpha$ has Gaussian distribution $P[\bar{X}_i^\alpha] \propto \exp[(1 - \exp(\beta\omega_0))M\omega_0(\bar{X}_i^\alpha)^2/2]$ resulting from (17.16). Both for high temperature ($\beta \to 0$) and in the adiabatic limit ($\omega_0 \to 0$), the classical result $P[\bar{X}_i^\alpha] \propto \exp[-\beta k_s(\bar{X}_i^\alpha)^2/2]$ is obtained. Note that the Green function $G_i^\alpha(\omega)$ is evaluated in bosonic states that are not eigenstates of the bosonic number operator $b_i^\dagger b_i$, and therefore in principle is a non-equilibrium Green function with different analytical properties as retarded Green functions $G_i(\omega)$ used elsewhere in the text. On average however, i.e. for the disorder averaged Green function $\langle G_i^\alpha\rangle$ which is obtained as the average over α_i instead of ϵ_i as in the previous sections, the full analytical properties of a retarded Green function are recovered.

The self-energy $\Sigma_i^\alpha(\omega)$ can be expressed as a continued fraction like for the Holstein model at zero temperature. The expression is derived at considerably less ease than before – e.g. using Mori-Zwanzig projection techniques [22] – and acquires a less systematic form. From the top level of the continued fraction,

$$\Sigma_i^\alpha(\omega) = \cfrac{\omega_0(\varepsilon_p - 2\mathrm{i}\sqrt{\varepsilon_p\omega_0}\,\mathrm{Im}\,\alpha_i)}{\omega - \cfrac{2\sqrt{\varepsilon_p\omega_0}(\varepsilon_p+\omega_0)\,\mathrm{Re}\,\alpha_i + \varepsilon_p\omega_0(1 - 4\mathrm{i}\,\mathrm{Re}\,\alpha_i\;\mathrm{Im}\,\alpha_i)}{\varepsilon_p - 2\mathrm{i}\sqrt{\varepsilon_p\omega_0}\,\mathrm{Im}\,\alpha_i} - \ldots} \tag{17.18}$$

we deduce that $\Sigma_i^\alpha(\omega)$ is of order ω_0, while $\bar{X}_i^\alpha$ is of order 1. The expression for $G_i^\alpha(\omega)$ therefore acquires the correct form as an expansion in ω_0. Note that (17.16)–(17.18) hold for any parameters values, but are constructed to work in the limit of small ω_0. The continued fraction (17.15), which is straightforwardly generalized to arbitrary eigenstates $|n\rangle$ of $b^\dagger b$, is not applicable in this case: For $\omega_0 \to 0$ the number of bosons in the thermal trace becomes large, which renders an expansion in the number of excited bosons useless.

By (17.16)–(17.18) the Holstein model for small $\omega_0 \to 0$ and finite T is cast in a form that is amenable to the stochastic method explicated in the preceding sections. Here, we do not supply actual calculations based on that. The bottom line instead is the interpretation provided by our reformulation: Temperature induced spatial fluctuations act to a certain degree like (static) disorder. In (17.17) the main source of resistivity due to scattering off thermally excited phonons is translated to disorder scattering: With increasing T, the amount of fluctuations of the disorder potential $\sqrt{\varepsilon_p k_s}\bar{X}_i^\alpha$ increases, and electron motion is strongly suppressed. We know from disordered systems that the suppression is much larger than expected from first estimates based on uncorrelated scattering, which neglect correlations in the electron motion which eventually lead to Anderson localization. Will an electron subject to electron-phonon interaction ever be localized at finite T? Exactly for $\omega_0 = 0$, with $\Sigma_i^\alpha = 0$, we end up with a disorder problem, and localization can occur. But otherwise, surely not: For any $\omega_0 \neq 0$ the ionic potential seen by an electron is not static but changes on a timescale $\propto 1/\omega_0$. Anderson localization itself is then suppressed by incoherent scattering where the electron exchanges energy with ions, e.g. by absorption of thermally excited phonons.[6] Nevertheless strong suppression of electron transport at small ω_0 exists as a precursor of Anderson localization.

17.3 Summary

At the end of this tutorial we shall return to the initial question we raised: How to set up a kind of mean-field theory for spatial fluctuations and correlations. The essential idea argued for is to adopt a stochastic viewpoint: The mean-field in the theory has to be the distribution of a certain quantity – that is a stochastic mean-field theory which does not have a *mean*-field at all. We first had to convince ourselves – taking disordered systems as the example for fluctuations of a potential in space – that important quantities like the density of states are indeed best understood as random quantities which should be described by their distribution. The main effort was to construct a working scheme, the LD approach, out of this basic premise of the stochastic viewpoint. Technically that included the derivation of a closed set of stochastic equations for the distribution of the local density of states as the quantity of interest. In the derivation a complicated set of equations could be collapsed into a single equation if formulated with the help of distributions. For the solution of this stochastic equation we discussed the application of Monte Carlo sampling.

As much as we used disordered systems to motivate the central concepts leading to the LD approach we took them as the first example to demonstrate its application. Notably, even a complex non-local effect like Anderson localization is correctly described by distributions of a local quantity. This demonstrates how correlations turn up in local distributions. On the other hand we had to accept that a disordered system is always far from the limit $d = \infty$. Both the second example – Anderson localization of a Holstein polaron as an interacting disordered system – and the third example – the Holstein model at finite temperature – show that we generally cannot separate temporal fluctuations from spatial ones. The competition between the different physical mechanism present in these problems gives rise to very rich physical behavior. The central features of such systems become accessible only within a theory which accounts for both spatial and temporal fluctuations on an equal footing, as the combined LD+DMFT approach does.

[6] Remember the discussion in the previous section concerning the case of large ω_0, opposite to the adiabatic limit addressed here. There we noted that only in a coherent polaron band Anderson localization affects a polaron like a free – albeit heavy – particle.

There is a number of (open) questions we could not touch upon here. The calculation of transport properties is one important example, which is not really understood at the present stage of development. Taking the Holstein model at finite temperature as an example, we sketched how to address the issue of transport at $T > 0$ in the notoriously difficult limit of small ω_0 by means of a stochastic formulation. To actually resolve this issue within the LD approach we have to specify a way how to obtain the electric conductivity from local distributions, aside from the need to actually perform the numerical calculations. There is no definite answer yet, which is ready to be implemented. We nevertheless believe to have given arguments that thinking in terms of distributions can prove worthwhile also here. Maybe we should rephrase our introductory word of warning concerning the content of this tutorial: It's not just about a method, it's about a way of thinking!

References

1. N.W. Ashcroft, N.D. Mermin, *Solid State Physics* (Saunders College Publ., Philadelphia, 1976)
2. J. Hubbard, Phys. Rev. Lett. **3**, 77 (1959)
3. R.L. Stratonovich, Dokl. Akad. Nauk SSSR **115**, 1097 (1957)
4. A. Georges, G. Kotliar, W. Krauth, M.J. Rozenberg, Rev. Mod. Phys. **68**, 13 (1996)
5. P.W. Anderson, Phys. Rev. **109**, 1492 (1958)
6. R.J. Elliot, J.A. Krumhansl, P.L. Leath, Rev. Mod. Phys. **46**, 465 (1974)
7. P.A. Lee, T.V. Ramakrishnan, Rev. Mod. Phys. **57**, 287 (1985)
8. B. Kramer, A. Mac Kinnon, Rep. Prog. Phys. **56**, 1469 (1993)
9. R. Abou-Chacra, D.J. Thouless, P.W. Anderson, J. Phys. C **6**, 1734 (1973)
10. S.M. Girvin, M. Jonson, Phys. Rev. B **22**, 3583 (1980)
11. D.E. Logan, P.G. Wolynes, Phys. Rev. B **29**, 6560 (1984)
12. D.E. Logan, P.G. Wolynes, Phys. Rev. B **36**, 4135 (1987)
13. V. Dobrosavljević, G. Kotliar, Philos. Trans. Roy. Soc. Lond., Ser. A **356**, 57 (1998)
14. A. Alvermann, H. Fehske, Eur. Phys. J. B **48**, 205 (2005)
15. T. Holstein, Ann. Phys. (N.Y.) **8**, 343 (1959)
16. H. Sumi, J. Phys. Soc. Jpn. **36**, 770 (1974)
17. Y.A. Firsov, *Polarons* (Izd. Nauka, Moscow, 1975)
18. H. Fehske, A. Alvermann, M. Hohenadler, G. Wellein, in *Polarons in Bulk Materials and Systems With Reduced Dimensionality, International School of Physics Enrico Fermi*, Vol. 161, ed. by G. Iadonisi, J. Ranninger, G. De Filippis (IOS Press, Amsterdam, 2006), *International School of Physics Enrico Fermi*, Vol. 161, pp. 285–296
19. H. Fehske and S.A. Trugman in *Polarons in Advanced Materials*, Ed. A.S. Alexandrov, Springer Series in Material Sciences Vol. 103, pp. 393–461 (Canopus/Springer, Dordrecht 2007)
20. S. Ciuchi, F. de Pasquale, S. Fratini, D. Feinberg, Phys. Rev. B **56**, 4494 (1997)
21. F.X. Bronold, A. Alvermann, H. Fehske, Philos. Mag. **84**, 673 (2004)
22. P. Fulde, *Electron Correlation in Molecules and Solids* (Springer-Verlag, Berlin, 1991)

Part VIII

Iterative Methods for Sparse Eigenvalue Problems

18 Exact Diagonalization Techniques

Alexander Weiße and Holger Fehske

Institut für Physik, Universität Greifswald, 17487 Greifswald, Germany

In this chapter we show how to calculate a few eigenstates of the full Hamiltonian matrix of an interacting quantum system. Naturally, this implies that the Hilbert space of the problem has to be truncated, either by considering finite systems or by imposing suitable cut-offs, or both. All of the presented methods are iterative, i.e., the Hamiltonian matrix is applied repeatedly to a set of vectors from the Hilbert space. In addition, most quantum many-particle problems lead to a sparse matrix representation of the Hamiltonian, where only a very small fraction of the matrix elements is non-zero.

18.1 Basis Construction

18.1.1 Typical Quantum Many-Particle Models

Before we can start applying sparse matrix algorithms, we need to translate the considered many-particle Hamiltonian, given in the language of second quantization, into a sparse Hermitian matrix. Usually, this is the intellectually and technically challenging part of the project, in particular, if we want to take into account symmetries of the problem.

Typical lattice models in solid state physics involve electrons, spins and phonons. Within this part we will focus on the Hubbard model,

$$H = -t \sum_{\langle ij \rangle, \sigma} \left(c_{i\sigma}^{\dagger} c_{j\sigma} + \text{H.c.} \right) + U \sum_{i} n_{i\uparrow} n_{i\downarrow} , \tag{18.1}$$

which describes a single band of electrons $c_{i\sigma}^{(\dagger)}$ ($n_{i\sigma} = c_{i\sigma}^{\dagger} c_{i\sigma}$) with on-site Coulomb interaction U. Originally [1, 2, 3], it was introduced to study correlation effects and ferromagnetism in narrow band transition metals. After the discovery of high-T_C superconductors the model became very popular again, since it is considered as the simplest lattice model which, in two dimensions, may have a superconducting phase. In one dimension, the model is exactly solvable [4, 5], hence we can check our numerics for correctness. From the Hubbard model at half-filling, taking the limit $U \to \infty$, we can derive the Heisenberg model

$$H = \sum_{ij} J_{ij} \, \boldsymbol{S}_i \cdot \boldsymbol{S}_j , \tag{18.2}$$

A. Weiße and H. Fehske: *Exact Diagonalization Techniques*, Lect. Notes Phys. **739**, 529–544 (2008)
DOI 10.1007/978-3-540-74686-7_18

which accounts for the magnetic properties of insulating compounds that are governed by the exchange interaction $J \sim t^2/U$ between localized spins $\boldsymbol{S}_i$. In many solids the electronic degrees of freedom will interact also with vibrations of the crystal lattice, described in harmonic approximation by bosons $b_i^{(\dagger)}$ (phonons). This leads to microscopic models like the Holstein-Hubbard model

$$\begin{aligned} H = & -t \sum_{\langle ij\rangle,\sigma} (c_{i\sigma}^{\dagger} c_{j\sigma} + \text{H.c.}) + U \sum_{i} n_{i\uparrow} n_{i\downarrow} \\ & - g\omega_0 \sum_{i,\sigma} (b_i^{\dagger} + b_i) n_{i\sigma} + \omega_0 \sum_{i} b_i^{\dagger} b_i \,. \end{aligned} \tag{18.3}$$

With the methods described in this part, such models can be studied on finite clusters with a few dozen sites, both at zero and at finite temperature. In special cases, e.g., for the problem of few polarons, also infinite systems are accessible.

18.1.2 The Hubbard Model and its Symmetries

To be specific, let us derive all the general concepts of basis construction for the Hubbard model on an one-dimensional chain or ring. For a single site i, the Hilbert space of the model (18.1) consists of four states,

(i) $|0\rangle$ = no electron at site i,
(ii) $c_{i\downarrow}^{\dagger}|0\rangle$ = one down-spin electron at site i,
(iii) $c_{i\uparrow}^{\dagger}|0\rangle$ = one up-spin electron at site i, and
(iv) $c_{i\uparrow}^{\dagger} c_{i\downarrow}^{\dagger}|0\rangle$ = two electrons at site i.

Consequently, for a finite cluster of L sites, the full Hilbert space has dimension 4^L. This is a rapidly growing number, and without symmetrization we could not go beyond $L \approx 16$ even on the biggest supercomputers.

Given a symmetry of the system, i.e. an operator A that commutes with H, the Hamiltonian will not mix states from different eigenspaces of A. Therefore, the matrix representing H will acquire a block structure, and we can handle each block separately (see Fig. 18.1). The Hubbard Hamiltonian (18.1) has a number of symmetries:

- Particle number conservation: H commutes with total particle number

$$N_{\text{e}} = \sum_{i,\sigma} n_{i\sigma} \,. \tag{18.4}$$

- $SU(2)$ spin symmetry: H commutes with all components of the total spin

$$S^{\alpha} = \frac{1}{2} \sum_{i} \sum_{\mu,\nu} c_{i\mu}^{\dagger} \sigma_{\mu\nu}^{\alpha} c_{i\nu} \,, \tag{18.5}$$

where σ^{α} denotes the Pauli matrices, and $\mu, \nu \in \{\uparrow, \downarrow\}$.

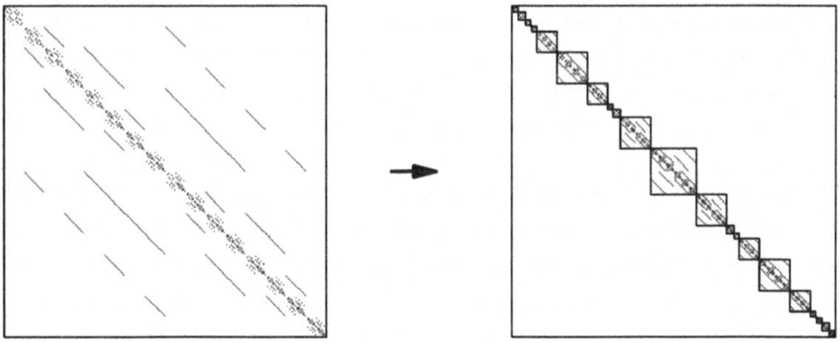

Fig. 18.1. With the use of symmetries the Hamiltonian matrix acquires a block structure. Here: The matrix for the Hubbard model when particle number conservation is neglected (**left**) or taken into account (**right**)

- Particle-hole symmetry: For an even number of lattice sites H is invariant under the transformation

$$Q: \; c_{i,\sigma} \to (-1)^i c^\dagger_{i,-\sigma} \,, \qquad c^\dagger_{i,\sigma} \to (-1)^i c_{i,-\sigma} \,, \tag{18.6}$$

  except for a constant.
- Translational invariance: Assuming periodic boundary conditions, i.e., $c^{(\dagger)}_{L,\sigma} = c^{(\dagger)}_{0,\sigma}$, H commutes with the translation operator

$$T: \; c^{(\dagger)}_{i,\sigma} \to c^{(\dagger)}_{i+1,\sigma} \,. \tag{18.7}$$

  Here L is the number of lattice sites.
- Inversion symmetry: H is symmetric with respect to the inversion

$$I: \; c^{(\dagger)}_{i,\sigma} \to c^{(\dagger)}_{L-i,\sigma} \,. \tag{18.8}$$

For the basis construction the most important of these symmetries are the particle number conservation, the spin-S^z conservation and the translational invariance. Note that the conservation of both $S^z = (N_\uparrow - N_\downarrow)/2$ and $N_e = N_\uparrow + N_\downarrow$ is equivalent to the conservation of the total number of spin-$\uparrow$ and of spin-$\downarrow$ electrons, $N_\uparrow$ and $N_\downarrow$, respectively. In addition to S^z we could also fix the total spin S^2, but the construction of the corresponding eigenstates is too complicated for most practical computations.

18.1.3 A Basis for the Hubbard Model

Let us start with building the basis for a system with L sites and fixed electron numbers $N_\uparrow$ and $N_\downarrow$. Each element of the basis can be identified by the positions of the up and down electrons, but for uniqueness we also need to define some normal order. For the Hubbard model it is convenient to first sort the electrons by the spin index, then by the lattice index, i.e.,

$$c_{3\uparrow}^{\dagger} c_{2\uparrow}^{\dagger} c_{0\uparrow}^{\dagger} c_{3\downarrow}^{\dagger} c_{1\downarrow}^{\dagger} |0\rangle \tag{18.9}$$

is a valid ordered state. This ordering has the advantage that the nearest-neighbor hopping in the Hamiltonian does not lead to complicated phase factors, when applied to our basis states. Finding all the basis states is a combinatorics problem: There are $\binom{L}{N_\uparrow}$ ways of distributing $N_\uparrow$ (indistinguishable) up-spin electrons on L sites, and similarly, $\binom{L}{N_\downarrow}$ ways of distributing $N_\downarrow$ down-spin electrons on L sites. Hence, the total number of states in our basis is $\binom{L}{N_\uparrow}\binom{L}{N_\downarrow}$. If we sum up the dimensions of all $(N_\uparrow, N_\downarrow)$-blocks, we obtain

$$\sum_{N_\uparrow=0}^{L} \sum_{N_\downarrow=0}^{L} \binom{L}{N_\uparrow}\binom{L}{N_\downarrow} = 2^L 2^L = 4^L \,, \tag{18.10}$$

which is the total Hilbert space dimension we derived earlier. The biggest block in our symmetrized Hamiltonian has $N_\uparrow = N_\downarrow = L/2$ and dimension $\binom{L}{L/2}^2$. This is roughly a factor of $\pi L/2$ smaller than the original 4^L. Below we will reduce the dimension of the biggest block by another factor of L using translational invariance.

Knowing the basic structure and the dimension of the Hilbert space with fixed particle numbers, how can we implement it on a computer? An efficient way to do so, is using integer numbers and bit operations that are available in many programming languages. Assume, we work with a lattice of $L = 4$ sites and $N_\uparrow = 3$, $N_\downarrow = 2$. We can then translate the state of (18.9) into a bit pattern,

$$c_{3\uparrow}^{\dagger} c_{2\uparrow}^{\dagger} c_{0\uparrow}^{\dagger} c_{3\downarrow}^{\dagger} c_{1\downarrow}^{\dagger} |0\rangle \to (\uparrow,\uparrow,0,\uparrow) \times (\downarrow,0,\downarrow,0) \to 1101 \times 1010 \,. \tag{18.11}$$

To build the other basis states, we need all four-bit integers with three bits set to one, as well as all four-bit integers with two bits set. We leave this to the reader as a little programming exercise, and just quote the result in Table 18.1.

The complete basis is given by all 24 pairs of the four up-spin and the six down-spin states. Having ordered the bit patterns by the integer values they correspond to,

Table 18.1. Basis states of the Hubbard model on four sites with three up- and two down-spin electrons

no.	↑-patterns	no.	↓-patterns
0	0111 = 7	0	0011 = 3
1	1011 = 11	1	0101 = 5
2	1101 = 13	2	0110 = 6
3	1110 = 14	3	1001 = 9
		4	1010 = 10
		5	1100 = 12

we can label each state by its indices (i, j) in the list of up and down patterns, or combine the two indices to an overall index $n = i \cdot 6 + j$. Our sample state (18.9) corresponds to the index pair $(2, 4)$, which is equivalent to the state $2 \cdot 6 + 4 = 16$ of the total 24 states.

18.1.4 The Hamiltonian Matrix

Having found all basis states, we can now apply the Hamiltonian (18.1) to each of them, to obtain the matrix elements. The hopping term corresponds to the left or right shift of single bits. For periodic boundary conditions we need to take care of potential minus signs, whenever an electron is wrapped around the boundary and the number of electrons it commutes through is odd. The Coulomb interaction merely counts double occupancy, i.e. bits which are set in both the up and down spin part of the basis state. For our sample state (18.9) we obtain:

$$\begin{aligned} \uparrow \text{-hopping}: &\ 1101 \times 1010 \rightarrow -t\,(1011 + 1110) \times 1010\,, \\ \downarrow \text{-hopping}: &\ 1101 \times 1010 \rightarrow -t\,1101 \times (0110 + 1100 + 1001 - 0011)\,, \\ U\text{-term}: &\ 1101 \times 1010 \rightarrow U\,1101 \times 1010\,. \end{aligned} \tag{18.12}$$

Now we need to find the indices of the resulting states on the right. For the Hubbard model with its decomposition into two spin channels, we can simply use a table which translates the integer value of the bit pattern into the index in the list of up and down spin states (see Table 18.1). Note, however, that this table has a length of 2^L. When simulating spin or phonon models such a table would easily exceed all available memory. For finding the index of a given basis state we then need to resort to other approaches, like hashing, fast search algorithms or some decomposition of the state [6]. Having found the indices and denoting our basis in a ket-notation, $|n\rangle$, (18.12) reads

$$\begin{aligned} \uparrow \text{-hopping}: &\quad |16\rangle \rightarrow -t\,(|10\rangle + |22\rangle)\,, \\ \downarrow \text{-hopping}: &\quad |16\rangle \rightarrow -t\,(|14\rangle + |17\rangle + |15\rangle - |12\rangle)\,, \\ U\text{-term}: &\quad |16\rangle \rightarrow U\,|16\rangle\,. \end{aligned} \tag{18.13}$$

To obtain the complete Hamiltonian matrix we have to repeat this procedure for all 24 basis states. In each case we obtain a maximum of $2L = 8$ off-diagonal non-zero matrix elements. Thus, the matrix is indeed very sparse (see Fig. 18.2). The generalization of the above considerations to arbitrary values of L, $N_\uparrow$, and $N_\downarrow$ is straight-forward. For spatial dimensions larger than one we need to be a bit more careful with fermionic phase factors. In general, minus signs will occur not only at the boundaries, but also for other hopping processes.

18.1.5 Using Translation Symmetry

We mentioned earlier that the translation symmetry of the Hubbard model (or any other lattice model) can be used for a further reduction of the Hilbert space dimension. What we need are the eigenstates of the translation operator T, which can be

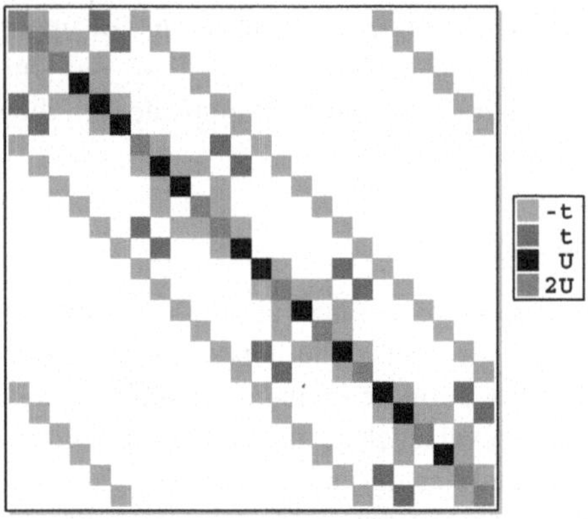

Fig. 18.2. Schematic representation of the Hamiltonian matrix of the Hubbard model with $L = 4$, $N_\uparrow = 3$, $N_\downarrow = 2$, and periodic boundary conditions

constructed using the projector

$$P_k = \frac{1}{L} \sum_{j=0}^{L-1} \mathrm{e}^{2\pi \mathrm{i} jk/L} T^j \,. \tag{18.14}$$

Clearly, for a given (unsymmetrized) state $|n\rangle$, the state $P_k|n\rangle$ is an eigenstate of T,

$$T P_k |n\rangle = \frac{1}{L} \sum_{j=0}^{L-1} \mathrm{e}^{2\pi \mathrm{i} jk/L} T^{j+1} |n\rangle = \mathrm{e}^{-2\pi \mathrm{i} k/L} P_k |n\rangle \,, \tag{18.15}$$

where the corresponding eigenvalue is $\exp(-2\pi \mathrm{i} k/L)$ and $2\pi k/L$ is the discrete lattice momentum. Here we made use of the fact that $T^L = 1$ (on a ring with L sites, L translations by one site let you return to the origin). This property also implies $\exp(-2\pi \mathrm{i} k) = 1$, hence k has to be an integer. Due to the periodicity of the exponential, we can restrict ourselves to $k = 0, 1, \ldots, (L-1)$.

The normalization of the state $P_k|n\rangle$ requires some care. We find

$$\begin{aligned} P_k^\dagger &= \frac{1}{L} \sum_{j=0}^{L-1} \mathrm{e}^{-2\pi \mathrm{i} jk/L} T^{-j} = \frac{1}{L} \sum_{j'=0}^{L-1} \mathrm{e}^{2\pi \mathrm{i} j'k/L} T^{j'} = P_k \\ P_k^2 &= \frac{1}{L^2} \sum_{i,j=0}^{L-1} \mathrm{e}^{2\pi \mathrm{i}(i-j)k/L} T^{i-j} = \frac{1}{L} \sum_{j'=0}^{L-1} \mathrm{e}^{2\pi \mathrm{i} j'k/L} T^{j'} = P_k \,, \end{aligned} \tag{18.16}$$

as we expect for a projector. Hence, $\langle n|P_k^\dagger P_k|n\rangle = \langle n|P_k^2|n\rangle = \langle n|P_k|n\rangle$. For most $|n\rangle$ the states $T^j|n\rangle$ with $j = 0, 1, \ldots, (L-1)$ will differ from each other, therefore $\langle n|P_k|n\rangle = 1/L$. However, some states are mapped onto themselves by a translation T^{ν_n} with $\nu_n < L$, i.e., $T^{\nu_n}|n\rangle = \mathrm{e}^{\mathrm{i}\phi_n}|n\rangle$ with a phase ϕ_n (usually 0 or

π). Nevertheless $T^L|n\rangle = |n\rangle$, therefore ν_n has to be a divider of L with $q_n = L/\nu_n$ an integer. Calculating the norm then gives

$$\langle n|P_k|n\rangle = \frac{1}{L}\sum_{j=0}^{q_n} \mathrm{e}^{\mathrm{i}(2\pi k/q_n + \phi_n)\,j} \,, \tag{18.17}$$

which equals $q_n/L = 1/\nu_n$ or 0 depending on k and ϕ_n.

How do the above ideas translate into a reduced dimension of our Hilbert space? Let us first consider the $\uparrow$-patterns from Table 18.1: All four patterns (states) are connected with a translation by one site, i.e., starting from the pattern $|0_\uparrow\rangle = 0111$ the other patterns are obtained through $|n_\uparrow\rangle = T^{-n}|0_\uparrow\rangle$,

$$\begin{aligned}
|0_\uparrow\rangle &= T^0|0_\uparrow\rangle = 0111 \,, \\
|1_\uparrow\rangle &= T^{-1}|0_\uparrow\rangle = 1011 \,, \\
|2_\uparrow\rangle &= T^{-2}|0_\uparrow\rangle = 1101 \,, \\
|3_\uparrow\rangle &= T^{-3}|0_\uparrow\rangle = 1110 \,.
\end{aligned} \tag{18.18}$$

We can call this group of connected states a cycle, which is completely described by knowing one of its members. It is convenient to use the pattern with the smallest integer value to be this special member of the cycle, and we call it the representative of the cycle.

Applying the projector to the representative of the cycle, $P_k|0_\uparrow\rangle$, we can generate L linearly independent states, which in our case reads

$$\begin{aligned}
P_0|0_\uparrow\rangle &= (0111 + 1011 + 1101 + 1110)/L \,, \\
P_1|0_\uparrow\rangle &= (0111 - \mathrm{i}\,1011 - 1101 + \mathrm{i}\,1110)/L \,, \\
P_2|0_\uparrow\rangle &= (0111 - 1011 + 1101 - 1110)/L \,, \\
P_3|0_\uparrow\rangle &= (0111 + \mathrm{i}\,1011 - 1101 - \mathrm{i}\,1110)/L \,.
\end{aligned} \tag{18.19}$$

The advantage of these new states, which are linear combinations of all members of the cycle in a spirit similar to discrete Fourier transformation, becomes clear when we apply the Hamiltonian: Whereas the Hamiltonian mixes the states in (18.18), all matrix elements between the states in (18.19) vanish. Hence, we have decomposed the four-dimensional Hilbert space into four one-dimensional blocks.

In a next step we repeat this procedure for the $\downarrow$-patterns of Table 18.1. These can be decomposed into two cycles represented by the states $|0_\downarrow\rangle = 0011$ and $|1_\downarrow\rangle = 0101$, where due to $T^2|1_\downarrow\rangle = -|1_\downarrow\rangle$ the second cycle has size $\nu_1 = 2$. Note, that we also have phase factors here, since the number of fermions is even. To get the complete symmetrized basis, we need to combine the up and down spin representatives, thereby taking into account relative shifts between the states. For our sample case the combined representatives,

$$|r\rangle = |n_\uparrow\rangle T^j |m_\downarrow\rangle \tag{18.20}$$

no.	patterns
0	0111×0011
1	0111×0110
2	0111×1100
3	0111×1001
4	0111×0101
5	0111×1010

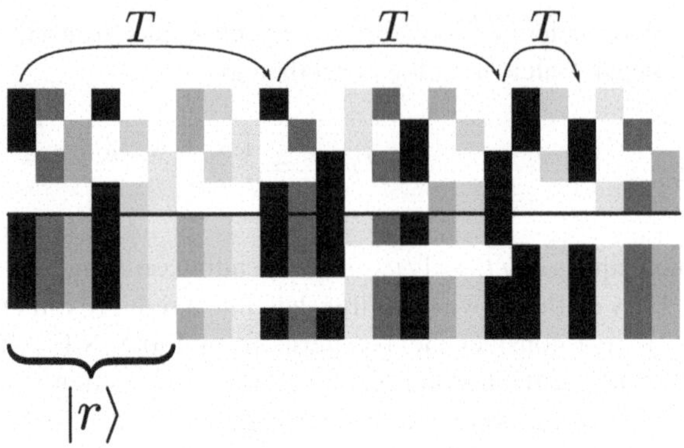

Fig. 18.3. Decomposition of the basis for $L = 4$, $N_\uparrow = 3$, $N_\downarrow = 2$ into six cycles

with $j = 0, 1, \ldots, \min(\nu_n, \nu_m) - 1$, are given in Fig. 18.3.

The basis of each of the L fixed-k (fixed-momentum) Hilbert spaces is then given by the states

$$|r_k\rangle = \frac{P_k|r\rangle}{\sqrt{\langle r|P_k|r\rangle}}, \tag{18.21}$$

where we discard those $|r\rangle$ with $\langle r|P_k|r\rangle = 0$. In our example all six states have $\langle r|P_k|r\rangle = 1/4 \,\forall k$ and no state is discarded. Therefore the dimension of each fixed-k space is six, and summing over all four k we obtain the original number of states, 24. For other particle numbers or lattice sizes we may obtain representatives $|r\rangle$ with $\langle r|P_k|r\rangle = 0$ for certain k. An example is the case $N_\uparrow = N_\downarrow = 2$, $L = 4$ which leads to ten representatives, but two of them have $\langle r|P_k|r\rangle = 0$ for $k = 1$ and $k = 3$. Adding the dimensions of the four k-subspaces, we find $10+8+10+8 = 36$, which agrees with $\binom{L}{N_\downarrow}\binom{L}{N_\uparrow} = 6^2$.

When calculating the Hamiltonian matrix for a given k-sector, we can make use of the fact that H commutes with T, and therefore also with P_k. Namely, the matrix element between two states $|r_k\rangle$ and $|r'_k\rangle$ is simply given by

$$\langle r'_k|H|r_k\rangle = \frac{\langle r'|P_k H P_k|r\rangle}{\sqrt{\langle r'|P_k|r'\rangle\langle r|P_k|r\rangle}} = \frac{\langle r'|P_k H|r\rangle}{\sqrt{\langle r'|P_k|r'\rangle\langle r|P_k|r\rangle}}, \tag{18.22}$$

i.e., we need to apply the projector only once after we applied H to the representative $|r\rangle$. Repeating the procedure for all representatives, we obtain the matrix for a given k. The full matrix with fixed particle numbers $N_\uparrow$ and $N_\downarrow$ is decomposed into L blocks with fixed k. For example, the 24×24 matrix from Fig. 18.2 is decomposed into the four 6×6 matrices.

$$H_{k=0} = \begin{pmatrix} 2U & -t & -t & t & t & 0 \\ -t & 2U & -t & -t & 0 & t \\ -t & -t & 2U & 0 & -t & -t \\ t & -t & 0 & U & -t & t \\ t & 0 & -t & -t & U & -t \\ 0 & t & -t & t & -t & U \end{pmatrix} \quad H_{k=1} = \begin{pmatrix} 2U & -t & -\mathrm{i}t & -\mathrm{i}t & t & 0 \\ -t & 2U & -t & -t & -2\mathrm{i}t & t \\ \mathrm{i}t & -t & 2U & 0 & -t & -\mathrm{i}t \\ \mathrm{i}t & -t & 0 & U & -t & -\mathrm{i}t \\ t & 2\mathrm{i}t & -t & -t & U & -t \\ 0 & t & \mathrm{i}t & \mathrm{i}t & -t & U \end{pmatrix}$$

$$H_{k=2} = \begin{pmatrix} 2U & -t & t & -t & t & 0 \\ -t & 2U & -t & -t & 0 & t \\ t & -t & 2U & 0 & -t & t \\ -t & -t & 0 & U & -t & -t \\ t & 0 & -t & -t & U & -t \\ 0 & t & t & -t & -t & U \end{pmatrix} \quad H_{k=3} = \begin{pmatrix} 2U & -t & \mathrm{i}t & \mathrm{i}t & t & 0 \\ -t & 2U & -t & -t & 2\mathrm{i}t & t \\ -\mathrm{i}t & -t & 2U & 0 & -t & \mathrm{i}t \\ -\mathrm{i}t & -t & 0 & U & -t & \mathrm{i}t \\ t & -2\mathrm{i}t & -t & -t & U & -t \\ 0 & t & -\mathrm{i}t & -\mathrm{i}t & -t & U \end{pmatrix} \tag{18.23}$$

Note that except for $k = 0$ and $k = 2$, which correspond to the momenta zero and π, the matrices H_k are complex. Their dimension, however, is a factor of L smaller than the dimension of the initial space with fixed particle numbers. At first glance, the above matrices look rather dense. This is due to the small dimension of our sample system. For larger L and N_e the Hamiltonian is as sparse as the example of Fig. 18.1.

18.1.6 A few Remarks about Spin Systems

We mentioned earlier that the Heisenberg model (18.2) can be derived from the Hubbard model (18.1) considering the limit $U \to \infty$. Consequently, the numerical setup for both models is very similar. For a model with $|\boldsymbol{S}_i| = 1/2$, we can choose the z-axis as the quantization axis and encode the two possible spin directions $\downarrow$ and $\uparrow$ into the bit values zero and one, e.g., $\downarrow\uparrow\downarrow\downarrow \to 0100$. If applicable, the conservation of the total spin $S^z = \sum_i S^z_i$ is similar to a particle number conservation, i.e., we can easily construct all basis states with fixed S^z using the ideas described earlier. The same holds for translational invariance, where now the construction of a symmetric basis is made easier by the lack of fermionic phase factors (spin operators at different sites commute). When calculating matrix elements it is convenient to rewrite the exchange interaction as

$$\boldsymbol{S}_i \boldsymbol{S}_j = \frac{1}{2}\left(S^+_i S^-_j + S^-_i S^+_j\right) + S^z_i S^z_j\,, \tag{18.24}$$

where the operators $S^\pm_i = S^x_i \pm \mathrm{i} S^y_i$ rise or lower the S^z_i value at site i, which is easy to implement in our representation. Note also, that from this equation the conservation of the total S^z is obvious.

If the considered solid consists of more complex ions with partially filled shells, we may also arrive at Heisenberg type models with $|\boldsymbol{S}_i| > 1/2$. In this case we need $2|\boldsymbol{S}_i| + 1$ states per site to describe all possible S^z-orientations and, of course, this requires more than one bit per site. Numbering all possible states with a given total S^z is slightly more complicated. For instance, we can proceed recursively adding one site at each time.

18.1.7 Phonon Systems

Having constructed a symmetrized basis for the Hubbard and Heisenberg type models, let us now comment on bosonic models and phonons, in particular. For such systems the particle number is usually not conserved, and the accessible Hilbert space is infinite even for a single site. For numerical studies we therefore need an appropriate truncation scheme, which preserves enough of the Hilbert space to describe the considered physics, but restricts the dimension to manageable values. Assume we are studying a model like the Holstein-Hubbard model (18.3), where the pure phonon part is described by a set of harmonic Einstein oscillators, one at each site. For an L-site lattice the eigenstates of this phonon system are given by the Fock states

$$|m_0, \ldots, m_{L-1}\rangle = \prod_{i=0}^{L-1} \frac{(b_i^\dagger)^{m_i}}{\sqrt{m_i!}} |0\rangle \tag{18.25}$$

and the corresponding eigenvalue is

$$E_\mathrm{p} = \omega_0 \sum_{i=0}^{L-1} m_i \,. \tag{18.26}$$

If we are interested in the ground state or the low energy properties of the interacting electron-phonon model (18.3), certainly only phonon states with a rather low energy will contribute. Therefore, a good truncated basis for the phonon Hilbert space is given by the states

$$|m_0, \ldots, m_{L-1}\rangle \quad \text{with} \quad \sum_{i=0}^{L-1} m_i \le M \,, \tag{18.27}$$

which include all states with $E_\mathrm{p} \le \omega_0 M$. The dimension of the resulting Hilbert space is $\binom{L+M}{M}$.

To keep the required M small, we apply another trick [7]. After Fourier transforming the phonon subsystem,

$$b_i = \frac{1}{\sqrt{L}} \sum_{k=0}^{L-1} \mathrm{e}^{2\pi \mathrm{i} i k/L} \widetilde{b}_k \,, \tag{18.28}$$

we observe that the phonon mode with $k = 0$ couples to a conserved quantity: The total number of electrons N_e,

$$\begin{aligned} H = {} & -t \sum_{\langle ij\rangle,\sigma} (c_{i\sigma}^\dagger c_{j\sigma} + \mathrm{H.c.}) + U \sum_i n_{i\uparrow} n_{i\downarrow} + \omega_0 \sum_k \widetilde{b}_k^\dagger \widetilde{b}_k \\ & - \frac{g\omega_0}{\sqrt{L}} \sum_{i,\sigma} \sum_{k\neq 0} \mathrm{e}^{-2\pi \mathrm{i} i k/L} (\widetilde{b}_k^\dagger + \widetilde{b}_{-k}) n_{i\sigma} - \frac{g\omega_0}{\sqrt{L}} (\widetilde{b}_0^\dagger + \widetilde{b}_0) N_\mathrm{e} \,. \end{aligned} \tag{18.29}$$

With a constant shift $\widetilde{b}_0 = \widehat{b}_0 + gN_\mathrm{e}/\sqrt{L}$ this part of the model can thus be solved analytically. Going back to real space and using the equivalently shifted phonons $b_i = \bar{b}_i + gN_\mathrm{e}/L$, the transformed Hamiltonian reads

$$
\begin{aligned}
H = -t \sum_{\langle ij\rangle,\sigma} (c_{i\sigma}^{\dagger} c_{j\sigma} + \text{H.c.}) + U \sum_i n_{i\uparrow} n_{i\downarrow} + \omega_0 \sum_i \bar{b}_i^{\dagger} \bar{b}_i \\
- g\omega_0 \sum_i (\bar{b}_i^{\dagger} + \bar{b}_i)(n_{i\uparrow} + n_{i\downarrow} - N_e/L) - \omega_0 (g N_e)^2 / L \,. \quad (18.30)
\end{aligned}
$$

Since the shifted phonons $\bar{b}_i^{(\dagger)}$ couple only to the local charge fluctuations, in a simulation the same accuracy can be achieved with a much smaller cutoff M, compared to the original phonons $b_i^{(\dagger)}$. This is particularly important in the case of strong interaction g.

As in the electronic case, we can further reduce the basis dimension using the translational symmetry of our lattice model. Under periodic boundary conditions, the translation operator T transforms a given basis state like

$$
T|m_0, \ldots, m_{L-1}\rangle = |m_{L-1}, m_0, \ldots, m_{L-2}\rangle \,. \quad (18.31)
$$

Since we are working with bosons, no additional phase factors can occur, and everything is a bit easier. As before, we need to find the representatives $|r_{\mathrm{p}}\rangle$ of the cycles generated by T, and then construct eigenstates of T with the help of the projection operator P_k. When combining the electronic representatives $|r_{\mathrm{e}}\rangle$ from (18.20) with the phonon representatives $|r_{\mathrm{p}}\rangle$, we proceed in the same way, as we did for the up and down spin channels, $|r\rangle = |r_{\mathrm{e}}\rangle T^j |r_{\mathrm{p}}\rangle$. A full symmetrized basis state of the interacting electron-phonon model is then given by $P_k|r\rangle$. Note that the product structure of the electron-phonon basis is preserved during symmetrization, which is a big advantage for parallel implementations [8].

Having explained the construction of a symmetrized basis and of the corresponding Hamiltonian matrix for both electron and phonon systems, we are now ready to work with these matrices. In particular, we will show how to calculate eigenstates and dynamic correlations of our physical systems.

18.2 Eigenstates of Sparse Matrices

18.2.1 The Lanczos Algorithm

The Lanczos algorithm is one of the simplest methods for the calculation of extremal (smallest or largest) eigenvalues of sparse matrices [9]. Initially it was developed for the tridiagonalization of Hermitian matrices [10], but it turned out, not to be particularly successful for this purpose. The reason for its failure as a tridiagonalization algorithm is the underlying recursion procedure, which rapidly converges to eigenstates of the matrix and therefore looses the orthogonality between subsequent vectors that is required for tridiagonalization. Sometimes, however, deficiencies turn into advantages, and the Lanczos algorithm made a successful career as an eigenvalue solver.

The basic structure and the implementation of the algorithm is very simple. Starting from a random initial state (vector) $|\phi_0\rangle$, we construct the series of states

$H^n|\phi_0\rangle$ by repeatedly applying the matrix H (i.e., the Hamiltonian). This series of states spans what is called a Krylov space in the mathematical literature, and the Lanczos algorithm therefore belongs to a broader class of algorithms that work on Krylov spaces [11]. Next we orthogonalize these states against each other to obtain a basis of the Krylov space. Expressed in terms of this basis, the matrix turns out to be tridiagonal. We can easily perform these two steps in parallel, and obtain the following recursion relation:

$$\begin{aligned} |\phi'\rangle &= H|\phi_n\rangle - \beta_n|\phi_{n-1}\rangle \,, \\ \alpha_n &= \langle\phi_n|\phi'\rangle \,, \\ |\phi''\rangle &= |\phi'\rangle - \alpha_n|\phi_n\rangle \,, \\ \beta_{n+1} &= ||\phi''|| = \sqrt{\langle\phi''|\phi''\rangle} \,, \\ |\phi_{n+1}\rangle &= |\phi''\rangle/\beta_{n+1} \,, \end{aligned} \tag{18.32}$$

where $|\phi_{-1}\rangle = 0$ and $|\phi_0\rangle$ is a random normalized state, $||\phi_0|| = 1$.

The coefficients α_n and β_n form the tridiagonal matrix, which we are looking for,

$$\widetilde{H}_N = \begin{pmatrix} \alpha_0 & \beta_1 & 0 & \dots\dots\dots & & 0 \\ \beta_1 & \alpha_1 & \beta_2 & 0 & \dots\dots & 0 \\ 0 & \beta_2 & \alpha_2 & \beta_3 & 0 & 0 \\ & & \ddots & \ddots & \ddots & \\ 0 & \dots & 0 & \beta_{N-2} & \alpha_{N-2} & \beta_{N-1} \\ 0 & \dots\dots & & 0 & \beta_{N-1} & \alpha_{N-1} \end{pmatrix} . \tag{18.33}$$

With increasing recursion order N the eigenvalues of $\widetilde{H}_N$ – starting with the extremal ones – converge to the eigenvalues of the original matrix H. In Fig. 18.4 we illustrate this for the ground-state energy of the one-dimensional Hubbard model (18.1) on a ring of 12 and 14 sites. Using only particle number conservation, the corresponding matrix dimensions are $D = \binom{12}{6}^2 = 853776$ and $D = \binom{14}{7}^2 = 11778624$, respectively. With about 90 iterations the precision of the lowest eigenvalue is better than 10^{-13}, where we compare with the exact result obtained with Bethe ansatz [4]. The eigenvalues of the tridiagonal matrix were calculated with standard library functions from the LAPACK collection [12]. Since $N \ll D$, this accounts only for a tiny fraction of the total computation time, which is governed by the application of H on $|\phi_n\rangle$.

Having found the extremal eigenvalues, we can also calculate the corresponding eigenvectors of the matrix. If the eigenvector $|\psi\rangle$ of the tridiagonal matrix $\widetilde{H}_N$ has the components ψ_j, i.e., $|\psi\rangle = \{\psi_0, \psi_1, \dots, \psi_{N-1}\}$, the eigenvector $|\Psi\rangle$ of the original matrix H is given by

$$|\Psi\rangle = \sum_{j=0}^{N-1} \psi_j|\phi_j\rangle \,. \tag{18.34}$$

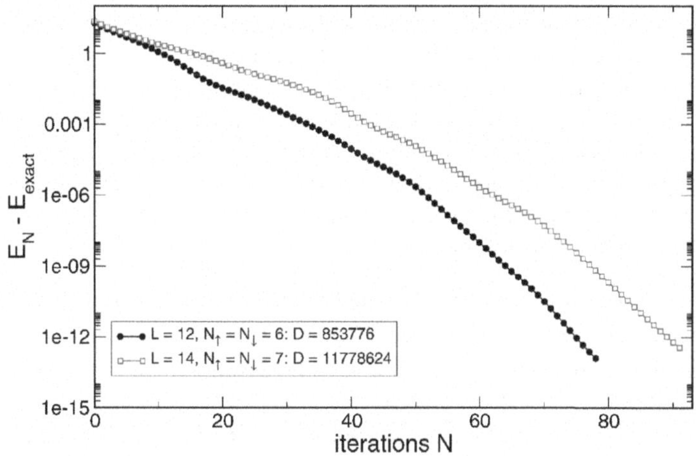

Fig. 18.4. Convergence of the Lanczos recursion for the ground-state energy of the Hubbard model on a ring of $L = 12$ and $L = 14$ sites

To calculate this sum we simply need to repeat the above Lanczos recursion with the same start vector $|\phi_0\rangle$, thereby omitting the scalar products for α_j and β_j, which we know already.

The efficiency of the Lanczos algorithm is based on three main properties:

(i) It relies only on matrix vector multiplications (MVM) of the matrix H with a certain vector $|\phi_n\rangle$. If H is sparse, this requires only of the order of D operations, where D is the dimension of H.

(ii) When calculating eigenvalues, the algorithm requires memory only for two vectors of dimension D and for the matrix H. For exceptionally large problems, the matrix can be re-constructed on-the-fly for each MVM, and the memory consumption is determined by the vectors. When calculating eigenvectors we need extra memory.

(iii) The first few eigenvalues on the upper and lower end of the spectrum of H usually converge very quickly. In most cases $N \lesssim 100$ iterations are sufficient.

Extensions of the Lanczos algorithm can also be used for calculating precise estimates of the full spectral density of H, or of dynamical correlation functions that depend on the spectrum of H and on the measured operators. We will discuss more details in Chap. 19 when we describe Chebyshev expansion based methods, such as the Kernel Polynomial Method.

18.2.2 The Jacobi-Davidson Algorithm

The Jacobi-Davidson method is a recent, more involved approach to the sparse eigenvalue problem, which was suggested by Sleijpen and van der Vorst [13] as a combination of Davidson's method [14] and a procedure described by Jacobi [15].

It has the advantage that not only the lowest eigenstates but also excitations converge rapidly. In addition, it can correctly resolve degeneracies.

In the Jacobi-Davidson algorithm, like in the Lanczos algorithm, a set of vectors $V_N = \{|v_0\rangle, \ldots, |v_{N-1}\rangle\}$ is constructed iteratively, and the eigenvalue problem for the Hamiltonian H is solved within this subspace. However, in contrast to the Lanczos algorithm, we do not work in the Krylov space of H, but instead expand V_N with a vector that is orthogonal to our current approximate eigenstates. In more detail, the procedure is as follows:

(i) Initialize the set V with a random normalized start vector, $V_1 = \{|v_0\rangle\}$.
(ii) Compute all unknown matrix elements $\langle v_i|H|v_j\rangle$ of $\widetilde{H}_N$ with $|v_i\rangle \in V_N$.
(iii) Compute an eigenstate $|s\rangle$ of $\widetilde{H}_N$ with eigenvalue θ, and express $|s\rangle$ in the original basis, $|u\rangle = \sum_i |v_i\rangle\langle v_i|s\rangle$.
(iv) Compute the associated residual vector $|r\rangle = (H-\theta)|u\rangle$ and stop the iteration, if its norm is sufficiently small.
(v) Otherwise, (approximately) solve the linear equation

$$(1 - |u\rangle\langle u|)(H - \theta)(1 - |u\rangle\langle u|)|t\rangle = -|r\rangle\,. \tag{18.35}$$

(vi) Orthogonalize $|t\rangle$ against V_N with the modified Gram-Schmidt method and append the resulting vector $|v_N\rangle$ to V_N, obtaining the set V_{N+1}.
(vii) Return to step (ii).

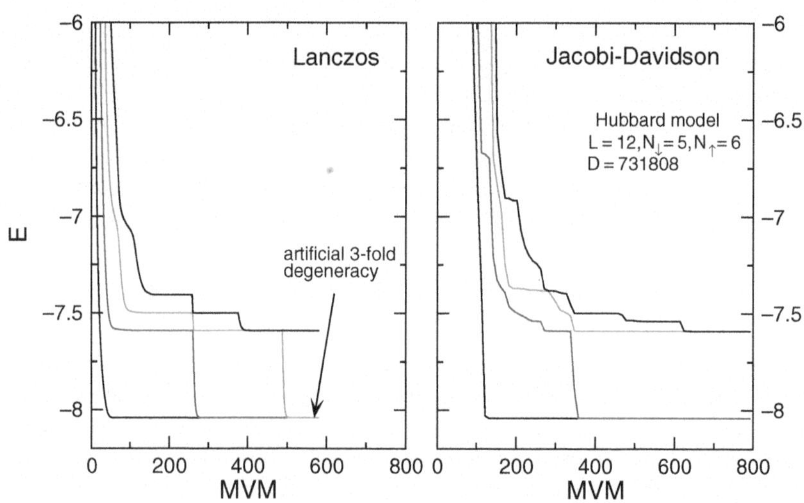

Fig. 18.5. Comparison of the Jacobi-Davidson algorithm and the Lanczos algorithm applied to the four lowest eigenstates of the Hubbard model with $L = 12$, $N_\downarrow = 5$, $N_\uparrow = 6$. Jacobi-Davidson correctly resolves the two-fold degeneracy, standard Lanczos (although faster) cannot distinguish true and artificial degeneracy

For (18.35) we only need an approximate solution, which can be obtained, for instance, with a few steps of the Generalized Minimum Residual Method (GMRES) or the Quasi Minimum Residual Method (QMR) [16]. If more than one eigenstate is desired, the projection operator $(1 - |u\rangle\langle u|)$ needs to be extended by the already converged eigenstates, $(1 - \sum_k |u_k\rangle\langle u_k|)$, such that the search continues in a new, yet unexplored direction. Since the Jacobi-Davidson algorithm requires memory for all the vectors in V_N, it is advisable to restart the calculation after a certain number of steps. There are clever strategies for this restart, and also for the calculation of interior eigenstates, which are hard to access with Lanczos. More details can be found in the original papers [13, 17] or in text books [18].

In Fig. 18.5 we give a comparison of the Lanczos and the Jacobi-Davidson algorithms, calculating the four lowest eigenstates of the Hubbard model on a ring of $L = 12$ sites with $N_\downarrow = 5$ and $N_\uparrow = 6$ electrons. The matrix dimension is $D = 731808$, and each of the lowest states is two-fold degenerate. In terms of speed and memory consumption the Lanczos algorithm has a clear advantage, but with the standard setup we have difficulties resolving the degeneracy. The method tends to create artificial copies of well converged eigenstates, which are indistinguishable from the true degenerate states. The problem can be circumvented with more advanced variants of the algorithm, such as Block or Band Lanczos [9, 18], but we loose the simplicity of the method and part of its speed. Jacobi-Davidson then is a strong competitor. It is not much slower and it correctly detects the two-fold degeneracy, since the converged eigenstates are explicitly projected out of the search space.

References

1. J. Hubbard, Proc. Roy. Soc. London, Ser. A **276**, 238 (1963)
2. M.C. Gutzwiller, Phys. Rev. Lett. **10**, 159 (1963)
3. J. Kanamori, Prog. Theor. Phys. **30**, 275 (1963)
4. E.H. Lieb, F.Y. Wu, Phys. Rev. Lett. **20**, 1445 (1968)
5. F.H.L. Essler, H. Frahm, F. Göhmann, A. Klümper, V.E. Korepin, *The One-Dimensional Hubbard Model* (Cambridge University Press, Cambridge, 2005)
6. R. Sedgewick, *Algorithmen* (Addison-Wesley, Bonn, 1992)
7. S. Sykora, A. Hübsch, K.W. Becker, G. Wellein, H. Fehske, Phys. Rev. B **71**, 045112 (2005)
8. B. Bäuml, G. Wellein, H. Fehske, Phys. Rev. B **58**, 3663 (1998)
9. J.K. Cullum, R.A. Willoughby, *Lanczos Algorithms for Large Symmetric Eigenvalue Computations*, vol. I & II (Birkhäuser, Boston, 1985)
10. C. Lanczos, J. Res. Nat. Bur. Stand. **45**, 255 (1950)
11. Y. Saad, *Numerical Methods for Large Eigenvalue Problems* (University Press, Manchester, 1992). URL `http://www-users.cs.umn.edu/~saad/books.html`
12. Linear Algebra PACKage. URL `http://www.netlib.org`
13. G.L.G. Sleijpen, H.A. van der Vorst, SIAM J. Matrix Anal. Appl. **17**, 401 (1996)
14. E.R. Davidson, J. Comput. Phys. **17**, 87 (1975)
15. C.G.J. Jacobi, J. Reine und Angew. Math. **30**, 51 (1846)

16. Y. Saad, *Iterative Methods for Sparse Linear Systems*, 2nd edn. (SIAM, Philadelphia, 2003). URL `http://www-users.cs.umn.edu/~saad/books.html`
17. D.R. Fokkema, G.L.G. Sleijpen, H.A. van der Vorst, SIAM J. Sci. Comp. **20**, 94 (1998)
18. Z. Bai, J. Demmel, J. Dongarra, A. Ruhe, H. van der Vorst (eds.), *Templates for the Solution of Algebraic Eigenvalue Problems: A Practical Guide* (SIAM, Philadelphia, 2000). URL `http://www.cs.utk.edu/~dongarra/etemplates/`

19 Chebyshev Expansion Techniques

Alexander Weiße and Holger Fehske

Institut für Physik, Universität Greifswald, 17487 Greifswald, Germany

With the Lanczos and the Jacobi-Davidson algorithm we are able to calculate a few of the many eigenstates of a sparse matrix. However, it is hardly feasible to calculate all eigenstates of matrices with dimensions larger than a million, not to speak of dimensions like 10^9. Nevertheless, we are interested in dynamic correlation functions and finite temperature properties, which depend on the complete spectrum of the Hamiltonian.

In this chapter we introduce the Kernel Polynomial Method (KPM), a numerical approach that on the basis of Chebyshev expansion allows a very precise calculation of the spectral properties of large sparse matrices and of the static and dynamic correlation functions, which depend on them. In addition, we show how the KPM successfully competes against the very popular Lanczos Recursion and Maximum Entropy Method and can be easily embedded into other numerical techniques, such as Cluster Perturbation Theory or Monte Carlo simulation. Characterized by a resource consumption that scales linearly with the problem dimension the KPM enjoyed growing popularity over the last decade and found broad application not only in physics (for a recent more detailed review see [1]).

19.1 Chebyshev Expansion and Kernel Polynomial Approximation

19.1.1 General Aspects

Let us first recall the basic properties of expansions in orthogonal polynomials and of Chebyshev expansion in particular. Given a positive weight function $w(x)$ defined on the interval $[a, b]$ we can introduce a scalar product

$$\langle f|g\rangle = \int_a^b w(x) f(x) g(x)\,\mathrm{d}x \tag{19.1}$$

between two integrable functions $f, g{:}[a, b] \to \mathbb{R}$. With respect to this scalar product there exists a complete set of polynomials $p_n(x)$, which fulfil the orthogonality relations $\langle p_n|p_m\rangle = \delta_{n,m}/h_n$, where $h_n = 1/\langle p_n|p_n\rangle$ denotes the inverse of the squared norm of $p_n(x)$. These orthogonality relations allow for an easy expansion

A. Weiße and H. Fehske: *Chebyshev Expansion Techniques*, Lect. Notes Phys. **739**, 545–577 (2008)
DOI 10.1007/978-3-540-74686-7_19

of a given function $f(x)$ in terms of the $p_n(x)$, since the expansion coefficients are proportional to the scalar products of f and p_n,

$$f(x) = \sum_{n=0}^{\infty} \alpha_n \, p_n(x) \tag{19.2}$$

with $\alpha_n = \langle p_n|f\rangle \, h_n$.

In general, all types of orthogonal polynomials can be used for such an expansion and for the KPM approach which we discuss in this chapter (see e.g. [2]). However, as we frequently observe whenever we work with polynomial expansions [3], Chebyshev polynomials [4, 5] of first and second kind turn out to be the best choice for most applications, mainly due to the good convergence properties of the corresponding series and the close relation to Fourier transform [6, 7]. The latter is also an important prerequisite for the derivation of optimal kernels (see below), which are required for the regularization of finite-order expansions, and which so far have not been derived for other sets of orthogonal polynomials.

There are two sets of Chebyshev polynomials, both defined on the interval $[a, b] = [-1, 1]$: The weight function $w(x) = (\pi\sqrt{1-x^2})^{-1}$ yields the polynomials of first kind, T_n, and the weight function $w(x) = \pi\sqrt{1-x^2}$ those of second kind, U_n. In what follows we focus on the $T_n = \cos(n \arccos(x))$, which after substituting $x = \cos(\varphi)$ can be shown to fulfil the orthogonality relation $\langle T_n|T_m\rangle = \delta_{n,m}\,(1+\delta_{n,0})/2$. Moreover, we can easily prove the recursion relation

$$T_{m+1}(x) = 2\,x\,T_m(x) - T_{m-1}(x)\;, \tag{19.3}$$

and the addition formula

$$2\,T_m(x)T_n(x) = T_{m+n}(x) + T_{m-n}(x)\;, \tag{19.4}$$

where $T_{-n}(x) = T_n(x)$ and $T_0(x) = 1$.

Expanding a function f in the standard way of (19.2), the determination of the coefficients $\langle T_n|f\rangle$ requires integrations over the weight function $w(x)$, see (19.1). In practical applications to matrix problems this prohibits a simple iterative scheme, but a solution follows from a slight rearrangement of the expansion, namely

$$f(x) = \frac{1}{\pi\sqrt{1-x^2}} \left(\mu_0 + 2\sum_{n-1}^{\infty} \mu_n \, T_n(x) \right) \tag{19.5}$$

with the modified coefficients (moments)

$$\mu_n = \int_{-1}^{1} f(x) T_n(x) \, \mathrm{d}x\;. \tag{19.6}$$

These two equations are the general basis for the Chebyshev expansion. In the remaining sections we will explain how to translate physical quantities into polynomial expansions of the form of (19.5), how to calculate the moments μ_n in practice, and how to improve the convergence of the approach.

19.1.2 Calculation of Moments

A common feature of basically all Chebyshev expansions is the requirement for a rescaling of the underlying matrix or Hamiltonian H. While Chebyshev polynomials are defined on the real interval $[-1, 1]$, the quantities we are interested in usually depend on the eigenvalues $\{E_k\}$ of the considered (finite-dimensional) matrix. To fit this spectrum into the interval $[-1, 1]$ we apply a simple linear transformation to the Hamiltonian and all energy scales,

$$\widetilde{H} = \frac{H - b}{a} , \qquad \widetilde{E} = \frac{E - b}{a} , \tag{19.7}$$

and denote all rescaled quantities with a tilde hereafter. Given the extremal eigenvalues of the Hamiltonian, E_{min} and E_{max}, which can be calculated, e.g. with the Lanczos algorithm [8], or for which bounds may be known analytically, the scaling factors a and b read $a = (E_{\mathrm{max}} - E_{\mathrm{min}})/(2-\epsilon)$, $b = (E_{\mathrm{max}} + E_{\mathrm{min}})/2$. The parameter ϵ is a small cut-off introduced to avoid stability problems that arise if the spectrum includes or exceeds the boundaries of the interval $[-1, 1]$. It can be fixed, e.g. to $\epsilon = 0.01$, or adapted to the resolution of the calculation, which for an expansion of finite order N is proportional $1/N$ (see below).

The next similarity of most Chebyshev expansions is the form of the moments, namely their dependence on the matrix or Hamiltonian $\widetilde{H}$. In general, we find two types of moments: Simple expectation values of Chebyshev polynomials in $\widetilde{H}$,

$$\mu_n = \langle\beta|T_n(\widetilde{H})|\alpha\rangle , \tag{19.8}$$

where $|\alpha\rangle$ and $|\beta\rangle$ are certain states of the system, or traces over such polynomials and a given operator A,

$$\mu_n = \mathrm{Tr}[A\, T_n(\widetilde{H})] . \tag{19.9}$$

Handling the first case is rather straightforward. Starting from the state $|\alpha\rangle$ we can iteratively construct the states $|\alpha_n\rangle = T_n(\widetilde{H})|\alpha\rangle$ by using the recursion relations for the T_n (see (19.3)),

$$|\alpha_0\rangle = |\alpha\rangle , \qquad |\alpha_1\rangle = \widetilde{H}|\alpha_0\rangle , \qquad |\alpha_{n+1}\rangle = 2\widetilde{H}|\alpha_n\rangle - |\alpha_{n-1}\rangle . \tag{19.10}$$

Scalar products with $|\beta\rangle$ then directly yield $\mu_n = \langle\beta|\alpha_n\rangle$.

The iterative calculation of the moments, in particular the application of $\widetilde{H}$ to the state $|\alpha_n\rangle$, represents the most time consuming part of the whole expansion approach and determines its performance. If $\widetilde{H}$ is a sparse matrix of dimension D the MVM is an order $O(D)$ process and the calculation of N moments therefore requires $O(ND)$ operations and time. The memory consumption depends on the implementation. For moderate problem dimension we can store the matrix and, in addition, need memory for two vectors of dimension D. For very large D the matrix certainly does not fit into the memory and has to be reconstructed on-the-fly in each iteration or retrieved from disc. The two vectors then determine the memory consumption of the calculation. Overall, the resource consumption of the moment iteration is similar or even slightly better than that of the Lanczos algorithm, which requires a few more vector operations (see our comparison in Sect. 19.3). In contrast to Lanczos, Chebyshev iteration is completely stable and can be carried out to arbitrary high order.

The moment iteration can be simplified even further, if $|\beta\rangle = |\alpha\rangle$. In this case the product relation (19.4) allows for the calculation of two moments from each new $|\alpha_n\rangle$

$$\mu_{2n} = 2\langle\alpha_n|\alpha_n\rangle - \mu_0 \,, \qquad \mu_{2n+1} = 2\langle\alpha_{n+1}|\alpha_n\rangle - \mu_1 \,, \tag{19.11}$$

which is equivalent to two moments per MVM. The numerical effort for N moments is thus reduced by a factor of two. In addition, like many other numerical approaches KPM benefits considerably from the use of symmetries that reduce the Hilbert space dimension.

The second case where the moments depend on a trace over the whole Hilbert space, at first glance, looks far more complicated. Based on the previous considerations we would estimate the numerical effort to be proportional to D^2, because the iteration needs to be repeated for all D states of a given basis. It turns out, however, that extremely good approximations of the moments can be obtained with a much simpler approach: The stochastic evaluation of the trace [2, 9, 10], i.e., an estimate of μ_n based on the average over only a small number $R \ll D$ of randomly chosen states $|r\rangle$

$$\mu_n = \mathrm{Tr}[A\, T_n(\widetilde{H})] \approx \frac{1}{R}\sum_{r=0}^{R-1}\langle r|A\, T_n(\widetilde{H})|r\rangle \,. \tag{19.12}$$

The number of random states R does not scale with D. It can be kept constant or even reduced with increasing D. To understand this, let us consider the convergence properties of the above estimate. Given an arbitrary basis $\{|i\rangle\}$ and a set of independent identically distributed random variables $\xi_{ri} \in \mathbb{C}$, which in terms of the statistical average $\langle\langle\ldots\rangle\rangle$ fulfil

$$\langle\langle\xi_{ri}\rangle\rangle = 0 \,, \qquad \langle\langle\xi_{ri}\xi_{r'j}\rangle\rangle = 0 \,, \qquad \langle\langle\xi^*_{ri}\xi_{r'j}\rangle\rangle = \delta_{rr'}\delta_{ij} \,, \tag{19.13}$$

a random vector is defined through $|r\rangle = \sum_{i=0}^{D-1}\xi_{ri}|i\rangle$. We can now calculate the statistical expectation value of the trace estimate $\Theta = \frac{1}{R}\sum_{r=0}^{R-1}\langle r|B|r\rangle$ for some Hermitian operator B with matrix elements $B_{ij} = \langle i|B|j\rangle$, and indeed find,

$$\langle\langle\Theta\rangle\rangle = \langle\langle\frac{1}{R}\sum_{r=0}^{R-1}\langle r|B|r\rangle\,\rangle\rangle = \frac{1}{R}\sum_{r=0}^{R-1}\sum_{i,j=0}^{D-1}\langle\langle\xi^*_{ri}\xi_{rj}\rangle\rangle B_{ij} = \sum_{i=0}^{D-1}B_{ii} = \mathrm{Tr}(B). \tag{19.14}$$

Of course, this only shows that we obtain the correct result on average. To assess the associated error we also need to study the fluctuation of Θ, which is characterized by $(\delta\Theta)^2 = \langle\langle\Theta^2\rangle\rangle - \langle\langle\Theta\rangle\rangle^2$. Evaluating $\langle\langle\Theta^2\rangle\rangle$, we get for the fluctuation

$$(\delta\Theta)^2 = \frac{1}{R}\left[\mathrm{Tr}(B^2) + (\langle\langle|\xi_{ri}|^4\rangle\rangle - 2)\sum_{j=0}^{D-1} B_{jj}^2\right] . \tag{19.15}$$

The trace of B^2 will usually be of order $O(D)$, and the relative error of the trace estimate, $\delta\Theta/\Theta$, is thus of order $O(1/\sqrt{RD})$. It is this favorable behavior, which ensures the convergence of the stochastic approach, and which was the basis for our initial statement that the number of random states $R \ll D$ can be kept small or even be reduced with the problem dimension D.

19.1.3 Damping of Gibbs Oscillations – Kernel Polynomials

In the preceding sections we introduced the basic ideas underlying the expansion of a function $f(x)$ in an infinite series of Chebyshev polynomials, and gave a few hints for the numerical calculation of the expansion coefficients μ_n. For a numerical approach, however, the total number of moments will remain finite, and we have to look for the best (uniform) approximation to $f(x)$ by polynomials of given maximal degree N. Introducing the concept of kernels, we will investigate and optimize the convergence properties of the mapping $f(x) \to f_{\mathrm{KPM}}(x)$ from the considered function $f(x)$ to our approximation $f_{\mathrm{KPM}}(x)$.

Experience shows that a simple truncation of an infinite series,

$$f(x) \approx \frac{1}{\pi\sqrt{1-x^2}}\left(\mu_0 + 2\sum_{n=1}^{N-1} \mu_n T_n(x)\right) , \tag{19.16}$$

leads to poor precision and fluctuations – also known as Gibbs oscillations – near points where the function $f(x)$ is not continuously differentiable. The situation is even worse for discontinuities or singularities of $f(x)$, as we illustrate in Fig. 19.1. A common procedure to damp these oscillations relies on an appropriate modification of the expansion coefficients, $\mu_n \to g_n\mu_n$, which depends on the order of the approximation N,

$$f_{\mathrm{KPM}}(x) = \frac{1}{\pi\sqrt{1-x^2}}\left(g_0\mu_0 + 2\sum_{n=1}^{N-1} g_n\mu_n T_n(x)\right) . \tag{19.17}$$

This truncation of the infinite series to order N together with the corresponding modification of the coefficients is equivalent to the convolution of $f(x)$ with a kernel $K_N(x,y)$,

$$f_{\mathrm{KPM}}(x) = \int_{-1}^{1} \pi\sqrt{1-y^2} K_N(x,y) f(y)\,\mathrm{d}y , \tag{19.18}$$

where

$$K_N(x,y) = g_0\phi_0(x)\phi_0(y) + 2\sum_{n=1}^{N-1} g_n\phi_n(x)\phi_n(y) , \tag{19.19}$$

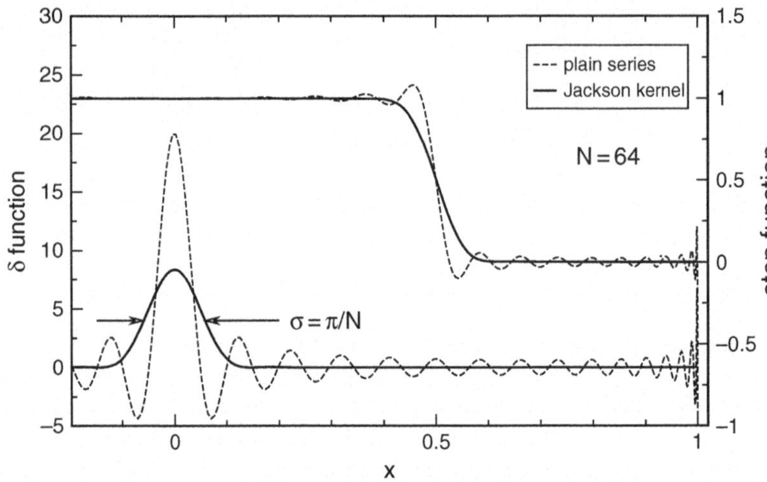

Fig. 19.1. Order $N = 64$ expansions of $\delta(x)$ and a step. Whereas the truncated series (Dirichlet kernel) strongly oscillate, the Jackson results smoothly converge to the expanded functions

and $\phi_n(x) = T_n(x)/(\pi\sqrt{1-x^2})$. This way the problem translates into finding an optimal kernel $K_N(x, y)$, i.e., coefficients g_n. Clearly the notion of *optimal* depends on the application considered.

The standard truncated series corresponds to the choice $g_n^{\mathrm{D}} = 1$, which leads to what is usually called the Dirichlet kernel,

$$K_N^{\mathrm{D}}(x, y) = [\phi_N(x)\phi_{N-1}(y) - \phi_{N-1}(x)\phi_N(y)]/(x - y) \,. \tag{19.20}$$

An approximation based on this kernel for $N \to \infty$ converges within the integral norm $||f||_2 = \sqrt{\langle f|f\rangle}$, i.e. we have

$$||f - f_{\mathrm{KPM}}||_2 \xrightarrow{N\to\infty} 0 \,. \tag{19.21}$$

This is, of course, not particularly restrictive and leads to the disadvantages we mentioned earlier.

A much better criterion would be uniform convergence,

$$||f - f_{\mathrm{KPM}}||_\infty = \max_{-1<x<1} |f(x) - f_{\mathrm{KPM}}(x)| \xrightarrow{N\to\infty} 0 \,, \tag{19.22}$$

and, indeed, this can be achieved for continuous functions f under very general conditions. Specifically, it suffices to demand that:

(i) The kernel is positive: $K_N(x, y) > 0 \; \forall x, y \in [-1, 1]$.
(ii) The kernel is normalized, $\int_{-1}^{1} K(x, y)\,\mathrm{d}x = \phi_0(y)$, which is equivalent to $g_0 = 1$.
(iii) The second coefficient g_1 approaches 1 as $N \to \infty$.

The conditions (i) and (ii) are very useful for practical applications: The first ensures that approximations of positive quantities become positive, the second conserves the integral of the expanded function, $\int_{-1}^{1} f_{\mathrm{KPM}}(x)\,\mathrm{d}x = \int_{-1}^{1} f(x)\,\mathrm{d}x$. Applying the kernel, for example, to a density of states thus yields an approximation which is strictly positive and normalized.

The simplest kernel which fulfils all three conditions is the Fejér kernel [11],

$$K_N^{\mathrm{F}}(x,y) = \frac{1}{N}\sum_{\nu=1}^{N} K_\nu^{\mathrm{D}}(x,y)\,, \tag{19.23}$$

i.e., $g_n^{\mathrm{F}} = 1 - n/N$, which is the arithmetic mean of all Dirichlet approximations of order less or equal N. However, with the coefficients g_n^{F} of the Fejér kernel we have not fully exhausted the freedom offered by the above conditions. We can hope to further improve the kernel by optimizing the g_n in some sense, which will lead us to recover old results by Jackson [12, 13]. In particular, let us tighten the third condition by demanding that the kernel has optimal resolution in the sense that

$$Q := \int_{-1}^{1}\int_{-1}^{1} (x-y)^2 K_N(x,y)\,\mathrm{d}x\,\mathrm{d}y \tag{19.24}$$

is minimal. Since $K_N(x,y)$ is peaked at $x = y$, Q is basically the squared width of this peak and a measure for the resolution of the kernel. For sufficiently smooth functions this more stringent condition will minimize the error $||f - f_{\mathrm{KPM}}||_\infty$, and in all other cases lead to optimal resolution and smallest broadening of sharp features.

The optimization [1, 12, 13] leads to a kernel first described by Jackson, $K_N^{\mathrm{J}}(x,y)$ with

$$g_n^{\mathrm{J}} = \frac{(N-n+1)\cos\left(\pi n/(N+1)\right) + \sin(\pi n/(N+1))\cot(\pi/(N+1))}{N+1}\,, \tag{19.25}$$

which yields the minimal value of Q,

$$Q_{\mathrm{min}} = 1 - \cos\frac{\pi}{N+1} \simeq \frac{1}{2}\left(\frac{\pi}{N}\right)^2\,. \tag{19.26}$$

This shows that for large N the resolution $\sqrt{Q}$ of the new kernel is proportional to $1/N$.

The quantity $\sqrt{Q_{\mathrm{min}}}$ obtained in (19.26) is mainly a measure for the spread of the kernel $K_N^{\mathrm{J}}(x,y)$ in the x-y-plane. For practical calculations, which may also involve singular functions, it is reasonable to ask for the broadening of a δ-function under convolution with the kernel, $\delta_{\mathrm{KPM}}(x-a) = g_0\phi_0(x)T_0(a) + 2\sum_{n=1}^{N-1} g_n\phi_n(x)T_n(a)$. It can be characterized by the variance $\sigma^2 = \langle\langle x^2\rangle\rangle - \langle\langle x\rangle\rangle^2$, which after a short calculation is found to be

$$\sigma^2 \simeq \left(\frac{\pi}{N}\right)^2\left(1 - a^2 + \frac{4a^2-3}{N}\right)\,. \tag{19.27}$$

Using the Jackson kernel, an order N expansion of a δ-function at $x = 0$ thus results in a broadened peak of width $\sigma = \pi/N$, whereas close to the boundaries, $a = \pm 1$, we find $\sigma = \pi/N^{3/2}$. It turns out that this peak is a good approximation to a Gaussian (see Fig. 19.1),

$$\delta^{\mathrm{J}}_{\mathrm{KPM}}(x) \approx \frac{1}{\sqrt{2\pi\sigma^2}} \mathrm{e}^{-x^2/(2\sigma^2)} . \tag{19.28}$$

The Jackson kernel is the best choice for most of the applications we discuss below. In some situations, however, special analytical properties of the expanded functions become important, which only other kernels can account for. Single-particle Green functions that appear in the Cluster Perturbation Theory (see Sect. 19.3), are an example. Considering the imaginary part of the Plemelj-Dirac formula, $\lim_{\epsilon\to 0} 1/(x + \mathrm{i}\epsilon) = \mathcal{P}(1/x) - \mathrm{i}\pi\delta(x)$ (here $\mathcal{P}$ denotes the principal value), which frequently occurs in connection with Green functions, the δ-function on the right hand side is approached in terms of a Lorentz curve,

$$\delta(x) = -\frac{1}{\pi} \lim_{\epsilon\to 0} \mathrm{Im} \frac{1}{x + \mathrm{i}\epsilon} = \lim_{\epsilon\to 0} \frac{\epsilon}{\pi(x^2 + \epsilon^2)} . \tag{19.29}$$

It has a different and broader shape compared to the approximations of $\delta(x)$ we get with the Jackson kernel. We can construct [1] a positive normalized kernel which perfectly mimics the above behavior, and consequently call it the Lorentz kernel $K^{\mathrm{L}}_N(x, y)$ with

$$g^{\mathrm{L}}_n = \frac{\sinh[\lambda(1 - n/N)]}{\sinh(\lambda)} . \tag{19.30}$$

Here, λ is a free parameter which as a compromise between good resolution and sufficient damping of the Gibbs oscillations we empirically choose in order of four. It is related to the ϵ-parameter of the Lorentz curve, i.e. to its resolution, via $\epsilon = \lambda/N$. Note also, that in the limit $\lambda \to 0$ we recover the Fejér kernel $K^{\mathrm{F}}_N(x, y)$, suggesting that both kernels share many of their convergence properties.

19.1.4 Multi-Dimensional Expansions

For the calculation of finite-temperature dynamical correlation functions we will later need expansions of functions of two variables. Let us therefore briefly comment on the generalization of the above considerations to d-dimensional space, which is easily obtained by extending the scalar products $\langle .|. \rangle$ to functions $f, g : [-1, 1]^d \to \mathbb{R}$. As in the one-dimensional case, a simple truncation of the infinite series will lead to Gibbs oscillations and poor convergence. Fortunately, we can easily generalize our results for kernel approximations. In particular, we find that the extended kernel $K_N(\boldsymbol{x}, \boldsymbol{y}) = \prod_{j=1}^{d} K_N(x_j, y_j)$ maps an infinite series onto a truncated series

$$f_{\mathrm{KPM}}(\boldsymbol{x}) = \frac{\sum\limits_{\boldsymbol{n}=0}^{N-1} \mu_{\boldsymbol{n}} h_{\boldsymbol{n}} \prod\limits_{j=1}^{d} g_{n_j} T_{n_j}(x_j)}{\prod\limits_{j=1}^{d} \pi\sqrt{1 - x_j^2}} , \tag{19.31}$$

where we can take the g_n of any of the previously discussed kernels. If we use the g_n^{J} of the Jackson kernel, $K_N^{\mathrm{J}}(\boldsymbol{x},\boldsymbol{y})$ fulfils generalizations of our conditions for an optimal kernel, namely

(i) $K_N^{\mathrm{J}}(\boldsymbol{x},\boldsymbol{y})$ is positive $\forall\, \boldsymbol{x},\boldsymbol{y} \in [-1,1]^d$.
(ii) $K_N^{\mathrm{J}}(\boldsymbol{x},\boldsymbol{y})$ is normalized with

$$\int_{-1}^{1} \cdots \int_{-1}^{1} f_{\mathrm{KPM}}(\boldsymbol{x})\, \mathrm{d}x_1 \ldots \mathrm{d}x_d = \int_{-1}^{1} \cdots \int_{-1}^{1} f(\boldsymbol{x})\, \mathrm{d}x_1 \ldots \mathrm{d}x_d\,. \tag{19.32}$$

(iii) $K_N^{\mathrm{J}}(\boldsymbol{x},\boldsymbol{y})$ has optimal resolution in the sense that

$$Q = \int_{-1}^{1} \cdots \int_{-1}^{1} (\boldsymbol{x}-\boldsymbol{y})^2 K_N(\boldsymbol{x},\boldsymbol{y})\, \mathrm{d}x_1 \ldots \mathrm{d}x_d\, \mathrm{d}y_1 \ldots \mathrm{d}y_d = d(g_0 - g_1) \tag{19.33}$$

is minimal.

Note that for simplicity the order of the expansion, N, was chosen to be the same for all spatial directions. Of course, we could also define more general kernels, $K_{\boldsymbol{N}}(\boldsymbol{x},\boldsymbol{y}) = \prod_{j=1}^{d} K_{N_j}(x_j,y_j)$, where the vector $\boldsymbol{N}$ denotes the orders of expansion for the different spatial directions.

19.1.5 Numerical Implementation

Having discussed the theory behind Chebyshev expansion, the calculation of moments, and the various kernel approximations, let us now come to the practical issues of the implementation of KPM, namely to the reconstruction of the expanded function $f(x)$ from its moments μ_n. Knowing a finite number N of coefficients μ_n, we usually want to reconstruct $f(x)$ on a finite set of abscissas x_k. Naively we could sum up (19.17) separately for each point, thereby making use of the recursion relations for T_n, i.e., $f(x_k) = (g_0\mu_0 + 2\sum_{n=1}^{N-1} g_n\mu_n T_n(x_k))/(\pi\sqrt{1-x_k^2})$. For a set $\{x_k\}$ containing $\tilde{N}$ points these summations would require of the order of $N\tilde{N}$ operations. We can do much better, remembering the definition $T_n(x) = \cos(n \arccos(x))$ and the close relation between KPM and Fourier expansion: First, we may introduce the short-hand notation $\tilde{\mu}_n = \mu_n g_n$ for the kernel improved moments. Second and more important, we make a special choice for our data points,

$$x_k = \cos\frac{\pi(k+1/2)}{\tilde{N}} \tag{19.34}$$

with $k = 0,\ldots,(\tilde{N}-1)$, which coincides with the abscissas of Chebyshev numerical integration [4]. The number $\tilde{N}$ of points in the set $\{x_k\}$ is not necessarily the same as the number of moments N. Usually we will consider $\tilde{N} \geq N$ and a reasonable choice is, e.g. $\tilde{N} = 2N$. All values $f(x_k)$ can now be obtained through a discrete cosine transform,

$$\gamma_k = \pi\sqrt{1-x_k^2}\, f(x_k) = \tilde{\mu}_0 + 2\sum_{n=1}^{N-1} \tilde{\mu}_n \cos\left(\frac{\pi n(k+1/2)}{\tilde{N}}\right) \qquad (19.35)$$

which allows for the use of divide-and-conquer type algorithms that require only $\tilde{N}\log\tilde{N}$ operations – a clear advantage over the above estimate $N\tilde{N}$.

Routines for fast discrete cosine transform are implemented in many mathematical libraries or Fast Fourier Transform (FFT) packages, for instance, in FFTW [14, 15] that ships with most Linux distributions. If no direct implementation is at hand we may also use fast discrete Fourier transform. With

$$\lambda_n = \begin{cases} (2-\delta_{n,0})\,\tilde{\mu}_n\, e^{i\pi n/(2\tilde{N})} & 0 < n < N \\ 0 & \text{otherwise} \end{cases} \qquad (19.36)$$

and the standard definition of discrete Fourier transform,

$$\tilde{\lambda}_k = \sum_{n=0}^{\tilde{N}-1} \lambda_n e^{2\pi i n k/\tilde{N}}\,, \qquad (19.37)$$

after some reordering we find for an even number of data points

$$\gamma_{2j} = \mathrm{Re}(\tilde{\lambda}_j)\,, \qquad \gamma_{2j+1} = \mathrm{Re}(\tilde{\lambda}_{\tilde{N}-1-j})\,, \qquad (19.38)$$

with $j = 0,\ldots,\tilde{N}/2-1$. If we need only a discrete cosine transform this setup is not optimal, as it makes no use of the imaginary part which the complex FFT calculates. It turns out, however, that the wasted imaginary part is exactly what we need when we later calculate Green functions and other complex quantities, i.e., we can use the setup

$$\gamma_{2j} = \tilde{\lambda}_j\,, \qquad \gamma_{2j+1} = \tilde{\lambda}^*_{\tilde{N}-1-j}\,, \qquad (19.39)$$

to evaluate (19.58).

19.2 Applications of the Kernel Polynomial Method

Having described the mathematical background of the KPM, we are now in the position to present practical applications of the approach. KPM can be used whenever we are interested in the spectral properties of large matrices or in correlation functions that can be expressed through the eigenstates of such matrices. In what follows, we try to cover all types of accessible quantities, focusing on lattice models from solid state physics.

19.2.1 Density of States

The first and basic application of Chebyshev expansion and KPM is the calculation of the spectral density of Hermitian matrices, which could correspond to the densities of states of both interacting or non-interacting quantum models [2, 9, 16, 17]. To be specific, let us consider a D-dimensional matrix M with eigenvalues E_k, whose spectral density is defined as

$$\rho(E) = \frac{1}{D} \sum_{k=0}^{D-1} \delta(E - E_k) \,. \tag{19.40}$$

As described earlier, the expansion of $\rho(E)$ in terms of Chebyshev polynomials requires a rescaling of $M \to \widetilde{M}$, such that the spectrum of $\widetilde{M} = (M - b)/a$ fits the interval $[-1, 1]$. Given the eigenvalues $\widetilde{E}_k$ of $\widetilde{M}$ the rescaled density $\widetilde{\rho}(\widetilde{E})$ reads $\widetilde{\rho}(\widetilde{E}) = D^{-1} \sum_{k=0}^{D-1} \delta(\widetilde{E} - \widetilde{E}_k)$, and according to (19.6) the expansion coefficients become

$$\begin{aligned} \mu_n &= \int_{-1}^{1} \widetilde{\rho}(\widetilde{E})\, T_n(\widetilde{E})\, \mathrm{d}\widetilde{E} = \frac{1}{D} \sum_{k=0}^{D-1} T_n(\widetilde{E}_k) \\ &= \frac{1}{D} \sum_{k=0}^{D-1} \langle k | T_n(\widetilde{M}) | k \rangle = \frac{1}{D} \,\mathrm{Tr}(T_n(\widetilde{M})) \,. \end{aligned} \tag{19.41}$$

This is exactly the trace form that we introduced in Sect. 19.1, and we can immediately calculate the μ_n using the stochastic techniques described before. Knowing the moments we can reconstruct $\widetilde{\rho}(\widetilde{E})$ for the whole range $[-1, 1]$, and a final rescaling yields $\rho(E)$.

As the first physical example let us consider percolation of non-interacting fermions in disordered solids. The percolation problem is characterized by the interplay of pure classical and quantum effects. Besides the question of finding a percolating path of accessible sites through a given lattice the quantum nature of the electrons imposes further restrictions on the existence of extended states and, consequently, of a finite dc-conductivity. As a particularly simple model describing this situation we consider a tight-binding one-electron Hamiltonian

$$H = \sum_{i=1} \epsilon_i c_i^\dagger c_i - t \sum_{\langle ij \rangle} \left(c_i^\dagger c_j + \mathrm{H.c.} \right) \tag{19.42}$$

on a simple cubic lattice with L^3 sites and random on-site energies ϵ_i drawn from the bimodal distribution $p(\epsilon_i) = p\,\delta(\epsilon_i - \epsilon_A) + (1 - p)\,\delta(\epsilon_i - \epsilon_B)$, also known as the binary alloy model (see Chap. 17). In the limit $\Delta = (\epsilon_B - \epsilon_A) \to \infty$ the wave-function of the A sub-band vanishes identically on the B-sites, making them completely inaccessible for the quantum particles. We then arrive at a situation where non-interacting electrons move on a random ensemble of lattice points, which, depending on p, may span the entire lattice or not. The corresponding Hamiltonian reads $H = -t\sum_{\langle ij\rangle\in A}(c_i^\dagger c_j + \text{H.c.})$, where the summation extends over nearest-neighbor A-sites only and, without loss of generality, ϵ_A is chosen to be zero.

In the theoretical investigation of disordered systems it turned out that distribution functions for the random quantities take the center stage [18, 19]. The distribution $f(\rho_i(E))$ of the local density of states (LDOS)

$$\rho_i(E) = \sum_{n=1}^{N} |\psi_n(\boldsymbol{r}_i)|^2\, \delta(E - E_n) \tag{19.43}$$

is particularly suited because $\rho_i(E)$ measures the local amplitude of the wavefunction at site $\boldsymbol{r}_i$. It therefore contains direct information about the localization properties. In contrast to the (arithmetically averaged) *mean* DOS, $\rho_{\text{me}}(E) = \langle\rho_i(E)\rangle$, the LDOS becomes critical at the localization transition [20, 21]. Therefore the (geometrically averaged) so-called *typical* DOS, $\rho_{\text{ty}}(E) = \exp(\langle\ln\rho_i(E)\rangle)$, is frequently used to monitor the transition from extended to localized states. The typical DOS puts sufficient weight on small values of ρ_i and a comparison to $\rho_{\text{me}}(E)$ allows to detect the localization transition.

Using the KPM the LDOS can be easily calculated for a large number of samples, K_r, and sites, K_s. The mean and typical DOS are then simply obtained from

$$\rho_{\text{me}}(E) = \frac{1}{K_r K_s}\sum_{k=1}^{K_r}\sum_{i=1}^{K_s}\rho_i(E)\,, \quad \rho_{\text{ty}}(E) = \exp\left[\frac{1}{K_r K_s}\sum_{k=1}^{K_r}\sum_{i=1}^{K_s}\ln(\rho_i(E))\right]\,, \tag{19.44}$$

respectively. We classify a state at energy E with $\rho_{\text{me}}(E) \neq 0$ as localized if $\rho_{\text{ty}}(E) = 0$ and as extended if $\rho_{\text{ty}}(E) \neq 0$.

In order to discuss possible localization phenomena let us investigate the behavior of the mean DOS for the quantum percolation models (19.42). As long as ϵ_A and ϵ_B do not differ too much there exists an asymmetric (if $p \neq 0.5$) but still connected electronic band [22]. At about $\Delta \simeq 4tD$ this band separates into two sub-bands centered at ϵ_A and ϵ_B, respectively. The most prominent feature in the split-band regime is the series of spikes at discrete energies within the band. As an obvious guess, we might attribute these spikes to eigenstates on islands of A or B sites being isolated from the main cluster [23, 24]. It turns out, however, that some of the spikes persist, even if we neglect all finite clusters and restrict the calculation to the spanning cluster of A sites, A_∞. This is illustrated in the upper panels of Fig. 19.2, where we compare the DOS of the model (19.42) (at $\Delta \to \infty$) to that of the spanning cluster only Hamiltonian. Increasing the concentration of accessible sites the mean DOS of the spanning cluster is evocative of the DOS of the simple cubic lattice, but even at large values of p a sharp peak structure remains at $E = 0$ (cf. Fig. 19.2, lower panels). Note that the most dominant peaks at $E/t = 0, \pm1, \pm\sqrt{2}$, $\left(\pm1 \pm \sqrt{5}\right)/2, \ldots$ correspond to eigenvalues of the tight-binding model on small clusters with different geometries. We can thus argue that the wavefunctions, which belong to these special energies, are localized on some dead ends of the spanning cluster. The assumption that the distinct peaks correspond to localized wavefunctions is corroborated by the fact that the typical DOS vanishes or, at least, shows

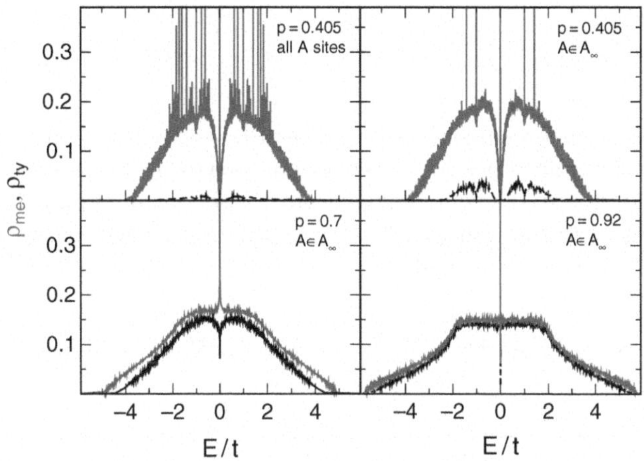

Fig. 19.2. Mean (*upper curves*) and typical (*lower curves*) DOS for the quantum percolation model in the limit $\Delta \to \infty$. While in the upper left panel all A-sites are taken into account, the other three panels show data for the restricted model on the spanning cluster A_∞ only (note that ρ_{ty} is smaller in the former case because there are more sites with vanishing amplitude of the wavefunction). System sizes were adapted to ensure that A_∞ always contains the same number of sites, i.e., 57^3 for $p = 0.405$, 46^3 for $p = 0.70$, and 42^3 for $p = 0.92$. In order to obtain these high-resolution date we used $N = 32768$ Chebyshev moments and $K_s \times K_r = 32 \times 32$

a dip at these energies. Occurring also for finite Δ, this effect becomes more pronounced as $\Delta \to \infty$ and in the vicinity of the classical percolation threshold p_c. For a more detailed discussion see [25].

19.2.2 Correlation Functions at Finite Temperature

Densities of states provide only the most basic information about a given quantum system, and much more details can usually be learned from the study of correlation functions.

Given the eigenstates $|k\rangle$ of an interacting quantum system the thermodynamic expectation value of an operator A reads

$$\langle A \rangle = \frac{1}{ZD} \operatorname{Tr}(A e^{-\beta H}) = \frac{1}{ZD} \sum_{k=0}^{D-1} \langle k|A|k\rangle \, e^{-\beta E_k} , \qquad (19.45)$$

where H is the Hamiltonian of the system, E_k the energy of the eigenstate $|k\rangle$, and $Z = \operatorname{Tr}(\exp(-\beta H))/D = D^{-1} \sum_{k=0}^{D-1} \exp(-\beta E_k)$ the partition function. Using the function $a(E) = D^{-1} \sum_{k=0}^{D-1} \langle k|A|k\rangle \, \delta(E - E_k)$ and the (canonical) density of states $\rho(E)$, we can express the thermal expectation value in terms of integrals over the Boltzmann weight,

$$\langle A\rangle = \frac{1}{Z}\int\limits_{-\infty}^{\infty} a(E)\,\mathrm{e}^{-\beta E}\,\mathrm{d}E\,, \qquad Z = \int\limits_{-\infty}^{\infty} \rho(E)\,\mathrm{e}^{-\beta E}\,\mathrm{d}E\,. \tag{19.46}$$

Of course, similar relations hold also for non-interacting fermion systems, where the Boltzmann weight $\exp(-\beta E)$ has to be replaced by the Fermi function $f(E) = 1/(1+\exp(\beta(E-\mu)))$ and the single-electron wave functions play the role of $|k\rangle$.

Again, the particular form of $a(E)$ suggests an expansion in Chebyshev polynomials, and after rescaling we find

$$\mu_n = \int\limits_{-1}^{1} \tilde{a}(E)\,T_n(E)\,\mathrm{d}E = \frac{1}{D}\sum_{k=0}^{D-1}\langle k|A|k\rangle\,T_n(\widetilde{E}_k) = \frac{1}{D}\,\mathrm{Tr}\left(AT_n(\widetilde{H})\right)\,, \tag{19.47}$$

which can be evaluated employing the stochastic approach, outlined in Sect. 19.1.

For interacting systems at low temperature the expression in (19.46) is a bit problematic, since the Boltzmann factor puts most of the weight on the lower end of the spectrum and heavily amplifies small numerical errors in $\rho(E)$ and $a(E)$. We can avoid these problems by calculating the ground state and some of the lowest excitations exactly, using standard iterative diagonalization methods like Lanczos or Jacobi-Davidson (see Sect. 18.2). Then we split the expectation value of A and the partition function Z into contributions from the exactly known states and contributions from the rest of the spectrum,

$$\langle A\rangle = \frac{1}{ZD}\sum_{k=0}^{C-1}\langle k|A|k\rangle\,\mathrm{e}^{-\beta E_k} + \frac{1}{Z}\int\limits_{-\infty}^{\infty} a_s(E)\mathrm{e}^{-\beta E}\,\mathrm{d}E\,,$$

$$Z = \frac{1}{D}\sum_{k=0}^{C-1}\mathrm{e}^{-\beta E_k} + \int\limits_{-\infty}^{\infty}\rho_s(E)\mathrm{e}^{-\beta E}\,\mathrm{d}E\,. \tag{19.48}$$

Here $a_s(E) = D^{-1}\sum_{k=C}^{D-1}\langle k|A|k\rangle\,\delta(E-E_k)$ and $\rho_s(E) = D^{-1}\sum_{k=C}^{D-1}\delta(E-E_k)$ describe the rest of the spectrum and can be expanded in Chebyshev polynomials easily. Based on the known states we can introduce the projection operator $P = 1 - \sum_{k=0}^{C-1}|k\rangle\langle k|$ and find for the expansion coefficients of $\tilde{a}_s(E)$

$$\mu_n = \frac{1}{D}\,\mathrm{Tr}(PAT_n(\widetilde{H})) \approx \frac{1}{RD}\sum_{r=0}^{R-1}\langle r|PAT_n(\widetilde{H})P|r\rangle\,, \tag{19.49}$$

and similarly for those of $\widetilde{\rho}_s(E)$:

$$\mu_n = \frac{1}{D}\,\mathrm{Tr}(PT_n(\widetilde{H})) \approx \frac{1}{RD}\sum_{r=0}^{R-1}\langle r|PT_n(\widetilde{H})P|r\rangle\,. \tag{19.50}$$

Note, that in addition to the two vectors for the Chebyshev recursion we now need memory also for the eigenstates $|k\rangle$. Otherwise the resource consumption is the same as in the standard scheme.

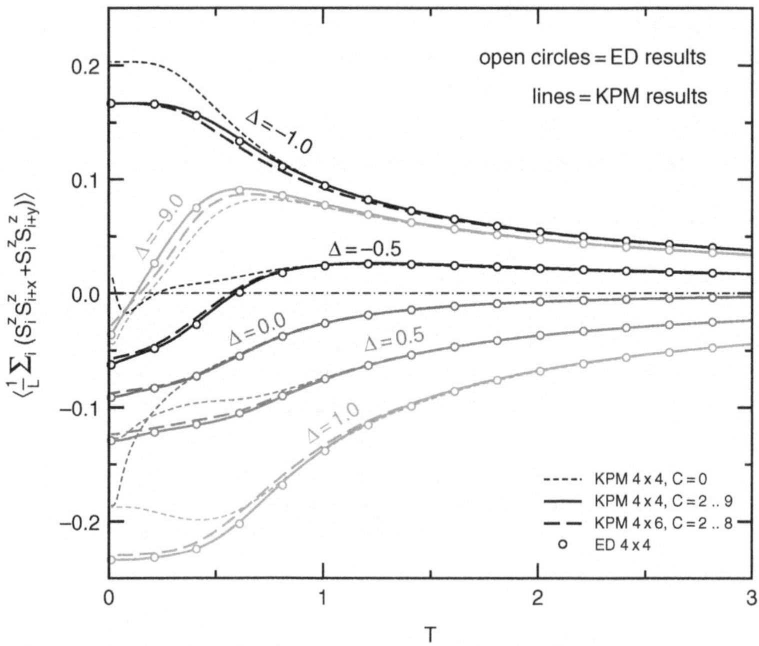

Fig. 19.3. Nearest-neighbor S^z-S^z correlations of the XXZ model on a square lattice. Lines represent the KPM results with separation of low-lying eigenstates (*bold solid* and *bold dashed*) and without (*thin dashed*), open symbols denote exact results from a complete diagonalization of a 4×4 system

We illustrate the accuracy of this approach in Fig. 19.3 considering the nearest-neighbor S^z-S^z correlations of the square-lattice spin-$1/2$ XXZ model as an example,

$$H = \sum_{i,\delta} (S_i^x S_{i+\delta}^x + S_i^y S_{i+\delta}^y + \Delta S_i^z S_{i+\delta}^z) \,. \tag{19.51}$$

As a function of temperature and for an anisotropy $-1 < \Delta < 0$ this model shows a quantum to classical crossover in the sense that the correlations are antiferromagnetic at low temperature (quantum effect) and ferromagnetic at high temperature (as expected for the classical model) [26, 27, 28]. Comparing the KPM results with the exact correlations of a 4×4 system, which were obtained from a complete diagonalization of the Hamiltonian, the improvement due to the separation of only a few low-lying eigenstates is obvious. Whereas for $C = 0$ the data is more or less random below $T \approx 1$, the agreement with the exact data is perfect, if the ground state and one or two excitations are considered separately. The numerical effort required for these calculations differs largely between complete diagonalization and the KPM method. For the former, 18 or 20 sites are practically the limit, whereas the latter can easily handle 30 sites or more.

Note that for non-interacting systems the above separation of the spectrum is not required, since for $T \to 0$ the Fermi function converges to a simple step function without causing any numerical problems.

19.2.3 Spectral Functions and Dynamical Response

19.2.3.1 General Considerations

Having discussed simple expectation values and static correlations, the calculation of time dependent quantities is the natural next step in the study of complex quantum models. This is motivated also by many experimental setups, which probe the response of a physical system to time dependent external perturbations. Examples are inelastic scattering experiments or measurements of transport coefficients. In the framework of linear response theory and the Kubo formalism the system's response is expressed in terms of dynamical correlation functions, which can also be calculated efficiently with Chebyshev expansion and KPM.

Given two operators A and B a general dynamical correlation function can be defined through

$$\langle A; B\rangle^{\pm}_{\omega} = \lim_{\epsilon\to 0} \langle 0|A \frac{1}{\omega + \mathrm{i}\epsilon \mp H} B|0\rangle = \lim_{\epsilon\to 0} \sum_{k=0}^{D-1} \frac{\langle 0|A|k\rangle\langle k|B|0\rangle}{\omega + \mathrm{i}\epsilon \mp E_k} , \qquad (19.52)$$

where E_k is the energy of the many-particle eigenstate $|k\rangle$ of the Hamiltonian H, $|0\rangle$ its ground state, and $\epsilon > 0$.

If we assume that the product $\langle 0|A|k\rangle\langle k|B|0\rangle$ is real the imaginary part

$$\mathrm{Im}\langle A; B\rangle^{\pm}_{\omega} = -\pi \sum_{k=0}^{D-1} \langle 0|A|k\rangle\langle k|B|0\rangle\, \delta(\omega \mp E_k) \qquad (19.53)$$

has a similar structure as, e.g., the local density of states in (19.43), and in fact, with $\rho_i(E)$ we already calculated a dynamical correlation function. Rescaling the Hamiltonian $H \to \widetilde{H}$ and all energies $\omega \to \widetilde{\omega}$ we can proceed as usual and expand $\mathrm{Im}\langle A; B\rangle^{\pm}_{\omega}$ in Chebyshev polynomials,

$$\mathrm{Im}\langle A; B\rangle^{\pm}_{\widetilde{\omega}} = -\frac{1}{\sqrt{1-\widetilde{\omega}^2}} \left(\mu_0 + 2\sum_{n=1}^{\infty} \mu_n T_n(\widetilde{\omega}) \right) . \qquad (19.54)$$

Again, the moments are obtained from expectation values

$$\mu_n = \frac{1}{\pi} \int_{-1}^{1} \mathrm{Im}\langle A; B\rangle^{\pm}_{\widetilde{\omega}}\, T_n(\widetilde{\omega})\, \mathrm{d}\widetilde{\omega} = \langle 0|A T_n(\mp\widetilde{H}) B|0\rangle . \qquad (19.55)$$

In many cases, especially for the spectral functions and optical conductivities studied below, only the imaginary part of $\langle A; B\rangle^{\pm}_{\omega}$ is of interest, and the above setup is all we need. Sometimes however – e.g., within the cluster perturbation theory discussed in Sect. 19.3 – also the real part of a general correlation function $\langle A; B\rangle_\omega^\pm$ is required. Fortunately it can be calculated with almost no additional effort: The analytical properties of $\langle A; B\rangle_\omega^\pm$ arising from causality imply that its real part is fully determined by the imaginary part. Indeed, using the Hilbert transforms of the Chebyshev polynomials,

$$\mathcal{P}\int_{-1}^{1} \frac{T_n(y)\,\mathrm{d}y}{(y-x)\sqrt{1-y^2}} = \pi\, U_{n-1}(x)\,,$$
$$\mathcal{P}\int_{-1}^{1} \frac{\sqrt{1-y^2}\,U_{n-1}(y)\,\mathrm{d}y}{(y-x)} = -\pi\, T_n(x)\,, \tag{19.56}$$

we obtain

$$\begin{aligned}\mathrm{Re}\langle A; B\rangle_{\tilde\omega}^\pm &= \sum_{k=0}^{D-1} \langle 0|A|k\rangle\langle k|B|0\rangle\; \mathcal{P}\left(\frac{1}{\tilde\omega \mp \tilde E_k}\right)\\ &= -\frac{1}{\pi}\mathcal{P}\int_{-1}^{1} \frac{\mathrm{Im}\langle A; B\rangle_{\tilde\omega'}^\pm}{\tilde\omega - \tilde\omega'}\,\mathrm{d}\omega' = -2\sum_{n=1}^{\infty}\mu_n\, U_{n-1}(\tilde\omega)\,.\end{aligned} \tag{19.57}$$

The full correlation function

$$\begin{aligned}\langle A; B\rangle_{\tilde\omega}^\pm &= \frac{-\mathrm{i}\mu_0}{\sqrt{1-\tilde\omega^2}} - 2\sum_{n=1}^{\infty}\mu_n\left(U_{n-1}(\tilde\omega) + \frac{\mathrm{i}\,T_n(\tilde\omega)}{\sqrt{1-\tilde\omega^2}}\right)\\ &= \frac{-\mathrm{i}}{\sqrt{1-\tilde\omega^2}}\left(\mu_0 + 2\sum_{n=1}^{\infty}\mu_n \mathrm{e}^{-\mathrm{i}n\arccos\tilde\omega}\right)\end{aligned} \tag{19.58}$$

can thus be reconstructed from the same moments μ_n that we derived for its imaginary part (19.55). In contrast to the real quantities we considered so far, the reconstruction merely requires complex Fourier transform (see (19.39)). If only the imaginary or real part of $\langle A; B\rangle_\omega^\pm$ is needed, a cosine or sine transform, respectively, is sufficient.

Note that the calculation of dynamical correlation functions for non-interacting electron systems is not possible with the scheme discussed in this section, not even at zero temperature. At finite band filling (finite chemical potential) the ground state consists of a sum over occupied single-electron states, and dynamical correlation functions thus involve a double summation over matrix elements between all single-particle eigenstates, weighted by the Fermi function. See the section on the optical conductivity for a discussion of this case, which covers also the calculation of dynamical correlation functions at finite temperature.

19.2.3.2 One-Particle Spectral Function

An important example of a dynamical correlation function is the (retarded) Green function in momentum space,

$$G_\sigma(\boldsymbol{k},\omega) = \langle c_{\boldsymbol{k},\sigma}; c^\dagger_{\boldsymbol{k},\sigma}\rangle^+_\omega + \langle c^\dagger_{\boldsymbol{k},\sigma}; c_{\boldsymbol{k},\sigma}\rangle^-_\omega \,, \tag{19.59}$$

and the associated spectral function

$$A_\sigma(\boldsymbol{k},\omega) = -\frac{1}{\pi}\,\mathrm{Im}\; G_\sigma(\boldsymbol{k},\omega) = A^+_\sigma(\boldsymbol{k},\omega) + A^-_\sigma(\boldsymbol{k},\omega) \,, \tag{19.60}$$

which characterizes the electron absorption or emission of an interacting system. For instance, A^- can be measured experimentally in angle resolved photo-emission spectroscopy (ARPES).

Exemplarily let us consider the one-dimensional Holstein model

$$H = -t\sum_i (c^\dagger_i c_{i+1} + \mathrm{H.c.}) - g\omega_0 \sum_{i,\sigma} (b^\dagger_i + b_i) n_i + \omega_0 \sum_i b^\dagger_i b_i \,, \tag{19.61}$$

which is one of the basic models for the study of electron-lattice interaction in electronically low-dimensional solids. In (19.61), the electrons are approximated by spinless fermions $c_i^{(\dagger)}$, the density of which couples to the local lattice distortion described by dispersionless phonons $b_i^{(\dagger)}$. At half-filling, i.e., 0.5 fermions per site, the model allows for the study of quantum effects at the transition from a (Luttinger liquid) metal to a (Peierls) insulator, marked by the opening of a gap at the Fermi wave vector and the development of charge-density-wave (CDW) long-range order and a matching lattice distortion [29, 30, 31]. The Peierls insulator can be classified as traditional band insulator and polaronic superlattice in the strong electron-phonon coupling adiabatic ($\omega_0/t \ll 1$) and anti-adiabatic ($\omega_0/t \gg 1$) regimes, respectively.

Figure 19.4 shows KPM data for the spectral function of the half-filled Holstein model and assesses its quality by comparing with results from Dynamical Density Matrix Renormalization Group (DDMRG) [32] calculations. In the spinless case, the photo-emission (A^-) and inverse photo-emission (A^+) parts read

$$\begin{aligned} A^-(k,\omega) &= \sum_l |\langle l, N_\mathrm{e}-1|\, c_k\, |0, N_\mathrm{e}\rangle|^2\, \delta[\omega + (E_{l,N_\mathrm{e}-1} - E_{0,N_\mathrm{e}})] \,, \\ A^+(k,\omega) &= \sum_l |\langle l, N_\mathrm{e}+1|\, c^\dagger_k\, |0, N_\mathrm{e}\rangle|^2\, \delta[\omega - (E_{l,N_\mathrm{e}+1} - E_{0,N_\mathrm{e}})] \,, \end{aligned} \tag{19.62}$$

where $|l, N_\mathrm{e}\rangle$ denotes the lth eigenstate with N_e electrons and energy E_{l,N_e}. For the parameters of Fig. 19.4 the system is in an insulating phase with a finite charge excitation gap at the Fermi momentum $k = \pm\pi/2$. Below and above the gap the spectrum is characterized by broad multi-phonon absorption, reflecting the Poisson-like phonon distribution in the ground state. Compared to DDMRG, KPM offers the better resolution and unfolds all the discrete phonon sidebands. Concerning numerical performance DDMRG has the advantage of a small optimized Hilbert space

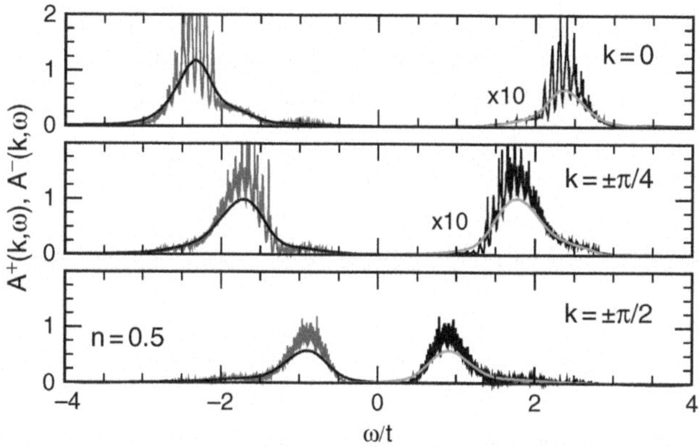

Fig. 19.4. Single-particle spectral functions $A(k,\omega)$ (for electron removal, $\omega < 0$, and electron injection, $\omega > 0$) of the spinless Holstein model at half-filling on an eight-site lattice with periodic boundary conditions. The system is in the Peierls/CDW insulating phase ($\omega_0/t = 0.1$ and $g = 4$). The rapidly oscillating thin lines are the KPM results ($M = 32$) while the smooth thick line are the DDMRG data ($M = 16$) obtained with the pseudo-site method for the same lattice size

[33, 34], which can be handled with standard workstations. However, the basis optimization is rather time consuming and, in addition, each frequency value ω requires a new simulation. The KPM calculations, on the other hand, involved matrix dimensions between 10^8 and 10^{10}, and we therefore used high-performance computers such as Hitachi SR8000-F1 or IBM p690 for the moment calculation. For the reconstruction of the spectra, of course, a desktop computer is sufficient.

19.2.3.3 Optical Conductivity

The next example of a dynamical correlation function is the optical conductivity. Here the imaginary and real parts of our general correlation functions $\langle A; B\rangle_\omega$ change their roles due to an additional frequency integration. The so-called regular contribution to the real part of the optical conductivity is thus given by,

$$\sigma^{\mathrm{reg}}(\omega) = \frac{1}{\omega} \sum_{E_k > E_0} |\langle k|J|0\rangle|^2 \, \delta(\omega - (E_k - E_0)) \, , \tag{19.63}$$

with the current operator $J = -\mathrm{i}qt\sum_{i,\sigma}(c^\dagger_{i,\sigma}c_{i+1,\sigma} - \mathrm{H.c.})$. The latter follows from the continuity equation $\dot{n}_{i\sigma} = \mathrm{i}[H, n_{i\sigma}] = j_{i-1,\sigma} - j_{i\sigma}$, where $j_{i\sigma}$ is the local particle current. After rescaling the energy and shifting the frequency, $\omega = \widetilde{\omega} + \widetilde{E}_0$, the sum can be expanded as described earlier, now with $J|0\rangle$ as the initial state for the Chebyshev recursion. Back-scaling and dividing by ω then yields the final result.

The finite-temperature extension of (19.63) is given by

$$\sigma^{\mathrm{reg}}(\omega) = \sum_{k,q} \frac{|\langle k|J|q\rangle|^2 (\mathrm{e}^{-\beta E_k} - \mathrm{e}^{-\beta E_q})}{ZD\,\omega} \, \delta(\omega - \omega_{qk}) \, , \tag{19.64}$$

with $\omega_{qk} = E_q - E_k$. Compared to (19.63) a straight-forward expansion of the finite temperature conductivity is spoiled by the presence of the Boltzmann weighting factors. A solution comes from the current matrix element density

$$j(x,y) = \frac{1}{D}\sum_{k,q} |\langle k|J|q\rangle|^2 \, \delta(x - E_k)\, \delta(y - E_q) \, . \tag{19.65}$$

Being a function of two variables, $j(x,y)$ can be expanded with two-dimensional KPM,

$$\tilde{j}(x,y) = \sum_{n,m=0}^{N-1} \frac{\mu_{nm} h_{nm} g_n g_m T_n(x) T_m(y)}{\pi^2 \sqrt{(1-x^2)(1-y^2)}} \, , \tag{19.66}$$

where $\tilde{j}(x,y)$ refers to the rescaled $j(x,y)$, g_n are the usual kernel damping factors, and h_{nm} account for the correct normalization. The moments μ_{nm} are obtained from

$$\mu_{nm} = \int_{-1}^{1}\int_{-1}^{1} \tilde{j}(x,y) T_n(x) T_m(y) \, \mathrm{d}x \, \mathrm{d}y = \frac{1}{D} \operatorname{Tr}\left(T_n(\tilde{H}) J T_m(\tilde{H}) J\right) \, , \tag{19.67}$$

and again the trace can be replaced by an average over a relatively small number R of random vectors $|r\rangle$. The numerical effort for an expansion of order $n, m < N$ ranges between $2RDN$ and RDN^2 operations, depending on whether memory is available for up to N vectors of the Hilbert space dimension D or not. Given the operator density $j(x,y)$ we find the optical conductivity by integrating over Boltzmann factors,

$$\begin{aligned} \sigma^{\mathrm{reg}}(\omega) &= \frac{1}{Z\omega} \int_{-\infty}^{\infty} j(y+\omega, y) \left(\mathrm{e}^{-\beta y} - \mathrm{e}^{-\beta(y+\omega)}\right) \mathrm{d}y \\ &= \sum_{k,q} \frac{|\langle k|J|q\rangle|^2 (\mathrm{e}^{-\beta E_k} - \mathrm{e}^{-\beta E_q})}{ZD\omega} \, \delta(\omega - \omega_{qk}) \, , \end{aligned} \tag{19.68}$$

and, as above, we get the partition function Z from an integral over the density of states $\rho(E)$. The latter can be expanded in parallel to $j(x,y)$. Note that the calculation of the conductivity at different temperatures is based on the same operator density $j(x,y)$, i.e., it needs to be expanded only once for all temperatures.

As a physical example, we consider the conductivity for the Anderson model of non-interacting fermions moving in a random potential [18],

$$H = -t \sum_{\langle ij \rangle} c_i^\dagger c_j + \sum_i \epsilon_i c_i^\dagger c_i \, . \tag{19.69}$$

Here hopping occurs along nearest neighbor bonds $\langle ij \rangle$ on a simple cubic lattice and the local potential ϵ_i is chosen randomly with uniform distribution in the interval $[-\gamma/2, \gamma/2]$. With increasing strength of disorder, γ, the single-particle eigenstates of the model tend to become localized in the vicinity of a particular lattice site, which excludes these states from contributing to electronic transport. Disorder can therefore drive a transition from metallic behavior with delocalized fermions to insulating behavior with localized fermions [35, 36, 37].

Since the Anderson model describes non-interacting fermions, the eigenstates $|k\rangle$ occurring in $\sigma(\omega)$ now denote single-particle wave functions and the Boltzmann weight has to be replaced by the Fermi function,

$$\sigma^{\text{reg}}(\omega) = \sum_{k,q} \frac{|\langle k|J|q\rangle|^2 (f(E_k) - f(E_q))}{\omega} \delta(\omega - \omega_{qk}) . \qquad (19.70)$$

Clearly, from a computational point of view this expression is of the same complexity for both, zero and finite temperature, i.e. we need the more advanced 2D KPM approach [38].

Figure 19.5 shows the optical conductivity of the Anderson model at $\gamma/t = 12$ for different inverse temperatures $\beta = 1/T$. The chemical potential is chosen as $\mu = 0$, i.e., the system is still in the metallic phase. However, the conductivity shows a pronounced dip near $\omega = 0$ with the functional form $\sigma(\omega) \sim \sigma_0 + |\omega|^\alpha$. For stronger disorder γ or a different chemical potential μ, the system will become insulating and the dc-conductivity σ_0 will vanish. The role of temperature, in this example, is limited to suppressing $\sigma(\omega)$, mainly through the $(f(E_k) - f(E_q))$ term in (19.70). The model (19.69) does not describe thermally activated hopping, since there are no phonons included.

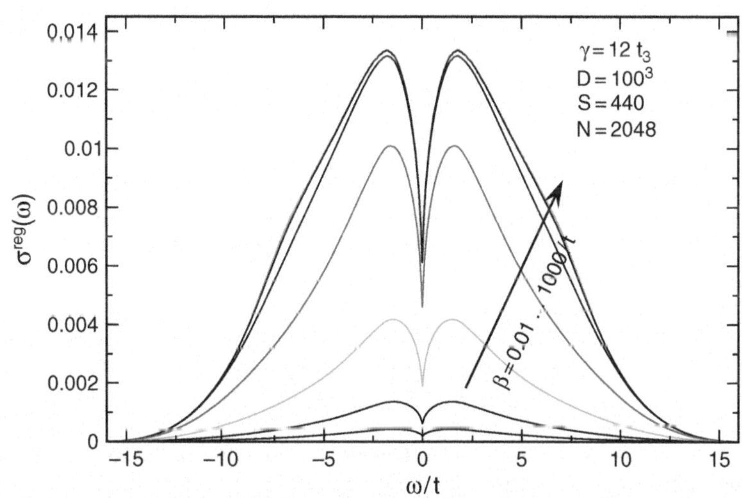

Fig. 19.5. Optical conductivity of the 3D Anderson model with $\gamma = 12$. Note that all curves are derived from the same matrix element density $j(x, y)$, which was calculated for a 100^3 site cluster with expansion order $N = 2048$ and averaged over $K_r = 440$ samples

19.2.4 Time Evolution of Quantum Systems

Dynamical correlation functions are an important aspect in the description of interacting quantum systems and, in many cases, are directly related to experimental results, in particular spectroscopy data. On the other hand, new experimental setups and techniques led to an increased interest in the real time dynamics of quantum systems. Chebyshev expansion is applicable also in this situation.

Starting from the time dependent Schrödinger equation,

$$\mathrm{i}\partial_t|\psi\rangle = H|\psi\rangle\,, \tag{19.71}$$

the approach is surprisingly simple: Assuming that at time $t = 0$ the system is in the state $|\psi_0\rangle$, its state at a later time is

$$|\psi_t\rangle = \mathrm{e}^{-\mathrm{i}Ht}|\psi_0\rangle\,, \tag{19.72}$$

and the problem translates into calculating the time evolution operator $U(t) = \exp(-\mathrm{i}Ht)$ for a given Hamiltonian H and time t. Using the rescaling introduced in (19.7), we can expand $U(t)$ in a series of Chebyshev polynomials [39, 40, 41],

$$U(t) = \mathrm{e}^{-\mathrm{i}(a\widetilde{H}+b)t} = \mathrm{e}^{-\mathrm{i}bt}\left(c_0 + 2\sum_{k=1}^{N} c_k T_k(\widetilde{H})\right)\,, \tag{19.73}$$

where the expansion coefficients c_k are given by

$$c_k = \int_{-1}^{1} \frac{T_k(x)\mathrm{e}^{-\mathrm{i}axt}}{\pi\sqrt{1-x^2}}\,\mathrm{d}x = (-\mathrm{i})^k J_k(at)\,, \tag{19.74}$$

and $J_k(at)$ denotes the Bessel function of order k. The Chebyshev polynomials of the Hamiltonian, $T_k(\widetilde{H})$, are calculated with the recursion we introduced earlier, see (19.3). Thus, the wave function at a later time is obtained simply through a set of MVMs with the Hamiltonian.

Asymptotically the Bessel function behaves as

$$J_k(z) \sim \frac{1}{k!}\left(\frac{z}{2}\right)^k \sim \frac{1}{\sqrt{2\pi k}}\left(\frac{\mathrm{e}z}{2k}\right)^k \tag{19.75}$$

for $k \to \infty$, hence for $k \gg at$ the expansion coefficients c_k decay superexponentially and the series can be truncated with negligible error. With an expansion order of $N \gtrsim 1.5at$ we are usually on the safe side. Moreover, we can check the quality of our approximation by comparing the norms of $|\psi_t\rangle$ and $|\psi_0\rangle$. For sparse matrices the whole time evolution scheme is therefore linear in both, the matrix dimension and the time.

The Chebyshev expansion method converges much faster than other time integration methods, in particular, it is faster than the popular Crank-Nicolson method

[42]. Within this approach the time interval t is divided into small steps $\Delta t = t/N$, and the wave function is propagated in a mixed explicit/implicit manner,

$$(1 + \tfrac{1}{2}\mathrm{i}H\Delta t)|\psi_{n+1}\rangle = (1 - \tfrac{1}{2}\mathrm{i}H\Delta t)|\psi_n\rangle \, . \tag{19.76}$$

Thus, each step requires both a sparse MVM and the solution of a sparse linear system. Obviously, this is more complicated than the Chebyshev recursion, which requires only MVMs. In the Crank-Nicolson method the time evolution operator is approximated as

$$U(t) = \left(\frac{1 - \mathrm{i}Ht/(2N)}{1 + \mathrm{i}Ht/(2N)}\right)^N \, . \tag{19.77}$$

In Fig. 19.6 we compare this approximation with the Chebyshev approximation by replacing H with the real variable x (this is equivalent to working with a diagonal matrix H). In both cases we consider time $t = 10$ and expansion order $N = 15$. Whereas the Chebyshev result agrees perfectly with the exact result $\exp(\mathrm{i}xt)$, the Crank-Nicolson approximation needs much higher N to achieve the same accuracy ($N \approx 90$).

Having explained the time evolution algorithm, let us now consider a specific example: the formation of a polaron on an one-dimensional lattice. The Hamiltonian for this problem was introduced at the beginning of this chapter, see (18.3). The polaron problem corresponds to the case of a single electron interacting with finite frequency lattice vibrations, i.e., we can omit the spin indices and the Hubbard term does not contribute. Bonča, Trugman and co-workers [43, 44] introduced a highly efficient variational basis for the polaron problem, which can be used to study its ground-state properties and lowest excitations on an infinite lattice, as well as the

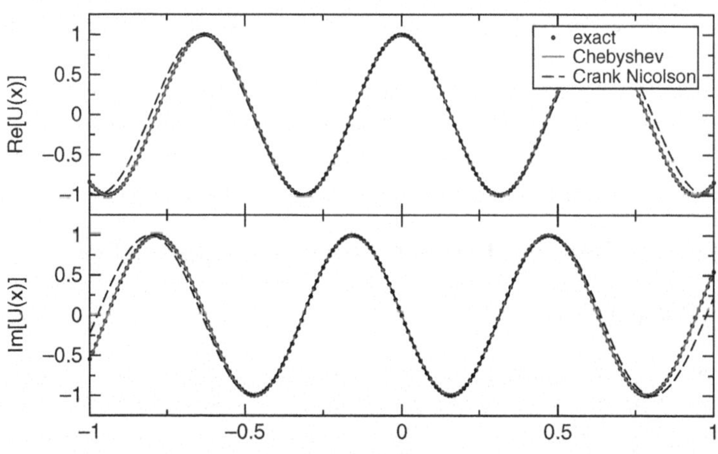

Fig. 19.6. Comparison of the Chebyshev and the Crank-Nicolson approximation of the function $U(t) = \exp(\mathrm{i}xt)$ with $t = 10$ and expansion order $N = 15$

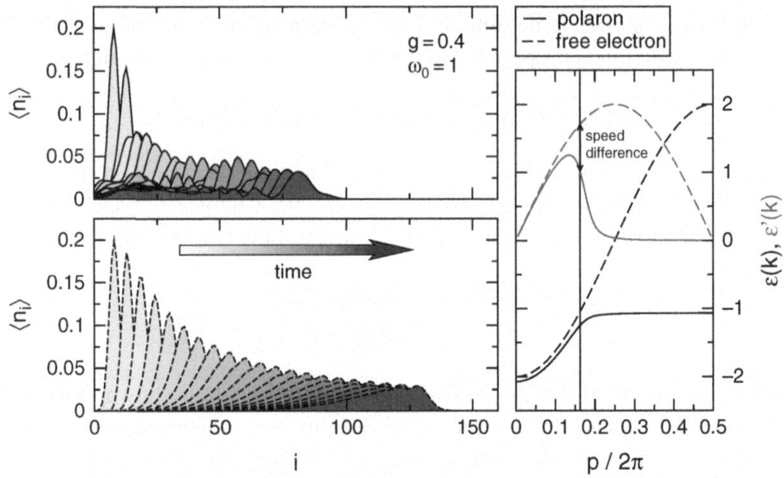

Fig. 19.7. Formation of a polaron for electron-lattice coupling $g = 0.4$ and phonon frequency $\omega_0 = 1$ (**upper panel**), compared to the motion of a non-interacting wave packet (**lower panel**). The right panel shows the underlying dispersions (*lower curves*) and velocities (*upper curves*)

quantum dynamics of such a system (for a recent review see also [45]). In Fig. 19.7 we show the time evolution of a single-electron wave packet

$$|\psi_0\rangle = \sum_j e^{ipj-(j-j_0)^2/(2\sigma^2)}\, c_j^\dagger|0\rangle\,, \tag{19.78}$$

where in the upper and lower panels the electron-phonon coupling g is finite or zero, respectively. For finite g, within the first few time steps a polaron is formed, which then travels at lower speed, compared to the non-interacting wave packet. The speed difference is given by the difference of the derivatives $\varepsilon'(k)$ of the underlying dispersions $\varepsilon(k)$ at the mean momentum p, see right hand panel. The Chebyshev expansion method allows for a fast and reliable simulation of this interesting problem.

19.3 KPM in Relation to other Numerical Approaches

19.3.1 KPM and CPT

The spectrum of a finite system of L sites, which we obtain through KPM, differs in many respects from that of an infinite system, $L \to \infty$, especially since for a finite system the lattice momenta $K = \pi m/L$ and the energy levels are discrete. While we cannot easily increase L without reaching computationally inaccessible Hilbert space dimensions, we can try to extrapolate from a finite to the infinite system.

With the Cluster Perturbation Theory (CPT) [46, 47, 48] a straightforward way to perform this task approximatively has recently been devised. In this scheme one first calculates the Green function $G^c_{ij}(\omega)$ for all sites $i, j = 1, \ldots, L$ of a L-size cluster with open boundary conditions, and then recovers the infinite lattice by pasting identical copies of this cluster at their edges. The glue is the hopping V between these clusters, where $V_{mn} = t$ for $|m - n| = 1$ and $m, n \equiv 0, 1(\mathrm{mod} L)$, which is dealt with in first order perturbation theory. Then the Green function $G_{ij}(\omega)$ of the infinite lattice is given through a Dyson equation

$$G_{ij}(\omega) = G^c_{ij}(\omega) + \sum_{mn} G^c_{ik}(\omega) V_{mn} G_{nj}(\omega) \;, \tag{19.79}$$

where indices of $G^c(\omega)$ are counted modulo L. Obviously this order of perturbation in V is exact for the non-interacting system. The Dyson equation is solved by Fourier transformation over momenta $K = kL$ corresponding to translations by L sites

$$G_{ij}(K, \omega) = \left[\frac{G^c(\omega)}{1 - V(K) G^c(\omega)} \right]_{ij} \;. \tag{19.80}$$

from which one finally obtains

$$G(k, \omega) = \frac{1}{L} \sum_{i,j=1}^{L} G^c_{ij}(Lk, \omega) \mathrm{e}^{-\mathrm{i}k(i-j)} \;. \tag{19.81}$$

Hence, from the Green function $G^c_{ij}(\omega)$ on a finite cluster we construct a Green function $G(k, \omega)$ with continuous momenta k.

Two approximations are made, one by using first order perturbation theory in $V = t$, the second on assuming translational symmetry in $G_{ij}(\omega)$ which is satisfied only approximately. In principle, the CPT spectral function $G(k, \omega)$ does not contain any more information than the cluster Green function $G^c_{ij}(\omega)$ already does. But extrapolating to the infinite system it gives a first hint at the scenario in the thermodynamic limit. Providing direct access to spectral functions, still without relying on possibly erroneous approximations, CPT occupies a niche between variational approaches like (D)DMRG [32, 49] and methods directly working in the thermodynamic limit like the variational ED method [43].

On applying the CPT crucial attention has to be paid to the kernel used in the reconstruction of $G^c_{ij}(\omega)$. As it turns out, the Jackson kernel is an inadequate choice here, since already for the non-interacting tight-binding model it introduces spurious structures into the spectra [1]. The failure can be attributed to the shape of the Jackson kernel: Being optimized for high resolution, a pole in the Green function will give a sharp peak with most of its weight concentrated at the center, and rapidly decaying tails. The reconstructed (cluster) Green function therefore does not satisfy the correct analytical properties required in the CPT step. To guarantee these properties, instead, we use the Lorentz kernel, which is constructed in order to mimic the effect of a finite imaginary part in the energy argument of a Green function.

Using $G^c_{ij}(\omega) = G^c_{ji}(\omega)$ (no magnetic field), for a L-site chain L diagonal and $L(L-1)/2$ off-diagonal elements of $G^c_{ij}(\omega)$ have to be calculated. The latter can be reduced to Chebyshev iterations for the operators $c_i^{(\dagger)} + c_j^{(\dagger)}$. The numerical effort can be further reduced by a factor $1/L$: If we keep the ground state $|0\rangle$ of the system we can calculate the moments $\mu_n^{ij} = \langle 0|c_i T_n(\widetilde{H}) c_j^\dagger|0\rangle$ for L elements $i = 1, \ldots, L$ of $G_{ij}^c(\omega)$ in a single Chebyshev iteration. To achieve a similar reduction within the Lanczos recursion we had to explicitly construct the eigenstates to the Lanczos eigenvalues. Then the factor $1/L$ is exceeded by at least ND additional operations for the construction of N eigenstates of a D-dimensional sparse matrix. Hence using KPM for the CPT cluster diagonalization the numerical effort can be reduced by a factor of $1/L$ in comparison to the Lanczos recursion.

As an example we consider the 1D Hubbard model

$$H = -t \sum_{i,\sigma} (c_{i,\sigma}^\dagger c_{i+1,\sigma} + \text{H.c.}) + U \sum_i n_{i\uparrow} n_{i\downarrow} , \tag{19.82}$$

which is exactly solvable by Bethe ansatz [50] and was also extensively studied with DDMRG [51]. It thus provides the opportunity to assess the precision of the KPM-based CPT. The top left panel of Fig. 19.8 shows the one-particle spectral function at half-filling, calculated on the basis of $L = 16$ site clusters and an expansion order of $N = 2048$. The matrix dimension is $D \approx 1.7 \cdot 10^8$. Remember that the cluster Green function is calculated for a chain with open boundary conditions. The reduced symmetry compared to periodic boundary conditions results in a larger dimension of the Hilbert space that has to be dealt with numerically.

In the top right panel the dots show the Bethe ansatz results for a $L = 64$ site chain, and the lines denote the $L \to \infty$ spinon and holon excitations each electron separates into (spin-charge separation). So far the Bethe ansatz does not allow for a direct calculation of the structure factor, the data thus represents only the position and density of the eigenstates, but is not weighted with the matrix elements of the operators $c_{k\sigma}^{(\dagger)}$. Although for an infinite system we would expect a continuous response, the CPT data shows some faint fine-structure. A comparison with the finite-size Bethe ansatz data suggests that these features are an artifact of the finite-cluster Greens function which the CPT spectral function is based on. The fine-structure is also evident in the lower panel of Fig. 19.8, where we compare with DDMRG data for a $L = 128$ site system. Otherwise the CPT nicely reproduces all expected features, like the excitation gap, the two pronounced spinon and holon branches, and the broad continuum. Note also, that CPT is applicable to all spatial dimensions, whereas DDMRG works well only for 1D models.

19.3.2 Chebyshev Expansion and Maximum Entropy

Having demonstrated the wide applicability of KPM, let us now discuss some direct competitors of KPM, i.e., methods that share the broad application range and some of its general concepts.

The first of these approaches, the combination of Chebyshev expansion and Maximum Entropy Method (MEM), is basically an alternative procedure to transform moment data μ_n into convergent approximations of the considered function

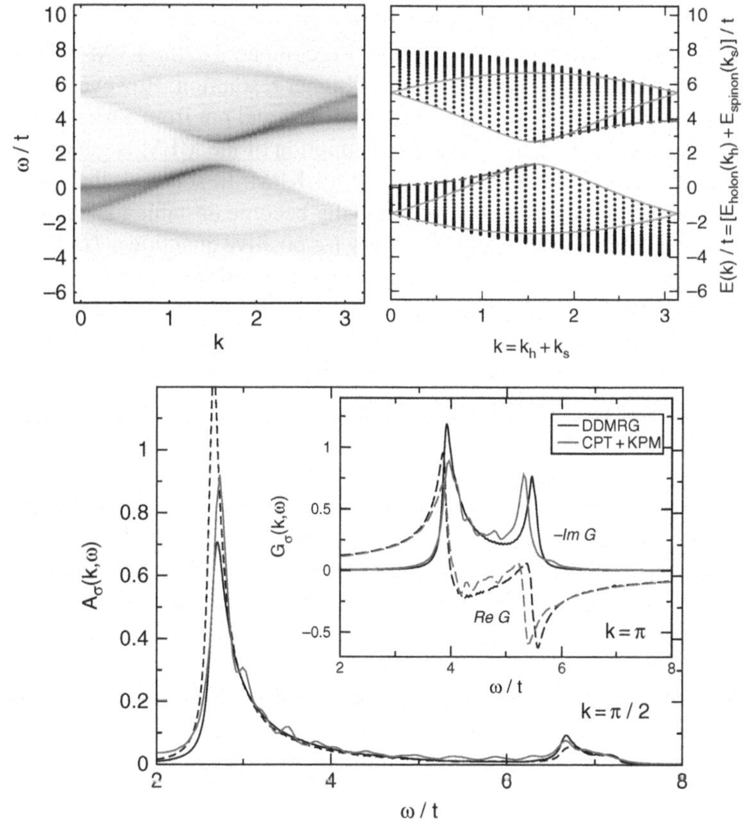

Fig. 19.8. Spectral function of the 1D Hubbard model for half-filling and $U = 4t$. **Top left**: CPT result with cluster size $L = 16$ and expansion order $N = 2048$. For similar data based on Lanczos recursion see [47]. **Top right**: Within the exact Bethe ansatz solution each electron separates into the sum of independent spinon (red dashed) and holon (green) excitations. The dots mark the energies of a 64-site chain. **Bottom**: CPT data compared to selected DDMRG results for a system with $L = 128$ sites, open boundary conditions and a broadening of $\epsilon = 0.0625t$. Note that in DDMRG the momenta are approximate

$f(x)$. To achieve this, instead of (or in addition to) applying kernel polynomials, an entropy

$$S(f, f_0) = \int_{-1}^{1} \left[f(x) - f_0(x) - \log\left(\frac{f(x)}{f_0(x)}\right) \right] \mathrm{d}x \tag{19.83}$$

is maximized under the constraint that the moments of the estimated $f(x)$ agree with the given data. The function $f_0(x)$ describes our initial knowledge about $f(x)$, and may in the worst case just be a constant. Being related to Maximum Entropy approaches to the classical moment problem [52, 53], for the case of Chebyshev moments different implementations of MEM have been suggested [9, 54, 55]. Since for a given set of N moments μ_n the approximation to the function $f(x)$ is usually not restricted to a polynomial of degree $N-1$, compared to the KPM with Jackson kernel the MEM usually yields estimates of higher resolution. However, this higher resolution results from adding a priori assumptions and not from a true information gain (see also Fig. 19.9). The resource consumption of the MEM is generally much higher than the $N \log N$ behavior we found for KPM. In addition, the approach is non-linear in the moments and can occasionally become unstable for large N. Note also that as yet MEM have been derived only for positive quantities, $f(x) > 0$, such as densities of states or strictly positive correlation functions.

MEM, nevertheless, is a good alternative to KPM, if the calculation of the μ_n is particularly time consuming. Based on only a moderate number of moments it yields very detailed approximations of $f(x)$, and we obtained very good results for some computationally demanding problems [56].

19.3.3 Lanczos Recursion

The Lanczos recursion technique [57] is certainly the most capable competitor of KPM. The use of the Lanczos algorithm [8, 58] for the characterization of spectral densities [59, 60] was first proposed at about the same time as the Chebyshev expansion approaches, and in principle Lanczos recursion is also a kind of modified moment expansion [61, 62]. Its generalization from spectral densities to zero-temperature dynamical correlation functions was first given in terms of continued fractions [63], and later also an approach based on the eigenstates of the tridiagonal matrix was introduced and termed Spectral Decoding Method [64]. This technique

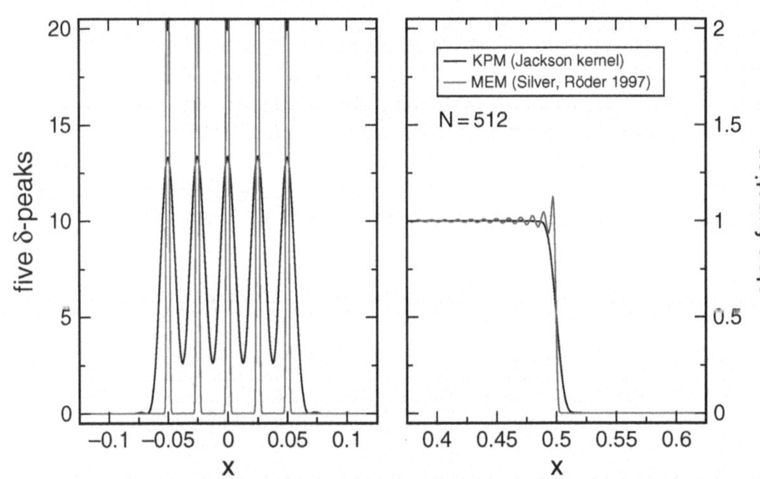

Fig. 19.9. Comparison of a KPM and a MEM approximation to a spectrum consisting of five isolated δ-peaks, and to a step function. The expansion order is $N = 512$. Clearly, for the δ-peaks MEM yields a higher resolution, but for the step function the Gibbs oscillations return (algorithm of [54])

was then generalized to finite temperature [65, 66], and, in addition, some variants of the approach for low temperature [67] and based on the micro-canonical ensemble [68] have been proposed recently.

To give an impression, in Table 19.1 we compare the setup for the calculation of a zero-temperature dynamical correlation function within the Chebyshev and the Lanczos approach. The most time consuming step for both methods is the recursive construction of a set of vectors $|\phi_n\rangle$, which in terms of scalar products yield the moments μ_n of the Chebyshev series or the elements α_n, β_n of the Lanczos tridiagonal matrix. In terms of the number of operations the Chebyshev recursion has a

Table 19.1. Comparison of Chebyshev expansion and Lanczos recursion for the calculation of a zero-temperature dynamical correlation function $f(\omega) = \sum_n |\langle n|A|0\rangle|^2 \delta(\omega - \omega_n)$. We assume N MVMs with a D-dimensional sparse matrix H, and a reconstruction of $f(\omega)$ at M points ω_i

Chebyshev / KPM	Lanczos recursion
Initialization:	Initialization:
$\widetilde{H} = (H - b)/a$ $\lvert\phi_0\rangle = A\lvert 0\rangle, \quad \lvert\phi_1\rangle = \widetilde{H}\lvert\phi_0\rangle$ $\mu_0 = \langle\phi_0\lvert\phi_0\rangle, \quad \mu_1 = \langle\phi_1\lvert\phi_0\rangle$	$\beta_0 = \sqrt{\langle 0\lvert A^\dagger A\rvert 0\rangle}$ $\lvert\phi_0\rangle = A\lvert 0\rangle/\beta_0, \quad \lvert\phi_{-1}\rangle = 0$
$O(ND)$	$O(ND)$
Recursion for $2N$ moments μ_n:	Recursion for N coefficients α_n, β_n:
$\lvert\phi_{n+1}\rangle = 2\widetilde{H}\lvert\phi_n\rangle - \lvert\phi_{n-1}\rangle$ $\mu_{2n+2} = 2\langle\phi_{n+1}\lvert\phi_{n+1}\rangle - \mu_0$ $\mu_{2n+1} = 2\langle\phi_{n+1}\lvert\phi_n\rangle - \mu_1$	$\lvert\phi'\rangle = H\lvert\phi_n\rangle - \beta_n\lvert\phi_{n-1}\rangle, \ \alpha_n = \langle\phi_n\lvert\phi'\rangle$ $\lvert\phi''\rangle = \lvert\phi'\rangle - \alpha_n\lvert\phi_n\rangle, \ \beta_{n+1} = \sqrt{\langle\phi''\lvert\phi''\rangle}$ $\lvert\phi_{n+1}\rangle = \lvert\phi''\rangle/\beta_{n+1}$
→ very stable	→ tends to lose orthogonality
$O(M \log M)$	$O(NM)$
Reconstruction in three simple steps:	Reconstruction via continued fraction:
Apply kernel: $\widetilde{\mu}_n = g_n\mu_n$ Fourier transform: $\widetilde{\mu}_n \to \widetilde{f}(\widetilde{\omega}_i)$ Rescale: $f(\omega_i) = \dfrac{\widetilde{f}[(\omega_i - b)/a]}{\pi\sqrt{a^2 - (\omega_i - b)^2}}$	$f(z) = -\dfrac{1}{\pi}\,\mathrm{Im}\,\dfrac{\beta_0^2}{z - \alpha_0 - \dfrac{\beta_1^2}{z - \alpha_1 - \dots}}$ where $z = \omega_i + i\epsilon$
→ procedure is linear in μ_n → well defined resolution $\propto 1/N$	→ procedure is non-linear in α_n, β_n → ϵ is somewhat arbitrary

small advantage, but, of course, the application of the Hamiltonian as the dominant factor is the same for both methods. As a drawback, at high expansion order the Lanczos iteration tends to lose the orthogonality between the vectors $|\phi_n\rangle$, which it intends to establish by construction. When the Lanczos algorithm is applied to eigenvalue problems this loss of orthogonality usually signals the convergence of extremal eigenstates, and the algorithm then starts to generate artificial copies of the converged states (see Fig. 18.5). For the calculation of spectral densities or correlation functions this means that the information content of the α_n and β_n does no longer increase proportionally to the number of iterations. Unfortunately, this deficiency can only be cured with more complex variants of the algorithm, which also increase the resource consumption. Chebyshev expansion is free from such defects, as there is a priori no orthogonality between the $|\phi_n\rangle$.

The reconstruction of the considered function from its moments μ_n or coefficients α_n, β_n, respectively, is also faster and simpler within the KPM, as it makes use of FFT. In addition, the KPM is a linear transformation of the moments μ_n, a property we used extensively above when averaging moment data instead of the corresponding functions. Continued fractions, in contrast, are non-linear in the coefficients α_n, β_n. A further advantage of KPM is our good understanding of its convergence and resolution as a function of the expansion order N. For the Lanczos algorithm these issues have not been worked out with the same rigor.

In Fig. 19.10 we compare KPM and Lanczos recursion, calculating the spectral function $-\pi^{-1}\,\mathrm{Im}\langle 0|c_{0\uparrow}(\omega - H)^{-1}c_{0\uparrow}^{\dagger}|0\rangle$ for the Hubbard model on a $L = 12$ site ring and half-filling. With the Jackson kernel all features of the dynamical correlation function are resolved sharply, whereas with Lanczos recursion, by construction, we observe Lorentzian broadening. The Lanczos recursion data therefore is

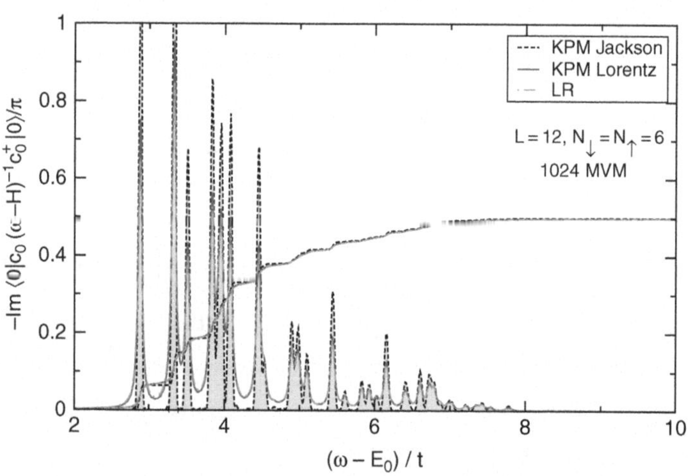

Fig. 19.10. The spectral function $-\pi^{-1}\,\mathrm{Im}\langle 0|c_{0\uparrow}(\omega - H)^{-1}c_{0\uparrow}^{\dagger}|0\rangle$ calculated for the Hubbard model with $L = 12$, $N_\downarrow = N_\uparrow = 6$ using KPM and Lanczos recursion (LR). Lanczos recursion closely matches KPM with Lorentz kernel

comparable to KPM with Lorentz kernel, except that the calculation takes a little bit longer (about 10% in this simple case). Note also, that within KPM the calculation of non-diagonal correlation functions, like $\langle 0|c_i(\omega - H)^{-1}c_j^\dagger|0\rangle$ with $i \neq j$, is much easier – see our discussion in Sect. 19.3.1.

In conclusion, we think that the Lanczos algorithm is an excellent tool for the calculation of extremal eigenstates of large sparse matrices, but for spectral densities and correlation functions the KPM (MEM) is the better choice. Of course, the advantages of both algorithms can be combined, e.g. when the Chebyshev expansion starts from an exact eigenstate that was calculated with the Lanczos algorithm.

Acknowledgements

We would like to thank A. Alvermann, B. Bäuml, G. Hager, M. Hohenadler, E. Jeckelmann, M. Kinateder, G. Schubert, R.N. Silver, and G. Wellein for valuable discussions and technical support. This work was supported by Deutsche Forschungsgemeinschaft through SFB TR24 and SFB 512. Furthermore, we acknowledge generous computer granting by John von Neumann-Institut für Computing Jülich (NIC), Leibniz-Rechenzentrum München (LRZ) and Norddeutscher Verbund für Hoch- und Höchstleistungsrechnen (HLRN).

References

1. A. Weiße, G. Wellein, A. Alvermann, H. Fehske, Rev. Mod. Phys. **78**, 275 (2006)
2. R.N. Silver, H. Röder, Int. J. Mod. Phys. C **5**, 935 (1994)
3. J.P. Boyd, *Chebyshev and Fourier Spectral Methods*. No. 49 in Lecture Notes in Engineering (Springer-Verlag, Berlin, 1989)
4. M. Abramowitz, I.A. Stegun (eds.), *Handbook of Mathematical Functions with formulas, graphs, and mathematical tables* (Dover, New York, 1970)
5. T.J. Rivlin, *Chebyshev polynomials: From Approximation Theory to Algebra and Number Theory*, 2nd edn. Pure and Applied Mathematics (John Wiley & Sons, New York, 1990)
6. E.W. Cheney, *Introduction to Approximation Theory* (McGraw-Hill, New York, 1966)
7. G.G. Lorentz, *Approximation of Functions* (Holt, Rinehart and Winston, New York, 1966)
8. C. Lanczos, J. Res. Nat. Bur. Stand. **45**, 255 (1950)
9. J. Skilling, in *Maximum Entropy and Bayesian Methods*, ed. by J. Skilling (Kluwer, Dordrecht, 1988), Fundamental Theories of Physics, pp. 455–466
10. D.A. Drabold, O.F. Sankey, Phys. Rev. Lett. **70**, 3631 (1993)
11. L. Fejér, Math. Ann. **58**, 51 (1904)
12. D. Jackson, Über die Genauigkeit der Annäherung stetiger Funktionen durch ganze rationale Funktionen gegebenen Grades und trigonometrische Summen gegebener Ordnung. Ph.D. thesis, Georg-August-Universität Göttingen (1911)
13. D. Jackson, T. Am. Math. Soc. **13**, 491 (1912)

14. M. Frigo, S.G. Johnson, Proceedings of the IEEE **93**(2), 216 (2005). Special issue on "Program Generation, Optimization, and Platform Adaptation"
15. M. Frigo, S.G. Johnson. FFTW fast fourier transform library. URL `http://www.fftw.org/`
16. J.C. Wheeler, Phys. Rev. A **9**, 825 (1974)
17. R.N. Silver, H. Röder, A.F. Voter, D.J. Kress, J. Comput. Phys. **124**, 115 (1996)
18. P.W. Anderson, Phys. Rev. **109**, 1492 (1958)
19. R. Abou-Chacra, D.J. Thouless, P.W. Anderson, J. Phys. C Solid State **6**, 1734 (1973)
20. R. Haydock, R.L. Te, Phys. Rev. B **49**, 10845 (1994)
21. V. Dobrosavljević, A.A. Pastor, B.K. Nikolić, Europhys. Lett. **62**, 76 (2003)
22. C.M. Soukoulis, Q. Li, G.S. Grest, Phys. Rev. B **45**, 7724 (1992)
23. S. Kirkpatrick, T.P. Eggarter, Phys. Rev. B **6**, 3598 (1972)
24. R. Berkovits, Y. Avishai, Phys. Rev. B **53**, R16125 (1996)
25. G. Schubert, A. Weiße, H. Fehske, Phys. Rev. B **71**, 045126 (2005)
26. K. Fabricius, B.M. McCoy, Phys. Rev. B **59**, 381 (1999)
27. C. Schindelin, H. Fehske, H. Büttner, D. Ihle, Phys. Rev. B **62**, 12141 (2000)
28. H. Fehske, C. Schindelin, A. Weiße, H. Büttner, D. Ihle, Braz. J. Phys. **30**, 720 (2000)
29. A. Weiße, H. Fehske, Phys. Rev. B **58**, 13526 (1998)
30. M. Hohenadler, G. Wellein, A.R. Bishop, A. Alvermann, H. Fehske, Phys. Rev. B **73**, 245120 (2006)
31. H. Fehske, E. Jeckelmann, in *Polarons in Bulk Materials and Systems With Reduced Dimensionality, International School of Physics Enrico Fermi*, Vol. 161, ed. by G. Iadonisi, J. Ranninger, G. De Filippis (IOS Press, Amsterdam, 2006), *International School of Physics Enrico Fermi*, Vol. 161, pp. 297–311
32. E. Jeckelmann, Phys. Rev. B **66**, 045114 (2002)
33. A. Weiße, H. Fehske, G. Wellein, A.R. Bishop, Phys. Rev. B **62**, R747 (2000)
34. E. Jeckelmann, H. Fehske, in *Polarons in Bulk Materials and Systems With Reduced Dimensionality, International School of Physics Enrico Fermi*, Vol. 161, ed. by G. Iadonisi, J. Ranninger, G. De Filippis (IOS Press, Amsterdam, 2006), *International School of Physics Enrico Fermi*, Vol. 161, pp. 247–284
35. D.J. Thouless, Phys. Rep. **13**, 93 (1974)
36. P.A. Lee, T.V. Ramakrishnan, Rev. Mod. Phys. **57**, 287 (1985)
37. B. Kramer, A. Mac Kinnon, Rep. Prog. Phys. **56**, 1469 (1993)
38. A. Weiße, Eur. Phys. J. B **40**, 125 (2004)
39. H. Tal-Ezer, R. Kosloff, J. Chem. Phys. **81**, 3967 (1984)
40. J.B. Wang, T.T. Scholz, Phys. Rev. A **57**, 3554 (1998)
41. V.V. Dobrovitski, H. De Raedt, Phys. Rev. E **67**, 056702 (2003)
42. W.H. Press, S.A. Teukolsky, W.T. Vetterling, B.P. Flannery, *Numerical Recipes in C: The Art of Scientific Computing*, 2nd edn. (Cambridge University Press, Cambridge, 1992)
43. J. Bonča, S.A. Trugman, I. Batistić, Phys. Rev. B **60**, 1633 (1999)
44. S.A. Trugman, L.C. Ku, J. Bonča, J. Supercond. **17**, 193 (2004)
45. H. Fehske and S.A. Trugman in *Polarons in Advanced Materials*, Ed. A.S. Alexandrov, Springer Series in Material Sciences Vol. 103, pp. 393–461 (Canopus/Springer, Dordrecht 2007)
46. C. Gros, R. Valentí, Ann. Phys. (Leipzig) **3**, 460 (1994)
47. D. Sénéchal, D. Perez, M. Pioro-Ladrière, Phys. Rev. Lett. **84**, 522 (2000)
48. D. Sénéchal, D. Perez, D. Plouffe, Phys. Rev. B **66**, 075129 (2002)
49. S.R. White, Phys. Rev. Lett. **69**, 2863 (1992)
50. F.H.L. Essler, H. Frahm, F. Göhmann, A. Klümper, V.E. Korepin, *The One-Dimensional Hubbard Model* (Cambridge University Press, Cambridge, 2005)

51. E. Jeckelmann, F. Gebhard, F.H.L. Essler, Phys. Rev. Lett. **85**, 3910 (2000)
52. L.R. Mead, N. Papanicolaou, J. Math. Phys. **25**, 2404 (1984)
53. I. Turek, J. Phys. C Solid State **21**, 3251 (1988)
54. R.N. Silver, H. Röder, Phys. Rev. E **56**, 4822 (1997)
55. K. Bandyopadhyay, A.K. Bhattacharya, P. Biswas, D.A. Drabold, Phys. Rev. E **71**, 057701 (2005)
56. B. Bäuml, G. Wellein, H. Fehske, Phys. Rev. B **58**, 3663 (1998)
57. E. Dagotto, Rev. Mod. Phys. **66**, 763 (1994)
58. J.K. Cullum, R.A. Willoughby, *Lanczos Algorithms for Large Symmetric Eigenvalue Computations*, Vol. I & II (Birkhäuser, Boston, 1985)
59. R. Haydock, V. Heine, M.J. Kelly, J. Phys. C Solid State **5**, 2845 (1972)
60. R. Haydock, V. Heine, M.J. Kelly, J. Phys. C Solid State **8**, 2591 (1975)
61. P. Lambin, J.P. Gaspard, Phys. Rev. B **26**, 4356 (1982)
62. C. Benoit, E. Royer, G. Poussigue, J. Phys. Cond. Mat. **4**, 3125 (1992)
63. E. Gagliano, C. Balseiro, Phys. Rev. Lett. **59**, 2999 (1987)
64. Q. Zhong, S. Sorella, A. Parola, Phys. Rev. B **49**, 6408 (1994)
65. J. Jaklič, P. Prelovšek, Phys. Rev. B **49**, 5065 (1994)
66. J. Jaklič, P. Prelovšek, Adv. Phys. **49**, 1 (2000)
67. M. Aichhorn, M. Daghofer, H.G. Evertz, W. von der Linden, Phys. Rev. B **67**, 161103 (2003)
68. M.W. Long, P. Prelovšek, S. El Shawish, J. Karadamoglou, X. Zotos, Phys. Rev. B **68**, 235106 (2003)

Part IX

The Density Matrix Renormalisation Group: Concepts and Applications

20 The Conceptual Background of Density-Matrix Renormalization

Ingo Peschel and Viktor Eisler

Fachbereich Physik, Freie Universität Berlin, 14195 Berlin, Germany

In the treatment of many-particle quantum systems, one approach is to work with the wave function and to look for an approximation which is as good as possible. The density-matrix renormalization group method (DMRG) is a numerical procedure which does that by selecting an optimal subspace of the complete Hilbert space in a systematic way. It was developed in the early nineties by Steven White [1, 2] and has since then become the most powerful tool for treating one-dimensional quantum systems [3, 4, 5]. This is due to the fact that it combines spectacular accuracies like ten decimal places for ground-state energies, with the possibility to treat large systems with e.g. hundreds of spins. Recently it has also been extended to time-dependent problems. All this will be described in more detail in the following contributions.

20.1 Introduction

In this introductory chapter, we want to give a general background for the method and discuss some concepts which arise in the characterization and description of quantum states. These are not only relevant for the DMRG but appear also in other contexts and have a basic interest in themselves. Specifically, this will be entangled states, reduced density matrices, entanglement entropies and matrix-product states. The emphasis will be on reduced density matrices and their features. These are crucial for the performance of the DMRG but they also arise naturally if one wants to quantify entanglement properties. The latter have been the topic of many recent studies and we will also give a brief account of that.

20.2 Entangled States

The notion of entanglement (in German "Verschränkung") was introduced by Schrödinger in 1935 [6] and plays a central role in the discussion of fundamental aspects of quantum mechanics [7]. It is usually illustrated with the example of two spins one-half with basis states $|+\rangle$ and $|-\rangle$. The simplest states of the composite system have product form, e.g.

$$|\Psi\rangle = |+\rangle|-\rangle \,, \tag{20.1}$$

I. Peschel and V. Eisler: *The Conceptual Background of Density-Matrix Renormalization*, Lect. Notes Phys. **739**, 581–596 (2008)
DOI 10.1007/978-3-540-74686-7_20

or, in general,

$$|\Psi\rangle = [a\,|+\rangle + b\,|-\rangle][c\,|+\rangle + d\,|-\rangle]\,. \tag{20.2}$$

In this case the two spins are independent of each other and all expectation values factorize. By contrast, an entangled state is

$$|\Psi\rangle = \frac{1}{\sqrt{2}}\,[|+\rangle|+\rangle + |-\rangle|-\rangle]\,. \tag{20.3}$$

This cannot be written in product form and expectation values do not factorize. The parts of the composite system are interwoven in the wave function. This is typical for interacting systems and is the situation one normally encounters, and has to deal with, in many-particle problems.

In the two-spin case it is relatively easy to check whether a state has product form or not. In the general case, one proceeds as follows. One divides the system into two parts 1 and 2.

1	2

Then a state $|\Psi\rangle$ of the total system can be written

$$|\Psi\rangle = \sum_{m,n} A_{mn}|\Psi_m^1\rangle|\Psi_n^2\rangle\,, \tag{20.4}$$

where $|\Psi_m^1\rangle$ and $|\Psi_n^2\rangle$ are orthonormal basis functions in the two Hilbert spaces. But a rectangular matrix $\mathbf{A}$ can always be written in the form $\mathbf{UDV}'$ where $\mathbf{U}$ is unitary, $\mathbf{D}$ is diagonal and the rows of $\mathbf{V}$ are orthonormal. This is called the singular-value decomposition and similar to the principal-axis transformation of a symmetric square matrix [8]. Using this in (20.4) and forming new bases by combining the $|\Psi_m^1\rangle$ with $\mathbf{U}$ and the $|\Psi_n^2\rangle$ with $\mathbf{V}'$, one obtains the Schmidt decomposition [9]

$$|\Psi\rangle = \sum_{n} \lambda_n\,|\Phi_n^1\rangle|\Phi_n^2\rangle \tag{20.5}$$

which gives the total wave function as a single sum of products of orthonormal functions. Here the number of terms is limited by the smaller of the two Hilbert spaces and the weight factors λ_n are the elements of the diagonal matrix $\mathbf{D}$. If $|\Psi\rangle$ is normalized, their absolute magnitudes squared sum to one. The entanglement properties are encoded in the set of λ_n. Only if all except one are zero, the sum reduces to a single term and $|\Psi\rangle$ is a product state. On the other hand, if all λ_n are equal in size, one would call the state maximally entangled. Of course, this refers to a particular bipartition and one should investigate different partitions to obtain a complete picture. One could also ask for the entanglement of more than two parts but it turns out that there is no general extension of the Schmidt decomposition.

20.3 Reduced Density Matrices

The entanglement structure just discussed can also be found from the density matrices associated with the state $|\Psi\rangle$. This is, in fact, the standard way to obtain it.

Starting from the total density matrix

$$\rho = |\Psi\rangle\langle\Psi| \;, \tag{20.6}$$

one can, for a chosen division, take the trace over the degrees of freedom in one part of the system. This gives the reduced density matrix for the other part, i.e.

$$\rho_1 = \mathrm{Tr}_2(\rho) \;\;,\;\; \rho_2 = \mathrm{Tr}_1(\rho) \;. \tag{20.7}$$

These Hermitian operators can be used to calculate arbitrary expectation values in the subsystems, but this is not all. From (20.5) it follows that their diagonal forms are

$$\rho_\alpha = \sum_n |\lambda_n|^2 \; |\Phi_n^\alpha\rangle\langle\Phi_n^\alpha| \;\;,\;\; \alpha = 1, 2 \;. \tag{20.8}$$

This means that

- ρ_1 and ρ_2 have the same non-zero eigenvalues,
- these eigenvalues are given by $w_n = |\lambda_n|^2$.

Therefore the eigenvalue spectrum of the ρ_α gives directly the weights in the Schmidt decomposition and a glance at this spectrum shows the basic entanglement features of the state, for the chosen bipartition. One also sees that the $|\Phi_n^\alpha\rangle$ appearing in (20.5) are the eigenfunctions of ρ_α.

In the DMRG algorithm, these properties are used to truncate the Hilbert space by calculating the ρ_α, selecting the m states $|\Phi_n^\alpha\rangle$ with largest weights w_n and deleting the rest. This procedure is expected to work well if the total weight of the discarded states is sufficiently small. Therefore the form of the density-matrix spectra is decisive for the success of the method and will be discussed in the following.

20.4 Solvable Models

The reduced density matrices can be determined for the ground states of a number of standard systems. These are integrable spin chains like the XY model and the XXZ model, free bosons like coupled oscillators and free fermions like hopping models. In all these cases the reduced density matrices are found to have the form

$$\rho_\alpha = K \exp\Big(-\sum_l \varepsilon_l \, c_l^\dagger \, c_l\Big) = K\,\mathrm{e}^{-H} \;, \tag{20.9}$$

where, depending on the problem, the $c_l^\dagger, c_l$ are fermionic or bosonic creation and annihilation operators and the ε_l are the corresponding single-particle eigenvalues. Before we discuss (20.9) further, let us describe briefly how one can derive this result. Basically, there are three methods to obtain the ρ_α.

(1) Integration over part of the variables according to the definition. This can be done e.g. for coupled harmonic oscillators [10, 11]. In this case the ground state is a Gaussian in the normal coordinates and has the general form

$$\Phi(u_1, u_2, \ldots, u_N) = C \exp\Big(-\frac{1}{2} \sum_{m,n}^{N} B_{m,n}\, u_m\, u_n \Big) , \tag{20.10}$$

in terms of the original coordinates u_n of the N oscillators, where C is a normalization constant. By forming ρ and integrating out e.g. the variables $u_{M+1}, \ldots, u_N$ one obtains $\rho_1(u_1, u_2, \ldots, u_M \,|\, u'_1, u'_2, \ldots, u'_M)$ which is again a Gaussian. As it stands, this is an integral operator but one can convert the terms involving $(u_n - u'_n)^2$ into derivatives $\partial^2/\partial u_n^2$ and thereby obtain a differential operator in the exponent. This leads to a quadratic expression in terms of boson operators and gives (20.9) after diagonalization. The single-particle eigenvalues follow from a combination of submatrices of $\mathbf{B}$. The method can also be used for systems of non-interacting fermions. In this case one first has to write the ground state in exponential form and then use Grassmann variables for the integration [12].

(2) Via correlation functions [13]. Consider a system of free electrons hopping on a lattice in a state described by a Slater determinant. In such a state, all many-particle correlation functions factorize into products of one-particle functions. For example,

$$\langle c_m^\dagger c_n^\dagger c_k c_l \rangle = \langle c_m^\dagger c_l \rangle \langle c_n^\dagger c_k \rangle - \langle c_m^\dagger c_k \rangle \langle c_n^\dagger c_l \rangle \, . \tag{20.11}$$

If all sites are in the same subsystem, a calculation using the reduced density matrix must give the same result. This is guaranteed by Wick's theorem if ρ_α is the exponential of a free-fermion operator

$$\rho_\alpha = K \exp\Big(-\sum_{i,j} H_{ij}\, c_i^\dagger\, c_j \Big) \, . \tag{20.12}$$

The matrix H_{ij}, where i and j are sites in the subsystem, is determined by the one-particle correlation function $C_{ij} = \langle c_i^\dagger c_j \rangle$ via

$$\mathbf{H} = \ln \left[\frac{\mathbf{1} - \mathbf{C}}{\mathbf{C}} \right] . \tag{20.13}$$

The method has been used in various fermionic problems [14, 15, 16, 17, 18, 19, 20, 21, 22]. If there is pair creation and annihilation, one has to include the anomalous correlation functions $\langle c_i^\dagger c_j^\dagger \rangle$ and $\langle c_i c_j \rangle$. The approach works for arbitrary dimensions and also for bosonic systems [22, 23, 24].

(3) Via the connection to two-dimensional classical models. Consider a quantum chain of finite length and imagine that one can obtain its state $|\Psi\rangle$ from an initial state $|\Psi_s\rangle$ by applying a proper operator $\mathbf{T}$ many times. If $\mathbf{T}$ is the row-to-row transfer matrix of a classical model, one has thereby related $|\Psi\rangle$ to the partition function of a two-dimensional semi-infinite strip of that system. The total

density matrix $|\Psi\rangle\langle\Psi|$ is then given by two such strips. This is sketched on the left of Fig. 20.1. The reduced density matrix, e.g. for the left part of the chain, follows by identifying the variables along the right part of the horizontal edges and summing them, which means mending the two half-strips together. In this way, ρ_α is expressed as the partition function of a full strip with a perpendicular cut, as shown on the right of Fig. 20.1.

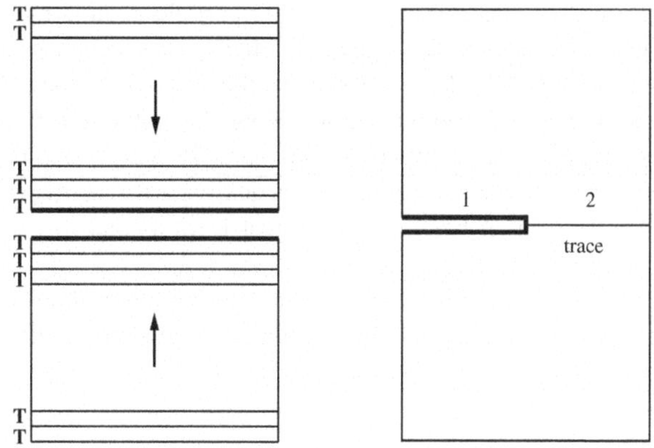

Fig. 20.1. Density matrices for a quantum chain as two-dimensional partition functions. **Left**: Expression for ρ. **Right**: Expression for ρ_1. The matrices are defined by the variables along the thick lines

This approach works for the ground state of a number of integrable quantum chains [11, 25, 26]. For example, the Ising chain in a transverse field can in this way be related to a two-dimensional Ising model where the lattice is rotated by 45° with respect to the horizontal. However, to actually calculate such a partition function and thus ρ_α, one needs a further ingredient, namely the corner transfer matrices introduced by Baxter [27]. These are partition functions for a whole quadrant as shown in Fig. 20.2. For some non-critical integrable models, they are known in the thermodynamic limit and have exponential form. By multiplying four of them as in

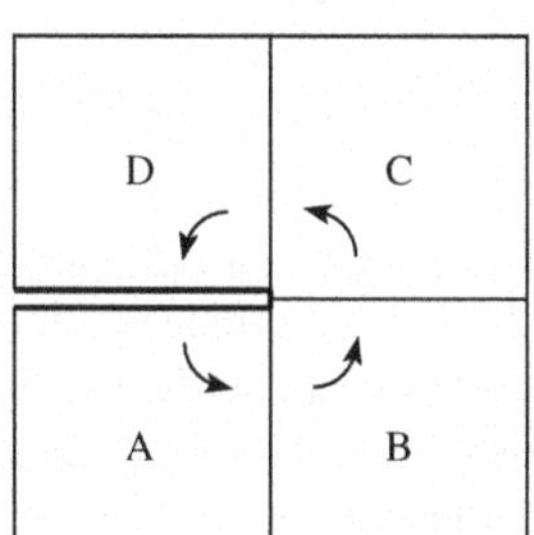

Fig. 20.2. Two-dimensional system built from four quadrants with corresponding corner transfer matrices $\mathbf{A}, \mathbf{B}, \mathbf{C}, \mathbf{D}$. The arrows indicate the direction of transfer

the figure one can obtain the reduced density matrix for a half-chain which is much longer than the correlation length.

For a continuum model, the representation just described can be viewed as a path-integral picture. This can be utilized in particular if the two-dimensional system is critical and conformally invariant [28, 29].

Returning to (20.9), one sees that ρ_α has a thermal form with some effective free-particle Hamiltonian H appearing in the exponent. The eigenstates $|\Phi_n^\alpha\rangle$ and their eigenvalues w_n are then specified by the single-particle occupation numbers and the values of the ε_l. The latter can be given explicitly in a few cases but otherwise have to be found numerically. Degeneracies in the w_n will occur either if one of the ε_l is zero or if they are commensurate. Note that although the ρ_α look like thermal density operators, no temperature appears. However, one can ascribe an effective temperature to the subsystem if one is dealing with a critical model where the low-lying spectrum of H has the same linear form as that of the Hamiltonian itself [30, 31].

For completeness, we mention that ρ_α can also be determined for some other states with high symmetry [32] and for a number of systems with infinite-range interactions [33].

20.5 Spectra

The free-particle models discussed above can be used to calculate the density-matrix spectra and to show their typical features. It turns out that there are differences between critical and non-critical systems and also between one and two dimensions. We will present results for two particular models in their ground states. One is the Ising chain in a transverse field with Hamiltonian

$$H = -\sum_n \sigma_n^z - \lambda \sum_n \sigma_n^x \sigma_{n+1}^x \,, \tag{20.14}$$

which has a non-degenerate ground state without long-range order for $\lambda < 1$, a two-fold degenerate one for $\lambda > 1$ and a quantum critical point at $\lambda = 1$. It can be viewed also as a fermionic model with pair creation and annihilation terms. The other one is a fermionic hopping model which in one dimension has the Hamiltonian

$$H = -\sum_n t_n (c_n^\dagger c_{n+1} + c_{n+1}^\dagger c_n) \,. \tag{20.15}$$

The homogeneous system with $t_n = 1$ is a critical model where the ground-state correlations decay algebraically. If t_n alternates between $1+\delta$ and $1-\delta$ one has a dimerized chain with finite correlation length. The homogeneous model will also be considered in two dimensions.

Figure 20.3 shows the spectra for a transverse Ising chain with open ends which is divided in the middle. On the left, the single-particle eigenvalues ε_l are plotted for three values of λ. They show the following features:

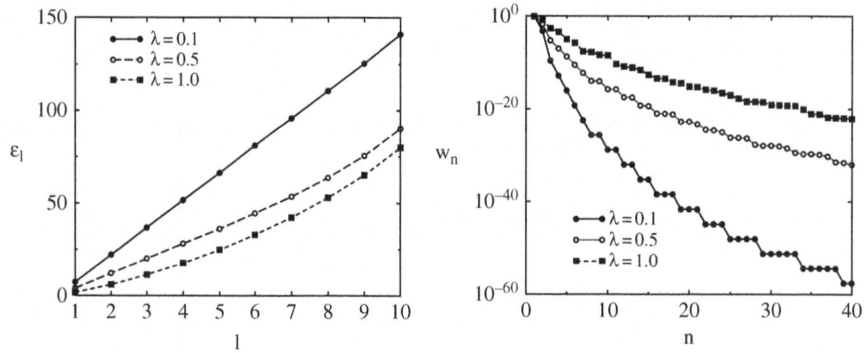

Fig. 20.3. Density-matrix spectra for one-half of a transverse Ising chain with $N = 20$ sites in its ground state. **Left**: Single-particle eigenvalues ε_l. **Right**: Total eigenvalues w_n. After [12]

- If the system is non-critical, the dispersion is linear for the lowest ε_l, i.e. they are equally spaced;
- The spacing becomes smaller and the linear region shrinks as one approaches the critical point;
- At the critical point, the linear region of the dispersion curve is no longer visible.

The equidistance of the levels becomes exact in the limit of an infinite system where it follows from the corresponding corner transfer matrix spectrum. The explicit formula in this case is, for $\lambda < 1$

$$\varepsilon_l = \varepsilon\,(2l - 1)\,, \quad l = 1, 2, 3 \ldots\,, \tag{20.16}$$

where $\varepsilon = \pi\ I(k')/I(k)$. Here $I(k)$ denotes the complete elliptic integral of the first kind, $k = \lambda$ and $k' = \sqrt{1 - k^2}$ [25]. The deviations from the linear law are therefore finite-size effects which, for fixed system size, increase near the critical point.

The eigenvalues w_n of ρ_1 which follow from the single-particle spectrum, are displayed in the right part of Fig. 20.3. One sees an extremely rapid decrease (please note the vertical scale), because the ε_l appearing in the exponent are all rather large. This is a typical property of non-critical quantum chains. For the equidistant levels (20.16) one can also determine the asymptotic form of the w_n [34]. The decay becomes slower near the critical point, but is still impressive even for $\lambda = 1$.

A closer look at critical systems, however, shows an important difference. The spectra then depend on the size of the subsystem in an essential way. Specifically, the single-particle dispersion becomes flatter and flatter as the size increases, and correspondingly also the w_n-curves become flatter. This is shown in Fig. 20.4 for a segment of L sites in an infinite homogeneous hopping model. For very large L, the ε_l are in this case predicted to have again a linear dispersion as in (20.16)

$$\varepsilon_l = \pm\,\frac{\pi^2}{2\ln L}(2l - 1)\,, \quad l = 1, 2, 3 \ldots\,, \tag{20.17}$$

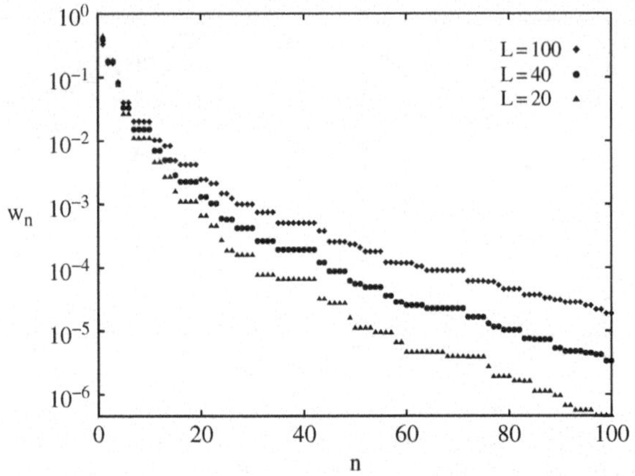

Fig. 20.4. Size dependence of the density-matrix spectrum in a critical system. Shown are the largest w_n for segments of different length in an infinite hopping model

which is also a conformal result [15]. Although in practical numerical calculations one always finds some curvature in the dispersion, the weak L-dependence indicated by (20.17) is what one sees in the figure. Thus for systems of conventional size ($L \sim 100$) the w_n-spectra still decay rather rapidly.

It is also interesting to look at the single-particle eigenfunctions ϕ_l associated with the ε_l. The ones for the lowest positive ε_l are shown in Fig. 20.5 for a segment of an infinite hopping model. One sees that they are concentrated near the two boundaries, i.e. near the interfaces with the rest of the chain. The difference is that in the critical case, shown on the left, the amplitudes decay slowly into the interior (actually with a power $x = -1/2$), while in the non-critical dimerized system, shown on the right, the decay is exponential and reflects the finite correlation length. The concentration near the boundary is typical for all low-lying single-particle states. In non-critical chains, it has an interesting consequence for the spectrum because if the

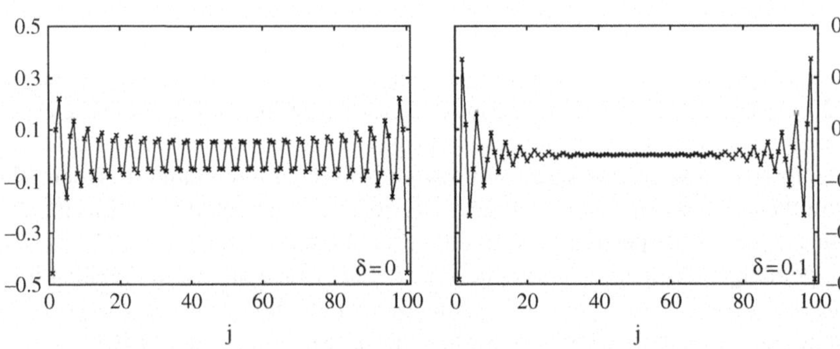

Fig. 20.5. Lowest lying single-particle eigenstates in a simple ($\delta = 0$) and a dimerized ($\delta = 0.1$) hopping model for a segment of $L = 100$ sites

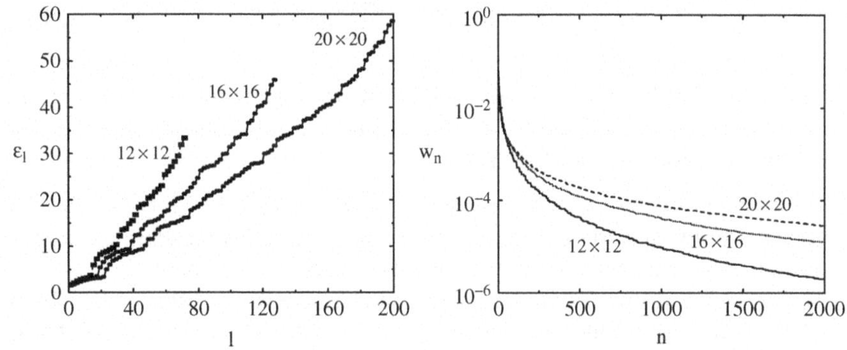

Fig. 20.6. Density-matrix spectra for halves of $N \times N$ hopping models in their ground states. The sizes are indicated in the figures. **Left**: Single-particle eigenvalues ε_l. **Right**: Total eigenvalues w_n. After [12]

subsystem has two boundaries with the rest, one also finds two such states which are practically degenerate and only differ in their reflection symmetry. As remarked above, this leads to degeneracies in the w_n and thus to a considerably slower decay of the spectrum than if one has only one boundary. Such a feature was noted early on when comparing DMRG calculations for open chains and rings. It also gives an indication to what happens in two dimensions.

Spectra for homogeneous two-dimensional hopping models in the form of $N \times N$ squares which are divided into two halves of size $N \times N/2$ are shown in Fig. 20.6. The lowest ε_l now have a kind of band structure with about N states in the lowest band. These can be associated with the interface. The picture would be even clearer if one considered a non-critical system where these states are more localized. This band structure has drastic consequences for the w_n, as seen on the right. After an initial decay, the spectrum flattens extremely, because the corresponding w_n can be generated by a large number of different single-particle combinations. This indicates that a DMRG calculation will not be successful in this case. Due to the long interface one has a much higher entanglement in the wave function than in one dimension. This feature will be discussed again in the next section in a somewhat different way.

20.6 Entanglement Entropy

In the previous section several examples have been given on how entangled states of bipartite systems can be fully characterized by means of reduced density matrices. However, in general this involves a large number of parameters and it would be useful to have a simple measure that allows for an easy quantification and comparison of entanglement content. This can be achieved by a generalization of the usual entropy definition to reduced density matrices. The entanglement (also known as von Neumann) entropy therefore reads

$$S_1 = -\mathrm{Tr}(\rho_1 \ln \rho_1) = -\sum_n w_n \ln w_n \,, \tag{20.18}$$

where the trace has been rewritten as a sum using the eigenvalues w_n. The entropy is defined in a way that certain basic requirements are automatically fulfilled. The most important properties are as follows:

- The entropy is determined purely by the spectrum of ρ_1, which is known to be identical to the spectrum of ρ_2, therefore $S_1 = S_2$ holds for arbitrary bipartitions, thus giving a measure of the mutual connection of the parts;
- The entropy vanishes for product states, and has a maximal value of $S = \ln M$ when all the eigenvalues are equal, $w_n = 1/M$ for $n = 1, 2, \ldots, M$. Using this one can write in general $S = \ln M_{\mathrm{eff}}$, where M_{eff} is the effective number of coupled states in parts 1 and 2.

Apart from these basic properties, the entanglement entropy shows features which result from the specific underlying density-matrix spectra. Correspondingly, they are different for critical and non-critical systems and depend on the dimensionality. We discuss this again for solvable models.

Consider the case of free fermions or bosons where the reduced density matrix has the exponential form (20.9). Then the entanglement entropy is given by the same expression as in thermodynamics, namely

$$S = \pm \sum_l \ln(1 \pm \mathrm{e}^{-\varepsilon_l}) + \sum_l \frac{\varepsilon_l}{\mathrm{e}^{\varepsilon_l} \pm 1} \,, \tag{20.19}$$

where the upper (lower) sign refers to fermions (bosons), respectively. In one dimension, these sums can be evaluated analytically in terms of elliptic integrals, if the ε_l have a linear dispersion as in (20.16) [15]. In this way, one can obtain S for the non-critical transverse Ising chain, the XY chain or a chain of harmonic oscillators and finds that it is finite and typically of the order one. Thus the corresponding ground states have $M_{\mathrm{eff}} \sim 1 - 10$ and are only weakly entangled as can be seen also from the density-matrix spectra.

The critical case is different, however, since as shown above the spectra then vary with the size of the subsystem. Using the asymptotic form (20.17) for a segment in a hopping model, one can evaluate S for large $\ln L$ by converting the sums into integrals. This gives

$$S = \frac{2 \ln L}{\pi^2} \left[\int_0^\infty \mathrm{d}\varepsilon \, \ln(1 + \mathrm{e}^{-\varepsilon}) + \int_0^\infty \mathrm{d}\varepsilon \, \frac{\varepsilon}{\mathrm{e}^{\varepsilon} + 1} \right] , \tag{20.20}$$

and since both integrals equal $\pi^2/12$ one obtains

$$S = \frac{1}{3} \ln L \,. \tag{20.21}$$

This logarithmic behavior can already be observed in numerical calculations for relatively small systems, where the law (20.17) is not yet strictly obeyed. It has been found for a number of one-dimensional models which indicates that it is a universal feature of the critical state. In fact, the result can be derived from conformal invariance using the path-integral representation of the reduced density matrix [28, 29]. The prefactor of the logarithm is then seen to involve the so-called central charge c which classifies the conformally invariant models. Besides that, it only depends on the number of contact points $\nu = 1, 2$ between the (singly connected) subsystem and the rest of the chain. Thus one has for large L

$$S = \nu \frac{c}{6} \ln L + k \,, \tag{20.22}$$

where k is a non-universal constant depending on the the model parameters and the geometry. Comparing (20.21) and (20.22), one sees that the hopping model corresponds to $c = 1$. The effective number of entangled states in a critical chain therefore increases as a power of the subsystem size

$$M_{\text{eff}} \sim L^{\nu\, c/6} \,. \tag{20.23}$$

These results also show that the entanglement entropy belongs to the quantities displaying critical behavior at a quantum phase transition. This is illustrated in Fig. 20.7 for the dimerized hopping model introduced in (20.15). The entropy is plotted there against the dimerization parameter δ, which measures the distance from the critical point $\delta = 0$. With increasing subsystem size, the curves become more and more peaked, signaling a singularity in the thermodynamic limit. One can also verify that the entropy has the usual finite-size scaling properties [29]. These features were also found in hopping models with an energy current [20].

For higher-dimensional systems, the spectra in Fig. 20.6 give some indication on the behavior of the entropy. The low-lying band of ε_l roughly has the effect of multiplying the contribution of one eigenvalue by the length of the interface. Indeed, there is a long-standing conjecture, called the "area law", which originated in

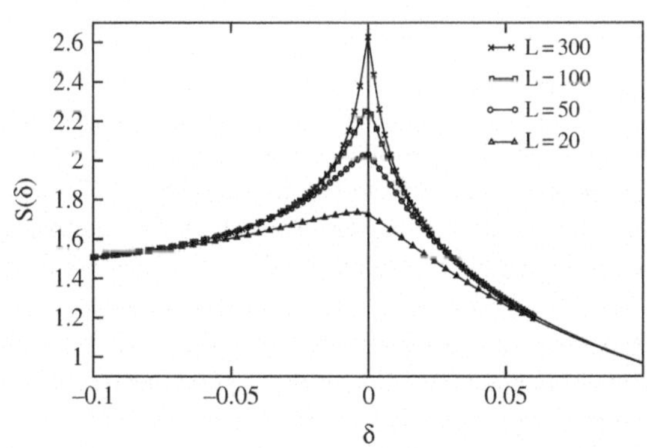

Fig. 20.7. Entanglement entropy for segments of different size L in a one-dimensional hopping model as a function of the dimerization parameter δ. The development of a singularity in case of vanishing dimerization is clearly visible

the context of black-hole physics [35, 36]. It states, that the entropy of an entangled state obtained by tracing out the degrees of freedom inside a given region in space (the black hole) should scale with the surface area of that region (instead of its volume). It was first checked numerically for massless bosonic fields in three spatial dimensions [36] and has recently been proven for non-critical harmonic lattice systems in arbitrary dimensions [24].

The idea of an area law is very plausible given the fact, that the entanglement entropy measures mutual connections in a wave function. However, it is not universally valid. In one dimension, the surface area of a subsystem is just the number of contact points with the rest of the system, which would lead to a constant entropy. This is indeed the case for non-critical systems, but as the results presented above show, not at criticality. It is therefore an intriguing question whether this is only a peculiarity of these one dimensional systems. Several studies in this direction have shown, that in the fermionic case the violation of the area law carries on to higher dimensional critical systems [22, 37, 38], if the Fermi surface is finite [39]. Thus, to leading order the behavior of the entropy for fermionic systems is given by

$$S \sim \begin{cases} L^{d-1} & \text{non-critical case} \\ L^{d-1} \ln L & \text{critical case} \end{cases} \tag{20.24}$$

where L characterizes the linear dimension. By contrast, no logarithmic corrections were found for bosonic systems in the case of a vanishing gap [22]. In terms of M_{eff}, the area dependence leads to $M_{\text{eff}} \sim \exp(L^{d-1})$ and thus to an exponentially large number of coupled states. This is another way of formulating the difficulty treating higher-dimensional systems with the DMRG algorithm.

The entanglement entropy has an interesting history. The first study dates back to 1986 [35] and contained the postulate of the area law in an astrophysical context. The idea was to interpret the black-hole entropy, originally coming from an effective thermodynamic description and proportional to the area of the event horizon, as an entanglement entropy, thus ascribing a quantum mechanical origin to it. Some years later, it was studied within field theory and called "geometric entropy" [28]. Its significance in connection with quantum critical phenomena was noted around 2003, with input from the field of quantum information. This triggered a large number of studies related in particular to its critical behavior. The most recent direction concerns its time dependence after a quench in the system [21, 40].

As noted above, the entanglement entropy explains the much larger effort which is necessary for treating critical or higher dimensional systems with DMRG. However, it has also been used in a constructive way to find an optimal order of the single-particle basis states and an optimal size of the blocks in quantum-chemical applications of the DMRG, where the orbitals play the role of the sites in spin chains [41].

20.7 Matrix-Product States

In this last section we want to discuss a particular class of entangled states which occurs in certain spin systems and also in the DMRG algorithm.

Consider, for example, a chain of N spins one-half. A general product state analogous to (20.2) is then, in a slightly different notation,

$$|\Psi\rangle = \prod_{n=1}^{N} \left[a_n(+)\,|+\rangle_n + a_n(-)\,|-\rangle_n \right] . \tag{20.25}$$

Thus at each site one has two coefficients for the two spin directions. Multiplying out the product, one can write this as

$$|\Psi\rangle = \sum_{\mathbf{s}} c(\mathbf{s})\,|\mathbf{s}\rangle , \tag{20.26}$$

where $\mathbf{s} = (s_1, s_2, \ldots, s_N)$ denotes a configuration of all spins and the coefficient $c(\mathbf{s})$ is the product

$$c(\mathbf{s}) = a_1(s_1)\, a_2(s_2) \ldots a_N(s_N) . \tag{20.27}$$

This can be generalized in the following way. Instead of two numbers $a_n(s_n)$ one associates two *matrices* $\mathbf{A}_n(s_n)$ with each site n. These matrices operate in an auxiliary space. The weight of a configuration is then calculated by forming a product as in (20.27). The result is now a matrix from which one still has to obtain a number. This can be done in two obvious ways. For a ring, one simply takes the trace

$$c(\mathbf{s}) = \mathrm{Tr}\left(\mathbf{A}_1(s_1)\,\mathbf{A}_2(s_2)\ldots\mathbf{A}_N(s_N)\right) , \tag{20.28}$$

whereas, for an open chain, one uses boundary vectors in the auxiliary space

$$c(\mathbf{s}) = \mathbf{u}'\,\mathbf{A}_1(s_1)\,\mathbf{A}_2(s_2)\ldots\mathbf{A}_N(s_N)\,\mathbf{v} . \tag{20.29}$$

The simplest case is a homogeneous state where the matrices are the same for all sites. Such states were first considered in the eighties [42, 43] and occur as ground states of certain spin chains with competing interactions [44]. The best-known example is the spin-one chain with bilinear and biquadratic interactions and a certain ratio of the couplings, where the valence-bond ground state [45] can be written in this form using 2×2 matrices. They also appear in non-equilibrium models describing, for example, the diffusion of hard-core particles between two reservoirs. This case corresponds to (20.29) and, depending on the parameters, the dimension m of the matrices can be finite or infinite [46].

These states have two important properties:

- They have a finite entanglement governed by the dimension of the matrices;
- For homogeneous states, the correlation functions are sums of $m^2 - 1$ exponentials, unless the matrices have special features, and the correlation length is finite.

The first property can be seen very easily. For an open chain which is divided into two parts, there are m connections between the matrix product to the left and to the right of the interface. Thus

$$|\Psi\rangle = \sum_{n=1}^{m} \beta_n \, |\phi_n^1\rangle|\phi_n^2\rangle \,. \tag{20.30}$$

This is not yet the Schmidt decomposition (20.5) because in general the states $|\phi_n^\alpha\rangle$ are not orthogonal. Nevertheless, the number of terms in the Schmidt decomposition is limited by m, the dimension of the matrices. For a ring, where one has two interfaces, it is limited by m^2. Correspondingly, the reduced density matrices have up to m resp. m^2 non-zero eigenvalues. If m is small, this gives the possibility to detect such states by investigating the density-matrix spectrum [47].

The second property excludes in principle the description of critical systems by such a state. However, taking the matrices large enough, one may still obtain a very good approximation for a system of finite size. The question of representing a quantum state in terms of a matrix product has recently been investigated in detail [48]. This was motivated partly by the fact that the DMRG produces its approximate wave function in the form of an (inhomogeneous) matrix product [49], as will be discussed in the next contribution. An alternative to the usual DMRG procedure could then be to start with a matrix-product Ansatz from the beginning and to find the matrices for the ground state by minimizing the energy [50]. This idea can be extended to higher dimensions [51]. For example, in a square lattice the analogue of the matrices would be fourth-order tensors which permit to connect each site to its four neighbors.

20.8 Summary

In this contribution we have discussed quantum states in terms of their entanglement properties. This approach is an alternative to the conventional characterization via correlation functions and the topic of intense current research. It also provides the framework in which the DMRG operates. Some knowledge of it is therefore indispensable for a deeper understanding and an appreciation of the nature of this intriguing numerical method. We have dealt with particular many-body states in order to illustrate basic features of entanglement. The DMRG is also an ideal tool if one wants to study these features for more complicated systems, because the algorithm is based on density-matrix spectra and determines them routinely. However, it has much wider applications as will be described in the following chapters.

References

1. S.R. White, Phys. Rev. Lett. **69**, 2863 (1992)
2. S.R. White, Phys. Rev. B **48**, 10345 (1993)

3. I. Peschel, X. Wang, M. Kaulke, K. Hallberg (eds.), *Density-Matrix Renormalization*. Lecture Notes in Physics 528 (Springer, Berlin, 1999)
4. U. Schollwöck, Rev. Mod. Phys. **77**, 259 (2005)
5. K.A. Hallberg, Adv. Phys. **55**, 477 (2006)
6. E. Schrödinger, Naturwissenschaften **23**, 807 (1935)
7. A. Ekert, P.L. Knight, Am. J. Phys **63**, 415 (1995)
8. R. Horn, C. Johnson, *Topics in Matrix Analysis* (Cambridge University Press, 1991), Chap. 3
9. E. Schmidt, Math. Annalen **63**, 433 (1907)
10. M.-C. Chung, I. Peschel, J. Phys. A: Math. Gen. **32**, 8419 (1999)
11. M.-C. Chung, I. Peschel, Phys. Rev. B **62**, 4191 (2000)
12. M.-C. Chung, I. Peschel, Phys. Rev. B **64**, 064412 (2001)
13. I. Peschel, J. Phys. A: Math. Gen. **36**, L205 (2003)
14. G. Vidal, J.I. Latorre, E. Rico, A. Kitaev, Phys. Rev. Lett. **90**, 227902 (2003)
15. I. Peschel, J. Stat. Mech. P06004 (2004)
16. B.Q. Jin, V.E. Korepin, J. Stat. Phys. **116**, 79 (2004)
17. J.P. Keating, F. Mezzadri, Phys. Rev. Lett. **94**, 050501 (2005)
18. J. Eisert, M. Cramer, Phys. Rev. A **72**, 042112 (2005)
19. N. Laflorencie, Phys. Rev. B **72**, 140408 (2005)
20. V. Eisler, Z. Zimborás, Phys. Rev. A **71**, 042318 (2005)
21. P. Calabrese, J.L. Cardy, J. Stat. Mech. P04010 (2005)
22. T. Barthel, M.-C. Chung, U. Schollwöck, Phys. Rev. A **74**, 022329 (2006)
23. H. Casini, M. Huerta, J. Stat. Mech. P12012 (2005)
24. M. Cramer, J. Eisert, M.B. Plenio, J. Dreiig, Phys. Rev. A **73**, 012309 (2006)
25. I. Peschel, M. Kaulke, Ö. Legeza, Ann. Physik (Leipzig) **8**, 153 (1999)
26. I. Peschel, J. Stat. Mech. P12005 (2004)
27. R.J. Baxter, *Exactly Solved Models in Statistical Mechanics* (Academic Press, London, 1982)
28. C. Holzhey, F. Larsen, F. Wilczek, Nucl. Phys. B **424**, 443 (1994)
29. P. Calabrese, J.L. Cardy, J. Stat. Mech. P06002 (2004)
30. S.A. Cheong, C.L. Henley, Phys. Rev. B **69**, 075112 (2004)
31. V. Eisler, Ö. Legeza, Z. Rácz, J. Stat. Mech. P11013 (2006)
32. V. Popkov, M. Salerno, G. Schütz, Phys. Rev. A **72**, 032327 (2005)
33. J. Vidal, S. Dusuel, T. Barthel, J. Stat. Mech.: Th. Exp. P01015 (2007)
34. K. Okunishi, Y. Hieida, Y. Akutsu, Phys. Rev. E **59**, R6227 (1999)
35. L. Bombelli, R.K. Koul, J. Lee, R.D. Sorkin, Phys. Rev. D **34**, 373 (1986)
36. M. Srednicki, Phys. Rev. Lett. **71**, 666 (1993)
37. M.M. Wolf, Phys. Rev. Lett. **96**, 010404 (2006)
38. D. Gioev, I. Klich, Phys. Rev. Lett. **96**, 100503 (2006)
39. W. Li, L. Ding, R. Yu, T. Roscilde, S. Haas, Phys. Rev. B **74**, 073103 (2006)
40. G. De Chiara, S. Montangero, P. Calabrese, R. Fazio, J. Stat. Mech. P03001 (2006)
41. Ö. Legeza, J. Sólyom, Phys. Rev. B **70**, 205118 (2004)
42. V. Hakim, J.P. Nadal, J. Phys. A: Math. Gen. **16**, L213 (1983)
43. M. Fannes, B. Nachtergaele, R.F. Werner, Europhys. Lett. **10**, 633 (1989)
44. A. Klümper, A. Schadschneider, J. Zittartz, Z. Physik B **87**, 281 (1992)
45. I. Affleck, T. Kennedy, E.H. Lieb, H. Tasaki, Phys. Rev. Lett. **59**, 799 (1987)
46. B. Derrida, M.R. Evans, V. Hakim, V. Pasquier, J. Phys. A: Math. Gen. **26**, 1493 (1993)
47. I. Peschel, M. Kaulke, [3], Chap. II, § 3.1, p. 279

48. F. Verstraete, J.I. Cirac, Phys. Rev. B **73**, 094423 (2006)
49. S. Östlund, S. Rommer, Phys. Rev. Lett. **75**, 3537 (1995)
50. F. Verstraete, D. Porras, J.I. Cirac, Phys. Rev. Lett. **93**, 227205 (2004)
51. F. Verstraete, J.I. Cirac (2004). URL http://arxiv.org/abs/cond-mat/0407066. Preprint

21 Density-Matrix Renormalization Group Algorithms

Eric Jeckelmann

Institut für Theoretische Physik, Leibniz Universität Hannover, 30167 Hannover, Germany

In this chapter I will introduce the basic Density Matrix Renormalization Group (DMRG) algorithms for calculating ground states in quantum lattice many-body systems using the one-dimensional spin-$\frac{1}{2}$ Heisenberg model as illustration. I will attempt to present these methods in a manner which combines the advantages of both the traditional formulation in terms of renormalized blocks and superblocks and the new description based on matrix-product states. The latter description is physically more intuitive but the former description is more appropriate for writing an actual DMRG program. Pedagogical introductions to DMRG which closely follow the original formulation are available in [2, 1]. The conceptual background of DMRG and matrix-product states is discussed in the previous chapter and should be read before. Extensions of the basic DMRG algorithms are presented in the chapters that follow this one.

21.1 Introduction

The DMRG was developed by White [3, 4] in 1992 to overcome the problems arising in the application of real-space renormalization groups to quantum lattice many-body systems in solid-state physics. Since then the approach has been extended to a great variety of problems in all fields of physics and even in quantum chemistry. The numerous applications of DMRG are summarized in two recent review articles [5, 6]. Additional information about DMRG can be found at `http://www.dmrg.info`.

Originally, DMRG has been considered as an extension of real-space renormalization group methods. The key idea of DMRG is to renormalize a system using the information provided by a reduced density matrix rather than an effective Hamiltonian (as done in most renormalization groups), hence the name density-matrix renormalization. Recently, the connection between DMRG and matrix-product states has been emphasized (for a recent review, see [7]) and has lead to significant extensions of the DMRG approach. From this point of view, DMRG is an algorithm for optimizing a variational wavefunction with the structure of a matrix-product state.

The outline of this chapter is as follows: First I briefly introduce the DMRG matrix-product state and examine its relation to the traditional DMRG blocks and

E. Jeckelmann: *Density-Matrix Renormalization Group Algorithms*, Lect. Notes Phys. **739**, 597–619 (2008)
DOI 10.1007/978-3-540-74686-7_21

superblocks in Sect. 1. In the next three Sect. I present a numerical renormalization group method, then the infinite-system DMRG algorithm, and finally the finite-system DMRG algorithm. In Sect. 5 the use of additive quantum numbers is explained. In the next two sections the estimation of numerical errors and code optimization are discussed. In the last section some extensions of DMRG are presented.

21.2 Matrix-Product States and (Super-)Blocks

We consider a quantum lattice system with N sites $n = 1, \ldots, N$. Let $\mathcal{B}(n) = \{|s_n\rangle; s_n = 1, \ldots, d_n\}$ denote a complete basis of the Hilbert space for site n (all bases used here are orthonormal). The tensor product of these bases yields a complete basis of the system Hilbert space $\mathcal{H}$

$$\{|\boldsymbol{s} = (s_1, \ldots, s_N)\rangle = |s_1\rangle \otimes \cdots \otimes |s_N\rangle; s_n = 1, \ldots, d_n; n = 1, \ldots, N\} \,. \tag{21.1}$$

For instance, for the spin-$\frac{1}{2}$ Heisenberg model $d_n = 2$ and $\mathcal{B}(n) = \{|\uparrow\rangle, |\downarrow\rangle\}$.

Any state $|\psi\rangle$ of $\mathcal{H}$ can be expanded in this basis: $|\psi\rangle = \sum_{\boldsymbol{s}} c(\boldsymbol{s})|\boldsymbol{s}\rangle$. As explained in Chap. 20, the coefficients $c(\boldsymbol{s})$ can take the form of a matrix product. Here we consider a particular matrix-product state

$$c(\boldsymbol{s}) = \mathsf{A}_1(s_1) \ldots \mathsf{A}_j(s_j) \mathsf{C}_j \mathsf{B}_{j+1}(s_{j+1}) \ldots \mathsf{B}_N(s_N) \,, \tag{21.2}$$

where C_j is a $(a_j \times b_{j+1})$-matrix (i.e., with a_j rows and b_{j+1} columns). The $(a_{n-1} \times a_n)$-matrices $\mathsf{A}_n(s_n)$ (for $s_n = 1, \ldots, d_n; n = 1, \ldots, j$) and the $(b_n \times b_{n+1})$-matrices $\mathsf{B}_n(s_n)$ (for $s_n = 1, \ldots, d_n; n = j+1, \ldots, N$) fulfill the orthonormalization conditions

$$\sum_{s_n=1}^{d_n} (\mathsf{A}_n(s_n))^\dagger \, \mathsf{A}_n(s_n) = \mathsf{I} \quad \text{and} \quad \sum_{s_n=1}^{d_n} \mathsf{B}_n(s_n) \, (\mathsf{B}_n(s_n))^\dagger = \mathsf{I} \,, \tag{21.3}$$

(I is the identity matrix), and the boundary condition $a_0 = b_{N+1} = 1$. These conditions imply that $a_n \le d_n a_{n-1} \le \prod_{k=1}^{n} d_k$ and $b_n \le d_n b_{n+1} \le \prod_{k=n}^{N} d_k$.

Obviously, this matrix-product state splits the lattice sites in two groups. The sites $n = 1, \ldots, j$ make up a left block $L(j)$ and the sites $n = j+1, \ldots, N$ constitute a right block $R(j+1)$. From the matrix elements of $\mathsf{A}_n(s_n)$ and $\mathsf{B}_n(s_n)$ we can define third-rank tensors

$$\phi_\alpha^{L(n)}(\alpha', s_n) = [\mathsf{A}_n(s_n)](\alpha', \alpha) \,, \tag{21.4}$$

for $s_n = 1, \ldots, d_n$; $\alpha = 1, \ldots, a_n$, and $\alpha' = 1, \ldots, a_{n-1}$, and

$$\phi_\beta^{R(n)}(s_n, \beta') = [\mathsf{B}_n(s_n)](\beta, \beta') \,, \tag{21.5}$$

for $s_n = 1, \ldots, d_n$; $\beta = 1, \ldots, b_n$, and $\beta' = 1, \ldots, b_{n+1}$. Using these tensors one can iteratively define a set of orthonormal states in the Hilbert space associated with each left block

$$\left|\phi_\alpha^{L(1)}\right\rangle = |s_1\rangle \; ; \; \alpha = s_1 = 1, \ldots, d_1;$$

$$\left|\phi_\alpha^{L(n)}\right\rangle = \sum_{\alpha'=1}^{a_{n-1}} \sum_{s_n=1}^{d_n} \phi_\alpha^{L(n)}(\alpha', s_n) \left|\phi_{\alpha'}^{L(n-1)}\right\rangle \otimes |s_n\rangle \; ; \; \alpha = 1, \ldots, a_n \; , \tag{21.6}$$

and each right block

$$\left|\phi_\beta^{R(N)}\right\rangle = |s_N\rangle \; ; \; \beta = s_N = 1, \ldots, d_N;$$

$$\left|\phi_\beta^{R(n)}\right\rangle = \sum_{\beta'=1}^{b_{n+1}} \sum_{s_n=1}^{d_n} \phi_\beta^{R(n)}(s_n, \beta') \, |s_n\rangle \otimes \left|\phi_{\beta'}^{R(n+1)}\right\rangle \; ; \; \beta = 1, \ldots, b_n \; . \tag{21.7}$$

The orthonormality of each set of block states (i.e., the states belonging to the same block Hilbert space) follows directly from the orthonormalization conditions for the matrices $\mathsf{A}_n(s_n)$ and $\mathsf{B}_n(s_n)$.

Every set of block states spans a subspace of the Hilbert space associated with the block. Using these states one can build an effective or renormalized (i.e., approximate) representation of dimension a_n or b_n for every block. By definition, an effective representation of dimension a_n for the block $L(n)$ is made of vector and matrix representations in a subspace basis $\mathcal{B}(L, n)$ for every state and operator (acting on sites in $L(n)$) which our calculation requires. Note that if $a_n = \prod_{k=1}^{n} d_k$, the block state set is a complete basis of the block Hilbert space and the "effective" representation is actually exact. An effective representation of dimension b_n for a right block $R(n)$ is defined similarly using a subspace basis $\mathcal{B}(R, n)$.

If we combine the left block $L(j)$ with the right block $R(j+1)$, we obtain a so-called superblock $\{L(j) + R(j+1)\}$ which contains the sites 1 to N. The tensor-product basis $\mathcal{B}(SB, j) = \mathcal{B}(L, j) \otimes \mathcal{B}(R, j+1)$ of the block bases is called a superblock basis and spans a $(a_j b_{j+1})$-dimensional subspace of the system Hilbert space $\mathcal{H}$. The matrix-product state given by (21.2) can be expanded in this basis

$$|\psi\rangle = \sum_{\alpha=1}^{a_j} \sum_{\beta=1}^{b_{j+1}} [\mathsf{C}_j](\alpha, \beta) \, |\phi_\alpha^{L(j)} \phi_\beta^{R(j+1)}\rangle \; , \tag{21.8}$$

where $[\mathsf{C}_j](\alpha, \beta)$ denotes the matrix elements of C_j and

$$|\phi_\alpha^{L(j)} \phi_\beta^{R(j+1)}\rangle = |\phi_\alpha^{L(j)}\rangle \otimes |\phi_\beta^{R(j+1)}\rangle \in \mathcal{B}(SB, j) \; . \tag{21.9}$$

We note that the square norm of $|\psi\rangle$ is given by $\langle\psi|\psi\rangle = \mathrm{Tr}\, \mathsf{C}_j^\dagger \mathsf{C}_j$.

If $a_j = \prod_{n=1}^{j} d_n$ and $b_{j+1} = \prod_{n=j+1}^{N} d_n$, the superblock basis $\mathcal{B}(SB, j)$ is a complete basis of $\mathcal{H}$ and any state $|\psi\rangle \in \mathcal{H}$ can be written in the form (21.8). For a large lattice these conditions mean that some matrix dimensions are very large (at least $2^{N/2}$ for a spin-$\frac{1}{2}$ model). However, a matrix-product state is numerically tractable only if all matrix dimensions are kept small, for instance $a_n, b_k \leq m$ with m up to a few thousands. A matrix-product state with restricted matrix sizes can be considered as an approximation for states in $\mathcal{H}$. In particular, it can be used as a variational ansatz for the ground state of the system Hamiltonian H. Thus the system energy $E = \langle\psi|H|\psi\rangle/\langle\psi|\psi\rangle$ is a function of the matrices $\mathsf{A}_n(s_n)$, $\mathsf{B}_n(s_n)$, and C_j. It has to be minimized with respect to these variational parameters subject to the constraints (21.3) to determine the ground state. In the following sections I will present three algorithms (a numerical renormalization group, the infinite-system DMRG method, and the finite-system DMRG method) for carrying out this minimization.

21.3 Numerical Renormalization Group

The Numerical Renormalization Group (NRG) method was developed by Wilson a few decades ago to solve the Kondo impurity problem [8]. The key idea is a decomposition of the lattice into subsystems (blocks) of increasing size. To calculate the ground state properties of a large lattice one starts from an exact representation of a small subsystem and builds effective representations of larger subsystems iteratively, adding one site at every iteration as illustrated in Fig. 21.1. Here I formulate this procedure for a quantum lattice system in the framework of a matrix-product state (21.2). To find a fixed point in an infinite chain, we consider that $j \equiv N$ in (21.2) while for a finite lattice size N we set $b_n = 1$ for all sites n. In both cases the right blocks do not play any role and j is increased by one in every iteration using the following procedure.

We want to calculate an effective representation of dimension a_{j+1} for the left block $L(j+1)$ assuming that we know an effective representation of dimension a_j for the left block $L(j)$. First, from the known bases $\mathcal{B}(L,j)$ of $L(j)$ and $\mathcal{B}(j+1)$ for the site $j+1$ we can define a tensor-product basis of dimension $a_j d_{j+1}$ for $L(j+1)$

$$\left|\phi_\alpha^{L(j)}\, s_{j+1}\right\rangle = \left|\phi_\alpha^{L(j)}\right\rangle \otimes |s_{j+1}\rangle \,, \tag{21.10}$$

with $\alpha = 1, \ldots, a_j$ and $s_{j+1} = 1, \ldots, d_{j+1}$. Second, every operator acting on sites in $L(j+1)$ can be decomposed into a sum of operator pairs

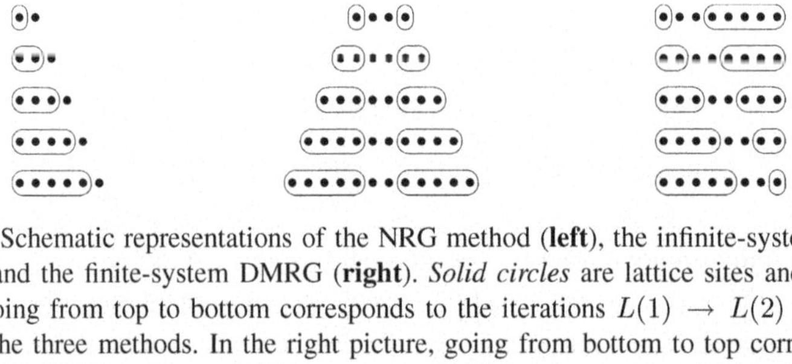

Fig. 21.1. Schematic representations of the NRG method (**left**), the infinite-system DMRG (**center**), and the finite-system DMRG (**right**). *Solid circles* are lattice sites and ovals are blocks. Going from top to bottom corresponds to the iterations $L(1) \rightarrow L(2) \rightarrow \cdots \rightarrow L(5)$ for the three methods. In the right picture, going from bottom to top corresponds to the iterations $R(N = 8) \rightarrow R(7) \rightarrow \cdots \rightarrow R(4)$ in a sweep from right to left of the finite-system DMRG

$$\mathcal{O} = \sum_{k} \mathcal{O}^{L,k}\, \mathcal{O}^{S,k} , \tag{21.11}$$

where the operators $\mathcal{O}^{L,k}$ act only on sites in $L(j)$ and the operators $\mathcal{O}^{S,k}$ act only on the site $j+1$. For instance, the (one-dimensional) Heisenberg Hamiltonian on the block $L(j+1)$

$$H = \sum_{n=1}^{j} \boldsymbol{S}_n \boldsymbol{S}_{n+1} , \tag{21.12}$$

can be decomposed as

$$H = \sum_{n=1}^{j-1} \boldsymbol{S}_n \boldsymbol{S}_{n+1} \otimes I + S_j^z \otimes S_{j+1}^z + \frac{1}{2}\left(S_j^+ \otimes S_{j+1}^- + S_j^- \otimes S_{j+1}^+\right) , \tag{21.13}$$

where I is the identity operator and $\boldsymbol{S}_n, S_n^z, S_n^+, S_n^-$ are the usual spin operators for the site n. As a result the matrix representation of $\mathcal{O}$ in the basis (21.10)

$$\mathcal{O}(\alpha, s_{j+1}, \alpha', s'_{j+1}) = \left\langle \phi_\alpha^{L(j)}\; s_{j+1} \middle| \mathcal{O} \middle| \phi_{\alpha'}^{L(j)}\; s'_{j+1} \right\rangle , \tag{21.14}$$

is given by

$$\mathcal{O}(\alpha, s_{j+1}, \alpha', s'_{j+1}) = \sum_k \mathcal{O}^{L,k}(\alpha, \alpha')\, \mathcal{O}^{S,k}(s_{j+1}, s'_{j+1}) , \tag{21.15}$$

where

$$\mathcal{O}^{L,k}(\alpha, \alpha') = \left\langle \phi_\alpha^{L(j)} \middle| \mathcal{O}^{L,k} \middle| \phi_{\alpha'}^{L(j)} \right\rangle \tag{21.16}$$

denotes the known matrix representations of $\mathcal{O}^{L,k}$ in the basis $\mathcal{B}(L,j)$ of the block $L(j)$. The matrix representations of the site operators

$$\mathcal{O}^{S,k}(s_{j+1}, s'_{j+1}) = \langle s_{j+1} | \mathcal{O}^{S,k} | s'_{j+1} \rangle , \tag{21.17}$$

can be calculated exactly. For instance, they correspond to the Pauli matrices for the spin operators S_n^x, S_n^y, S_n^z in the spin-$\frac{1}{2}$ basis $\mathcal{B}(n) = \{|\uparrow\rangle, |\downarrow\rangle\}$.

Using this procedure we can construct the matrix representation (21.14) of the Hamiltonian (restricted to the block $L(j+1)$) in the basis (21.10). This matrix can be fully diagonalized numerically. In practice, this sets an upper limit of a few thousands on $a_j d_{j+1}$. The eigenvectors are denoted $\phi_\mu^{L(j+1)}(\alpha, s_{j+1})$ for $\mu = 1, \ldots, a_j d_{j+1}$ and are ordered by increasing eigenenergies $\epsilon_\mu^{L(j+1)}$. The a_{j+1} eigenvectors with the lowest eigenenergies are used to define a new basis $\mathcal{B}(L, j+1)$ of $L(j+1)$ through (21.6) and the other eigenvectors are discarded. The matrix representation in $\mathcal{B}(L, j+1)$ for any operator acting in $L(j+1)$

$$\mathcal{O}(\mu, \mu') = \left\langle \phi_\mu^{L(j+1)} \middle| \mathcal{O} \middle| \phi_{\mu'}^{L(j+1)} \right\rangle \; ; \; \mu, \mu' = 1, \ldots, a_{j+1} , \tag{21.18}$$

can be calculated using the orthogonal transformation and projection defined by the reduced set of eigenvectors. Explicitly, we have to perform two successive matrix products

$$M(\alpha, s_{j+1}, \mu') = \sum_{\alpha'=1}^{a_j} \sum_{s'_{j+1}=1}^{d_{j+1}} \mathcal{O}(\alpha, s_{j+1}, \alpha', s'_{j+1})\, \phi_{\mu'}^{L(j+1)}(\alpha', s'_{j+1}) \, ,$$

$$\mathcal{O}(\mu, \mu') = \sum_{\alpha=1}^{a_j} \sum_{s_{j+1}=1}^{d_{j+1}} \left(\phi_{\mu}^{L(j+1)}(\alpha, s_{j+1})\right)^{*} M(\alpha, s_{j+1}, \mu') \, . \quad (21.19)$$

Vector representations of states in $L(j+1)$ can be obtained using the same principles. Therefore, we have obtained an effective representation of dimension a_{j+1} for the block $L(j+1)$. We note that the block states (21.6) are not explicitly calculated. Only matrix and vector representations for operators and states in that basis and the transformation from a basis to the next one need to be calculated explicitly.

Once the effective representation of $L(j+1)$ has be determined, the procedure can be repeated to obtain the effective representation of the next larger block. This procedure has to be iterated until $j+1 = N$ for a finite system or until a fixed point is reached if one investigates an infinite system. After the last iteration physical quantities for the (approximate) ground state and low-energy excitations can be calculated using the effective representation of $L(N)$. For instance, expectation values are given by

$$\langle \psi | \mathcal{O} | \psi \rangle = \sum_{\mu,\mu'=1}^{a_N} [\mathsf{C}_N^{\dagger}](\mu)\, \mathcal{O}(\mu, \mu')\, [\mathsf{C}_N](\mu') \, , \quad (21.20)$$

where $\mathcal{O}(\mu, \mu')$ is the matrix representation of $\mathcal{O}$ in the basis $\mathcal{B}(L, N)$ and C_N is the $(a_N \times 1)$-matrix corresponding to the state $|\psi\rangle$ in (21.2) and (21.8). For the ground state we obviously have $[\mathsf{C}_N](\mu) = \delta_{\mu,1}$.

The NRG method is efficient and accurate for quantum impurity problems such as the Kondo model but fails utterly for quantum lattice problems such as the Heisenberg model. One reason is that in many quantum systems the exact ground state can not be represented accurately by a matrix-product state (21.2) with restricted matrix sizes. However, another reason is that in most cases the NRG algorithm does not generate the optimal block representation for the ground state of a quantum lattice system and thus does not even find the matrix-product state (21.2) with the minimal energy for given matrix sizes.

21.4 Infinite-System DMRG Algorithm

The failure of the NRG method for quantum lattice problems can be understood qualitatively. The subsystem represented by a block $L(n)$ always has an artificial boundary at which the low-energy eigenstates of a quantum lattice Hamiltonian tend to vanish. Thus at later iterations the low-energy eigenstates of the effective Hamiltonian in larger subsystems have unwanted features like nodes where the artificial boundaries of the previous subsystems were located. White and Noack [9] have shown that this difficulty can be solved in single-particle problems if the effects of the subsystem environment are taken into account self-consistently. DMRG is the application of this idea to many-particle problems. In his initial papers [3, 4], White described two DMRG algorithms: The infinite-system method presented in this section and the finite-system method discussed in the next section.

The infinite-system method is certainly the simplest DMRG algorithm and is the starting point of many other DMRG methods. In this approach the system size increases by two sites in every iteration, $N \to N+2$, as illustrated in Fig. 21.1. The right block $R(j+1)$ is always an image (reflection) of the left block $L(j)$, which implies that $j \equiv N/2$ in (21.2). Therefore, the superblock structure is $\{L(N/2) + R(N/2+1)\}$ and an effective representation for the N-site system is known if we have determined one for $L(N/2)$.

As in the NRG method an iteration consists in the calculation of an effective representation of dimension a_{j+1} for the block $L(j+1)$ assuming that we already know an effective representation of dimension a_j for the block $L(j)$. First, we proceed as with the NRG method and determine an effective representation of dimension $a_j d_{j+1}$ for $L(j+1)$ using the tensor product basis (21.10). Next, the effective representation of $R(j+2)$ is chosen to be an image of $L(j+1)$. The quantum system is assumed to be homogeneous and symmetric (invariant under a reflection $n \to n' = N - n + 3$ through the middle of the (N+2)-site lattice) to allow for this operation. Therefore, one can define a one-to-one mapping between the site and block bases on the left- and right-hand sides of the superblock. We consider a mapping between the tensor product bases for $L(j+1)$ and $R(j+2)$

$$\left|\phi_\alpha^{L(j)} s_{j+1}\right\rangle \leftrightarrow \left|s_{j+2}\, \phi_\beta^{R(j+3)}\right\rangle . \tag{21.21}$$

Thus, the matrix representation of any operator acting in $R(j+2)$

$$\mathcal{O}(s_{j+2}, \beta, s'_{j+2}, \beta') = \left\langle s_{j+2}\, \phi_\beta^{R(j+3)} \middle| \mathcal{O} \middle| s'_{j+2}\, \phi_{\beta'}^{R(j+3)} \right\rangle , \tag{21.22}$$

is given by the matrix representation (21.14) of the corresponding (reflected) operator in $L(j+1)$ through the basis mapping.

A superblock basis $\mathcal{B}(SB, j+1)$ of dimension $\mathcal{D}_{j+1} = a_j d_{j+1} d_{j+2} b_{j+3}$ can be defined using the tensor product of the block bases

$$\left|\phi_\alpha^{L(j)} s_{j+1}\, s_{j+2}\, \phi_\beta^{R(j+3)}\right\rangle = \left|\phi_\alpha^{L(j)} s_{j+1}\right\rangle \otimes \left|s_{j+2}\, \phi_\beta^{R(j+3)}\right\rangle , \tag{21.23}$$

for $\alpha = 1, \ldots, a_j$; $s_{j+1} = 1, \ldots, d_{j+1}$; $s_{j+2} = 1, \ldots, d_{j+2}$; and $\beta = 1, \ldots, b_{j+3}$. Every operator acting on the superblock (i.e., the $(N+2)$-site lattice) can be decomposed in a sum of operator pairs

$$\mathcal{O} = \sum_{k=1}^{n_k} \mathcal{O}^{L,k} \, \mathcal{O}^{R,k} \, , \qquad (21.24)$$

where the operator parts $\mathcal{O}^{L,k}$ and $\mathcal{O}^{R,k}$ act on sites in $L(j+1)$ and $R(j+2)$, respectively. As an example, the Heisenberg Hamiltonian on a $(N+2)$-site chain can be written

$$H = \sum_{n=1}^{j} \boldsymbol{S}_n \boldsymbol{S}_{n+1} \otimes I + I \otimes \sum_{n=j+2}^{N+1} \boldsymbol{S}_n \boldsymbol{S}_{n+1} + \\ S^z_{j+1} \otimes S^z_{j+2} + \frac{1}{2} \left(S^+_{j+1} \otimes S^-_{j+2} + S^-_{j+1} \otimes S^+_{j+2} \right) \, , \qquad (21.25)$$

where I is the identity operator. Therefore, the matrix representation of any operator in the superblock basis

$$\mathcal{O}(\alpha, s_{j+1}, s_{j+2}, \beta, \alpha', s'_{j+1}, s'_{j+2}, \beta') = \\ \left\langle \phi^{L(j)}_{\alpha} \, s_{j+1} \, s_{j+2} \, \phi^{R(j+3)}_{\beta} \middle| \mathcal{O} \middle| \phi^{L(j)}_{\alpha'} \, s'_{j+1} \, s'_{j+2} \, \phi^{R(j+3)}_{\beta'} \right\rangle \, , \qquad (21.26)$$

(for $\alpha, \alpha' = 1, \ldots, a_j$; $s_{j+1}, s'_{j+1} = 1, \ldots, d_{j+1}$; $s_{j+2}, s'_{j+2} = 1, \ldots, d_{j+2}$; and $\beta, \beta' = 1, \ldots, b_{j+3}$) is given by the sum of the tensor products of the matrix representations (21.14) and (21.22) for the block operators

$$\mathcal{O}(\alpha, s_{j+1}, s_{j+2}, \beta, \alpha', s'_{j+1}, s'_{j+2}, \beta') = \\ \sum_{k=1}^{n_k} \mathcal{O}^{L,k}(\alpha, s_{j+1}, \alpha', s'_{j+1}) \, \mathcal{O}^{R,k}(s_{j+2}, \beta, s'_{j+2}, \beta') \, . \qquad (21.27)$$

Storing the matrix representations (21.14) and (21.22) for the block operators requires a memory amount $\propto n_k[(a_j d_{j+1})^2 + (d_{j+2} b_{j+3})^2]$, but calculating and storing the superblock matrix (21.26) require $n_k(\mathcal{D}_{j+1})^2$ additional operations and a memory amount $\propto (\mathcal{D}_{j+1})^2$. As the number of operator pairs n_k is typically much smaller than the matrix dimensions a_j, b_{j+3} ($n_k = 5$ in the Heisenberg model on a open chain), one should not calculate the superblock matrix representation (21.26) explicitly but work directly with the right-hand side of (21.27). For instance, the application of the operator $\mathcal{O}$ to a state $|\psi\rangle \subset \mathcal{H}$ yields a new state $|\psi'\rangle = \mathcal{O}|\psi\rangle$, which can be calculated without computing the superblock matrix (21.26) explicitly. If

$$[\mathsf{C}_{j+1}](\alpha, s_{j+1}, s_{j+2}, \beta) = \left\langle \phi^{L(j)}_{\alpha} \, s_{j+1} \, s_{j+2} \, \phi^{R(j+3)}_{\beta} \middle| \psi \right\rangle \, , \qquad (21.28)$$

is the vector representation of $|\psi\rangle$ in the superblock basis (21.23), the vector representation C'_{j+1} of $|\psi'\rangle$ in this basis is obtained through double matrix products with the block operator matrices in (21.27)

$$
\begin{aligned}
&V_k(\alpha', s'_{j+1}, s_{j+2}, \beta) \\
&\quad = \sum_{\beta'=1}^{b_{j+3}} \sum_{s'_{j+2}=1}^{d_{j+2}} [\mathsf{C}_{j+1}](\alpha', s'_{j+1}, s'_{j+2}, \beta')\, \mathcal{O}^{R,k}(s_{j+2}, \beta, s'_{j+2}, \beta') , \\
&[\mathsf{C}'_{j+1}](\alpha, s_{j+1}, s_{j+2}, \beta) \\
&\quad = \sum_{k=1}^{n_k} \sum_{\alpha'=1}^{a_j} \sum_{s'_{j+1}=1}^{d_{j+1}} \mathcal{O}^{L,k}(\alpha, s_{j+1}, \alpha', s'_{j+1}) V_k(\alpha', s'_{j+1}, s_{j+2}, \beta) .
\end{aligned}
\tag{21.29}
$$

Performing these operations once requires only $n_k \mathcal{D}_{j+1}(a_j d_{j+1} + d_{j+2} b_{j+3})$ operations, while computing a matrix-vector product using the superblock matrix (21.26) would require $(\mathcal{D}_{j+1})^2$ operations. In practice, this sets an upper limit of the order of a few thousands for the matrix dimensions a_n, b_n.

As we want to calculate the ground state of the system Hamiltonian H, the next task is to set up the superblock representation (21.27) of H and then to determine the vector representation (21.28) of its ground state in the superblock basis. To determine the ground state without using the superblock matrix (21.26) of H we use iterative methods such as the Lanczos algorithm or the Davidson algorithm, see Chap. 18. These algorithms do not require an explicit matrix for H but only the operation $|\psi'\rangle = H|\psi\rangle$, which can be performed very efficiently with (21.29) as discussed above.

Once the superblock ground state C_{j+1} has been determined, the next step is finding an effective representation of dimension $a_{j+1} < a_j d_{j+1}$ for $L(j+1)$ which described this ground state as closely as possible. Thus we look for the best approximation $\widetilde{\mathsf{C}}_{j+1}$ of the superblock ground state C_{j+1} with respect to a new basis $\mathcal{B}(L, j+1)$ of dimension a_{j+1} for $L(j+1)$. As discussed in Chap. 20 this can be done using the Schmidt decomposition or more generally reduced density matrices. Choosing the density-matrix eigenvectors with the highest eigenvalues is an optimal choice for constructing a smaller block basis (see Sect. 21.7). Therefore, if the DMRG calculation targets a state with a vector representation $[\mathsf{C}_{j+1}](\alpha, s_{j+1}, s_{j+2}, \beta)$ in the superblock basis (21.23), we calculate the reduced density matrix for the left block $L(j+1)$

$$
\begin{aligned}
&\rho(\alpha, s_{j+1}, \alpha', s'_{j+1}) \\
&\quad = \sum_{s_{j+2}=1}^{d_{j+2}} \sum_{\beta=1}^{b_{j+3}} \left([\mathsf{C}_{j+1}](\alpha, s_{j+1}, s_{j+2}, \beta)\right)^* [\mathsf{C}_{j+1}](\alpha', s'_{j+1}, s_{j+2}, \beta)
\end{aligned}
\tag{21.30}
$$

for $\alpha, \alpha' = 1, \ldots, a_j$ and $s_{j+1}, s'_{j+1} = 1, \ldots, d_{j+1}$. This density matrix has $a_j d_{j+1}$ eigenvalues $w_\mu \geq 0$ with

$$
\sum_{\mu=1}^{a_j d_{j+1}} w_\mu = 1 . \tag{21.31}
$$

We note $\phi_{\mu}^{L(j+1)}(\alpha, s_{j+1})$ the corresponding eigenvectors. The a_{j+1} eigenvectors with the largest eigenvalues are used to define a new basis $\mathcal{B}(L, j+1)$ of $L(j+1)$ through (21.6) and the other eigenvectors are discarded. As done in the NRG method, the matrix representation of any operator in $L(j+1)$ can be calculated using the orthogonal transformation and projection (21.19) defined by the reduced set of eigenvectors. If necessary, vector representations of states in $L(j+1)$ can be obtained using the same principles.

Thus, we have obtained an effective representation of dimension a_{j+1} for the block $L(j+1)$. We note that as with the NRG method the block states (21.6) are not explicitly calculated. Only matrix and vector representations of operators and states in that basis and the transformation from a basis to the next one need to be calculated explicitly. The procedure can be repeated to obtain an effective representation of the next larger blocks (i.e., for the next larger lattice size). Iterations are continued until a fixed point has been reached.

As an illustration Fig. 21.2 shows the convergence of the ground state energy per site as a function of the superblock size N in the one-dimensional spin-$\frac{1}{2}$ Heisenberg model. The energy per site $E_{\mathrm{DMRG}}(N)$ is calculated from the total energy E_0 for two consecutive superblocks $E_{\mathrm{DMRG}}(N) = [E_0(N) - E_0(N-2)]/2$. The exact result for an infinite chain is $E_{\mathrm{exact}} = \frac{1}{4} - \ln(2)$ according to the Bethe ansatz solution [10]. The matrix dimensions a_n, b_n are chosen to be not greater than a number m which is the maximal number of density-matrix eigenstates kept at each iteration. As N increases, $E_{\mathrm{DMRG}}(N)$ converges to a limiting value $E_{\mathrm{DMRG}}(m)$ which is the minimal energy for a matrix-product state (21.2) with matrix dimensions up to m. This energy minimum $E_{\mathrm{DMRG}}(m)$ is always higher than the exact ground state energy E_{exact} as expected for a variational method. The error in $E_{\mathrm{DMRG}}(m)$ is dominated by truncation errors, which decrease rapidly as the number m increases (see the discussion of truncation errors in Sect. 21.7).

Once a fixed point has been reached, ground state properties can be calculated. For instance, a ground state expectation value $\bar{O} = \langle\psi|O|\psi\rangle$ is obtained in two

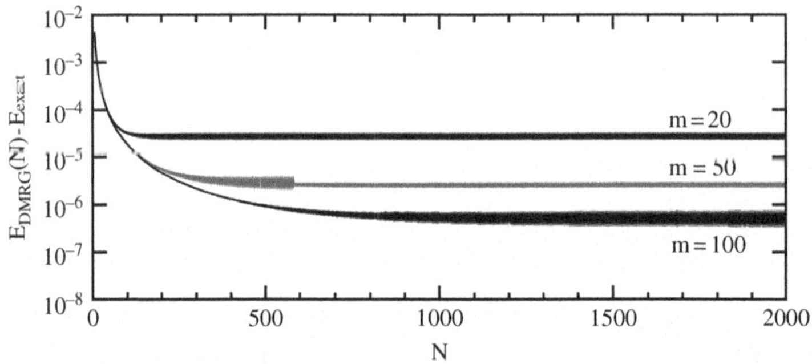

Fig. 21.2. Convergence of the ground state energy per site calculated with the infinite-system DMRG algorithm in a spin-$\frac{1}{2}$ Heisenberg chain as a function of the superblock size N for three different numbers m of density-matrix eigenstates kept

steps: First, one calculates $|\psi'\rangle = \mathcal{O}|\psi\rangle$ using (21.29), then the expectation value is computed as a scalar product $\bar{\mathcal{O}} = \langle\psi|\psi'\rangle$. Explicitly,

$$\langle\psi|\psi'\rangle = \sum_{\alpha=1}^{a_j}\sum_{s_{j+1}=1}^{d_{j+1}}\sum_{s_{j+2}=1}^{d_{j+2}}\sum_{\beta=1}^{b_{j+3}} \left([\mathsf{C}_{j+1}](\alpha, s_{j+1}, s_{j+2}, \beta)\right)^* [\mathsf{C}'_{j+1}](\alpha, s_{j+1}, s_{j+2}, \beta) \,. \tag{21.32}$$

Experience shows that the infinite-system DMRG algorithm yields accurate results for local physical quantities such as spin density and short-range spin correlations in quantum lattice systems with "good" features (infinite homogeneous one-dimensional systems with short-range interactions such as the Heisenberg model on an open chain). These local quantities are calculated using operators $\mathcal{O}$ acting only on sites around the middle of the last (i.e., longest) lattice (more precisely, in an interval of length $N - N'$ around the center of a N-site lattice if the fixed point has been reached after $N'/2 < N/2$ iterations). For other types of systems and other physical quantities the infinite-system algorithm fails in most cases. The reasons for these failures are the same as for NRG. First, the exact ground state may not be represented accurately by a matrix-product state (21.2) with restricted matrix sizes. For instance, the matrix-product state cannot reproduce long-range power-law correlations, see Chap. 20. Second, very often the infinite-system DMRG algorithm does not generate the optimal block representation for the ground state (i.e., does not find the best possible matrix-product state (21.2) for preset matrix sizes) when the system does not have the good features mentioned above.

21.5 Finite-System DMRG Algorithm

The finite-system method is a more versatile DMRG algorithm than the infinite-system method as it can be applied to almost any quantum lattice problem. It is also more reliable as it always finds the best possible matrix-product representation (21.2) for a given quantum state. In the finite-system DMRG method the lattice size N is kept constant. The superblock structure is $\{L(j) + R(j+1)\}$, where j is varied iteratively by one site from $N-2$ to 2 in a sweep from right to left and from 2 to $N-2$ in a sweep from left to right, see Fig. 21.1. If the system has the reflection symmetry used in the infinite-system algorithm, j need to be varied from $N/2$ to 2 and back only. At the start of the finite-system algorithm, one calculates effective representations for the left blocks $L(1)$ to $L(N-3)$ using the NRG method, the infinite-system DMRG algorithm, or other methods, even using random transformations in (21.6), as they can be poor approximations of the optimal representations. This initial calculation is called the warmup sweep.

We first proceed with a sweep through the lattice from right to left, reducing j by one at every iteration starting from $j = N - 2$. For this purpose, we have to compute an effective representation of dimension b_{j+1} for $R(j+1)$ using the effective representation of dimension b_{j+2} for the right block $R(j+2)$ calculated in the previous iteration. For the first iteration $j = N - 2$, the exact representation of $R(N)$ is used. As done for left blocks in the NRG and infinite-system DMRG algorithm, we first define a tensor-product basis of dimension $d_{j+1}b_{j+2}$ for the new right block using the site basis $\mathcal{B}(j+1)$ and the subspace basis $\mathcal{B}(R, j+2)$ of $R(j+2)$

$$\left|s_{j+1}\, \phi_{\beta}^{R(j+2)}\right\rangle = |s_{j+1}\rangle \otimes \left|\phi_{\beta}^{R(j+2)}\right\rangle , \tag{21.33}$$

for $s_{j+1} = 1, \ldots, d_{j+1}$ and $\beta = 1, \ldots, b_{j+2}$. The matrix representation (21.22) of any operator $\mathcal{O}$ acting in $R(j+1)$ can be calculated similarly to (21.15)

$$\mathcal{O}(s_{j+1}, \beta, s'_{j+1}, \beta') = \sum_{k} \mathcal{O}^{S,k}(s_{j+1}, s'_{j+1})\, \mathcal{O}^{R,k}(\beta, \beta') , \tag{21.34}$$

where the $\mathcal{O}^{S,k}(s_{j+1}, s'_{j+1})$ are site-operator matrices (21.17) and $\mathcal{O}^{R,k}(\beta, \beta')$ denotes the known matrix representations of operators acting on sites of $R(j+2)$ in the basis $\mathcal{B}(R, j+2)$. Thus we obtain an effective representation of dimension $d_{j+1}b_{j+2}$ for $R(j+1)$. Next, we use the available effective representation of dimension a_{j-1} for the left block $L(j-1)$, which has been obtained during the previous sweep from left to right (or the result of the warmup sweep if this is the first sweep from right to left). With this block $L(j-1)$ we build an effective representation of dimension $a_{j-1}d_j$ for $L(j)$ using a tensor-product basis (21.10) as done in the NRG and infinite-system DMRG methods.

Now we consider the superblock $\{L(j) + R(j+1)\}$ and its tensor-product basis analogue to (21.23) and set up the representation of operators in this basis, especially the Hamiltonian, similarly to (21.27). As for the infinite-system algorithm we determine the ground state C_j of the superblock Hamiltonian in the superblock basis using the Lanczos or Davidson algorithm and the efficient implementation of the matrix-vector product (21.29). Typically, we have already obtained a representation of the ground state C_{j+1} for the superblock configuration $\{L(j+1) + R(j+2)\}$ in the previous iteration. This state can be transformed exactly in the superblock basis for $\{L(j) + R(j+1)\}$ using

$$[\mathsf{C}_j^{\mathrm{G}}](\alpha, s_j, s_{j+1}, \beta) = \sum_{\alpha'=1}^{a_j} \phi_{\alpha'}^{L(j)}(\alpha, s_j) \sum_{s_{j+2}=1}^{d_{j+2}} \sum_{\beta'=1}^{b_{j+3}} [\mathsf{C}_{j+1}](\alpha', s_{j+1}, s_{j+2}, \beta') \left(\phi_{\beta}^{R(j+2)}(s_{j+2}, \beta')\right)^{*} , \tag{21.35}$$

for $\alpha = 1, \ldots, a_{j-1}$; $s_j = 1, \ldots, d_j$; $s_{j+1} = 1, \ldots, d_{j+1}$; and $\beta = 1, \ldots, b_{j+2}$. The functions $\phi_{\beta}^{R(j+2)}(s_{j+2}, \beta')$ are the density-matrix eigenvectors of $R(j+2)$ calculated in the previous iteration while the functions $\phi_{\alpha'}^{L(j)}(\alpha, s_j)$ are the density-matrix eigenvectors of $L(j)$ calculated during the previous sweep from left to right (or during the warmup sweep if this is the first sweep from right to left). The state

$\mathsf{C}_j^{\mathrm{G}}$ can be used as the initial vector for the iterative diagonalization routine. When the finite-system DMRG algorithm has already partially converged, this initial state $\mathsf{C}_j^{\mathrm{G}}$ is a good guess for the exact ground state C_j of the superblock Hamiltonian in the configuration $\{L(j) + R(j+1)\}$ and thus the iterative diagonalization method converges in a few steps. This can result in a speed up of one or two orders of magnitude compared to a diagonalization using a random initial vector $\mathsf{C}_j^{\mathrm{G}}$.

Once the superblock representation C_j of the targeted ground state has been obtained, we calculate the reduced density matrix for the right block $R(j+1)$

$$\rho(s_{j+1}, \beta, s'_{j+1}, \beta') = \sum_{s_j=1}^{d_j} \sum_{\alpha=1}^{a_{j-1}} \left([\mathsf{C}_j](\alpha, s_j, s_{j+1}, \beta)\right)^* [\mathsf{C}_j](\alpha, s_j, s'_{j+1}, \beta') , \quad (21.36)$$

for $\beta, \beta' = 1, \ldots, b_{j+2}$ and $s_{j+1}, s'_{j+1} = 1, \ldots, d_{j+1}$. We denote the eigenvectors of this density matrix $\phi_\mu^{R(j+1)}(s_{j+1}, \beta)$ with $\mu = 1, \ldots, d_{j+1}b_{j+2}$. The b_{j+1} eigenvectors with the largest eigenvalues are chosen to define a new basis $\mathcal{B}(R, j+1)$ of $R(j+1)$ through (21.7) and the other eigenvectors are discarded. As already mentioned in the previous section, this is the optimal choice for preserving C_j while reducing the basis dimension from $d_{j+1}b_{j+2}$ to b_{j+1} (see Chap. 20 and Sect. 21.7 for more detail). The matrix representation of any operator acting only on sites in $R(j+1)$ can be calculated in the new basis with two successive matrix products

$$M(s_{j+1}, \beta, \mu') = \sum_{\beta'=1}^{b_{j+2}} \sum_{s'_{j+1}=1}^{d_{j+1}} O(s_{j+1}, \beta, s'_{j+1}, \beta') \, \phi_{\mu'}^{R(j+1)}(s'_{j+1}, \beta') ,$$

$$O(\mu, \mu') = \sum_{\beta=1}^{b_{j+2}} \sum_{s_{j+1}=1}^{d_{j+1}} \left(\phi_\mu^{R(j+1)}(s_{j+1}, \beta)\right)^* M(s_{j+1}, \beta, \mu') , \quad (21.37)$$

for $\mu, \mu' = 1, \ldots, b_{j+1}$ as done for a right block in (21.19).

Thus we have obtained an effective representation of dimension b_{j+1} for the right block $R(j+1)$. We note that the block states (21.7) are not explicitly calculated. Only matrix and vector representations of operators and states in that basis and the transformation from a basis to the next one need to be calculated explicitly. The procedure is repeated in the next iteration to obtain an effective representation for the next larger right block. Iterations are continued until the sweep from right to left is completed ($j + 1 = 3$). This right-to-left sweep is illustrated in the right picture of Fig. 21.1 going from bottom to top.

Then we exchange the roles of the left and right blocks and perform a sweep from left to right. Effective representations for the left blocks $L(j), j = 2, \ldots, N-3$, are built iteratively. The effective representation for $R(j+3)$ which has been calculated during the last right-to-left sweep is used to make an effective tensor-product-basis representation of $R(j+2)$ and thus to complete the superblock $\{L(j+1) + R(j+2)\}$. This left-to-right sweep corresponds to the right picture of Fig. 21.1 going from top to bottom.

When this left-to-right sweep is done, one can start a new couple of sweeps back and forth. The ground state energy calculated with the superblock Hamiltonian decreases progressively as the sweeps are performed. This results from the progressive optimization of the matrix-product state (21.2) for the ground state. Figure 21.3 illustrates this procedure for the total energy of a 400-site Heisenberg chain. The matrix dimensions a_n, b_n are chosen to be not greater than $m = 20$ (maximal number of density-matrix eigenstates kept at each iteration). The sweeps are repeated until the procedure converges (i.e., the ground state energy converges). In Fig. 21.3 the DMRG energy converges to a value $E_{\mathrm{DMRG}}(m = 20)$ which lies about 0.008 above the exact result for the 400-site Heisenberg chain. As it corresponds to a variational wavefunction (21.2) the DMRG energy $E_{\mathrm{DMRG}}(m)$ always lies above the exact ground state energy and decreases as m increases.

Once convergence is achieved, ground state properties can be calculated with (21.29) and (21.32) as explained in the previous section. Contrary to the infinite-system algorithm, however, the finite-system algorithm yields consistent results for the expectation values of operators acting on any lattice site. For example, we show in Fig. 21.4 the staggered spin bond order $(-1)^n(\langle \boldsymbol{S}_n \boldsymbol{S}_{n+1}\rangle + \ln(2) - 1/4)$ and the staggered spin-spin correlation function $C(r) = (-1)^r \langle \boldsymbol{S}_n \boldsymbol{S}_{n+r}\rangle$ obtained in the 400-site Heisenberg chain using up to $m = 200$ density-matrix eigenstates. A strong staggered spin bond order is observed close to the chain edges (Friedel oscillations) while a smaller one is still visible in the middle of the chain because of its finite size. For a distance up to $r \approx 100$ the staggered spin-spin correlation function $C(r)$ decreases approximately as a power-law $1/r$ as expected but a deviation from this behavior occurs for larger r because of the chain edges. Finite-size

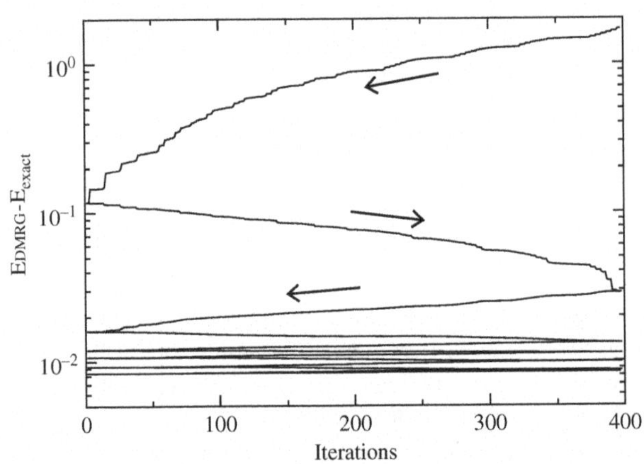

Fig. 21.3. Convergence of the ground state energy calculated with the finite-system DMRG algorithm using $m = 20$ density-matrix eigenstates as a function of the iterations in a 400-site spin-$\frac{1}{2}$ Heisenberg chain. Arrows show the sweep direction for the first three sweeps starting from the top

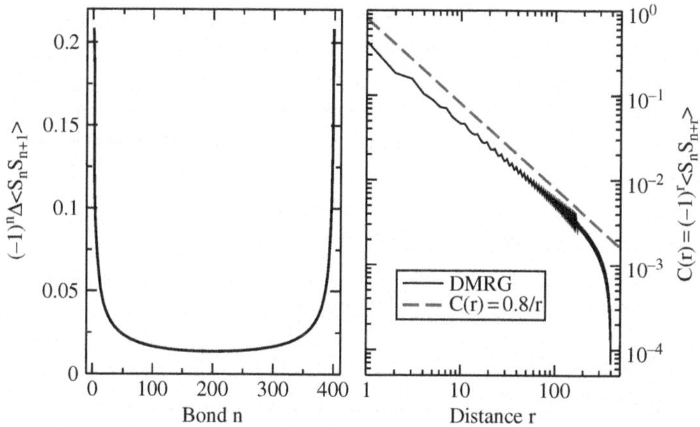

Fig. 21.4. Staggered spin bond order (**left**) $(-1)^n(\langle \boldsymbol{S}_n \boldsymbol{S}_{n+1}\rangle - \frac{1}{4} + \ln(2))$ and staggered spin-spin correlation function (**right**) $C(r) = (-1)^r \langle \boldsymbol{S}_n \boldsymbol{S}_{n+r}\rangle$. Both quantities have been calculated using the finite-system DMRG algorithm with $m = 200$ in a 400-site spin-$\frac{1}{2}$ Heisenberg chain. The *dashed line* is a guide for the eye

and chain-end effects are unavoidable and sometimes troublesome features of the finite-size DMRG method.

Contrary to the infinite-system algorithm the finite-system algorithm always finds the optimal matrix-product state (21.2) with restricted matrix sizes. Nevertheless, experience shows that the accuracy of DMRG calculations depends significantly on the system investigated because the matrix-product state (21.2) with restricted matrix sizes can be a good or a poor approximation of the true ground state. In practice, this implies that physical quantities calculated with DMRG can approach the exact results rapidly or slowly for an increasing number m of density-matrix eigenstates kept. This so-called truncation error is discussed in Sect. 21.7. For instance, the finite-system DMRG method yields excellent results for gapped one-dimensional systems but is less accurate for critical systems or in higher dimensions for the reason discussed in Chap. 20.

21.6 Additive Quantum Numbers

Symmetries and quantum numbers play an important role in the solution and analysis of quantum many-body systems. Here I discuss additive quantum numbers, which constitute the simplest implementation of quantum numbers within the basic DMRG methods. A quantum number is additive when the quantum number of the tensor product of two states is given by the sum of the quantum numbers of both states. The use of other symmetries and quantum numbers is described in [7, 11].

We consider an operator Q acting in $\mathcal{H}$ which is the sum of Hermitian site operators, $Q = \sum_{n=1}^{N} Q_n$, where the operator Q_n acts only on the site n. A typical example is the z-component of the total spin $S^z = \sum_{n=1}^{N} S_n^z$ in a spin system such as the Heisenberg model. If Q commutes with the system Hamiltonian H, eigenstates of H can be chosen so that they are also eigenstates of Q. If the target of the DMRG calculation is an eigenstate of Q (for instance, the ground state of H), one can show that the reduced density operators for left and right blocks commute with the operators

$$Q^{L(j)} = \sum_{n=1}^{j} Q_n \text{ and } Q^{R(j+1)} = \sum_{n=j+1}^{N} Q_n \,, \tag{21.38}$$

respectively. As a consequence, the density-operator eigenstates (21.6) and (21.7) can be chosen to be eigenstates of $Q^{L(j)}$ or $Q^{R(j+1)}$ and the block basis states can be labeled with an index identifying their quantum number (the corresponding eigenvalue of $Q^{L(j)}$ or $Q^{R(j+1)}$). For instance, the left block basis becomes

$$\mathcal{B}(L,j) = \left\{ \left| \phi_{r,\alpha}^{L(j)} \right\rangle ; \, r = 1,2,\ldots ; \alpha = 1,\ldots,a_{r,j} \right\} \,, \tag{21.39}$$

where the index r numbers the possible quantum numbers $q_r^{L(j)}$ of $Q^{L(j)}$, α numbers $a_{r,j}$ basis states with the same quantum number, and $\sum_r a_{r,j} = a_j$.

We note that $Q^{L(j+1)} = Q^{L(j)} + Q_{j+1}$. Thus if we choose the site basis states in $\mathcal{B}(j+1)$ to be eigenstates of the site operator Q_{j+1} and denote $|t, s_{j+1}\rangle$ a basis state with quantum number $q_t^{S(j+1)}$, the tensor product state (21.10) becomes

$$\left| \phi_{r,\alpha}^{L(j)} ; t, s_{j+1} \right\rangle = \left| \phi_{r,\alpha}^{L(j)} \right\rangle \otimes |t, s_{j+1}\rangle \,, \tag{21.40}$$

and its quantum number (eigenvalue of $Q^{L(j+1)}$) is given by $q_p^{L(j+1)} = q_r^{L(j)} + q_t^{S(j+1)}$. Therefore, the corresponding density-matrix eigenstates take the form $\phi_{p,\alpha}^{L(j+1)}(r, \alpha', t, s_{j+1})$ and vanish if $q_p^{L(j+1)} \neq q_r^{L(j)} + q_t^{S(j+1)}$, see (21.6). Similarly, the density-matrix eigenstates for a right block are noted $\phi_{p,\beta}^{R(j+1)}(t, s_{j+1}, r, \beta')$ and vanish if $q_p^{R(j+1)} \neq q_r^{R(j+2)} + q_t^{S(j+1)}$. We can save computer time and memory if we use this rule to compute and store only the terms which do not identically vanish.

Furthermore, as $Q = Q^{L(j)} + Q_{j+1} + Q_{j+2} + Q^{R(j+3)}$, a superblock basis state (21.23) can be written

$$\left| \phi_{p,\alpha}^{L(j)} ; r, s_{j+1} ; t, s_{j+2} ; \phi_{v,\beta}^{R(j+3)} \right\rangle = \left| \phi_{p,\alpha}^{L(j)} ; r, s_{j+1} \right\rangle \otimes \left| t, s_{j+2} ; \phi_{v,\beta}^{R(j+3)} \right\rangle \,, \tag{21.41}$$

and its quantum number (eigenvalue of Q) is given by $q = q_p^{L(j)} + q_r^{S(j+1)} + q_t^{S(j+2)} + q_v^{R(j+3)}$. Therefore, the superblock representation (21.28) of a state $|\psi\rangle$ with a quantum number q can be written $[\mathsf{C}_{j+1}](p, \alpha, r, s_{j+1}, t, s_{j+2}, v, \beta)$ and vanishes if $q \neq q_p^{L(j)} + q_r^{S(j+1)} + q_t^{S(j+2)} + q_v^{R(j+2)}$. Here again we can save computer time and memory if we use this rule to compute and store only the components of C_{j+1} which do not identically vanish.

If an operator $\mathcal{O}$ has a simple commutation relation with Q of the form $[Q, \mathcal{O}] = \Delta q\ \mathcal{O}$, where Δq is a number, the matrix elements $\langle \mu|\mathcal{O}|\nu\rangle$ of $\mathcal{O}$ in the eigenbasis of Q vanish but for special combinations of the eigenstates $|\mu\rangle$ and $|\nu\rangle$. Similar rules apply for the related operators $Q^{L(n)}$, $Q^{R(n)}$, and Q_n. Explicitly, for the matrices (21.17) of site operators one finds that $\langle p, s_n|\mathcal{O}|p', s'_n\rangle = 0$ for $q_p^{S(n)} \neq q_{p'}^{S(n)} + \Delta q$ if $[Q_n, \mathcal{O}] = \Delta q\ \mathcal{O}$. For instance, for the spin operator S_n^+ with $[S^z, S_n^+] = \hbar S_n^+$ only $\langle \uparrow |S_n^+| \downarrow\rangle$ does not vanish. For the matrix representation of left block operators (21.14) one finds that

$$\left\langle \phi_{p\alpha}^{L(j)}; r, s_{j+1} \middle| \mathcal{O} \middle| \phi_{p'\alpha'}^{L(j)}; r', s'_{j+1} \right\rangle = 0 \,, \tag{21.42}$$

for $q_p^{L(j)} + q_r^{S(j+1)} \neq q_{p'}^{L(j)} + q_{r'}^{S(j+1)} + \Delta q$ if $[Q^{L(j+1)}, \mathcal{O}] = \Delta q\ \mathcal{O}$. A similar rule, applies to matrix representations (21.22) in a right block

$$\left\langle t, s_{j+2}; \phi_{v\beta}^{R(j+3)} \middle| \mathcal{O} \middle| t', s'_{j+2}; \phi_{v'\beta'}^{R(j+3)} \right\rangle = 0 \,, \tag{21.43}$$

for $q_v^{R(j+3)} + q_t^{S(j+2)} \neq q_{v'}^{R(j+3)} + q_{t'}^{S(j+2)} + \Delta q$ if $[Q^{R(j+2)}, \mathcal{O}] = \Delta q\ \mathcal{O}$. Therefore, we can reduce the computer time and memory used if we compute and save only the matrix elements which are not identically zero because of the conservation of additive quantum numbers. Moreover, if we implement these rules, the computational cost of the operations (21.15), (21.19), (21.29), (21.30), (21.32), and (21.34) to (21.37) is also substantially reduced.

In summary, using additive quantum numbers increases the complexity of a DMRG program but can reduce the computational effort significantly. In Chap. 22 it is shown that quantum numbers and symmetries can also be used with DMRG to investigate additional properties such as excited states.

21.7 Truncation Errors

There are three main sources of numerical errors in the finite-system DMRG method:

- The iterative diagonalization algorithm used to find the ground state of the superblock Hamiltonian (diagonalization error),
- the iterative optimization of matrices in the matrix-product state (21.2) (convergence error), and
- the restrictions put on the matrix dimensions a_n and b_n (truncation error).

Diagonalization errors originate from errors in the calculation of the matrix C_j in (21.2) but they propagate to the other matrices through the density-matrix based selection of the block basis states. These errors can always be made negligible compared to the other two error sources in ground state calculations. However, as the superblock diagonalization is the most time-consuming task and the other two error sources limit the overall accuracy anyway, one should not determine the superblock ground state with too much precision but strike a balance between accuracy and computational cost. In DMRG algorithms that target other states than the ground state (for instance, dynamical correlation functions, see Chap. 22), the diagonalization error may become relevant.

Convergence errors corresponds to non-optimal matrices $\mathsf{A}_n(s_n)$ and $\mathsf{B}_n(s_n)$ in the matrix-product state (21.2). They are negligible in DMRG calculations for ground state properties in non-critical one-dimensional open systems with nearest-neighbor interactions. For such cases DMRG converges after very few sweeps through the lattice. Convergence problems occur frequently in critical or inhomogeneous systems and in systems with long-range interactions (this effectively includes all systems in dimension larger than one, see the last section). However, if one performs enough sweeps through the lattice (up to several tens in hard cases), these errors can always be made smaller than truncation errors (i.e., the finite-system DMRG algorithm always finds the optimal matrices for a matrix-product state (21.2) with restricted matrix sizes).

Truncation errors are usually the dominant source of inaccuracy in the finite-system DMRG method. They can be systematically reduced by increasing the matrix dimensions a_n, b_n used in (21.2). In actual computations, however, they can be significant and it is important to estimate them reliably. In the finite-system DMRG algorithm a truncation error is introduced at every iteration when a tensor-product basis of dimension $a_j d_{j+1}$ for the left block $L(j+1)$ is reduced to a basis of dimension a_{j+1} during a sweep from left to right and, similarly, when a tensor-product basis of dimension $b_{j+2} d_{j+1}$ for the right block $R(j+1)$ is reduced to a basis of dimension b_{j+1} during a sweep from right to left. Each state $|\psi\rangle$ which is defined using the original tensor-product basis (usually, the superblock ground state) is replaced by an approximate state $|\widetilde{\psi}\rangle$ which is defined using the truncated basis. It has been shown [1] that the optimal choice for constructing a smaller block basis for a given target state $|\psi\rangle$ consists in choosing the eigenvectors with the highest eigenvalues w_μ from the reduced density-matrix (21.30) or (21.36) of $|\psi\rangle$ for this block. More precisely, this choice minimizes the differences $= \left| |\psi\rangle - |\widetilde{\psi}\rangle \right|^2$ between the target state $|\psi\rangle$ and its approximation $|\widetilde{\psi}\rangle$.

The minimum of S is given by the weight P of the discarded density-matrix eigenstates. With $w_1 \gtrsim w_2 \gtrsim \cdots \gtrsim w_{a_j d_{j+1}}$ we can write

$$S_{\text{min}} = P(a_{j+1}) = \sum_{\mu=1+a_{j+1}}^{a_j d_{j+1}} w_\mu = 1 - \sum_{\mu=1}^{a_{j+1}} w_\mu \tag{21.44}$$

for the left block $L(j+1)$ and similarly $S_{\text{min}} = P(b_{j+1}) = 1 - \sum_{\mu=1}^{b_{j+1}} w_\mu$ for the right block $R(j+1)$. It can be shown that errors in physical quantities depend directly on the discarded weight. For the ground-state energy the truncation introduces an error

$$\frac{\langle\widetilde{\psi}|H|\widetilde{\psi}\rangle}{\langle\widetilde{\psi}|\widetilde{\psi}\rangle} - \frac{\langle\psi|H|\psi\rangle}{\langle\psi|\psi\rangle} \propto P(a_{j+1}) \text{ or } P(b_{j+1}) , \tag{21.45}$$

while for other expectation values the truncation error is

$$\frac{\langle\widetilde{\psi}|O|\widetilde{\psi}\rangle}{\langle\widetilde{\psi}|\widetilde{\psi}\rangle} - \frac{\langle\psi|O|\psi\rangle}{\langle\psi|\psi\rangle} \propto \sqrt{P(a_{j+1})} \text{ or } \sqrt{P(b_{j+1})} \tag{21.46}$$

for $P(a_{j+1}), P(b_{j+1}) \ll 1$, respectively. Therefore, truncation errors for physical quantities are small when the discarded weight is small. Clearly, the discarded weight is small when the eigenvalues w_μ of the reduced density-matrices (21.30) and (21.36) decrease rapidly with increasing index μ. As discussed in Chap. 20, there are various quantum systems for which the spectrum of reduced density-matrices for subsystems has this favorable property. For such quantum systems the matrix-product state (21.2) is a good approximation and DMRG truncation errors decrease rapidly with increasing matrix sizes a_{j+1} and b_{j+1}.

In practice, there are two established methods for choosing the matrix dimensions in a systematic way in order to cope with truncation errors. First, we can perform DMRG sweeps with matrix dimensions not greater than a fixed number m of density-matrix eigenstates kept, $a_n, b_n \lesssim m$. In that approach, physical quantities [ground state energy $E_{\mathrm{DMRG}}(m)$ and other expectation values $\bar{O}_{\mathrm{DMRG}}(m)$] are calculated for several values of m and their scaling with increasing m is analyzed. Usually, one finds a convergence to a fixed value with corrections that decreases monotonically with m. This decrease is exponential in favorable cases (gapped one-dimensional systems with short-range interactions) but can be as slow as m^{-2} for systems with non-local Hamiltonians. As an example, we show in Fig. 21.5 the truncation error in the ground state energy for a 100-site Heisenberg chain. For open boundary conditions (a favorable case for a matrix product state (21.2) and thus for DMRG) the error decreases very rapidly with m until it reaches the order of

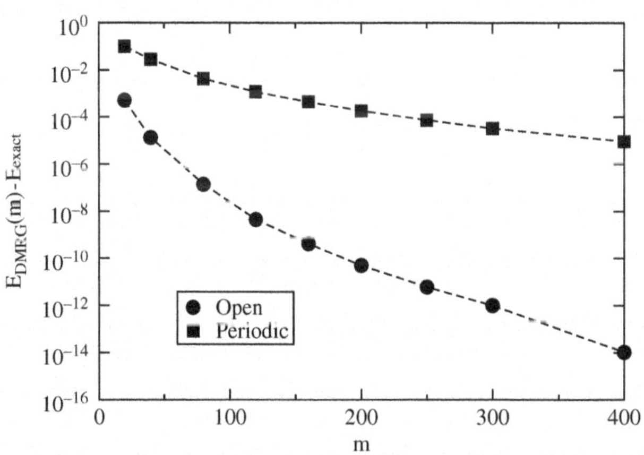

Fig. 21.5. Error in the ground state energy calculated with the finite-system DMRG algorithm as a function of the number m of density-matrix eigenstates kept for the spin-$\frac{1}{2}$ Heisenberg Hamiltonian on a one-dimensional 100-site lattice with open (*circles*) and periodic (*squares*) boundary conditions

magnitude of round-off errors in the computer system used. For periodic boundary conditions (a less favorable case) the error decreases slowly with m and is still significant for the largest number of density-matrix eigenstates considered $m = 400$.

In the second approach the density-matrix eigenbasis is truncated so that the discarded weight is approximately constant, $P(a_{j+1}), P(b_{j+1}) \lesssim P$, and thus a variable number of density-matrix eigenstates is kept at every iteration. The physical quantities obtained with this procedure depends on the chosen discarded weight P. Empirically, one finds that the relations (21.45) and (21.46) hold for DMRG results calculated with various P. For the energy one has $E_{\text{DMRG}}(P) \approx E(P = 0) + cP$ and for other expectation values $\bar{O}_{\text{DMRG}}(P) \approx \bar{O}(P = 0) + c'\sqrt{P}$ if P is small enough. Therefore, we can carry out DMRG calculations for several values of the discarded weight P and obtain results $E(P = 0)$ and $\bar{O}(P = 0)$ in the limit of vanishing discarded weight $P \to 0$ using an extrapolation. In practice, this procedure yields reliable estimations of the truncation errors and often the extrapolated results are more accurate than those obtained directly with DMRG for the smallest value of P used in the extrapolation.

It should be noted that if one works with a fixed number m of density-matrix eigenstates kept, it is possible to calculate an average discarded weight $P(m) = \sum_j P(b_{j+1})$ over a sweep. In many cases, the physical quantities $E_{\text{DMRG}}(m)$ and $\bar{O}_{\text{DMRG}}(m)$ scale with $P(m)$ as in (21.45) and (21.46), respectively. Therefore, an extrapolation to the limit of vanishing discarded weight $P(m) \to 0$ is also possible (see [12] for some examples).

21.8 Computational Cost and Optimization

Theoretically, the computational cost for one calculation with the infinite-system algorithm or for one sweep of the finite-system algorithm is proportional to $Nn_k m^3 d^3$ for the number of operations and to $Nn_k m^2 d^2$ for the memory if one assumes that about m density-matrix eigenstates are kept at every iteration and the site Hilbert space dimension is d. In practice, various optimization techniques such as the additive quantum numbers of Sect. 21.6 lead to a more favorable scaling with m. The actual computational effort varies greatly with the model investigated and the physical quantities which are computed. Using a highly optimized code the infinite-system DMRG simulations shown in Fig. 21.2 take from 30 seconds for $m = 20$ to 7 minutes for $m = 100$ on 3 GHz Pentium 4 processor while the finite-system DMRG calculations shown in Figs. 21.3 and 21.4 take about 20 minutes. These calculations use less than to 300 MBytes of memory. For more difficult problems with $m \lesssim 10^4$, the computational cost can reach thousands of CPU hours and hundreds of GBytes of memory.

Even for less challenging problems, it is useful to consider some basic optimization issues for a DMRG code. First of all, an efficient dynamical memory management should be used because many vectors, matrices, and tensors of higher ranks with variable sizes have too be stored temporarily. Even for the simplest applications this amounts to the allocation and release of GBytes of memory during a DMRG simulation. Second, processor-optimized linear algebra routines should be used because most of the CPU time is spent for linear algebra operations such as products of matrices. Generic BLAS routines [13] can be as much as two orders of magnitude slower than processor-optimized ones. Third, the most expensive part of a DMRG iteration is usually the calculation of the superblock representation (21.28) of the target state, typically the ground state of the superblock Hamiltonian. Thus one should use the procedures (21.29) and (21.35) and the efficient iterative algorithms described in Chap. 18 to perform this task. Finally, one should also consider using parallelization and optimization techniques for high-performance computers (see Chap. 27). Unfortunately, the basic DMRG algorithms presented in Sects. 21.4 and 21.5 are inherently sequential and a parallelization is possible only at a low level. For instance, the superblock product (21.29) or, at an even lower level, matrix products (i.e., the BLAS routines) can be parallelized. The parallelization of a DMRG code is discussed in more detail in [14].

21.9 Basic Extensions

The finite-system DMRG algorithm described in Sect. 21.5 can readily be applied to one-dimensional quantum spin chains. It can also be used to study fermions, bosons, and systems in higher dimensions without much difficulty.

To apply the DMRG algorithm to fermion systems we just have to take into account the fermion commutation sign in the operator decomposition (21.11) and (21.24) and in the tensor product of their matrix representations (21.15), (21.27), and (21.34). Using the total number of fermion as an additive quantum number is very helpful for that purpose. For instance, if $|\alpha\rangle, |\alpha'\rangle, |\beta\rangle, |\beta'\rangle$ denote states with a fixed number of fermions and $\mathcal{O}_1$,$\mathcal{O}_2$ are operators, the matrix element of the tensor-product operator $\mathcal{O}_1 \otimes \mathcal{O}_2$ for the tensor-product states $|\alpha\beta\rangle = |\alpha\rangle \otimes |\beta\rangle$ and $|\alpha'\beta'\rangle = |\alpha'\rangle \otimes |\beta'\rangle$ is

$$\langle \alpha\beta \,|\mathcal{O}_1 \otimes \mathcal{O}_2|\, \alpha'\beta' \rangle = (-1)^{q|\Delta q|} \langle \alpha \,|\mathcal{O}_1|\, \alpha' \rangle \langle \beta \,|\mathcal{O}_2|\, \beta' \rangle \ , \tag{21.47}$$

where q is the number of fermions in the state $|\alpha'\rangle$ and Δq is the difference between the number of fermions in the states $|\beta\rangle$ and $|\beta'\rangle$.

To apply the DMRG method to boson systems such as electron-phonon models, we must first choose an appropriate finite basis for each boson site to represent the infinite Hilbert space of a boson as best as possible, which is done also in exact diagonalization methods [12]. Then the finite-system DMRG algorithm can be used without modification. However, the computational cost scales as d^3 for the CPU time and as d^2 for the memory if d states are used to represent each boson site. Typically, $d = 10-100$ is required for accurate computations in electron-phonon models. Therefore, simulating boson systems with the standard DMRG algorithms is significantly more demanding than spin systems. More sophisticated DMRG algorithms have been developed to reduce the computational effort involved in solving boson systems. The best algorithms scale as d or $d\ln(d)$ and are presented in [12].

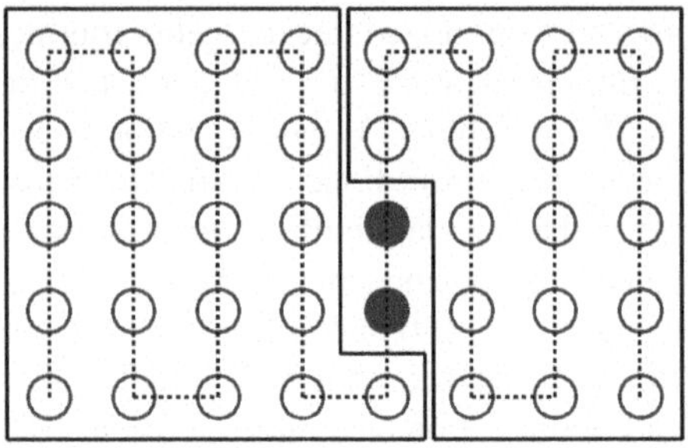

Fig. 21.6. Schematic representations of the site sequence (*dashed line*) in a two-dimensional lattice. The site in the bottom left corner is site 1. The superblock structure $\{L(21)+$ site 22 + site 23 $+R(24)\}$ is shown with *solid lines* delimiting the left and right blocks and full circles indicating both sites

The finite-system DMRG method can be applied to quantum systems with various degrees of freedom, on lattices in dimension larger than one, and to the non-local Hamiltonians considered in quantum chemistry and momentum space, see Chap. 24. We just have to order the lattice sites from 1 to N in some way to be able to carry out the algorithm described in Sect. 21.5. For instance, Fig. 21.6 shows one possible site sequence for a two-dimensional cluster. It should be noted that sites which are close in the two-dimensional lattice are relatively far apart in the sequence. This corresponds to an effective long-range interaction between the sites even if the two-dimensional system includes short-range interactions only, and results in a slower convergence and larger truncation errors than in truly one-dimensional systems with short-range interactions. As a consequence, reordering of the lattice sites can significantly modify the accuracy of a DMRG calculation and various site sequences should be considered for those systems which do not have a natural order. The difficulty with DMRG simulations and more generally with matrix-product states in dimensions larger than one is discussed fully in Chap. 20.

References

1. I. Peschel, X. Wang, M. Kaulke, K. Hallberg (eds.), *Density-Matrix Renormalization, A New Numerical Method in Physics* (Springer, Berlin, 1999)
2. R.M. Noack, S.R. Manmana, in *Lectures on the Physics of Highly Correlated Electron Systems IX: Ninth Training Course in the Physics of Correlated Electron Systems and High-Tc Superconductors, AIP Conf. Proc.*, Vol. 789, ed. by A. Avella, F. Mancini (AIP, 2005), pp. 93–163
3. S.R. White, Phys. Rev. Lett. **69**(19), 2863 (1992)
4. S.R. White, Phys. Rev. B **48**(14), 10345 (1993)
5. U. Schollwock, Rev. Mod. Phys. **77**(1), 259 (2005)
6. K. Hallberg, Adv. Phys. **55**, 477 (2006)
7. I.P. McCulloch (2007). URL `http://arxiv.org/abs/cond-mat/0701428`. Preprint

8. K.G. Wilson, Rev. Mod. Phys. **47**(4), 773 (1975)
9. S.R. White, R.M. Noack, Phys. Rev. Lett. **68**(24), 3487 (1992)
10. L. Hulthén, Arkiv Mat. Astr. Fysik **26A**(11), 1 (1938)
11. I.P. McCulloch, M. Gulàcsi, Europhys. Lett. **57**(6), 852 (2002)
12. E. Jeckelmann, H. Fehske, in *Proceedings of the International School of Physics "Enrico Fermi" - Course CLXI Polarons in Bulk Materials and Systems with Reduced Dimensionality* (IOS Press, Amsterdam, 2006), pp. 247–284
13. Basic Linear Algebra Subprograms (BLAS). URL `http://www.netlib.org/blas/`
14. G. Hager, E. Jeckelmann, H. Fehske, G. Wellein, J. Comput. Phys. **194**, 795 (2004)

22 Dynamical Density-Matrix Renormalization Group

Eric Jeckelmann[1] and Holger Benthien[2]

[1] Institut für Theoretische Physik, Leibniz Universität Hannover, 30167 Hannover, Germany

[2] Physical and Theoretical Chemistry Laboratory, Oxford University, Oxford OX1 3QZ, United Kingdom

The dynamical density-matrix renormalization group (DDMRG) method is a numerical technique for calculating the zero-temperature dynamical properties in low-dimensional quantum many-body systems. For the one-dimensional Hubbard model and its extensions, DDMRG allows for accurate calculations of these properties for lattices with hundreds of sites and particles and for any excitation energy. The key idea of this approach is a variational principle for dynamical correlation functions. The variational problem can be solved with a standard density-matrix renormalization group (DMRG) method. Combined with a finite-size-scaling analysis for dynamical spectra, the DDMRG method enables us to study dynamical properties in the thermodynamic limit. An efficient calculation of momentum-dependent quantities with DMRG is achieved using open boundary conditions and quasi-momenta. These techniques are illustrated with the photoemission spectral function of the half-filled one-dimensional Hubbard model.

22.1 Introduction

Calculating the dynamical correlation functions of quantum many-body systems has been a long-standing problem of theoretical physics because many experimental techniques probe these properties. For instance, solid-state spectroscopy experiments, such as optical absorption, photoemission, or nuclear magnetic resonance, measure the dynamical correlations between an external time-dependent perturbation and the response of electrons and phonons in solids [1]. Typically, the zero-temperature dynamic response of a quantum system is given by a dynamical correlation function (with $\hbar = 1$)

$$G_X(\omega + i\eta) = -\frac{1}{\pi} \left\langle \psi_0 \left| X^\dagger \frac{1}{E_0 + \omega + i\eta - H} X \right| \psi_0 \right\rangle , \qquad (22.1)$$

where H is the time-independent Hamiltonian of the system, E_0 and $|\psi_0\rangle$ are its ground-state energy and wavefunction, X is the quantum operator corresponding to the physical quantity which is analyzed, and $X^\dagger$ is the Hermitian conjugate of X. A small real number $\eta > 0$ is used to shift the poles of the correlation function into the complex plane. The spectral function $G_X(\omega + i\eta)$ is also the Laplace transform (up to a constant prefactor) of the zero-temperature time-dependent correlation function

E. Jeckelmann and H. Benthien: *Dynamical Density-Matrix Renormalization Group*, Lect. Notes Phys. **739**, 621–635 (2008)
DOI 10.1007/978-3-540-74686-7_22

$$G_X(t \geq 0) = \langle \psi_0 | X^\dagger(t) X(0) | \psi_0 \rangle , \tag{22.2}$$

where $X(t)$ is the Heisenberg representation of the operator X. In general, we are interested in the imaginary part of the correlation function for $\eta \to 0$

$$I_X(\omega + \mathrm{i}\eta) = \mathrm{Im}\, G_X(\omega + \mathrm{i}\eta) = \frac{1}{\pi} \left\langle \psi_0 \left| X^\dagger \frac{\eta}{(E_0 + \omega - H)^2 + \eta^2} X \right| \psi_0 \right\rangle . \tag{22.3}$$

A fundamental model for one-dimensional correlated electron systems is the Hubbard model [2] defined by the Hamiltonian

$$H = -t \sum_{j;\sigma} \left(c^\dagger_{j\sigma} c_{j+1\sigma} + c^\dagger_{j+1\sigma} c_{j\sigma} \right) + U \sum_j n_{j\uparrow} n_{j\downarrow} - \frac{U}{2} \sum_j n_j . \tag{22.4}$$

It describes electrons with spin $\sigma = \uparrow, \downarrow$ which can hop between neighboring sites on a lattice. Here $c^\dagger_{j\sigma}$ and $c_{j\sigma}$ are creation and annihilation operators for electrons with spin σ at site j $(= 1, \ldots, N)$, $n_{j\sigma} = c^\dagger_{j\sigma} c_{j\sigma}$ are the corresponding density operators, and $n_j = n_{j\uparrow} + n_{j\downarrow}$. The hopping integral t gives rise to a single-electron band of width $4t$. The Coulomb repulsion between electrons is mimicked by a local Hubbard interaction $U \geq 0$. The chemical potential has been chosen $\mu = U/2$ so that the number of electrons is equal to the number of sites N (half-filled band) in the grand-canonical ground state and the Fermi energy is $\varepsilon_\mathrm{F} = 0$ in the thermodynamic limit. The photoemission spectral function $A(k, \omega)$ is the imaginary part of the one-particle Green's function

$$A_\sigma(k, \omega \leq 0) = \lim_{\eta \to 0} I_X(-\omega + \mathrm{i}\eta) , \tag{22.5}$$

for the operator $X = c_{k\sigma}$ which annihilates an electron with spin σ in the Bloch state with wavevector $k \in (-\pi, \pi]$. This spectral function corresponds to the spectrum measured in angle-resolved photoemission spectroscopy experiments. We note that the spectral function of the Hubbard model is symmetric with respect to spatial reflection $A_\sigma(-k, \omega) = A_\sigma(k, \omega)$ and spin-reflection $A_\uparrow(k, \omega) = A_\downarrow(k, \omega)$. The one-particle density of states (DOS) is

$$n_\sigma(\omega \leq 0) = \frac{1}{N} \sum_k A_\sigma(k, \omega) . \tag{22.6}$$

At half-filling the inverse photoemission spectral function $B_\sigma(k, \omega \geq 0)$ is related to $A_\sigma(k, \omega)$ through the relation $B_\sigma(k, \omega) = A_\sigma(k + \pi, -\omega)$ and thus $n_\sigma(\omega \geq 0) = \frac{1}{N} \sum_k B_\sigma(k, \omega)$ is equal to $n_\sigma(-\omega)$.

Since its invention in 1992 the DMRG method [3, 4] has established itself has the most powerful numerical method for studying the properties of one-dimensional lattice models such as the Hubbard model (for reviews, see [5, 6, 7]). Several approaches have been developed to obtain excited states or to calculate dynamical quantities with DMRG. We will first discuss a few approaches which are both simple and efficient but do not allow for the calculation of continuous or complicated spectra. Then we will present the dynamical DMRG method, which is presently the best frequency-space DMRG approach for calculating zero-temperature dynamical correlation functions when the spectrum is complex or continuous and allows us to determine spectral properties in the thermodynamic limit (i.e., for infinitely large lattices). The basic principles of the DMRG method are described in the Chaps. 20 and 21 of this book and are assumed to be known. The direct calculation of time-dependent quantities (22.2) within DMRG is explained in Chap. 23 while methods for computing dynamical quantities at finite temperature are described in Chap. 25.

22.2 Methods for Simple Discrete Spectra

22.2.1 Direct Evaluation of Excited States

Let $\{|n\rangle, n = 0, 1, 2, \dots\}$ be the complete set of eigenstates of H with eigenenergies E_n ($|n = 0\rangle$ corresponds to the ground state $|\psi_0\rangle$). The spectrum (22.3) can be written

$$I_X(\omega + \mathrm{i}\eta) = \frac{1}{\pi} \sum_n |\langle n|X|0\rangle|^2 \frac{\eta}{(E_n - E_0 - \omega)^2 + \eta^2} . \tag{22.7}$$

$E_n - E_0$ is the excitation energy and $|\langle n|X|0\rangle|^2$ the spectral weight of the n-th excited state. Obviously, only states with a finite spectral weight contribute to the dynamical correlation function. Typically, the number of contributing excited states scales as a power of the system size N (while the Hilbert space dimension increases exponentially with N). In principle, one can calculate the contributing excited states only and reconstruct the spectrum from the sum over these states (22.7).

The simplest method for computing excited states within DMRG is to target the lowest M eigenstates $|\psi_s\rangle$ instead of the sole ground state using the standard algorithm. In that case, the density matrix is formed as the sum

$$\rho = \sum_{s=1}^{M} c_s \rho_s \tag{22.8}$$

of the density matrices $\rho_s = |\psi_s\rangle\langle\psi_s|$ for each target state [8]. As a result the DMRG algorithm produces an effective Hamiltonian describing these M states accurately. Here the coefficients $c_s > 0$ are normalized weighting factors ($\sum_s c_s = 1$), which allow us to vary the influence of each target state in the formation of the density matrix. This approach yields accurate results for some problems such as the Holstein polaron [9]. In most cases, however, this approach is limited to a small number M of excited states (of the order of ten) because DMRG truncation errors grow rapidly with the number of targeted states (for a fixed number of density-matrix eigenstates kept). This is not sufficient for calculating a complete spectrum for a large system and often does not even allow for the calculation of low-energy excitations. For instance, in the strong-coupling regime $U \gg t$ of the half-filled one-dimensional

Hubbard model, the lowest excitation contributing to the photoemission spectral function $A(k,\omega)$ has an energy $-\omega = E_n - E_0 \approx U/2$ while there are many spin excitations with energies smaller than $U/2$. Therefore, in order to obtain the first contributing excitation, one would have to target an extremely large number M of eigenstates with DMRG.

22.2.2 Quantum Numbers and Symmetries

The simplest DMRG method for calculating specific excited states (rather than the lowest eigenstates) uses the conserved quantum numbers and the symmetries of the system. If quantum numbers and symmetry operators are well-defined in every subsystem, DMRG calculations can be carried out to obtain the lowest eigenstates in a specific symmetry subspace and for specific quantum numbers. As an example, the total number of particles is conserved in the Hubbard model (i.e., the particle number operator $N = \sum_j n_j$ commutes with the Hamilton operator H). Thus one can target the M lowest eigenstates for a given number of particles. This yields useful information about excitations contributing to the spectral functions $A(k,\omega)$ and $B(k,\omega)$. For instance, a gap in the density of states $n_\sigma(\omega)$ is given by

$$E_c = E_0(+1) + E_0(-1) - 2E_0(0) \; , \tag{22.9}$$

where $E_0(z)$ is the lowest eigenenergy for a system with z electrons added or removed from the half-filled band.

It is also possible to target simultaneously the lowest eigenstates for two different sets of quantum numbers or in two different symmetry subspaces using (22.8) and thus to calculate matrix elements $\langle n|X|0\rangle$ between these states. This allows one to reconstruct the dynamical correlation function at low energy using the Lehmann representation (22.7). For instance, to calculate the photoemission spectral function (22.5) of the Hubbard model we would target the ground state for N electrons and the lowest M eigenstates with $N-1$ electrons as only those states contribute to $A(k,\omega)$. This way we circumvent the many low-energy spin excitations in the N-electron subspace. In practice, this method works only for the onset of the spectrum because there are still a large number of weightless spin excitations between successive $(N-1)$-electron states contributing to $A(k,\omega)$ and one would again have to target a very large number M of eigenstates to access the relevant high-energy excitations.

Using quantum numbers and symmetries is the most efficient and accurate method for calculating specific low-lying excited states with DMRG. For instance, symmetries and quantum numbers have been used successfully to study optical excitations and the low-energy optical conductivity spectrum in various extended one-dimensional Hubbard models describing conjugated polymers [10, 11]. However, this approach is obviously restricted to those problems which have relevant symmetries and quantum numbers and provides at most the lowest M eigenstates with the chosen symmetries and quantum numbers, where M is at most a few tens for realistic applications. Thus it is not appropriate for high-energy excitations and for complex or continuous spectra.

22.2.3 Lanczos-DMRG Method

The Lanczos-DMRG method [12, 13] combines DMRG with the Lanczos algorithm [14] to compute dynamical correlation functions. Starting from the states $|\phi_{-1}\rangle = 0$ and $|\phi_0\rangle = X|\psi_0\rangle$, the Lanczos algorithm recursively generates a set of so-called Lanczos vectors:

$$|\phi_{n+1}\rangle = H|\phi_n\rangle - a_n|\phi_n\rangle - b_n^2|\phi_{n-1}\rangle \, , \tag{22.10}$$

where $a_n = \langle\phi_n|H|\phi_n\rangle/\langle\phi_n|\phi_n\rangle$ and $b_{n+1}^2 = \langle\phi_{n+1}|\phi_{n+1}\rangle/\langle\phi_n|\phi_n\rangle$ for $n = 0, \dots, L-1$. These Lanczos vectors span a Krylov subspace containing the excited states contributing to the dynamical correlation function (22.1). Calculating L Lanczos vectors gives the first $2L-1$ moments of a spectrum and up to L excited states contributing to it. The spectrum can be obtained from the continued fraction expansion

$$-\pi G_X(z - E_0) = \cfrac{\langle\psi_0|X^\dagger X|\psi_0\rangle}{z - a_0 - \cfrac{b_1^2}{z - a_1 - \cfrac{b_2^2}{z - \dots}}} \, . \tag{22.11}$$

This procedure has proved to be efficient and reliable in the context of exact diagonalizations (see Chap. 18).

Within a DMRG calculation the Lanczos algorithm is applied to the effective superblock operators H and X and serves two purposes. Firstly, it is used to compute the full dynamical spectrum. Secondly, in addition to the ground state $|\psi_0\rangle$ some Lanczos vectors $\{|\phi_n\rangle, n = 0, \dots, M \leq L\}$ are used as target (22.8) to construct an effective representation of the Hamiltonian which describes both ground state and excited states accurately. Be reminded that a target state does not need to be an eigenstate of the Hamiltonian but can be any quantum state which is well-defined and can be computed in every superblock during a DMRG sweep through the lattice. Unfortunately, DMRG truncation errors increase rapidly with the number M of target Lanczos vectors for a fixed number of density-matrix eigenstates kept and the method becomes numerically unstable. Therefore, only the first few Lanczos vectors (often only the first one $|\phi_0\rangle$) are included as target in most applications of Lanczos DMRG. In that case, the density-matrix renormalization does not necessarily converge to an optimal representation of H for all excited states contributing to a dynamical correlation function and the calculated spectrum is not always reliable. For instance, the shape of continuous spectra (for very large systems $N \gg 1$) can not be determined accurately with the Lanczos-DMRG method [13]. Nevertheless, Lanczos DMRG is a relatively simple and quick method for calculating dynamical properties within DMRG. In practice, it gives reliable and accurate results for simple discrete spectra made of (or dominated by) a few peaks only and it has been used successfully in several studies of low-dimensional correlated systems (see [6, 7]).

22.3 Dynamical DMRG

22.3.1 Correction Vector

The correction vector associated with the dynamical correlation function $G_X(\omega + i\eta)$ is defined by [15]

$$|\psi_X(\omega + i\eta)\rangle = \frac{1}{E_0 + \omega + i\eta - H}|X\rangle \,, \tag{22.12}$$

where $|X\rangle = X|\psi_0\rangle$ is identical to the first Lanczos vector. If the correction vector is known, the dynamical correlation function can be calculated directly

$$G_X(\omega + i\eta) = -\frac{1}{\pi}\langle X|\psi_X(\omega + i\eta)\rangle \,. \tag{22.13}$$

To calculate a correction vector an inhomogeneous linear equation system

$$(E_0 + \omega + i\eta - H)|\psi\rangle = |X\rangle \,, \tag{22.14}$$

has to be solved for the unknown state $|\psi\rangle$. Typically, the vector space dimension is very large and the equation system is solved with the conjugate gradient method [16] or other iterative methods [17].

The distinctive characteristic of a correction vector approach to the calculation of dynamical properties is that a specific quantum state (22.12) is constructed to compute the dynamical correlation function at each frequency ω. To obtain a complete dynamical spectrum, the procedure has to be repeated for many different frequencies. Therefore, in the context of exact diagonalizations the correction-vector approach is less efficient than the Lanczos technique (22.10) and (22.11). For DMRG calculations, however, this is a highly favorable characteristic. The dynamical correlation function can be determined for each frequency ω separately using effective representations of the system Hamiltonian H and operator X which describe a single excitation energy accurately. The approach can be extended to higher-order dynamic response functions such as third-order optical polarizabilities [18].

In practice, in a correction-vector DMRG calculation [13] two correction vectors with close frequencies ω_1 and ω_2 and finite broadening $\eta \sim \omega_2 - \omega_1 > 0$ are calculated from the effective superblock operators H and X and used as target (22.8) beside the ground state $|\psi_0\rangle$ and the first Lanczos vector $|X\rangle$. This is sufficient to obtain an accurate effective representation of the system excitations for frequencies $\omega_1 \lesssim \omega \lesssim \omega_2$. The spectrum is then calculated for this frequency interval using (22.13). The calculation is repeated for several (possibly overlapping) intervals to determine the spectral function over a large frequency range. This correction-vector DMRG method allows one to perform accurate calculations of complex or continuous spectra for all frequencies in large lattice quantum many-body systems [6, 7, 13].

22.3.2 Variational Principle

The success of the correction-vector DMRG method for calculating dynamical properties show that using specific target states for each frequency is the right approach. This idea can be further improved using a variational formulation of the problem [19]. Consider the functional

$$W_{X,\eta}(\omega, \psi) = \langle\psi|(E_0 + \omega - H)^2 + \eta^2|\psi\rangle + \eta\langle X|\psi\rangle + \eta\langle\psi|X\rangle \ . \quad (22.15)$$

For any $\eta \neq 0$ and a fixed frequency ω this functional has a well-defined and non-degenerate minimum $|\psi_{\text{min}}\rangle$. This state is related to the correction vector (22.12) by

$$(H - E_0 - \omega + \text{i}\eta)|\psi_{\text{min}}\rangle = \eta|\psi_X(\omega + \text{i}\eta)\rangle \ . \quad (22.16)$$

The value of the minimum yields the imaginary part of the dynamical correlation function

$$W_{X,\eta}(\omega, \psi_{\text{min}}) = -\pi\eta I_X(\omega + \text{i}\eta) \ . \quad (22.17)$$

Therefore, the calculation of spectral functions can be formulated as a minimization problem.

This variational formulation is completely equivalent to the correction-vector method if we can calculate $|\psi_{\text{min}}\rangle$ and $|\psi_X(\omega + \text{i}\eta)\rangle$ exactly. However, if we can only calculate approximate states with an error of the order $\varepsilon \ll 1$, the variational formulation (22.17) gives the imaginary part $I_X(\omega + \text{i}\eta)$ of the correlation function with an accuracy of the order of ε^2, while the correction-vector approach (22.13) yields results with an error of the order of ε.

22.3.3 DDMRG Algorithm

The DMRG method can be used to minimize the functional $W_{X,\eta}(\omega, \psi)$ and thus to calculate the dynamical correlation function $G_X(\omega + \text{i}\eta)$. This approach is called the dynamical DMRG method. The minimization of the functional is easily integrated into the standard DMRG algorithm. At every step of a sweep through the system lattice, the following calculations are performed for the effective superblock operators H and X:

(i) The ground state vector $|\psi_0\rangle$ of H and its energy E_0 are calculated as in the standard DMRG method.
(ii) The state $|X\rangle = X|\psi_0\rangle$ is calculated.
(iii) The functional $W_{X,\eta}(\omega, \psi)$ is minimized using an iterative minimization algorithm. This yields the imaginary part $I_X(\omega + \text{i}\eta)$ of the dynamical correlation function and the state $|\psi_{\text{min}}\rangle$.
(iv) The correction vector is calculated using (22.16).
(v) The states $|\psi_0\rangle$, $|X\rangle$, and $|\psi_X(\omega + \text{i}\eta)\rangle$ are used as target (22.8) of the density-matrix renormalization process.

The robust finite-system DMRG algorithm must be used to perform several sweeps through a lattice of fixed size. Sweeps are repeated until the procedure has converged to the minimum of $W_{X,\eta}(\omega, \psi)$.

To obtain the spectrum $I_X(\omega + i\eta)$ over a range of frequencies, one has to repeat this calculation for several values of ω. The computational effort is thus roughly proportional to the number of frequencies. As with the correction-vector approach, one can perform a DDMRG calculation for two close frequencies ω_1 and ω_2 simultaneously, and then calculate the dynamical correlation function for frequencies ω between ω_1 and ω_2 without targeting the corresponding correction vectors. This approach can significantly reduce the computer time necessary to determine the spectrum over a frequency range but the results obtained for $\omega \neq \omega_1, \omega_2$ are less accurate than for the targeted frequencies $\omega = \omega_1$ and $\omega = \omega_2$.

22.3.4 Accuracy of DDMRG

First, it should be noted that DDMRG calculations are always performed for a finite parameter η. The spectrum $I_X(\omega + i\eta)$ is equal to the convolution of the true spectrum $I_X(\omega)$ with a Lorentzian distribution of width η

$$I_X(\omega + i\eta) = \int_{-\infty}^{+\infty} d\omega' I_X(\omega') \frac{1}{\pi} \frac{\eta}{(\omega - \omega')^2 + \eta^2} . \tag{22.18}$$

Therefore, DDMRG spectra are always broadened and it is sometimes necessary to perform several calculations for various η to determine $I_X(\omega)$ accurately. In most cases, however, the appropriate broadening for DDMRG calculations has merely the positive side effects of smoothing out numerical errors and hiding the discreteness of the spectrum, which is a finite-size effect (see the next section).

If a complete spectrum $I_X(\omega + i\eta)$ has been obtained, it is possible to calculate the total spectral weight by integration and to compare it to ground state expectation values using sum rules. This provides an independent check of DDMRG results. For instance, the total weight of the photoemission spectral function must fulfill the relation

$$\int_{-\infty}^{0} d\omega A_\sigma(k, \omega) = n_\sigma(k) , \tag{22.19}$$

where $n_\sigma(k) = \langle \psi_0 | c^\dagger_{k\sigma} c_{k\sigma} | \psi_0 \rangle$ is the ground state momentum distribution.

Numerous comparisons with exact analytical results and accurate numerical simulations have demonstrated the unprecedented accuracy and reliability of the dynamical DMRG method for calculating dynamical correlation functions in one-dimensional correlated systems [6, 9, 19, 20, 21, 22] and quantum impurity problems [23, 24, 25]. As an example, we show in Fig. 22.1 the local DOS of the half-filled one-dimensional Hubbard model calculated with DDMRG for two values of U. The local DOS is obtained by substituting $X = c_{j\sigma}$ and $X = c^\dagger_{j\sigma}$ for $c_{k\sigma}$ and $c^\dagger_{k\sigma}$ in the definition of the spectral functions $A(k, \omega)$ and $B(k, \omega)$, respectively.

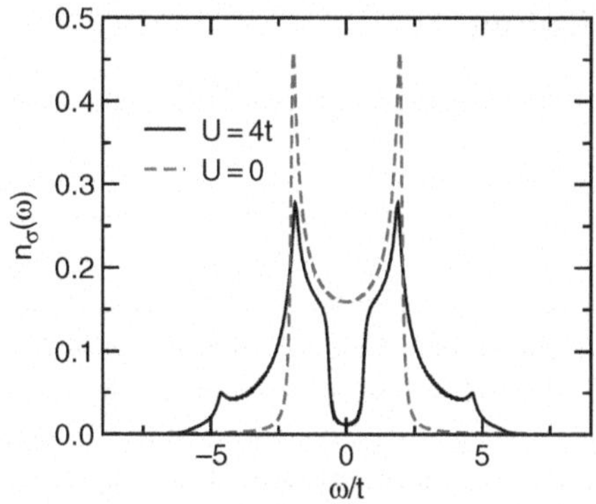

Fig. 22.1. Local density of states of the half-filled one-dimensional Hubbard model for $U = 0$ and $U = 4t$ calculated in the middle of an open chain with 128 sites using DDMRG and a broadening $\eta = 0.08t$

The local DOS does not depend on the site j for periodic boundary conditions and is equal to the integrated DOS defined in Sect. 22.1. For open boundary conditions we have checked that the local DOS in the middle of the chain is indistinguishable from the integrated DOS (22.6) for the typical broadening η used in DDMRG calculations [22]. On the scale of Fig. 22.1 the DDMRG DOS for the metallic regime ($U = 0$) is indistinguishable from the exact result (with the same broadening η), which illustrates the accuracy of DDMRG. For the insulating regime $U = 4t$, one clearly sees the opening of the Mott-Hubbard gap in Fig. 22.1. The width of the gap agrees with the exact result $E_c \approx 1.286t$ calculated with the Bethe Ansatz method [2]. The shape of the spectrum around the spectrum onsets at $\omega \approx \pm E_c/2 \approx 0.643t$ also agrees with field-theoretical predictions as discussed in the next section. The effects of the broadening $\eta = 0.08t$ are also clearly visible in Fig. 22.1: For $U = 4t$ spectral weight is seen inside the Mott-Hubbard gap and for $U = 0$ the DOS divergences at $\omega = \pm 2t$ have been broadened into two sharp peaks.

The numerical errors in the DDMRG method are dominated by the truncation of the Hilbert space. As in a ground state DMRG calculation, this truncation error decreases (and thus the accuracy of DDMRG target states and physical results increases) when more density-matrix eigenstates are kept. As the variational approach yields a smaller error in the spectrum than the correction-vector approach for the same accuracy in the targeted states, the DDMRG method is usually more accurate than the correction-vector DMRG method for the same number of density-matrix eigenstates kept or, equivalently, the DDMRG method is faster than the correction-vector DMRG method for a given accuracy.

22.4 Finite-Size Scaling

If only a finite number of eigenstates contributes to a dynamical correlation function, the spectrum (22.7) is discrete in the limit $\eta \to 0$

$$I_X(\omega) = \sum_n |\langle n|X|0\rangle|^2 \delta(E_n - E_0 - \omega) . \qquad (22.20)$$

If the number of contributing eigenstates is infinite (for instance, in the thermodynamic limit $N \to \infty$ of the Hubbard model), the spectrum $I_X(\omega)$ may also include continuous structures. DDMRG allows us to calculate the spectrum of a large but finite system with a broadening given by the parameter η. To determine the properties of a dynamical spectrum in the thermodynamic limit, one has to calculate

$$I_X(\omega) = \lim_{\eta \to 0} \lim_{N \to \infty} I_X(\omega + i\eta) . \qquad (22.21)$$

It should be noted that the order of limits in the above formula is important. Computing both limits from numerical results requires a lot of accurate data for different values of η and N and can be the source of large extrapolation errors. A better approach is to use a broadening $\eta(N) > 0$ which decreases with increasing N and vanishes in the thermodynamic limit [19]:

$$I(\omega) = \lim_{N \to \infty} I_X(\omega + i\eta(N)) . \qquad (22.22)$$

The function $\eta(N)$ depends naturally on the specific problem studied and can also vary for each frequency ω considered. For one-dimensional correlated electron systems such as the Hubbard model, one finds empirically that the optimal scaling is

$$\eta(N) = \frac{c}{N} , \qquad (22.23)$$

where the constant c is comparable to the effective band width of the excitations contributing to the spectrum around ω.

In Fig. 22.2 we see that the DOS of the half-filled one-dimensional Hubbard model becomes progressively step-like around $\omega \approx 0.643t$ as the system size is increased using a size-dependent broadening $\eta = 10.24t/N$. The slope of $n_\sigma(\omega)$ has a maximum at a frequency which tends to half the value of the Mott-Hubbard gap $E_c \approx 1.286t$ for $N \to \infty$. The height of the maximum diverges as $\eta^{-1} \sim N$ for increasing N (see the inset in Fig. 22.2). This demonstrates the presence of a Dirac-function peak $\delta(\omega - E_c/2)$ in the derivative of $n_\sigma(\omega)$ [19] or, equivalently, a step increase of the DOS at the spectrum onset in the thermodynamic limit, in agreement with the field-theoretical result for a one-dimensional Mott insulator [26]. Thus the features of the infinite-system spectrum can be determined accurately from DDMRG data for finite systems using a finite-size scaling analysis with a size-dependent broadening $\eta(N)$.

It should be noted that a good approximation for a continuous infinite-system spectrum can sometimes be obtained at a much lower computational cost than this scaling analysis by solving the convolution equation (22.18) numerically for an unknown smooth function $I_X(\omega')$ using DDMRG data for a finite system on the left-hand side (deconvolution) [9, 24, 27].

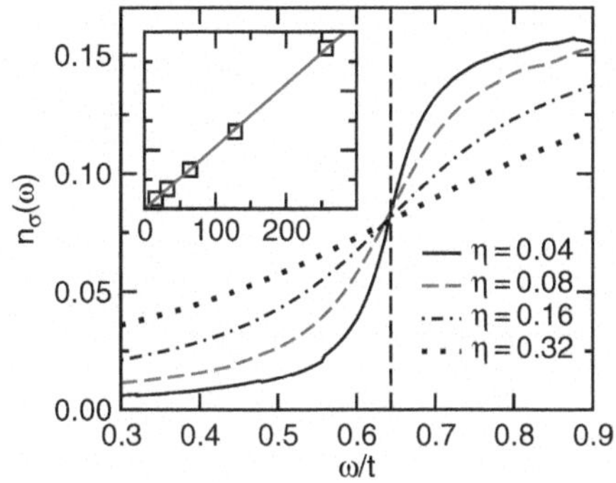

Fig. 22.2. Expanded view of the DOS around the spectrum onset at $\omega = E_c/2 \approx 0.643t$ (*vertical dashed line*) in the the half-filled one-dimensional Hubbard model for $U = 4t$. The data have been obtained with DDMRG for various system sizes from $N = 32$ to $N = 256$ with a broadening $\eta = 10.24t/N$. The inset shows the slope of $n_\sigma(\omega)$ at $\omega = E_c/2$ as a function of the system size

22.5 Momentum-Dependent Quantities

The DMRG method is usually implemented in real space where it performs optimally for one-dimensional systems with open boundary conditions and short-range interactions only [5, 6]. If periodic boundary conditions are used, momentum dependent operators, such as the operators $c_{k\sigma}$ used in the definition of the spectral functions $A(k,\omega)$, can be readily expanded as a function of local (real space) operators using plane waves (or Bloch states) [12]

$$c_{k\sigma} = \frac{1}{\sqrt{N}} \sum_{j=1}^{N} \mathrm{e}^{-\mathrm{i}kj} c_{j\sigma} , \tag{22.24}$$

with wavevectors $k = 2\pi z/N$ (momentum $p = \hbar k$) for integers $-N/2 < z \leq N/2$. These plane waves are the one-electron eigenstates of the Hamiltonian (22.4) in the non-interacting limit ($U = 0$) for periodic boundary conditions.

Since DMRG calculations can be performed for much larger systems using open boundary conditions, it is desirable to extend the definition of the spectral function $A(k,\omega)$ to that case. Combining plane waves with filter functions to reduce boundary effects is a possible approach [13] but this method is complicated and does not always yield good results [22]. A simple and efficient approach is based on the eigenstates of the particle-in-a-box problem [i.e., the one-electron eigenstates of the Hamiltonian (22.4) with $U = 0$ on an open chain]. The operators are defined for quasi-wavevectors $k = \pi z/(N+1)$ (quasi-momenta $p = \hbar k$) with integers $1 \leq z \leq N$ by

$$c_{k\sigma} = \sqrt{\frac{2}{N+1}} \sum_{j=1}^{N} \sin(kj) c_{j\sigma} . \tag{22.25}$$

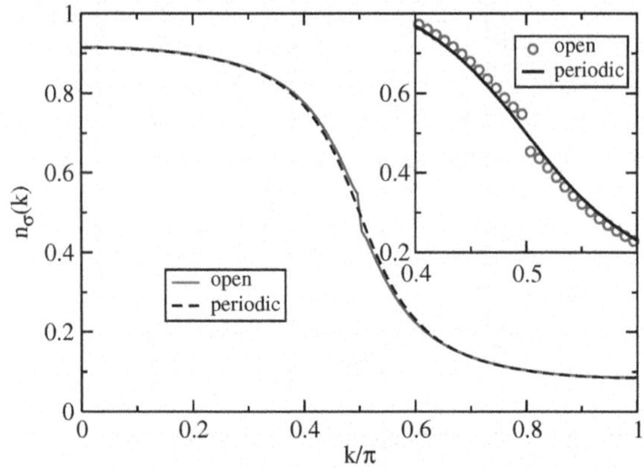

Fig. 22.3. Momentum and quasi-momentum distribution in the half-filled one-dimensional Hubbard model for $U = 4t$ calculated using DMRG on a 128-site lattice with periodic and open boundary conditions, respectively. The inset shows an expanded view of the same data around the Fermi point $k_\mathrm{F} = \pi/2$

Both definitions of $c_{k\sigma}$ are equivalent in the thermodynamic limit $N \to \infty$. Numerous test for momentum-dependent quantities [such as the spectral function $A(k,\omega)$] have shown that both approaches are also consistent in the entire Brillouin zone for finite systems [21, 22]. For instance, in Fig. 22.3 we show the ground state momentum distribution $n_\sigma(k)$ of the half-filled one-dimensional Hubbard model calculated with DMRG for both periodic and open boundary conditions. Small quantitative differences are observed only for a few special k-points corresponding to low-energy excitations, close to the Fermi wavevector $k_\mathrm{F} = \pi/2$. Therefore, open chains and the definition (22.25) can be used to investigate momentum-dependent quantities such as spectral functions $A(k,\omega)$.

22.6 Application: Spectral Function of the Hubbard Model

The DDMRG method and the quasi-momentum technique allow us to calculate the spectral properties of one-dimensional correlated systems on large lattices. To illustrate the capability of this approach we have calculated the photoemission spectral function $A_\sigma(k,\omega)$ of the half-filled one-dimensional Hubbard model. In Fig. 22.4 we show a density plot of this spectral function for $U = 4t$ on a 128-site lattice. Results for stronger coupling U/t are qualitatively similar [22]. In Fig. 22.4 we observe dispersive structures which correspond well to the excitation branches (spinon and holon) predicted by field theory for one-dimensional Mott insulators in the weak coupling regime (i.e., $U/t \ll 1$ in the Hubbard model) [26]. The DDMRG results can also be compared to those obtained with other numerical methods (see Chap. 19).

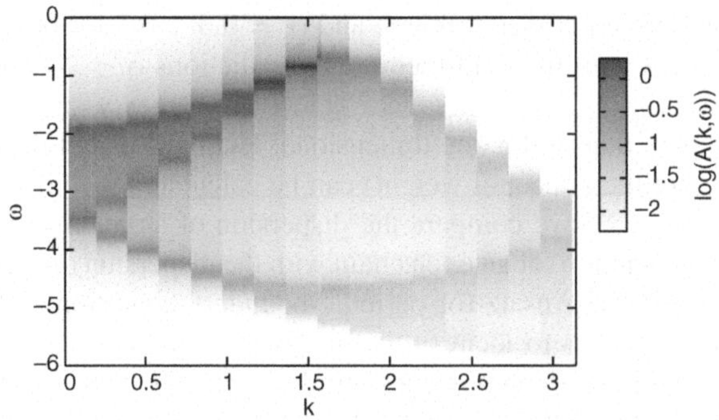

Fig. 22.4. Density plot of the spectral function $A_\sigma(k,\omega)$ in the half-filled one-dimensional Hubbard model for $U = 4t$ calculated on a 128-site open chain using DDMRG with $\eta = 0.0625t$ and quasi-momenta

An advantage of the DDMRG approach over other numerical techniques is that it allows for the simulation of systems large enough to obtain information on the spectrum in the thermodynamic limit. For instance, using the scaling analysis of Sect. 22.4 we have found that for a given k-point the spectrum maximum diverges as a power-law $\eta^{-\alpha}$ ($\sim N^\alpha$) for $\eta \to 0$ ($N \to \infty$) at the spectrum onset (i.e. on the low-energy holon and spinon excitation branches). This indicates a power-law divergence of the spectral weight in the thermodynamic limit [19]: $A_\sigma(k,\omega) \sim (\varepsilon(k) - \omega)^{-\alpha}$ for $\omega < \varepsilon(k)$ and $|k| \leq k_{\mathrm{F}}$, where $\varepsilon(k)$ is the gap dispersion set by the spinon branch for $|k| < Q \approx 0.4\pi$ and by the holon branch for $|k| > Q$. This behavior has been predicted for generic one-dimensional Mott insulators using field theory [26]. From the DDMRG data, we obtain $\alpha = 0.79 \pm 0.05$ for the Q-point

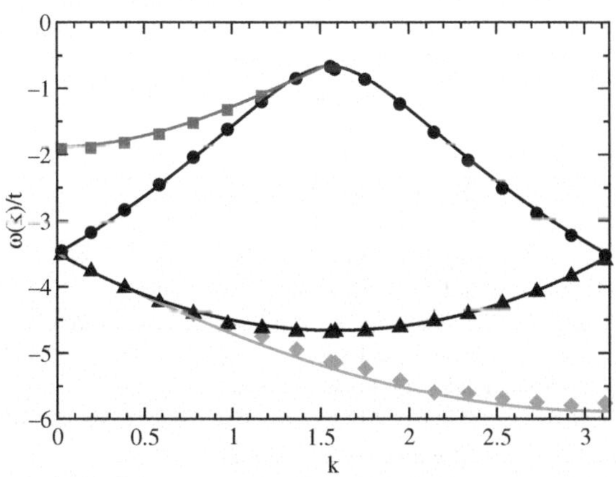

Fig. 22.5. Dispersion of structures found in the DDMRG spectral function of Fig. 22.4 (symbols). Lines show the dispersion of corresponding excitation branches calculated with the Bethe Ansatz for periodic boundary conditions

where spinon and holon branches merge and $\alpha = 0.5 \pm 0.1$ for other $|k| \leq k_{\mathrm{F}}$ in excellent agreement with the field-theoretical predictions $\alpha = 3/4$ and $\alpha = 1/2$, respectively.

Finally, we note that in the one-dimensional Hubbard model the dispersion of excitations (but not their spectral weight) can be calculated with the Bethe Ansatz method [2]. In Fig. 22.5 we compare the dispersion of structures observed in the DDMRG spectral function for an open chain with the dispersion of some excitations obtained with the Bethe Ansatz for periodic boundary conditions. The agreement is excellent and allows us to identify the dominant structures, such as the spinon branch (squares) and the holon branches (circles) [21, 22]. This demonstrates once again the accuracy of the DDMRG method combined with the quasi-momenta technique. In summary, DDMRG provides a powerful and versatile approach for investigating the dynamical properties in low-dimensional lattice quantum many-body systems.

References

1. H. Kuzmany, *Solid-State Spectroscopy* (Springer, Berlin, 1998)
2. F. Essler, H. Frahm, F. Göhmann, A. Klümper, V. Korepin, *The One-Dimensional Hubbard Model* (Cambridge University Press, Cambridge, 2005)
3. S.R. White, Phys. Rev. Lett. **69**(19), 2863 (1992)
4. S.R. White, Phys. Rev. B **48**(14), 10345 (1993)
5. I. Peschel, X. Wang, M. Kaulke, K. Hallberg (eds.), *Density-Matrix Renormalization, A New Numerical Method in Physics* (Springer, Berlin, 1999)
6. U. Schollwöck, Rev. Mod. Phys. **77**(1), 259 (2005)
7. K. Hallberg, Adv. Phys. **55**, 477 (2006)
8. S.R. White, D.A. Huse, Phys. Rev. B **48**(6), 3844 (1993)
9. E. Jeckelmann, H. Fehske, in *Proceedings of the International School of Physics "Enrico Fermi" - Course CLXI Polarons in Bulk Materials and Systems with Reduced Dimensionality* (IOS Press, Amsterdam, 2006), pp. 247–284
10. S. Ramasesha, S.K. Pati, H.R. Krishnamurthy, Z. Shuai, J.L. Brédas, Phys. Rev. B **54**(11), 7598 (1996)
11. M. Boman, R.J. Bursill, Phys. Rev. B **57**(24), 15167 (1998)
12. K.A. Hallberg, Phys. Rev. B **52**(14), R9827 (1995)
13. T.D. Kühner, S.R. White, Phys. Rev. B **60**(1), 335 (1999)
14. E.R. Gagliano, C.A. Balseiro, Phys. Rev. Lett. **59**(26), 2999 (1987)
15. Z.G. Soos, S. Ramasesha, J. Chem. Phys. **90**(2), 1067 (1989)
16. W. Press, S. Teukolsky, W. Vetterling, B. Flannery, *Numerical Recipes in C++. The Art of Scientific Computing* (Cambridge University Press, Cambridge, 2002)
17. S. Ramasesha, J. Comp. Chem. **11**(5), 545 (1990)
18. S.K. Pati, S. Ramasesha, Z. Shuai, J.L. Brédas, Phys. Rev. B **59**(23), 14827 (1999)
19. E. Jeckelmann, Phys. Rev. B **66**(4), 045114 (2002)
20. E. Jeckelmann, F. Gebhard, F.H.L. Essler, Phys. Rev. Lett. **85**(18), 3910 (2000)
21. H. Benthien, F. Gebhard, E. Jeckelmann, Phys. Rev. Lett. **92**(25), 256401 (2004)
22. H. Benthien, Dynamical properties of quasi one-dimensional correlated electron systems. Ph.D. thesis, Philipps-Universität, Marburg, Germany (2005)

23. F. Gebhard, E. Jeckelmann, S. Mahlert, S. Nishimoto, R. Noack, Eur. Phys. J. B **36**, 491 (2003)
24. S. Nishimoto, E. Jeckelmann, J. Phys. Condens. Matter **16**, 613 (2004)
25. C. Raas, G.S. Uhrig, F.B. Anders, Phys. Rev. B **69**(4), 041102 (2004)
26. F.H.L. Essler, A.M. Tsvelik, Phys. Rev. B **65**(11), 115117 (2002)
27. C. Raas, G. Uhrig, Eur. Phys. J. B **45**, 293 (2005)

23 Studying Time-Dependent Quantum Phenomena with the Density-Matrix Renormalization Group

Reinhard M. Noack[1], Salvatore R. Manmana[2], Stefan Wessel[3], and Alejandro Muramatsu[3]

[1] Fachbereich Physik, Philipps-Universität Marburg, 35032 Marburg, Germany
[2] Institute of Theoretical Physics, École Polytechnique Fédérale de Lausanne, CH-1015 Lausanne, Switzerland
[3] Institut für Theoretische Physik III, Universität Stuttgart, 70550 Stuttgart, Germany

Recently, the Density Matrix Renormalization Group (DMRG) has been extended to calculate the time evolution of an arbitrary state. Here, we will discuss this extension of the DMRG method, in particular, the general properties of the DMRG that are relevant to the extension, the basic issues that are involved in calculating time-dependence within the DMRG, and the first attempts at formulating time-dependent DMRG (t-DMRG) algorithms. Moreoever, we describe adaptive t-DMRG methods, which tailor the reduced Hilbert space to one particular time step and which are therefore the most efficient algorithms for the majority of applications. Finally, we discuss in detail the application of the t-DMRG to a system of interacting spinless fermions which are quenched by suddenly changing the interaction strength. This system provides a very useful test bed for the method, but also raises physical issues which are illustrative of the general behavior of quenched interacting quantum systems.

23.1 Time Dependence in Interacting Quantum Systems

The calculation of the time dependence of interacting quantum mechanical systems is, generally, a difficult problem. Although the time dependence of the wave vector $|\psi\rangle$ is governed by the time-dependent Schrödinger equation

$$\mathrm{i}\hbar \frac{\partial |\psi\rangle}{\partial t} = H|\psi\rangle \,, \tag{23.1}$$

with formal solution

$$|\psi(t)\rangle = \mathrm{e}^{\ \mathrm{i}Ht/\hbar}|\psi(t_0)\rangle \tag{23.2}$$

for a time-independent Hamiltonian H given an initial state at time $t = t_0$, $|\psi(t_0)\rangle$, this formal expression does not help very much in finding an actual solution: Calculating the exponential of the Hamiltonian applied to an arbitrary state is, in general, a quite difficult problem.

Here we will concern ourselves primarily with the case of systems undergoing a sudden change or quench, as formulated above, i.e., the system is prepared in an initial state at time $t_0 \equiv 0$ and evolves via a Hamiltonian that is time-independent

R. M. Noack et al.: *Studying Time-Dependent Quantum Phenomena with the Density-Matrix Renormalization Group*, Lect. Notes Phys. **739**, 637–652 (2008)
DOI 10.1007/978-3-540-74686-7_23

for $t > 0$. In order to simplify the notation, we will take $\hbar = 1$ and define $t_0 \equiv 0$ in the following. This physical situation is interesting in a number of experimental contexts. Examples include experiments in which the depth of an optical lattice containing trapped cold atoms is suddenly changed, leading to the collapse and revival of a Bose-Einstein condensate [1], the realization of a quantum version of Newton's cradle [2], the quenching of a ferromagnetic spinor Bose-Einstein condensate [3], and transport across quantum dots [4, 5] and other nanostructures. One should also consider what aspects of time-dependent behavior are interesting. In these systems, the detailed time evolution of various observables can be followed experimentally on short to intermediate time scales. For example, for the system of ^{87}Rb atoms on an optical lattice, snapshots of the momentum distribution can be obtained by releasing the condensate at different times after the quench and then performing time-of-flight measurements [1]. What is interesting is, first of all, the transient behavior, in this case, oscillations between a momentum distribution characteristic of a Bose-Einstein condensate and that of a bosonic Mott insulator. After a somewhat longer period of time, one can ask the question of whether there is relaxation of these oscillations to stationary or quasi-stationary behavior, and, if so, how can this behavior be characterized.

Numerically, the way to proceed, given an initial state $|\psi(0)\rangle$, is to propagate through a succession of discrete time intervals of size Δt. The time interval Δt is chosen to be sufficiently small so that $|\psi(t+\Delta t)\rangle$ can be calculated to the desired accuracy given $|\psi(t)\rangle$. For the single-particle Schrödinger equation, an appropriately chosen discretization in time and space leads to finite difference equations which can be iterated numerically; the most well-known variants are the Crank-Nicolson method and the Runge-Kutta method. For interacting many-particle systems, it is less evident how to formulate a well-behaved and efficient algorithm, but a discretization in time nevertheless forms the basis for most tenable algorithms.

One class of such algorithms involves projecting the time-propagation operator over a finite interval, $\exp(-\mathrm{i}H\Delta t)$, onto the Krylov subspace, the subspace generated by applying the Hamiltonian n times to an arbitrary initial vector, $|u_0\rangle$,

$$\left\{|u_0\rangle, H|u_0\rangle, H^2|u_0\rangle, ..., H^n|u_0\rangle\right\} .$$

The Lanczos and the related Arnoldi method involve projecting an operator onto an orthogonalized version of this Krylov subspace, where n is typically chosen to be much smaller than the total dimension of the Hilbert space (see also Chaps. 18 and 19). In the original methods, the operator projected is the Hamiltonian, and the lowest (or highest) few eigenstates are good variational (anti-variational) approximations to the exact eigenstates. However, variants of these methods can also be used to approximate the unitary time-evolution operator. In the Lanczos procedure, the Hamiltonian becomes tridiagonal in the Lanczos basis, a basis for the Krylov subspace orthogonalized via the Lanczos recursion. The time evolution operator is then the exponential of a tridiagonal matrix, which can be formed explicitly and efficiently. For a given time interval Δt and bandwidth of the matrix representation of H, explicit error bounds can be given for the Euclidean norm of the wave function

[6]. In practice, very good approximations for an appropriately chosen Δt can be achieved by taking $n \approx 10$–20 [7]. These algorithms can be applied to almost any system that can be treated with exact diagonalization to find ground-state properties and provide a very useful numerically exact (or at least very controllable) method for small system sizes. However, these methods are limited in the same way that exact diagonalization for the ground-state properties are limited: Since the entire Hilbert space, which grows exponentially in system size, must be treated, the computational complexity and the memory required also grow exponentially.

One class of algorithms that overcomes this exponential limitations, at least for a certain class of low-dimensional interacting quantum lattice models with short-range interactions, is the density-matrix renormalization group. It is related to exact diagonalization in that it carries out a series of iterative diagonalizations in order to form a good variational approximation to particular states of a system in a reduced Hilbert space.

23.1.1 Calculating Time Evolution Within the DMRG

In its original formulation, the density-matrix renormalization group algorithm is a method for calculating the properties of extremal eigenstates (e.g., the ground state and low-lying excited states) of an interacting quantum system on a finite lattice in a truncated basis, i.e., on a carefully chosen subspace of the complete Hilbert space. This is done by iteratively building up a variational state, a particular case of a matrix-product state. For details, see Chap. 21. The fundamental idea behind the approximation is to divide the complete system (superblock) into two parts: a system block and an environment block; see Fig. 23.1. Once such a division is made, the basis is tailored to suit a particular state or set of states $|\psi_\alpha\rangle$ using the reduced density matrix for the system block which has the form

$$\hat{\rho}_{\text{sub}} = \sum_\alpha W_\alpha \sum_j \langle j|\psi_\alpha\rangle\langle\psi_\alpha|j\rangle \,, \tag{23.3}$$

where the states $|j\rangle$ are a basis for the environment block, and the W_α are positive semi-definite weights, which must sum to one. When only one state enters into the sum (i.e., only one $w_\alpha = 1$), the superblock is in a pure state, otherwise it is in a mixed state. The states $|\psi_\alpha\rangle$ are called target states.

In order to truncate the basis, a given number, m, states with the largest weights, i.e., the largest density-matrix eigenvalues, are retained. For the case of a pure state, this is equivalent to representing the wave function of the superblock in a reduced basis by truncating the Schmidt or singular-value decomposition:

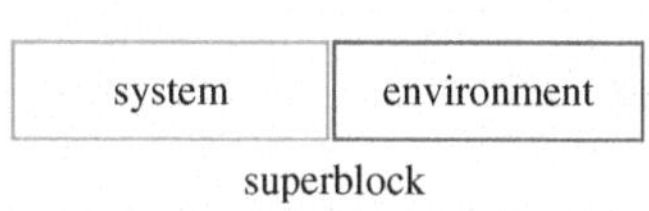

Fig. 23.1. Decomposition used in the DMRG: The superblock, which encompasses the entire system studied, is divided into a system block and an environment block

$$|\psi_0\rangle \approx \sum_{\gamma}^{m \leq \dim(\gamma)} \sqrt{w_\gamma}\, |\phi_\gamma\rangle\, |\chi_\gamma\rangle\,, \qquad (23.4)$$

where the w_γ are the nonzero eigenvalues of the reduced density matrices of either the system or the environment blocks (which are identical), and the $|\phi_\gamma\rangle$ and $|\chi_\gamma\rangle$ are the eigenstates of the reduced density matrices of the system and the environment blocks, respectively. This expression can straightforwardly be generalized to the case of a mixed state. A matrix-product state is built up out of a succession of such approximations.

In order to accurately calculate a state that evolves in time, the DMRG algorithm must be extended in two ways: First, states other than extremal eigenstates must be generated, and second, the basis must be adapted to the time-evolving state. Different choices can be made in how these extensions are carried out; these choices can be used to classify the various algorithms.

The simplest and earliest algorithm, formulated by Cazalilla and Marston [8], adapts the basis for the initial state only. More specifically, the initial state $|\psi(0)\rangle$ is determined using a ground-state DMRG calculation, carried out with a Hamiltonian H_0. The wave vector is then propagated through a set of time steps without further changing the basis, i.e., the basis is adapted to the initial state only and is not changed. The accuracy of this method clearly depends on how well the basis adapted for the initial state represents the time-evolved state. Since one is, in most cases, interested in a time-evolved state which is significantly different from the initial state, this method, will not, in general, provide an accurate approximation for the time-evolved behavior.

Luo, Xiang and Wang [9] subsequently pointed out that better accuracy could be achieved for the test quantity calculated in [8], the tunnel current across a quantum dot, when information on *all* relevant time scales is included in the DMRG procedure. They did this by including in the density matrix (23.3) states at all discrete time steps,

$$|\psi(0)\rangle, |\psi(\Delta t)\rangle, |\psi(2\Delta t)\rangle, \ldots, |\psi(T)\rangle \qquad (23.5)$$

up to a maximum time T. This scheme is illustrated conceptually in Fig. 23.2(a). While this technique should evidently improve the accuracy at times removed from $t = 0$, the penalty that must be paid is that the set of bases built up by the DMRG procedure, i.e., the matrix-product state that is generated, is adapted for a set of states rather than for a single state. Therefore, for a fixed number of states kept at each step, the accuracy of the representation of each particular state suffers. In other words, the longer the desired maximum time T, the more poorly the matrix-product state is adapted for a given time, at least at fixed numerical effort.

23.1.2 Adaptive Algorithm

Generally, procedures for propagating differential equations forward through a finite difference in time depend only on the previous time step, or, at most, on a small number of previous time steps. Therefore, it seems natural to address the problem

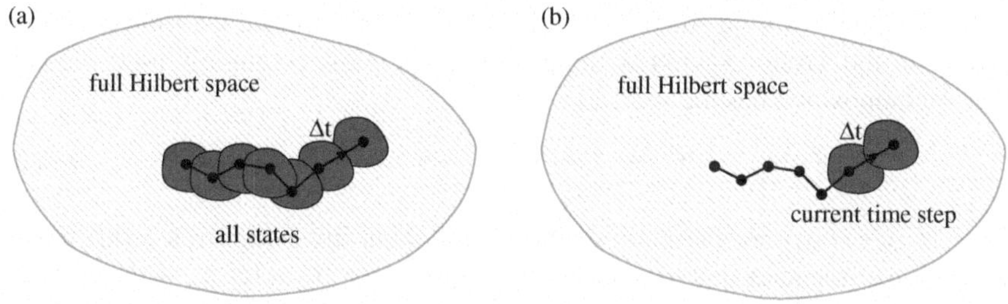

Fig. 23.2. Schematic sketch of the portions of the complete Hilbert space for which the matrix-product state is adapted for **(a)** the complete t-DMRG and **(b)** the adaptive t-DMRG

described at the end of the last section by targeting only states associated with the previous and the current time step. While this statement seems straightforward at first glance, the problem of how to formulate a controlled, efficient algorithm incorporating this strategy is less straightforward. In particular, the original DMRG algorithm targets extremal eigenstates of the Hamiltonian, i.e., the ground state and low-lying excited states within a particular symmetry sector. Additional states can also be targeted, such as the correction vector when dynamical quantities are desired (see Chap. 22), but they are generally generated by applying an operator to one of the extremal eigenstates, or by minimizing an additional functional. For an arbitrary time step, however, the only information available is $|\psi(t)\rangle$, which is not an extremal state of a particular functional. Information on this state is encoded as a matrix-product state, i.e., a series of transformations to the basis of the reduced density matrix for successive partitions of the system. Given this state and the Hamiltonian H that determines the time evolution, the state $|\psi(t+\Delta t)\rangle = \exp(-\mathrm{i}H\Delta t)|\psi(t)\rangle$ must be calculated. This must be done by re-adapting the basis to $|\psi(t+\Delta t)\rangle$.

In general, such a re-adaption is carried out by performing a finite-system DMRG sweep in which the state $|\psi(t+\Delta t)\rangle$ is targeted at each step. By doing this for every bipartite decomposition of the system, the matrix-product state is optimized for the new state. Note, however, that in order to calculate $|\psi(t+\Delta t)\rangle$ accurately, an accurate representation of $|\psi(t)\rangle$ must also be available at each step. Therefore, the basis must simultaneous be re-adapted for $|\psi(t)\rangle$. However, $|\psi(t)\rangle$ cannot be explicitly recalculated because the previous time step is not known. This technical problem is the reason that the adaptive method was not developed earlier. The solution is to transform the wave function $|\psi(t)\rangle$ from the last step using the so-called wave-function transformation; for details, see Chap. 21, Sect. 4, and, in particular, (35). Note that such a transformation is not exact; it introduces an additional error that is the truncation error of the particular finite-system step into the representation of $|\psi(t)\rangle$. Therefore, one should avoid performing superfluous finite-system sweeps in the time-dependent DMRG; unlike in the ground-state DMRG, additional sweeps are not guaranteed to always improve the wave function.

The original work on adaptive t-DMRG [10, 11, 12] treated the time evolution operator in the Trotter-Suzuki decomposition. The most commonly used second-order decomposition has the form

$$e^{-iH\Delta t} \approx e^{-iH_{\text{even}}\Delta t/2} e^{-iH_{\text{odd}}\Delta t} e^{-iH_{\text{even}}\Delta t} + \mathcal{O}((\Delta t)^3) , \qquad (23.6)$$

where H_{even} (H_{odd}) are the parts of the Hamiltonian involving even (odd) bonds and we have assumed that H can be decomposed as $H = H_{\text{even}} + H_{\text{odd}}$. Here $H_{\text{even}} = \sum_{i=1}^{L} H_{2i,2i+1}$ is a sum over the even bond operators and, similarly, $H_{\text{odd}} = \sum_{i=1}^{L} H_{2i-1,2i}$. Note that only Hamiltonians composed solely of nearest-neighbor connections can be decomposed in this way. For one-dimensional lattices, this decomposition can be integrated quite readily into the usual finite-system algorithm. Since the exponentials of the individual bond operators within the even or odd bonds commute with one another, the terms can be ordered so that only one bond term is applied at each finite-system step. This bond term is chosen so that it corresponds to the two exactly treated sites in the finite-system superblock configuration, as depicted in Fig. 23.3. The advantage of this scheme is that the complete Hilbert space of the two sites is present, so that this piece of the time-evolution operator can be applied *exactly* and very efficiently. A complete sweep back and forth then successively applies all the bond operators and, at the end of the sweep, the propagation through the time step is complete. For more detailed descriptions of the algorithms, see [11, 12].

Feiguin and White [13] subsequently pointed out that an adaptive t-DMRG algorithm can also be formulated without carrying out a Trotter-Suzuki decomposition. Instead, the complete time evolution operator is applied at each step of the finite-system procedure, and sweeping is carried out only to adapt the matrix-product state. Different schemes are then possible to carry out the propagation through a time step; in [13] the Runge-Kutta method was used. However, if an integrator is used, it would clearly be better to use one that preserves unitarity, such as Crank-Nicholson. The most accurate and efficient scheme seems to be to decompose $\exp(-iH\Delta t)$ in a Lanczos basis [7] or using the Arnoldi method [14], just as is done in the exact diagonalization method discussed above in Sect. 23.1. This scheme has the advantage of preserving unitarity and converges exponentially in the number of applications of H.

Another crucial issue in the general adaptive algorithm is which states to target in the density matrix. If one considers the time evolution of the density matrix through one time step

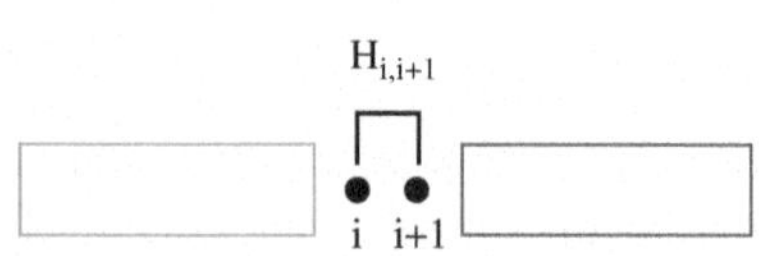

Fig. 23.3. Four-site superblock decomposition showing how an individual bond operator is applied in the Trotter-Suzuki decomposition-based variant of the adaptive t-DMRG

$$\rho(t+\Delta t) = \int_t^{t+\Delta t} \mathrm{d}t' |\psi(t')\rangle\langle\psi(t')| \,, \tag{23.7}$$

it is clear that targeting additional states within the time interval $[t, t+\Delta t]$ could be helpful [13]. Just how many intermediate time steps should be targeted depends on the overall time step Δt and the details of the system studied. In practice, we target one to a few states at intermediate times in the calculations presented in Sect. 23.2; this issue is illustrated numerically there.

In general, which variant of the adaptive t-DMRG to use will depend on the problem treated. First, the Lanczos (and related variants of the general adaptive scheme) can be applied to a more general class of systems than the Trotter method. When the Trotter method can be applied, it is generally computationally more efficient for a given formal accuracy, i.e., equal number of states kept m or equal cutoff in discarded weight or quantum information loss. However, in general, the different methods should be compared in test runs for particular systems in order determine which one yields the most accurate and stable results for given computational effort.

23.2 Sudden Quench of Interacting Fermions

In order to illustrate and test the adaptive t-DMRG algorithm as well as to explore typical physical issues that crop up in suddenly perturbed strongly interacting systems, we consider a system of spinless fermions with nearest-neighbor Coulomb repulsion

$$H = -t_\mathrm{h} \sum_j \left(c_{j+1}^\dagger c_j + \mathrm{H.c.}\right) + V \sum_j n_j n_{j+1} \,. \tag{23.8}$$

Here t_h is the hopping integral, which we take to be the unit of energy, $t_h = 1$, unless explicitly stated otherwise, $c_j^\dagger$ (c_j) creates (annihilates) a fermion on site j, V denotes the strength of the Coulomb repulsion, and $n_j = c_j^\dagger c_j$ is the local density operator. In addition to being one of the simplest models of interacting fermions, this system is well-suited as a test bed for various reasons. First, the system can be mapped onto the anisotropic (XXZ) Heisenberg chain, which is exactly solvable. Therefore, the ground-state phase diagram, as well as many aspects of the excitation spectrum, are known. Second, at half-filling, there is a phase transition between two qualitatively different phases as a function of V. In the fermionic language, the transition is between a metallic (more precisely, Luttinger liquid) phase for $V < V_\mathrm{c} = 2$ and a charge-density-wave (CDW) insulator for $V > 2$. In the spin language, this corresponds to a transition between XY symmetry and Ising symmetry and the corresponding phases. Third, as we will see, the atomic limit of this model leads to an exactly understandable, but non-trivial time evolution. In order to take advantage of these features of the model, we will treat exclusively the half-filled ($\langle n\rangle = 0.5$) system here.

Our physical motivation for considering this system comes from the "collapse and revival" phenomena observed in experiments with atoms trapped in optical lattices [1]. When the depth of the optical lattice is suddenly changed, the effective hopping and interaction strength of the corresponding model are suddenly changed; this can be parameterized as a change of their ratio. In the bosonic systems treated in [1], the parameters were changed in such a way that a transition from a superfluid to a bosonic Mott insulator was induced. Although more difficult to realize experimentally, trapping fermionic atoms is also possible [15, 16]. As we will see, the phenomena observed when the model parameters of fermionic systems are suddenly changed is reminiscent of that found in the bosonic systems. In view of this, we will treat a system initially prepared to be in the ground state of Hamiltonian (23.8) with a particular value of the interaction V_0, i.e., $|\psi(0)\rangle = |\psi_0(V_0)\rangle$, the ground state of $H(V_0)$. At time $t = 0$, the interaction strength will be suddenly changed to a value V and the time evolution will be subsequently carried out using $H(V)$.

In order to investigate the single-particle properties of the system, which are related to its metallic or insulating nature, we examine the momentum distribution function

$$\langle n_k \rangle(t) = \frac{1}{L} \sum_{l,m=1}^{L} \mathrm{e}^{\mathrm{i}k(l-m)} \langle c_l^\dagger c_m \rangle(t) \,, \tag{23.9}$$

i.e., the Fourier transform of the one-particle density matrix, $\rho_{lm} = \langle c_l^\dagger c_m \rangle$, as a function of time. In an insulator, $\langle n_k \rangle$ has a finite slope at the Fermi wave vector, $k = k_\mathrm{F}$, while for a conventional (Fermi liquid) metal, there is a jump discontinuity at k_F. For a one-dimensional interacting metal, a Luttinger liquid, the jump is replaced by a power-law singularity in the slope at k_F [17, 18]. Note that the behavior of the density-density correlation function is also interesting for characterizing the CDW insulating phase [19, 20]; however, for the sake of compactness, we will consider only single-particle properties here.

We use the adaptive t-DMRG method described in Sect. 23.1.2, using both the Lanczos and the Trotter treatment of the time step. In all cases, we set a fixed threshold of discarded weight as well as a limit on the maximum number of states kept; we set the number of states limit to be appropriate for the weight cutoff and the system parameters. Typical values for this system are a weight cutoff of 10^{-9} and a maximum of 1500 states kept.

We have carried out extensive tests, comparing both variants of the adaptive t-DMRG algorithm with each other and with control results where available. Unfortunately, there are not many interacting quantum systems for which exact results can be obtained. In order to calculate the full time evolution, all eigenstates of the system must be obtained; exact methods for the ground state, such as the Bethe ansatz, are generally not powerful enough to obtain the full time evolution.[4] For spinless fermions, exact results for the time evolution are available for zero interaction $V = 0$ and in the atomic limit, $t_h = 0$. In addition, on sufficiently small

[4] There have been recent advances for single-impurity systems using the Bethe ansatz; see [21, 22].

systems, we can compare with time evolution calculated using the Lanczos method, for which the numerical errors are well-controlled enough so that the numerical error can be made arbitrarily small. The behavior of various quantities can be considered. Since the time evolution is unitary, the expectation value of the Hamiltonian $\langle H(V) \rangle$ and all higher powers of H, $\langle H^2 \rangle$, $\langle H^3 \rangle, \ldots$, will be conserved. Any appreciably change in these expectation values with time then signifies a breakdown in accuracy. Since the average energy is not particularly important physically, the accuracy of the relevant observables is more important. Here the most important observable is the momentum distribution $\langle n_k \rangle$; other useful quantities include the local density and the density-density correlation function.

In Fig. 23.4, we compare the maximum error over k in the momentum distribution of various t-DMRG calculations for the same system, an $L = 18$ site chain with open boundary conditions in which the interaction is changed from $V_0 = 0.5$ to $V = 10$. The numerically exact (for the time range shown) benchmark is provided by a Lanczos time evolution calculation. In the Lanczos t-DMRG method, it is important to optimize the number of intermediate time steps targeted. As can be seen the accuracy of the Lanczos t-DMRG method depends strongly on the number of intermediate time steps taken. For the time step taken here, the best accuracy occurs

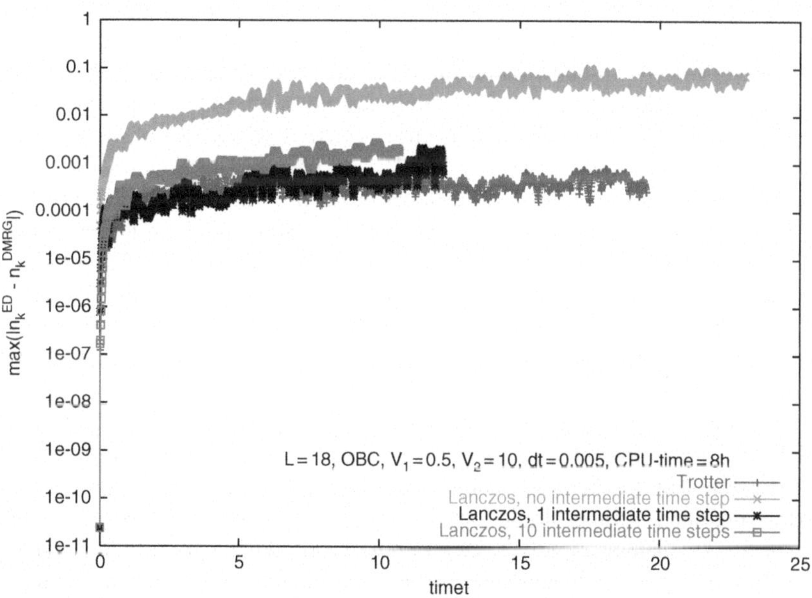

Fig. 23.4. Maximum value of the deviation of the momentum distribution $\langle n_k \rangle$ obtained with the Lanczos and Trotter adaptive t-DMRG methods from a numerically exact Lanczos time-evolution calculation for a system of interacting spinless fermions with $L = 18$ sites pushed out of equilibrium by changing the interaction from $V_0 = 0.5$ to $V = 10$ at time $t = 0$. the time step is $\Delta t = 5 \times 10^{-3}$ and the calculations were all limited to 8 CPU hours with fixed discarded weight

when one intermediate time step is taken, with zero and ten intermediate time steps significantly less accurate. The Trotter method yields the most accurate result for times less than approximately one, whereas the Lanczos t-DMRG with one intermediate time step yields somewhat more accurate results for times between one and five. Note that the CPU time has been held to the same value for all the runs, so that the length of the curves in time indicate the relative efficiency. For example, the Trotter method uses about 2/3 the CPU time of the comparably accurate Lanczos t-DMRG with one intermediate time step. Therefore, for fixed CPU time, one could gain better accuracy for the same time range by taking a larger m. This result is typical for the system of spinless fermions treated here. We note, however, that a larger time step can be taken in the Lanczos t-DMRG than the Trotter variant to obtain the same accuracy. We nevertheless find that the Trotter method with an optimal choice of parameters yields the most accurate results for a given computational effort for the results shown here; the majority of the results shown are therefore calculated using it. A more extensive analysis of the errors can be found in [19].

One useful limit to consider is the atomic limit, $t_h = 0$. With no hopping, the particle number can be treated as classical variable and the Hamiltonian, which consists of only a Coulomb repulsion, corresponds to the classical Ising model. In the Ising language, the ground state is an unmagnetized antiferromagnetic state, which corresponds to a CDW state at $q = \pi$ site and is two-fold degenerate. Excitations out of the ground state involve forming at least one domain wall, each of which has an energy cost V. Such excited states are highly degenerate because the number of ways of making such an excitation is at least of the order of the system size. Therefore, the complete excitation spectrum consists of a series of highly degenerate, dispersionless levels at energy V, $2V, \ldots$ The time dependence of the relevant observables can be calculated explicitly. Any observable composed of the local density operator $n_{i\sigma}$, such as the density-density or spin-spin correlation function, is time-independent because $n_{i\sigma}$ commutes with H when $t_h = 0$. The functional dependence of the single-particle density matrix on time and thus the frequencies that enter into its Fourier transform $\langle n_k \rangle$ can be easily obtained. It consists of two cosine terms with frequencies $\omega_1 = V$ and $\omega_2 = 2V$ is therefore periodic with period $T = 2\pi/V$ [19].

We display the behavior of the momentum distribution $\langle n_k \rangle$ Fig. 23.5(a). The initial state is the ground state of $H(V_0 = 0.5)$ with $t_h = 1$, i.e., an interacting metallic state. (In the thermodynamic limit, the jump at $k_\mathrm{F} = \pi/2$ would develop into the singular, Luttinger-liquid form.) As $\langle n_k \rangle$ develops from the pseudo-metallic form at $t = 0$, changes rapidly, even attaining inverted behavior as a function of k at $t = 0.3$. At $t \approx 0.62$, in agreement with the argument above, there is a complete revival of the momentum distribution. The Fourier transform in the time domain, Fig. 23.5(b), clearly shows the expected sharp peaks at $\omega_1 = V$ and $\omega_2 = 2V$.

We now turn to the case of finite t_h, treating first time evolution with large V/t_h, $V = 40$ (with $t_h = 1$), which is well into the CDW insulating phase for the ground state. The initial state is the ground state of $H(V_0 = 0.5)$, which has distinctly metallic character. As can be seen from the surface plot in Fig. 23.6, the behavior

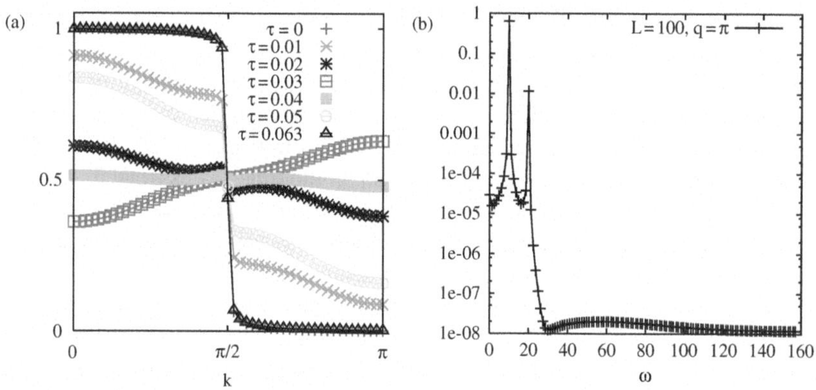

Fig. 23.5. **(a)** Momentum distribution $\langle n_k \rangle$ in the atomic limit, $t_h = 0$, $V = 10$ on a $L = 100$ site system at the indicated times. **(b)** Fourier transform in the time domain of the $k = \pi$ component from **(a)**. The two sharp peak occur at angular frequencies $\omega_1 = 10$ and $\omega_2 = 20$

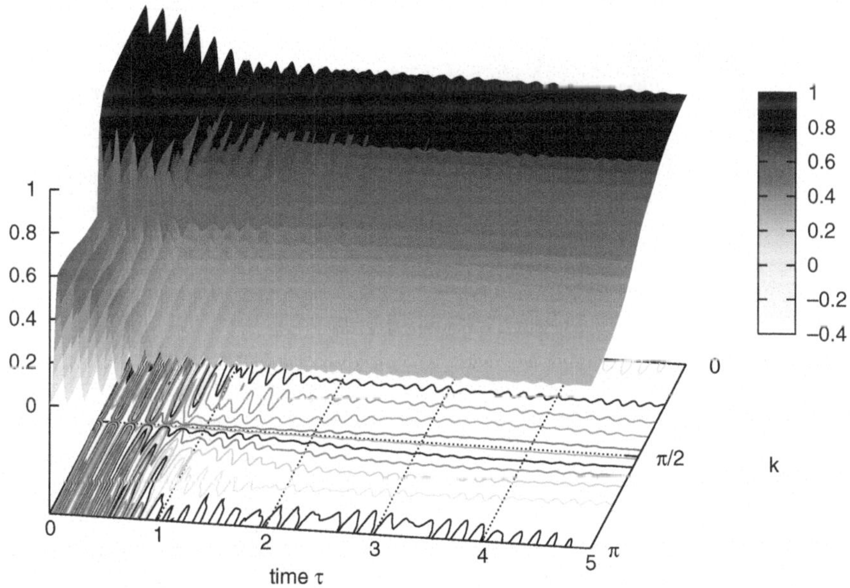

Fig. 23.6. Momentum distribution $\langle n_k \rangle$ plotted as a surface depending on momentum k and time t when the initial ground state of $H(V_0 = 0.5)$ is time-evolved with Hamiltonian $H(V = 40)$. The system size is $L = 50$, the time step $\Delta t = 0.001$, and up to 1500 states are kept with a discarded weight cutoff of 10^{-9}

of the momentum distribution at short time is similar to that in the atomic limit. There are strong oscillations at all k with a period $T = 0.157$, that is shorter due to the larger value of V. However, the revival is not complete, and, after a number of oscillations and a time of the order of $1/t_h$, the oscillations damp out. At larger times, there are still residual oscillations which do not become smaller, but also show no significant drift or revival phenomena on the time scales treated. We argue that this indicates that a quasi-stationary state has been reached. Note that, although $\langle n_k \rangle$ is still relatively steeply changing near the Fermi wave vector $k_\mathrm{F} = \pi/2$, the slope is actually finite at k_F, characteristic of insulating behavior.

When the time evolution for the same initial state is carried out with the smaller interaction $V = 10$, Fig. 23.7, oscillations as a function of time are still evident. However, the period is significantly longer, as would be expected from the smaller value of V ($T = 0.628$). However, the time over which the oscillations decay is still of the order of $1/t_h$. Therefore, only two distinct oscillations are evident before quasi-stationary behavior is reached. The relatively steeply changing portion of the momentum distribution function is somewhat more pronounced than in the $V = 40$ case, but it still has insulating character.

For a much smaller interaction, $V = 2.5$, Fig. 23.8, no oscillations occur; the metallic quasi-jump decays smoothly to an insulating form that has a somewhat larger rapidly changing region than for the larger values of V. There still seems

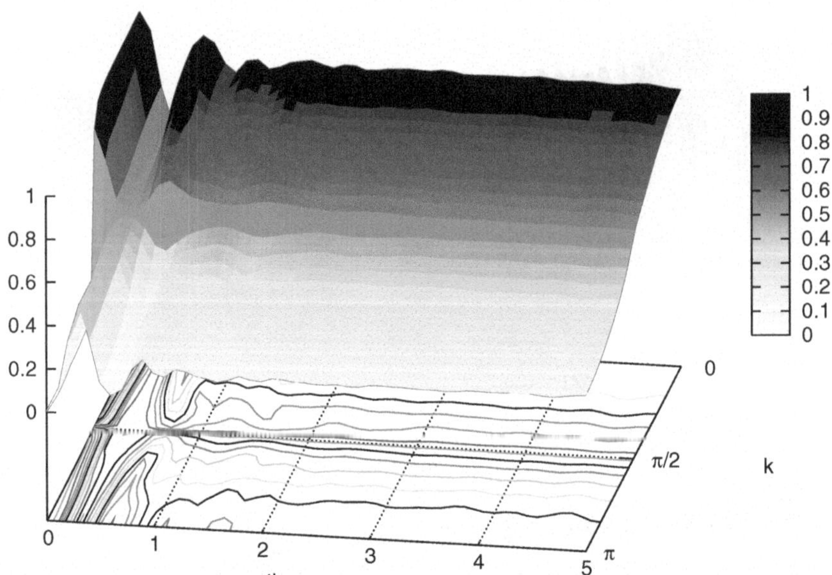

Fig. 23.7. Momentum distribution $\langle n_k \rangle$ plotted as a surface depending on momentum k and time t when the initial ground state of $H(V_0 = 0.5)$ is time-evolved with Hamiltonian $H(V = 10)$. The system size is $L = 50$, the time step $\Delta t = 0.005$, and up to 1000 states are kept with a discarded weight cutoff of 10^{-9}

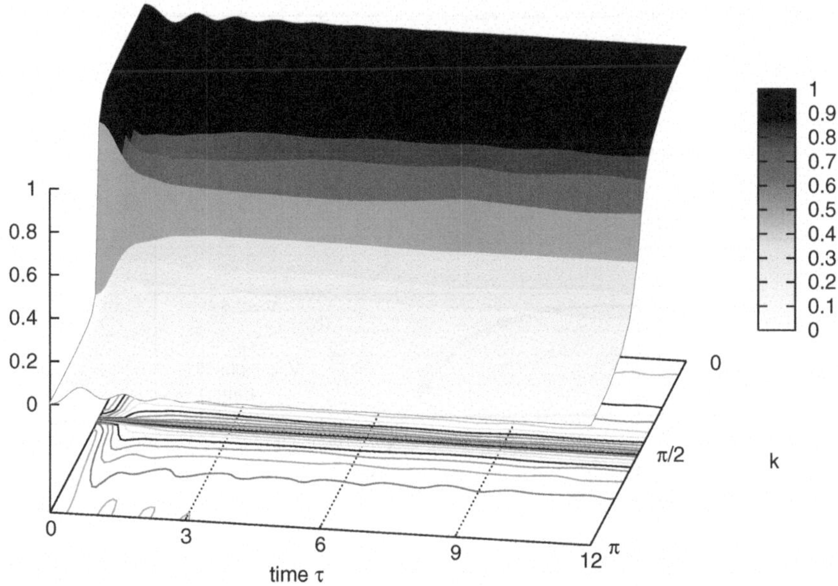

Fig. 23.8. Momentum distribution $\langle n_k \rangle$ plotted as a function of k and when the initial ground state of is time-evolved with Hamiltonian $H(V = 2.5)$. The system size is $L = 50$, the time step $\Delta t = 0.005$, and up to 1000 states are kept with a discarded weight cutoff of 10^{-9}

to be convergence to quasi-stationary behavior for large t. The change is relatively rapid until a time $t = 1/t_h$, and then somewhat more gradual between $t = 1/t_h$ and $t = 2/t_h$. We have also investigated evolution with smaller values of V. At the critical point $V = V_c = 2$ and slightly below, the behavior does not change significantly from that shown in Fig. 23.8. This is an indication that carrying out the evolution at a critical point rather than slightly away from it has no large effect. In addition, we investigated the case of a non-integrable system by turning on an addition next-nearest-neighbor Coulomb repulsion. This also had no qualitative effect on the behavior of $\langle n_k \rangle$ with time [23].

At all three parameter values discussed until now, we have observed relaxation to quasi-stationary behavior in the momentum distribution, and all on a similar time scale. In order to ascertain that this behavior is generic, we can examine the dependence on the initial state. For a given V, the average energy $\langle H \rangle$ will increase as the interaction for the initial state V_0 is moved away from V in either direction. Therefore, it is often possible to find two different initial states with the same average energy. The quasi-stationary behavior at sufficiently long times can then be compared to see if it is generic. We find that unless the initial states are very far apart, the momentum distribution does converge, to good quantitative agreement, to the same quasi-stationary behavior. This is illustrated for a particular parameter value, $V = 2.5$, in Fig. 23.9. As can be seen in Fig. 23.9(a), the two initial states, at $V_0 = 0.5$ and $V_0 = 5.0086$ are qualitatively quite different: the state with the lower

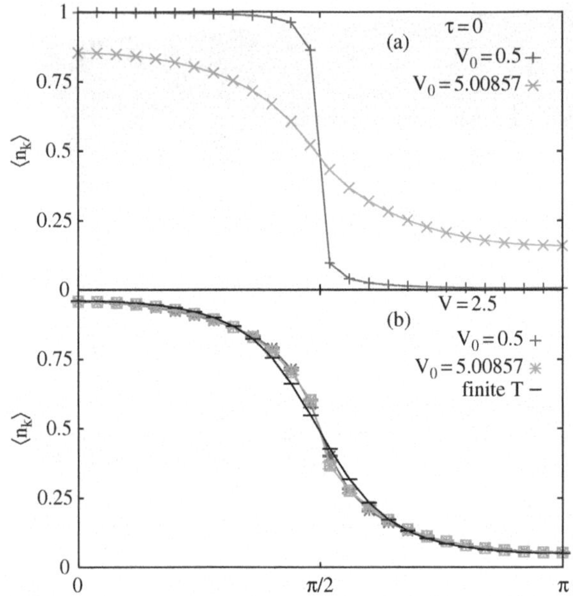

Fig. 23.9. **(a)** Momentum distribution for two initial states at $V_0 = 0.5$ and $V_0 = 5.0086$ with the same energy expectation value $\langle H(V = 2.5)\rangle$ for the time-evolving Hamiltonian. **(b)** Momentum distribution of the two initial states of **(a)** after being time-evolved with $H(V = 2.5)$ to a time $T = 4.5$. Also shown is the momentum distribution for a thermal state with the same average energy calculated using the quantum Monte Carlo method

V_0 has a clearly metallic initial momentum distribution, while the $V_0 = 5.0086$ state has clearly insulating character. Nevertheless, after a time $t = 4.5/t_h$, Fig. 23.9(b), they agree almost exactly, with $\langle n_k \rangle$ showing insulating behavior, but with a somewhat steeper slope at k_F than the insulating initial state. Also depicted in Fig. 23.9(b) is $\langle n_k \rangle$ for a thermal state with the same average energy as both initial states. The quasi-stationary state shows a small, but appreciable difference with the thermal state. Such a difference becomes larger when the time evolution is carried out with larger values of V. Therefore, we conclude that there is a generic quasi-stationary momentum distribution for a wide range of initial states and time-evolving parameter values, but that this state is almost always significantly different from the thermal state with the same average energy and the same interaction strength. We have also studied the density-density correlation function and have come to analogous conclusions [19, 20].

23.3 Discussion

In this chapter, we have given an outline of the method by which the DMRG technique can be used to calculate the time dependence of interacting quantum systems. For most applications, some version of the adaptive t-DMRG will be the best-suited method. Note, however, that it is possible that a system can have a strong enough dependence on a wide range of earlier times so that the complete t-DMRG method (i.e., targeting all time steps) could be advantageous in relatively rare circumstances [14, 24].

Within the adaptive t-DMRG, there are two major variants. The first, the Trotter method [10, 11, 12], is based on a Trotter-Suzuki decomposition which allows one to decompose the time evolution operator into pieces that can be treated efficiently and exactly within a DMRG sweeping procedure. While this method is generally quite efficient, it is limited, at least in its simplest form, to one-dimensional systems with nearest-neighbor interactions and also suffering from a systematic error in the size of the time step. The second variant treats the evolution through a time step directly [13]. The most effective way to do this seems to be to treat the exponential time evolution operator in a Lanczos expansion or using the closely related Arnoldi method [7, 14]. This method can treat more general Hamiltonians and seems to be more stable and, in some cases, more accurate for some systems, but is usually computationally more expensive than the Trotter method for similar accuracy.

As an example, we have applied the adaptive t-DMRG to a one-dimensional system of interacting spinless fermions. By starting with a metallic state and time-evolving with a Hamiltonian with CDW insulating ground state, we find oscillations in the single-particle momentum distribution that are reminiscent of collapse and revival phenomena found in bosonic systems on an optical lattice. These oscillations are damped out on the scale of the inverse hopping and attain quasi-stationary behavior for a wide range of interaction strengths. Different initial states with the same average energy lead to very similar quasi-stationary behavior, indicating that this behavior is generic. However, the quasi-stationary behavior cannot be easily characterized as a thermal distribution, at least when the temperature is fixed by the average energy. One possibility to describe this behavior is to use a more general ensemble such as the generalized Gibbs ensemble rather than the Boltzmann ensemble [23, 25, 26]. Since the generalized Gibbs ensemble used in [23, 25, 26] is parameterized by an indefinite number of parameters, each coupled to a successively higher power of the Hamiltonian H, any distribution can, in principle, be described. What is required then is a simple physical description using a small number of parameters. How to do this, and how to describe the long-time behavior of suddenly perturbed interacting quantum systems in general, is clearly a very interesting area for further research.

References

1. M. Greiner, O. Mandel, T. Hänsch, I. Bloch, Nature **419**, 51 (2002)
2. T. Kinoshita, T. Wenger, D. Weiss, Nature **440**, 900 (2006)
3. L. Sadler, J. Higbie, S. Leslie, M. Vengalattore, D. Stamper-Kurn, Nature **443**, 312 (2006)
4. Z. Yao, H. Postma, L. Balents, C. Dekker, Nature **402**, 273 (1999)
5. O. Auslaender, A. Yacoby, R.d. et al., Phys. Rev. Lett. **84**, 1764 (2000)

6. M. Hochbruck, C. Lubich, BIT **39**, 620 (1999)
7. S. Manmana, A. Muramatsu, R. Noack, in *AIP Conf. Proc.*, Vol. 789 (2005), Vol. 789, p. 269
8. M. Cazalilla, J. Marston, Phys. Rev. Lett. **88**, 256403 (2002)
9. H. Luo, T. Xiang, X. Wang, Phys. Rev. Lett. **91**, 049701 (2003)
10. G. Vidal, Phys. Rev. Lett. **93**, 040502 (2004)
11. A. Daley, C. Kollath, U. Schollwöck, G. Vidal, J. Stat. Mech.: Theor. Exp. p. P04005 (2004)
12. S. White, A. Feiguin, Phys. Rev. Lett. **93**, 076401 (2004)
13. A. Feiguin, S. White, Phys. Rev. B **72**, 020404(R) (2005)
14. P. Schmitteckert, Phys. Rev. B **70**, 121302(R) (2004)
15. L. Pezzé, L. Pitaevskii, A.S. et al., Phys. Rev. Lett. **93**, 120401 (2004)
16. M. Köhl, H. Moritz, T.S. et al., Phys. Rev. Lett. **94**, 080403 (2004)
17. K. Schönhammer, in *Physics and Chemstry of Materials with Low-Dimensional Structures*, Vol. 25, ed. by D. Baeriswyl, L. Degiorgi (Kluwer Academic Publishers, Norwell Dordrecht, 2004), Vol. 25
18. T. Giamarchi, *Quantum Physics in One Dimension* (Oxford University Press, Oxford, 2004)
19. S. Manmana, Nonequilibrium dynamics of strongly correlated quantum systems. Ph.D. thesis, University of Stuttgart, Stuttgart (2006)
20. S. Manmana, S. Wessel, R. Noack, A. Muramatsu. In preparation
21. P. Mehta, N. Andrei, Phys. Rev. Lett. **96**, 216802 (2006)
22. P. Mehta, S.P. Chao, N. Andrei. URL `http://arxiv.org/abs/cond-mat/0703426`. Erratum
23. S. Manmana, S. Wessel, R. Noack, A. Muramatsu, Phys. Rev. Lett. **98**, 210405 (2007)
24. P. Schmitteckert. Private communication
25. M. Rigol, V. Dunjko, V. Yurovsky, M. Olshanii, Phys. Rev. Lett. **98**, 050405 (2007)
26. M.A. Cazalilla, Phys. Rev. Lett. **97**, 156403 (2006)

24 Applications of Quantum Information in the Density-Matrix Renormalization Group

Ö. Legeza[1], R.M. Noack[2], J. Sólyom[1], and L. Tincani[2]

[1] Research Institute for Solid State Physics and Optics, H-1525 Budapest, Hungary
[2] Fachbereich Physik, Philipps-Universität Marburg, 35032 Marburg, Germany

In the past few years, there has been an increasingly active exchange of ideas and methods between the formerly rather disjunct fields of quantum information and many-body physics. This has been due, on the one hand, to the growing sophistication of methods and the increasing complexity of problems treated in quantum information theory, and, on the other, to the recognition that a number of central issues in many-body quantum systems can fruitfully be approached from the quantum information point of view. Nowhere has this been more evident than in the context of the family of numerical methods that go under the rubric density-matrix renormalization group. In particular, the concept of entanglement and its definition, measurement, and manipulation lies at the heart of much of quantum information theory [1]. The density-matrix renormalization group (DMRG) methods use properties of the entanglement of a bipartite system to build up an accurate approximation to particular many-body wave functions. The cross-fertilization between the two fields has led to improvements in the understanding of interacting quantum systems in general and the DMRG method in particular, has led to new algorithms related to and generalizing the DMRG, and has opened up the possibility of studying many new physical problems, ones of interest both for quantum information theory and for understanding the behavior of strongly correlated quantum systems [2].

In this line, we discuss some relevant concepts in quantum information theory, including the relation between the DMRG and data compression and entanglement. As an application, we will use the quantum information entropy calculated with the DMRG to study quantum phase transitions, in particular in the bilinear-biquadratic spin-one chain and in the frustrated spin-1/2 Heisenberg chain.

24.1 Basic Concepts of Quantum Information Theory

Perhaps the most fundamental measure in quantum information is the von Neumann entropy, which quantifies the quantum information or entanglement between two parts of a bipartite system. For a system of size N, it is defined as

$$s(N) = -\mathrm{Tr}\left[\rho^{(N)} \ln \rho^{(N)}\right] . \tag{24.1}$$

Here $\rho^{(N)}$ is the density matrix for the system and the trace is over the degrees of freedom of the system. Implicit in this description is that the system can be thought

Ö. Legeza et al.: *Applications of Quantum Information in the Density-Matrix Renormalization Group*, Lect. Notes Phys. **739**, 653–664 (2008)
DOI 10.1007/978-3-540-74686-7_24

of as forming one part of a larger, bipartite system which can always be constructed to be in a pure state.

The von Neumann entropy has been found to be intimately connected to many-body properties of a quantum system such as the quantum criticality. In one dimension, $s(N)$ will increase logarithmically with N if the system is quantum critical, but will saturate with N if the system is not [3, 4]. If a quantum critical system is also conformally invariant, additional, specific statements can be made about the entropy (see below) [5]. In higher dimensions, the von Neumann entropy will be bounded from below by a number proportional to the area (or length or volume, as appropriate) of the interface between the two parts of the system [6].

Since the von Neumann entropy is also a quantification of the fundamental approximation in the DMRG, a number of entanglement-based approaches to improve the performance and to extend the applicability of DMRG [2, 7, 8, 9], have been developed in the past few years [10, 11, 12, 13, 14, 15, 16].

For a more extensive discussion of the relationship of entanglement and von Neumann entropy with the fundamentals of the DMRG, see Chap. 20 of this volume, especially Sects. 2 and 6.

24.1.1 DMRG and Quantum Data Compression

The reduction of the Hilbert space carried out in the DMRG method is closely related to the problem of quantum data compression [17, 18]. In quantum data compression, the Hilbert space of the system Λ is divided into two parts: The "typical subspace" Λ_{typ}, which is retained, and the "atypical subspace" Λ_{atyp}, which is discarded. For pure states, there is a well defined relationship between Λ_{typ} and the von Neumann entropy of the corresponding ensemble. In general, it has been shown that

$$\beta \equiv \ln\left(\dim \Lambda_{\text{typ}}\right) - s\ , \tag{24.2}$$

is independent of the system size for large enough systems [11, 19].

Since one fundamentally treats a bipartite system in the DMRG, each subsystem is, in general, in a mixed state. In the context of the DMRG, the accessible information [20, 21] of mixed-state ensembles can be interpreted as the information loss due to the truncation procedure. This information loss is a better measure of the error than the discarded weight of the reduced density matrix

$$\varepsilon_{\text{TE}} = 1 - \sum_{\alpha=1}^{m} w_\alpha\ , \tag{24.3}$$

(also called the truncation error). Here the w_α are the eigenvalues of the reduced density matrix ρ of either subsystem; both must have the same nonzero eigenvalue spectrum.

Based on these considerations, the convergence of DMRG can be improved significantly by selecting the states kept using a criterion related to the accessible information. In general, the accessible information must be less than the Kholevo bound [20]

$$I \leq s(\rho) - p_{\text{typ}} s(\rho_{\text{typ}}) - (1 - p_{\text{typ}})\, s(\rho_{\text{atyp}}) \,, \tag{24.4}$$

where ρ_{typ} or ρ_{atyp} are the portions of the density matrix formed from the basis states that are kept and discarded, respectively. The probability p_{typ} is chosen to be appropriate for the corresponding binary channel. The behavior of the mutual information for particular ensembles as a function of p_{typ}, including various bounds on the mutual information can be found in [21]. For the DMRG, the atypical subspace should contain as little information as possible if the approximation is to be accurate. Assuming that this is the case, we take $p_{\text{typ}} = 1$, and the number of block states are selected so that $s(\rho) - s(\rho_{\text{typ}}) \leq \chi$. This *a priori*-defined χ satisfies a well-defined entropy sum rule which is related to the total quantum information generated by the DMRG. Deviations from this sum rule provide a measure of the error of the DMRG calculation. Therefore, χ can be chosen to control its accuracy.

Figure 24.1 shows the relative error of the ground-state energy, defined as $(E_{\text{DMRG}} - E_{\text{exact}})/E_{\text{exact}}$, plotted on a logarithmic scale for various values of the Coulomb interaction U for the one-dimensional Hubbard model. In Fig. 24.1(a), it is plotted as a function of ε_{TE}, whereas in Fig. 24.1(b), it is plotted as a function of χ. As can be seen, the error in the latter plot behaves very stably as a function of χ, even for very small values of the retained eigenvalues of ρ_{typ}. On the other hand, the error in the energy behaves somewhat less regularly as a function of ε_{TE}. Therefore, an extrapolation of the energy as a function of χ would be significantly better behaved than one as a function of ε_{TE}. We find that such behavior is representative; generically, extrapolation with χ is as stable or more stable than extrapolation with ε_{TE} for a wide variety of systems [11].

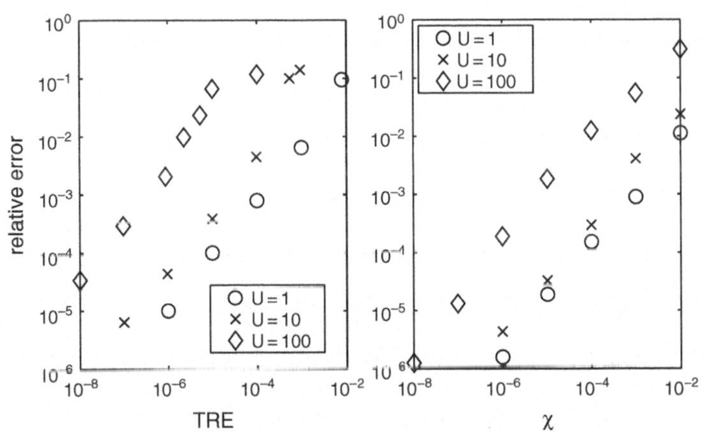

Fig. 24.1. The relative error of the ground-state energy for the half-filled Hubbard chain for various values of the on-site Coulomb interaction U on an $N = 80$-site lattice with periodic boundary conditions as a function of **(a)** the truncation error and **(b)** the threshold value of the Kholevo bound on accessible information, see (24.4). Taken from [11]

24.1.2 DMRG and Non-Local Models

Another application of quantum information to the DMRG is to non-local quantum lattice models that occur in quantum chemical applications [22] or when the Hubbard and related models are represented in momentum space [23, 24]. A general non-local fermionic Hamiltonian has the form

$$\widehat{H} = \sum_{p,q,\sigma} T^{\sigma}_{p,q}\, \widehat{c}^{\dagger}_{p,\sigma} \widehat{c}_{q,\sigma} + \sum_{p,q,r,s,\sigma,\sigma'} V^{\sigma,\sigma'}_{p,q,r,s}\, \widehat{c}^{\dagger}_{p,\sigma} \widehat{c}^{\dagger}_{q,\sigma'} \widehat{c}_{r,\sigma'} \widehat{c}_{s,\sigma} \,. \tag{24.5}$$

Here $\widehat{c}^{\dagger}_{p,\sigma}$ creates a fermion with spin σ in single-particle orbital p. In quantum chemistry, the $T^{\sigma}_{p,q}$ represent the one-particle overlap integrals, while in momentum-space models only the diagonal elements, which contain the dispersion, are nonzero. The $V^{\sigma,\sigma'}_{p,q,r,s}$ are two-particle overlap integrals, which are related to the Coulomb interaction, in quantum chemistry. For momentum-space models, $V^{\sigma,\sigma'}_{p,q,r,s}$ contains the Fourier-transformed Coulomb interaction; additional symmetries (momentum conservation, interaction only between opposite spin species, etc.) generally simplify its structure significantly.

In contrast to short-range models in real space, the optimal ordering of the "lattice sites", in such models, i.e., the orbitals of the single-particle basis, is not evident. However, finding a sufficiently good ordering seems to be crucial to formulating efficient DMRG algorithms for such systems [10, 25]. From a quantum information point of view, the lattice sites are inequivalent in general; their relative importance only becomes clear when the interaction is turned on. Therefore, the entropy profile of the bipartite partitioning of the finite system depends very much on the ordering of the lattice sites. The question, then, is how to quantify their importance in terms of quantum information. Unfortunately, there is, as yet, no clear-cut solution to this problem. However, various quantum-information-based quantities lend insight. One such quantity is the single-site entropy s^p, which is formed by taking a single site as one part of a bipartite system, and then calculating the entropy for this subsystem in the usual way. This quantity encodes the entanglement between the site and the remainder of the system, i.e., the extent to which the site shares quantum information with the rest of the system. Heuristic schemes to order sites based on this site entropy have been proposed [10]: Generally, sites with the largest single-site entropy should be placed close together in the middle of the order. For such cases the size of the typical subspace can also be reduced using entanglement-based optimization of the ordering of lattice sites [10, 11, 16, 22]. The problem, however, is that the single-site entropy does not encode the quantum information exchange between *pairs* of sites, i.e., how important it is to place particular sites in proximity to each other.

One quantity that can overcome this problem is the mutual two-site information

$$I_{p,q} \equiv \frac{1}{2}\left(s^p + s^q - s^{pq}\right)\left(1 - \delta_{pq}\right) \geq 0 \,, \tag{24.6}$$

where s^{pq} is the entropy of a subsystem consisting of two (not necessarily adjacent) lattice sites p and q. This quantity has proven to be especially useful when

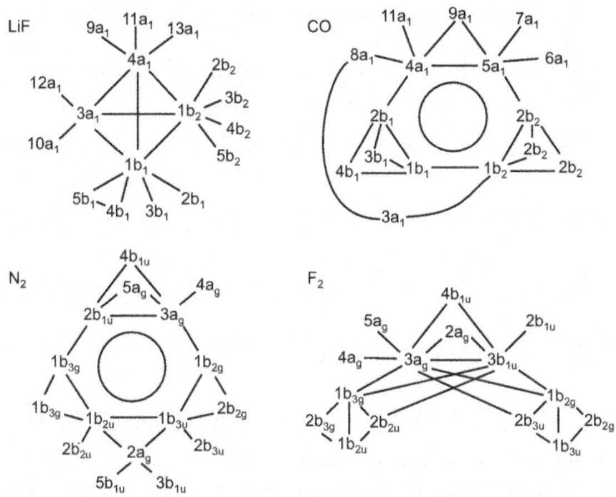

Fig. 24.2. Diagram of $I_{p,q}$ for the molecules LiF, CO, N_2, and F_2 calculated at Hartree-Fock ordering with $m = 200$: Lines connect orbital labels with $I_{p,q} > 0.01$. The circle for CO and N_2 denotes that the surrounding orbitals are all connected with each other. Taken from [22]

applied to quantum chemical systems. In Fig. 24.2, we show the topology of $I_{p,q}$ for four prototypical small molecules, LiF, CO, N_2, and F_2, with a particular basis set; for details, see [22]. As can be seen, the mutual two-site information yields a picture of the detailed connectivity of the orbitals, which is different for each molecule. An attempt to optimize ordering of orbitals using a cost function based on this information has led to moderate success [22]. However, more work needs to be done both on defining a meaningful measure of mutual two-site information, and in developing heuristics to optimize ordering based on this measure. A related problem has cropped up in an attempt to map the one-dimensional Hubbard model with periodic boundary condition to a model with open boundary conditions [16]. The transformed effective interaction, which has the form $V_{p,q,r,s}^{\sigma,\sigma'}$ (see Hamiltonian (24.5)), is then nonlocal. An analysis of the entanglement generated by these nonlocal terms has been used to optimize the site ordering. Such insights are also relevant to quantum chemical problems.

24.2 Entropic Analysis of Quantum Phase Transitions

The local measure of entanglement, the ℓ-site entropy with $\ell = 1, 2, ...N$, which is obtained from the reduced density matrix ρ, can be used to detect and locate quantum phase transitions (QPTs) [26, 27, 28, 29]. As an example, Fig. 24.3 shows the block entropy for $\ell = N/2$ for the most general isotropic spin-one chain model described by the Hamiltonian

$$H = \sum_i \left[\cos\theta \, (\boldsymbol{S}_i \cdot \boldsymbol{S}_{i+1}) + \sin\theta \, (\boldsymbol{S}_i \cdot \boldsymbol{S}_{i+1})^2 \right] , \qquad (24.7)$$

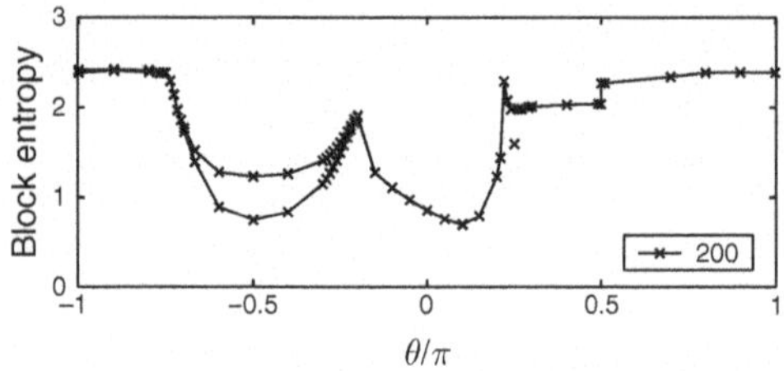

Fig. 24.3. Entropy of blocks of $\ell = N/2$ and $\ell = N/2 + 1$ sites of the bilinear-biquadratic spin $S = 1$ model for a chain with $N = 200$ sites

as a function of θ, where we take $-\pi < \theta \leq \pi$. In one dimension, the model is known to have at least four different phases [30, 31, 32]. The ground state is ferromagnetic for $\theta < -3\pi/4$ and $\theta > \pi/2$. For $-3\pi/4 < \theta < -\pi/4$, the ground state is dimerized; the point $\theta = -\pi/4$ is exactly solvable [33, 34]. In the range $-\pi/4 < \theta < \pi/4$, the system is in the Haldane phase, there is an exact solution at $\theta = \pi/4$ [35, 36, 37], and for $\pi/4 < \theta < \pi/2$, the phase is spin nematic (trimerized). The issue of whether a quantum quadrupolar phase [38, 39, 40], exists near $\theta = -3\pi/4$ has not yet been settled [41]. These phases and the corresponding QPTs are reflected in the block entropy, Fig. 24.3. The jump in the entropy at $\pi/2$ indicates a first-order transition. At $\theta = -3\pi/4$, there is only a cusp in the block entropy, but a jump in the single-entropy s^p (as defined above) indicates that this transition is first order [42]. The cusps at $\theta = -\pi/4$ and $\pi/4$ indicate second-order transitions, and the bifurcation of the entropy curves for $\ell = N/2$ and $\ell = N/2 + 1$ indicates that there is a spatially inhomogeneous dimerized phase between $-3\pi/4 < \theta < -\pi/4$.

Note that the entropy has a minimum at $\theta = \arctan 1/3 \simeq 0.1024\pi$, which is at the valence-bond-solid (VBS) point [43], but that it remains a continuous curve. The extremum of the entropy indicates a change in the wave function and can also signal a phase transition even if it remains a continuous curve. Such behavior has also been found in the $1/n$-filled SU(n) $n = 2, 3, 4, 5$ Hubbard model at $U = 0$, where an infinite-order (Kosterlitz-Thouless-like) phase transition takes place [27, 44, 45, 46]. Since there is no sharply defined transition in the entropy, however, additional methods must be used to classify the ground-state properties on either side of an extremum. One possibility is an analysis of the entropy profile $s(\ell)$ as the subsystem size ℓ is changed from $\ell = 0$ to N for fixed model parameters; see below. Note that there is also another minimum in the block entropy at $\theta = -\pi/2$; this corresponds to a point where the model can be partially mapped to the nine-state quantum Potts model whose ground state is exactly known [47, 48]. However, there is no known phase transition at this point.

For models that map to a conformal field theory [49], an analytic expression for the entropy profile has been derived, and this form has been shown to be satisfied by critical spin models. The entropy for a subsystem of length ℓ in a finite system of length N with open boundary conditions within conformal field theory has the form [5]

$$s(\ell) = \frac{c}{6} \ln \left[\frac{2N}{\pi} \sin \left(\frac{\pi \ell}{N} \right) \right] + g \,, \tag{24.8}$$

where c is the central charge. This quantity contains a constant term which depends on the ground-state degeneracy and a constant shift g which depends on the boundary conditions. As will be shown below, there can be an additional oscillatory term which decays with system size and distance from the boundary as a power law [50, 51].

A new method to analyze the oscillatory nature of the finite subsystem entropy $s(\ell)$, is based on the Fourier spectrum of $s(\ell)$,

$$\widetilde{s}(q) = \frac{1}{N} \sum_{\ell=0}^{N} \mathrm{e}^{-\mathrm{i}q\ell} s(\ell) \,, \tag{24.9}$$

with $s(0) = s(N) = 0$, where $q = 2\pi n/N$ and $n = 0, \ldots, N-1$, is appropriate to study cases when no true phase transition takes place, i.e., when only the character of the decaying correlation function changes.

Figure 24.4(a) shows the block entropy at $\theta = \pi/4$ for the so-called trimerized phase, a phase characterized by three soft modes in the energy spectrum at $k = 0, \pm 2\pi/3$ [35, 36, 37]. The solid line is a fit using (24.8), which yields $c = 2$ in agreement with [30, 31, 32] and [52]. The oscillation in the entropy with a period of three is related to these three soft modes, as is apparent in Fig. 24.4(b) in which the peaks in the Fourier spectrum appear at $k = 0, \pm 2\pi/3$ [53]. A similar analysis at $\theta = -\pi/4$ yields $c = 3/2$ in the thermodynamic limit. The corresponding $\widetilde{s}(q)$ has

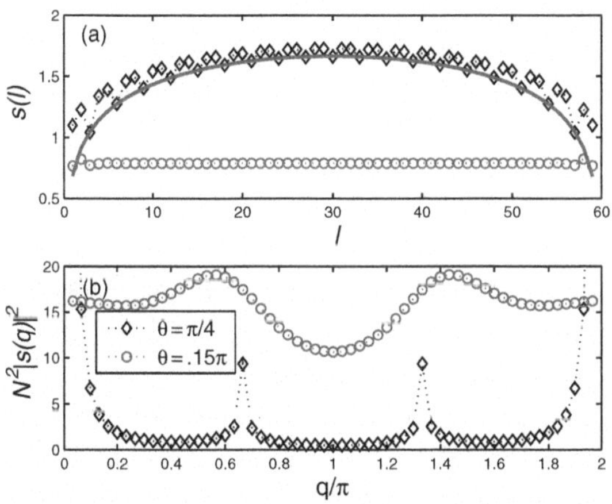

Fig. 24.4. **(a)** Von Neumann entropy of a subsystem of size ℓ on an $L = 60$ bilinear-biquadratic spin chain for $\theta = \pi/4$ and $\theta = 0.15\pi$. The fit to the upper, $\theta = \pi/4$, curve has been carried out using (24.8), and yields $c = 2$, $q^* = 2\pi/3$. The lower curve, for $\theta = 0.15\pi$, has a small value and the entropy saturates because the phase is gapped. **(b)** Power spectrum $N^2|\widetilde{s}(q)|^2$ of the data of **(a)**. The curve for $\theta = 0.15\pi$ has been multiplied by a factor of 5 to enhance its readability on this scale

peaks at $q = 0$ and π for finite systems. It is known that for $\theta < \pi/4$ the soft modes become gapped and the minimum of the energy spectrum moves from $q = 2\pi/3$ toward $q = \pi$ as θ approaches the VBS point [43].

In order to characterize the various phases in the thermodynamic limit, a finite-size extrapolation must be carried out. Fig. 24.5 displays the behavior of $\widetilde{s}(q^*)$ with system size for a number of values of θ that are representative of the different phases. The wave vector q^* is chosen to be appropriate for the corresponding phase, for example, $q^* = 2\pi/3$ in the trimerized phase. The value $q^* = 0.53$ for $\theta = 0.15\pi$ (in the incommensurate phase) is the location of the incommensurate peak; see Fig. 24.4(b). As can be seen, all $\widetilde{s}(q^*) \to 0$ for $N \to \infty$, except in the range $-3\pi/4 < \theta < -\pi/4$ where $\widetilde{s}(q = \pi)$ remains finite, signaling the bond-ordered nature of the dimerized phase. Note that the $q^* = 0$ peak (not shown) also scales to a finite value in much of the phase diagram.

In Fig. 24.6, we summarize the behavior of $\widetilde{s}(q)$ for finite systems and in the $N \to \infty$ limit. We determine the position of the peaks in $\widetilde{s}(q)$ on finite systems by finding the maxima in splines fit through the discrete allowed q points. Infinite-system behavior, obtained from extrapolations (see Fig. 24.5), is also depicted. In the ferromagnetic phase, $\theta < -3\pi/4, \theta > \pi/2$, there is a sole peak at $q^* = 0$, as expected. The $q^* = 0$ peak is present for all θ and persists in the thermodynamic limit. In the dimer phase, $-3\pi/4 < \theta < -\pi/4$, the $q^* = \pi$ peak persists in the thermodynamic limit (see Fig. 24.5). Two different behaviors can be seen in the Haldane phase, $-\pi/4 < \theta < \pi/4$; for $\theta < \theta_{\mathrm{VBS}}$, the $q^* = \pi$ peak present in finite-size systems vanishes in the thermodynamic limit. For $\theta > \theta_{\mathrm{VBS}}$, the incommensurate peak present only in finite systems can be seen to move from $q = 0$ to $2\pi/3$ as θ goes towards $\pi/4$, as also seen in Fig. 24.4. Finally, in the spin nematic phase, $\pi/4 < \theta < \pi/2$, there is a peak at $q^* = 2\pi/3$ which scales to zero as $N \to \infty$.

Therefore, incommensurability can be detected by the entropy analysis as well. It is known [54] that the VBS point is a disorder point, where incommensurate

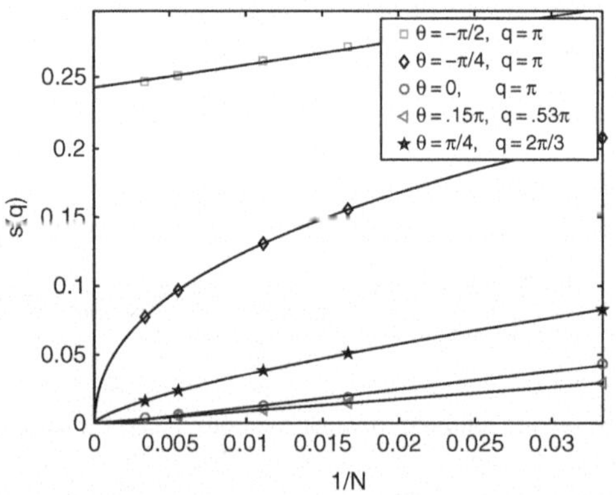

Fig. 24.5. Finite-size scaling of $\widetilde{s}(q)$ for a number of representative values of θ at the appropriate wave vector q. The continuous lines are fits to a form $AN^{-\alpha} + B$

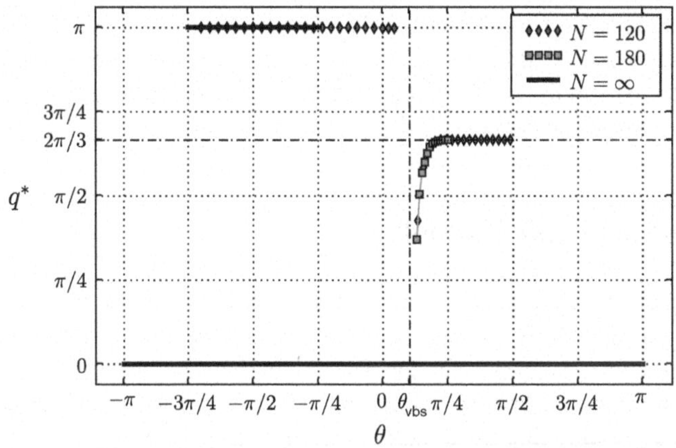

Fig. 24.6. Position of the peak q^* in the Fourier-transformed block entropy $|\tilde{s}(q)|^2$ plotted as a function of the parameter θ for the bilinear-biquadratic spin chain on system sizes of $N = 120$ and $N = 180$ (for higher resolution near θ_{VBS}), as well as in the thermodynamic limit. The peak at $q^* = 0$ on finite systems, which is present for all θ, has been removed for readability

oscillations appear in the decaying correlation function; however, the shift of the minimum of the static structure factor appears only at a larger value, $\theta_{\mathrm{L}} = 0.138\pi$, the Lifshitz point. In contrast to this, the minimum of the block entropy shown in Fig. 24.3 is exactly at the VBS point, and therefore indicates the location of the commensurate-incommensurate transition correctly.

A similar analysis can be carried out for the frustrated $J - J'$ Heisenberg spin $1/2$ chain with Hamiltonian

$$H = \sum_i \left[J\left(\boldsymbol{S}_i \cdot \boldsymbol{S}_{i+1}\right) + J'\left(\boldsymbol{S}_i \cdot \boldsymbol{S}_{i+2}\right) \right], \tag{24.10}$$

with the ratio J'/J ($J', J > 0$) playing the role of the parameter θ in the bilinear-biquadratic model. For $J'/J < J_c \approx 0.2411$, the model is in a critical Heisenberg phase, while a spin gap develops for $J'/J > J_c$. At $J'/J = 0.5$, the Majumdar-Ghosh point, the model is exactly solvable and the ground state is a product of

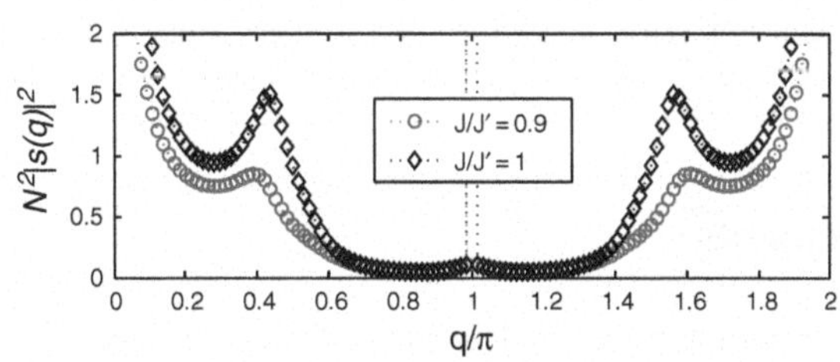

Fig. 24.7. Power spectrum of the block entropy $N^2|\tilde{s}(q)|^2$ for the frustrated Heisenberg chain at $J/J' = 1$, calculated on a chain of length $N = 128$

local dimers [55]. As a function of J'/J, the block entropy is continuous, but has a minimum at $J/J' = 0.5$. For $J/J' > 0.5$ an extra peak appears in the Fourier spectrum of $\widetilde{s}(q)$ and moves from 0 to $\pi/2$ as J/J' gets larger. The development of the incommensurate peaks near $J/J' = 1$ can be seen in Fig. 24.7.

24.3 Discussion and Outlook

In this chapter, we have sketched the intimate relationship between quantum information and the family of density-matrix renormalization group methods. The fundamental approximation in the DMRG can perhaps be best understood in quantum information terms: The wave function of a bipartite system is most accurately represented by minimizing the quantum information loss, or by carrying out an optimal lossy quantum data compression. The quantum information loss can be used within the DMRG as a measure of the accuracy that is alternate to the discarded weight of density matrix eigenvalues and has a number of advantages. We have outlined some efforts to use quantum information quantities, specifically, the one-site entropy and the mutual quantum information for two sites, to optimize the ordering of single-particle orbitals in non-local Hamiltonians. Finally, we have discussed how the von Neumann entropy calculated during the DMRG procedure can be used to study quantum phase transitions. Jumps, cusps, and minima or maxima in the mid-block entropy signal first, second, and infinite-order phase transitions, while information about the spatial structure of phase can be gleaned from dependence of the entropy on the position of the partition of the bipartite system and from its Fourier transform.

There are a number of possibilities to further apply quantum information theory within the DMRG approach. For example, the possibility of using various quantum information entropies, the entropy reduction by basis-state transformations, entanglement localization, bounds on accessible information of mixed states, and the description of the dynamics of mixed states in terms of an effective temperature have not yet been fully explored [1, 56]. These aspects of quantum information are also closely related to a more quantum-information oriented formulation of the DMRG which has led to more general algorithms based on matrix-product states and their generalizations. Application of these generalizations include systems at finite temperature, systems with dissipation, the calculation of dynamical and time-dependent behavior, and more efficient treatment of higher-dimensional systems [2].

Acknowledgements

This work was supported in part by Hungarian Research Fund (OTKA) Grants No. K 68340 and NF 61726 and by the János Bolyai Research Fund.

References

1. A. Galindo, M. Martin-Delgado, Rev. Mod. Phys. **74**, 347 (2002)
2. U. Schollwöck, Rev. Mod. Phys. **77**, 259 (2005)
3. G. Vidal, J. Latorre, E. Rico, A. Kitaev, Phys. Rev. Lett. **90**, 227902 (2003)
4. J. Latorre, E. Rico, G. Vidal, Quant. Inf. and Comp. **4**, 48 (2004)
5. P. Calabrese, J. Cardy, J. Stat. Mech.: Theor. Exp. (2004)
6. M. Srednicki, Phys. Rev. Lett. **71**, 666 (1993)
7. S. White, Phys. Rev. Lett. **69**, 2863 (1992)
8. S. White, Phys. Rev. B **48**, 10345 (1993)
9. R. Noack, S. Manmana, in *Lectures on the physics of highly correlated electron systems IX, AIP Conference proceedings*, Vol. 789, ed. by A. Avella, F. Mancini (Melville, New York, 2005), *AIP Conference proceedings*, Vol. 789, p. 93
10. Ö. Legeza, J. Sólyom, Phys. Rev. B **68**, 195116 (2003)
11. O. Legeza, J. Sólyom, Phys. Rev. B **70**, 205118 (2004)
12. F. Verstraete, D. Porras, J. Cirac, Phys. Rev. Lett. **93**, 227205 (2004)
13. F. Verstraete, J. Cirac. URL `http://arxiv.org/abs/cond-mat/0407066`. Preprint
14. S.R. White, A. Feiguin, Phys. Rev. Lett. **93**, 076401 (2004)
15. A.J. Daley, C. Kollath, U. Schollwöck, G. Vidal, J. Stat. Mech.: Theor. Exp. P04005 (2004)
16. O. Legeza, F. Gebhard, J. Rissler, Phys. Rev. B **74**, 195112 (2006)
17. B. Schumacher, Phys. Rev. A **51**, 2738 (1995)
18. R. Jozsa, J. Mod. Opt. **41**, 2315 (1994)
19. G. Vidal, Phys. Rev. Lett. **91**, 147902 (2003)
20. A. Kholevo, Probl. Inf. Transm.(USSR) **177**, 9 (1973)
21. C. Fuchs, C. Caves, Phys. Rev. Lett. **73**, 3047 (1994)
22. J. Rissler, R. Noack, S. White, Chem. Phys. **323**, 519 (2006)
23. T. Xiang, Phys. Rev. B **53**, 10445 (1996)
24. S. Nishimoto, E. Jeckelmann, F. Gebhard, R. Noack, Phys. Rev. B **65**, 165114 (2002)
25. G.L. Chan, M. Head-Gordon, J. Chem. Phys. **116**, 4462 (2002)
26. P. Zanardi, Phys. Rev. A **65**, 42101 (2002)
27. S.J. Gu, S.S. Deng, Y.Q. Li, H.Q. Lin, Phys. Rev. Lett. **93**, 86402 (2004)
28. J. Vidal, G. Palacios, R. Mosseri, Phys. Rev. A **69**, 022107 (2004)
29. J. Vidal, R. Mosseri, J. Dukelsky, Phys. Rev. A **69**, 054101 (2004)
30. G. Fath, J. Sólyom, Phys. Rev. B **44**, 11836 (1991)
31. G. Fath, J. Sólyom, Phys. Rev. B **47**, 872 (1993)
32. G. Fath, J. Sólyom, Phys. Rev. B **51**, 3620 (1995)
33. L. Takhtajan, Phys. Lett. A **87**, 479 (1982)
34. H.M. Babujian, Phys. Lett. A **90**, 479 (1982)
35. G. Uimin, JETP Lett. **12**, 225 (1970)
36. C. Lai, J. Math. Phys. **15**, 1675 (1974)
37. B. Sutherland, Phys. Rev. B **12**, 3795 (1975)
38. A. Chubukov, J. Phys. Condens. Matter **2**, 1593 (1990)
39. A. Chubukov, Phys. Rev. B **43**, 3337 (1991)
40. A. Läuchli, G. Schmid, S. Trebst, Phys. Rev. B **74**, 144426 (2006)
41. K. Buchta, G. Fath, Ö. Legeza, J. Sólyom, Phys. Rev. B **72**, 054433 (2005)
42. Ö. Legeza, J. Sólyom, Phys. Rev. Lett. **96**, 116401 (2006)
43. I. Affleck, T. Kennedy, E. Lieb, H. Tasaki, Phys. Rev. Lett. **59**, 799 (1987)

44. D. Larsson, H. Johannesson, Phys. Rev. Lett. **95**, 196406 (2005)
45. D. Larsson, H. Johannesson, Phys. Rev. A **73**, 155108 (2007)
46. K. Buchta, Ö. Legeza, E.S.J. Sólyom, Phys. Rev. B **75**, 155108 (2007)
47. J. Parkinson, J. Phys. C **20**, L1029 (1987)
48. J. Parkinson, J. Phys. C **21**, 3793 (1988)
49. C. Holzhey, F. Larsen, F. Wilczek, Nucl. Phys. B **424**, 443 (1994)
50. I. Affleck, A.W.W. Ludwig, Phys. Rev. Lett. **67**, 161 (1991)
51. N. Laflorencie, E.S. Sørensen, M.S. Chang, I. Affleck, Phys. Rev. Lett. **96**, 100603 (2006)
52. C. Itoi, M.H. Kato, Phys. Rev. B **55**, 8295 (1997)
53. Ö. Legeza, J. Sólyom, L. Tincani, R.M. Noack, Phys. Rev. Lett. **99**, 087203 (2007)
54. U. Schollwöck, T. Jolicoeur, T. Garel, Phys. Rev. B **53**, 3304 (1996)
55. C.K. Majumdar, D.K. Ghosh, J. Mat. Phys. **10**, 1388, 1399 (1969)
56. L. Amico, R. Fazio, A. Osterloh, V. Vedral. URL `http://arxiv.org/abs/quant-ph/0703044`. Preprint

25 Density-Matrix Renormalization Group for Transfer Matrices: Static and Dynamical Properties of 1D Quantum Systems at Finite Temperature

Stefan Glocke[1], Andreas Klümper[1], and Jesko Sirker[2,3]

[1] Fachbereich Physik, Bergische Universität Wuppertal, 42097 Wuppertal, Germany
[2] Department of Physics and Astronomy, University of British Columbia, Vancouver, BC, V6T 1Z1, Canada
[3] Max-Planck Institute for Solid State Research, 70569 Stuttgart, Germany

The density-matrix renormalization group (DMRG) applied to transfer matrices allows it to calculate static as well as dynamical properties of one-dimensional (1D) quantum systems at finite temperature in the thermodynamic limit. To this end the quantum system is mapped onto a 2D classical system by a Trotter-Suzuki decomposition. Here we discuss two different mappings: The standard mapping onto a 2D lattice with checkerboard structure as well as an alternative mapping introduced by two of us. For the classical system an appropriate quantum transfer matrix is defined which is then treated using a DMRG scheme. As applications, the calculation of thermodynamic properties for a spin-1/2 Heisenberg chain in a staggered magnetic field and the calculation of boundary contributions for open spin chains are discussed. Finally, we show how to obtain real-time dynamics from a classical system with complex Boltzmann weights and present results for the autocorrelation function of the XXZ-chain.

25.1 Introduction

Several years after the invention of the DMRG method to study ground-state properties of 1D quantum systems [1], Nishino showed that the same method can also be applied to the transfer matrix of a 2D classical system hence allowing to calculate its partition function at finite temperature [2]. The same idea can also be used to calculate the thermodynamic properties of a 1D quantum system after mapping it to a 2D classical one with the help of a Trotter-Suzuki decomposition [3, 4, 5]. Bursill et al. [6] then presented the first application but the density matrix chosen in this work to truncate the Hilbert space was not optimal so that the true potential of this new numerical method was not immediately clear. This changed when Wang and Xiang [7] and Shibata [8] presented an improved algorithm and showed that the density-matrix renormalization group applied to transfer matrices (which we will denote as TMRG from hereon) is indeed a serious competitor to other numerical methods as for example Quantum-Monte-Carlo (QMC). Since then, the TMRG method has been successfully applied to a number of systems including various spin

S. Glocke et al.: *Density-Matrix Renormalization Group for Transfer Matrices: Static and Dynamical Properties of 1D Quantum Systems at Finite Temperature*, Lect. Notes Phys. **739**, 665–677 (2008)
DOI 10.1007/978-3-540-74686-7_25

chains, the Kondo lattice model, the $t - J$ chain and ladder and also spin-orbital models [9, 10, 11, 12, 13, 14, 15, 16, 17].

The main advantage of the TMRG algorithm is that the thermodynamic limit can be performed exactly thus avoiding an extrapolation in system size. Furthermore, there are no statistical errors and results can be obtained with an accuracy comparable to $(T = 0)$ DMRG calculations. Similar to the $(T = 0)$ DMRG algorithms, the method is best suited for 1D systems with short range interactions. These systems can, however, be either bosonic or fermionic because no negative sign problem as in QMC exists. Most important, there are two areas where TMRG seems to have an edge over any other numerical methods known today. These are:

(i) Impurity or boundary contributions, and
(ii) real-time dynamics at finite temperature.

As first shown by Rommer and Eggert [18], the TMRG method allows it to separate an impurity or boundary contribution from the bulk part thus giving direct access to quantities which are of order $\mathcal{O}(1/L)$ compared to the $\mathcal{O}(1)$ bulk contribution (here L denotes the length of the system). We will discuss this in more detail in Sect. 25.5. Calculating numerically the dynamical properties for large or even infinite 1D quantum systems constitutes a particularly difficult problem because QMC and TMRG algorithms can usually only deal with imaginary-time correlation functions. The analytical continuation of numerical data is, however, an ill-posed problem putting severe constraints on the reliability of results obtained this way. Very recently, two of us have presented a modified TMRG algorithm which allows for the direct calculation of real-time correlations [19]. This new algorithm will be discussed in Sect. 25.6.

Before coming to these more recent developments we will discuss the definition of an appropriate quantum transfer matrix for the classical system in Sect. 25.2 and describe how the DMRG algorithm is applied to this object in Sect. 25.3. Here we will follow in parts the article by Wang and Xiang in [20] but, at the same time, also discuss an alternative Trotter-Suzuki decomposition [15, 16].

25.2 Quantum Transfer Matrix Theory

The TMRG method is based on a Trotter-Suzuki decomposition of the partition function, mapping a 1D quantum system to a 2D classical one [3, 4, 5]. In the following, we discuss both the standard mapping introduced by Suzuki [5] as well as an alternative one [15, 16] starting from an arbitrary Hamiltonian H of a 1D quantum system with length L, periodic boundary conditions and nearest-neighbor interaction

$$H = \sum_{i=1}^{L} h_{i,i+1} \, . \tag{25.1}$$

The standard mapping, widely used in QMC and TMRG calculations, is described in detail in [20]. Therefore we only summarize it briefly here. First, the Hamiltonian is decomposed into two parts, $H = H_e + H_o$, where each part is a sum of commuting terms. Here H_e (H_o) contains the interactions $h_{i,i+1}$ with i even (odd). By discretizing the imaginary time, the partition function becomes

$$Z = \mathrm{Tr}\, \mathrm{e}^{-\beta H} = \lim_{M\to\infty} \mathrm{Tr}\left\{\left[\mathrm{e}^{-\epsilon H_e}\mathrm{e}^{-\epsilon H_o}\right]^M\right\} \quad (25.2)$$

with $\epsilon = \beta/M$, β being the inverse temperature and M an integer (the so called Trotter number). By inserting $2M$ times a representation of the identity operator, the partition function is expressed by a product of local Boltzmann weights

$$\tau_{k,k+1}^{i,i+1} = \left\langle s_k^i s_k^{i+1}\right| \mathrm{e}^{-\epsilon H_{e,o}} \left| s_{k+1}^i s_{k+1}^{i+1}\right\rangle , \quad (25.3)$$

denoted in a graphical language by a shaded plaquette (see Fig. 25.1). The subscripts i and k represent the spin coordinates in the space and the Trotter (imaginary time) directions, respectively. A column-to-column transfer matrix $\mathcal{T}_M$, the so called *quantum transfer matrix* (QTM), can now be defined using these local Boltzmann weights

$$\mathcal{T}_M = \left(\tau_{1,2}\tau_{3,4}\cdots\tau_{2M-1,2M}\right)\left(\tau_{2,3}\tau_{4,5}\cdots\tau_{2M,1}\right) . \quad (25.4)$$

and is shown in the left part of Fig. 25.1. The partition function is then simply given by

$$Z = \mathrm{Tr}\, \mathcal{T}_M^{L/2} . \quad (25.5)$$

The disadvantage of this Trotter-Suzuki mapping to a 2D lattice with checkerboard structure is that the QTM is two columns wide. This increases the amount of memory necessary to store it and also complicates the calculation of correlation functions.

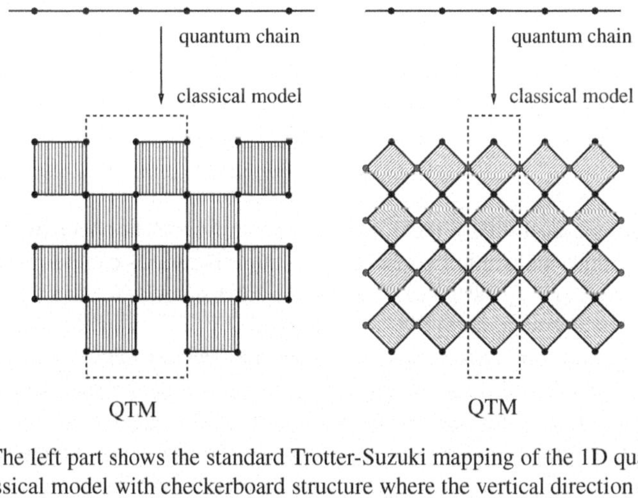

Fig. 25.1. The left part shows the standard Trotter-Suzuki mapping of the 1D quantum chain to a 2D classical model with checkerboard structure where the vertical direction corresponds to imaginary time. The QTM is two-column wide. The right part shows the alternative mapping. Here, the QTM is only one column wide

Alternatively, the partition function can also be expressed by [15, 16]

$$Z = \lim_{M\to\infty} \mathrm{Tr}\left\{[\mathcal{T}_1(\epsilon)\mathcal{T}_2(\epsilon)]^{M/2}\right\}, \tag{25.6}$$

with $\mathcal{T}_{1,2}(\epsilon) = T_{R,L}\exp[-\epsilon H + \mathcal{O}(\epsilon^2)]$. Here, $T_{R,L}$ are the right- and left-shift operators, respectively. The obtained classical lattice has alternating rows and additional points in a mathematical auxiliary space. Its main advantage is that it allows to formulate a QTM which is only one column wide (see right part of Fig. 25.1). The derivation of this QTM is completely analogous to the standard one, even the shaded plaquettes denote the same Boltzmann weight. Here, however, these weights are rotated by 45° clockwise and anti-clockwise in an alternating fashion from row to row. Using this transfer matrix, $\widetilde{\mathcal{T}}_M$, the partition function is given by $Z = \mathrm{Tr}\ \widetilde{\mathcal{T}}_M^L$.

25.2.1 Physical Properties in the Thermodynamic Limit

The reason why this transfer matrix formalism is extremely useful for numerical calculations has to do with the eigenspectrum of the QTM. At infinite temperature it is easy to show [21] that the largest eigenvalue of the QTM $\mathcal{T}_M(\widetilde{\mathcal{T}}_M)$ is given by S^2 (S) and all other eigenvalues are zero. Here S denotes the number of degrees of freedom of the physical system per lattice site. Decreasing the temperature, the gap between the leading eigenvalue Λ_0 and next-leading eigenvalues Λ_n $(n > 0)$ of the transfer matrix shrinks. The ratio between Λ_0 and each of the other eigenvalues Λ_n, however, defines a *correlation length* $1/\xi_n = \ln|\Lambda_0/\Lambda_n|$ [20, 21]. Because an 1D quantum system cannot order at finite temperature, any correlation length ξ_n will stay finite for $T > 0$, i.e., the gap between the leading and any next-leading eigenvalue stays finite. Therefore the calculation of the free energy in the *thermodynamic limit* boils down to the calculation of the largest eigenvalue Λ_0 of the QTM

$$\begin{aligned} f &= -\lim_{L\to\infty}\frac{1}{\beta L}\ln Z = -\lim_{L\to\infty}\lim_{\epsilon\to 0}\frac{1}{\beta L}\ln \mathrm{Tr}\,\widetilde{\mathcal{T}}_M^L \\ &= -\lim_{\epsilon\to 0}\lim_{L\to\infty}\frac{1}{\beta L}\ln\left\{\Lambda_0^L\left[1+\sum_{l>1}\underbrace{(\Lambda_l/\Lambda_0)^L}_{\xrightarrow{L\to\infty}0}\right]\right\} = -\lim_{\epsilon\to 0}\frac{\ln\Lambda_0}{\beta}\,. \end{aligned} \tag{25.7}$$

Here the interchangeability of the limits $L \to \infty$ and $\epsilon \to 0$ has been used [5]. Local expectation values and static two-point correlation functions can be calculated in a similar fashion (see e.g. [20] and [21]). In the next section, we are going to show how the eigenvalues of the QTM are computed by means of the density matrix renormalization group. This is possible since the transfer matrices are built from local objects. Instead of sums of local objects we are dealing with products, but this is not essential to the numerical method. However, there are a few important differences in treating transfer matrices instead of Hamiltonians. At first sight, these differences look technical, but at closer inspection they reveal a physical core.

The QTMs as introduced above are real valued, but not symmetric. This is not a serious drawback for numerical computations, but certainly inconvenient. So the first question that arises is whether the transfer matrices can be symmetrized. Unfortunately, this is not the case. If the transfer matrix were replaceable by a real symmetric (or a hermitean) matrix all eigenvalues would be real and the ratios of next-leading eigenvalues to the leading eigenvalue would be real, positive or negative. Hence all correlation functions would show commensurability with the lattice. However, we know that a generic quantum system at sufficiently low temperatures yields incommensurate oscillations with wave vectors being multiples of the Fermi vector taking rather arbitrary values.

Therefore we know that the spectrum of a QTM must consist of real eigenvalues or of complex eigenvalues organized in complex conjugate pairs. This opens the possibility to understand the QTM as a *normal matrix* upon a suitable choice of the underlying scalar product. Unfortunately, the above introduced matrices are not *normal* with respect to standard scalar products, i.e. we do not have $[\widetilde{\mathcal{T}}_M, \widetilde{\mathcal{T}}_M^\dagger] = 0$.

25.3 The Method – DMRG Algorithm for the QTM

Next, we describe how to increase the length of the transfer matrix in imaginary time, i.e. the inverse temperature, by successive DMRG steps. Like in the ordinary DMRG, we first divide the QTM into two parts, the system S and the environment block E. Using the QTM, $\widetilde{\mathcal{T}}_M$, the density matrix is defined by

$$\rho = \widetilde{\mathcal{T}}_M^L , \tag{25.8}$$

which reduces to $\rho = \left|\Psi_0^R\right\rangle \left\langle\Psi_0^L\right|$ up to a normalization constant in the thermodynamic limit. As in the zero-temperature DMRG algorithm, a reduced density matrix ρ_S is obtained by taking a partial trace over the environment

$$\rho_S = \mathrm{Tr}_E\{\left|\Psi_0^R\right\rangle \left\langle\Psi_0^L\right|\} . \tag{25.9}$$

Note that this matrix is real but non-symmetric, which complicates its numerical diagonalization. It also allows for complex conjugated pairs of eigenvalues which have to be treated separately (see [21] for details).

In actual computations, the Trotter-Suzuki parameter ϵ is fixed. Therefore the temperature $T \sim 1/\epsilon M$ is decreased by an iterative algorithm $M \to M+1$. In the following, the blocks of the QTM, $\widetilde{\mathcal{T}}_M$, are shown in a 90°-rotated view.

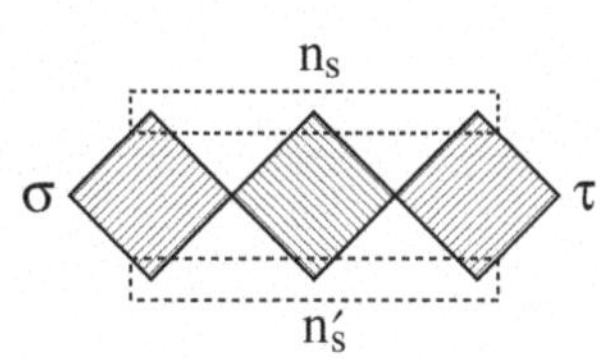

Fig. 25.2. The system block Γ. The plaquettes are connected by a summation over the adjacent corner spins

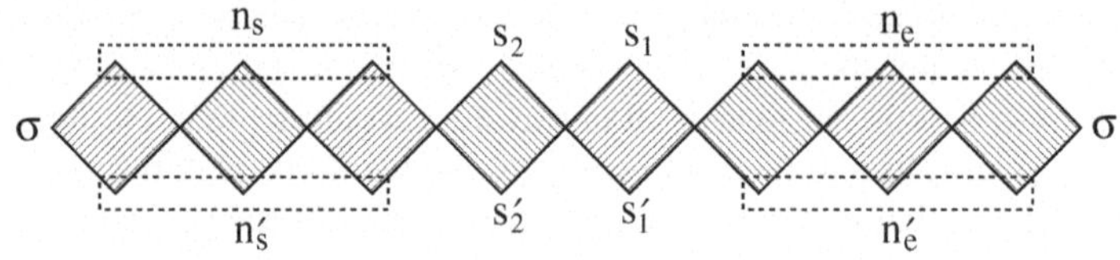

Fig. 25.3. The superblock is closed periodically by a summation over all σ states

(i) First we construct the initial system block Γ (see Fig. 25.2) consisting of M plaquettes so that $S^M \leq N < S^{M+1}$, where S is the dimension of the local Hilbert space and N is the number of states which we want to keep. n_s, n'_s are block-spin variables and contain $\widetilde{N} = S^M$ states. The $S^2 \cdot \widetilde{N}^2$-dimensional array $\Gamma(\sigma, n_s, \tau, n'_s)$ is stored.

(ii) The enlarged system block $\widetilde{\Gamma}(\sigma, n_s, s_2, \tau, s'_2, n'_s)$, a $(S^4 \cdot \widetilde{N}^2)$-dimensional array, is formed by adding a plaquette to the system block. If $h_{i,i+1}$ is real and translationally invariant, the environment block can be constructed by a 180°-rotation and a following inversion of the system block. Otherwise the environment block has to be treated separately like the system block. Together both blocks form the superblock (see Fig. 25.3).

(iii) The leading eigenvalue Λ_0 and the corresponding left and right eigenstates

$$\langle \Psi_0^L | = \Psi^L(s_1, n_s, s_2, n_e) \ , \quad |\Psi_0^R\rangle = \Psi^R(s'_1, n'_s, s'_2, n'_e)$$

are calculated and normalized $\langle \Psi_0^L | \Psi_0^R \rangle = 1$. Now thermodynamic quantities can be evaluated at the temperature $T = 1/(2\epsilon(M+1))$.

(iv) A reduced density matrix is calculated by performing the trace over the environment

$$\rho_s(n'_s, s'_2 | n_s, s_2) = \sum_{s_1, n_e} |\Psi_0^R\rangle \langle \Psi_0^L| = \sum_{s_1, n_e} \Psi^R(s_1, n'_s, s'_2, n_e) \Psi^L(s_1, n_s, s_2, n_e)$$

and the complete spectrum is computed. A $(N \times (S \cdot \widetilde{N}))$-matrix $V^L(\widetilde{n}_s | n_s, s_2)$ $(V^R(\widetilde{n}'_s | n'_s, s'_2))$ is constructed using the left (right) eigenstates belonging to the N largest eigenvalues, where $\widetilde{n}_s$ $(\widetilde{n}'_s)$ is a new renormalized block-spin variable with only N possible values.

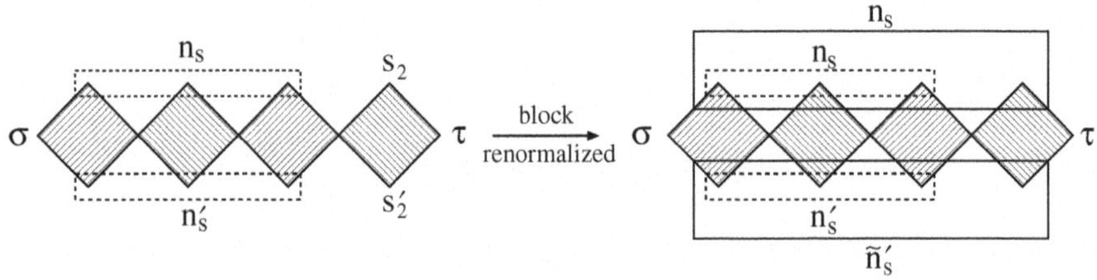

Fig. 25.4. The renormalization step for the system block

(v) Using V^L and V^R the system block is renormalized. The renormalization (see Fig. 25.4) is given by

$$\Gamma(\sigma, \widetilde{n}_s, \tau, \widetilde{n}'_s) = \sum_{n_s, s_2} \sum_{n'_s, s'_2} V^L(\widetilde{n}_s | n_s, s_2) \widetilde{\Gamma}(\sigma, n_s, s_2, \tau, s'_2, n'_s) V^R(\widetilde{n}'_s | n'_s, s'_2) .$$

Now the algorithm is repeated starting with step 2 using the new system block. However, the block-spin variables can now take N instead of $\widetilde{N}$ values.

25.4 An Example: The Spin-1/2 Heisenberg Chain with Staggered and Uniform Magnetic Fields

As example, we show here results for the magnetization of a spin-1/2 Heisenberg chain subject to a staggered magnetic field h_s and an uniform field $h_u = g\mu_B H_{ext}/J$

$$H = J \sum_i \left[\boldsymbol{S}_i \cdot \boldsymbol{S}_{i+1} - h_u S_i^z - (-1)^i h_s S_i^x \right] , \tag{25.10}$$

where H_{ext} is the external uniform magnetic field and g the Landé factor. An effective staggered magnetic field is realized in spin-chain compounds as for example copper pyrimidine dinitrate (CuPM) or copper benzoate if an external uniform magnetic field H_{ext} is applied [22]. For CuPM the magnetization as a function of applied

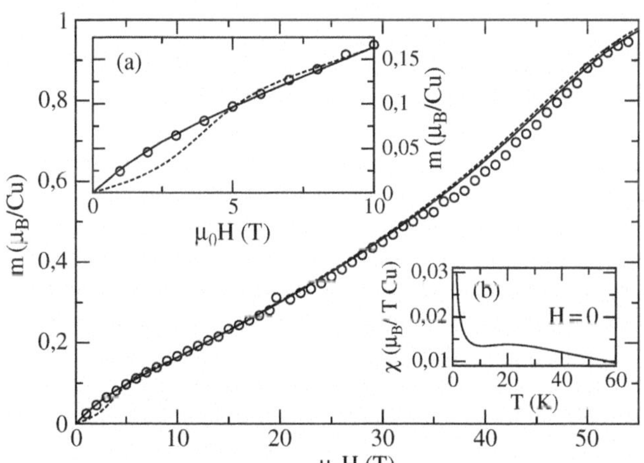

Fig. 25.5. TMRG data (solid line) and experimental magnetization curves (circles) for CuPM at a temperature $T = 1.6$ K with the magnetic field applied along the c'' axis. For comparison ED data for a system of 16 sites and $T = 0$ are shown (dashed lines). Here $J/k_B = 36.5$ K, $h_u = g\mu_B H_{ext}/J$, $h_s = 0.11\, h_u$ and $g = 2.19$. Inset **(a)**: Magnetization for small magnetic fields. Inset **(b)**: Susceptibility as a function of temperature T at $H_{ext} = 0$ calculated by TMRG

magnetic field H_{ext} has been measured experimentally. In Fig. 25.5 the excellent agreement between these experimental and TMRG data at a temperature $T = 1.6\,\text{K}$ with a magnetic field applied along the c'' axis is shown. Along the c'' axis the effect due to the induced staggered field is largest (see [23] for more details). Note that at low magnetic fields the TMRG data describe the experiment more accurately than the exact diagonalization (ED) data, because there are no finite size effects (see inset (a) of Fig. 25.5). For a magnetic field H_{ext} applied along the c'' axis a gap, $\Delta \propto H_{\text{ext}}^{2/3}$, is induced with multiplicative logarithmic corrections. For $H_{\text{ext}} \to 0$ and low T the susceptibility diverges $\chi \sim 1/T$ because of the staggered part [24] (see inset (b) of Fig. 25.5).

25.5 Impurity and Boundary Contributions

In recent years much interest has focused on the question how impurities and boundaries influence the physical properties of spin chains [25, 26, 27, 28, 29]. The doping level p defines an average chain length $\bar{L} = 1/p - 1$ and impurity or boundary contributions are of order $\sim \mathcal{O}\left(1/\bar{L}\right)$ compared to the bulk. This makes it very difficult to separate these contributions from finite-size corrections if numerical data for finite systems (e.g. from QMC calculations) are used. TMRG, on the other hand, allows to study directly impurities embedded into an infinite chain [18]. We will discuss here only the simplest case that a single bond or a single site is different from the rest. The partition function is then given by

$$Z = \text{Tr}\left(\widetilde{\mathcal{T}}_M^{L-1} \mathcal{T}_{\text{imp}}\right) , \tag{25.11}$$

where $\mathcal{T}_{\text{imp}}$ is the QTM describing the site impurity or the modified bond. In the thermodynamic limit the total free energy then becomes

$$F = -T \ln Z = L f_{\text{bulk}} + F_{\text{imp}} = -LT \ln \Lambda_0 - T \ln(\lambda_{\text{imp}}/\Lambda_0) , \tag{25.12}$$

with Λ_0 being the largest eigenvalue of the QTM, $\widetilde{\mathcal{T}}_M$, and $\lambda_{\text{imp}} = \langle \Psi_0^L | \mathcal{T}_{\text{imp}} | \Psi_0^R \rangle$.

As example, we want to consider a semi-infinite spin-1/2 XXZ-chain with an open boundary. In this case translational invariance is broken and field theory predicts Friedel-type oscillations in the local magnetization $\langle S^z(r) \rangle$ and susceptibility $\chi(r) = \partial \langle S^z(r) \rangle / \partial h$ near the boundary [30, 31]. Using the TMRG method the local magnetization can be calculated by

$$\langle S^z(r) \rangle = \frac{\langle \Psi_L^0 | \widetilde{\mathcal{T}}(S^z) \widetilde{\mathcal{T}}^{r-1} \mathcal{T}_{\text{imp}} | \Psi_R^0 \rangle}{\Lambda_0^r \lambda_{\text{imp}}} , \tag{25.13}$$

where $\widetilde{\mathcal{T}}(S^z)$ is the transfer matrix with the operator S^z included and $\mathcal{T}_{\text{imp}}$ is the transfer matrix corresponding to the bond with zero exchange coupling. Hence $\mathcal{T}_{\text{imp}} | \Psi_R^0 \rangle$ is nothing but the state describing the open boundary at the right. In Fig. 25.6 the susceptibility profile as a function of the distance r from the boundary for various temperatures as obtained by TMRG calculations [31] is shown. For more details the reader is referred to [18] and [31].

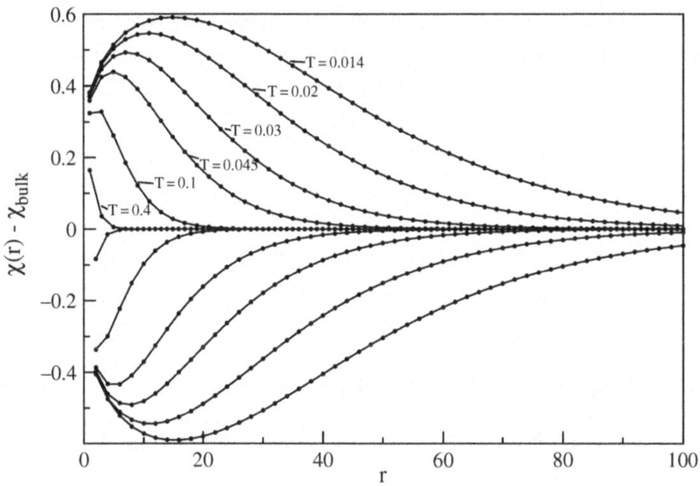

Fig. 25.6. Susceptibility profile for $\Delta = 0.6$ and different temperatures T. $N = 240$ states have been kept in the DMRG algorithm. The lines are a guide to the eye

25.6 Real-Time Dynamics

Finally, we want to discuss a very recent development in the TMRG method. The Trotter-Suzuki decomposition of a 1D quantum system yields a 2D classical model with one axis corresponding to imaginary time (inverse temperature). It is therefore straightforward to calculate imaginary-time correlation functions (CFs). Although the results for the imaginary-time CFs obtained by TMRG are very accurate, the results for real times (real frequencies) involve errors of unknown magnitude because the analytical continuation poses an ill-conditioned problem. In practice, the maximum entropy method is the most efficient way to obtain spectral functions from TMRG data. The combination of TMRG and maximum entropy has been used to calculate spectral functions for the XXZ-chain [17] and the Kondo-lattice model [14]. However, it is in principle impossible to say how reliable these results are because of the afore mentioned problems connected with the analytical continuation of numerical data. It is therefore desirable to avoid this step completely and to calculate real-time correlation functions directly.

A TMRG algorithm to do this has recently been proposed by two of us [19]. Starting point is an arbitrary two-point CF for an operator $\widehat{O}_r(t)$ at site r and time t

$$\langle \widehat{O}_r(t)\widehat{O}_0(0)\rangle = \frac{\mathrm{Tr}(\widehat{O}_r(t)\widehat{O}_0(0)\mathrm{e}^{-\beta H})}{\mathrm{Tr}(\mathrm{e}^{-\beta H})} = \frac{\mathrm{Tr}\left(\mathrm{e}^{-\beta H/2}\mathrm{e}^{itH}\widehat{O}_r\mathrm{e}^{-itH}\widehat{O}_0\mathrm{e}^{-\beta H/2}\right)}{\mathrm{Tr}\left(\mathrm{e}^{-\beta H/2}\mathrm{e}^{itH}\mathrm{e}^{-itH}\mathrm{e}^{-\beta H/2}\right)} . \tag{25.14}$$

Here we have used the cyclic invariance of the trace and have written the denominator in analogy to the numerator. In the following we will use the standard Trotter-Suzuki decomposition leading to a 2D checkerboard model.

The crucial step in our approach to calculate real-time dynamics directly is to introduce a second Trotter-Suzuki decomposition of $\exp(-\mathrm{i}\delta H)$ with $\delta = t/N$ in addition to the usual one for the partition function described in Sect. 25.2. We can then define a column-to-column transfer matrix

$$\begin{aligned}\mathcal{T}_{2N,M} &= (\tau_{1,2}\tau_{3,4}\cdots\tau_{2M-1,2M})(\tau_{2,3}\tau_{4,5}\cdots\tau_{2M,2M+1})\\ &\quad(\bar{v}_{2M+1,2M+2}\cdots\bar{v}_{2M+2N-1,2M+2N})\\ &\quad(\bar{v}_{2M+2,2M+3}\cdots\bar{v}_{2M+2N,2M+2N+1})\\ &\quad(v_{2M+2N+1,2M+2N+2}\cdots v_{2M+4N-1,2M+4N})\\ &\quad(v_{2M+2N+2,2M+2N+3}\cdots v_{2M+4N,1})\,,\end{aligned} \tag{25.15}$$

where the local transfer matrices have matrix elements

$$\begin{aligned}\tau(s_k^i s_k^{i+1}|s_{k+1}^i s_{k+1}^{i+1}) &= \langle s_k^i s_k^{i+1}|\mathrm{e}^{-\epsilon h_{i,i+1}}|s_{k+1}^i s_{k+1}^{i+1}\rangle\\ v(s_l^i s_l^{i+1}|s_{l+1}^i s_{l+1}^{i+1}) &= \langle s_l^i s_l^{i+1}|\mathrm{e}^{-i\delta h_{i,i+1}}|s_{l+1}^i s_{l+1}^{i+1}\rangle\end{aligned} \tag{25.16}$$

and $\bar{v}$ is the complex conjugate. Here $i = 1,\dots,L$ is the lattice site, $k = 1,\dots,2M$ ($l = 1,\dots,2N$) the index of the imaginary time (real time) slices and $s^i_{k(l)}$ denotes a local basis. The denominator in (25.14) can then be represented by $\mathrm{Tr}(\mathcal{T}_{2N,M}^{L/2})$ where $N, M, L \to \infty$. A similar path-integral representation holds for the numerator in (25.14). Here we have to introduce an additional modified transfer matrix $\mathcal{T}_{2N,M}(\widehat{O})$ which contains the operator $\widehat{O}$ at the appropriate position. For $r > 1$ we find

$$\begin{aligned}\langle\widehat{O}_r(t)\widehat{O}_0(0)\rangle &= \lim_{N,M\to\infty}\lim_{L\to\infty}\frac{\mathrm{Tr}(\mathcal{T}(\widehat{O})\mathcal{T}^{[r/2]-1}\mathcal{T}(\widehat{O})\mathcal{T}^{L/2-[r/2]-1})}{\mathrm{Tr}(\mathcal{T}^{L/2})}\\ &= \lim_{N,M\to\infty}\frac{\langle\Psi_0^L|\mathcal{T}(\widehat{O})\mathcal{T}^{[r/2]-1}\mathcal{T}(\widehat{O})|\Psi_0^R\rangle}{\Lambda_0^{[r/2]+1}\langle\Psi_0^L|\Psi_0^R\rangle}\,.\end{aligned} \tag{25.17}$$

Here $[r/2]$ denotes the first integer smaller than or equal to $r/2$ and we have set $\mathcal{T} \equiv \mathcal{T}_{2N,M}$. A graphical representation of the transfer matrices appearing in the numerator of (25.17) is shown in Fig. 25.7. This new transfer matrix can again be treated with the DMRG algorithm described in Sect. 25.3 where either a τ or v plaquette is added corresponding to a decrease in temperature T or an increase in real time t, respectively.

To demonstrate the method, results for the longitudinal spin-spin autocorrelation function of the XXZ-chain at infinite temperature are shown in Fig. 25.8. For $\Delta = 0$ the XXZ-model corresponds to free spinless fermions and is exactly solvable. We focus on the case of free fermions, as here the analysis of the dynamical TMRG (DTMRG) method, its results and numerical errors can be done to much greater extent than in the general case. The performance of the DTMRG itself is expected to be independent of the strength of the interaction. The comparison with the exact result in Fig. 25.8 shows that the maximum time before the DTMRG algorithm breaks down increases with the number of states. However, the improvement when taking

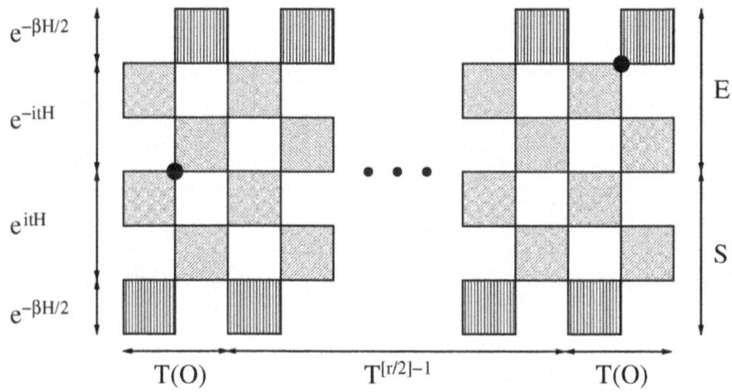

Fig. 25.7. Transfer matrices appearing in the numerator of (25.17) for $r > 1$ with r even. The two big black dots denote the operator $\widehat{O}$. $\mathcal{T}, \mathcal{T}(\widehat{O})$ consist of three parts: A part representing $\exp(-\beta H)$ (vertically striped plaquettes), another for $\exp(itH)$ (stripes from lower left to upper right) and a third part describing $\exp(-itH)$ (upper left to lower right). $\mathcal{T}, \mathcal{T}(\widehat{O})$ are split into system (S) and environment (E)

$N = 400$ instead of $N = 300$ states is marginal. The reason for the breakdown of the DTMRG computation can be traced back to an increase of the discarded weight (see inset of Fig. 25.9). Throughout the RG procedure we keep only N of the leading eigenstates of the reduced density matrix ρ_S. As long as the discarded states carry a total weight less than, say, 10^{-3} the results are faithful. For infinite temperature and $\Delta = 0$ we could explain the rapid increase of the discarded weight with time by deriving an explicit expression for the leading eigenstate of the QTM as well as for the corresponding reduced density matrix. At the free fermion point the spectrum of this density matrix is multiplicative. Hence, from the one-particle spectrum

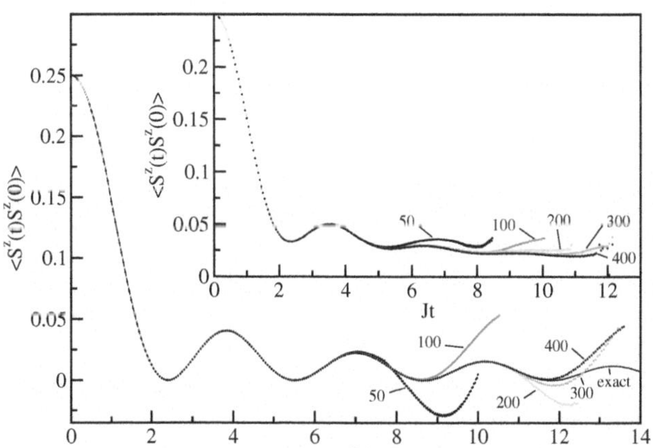

Fig. 25.8. Autocorrelation function for $\Delta = 0$ and $\Delta = 1$ (inset) at $T = \infty$, where $N = 50 - 400$ states have been kept and $\delta = 0.1$. The exact result is shown for comparison in the case $\Delta = 0$

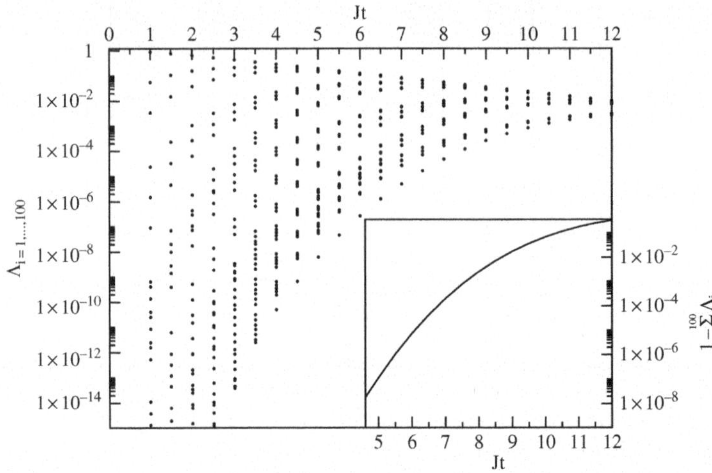

Fig. 25.9. Largest 100 eigenvalues Λ_i of ρ_S for $\Delta = 0$ and $T = \infty$ calculated exactly. The inset shows the discarded weight $1 - \sum_{i=1}^{100} \Lambda_i$

which is calculated by simple numerics we obtain the entire spectrum. As shown in Fig. 25.9 this spectrum becomes more dense with increasing time thus setting a characteristic time scale $t_c(N)$, quite independent of the discretization δ of the real time, where the algorithm breaks down. Despite these limitations, it is often possible to extrapolate the numerical data to larger times using physical arguments thus allowing to obtain frequency-dependent quantities by a direct Fourier transform. This way the spin-lattice relaxation rate for the Heisenberg chain has been successfully calculated [32].

Acknowledgement

S.G. acknowledges support by the DFG under contracts KL645/4-2 and GK1052 (Representation theory and its applications in mathematics and physics) and J.S. by the DFG and NSERC. The numerical calculations have been performed in part using the Westgrid Facility (Canada).

References

1. S. White, Phys. Rev. Lett. **69**, 2863 (1992)
2. T. Nishino, J. Phys. Soc. Jpn. **64**, 3598 (1995)
3. H. Trotter, Proc. Amer. Math. Soc. **10**, 545 (1959)
4. M. Suzuki, Commun. Math. Phys. **51**(2), 183 (1976)
5. M. Suzuki, Phys. Rev. B **31**, 2957 (1985)
6. R. Bursill, T. Xiang, G. Gehring, J. Phys. - Condens. Mat. **8**, L583 (1996)

7. X. Wang, T. Xiang, Phys. Rev. B **56**, 5061 (1997)
8. N. Shibata, J. Phys. Soc. Jpn. **66**, 2221 (1997)
9. S. Eggert, S. Rommer, Phys. Rev. Lett. **81**, 1690 (1998)
10. A. Klümper, R. Raupach, F. Schönfeld, Phys. Rev. B **59**, 3612 (1999)
11. B. Ammon, M. Troyer, T. Rice, N. Shibata, Phys. Rev. Lett. **82**, 3855 (1999)
12. J. Sirker, G. Khaliullin, Phys. Rev. B **67**, 100408(R) (2003)
13. J. Sirker, Phys. Rev. B **69**, 104428 (2004)
14. T. Mutou, N. Shibata, K. Ueda, Phys. Rev. Lett. **81**, 4939 (1998)
15. J. Sirker, A. Klümper, Europhys. Lett. **60**, 262 (2002)
16. J. Sirker, A. Klümper, Phys. Rev. B **66**, 245102 (2002)
17. F. Naef, X. Wang, X. Zotos, W. von der Linden, Phys. Rev. B **60**, 359 (1999)
18. S. Rommer, S. Eggert, Phys. Rev. B **59**, 6301 (1999)
19. J. Sirker, A. Klümper, Phys. Rev. B **71**, 241101(R) (2005)
20. I. Peschel, X. Wang, M. Kaulke, K. Hallberg (eds.), *Density-Matrix Renormalization, Lecture Notes in Physics*, Vol. 528 (Springer, Berlin, 1999). See also references therein
21. J. Sirker, Transfer matrix approach to thermodynamics and dynamics of one-dimensional quantum systems. Ph.D. thesis, Universität Dortmund (2002)
22. M. Oshikawa, I. Affleck, Phys. Rev. Lett. **79**, 2883 (1997)
23. S. Glocke, A. Klümper, H. Rakoto, J. Broto, A. Wolter, S. Süllow, Phys. Rev. B **73**, 220403(R) (2006)
24. M. Oshikawa, I. Affleck, Phys. Rev. B **60**, 1038 (1999)
25. S. Eggert, I. Affleck, Phys. Rev. B **46**, 10866 (1992)
26. S. Fujimoto, S. Eggert, Phys. Rev. Lett. **92**, 037206 (2004)
27. J. Sirker, M. Bortz, J. Stat. Mech. - Theor. Exp. p. P01007 (2006)
28. A. Furusaki, T. Hikihara, Phys. Rev. B **69**, 094429 (2004)
29. J. Sirker, N. Laflorencie, S. Fujimoto, S. Eggert, I. Affleck, Cond. Mat. **0610**, 0610165 (2006)
30. S. Eggert, I. Affleck, Phys. Rev. Lett. **75**, 934 (1995)
31. M. Bortz, J. Sirker, J. Phys. A - Math. Gen. **38**, 5957 (2005)
32. J. Sirker, Phys. Rev. B **73**, 224424 (2006)

Part X

Concepts of High Performance Computing

26 Architecture and Performance Characteristics of Modern High Performance Computers

Georg Hager and Gerhard Wellein

Regionales Rechenzentrum Erlangen der Friedrich-Alexander-Universität
Erlangen-Nürnberg, 91058 Erlangen, Germany

In the past two decades the accessible compute power for numerical simulations has increased by more than three orders of magnitude. Many-particle physics has largely benefited from this development because the complex particle-particle interactions often exceed the capabilities of analytical approaches and require sophisticated numerical simulations. The significance of these simulations, which may require large amounts of data and compute cycles, is frequently determined both by the choice of an appropriate numerical method or solver and the efficient use of modern computers. In particular, the latter point is widely underestimated and requires an understanding of the basic concepts of current (super) computer systems.

In this chapter we present a comprehensive introduction to the architectural concepts and performance characteristics of state-of-the art high performance computers, ranging from the "poor man's" Linux cluster to leading edge supercomputers with thousands of processors. In Sect. 26.1 we discuss basic features of modern commodity microprocessors with a slight focus on Intel and AMD products. Vector systems (NEC SX8) are briefly touched. The main emphasis is on the various approaches used for on-chip parallelism and data access, including cache design, and the resulting performance characteristics.

In Sect. 26.2 we turn to the fundamentals of parallel computing. First we explain the basics and limitations of parallelism without specialization to a concrete method or computer system. Simple performance models are established which help to understand the most severe bottlenecks that will show up with parallel programming.

In terms of concrete manifestations of parallelism we then cover the principles of distributed-memory parallel computers, of which clusters are a variant. These systems are programmed using the widely accepted *message passing* paradigm where processes running on the compute nodes communicate via a library that sends and receives messages between them and thus serves as an abstraction layer to the hardware interconnect. Whether the program is run on an inexpensive cluster with bare Gigabit Ethernet or on a special-purpose vector system featuring a high-performance switch like the NEC IXS does not matter as far as the parallel programming paradigm is concerned. The Message Passing Interface (MPI) has emerged as the quasi-standard for message passing libraries. We introduce the most important MPI functionality using some simple examples. As the network is often a performance-limiting aspect with MPI programming, some comments are made

G. Hager and G. Wellein: *Architecture and Performance Characteristics of Modern High Performance Computers*, Lect. Notes Phys. **739**, 681–730 (2008)
DOI 10.1007/978-3-540-74686-7_26

about basic performance characteristics of networks and the influence of bandwidth and latency on overall data transfer efficiency.

Price/performance considerations usually drive distributed-memory parallel systems into a particular direction of design. Compute nodes comprise multiple processors which share the same address space (shared memory). Two types of shared memory nodes are in wide use and will be discussed here: The uniform memory architecture (UMA) provides the same view/performance of physical memory for all processors and is used, e.g., in most current Intel-based systems. With the success of AMD Opteron CPUs in combination with Hypertransport technology the cache-coherent non-uniform memory architecture (ccNUMA) has gained increasing attention. The concept of having a single address space on a physically distributed memory (each processor can access local and remote memory) allows for scaling available memory bandwidth but requires special care in programming and usage.

Common to all shared-memory systems are mechanisms for establishing cache coherence, i.e. ensuring consistency of the different views to data on different processors in presence of caches. One possible implementation of a cache coherence protocol is chosen to illustrate the potential bottlenecks that coherence traffic may impose. Finally, an introduction to the current standard for shared-memory scientific programming, OpenMP, is given.

26.1 Microprocessors

In the "old days" of scientific supercomputing roughly between 1975 and 1995, leading-edge high performance systems were specially designed for the HPC market by companies like Cray, NEC, Thinking Machines, or Meiko. Those systems were way ahead of standard commodity computers in terms of performance and price. Microprocessors, which had been invented in the early 1970s, were only mature enough to hit the HPC market by the end of the 1980s, and it was not until the end of the 1990s that clusters of standard workstation or even PC-based hardware had become competitive at least in terms of peak performance. Today the situation has changed considerably. The HPC world is dominated by cost-effective, off-the-shelf systems with microprocessors that were not primarily designed for scientific computing. A few traditional supercomputer vendors act in a niche market. They offer systems that are designed for high application performance on the single CPU level as well as for highly parallel workloads. Consequently, the scientist is likely to encounter commodity clusters first and only advance to more specialized hardware as requirements grow. For this reason we will mostly be focused on microprocessor-based systems in this paper. Vector computers show a different programming paradigm which is in many cases close to the requirements of scientific computation, but they have become rare animals.

Microprocessors are probably the most complicated machinery that man has ever created. Understanding all inner workings of a CPU is out of the question for the scientist and also not required. It is helpful, though, to get a grasp of the high-level features in order to understand potential bottlenecks. Figure 26.1 shows a very

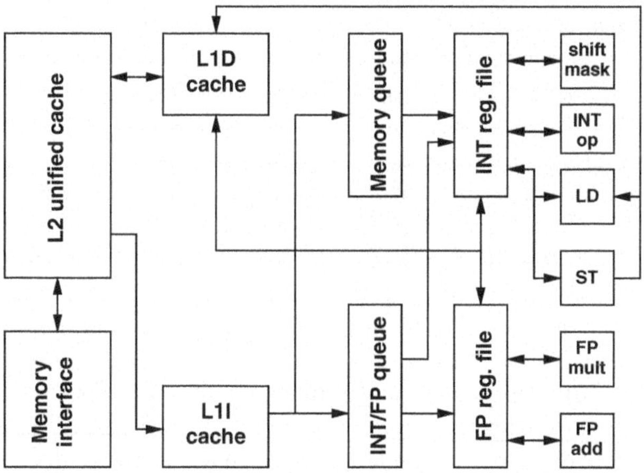

Fig. 26.1. Simplified block diagram of a typical microprocessor

simplified block diagram of a modern microprocessor. The components that actually do work for a running application are the arithmetic units for floating-point (FP) and integer (INT) operations and make up for only a very small fraction of chip area. The rest consists of administrative logic that helps to feed those units with operands. All operands must reside in CPU registers which are generally divided into floating-point and integer (or general purpose) varieties. Typical CPUs nowadays have between 16 and 128 registers of both kinds. Load (LD) and store (ST) units handle instructions that transfer data to and from registers. Instructions are sorted into several queues, waiting to be executed, probably not in the order they were issued (see below). Finally, caches hold data and instructions to be (re-)used soon. A lot of additional logic, i.e. branch prediction, reorder buffers, data shortcuts, transaction queues etc. that we cannot touch upon here is built into modern processors. Vendors provide extensive documentation about those details [1, 2].

26.1.1 Performance Metrics and Benchmarks

All those components can operate at some maximum speed called *peak performance*. Whether this limit can be reached with a specific application code depends on many factors and is one of the key topics of Chap. 27. Here we would like to introduce some basic performance metrics that can quantify the speed of a CPU. Scientific computing tends to be quite centric to floating-point data, usually with double precision (DP). The performance at which the FP units generate DP results for multiply and add operations is measured in floating-point operations per second (Flops/sec). The reason why more complicated arithmetic (divide, square root, trigonometric functions) is not counted here is that those are executed so slowly compared to add and multiply as to not contribute significantly to overall performance in most cases (see also Sect. 27.1). At the time of writing, standard microprocessors feature a peak performance between 4 and 12 GFlops/sec.

As mentioned above, feeding arithmetic units with operands is a complicated task. The most important data paths from the programmer's point of view are those to and from the caches. The speed, or bandwidth of those paths is quantified in GBytes/sec. The GFlops/sec and GBytes/sec metrics usually suffice for explaining most relevant performance features of microprocessors.[1]

Fathoming the chief performance characteristics of a processor is one of the purposes of low-level benchmarking. A low-level benchmark is a program that tries to test some specific feature of the architecture like, e.g., peak performance or memory bandwidth. One of the most prominent examples is the vector triad. It comprises a nested loop, the inner level executing a combined vector multiply-add operation (see Listing 26.1). The purpose of this benchmark is to measure the performance of data transfers between memory and arithmetic units of a microprocessor. On the inner level, three load streams for arrays `B`, `C` and `D` and one store stream for `A` are active. Depending on `N`, this loop might execute in a very small time, which would be hard to measure. The outer loop thus repeats the triad `R` times so that execution time becomes large enough to be accurately measurable. In a real benchmarking situation one would choose `R` according to `N` so that the overall execution time stays roughly constant for different `N`.

Still the outer loop serves another purpose. In situations where `N` is small enough to fit into some processor cache, one would like the benchmark to reflect the performance of this cache. With `R` suitably chosen, startup effects become negligible and this goal is achieved.

The aim of the `dummy()` subroutine is to prevent the compiler from doing an obvious optimization: Without the call, the compiler might discover that the inner loop does not depend at all on the outer loop index `j` and drop the outer loop right away. The call to `dummy()`, which should reside in another compilation unit, fools the compiler into believing that the arrays may change between outer loop iterations.

Listing 26.1. Basic code fragment for the vector triad benchmark, including performance measurement

```
double precision A(N),B(N),C(N),D(N),S,E,MFLOPS
S = get_walltime()
do j=1,R
  do i=1,N
    A(i) = B(i) + C(i) * D(i)         ! 3 loads, 1 store
  enddo
  call dummy(A,B,C,D)                 ! prevent loop interchange
enddo
E = get_walltime()
MFLOPS = R*N*2.d0/((E-S)*1.d6)        ! compute MFlop/sec rate
```

[1] Please note that the giga and mega prefixes refer to a factor of 10^9 and 10^6, respectively, when used in conjunction with ratios like bandwidth or performance.

This effectively prevents the optimization described, and the cost for the call are negligible as long as `N` is not too small. Optionally, the call can be masked by an `if` statement whose condition is never true (a fact that must of course also be hidden from the compiler).

The `MFLOPS` variable is computed to be the MFlops/sec rate for the whole loop nest. Please note that the most sensible time measure in benchmarking is wallclock time. Any other "time" that the runtime system may provide, first and foremost the often-used CPU time, is prone to misinterpretation because there might be contributions from I/O, context switches, other processes etc. that CPU time cannot encompass. This is even more true for parallel programs (see Sect. 26.2).

Figure 26.2 shows performance graphs for the vector triad obtained on current microprocessor and vector systems. For very small loop lengths we see poor performance no matter which type of CPU or architecture is used. On standard microprocessors, performance grows with `N` until some maximum is reached, followed by several sudden breakdowns. Finally, performance stays constant for very large loops. Those characteristics will be analyzed and explained in the following sections.

Vector processors (dotted line in Fig. 26.2) show very contrasting features. The low-performance region extends much farther than on microprocessors, but after saturation at some maximum level there are no breakdowns any more. We conclude that vector systems are somewhat complementary to standard CPUs in that they meet different domains of applicability. It may, however, be possible to optimize real-world code in a way that circumvents the low-performance regions. See Sect. 27.1 for details.

Low-level benchmarks are powerful tools to get information about the basic capabilities of a processor. However, they often cannot accurately predict the behavior of real application code. In order to decide whether some CPU or architecture is

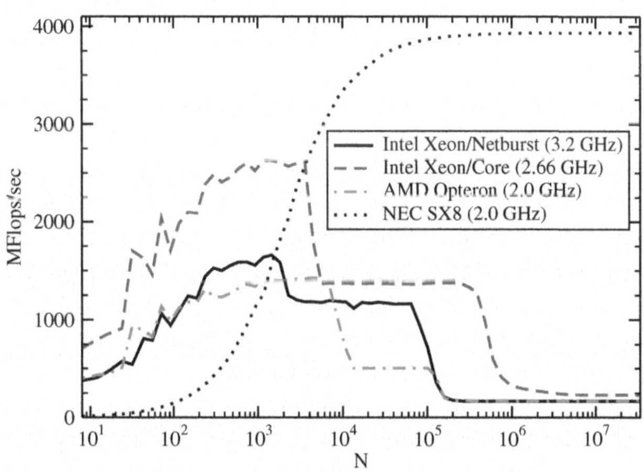

Fig. 26.2. Serial vector triad performance data for different architectures. Note the entirely different performance characteristics of the vector processor (NEC SX8)

well-suited for some application (e.g., in the run-up to a procurement), the only safe way is to prepare application benchmarks. This means that an application code is used with input parameters that reflect as closely as possible the real requirements of production runs but lead to a runtime short enough for testing (no more than a few minutes). The decision for or against a certain architecture should always be heavily based on application benchmarking. Standard benchmark collections like the SPEC suite [3] can only be rough guidelines.

26.1.2 Moore's Law

Computer technology had been used for scientific purposes and, more specifically, for numerical calculations in physics long before the dawn of the desktop PC. For more than 30 years scientists could rely on the fact that no matter which technology was implemented to build computer chips, their complexity or general capability doubled about every 24 months. In its original form, Moore's law stated that the number of components (transistors) on a chip required to hit the "sweet spot" of minimal manufacturing cost per component would increase at the indicated rate [4]. This has held true since the early 1960s despite substantial changes in manufacturing technologies that have happened over the decades. Amazingly, the growth in complexity has always roughly translated to an equivalent growth in compute performance, although the meaning of performance remains debatable as a processor is not the only component in a computer (see below for more discussion regarding this point).

Increasing chip transistor counts and clock speeds have enabled processor designers to implement many advanced techniques that lead to improved application performance. A multitude of concepts have been developed, including the following:

(i) *Pipelined functional units.* Of all innovations that have entered computer design, pipelining is perhaps the most important one. By subdividing complex operations (like, e.g., floating point addition and multiplication) into simple components that can be executed using different functional units on the CPU, it is possible to increase instruction throughput, i.e. the number of instructions executed per clock cycle. Optimally pipelined execution leads to a throughput of one instruction per cycle. At the time of writing, processor designs exist that feature pipelines with more than 30 stages. See the next section for details.

(ii) *Superscalar architecture.* Superscalarity provides for an instruction throughput of more than one per cycle by using multiple, identical functional units concurrently. This is also called instruction-level parallelism (ILP). Modern microprocessors are up to six-way superscalar.

(iii) *Out-of-order execution.* If arguments to instructions are not available on time, e.g. because the memory subsystem is too slow to keep up with processor speed, out-of-order execution can avoid pipeline bubbles by executing instructions that appear later in the instruction stream but have their parameters available. This improves instruction throughput and makes it easier for compilers to arrange machine code for optimal performance. Current out-of-order designs can keep hundreds of instructions in flight at any time, using a reorder buffer that stores instructions until they become eligible for execution.

(iv) *Larger caches*. Small, fast, on-chip memories serve as temporary data storage for data that is to be used again soon or that is close to data that has recently been used. This is essential due to the increasing gap between processor and memory speeds (see Sect. 26.1.5). Enlarging the cache size is always good for application performance.

(v) *Advancement of instruction set design*. In the 1980s, a general move from the Complex Instruction Set Computing (CISC) to the Reduced Instruction Set Computing (RISC) paradigm took place. In CISC, a processor executes very complex, powerful instructions, requiring a large effort for decoding but keeping programs small and compact, lightening the burden on compilers. RISC features a very simple instruction set that can be executed very rapidly (few clock cycles per instruction; in the extreme case each instruction takes only a single cycle). With RISC, the clock rate of microprocessors could be increased in a way that would never have been possible with CISC. Additionally, it frees up transistors for other uses. Nowadays, most computer architectures significant for scientific computing use RISC at the low level. Recently, Intel's Itanium line of processors have introduced Explicitly Parallel Instruction Computing (EPIC) which extends the RISC idea to incorporate information about parallelism in the instruction stream, i.e. which instructions can be executed in parallel. This reduces hardware complexity because the task of establishing instruction-level parallelism is shifted to the compiler, making out-of-order execution obsolete.

In spite of all innovations, processor vendors have recently been facing high obstacles in pushing performance limits to new levels. It becomes more and more difficult to exploit the potential of ever-increasing transistor numbers with standard, monolithic RISC processors. Consequently, there have been some attempts to simplify the designs by actually giving up some architectural complexity in favor of more straightforward ideas like larger caches, multi-core chips (see below) and even heterogeneous architectures on a single chip.

26.1.3 Pipelining

Pipelining in microprocessors serves the same purpose as assembly lines in manufacturing: Workers (functional units) do not have to know all details about the final product but can be highly skilled and specialized for a single task. Each worker executes the same chore over and over again on different objects, handing the half-finished product to the next worker in line. If it takes m different steps to finish the product, m products are continually worked on in different stages of completion. If all tasks are carefully tuned to take the same amount of time (the time step), all workers are continuously busy. At the end, one finished product per time step leaves the assembly line.

Complex operations like loading and storing data or floating-point arithmetic cannot be executed in a single cycle without excessive hardware requirements. Fortunately, the assembly line concept is applicable here. The most simple setup is a fetch-decode-execute pipeline, in which each stage can operate independently of the others. While an instruction is being executed, another one is being decoded and a third one is being fetched from instruction (L1I) cache. These still complex tasks are usually broken down even further. The benefit of elementary subtasks is the potential for a higher clock rate as functional units can be kept simple. As an example, consider floating-point multiplication for which a possible division in to five subtasks is depicted in Fig. 26.3. For a vector product `A(:)=B(:)*C(:)`, execution begins with the first step, separation of mantissa and exponent, on elements `B(1)` and `C(1)`. The remaining four functional units are idle at this point. The intermediate result is then handed to the second stage while the first stage starts working on `B(2)` and `C(2)`. In the second cycle, only three out of five units are still idle. In the fifth cycle the pipeline has finished its so-called wind-up phase (in other words, the multiply pipeline has a latency of five cycles). From then on, all units are continuously busy, generating one result per cycle (having a pipeline throughput of one). When the first pipeline stage has finished working on `B(N)` and `C(N)`, the wind-down phase starts. Four cycles later, the loop is finished and all results have been produced.

In general, for a pipeline of depth (or latency) m, executing N independent, subsequent operations takes $N + m - 1$ steps. We can thus calculate the expected speedup versus a general-purpose unit that needs m cycles to generate a single result,

$$\frac{T_{\text{seq}}}{T_{\text{pipe}}} = \frac{mN}{N + m - 1}, \tag{26.1}$$

which is proportional to m for large N. The throughput is

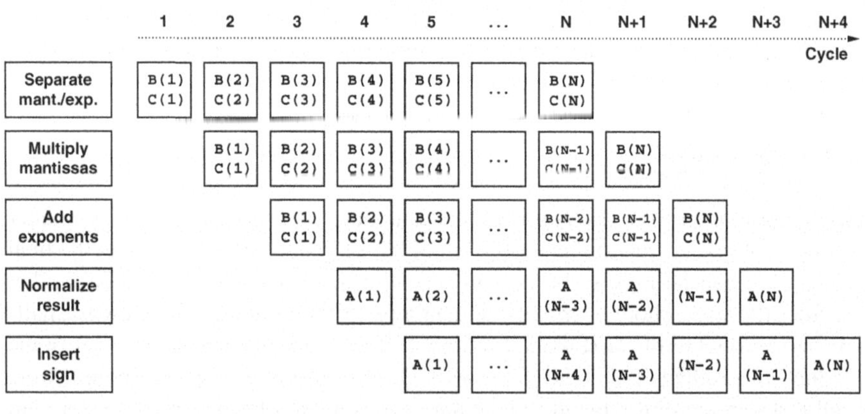

Fig. 26.3. Timeline for a simplified floating-point multiplication pipeline that executes `A(:)=B(:)*C(:)`. One result is generated on each cycle after a five-cycle wind-up phase

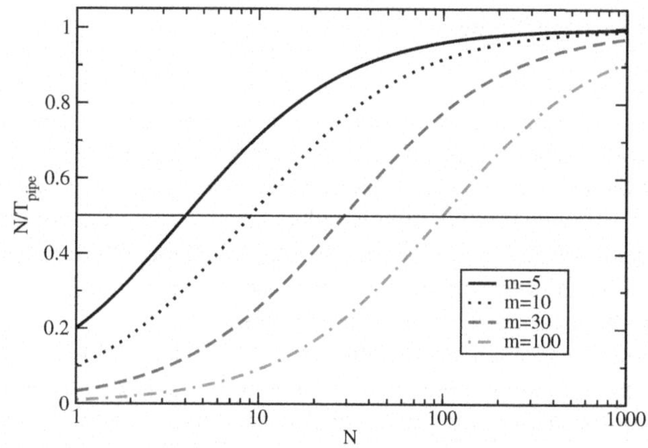

Fig. 26.4. Pipeline throughput as a function of the number of independent operations. m is the pipeline depth

$$\frac{N}{T_{\text{pipe}}} = \frac{1}{1 + \frac{m-1}{N}} , \qquad (26.2)$$

approaching one for large N (see Fig. 26.4). It is evident that the deeper the pipeline the larger the number of independent operations must be to achieve reasonable throughput because of the overhead incurred by wind-up and wind-down phases.

One can easily determine how large N must be in order to get at least p results per cycle ($0 < p \leq 1$):

$$p = \frac{1}{1 + \frac{m-1}{N_c}} \quad \Longrightarrow \quad N_c = \frac{(m-1)p}{1-p} . \qquad (26.3)$$

For $p = 0.5$ we arrive at $N_c = m - 1$. Taking into account that present-day microprocessors feature overall pipeline lengths between 10 and 35 stages, we can immediately identify a potential performance bottleneck in codes that use short, tight loops. In superscalar or even vector processors the situation becomes even worse as multiple identical pipelines operate in parallel, leaving shorter loop lengths for each pipe.

Another problem connected to pipelining arises when very complex calculations like FP divide or even transcendental functions must be executed. Those operations tend to have very long latencies (several tens of cycles for square root or divide, often more than 100 for trigonometric functions) and are only pipelined to a small level or not at all so that stalling the instruction stream becomes inevitable (this leads to so-called pipeline bubbles). Avoiding such functions is thus a primary goal of code optimization. This and other topics related to efficient pipelining will be covered in Sect. 27.1.

26.1.3.1 Software Pipelining

Note that although a depth of five is not unrealistic for a FP multiplication pipeline, executing a real code involves more operations like, e.g., loads, stores, address calculations, opcode fetches etc. that must be overlapped with arithmetic. Each operand of an instruction must find its way from memory to a register, and each result must be written out, observing all possible interdependencies. It is the compiler's job to arrange instructions in a way to make efficient use of all the different pipelines. This is most crucial for in-order architectures, but also required on out-of-order processors due to the large latencies for some operations.

As mentioned above, an instruction can only be executed if its operands are available. If operands are not delivered on time to execution units, all the complicated pipelining mechanisms are of no use. As an example, consider a simple scaling loop:

```
do i=1,N
  A(i) = s * A(i)
enddo
```

Seemingly simple in a high-level language, this loop transforms to quite a number of assembly instructions for a RISC processor. In pseudo-code, a naive translation could look like this:

```
loop:   load A(i)
        mult A(i) = A(i) * s
        store A(i)
        branch -> loop
```

Although the multiply operation can be pipelined, the pipeline will stall if the load operation on `A(i)` does not provide the data on time. Similarly, the store operation can only commence if the latency for `mult` has passed and a valid result is available. Assuming a latency of four cycles for `load`, two cycles for `mult` and two cycles for `store`, it is clear that above pseudo-code formulation is extremely inefficient. It is indeed required to interleave different loop iterations to bridge the latencies and avoid stalls:

```
loop:   load A(i+6)
        mult A(i+2) = A(i+2) * s
        store A(i)
        branch -> loop
```

Here we assume for simplicity that the CPU can issue all four instructions of an iteration in a single cycle and that the final branch and loop variable increment comes at no cost. Interleaving of loop iterations in order to meet latency requirements is called *software pipelining*. This optimization asks for intimate knowledge about processor architecture and insight into application code on the side of compilers. Often, heuristics are applied to arrive at optimal code.

It is, however, not always possible to optimally software pipeline a sequence of instructions. In the presence of dependencies, i.e., if a loop iteration depends on the result of some other iteration, there are situations when neither the compiler nor the processor hardware can prevent pipeline stalls. For instance, if the simple scaling loop from the previous example is modified so that computing `A(i)` requires `A(i+offset)`, with `offset` being either a constant that is known at compile time or a variable:

real dependency	pseudo-dependency	general version

`do i=2,N`
`  A(i)=s*A(`**`i-1`**`)`
`enddo`
	`do i=1,N-1`
	`  A(i)=s*A(`**`i+1`**`)`
	`enddo`
		`start=max(1,1-offset)`
		`end=min(N,N-offset)`
		`do i=start,end`
		`  A(i)=s*A(`**`i+offset`**`)`
		`enddo`

As the loop is traversed from small to large indices, it makes a huge difference whether the offset is negative or positive. In the latter case we speak of a pseudo-dependency, because `A(i+1)` is always available when the pipeline needs it for computing `A(i)`, i.e. there is no stall. In case of a real dependency, however, the pipelined computation of `A(i)` must stall until the result `A(i-1)` is completely finished. This causes a massive drop in performance as can be seen on the left of Fig. 26.5. The graph shows the performance of the above scaling loop in MFlops/sec versus loop length. The drop is clearly visible only in cache because of the small latencies of on-chip caches. If the loop length is so large that all data has to be fetched from memory, the impact of pipeline stalls is much less significant.

Although one might expect that it should make no difference whether the offset is known at compile time, the right graph in Fig. 26.5 shows that there is a dramatic performance penalty for a variable offset. Obviously the compiler cannot optimally software pipeline or otherwise optimize the loop in this case. This is actually a common phenomenon, not exclusively related to software pipelining; any obstruction that hides information from the compiler can have a substantial performance impact.

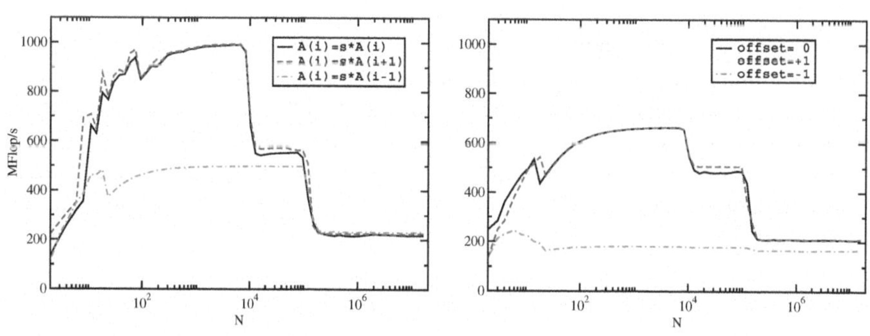

Fig. 26.5. Influence of constant (**left**) and variable (**right**) offsets on the performance of a scaling loop. (AMD Opteron 2.0 GHz)

There are issues with software pipelining linked to the use of caches. See below for details.

26.1.4 Superscalar Processors

If a processor is designed to be capable of executing more than one instruction or, more generally, producing more than one result per cycle, this goal is reflected in many of its design details:

- Multiple instructions can be fetched and decoded concurrently (4–6 nowadays).
- Address and other integer calculations are performed in multiple integer (add, mult, shift, mask) units (2–6).
- Multiple DP floating-point pipelines can run in parallel. Often there are one or two combined mult-add pipes that perform `a=b+c*d` with a throughput of one each.
- Single Instruction Multiple Data (SIMD) extensions are special instructions that issue identical operations on a whole array of integer or FP operands, probably in special registers. Whether SIMD will pay off on a certain code depends crucially on its recurrence structure and cache reuse. Examples are Intel's SSE and successors, AMD's 3dNow! and the AltiVec extensions in Power and PowerPC processors.
- Caches are fast enough to sustain more than one DP load or store operation per cycle, and there are as many execution units for loads and stores available (2–4).

Out-of-order execution and compiler optimization must work together in order to fully exploit superscalarity. However, even on the most advanced architectures it is extremely hard for compiler-generated code to achieve a throughput of more than 2–3 instructions per cycle. This is why programmers with very high demands for performance sometimes still resort to the use of assembly language.

26.1.5 Memory Hierarchies

Data can be stored in a computer system in a variety of ways. As described above, CPUs feature a set of registers for instruction arguments that can be accessed without any delays. In addition there are one or more small but very fast caches that hold data items that have been used recently. Main memory is much slower but also much larger than cache. Finally, data can be stored on disk and copied to main memory as needed. This a is a complex memory hierarchy, and it is vital to understand how data transfer works between the different levels in order to identify performance bottlenecks. In the following we will concentrate on all levels from CPU to main memory (see Fig. 26.6).

26.1.5.1 Cache

Caches are low-capacity, high-speed memories that are nowadays usually integrated on the CPU die. The need for caches can be easily understood by the fact that data

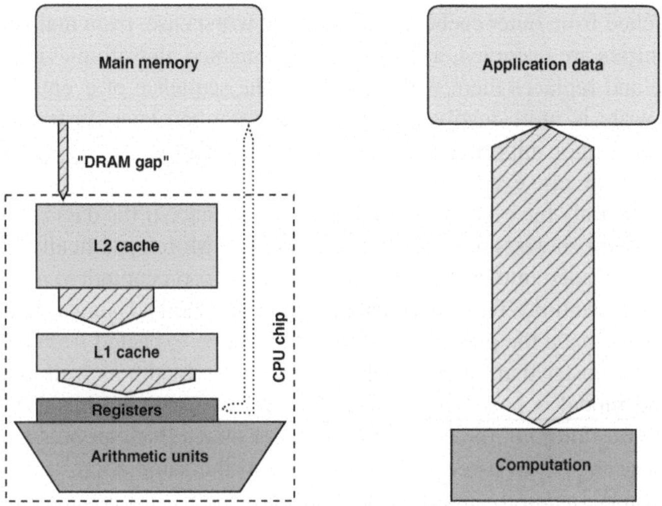

Fig. 26.6. **Left**: simplified data-centric memory hierarchy in a cache-based microprocessor (direct access paths from registers to memory are not available on all architectures). There is usually a separate L1 cache for instructions. This model must be mapped to the data access requirements of an application (**right**)

transfer rates to main memory are painfully slow compared to the CPU's arithmetic performance. At a peak performance of several GFlops/sec, memory bandwidth, i.e. the rate at which data can be transferred from memory to the CPU, is still stuck at a couple of GBytes/sec, which is entirely insufficient to feed all arithmetic units and keep them busy continuously (see Sect. 27.1 for a more thorough analysis). To make matters worse, in order to transfer a single data item (usually one or two DP words) from memory, an initial waiting time called *latency* occurs until bytes can actually flow. Often, latency is defined as the time it takes to transfer a zero-byte message. Memory latency is usually of the order of several hundred CPU cycles and is composed of different contributions from memory chips, the chipset and the processor. Although Moore's law still guarantees a constant rate of improvement in chip complexity and (hopefully) performance, advances in memory performance show up at a much slower rate. The term *DRAM gap* has been coined for the increasing distance between CPU and memory in terms of latency and bandwidth.

Caches can alleviate the effects of the DRAM gap in many cases. Usually there are at least two levels of cache (see Fig. 26.6), and there are two L1 caches, one for instructions (I-cache) and one for data. Outer cache levels are normally unified, storing data as well as instructions. In general, the closer a cache is to the CPU's registers, i.e. the higher its bandwidth and the lower its latency, the smaller it must be to keep administration overhead low. Whenever the CPU issues a read request (load) for transferring a data item to a register, first-level cache logic checks whether this item already resides in cache. If it does, this is called a cache hit and the request can be satisfied immediately, with low latency. In case of a cache miss, however, data must be fetched from outer cache levels or, in the worst case, from main memory. If all cache entries are occupied, a hardware-implemented algorithm evicts old items from cache and replaces them with new data. The sequence of events for a cache miss on a write is more involved and will be described later. Instruction caches are usually of minor importance as scientific codes tend to be largely loop-based; I-cache misses are rare events.

Caches can only have a positive effect on performance if the data access pattern of an application shows some *locality of reference*. More specifically, data items that have been loaded into cache are to be used again soon enough to not have been evicted in the meantime. This is also called temporal locality. Using a simple model, we will now estimate the performance gain that can be expected from a cache that is a factor of τ faster than memory (this refers to bandwidth as well as latency; a more refined model is possible but does not lead to additional insight). Let β be the cache reuse ratio, i.e. the fraction of loads or stores that can be satisfied from cache because there was a recent load or store to the same address. Access time to main memory (again this includes latency and bandwidth) is denoted by T_m. In cache, access time is reduced to $T_\text{c} = T_\text{m}/\tau$. For some finite β, the average access time will thus be $T_\text{av} = \beta T_\text{c} + (1-\beta)T_\text{m}$, and we calculate an access performance gain of

$$G(\tau, \beta) = \frac{T_\text{m}}{T_\text{av}} = \frac{\tau T_\text{c}}{\beta T_\text{c} + (1-\beta)\tau T_\text{c}} = \frac{\tau}{\beta + \tau(1-\beta)} \, . \qquad (26.4)$$

As Fig. 26.7 shows, a cache can only lead to a significant performance advantage if the hit ratio is relatively close to one.

However, many applications use streaming patterns where large amounts of data are loaded to the CPU, modified and written back, without the potential of reuse in

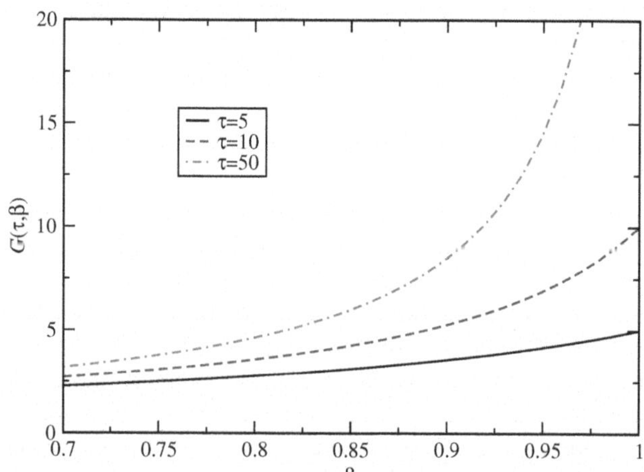

Fig. 26.7. Performance gain vs. cache reuse ratio. τ parametrizes the speed advantage of cache vs. main memory

time. For a cache that only supports temporal locality, the reuse ratio β (see above) is zero for streaming. Each new load is expensive as an item has to be evicted from cache and replaced by the new one, incurring huge latency. In order to reduce the latency penalty for streaming, caches feature a peculiar organization into cache lines. All data transfers between caches and main memory happen on the cache line level. The advantage of cache lines is that the latency penalty of a cache miss occurs only on the first miss on an item belonging to a line. The line is fetched from memory as a whole; neighboring items can then be loaded from cache with much lower latency, increasing the cache hit ratio γ, not to be confused with the reuse ratio β. So if the application shows some spatial locality, i.e. if the probability of successive accesses to neighboring items is high, the latency problem can be significantly reduced. The downside of cache lines is that erratic data access patterns are not supported. On the contrary, not only does each load incur a miss and subsequent latency penalty, it also leads to the transfer of a whole cache line, polluting the memory bus with data that will probably never be used. The effective bandwidth available to the application will thus be very low. On the whole, however, the advantages of using cache lines prevail, and very few processor manufacturers have provided means of bypassing the mechanism.

Assuming a streaming application working on DP floating point data on a CPU with a cache line length of $L_c = 16$ words, spatial locality fixes the hit ratio at $\gamma = (16-1)/16 = 0.94$, a seemingly large value. Still it is clear that performance is governed by main memory bandwidth and latency – the code is memory-bound. In order for an application to be truly cache-bound, i.e. decouple from main memory so that performance is not governed by bandwidth or latency any more, γ must be large enough that the time it takes to process in-cache data becomes larger than the time for reloading it. If and when this happens depends of course on the details of the operations performed.

By now we can interpret the performance data for cache-based architectures on the vector triad in Fig. 26.2. At very small loop lengths, the processor pipeline is too long to be efficient. Wind-up and wind-down phases dominate and performance is poor. With growing `N` this effect becomes negligible, and as long as all four arrays fit into the innermost cache, performance saturates at a high value that is set by cache bandwidth and the ability of the CPU to issue load and store instructions. Increasing `N` a little more gives rise to a sharp drop in performance because the innermost cache is not large enough to hold all data. Second-level cache has usually larger latency but similar bandwidth to L1 so that the penalty is larger than expected. However, streaming data from L2 has the disadvantage that L1 now has to provide data for registers as well as continuously reload and evict cache lines from/to L2, which puts a strain on the L1 cache's bandwidth limits. This is why performance is usually hard to predict on all but the innermost cache level and main memory. For each cache level another performance drop is observed with rising `N`, until finally even the large outer cache is too small and all data has to be streamed from main memory. The sizes of the different caches are directly related the locations of the bandwidth breakdowns. Section 27.1 will describe how to predict performance for simple loops from basic parameters like cache or memory bandwidths and the data demands of the application.

Storing data is a little more involved than reading. In presence of caches, if data to be written out already resides in cache, a write hit occurs. There are several possibilities for handling this case, but usually outermost caches work with a write-back strategy: The cache line is modified in cache and written to memory as a whole when evicted. On a write miss, however, cache-memory consistency dictates that the cache line in question must first be transferred from memory to cache before it can be modified. This is called read for ownership (RFO) and leads to the situation that a data write stream from CPU to memory uses the bus twice, once for all the cache line RFOs and once for evicting modified lines (the data transfer requirement for the triad benchmark code is increased by 25 % due to RFOs). Consequently, streaming applications do not usually profit from write-back caches and there is often a wish for avoiding RFO transactions. Some architectures provide this option, and there are generally two different strategies:

- *Non-temporal stores*. These are special store instructions that bypass all cache levels and write directly to memory. Cache does not get polluted by store streams that do not exhibit temporal locality anyway. In order to prevent excessive latencies, there is usually a write combine buffer of sorts that bundles a number of successive stores.
- *Cache line zero*. Again, special instructions serve to zero out a cache line and mark it as modified without a prior read. The data is written to memory when evicted. In comparison to non-temporal stores, this technique uses up cache space for the store stream. On the other hand it does not slow down store operations in cache-bound situations.

Both can be applied by the compiler and hinted at by the programmer by means of directives. In very simple cases compilers are able to apply those instructions automatically in their optimization stages, but one must take care to not slow down a cache-bound code by using non-temporal stores, rendering it effectively memory-bound.

26.1.5.2 Cache Mapping

So far we have implicitly assumed that there is no restriction on which cache line can be associated with which memory locations. A cache design that follows this rule is called fully associative. Unfortunately it is quite hard to build large, fast and fully associative caches because of large bookkeeping overhead: For each cache line the cache logic must store its location in the CPU's address space, and each memory access must be checked against the list of all those addresses. Furthermore, the decision which cache line to replace next if the cache is full is made by some algorithm implemented in hardware. Usually, there is a least-recently-used (LRU) strategy that makes sure only the oldest items are evicted.

The most straightforward simplification of this expensive scheme consists in a direct-mapped cache which maps the full cache size repeatedly into memory (see

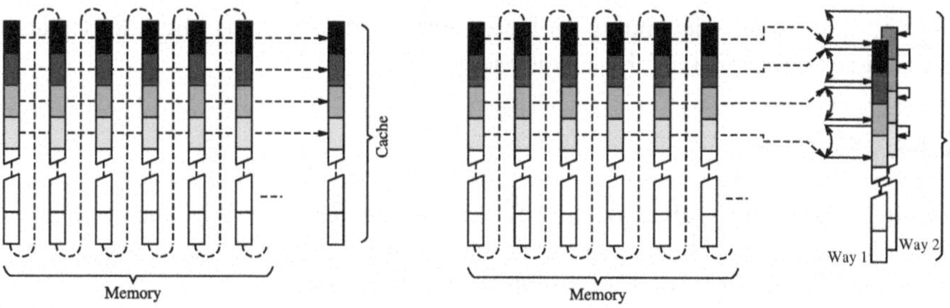

Fig. 26.8. Direct-mapped (**left**) and two-way set-associative cache (**right**). Shaded boxes indicate cache lines

Fig. 26.8 (left)). Memory locations that lie a multiple of the cache size apart are always mapped to the same cache line, and the cache line that corresponds to some address can be obtained very quickly by masking out the most significant bits. Moreover, an algorithm to select which cache line to evict is pointless. No hardware and no clock cycles need to be spent for it.

The downside of a direct-mapped cache is that it is disposed toward cache thrashing, which means that cache lines are loaded into and evicted from cache in rapid succession. This happens when an application uses many memory locations that get mapped to the same cache line. A simple example would be a strided triad code for DP data:

```
do i=1,N,CACHE_SIZE/8
  A(i) = B(i) + C(i) * D(i)
enddo
```

By using the cache size in units of DP words as a stride, successive loop iterations hit the same cache line so that *every* memory access generates a cache miss. This is different from a situation where the stride is equal to the line length; in that case, there is still some (albeit small) N for which the cache reuse is 100 %. Here, the reuse fraction is exactly zero no matter how small N may be.

To keep administrative overhead low and still reduce the danger of cache thrashing, a set-associative cache is divided into m direct-mapped caches of equal size, so-called ways. The number of ways m is the number of different cache lines a memory address can be mapped to (see Fig. 26.8 (right) for an example of a two-way set-associative cache). On each memory access, the hardware merely has to determine which way the data resides in or, in the case of a miss, which of the m possible cache lines should be evicted.

For each cache level the tradeoff between low latency and prevention of thrashing must be considered by processor designers. Innermost (L1) caches tend to be less set-associative than outer cache levels. Nowadays, set-associativity varies between two- and 16-way. Still, the effective cache size, i.e. the part of the cache that is actually useful for exploiting spatial and temporal locality in an application code could be quite small, depending on the number of data streams, their strides and mutual offsets. See Sect. 27.1 for examples.

26.1.5.3 Prefetch

Although exploiting spatial locality by the introduction of cache lines improves cache efficiency a lot, there is still the problem of latency on the first miss. Figure 26.9 visualizes the situation for a simple vector norm kernel:

```
do i=1,N
  S = S + A(i)*A(i)
enddo
```

There is only one load stream in this code. Assuming a cache line length of four elements, three loads can be satisfied from cache before another miss occurs. The long latency leads to long phases of inactivity on the memory bus.

Making the lines very long will help, but will also slow down applications with erratic access patterns even more. As a compromise one has arrived at typical cache line lengths between 64 and 128 bytes (8–16 DP words). This is by far not big enough to get around latency, and streaming applications would suffer not only from insufficient bandwidth but also from low memory bus utilization. Assuming a typical commodity system with a memory latency of 100 ns and a bandwidth of 4GBytes/sec, a single 128-byte cache line transfer takes 32 ns, so 75 % of the potential bus bandwidth is unused. Obviously, latency has an even more severe impact on performance than bandwidth.

The latency problem can be solved in many cases, however, by *prefetching*. Prefetching supplies the cache with data ahead of the actual requirements of an application. The compiler can do this by interleaving special instructions with the software pipelined instruction stream that touch cache lines early enough to give the hardware time to load them into cache (see Fig. 26.10). This assumes there is the potential of asynchronous memory operations, a prerequisite that is to some extent true for current architectures. As an alternative, some processors feature a

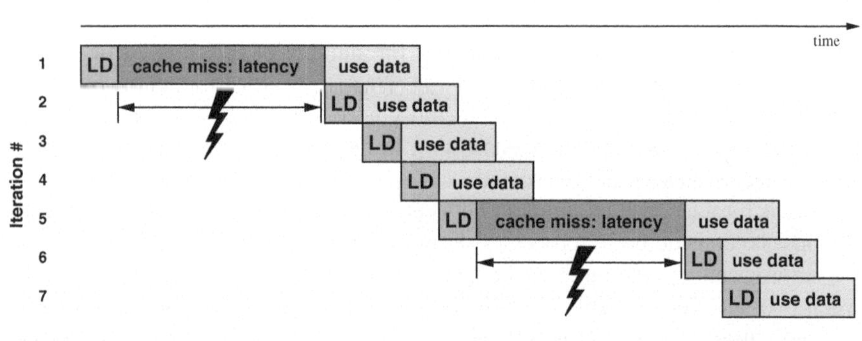

Fig. 26.9. Timing diagram on the influence of cache misses and subsequent latency penalties for a vector norm loop. The penalty occurs on each new miss

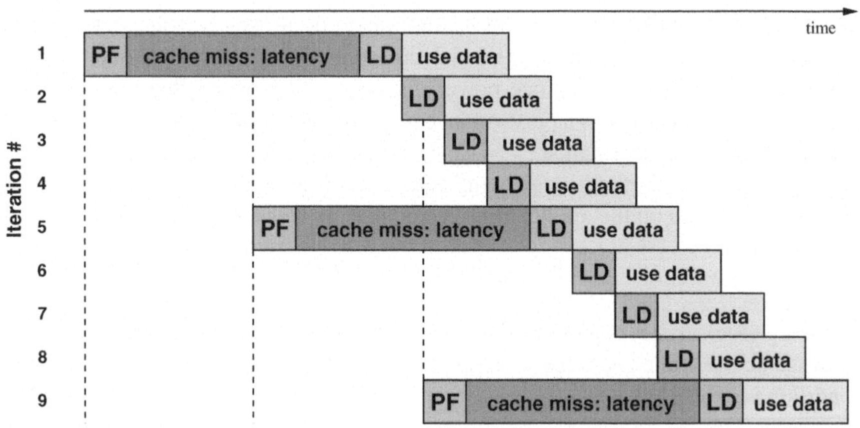

Fig. 26.10. Calculation and data transfer can be overlapped much better with prefetching. In this example, two outstanding prefetches are required to hide latency completely

hardware prefetcher that can detect regular access patterns and tries to read ahead application data, keeping up the continuous data stream and hence serving the same purpose as prefetch instructions. Whichever strategy is used, it must be emphasized that prefetching requires resources that are limited by design. The memory subsystem must be able to sustain a certain number of outstanding prefetch operations, i.e. pending prefetch requests, or else the memory pipeline will stall and latency cannot be hidden completely. Applications with many data streams can easily overstrain the prefetch mechanism. Nevertheless, if main memory access is unavoidable, a good programming guideline is to try to establish long continuous data streams.

Figs. 26.9 and 26.10 stress the role of prefetching for hiding latency, but the effects of bandwidth limitations are ignored. It should be clear that prefetching cannot enhance available memory bandwidth, although the transfer time for a single cache line is dominated by latency.

26.1.6 Multi-Core Processors

In recent years it has become increasingly clear that, although Moore's law is still valid and will be at least for the next decade, standard microprocessors are starting to hit the "heat barrier": Switching and leakage power of several-hundred-million-transistor chips are so large that cooling becomes a primary engineering effort and a commercial concern. On the other hand, the necessity of an ever-increasing clock frequency is driven by the insight that architectural advances and growing cache sizes alone will not be sufficient to keep up the one-to-one correspondence of Moore's law with application performance.

Processor vendors are looking for a way out of this dilemma in the form of *multi-core* designs. The technical motivation behind multi-core is based on the observation that power dissipation of modern CPUs is proportional to the third power of clock frequency f_c (actually it is linear in f_c and quadratic in supply voltage V_{cc}, but a decrease in f_c allows for a proportional decrease in V_{cc}). Lowering f_c and thus V_{cc} can therefore dramatically reduce power dissipation. Assuming that a single core with clock frequency f_c has a performance of p and a power dissipation of W, some relative change in performance $\varepsilon_p = \Delta p/p$ will emerge for a relative clock change of $\varepsilon_f = \Delta f_c/f_c$. All other things being equal, $|\varepsilon_f|$ is an upper limit for $|\varepsilon_p|$, which in turn will depend on the applications considered. Power dissipation is

$$W + \Delta W = (1 + \varepsilon_f)^3 W . \tag{26.5}$$

Reducing clock frequency opens the possibility to place more than one CPU core on the same die while keeping the same power envelope as before. For m cores, this condition is expressed as

$$(1 + \varepsilon_f)^3 m = 1 \quad \Longrightarrow \quad \varepsilon_f = m^{-1/3} - 1 \tag{26.6}$$

Figure 26.11 shows the required relative frequency reduction with respect to the number of cores. The overall performance of the multi-core chip,

$$p_m = (1 + \varepsilon_p) p m , \tag{26.7}$$

should at least match the single-core performance so that

$$\varepsilon_p > \frac{1}{m} - 1 \tag{26.8}$$

is a limit on the performance penalty for a relative clock frequency reduction of ε_f that should be observed for multi-core to stay useful.

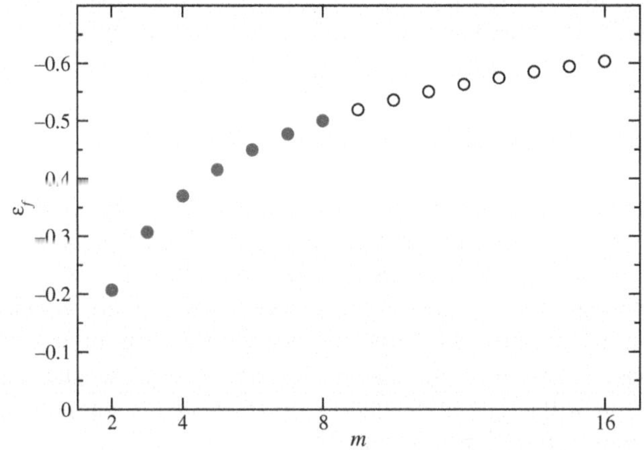

Fig. 26.11. Relative frequency reduction required to keep a given power envelope versus number of cores on a multi-core chip. The filled dots represent available technology at the time of writing

Of course it is not easy to grow the CPU die by a factor of m with a given manufacturing technology. Hence the most simple way to multi-core is to place separate CPU dies in a common package. At some point advances in manufacturing technology, i.e. smaller structure lengths, will then enable the integration of more cores on a single die. Additionally, some compromises regarding the single-core performance of a multi-core chip with respect to the previous generation will be made so that the number of transistors per core will go down as will the clock frequency. Some manufacturers have even adopted a more radical approach by designing new, much simpler cores, albeit at the cost of introducing new programming paradigms.

Finally, the over-optimistic assumption (26.7) that m cores show m times the performance of a single core will only be valid in the rarest of cases. Nevertheless, multi-core has by now been adopted by all major processor manufacturers. There are, however, significant differences in how the cores in a package can be arranged to get good performance. Caches can be shared or exclusive to each core, the memory interface can be on- or off-chip, fast data paths between the cores' caches may or may not exist, etc.

The most important conclusion one must draw from the multi-core transition is the absolute demand for parallel programming. As the single core performance will at best stagnate over the years, getting more speed for free through Moore's law just by waiting for the new CPU generation does not work any more. The following section outlines the principles and limitations of parallel programming. More details on dual- and multi-core designs will be discussed in the section on shared-memory programming Sect. 26.2.4.

In order to avoid any misinterpretation we will always use the terms core, CPU and processor synonymously.

26.2 Parallel Computing

We speak of *parallel computing* whenever a number of processors (cores) solve a problem in a cooperative way. All modern supercomputer architectures depend heavily on parallelism, and the number of CPUs in large-scale supercomputers increases steadily. A common measure for supercomputer speed has been established by the Top 500 list [5] that is published twice a year and ranks parallel computers based on their performance in the LINPACK benchmark that solves a dense system of linear equations of unspecified size. Although LINPACK is not generally accepted as a good metric because it covers only a single architectural aspect (peak performance), the list can still serve as an important indicator for trends in supercomputing. The main tendency is clearly visible from a comparison of processor number distributions in Top 500 systems (see Fig. 26.12): Top of the line HPC systems do not rely on Moore's law alone for performance but parallelism becomes more important every year. This trend will accelerate even more by the advent of multi-core processors – the June 2006 list contains only very few dual-core systems (see also Sect. 26.1.6).

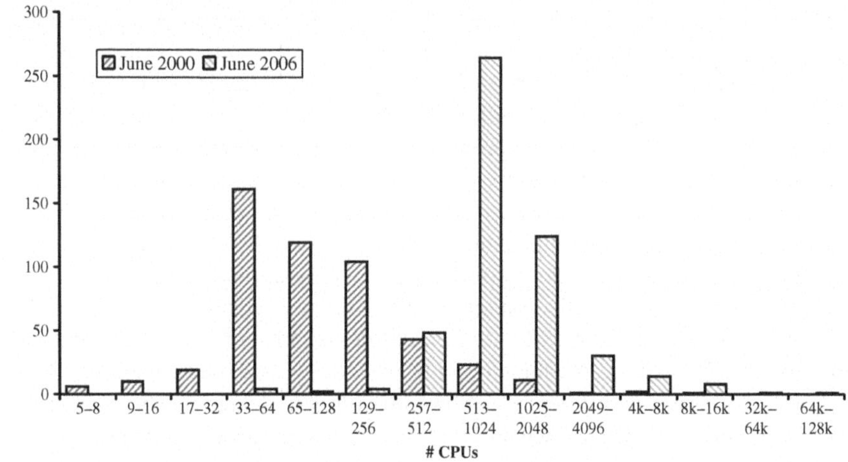

Fig. 26.12. Number of systems vs. processor count in the June 2000 and June 2006 Top 500 lists. The average number of CPUs has grown 16-fold in six years

26.2.1 Basic Principles of Parallelism

Parallelization is the process of formulating a problem in a way that lends itself to concurrent execution by several execution units of some kind. This is not only a common problem in computing but also in many other areas like manufacturing, traffic flow and even business processes. Ideally, the execution units (workers, assembly lines, border crossings, CPUs,...) are initially given some amount of work to do which they execute in exactly the same amount of time. Therefore, using N workers, a problem that takes a time T to be solved sequentially will now take only T/N. We call this a speedup of N.

Of course, reality is not perfect and some concessions will have to be made. Not all workers might execute at the same speed (see Fig. 26.13), and the tasks might not be easily partitionable into N equal chunks. Moreover there might be shared resources like, e.g., tools that only exist once but are needed by all workers. This will effectively serialize part of the concurrent execution (Fig. 26.14). Finally, the parallel work-flow may require some communication between workers, adding some overhead that would not be present in the serial case (Fig. 26.15). All these effects can impose limits on speedup. How well a task can be parallelized is usually quantified by some scalability metric.

26.2.2 Performance Models for Parallel Scalability

In order to be able to define scalability we first have to identify the basic measurements on which derived performance metrics are built. In a simple model, the overall problem size (amount of work) shall be $s + p = 1$, where s is the serial (non-parallelizable) and p is the perfectly parallelizable fraction. The 1-CPU (serial) runtime for this case,

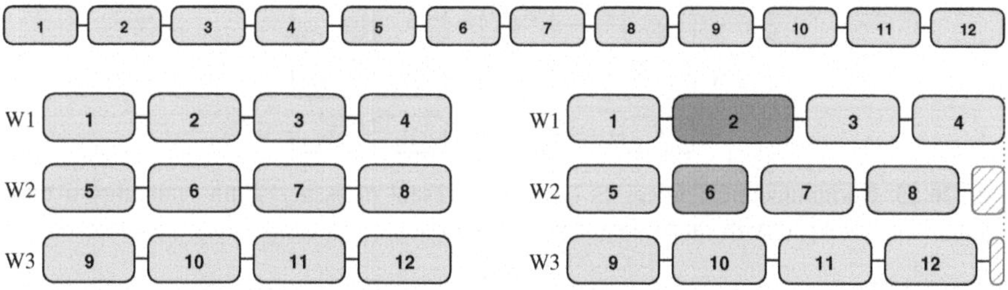

Fig. 26.13. Parallelizing a sequence of tasks (**top**) using three workers (W1... W3). **Left bottom**: perfect speedup. **Right bottom**: some tasks executed by different workers at different speeds lead to load imbalance. Hatched regions indicate unused resources

$$T_{\mathrm{f}}^{\mathrm{s}} = s + p\,, \tag{26.9}$$

is thus normalized to one. Solving the same problem on N CPUs will require a runtime of

$$T_{\mathrm{f}}^{\mathrm{p}} = s + \frac{p}{N}\,. \tag{26.10}$$

This is called *strong scaling* because the amount of work stays constant no matter how many CPUs are used. Here the goal of parallelization is minimization of time to solution for a given problem.

If time to solution is not the primary objective because larger problem sizes (for which available memory is the limiting factor) are of interest, it is appropriate to scale the problem size with some power of N so that the total amount of work is $s+pN^{\alpha}$, where α is a positive but otherwise free parameter. Here we use the implicit assumption that the serial fraction s is a constant. We define the serial runtime for the scaled problem as

$$T_{\mathrm{v}}^{\mathrm{s}} = s + pN^{\alpha}\,. \tag{26.11}$$

Consequently, the parallel runtime is

$$T_{\mathrm{v}}^{\mathrm{p}} = s + pN^{\alpha-1}\,. \tag{26.12}$$

The term *weak scaling* has been coined for this approach.

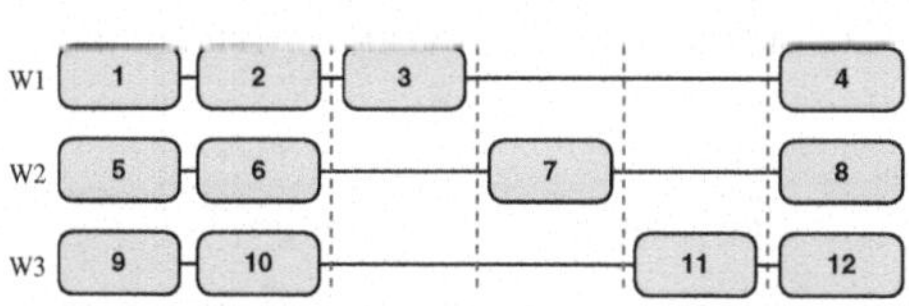

Fig. 26.14. Parallelization in presence of a bottleneck that effectively serializes part of the concurrent execution. Tasks 3, 7 and 11 cannot overlap across the dashed barriers

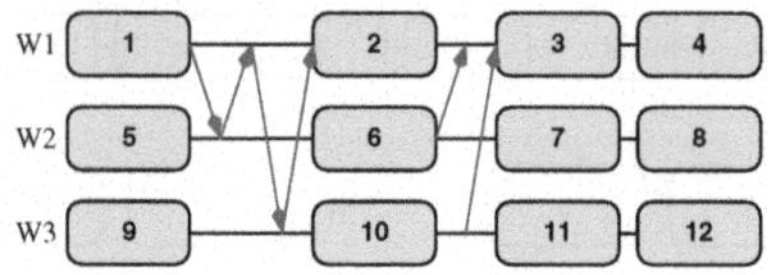

Fig. 26.15. Communication processes (arrows represent messages) limit scalability if they cannot be overlapped with each other or with calculation

26.2.2.1 Scalability Limitations

In a simple Ansatz, application speedup can be defined as the quotient of parallel and serial performance for fixed problem size. In the following we will define performance as work over time, unless otherwise noted. Serial performance for fixed problem size (work) $s + p$ is thus

$$P_{\mathrm{f}}^{\mathrm{s}} = \frac{s+p}{T_{\mathrm{f}}^{\mathrm{s}}} = 1 \,, \tag{26.13}$$

as expected. Parallel performance is in this case

$$P_{\mathrm{f}}^{\mathrm{p}} = \frac{s+p}{T_{\mathrm{f}}^{\mathrm{p}}(N)} = \frac{1}{s + \frac{1-s}{N}} \,, \tag{26.14}$$

and application speedup (scalability) is

$$S_{\mathrm{f}} = \frac{P_{\mathrm{f}}^{\mathrm{p}}}{P_{\mathrm{f}}^{\mathrm{s}}} = \frac{1}{s + \frac{1-s}{N}} \,. \tag{26.15}$$

With (26.15) we have derived the well-known *Amdahl law* which limits application speedup for large N to $1/s$. It answers the question "How much faster (in terms of runtime) does my application run when I put the same problem on N CPUs?" On the other hand, in the case of weak scaling where workload grows with CPU count, the question to ask is "How much more work can my program do in a given amount of time when I put a larger problem on N CPUs?" Serial performance as defined above is again

$$P_{\mathrm{v}}^{\mathrm{s}} = \frac{s+p}{T_{\mathrm{f}}^{\mathrm{s}}} = 1 \,, \tag{26.16}$$

as $N = 1$. Based on (26.11) and (26.12), Parallel performance (work over time) is

$$P_{\mathrm{v}}^{\mathrm{p}} = \frac{s+pN^{\alpha}}{T_{\mathrm{v}}^{\mathrm{p}}(N)} = \frac{s+(1-s)N^{\alpha}}{s+(1-s)N^{\alpha-1}} = S_{\mathrm{v}} \,, \tag{26.17}$$

again identical to application speedup. In the special case $\alpha = 0$ (strong scaling) we recover Amdahl's law. With $0 < \alpha < 1$, we get for large CPU counts

$$S_{\mathrm{v}} \xrightarrow{N \gg 1} \frac{s+(1-s)N^{\alpha}}{s} = 1 + \frac{p}{s} N^{\alpha} \,, \tag{26.18}$$

which is linear in N^{α}. As a result, weak scaling allows us to cross the Amdahl Barrier and get unlimited performance, even for small α. In the ideal case $\alpha = 1$, (26.17) simplifies to

$$S_{\mathrm{v}}(\alpha = 1) = s + (1 - s)N \,, \tag{26.19}$$

and speedup is linear in N, even for small N. This is called *Gustafson's law*. Keep in mind that the terms with N or N^{α} in the previous formulas always bear a prefactor that depends on the serial fraction s, thus a large serial fraction can lead to a very small slope.

26.2.2.2 Parallel Efficiency

In the light of the considerations about scalability, one other point of interest is the question how effectively a given resource, i.e., CPU power, can be used in a parallel program (in the following we assume that the serial part of the program is executed on one single worker while all others have to wait). Usually, parallel efficiency is then defined as

$$\varepsilon = \frac{\text{performance on } N \text{ CPUs}}{N \times \text{ performance on one CPU}} = \frac{\text{speedup}}{N} . \tag{26.20}$$

We will only consider weak scaling, as the limit $\alpha \rightarrow 0$ will always recover the Amdahl case. We get

$$\varepsilon = \frac{S_{\mathrm{v}}}{N} = \frac{sN^{-\alpha} + (1 - s)}{sN^{1-\alpha} + (1 - s)} \,. \tag{26.21}$$

For $\alpha = 0$ this yields $1/(sN + (1 - s))$, which is the expected ratio for the Amdahl case and approaches zero with large N. For $\alpha = 1$ we get $s/N + (1 - s)$, which is also correct because the more CPUs are used the more CPU cycles are wasted, and, starting from $\varepsilon = s + p = 1$ for $N = 1$, efficiency reaches a limit of $1 - s = p$ for large N. Weak scaling enables us to use at least a certain fraction of CPU power, even when the CPU count is very large. Wasted CPU time grows linearly with N, though, but this issue is clearly visible with the definitions used.

26.2.2.3 Refined Performance Models

There are situations where Amdahl's and Gustafson's laws are not appropriate because the underlying model does not encompass components like communication, load imbalance, parallel startup overhead etc. As an example, we will include a simple communication model. For simplicity we presuppose that communication cannot be overlapped with computation (see Fig. 26.15), an assumption that is actually true for many parallel architectures. In a parallel calculation, communication must thus be accounted for as a correction term in parallel runtime (26.12):

$$T_{\mathrm{v}}^{\mathrm{pc}} = s + pN^{\alpha-1} + c_{\alpha}(N) \,. \tag{26.22}$$

The communication overhead $c_\alpha(N)$ must not be included into the definition of work that is used to derive performance as it emerges from processes that are solely a result of the parallelization. Parallel speedup is then

$$S_{\text{v}}^{\text{c}} = \frac{s + pN^\alpha}{T_{\text{v}}^{\text{pc}}(N)} = \frac{s + (1-s)N^\alpha}{s + (1-s)N^{\alpha-1} + c_\alpha(N)} \,. \tag{26.23}$$

The functional dependence $c_\alpha(N)$ can have a variety of forms; the dependency on α is sometimes functional, sometimes conceptual. Furthermore we assume that the amount of communication is the same for all workers. A few special cases are described below:

- $\alpha = 0$, *blocking network:* If the communication network has a bus-like structure, i.e. only one message can be in flight at any time, and the communication overhead per CPU is independent of N then $c_\alpha(N) = (\kappa + \lambda)N$, where κ is message transfer time and λ is latency. Thus,

$$S_{\text{v}}^{\text{c}} = \frac{1}{s + \frac{1-s}{N} + (\kappa + \lambda)N} \xrightarrow{N \gg 1} \frac{1}{(\kappa + \lambda)N} \,, \tag{26.24}$$

  i.e. performance is dominated by communication and even goes to zero for large CPU numbers. This is a very common situation as it also applies to the presence of shared resources like memory paths, I/O devices and even on-chip arithmetic units.
- $\alpha = 0$, *non-blocking network:* If the communication network can sustain $N/2$ concurrent messages with no collisions, $c_\alpha(N) = \kappa + \lambda$ and

$$S_{\text{v}}^{\text{c}} = \frac{1}{s + \frac{1-s}{N} + \kappa + \lambda} \xrightarrow{N \gg 1} \frac{1}{s + \kappa + \lambda} \,. \tag{26.25}$$

  In this case the situation is quite similar to the Amdahl case and performance will saturate at a lower value than without communication.
- $\alpha = 0$, *non-blocking network, 3D domain decomposition:* There are also cases where communication overhead decreases with N for strong scaling, e.g. like $c_\alpha(N) = \kappa N^{-\beta} + \lambda$. For any $\beta > 0$ performance at large N will be dominated by s and the latency:

$$S_{\text{v}}^{\text{c}} = \frac{1}{s + \frac{1-s}{N} + \kappa N^{-\beta} + \lambda} \xrightarrow{N \gg 1} \frac{1}{s + \lambda} \,. \tag{26.26}$$

  This arises, e.g., when domain decomposition (see Sect. 26.2.3) is employed on a computational domain along all coordinate axes. In this case $\beta = 2/3$.
- $\alpha = 1$, *non-blocking network, 3D domain decomposition:* Finally, when the problem size grows linearly with N, one may end up in a situation where communication per CPU stays independent of N. As this is weak scaling, the numerator leads to linear scalability with an overall performance penalty (prefactor):

$$S_{\text{v}}^{\text{c}} = \frac{s + pN}{s + p + \kappa + \lambda} \xrightarrow{N \gg 1} \frac{(1-s)N}{1 + \kappa + \lambda} \,. \tag{26.27}$$

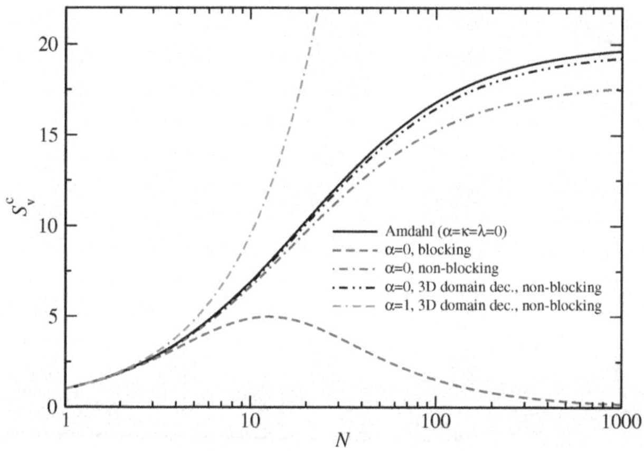

Fig. 26.16. Predicted parallel scalability for different models at $s = 0.05$. In general, $\kappa = 0.005$ and $\lambda = 0.001$ except for the Amdahl case which is shown for reference

Figure 26.16 illustrates the four cases at $\kappa = 0.005$, $\lambda = 0.001$ and $s = 0.05$ and compares with Amdahl's law. Note that the simplified models we have covered in this section are far from accurate for many applications. In order to check whether some performance model is appropriate for the code at hand, one should measure scalability for some processor numbers and fix the free model parameters by least-squares fitting.

26.2.3 Distributed-Memory Computing

After covering the principles and limitations of parallelization we will now turn to the concrete architectures that are at the programmer's disposal to implement a parallel algorithm on. Two primary paradigms have emerged, and each features a dominant and standardized programming model: *Distributed-memory* and *shared-memory* systems. In this section we will be concerned with the former while the next section covers the latter.

Figure 26.17 shows a simplified block diagram, or programming model, of a distributed-memory parallel computer. Each processor P (with its own local cache C) is connected to exclusive local memory, i.e. no other CPU has direct access to it. Although many parallel machines today, first and foremost the popular PC clusters, consist of a number of shared-memory compute nodes with two or more CPUs for price/performance reasons, the programmer's view does not reflect that (it is even possible to use distributed-memory programs on machines that feature shared memory only). Each node comprises at least one network interface (NI) that mediates the connection to a communication network. On each CPU runs a serial process that can communicate with other processes on other CPUs by means of the network. In the simplest case one could use standard switched Ethernet, but a number of more advanced technologies have emerged that can easily have ten times the bandwidth

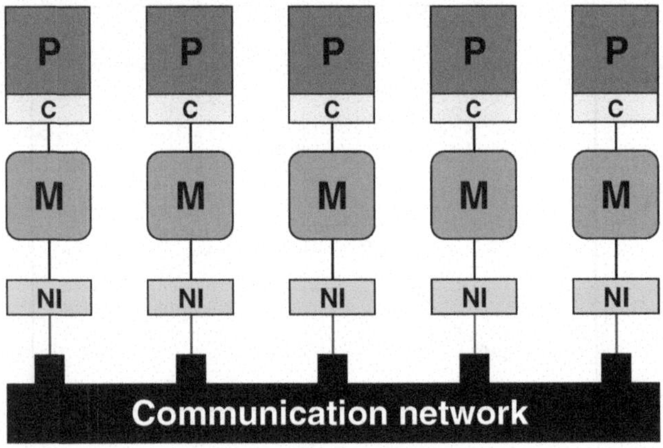

Fig. 26.17. Simplified programmer's view, or programming model, of a distributed-memory parallel computer

and 1/10 th of the latency of Gbit Ethernet. As shown in the section on performance models, the exact layout and speed of the network has considerable impact on application performance. The most favorable design consists of a non-blocking wire-speed network that can switch $N/2$ connections between its N participants without any bottlenecks. Although readily available for small systems with tens to a few hundred nodes, non-blocking switch fabrics become vastly expensive on very large installations and some compromises are usually made, i.e. there will be a bottleneck if all nodes want to communicate concurrently.

26.2.3.1 Domain Decomposition

On a distributed-memory system it is the programmer's responsibility to divide the problem into pieces in an appropriate way and distribute data across the processes. All process-to-process communication is explicit in the program. A very common method in parallel programming is domain decomposition. As an example consider a two-dimensional simulation code that updates physical variables on a $n \times n$ grid. Domain decomposition subdivides the computational domain into N subdomains. How exactly this is to be done is the choice of the programmer, but some guidelines should be observed (see Fig. 26.18). First, the computational effort should be equal for all domains to avoid load imbalance. Second, next-neighbor interactions require communication across domain boundaries. The data volume to be considered here is proportional to the overall length of the cuts. Comparing the two alternatives in Fig. 26.18, one arrives at a communication cost of $n(N-1)$ for stripe domains, whereas an optimal decomposition into square subdomains leads to a cost of $2n(\sqrt{N}-1)$. Hence for large N the optimal decomposition has an advantage in communication cost of $2/\sqrt{N}$. Whether this difference is significant or not in reality depends on the problem size and other factors, of course.

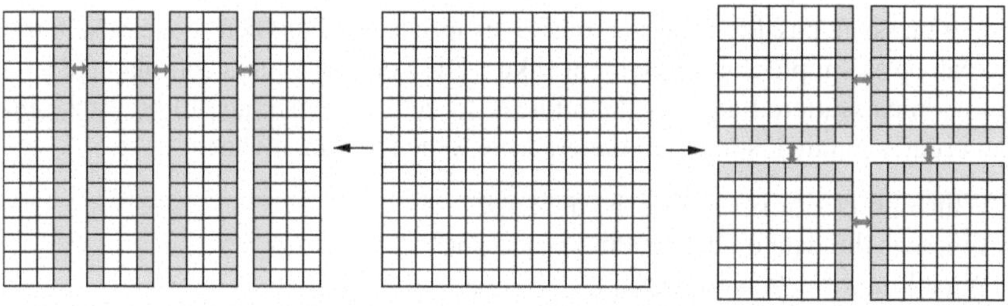

Fig. 26.18. Domain decomposition of a two-dimensional simulation with next-neighbor interactions. Cutting into stripes (**left**) is simple but incurs more communication than optimal decomposition (**right**). Shaded cells participate in network communication

Note that domain decomposition has the attractive property that domain boundary area grows more slowly than volume if the problem size increases with N constant. Therefore one can alleviate communication bottlenecks just by choosing a larger problem size. The expected effects of strong and weak scaling with optimal domain decomposition in three dimensions have been discussed in (26.26) and (26.27).

26.2.3.2 The Message Passing Paradigm and MPI

As mentioned above, distributed-memory parallel programming requires the use of explicit message passing (MP), i.e. communication between processes. This is surely the most tedious and complicated but also the most flexible parallelization method. Nowadays there is an established standard for message passing called MPI (Message Passing Interface) that is supported by all vendors [6]. MPI conforms to the following rules:

- The same program runs on all processes (Single Program Multiple Data, SPMD). This is no restriction compared to the more general MPMD (Multiple Program Multiple Data) model as all processes taking part in a parallel calculation can be distinguished by a unique identifier called *rank* (see below).
- The program is written in a sequential language like Fortran, C or C++. Data exchange, i.e. sending and receiving of messages, is done via calls to an appropriate library.
- All variables in a process are local to this process. There is no concept of shared memory.

One should add that message passing is not the only possible programming paradigm for distributed-memory machines. Specialized languages like High Performance Fortran (HPF), Unified Parallel C (UPC) etc. have been created with support for distributed-memory parallelization built in, but they have not developed a broad user community and it is as yet unclear whether those approaches can match the efficiency of MPI.

In a message passing program, messages move data between processes. A message can be as simple as a single item (like a DP word) or even a complicated structure, perhaps scattered all over the address space. For a message to be transmitted in an orderly manner, some parameters have to be fixed in advance:

- Which processor is sending the message?
- Where is the data on the sending processor?
- What kind of data is being sent?
- How much data is there?
- Which process/es is/are going to receive the message?
- Where should the data be left on the receiving process(es)?
- How much data are the receiving processes prepared to accept?

As we will see, all MPI calls that actually transfer data have to specify those parameters in some way. MPI is a very broad standard with (in its latest version) over 500 library routines. Fortunately, most applications merely require less than twenty of those to work.

26.2.3.3 A Brief Glance on MPI

In order to compile and link MPI programs, compilers and linkers need options that specify where include files and libraries can be found. As there is considerable variation in those locations across installations, most MPI implementations provide compiler wrapper scripts (often called `mpicc`, `mpif77`, etc.) that supply the required options automatically but otherwise behave like normal compilers. Note that the way that MPI programs should be compiled and started is not fixed by the standard, so please consult your system documentation.

Listing 26.2. A very simple, fully functional "Hello World" MPI program

```
program mpitest
use MPI

integer rank, size, ierror

call MPI_Init(ierror)
call MPI_Comm_size(MPI_COMM_WORLD, size, ierror)
call MPI_Comm_rank(MPI_COMM_WORLD, rank, ierror)

write(*,*) 'Hello World, I am ',rank,' of ',size

call MPI_Finalize(ierror)

end
```

Listing 26.2 shows a simple "Hello World" type MPI program in Fortran 90. In line 2, the `MPI` module is loaded which provides required globals and definitions (in Fortran 77 and C/C++ one would use the preprocessor to read in the `mpif.h` or `mpi.h` header files, respectively). All MPI calls take an `INTENT(OUT)` argument, here called `ierror`, that transports information about the success of the MPI operation to the user code (in C/C++, the return code is used for that). As failure resiliency is not built into the MPI standard today and checkpoint/restart features are usually implemented by the user code anyway, the error code is rarely checked at all.

The first call in every MPI code should go to `MPI_Init` and initializes the parallel environment (line 6). In C/C++, `&argc` and `&argv` are passed to `MPI_Init` so that the library can evaluate and remove any additional command line arguments that may have been added by the MPI startup process. After initialization, MPI has set up a so-called communicator, called `MPI_COMM_WORLD`. A communicator defines a group of MPI processes that can be referred to by a communicator handle. The `MPI_COMM_WORLD` handle describes all processes that have been started as part of the parallel program. If required, other communicators can be defined as subsets of `MPI_COMM_WORLD`. Nearly all MPI calls require a communicator as an argument.

The calls to `MPI_Comm_size` and `MPI_Comm_rank` in lines 7 and 8 serve to determine the number of processes (`size`) in the parallel program and the unique identifier (the rank) of the calling process, respectively. The ranks in a communicator, in this case `MPI_COMM_WORLD`, are numbered starting from zero up to $N-1$. In line 12, the parallel program is shut down by a call to `MPI_Finalize`. Note that no MPI process except rank 0 is guaranteed to execute any code beyond `MPI_Finalize`.

In order to compile and run the source code in Listing 26.2, a common implementation would require the following steps:

```
$ mpif90 -O3 -o hello.exe hello.F90
$ mpirun -np 4 ./hello.exe
```

This would compile the code and start it with four processes. Be aware that processors may have to be allocated from some batch system before parallel programs can be launched. How MPI processes are mapped to actual processors is entirely up to the implementation. The output of this program could look like the following:

```
Hello World, I am 3 of 4
Hello World, I am 0 of 4
Hello World, I am 2 of 4
Hello World, I am 1 of 4
```

Although the `stdout` and `stderr` streams of MPI programs are usually redirected to the terminal where the program was started, the order in which outputs from different ranks will arrive is undefined.

This example did not contain any real communication apart from starting and stopping processes. An MPI message is defined as an array of elements of a particular MPI datatype. Data types can either be basic types (corresponding to the standard types that every programming language knows) or derived types that must be defined by appropriate MPI calls. The reason why MPI needs to know the data types of messages is that it supports heterogeneous environments where it may be necessary to do on-the-fly data conversions. For some message transfer to take place, the data types on sender and receiver sides must match. If there is exactly one sender and one receiver we speak of point-to-point communication. Both ends are identified uniquely by their ranks. Each message can carry an additional integer label, the so-called tag that may be used to identify the type of a message, as a sequence number or any other accompanying information. In Listing 26.3 we show an MPI program fragment that computes an integral over some function `f(x)` in parallel. Each MPI process gets assigned a subinterval of the integration domain (lines 9 and 10), and some other function can then perform the actual integration (line 12). After that each process holds its own partial result, which should be added to get the final integral. This is done at rank 0, who executes a loop over all ranks from 1 to `size` − 1, receiving the local integral from each rank in turn via `MPI_Recv` and accumulating the result in `res`. Each rank apart from 0 has to call `MPI_Send` to transmit the data. Hence there are `size` − 1 send and `size` − 1 matching receive operations. The data types on both sides are specified to be `MPI_DOUBLE_PRECISION`, which corresponds to the usual `double precision` type in Fortran (be aware that MPI types are named differently in C/C++ than in Fortran). The message tag is not used here, so we set it to 0 because identical tags are required for message matching as well.

While all parameters are necessarily fixed on `MPI_Send`, there is some more variability on the receiver side. `MPI_Recv` allows wildcards so that the source rank and the tag do not have to be specified. Using `MPI_ANY_SOURCE` as source rank and `MPI_ANY_TAG` as tag will match any message, from any source, with any tag as long as the other matching criteria like data type and communicator are met (this would have been possible in the integration example without further code changes). After `MPI_Recv` has returned to the user code, the `status` array can be used to extract the missing pieces of information, i.e. the actual source rank and message tag, and also the length of the message as the array size specified in `MPI_Recv` is only an upper limit.

The accumulation of partial results as shown above is an example for a *reduction* operation, performed on all processes in the communicator. MPI has mechanisms that make reductions much simpler and in most cases more efficient than looping over all ranks and collecting results. As reduction is a procedure that all ranks in a communicator participate in, it belongs to the so-called collective communication operations in MPI. Collective communication, as opposed to point-to-point communication, requires that every rank calls the same routine, so it is impossible for a message sent via point-to-point to match a receive that was initiated using a collective

Listing 26.3. Program fragment for parallel integration in MPI

```
  integer stat(MPI_STATUS_SIZE)
  call MPI_Comm_size(MPI_COMM_WORLD, size, ierror)
  call MPI_Comm_rank(MPI_COMM_WORLD, rank, ierror)
! integration limits
  a=0.d0
  b=2.d0
  res=0.d0
! limits for "me"
  mya=a+rank*(b-a)/size
  myb=mya+(b-a)/size
! integrate f(x) over my own chunk - actual work
  psum = integrate(mya,myb)
! rank 0 collects partial results
  if(rank.eq.0) then
     res=psum
     do i=1,size-1
        call MPI_Recv(tmp, &    ! receive buffer
                      1,   &    ! array length
                                ! datatype
                      MPI_DOUBLE_PRECISION,&
                      i,   &    ! rank of source
                      0,   &    ! tag (additional label)
                                ! communicator
                      MPI_COMM_WORLD,&
                      stat,&    ! status array (msg info)
                      ierror)
        res=res+tmp
     enddo
     write (*,*) 'Result: ',res
! ranks != 0 send their results to rank 0
  else
     call MPI_Send(psum,   &    ! send buffer
                   1,      &    ! array length
                   MPI_DOUBLE_PRECISION,&
                   0,      &    ! rank of destination
                   0,      &    ! tag
                   MPI_COMM_WORLD,ierror)
  endif
```

call. The whole `if...else...endif` construct (apart from printing the result) in Listing 26.3 could have been written as a single call:

```
call MPI_Reduce(psum,     & ! send buffer
                res,      & ! receive buffer
                1,        & ! array length
                MPI_DOUBLE_PRECISION,&
                MPI_SUM,& ! type of operation
                0,        & ! root (accumulate res there)
                MPI_COMM_WORLD,ierror)
```

Most collective routines define a root rank at which some general data source or sink is located. Although rank 0 is a natural choice for root, it is in no way different from other ranks.

There are collective routines not only for reduction but also for barriers (each process stops at the barrier until all others have reached the barrier as well), broadcasts (the root rank transmits some data to everybody else), scatter/gather (data is distributed from root to all others or collected at root from everybody else), and complex combinations of those. Generally speaking, it is a good idea to prefer collectives over point-to-point constructs that emulate the same semantics. Good MPI implementations are optimized for data flow on collective operations and also have some knowledge about network topology built in.

All MPI functionalities described so far have the property that the call returns to the user program only after the message transfer has progressed far enough so that the send/receive buffer can be used without problems. That means, received data has arrived completely and sent data has left the buffer so that it can be safely modified without inadvertently changing the message. In MPI terminology, this is called blocking communication. Although collective operations are always blocking, point-to-point communication can be performed with non-blocking calls as well. A non-blocking point-to-point call merely initiates a message transmission and returns very quickly to the user code. In an efficient implementation, waiting for data to arrive and the actual data transfer occur in the background, leaving resources free for computation. In other words, non-blocking MPI is a way in which computation and communication may be overlapped. As long as the transfer has not finished (which can be checked by suitable MPI calls), the message buffer must not be used. Non-blocking and blocking MPI calls are mutually compatible, i.e. a message sent via a blocking send can be matched by a non-blocking receive. Table 26.1 gives a rough overview of available communication modes in MPI.

26.2.3.4 Basic Performance Characteristics of Networks

As mentioned before, there are various options for the choice of a network in a distributed-memory computer. The simplest and cheapest solution to date is Gbit Ethernet, which will suffice for many throughput applications but is far too slow – in terms of bandwidth and latency – for parallel code with any need for fast communication. Assuming that the total transfer time for a message of size N [bytes] is

Table 26.1. Non-exhaustive overview on MPI's communication modes

	Point-to-point	Collective

Blocking	`MPI_Send(buf,...)`
	`MPI_Ssend(buf,...)`
	`MPI_Bsend(buf,...)`
	`MPI_Recv(buf,...)`
	(`buf` can be used after call returns)
		`MPI_Barrier(...)`
		`MPI_Bcast(...)`
		`MPI_Reduce(...)`
		(all processes in communicator must call)

Non-blocking	`MPI_Isend(buf,...)`
	`MPI_Irecv(buf,...)`
	(`buf` can not be used or modified after call returns; check for completion with `MPI_Wait(...)` or `MPI_Test(...)`)
		N/A

composed of latency and streaming parts,

$$T = T_l + \frac{N}{B} \tag{26.28}$$

and B being the maximum network bandwidth in MBytes/sec, the effective bandwidth is

$$B_{\text{eff}} = \frac{N}{T_l + \frac{N}{B}} \,. \tag{26.29}$$

In Fig. 26.19, the model parameters in (26.29) are fitted to real data obtained on a Gbit Ethernet network. Obviously this simple model is able to describe the gross features well.

The measurement of the effective bandwidth is frequently done with the Ping-Pong benchmark. The basic code sends a message of size N [bytes] once back and forth between two nodes:

```
S = get_walltime()
if(rank.eq.0) then
  call MPI_Send(buf,N,MPI_BYTE,1,0,...)
  call MPI_Recv(buf,N,MPI_BYTE,1,0,...)
else
  call MPI_Recv(buf,N,MPI_BYTE,0,0,...)
  call MPI_Send(buf,N,MPI_BYTE,0,0,...)
endif
E = get_walltime()
MBYTES   = 2*N/(E-S)/1.d6      ! MByte/sec rate
TIME     = (E-S)/2*1.d6        ! transfer time in microsecs
                               ! for single message
```

Bandwidth in MBytes/sec is then reported for different N (see Fig. 26.20). Common to all interconnects, we observe very low bandwidth for small message sizes as expected from the model (26.29). Latency can be measured directly by taking the

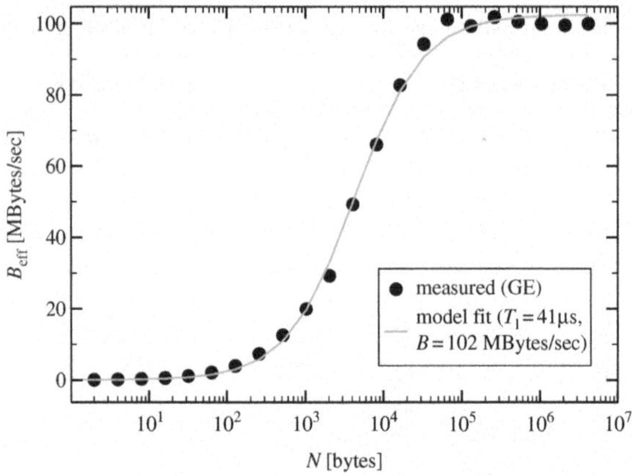

Fig. 26.19. Fit of the model for effective bandwidth (26.29) to data measured on a Gbit Ethernet network

$N = 0$ limit of transfer time (inset in Fig. 26.20). The reasons for latency can be diverse:

- All data transmission protocols have some overhead in the form of administrative data like message headers etc.
- Some protocols (like, e.g., TCP/IP as used over Ethernet) define minimum message sizes, so even if the application sends a single byte, a small frame of $N > 1$ bytes is transmitted.
- Initiating a message transfer is a complicated process that involves multiple software layers, depending on the complexity of the protocol. Each software layer adds to latency.
- Standard PC hardware as frequently used in clusters is not optimized towards low-latency I/O.

In fact, high-performance networks try to improve latency by reducing the influence of all of the above. Lightweight protocols, optimized drivers and communication devices directly attached to processor buses are all used by vendors to provide low MPI latency.

For large messages, effective bandwidth saturates at some maximum value. Structures like local minima etc. frequently occur but are very dependent on hardware and software implementations (e.g., the MPI library could decide to switch to a different buffering algorithm beyond some message size). Although saturation bandwidths can be quite high (there are systems where achievable MPI bandwidths are comparable to the local memory bandwidth of the processor), many applications work in a region on the bandwidth graph where latency effects still play a dominant role. To quantify this problem, the $N_{1/2}$ value is often reported. This is the message size at which $B_{\mathrm{eff}} = B/2$ (see Fig. 26.20). In the model (26.29), $N_{1/2} = BT_1$. From this point of view it makes sense to ask whether an increase in maximum network bandwidth by a factor of β is really beneficial for all messages. At message size N,

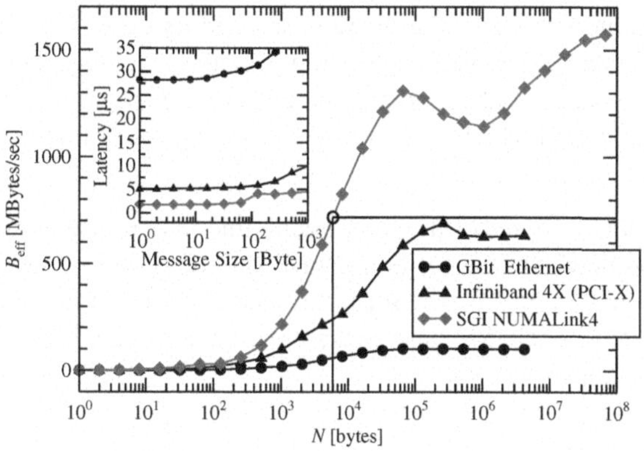

Fig. 26.20. Result of the PingPong benchmark for three different networks. The $N_{1/2}$ point is marked for the NumaLink4 data. Inset: Latencies can be deduced by extrapolating to zero message length

the improvement in effective bandwidth is

$$\frac{B_{\text{eff}}(\beta B, T_1)}{B_{\text{eff}}(B, T_1)} = \frac{1 + N/N_{1/2}}{1 + N/\beta N_{1/2}}, \tag{26.30}$$

so that for $N = N_{1/2}$ and $\beta = 2$ the gain is only 33 %. In case of a reduction of latency by a factor of β, the result is the same. Hence it is desirable to improve on both latency and bandwidth to make an interconnect more efficient for all applications.

Please note that the simple PingPong algorithm described above cannot pinpoint saturation effects: If the network fabric is not completely non-blocking and all nodes transmit or receive data (as is often the case with collective MPI operations), aggregated bandwidth, i.e. the sum over all effective bandwidths for all point-to-point connections, is lower than the theoretical limit. This can severely throttle the performance of applications on large CPU numbers as well as overall throughput of the machine.

26.2.4 Shared-Memory Computing

A shared-memory parallel computer is a system in which a number of CPUs work on a common, shared physical address space. This is fundamentally different from the distributed-memory paradigm as described in the previous section. Although transparent to the programmer as far as functionality is concerned, there are two varieties of shared-memory systems that have very different performance characteristics:

- *Uniform Memory Access* (UMA) systems feature a flat memory model: Memory bandwidth and latency are the same for all processors and all memory locations. This is also called symmetric multiprocessing (SMP).
- On *cache-coherent Non-Uniform Memory Access* (ccNUMA) machines, memory is physically distributed but logically shared. The physical layout of such

systems is quite similar to the distributed-memory case (Fig. 26.17), but network logic makes the aggregated memory of the whole system appear as one single address space. Due to the distributed nature, memory access performance varies depending on which CPU accesses which parts of memory (local vs. remote access).

With multiple CPUs, copies of the same cache line may reside in different caches, probably in modified state. So for both above varieties, *cache coherence protocols* must guarantee consistency between cached data and data in memory at all times. Details about UMA, ccNUMA and cache coherence mechanisms are provided in the following sections.

26.2.4.1 UMA

The simplest implementation of a UMA system is a dual-core processor in which two CPUs share a single path to memory. Technical details vary among vendors, and it is very common in high performance computing to use more than one chip in a compute node (be they single-core or multi-core), which adds to diversity. In Figs. 26.21 and 26.22, two typical representatives of UMA systems used in HPC are shown.

In Fig. 26.21 two (single-core) processors, each in its own socket, communicate and access memory over a common bus, the so-called front-side bus (FSB). All arbitration protocols required to make this work are already built into the CPUs. The chipset (often termed north-bridge) is responsible for driving the memory modules and connects to other parts of the node like I/O subsystems.

In Fig. 26.22, two dual-core chips connect to the chipset, each with its own FSB. The chipset plays an important role in enforcing cache coherence and also mediates the connection to memory. In principle, a system like this could be designed so

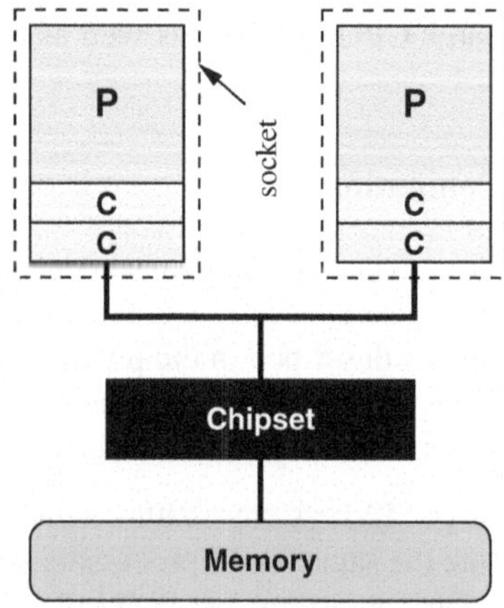

Fig. 26.21. A UMA system with two single-core CPUs that share a common front-side bus (FSB)

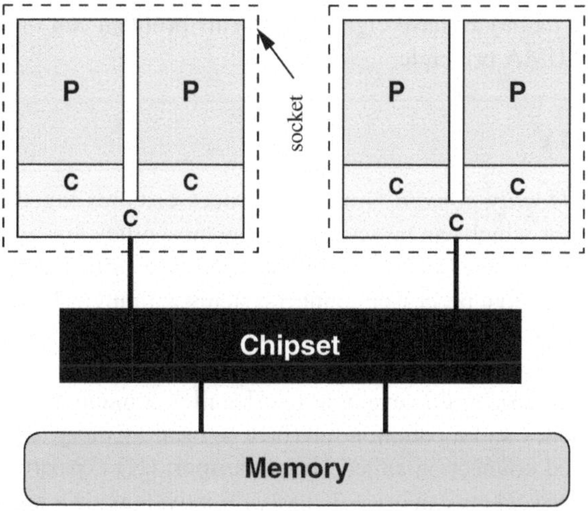

Fig. 26.22. A UMA system in which the FSBs of two dual-core chips are connected separately to the chipset

that the bandwidth from chipset to memory matches the aggregated bandwidth of the front-side buses. Each dual-core chip features a separate L1 on each CPU but a shared L2 cache for both. The advantage of a shared cache is that, to an extent limited by cache size, data exchange between cores can be done there and does not have to resort to the slow front-side bus. Of course, a shared cache should also meet the bandwidth requirements of all connected cores, which might not be the case. Due to the shared caches and FSB connections this kind of node is, while still a UMA system, quite sensitive to the exact placement of processes or threads on its cores. For instance, with only two processes it may be desirable to keep (pin) them on separate sockets if the memory bandwidth requirements are high. On the other hand, processes communicating a lot via shared memory may show more performance when placed on the same socket because of the shared L2 cache. Operating systems as well as some modern compilers usually have tools or library functions for observing and implementing thread or process pinning.

The general problem of UMA systems is that bandwidth bottlenecks are bound to occur when the number of sockets, or FSBs, is larger than a certain limit. In very simple designs like the one in Fig. 26.21, a common memory bus is used that can only transfer data to one CPU at a time (this is also the case for all multi-core chips available today).

In order to maintain scalability of memory bandwidth with CPU number, non-blocking crossbar switches can be built that establish point-to-point connections between FSBs and memory modules (similar to the chip set in Fig. 26.22). Due to the very large aggregated bandwidths those become very expensive for a larger number of sockets. At the time of writing, the largest UMA systems with scalable bandwidth (i.e. the memory bandwidth matches the aggregated FSB bandwidths of all processors in the node) have eight CPUs. This problem can only be solved by giving up on the UMA principle.

26.2.4.2 ccNUMA

In ccNUMA, a *locality domain* is a set of processor cores together with locally connected memory, which can be accessed in the most efficient way, i.e. without resorting to a network of any kind. Although the ccNUMA principle provides scalable bandwidth for very large processor counts (systems with up to 1024 CPUs in a single address space with a single OS instance are available today), it is also found in inexpensive two- or four-socket AMD Opteron nodes frequently used for HPC clustering (see Fig. 26.23). In this example two locality domains, i.e. dual-core chips with separate caches and a common interface to local memory, are linked using a special high-speed connection called Hypertransport (HT). Apart from the minor peculiarity that the sockets can drive memory directly, making a north-bridge obsolete, this system differs substantially from networked UMA designs in that the HT link can mediate direct *coherent* access from one processor to another processor's memory. From the programmer's point of view this mechanism is transparent: All the required protocols are handled by the HT hardware.

In Fig. 26.24 another approach to ccNUMA is shown, which is flexible enough to scale to large machines, and used, e.g., in SGI Altix systems. Each processor socket connects to a communication interface (S) that provides memory access as well as connectivity to the proprietary NUMALink (NL) network. The NL network relies on routers (R) to switch connections for non-local access. As with HT, the NL hardware allows for transparent access to the whole address space of the machine from all CPUs. Although shown here only with four sockets, multi-level router fabrics can be built that scale up to hundreds of CPUs. It must, however, be noted that each piece of hardware inserted into a data connection (communication interfaces, routers) adds to latency, making access characteristics very inhomogeneous across

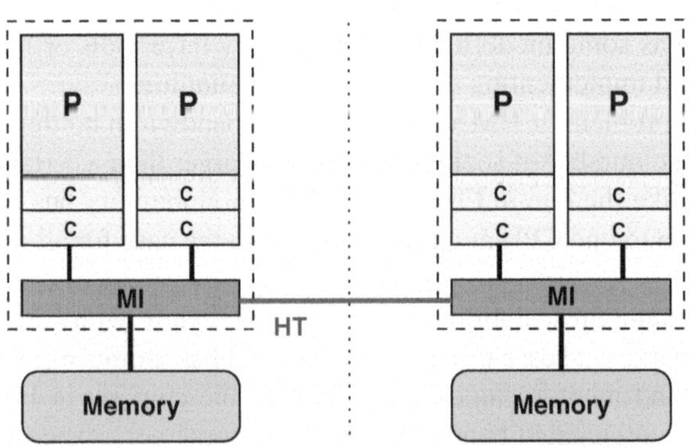

Fig. 26.23. Hypertransport-based ccNUMA system with two locality domains (one per socket) and four cores

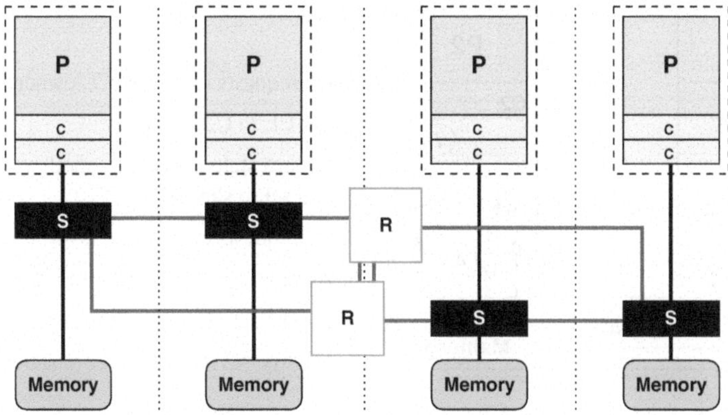

Fig. 26.24. ccNUMA system with routed NUMALink network and four locality domains

the system. Furthermore, as is the case with networks for distributed-memory computers, providing wire-equivalent speed, non-blocking bandwidth in large systems is extremely expensive.

In all ccNUMA designs network connections must have bandwidth and latency characteristics comparable to those of local memory. Although this is the case for all contemporary systems, even a penalty factor of two for non-local transfers can badly hurt application performance if access can not be restricted inside locality domains. This locality problem is the first of two obstacles to take with high performance software on ccNUMA. It occurs even if there is only one serial program running on a ccNUMA machine. The second problem is potential congestion if two processors from different locality domains access memory in the same locality domain, fighting for memory bandwidth. Even if the network is non-blocking and its performance matches the bandwidth and latency of local access, congestion can occur. Both problems can be solved by carefully observing the data access patterns of an application and restricting data access of each processor to its own locality domain. Section 27.2.3 will elaborate on this topic.

In inexpensive ccNUMA systems I/O interfaces are often connected to a single locality domain. Although I/O transfers are usually slow compared to memory bandwidth, there are, e.g., high-speed network interconnects that feature multi-GB bandwidths between compute nodes. If data arrives at the wrong locality domain, written by an I/O driver that has positioned its buffer space disregarding any ccNUMA constraints, it should be copied to its optimal destination, reducing effective bandwidth by a factor of four (three if RFOs can be avoided, see Sect 26.1.5.2. In this case even the most expensive interconnect hardware is wasted. In truly scalable ccNUMA designs this problem is circumvented by distributing I/O connections across the whole machine and using ccNUMA-aware drivers.

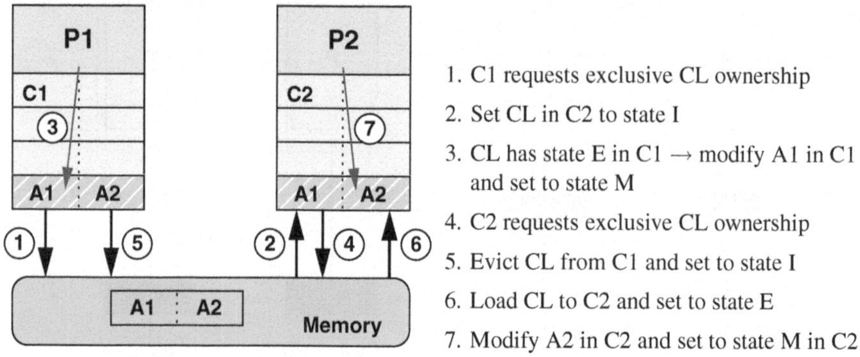

Fig. 26.25. Two processors P1, P2 modify the two parts A1, A2 of the same cache line in caches C1 and C2. The MESI coherence protocol ensures consistency between cache and memory

26.2.4.3 Cache Coherence

Cache coherence mechanisms are required in all cache-based multiprocessor systems, UMA as well as ccNUMA. This is because potentially copies of the same cache line could reside in several CPU caches. If, e.g., one of those gets modified and evicted to memory, the other caches' contents reflect outdated data. Cache coherence protocols ensure a consistent view of memory under all circumstances.

Figure 26.25 shows an example on two processors P1 and P2 with respective caches C1 and C2. Each cache line holds two items. Two neighboring items A1 and A2 in memory belong to the same cache line and are modified by P1 and P2, respectively. Without cache coherence, each cache would read the line from memory, A1 would get modified in C1, A2 would get modified in C2 and some time later both modified copies of the cache line would have to be evicted. As all memory traffic is handled in chunks of cache line size, there is no way to determine the correct values of A1 and A2 in memory.

Under control of cache coherence logic this discrepancy can be avoided. As an example we pick the MESI protocol, which draws its name from the four possible states a cache line can take:

M *modified:* The cache line has been modified in this cache, and it resides in no other cache than this one. Only upon eviction will memory reflect the most current state.

E *exclusive:* The cache line has been read from memory but not (yet) modified. However, it resides in no other cache.

S *shared:* The cache line has been read from memory but not (yet) modified. There may be other copies in other caches of the machine.

I *invalid:* The cache line does not reflect any sensible data. Under normal circumstances this happens if the cache line was in shared state and another processor has requested exclusive ownership. A cache miss occurs if and only if the chosen line is invalid.

The order of events is depicted in Fig. 26.25. The question arises how a cache line in state M is notified when it should be evicted because another cache needs to read the most current data. Similarly, cache lines in state S or E must be invalidated if another cache requests exclusive ownership. In small systems a *bus snoop* is used to achieve this: Whenever notification of other caches seems in order, the originating cache broadcasts the corresponding cache line address through the system, and all caches "snoop" the bus and react accordingly. While simple to implement, this method has the crucial drawback that address broadcasts pollute the system buses and reduce available bandwidth for useful memory accesses. A separate network for coherence traffic can alleviate this effect but is not always practicable.

A better alternative, usually applied in larger ccNUMA machines, is a directory-based protocol where bus logic like chip sets or memory interfaces keep track of the location and state of each cache line in the system. This uses up some small part of main memory (usually far less than 10 %), but the advantage is that state changes of cache lines are transmitted only to those caches that actually require them. This greatly reduces coherence traffic through the system. Today even workstation chip sets implement snoop filters that serve the same purpose.

Coherence traffic can severely hurt application performance if the same cache line is written to frequently by different processors (false sharing). In Sect. 27.2.1.2 we will give hints for avoiding false sharing in user code.

26.2.4.4 Short Introduction to Shared-Memory Programming with OpenMP

As mentioned before, programming shared-memory systems can be done in an entirely distributed-memory fashion, i.e. the processes making up an MPI program can run happily on a UMA or ccNUMA machine, not knowing that the underlying hardware provides more efficient means of communication. In fact, even on large constellation clusters (systems where the number of nodes is smaller than the number of processors per node), the dominant parallelization method is often MPI due to its efficiency and flexibility. After all, an MPI code can run on shared- as well as distributed-memory systems, and efficient MPI implementations transparently use shared memory for communication if available.

However, MPI is not only the most flexible but also the most tedious way of parallelization. Shared memory opens the possibility to have immediate access to all data from all processors without explicit message passing. The established standard in this field is OpenMP [7]. OpenMP is a set of compiler directives that a non-OpenMP-capable compiler would just regard as comments and ignore. Thus, a well-written parallel OpenMP program is also a valid serial program (of course it is possible to write OpenMP code that will not run sequentially, but this is not the intention of the method). In contrast to MPI, the central entity in OpenMP is not a process but a thread. Threads are also called lightweight processes because several of them can share a common address space and mutually access data. Spawning a thread is much less costly than forking a new process, because threads share everything but instruction pointer (the address of the next instruction to be executed), stack pointer and register state. Each thread can, by means of its local stack pointer, also have private variables, but as all data is accessible via the common address space, it is only a matter of taking the address of an item to make it accessible to all other threads as well: Thread-private data is for convenience, not for protection.

It is indeed possible to use operating system threads (POSIX threads) directly, but this option is seldom used with numerical software. OpenMP is a layer that adapts the raw OS thread interface to make it more usable with the typical loop structures that numerical software tends to show. As an example, consider a parallel version of a simple integration program (Listing 26.4). This is valid serial code, but equipping it with the comment lines starting with the sequence `!$OMP` (called a sentinel) and using an OpenMP-capable compiler makes it shared-memory parallel. The `PARALLEL` directive instructs the compiler to start a parallel region (see Fig. 26.26). A team of threads is spawned that executes identical copies of everything up to `END PARALLEL` (the actual number of threads is unknown at compile time as it is set by an environment variable). By default, all variables which were present in the program before the parallel region are shared among all threads. However, that would include `x` and `sum` of which we later need private versions for each thread. OpenMP provides a way to make existing variables private by means of the `PRIVATE` clause. If, in the above example, any thread in a parallel region writes to `sum` (see line 4), it will update its own private copy, leaving the other threads' untouched. Therefore, before the loop starts each thread's copy of `sum` is set to zero.

In order to share some amount of work between threads and actually reduce wallclock time, work sharing directives can be applied. This is done in line 5 using the `DO` directive with the optional `SCHEDULE` clause. The `DO` directive is always related to the immediately following loop (line 6) and generates code that distributes

Listing 26.4. A simple program for numerical integration of a function $f(x)$ in OpenMP

```
1     pi=0.d0
2     w=1.d0/n
3  !$OMP PARALLEL PRIVATE(x,sum)
4     sum=0.d0
5  !$OMP DO SCHEDULE(STATIC)
6     do i=1,n
7        x=w*(i-0.5d0)
8        sum=sum+f(x)
9     enddo
10 !$OMP END DO
11 !$OMP CRITICAL
12    pi=pi+w*sum
13 !$OMP END CRITICAL
14 !$OMP END PARALLEL
```

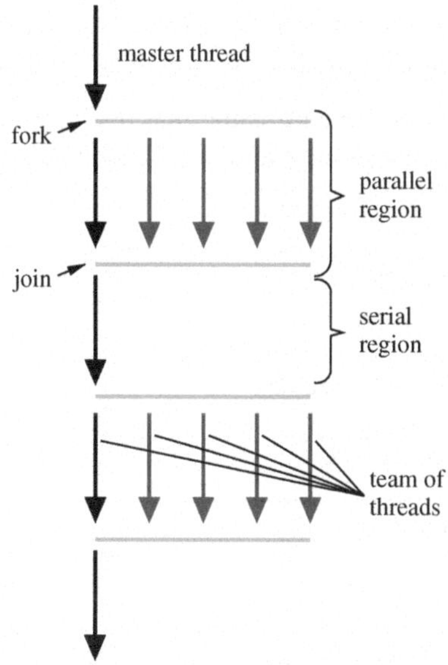

Fig. 26.26. Model for OpenMP thread operations: The master thread forks a thread team that work on shared memory in a parallel region. After the parallel region, the threads are joined or put to sleep until the next parallel region starts

the loop iterations among the team of threads (please note that the loop counter variable is automatically made private). How this is done is controlled by the argument of `SCHEDULE`. The simplest possibility is `STATIC` which divides the loop in chunks of (roughly) equal size and executes each thread on a chunk. If for some reason the amount of work per loop iteration is not constant but, e.g., decreases with loop count, this strategy is suboptimal because different threads will get vastly different workloads, which leads to load imbalance. One solution would be to use a chunk size like in "`STATIC,1`" that dictates that chunks of size one should be distributed across threads in a round-robin manner. There are alternatives to static schedule for other types of workload (`DYNAMIC`, `GUIDED`).

The parallelized loop computes a partial sum in each thread's private `sum` variable. To get the final result, all the partial sums must be accumulated in the global `pi` variable (line 12), but `pi` is shared so that uncontrolled updates would lead to a race condition, i.e. the exact order and timing of operations will influence the result. In OpenMP, *critical sections* solve this problem by making sure that at most one thread at a time executes some piece of code. In the example, the `CRITICAL` and `END CRITICAL` directives bracket the update to `pi` so that a correct result emerges at all times.

Critical sections hold the danger of deadlocks when used inappropriately. A deadlock arises when one or more threads wait for resources that will never become available, a situation that is easily generated with badly arranged `CRITICAL` directives. When a thread encounters a `CRITICAL` directive inside a critical region, it will block forever. OpenMP provides two solutions for this problem:

- A critical section may be given a name that distinguishes it from others. The name is specified in parentheses after the `CRITICAL` directive:

```
!$OMP PARALLEL DO PRIVATE(x)
  do i=1,N
    x=sin(2*PI*x/N)
!$OMP CRITICAL (psum)
    sum=sum+func(x)
!$OMP END CRITICAL (psum)
  enddo
!$OMP END PARALLEL DO
  ...
  SUBROUTINE func(v)
  double precision v
!$OMP CRITICAL (prand)
  v=v+random_func()
!$OMP END CRITICAL (prand)
  END SUBROUTINE func
```

  Without the names on the two different critical sections in this code would deadlock.
- There are OpenMP API functions (see below) that support the use of locks for protecting shared resources. The advantage of locks is that they are ordinary variables that can be arranged as arrays or in structures. That way it is possible to protect each single element of an array of resources individually.

Whenever there are different shared resources in a program that must be protected from concurrent access each for its own but are otherwise unconnected, named critical sections or OpenMP locks should be used both for correctness and performance reasons.

In some cases it may be useful to write different code depending on OpenMP being enabled or not. The directives themselves are no problem here because they will be ignored gracefully. Conditional compilation however is supported by the preprocessor symbol `_OPENMP` which is defined only if OpenMP is available and (in Fortran) the special sentinel `!$` that acts as a comment only if OpenMP is not enabled (see Listing 26.5). Here we also see a part of OpenMP that is not concerned with directives. The `use omp_lib` declaration loads the OpenMP API function prototypes (in C/C++, `#include <omp.h>` serves the same purpose). The `omp_get_thread_num()` function determines the thread ID, a number between zero and the number of threads minus one, while `omp_get_num_threads()` returns the number of threads in the current team. So if the general disposition of OpenMP towards loop-based code is not what the programmer wants, one can easily switch to an MPI-like style where thread ID determines the tasks of each thread.

In above example the second API call (line 8) is located in a `SINGLE` region, which means that it will be executed by exactly one thread, namely the one that

Listing 26.5. Fortran sentinels and conditional compilation with OpenMP

```
1  !$ use omp_lib
2    myid=0
3    numthreads=1
4  #ifdef _OPENMP
5  !$OMP PARALLEL PRIVATE(myid)
6    myid = omp_get_thread_num()
7  !$OMP SINGLE
8    numthreads = omp_get_num_threads()
9  !$OMP END SINGLE
10 !$OMP CRITICAL
11   write(*,*) 'Parallel program - this is thread ',myid,&
12                                               ' of ',numthreads
13 !$OMP END CRITICAL
14 !$OMP END PARALLEL
15 #else
16   write(*,*) 'Serial program'
17 #endif
```

reaches the SINGLE directive first. This is done because numthreads is global and should be written to only by one thread. In the critical region each thread just prints a message, but a necessary requirement for the numthreads variable to have the updated value is that no thread leaves the SINGLE region before update has been promoted to memory. The END SINGLE directive acts as an implicit barrier, i.e. no thread can continue executing code before all threads have reached the same point. The OpenMP memory model ensures that barriers enforce memory consistency: Variables that have been held in registers are written out so that cache coherence can make sure that all caches get updated values. This can also be initiated under program control via the FLUSH directive, but most OpenMP work-sharing and synchronization constructs perform implicit barriers and hence flushes at the end.

There is an important reason for serializing the write statements in line 10. As a rule, I/O operations and general OS functionality, but also common library functions should be serialized because they are usually not *thread-safe*, i.e. calling them in parallel regions from different threads at the same time may lead to errors. A prominent example is the rand() function from the C library as it uses a static variable to store its hidden state (the seed). Although local variables in functions are private to the calling thread, static data is shared by definition. This is also true for Fortran variables with a SAVE attribute.

One should note that the OpenMP standard gives no hints as to how threads are to be distributed among the processors, let alone observe locality constraints. Usually the OS makes a good choice regarding placement of threads, but sometimes (especially on multi-core architectures and ccNUMA systems) it makes sense to

Listing 26.6. C/C++ example with reduction clause for adding noise to the elements of an array and calculating its vector norm. `rand()` is not thread-safe so it must be protected by a critical region

```
1    double r,s;
2  #pragma omp parallel for private(r) reduction(+:s)
3    for(i=0; i<N; ++i) {
4  #pragma omp critical
5      {
6        r = rand();                // not thread-safe
7      }
8      a[i] += func(r/RAND_MAX); // func() is thread-safe
9      s = s + a[i] * a[i];       // calculate norm
10   }
```

use OS-level tools, compiler support or library functions to explicitly pin threads to cores. See Sect. 27.2.3 for details.

So far, all OpenMP examples were concerned with the Fortran bindings. Of course there is also a C/C++ interface that has the same functionality. The C/C++ sentinel is called `#pragma omp`, and the only way to do conditional compilation is to use the `_OPENMP` symbol. Loop parallelization only works for canonical `for` loops that have standard integer-type loop counters (i.e., no STL[2]-style iterator loops) and is done via `#pragma omp for`. All directives that act on code regions apply to compound statements and an explicit ending directive is not required.

The example in Listing 26.6 shows a C code that adds some random noise to the elements of an array `a[]` and calculates its vector norm. As mentioned before, `rand()` is not thread-safe and must be protected with a critical region. The function `func()`, however, is assumed to be thread-safe as it only uses automatic (stack) variables and can thus be called safely from a parallel region (line 8). Another peculiarity in this example is the fusion of the `parallel` and `for` directives to `parallel for`, which allows for more compact code. Finally, the reduction operation is not performed using critical updates as in the integration example. Instead, an OpenMP `reduction` clause is used (end of line 2) that automatically initializes the summation variable `s` with a sensible starting value, makes it private and accumulates the partial results to it.

A word of caution is in order concerning thread-local variables. Usually the OS shell restricts the maximum size of all stack variables of its processes. This limit can often be adjusted by the user or the administrators. However, in a threaded program there are as many stacks as there are threads, and the way the thread-local stacks get their limit set is not standardized at all. Please consult OS and compiler

[2] Standard template library

documentation as to how thread-local stacks are limited. Stack overflows are a frequent source of problems with OpenMP programs.

Running an OpenMP program is as simple as starting the executable binary just like in the serial case. The number of threads to be used is determined by an environment variable called `OMP_NUM_TREADS`. There may be other means to influence the way the program is running, e.g. OS scheduling of threads, pinning, getting debug output etc., but those are not standardized.

26.3 Conclusion and Outlook

We have presented architectural characteristics of current cache-based microprocessors and the systems they are used in. The dominant parallel architectures (distributed and shared memory) have been outlined and their main programming methodologies explained. We have deliberately focused on mainstream technology because it is yet unclear whether any of the new approaches to computing currently put forward by the HPC industry will prevail.

For processor manufacturers, the multi-core path is surely the way to go for the next decade. Moore's law will be valid for a long time, and in a few years we will see tens or even hundreds of cores on a single die. The bandwidth bottlenecks implied will require some new approaches to high-performance programming, other than just putting all cores to use with plain OpenMP or MPI. This problem is under intense discussion in the HPC community today.

As far as more radical approaches are concerned, there is a clear tendency towards the use of building blocks for special-purpose computing. One might argue that vector systems have concentrated on special-purpose computing for the last 30 years, but today the aim is different. Approaches like FPGAs (Field Programmable Gate Arrays), computing with graphics hardware (GPUs) and new ideas like the Cell processor with vector-like slave units currently show benefits only for very narrow fields of applications. Some trends indicate a move to hybrid architectures, turning away from homogeneous parallelism to a diversity of programming models and specialized functionality in a single machine. How programming languages and tools in such an environment should look like is, even with some systems already in the planning stage, as yet largely unknown. Given the considerable inertia of the scientific computing community when it comes to adopting new standards, MPI and OpenMP can be expected to be around for a very long time, though.

References

1. Intel 64 and IA-32 Architectures Optimization Reference Manual (2006). URL `http://developer.intel.com/design/processor/manuals/248966.pdf`
2. Software Optimization Guide for AMD64 Processors (2005). URL `http://www.amd.com/us-en/assets/content_type/white_papers_and_tech_docs/25112.PDF`

3. URL `http://www.spec.org/`
4. G.E. Moore, Electronics **38**(8) (1965)
5. URL `http://www.top500.org`
6. URL `http://www.mpi-forum.org`
7. URL `http://www.openmp.org`

27 Optimization Techniques for Modern High Performance Computers

Georg Hager and Gerhard Wellein

Regionales Rechenzentrum Erlangen der Friedrich-Alexander-Universität Erlangen-Nürnberg, 91058 Erlangen, Germany

The rapid development of faster and more capable processors and architectures has often led to the false conclusion that the next generation of hardware will easily meet the scientist's requirements. This view is at fault for two reasons: First, utilizing the full power of existing systems by proper parallelization and optimization strategies, one can gain a competitive advantage without waiting for new hardware. Second, computer industry has now reached a turning point where exponential growth of compute power has ended and single-processor performance will stagnate at least for the next couple of years. The advent of multi-core CPUs was triggered by this development, making the need for more advanced, parallel, and well-optimized algorithms imminent.

This chapter describes different ways to write efficient code on current supercomputer systems. In Sect. 27.1, simple common sense optimizations for scalar code like strength reduction, correct layout of data structures and tabulation are covered first. Many scientific programs are limited by the speed of the computer system's memory interface, so it is vital to avoid slow data paths or, if this is not possible, at least use them efficiently. After some theoretical considerations on data access and performance estimates based on code analysis and hardware characteristics, techniques like loop transformations and cache blocking are explained using examples from linear algebra (matrix-vector multiplication, matrix transpose). The importance of interpreting compiler logs is emphasized. Along the discussion of performance measurements for vanilla and optimized codes we introduce peculiarities like cache thrashing and translation look-aside buffer misses, both potential showstoppers for compute performance. In a case study we apply the acquired knowledge on sparse matrix-vector multiplication, a performance-determining operation required for practically all sparse diagonalization algorithms.

Turning to shared-memory parallel programming in Sect. 27.2, we identify typical pitfalls (OpenMP loop overhead and false sharing) that can severely limit parallel scalability, and show some ways to circumvent them. The abundance of AMD Opteron nodes in clusters has initiated the necessity for optimizing memory locality. ccNUMA can lead to diverse bandwidth bottlenecks, and few compilers support special features for ensuring memory locality. Programming techniques which can alleviate ccNUMA effects are therefore described in detail using a parallelized sparse matrix-vector multiplication as a nontrivial but instructive example.

G. Hager and G. Wellein: *Optimization Techniques for Modern High Performance Computers*, Lect. Notes Phys. **739**, 731–767 (2008)
DOI 10.1007/978-3-540-74686-7_27

27.1 Optimizing Serial Code

In the age of multi-1000-processor parallel computers, writing code that runs efficiently on a single CPU has grown slightly old-fashioned in some circles. The argument for this point of view is derived from the notion that it is easier to add more CPUs and boasting massive parallelism instead of investing effort into serial optimization.

Nevertheless there can be no doubt that single-processor optimizations are of premier importance. If a speedup of two can be gained by some straightforward common sense optimization as described in the following section, the user will be satisfied with half the number of CPUs in the parallel case. In the face of Amdahl's law the benefit will usually be even larger. This frees resources for other users and projects and puts the hardware that was often acquired for considerable amounts of money to better use. If an existing parallel code is to be optimized for speed, it must be the first goal to make the single-processor run as fast as possible.

27.1.1 Common Sense Optimizations

Often very simple changes to code can lead to a significant performance boost. The most important common sense guidelines regarding the avoidance of performance pitfalls are summarized in the following. Those may seem trivial, but experience shows that many scientific codes can be improved by the simplest of measures.

27.1.1.1 Do Less Work!

In all but the rarest of cases, rearranging the code such that less work than before is being done will improve performance. A very common example is a loop that checks a number of objects to have a certain property, but all that matters in the end is that *any* object has the property at all:

```
logical FLAG
FLAG = .false.
do i=1,N
  if(complex_func(A(i)) < THRESHOLD) then
    FLAG = .true.
  endif
enddo
```

If `complex_func()` has no side effects, the only information that gets communicated to the outside of the loop is the value of `FLAG`. In this case, depending on the probability for the conditional to be true, much computational effort can be saved by leaving the loop as soon as `FLAG` changes state:

```
logical FLAG
FLAG = .false.
do i=1,N
  if(complex_func(A(i)) < THRESHOLD) then
    FLAG = .true.
    exit
  endif
enddo
```

27.1.1.2 Avoid Expensive Operations!

Sometimes, implementing an algorithm is done in a thoroughly one-to-one way, translating formulae to code without any reference to performance issues. While this is actually good (performance optimization always bears the slight danger of changing numerics, if not results), in a second step all those operations should be eliminated that can be substituted by cheaper alternatives. Prominent examples for such strong operations are trigonometric functions or exponentiation. Bear in mind that an expression like `x**2.0` is often not optimized by the compiler to become `x*x` but left as it stands, resulting in the evaluation of an exponential and a logarithm. The corresponding optimization is called strength reduction. Apart from the simple case described above, strong operations often appear with a limited set of fixed arguments. This is an example from a simulation code for non-equilibrium spin systems:

```
integer iL,iR,iU,iO,iS,iN,edelz
double precision tt
...                                 ! load spin orientations
edelz=iL+iR+iU+iO+iS+iN             ! loop kernel
BF= 0.5d0*(1.d0+tanh(edelz/tt))
```

The last two lines are executed in a loop that accounts for nearly the whole runtime of the application. The integer variables store spin orientations (up or down, i.e. -1 or $+1$, respectively), so the `edelz` variable only takes integer values in the range $\{-6,\ldots,+6\}$. The `tanh()` function is one of those operations that take vast amounts of time (at least tens of cycles), even if implemented in hardware. In the case described, however, it is easy to eliminate the `tanh()` call completely by tabulating the function over the range of arguments required, assuming that `tt` does not change its value so that the table does only have to be set up once:

```
double precision tanh_table(-6:6)
integer iL,iR,iU,iO,iS,iN, edelz
double precision, tt
...
do i=-6,6                            ! do this once
  tanh_table(i) = tanh(dble(i)/tt)
```

```
enddo
...
edelz=iL+iR+iU+iO+iS+iN            ! loop kernel
BF= 0.5d0*(1.d0+tanh_table(edelz))
```

The table lookup is performed at virtually no cost compared to the `tanh()` evaluation since the table will, due to its small size and frequent use, be available in L1 cache at access latencies of a few CPU cycles.

27.1.1.3 Shrink the Working Set!

The working set of a code is the amount of memory it uses (i.e. actually touches) in the course of a calculation. In general, shrinking the working set by whatever means is a good thing because it raises the probability for cache hits. If and how this can be achieved and whether it pays off performance-wise depends heavily on the algorithm and its implementation, of course. In the above example, the original code used standard four-byte integers to store the spin orientations. The working set was thus much larger than the L2 cache of any processor. By changing the array definitions to use `integer*1` for the spin variables, the working set could be reduced by nearly a factor of four, and became comparable to cache size.

Many recent microprocessor designs have instruction set extensions for integer and floating-point SIMD operations (see also Sect. 26.1.4) that allow the concurrent execution of arithmetic operations on a wide register that can hold, e.g., two DP or four SP floating-point words. Although vector processors also use SIMD instructions and the use of SIMD in microprocessors is often coined vectorization, it is more similar to the multi-track property of modern vector systems. Generally speaking, a vectorizable loop in this context will run faster if more operations can be performed with a single instruction, i.e. the size of the data type should be as small as possible. Switching from DP to SP data could result in up to a twofold speedup, with the additional benefit that more items fit into the cache.

Consider, however, that not all microprocessors can handle small types efficiently. Using byte-size integers for instance could result in very ineffective code that actually works on larger word sizes but extracts the byte-sized data by mask and shift operations.

27.1.1.4 Eliminate Common Subexpressions!

Common subexpression elimination is an optimization that is often considered a task for compilers. Basically one tries to save time by pre-calculating parts of complex expressions and assigning them to temporary variables before a loop starts:

```
! inefficient
do i=1,N
  A(i)=A(i)+s+r*sin(x)
enddo
```

$\longrightarrow$

```
tmp=s+r*sin(x)
do i=1,N
  A(i)=A(i)+tmp
enddo
```

A lot of compute time can be saved by this optimization, especially where strong operations (like `sin()`) are involved. Although it may happen that subexpressions are obstructed by other code and not easily recognizable, compilers are in principle able to detect this situation. They will however often refrain from pulling the subexpression out of the loop except with very aggressive optimizations turned on. The reason for this is the well-known non-associativity of FP operations: If floating-point accuracy is to be maintained compared to non-optimized code, associativity rules must not be used and it is left to the programmer to decide whether it is safe to regroup expressions by hand.

27.1.1.5 Avoid Conditionals in Tight Loops!

Tight loops, i.e. loops that have few operations in them, are typical candidates for software pipelining (see Sect. 26.1.3.1), loop unrolling and other optimization techniques (see below). If for some reason compiler optimization fails or is inefficient, performance will suffer. This can easily happen if the loop body contains conditional branches:

```
do j=1,N
  do i=1,N
    if(i.ge.j) then
      sign=1.d0
    else if(i.lt.j) then
      sign=-1.d0
    else
      sign=0.d0
    endif
    C(j) = C(j) + sign * A(i,j) * B(i)
  enddo
enddo
```

In this multiplication of a matrix with a vector, the upper and lower triangular parts get different signs and the diagonal is ignored. The `if` statement serves to decide about which factor to use. Each time a corresponding conditional branch is encountered by the processor, some branch prediction logic tries to guess the most probable outcome of the test before the result is actually available, based on statistical methods. The instructions along the chosen path are then fetched, decoded, and generally fed into the pipeline. If the anticipation turns out to be false (this is called a mispredicted branch or branch miss), the pipeline has to be flushed back to the position of the branch, implying many lost cycles. Furthermore, the compiler refrains from doing advanced optimizations like loop unrolling (see Sect. 27.1.3.2).

Fortunately the loop nest can be transformed so that all `if` statements vanish:

```
do j=1,N
  do i=j+1,N
    C(j) = C(j) + A(i,j) * B(i)
  enddo
```

```
enddo
do j=1,N
  do i=1,j-1
    C(j) = C(j) - A(i,j) * B(i)
  enddo
enddo
```

By using two different variants of the inner loop, the conditional has virtually been moved outside. One should add that there is more optimization potential in this loop nest. Please consider the section on data access below for more information.

27.1.1.6 Use Compiler Logs!

The previous sections have pointed out that the compiler is a crucial component in writing efficient code. It is very easy to hide important information from the compiler, forcing it to give up optimization at an early stage. In order to make the decisions of the compiler's intelligence available to the user, many compilers offer options to generate annotated source code listings or at least logs that describe in some detail what optimizations were performed. Listing 27.1 shows an example for a compiler annotation regarding a standard vector triad loop as in listing 26.1. Unfortunately, not all compilers have the ability to write such comprehensive code annotations and users are often left with guesswork.

27.1.2 Data Access

Of all possible performance-limiting factors in HPC, the most important one is data access. As explained earlier, microprocessors tend to be inherently unbalanced with respect to the relation of theoretical peak performance versus memory bandwidth. As many applications in science and engineering consist of loop-based code that

Listing 27.1. Compiler log for a software pipelined triad loop

```
#<swps> 16383 estimated iterations before pipelining
#<swps>     4 unrollings before pipelining
#<swps>    20 cycles per 4 iterations
#<swps>     8 flops        ( 20% of peak) (madds count as 2)
#<swps>     4 flops        ( 10% of peak) (madds count as 1)
#<swps>     4 madds        ( 20% of peak)
#<swps>    16 mem refs     ( 80% of peak)
#<swps>     5 integer ops  ( 12% of peak)
#<swps>    25 instructions ( 31% of peak)
#<swps>     2 short trip threshold
#<swps>    13 integer registers used.
#<swps>    17 float registers used.
```

moves large amounts of data in and out of the CPU, on-chip resources tend to be underutilized and performance is limited only by the relatively slow data paths to memory or even disks. Any optimization attempt should therefore aim at reducing traffic over slow data paths, or, should this turn out to be infeasible, at least make data transfer as efficient as possible.

27.1.2.1 Balance and Lightspeed Estimates

Some programmers go to great lengths trying to improve the efficiency of code. In order to decide whether this makes sense or if the program at hand is already using the resources in the best possible way, one can often estimate the theoretical performance of loop-based code that is bound by bandwidth limitations by simple rules of thumb. The central concept to introduce here is balance. For example, the machine balance B_{m} of a processor is the ratio of possible memory bandwidth in GWords/sec to peak performance in GFlops/sec:

$$B_{\mathrm{m}} = \frac{\text{memory bandwidth [GWords/sec]}}{\text{peak performance [GFlops/sec]}} . \tag{27.1}$$

Memory bandwidth could also be substituted by the bandwidth to caches or even network bandwidths, although the metric is generally most useful for codes that are really memory-bound. Access latency is assumed to be hidden by techniques like prefetching and software pipelining. As an example, consider a processor with a clock frequency of 3.2 GHz that can perform at most two flops per cycle and has a memory bandwidth of 6.4 GBytes/sec. This processor would have a machine balance of 0.125 W/F. At the time of writing, typical values of B_{m} lie in the range between 0.1 W/F for commodity microprocessors and 0.5 W/F for top of the line vector computers. Due to the continuously growing DRAM gap and the advent of multi-core designs, machine balance for standard architectures will presumably decrease further in the future. Table 27.1 shows typical balance values for several possible transfer paths.

In order to quantify the requirements of some code that runs on a machine with a certain balance, we further define the code balance of a loop to be

$$B_{\mathrm{c}} = \frac{\text{data traffic volume [Words]}}{\text{floating point operations [Flops]}} . \tag{27.2}$$

Table 27.1. Typical balance values for operations limited by different transfer paths

data path	balance
cache	0.5–1.0
machine (memory)	0.05–0.5
interconnect (high speed)	0.01–0.04
interconnect (GBit ethernet)	0.001–0.003
disk	0.001–0.02

Now it is obvious that the expected maximum fraction of peak performance one can expect from a code with balance B_c on a machine with balance B_m is

$$l = \min\left(1, \frac{B_m}{B_c}\right) . \tag{27.3}$$

We call this fraction the lightspeed of a code. If $l \simeq 1$, loop performance is not limited by bandwidth but other factors, either inside the CPU or elsewhere. Note that this simple performance model is based on some crucial assumptions:

- The loop code makes use of all arithmetic units (multiplication and addition) in an optimal way. If this is not the case, e.g., when only additions are used, one must introduce a correction term that reflects the ratio of MULT to ADD operations.
- Code is based on double precision floating-point arithmetic. In cases where this is not true, one can easily derive similar, more appropriate metrics (e.g., words per instruction).
- Data transfer and arithmetic overlap perfectly.
- The system is in throughput mode, i.e. latency effects are negligible.

We must emphasize that more advanced strategies for performance modeling do exist and refer to the literature [1, 2].

As an example consider the standard vector triad benchmark introduced in Sect. 26.1.5. The kernel loop,

```
do i=1,N
  A(i) = B(i) + C(i) * D(i)
enddo
```

features two flops per iteration, for which three loads (to elements `B(i)`, `C(i)`, and `D(i)`) and one store operation (to `A(i)`) provide the required input data. The code balance is thus $B_c = (3 + 1)/2 = 2$. On a CPU with machine balance $B_m = 0.1$, we can then expect a lightspeed ratio of 0.05, i.e. 5 % of peak.

Standard cache-based microprocessors usually feature an outermost cache level with write-back strategy. As explained in Sect. 26.1.5, cache line read for ownership (RFO) is then required to ensure cache-memory coherence if nontemporal stores or cache line zero is not used. Under such conditions, the store stream to array `A` must be counted twice in calculating the code balance, and we would end up with a lightspeed estimate of $l_{RFO} = 0.04$.

27.1.2.2 Storage Order of Multi-Dimensional Arrays

Multi-dimensional arrays, first and foremost matrices or matrix-like structures, are omnipresent in scientific computing. Data access is a crucial topic here as the mapping between the inherently one-dimensional, cache line based memory layout of standard computers and any multi-dimensional data structure must be matched to the order in which code loads and stores data so that spatial and temporal locality can be employed. In Sect. 26.1.5 it was shown that strided access to a one-dimensional array reduces spatial locality, leading to low utilization of the available bandwidth. When dealing with multi-dimensional arrays, those access patterns can be generated quite naturally:

Stride-N access

```
do i=1,N
  do j=1,N
    A(i,j) = i*j
  enddo
enddo
```

Stride-1 access

```
for(i=0; i<N; ++i) {
  for(j=0; j<N; ++j) {
    a[i][j] = i*j;
  }
}
```

These Fortran and C codes perform exactly the same task, and the second array index is the fast (inner loop) index both times, but the memory access patterns are quite distinct. In the Fortran example, the memory address is incremented in steps of `N*sizeof(double)`, whereas in the C example the stride is optimal. This is because Fortran follows the so-called column major order whereas C follows row major order for multi-dimensional arrays (see Fig. 27.1). Although mathematically insignificant, the distinction must be kept in mind when optimizing for data access.

27.1.2.3 Case Study: Dense Matrix Transpose

For the following example we assume column major order as implemented in Fortran. Calculating the transpose of a dense matrix, $A = B^{\mathrm{T}}$, involves strided memory access to `A` or `B`, depending on how the loops are ordered. The most unfavorable way of doing the transpose is shown here:

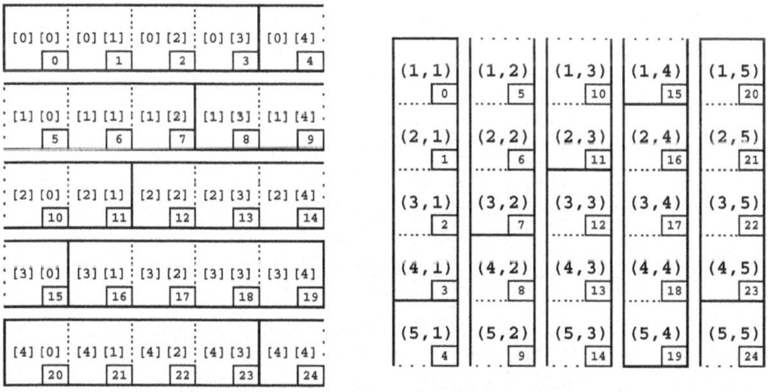

Fig. 27.1. Row major order (**left**) and column major order (**right**) storage schemes for matrices. The small numbers indicate the offset of the element with respect to the starting address of the array. Solid frames symbolize cache lines

```
do i=1,N
  do j=1,N
    A(i,j) = B(j,i)
  enddo
enddo
```

Write access to matrix A is strided (see Fig. 27.2). Due to RFO transactions, strided writes are more expensive than strided reads. Starting from this worst possible code we can now try to derive expected performance features. As matrix transpose does not perform any arithmetic, we will use effective bandwidth (i.e., GBytes/sec available to the application) to denote performance.

Let C be the cache size and L_c the cache line size, both in DP words. Depending on the size of the matrices we can expect three primary performance regimes:

- In case the two matrices fit into a CPU cache ($2N^2 \lesssim C$), we expect effective bandwidths of the order of cache speeds. Spatial locality is of importance only between different cache levels; optimization potential is limited.
- If the matrices are too large to fit into cache but still

$$NL_c \lesssim C\,, \tag{27.4}$$

  the strided access to A is insignificant because all stores performed during a complete traversal of a row that cause a write miss start a cache line RFO. Those lines are most probably still in cache for the next $L_c - 1$ rows, alleviating the effect of strided write (spatial locality). Effective bandwidth should be of the order of the processor's memory bandwidth.
- If N is even larger so that $NL_c \gtrsim C$, each store to A causes a cache miss and a subsequent RFO. A sharp drop in performance is expected at this point as only one out of L_c cache-line entries is actually used for the store stream and any spatial locality is suddenly lost.

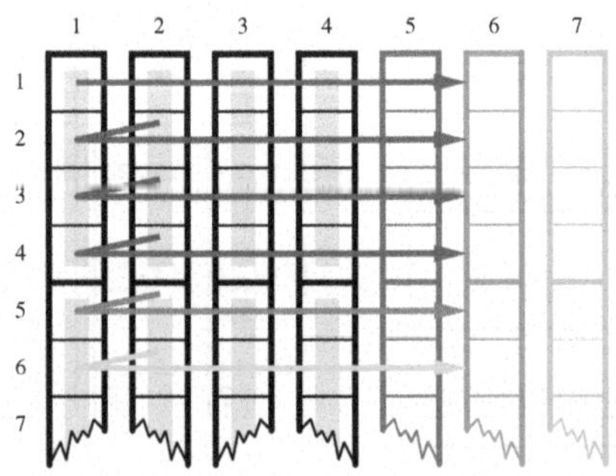

Fig. 27.2. Cache line traversal for vanilla matrix transpose (strided store stream, column major order). If the leading matrix dimension is a multiple of the cache line size, each column starts on a line boundary

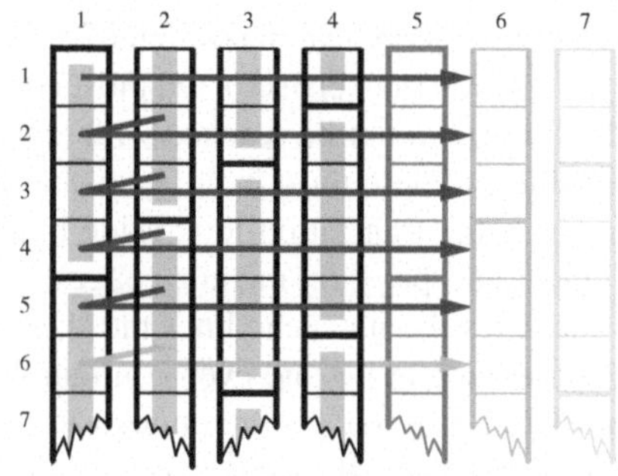

Fig. 27.3. Cache line traversal for padded matrix transpose. Successive iterations hit different cache lines

The vanilla graph in Fig. 27.4 shows that the assumptions described above are essentially correct, although the strided write seems to be very unfavorable even when the whole working set fits into cache. This is because the L1 cache on the considered architecture is of write-through type, i.e. the L2 cache is always updated on a write, regardless whether there was an L1 hit or miss. The RFO transactions between the two caches hence waste the major part of available internal bandwidth.

In the second regime described above, performance stays roughly constant up to a point where the fraction of cache used by the store stream for N cache lines becomes comparable to the L2 size. Effective bandwidth is around 1.8 GBytes/sec, a

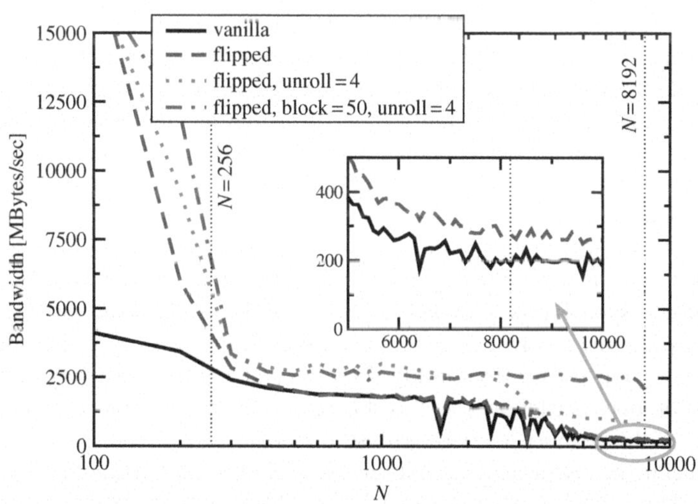

Fig. 27.4. Performance (effective bandwidth) for different implementations of the dense matrix transpose on a modern microprocessor with 1 MByte of L2 cache. The $N = 256$ and $N = 8192$ lines indicate the positions where the matrices fit into cache and where N cache lines fit into cache, respectively. (Intel Xeon/Nocona 3.2 Ghz)

mediocre value compared to the theoretical maximum of 5.3 GBytes/sec (delivered by two-channel memory at 333 MTransfers/sec). On most commodity architectures the theoretical bandwidth limits can not be reached with compiler-generated code, but 50 % is usually attainable, so there must be a factor that further reduces available bandwidth. This factor is the translation look-aside buffer (TLB) that caches the mapping between logical and physical memory pages. The TLB can be envisioned as an additional cache level with cache lines the size of memory pages (the page size is often 4 kB, sometimes 16 kB and even configurable on some systems). On the architecture considered, it is only large enough to hold 64 entries, which corresponds to 256 kBytes of memory at a 4 kB page size. This is smaller than the whole L2 cache, so it must be expected that this cache level cannot be used with optimal performance. Moreover, if N is larger than 512, i.e. if one matrix row exceeds the size of a page, every single access in the strided stream causes a TLB miss. Even if the page tables reside in L2 cache, this penalty reduces effective bandwidth significantly because every TLB miss leads to an additional access latency of at least 57 processor cycles. At a core frequency of 3.2 GHz and a bus transfer rate of 666 MWords/sec, this matches the time needed to transfer more than half a cache line!

At $N \gtrsim 8192$, performance has finally arrived at the expected low level. The machine under investigation has a theoretical memory bandwidth of 5.3 GBytes/sec of which around 200 MBytes/sec actually "hit the floor".

At a cache line length of 16 words (of which only one is used for the strided store stream), three words per iteration are read or written in each loop iteration for the in-cache case whereas 33 words are read or written for the worst case. We thus expect a 1 : 11 performance ratio, roughly the value observed.

We must stress here that performance predictions based on architectural specifications do work in many, but not in all cases, especially on commodity systems where factors like chip sets, memory chips, interrupts etc. are basically uncontrollable. Sometimes only a qualitative understanding of the reasons for some peculiar performance behavior can be developed, but this is often enough to derive the next logical optimization steps.

The first and most simple optimization for dense matrix transpose would consist in interchanging the order of the loop nest, i.e. pulling the `i` loop inside. This would render the access to matrix `B` strided but eliminate the strided write for `A`, thus saving roughly half the bandwidth (5/11, to be exact) for very large N. The measured performance gain (see the inset in Fig. 27.4, flipped graph), albeit very noticeable, falls short of this expectation. One possible reason for this could be a slightly better effectivity of the memory interface with strided writes.

In general, the performance graphs in Fig. 27.4 look quite erratic at some points. At first sight it is unclear whether some N should lead to strong performance penalties as compared to neighboring values. A closer look (vanilla graph in Fig. 27.5) reveals that powers of two in array dimensions seem to be quite unfavorable (the benchmark program allocates new matrices with appropriate dimensions for each new N). As mentioned in Sect. 26.1.5.2, strided memory access leads to thrashing

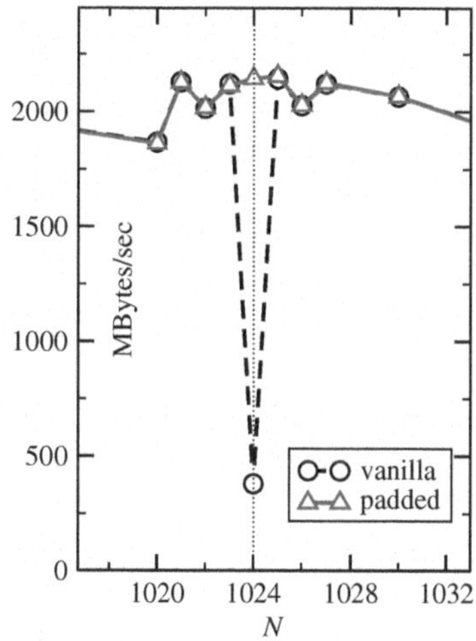

Fig. 27.5. Cache thrashing for unfavorable choice of array dimensions (dashed). Padding removes thrashing completely (solid)

when successive iterations hit the same (set of) cache line(s) because of insufficient associativity. Fig. 27.2 shows clearly that this can easily happen with matrix transpose if the leading dimension is a power of two. On a direct-mapped cache of size C, every C/N-th iteration hits the same cache line. At a line length of L_c words, the effective cache size is

$$C_\mathrm{eff} = L_\mathrm{c} \max\left(1, \frac{C}{N}\right) . \tag{27.5}$$

It is the number of cache words that are actually usable due to associativity constraints. On an m-way set-associative cache this number is merely multiplied by m. Considering a real-world example with $C = 2^{17}$ (1 MByte), $L_\mathrm{c} = 16$, $m = 8$ and $N = 1024$ one arrives at $C_\mathrm{eff} = 2^{11}$ DP words, i.e. 16 kBytes. So $NL_\mathrm{c} \gg C_\mathrm{eff}$ and performance should be similar to the very large N limit described above, which is roughly true.

A simple code modification, however, eliminates the thrashing effect: Assuming that matrix `A` has dimensions 1024×1024, enlarging the leading dimension by p (called padding) to get `A(1024+`p`,1024)` leads to a fundamentally different cache use pattern. After L_c/p iterations, the address belongs to another set of m cache lines and there is no associativity conflict if $Cm/N > L_\mathrm{c}/p$ (see Fig. 27.3). In Fig. 27.5 the striking effect of padding the leading dimension by $p = 1$ is shown with the padded graph.

Generally speaking, one should by all means stay away from powers of two in array dimensions. It is clear that different dimensions may require different paddings to get optimal results, so sometimes a rule of thumb is applied: Try to make leading array dimensions odd multiples of 16.

Further optimization approaches will be considered in the following sections.

27.1.3 Data Access Optimizations and Classification of Algorithms

The optimization potential of many loops on cache-based processors can easily be estimated just by looking at basic parameters like the scaling behavior of data transfers and arithmetic operations versus problem size. It can then be decided whether investing optimization effort would make sense.

27.1.3.1 $O(N)/O(N)$

If both the number of arithmetic operations and the number of data transfers (loads/stores) are proportional to the problem size (or loop length) N, optimization potential is usually very limited. Scalar products, vector additions and sparse matrix-vector multiplication are examples for this kind of problems. They are inevitably memory-bound for large N, and compiler-generated code achieves good performance because $O(N)/O(N)$ loops tend to be quite simple and the correct software pipelining strategy is obvious. Loop nests, however, are a different matter (see below).

But even if loops are not nested there is sometimes room for improvement. As an example, consider the following vector additions:

```
do i=1,N
  A(i) = B(i) + C(i)
enddo
do i=1,N
  Z(i) = B(i) + E(i)
enddo
```

loop fusion →

```
! optimized
do i=1,N
  A(i) = B(i) + C(i)
! save a load for B(i)
  Z(i) = B(i) + E(i)
enddo
```

Each of the loops on the left has no options left for optimization. The code balance is 3/1 as there are two loads, one store and one addition per loop (not counting RFOs). Array B, however, is loaded again in the second loop, which is unnecessary: Fusing the loops into one has the effect that each element of B only has to be loaded once, reducing code balance to 5/2. All else being equal, performance in the memory-bound case will improve by a factor of 6/5 (if RFO cannot be avoided, this will be 8/7).

Loop fusion has achieved an $O(N)$ data reuse for the two-loop constellation so that a complete load stream could be eliminated. In simple cases like the one above, compilers can often apply this optimization by themselves.

27.1.3.2 $O(N^2)/O(N^2)$

In typical two-level loop nests where each loop has a trip count of N, there are $O(N^2)$ operations for $O(N^2)$ loads and stores. Examples are dense matrix-vector multiplication, matrix transpose, matrix addition etc., Although the situation on the inner level is similar to the $O(N)/O(N)$ case and the problems are generally memory-bound, the nesting opens new opportunities. Optimization, however, is again usually limited to a constant factor of improvement. Consider dense matrix-vector multiplication (MVM):

```
do i=1,N
  tmp = C(i)
  do j=1,N
    tmp = tmp + A(j,i) * B(j)
  enddo
  C(i) = tmp
enddo
```

This code has a balance of 1 (two loads for A and B and two flops). Array C is indexed by the outer loop variable, so updates can go to a register (here clarified through the use of the scalar tmp although compilers can do this transformation automatically) and do not count as load or store streams. Matrix A is only loaded once, but B is loaded N times, once for each outer loop iteration. One would like to apply the same fusion trick as above, but there are not just two but N inner loops to fuse. The solution is loop unrolling: The outer loop is traversed with a stride m and the inner loop is replicated m times. Obviously, one has to deal with the situation that the outer loop count might not be a multiple of m. This case has to be handled by a remainder loop:

```
! remainder loop
do r=1,mod(N,m)
  do j=1,N
    C(r) = C(r) + A(j,r) * B(j)
  enddo
enddo
! main loop
do i=r,N,m
  do j=1,N
    C(i) = C(i) + A(j,i) * B(j)
  enddo
  do j=1,N
    C(i+1) = C(i+1) + A(j,i+1) * B(j)
  enddo
  ! m times
  ...
  do j=1,N
    C(i+m-1) = C(i+m-1) + A(j,i+m-1) * B(j)
  enddo
enddo
```

The remainder loop is obviously subject to the same optimization techniques as the original loop, but otherwise unimportant. For this reason we will ignore remainder loops in the following.

By just unrolling the outer loop we have not gained anything but a considerable code bloat. However, loop fusion can now be applied easily:

```
! remainder loop ignored
do i=1,N,m
  do j=1,N
    C(i) = C(i) + A(j,i) * B(j)
    C(i+1) = C(i+1) + A(j,i+1) * B(j)
    ! m times
    ...
    C(i+m-1) = C(i+m-1) + A(j,i+m-1) * B(j)
  enddo
enddo
```

The combination of outer loop unrolling and fusion is often called unroll and jam. By m-way unroll and jam we have achieved an m-fold reuse of each element of B from register so that code balance reduces to $(m + 1)/(2m)$ which is clearly smaller than one for $m > 1$. If m is very large, the performance gain can get close to a factor of two. In this case array B is only loaded a few times or, ideally, just once from memory. As A is always loaded exactly once and has size N^2, the total memory traffic with m-way unroll and jam amounts to $N^2(1+1/m)+N$. Fig. 27.6 shows the memory access pattern for vanilla and 2-way unrolled dense MVM.

All this assumes, however, that register pressure is not too large, i.e. the CPU has enough registers to hold all the required operands used inside the now quite sizeable loop body. If this is not the case, the compiler must spill register data to cache, slowing down the computation. Again, compiler logs can help identify such a situation.

Unroll and jam can be carried out automatically by some compilers at high optimization levels. Be aware though that a complex loop body may obscure important information and manual optimization could be necessary, either – as shown above – by hand-coding or compiler directives that specify high-level transformations like unrolling. Directives, if available, are the preferred alternative as they are much easier to maintain and do not lead to visible code bloat. Regrettably, compiler directives are inherently non-portable.

The matrix transpose code from the previous section is another example for a problem of $O(N^2)/O(N^2)$ type, although in contrast to dense MVM there is no direct opportunity for saving on memory traffic; both matrices have to be read

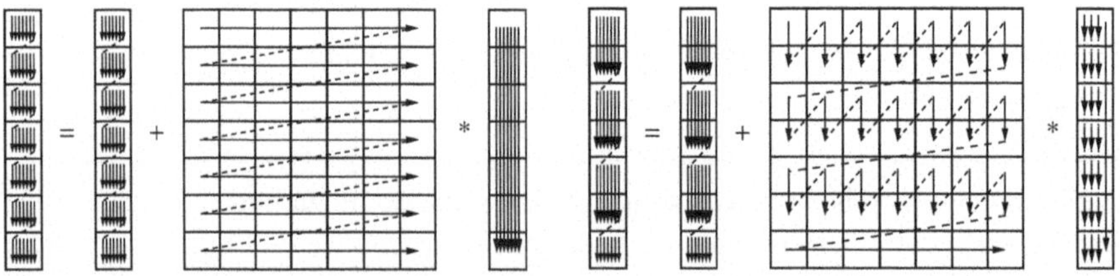

Fig. 27.6. Vanilla (**left**) and 2-way unrolled (**right**) dense matrix vector multiplication. The remainder loop is only a single (outer) iteration in this example

or written exactly once. Nevertheless, by using unroll and jam on the flipped version a significant performance boost of nearly 50 % is observed (see dotted line in Fig. 27.4):

```
do j=1,N,m
  do i=1,N
    A(i,j)       = B(j,i)
    A(i,j+1)     = B(j+1,i)
    ...
    A(i,j+m-1)  = B(j+m-1,i)
  enddo
enddo
```

Naively one would not expect any effect at $m = 4$ because the basic analysis stays the same: In the mid-N region the number of available cache lines is large enough to hold up to L_c columns of the store stream. The left picture in Fig. 27.7 shows the situation for $m = 2$. However, the fact that m words in each of the load stream's cache lines are now accessed in direct succession reduces the TLB misses by a factor of m, although the TLB is still way too small to map the whole working set.

Even so, cutting down on TLB misses does not remedy the performance breakdown for large N when the cache gets too small to hold N cache lines. It would be nice to have a strategy which reuses the remaining $L_c - m$ words of the strided stream's cache lines right away so that each line may be evicted soon and would not have to be reclaimed later. A brute force method is L_c-way unrolling, but this approach leads to large-stride accesses in the store stream and is not a general solution as large unrolling factors raise register pressure in loops with arithmetic operations. Loop blocking can achieve optimal cache line use without additional register pressure. It does not save load or store operations but increases the cache hit ratio. For a loop nest of depth d, blocking introduces up to d additional outer loop levels that cut the original inner loops into chunks:

```
do jj=1,N,b
  jstart=jj; jend=jj+b-1
  do ii=1,N,b
    istart=ii; iend=ii+b-1
    do j=jstart,jend,m
      do i=istart,iend
        a(i,j)   = b(j,i)
        a(i,j+1) = b(j+1,i)
        ...
        a(i,j+m-1) = b(j+m-1,i)
      enddo
    enddo
  enddo
enddo
```

In this example we have used 2D blocking with identical blocking factors b for both loops in addition to m-way unroll and jam. Obviously, this change does not

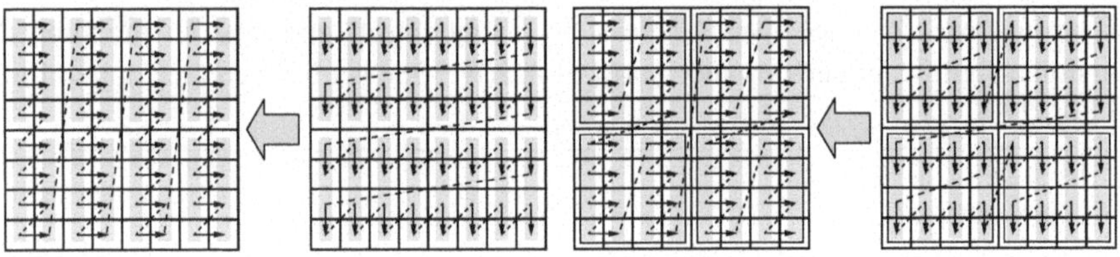

Fig. 27.7. Two-way unrolled (**left**) and blocked/unrolled (**right**) flipped matrix transpose, i.e. with strided load

alter the loop body so the number of registers needed to hold operands stays the same. However, the cache line access characteristics are much improved (see the right picture in Fig. 27.7 which shows a combination of two-way unrolling and 4×4 blocking). If the blocking factors are chosen appropriately, the cache lines of the strided stream will have been used completely at the end of a block and can be evicted soon. Hence we expect the large-N performance breakdown to disappear. The dotted-dashed graph in Fig. 27.4 demonstrates that 50×50 blocking combined with 4-way unrolling alleviates all memory access problems induced by the strided stream.

Loop blocking is a very general and powerful optimization that can often not be performed by compilers. The correct blocking factor to use should be determined experimentally through careful benchmarking, but one may be guided by typical cache sizes, i.e. when blocking for L1 cache the aggregated working set size of all blocked inner loop nests should not be much larger than half the cache. Which cache level to block for depends on the operations performed and there is no general recommendation.

27.1.3.3 $O(N^3)/O(N^2)$

If the number of operations is larger than the number of data items by a factor that grows with problem size, we are in the very fortunate situation to have tremendous optimization potential. By the techniques described above (unroll and jam, loop blocking) it is usually possible for these kinds of problems to render the implementation cache-bound. Examples for algorithms that show $O(N^3)/O(N^2)$ characteristics are dense matrix-matrix multiplication (MMM) and dense matrix diagonalization. It is beyond the scope of this contribution to develop a well-optimized MMM, let alone eigenvalue calculation, but we can demonstrate the basic principle by means of a simpler example which is actually of the $O(N^2)/O(N)$ type:

```
do i=1,N
  do j=1,N
    sum = sum + foo(A(i),B(j))
  enddo
enddo
```

The complete data set is $O(N)$ here but $O(N^2)$ operations (calls to `foo()`, additions) are performed on it. In the form shown above, array `B` is loaded from memory N times, so the total memory traffic amounts to $N(N+1)$ words. m-way unroll and jam is possible and will immediately reduce this to $N(N/m+1)$, but the disadvantages of large unroll factors have been pointed out already. Blocking the inner loop with a block size of b, however,

```
do jj=1,N,b
  jstart=jj; jend=jj+b-1
  do i=1,N
    do j=jstart,jend
      sum = sum + foo(A(i),B(j))
    enddo
  enddo
enddo
```

has two effects:

- Array `B` is now loaded only once from memory, provided that b is small enough so that b elements fit into cache and stay there as long as they are needed.
- Array `A` is loaded from memory N/b times instead of once.

Although `A` is streamed through cache N/b times, the probability that the current block of `B` will be evicted is quite low, the reason being that those cache lines are used very frequently and thus kept by the LRU replacement algorithm. This leads to an effective memory traffic of $N(N/b+1)$ words. As b can be made much larger than typical unrolling factors, blocking is the best optimization strategy here. Unroll and jam can still be applied to enhance in-cache code balance. The basic N^2 dependence is still there, but with a prefactor that can make the difference between memory-bound and cache-bound behavior. A code is cache-bound if main memory bandwidth and latency are not the limiting factors for performance any more. Whether this goal is achievable on a certain architecture depends on the cache size, cache and memory speeds, and the algorithm, of course.

Algorithms of the $O(N^3)/O(N^2)$ type are typical candidates for optimizations that can potentially lead to performance numbers close to the theoretical maximum. If blocking and unrolling factors are chosen appropriately, dense MMM, e.g., is an operation that usually achieves over 90 % of peak for $N \times N$ matrices if N is not too small. It is provided in highly optimized versions by system vendors as, e.g., contained in the BLAS (Basic Linear Algebra Subsystem) library. One might ask why unrolling should be applied at all when blocking already achieves the most important task of making the code cache-bound. The reason is that even if all the data resides in cache, many processor architectures do not have the capability for sustaining enough loads and stores per cycle to feed the arithmetic units continuously. The once widely used but now outdated MIPS R1X000 family of processors for instance could only sustain one load *or* store operation per cycle, which makes unroll and jam mandatory if the kernel of a loop nest uses more than one stream, especially in cache-bound situations like the blocked $O(N^2)/O(N)$ example above.

Although demonstrated here for educational purpose, there is no need to hand-code and optimize standard linear algebra and matrix operations. They should always be used from optimized libraries, if available. Nevertheless the techniques described can be applied in many real-world codes. An interesting example with some complications is sparse MVM (see next section).

27.1.4 Case Study: Sparse Matrix-Vector Multiplication

An interesting real-world application of the blocking and unrolling strategies discussed in the previous sections is the multiplication of a sparse matrix with a vector. It is a key ingredient in most iterative matrix diagonalization algorithms (Lanczos, Davidson, Jacobi-Davidson; see Chap. 18) and usually a performance-limiting factor. A matrix is called sparse if the number of non-zero entries N_{nz} grows linearly with the number of matrix rows N_{r}. Of course, only the non-zeroes are stored at all for efficiency reasons. Sparse MVM (sMVM) is hence an $O(N_{\mathrm{r}})/O(N_{\mathrm{r}})$ problem and inherently memory-bound if N_{r} is reasonably large. Nevertheless, the presence of loop nests enables some significant optimization potential. Fig. 27.8 shows that sMVM generally requires some strided or even indirect addressing of the r.h.s. vector, although there exist matrices for which memory access patterns are much more favorable. In the following we will keep at the general case.

27.1.4.1 Sparse Matrix Storage Schemes

Several different storage schemes for sparse matrices have been developed, some of which are suitable only for special kinds of matrices [3]. Of course, memory access patterns and thus performance characteristics of sMVM depend heavily on the storage scheme used. The two most important and also general formats are CRS (Compressed Row Storage) and JDS (Jagged Diagonals Storage). We will see that CRS is well-suited for cache-based microprocessors while JDS supports dependency and loop structures that are favorable on vector systems.

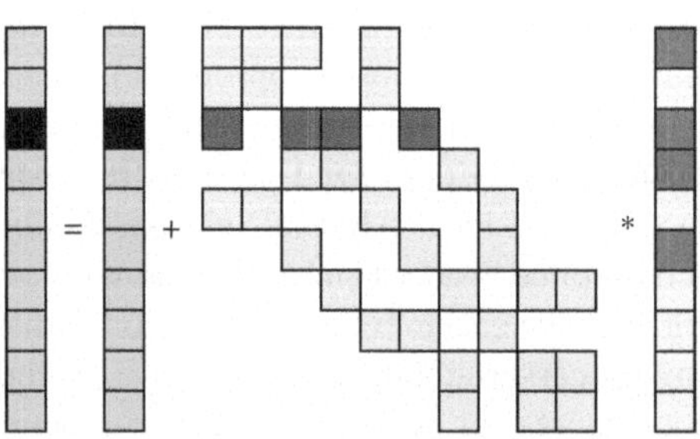

Fig. 27.8. Sparse matrix-vector multiplication. Dark elements visualize entries involved in updating a single l.h.s. element. Unless the sparse matrix rows have no gaps between the first and last non-zero elements, some indirect addressing of the r.h.s. vector is inevitable

In CRS, an array `val` of length N_{nz} is used to store all non-zeroes of the matrix, row by row, without any gaps, so some information about which element of `val` originally belonged to which row and column must be supplied. This is done by two additional integer arrays, `col_idx` of length N_{nz} and `row_ptr` of length N_r. `col_idx` stores the column index of each non-zero element in `val`, and `row_ptr` contains the indices at which new rows start in `val` (see Fig. 27.9). The basic code to perform a MVM using this format is quite simple:

```
do i = 1,Nr
  do j = row_ptr(i), row_ptr(i+1) - 1
    c(i) = c(i) + val(j) * b(col_idx(j))
  enddo
enddo
```

The following points should be noted:

- There is a long outer loop (length N_r).
- The inner loop may be short compared to typical microprocessor pipeline lengths.
- Access to result vector `c` is well optimized: It is only loaded once from memory.
- The non-zeroes in `val` are accessed with stride one.
- As expected, the r.h.s. vector `b` is accessed indirectly. This may however not be a serious performance problem depending on the exact structure of the matrix. If the non-zeroes are concentrated mainly around the diagonal, there will even be considerable spatial and/or temporal locality.
- $B_c = 5/4$ if the integer load to `col_idx` is counted with four bytes.

Some of those points will be of importance later when we demonstrate parallel sMVM (see Sect. 27.2.2).

JDS requires some rearrangement of the matrix entries beyond simple zero elimination. First, all zeroes are eliminated from the matrix rows and the non-zeroes are shifted to the left. Then the matrix rows are sorted by descending number of non-zeroes so that the longest row is at the top and the shortest row is at the bottom. The permutation map generated during the sorting stage is stored in array `perm` of length N_r. Finally, the now established columns are stored in array `val` consecutively. These columns are also called jagged diagonals as they traverse the original

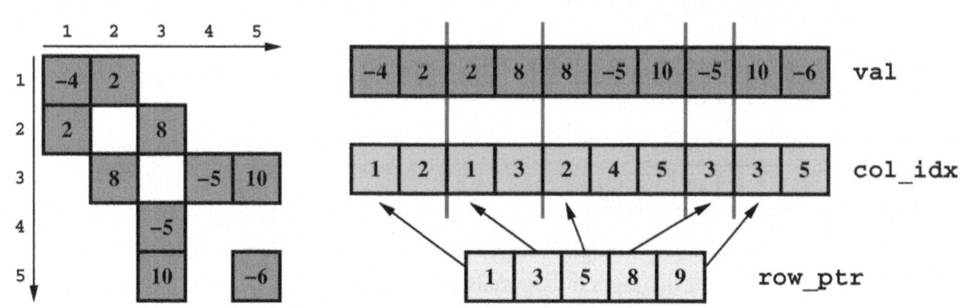

Fig. 27.9. CRS sparse matrix storage format

sparse matrix from left top to right bottom (see Fig. 27.10). For each non-zero the original column index is stored in `col_idx` just like in the CRS. In order to have the same element order on the r.h.s. and l.h.s. vectors, the `col_idx` array is subject to the above-mentioned permutation as well. Array `jd_ptr` holds the start indices of the N_j jagged diagonals. A standard code for sMVM in JDS format is only slightly more complex than with CRS:

```
do diag=1, Nj
  diagLen = jd_ptr(diag+1) - jd_ptr(diag)
  offset = jd_ptr(diag)
  do i=1, diagLen
    c(i) = c(i) + val(offset+i) * b(col_idx(offset+i))
  enddo
enddo
```

The `perm` array storing the permutation map is not required here; usually, all sMVM operations are done in permuted space. These are the notable properties of this loop:

- There is a long inner loop without dependencies, which makes JDS a much better storage format for vector processors than CRS.
- The outer loop is short (number of jagged diagonals).

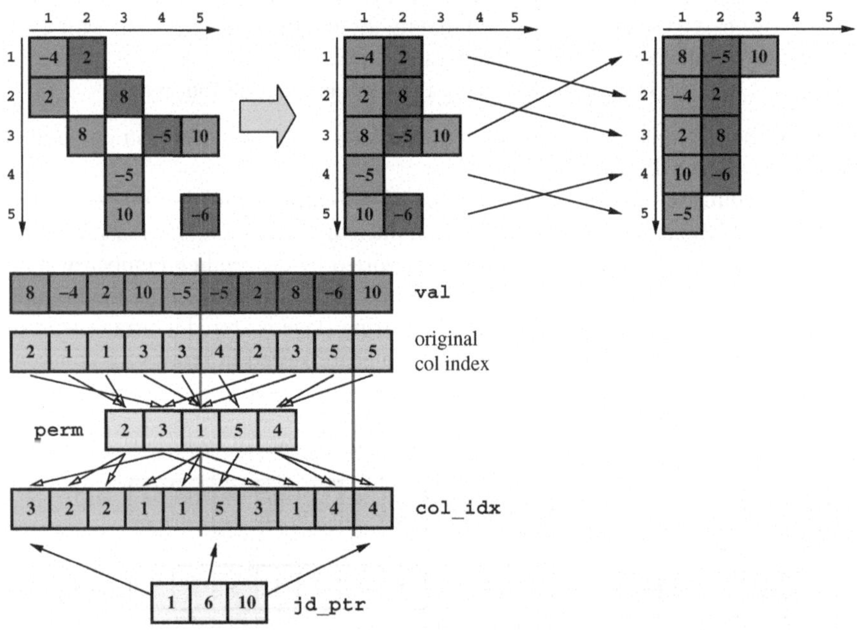

Fig. 27.10. JDS sparse matrix storage format. The permutation map is also applied to the column index array. One of the jagged diagonals is marked

- The result vector is loaded multiple times (at least partially) from memory, so there might be some optimization potential.
- The non-zeroes in `val` are accessed with stride one.
- The r.h.s. vector is accessed indirectly, just as with CRS. The same comments as above do apply, although a favorable matrix layout would feature straight diagonals, not compact rows. As an additional complication the matrix rows as well as the r.h.s. vector are permuted.
- $B_c = 9/4$ if the integer load to `col_idx` is counted with four bytes.

The code balance numbers of CRS and JDS sMVM seem to be quite in favor of CRS.

27.1.4.2 Optimizing JDS Sparse MVM

Unroll and jam should be applied to the JDS sMVM, but it usually requires the length of the inner loop to be independent of the outer loop index. Unfortunately, the jagged diagonals are generally not all of the same length, violating this condition. However, an optimization technique called *loop peeling* can be employed which, for m-way unrolling, cuts rectangular $m \times x$ chunks and leaves $m-1$ partial diagonals over for separate treatment (see Fig. 27.11; the remainder loop is omitted as usual):

```
do diag=1,Nj,2   ! 2-way unroll & jam
  diagLen = min( (jd_ptr(diag+1)-jd_ptr(diag)) ,\
                 (jd_ptr(diag+2)-jd_ptr(diag+1)) )
  offset1 = jd_ptr(diag)
  offset2 = jd_ptr(diag+1)
  do i=1, diagLen
    c(i) = c(i)+val(offset1+i)*b(col_idx(offset1+i))
    c(i) = c(i)+val(offset2+i)*b(col_idx(offset2+i))
  enddo
  ! peeled-off iterations
  offset1 = jd_ptr(diag)
  do i=(diagLen+1),(jd_ptr(diag+1)-jd_ptr(diag))
    c(i) = c(i)+val(offset1+i)*b(col_idx(offset1+i))
  enddo
enddo
```

Assuming that the peeled-off iterations account for a negligible contribution to CPU time, m-way unroll and jam reduces code balance to

$$B_c = \frac{1}{m} + \frac{5}{4} .$$

If m is large enough, this can get close to the CRS balance. However, as explained before large m leads to strong register pressure and is not always desirable. Generally, a sensible combination of unrolling and blocking is employed to reduce memory traffic and enhance in-cache performance at the same time. Blocking is indeed possible for JDS sMVM as well (see Fig. 27.12):

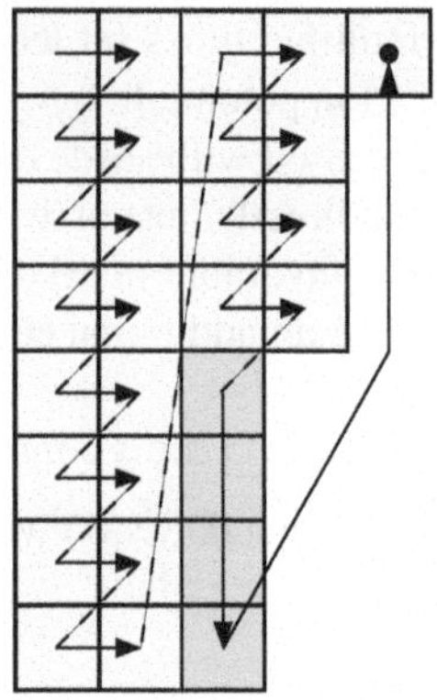

Fig. 27.11. JDS matrix traversal with 2-way unroll and jam and loop peeling. The peeled iterations are marked

```
! loop over blocks
do ib=1, Nr, bl
  block_start = ib
  block_end   = min(ib+bl-1, Nr)
  ! loop over diagonals in one block
  do diag=1, Nj
    diagLen = jd_ptr(diag+1)-jd_ptr(diag)
    offset = jd_ptr(diag)
    if(diagLen .ge. block_start) then
      ! standard JDS sMVM kernel
      do i=block_start, min(block_end,diagLen)
        c(i) = c(i)+val(offset+i)*b(col_idx(offset+i))
      enddo
    endif
  enddo
enddo
```

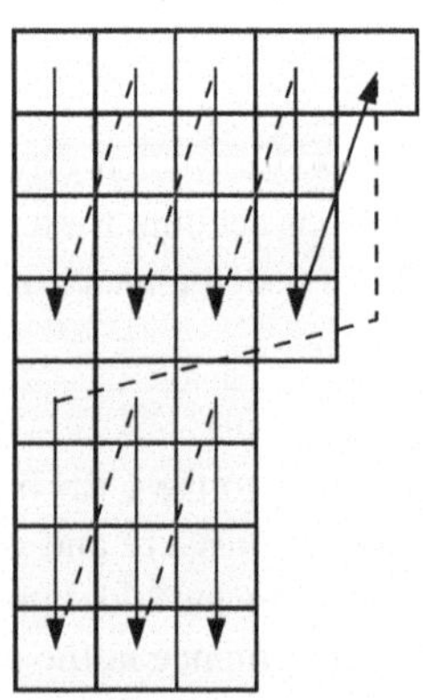

Fig. 27.12. JDS matrix traversal with 4-way loop blocking

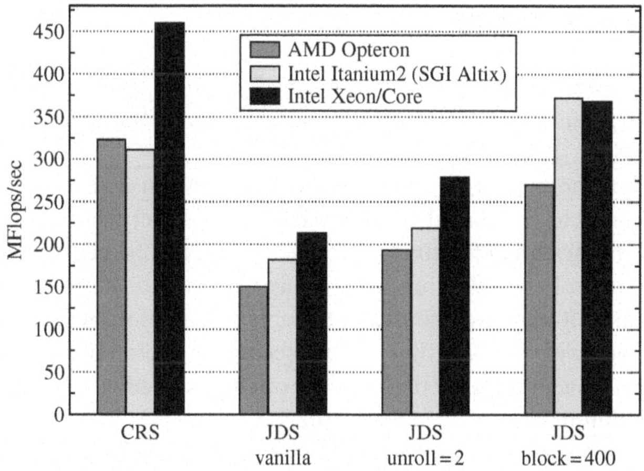

Fig. 27.13. Performance comparison of sparse MVM codes with different optimizations. A matrix with 1.7×10^7 unknowns and 20 jagged diagonals was chosen. The blocking size of 400 has proven to be optimal for a wide range of architectures

With this optimization the result vector is effectively loaded only once from memory if the block size `bl` is not too large. The code should thus get similar performance as the CRS version, although code balance has not been changed. As anticipated above with dense matrix transpose, blocking does not optimize for register reuse but for cache utilization.

Fig. 27.13 shows a performance comparison of CRS and plain, 2-way unrolled and blocked ($b = 400$) JDS sMVM on three different architectures. The CRS variant seems to be preferable for standard AMD and Intel microprocessors, which is not surprising because it features the lowest code balance right away without any subsequent manual optimizations and the short inner loop length is less unfavorable on CPUs with out-of-order capabilities. The Intel Itanium2 processor with its EPIC architecture, however, shows mediocre performance for CRS and tops at the blocked JDS version. This architecture can not cope very well with the short loops of CRS due to the absence of out-of-order processing and the compiler, despite detecting all instruction-level parallelism on the inner loop level, not being able to overlap the wind-down of one row with the wind-up phase of the next.

27.2 Shared-Memory Parallelization

OpenMP seems to be the easiest way to write parallel programs as it features a simple, directive-based interface and incremental parallelization, meaning that the loops of a program can be tackled one by one without major code restructuring. It turns out, however, that getting a truly scalable OpenMP program is a significant undertaking in all but the most trivial cases. This section pinpoints some of the performance problems that can arise with shared-memory programming and how they can be circumvented. We then turn to the OpenMP parallelization of the sparse MVM code that has been demonstrated in the previous sections.

27.2.1 Performance Pitfalls

Like any other parallelization method, OpenMP is prone to the standard problems of parallel programming: Serial fraction (Amdahl's law) and load imbalance, both introduced in Sect. 26.2.

An overabundance of serial code can easily arise when critical sections become out of hand. If all but one threads continuously wait for a critical section to become available, the program is effectively serialized. This can be circumvented by employing finer control on shared resources using named critical sections or OpenMP locks. Sometimes it may even be useful to supply thread-local copies of otherwise shared data that may be pulled together by a reduction operation at the end of a parallel region. The load imbalance problem can often be solved by choosing a different OpenMP scheduling strategy (see Sect. 26.2.4.4).

There are, however, very specific performance problems that are inherently connected to shared-memory programming in general and OpenMP in particular.

27.2.1.1 OpenMP Overhead

Whenever a parallel region is started or stopped or a parallel loop is initiated or ended, there is some non-negligible overhead involved. Threads must be spawned or at least woken up from an idle state, the size of the work packages (chunks) for each thread must be determined, and in the case of dynamic or guided scheduling schemes each thread that becomes available must be supplied with a new chunk to work on. Generally, the overhead caused by the start of a parallel region consists of a (large) constant part and a part that is proportional to the number of threads. There are vast differences from system to system as to how large this overhead can be, but it is generally of the order of at least hundreds if not thousands of CPU cycles. If the programmer follows some simple guidelines, the adverse effects of OpenMP overhead can be much reduced:

– Avoid parallelizing short, tight loops. If the loop body does not contain much work, i.e. if each iteration executes in a very short time, OpenMP loop overhead will lead to very bad performance. It is often beneficial to execute a serial version if the loop count is below some threshold. The OpenMP `IF` clause helps with this:

```
!$OMP PARALLEL DO IF(N>10000)
  do i=1,N
    A(i) = B(i) + C(i) * D(i)
  enddo
!$OMP END PARALLEL DO
```

Fig. 27.14 shows a comparison of vector triad data in the purely serial case and with one and four OpenMP threads, respectively. The presence of OpenMP causes overhead at small N even if only a single thread is used. Using the `IF` clause leads to an optimal combination of threaded and serial loop versions if

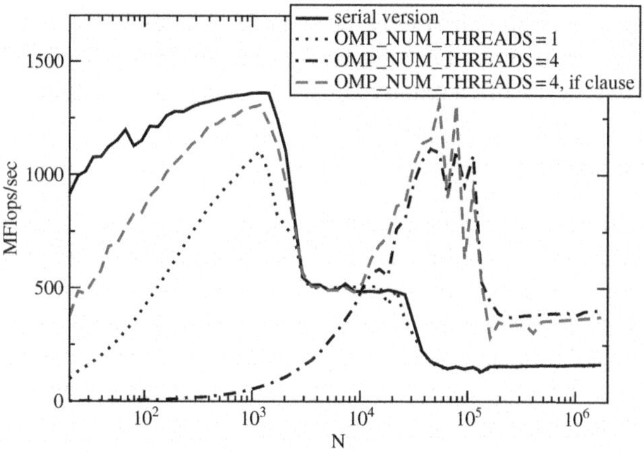

Fig. 27.14. OpenMP overhead and the benefits of the `IF(N>10000)` clause for the vector triad benchmark. Note the impact of aggregate cache size on the position of the performance breakdown from L2 to memory. (AMD Opteron 2.0 GHz)

the threshold is chosen appropriately, and is hence mandatory when large loop lengths cannot be guaranteed.

As a side-note, there is another harmful effect of short loop lengths: If the number of iterations is comparable to the number of threads, load imbalance may cause bad scalability.

- In loop nests, parallelize on a level as far out as possible. This is inherently connected to the previous advice. Parallelizing inner loop levels leads to increased OpenMP overhead because a team of threads is spawned or woken up multiple times.
- Be aware that most OpenMP work-sharing constructs (including `OMP DO` and `END DO`) insert automatic barriers at the end so that all threads have completed their share of work before anything after the construct is executed. In cases where this is not required, a `NOWAIT` clause removes the implicit barrier:

```
!$OMP PARALLEL
!$OMP DO
  do i=1,N
    A(i) = func1(B(i))
  enddo
!$OMP END DO NOWAIT
! still in parallel region here. do more work:
!$OMP CRITICAL
  CNT = CNT + 1
!$OMP END CRITICAL
!$OMP END PARALLEL
```

There is also an implicit barrier at the end of a parallel region that cannot be removed. In general, implicit barriers add to synchronization overhead like critical regions, but they are often required to protect from race conditions.

27.2.1.2 False Sharing

The hardware-based cache coherence mechanisms described in Sect. 26.2.4 make the use of caches in a shared-memory system transparent to the programmer. In some cases, however, cache coherence traffic can throttle performance to very low levels. This happens if the same cache line is modified continuously by a group of threads so that the cache coherence logic is forced to evict and reload it in rapid succession. As an example, consider a program fragment that calculates a histogram over the values in some large integer array A that are all in the range $\{1,\ldots,8\}$:

```
integer, dimension(8) :: S
integer IND
S = 0
do i=1,N
  IND = A(i)
  S(IND) = S(IND) + 1
enddo
```

In a straightforward parallelization attempt one would probably go about and make S two-dimensional, reserving space for the local histogram of each thread:

```
  integer, dimension(:,:), allocatable :: S
  integer IND,ID,NT
!$OMP PARALLEL PRIVATE(ID,IND)
!$OMP SINGLE
  NT = omp_get_num_threads()
  allocate(S(0:NT,8))
  S = 0
!$OMP END SINGLE
 ID = omp_get_thread_num() + 1
!$OMP DO
  do i=1,N
   IND = A(i)
   S(ID,IND) = S(ID,IND) + 1
  enddo
!$OMP END DO NOWAIT
  ! calculate complete histogram
!$OMP CRITICAL
  do j=1,8
  S(0,j) = S(0,j) + S(ID,j)
  enddo
!$OMP END CRITICAL
!$OMP END PARALLEL
```

The loop starting at line 18 collects the partial results of all threads. Although this is a valid OpenMP program, it will not run faster but much more slowly when using four threads instead of one. The reason is that the two-dimensional array `S` contains all the histogram data from all threads. With four threads these are 160 bytes, less than two cache lines on most processors. On each histogram update to `S` in line 10, the writing CPU must gain exclusive ownership of one of the two cache lines, i.e. every write leads to a cache miss and subsequent coherence traffic. Compared to the situation in the serial case where `S` fits into the cache of a single CPU, this will result in disastrous performance.

One should add that false sharing can be eliminated in simple cases by the standard register optimizations of the compiler. If the crucial update operation can be performed to a register whose contents are only written out at the end of the loop, no write misses turn up. This is not possible in the above example, however, because of the computed second index to `S` in line 10.

Getting rid of false sharing by manual optimization is often a simple task once the problem has been identified. A standard technique is array padding, i.e. insertion of a suitable amount of space between memory locations that get updated by different threads. In the histogram example above, an even more painless solution exists in the form of data privatization: On entry to the parallel region, each thread gets its own *local* copy of the histogram array in its own stack space. It is very unlikely that those different instances will occupy the same cache line, so false sharing is not a problem. Moreover, the code is simplified and made equivalent with the serial version by using the `REDUCTION` clause introduced in Sect. 26.2.4.4:

```
  integer, dimension(0) :: S
  integer IND
  S=0
!$OMP PARALLEL DO PRIVATE(IND) REDUCTION(+:S)
  do i=1,N
    IND = A(i)
    S(IND) = S(IND) + 1
  enddo
!$OMP EMD PARALLEL DO
```

Setting `S` to zero is only required for serial equivalence as the reduction clause automatically initializes the variables in question with appropriate starting values. We must add that OpenMP reduction to arrays in Fortran does not work for allocatable, pointer or assumed size types.

27.2.2 Case Study: Parallel Sparse Matrix-Vector Multiplication

As an interesting application of OpenMP to a nontrivial problem we now extend the considerations on sparse MVM data layout and optimization by parallelizing the CRS and JDS matrix-vector multiplication codes from Sect. 27.1.4.

No matter which of the two storage formats is chosen, the general parallelization approach is always the same: In both cases there is a parallelizable loop that calculates successive elements (or blocks of elements) of the result vector (see Fig. 27.15). For the CRS matrix format, this principle can be applied in a straightforward manner:

```
!$OMP PARALLEL DO PRIVATE(j)¹
  do i = 1,Nr
    do j = row_ptr(i), row_ptr(i+1) - 1
      c(i) = c(i) + val(j) * b(col_idx(j))
    enddo
  enddo
!$OMP END PARALLEL DO
```

Due to the long outer loop, OpenMP overhead is usually not a problem here. Depending on the concrete form of the matrix, however, some loop imbalance might occur if very short or very long matrix rows are clustered at some regions. A different kind of OpenMP scheduling strategy like DYNAMIC or GUIDED might help in this situation.

The vanilla JDS sMVM is also parallelized easily:

```
!$OMP PARALLEL PRIVATE(diag,diagLen,offset)
  do diag=1, Nj
    diagLen = jd_ptr(diag+1) - jd_ptr(diag)
    offset = jd_ptr(diag)
!$OMP DO
    do i=1, diagLen
      c(i) = c(i) + val(offset+i) * b(col_idx(offset+i))
    enddo
!$OMP END DO
  enddo
```

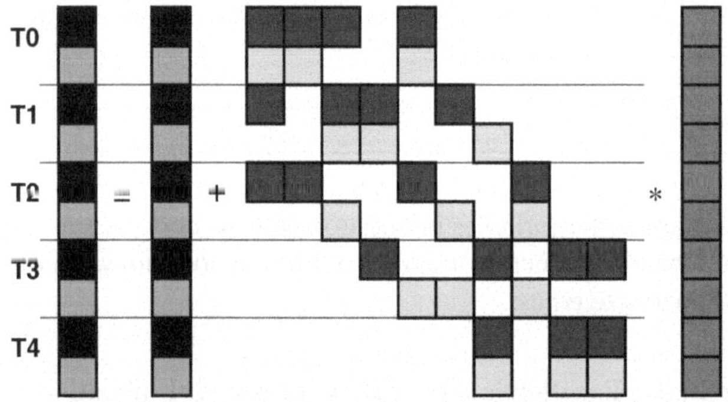

Fig. 27.15. Parallelization approach for sparse MVM (five threads). All marked elements are handled in a single iteration of the parallelized loop. The r.h.s. vector is accessed by all threads

[1] The privatization of inner loop indices in the lexical extent of a parallel outer loop is not required in Fortran, but it is in C/C++ [4].

```
!$OMP END PARALLEL
```

The parallel loop is the inner loop in this case, but there is no OpenMP overhead problem as the loop count is large. Moreover, in contrast to the parallel CRS version, there is no load imbalance because all inner loop iterations contain the same amount of work. All this would look like an ideal situation were it not for the bad code balance of vanilla JDS sMVM. However, the unrolled and blocked versions can be equally well parallelized. For the blocked code (see Fig. 27.12), the outer loop over all blocks is a natural candidate:

```
!$OMP DO PARALLEL DO PRIVATE(block_start,block_end,i,diag,
!$OMP& diagLen,offset)
  do ib=1,Nr,b
    block_start = ib
    block_end   = min(ib+b-1,Nr)
    do diag=1,Nj
      diagLen = jd_ptr(diag+1)-jd_ptr(diag)
      offset  = jd_ptr(diag)
      if(diagLen .ge. block_start) then
        do i=block_start, min(block_end,diagLen)
          c(i) = c(i)+val(offset+i)*b(col_idx(offset+i))
        enddo
      endif
    enddo
  enddo
!$OMP END PARALLEL DO
```

This version has even got less OpenMP overhead because the DO directive is on the outermost loop. Unfortunately, there is more potential for load imbalance because of the matrix rows being sorted for size. But as the dependence of workload on loop index is roughly predictable, a static schedule with a chunk size of one can remedy most of this effect.

Fig. 27.16 shows performance and scaling behavior of the parallel CRS and blocked JDS versions on three different architectures. In all cases, the code was run on as few locality domains or sockets as possible, i.e. first filling one locality domain or socket before going to the next. On the ccNUMA systems (Altix and Opterons, equivalent to the block diagrams in Figs. 26.23 and 26.24), the performance characteristics with growing CPU number is obviously fundamentally different from the UMA system (Xeon/Core node like in Fig. 26.22). Both code versions seem to be extremely unsuitable for ccNUMA. Only the UMA node shows the expected behavior of strong bandwidth saturation at 2 threads and significant speedup when the second socket gets used (additional bandwidth due to second FSB).

The reason for the failure of ccNUMA to deliver the expected bandwidth lies in our ignorance of a necessary prerequisite for scalability that we have not honored yet: Correct data and thread placement for access locality.

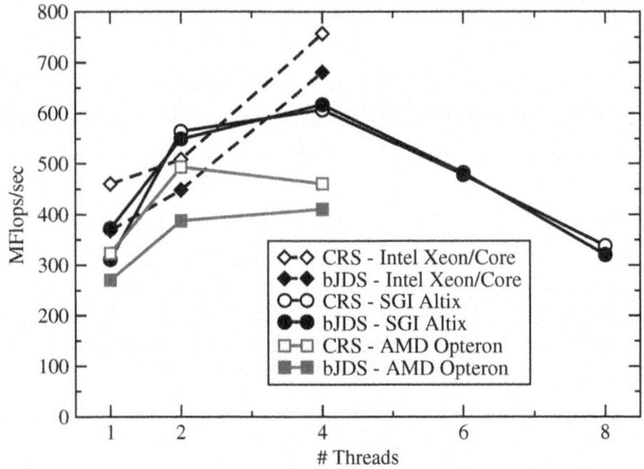

Fig. 27.16. Performance and strong scaling for straightforward OpenMP parallelization of sparse MVM on three different architectures, comparing CRS (open symbols) and blocked JDS (closed symbols) variants. The Intel Xeon/Core system (dashed) is of UMA type, the other two systems are ccNUMA

27.2.3 Locality of Access on ccNUMA

It was mentioned already in the section on ccNUMA architecture that locality and congestion problems (see Figs. 27.17 and 27.18) tend to turn up when threads/processes and their data are not carefully placed across the locality domains of a ccNUMA system. Unfortunately, the current OpenMP standard does not refer to placement at all and it is up to the programmer to use the tools that system builders provide.

The placement problem has two dimensions: First, one has to make sure that memory gets mapped into the locality domains of processors that actually access them. This minimizes NUMA traffic across the network. Second, threads or processes must be "pinned" to those CPUs which had originally mapped their memory regions in order not to lose locality of access. In this context, *mapping* means that a page table entry is set up which describes the association of a physical with a vir-

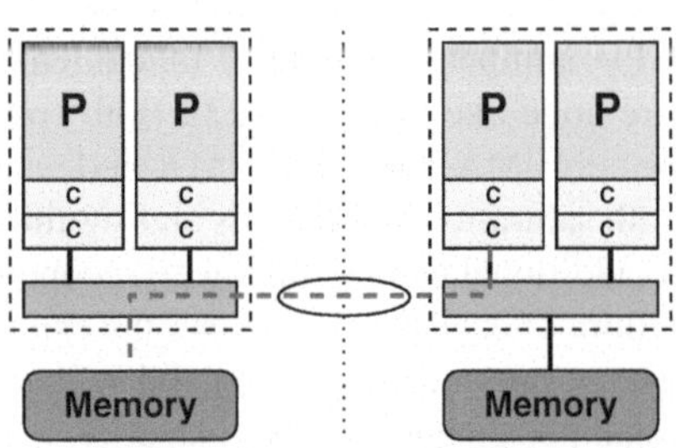

Fig. 27.17. Locality problem on a ccNUMA system. Memory pages got mapped into a locality domain that is not connected to the accessing processor, leading to NUMA traffic

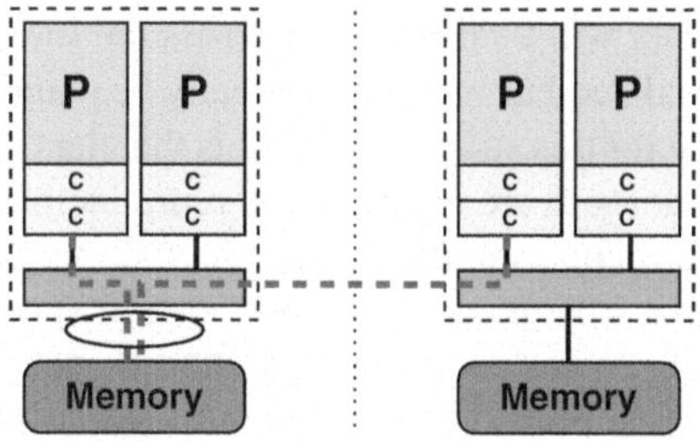

Fig. 27.18. Congestion problem on a ccNUMA system. Even if the network is very fast, a single locality domain can usually not saturate the bandwidth demands from concurrent local and non-local accesses

tual memory page. Consequently, locality of access in ccNUMA systems is always followed on the page level, with typical page sizes of (commonly) 4 kB or (more rarely) 16 kB, sometimes larger. Hence strict locality may be hard to implement with working sets that only encompass a few pages.

27.2.3.1 Ensuring Locality of Memory Access

Fortunately, the initial mapping requirement can be enforced in a portable manner on all current ccNUMA architectures. They support a first touch policy for memory pages: A page gets mapped into the locality domain of the processor that first reads or writes to it. Merely allocating memory is not sufficient (and using `calloc()` in C will most probably be counterproductive). It is therefore the data initialization code that deserves attention on ccNUMA:

```
integer,parameter::N=1000000
double precision A(N), B(N)


! executed on single
! locality domain
READ(1000) A
! congestion problem
!$OMP PARALLEL DO
  do i = 1, N
    B(i) = func(A(i))
  enddo
!$OMP END PARALLEL DO
```

```
integer,parameter::N=1000000
double precision A(N), B(N)
!$OMP PARALLEL DO
  do i=1,N
    A(i) = 0.d0
!$OMP END PARALLEL DO
! A is mapped now
READ(1000) A
!$OMP PARALLEL DO
  do i = 1, N
    B(i) = func(A(i))
  enddo
!$OMP END PARALLEL DO
```

On the left, initialization of `A` is done in a serial region using a `READ` staement, so the array data gets mapped to a single locality domain (maybe more if the array is very large). The access to `A` in the parallel loop will then lead to congestion. The version on the right corrects this problem by initializing `A` in parallel, first-touching its elements in the same way they are accessed later. Although the `READ` operation is still sequential, the data will be distributed across the locality domains. Array B does not have to be initialized but will automatically be mapped correctly.

A required condition for this strategy to work is that the OpenMP loop schedules of initialization and work loops are identical and reproducible, i.e. the only possible choice is STATIC with a constant chunk size. As the OpenMP standard does not define a default schedule, it is generally a good idea to specify it explicitly on all parallel loops. All current compilers choose STATIC by default, though. Of course, the use of a static schedule poses some limits on possible optimizations for eliminating load imbalance. One option is the choice of an appropriate chunk size (as small as possible, but at least several pages).

Unfortunately it is not always at the programmer's discretion how and when data is touched first. In C/C++, global data (including objects) is initialized before the main() function even starts. If globals cannot be avoided, properly mapped local copies of global data may be a possible solution, code characteristics in terms of communication vs. calculation permitting [5]. A discussion of some of the problems that emerge from the combination of OpenMP with C++ can be found in [6].

27.2.3.2 ccNUMA Optimization of Sparse MVM

It should now be obvious that the bad scalability of OpenMP-parallelized sparse MVM codes on ccNUMA systems (see Fig. 27.16) is due to congestion that arises because of wrong data placement. By writing parallel initialization loops that exploit first touch mapping policy, scaling can be improved considerably. We will restrict ourselves to CRS here as the strategy is basically the same for JDS. Arrays c, val, col_idx, row_ptr and b must be initialized in parallel:

```
!$OMP PARALLEL DO
  do i=1,Nr
    row_ptr(i) = 0 ; c(i) = 0.d0 ; b(i) = 0.d0
  enddo
!$OMP END PARALLEL DO
.... ! preset row_ptr array
!$OMP PARALLEL DO PRIVATE(start,end,j)
  do i=1,Nr
    start = row_ptr(i) ; end = row_ptr(i+1)
    do j=start,end-1
      val(j) = 0.d0 ; col_idx(j) = 0
    enddo
  enddo
!$OMP END PARALLEL DO
```

The initialization of b is based on the assumption that the non-zeroes of the matrix are roughly clustered around the main diagonal. Depending on the matrix structure it may be hard in practice to perform proper placement for the r.h.s. vector at all.

Fig. 27.19 shows performance data for the same architectures and sMVM codes as in Fig. 27.16 but with appropriate ccNUMA placement. There is no change in

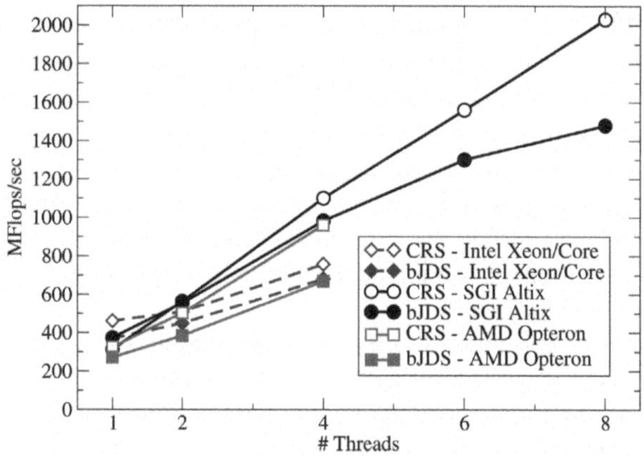

Fig. 27.19. Performance and strong scaling for ccNUMA-optimized OpenMP parallelization of sparse MVM on three different architectures, comparing CRS (open symbols) and blocked JDS (closed symbols) variants. Cf. Fig. 27.16 for performance without proper placement

scalability for the UMA platform, which was to be expected, but also on the cc-NUMA systems for up to two threads. The reason is of course that both architectures feature two-processor locality domains which are of UMA type. On four threads and above, the locality optimizations yield dramatically improved performance. Especially for the CRS version scalability is nearly perfect when going from $2n$ to $2(n+1)$ threads (due to bandwidth limitations inside the locality domains, scalability on ccNUMA systems should always be reported with reference to performance on all cores of a locality domain). The JDS variant of the code benefits from the optimizations as well, but falls behind CRS for larger thread numbers. This is because of the permutation map for JDS which makes it hard to place larger portions of the r.h.s. vector into the correct locality domains, leading to increased NUMA traffic.

It should be obvious by now that data placement is of premier importance on cc-NUMA architectures, including commonly used two-socket cluster nodes. In principle, ccNUMA features superior scalability for memory-bound codes, but UMA systems are much easier to handle and require no code optimization for locality of access. It is to be expected, though, that ccNUMA designs will prevail in the mid-term future.

27.2.3.3 Pinning

One may speculate that the considerations about locality of access on ccNUMA systems from the previous section do not apply for MPI-parallelized code. Indeed, MPI processes have no concept of shared memory. They allocate and first-touch memory pages in their own locality domain *by default*. Operating systems are nowadays capable of maintaining strong affinity between threads and processors, meaning that a thread (or process) will be reluctant to leave the processor it was initially started on. However, it might happen that system processes or interactive load push threads off their original CPUs. It is not guaranteed that the previous state will be re-established after the disturbance. One indicator for insufficient thread affinity are erratic performance numbers (i.e., varying from run to run). Even on UMA systems insufficient affinity can lead to problems if the UMA node is divided into sections (e.g., sockets with dual-core processors like in Fig. 26.22) that have separate paths to memory and internal shared caches. It may be of advantage to keep neighboring thread IDs on the cores of a socket to exploit the advantage of shared caches. If only one core per socket is used, migration of both threads to the same socket should be avoided if the application is bandwidth-bound.

The programmer can avoid those effects by pinning threads to CPUs. Every operating system has ways of limiting the mobility of threads and processes. Unfortunately, these are by no means portable, but there is always a low-level interface with library calls that access the basic functionality. Under the Linux OS, PLPA [7] can be used for that purpose. The following is a C example that pins each thread to a CPU whose ID corresponds to the thread ID:

```
#include <plpa.h>
...
#pragma omp parallel
{
  plpa_cpu_set_t mask;
  PLPA_CPU_ZERO(&mask);
  int id = omp_get_thread_num();
  PLPA_CPU_SET(id,&mask);
  PLPA_NAME(sched_setaffinity)((pid_t)0, (size_t)32, &mask);
}
```

The mask variable is used as a bit mask to identify those CPUs the thread should be restricted to by setting the corresponding bits to one (this could be more than one bit, a feature often called *CPU set*). After this code has executed, no thread will be able to leave its CPU any more.

System vendors often provide high-level interfaces to the pinning or CPU set mechanism. Please consult the system documentation for details.

27.3 Conclusion and Outlook

In this chapter we have presented basic optimization techniques on the processor and the shared-memory level. Although we have mainly used examples from linear algebra for clarity of presentation, the concepts can be applied to all numerical program codes. Although compilers are often surprisingly smart in detecting optimization opportunities, they are also easily deceived by the slightest obstruction of their view on program source. Regrettably, compiler vendors are very reluctant to build tools into their products that facilitate the programmer's work by presenting a clear view on optimizations performed or dropped.

There is one important topic in code optimization that we have neglected for brevity: The start of any serious optimization attempt on a nontrivial application should be the production of a *profile* that identifies the hot spots, i.e. the parts of the code that take the most time to execute. Many tools, free and commercial, exist in this field and more are under development. In which form a programmer should be presented performance data for a parallel run with thousands of processors and how the vast amounts of data can be filtered to extract the important insights is the subject of intense research. Multi-core technologies are adding another dimension to this problem.

References

1. A. Hoisie, O. Lubeck, H. Wassermann, Int. J. High Perform. Comp. Appl. **14**, 330 (2000)
2. P.F. Spinnato, G. van Albada, P.M. Sloot, IEEE Trans. Parallel Distrib. Systems **15**(1), 81 (2004)
3. R. Barrett, M. Berry, T. Chan, J. Demmel, J. Donato, J. Dongarra, V. Eijkhout, R. Pozo, C. Romine, H. van der Vorst, *Templates for the Solution of Linear Systems: Building Blocks for Iterative Methods* (SIAM, 1994)
4. URL `http://www.openmp.org`
5. B. Chapman, F. Bregier, A. Patil, A. Prabhakar, Concurrency Comput.: Pract. Exper. **14**, 713 (2002)
6. C. Terboven, D. an Mey, in *Proceedings of IWOMP2006 — International Workshop on OpenMP, Reims, France, June 12–15, 2006.* (2006). URL `http://iwomp.univ-reims.fr/cd/papers/TM06.pdf`
7. URL `http://www.open-mpi.org/software/plpa/`

A Appendix: Abbreviations

Abbreviation	Explanation
1D,2D,3D	one-, two-, three-dimensional
1D3V	**1D** in a usual space and **3D** in a velocity space
ABINIT	DFT software package (open source)
AIREBO	**A**daptive **I**ntermolecular **R**eactive **B**ond **O**rder
API	**A**pplication **P**rogramming **I**nterface
ARPES	**A**ngle-**R**esolved **P**hoto-**E**mission **S**pectroscopy
BE	**B**oltzmann **E**quation
BIT1	1D3V PIC code
BLAS	**B**asic **L**inear **A**lgebra **S**ubsystem
BO	**B**orn-**O**ppenheimer
CASTEP	DFT software package (commercial)
ccNUMA	**c**ache-**c**oherent **N**on-**U**niform **M**emory **A**rchitecture
CDW	**C**harge **D**ensity **W**ave
CF	**C**orrelation **F**unctions
CI	**C**onfiguration **I**nteraction
CIC	**C**loud **i**n **C**ell
CISC	**C**omplex **I**nstruction **S**et **C**omputing
CO	**C**omplex **O**bject
CP	**C**ar-**P**arrinello
CPA	**C**oherent **P**otential **A**pproximation
CP-PAW	Car-Parrinello software package
CPT	**C**luster **P**erturbation **T**heory
CPU	**C**entral **P**rocessing **U**nit
CRS	**C**ompressed **R**ow **S**torage
dc	**d**irect **c**urrent
DDCF	**D**ensity-**D**ensity time **C**orrelation **F**unctions
DDMRG	**D**ynamical **D**ensity **M**atrix **R**enormalization **G**roup
DFT	**D**ensity **F**unctional **T**heory
DMC	**D**iagrammatic **M**onte **C**arlo
DMFT	**D**ynamical **M**ean-**F**ield **T**heory
DMRG	**D**ensity-**M**atrix **R**enormalization **G**roup

(continued)

(continued)

Abbreviation	Explanation
DOS	**D**ensity **o**f **S**tates
DP	**D**ouble **P**recision
DRAM	**D**ynamic **R**andom **A**ccess **M**emory
DTMRG	**D**ynamical **TMRG**
ED	**E**xact **D**iagonalization
EDIP	**E**nvironment-**D**epedent **I**nteraction **P**otential
EIRENE	A Monte Carlo linear transport solver
EPIC	**E**xplicitly **P**arallel **I**nstruction **C**omputing
FD	**F**eynman **D**iagram
FFT	**F**ast **F**ourier **T**ransform
FFTW	**F**astest **F**ourier **T**ransform in the **W**est (FFT library)
FHI98md	DFT software package
FMM	**F**ast **M**ultipole **M**ethod
FP	**F**loating **P**oint
FPGA	**F**ield **P**rogrammable **G**ate **A**rrays
FSB	**F**ront **S**ide **B**us
GAUSSIAN	computational chemistry software program
GF	**G**reen **F**unction
GTO	**G**aussian **T**ype **O**rbitals
GMRES	**G**eneralized **M**inimum **Res**idual Method
GPU	**G**raphics **P**rocessing **U**nit
HF	**H**artree-**F**ock
HPC	**H**igh **P**erformance **C**omputing
HPF	**H**igh **P**erformance **F**ortran
HT	**H**yper**t**ransport
IKP	**I**mproved **K**elbg **P**otential
ILP	**I**nstruction-**L**evel **P**arallelism
JDS	**J**agged **D**iagonals **S**torage
KPM	**K**ernel **P**olynomial **M**ethod
LAPACK	**L**inear **A**lgebra **P**ackage
LD	**L**ocal **D**istribution
LDA	**L**ocal **D**ensity **A**pproximation
LDA-KS	**L**ocal **D**ensity **A**pproximation in the **K**ohn-**S**ham scheme
LDOS	**L**ocal **D**ensity **o**f **S**tates
LINPACK	**Lin**ear Algebra **Pack**age (superseeded by LAPACK)
LJ	**L**ennard-**J**ones
LR	**L**anczos **R**ecursion
LRU	**L**east **R**ecently **U**sed
MC	**M**onte **C**arlo
MD	**M**olecular **D**ynamics
MEM	**M**aximum **E**ntropy **M**ethod
MESI	**M**odified/**E**xclusive/**S**hared/**I**nvalid protocol
MIPS	**M**icroprocessor without **I**nterlocked **P**ipeline **S**tages
MIT	**M**etal-**I**nsulator **T**ransition
MMM	**M**atrix **M**atrix **M**ultiplication

Abbreviation	Explanation
MOLPRO	quantum chemistry software package
MP	**M**essage **P**assing
MPI	**M**essage **P**assing **I**nterface
MPMD	**M**ultiple **P**rogram **M**ultiple **D**ata
MVM	**M**atrix **V**ector **M**ultiplication
NGP	**N**earest **G**rid **P**oint
NI	**N**etwork **I**nterface
NL	**N**UMA **L**ink
NRG	**N**umerical **R**enomalization **G**roup
NUMA	**N**on-**U**niform **M**emory **A**rchitecture
NWChem	computational chemistry software package
OpenMP	**Open** **M**ulti-**P**rocessing
OS	**O**perating **S**ystem
PDP1	**P**rogrammed **D**ata **P**rocessor **1**
PES	**P**otential **E**nergy **S**urface
PIC	**P**article-**i**n-**C**ell
PIC-MCC	**P**article-**i**n-**C**ell **M**onte **C**arlo **C**ollision
PIMC	**P**ath **I**ntegral **M**onte **C**arlo
PJT	**P**seudo **J**ahn-**T**eller
PLPA	**P**ageable **L**ink **P**ack **A**rea
POSIX	**P**ortable **O**perating **S**ystem **I**nterface
QMC	**Q**uantum **M**onte **C**arlo
QMD	**Q**uantum **M**olecular **D**ynamics
QMR	**Q**uasi **M**inimum **R**esidual Method
QP	**Q**uantum **P**article
QPT	**Q**uantum **P**hase **T**ransition
REBO	**R**eactive **E**mpirical **B**ond **O**rder
RFO	**R**ead **F**or **O**wnership
RG	**R**enormalization **G**roup
RISC	**R**educed **I**nstruction **S**et **C**omputing
RKHS	**R**eproducing **K**ernel **H**ilbert **S**pace
SIAM	**S**ingle **I**mpurity **A**nderson **M**odel; **S**ociety for **I**ndustrial and **A**pplied **M**athematics
SIMD	**S**ingle **I**nstruction **M**ultiple **D**ata
SMP	**S**ymmetric **M**ulti-**P**rocessing
sMVM	**s**parse **M**atrix **V**ector **M**ultiplication
SO	**S**tochastic **O**ptimization
SPEC	**S**tandard **P**erformance **E**valuation **C**orporation
SP	**S**ingle **P**recision
SPMD	**S**ingle **P**rogram **M**ultiple **D**ata
STL	**S**tandard **T**emplate **L**ibrary
STM	**S**canning **T**unneling **M**icroscopy
STO	**S**later **T**ype **O**rbitals
TCP/IP	**T**ransmission **C**ontrol **P**rotocol / **I**nternet **P**rotocol
TLB	**T**ranslation **L**ook-aside **B**uffer

(continued)

(continued)

Abbreviation	Explanation
TMRG	**T**ransfer **M**atrix **R**enormalization **G**roup
UMA	**U**niform **M**emory **A**rchitecture
UPC	**U**nified **P**arallel **C**
VASP	ab initio molecular dynamics software package
VBS	**V**alence **B**ond **S**olid
WF	**W**ave **F**unction
XOOPIC	**X**-windows **O**bject **O**riented **PIC**
XPDP1	**X**-windows **PDP1** plasma code

Index

The manufacturer's authorised representative in the EU is Springer Nature Customer Service Centre GmbH, Europaplatz 3, 69115 Heidelberg, Germany. If you have any concerns regarding our products, please contact ProductSafety@springernature.com

Printed and bound by CPI Group (UK) Ltd, Croydon, CR0 4YY
22/09/2026
02220988-0011